Intermediate Algebra

Intermediate Algebra

EDITION 3

Mark Dugopolski

Southeastern Louisiana University

Boston   Burr Ridge, IL   Dubuque, IA   Madison, WI
New York   San Francisco   St. Louis   Bangkok
Bogotá   Caracas   Lisbon   London   Madrid   Mexico City   Milan
New Delhi   Seoul   Singapore   Sydney
Taipei   Toronto

McGraw-Hill Higher Education

A Division of The **McGraw-Hill** *Companies*

INTERMEDIATE ALGEBRA, THIRD EDITION

This book is printed on acid-free paper.

1 2 3 4 5 6 7 8 9 0 VNH/VNH 0 9 8 7 6 5 4 3 2 1 0

ISBN 0–07–229466–3
ISBN 0–07–229484–1 (AIE)

Vice president and editorial director: *Kevin T. Kane*
Publisher: *JP Lenney*
Sponsoring editor: *William K. Barter*
Developmental editor: *Erin Brown*
Marketing manager: *Mary K. Kittell*
Project manager: *Vicki Krug*
Production supervisor: *Enboge Chong*
Senior designer: *Sabrina Dupont*
Photo research coordinator: *Jodi Banowetz*
Senior supplement coordinator: *Audrey A. Reiter*
Compositor: *Interactive Composition Corporation*
Typeface: *10 1/2/12 Times Roman*
Printer: *Von Hoffmann Press, Inc.*

Cover/interior designer: *Amanda Kavanagh*
Photo credits: Page 27, p. 53, p. 73: *US Army Corps of Engineers;* p. 121: *Paul Conklin;* p. 199, p. 206, p. 333, p. 387, p. 400, p. 447, p. 627: *Susan Van Etten;* p. 257: *Herb Snitzer/Stock Boston;* p. 537: *Molly McCallister;* p. 571: *Courtesy of British Airways;* p. 596: *Frederick von Huene/ von Huene Workshop. All other photos © PhotoDisc, Inc.* (with exception of pages 158, 317, 369, 460, 504, 631).

Library of Congress has cataloged Algebra for College Students as follows:

Dugopolski, Mark.
    Algebra for college students / Mark Dugopolski. — 2nd ed.
        p.   cm.
    Includes index.
        ISBN 0–07–232399–X
        1. Algebra.   I. Title.
    QA152.2.D828   2000
        512.9—dc21   99–40279
                                                          CIP

To my wife and daughters,
Cheryl, Sarah, and Alisha

CONTENTS

Chapter 4 — Systems of Linear Equations   199

Chapter 5 — Exponents and Polynomials   257

Appendix   A-1

Index   I-1

PREFACE

*I*ntermediate Algebra is designed to provide students with the algebra background needed for further college-level mathematics courses. The unifying theme of this text is the development of the skills necessary for solving equations and inequalities, followed by the application of those skills to solving applied problems. My primary goal in writing the third edition of *Intermediate Algebra* has been to retain the features that made the second edition so successful while incorporating the comments and suggestions of second-edition users. In addition, I have provided many new features that will help instructors to reach the goals that they have set for their students. As always, I endeavor to write texts that students can read, understand, and enjoy, and at the same time gain confidence in their ability to use mathematics. Although a complete development of each topic is provided in *Intermediate Algebra,* the text *Elementary Algebra* in this series would be more appropriate for students with no prior experience in algebra.

Content Changes

While the essence of the text remains, the topics have been rearranged and new features added to reflect the current needs of instructors who are teaching intermediate algebra courses.

- Graphing is covered earlier in the text. Graphing linear equations in two variables is now in Chapter 3 immediately following linear equations in one variable. An introduction to functions and their graphs is also included in Chapter 3.

- Functions are also covered earlier in the text. The phrase, "is a function of" can be found in Section 3.3. The definition of a function as an object is then given in Section 3.5.

- Systems of equations and inequalities (Chapter 4) is also covered earlier in the third edition.

- More emphasis is given to reading and understanding graphs. Exercises that involve graphs have been added to many sections of the text along with conceptual questions relating to the graphs.

In addition to these changes, the text and exercise sets have been carefully revised where necessary. Many new, applied examples have been added to the text and a large number of new, applied exercises included in the exercise sets. Particular care has been given to achieving an appropriate balance of problems that progressively increase in difficulty from routine exercises in the beginning of the set to more challenging exercises at the end of the set. As in earlier editions, fractions and decimals are used in the exercises and throughout the text discussions to help reinforce the basic arithmetic skills that are required for success in algebra.

Features

- Each chapter begins with a Chapter Opener that discusses a real application of algebra. The discussion is accompanied by a photograph, and in most cases by a real-data application graph that helps students to visualize algebra and more fully to understand the concepts discussed in the chapter. In addition, each chapter contains a Math at Work feature, which profiles a real person and the mathematics that he or she uses on the job. These two features have corresponding real-data exercises.

- **NEW!** An increased emphasis on real-data applications that involve graphs is a focus for the third edition. Applications have been added throughout the text to help demonstrate concepts, to motivate students, and to give students practice using new skills. Many of the real-data exercises contain data obtained from the Internet. Internet addresses are provided as a resource for both students and teachers. Because Internet addresses frequently change, a list of addresses will also be available on the web site. An Index of Applications listing applications by subject matter is included at the front of the text.

- Every section begins with In This Section, a list of topics that shows the student what will be covered. Because the topics correspond to the headings within each section, students will find it easy to locate and study specific concepts.

- Important ideas, such as definitions, rules, summaries, and strategies, are set apart in boxes for quick reference. Color is used to highlight these boxes as well as other significant points in the text.

- **NEW!** The third edition contains three new margin features that appear throughout the text:

  Calculator Close-ups give students an idea of how and when to use a graphing calculator. Some Calculator Close-ups simply introduce the features of a graphing calculator, whereas others enhance the understanding of algebraic concepts. For this reason many of the Calculator Close-ups will benefit even those students who do not use a graphing calculator. A graphing calculator is not required for studying from this text.

  Study Tips are included in the margins throughout the text. These short tips are meant to reinforce continually good study habits and to keep reminding students that it is never too late to make improvements in the manner in which they study.

  Helpful Hints are short comments that enhance the material in the text, provide another way of approaching a problem, or clear up misconceptions.

- At the end of every section there are Warm-up exercises, a set of ten simple statements that are to be answered true or false. These exercises are designed to provide a smooth transition between the ideas and the exercise sets. They help students to understand that every statement in mathematics is either true or false. They are also good for discussion or group work.

- **NEW!** Every section-ending exercise set in the third edition generally begins with six simple writing exercises. These exercises are designed to get students to review the definitions and rules of the section before doing more traditional exercises. For example, the student might be simply asked what properties of equality were discussed in this section.

- The end-of-section Exercises follow the same order as the textual material and contain exercises that are keyed to examples, as well as numerous exercises that are not keyed to examples. This organization allows the instructor to deal with only part of a section if necessary and to easily determine which exercises are appropriate to assign. The keyed exercises give the student a place to start practicing and building confidence, whereas the nonkeyed exercises are designed to "wean" the student from following examples in a step-by-step manner. Getting More Involved exercises are designed to encourage writing, discussion, exploration, and cooperative learning. Graphing Calculator Exercises require a graphing calculator and are identified with a graphing calculator logo. Exercises for which a scientific calculator would be helpful are identified with a scientific calculator logo.

- Every chapter ends with a four-part Wrap-up, which includes the following:

  The Chapter Summary lists important concepts along with brief illustrative examples.

  NEW! Enriching Your Mathematical Word Power appears at the end of each chapter and consists of multiple-choice questions in which the important terms are to be matched with their meanings. This feature emphasizes the importance of proper terminology.

  The Review Exercises contain problems that are keyed to the sections of the chapter as well as numerous miscellaneous exercises.

  The Chapter Test is designed to help the student assess his or her readiness for a test. The Chapter Test has no keyed exercises, thus enabling the student to work independently of the sections and examples.

- At the end of each chapter there is a Collaborative Activities feature that is designed to encourage interaction and learning in groups. Instructions and suggestions for using these activities and answers to all problems can be found in the Instructor's Solutions Manual.

- The Making Connections exercises at the end of Chapters 2–12 are designed to help students review and synthesize the new material with ideas from previous chapters, and in some cases to review material necessary for success in the upcoming chapter. Every Making Connections exercise set includes at least one applied exercise that requires ideas from one or more of the previous chapters.

Coverage

For those who wish to cover more on functions, Chapter 9 can be covered after functions are introduced in Chapter 3. For those who wish to cover less on functions, Sections 3.5 and 3.6 can be omitted. Some or all of Chapter 4 can be omitted for those who desire a less extensive treatment of systems of linear equations. However, if you have a graphing calculator to do the determinants, Cramer's rule with three variables is rather fun.

Supplements for the Instructor

ANNOTATED INSTRUCTOR'S EDITION

This ancillary includes answers to all exercises and tests. Each answer is printed next to each problem on the page where the problem appears. The answers are printed in a second color for ease of use by instructors.

PRINT AND COMPUTERIZED TEST BANK

The testing materials provide an array of formats that allow the instructor to create tests using both algorithmically generated test questions and those from a standard test bank. This testing system enables the instructor to choose questions either manually or randomly by section, question type, difficulty level, and other criteria. Testing is available for IBM, IBM compatible, and Macintosh computers. Instructors can edit questions in the testing system as well if they seek a degree of customization. The print version of the test bank is softcover and provides questions found in the computerized version along with answer keys. Each chapter of the print version contains three different tests. Additionally, the print test bank contains four different, comprehensive final exams.

INSTRUCTOR'S SOLUTIONS MANUAL

Prepared by Mark Dugopolski, this supplement contains detailed, worked solutions to all of the exercises in the text. The solutions are done by the techniques used in the text. Instructions and suggestions for using the Collaborative Activities feature in the text are also included in the Instructor's Solutions Manual.

Supplements for the Student

STUDENT'S SOLUTIONS MANUAL

Prepared by Mark Dugopolski, the Student's Solutions Manual contains complete worked-out solutions to all of the odd-numbered exercises in the text. It also contains solutions for all exercises in the Chapter Tests. It may be purchased by your students from McGraw-Hill.

DUGOPOLSKI VIDEO SERIES

The video tape series contains instructional material and presents opportunities for students to work problems and to check their results. The tapes are text-specific and cover all chapters of the text. The tapes are facilitated by instructors who introduce topics and work through examples. Students are encouraged to work examples on their own and to check their results with those provided.

DUGOPOLSKI TUTORIAL CD-ROM

This interactive CD-ROM is a self-paced tutorial specifically linked to the text and reinforces topics through unlimited opportunities to review concepts and to practice problem solving. The CD-ROM contains text-, chapter-, and section-specific tutorials, multiple-choice questions with feedback, as well as algorithmically-generated questions. It requires virtually no computer training on the part of students and supports IBM and Macintosh computers.

In addition, a number of other technology and Web-based ancillaries are under development; they will support the ever-changing technology needs in developmental mathematics. For further information about these or any supplements, please contact your local McGraw-Hill sales representative.

Acknowledgments

First I thank all of the students and professors who used the previous editions of this text, for without their support there would not be a third edition. I sincerely appreciate the efforts of the reviewers who made many helpful suggestions:

Gisela Acosta, *Valencia Community College, East*
Francisco E. Alarcon, *Indiana University of Pennsylvania*

Dimos Arsenidis, *C.S.U.L.B.*
Luis A. Beltran, *Miami-Dade Community College*
Don Bigwood, *Bismarck State College*
Ray F. Brinker, *Western Illinois University*
Mary Jean Brod, *University of Montana*
Deborah Bryant, *Santa Rosa Junior College*
Kathleen A. Cantone, *Onondaga Community College*
Jack Carson, Jr., *Madison Area Technical College*
Oiyin Pauline Chow, *Harrisburg Area Community College*
Al Coons, *Pima Community College*
Victor M. Cornell, *Mesa Community College*
Diane Daniels, *Mississippi State University*
Gregory Davis, *University of Wisconsin, Green Bay*
Joe DiCostanzo, *Johnson County Community College*
Elizabeth L. Doane, *Housatonic Community Technical College*
Lacey P. Echols, *Butler University*
Sue W. Fader, *Delaware Technical and Community College, Stanton Campus*
Cynthia Fleck, *Wright State University*
Dr. Jeanette W. Glover, *University of Memphis*
Jacqueline R. Grace, *SUNY New Paltz*
Roberta Grenz, *Community College of Southern Nevada*
Patricia L. Hirschy, *Asnuntuck Comm-Tech College*
Daniel L. Hostetler, *Univ. College/University of Cincinnati*
Heidi A. Howard, *Florida Community College at Jacksonville*
Milia Ison, *DeAnza College*
Doris J. Jones, *Langston University*
Giles Wilson Maloof, *Boise State University*
John Martin, *Santa Rosa Junior College*
Kenneth J. Mead, *Genesee Community College*
Aaron Montgomery, *Purdue University, North Central*
Christina Morian, *Lincoln University*
Daniel P. Munton, *Santa Rosa Junior College*
Jim Neary, *Ivy Tech State College*
Masood Poorandi, *Bethune Cookman College*
Christopher P. Reisch, *SUNY Buffalo*
Togba C. Sapolucia, *Northeast Houston Community College*
Fred Schineller, *Arizona State University*
Patricia L. Schulte, *Penn State University*
Mark Serebransky, *Camden County College*
Ann Sitomer, *Grossmont College*
Lourdes Triana, *Humboldt State University*
Paul J. Welsh, *Pima Community College-East Campus*
Albert E. White, *Lincoln University*
Jackie Wing, *Angelina Jr. College*

Dr. Judith B. Wood, *Central Florida Community College*
Vivian J. Zabrocki, *Montana State University, Billings.*

I would also like to thank those who reviewed the second edition.

Nancy Angle, *Cerritos College*
Chris Barker, *DeAnza College*
Richard Basic, *Lakeland Community College*
Richard A. Butterworth, *Massasoit Community College*
Nancy Carpenter, *Johnson County Community College*
Florence Chambers, *Joliet Junior College*
Irene Doo, *Austin Community College*
David Dudley, *Phoenix College*
James Fryxell, *College of Lake County*
Terry Fung, *Kean College*
Jane Hammontree, *Tulsa Junior College*
Robert A. Hawes, *Northern Essex Community College*
Dale Hughes, *Johnson County Community College*
Kathy Kepner, *Paducah Community College*
Judith Lenk, *Ocean County College*
Mitchel Levy, *Broward Community College*
Joy McMullen, *Lakeland Community College*
Gael Mericle, *Mankato State University*
Jane Morrison, *South Suburban Community College*
Linda Padilla, *Joliet Junior College*
Sue Parsons, *Cerritos College*
Joanne Peeples, *El Paso Community College*
Rose L. Pugh, *Bellevue Community College*
Scott Reed, *College of Lake County*
Fred Russell, *Charles County Community College*
Howard Sorken, *Broward Community College*
Patricia Stanley, *Ball State University*
Eric Stietzel, *Foothill College*
Diane Tesar, *South Suburban College*
Charles Waiveris, *Central Connecticut State University*

I also thank Edgar Reyes of Southeastern Louisiana University for his help with the CD-ROM and Rebecca Muller of Southeastern Louisiana University for her work on the printed test bank. Finally, my thanks go to Laurel Technical Services for error checking the manuscript. I thank the staff at McGraw-Hill for all of their help and encouragement throughout the revision process. Special thanks go to Bill Barter and Erin Brown. I also want to express my sincere appreciation to my wife, Cheryl, for her invaluable patience and support.

Hammond, Louisiana   M.D.

Rational Expressions

I nformation is everywhere—in the newspapers and magazines we read, the televisions we watch, and the computers we use. And now people are talking about the Information Superhighway, which will deliver vast amounts of information directly to consumers' homes. In the future the combination of telephone, television, and computer will give us on-the-spot health care recommendations, video conferences, home shopping, and perhaps even electronic voting and driver's license renewal, to name just a few. There is even talk of 500 television channels!

Some experts are concerned that the consumer will give up privacy for this technology. Others worry about regulation, access, and content of the enormous international computer network.

Whatever the future of this technology, few people understand how all their electronic devices work. However, this vast array of electronics rests on physical principles, which are described by mathematical formulas. In Exercises 49 and 50 of Section 6.6 we will see that the formula governing resistance for receivers connected in parallel involves rational expressions, which are the subject of this chapter.

Chapter Opener

Each **chapter opener** features a real-world situation that can be modeled using mathematics. Each chapter contains exercises that relate back to the chapter opener.

---

42. Trimming hedges. Lourdes can trim the hedges around her property in 8 hours by using an electric hedge trimmer. Rafael can do the same job in 15 hours by using a manual trimmer. How long would it take them to trim the hedges working together?

43. Filling the tub. It takes 10 minutes to fill Alisha's bathtub and 12 minutes to drain the water out. How long would it take to fill it with the drain accidentally left open?

FIGURE FOR EXERCISE 43

44. Eating machine. Charles can empty the cookie jar in $1\frac{1}{2}$ hours. It takes his mother 2 hours to bake enough cookies to fill it. If the cookie jar is full when Charles comes home from school, and his mother continues baking and restocking the cookie jar, then how long will it take him to empty the cookie jar?

45. Filing the invoices. It takes Gina 90 minutes to file the monthly invoices. If Hilda files twice as fast as Gina does, how long will it take them working together?

46. Painting alone. Julie can paint a fence by herself in 12 hours. With Betsy's help, it takes only 5 hours. How long would it take Betsy by herself?

47. Buying fruit. Molly bought $5.28 worth of oranges and $8.80 worth of apples. She bought 2 more pounds of oranges than apples. If apples cost twice as much per pound as oranges, then how many pounds of each did she buy?

48. Raising rabbits. Luke raises rabbits and raccoons to sell for meat. The price of raccoon meat is three times the price of rabbit meat. One day Luke sold 160 pounds of meat, $72 worth of each type. What is the price per pound of each type of meat?

49. Total resistance. If two receivers with resistances R_1 and R_2 are connected in parallel, then the formula

$$\frac{1}{R} = \frac{1}{R_1} + \frac{1}{R_2}$$

relates the total resistance for the circuit R with R_1 and R_2. Given that R_1 is 3 ohms and R is 2 ohms, find R_2.

FIGURE FOR EXERCISE 49

50. More resistance. Use the formula from Exercise 49 to find R_1 and R_2 given that the total resistance is 1.2 ohms and R_1 is 1 ohm larger than R_2.

51. Las Vegas vacation. Brenda of Horizon Travel has arranged for a group of gamblers to share the $24,000 cost of a charter flight to Las Vegas. If Brenda can get 40 more people to share the cost, then the cost per person will decrease by $100.
   a) How many people were in the original group?
   b) Write the cost per person as a function of the number of people sharing the cost.

FIGURE FOR EXERCISE 51

52. White-water rafting. Adventures, Inc. has a $1,500 group rate for an overnight rafting trip on the Colorado River. For the last trip five people failed to show, causing the price per person to increase by $25. How many were originally scheduled for the trip?

53. Doggie bag. Muffy can eat a 25-pound bag of dog food in 28 days, whereas Missy eats a 25-pound bag in 23 days. How many days would it take them together to finish a 50-pound bag of dog food.

54. Rodent food. A pest control specialist has found that 6 rats can eat an entire box of sugar-coated breakfast cereal in 13.6 minutes, and it takes a dozen mice 34.7 minutes to devour the same size box of cereal. How long would it take all 18 rodents, in a cooperative manner, to finish off a box of cereal?

Margin Notes

Margin notes include **Helpful Hints, Study Tips,** and **Calculator Close-ups.** The *Helpful Hints* point out common errors or reminders. The *Study Tips* provide practical suggestions for improving study habits. The optional *Calculator Close-ups* provide tips on using a graphing calculator to aid in your understanding of the material. They also include insightful suggestions for increasing calculator proficiency.

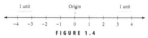

These Calculator Close-ups are designed to help reinforce the concepts of algebra, not replace them. Do not rely too heavily on your calculator or use it to replace the algebraic methods taught in this course.

Graphing on the Number Line

To construct a number line, we draw a straight line and label any convenient point with the number 0. Now we choose any convenient length and use it to locate points to the right of 0 as points corresponding to the positive integers and points to the left of 0 as points corresponding to the negative integers. See Fig. 1.4. The numbers corresponding to the points on the line are called the **coordinates** of the points. The distance between two consecutive integers is called a **unit**, and it is the same for any two consecutive integers. The point with coordinate 0 is called the **origin.** The numbers on the number line increase in size from left to right. When we compare the size of any two numbers, the larger number lies to the right of the smaller one on the number line.

FIGURE 1.4

It is often convenient to illustrate sets of numbers on a number line. The set of integers, J, is illustrated or **graphed** as in Fig. 1.5. The three dots to the right and left on the number line indicate that the integers go on indefinitely in both directions.

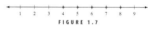

FIGURE 1.5

EXAMPLE 2   **Graphing on the number line**
List the elements of each set and graph each set on a number line.
a) $\{x \mid x$ is a whole number less than 4$\}$
b) $\{a \mid a$ is an integer between 3 and 9$\}$
c) $\{y \mid y$ is an integer greater than $-3\}$

Solution

a) The whole numbers less than 4 are 0, 1, 2, and 3. Figure 1.6 shows the graph of this set.

FIGURE 1.6

b) The integers between 3 and 9 are 4, 5, 6, 7, and 8. The graph is shown in Fig. 1.7.

FIGURE 1.7

c) The integers greater than -3 are -2, -1, 0, 1, and so on. To indicate the continuing pattern, we use a series of dots on the graph in Fig. 1.8.

FIGURE 1.8

---

38. *Three-digit number.* The sum of the digits of a three-digit number is 11. If the digits are reversed, the new number is 46 more than five times the old number. If the hundreds digit plus twice the tens digit is equal to the units digit, then what is the number?

39. *Working overtime.* To make ends meet, Ms. Farnsby works three jobs. Her total income last year was $48,000. Her income from teaching was just $6,000 more than her income from house painting. Royalties from her textbook sales were one-seventh of the total money she received from teaching and house painting. How much did she make from each source last year?

40. *Pocket change.* Harry has $2.25 in nickels, dimes, and quarters. If he had twice as many nickels, half as many dimes, and the same number of quarters, he would have $2.50. If he has 27 coins altogether, then how many of each does he have?

GETTING MORE INVOLVED

41. *Exploration.* Draw diagrams showing the possible ways to position three planes in three-dimensional space.

42. *Discussion.* Make up a system of three linear equations in three variables for which the solution set is $\{(0, 0, 0)\}$. A system with this solution set is called a *homogeneous* system. Why do you think it is given that name?

43. *Cooperative learning.* Working in groups, do parts (a)–(d) below. Then write a report on your findings.
a) Find values of a, b, and c so that the graph of $y = ax^2 + bx + c$ goes through the points $(-1, -2)$, $(1, 0)$, and $(2, 7)$.
b) Arbitrarily select three ordered pairs and find the equation of the parabola that goes through the three points.
c) Could more than one parabola pass through three given points? Give reasons for your answer.
d) Explain how to pick three points for which no parabola passes through all of them.

 **4.4   SOLVING LINEAR SYSTEMS USING MATRICES**

In this section

- Matrices
- The Augmented Matrix
- The Gaussian Elimination Method
- Inconsistent and Dependent Equations

You solved linear systems in two variables by substitution and addition in Sections 4.1 and 4.2. Those methods are done differently on each system. In this section you will learn the Gaussian elimination method, which is related to the addition method. The Gaussian elimination method is performed in the same way on every system. We first need to introduce some new terminology.

Matrices

A **matrix** is a rectangular array of numbers. The **rows** of a matrix run horizontally, and the **columns** of a matrix run vertically. A matrix with m rows and n columns has **order** $m \times n$ (read "m by n"). Each number in a matrix is called an **element** or **entry** of the matrix.

EXAMPLE 1   **Order of a matrix**
Determine the order of each matrix.

a) $\begin{bmatrix} -1 & 2 \\ 5 & \sqrt{2} \\ 0 & 3 \end{bmatrix}$ b) $\begin{bmatrix} 2 & 3 \\ -1 & 5 \end{bmatrix}$ c) $\begin{bmatrix} 1 & 2 & 3 \\ 4 & 5 & 6 \\ -1 & 0 & 2 \end{bmatrix}$ d) $[1 \ \ 3 \ \ 6]$

Solution
Because matrix (a) has 3 rows and 2 columns, its order is 3×2. Matrix (b) is a 2×2 matrix, matrix (c) is a 3×3 matrix, and matrix (d) is a 1×3 matrix.

The Augmented Matrix

The solution to a system of linear equations such as

$$x - 2y = -5$$
$$3x + y = 6$$

8.1 FACTORING AND COMPLETING THE SQUARE

- Review of Factoring
- Review of the Even-Root Property
- Completing the Square
- Miscellaneous Equations
- Imaginary Solutions

Factoring and the even-root property were used to solve quadratic equations in Chapters 5, 6, and 7. In this section we first review those methods. Then you will learn the method of completing the square, which can be used to solve any quadratic equation.

Review of Factoring

A quadratic equation is a second-degree polynomial equation of the form

$$ax^2 + bx + c = 0,$$

where a, b, and c are real numbers with $a \neq 0$. If the second-degree polynomial on the left-hand side can be factored, then we can solve the equation by breaking it into two first-degree polynomial equations (linear equations) using the following strategy.

Strategy for Solving Quadratic Equations by Factoring

1. Write the equation with 0 on the right-hand side.
2. Factor the left-hand side.
3. Use the zero factor property to set each factor equal to zero.
4. Solve the simpler equations.
5. Check the answers in the original equation.

EXAMPLE 1

Solving a quadratic equation by factoring

Solve $3x^2 - 4x = 15$ by factoring.

Solution

Subtract 15 from each side to get 0 on the right-hand side:

$$3x^2 - 4x - 15 = 0$$
$$(3x + 5)(x - 3) = 0 \quad \text{Factor the left-hand side.}$$
$$3x + 5 = 0 \quad \text{or} \quad x - 3 = 0 \quad \text{Zero factor property}$$
$$3x = -5 \quad \text{or} \quad x = 3$$
$$x = -\frac{5}{3}$$

The solution set is $\left\{-\frac{5}{3}, 3\right\}$. Check the solutions in the original equation. ■

Review of the Even-Root Property

In Chapter 7 we solved quadratic equations by using the even-root property.

Strategy Boxes

The **strategy boxes** generally provide a numbered list of concepts from a section or a set of steps to follow in problem solving. They can be used by students who prefer a more structured approach to problem solving or used as a study tool to review important points within sections.

MATH AT WORK

Cargo has been lost, or the hull of a ship has been damaged. What is the amount of money that should be paid to the insured party? Lisa M. Paccione, Ocean Marine Claim Representative for the St. Paul Insurance Company, investigates, evaluates, resolves, and pays these types of claims. Ms. Paccione does this by gathering data, occasionally doing a visual inspection, interviewing witnesses, and negotiating with attorneys.

MARINE INSURANCE AGENT

Decisions about losses are based on the insured party's individual policy as well as traditional marine practices and maritime law. When consignees suffer a cargo loss, they not only are compensated for the actual amount of the damaged goods, but also receive an additional "advance" in the settlement. Customarily, the advance is 10% over the value of the goods. The amount that St. Paul pays the insured party for a valid claim is computed by using a proportion. In Exercises 59 and 60 of this section you will solve problems involving this proportion.

EXAMPLE 8

Ratios and proportions

The ratio of men to women at a football game was 4 to 3. If there were 12,000 more men than women in attendance, then how many men and how many women were in attendance?

Solution

Let x represent the number of men in attendance and $x - 12,000$ represent the number of women in attendance. Because the ratio of men to women was 4 to 3, we can write the following proportion:

$$\frac{4}{3} = \frac{x}{x - 12,000}$$
$$4x - 48,000 = 3x$$
$$x = 48,000$$

So there were 48,000 men and 36,000 women at the game. ■

WARM-UPS

True or false? Explain.

1. In solving an equation involving rational expressions, multiply each side by the LCD for all of the denominators.
2. To solve $\frac{1}{x} + \frac{1}{2x} = \frac{1}{3}$, first change each rational expression to an equivalent rational expression with a denominator of $6x$.
3. Extraneous roots are not real numbers.
4. To solve $\frac{1}{x-2} + 3 = \frac{1}{x+2}$, multiply each side by $x^2 - 4$.
5. The solution set to $\frac{x}{3x+4} - \frac{6}{2x+1} = \frac{7}{5}$ is $\left\{-\frac{4}{3}, -\frac{1}{2}\right\}$.

Math at Work

The **Math at Work** feature that appears in each chapter explores the careers of individuals who use the mathematics presented in the chapter in their work. Students are referred to exercises that directly relate to the occupation highlighted in Math at Work.

Warm-ups

Warm-ups appear before each set of exercises at the end of every section. They are true or false statements that can be used to check conceptual understanding of material within each section.

Because there are no real even roots of negative numbers, the expressions

$$a^{1/2}, \quad x^{-3/4}, \quad \text{and} \quad y^{1/6}$$

are not real numbers if the variables have negative values. To simplify matters, we sometimes assume the variables represent only positive numbers when we are working with expressions involving variables with rational exponents. That way we do not have to be concerned with undefined expressions and absolute value.

E X A M P L E   8

Expressions involving variables with rational exponents
Use the rules of exponents to simplify the following. Write your answers with positive exponents. Assume all variables represent *positive* real numbers.

a) $x^{2/3}x^{4/3}$ b) $\dfrac{a^{1/2}}{a^{1/4}}$

c) $(x^{1/2}y^{-3})^{1/2}$ d) $\left(\dfrac{x^2}{y^{1/3}}\right)^{-1/2}$

Solution

a) $x^{2/3}x^{4/3} = x^{6/3}$ Use the product rule to add the exponents.

           $= x^2$ Reduce the exponent.

b) $\dfrac{a^{1/2}}{a^{1/4}} = a^{1/2-1/4}$ Use the quotient rule to subtract the exponents.

        $= a^{1/4}$ Simplify.

c) $(x^{1/2}y^{-3})^{1/2} = (x^{1/2})^{1/2}(y^{-3})^{1/2}$ Power of a product rule

               $= x^{1/4}y^{-3/2}$ Power of a power rule

               $= \dfrac{x^{1/4}}{y^{3/2}}$ Definition of negative exponent

d) Because this expression is a negative power of a quotient, we can first find the reciprocal of the quotient, then apply the power of a power rule:

$$\left(\frac{x^2}{y^{1/3}}\right)^{-1/2} = \left(\frac{y^{1/3}}{x^2}\right)^{1/2} = \frac{y^{1/6}}{x} \quad \frac{1}{3}\cdot\frac{1}{2}=\frac{1}{6}$$

WARM-UPS

True or false? Explain.

1. $4^{-1/2} = \dfrac{1}{2}$ 2. $16^{1/2} = 8$

3. $(3^{2/3})^3 = 9$ 4. $8^{-2/3} = -4$

5. $2^{1/2} \cdot 2^{1/2} = 2$ 6. $\left(\dfrac{1}{4}\right)^{1/2} = \dfrac{1}{2}$

7. $\dfrac{3}{3^{1/2}} = 3^{1/2}$ 8. $(2^9)^{1/2} = 2^3$

9. $3^{1/3} \cdot 6^{1/3} = 18^{2/3}$ 10. $2^{3/4} \cdot 2^{1/4} = 4$

COLLABORATIVE ACTIVITIES

Beorg's Business

In manufacturing or other businesses in which time is money and tasks are easily shared, problems involving work appear. An owner or manager who wants to know how to bid a job often develops a table of times needed to complete the job as determined by how much work is required and who could be assigned to the job.

Beorg owns a kaleidoscope-manufacturing company with two employees, Scott and Salina. It takes Scott one hour to make one kaleidoscope, and it takes Salina $\frac{1}{2}$ hour to make one kaleidoscope. Beorg wants to know how long it would take to complete a certain number of kaleidoscopes. Using the information given and answering the questions below, fill in the following table for Beorg.

Name of Employee	Time for one kaleidoscope	Time for 20 kaleidoscopes
Scott	1 hr	
Salina	$\frac{1}{2}$ hr	
Scott & Salina		
Sammy		
Scott & Sammy	$\frac{3}{4}$ hr	
Salina & Sammy		
Scott, Salina, & Sammy		

Grouping: Four students per group
Topic: Applications of work problems

1. How long will it take Scott and Salina working together to make one kaleidoscope?

2. Beorg hires a third person, Sammy, and has him and Scott make one kaleidoscope. Working together, it takes them $\frac{3}{4}$ hour to make one kaleidoscope. How long would it take Sammy by himself to make one kaleidoscope?

3. How long would it take Salina and Sammy working together to make one kaleidoscope? How long would it take for all three working together?

Now Beorg wants to finish his time table. He would like to have 20 kaleidoscopes completed each day.

4. Finish the preceding table, and find the best combination or combinations of employees to use to have 20 kaleidoscopes at the end of an 8-hour day.

Extension: Is Sammy in the combination or combinations you found in the last question? Is it worth having Sammy work? Remember that when someone is starting a new job, he or she may work more slowly until he or she learns how to do the job more efficiently. Find out how fast Sammy would need to work for production to double (40 kaleidoscopes in an 8-hour day).

WRAP-UP   CHAPTER 6

SUMMARY

Rational Expressions   **Examples**

Rational expression	The ratio of two polynomials with the denominator not equal to zero	$\dfrac{x^2 - 1}{2x - 3}$
Domain of a rational expression	The set of all possible numbers that can be used as replacements for the variable	$D = \left\{x \mid x \neq \dfrac{3}{2}\right\}$

Collaborative Activities

Collaborative Activities appear at the end of each chapter. The activities are designed to encourage interaction and learning in a group setting.

Solve each equation. Practice combining some steps. Look for more efficient ways to solve each equation. See Example 8.

81. $3x - 9 = 0$

82. $5x + 1 = 0$

83. $7 - z = -9$

84. $-3 - z = 3$

85. $\frac{2}{3}x = \frac{1}{2}$

86. $\frac{3}{2}x = -\frac{9}{5}$

87. $-\frac{3}{5}y = 9$

88. $-\frac{2}{7}w = 4$

89. $3y + 5 = 4y - 1$

90. $2y - 7 = 3y + 1$

91. $5x + 10(x + 2) = 110$

92. $1 - 3(x - 2) = 4(x - 1) - 3$

Solve each equation.

93. $\frac{P + 7}{3} - \frac{P - 2}{5} = \frac{7}{3} - \frac{P}{15}$

94. $\frac{w - 3}{8} - \frac{5 - w}{4} = \frac{4w - 1}{8} - 1$

95. $x - 0.06x = 50,000$

96. $x - 0.05x = 800$

97. $2.365x + 3.694 = 14.8095$

98. $-3.48x + 6.981 = 4.329x - 6.851$

Solve each problem. See Example 9.

99. *Public school enrollment.* The expression

$$0.45x + 39.05$$

can be used to approximate the total enrollment in millions in public elementary and secondary schools in the year 1985 + x (National Center for Education Statistics, www.nces.ed.gov).
a) What was the public school enrollment in 1992?
b) In which year will enrollment reach 50 million students?
c) Judging from the accompanying graph, is enrollment increasing or decreasing?

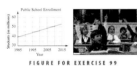

FIGURE FOR EXERCISE 99

100. *Teacher's average salary.* The expression

$$553.7x + 27,966$$

can be used to approximate the average annual salary in dollars of public school teachers in the year 1985 + x (National Center for Education Statistics, www.nces.ed.gov).

a) What was the average teacher's salary in 1993?
b) In which year will the average salary reach $40,000?

101. *Solid waste recovery.* In 1960 the United States generated 87.1 million tons of municipal solid waste and recovered (or recycled) only 4.3% of it (U.S. Department of Energy, www.doe.gov). The amount of solid waste generated in the United States in the year 1960 + n is given by

$$w = 3.14n + 87.1,$$

whereas the amount recovered is given by

$$w = 0.576n + 3.78,$$

where w is in millions of tons.
a) Use the accompanying graph to estimate the first year in which the United States generated over 100 million tons of municipal solid waste.
b) Find the year in which 13% of the municipal solid waste will be recovered by solving

$$0.576n + 3.78 = 0.13(3.14n + 87.1).$$

FIGURE FOR EXERCISE 101

102. *Recycling progress.* Find the year in which 14% of the municipal solid waste will be recovered by solving

$$0.576n + 3.78 = 0.14(3.14n + 87.1).$$

See the previous exercise.

GETTING MORE INVOLVED

103. *Exploration.* If you solved the equations in the two previous exercises, then you found the years in which recovery of municipal solid waste will reach 13% and 14%.
a) Write equations corresponding to recovery rates of 15%, 16%, 17%, 18%, and 19%.
b) Solve your equations to find the years in which those recovery rates will be achieved.
c) Use your results to judge whether we are making real progress in recovery of municipal solid waste.

104. *Writing.* Explain how to eliminate the decimals in an equation that involves numbers with decimal points. Would you use the same technique when using a calculator?

105. *Discussion.* Explain why the multiplication property of equality does not allow us to multiply each side of an equation by zero.

Exercises

The theme of mathematics in everyday situations is carried over to the exercise sets. Applications based on real-world data are included in each set. The **Index of Applications** can help students to quickly identify exercises that associate the mathematics that may be used in their areas of interest.

Find the complex solutions to each equation. See Example 10.

67. $x^2 + 2x + 5 = 0$

68. $x^2 + 4x + 5 = 0$

69. $x^2 + 12 = 0$

70. $-3x^2 - 21 = 0$

71. $5z^2 - 4z + 1 = 0$

72. $2w^2 - 3w + 2 = 0$

Find all real or imaginary solutions to each equation. Use the method of your choice.

73. $4x^2 + 25 = 0$

74. $5w^2 - 3 = 0$

75. $\left(p + \frac{1}{2}\right)^2 = \frac{9}{4}$

76. $\left(y - \frac{2}{3}\right)^2 = \frac{4}{9}$

77. $5t^2 + 4t - 3 = 0$

78. $3v^2 + 4v - 1 = 0$

79. $m^2 + 2m - 24 = 0$

80. $q^2 + 6q - 7 = 0$

81. $\left(a + \frac{2}{3}\right)^2 = -\frac{32}{9}$

82. $\left(w + \frac{1}{2}\right)^2 = -6$

83. $-x^2 + x + 6 = 0$

84. $-x^2 + x + 12 = 0$

85. $x^2 - 6x + 10 = 0$

86. $x^2 - 8x + 17 = 0$

87. $2x - 5 = \sqrt{7x + 7}$

88. $\sqrt{7x + 29} = x + 3$

89. $\frac{1}{x} + \frac{1}{x - 1} = \frac{1}{4}$

90. $\frac{1}{x} - \frac{2}{1 - x} = \frac{1}{2}$

If the solution to an equation is imaginary or irrational, it takes a bit more effort to check. Replace x by each given number to verify each statement.

91. Both $2 + \sqrt{3}$ and $2 - \sqrt{3}$ satisfy $x^2 - 4x + 1 = 0$.

92. Both $1 + \sqrt{2}$ and $1 - \sqrt{2}$ satisfy $x^2 - 2x - 1 = 0$.

93. Both $1 + i$ and $1 - i$ satisfy $x^2 - 2x + 2 = 0$.

94. Both $2 + 3i$ and $2 - 3i$ satisfy $x^2 - 4x + 13 = 0$.

Solve each problem.

95. *Approach speed.* The formula $1211.1L = CA^2S$ is used to determine the approach speed for landing an aircraft, where L is the gross weight of the aircraft in pounds, C is the coefficient of lift, S is the surface area of the wings in square feet (ft²), and A is approach speed in feet per second. Find A for the Piper Cheyenne, which has a gross weight of 8,700 lb, a coefficient of lift of 2.81, and wing surface area of 200 ft².

96. *Time to swing.* The period T (time in seconds for one complete cycle) of a simple pendulum is related to the length L (in feet) of the pendulum by the formula $8T^2 = \pi^2 L$. If a child is on a swing with a 10-foot chain, then how long does it take to complete one cycle of the swing?

97. *Time for a swim.* Tropical Pools figures that its monthly revenue in dollars on the sale of x above-ground pools is given by $R = 1500x - 3x^2$, where x is less than 25. What number of pools sold would provide a revenue of $17,568?

98. *Pole vaulting.* In 1981 Vladimir Poliakov (USSR) set a world record of 19 ft $\frac{3}{4}$ in. for the pole vault (Doubleday Almanac). To reach that height, Poliakov obtained a speed of approximately 36 feet per second on the runway. The function $h = -16t^2 + 36t$ gives his height t seconds after leaving the ground.
a) Use the formula to find the exact values of t for which his height was 18 feet.
b) Use the accompanying graph to estimate the value of t for which he was at his maximum height.
c) Approximately how long was he in the air?

FIGURE FOR EXERCISE 98

GETTING MORE INVOLVED

99. *Discussion.* Which of the following equations is not a quadratic equation?
a) $\pi x^2 - \sqrt{5}x - 1 = 0$ b) $3x^2 - 1 = 0$
c) $4x + 5 = 0$ d) $0.009x^2 = 0$

100. *Exploration.* Solve $x^2 - 4x + k = 0$ for k = 0, 4, 5, and 10.
a) When does the equation have only one solution?
b) For what values of k are the solutions real?
c) For what values of k are the solutions imaginary?

Getting More Involved appears within selected exercise sets. This feature may contain

 Writing,

 Cooperative Learning,

 Exploration, and/or

Discussion exercises. Each of these components is designed to give students an opportunity to improve and develop the ways in which they express mathematical ideas.

The exercise sets contain exercises that are keyed to examples, as well as exercises that are not keyed to examples.

Calculator Exercises

Calculator exercises are optional. They provide an opportunity for students to learn how a scientific or graphing calculator may be useful in solving various problems.

---

deducting the bonus. So

$$T = 0.40(100,000 - B).$$

a) Use the accompanying graph to estimate the values of T and B that satisfy both equations.

b) Solve the system algebraically to find the bonus and the amount of tax.

57. Textbook case. The accompanying graph shows the cost of producing textbooks and the revenue from the sale of those textbooks.

a) What is the cost of producing 10,000 textbooks?

b) What is the revenue when 10,000 textbooks are sold?

c) For what number of textbooks is the cost equal to the revenue?

d) The cost of producing zero textbooks is called the *fixed cost*. Find the fixed cost.

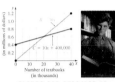

FIGURE FOR EXERCISE 57

58. Free market. The function $S = 5000 + 200x$ and $D = 9500 - 100x$ express the supply S and the demand D, respectively, for a popular compact disk brand as a function of its price x (in dollars).

a) Graph the functions on the same coordinate system.

b) What happens to the supply as the price increases?

c) What happens to the demand as the price increases?

d) The price at which supply and demand are equal is called the *equilibrium price*. What is the equilibrium price?

GETTING MORE INVOLVED

59. Discussion. Which of the following equations is not equivalent to $2x - 3y = 6$?

a) $3y - 2x = 6$ b) $y = \frac{2}{3}x - 2$

c) $x = \frac{3}{2}y + 3$ d) $2(x - 5) = 3y - 4$

60. Discussion. Which of the following equations is inconsistent with the equation $3x + 4y = 8$?

a) $y = \frac{3}{4}x + 2$ b) $6x + 8y = 16$

c) $y = -\frac{3}{4}x + 8$ d) $3x - 4y = 8$

GRAPHING CALCULATOR EXERCISES

61. Solve each system by graphing each pair of equations on a graphing calculator and using the trace feature or intersect feature to estimate the point of intersection. Find the coordinates of the intersection to the nearest tenth.

a) $y = 3.5x - 7.2$ b) $2.3x - 4.1y = 3.3$
$\ \ \ y = -2.3x + 9.1$ $3.4x + 9.2y = 1.3$

In this section

- The Addition Method
- Equations Involving Fractions or Decimals
- Applications

4.2 THE ADDITION METHOD

In Section 4.1 you used substitution to eliminate a variable in a system of equations. In this section we see another method for eliminating a variable in a system of equations.

The Addition Method

In the **addition method** we eliminate a variable by adding the equations.

EXAMPLE 1   **An independent system solved by addition**

Solve the system by the addition method:

$$3x - 5y = -9$$
$$4x + 5y = 23$$

---

95. $\left(-\frac{1}{16}\right)^{-3/4}$ **96.** $\left(\frac{9}{16}\right)^{-1/2}$

97. $(9x^8)^{1/2}$ **98.** $(-27x^9)^{1/3}$

99. $(3a^{-2/3})^{-3}$ **100.** $(5x^{-1/2})^{-2}$

101. $(a^{1/2}b)^{1/2}(ab^{1/2})$ **102.** $(m^{1/4}n^{1/2})^2(m^2n^3)^{1/2}$

103. $(km^{1/2})^3(k^3m^5)^{1/2}$ **104.** $(tv^{1/3})^2(t^2v^{-3})^{-1/2}$

 Use a scientific calculator with a power key (x^y) to find the decimal value of each expression. Round answers to four decimal places.

105. $2^{1/3}$ **106.** $5^{1/2}$

107. $-2^{1/2}$ **108.** $(-3)^{1/3}$

109. $1024^{1/10}$ **110.** $7776^{0.2}$

111. $8^{0.33}$ **112.** $289^{0.5}$

113. $\left(\frac{64}{15,625}\right)^{-1/6}$ **114.** $\left(\frac{32}{243}\right)^{-3/5}$

Simplify each expression. Assume a and b are positive real numbers and m and n are rational numbers.

115. $a^{m/2} \cdot a^{m/4}$ **116.** $b^{n/2} \cdot b^{-n/3}$

117. $\dfrac{a^{-m/5}}{a^{-m/3}}$ **118.** $\dfrac{b^{-n/4}}{b^{-n/3}}$

119. $(a^{-1/m}b^{-1/n})^{-mn}$ **120.** $(a^{-m/2}b^{-n/3})^{-6}$

121. $\left(\dfrac{a^{-3m}b^{-6m}}{a^{9m}}\right)^{-1/3}$ **122.** $\left(\dfrac{a^{-3/m}b^{9/n}}{a^{-6/m}b^{3/n}}\right)^{-1/3}$

In Exercises 123–130, solve each problem.

123. Diagonal of a box. The length of the diagonal of a box can be found from the formula

$$D = (L^2 + W^2 + H^2)^{1/2},$$

where L, W, and H represent the length, width, and height of the box, respectively. If the box is 12 inches long, 4 inches wide, and 3 inches high, then what is the length of the diagonal?

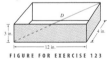

FIGURE FOR EXERCISE 123

124. Radius of a sphere. The radius of a sphere is a function of its volume, given by the formula

$$r = \left(\frac{0.75V}{\pi}\right)^{1/3}.$$

Find the radius of a spherical tank that has a volume of $\frac{32\pi}{3}$ cubic meters.

FIGURE FOR EXERCISE 124

125. Maximum sail area. According to the new International America's Cup Class Rules, the maximum sail area in square meters for a yacht in the America's Cup race is given by

$$S = (13.0368 + 7.84D^{1/3} - 0.8L)^2,$$

where D is the displacement in cubic meters (m^3), and L is the length in meters (m). (*Scientific American*, May 1992). Find the maximum sail area for a boat that has a displacement of 18.42 m^3 and a length of 21.45 m.

FIGURE FOR EXERCISE 125

126. Orbits of the planets. According to Kepler's third law of planetary motion, the average radius R of the orbit of a planet around the sun is determined by $R = T^{2/3}$, where T is the number of years for one orbit and R is measured in astronomical units or AUs (Windows to the Universe, www.windows.umich.edu).

a) It takes Mars 1.881 years to make one orbit of the sun. What is the average radius (in AUs) of the orbit of Mars?

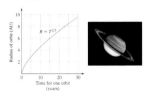

FIGURE FOR EXERCISE 126

WRAP-UP   **CHAPTER 5**

SUMMARY

Definitions		Examples
Definition of negative integral exponents	If a is a nonzero real number and n is a positive integer, then $$a^{-n} = \frac{1}{a^n}.$$	$2^{-3} = \frac{1}{2^3} = 8$
Definition of zero exponent	If a is any nonzero real number, then $a^0 = 1$. The expression 0^0 is undefined.	$3^0 = 1$

Rules of Exponents		Examples
If a and b are nonzero real numbers and m and n are integers, then the following rules hold.		
Negative exponent rules	$a^{-n} = \left(\frac{1}{a}\right)^n$, $a^{-1} = \frac{1}{a}$, and $\frac{1}{a^{-n}} = a^n$ Find the power and reciprocal in either order.	$5^{-1} = \frac{1}{5}$, $\frac{1}{5^{-3}} = 5^3$ $\left(\frac{2}{3}\right)^{-2} = \left(\frac{3}{2}\right)^2$
Product rule	$a^m \cdot a^n = a^{m+n}$	$3^5 \cdot 3^7 = 3^{12}$, $2^{-3} \cdot 2^{10} = 2^7$
Quotient rule	$\frac{a^m}{a^n} = a^{m-n}$	$\frac{x^8}{x^5} = x^3$, $\frac{5^4}{5^{-7}} = 5^{11}$
Power of a power rule	$(a^m)^n = a^{mn}$	$(5^2)^3 = 5^6$
Power of a product rule	$(ab)^n = a^n b^n$	$(2x)^3 = 8x^3$ $(2x^3)^4 = 16x^{12}$
Power of a quotient rule	$\left(\frac{a}{b}\right)^n = \frac{a^n}{b^n}$	$\left(\frac{x}{3}\right)^2 = \frac{x^2}{9}$

Scientific Notation		Examples
Converting from scientific notation	1. Determine the number of places to move the decimal point by examining the exponent on the 10. 2. Move to the right for a positive exponent and to the left for a negative exponent.	$4 \times 10^3 = 4000$ $3 \times 10^{-4} = 0.0003$

Wrap-up

Every chapter ends with a four-part **Wrap-up:**

The **Summary** lists important concepts along with brief illustrative examples.

ENRICHING YOUR MATHEMATICAL WORD POWER

For each mathematical term, choose the correct meaning.

1. nth root of a
   a. a square root
   b. the root of a^n
   c. a number b such that $a^n = b$
   d. a number b such that $b^n = a$

2. square of a
   a. a number b such that $b^2 = a$
   b. a^2
   c. $|a|$
   d. $\sqrt{a}$

3. cube root of a
   a. a^3
   b. a number b such that $b^3 = a$
   c. $a/3$
   d. a number b such that $b = a^3$

4. principal root
   a. the main root
   b. the positive even root of a positive number
   c. the positive odd root of a negative number
   d. the negative odd root of a negative number

5. odd root of a
   a. the number b such that $b^n = a$, where a is an odd number
   b. the opposite of the even root of a
   c. the nth root of a
   d. the number b such that $b^n = a$, where n is an odd number

6. index of a radical
   a. the number n in $n\sqrt{a}$
   b. the number n in $\sqrt[n]{a}$
   c. the number n in a^n
   d. the number n in $\sqrt{a^n}$

7. like radicals
   a. radicals with the same index
   b. radicals with the same radicand
   c. radicals with the same radicand and the same index
   d. radicals with even indices

8. integral exponent
   a. an exponent that is an integer
   b. a positive exponent
   c. a rational exponent
   d. a fractional exponent

9. rational exponent
   a. an exponent that produces a rational number
   b. an integral exponent
   c. an exponent that is a real number
   d. an exponent that is a rational number

10. radicand
   a. the expression $\sqrt[n]{a}$
   b. the expression $\sqrt{a}$
   c. the number a in $\sqrt[n]{a}$
   d. the number n in $\sqrt[n]{a}$

11. complex numbers
   a. $a + bi$, where a and b are real
   b. irrational numbers
   c. imaginary numbers
   d. $\sqrt{-1}$

12. imaginary unit
   a. 1
   b. -1
   c. i
   d. $\sqrt{1}$

13. imaginary number
   a. $a + bi$, where a and b are real
   b. i
   c. a complex number
   d. a complex number in which $b \neq 0$

14. complex conjugates
   a. i and $\sqrt{-1}$
   b. $a + bi$ and $a - bi$
   c. $(a + b)(a - b)$
   d. i and -1

REVIEW EXERCISES

7.1 *Simplify the expressions involving rational exponents. Assume all variables represent positive real numbers. Write your answers with positive exponents.*

1. $(-27)^{-2/3}$
2. $-25^{1/2}$
3. $(2^6)^{1/3}$
4. $(5^2)^{1/2}$
5. $100^{-3/2}$
6. $1000^{-2/3}$
7. $\frac{3x^{-1/2}}{3^{-2}x^{-1}}$
8. $\frac{(x^2y^{-3}z)^{1/2}}{x^{1/2}yz^{-1/2}}$
9. $(a^{1/2}b)^3(ab^{1/4})^2$
10. $(t^{-1/2})^{-2}(t^{-5}y^2)$

Enriching Your Mathematical Word Power enables students to review terms introduced in each chapter. It is intended to help reinforce students' command of mathematical terminology.

Review Exercises contain problems that are keyed to each section of the chapter as well as *miscellaneous exercises* that are not keyed to the sections. These *exercises* are designed to test the student's ability to synthesize various concepts.

The Chapter Test

The Chapter Test is designed to help the student assess his or her readiness for a test. Since the Chapter Test has no keyed exercises, students are given an opportunity to synthesize concepts found within the chapter.

---

49. Beets and beans. One serving of canned beets contains 1 gram of protein and 6 grams of carbohydrates. One serving of canned red beans contains 6 grams of protein and 20 grams of carbohydrates. How many servings of each would it take to get exactly 21 grams of protein and 78 grams of carbohydrates?

FIGURE FOR EXERCISE 48

CHAPTER 4 TEST

Solve the system by graphing.

1. $x + y = 4$
$y = 2x + 1$

Solve each system by substitution.

2. $y = 2x - 8$
$4x + 3y = 1$

3. $y = x - 5$
$3x - 4(y - 2) = 28 - x$

Solve each system by the addition method.

4. $3x + 2y = 3$
$4x - 3y = -13$

5. $3x - \quad = y = 5$
$-6x + 2y = 1$

Determine whether each system is independent, inconsistent, or dependent.

6. $y = 3x - 5$
$y = 3x + 2$

7. $2x + 2y = 8$
$x + \quad y = 4$

8. $y = 2x - 3$
$y = 5x - 14$

Solve the following system by elimination of variables.

9. $x + \quad y - \quad z = 2$
$2x - \quad y + 3z = -5$
$x - 3y + \quad z = 4$

Solve by the Gaussian elimination method.

10. $3x - \quad y = 1$
$x + 2y = 12$

11. $x - \quad y - \quad z = 1$
$-x - \quad y + 2z = -2$
$-x - 3y + \quad z = -5$

Evaluate each determinant.

12. $\begin{vmatrix} 2 & 3 \\ 4 & -3 \end{vmatrix}$

13. $\begin{vmatrix} 1 & -2 & -1 \\ 2 & 3 & 1 \\ 1 & 1 & 0 \end{vmatrix}$

study tip

Before you take an in-class exam on this chapter, work the sample test given here. Set aside one hour to work this test and use the answers in the back of this book to grade yourself. Even though your instructor might not ask exactly the same questions, you will get a good idea of your test readiness.

Solve each system by using Cramer's rule.

14. $2x - y = -4$
$3x + y = -1$

15. $x + y \quad = 0$
$x - y + 2z = 6$
$2x + y - \quad z = 1$

For each problem, write a system of equations in two or three variables. Use the method of your choice to solve each system.

16. One night the manager of the Sea Breeze Motel rented 5 singles and 12 doubles for a total of $390. The next night he rented 9 singles and 10 doubles for a total of $412. What is the rental charge for each type of room?

17. Jill, Karen, and Betsy studied a total of 93 hours last week. Jill's and Karen's study time totaled only one-half as much as Betsy's. If Jill studied 3 hours more than Karen, then how many hours did each one of the girls spend studying?

Solve the following problem by linear programming.

18. Find the maximum value of the function
$$P(x, y) = 8x + 10y$$
subject to the following constraints:
$$x \geq 0, y \geq 0$$
$$2x + 3y \leq 12$$
$$x + \quad y \leq 5$$

MAKING CONNECTIONS   CHAPTERS 1–8

Solve each equation.

1. $2x - 15 = 0$

2. $2x^2 - 15 = 0$

3. $2x^2 + x - 15 = 0$

4. $2x^2 + 4x - 15 = 0$

5. $|4x + 11| = 3$

6. $|4x^2 + 11x| = 3$

7. $\sqrt{x} = x - 6$

8. $(2x - 5)^{2/3} = 4$

Solve each inequality.

9. $1 - 2x < 5 - x$

10. $(1 - 2x)(5 - x) \leq 0$

11. $\dfrac{1 - 2x}{5 - x} \leq 0$

12. $|5 - x| < 3$

13. $3x - 1 < 5$ and $-3 \leq x$

14. $x - 3 < 1$ or $2x \geq 8$

Solve each equation for y.

15. $2x - 3y = 9$

16. $\dfrac{y - 3}{x + 2} = -\dfrac{1}{2}$

17. $3y^2 + cy + d = 0$

18. $my^2 - ny = w$

19. $\dfrac{1}{3}x - \dfrac{2}{5}y = \dfrac{5}{6}$

20. $y - 3 = -\dfrac{2}{3}(x - 4)$

Let $m = \dfrac{y_2 - y_1}{x_2 - x_1}$. Find the value of m for each of the following choices of x_1, x_2, y_1, and y_2.

21. $x_1 = 2, x_2 = 5, y_1 = 3, y_2 = 7$

22. $x_1 = -3, x_2 = 4, y_1 = 5, y_2 = -6$

23. $x_1 = 0.3, x_2 = 0.5, y_1 = 0.8, y_2 = 0.4$

24. $x_1 = \dfrac{1}{2}, x_2 = \dfrac{1}{3}, y_1 = \dfrac{3}{5}, y_2 = -\dfrac{4}{3}$

Solve each problem.

25. *Ticket prices.* In the summer of 1994 the rock group Pearl Jam testified before a congressional committee that Ticketmaster was unfairly raising the prices of the group's concert tickets. One member of the group stated that fans should not have to pay more than $20 to see Pearl Jam. Of course, for any concert, as ticket prices rise, the number of tickets sold decreases, as shown in the figure. If you use the formula $n = 48,000 - 400p$ to predict the number sold depending on the price p, then how many will be sold at $20 per ticket? How many will be sold at $25 per ticket? Use the bar graph to estimate the price if 35,000 tickets were sold.

FIGURE FOR EXERCISE 25

26. *Increasing revenue.* Even though the number of tickets sold for a concert decreases with increasing price, the revenue generated does not necessarily decrease. Use the formula $R = p(48,000 - 400p)$ to determine the revenue when the price is $20 and when the price is $25. What price would produce a revenue of $1.28 million? Use the graph to find the price that determines the maximum revenue.

FIGURE FOR EXERCISE 26

Making Connections

These nonkeyed exercises are designed to help students synthesize new material with ideas from previous chapters, and in some cases to review material necessary for success in the upcoming chapter. They may serve as a cumulative review.

MARK DUGOPOLSKI

INTERMEDIATE

Algebra

3RD EDITION

www.mhhe.com/dugopolski

FREE CD INCLUDED

McGRAW-HILL IS PROUD TO OFFER AN EXCITING NEW SUITE OF MULTIMEDIA PRODUCTS AND SERVICES CALLED COURSE SOLUTIONS.

Designed specifically to help you with your individual course needs, **Course Solutions** will assist you in integrating your syllabus with our premier titles and state-of-the-art new media tools that support them.

AT THE HEART OF COURSE SOLUTIONS YOU'LL FIND:

- Fully integrated multimedia
- A full-scale Online Learning Center
- A Course Integration Guide

AS WELL AS THESE UNPARALLELED SERVICES:

- McGraw-Hill Learning Architecture
- McGraw-Hill Course Consultant Service
- McGraw-Hill Student Tutorial Service
- McGraw-Hill Instructor Syllabus Service
- PageOut Lite
- PageOut: The Course Web Site Development Center
- Other Delivery Options

COURSE SOLUTIONS truly has the solutions to your every teaching need. Read on to learn how we can specifically help you with your classroom challenges.

SPECIAL ATTENTION

to your specific needs.

McGRAW-HILL LEARNING ARCHITECTURE

Each McGraw-Hill *Online Learning Center* is ready to be ported into our *McGraw-Hill Learning Architecture*—a full course management software system for Local Area Networks and Distance Learning Classes. Developed in conjunction with Top Class software, *McGraw-Hill Learning Architecture* is a powerful course management system available upon special request.

McGRAW-HILL COURSE CONSULTANT SERVICE

In addition to the *Course Integration Guide,* instructors using **Course Solutions** textbooks can access a special curriculum-based *Course Consultant Service* via a web-based threaded discussion list within each *Online Learning Center.* A **McGraw-Hill Course Solutions Consultant** will personally help you—as a text adopter—integrate this text and media into your course to fit your specific needs. This content-based service is offered in addition to our usual software support services.

McGRAW-HILL INSTRUCTOR SYLLABUS SERVICE

For *new* adopters of **Course Solutions** textbooks, McGraw-Hill will help correlate all text, supplement, and appropriate materials and services to your course syllabus. Simply call your McGraw-Hill sales representative for assistance.

PAGEOUT LITE

Free to **Course Solutions** textbook adopters, *PageOut Lite* is perfect for instructors who want to create their own Web site. In just a few minutes, even novices can turn their syllabus into a Web site using *PageOut Lite.*

PAGEOUT: THE COURSE WEB SITE DEVELOPMENT CENTER

For those that want the benefits of *PageOut Lite's* no-hassle approach to site development, but with even more features, we offer *PageOut: The Course Web Site Development Center.*

   PageOut shares many of *PageOut Lite's* features, but also enables you to create links that will take your students to your original material, other Web site addresses, and to *McGraw-Hill Online Learning Center* content. This means you can assign *Online Learning Center* content within your syllabus-based Web site. *PageOut's* gradebook function will tell you when each student has taken a quiz or worked through an exercise, automatically recording their scores for you. *PageOut* also features a discussion board list where you and your students can exchange questions and post announcements, as well as an area for students to build personal Web pages.

OTHER DELIVERY OPTIONS

Online Learning Centers are also compatible with a number of full-service online course delivery systems or outside educational service providers. For a current list of compatible delivery systems, contact your McGraw-Hill sales representative.

And for your students …

McGRAW-HILL STUDENT TUTORIAL SERVICE

Within each *Online Learning Center* resides a **FREE** *Student Tutorial Service.* This web-based "homework hotline"—available via a threaded discussion list—features guaranteed, 24-hour response time on weekdays.

www.mhhe.com/dugopolski

CHAPTER 1

The Real Numbers

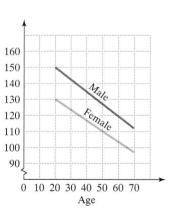

E verywhere you look people are running, riding, dancing, and exercising their way to fitness. In the past year more than $25 billion has been spent on sports equipment alone, and this amount is growing steadily.

Proponents of exercise claim that it can increase longevity, improve body image, decrease appetite, and generally enhance a person's health. While many sports activities can help you to stay fit, experts have found that aerobic, or dynamic, workouts provide the most fitness benefit. Some of the best aerobic exercises include cycling, running, and even jumping rope. Whatever athletic activity you choose, trainers recommend that you set realistic goals and work your way toward them consistently and slowly. To achieve maximum health benefits, experts suggest that you exercise three to five times a week for 15 to 60 minutes at a time.

There are many different ways to measure exercise. One is to measure the energy used, or the rate of oxygen consumption. Since heart rate rises as a function of increased oxygen, another easier measure of intensity of exercise is your heart rate during exercise. The desired heart rate, or target heart rate, for beneficial exercise varies for each individual depending on conditioning, age, and gender. In Exercises 101 and 102 of Section 1.4 you will see how an algebraic expression can determine your target heart rate for beneficial exercise.

1.1 SETS

Every subject has its own terminology, and **algebra** is no different. In this section we will learn the basic terms and facts about sets.

Set Notation

A **set** is a collection of objects. At home you may have a set of dishes and a set of steak knives. In algebra we generally discuss sets of numbers. For example, we refer to the numbers 1, 2, 3, 4, 5, and so on as the set of **counting numbers** or **natural numbers.** Of course, these are the numbers that we use for counting.

The objects or numbers in a set are called the **elements** or **members** of the set. To describe sets with a convenient notation, we use braces, { }, and name the sets with capital letters. For example,

$$A = \{1, 2, 3\}$$

means that set A is the set whose members are the natural numbers 1, 2, and 3. The letter N is used to represent the entire set of natural numbers.

A set that has a fixed number of elements such as $\{1, 2, 3\}$ is a **finite** set, whereas a set without a fixed number of elements such as the natural numbers is an **infinite** set. When listing the elements of a set, we use a series of three dots to indicate a continuing pattern. For example, the set of natural numbers is written as

$$N = \{1, 2, 3, \ldots\}.$$

The set of natural numbers *between* 4 and 40 can be written

$$\{5, 6, 7, 8, \ldots, 39\}.$$

Note that since the members of this set are *between* 4 and 40, it does not include 4 or 40.

Set-builder notation is another method of describing sets. In this notation we use a variable to represent the numbers in the set. A **variable** is a letter that is used to stand for some numbers. The set is then built from the variable and a description of the numbers that the variable represents. For example, the set

$$B = \{1, 2, 3, \ldots, 49\}$$

is written in set-builder notation as

$$B = \{x \mid x \text{ is a natural number less than } 50\}.$$

$\qquad\quad\uparrow\ \uparrow\qquad\qquad\qquad\qquad\quad\uparrow$

The set of numbers   such that   condition for membership

This notation is read as "B is the set of numbers x such that x is a natural number less than 50." Notice that the number 50 is not a member of set B.

The symbol $\in$ is used to indicate that a specific number is a member of a set, and $\notin$ indicates that a specific number is not a member of a set. For example, the statement $1 \in B$ is read as "1 is a member of B," "1 belongs to B," "1 is in B," or "1 is an element of B." The statement $0 \notin B$ is read as "0 is not a member of B," "0 does not belong to B," "0 is not in B," or "0 is not an element of B."

Two sets are **equal** if they contain exactly the same members. Otherwise, they are said to be not equal. To indicate equal sets, we use the symbol $=$. For sets that are not equal we use the symbol $\neq$. The elements in two equal sets do not need to be written in the same order. For example, $\{3, 4, 7\} = \{3, 4, 7\}$ and $\{2, 4, 1\} = \{1, 2, 4\}$, but $\{3, 5, 6\} \neq \{3, 5, 7\}$.

E X A M P L E   1

Set notation

Let $A = \{1, 2, 3, 5\}$ and $B = \{x \mid x$ is an even natural number less than 10$\}$. Determine whether each statement is true or false.

a) $3 \in A$ b) $5 \in B$ c) $4 \notin A$ d) $A = N$

e) $A = \{x \mid x$ is a natural number less than 6$\}$ f) $B = \{2, 4, 6, 8\}$

Solution

a) True, because 3 is a member of set A.

b) False, because 5 is not an even natural number.

c) True, because 4 is not a member of set A.

d) False, because A does not contain all of the natural numbers.

e) False, because 4 is a natural number less than 6, and $4 \notin A$.

f) True, because the even counting numbers less than 10 are 2, 4, 6, and 8.   ■

Union of Sets

Any two sets A and B can be combined to form a new set called their union that consists of all elements of A together with all elements of B.

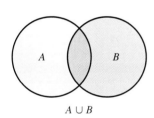

$A \cup B$

FIGURE 1.1

Union of Sets

If A and B are sets, the **union** of A and B, denoted $A \cup B$, is the set of all elements that are either in A, in B, or in both. In symbols,

$$A \cup B = \{x \mid x \in A \text{ or } x \in B\}.$$

In mathematics the word "or" is always used in an inclusive manner (allowing the possibility of both alternatives). The diagram in Fig. 1.1 can be used to illustrate $A \cup B$. Any point that lies within circle A, circle B, or both is in $A \cup B$. Diagrams (like Fig. 1.1) that are used to illustrate sets are called **Venn diagrams.**

E X A M P L E   2

helpful   hint

To remember what "union" means think of a labor union, which is a group formed by joining together many individuals.

Union of sets

Let $A = \{0, 2, 3\}$, $B = \{2, 3, 7\}$, and $C = \{7, 8\}$. List the elements in each of the following sets.

a) $A \cup B$ b) $A \cup C$

Solution

a) $A \cup B$ is the set of numbers that are in A, in B, or in both A and B.

$$A \cup B = \{0, 2, 3, 7\}$$

b) $A \cup C = \{0, 2, 3, 7, 8\}$ ■

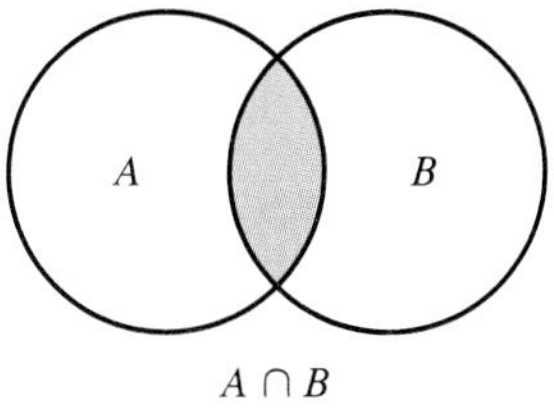

$A \cap B$

FIGURE 1.2

Intersection of Sets

Another way to form a new set from two known sets is by considering only those elements that the two sets have in common. The diagram shown in Fig. 1.2 illustrates the intersection of two sets A and B.

Intersection of Sets

If A and B are sets, the **intersection** of A and B, denoted $A \cap B$, is the set of all elements that are in both A and B. In symbols,

$$A \cap B = \{x \mid x \in A \text{ and } x \in B\}.$$

It is possible for two sets to have no elements in common. A set with no members is called the **empty set** and is denoted by the symbol $\varnothing$. Note that $A \cup \varnothing = A$ and $A \cap \varnothing = \varnothing$ for any set A.

CAUTION   The set $\{0\}$ is not the empty set. The set $\{0\}$ has one member, the number 0. Do not use the number 0 to represent the empty set.

E X A M P L E 3

Intersection of sets

Let $A = \{0, 2, 3\}$, $B = \{2, 3, 7\}$, and $C = \{7, 8\}$. List the elements in each of the following sets.

a) $A \cap B$ **b)** $B \cap C$ **c)** $A \cap C$

Solution

a) $A \cap B$ is the set of all numbers that are in both A and B. So $A \cap B = \{2, 3\}$.

b) $B \cap C = \{7\}$ **c)** $A \cap C = \varnothing$ ■

E X A M P L E 4

Membership and equality

Let $A = \{1, 2, 3, 5\}$, $B = \{2, 3, 7, 8\}$, and $C = \{6, 7, 8, 9\}$. Place one of the symbols $=$, $\neq$, $\in$, or $\notin$ in the blank to make each statement correct.

a) $5 \underline{\hspace{1cm}} A \cup B$ **b)** $5 \underline{\hspace{1cm}} A \cap B$

c) $A \cup B \underline{\hspace{1cm}} \{1, 2, 3, 5, 7, 8\}$ **d)** $A \cap B \underline{\hspace{1cm}} \{2\}$

Solution

a) $5 \in A \cup B$ because 5 is a member of A.

b) $5 \notin A \cap B$ because 5 must belong to *both* A and B to be a member of $A \cap B$.

c) $A \cup B = \{1, 2, 3, 5, 7, 8\}$ because the elements of A together with those of B are listed. Note that 2 and 3 are members of both sets but are listed only once.

d) $A \cap B \neq \{2\}$ because $A \cap B = \{2, 3\}$.   ■

Subsets

If every member of set A is also a member of set B, then we write $A \subseteq B$ and say that A is a **subset** of B. See Fig. 1.3. For example,

$$\{2, 3\} \subseteq \{2, 3, 4\}.$$

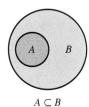

$A \subseteq B$

F I G U R E 1 . 3

If set A is not a subset of B, we write $A \nsubseteq B$.

CAUTION   To claim that $A \nsubseteq B$, there *must* be an element of A that does *not* belong to B. For example,

$$\{1, 2\} \nsubseteq \{2, 3, 4\}$$

because 1 is a member of the first set but not of the second.

Is the empty set $\varnothing$ a subset of $\{2, 3, 4\}$? If we say that $\varnothing$ is *not* a subset of $\{2, 3, 4\}$, then there must be an element of $\varnothing$ that does not belong to $\{2, 3, 4\}$. But that cannot happen because $\varnothing$ is empty. So $\varnothing$ is a subset of $\{2, 3, 4\}$. In fact, by the same reasoning, *the empty set is a subset of every set.*

EXAMPLE   5

Subsets

Determine whether each statement is true or false.

a) $\{1, 2, 3\}$ is a subset of the set of natural numbers.

b) The set of natural numbers is not a subset of $\{1, 2, 3\}$.

c) $\{1, 2, 3\} \not\subseteq \{2, 4, 6, 8\}$

d) $\{2, 6\} \subseteq \{1, 2, 3, 4, 5\}$

e) $\varnothing \subseteq \{2, 4, 6\}$

Solution

a) True, because 1, 2, and 3 are natural numbers.

b) True, because 5, for example, is a natural number and $5 \notin \{1, 2, 3\}$.

c) True, because 1 is in the first set but not in the second.

d) False, because 6 is in the first set but not in the second.

e) True, because we cannot find anything in $\varnothing$ that fails to be in $\{2, 4, 6\}$. ◼

Combining Three or More Sets

We know how to find the union and intersection of two sets. For three or more sets we use parentheses to indicate which pair of sets to combine first. In the following example, notice that different results are obtained from different placements of the parentheses.

EXAMPLE   6

Operations with three sets

Let $A = \{1, 2, 3, 4\}$, $B = \{2, 5, 6, 8\}$, and $C = \{4, 5, 7\}$. List the elements of each of the following sets.

a) $(A \cup B) \cap C$ **b)** $A \cup (B \cap C)$

Solution

a) The parentheses indicate that the union of A and B is to be found first and then the result, $A \cup B$, is to be intersected with C.

$$A \cup B = \{1, 2, 3, 4, 5, 6, 8\}$$

Now examine $A \cup B$ and C to find the elements that belong to both sets:

$$A \cup B = \{1, 2, 3, 4, 5, 6, 8\}$$
$$C = \{4, 5, 7\}$$

The only numbers that are members of $A \cup B$ and C are 4 and 5. Thus

$$(A \cup B) \cap C = \{4, 5\}.$$

b) In $A \cup (B \cap C)$, first find $B \cap C$:

$$B \cap C = \{5\}$$

Now $A \cup (B \cap C)$ consist of all members of A together with 5 from $B \cap C$:

$$A \cup (B \cap C) = \{1, 2, 3, 4, 5\}$$ ◼

True or false? Explain your answer.

Let $A = \{1, 2, 3, 4\}$, $B = \{3, 4, 5\}$, and $C = \{3, 4\}$.

1. $A = \{x \mid x$ is a counting number$\}$ False

2. The set B has an infinite number of elements.   False

3. The set of counting numbers less than 50 million is an infinite set.   False

4. $1 \in A \cap B$ False   **5.** $3 \in A \cup B$ True   **6.** $A \cap B = C$ True

7. $C \subseteq B$ True   **8.** $A \subseteq B$ False   **9.** $\varnothing \subseteq C$ True

10. $A \nsubseteq C$ True

1.1   EXERCISES

Reading and Writing   *After reading this section, write out the answers to these questions. Use complete sentences.*

1. What is a set?   A set is a collection of objects.

2. What is the difference between a finite set and an infinite set?
A finite set has a fixed number of elements and an infinite set does not.

3. What is a Venn diagram used for?
A Venn diagram is used to illustrate relationships between sets.

4. What is the difference between the intersection and the union of two sets?
The intersection of two sets consists of elements that are in both sets, whereas the union of two sets consists of elements that are in one, in the other, or in both sets.

5. What does it mean to say that set A is a subset of set B?
Every member of set A is also a member of set B.

6. Which set is a subset of every set?
The empty set is a subset of every set.

For Exercises 7–52 let N represent the natural numbers, $A = \{x \mid x$ is an odd counting number smaller than 10\}, $B = \{2, 4, 6, 8\}$, and $C = \{1, 2, 3, 4, 5\}$.

Determine whether each statement is true or false. Explain your answer. See Example 1.

7. $6 \in A$
False

8. $8 \in A$
False

9. $A \neq B$
True

10. $A = \{1, 3, 5, 7, \ldots\}$
False

11. $3 \in C$
True

12. $4 \notin B$
False

13. $A = \{1, 3, 7, 9\}$
False

14. $B \neq C$
True

15. $0 \in N$
False

16. $2.5 \in N$
False

17. $C = N$
False

18. $N = A$
False

List the elements in each set. If the set is empty, write $\varnothing$. See Examples 2 and 3.

19. $A \cap B$
$\varnothing$

20. $A \cup B$
$\{1, 2, 3, 4, 5, 6, 7, 8, 9\}$

21. $A \cap C$
$\{1, 3, 5\}$

22. $A \cup C$
$\{1, 2, 3, 4, 5, 7, 9\}$

23. $B \cup C$
$\{1, 2, 3, 4, 5, 6, 8\}$

24. $B \cap C$
$\{2, 4\}$

25. $A \cup \varnothing$ A **26.** $B \cup \varnothing$ B **27.** $A \cap \varnothing$ $\varnothing$

28. $B \cap \varnothing$ $\varnothing$ **29.** $A \cap N$ A **30.** $A \cup N$ N

Use one of the symbols $\in$, $\notin$, $=$, $\neq$, $\cup$, or $\cap$ in the blank of each statement to make it correct. See Example 4.

31. $A \cap B$ _____ $\varnothing$ $=$ **32.** $A \cap C$ _____ $\varnothing$ $\neq$

33. A _____ $B = \{1, 2, 3, 4, 5, 6, 7, 8, 9\}$ $\cup$

34. A _____ $B = \varnothing$ $\cap$ **35.** B _____ $C = \{2, 4\}$ $\cap$

36. B _____ $C = \{1, 2, 3, 4, 5, 6, 8\}$ $\cup$

37. 3 _____ $A \cap B$ $\notin$ **38.** 3 _____ $A \cap C$ $\in$

39. 4 _____ $B \cap C$ $\in$ **40.** 8 _____ $B \cup C$ $\in$

Determine whether each statement is true or false. Explain your answer. See Example 5.

41. $A \subseteq N$ True   **42.** $B \subseteq N$ True

43. $\{2, 3\} \subseteq C$ True   **44.** $C \subseteq A$ False

45. $B \nsubseteq C$ True   **46.** $C \nsubseteq A$ True

47. $\varnothing \subseteq B$ True   **48.** $\varnothing \subseteq C$ True

49. $A \subseteq \varnothing$ False   **50.** $B \subseteq \varnothing$ False

51. $A \cap B \subseteq C$ True   **52.** $B \cap C \subseteq \{2, 4, 6, 8\}$ True

For Exercises 53–78, let $D = \{3, 5, 7\}$, $E = \{2, 4, 6, 8\}$, and $F = \{1, 2, 3, 4, 5\}$.

List the elements in each set. If the set is empty, write $\varnothing$. See Example 6.

53. $D \cup E$
$\{2, 3, 4, 5, 6, 7, 8\}$

54. $D \cap E$
$\varnothing$

55. $D \cap F$
$\{3, 5\}$

56. $D \cup F$
$\{1, 2, 3, 4, 5, 7\}$

57. $E \cup F$
$\{1, 2, 3, 4, 5, 6, 8\}$

58. $E \cap F$
$\{2, 4\}$

59. $(D \cup E) \cap F$
$\{2, 3, 4, 5\}$

60. $(D \cup F) \cap E$
$\{2, 4\}$

61. $D \cup (E \cap F)$
$\{2, 3, 4, 5, 7\}$

62. $D \cup (F \cap E)$
$\{2, 3, 4, 5, 7\}$

63. $(D \cap F) \cup (E \cap F)$
$\{2, 3, 4, 5\}$

64. $(D \cap E) \cup (F \cap E)$
$\{2, 4\}$

65. $(D \cup E) \cap (D \cup F)$
$\{2, 3, 4, 5, 7\}$

66. $(D \cup F) \cap (D \cup E)$
$\{2, 3, 4, 5, 7\}$

Use one of the symbols $\in$, $\subseteq$, $=$, $\cup$, or $\cap$ in the blank of each statement to make it correct.

67. D _____ $\{x \mid x$ is an odd natural number$\}$ $\subseteq$

68. E _____ $\{x \mid x$ is an even natural number smaller than 9$\}$
$=$

69. 3 _____ D $\in$

70. $\{3\}$ _____ D $\subseteq$

71. D _____ $E = \varnothing$ $\cap$

72. $D \cap E$ _____ D $\subseteq$

73. $D \cap F$ _____ F $\subseteq$

74. $3 \notin E$ _____ F $\cap$

75. $E \not\subseteq E$ _____ F $\cap$

76. $E \subseteq E$ _____ F $\cup$

77. D _____ $F = F \cup D$ $\cup$

78. E _____ $F = F \cap E$ $\cap$

List the elements in each set.

79. $\{x \mid x$ is an even natural number less than 20$\}$
$\{2, 4, 6, \ldots, 18\}$

80. $\{x \mid x$ is a natural number greater than 6$\}$
$\{7, 8, 9, \ldots\}$

81. $\{x \mid x$ is an odd natural number greater than 11$\}$
$\{13, 15, 17, \ldots\}$

82. $\{x \mid x$ is an odd natural number less than 14$\}$
$\{1, 3, 5, \ldots, 13\}$

83. $\{x \mid x$ is an even natural number between 4 and 79$\}$
$\{6, 8, 10, \ldots, 78\}$

84. $\{x \mid x$ is an odd natural number between 12 and 57$\}$
$\{13, 15, 17, \ldots, 55\}$

Write each set using set-builder notation.

85. $\{3, 4, 5, 6\}$
$\{x \mid x$ is a natural number between 2 and 7$\}$

86. $\{1, 3, 5, 7\}$
$\{x \mid x$ is an odd natural number less than 8$\}$

87. $\{5, 7, 9, 11, \ldots\}$
$\{x \mid x$ is an odd natural number greater than 4$\}$

88. $\{4, 5, 6, 7, \ldots\}$
$\{x \mid x$ is a natural number greater than 3$\}$

89. $\{6, 8, 10, 12, \ldots, 82\}$
$\{x \mid x$ is an even natural number between 5 and 83$\}$

90. $\{9, 11, 13, 15, \ldots, 51\}$
$\{x \mid x$ is an odd natural number between 8 and 52$\}$

GETTING MORE INVOLVED

91. *Discussion.* If A and B are finite sets, could $A \cup B$ be infinite? Explain.   No

92. *Cooperative learning.* Work with a small group to answer the following questions. If $A \subseteq B$ and $B \subseteq A$, then what can you conclude about A and B? If $(A \cup B) \subseteq (A \cap B)$, then what can you conclude about A and B? $A = B, A = B$

93. *Discussion.* What is wrong with each statement? Explain.

a) $3 \subseteq \{1, 2, 3\}$ **a)** $3 \in \{1, 2, 3\}$

b) $\{3\} \in \{1, 2, 3\}$ **b)** $\{3\} \subseteq \{1, 2, 3\}$

c) $\varnothing = \{\varnothing\}$ **c)** $\varnothing \neq \{\varnothing\}$

94. *Exploration.* There are only two possible subsets of $\{1\}$, namely, $\varnothing$ and $\{1\}$.

a) List all possible subsets of $\{1, 2\}$. How many are there? $\varnothing, \{1\}, \{2\}, \{1, 2\}$

b) List all possible subsets of $\{1, 2, 3\}$. How many are there?   8

c) Guess how many subsets there are of $\{1, 2, 3, 4\}$. Verify your guess by listing all the possible subsets.   16

d) How many subsets are there for $\{1, 2, 3, \ldots, n\}$? 2^n

---

1.2   THE REAL NUMBERS

The set of real numbers is the basic set of numbers used in algebra. There are many different types of real numbers. To understand better the set of real numbers, we will study some of the subsets of numbers that make up this set.

The Rational Numbers

We have already discussed the set of counting or natural numbers. The set of natural numbers together with the number 0 is called the set of **whole numbers.** The whole numbers together with the negatives of the counting numbers form the set of

In this
section

• The Rational Numbers

• Graphing on the Number Line

• The Irrational Numbers

integers. We use the letters N, W, and J to name these sets:

$$N = \{1, 2, 3, \ldots\} \qquad \text{The natural numbers}$$
$$W = \{0, 1, 2, 3, \ldots\} \qquad \text{The whole numbers}$$
$$J = \{\ldots, -3, -2, -1, 0, 1, 2, 3, \ldots\} \qquad \text{The integers}$$

Rational numbers are numbers that are written as ratios or as quotients of integers. We use the letter Q (for quotient) to name the set of rational numbers and write the set in set-builder notation as follows:

$$Q = \left\{ \frac{a}{b} \middle| a \text{ and } b \text{ are integers, with } b \neq 0 \right\} \qquad \text{The rational numbers}$$

Examples of rational numbers are

$$7, \quad \frac{9}{4}, \quad -\frac{17}{10}, \quad 0, \quad \frac{0}{4}, \quad \frac{3}{1}, \quad -\frac{47}{3}, \quad \text{and} \quad \frac{-2}{-6}.$$

Note that the rational numbers are the numbers that can be expressed as a ratio (or quotient) of integers. The integer 7 is rational because we can write it as $\frac{7}{1}$.

Another way to describe rational numbers is by using their decimal form. To obtain the decimal form, we divide the denominator into the numerator. For some rational numbers the division terminates, and for others it continues indefinitely. The following examples show some rational numbers and their equivalent decimal forms:

$$\frac{26}{100} = 0.26 \qquad \text{Terminating decimal}$$

$$\frac{4}{1} = 4.0 \qquad \text{Terminating decimal}$$

$$\frac{1}{4} = 0.25 \qquad \text{Terminating decimal}$$

$$\frac{2}{3} = 0.6666 \ldots \qquad \text{The single digit 6 repeats.}$$

$$\frac{25}{99} = 0.252525 \ldots \qquad \text{The pair of digits 25 repeats.}$$

$$\frac{4177}{990} = 4.2191919 \ldots \qquad \text{The pair of digits 19 repeats.}$$

Rational numbers are defined as ratios of integers, but they can be described also by their decimal form. *The rational numbers are those decimal numbers whose digits either repeat or terminate.*

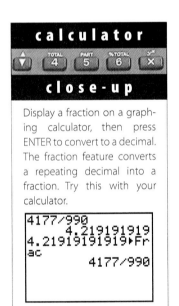

E X A M P L E   1

Subsets of the rational numbers

Determine whether each statement is true or false.

a) $0 \in W$ **b)** $N \subseteq J$ **c)** $0.75 \in J$ **d)** $J \subseteq Q$

Solution

a) True, because 0 is a whole number.

b) True, because every natural number is also a member of the set of integers.

c) False, because the rational number 0.75 is not an integer.

d) True, because the rational numbers include the integers.

Graphing on the Number Line

To construct a number line, we draw a straight line and label any convenient point with the number 0. Now we choose any convenient length and use it to locate points to the right of 0 as points corresponding to the positive integers and points to the left of 0 as points corresponding to the negative integers. See Fig. 1.4. The numbers corresponding to the points on the line are called the **coordinates** of the points. The distance between two consecutive integers is called a **unit,** and it is the same for any two consecutive integers. The point with coordinate 0 is called the **origin.** The numbers on the number line increase in size from left to right. When we compare the size of any two numbers, the larger number lies to the right of the smaller one on the number line.

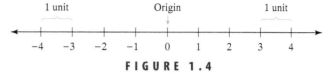

FIGURE 1.4

It is often convenient to illustrate sets of numbers on a number line. The set of integers, J, is illustrated or **graphed** as in Fig. 1.5. The three dots to the right and left on the number line indicate that the integers go on indefinitely in both directions.

FIGURE 1.5

E X A M P L E   2

Graphing on the number line

List the elements of each set and graph each set on a number line.

a) $\{x \mid x$ is a whole number less than $4\}$

b) $\{a \mid a$ is an integer between 3 and $9\}$

c) $\{y \mid y$ is an integer greater than $-3\}$

Solution

a) The whole numbers less than 4 are 0, 1, 2, and 3. Figure 1.6 shows the graph of this set.

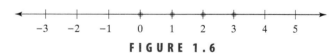

FIGURE 1.6

b) The integers between 3 and 9 are 4, 5, 6, 7, and 8. The graph is shown in Fig. 1.7.

FIGURE 1.7

c) The integers greater than -3 are -2, -1, 0, 1, and so on. To indicate the continuing pattern, we use a series of dots on the graph in Fig. 1.8.

FIGURE 1.8

When we draw a number line, we might label only the integers. But every point on the number line corresponds to a number. The set of all of these numbers is called the set of **real numbers,** *R*. It is easy to locate rational numbers on the number line. For example, the number $\frac{1}{2}$ corresponds to a point halfway between 0 and 1, and $\frac{3}{2}$ corresponds to a point halfway between 1 and 2, as shown in Fig. 1.9. However, on the number line, there are points such as $-\sqrt{2}$ that do not correspond to rational numbers. These points correspond to the set of **irrational numbers,** *I*. *The rational numbers Q and the irrational numbers I have no numbers in common, and together they form the set of real numbers R.*

FIGURE 1.9

The Irrational Numbers

Irrational numbers are those real numbers that cannot be expressed as a ratio of integers. To see an example of an irrational number, consider the positive square root of 2 (in symbols $\sqrt{2}$). The square root of 2 is a number that you can multiply by itself to get 2. So we can write (using a raised dot for times)

$$\sqrt{2} \cdot \sqrt{2} = 2.$$

If we look for $\sqrt{2}$ on a calculator or in Appendix B, we find 1.414. But if we multiply 1.414 by itself, we get

$$(1.414)(1.414) = 1.999396.$$

So $\sqrt{2}$ is not equal to 1.414 (in symbols, $\sqrt{2} \neq 1.414$). The square root of 2 is approximately 1.414 (in symbols, $\sqrt{2} \approx 1.414$). There is no terminating or repeating decimal that will give exactly 2 when multiplied by itself. So $\sqrt{2}$ is an irrational number. It can be shown that other square roots such as $\sqrt{3}$, $\sqrt{5}$, and $\sqrt{7}$ are also irrational numbers.

It is easy to write examples of irrational numbers in decimal form because *as decimal numbers they have digits that neither repeat nor terminate.* Each of the following numbers has a continuing pattern that guarantees that its digits will neither repeat nor terminate:

$$0.606000600000600000006 \ldots$$
$$0.15115111511115 \ldots$$
$$3.12345678910111213 \ldots$$

So each of these numbers is an irrational number.

Since we generally work with rational numbers, the irrational numbers may seem to be unnecessary. However, irrational numbers occur in some very real situations. Over 2000 years ago people in the Orient and Egypt observed that the ratio of the circumference and diameter is the same for any circle. This constant value was proven to be an irrational number by Johann Heinrich Lambert in 1767. Like other irrational numbers, it does not have any convenient representation as a decimal number. This number has been given the name π (Greek letter pi). See Fig. 1.10. The value of π rounded to nine decimal places is 3.141592654. When using π in computations, we frequently use the rational number 3.14 as an approximate value for π.

Figure 1.11 illustrates the subsets of the real numbers that we have been discussing.

calculator
close-up

A calculator gives a 10-digit rational approximation for $\sqrt{2}$. Note that if the approximate value is squared, you do not get 2.

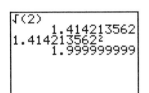

The screen shot that appears on this page and in succeeding pages may differ from the display on your calculator. You may have to consult your manual to get the desired results.

$$\pi = \frac{\text{Circumference}}{\text{Diameter}} \qquad \pi = \frac{C}{D}$$

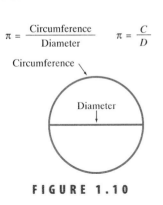

FIGURE 1.10

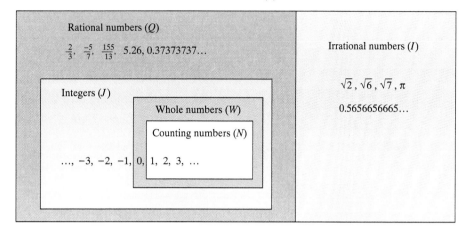

FIGURE 1.11

E X A M P L E   3

Classifying real numbers

Determine which elements of the set

$$\left\{ -\sqrt{7}, -\frac{1}{4}, 0, \sqrt{5}, \pi, 4.16, 12 \right\}$$

are members of each of the following sets.

a) Real numbers   **b)** Rational numbers   **c)** Integers

Solution

a) All of the numbers are real numbers.

b) The numbers $-\frac{1}{4}$, 0, 4.16, and 12 are rational numbers.

c) The only integers in this set are 0 and 12.

E X A M P L E   4

Subsets of the real numbers

Determine whether each of the following statements is true or false.

a) $\sqrt{7} \in Q$ **b)** $J \subseteq W$ **c)** $I \cap Q = \varnothing$ **d)** $-3 \in N$
e) $J \cap I = \varnothing$ **f)** $Q \subseteq R$ **g)** $R \subseteq N$ **h)** $\pi \in R$

Solution

a) False   **b)** False   **c)** True   **d)** False
e) True   **f)** True   **g)** False   **h)** True

WARM-UPS

True or false? Explain your answer.

1. The number π is a rational number.   False

2. The set of rational numbers is a subset of the set of real numbers.   True

3. Zero is the only number that is a member of both Q and I.   False

4. The set of real numbers is a subset of the set of irrational numbers.   False

(*continued*)

5. The decimal number 0.44444 . . . is a rational number.   True
6. The decimal number 4.212112111211112 . . . is a rational number.   False
7. Every irrational number corresponds to a point on the number line.   True
8. If a real number is not irrational, then it is rational.   True
9. The set of counting numbers from 1 through 5 trillion is an infinite set. False
10. There are infinitely many rational numbers.   True

1.2   EXERCISES

Reading and Writing   *After reading this section, write out the answers to these questions. Use complete sentences.*

1. What are the integers?
   The integers consist of the positive and negative counting numbers and zero.

2. What are the rational numbers?
   The rational numbers consist of all numbers that can be expressed as a ratio of integers.

3. What kinds of decimal numbers are rational numbers?
   The repeating or terminating decimal numbers are rational numbers.

4. What kinds of decimal numbers are irrational?
   Decimals that neither repeat nor terminate are irrational numbers.

5. What are the real numbers?
   The set of real numbers is the union of the rational and irrational numbers.

6. What is the ratio of the circumference and diameter of any circle?
   The ratio of the circumference and diameter of any circle is π.

Determine whether each statement is true or false. Explain your answer. See Example 1.

7. $-6 \in Q$ True
8. $\frac{2}{7} \in Q$ True
9. $0 \notin Q$ False
10. $0 \notin N$ True
11. $0.6666 \ldots \in Q$ True
12. $0.00976 \notin Q$ False
13. $N \subseteq Q$ True
14. $Q \subseteq J$ False

List the elements in each set and graph each set on a number line. See Example 2.

15. $\{x \mid x$ is a whole number smaller than 6$\}$
    $\{0, 1, 2, 3, 4, 5\}$

16. $\{x \mid x$ is a natural number less than 7$\}$
    $\{1, 2, 3, 4, 5, 6\}$
    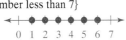

17. $\{a \mid a$ is an integer greater than $-5\}$
    $\{-4, -3, -2, -1, 0, 1, \ldots\}$

18. $\{z \mid z$ is an integer between 2 and 12$\}$
    $\{3, 4, 5, 6, 7, 8, 9, 10, 11\}$

19. $\{w \mid w$ is a natural number between 0 and 5$\}$
    $\{1, 2, 3, 4\}$

20. $\{y \mid y$ is a whole number greater than 0$\}$
    $\{1, 2, 3, 4, 5, \ldots\}$

21. $\{x \mid x$ is an integer between -3 and 5$\}$
    $\{-2, -1, 0, 1, 2, 3, 4\}$

22. $\{y \mid y$ is an integer between -4 and 7$\}$
    $\{-3, -2, -1, 0, 1, 2, 3, 4, 5, 6\}$

Determine which elements of the set

$$A = \left\{ -\sqrt{10}, -3, -\frac{5}{2}, -0.025, 0, \sqrt{2}, 3\frac{1}{2}, \frac{8}{2} \right\}$$

are members of the following sets. See Example 3.

23. Real numbers   All
24. Natural numbers $\left\{ \frac{8}{2} \right\}$
25. Whole numbers $\left\{ 0, \frac{8}{2} \right\}$
26. Integers $\left\{ -3, 0, \frac{8}{2} \right\}$
27. Rational numbers $\left\{ -3, -\frac{5}{2}, -0.025, 0, 3\frac{1}{2}, \frac{8}{2} \right\}$
28. Irrational numbers $\{ -\sqrt{10}, \sqrt{2} \}$

Determine whether each statement is true or false. Explain. See Example 4.

29. $Q \subseteq R$ True
30. $I \subseteq Q$ False
31. $I \cap Q = \{0\}$ False
32. $J \subseteq Q$ True

33. $I \cup Q = R$ True
34. $J \cap Q = \varnothing$ False
35. $0.2121121112 \ldots \in Q$
False
36. $0.3333 \ldots \in Q$
True
37. $3.252525 \ldots \in I$
False
38. $3.1010010001 \ldots \in I$
True
39. $0.999 \ldots \in I$ False
40. $0.666 \ldots \in Q$ True
41. $\pi \in I$ True
42. $\pi \in Q$ False

Place one of the symbols $\subseteq$, $\not\subseteq$, $\in$, or $\notin$ in each blank so that each statement is true.

43. N ___ W $\subseteq$
44. J ___ Q $\subseteq$
45. J ___ N $\not\subseteq$
46. Q ___ W $\not\subseteq$
47. Q ___ R $\subseteq$
48. I ___ R $\subseteq$
49. $\varnothing$ ___ I $\subseteq$
50. $\varnothing$ ___ Q $\subseteq$
51. N ___ R $\subseteq$
52. W ___ R $\subseteq$
53. 5 ___ J $\in$
54. -6 ___ J $\in$
55. 7 ___ Q $\in$
56. 8 ___ Q $\in$
57. $\sqrt{2}$ ___ R $\in$
58. $\sqrt{2}$ ___ I $\in$
59. 0 ___ I $\notin$
60. 0 ___ Q $\in$
61. $\{2, 3\}$ ___ Q $\subseteq$
62. $\{0, 1\}$ ___ N $\not\subseteq$
63. $\{3, \sqrt{2}\}$ ___ R $\subseteq$
64. $\{3, \sqrt{2}\}$ ___ Q $\not\subseteq$

GETTING MORE INVOLVED

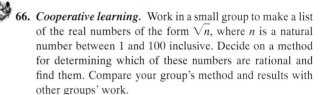

65. *Writing.* What is the difference between a rational and an irrational number? Why is $\sqrt{9}$ rational and $\sqrt{3}$ irrational?

66. *Cooperative learning.* Work in a small group to make a list of the real numbers of the form $\sqrt{n}$, where n is a natural number between 1 and 100 inclusive. Decide on a method for determining which of these numbers are rational and find them. Compare your group's method and results with other groups' work.

67. *Exploration.* Find the decimal representations of

$$\frac{2}{9}, \quad \frac{2}{99}, \quad \frac{23}{99}, \quad \frac{23}{999}, \quad \frac{234}{999}, \quad \frac{23}{9999}, \quad \text{and} \quad \frac{1234}{9999}.$$

a) What do these decimals have in common?
b) What is the relationship between each fraction and its decimal representation?

<table>
<tr><td>**1.3**</td><td># OPERATIONS ON THE SET OF REAL NUMBERS</td></tr>
</table>

Computations in algebra are performed with positive and negative numbers. In this section we will extend the basic operations of arithmetic to the negative numbers.

In this section

- Absolute Value
- Addition
- Subtraction
- Multiplication
- Division
- Division by Zero

Absolute Value

The real numbers are the coordinates of the points on the number line. However, we often refer to the points as numbers. For example, the numbers 5 and -5 are both five units away from 0 on the number line shown in Fig. 1.12. A number's *distance* from 0 on the number line is called the **absolute value** of the number. We write $|a|$ for "the absolute value of a." Therefore $|5| = 5$ and $|-5| = 5$.

FIGURE 1.12

EXAMPLE 1

Absolute value

Find the value of $|4|$, $|-4|$, and $|0|$.

Solution

Because both 4 and -4 are four units from 0 on the number line, we have $|4| = 4$ and $|-4| = 4$. Because the distance from 0 to 0 on the number line is 0, we have $|0| = 0$.

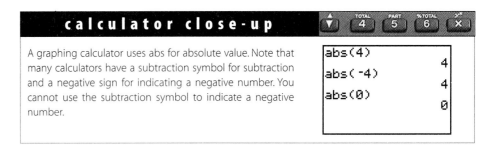

A graphing calculator uses abs for absolute value. Note that many calculators have a subtraction symbol for subtraction and a negative sign for indicating a negative number. You cannot use the subtraction symbol to indicate a negative number.

```
abs(4)
                    4
abs(-4)
                    4
abs(0)
                    0
```

Note that $|a|$ represents distance, and distance is never negative. So $|a|$ is greater than or equal to zero for any number a.

Two numbers that are located on opposite sides of zero and have the same absolute value are called **opposites** of each other. The opposite of zero is zero. Every number has a unique opposite. The numbers 9 and -9 are opposites of one another. The minus sign, $-$, is used to signify "opposite" in addition to "negative." When the minus sign is used in front of a number, it is read as "negative." When it is used in front of parentheses or a variable, it is read as "opposite." For example,

$$-(9) = -9$$

is read as "the opposite of 9 is negative 9," and

$$-(-9) = 9$$

is read as "the opposite of negative 9 is 9." In general, $-a$ is read "the opposite of a." If a is positive, $-a$ is negative. If a is negative, $-a$ is positive. Opposites have the following property.

Opposite of an Opposite

For any number a,
$$-(-a) = a.$$

E X A M P L E   2

Opposite of an opposite

Evaluate.

a) $-(-12)$ **b)** $-(-(-8))$

Solution

a) The opposite of negative 12 is 12. So $-(-12) = 12$.

b) The opposite of the opposite of -8 is -8. So $-(-(-8)) = -8$. ∎

Remember that we have defined $|a|$ to be the distance between 0 and a on the number line. Using opposites, we can give a symbolic definition of absolute value.

Absolute Value

$$|a| = \begin{cases} a & \text{if } a \text{ is positive or zero} \\ -a & \text{if } a \text{ is negative} \end{cases}$$

Using this definition, we write

$$|7| = 7$$

because 7 is positive. To find the absolute value of -7, we use the second line of the definition and write

$$|-7| = -(-7) = 7.$$

Addition

A good way to understand positive and negative numbers is to think of the *positive numbers as assets* and the *negative numbers as debts.* For this illustration we can think of assets simply as cash. Think of debts as unpaid bills such as the electric bill, the phone bill, and so on. If you have assets of $4 and $11 and no debts, then your net worth is $15. **Net worth** is the total of your debts and assets. If you have debts of $6 and $7 and no assets, then your net worth is $-$13. In symbols,

$$(-6) \qquad + \qquad (-7) \qquad = \qquad -13.$$

↑	↑	↑	↑
$6 debt	Added to	$7 debt	$13 debt

We can think of this addition as adding the absolute values of -6 and -7 (that is, $6 + 7 = 13$) and then putting a negative sign on that result to get -13. These examples illustrate the following rule.

Sum of Two Numbers with Like Signs

To find the sum of two numbers with the same sign, add their absolute values. The sum has the same sign as the original numbers.

If you have a debt of $5 and have only $5 in cash, then your debts equal your assets (in absolute value), and your net worth is $0. In symbols,

$$-5 \qquad + \qquad 5 \qquad = \qquad 0.$$

↑	↑	↑
Debt of $5	Asset of $5	Net worth

The number a and its opposite $-a$ have a sum of zero for any a. For this reason, a and $-a$ are called **additive inverses** of each other. Note that the words "negative," "opposite," and "additive inverse" are often used interchangeably.

Additive Inverse Property

For any real number a, there is a unique number $-a$ such that
$$a + (-a) = -a + a = 0.$$

To understand the sum of a positive and a negative number, consider the following situation. If you have a debt of $7 and $10 in cash, you may have $10 in hand, but your net worth is only $3. Your assets exceed your debts (in absolute value), and you have a positive net worth. In symbols,

$$-7 + 10 = 3.$$

Note that to get 3, we actually subtract 7 from 10. If you have a debt of $8 but have only $5 in cash, then your debts exceed your assets (in absolute value). You have a net worth of $-$3. In symbols,

$$-8 + 5 = -3.$$

Note that to get the 3 in the answer, we subtract 5 from 8.

As you can see from these examples, the sum of a positive number and a negative number (with different absolute values) may be either positive or negative. These examples illustrate the rule for adding numbers with unlike signs and different absolute values.

Sum of Two Numbers with Unlike Signs (and Different Absolute Values)

To find the sum of two numbers with unlike signs, subtract their absolute values.

The sum is positive if the number with the larger absolute value is positive. The sum is negative if the number with the larger absolute value is negative.

E X A M P L E   3   **Adding signed numbers**

Find each sum.

a) $-6 + 13$ b) $-9 + (-7)$ c) $2 + (-2)$

d) $-35.4 + 2.51$ e) $-7 + 0.05$ f) $\frac{1}{5} + \left(-\frac{3}{4}\right)$

Solution

a) The absolute values of -6 and 13 are 6 and 13. Subtract 6 from 13 to get 7. Because the number with the larger absolute value is 13 and it is positive, the result is 7.

b) $-9 + (-7) = -16$

c) $2 + (-2) = 0$

d) Line up the decimal points and subtract 2.51 from 35.40 to get 32.89. Because 35.4 is larger than 2.51 and 35.4 has a negative sign, the answer is negative.

$$-35.4 + 2.51 = -32.89$$

e) Line up the decimal points and subtract 0.05 from 7.00 to get 6.95. Because 7.00 is larger than 0.05 and 7.00 has a negative sign, the answer is negative.

$$-7 + 0.05 = -6.95$$

f) $\dfrac{1}{5} + \left(-\dfrac{3}{4}\right) = \dfrac{4}{20} + \left(-\dfrac{15}{20}\right) = -\dfrac{11}{20}$

Subtraction

Think of subtraction as removing debts or assets, and think of addition as receiving debts or assets. For example, if you have $10 in cash and $4 is taken from you, your resulting net worth is the same as if you have $10 and a water bill for $4 arrives in the mail. In symbols,

$$10 \quad - \quad 4 \quad = \quad 10 \quad + \quad (-4).$$

Remove Cash   Receive Debt

Removing cash is equivalent to receiving a debt.

Suppose that you have $17 in cash but owe $7 in library fines. Your net worth is $10. If the debt of $7 is canceled or forgiven, your net worth will increase to $17, the same as if you received \$7 in cash. In symbols,

$$10 \quad - \quad (-7) \quad = \quad 10 \quad + \quad 7.$$

Remove   Debt   Receive   Cash

Removing a debt is equivalent to receiving cash.

Notice that each preceding subtraction problem is equivalent to an addition problem in which we add the opposite of what we were going to subtract. These examples illustrate the definition of subtraction.

Subtraction of Real Numbers

For any real numbers a and b,

$$a - b = a + (-b).$$

E X A M P L E   4

c a l c u l a t o r

TOTAL	PART	%TOTAL	
4	5	6	×

c l o s e - u p

A graphing calculator can subtract signed numbers in any form. If your calculator has a subtraction symbol and a negative symbol, you will get an error message if you do not use them appropriately.

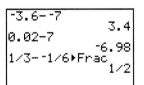

```
-3.6--7
           3.4
0.02-7
         -6.98
1/3--1/6▶Frac
           1/2
```

Subtracting signed numbers

Find each difference.

a) $-7 - 3$ **b)** $7 - (-3)$ **c)** $48 - 99$

d) $-3.6 - (-7)$ **e)** $0.02 - 7$ **f)** $\dfrac{1}{3} - \left(-\dfrac{1}{6}\right)$

Solution

a) To subtract 3 from -7, add the opposite of 3 and -7:

$$-7 - 3 = -7 + (-3) = -10$$

b) To subtract -3 from 7, add the opposite of -3 and 7. The opposite of -3 is 3:

$$7 - (-3) = 7 + (3) = 10$$

c) To subtract 99 from 48, add -99 and 48:

$$48 - 99 = 48 + (-99) = -51$$

d) $-3.6 - (-7) = -3.6 + 7 = 3.4$

e) $0.02 - 7 = 0.02 + (-7) = -6.98$

f) $\dfrac{1}{3} - \left(-\dfrac{1}{6}\right) = \dfrac{1}{3} + \dfrac{1}{6} = \dfrac{2}{6} + \dfrac{1}{6} = \dfrac{3}{6} = \dfrac{1}{2}$

Multiplication

The result of multiplying two numbers is called the **product** of the numbers. The numbers multiplied are **factors.** In algebra we use a raised dot to indicate multiplication, or we place symbols next to one another. For example, the product of a and b is written as $a \cdot b$ or ab. The product of 4 and x is $4x$. We also use parentheses to indicate multiplication. For example, the product of 4 and 3 is written as $4 \cdot 3$, $4(3)$, $(4)3$, or $(4)(3)$.

Multiplication is just a short way to do repeated additions. Adding five 2's gives

$$2 + 2 + 2 + 2 + 2 = 10.$$

So we have the multiplication fact $5 \cdot 2 = 10$. Adding together five negative 2's gives

$$(-2) + (-2) + (-2) + (-2) + (-2) = -10.$$

So we must have $5(-2) = -10$. We can think of $5(-2) = -10$ as saying that taking on five debts of \$2 each is equivalent to a debt of \$10. Losing five debts of \$2 each is equivalent to gaining \$10, so we must have $-5(-2) = 10$.

The rules for multiplying signed numbers are easy to state and remember.

Product of Signed Numbers

To find the product of two nonzero real numbers, multiply their absolute values.

     The product is *positive* if the numbers have the *same* sign.
     The product is *negative* if the numbers have *unlike* signs.

For example, to multiply -4 and -5, we multiply their absolute values ($4 \cdot 5 = 20$). Since -4 and -5 have the same sign, $(-4)(-5) = 20$. To multiply -6 and 3, we multiply their absolute values ($6 \cdot 3 = 18$). Since -6 and 3 have unlike signs, $-6 \cdot 3 = -18$.

E X A M P L E   5

Multiplying signed numbers

Find each product.

   a) $(-3)(-6)$ **b)** $-4(10)$ **c)** $(-0.01)(0.02)$ **d)** $\dfrac{4}{9} \cdot \left(-\dfrac{1}{5}\right)$

Solution

a) First multiply the absolute values ($3 \cdot 6 = 18$). Because -3 and -6 have the same sign, we get $(-3)(-6) = 18$.

b) $-4(10) = -40$ Opposite signs, negative result

c) When multiplying decimals, we total the number of decimal places used in the numbers multiplied to get the number of decimal places in the answer. Thus $(-0.01)(0.02) = -0.0002$.

d) $\dfrac{4}{9} \cdot \left(-\dfrac{1}{5}\right) = -\dfrac{4}{45}$ ■

c l o s e - u p

You can use parentheses or the times symbol to multiply on a graphing calculator. The answer for $(-0.01)(-0.02)$ is given in scientific notation. The -4 after the E means that the decimal point belongs four places to the left. So the answer is -0.0002. See Section 5.1 for more information on scientific notation.

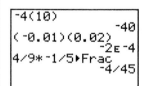

```
-4(10)
                 -40
(-0.01)(0.02)
               -2E-4
4/9*-1/5▶Frac
               -4/45
```

Division

Just as every real number has an additive inverse or opposite, every nonzero real number a has a **multiplicative inverse** or **reciprocal** $\dfrac{1}{a}$. The reciprocal of 3 is $\dfrac{1}{3}$, and

$$3 \cdot \frac{1}{3} = 1.$$

Multiplicative Inverse Property

For any nonzero real number a, there is a unique number $\dfrac{1}{a}$ such that

$$a \cdot \frac{1}{a} = \frac{1}{a} \cdot a = 1.$$

E X A M P L E   6

Finding multiplicative inverses

Find the multiplicative inverse (reciprocal) of each number.

   a) -2 **b)** $\dfrac{3}{8}$ **c)** -0.2

Solution

a) The multiplicative inverse (reciprocal) of -2 is $-\frac{1}{2}$ because

$$-2\left(-\frac{1}{2}\right) = 1.$$

b) The reciprocal of $\frac{3}{8}$ is $\frac{8}{3}$ because

$$\frac{3}{8} \cdot \frac{8}{3} = 1.$$

c) First convert the decimal number -0.2 to a fraction:

$$-0.2 = -\frac{2}{10}$$

$$= -\frac{1}{5}$$

So the reciprocal of -0.2 is -5 and $-0.2(-5) = 1$. ◼

Note that the reciprocal of any negative number is negative.

Earlier we defined subtraction for real numbers as addition of the additive inverse. We now define division for real numbers as multiplication by the multiplicative inverse (reciprocal).

Division of Real Numbers

For any real numbers a and b with $b \neq 0$,

$$a \div b = a \cdot \frac{1}{b}.$$

If $a \div b = c$, then a is called the **dividend,** b the **divisor,** and c the **quotient.** We also refer to $a \div b$ and $\frac{a}{b}$ as the quotient of a and b.

EXAMPLE 7

Dividing signed numbers

Find each quotient.

a) $-60 \div (-2)$ **b)** $-24 \div \frac{3}{8}$ **c)** $6 \div (-0.2)$

Solution

a) $-60 \div (-2) = -60 \cdot \left(-\frac{1}{2}\right) = 30$ Multiply by $-\frac{1}{2}$, the reciprocal of -2.

b) $-24 \div \frac{3}{8} = -24 \cdot \frac{8}{3} = -64$

c) $6 \div (-0.2) = 6(-5) = -30$ ◼

You can see from Examples 6 and 7 that a product or quotient is positive when the signs are the same and is negative when the signs are opposite:

same signs $\leftrightarrow$ positive result,

opposite signs $\leftrightarrow$ negative result.

Even though all division can be done as multiplication by a reciprocal, we generally use reciprocals only when dividing fractions. Instead, we find quotients using our knowledge of multiplication and the fact that

$$a \div b = c \qquad \text{if and only if} \qquad c \cdot b = a.$$

For example, $-72 \div 9 = -8$ because $-8 \cdot 9 = -72$. Using long division or a calculator, you can get

$$-43.74 \div 1.8 = -24.3$$

and check that you have it correct by finding $-24.3 \cdot 1.8 = -43.74$.

We use the same rules for division when division is indicated by a fraction bar. For example,

$$\frac{-6}{3} = -2, \qquad \frac{6}{-3} = -2, \qquad \frac{-1}{3} = \frac{1}{-3} = -\frac{1}{3}, \qquad \text{and} \qquad \frac{-6}{-3} = 2.$$

Note that if one negative sign appears in a fraction, the fraction has the same value whether the negative sign is in the numerator, in the denominator, or in front of the fraction. If the numerator and denominator of a fraction are both negative, then the fraction has a positive value.

Division by Zero

Why do we omit division by zero from the definition of division? If we write $10 \div 0 = c$, we need to find c such that $c \cdot 0 = 10$. But there is no such number. If we write $0 \div 0 = c$, we need to find c such that $c \cdot 0 = 0$. But $c \cdot 0 = 0$ is true for any number c. Having $0 \div 0$ equal to any number would be confusing. Thus $a \div b$ is defined only for $b \neq 0$. Quotients such as

$$5 \div 0, \qquad 0 \div 0, \qquad \frac{7}{0}, \qquad \text{and} \qquad \frac{0}{0}$$

are said to be *undefined*.

WARM-UPS

True or false? Explain your answer.

1. The additive inverse of -6 is 6.   True

2. The opposite of negative 5 is positive 5.   True

3. The absolute value of 6 is -6.   False

4. The result of a subtracted from b is the same as $b + (-a)$.   True

5. If a is positive and b is negative, then ab is negative.   True

6. If a is positive and b is negative, then $a + b$ is negative.   False

7. $(-3) - (-6) = -9$ False   **8.** $6 \div \left(-\frac{1}{2}\right) = -3$ False

9. $-3 \div 0 = 0$ False   **10.** $0 \div (-7) = 0$ True

1.3   EXERCISES

Reading and Writing   *After reading this section, write out the answers to these questions. Use complete sentences.*

1. What is absolute value?

The absolute value of a number is the number's distance from 0 on the number line.

2. How do you add two numbers with the same sign?

Add their absolute values, then affix the sign of the original numbers.

3. How do you add two numbers with unlike signs and different absolute values?

Subtract their absolute values and use the sign of the number with the larger absolute value.

4. What is the relationship between subtraction and addition?

The difference $a - b$ is defined as $a + (-b)$.

5. How do you multiply signed numbers?
Multiply their absolute values, then affix a positive sign if the original numbers have the same sign and a negative sign if the original numbers have opposite signs.

6. What is the relationship between division and multiplication?   The quotient $a \div b$ is defined as $a \cdot \frac{1}{b}$.

Evaluate. See Examples 1 and 2.

7. $\left|-34\right|$ 34

8. $\left|17\right|$ 17

9. $\left|0\right|$ 0

10. $\left|-15\right|$ 15

11. $\left|-6\right| - \left|-6\right|$ 0

12. $\left|8\right| - \left|-8\right|$ 0

13. $-\left|-9\right|$ -9

14. $-\left|-3\right|$ -3

15. $-(-9)$ 9

16. $-(-(8))$ 8

17. $-(-(-3))$ -3

18. $-(-(-2))$ -2

Find each sum. See Example 3.

19. $(-5) + 9$ 4

20. $(-3) + 10$ 7

21. $(-4) + (-3)$ -7

22. $(-15) + (-11)$ -26

23. $-6 + 4$ -2

24. $5 + (-15)$ -10

25. $7 + (-17)$ -10

26. $-8 + 13$ 5

27. $(-11) + (-15)$ -26

28. $-18 + 18$ 0

29. $18 + (-20)$ -2

30. $7 + (-19)$ -12

31. $-14 + 9$ -5

32. $-6 + (-7)$ -13

33. $-4 + 4$ 0

34. $-7 + 9$ 2

35. $-\frac{1}{10} + \frac{1}{5}$ $\frac{1}{10}$

36. $-\frac{1}{8} + \left(-\frac{1}{8}\right)$ $-\frac{1}{4}$

37. $\frac{1}{2} + \left(-\frac{2}{3}\right)$ $-\frac{1}{6}$

38. $\frac{3}{4} + \frac{1}{2}$ $\frac{5}{4}$

39. $-15 + 0.02$ -14.98

40. $0.45 + (-1.3)$ -0.85

41. $-2.7 + (-0.01)$ -2.71

42. $0.8 + (-1)$ -0.2

43. $47.39 + (-44.587)$ 2.803

44. $0.65357 + (-2.375)$ -1.72143

45. $0.2351 + (-0.5)$ -0.2649

46. $-1.234 + (-4.756)$ -5.99

Find each difference. See Example 4.

47. $7 - 10$ -3

48. $8 - 19$ -11

49. $-4 - 7$ -11

50. $-5 - 12$ -17

51. $7 - (-6)$ 13

52. $3 - (-9)$ 12

53. $-1 - 5$ -6

54. $-4 - 6$ -10

55. $-12 - (-3)$ -9

56. $-15 - (-6)$ -9

57. $20 - (-3)$ 23

58. $50 - (-70)$ 120

59. $\frac{9}{10} - \left(-\frac{1}{10}\right)$ 1

60. $\frac{1}{8} - \frac{1}{4}$ $-\frac{1}{8}$

61. $1 - \frac{3}{2}$ $-\frac{1}{2}$

62. $-\frac{1}{2} - \left(-\frac{1}{3}\right)$ $-\frac{1}{6}$

63. $2 - 0.03$ 1.97

64. $-0.02 - 3$ -3.02

65. $5.3 - (-2)$ 7.3

66. $-4.1 - 0.13$ -4.23

67. $-2.44 - 48.29$ -50.73

68. $-8.8 - 9.164$ -17.964

69. $-3.89 - (-5.16)$ 1.27

70. $0 - (-3.5)$ 3.5

Find each product. See Example 5.

71. $(25)(-3)$ -75

72. $(5)(-7)$ -35

73. $\left(-\frac{1}{3}\right)\left(-\frac{1}{2}\right)$ $\frac{1}{6}$

74. $\left(-\frac{1}{2}\right)\left(-\frac{6}{7}\right)$ $\frac{3}{7}$

75. $(0.3)(-0.3)$ -0.09

76. $(-0.1)(-0.5)$ 0.05

77. $(-0.02)(-10)$ 0.2

78. $(0.05)(-2.5)$ -0.125

Find the multiplicative inverse of each number. See Example 6.

79. 20 $\frac{1}{20}$ or 0.05

80. -5 $-\frac{1}{5}$ or -0.2

81. $-\frac{6}{5}$ $-\frac{5}{6}$

82. $-\frac{1}{8}$ -8

83. -0.3 $-\frac{10}{3}$

84. 0.125 8

Evaluate. See Example 7.

85. $-6 \div 3$ -2

86. $84 \div (-2)$ -42

87. $30 \div (-0.8)$ -37.5

88. $(-9)(-6)$ 54

89. $(-0.8)(0.1)$ -0.08

90. $7 \div (-0.5)$ -14

91. $(-0.1) \div (-0.4)$ 0.25

92. $(-18) \div (-0.9)$ 20

93. $9 \div \left(-\frac{3}{4}\right)$ -12

94. $-\frac{1}{3} \div \left(-\frac{5}{8}\right)$ $\frac{8}{15}$

95. $-\frac{2}{3}\left(-\frac{9}{10}\right)$ $\frac{3}{5}$

96. $\frac{1}{2}\left(-\frac{2}{5}\right)$ $-\frac{1}{5}$

97. $(0.25)(-365)$ -91.25

98. $7.5 \div (-0.15)$ -50

99. $(-51) \div (-0.003)$ 17,000

100. $(-2.8)(5.9)$ -16.52

Perform the following computations.

101. $-62 + 13$ -49

102. $-88 + 39$ -49

103. $-32 - (-25)$ -7

104. $-71 - (-19)$ -52

105. $\left|-15\right|$ 15

106. $-\left|-75\right|$ -75

107. $\frac{1}{2}(-684)$ -342

108. $\frac{1}{3}(-123)$ -41

109. $\frac{1}{2} - \left(-\frac{1}{4}\right)$ $\frac{3}{4}$

110. $\frac{1}{8} - \left(-\frac{1}{4}\right)$ $\frac{3}{8}$

111. $-57 \div 19$ -3

112. $0 \div (-36)$ 0

113. $\left|-17\right| + \left|-3\right|$ 20

114. $64 - \left|-12\right|$ 52

115. $0 \div (-0.15)$ 0

116. $-20 \div \left(-\frac{8}{3}\right)$ $\frac{15}{2}$

117. $27 \div (-0.15)$ -180

118. $33 \div (-0.2)$ -165

119. $-\dfrac{1}{3} + \dfrac{1}{6}$

$-\dfrac{1}{6}$

120. $-\dfrac{2}{3} + \dfrac{1}{6}$

$-\dfrac{1}{2}$

121. $-63 + |8|$

-55

122. $|-34| - 27$

7

123. $-\dfrac{1}{2} + \left(-\dfrac{1}{2}\right)$

-1

124. $-\dfrac{2}{3} + \left(-\dfrac{2}{3}\right)$

$-\dfrac{4}{3}$

125. $-\dfrac{1}{2} - 19$

$-\dfrac{39}{2}$

126. $-\dfrac{1}{3} - 22$

$-\dfrac{67}{3}$

127. $28 - 0.01$
27.99

128. $55 - 0.1$
54.9

129. $-29 - 0.3$
-29.3

130. $-0.241 - 0.3$
-0.541

131. $(-2)(0.35)$
-0.7

132. $(-3)(0.19)$
-0.57

133. $(-10)(-0.2)$
2

134. $\left(-\dfrac{1}{2}\right)(-50)$
25

Use an operation with signed numbers to solve each problem.

135. *Net worth of a family.* The average American family has an $85,000 house, a $45,000 mortgage, $2,300 in credit card debt, $1,500 in other debts, $1,200 in savings, and two cars worth $3,500 each. What is the net worth of the average American family?
$44,400

136. *Net worth of a bank.* Just before the recession, First Federal Homestead had $15.6 million in mortgage loans, had $23.3 million on deposit, and owned $8.5 million worth of real estate. After the recession started, the value of the real estate decreased to $4.8 million. What was the net worth of First Federal before the recession and after the recession started? (To a financial institution a loan is an asset and a deposit is a liability.)
$800,000, $-$2.9 million

137. *Warming up.* On January 11 the temperature at noon was 14°F in St. Louis and −6°F in Duluth. How much warmer was it in St. Louis?
20°

138. *Bitter cold.* On January 16 the temperature at midnight was −31°C in Calgary and −20°C in Toronto. How much warmer was it in Toronto?
11°C

139. *Below sea level.* The altitude of the floor of Death Valley is −282 feet (282 feet below sea level); the altitude of the shore of the Dead Sea is −1,296 feet (*Rand McNally World Atlas*). How many feet above the shore of the Dead Sea is the floor of Death Valley?
1,014 feet

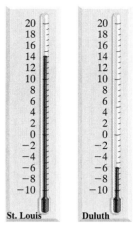

FIGURE FOR EXERCISE 137

140. *Highs and lows.* The altitude of the peak of Mt. Everest, the highest point on earth, is 29,028 feet. The world's greatest known ocean depth of −36,201 feet was recorded in the Marianas Trench (*Rand McNally World Atlas*). How many feet above the bottom of the Marianas Trench is a climber who has reached the top of Mt. Everest?
65,229 feet

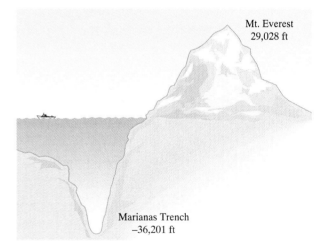

FIGURE FOR EXERCISE 140

GETTING MORE INVOLVED

141. *Discussion.* Why is it necessary to learn addition of signed numbers before learning subtraction of signed numbers and to learn multiplication of signed numbers before division of signed numbers?

142. *Writing.* Explain why 0 is the only real number that does not have a multiplicative inverse.

1.4   E V A L U A T I N G   E X P R E S S I O N S

In algebra you will learn to work with variables. However, there is often nothing more important than finding a numerical answer to a question. This section is concerned with computation.

Arithmetic Expressions

The result of writing numbers in a meaningful combination with the ordinary operations of arithmetic is called an **arithmetic expression** or simply an **expression.** An expression that involves more than one operation is called a **sum, difference, product,** or **quotient** if the last operation to be performed is addition, subtraction, multiplication, or division, respectively. Parentheses are used as **grouping symbols** to indicate which operations are performed first. The expression

$$5 + (2 \cdot 3)$$

is a sum because the parentheses indicate that the product of 2 and 3 is to be found before the addition is performed. So we evaluate this expression as follows:

$$5 + (2 \cdot 3) = 5 + 6 = 11$$

If we write $(5 + 2)3$, the expression is a product and it has a different value.

$$(5 + 2)3 = 7 \cdot 3 = 21$$

Brackets [ ] are also used to indicate grouping. If an expression occurs within absolute value bars | |, it is evaluated before the absolute value is found. So absolute value bars also act as grouping symbols. We perform first the operations within the innermost grouping symbols.

E X A M P L E   1

Grouping symbols

Evaluate each expression.

a) $5[(2 \cdot 3) - 8]$ **b)** $2[(4 \cdot 5) - |3 - 6|]$

Solution

a) $5[(2 \cdot 3) - 8] = 5[6 - 8]$ Innermost grouping first.

$$= 5[-2]$$
$$= -10$$

b) $2[(4 \cdot 5) - |3 - 6|] = 2[20 - |-3|]$

$$= 2[20 - 3]$$
$$= 2[17]$$
$$= 34$$

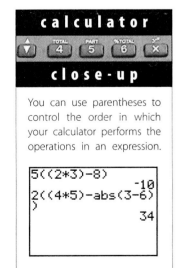

calculator

close-up

You can use parentheses to control the order in which your calculator performs the operations in an expression.

```
5((2*3)-8)
            -10
2((4*5)-abs(3-6)
)
            34
```

Exponential Expressions

We use the notation of exponents to simplify the writing of repeated multiplication. The product $5 \cdot 5 \cdot 5 \cdot 5$ is written as 5^4. The number 4 in 5^4 is called the exponent, and it indicates the number of times that the factor 5 occurs in the product.

Exponential Expression

For any natural number n and real number a,

$$a^n = \underbrace{a \cdot a \cdot a \cdot \ldots \cdot a}_{n \text{ factors of } a}.$$

We call a the **base,** n the **exponent,** and a^n an **exponential expression.**

We read a^n as "the nth power of a" or "a to the nth power." The exponential expressions 3^5 and 10^6 are read as "3 to the fifth power" and "10 to the sixth power." We can also use the words "squared" and "cubed" for the second and third powers, respectively. For example, 5^2 and 2^3 are read as "5 squared" and "2 cubed," respectively.

E X A M P L E   2

calculator
close-up

Powers are indicated on a graphing calculator using a caret (^). Most calculators also have an x^2-key for squaring. Note that parentheses are necessary in $(-3)^4$. Without parentheses, your calculator should get $-3^4 = -81$. Try it.

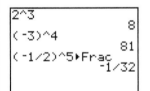

```
2^3
                    8
(-3)^4
                   81
(-1/2)^5▶Frac
                -1/32
```

Exponential expressions

Evaluate.

a) 2^3 b) $(-3)^4$ c) $\left(-\dfrac{1}{2}\right)^5$

Solution

a) $2^3 = 2 \cdot 2 \cdot 2 = 8$ b) $(-3)^4 = (-3)(-3)(-3)(-3) = 81$

c) $\left(-\dfrac{1}{2}\right)^5 = \left(-\dfrac{1}{2}\right)\left(-\dfrac{1}{2}\right)\left(-\dfrac{1}{2}\right)\left(-\dfrac{1}{2}\right)\left(-\dfrac{1}{2}\right) = -\dfrac{1}{32}$ ∎

Square Roots

Because $3^2 = 9$ and $(-3)^2 = 9$, both 3 and -3 are square roots of 9. We use the **radical symbol** $\sqrt{}$ to indicate the nonnegative or principal square root of 9. We write $\sqrt{9} = 3$.

Square Roots

If $a^2 = b$, then a is called a **square root** of b. If $a \geq 0$, then a is called the **principal square root** of b and we write $\sqrt{b} = a$.

The radical symbol is a grouping symbol. We perform all operations within the radical symbol before the square root is found.

E X A M P L E   3

Evaluating square roots

Evaluate.

a) $\sqrt{64}$ b) $\sqrt{9 + 16}$ c) $\sqrt{3(17 - 5)}$

Solution

a) Because $8^2 = 64$, we have $\sqrt{64} = 8$.

b) Because the radical symbol is a grouping symbol, add 9 and 16 before finding the square root:

$$\sqrt{9 + 16} = \sqrt{25} = 5$$

Note that $\sqrt{9} + \sqrt{16} = 3 + 4 = 7$. So $\sqrt{9 + 16} \neq \sqrt{9} + \sqrt{16}$.

c) $\sqrt{3(17 - 5)} = \sqrt{3(12)} = \sqrt{36} = 6$ ∎

calculator

close-up

Because the radical symbol on most calculators cannot be extended, parentheses are used to group the expression that is inside the radical.

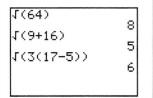

Order of Operations

To simplify the writing of expressions, we often omit some grouping symbols. If we saw the expression

$$5 + 2 \cdot 3$$

written without parentheses, we would not know how to evaluate it unless we had a rule for which operations to perform first. Expressions in which some or all grouping symbols are omitted, are evaluated consistently by using a rule called the **order of operations.**

Order of Operations

Evaluate inside any grouping symbols first. Where grouping symbols are missing use the following order.
1. Evaluate each exponential expression (in order from left to right).
2. Perform multiplication and division (in order from left to right).
3. Perform addition and subtraction (in order from left to right).

"In order from left to right" means that we evaluate the operations in the order in which they are written. For example,

$$20 \cdot 3 \div 6 = 60 \div 6 = 10 \qquad \text{and} \qquad 10 - 3 + 6 = 7 + 6 = 13.$$

If an expression contains grouping symbols, we evaluate within the grouping symbols using the order of operations.

E X A M P L E 4

calculator

close-up

When parentheses are omitted, most (but not all), calculators follow the same order of operations such as we do in this text. Try these computations on your calculator. To use a calculator effectively, you must practice with it.

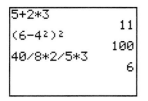

Order of operations
Evaluate each expression.

a) $5 + 2 \cdot 3$ **b)** $9 \cdot 2^3$ **c)** $(6 - 4^2)^2$ **d)** $40 \div 8 \cdot 2 \div 5 \cdot 3$

Solution

a) $5 + 2 \cdot 3 = 5 + 6$ Multiply first.

           $= 11$ Then add.

b) $9 \cdot 2^3 = 9 \cdot 8$ Evaluate the exponential expression first.

        $= 72$ Then multiply.

c) $(6 - 4^2)^2 = (6 - 16)^2$ Evaluate 4^2 within the parentheses first.

           $= (-10)^2$ Then subtract.

           $= 100$ $(-10)(-10) = 100$

d) Multiplication and division are done from left to right.

$$40 \div 8 \cdot 2 \div 5 \cdot 3 = 5 \cdot 2 \div 5 \cdot 3$$
$$= 10 \div 5 \cdot 3$$
$$= 2 \cdot 3$$
$$= 6$$

CAUTION Don't confuse -3^2 and $(-3)^2$. We interpret -3^2 as the opposite of 3^2. So $-3^2 = -(3^2) = -9$, whereas $(-3)^2 = (-3)(-3) = 9$.

E X A M P L E   5

The order of negative signs

Evaluate each expression.

a) -2^4 **b)** -5^2 **c)** $(3 - 5)^2$ **d)** $-(5^2 - 4 \cdot 7)^2$

Solution

a) To evaluate -2^4, find 2^4 first and then take the opposite. So $-2^4 = -16$.

b) $-5^2 = -(5^2) = -25$

c) Evaluate within the parentheses first, square that result, then take the opposite.

$$
\begin{aligned}
(3 - 5)^2 &= (-2)^2 &&\text{Evaluate within parentheses first.}\\
&= 4 &&\text{Square } -2 \text{ to get 4.}
\end{aligned}
$$

d)
$$
\begin{aligned}
-(5^2 - 4 \cdot 7)^2 &= -(25 - 28)^2 &&\text{Evaluate } 5^2 \text{ within the parentheses first.}\\
&= -(-3)^2 &&\text{Then subtract.}\\
&= -9 &&\text{Square } -3 \text{ to get 9, then take the opposite of}\\
& &&\text{9 to get } -9.
\end{aligned}
$$

> **helpful hint**
>
> "Everybody Loves My Dear Aunt Sally" is often used as a memory aid for the order of operations. Do **E**xponents and **L**ogarithms, **M**ultiplication and **D**ivision, and then **A**ddition and **S**ubtraction. Logarithms are discussed later in this text.

    When an expression involves a fraction bar, the numerator and denominator are each treated as if they are in parentheses. The next example illustrates how the fraction bar groups the numerator and denominator.

E X A M P L E   6

Order of operations in fractions

Evaluate each quotient.

a) $\dfrac{10 - 8}{6 - 8}$ **b)** $\dfrac{-6^2 + 2 \cdot 7}{4 - 3 \cdot 2}$

Solution

a)
$$
\begin{aligned}
\frac{10 - 8}{6 - 8} &= \frac{2}{-2} &&\text{Evaluate the numerator and denominator separately.}\\
&= -1 &&\text{Then divide.}
\end{aligned}
$$

b)
$$
\begin{aligned}
\frac{-6^2 + 2 \cdot 7}{4 - 3 \cdot 2} &= \frac{-36 + 14}{4 - 6}\\
&= \frac{-22}{-2} &&\text{Evaluate the numerator and denominator separately.}\\
&= 11 &&\text{Then divide.}
\end{aligned}
$$

> **calculator**
>
>
>
> **close-up**
>
> Some calculators use the built-up form for fractions $\left(\frac{1}{2}\right)$, but some do not (1/2). If your calculator does not use the built-up form, then you must enclose numerators and denominators (that contain operations) in parentheses as shown here.
>
> ```
> (10-8)/(6-8)
>                 -1
> ( -6²+2*7)/(4-3*2
> )
>                 11
> ```

Algebraic Expressions

The result of combining numbers and variables with the ordinary operations of arithmetic (in some meaningful way) is called an **algebraic expression.** For example,

$$2x - 5y, \quad 5x^2, \quad (x - 3)(x + 2), \quad b^2 - 4ac, \quad 5, \quad \text{and} \quad \frac{x}{2}$$

are algebraic expressions, or simply **expressions.** An expression such as $2x - 5y$ has no definite value unless we assign values to x and y. For example, if $x = 3$ and $y = 4$, then the value of $2x - 5y$ is found by replacing x with 3 and y with 4 and evaluating:

$$2x - 5y = 2(3) - 5(4) = 6 - 20 = -14$$

Nancy Gittins, Assistant Director of Finan-
cial Aid at Babson College, helps graduate
and undergraduate students to achieve their
goal of financing their educations. Because
recent tuition and fees can be as high as
$20,000 a year, many students need financial
aid to help defray these expenses. Federal
loans and state loans, as well as grants from

**FINANCIAL AID
DIRECTOR**

the federal and state levels, can be given to students who qualify. Ms. Gittins ad-
ministers many of these loans and grants and helps students to understand the different options that are available. The interest rate for these loans is now a variable
rate that is tied to treasury bills. The rate can be as high as 9.0% and as low as 7.3%.
In Exercise 107 of this section you will work with interest rates compounded
annually.

Note the importance of the order of operations in evaluating an algebraic expression.
To find the value of the difference $2x - 5y$ when $x = -2$ and $y = -3$, replace
x and y by -2 and -3, respectively, and then evaluate.

$$2x - 5y = 2(-2) - 5(-3) = -4 - (-15) = -4 + 15 = 11$$

E X A M P L E   7

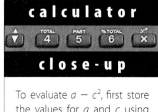

calculator

close-up

To evaluate $a - c^2$, first store
the values for a and c using
the STO key. Then enter the
expression.

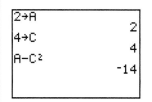

Value of an algebraic expression

Evaluate each expression for $a = 2$, $b = -3$, and $c = 4$.

a) $a - c^2$ **b)** $a - b^2$ **c)** $b^2 - 4ac$ **d)** $\dfrac{a - b}{c - b}$

Solution

a) Replace a by 2 and c by 4 in the expression $a - c^2$.

$$a - c^2 = 2 - 4^2 = 2 - 16 = -14$$

b) $a - b^2 = 2 - (-3)^2 = 2 - 9 = -7$

c) $b^2 - 4ac = (-3)^2 - 4(2)(4) = 9 - 32 = -23$

d) $\dfrac{a - b}{c - b} = \dfrac{2 - (-3)}{4 - (-3)} = \dfrac{5}{7}$

■

CAUTION When you replace a variable by a negative number, be sure to
use parentheses around the negative number. If we were to omit the parentheses in
Example 7(c), we would get $-3^2 - 4(2)(4) = -41$ instead of -23.

A symbol such as y_1 is treated like any other variable. We read y_1 as "y one" or
"y sub one." The 1 is called a **subscript.** We can think of y_1 as the "first y" and y_2 as
the "second y." We use the subscript notation in the following example.

E X A M P L E   8

An algebraic expression with subscripts

Let $y_1 = -12$, $y_2 = -5$, $x_1 = -3$, and $x_2 = 4$. Find the value of $\dfrac{y_2 - y_1}{x_2 - x_1}$.

Solution

Substitute the appropriate values into the expression

$$\frac{y_2 - y_1}{x_2 - x_1} = \frac{-5 - (-12)}{4 - (-3)} = \frac{7}{7} = 1.$$

When we evaluate an algebraic expression involving only one variable for many values of that variable, we get a collection of data. A graph (picture) of these data can give us useful information.

E X A M P L E   9

Reading a graph

The expression $0.85(220 - A)$ gives the target heart rate for beneficial exercise for an athlete who is A years old. Use the graph in Fig. 1.13 to estimate the target heart rate for a 40-year-old athlete. Use the graph to estimate the age of an athlete with a target heart rate of 170.

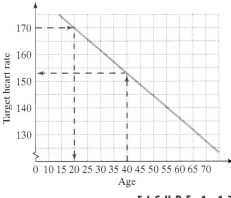

FIGURE 1.13

Solution

To find the target heart rate for a 40-year-old athlete, first draw a vertical line from age 40 up to the graph as shown in Fig. 1.13. From the point of intersection, draw a horizontal line to the heart rate scale. So the target heart rate for a 40-year-old athlete is about 153. To find the age corresponding to a heart rate of 170, first draw a horizontal line from heart rate 170 to the graph as shown in the figure. From the point of intersection, draw a vertical line down to the age scale. The heart rate of 170 corresponds to an age of about 20.

WARM-UPS

True or false? Explain your answer.

1. $2^3 = 6$ False

2. $-1 \cdot 2^2 = -4$ True

3. $-2^2 = -4$ True

4. $6 + 3 \cdot 2 = 18$ False

(continued)

5. $(6 + 3) \cdot 2 = 81$ False

6. $(6 + 3)^2 = 18$ False

7. $6 + 3^2 = 15$ True

8. $(-3)^3 = -3^3$ True

9. $|-3 - (-2)| = 5$ False

10. $|7 - 8| = |7| - |8|$ False

1.4   EXERCISES

Reading and Writing   *After reading this section, write out the answers to these questions. Use complete sentences.*

1. What is an arithmetic expression?

An arithmetic expression is the result of writing numbers in a meaningful combination with the ordinary operations of arithmetic.

2. How do you know whether to call an expression a sum, a difference, a product, or a quotient?

An expression is called a sum, a difference, a product, or a quotient if the last operation to be performed is addition, subtraction, multiplication, or division, respectively.

3. Why are grouping symbols used?

Grouping symbols are used to indicate the order in which operations are to be performed.

4. What is an exponential expression?

An exponential expression is an expression of the form a^n.

5. What is the purpose of the order of operations?

The order of operations tells us the order in which to perform operations when we omit grouping symbols.

6. What is the difference between -3^2 and $(-3)^2$?

The value of -3^2 is -9 and the value of $(-3)^2$ is 9.

Evaluate each expression. See Example 1.

7. $(-3 \cdot 4) - (2 \cdot 5)$ -22

8. $|-3 - 2| - |2 - 6|$ 1

9. $4[5 - |3 - (2 \cdot 5)|]$ -8

10. $-2|(-3 \cdot 4) - 6|$ -36

11. $(6 - 8)(|2 - 3| + 6)$ -14

12. $-5(6 + [(5 - 7) - 4])$ 0

Evaluate each exponential expression. See Example 2.

13. 2^5 32

14. 3^4 81

15. $(-1)^4$ 1

16. $(-1)^5$ -1

17. $\left(-\dfrac{1}{3}\right)^2$ $\dfrac{1}{9}$

18. $\left(-\dfrac{1}{2}\right)^6$ $\dfrac{1}{64}$

Evaluate each radical. See Example 3.

19. $\sqrt{49}$ 7

20. $\sqrt{100}$ 10

21. $\sqrt{36 + 64}$ 10

22. $\sqrt{25 - 9}$ 4

23. $\sqrt{4(7 + 9)}$ 8

24. $\sqrt{(11 + 2)(18 - 5)}$ 13

Evaluate each expression. See Examples 4 and 5.

25. $4 - 6 \cdot 2$ -8

26. $8 - 3 \cdot 9$ -19

27. $5 - 6(3 - 5)$ 17

28. $8 - 3(4 - 6)$ 14

29. $\left(\dfrac{1}{3} - \dfrac{1}{2}\right)\left(\dfrac{1}{4} - \dfrac{1}{2}\right)$ $\dfrac{1}{24}$

30. $\left(\dfrac{1}{2} - \dfrac{1}{4}\right)\left(\dfrac{1}{2} - \dfrac{3}{4}\right)$ $-\dfrac{1}{16}$

31. $-3^2 + (-8)^2 + 3$ 58

32. $-6^2 + (-3)^3$ -63

33. $-(2 - 7)^2$ -25

34. $-(1 - 3 \cdot 2)^3$ 125

35. $-5^2 \cdot 2^3$ -200

36. $2^4 - 4^2$ 0

37. $(-5)(-2)^3$ 40

38. $(-1)(2 - 8)^3$ 216

39. $-(3^2 - 4)^2$ -25

40. $-(6 - 2^3)^4$ -16

41. $-60 \div 10 \cdot 3 \div 2 \cdot 5 \div 6$ -7.5

42. $75 \div (-5)(-3) \div \dfrac{1}{2} \cdot 6$ 540

43. $5.5 - 2.3^4$ -22.4841

44. $5.3^2 - 4 \cdot 6.1$ 3.69

45. $(1.3 - 0.31)(2.9 - 4.88)$ -1.9602

46. $(6.7 - 9.88)^3$ -32.157432

47. $-388.8 \div (13.5)(9.6)$ -276.48

48. $(-4.3)(5.5) \div (3.2)(-1.2)$ 8.86875

Evaluate each expression. See Example 6.

49. $\dfrac{2 - 6}{9 - 7}$ -2

50. $\dfrac{9 - 12}{4 - 5}$ 3

51. $\dfrac{-3 - 5}{6 - (-2)}$ -1

52. $\dfrac{-14 - (-2)}{-3 - 3}$ 2

53. $\dfrac{4 + 2 \cdot 7}{3 \cdot 2 - 9}$ -6

54. $\dfrac{-6 - 2(-3)}{8 - 3(-3)}$ 0

55. $\dfrac{-3^2 - (-9)}{2 - 3^2}$ 0

56. $\dfrac{-2^4 - 5}{3^2 - 2^4}$ 3

Evaluate each expression for $a = -1$, $b = 3$, and $c = -4$. See Example 7.

57. $b^2 - 4ac$ -7

58. $\sqrt{a^2 - 4bc}$ 7

59. $\dfrac{a - b}{a - c}$ $-\dfrac{4}{3}$

60. $\dfrac{b - c}{b - a}$ $\dfrac{7}{4}$

61. $(a - b)(a + b)$ -8

62. $(a - c)(a + c)$ -15

63. $\sqrt{c^2 - 2c + 1}$ 5

64. $b^2 - 2b - 3$ 0

65. $\dfrac{2}{a} + \dfrac{b}{c} - \dfrac{1}{c}$ $-\dfrac{5}{2}$

66. $\dfrac{c}{a} + \dfrac{c}{b} - \dfrac{a}{b}$ 3

67. $|a - b|$ 4

68. $|b + c|$ 1

Find the value of $\dfrac{y_2 - y_1}{x_2 - x_1}$ for each choice of y_1, y_2, x_1, and x_2. See Example 8.

69. $y_1 = 4$, $y_2 = -6$, $x_1 = 2$, $x_2 = -7$ $\dfrac{10}{9}$

70. $y_1 = -3$, $y_2 = -3$, $x_1 = 4$, $x_2 = -5$ 0

71. $y_1 = -1$, $y_2 = 2$, $x_1 = -3$, $x_2 = 1$ $\dfrac{3}{4}$

72. $y_1 = -2$, $y_2 = 5$, $x_1 = 2$, $x_2 = 6$ $\dfrac{7}{4}$

73. $y_1 = 2.4$, $y_2 = 5.6$, $x_1 = 5.9$, $x_2 = 4.7$ -2.67

74. $y_1 = -5.7$, $y_2 = 6.9$, $x_1 = 3.5$, $x_2 = 4.2$ 18

Evaluate each expression without a calculator. Use a calculator to check.

75. $-2^2 + 5(3)^2$
41

76. $-3^2 + 3(6)^2$
99

77. $(-2 + 5)3^2$
27

78. $(-3 + 3)6^2$
0

79. $\sqrt{5^2 - 4(1)(6)}$
1

80. $\sqrt{6^2 - 4(2)(4)}$
2

81. $[13 + 2(-5)]^2$
9

82. $[6 + 2(-4)]^2$
4

83. $\dfrac{4 - (-1)}{-3 - 2}$
-1

84. $\dfrac{2 - (-3)}{3 - 5}$
$-\dfrac{5}{2}$

85. $3(-2)^2 - 5(-2) + 4$
26

86. $3(-1)^2 + 5(-1) - 6$
-8

87. $-4\left(\dfrac{1}{2}\right)^2 + 3\left(\dfrac{1}{2}\right) - 2$
$-\dfrac{3}{2}$

88. $8\left(\dfrac{1}{2}\right)^2 - 6\left(\dfrac{1}{2}\right) + 1$
0

89. $-\dfrac{1}{2}|6 - 2|$
-2

90. $-\dfrac{1}{3}|9 - 6|$
-1

91. $\dfrac{1}{2} - \dfrac{1}{3}\left|\dfrac{1}{4} - \dfrac{1}{2}\right|$
$\dfrac{5}{12}$

92. $\dfrac{1}{3} - \dfrac{1}{2}\left|\dfrac{1}{3} - \dfrac{1}{2}\right|$
$\dfrac{1}{4}$

93. $|6 - 3 \cdot 7| + |7 - 5|$
17

94. $|12 - 4| - |3 - 4 \cdot 5|$
-9

95. $3 - 7[4 - (2 - 5)]$
-46

96. $9 - 2[3 - (4 + 6)]$
23

97. $3 - 4(2 - |4 - 6|)$
3

98. $3 - (|-4| - |-5|)$
4

99. $4[2 - (5 - |-3|)^2]$
-8

100. $[5 - (-3)]^2 + [4 - (-2)]^2$
100

Solve each problem. See Example 9.

101. *Female target heart rate.* The algebraic expression $0.65(220 - A)$ gives the target heart rate for beneficial exercise for women, where A is the age of the woman.

How much larger is the target heart rate of a 25-year-old woman than that of a 65-year-old woman? Use the accompanying graph to estimate the age at which a woman's target heart rate is 115.   26 beats per minute, age 43

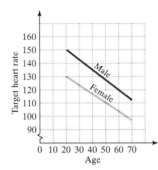

FIGURE FOR EXERCISES 101 AND 102

102. *Male target heart rate.* The algebraic expression $0.75(220 - A)$ gives the target heart rate for beneficial exercise for men, where A is the age of the man. Use the algebraic expression to find the target heart rate for a 20-year-old and a 50-year-old man. Use the accompanying graph to estimate the age at which a man's target heart rate is 115.   150, 127.5, age 67

Solve each problem.

103. *Perimeter of a pool.* The algebraic expression $2L + 2W$ gives the perimeter of a rectangle with length L and width W. Find the perimeter of a rectangular swimming pool that has length 34 feet and width 18 feet.   104 feet

104. *Area of a lot.* The algebraic expression for the area of a trapezoid, $0.5h(b_1 + b_2)$, gives the area of the property shown in the figure. Find the area if $h = 150$ feet, $b_1 = 260$ feet, and $b_2 = 220$ feet.
$36{,}000$ ft^2 (square feet)

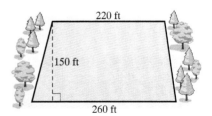

220 ft

150 ft

260 ft

FIGURE FOR EXERCISE 104

105. *Saving for retirement.* The expression $P(1 + r)^n$ gives the amount of an investment of P dollars invested for n years at interest rate r compounded annually. Long-term corporate bonds have had an average yield of 6.2% annually over the last 40 years (Fidelity Investments, www.fidelity.com).

a) Use the accompanying graph to estimate the amount of a $10,000 investment in corporate bonds after 30 years. $60,000

b) Use the given expression to calculate the value of a $10,000 investment after 30 years of growth at 6.2% compounded annually. $60,776.47

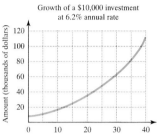
Growth of a $10,000 investment at 6.2% annual rate

FIGURE FOR EXERCISE 105

 106. *Saving for college.* The average cost of a B.A. in 2005 will be $252,000 at an Ivy League school (*Fortune Investors Guide*). The principal that must be invested at interest rate r compounded annually to have A dollars n years in the future is given by the algebraic expression

$$\frac{A}{(1 + r)^n}.$$

What investment in 1987 would amount to $252,000 in 2005 at 7% compounded annually? $74,557.71

 107. *Student loan.* A college student borrowed $4,000 at 8% compounded annually in her freshman year and did not have to make payments until 4 years later. Use the accompanying graph to estimate the amount that she owes at the time the payments start. Use the expression

$P(1 + r)^n$ to find the actual amount of the debt at the time the payments start. $5,500, $5,441.96

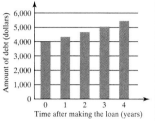 

FIGURE FOR EXERCISE 107

 108. *High cost of nursing care.* The average cost for a one-year stay in a nursing home in 1990 was $29,930 (*Fortune Investors Guide*). In n years from 1990 the average cost will be $29{,}930(1.05)^n$ dollars. Find the projected cost for a one-year stay in 2005. $62,222

 109. *Soaring cost of nursing care.* Some economists project that the average cost of a one-year stay in a nursing home n years from 1990 will be $29{,}930(1.08)^n$ dollars. How much more would you pay for a one-year stay in 2005 using this expression rather than the expression in the last exercise? $32,721

GETTING MORE INVOLVED

 110. *Discussion.* Evaluate $5(5(5 \cdot 3 + 6) + 4) + 7$ and $3 \cdot 5^3 + 6 \cdot 5^2 + 4 \cdot 5 + 7$. Explain why these two expressions must have the same value.

 111. *Cooperative learning.* Find some examples of algebraic expressions that are not mentioned in this text and explain to your class what they are used for.

1.5 **PROPERTIES OF THE REAL NUMBERS**

You know that the price of a hamburger plus the price of a Coke is the same as the price of a Coke plus the price of a hamburger. But, do you know which property of the real numbers is at work in this situation? In arithmetic we may be unaware when to use properties of the real numbers, but in algebra we need a better understanding of those properties. In this section we will study the properties of the basic operations on the set of real numbers.

Commutative Properties

We get the same result whether we evaluate $3 + 7$ or $7 + 3$. With multiplication, we have $4 \cdot 5 = 5 \cdot 4$. These examples illustrate the commutative properties.

Commutative Property of Addition

For any real numbers a and b,
$$a + b = b + a.$$

Commutative Property of Multiplication

For any real numbers a and b,
$$ab = ba.$$

In writing the product of a number and a variable, it is customary to write the number first. We write $3x$ rather than $x3$. In writing the product of two variables, it is customary to write them in alphabetical order. We write cd rather than dc.

Addition and multiplication are commutative operations, but what about subtraction and division? Because $7 - 3 = 4$ and $3 - 7 = -4$, subtraction is not commutative. To see that division is not commutative, consider the amount each person gets when a \$1 million lottery prize is divided between two people and when a \$2 prize is divided among 1 million people.

Associative Properties

Consider the expression $2 + 3 + 7$. Using the order of operations, we add from left to right to get 12. If we first add 3 and 7 to get 10 and then add 2 and 10, we also get 12. So
$$(2 + 3) + 7 = 2 + (3 + 7).$$
Now consider the expression $2 \cdot 3 \cdot 5$. Using the order of operations, we multiply from left to right to get 30. However, we can first multiply 3 and 5 to get 15 and then multiply by 2 to get 30. So
$$(2 \cdot 3) \cdot 5 = 2 \cdot (3 \cdot 5).$$
These examples illustrate the associative properties.

Associative Property of Addition

For any real numbers a, b, and c,
$$(a + b) + c = a + (b + c).$$

Associative Property of Multiplication

For any real numbers a, b, and c,
$$(ab)c = a(bc).$$

Consider the expression
$$4 - 9 + 8 - 5 - 8 + 6 - 13.$$
According to the accepted order of operations, we could evaluate this expression by computing from left to right. However, if we use the definition of subtraction, we can rewrite this expression as
$$4 + (-9) + 8 + (-5) + (-8) + 6 + (-13).$$
The commutative and associative properties of addition allow us to add these numbers in any order we choose. A good way to add them is to add the positive numbers,

add the negative numbers, and then combine the two totals:

$$4 + 8 + 6 + (-9) + (-5) + (-8) + (-13) = 18 + (-35) = -17$$

For speed we usually do not rewrite the expression. We just sum the positive numbers and sum the negative numbers, and then combine their totals.

E X A M P L E   1   **Using commutative and associative properties**

Evaluate.

a) $4 - 7 + 10 - 5$ **b)** $6 - 5 - 9 + 7 - 2 + 5 - 8$

Solution

a) $4 - 7 + 10 - 5 = 14 + (-12) = 2$

        ↑   ↑

Sum of the positive   Sum of the negative
    numbers   numbers

b) $6 - 5 - 9 + 7 - 2 + 5 - 8 = 18 + (-24) = -6$

Not all operations are associative. Using subtraction, for example, we have

$$(8 - 4) - 1 \neq 8 - (4 - 1)$$

because $(8 - 4) - 1 = 3$ and $8 - (4 - 1) = 5$. For division we have

$$(8 \div 4) \div 2 \neq 8 \div (4 \div 2)$$

because $(8 \div 4) \div 2 = 1$ and $8 \div (4 \div 2) = 4$. So subtraction and division are not associative.

Distributive Property

Imagine a parade in which 6 rows of horses are followed by 4 rows of horses with 3 horses in each row.

+ + + + + +   + + + +
+ + + + + +   + + + +
+ + + + + +   + + + +

There are 10 rows of 3 horses or 30 horses, or there are 18 horses followed by 12 horses for a total of 30 horses.

Using the order of operations, we evaluate the product $3(6 + 4)$ first by adding 6 and 4 and then multiplying by 3:

$$3(6 + 4) = 3 \cdot 10 = 30$$

Note that we also have

$$3 \cdot 6 + 3 \cdot 4 = 18 + 12 = 30.$$

Therefore

$$3(6 + 4) = 3 \cdot 6 + 3 \cdot 4.$$

Note that multiplication by 3 from outside the parentheses is *distributed* over each term inside the parentheses. This example illustrates the distributive property.

Distributive Property

For any real numbers a, b, and c,

$$a(b + c) = ab + ac.$$

Because subtraction is defined in terms of addition, multiplication distributes over subtraction as well as over addition. For example,

$$3(x - 2) = 3(x + (-2))$$
$$= 3x + (-6)$$
$$= 3x - 6.$$

Because multiplication is commutative, we can write the multiplication before or after the parentheses. For example,

$$(y + 6)3 = 3(y + 6)$$
$$= 3y + 18.$$

The distributive property is used in two ways. If we start with the product $5(x + 4)$ and write

$$5(x + 4) = 5x + 20,$$

we are writing a product as a sum. We are removing the parentheses. If we start with the difference $6x - 18$ and write

$$6x - 18 = 6(x - 3),$$

we are using the distributive property to write a difference as a product.

E X A M P L E   2

Using the distributive property

Use the distributive property to rewrite each sum or difference as a product and each product as a sum or difference.

a) $9x - 9$ **b)** $b(2 - a)$ **c)** $3a + ac$ **d)** $-2(x - 3)$

Solution

a) $9x - 9 = 9(x - 1)$

b) $b(2 - a) = 2b - ab$ Note that $b \cdot 2 = 2b$ by the commutative property.

c) $3a + ac = a(3 + c)$

d) $-2(x - 3) = -2x - (-2)(3)$
$$= -2x - (-6)$$
$$= -2x + 6$$

study tip

Get to class early so that you are relaxed and ready to go when class starts. Collect your thoughts and get your questions ready. If your instructor arrives early, you might be able to have your questions answered before class. Take responsibility for your education. Many come to learn, but not all learn.

Identity Properties

The numbers 0 and 1 have special properties. Addition of 0 to a number does not change the number, and multiplication of a number by 1 does not change the number. For this reason, 0 is called the **additive identity** and 1 is called the **multiplicative identity.**

Additive Identity Property

For any real number a,

$$a + 0 = 0 + a = a.$$

Multiplicative Identity Property

For any real number a,

$$a \cdot 1 = 1 \cdot a = a.$$

Inverse Properties

The ideas of *additive inverses* and *multiplicative inverses* were introduced in Section 1.3. Every real number a has a unique additive inverse or opposite, $-a$, such that $a + (-a) = 0$. Every nonzero real number a also has a unique multiplicative inverse (reciprocal), written $\frac{1}{a}$, such that $a\left(\frac{1}{a}\right) = 1$. For rational numbers the multiplicative inverse is easy to find. For example, the multiplicative inverse of $\frac{2}{5}$ is $\frac{5}{2}$ because

$$\frac{2}{5} \cdot \frac{5}{2} = \frac{10}{10} = 1.$$

calculator

close-up

Most scientific calculators have a key labeled $1/x$, which gives the reciprocal of the number on the display. Graphing calculators do not have a reciprocal key, but you can find reciprocals as shown here.

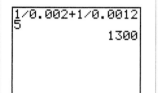

Additive Inverse Property

For any real number a, there is a unique number $-a$ such that
$$a + (-a) = -a + a = 0.$$

Multiplicative Inverse Property

For any nonzero real number a, there is a unique number $\frac{1}{a}$ such that
$$a \cdot \frac{1}{a} = \frac{1}{a} \cdot a = 1.$$

Reciprocals are used in problems involving rates. For example, if Brandon washes one car in $\frac{1}{3}$ of an hour, then he is washing cars at the rate of $1/\left(\frac{1}{3}\right)$ or 3 cars/hour (3 cars per hour). If Gilda washes one car in $\frac{1}{4}$ of an hour, then she is washing at the rate of $1/\left(\frac{1}{4}\right)$ or 4 cars/hour. In general, if one task is completed in x hours, then the rate is $\frac{1}{x}$ tasks/hour. If Brandon and Gilda maintain the same rates when working together, then their rate together is the sum of their individual rates, or 7 cars/hour.

EXAMPLE 3

Work rates

An old computer system can process one water bill in 0.002 hour. A new computer system can process one water bill in 0.00125 hour. If the old system is used simultaneously with the new one, then at what rate will the processing of the water bills be accomplished?

Solution

Since the old system does one bill in 0.002 hour, its rate is $\frac{1}{0.002}$ bills per hour. Since the new system does one bill in 0.00125 hour, its rate is $\frac{1}{0.00125}$ bills per hour. Their rate when working together is the sum of their individual rates:

$$\frac{1}{0.002} + \frac{1}{0.00125} = 1300$$

They are working together at the rate of 1300 bills per hour.

Multiplication Property of Zero

Zero has a property that no other number has. Multiplication involving zero always results in zero. It is the multiplication property of zero that prevents 0 from having a reciprocal.

Multiplication Property of Zero

For any real number a,

$$0 \cdot a = a \cdot 0 = 0.$$

E X A M P L E 4

Recognizing properties

Identify the property that is illustrated in each case.

a) $5 \cdot 9 = 9 \cdot 5$

b) $3 \cdot \dfrac{1}{3} = 1$

c) $1 \cdot 865 = 865$

d) $3 + (5 + a) = (3 + 5) + a$

e) $4x + 6x = (4 + 6)x$

f) $7 + (x + 3) = 7 + (3 + x)$

g) $4567 \cdot 0 = 0$

h) $239 + 0 = 239$

i) $-8 + 8 = 0$

j) $-4(x - 5) = -4x + 20$

Solution

a) Commutative property of multiplication

b) Multiplicative inverse property

c) Multiplicative identity property

d) Associative property of addition

e) Distributive property

f) Commutative property of addition

g) Multiplication property of zero

h) Additive identity property

i) Additive inverse property

j) Distributive property

WARM-UPS

True or false? Explain your answer.

1. Addition is a commutative operation.   True
2. $8 \div (4 \div 2) = (8 \div 4) \div 2$ False
3. $10 \div 2 = 2 \div 10$ False
4. $5 - 3 = 3 - 5$ False
5. $10 - (7 - 3) = (10 - 7) - 3$ False
6. $4(6 \div 2) = (4 \cdot 6) \div (4 \cdot 2)$ False
7. The multiplicative inverse of 0.02 is 50.   True
8. Division is not an associative operation.   True
9. $3 + 2x = 5x$ for any value of x.   False
10. A machine that washes one car in 0.04 hour is washing at the rate of 25 cars per hour.   True

1.5   EXERCISES

Reading and Writing   *After reading this section, write out the answers to these questions. Use complete sentences.*

1. What are the commutative properties?
The commutative property of addition says that $a + b = b + a$ and the commutative property of multiplication says that $a \cdot b = b \cdot a$.

2. What are the associative properties?
The associative property of addition says that $(a + b) + c = a + (b + c)$.

3. What is the difference between the commutative property of addition and the associative property of addition?
The commutative property of addition says that you get the same result when you add two numbers in either order. The associative property of addition deals with which two numbers are added first when adding three numbers.

4. What is the distributive property?
The distributive property says that $a(b + c) = ab + ac$.

5. Why is 0 called the additive identity?
Zero is the additive identity because adding zero to a number does not change the number.

6. Why is 1 called the multiplicative identity?
One is the multiplicative identity because multiplying a number by 1 does not change the number.

Evaluate. See Example 1.

7. $9 - 4 + 6 - 10$ 1
8. $-3 + 4 - 12 + 9$ -2
9. $6 - 10 + 5 - 8 - 7$ -14
10. $5 - 11 + 6 - 9 + 12 - 2$ 1
11. $-4 - 11 + 6 - 8 + 13 - 20$ -24
12. $-8 + 12 - 9 - 15 + 6 - 22 + 3$ -33
13. $-3.2 + 1.4 - 2.8 + 4.5 - 1.6$ -1.7
14. $4.4 - 5.1 + 3.6 - 2.3 + 8.1$ 8.7
15. $3.27 - 11.41 + 5.7 - 12.36 - 5$ -19.8
16. $4.89 - 2.1 + 7.58 - 9.06 - 5.34$ -4.03

Use the distributive property to rewrite each sum or difference as a product and each product as a sum or difference. See Example 2.

17. $4(x - 6)$ $4x - 24$
18. $5(a - 1)$ $5a - 5$
19. $2m + 10$ $2(m + 5)$
20. $3y + 9$ $3(y + 3)$
21. $a(3 + t)$ $3a + at$
22. $b(y + w)$ $by + bw$
23. $-2(w - 5)$ $-2w + 10$
24. $-4(m - 7)$ $-4m + 28$
25. $-2(3 - y)$ $-6 + 2y$
26. $-5(4 - p)$ $-20 + 5p$
27. $5x - 5$ $5(x - 1)$
28. $3y + 3$ $3(y + 1)$
29. $-1(-2x - y)$ $2x + y$
30. $-1(-4y - w)$ $4y + w$
31. $-3(-2w - 3y)$ $6w + 9y$
32. $-4(-x - 6)$ $4x + 24$
33. $3y - 15$ $3(y - 5)$
34. $5x + 10$ $5(x + 2)$
35. $3a + 9$ $3(a + 3)$
36. $7b - 49$ $7(b - 7)$

37. $\frac{1}{2}(4x + 8)$ $2x + 4$
38. $\frac{1}{3}(3x + 6)$ $x + 2$
39. $-\frac{1}{2}(2x - 4)$ $-x + 2$
40. $-\frac{1}{3}(9x - 3)$ $-3x + 1$

Find the multiplicative inverse (reciprocal) of each number.

41. $\frac{1}{2}$ 2
42. $\frac{1}{3}$ 3
43. 1   1
44. -1 -1
45. 6 $\frac{1}{6}$
46. 8 $\frac{1}{8}$
47. 0.25   4
48. 0.75 $\frac{4}{3}$
49. -0.7 $-\frac{10}{7}$
50. -0.9 $-\frac{10}{9}$
51. -1.8 $-\frac{5}{9}$
52. -2.6 $-\frac{5}{13}$

  Use a calculator to evaluate each expression. Round answers to four decimal places.

53. $\frac{1}{2.3} + \frac{1}{5.4}$ 0.6200
54. $\frac{1}{13.5} - \frac{1}{4.6}$ -0.1433
55. $\dfrac{\frac{1}{4.3}}{\frac{1}{5.6} + \frac{1}{7.2}}$ 0.7326
56. $\dfrac{\frac{1}{4.5} - \frac{1}{5.6}}{\frac{1}{3.2} + \frac{1}{2.7}}$ 0.0639

  Solve each problem. See Example 3.

57. *Fastest airliner.* The world's fastest airliner, the Concorde, travels one mile in 0.0006897 hour and carries 128 passengers (*The Doubleday Almanac*). Find its rate in miles per hour.   1,450 mph

58. *Fastest jet plane.* The U.S. Lockheed SR-71 is the world's fastest jet plane. The SR-71 can travel one mile in 0.000456 hour (*The Doubleday Almanac*). Find its rate in miles per hour.   2,193 mph

59. *Who's got the button.* A small clothing factory has three workers who attach buttons. Rita, Mary, and Sam can attach a single button in 0.01 hour, 0.02 hour, and 0.015 hour, respectively. At what hourly rate are they attaching buttons when working simultaneously?   217 buttons per hour

60. *Modern art.* Emilio can paint the exterior of a certain house in 36.5 hours. Alex can paint the same house in 30 hours. If they work together without interfering with each other, then at what hourly rate will the house be painted?   0.06 house per hour

Name the property that is illustrated in each case. See Example 4.

61. $3 + x = x + 3$ Commutative property of addition
62. $x \cdot 5 = 5x$ Commutative property of multiplication
63. $5(x - 7) = 5x - 35$ Distributive property
64. $a(3b) = (a \cdot 3)b$ Associative property of multiplication

65. $3(xy) = (3x)y$ Associative property of multiplication
66. $3(x - 1) = 3x - 3$ Distributive property
67. $4(0.25) = 1$ Multiplicative inverse property
68. $0.3 + 9 = 9 + 0.3$ Commutative property of addition
69. $y^3x = xy^3$ Commutative property of multiplication
70. $0 \cdot 52 = 0$ Multiplication property of zero
71. $1 \cdot x = x$ Multiplicative identity property
72. $(0.1)(10) = 1$ Multiplicative inverse property
73. $2x + 3x = (2 + 3)x$ Distributive property
74. $8 + 0 = 8$ Additive identity property
75. $7 + (-7) = 0$ Additive inverse property
76. $1 \cdot y = y$ Multiplicative identity property
77. $(36 + 79)0 = 0$ Multiplication property of zero
78. $5x + 5 = 5(x + 1)$ Distributive property
79. $xy + x = x(y + 1)$ Distributive property
80. $ab + 3ac = a(b + 3c)$ Distributive property

Complete each statement using the property named.

81. $5 + w =$ _____, commutative property of addition
$w + 5$
82. $2x + 2 =$ _____, distributive property
$2(x + 1)$
83. $5(xy) =$ _____, associative property of multiplication
$(5x)y$
84. $x + \dfrac{1}{2} =$ _____, commutative property of addition

$\dfrac{1}{2} + x$

85. $\dfrac{1}{2}x - \dfrac{1}{2} =$ _____, distributive property $\dfrac{1}{2}(x - 1)$
86. $3(x - 7) =$ _____, distributive property $3x - 21$
87. $6x + 9 =$ _____, distributive property $3(2x + 3)$
88. $(x + 7) + 3 =$ _____, associative property of addition
$x + (7 + 3)$
89. $8(0.125) =$ _____, multiplicative inverse property 1
90. $-1(a - 3) =$ _____, distributive property $-a + 3$
91. $0 = 5($_____$)$, multiplication property of zero 0
92. $8 \cdot ($_____$) = 8$, multiplicative identity property 1
93. $0.25 ($_____$) = 1$, multiplicative inverse property 4
94. $45(1) =$ _____, multiplicative identity property 45

GETTING MORE INVOLVED

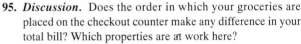

95. *Discussion.* Does the order in which your groceries are placed on the checkout counter make any difference in your total bill? Which properties are at work here?

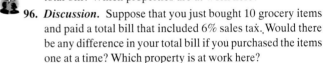

96. *Discussion.* Suppose that you just bought 10 grocery items and paid a total bill that included 6% sales tax. Would there be any difference in your total bill if you purchased the items one at a time? Which property is at work here?

1.6   USING THE PROPERTIES

The properties of the real numbers can be helpful when we are doing computations. In this section we will see how the properties can be applied in arithmetic and algebra.

In this section

- Using the Properties in Computation
- Like Terms
- Combining Like Terms
- Products and Quotients
- Removing Parentheses

Using the Properties in Computation

Consider the product of 36 and 200. Using the associative property of multiplication, we can write

$$(36)(200) = (36)(2 \cdot 100) = (36 \cdot 2)(100).$$

To find this product mentally, first multiply 36 by 2 to get 72, then multiply 72 by 100 to get 7200.

EXAMPLE 1

Using properties in computation

Evaluate each expression mentally by using an appropriate property.

a) $536 + 25 + 75$ **b)** $5 \cdot 426 \cdot \dfrac{1}{5}$ **c)** $7 \cdot 45 + 3 \cdot 45$

Solution

a) To perform this addition mentally, the associative property of addition can be applied as follows:

$$536 + (25 + 75) = 536 + 100 = 636$$

b) Use the commutative and associative properties of multiplication to rearrange mentally this product.

$$5 \cdot 426 \cdot \frac{1}{5} = 426 \cdot 5 \cdot \frac{1}{5} \qquad \text{Commutative property of multiplication}$$

$$= 426\left(5 \cdot \frac{1}{5}\right) \qquad \text{Associative property of multiplication}$$

$$= 426 \cdot 1 \qquad \text{Multiplicative inverse property}$$

$$= 426$$

c) Use the distributive property to rewrite the expression, then evaluate it.

$$7 \cdot 45 + 3 \cdot 45 = (7 + 3)45 = 10 \cdot 45 = 450$$

Like Terms

The properties of the real numbers are used also with algebraic expressions. Simple algebraic expressions such as

$$-2, \qquad 4x, \qquad -5x^2y, \qquad b, \qquad \text{and} \qquad -abc$$

are called terms. A **term** is a single number or the product of a number and one or more variables raised to powers. The number preceding the variables in a term is called the **coefficient.** In the term $4x$ the coefficient of x is 4. In the term $-5x^2y$ the coefficient of x^2y is -5. In the term b the coefficient of b is 1, and in the term $-abc$ the coefficient of abc is -1. If two terms contain the same variables with the same powers, they are called **like terms.** For example, $3x^2$ and $-5x^2$ are like terms, whereas $3x^2$ and $-2x^3$ are not like terms.

Combining Like Terms

We can combine any two like terms involved in a sum by using the distributive property. For example,

$$2x + 5x = (2 + 5)x \qquad \text{Distributive property}$$

$$= 7x \qquad \text{Add 2 and 5.}$$

Because the distributive property is valid for any real numbers, we have $2x + 5x = 7x$ for any real number x.

    We can also use the distributive property to combine any two like terms involved in a difference. For example,

$$-3xy - (-2xy) = [-3 - (-2)]xy \qquad \text{Distributive property}$$

$$= -1xy \qquad \text{Subtract.}$$

$$= -xy \qquad \text{Multiplying by } -1 \text{ is the same as taking the opposite.}$$

Of course, we do not want to write out these steps every time we combine like terms. We can combine like terms as easily as we can add or subtract their coefficients.

E X A M P L E　2

Combining like terms

Perform the indicated operation.

a) $b + 3b$　　　　　　　　　　　**b)** $5x^2 - 7x^2$

c) $5xy - (-13xy)$　　　　　　　**d)** $-2a + (-9a)$

Solution

a) $b + 3b = 1b + 3b = 4b$　　　　**b)** $5x^2 - 7x^2 = -2x^2$

c) $5xy - (-13xy) = 18xy$　　　　**d)** $-2a + (-9a) = -11a$　　　■

CAUTION　The distributive property allows us to combine only *like* terms. Expressions such as

$$3xw + 5, \qquad 7xy + 9t, \qquad 5b + 6a, \qquad \text{and} \qquad 6x^2 + 7x$$

do not contain like terms, so their terms cannot be combined.

Products and Quotients

We can use the associative property of multiplication to simplify the product of two terms. For example,

$$4(7x) = (4 \cdot 7)x \quad \text{Associative property of multiplication}$$
$$= (28)x$$
$$= 28x \qquad \text{Remove unnecessary parentheses.}$$

CAUTION　Multiplication does not distribute over multiplication. For example, $2(3 \cdot 4) \neq 6 \cdot 8$ because $2(3 \cdot 4) = 2(12) = 24$.

In the next example we use the fact that dividing by 3 is equivalent to multiplying by $\frac{1}{3}$, the reciprocal of 3:

$$3\left(\frac{x}{3}\right) = 3\left(x \cdot \frac{1}{3}\right) \quad \text{Definition of division}$$

$$= 3\left(\frac{1}{3} \cdot x\right) \quad \text{Commutative property of multiplication}$$

$$= \left(3 \cdot \frac{1}{3}\right)x \quad \text{Associative property of multiplication}$$

$$= 1 \cdot x \qquad 3 \cdot \frac{1}{3} = 1 \text{ (Multiplicative inverse property)}$$

$$= x \qquad \text{Multiplicative identity property}$$

To find the product $(3x)(5x)$, we use both the commutative and associative properties of multiplication:

$$(3x)(5x) = (3x \cdot 5)x \qquad \text{Associative property of multiplication}$$
$$= (3 \cdot 5x)x \qquad \text{Commutative property of multiplication}$$
$$= (3 \cdot 5)(x \cdot x) \qquad \text{Associative property of multiplication}$$
$$= (15)(x^2) \qquad \text{Simplify.}$$
$$= 15x^2 \qquad \text{Remove unnecessary parentheses.}$$

All of the steps in finding the product $(3x)(5x)$ are shown here to illustrate that every step is justified by a property. However, you should write $(3x)(5x) = 15x^2$ without doing any intermediate steps.

E X A M P L E   3

Multiplying terms

Find each product.

a) $(-5)(6x)$ b) $(-3a)(-8a)$ c) $(-4y)(-6)$ d) $(-5a)\left(\dfrac{b}{5}\right)$

Solution

a) $-30x$ b) $24a^2$ c) $24y$ d) $-ab$ ■

In the next example we use the properties to find quotients. Try to identify the property that is used at each step.

E X A M P L E   4

Dividing terms

Find each quotient.

a) $\dfrac{5x}{5}$

b) $\dfrac{4x + 8}{2}$

Solution

a) First use the definition of division to change the division by 5 to multiplication by $\frac{1}{5}$.

$$\frac{5x}{5} = 5x \cdot \frac{1}{5} = \left(\frac{1}{5} \cdot 5\right)x = 1 \cdot x = x$$

b) First use the definition of division to change division by 2 to multiplication by $\frac{1}{2}$.

$$\frac{4x + 8}{2} = (4x + 8) \cdot \frac{1}{2} = \frac{1}{2} \cdot (4x + 8) = 2x + 4$$

Since both $4x$ and 8 are divided by 2, we could have written

$$\frac{4x + 8}{2} = \frac{4x}{2} + \frac{8}{2} = 2x + 4.$$ ■

C A U T I O N   Do not divide a number into just one term of a sum. For example,

$$\frac{2 + 7}{2} \neq 1 + 7$$

because

$$\frac{2 + 7}{2} = \frac{9}{2} \qquad \text{and} \qquad 1 + 7 = 8.$$

Removing Parentheses

Multiplying a number by -1 merely changes the sign of the number. For example,

$$(-1)(6) = -6 \quad \text{and} \quad (-1)(-15) = 15.$$

Thus -1 times a number is the same as the *opposite* of the number. Using variables, we have

$$(-1)x = -x \quad \text{or} \quad -1(a + 2) = -(a + 2).$$

When a minus sign appears in front of a sum, we can think of it as multiplication by -1 and use the distributive property. For example,

$$\begin{aligned}
-(a + 2) &= -1(a + 2) \\
&= (-1)a + (-1)2 \quad \text{Distributive property} \\
&= -a + (-2) \\
&= -a - 2.
\end{aligned}$$

If a minus sign occurs in front of a difference, we can rewrite the expression as a sum. For example,

$$-(x - 5) = -1(x - 5) = (-1)x - (-1)5 = -x + 5.$$

Note that a minus sign in front of a set of parentheses affects each term in the parentheses, changing the sign of each term.

E X A M P L E   5   **Removing parentheses**

Simplify each expression.

a) $6 - (x + 8)$ **b)** $4x - 6 - (7x - 4)$ **c)** $3x - (-x + 7)$

Solution

a) $\begin{aligned}[t]
6 - (x + 8) &= 6 - x - 8 \quad \text{Change the sign of each term in parentheses.} \\
&= 6 - 8 - x \quad \text{Rearrange the terms.} \\
&= -2 - x \quad \text{Combine like terms.}
\end{aligned}$

b) $\begin{aligned}[t]
4x - 6 - (7x - 4) &= 4x - 6 - 7x + 4 \quad \text{Remove parentheses.} \\
&= 4x - 7x - 6 + 4 \quad \text{Rearrange the terms.} \\
&= -3x - 2 \quad \text{Combine like terms.}
\end{aligned}$

c) $\begin{aligned}[t]
3x - (-x + 7) &= 3x + x - 7 \quad \text{Remove parentheses.} \\
&= 4x - 7 \quad \text{Combine like terms.}
\end{aligned}$

The commutative and associative properties of addition allow us to rearrange the terms so that we may combine like terms. However, it is not necessary actually to write down the rearrangement. We can identify like terms and combine them without rearranging.

E X A M P L E   6   **More parentheses and like terms**

Simplify each expression.

a) $(-5x + 7) + (2x - 9)$ **b)** $-4x + 7x + 3(2 - 5x)$
c) $-3x(4x - 9) - (x - 5)$ **d)** $x - 0.03(x + 300)$

Solution

a) $(-5x + 7) + (2x - 9) = -3x - 2$ Combine like terms.

b) $-4x + 7x + 3(2 - 5x) = -4x + 7x + 6 - 15x$ Distributive property

$$= -12x + 6 \qquad \text{Combine like terms.}$$

c) $-3x(4x - 9) - (x - 5) = -12x^2 + 27x - x + 5$ Remove parentheses.

$$= -12x^2 + 26x + 5 \qquad \text{Combine like terms.}$$

d) $x - 0.03(x + 300) = 1x - 0.03x - 9$ Distributive property; $(-0.03)(300) = -9$

$$= 0.97x - 9 \qquad \text{Combine like terms: } 1.00 - 0.03 = 0.97$$

WARM-UPS

True or false? Explain your answer.

A statement involving variables should be marked true only if it is true for all values of the variable.

1. $5(x + 7) = 5x + 35$ True   **2.** $-4x + 8 = -4(x + 8)$ False

3. $-1(a - 3) = -(a - 3)$ True   **4.** $5y + 4y = 9y$ True

5. $(2x)(5x) = 10x$ False   **6.** $-2t(5t - 3) = -10t^2 + 6t$ True

7. $a + a = a^2$ False   **8.** $b \cdot b = 2b$ False

9. $1 + 7x = 8x$ False

10. $(3x - 4) - (8x - 1) = -5x - 3$ True

1.6   EXERCISES

Reading and Writing   *After reading this section, write out the answers to these questions. Use complete sentences.*

1. What is a term?

A term is a single number or a product of a number and one or more variables.

2. What are like terms?

Like terms contain the same variables with the same powers.

3. What is the coefficient of a term?

The coefficient of a term is the number preceding the variables.

4. Which property is used to combine like terms?

The distributive property is used to combine like terms.

5. What operations can you perform with unlike terms?

You can multiply and divide unlike terms.

6. How do you remove parentheses that are preceded by a negative sign?

You can remove parentheses preceded by a negative sign by taking the opposite of every term in the parentheses.

Perform each computation. Make use of appropriate rules to simplify each problem. See Example 1.

7. $45(200)$ 9,000   **8.** $25(300)$ 7,500

9. $\dfrac{4}{3}(0.75)$ 1   **10.** $5(0.2)$ 1

11. $(427 + 68) + 32$ 527   **12.** $(194 + 78) + 22$ 294

13. $47 \cdot 4 + 47 \cdot 6$ 470   **14.** $53 \cdot 3 + 53 \cdot 7$ 530

15. $19 \cdot 5 \cdot 2 \cdot \dfrac{1}{5}$ 38   **16.** $17 \cdot 4 \cdot 2 \cdot \dfrac{1}{4}$ 34

17. $(120)(400)$ 48,000   **18.** $150 \cdot 300$ 45,000

19. $13 \cdot 377(-5 + 5)$ 0   **20.** $(456 \cdot 8)\dfrac{1}{8}$ 456

21. $(348 + 5) + 45$ 398   **22.** $(135 + 38) + 12$ 185

23. $\dfrac{2}{3}(1.5)$ 1   **24.** $(1.25)(0.8)$ 1

25. $17 \cdot 101 - 17 \cdot 1$ 1,700

26. $33 \cdot 2 - 12 \cdot 33$ -330

27. $354 + 7 + 8 + 3 + 2$ 374

28. $564 + 35 + 65 + 72 + 28$ 764

29. $(567 + 874)(-2 \cdot 4 + 8)$ 0

30. $(567^2 + 48)[3(-5) + 15]$ 0

Combine like terms where possible. See Example 2.

31. $-4n + 6n$
   $2n$

32. $-3a + 15a$
   $12a$

33. $3w - (-4w)$
   $7w$

34. $3b - (-7b)$
   $10b$

35. $4mw^2 - 15mw^2$
$-11mw^2$

36. $2b^2x - 16b^2x$
$-14b^2x$

37. $-5x - (-2x)$
$-3x$

38. $-11 - 7t$
$-11 - 7t$

39. $-4 - 7z$
$-4 - 7z$

40. $-19m - (-3m)$
$-16m$

41. $4t^2 + 5t^2$
$9t^2$

42. $5a + 4a^2$
$5a + 4a^2$

43. $-4ab + 3a^2b$
$-4ab + 3a^2b$

44. $-7x^2y + 5x^2y$
$-2x^2y$

45. $9mn - mn$
$8mn$

46. $3cm - cm$
$2cm$

47. $x^3y - 3x^3y$
$-2x^3y$

48. $s^4t - 5s^4t$
$-4s^4t$

49. $-kz^6 - kz^6$
$-2kz^6$

50. $m^7w - m^7w$
0

Find each product or quotient. See Examples 3 and 4.

51. $4(7t)$
$28t$

52. $-3(4r)$
$-12r$

53. $(-2x)(-5x)$
$10x^2$

54. $(-3h)(-7h)$
$21h^2$

55. $(-h)(-h)$
h^2

56. $x(-x)$
$-x^2$

57. $7w(-4)$
$-28w$

58. $-5t(-1)$
$5t$

59. $-x(1 - x)$
$-x + x^2$

60. $-p(p - 1)$
$-p^2 + p$

61. $(5k)(5k)$
$25k^2$

62. $(-4y)(-4y)$
$16y^2$

63. $3\left(\dfrac{y}{3}\right)$
y

64. $5z\left(\dfrac{z}{5}\right)$
z^2

65. $9\left(\dfrac{2y}{9}\right)$
$2y$

66. $8\left(\dfrac{y}{8}\right)$
y

67. $\dfrac{6x^3}{2}$
$3x^3$

68. $\dfrac{-8x^2}{4}$
$-2x^2$

69. $\dfrac{3x^2y + 15x}{3}$
$x^2y + 5x$

70. $\dfrac{6xy^2 - 8w}{2}$
$3xy^2 - 4w$

71. $\dfrac{2x - 4}{-2}$
$-x + 2$

72. $\dfrac{-6x - 9}{-3}$
$2x + 3$

73. $\dfrac{-xt + 10}{-2}$
$\dfrac{1}{2}xt - 5$

74. $\dfrac{-2xt^2 + 8}{-4}$
$\dfrac{1}{2}xt^2 - 2$

Simplify each expression. See Example 5.

75. $a - (4a - 1)$ $-3a + 1$

76. $5x - (2x - 7)$ $3x + 7$

77. $6 - (x - 4)$ $10 - x$

78. $9 - (w - 5)$ $14 - w$

79. $4m + 6 - (m + 5)$ $3m + 1$

80. $5 - 6t - (3t + 4)$ $1 - 9t$

81. $-5b - (-at + 7b)$ $-12b + at$

82. $-4x^2 - (-7x^2 + 2y)$ $3x^2 - 2y$

83. $t^2 - 5w - (-2w - t^2)$ $2t^2 - 3w$

84. $n^2 - 6m - (-n^2 - 2m)$ $2n^2 - 4m$

85. $x^2 - (x^2 - y^2 - z)$ $y^2 + z$

86. $5w - (6w - 3xy - zy)$ $3xy + zy - w$

Simplify each expression. See Example 6.

87. $(2x + 7x) + (3 + 5)$ $9x + 8$

88. $(3x + 4x) + (5 + 12)$ $7x + 17$

89. $(-3x + 4) + (5x - 6)$ $2x - 2$

90. $(-4x + 11) + (6x - 8)$ $2x + 3$

91. $4a^2 - 5c - (6a^2 - 7c)$ $-2a^2 + 2c$

92. $3x^2 - 4 - (x^2 - 5)$ $2x^2 + 1$

93. $5(t^2 - 3w) - 2(-3w - t^2)$ $7t^2 - 9w$

94. $6(xy^2 + 2) - 5(-xy^2 - 1)$ $11xy^2 + 17$

95. $-7m + 3(m - 4) + 5m$ $m - 12$

96. $-6m + 4(m - 3) + 7m$ $5m - 12$

97. $8 - 7(k^3 + 3) - 4$ $-7k^3 - 17$

98. $6 + 5(k^3 - 2) - k^3 + 5$ $4k^3 + 1$

99. $x - 0.04(x + 50)$ $0.96x - 2$

100. $x - 0.03(x + 500)$ $0.97x - 15$

101. $0.1(x + 5) - 0.04(x + 50)$ $0.06x - 1.5$

102. $0.06x + 0.14(x + 200)$ $0.2x + 28$

103. $3k + 5 - 2(3k - 4) - k + 3$ $-4k + 16$

104. $5w - 2 + 4(w - 3) - 6(w - 1)$ $3w - 8$

105. $5.7 - 4.5(x - 3.9) - 5.42$ $-4.5x + 17.83$

106. $0.04(5.6x - 4.9) + 0.07(7.3x - 34)$ $0.735x - 2.576$

Simplify.

107. $3(1 - xy) - 2(xy - 5) - (35 - xy)$ $-4xy - 22$

108. $2(x^2 - 3) - (6x^2 - 2) + 2(-7x^2 - 4)$ $-18x^2 - 12$

109. $w \cdot 3w + 5w (-6w) - w(2w)$ $-29w^2$

110. $3w^3 + 5w^3 - 4w^3 + 12w^3 - 2w^2$ $16w^3 - 2w^2$

111. $3a^2w^2 - 5w^2 \cdot a^2 - 2aw \cdot 2aw$ $-6a^2w^2$

112. $-3(aw^2 + 5a^2w) - 2(-a^2w - a^2w)$ $-11a^2w - 3aw^2$

113. $\dfrac{1}{6} - \dfrac{1}{3}\left(-6x^2y - \dfrac{1}{2}\right)$ $2x^2y + \dfrac{1}{3}$

114. $-\dfrac{1}{2}bc - \dfrac{1}{2}bc(3 - a)$ $\dfrac{1}{2}abc - 2bc$

115. $-\dfrac{1}{2}m\left(-\dfrac{1}{2}m\right) - \dfrac{1}{2}m - \dfrac{1}{2}m$ $\dfrac{1}{4}m^2 - m$

116. $\dfrac{4wyt}{4} + \dfrac{-8wyt}{2} - \dfrac{-2wy}{2}$ $-3wyt + wy$

117. $\dfrac{-8t^3 - 6t^2 + 2}{-2}$ $4t^3 + 3t^2 - 1$

118. $\dfrac{7x^3 - 5x^3 - 4}{-2}$ $-x^3 + 2$

119. $\dfrac{-6xyz - 3xy + 9z}{-3}$ $2xyz + xy - 3z$

120. $\dfrac{20a^2b^4 - 10a^2b^4 + 5}{-5}$ $-2a^2b^4 - 1$

Write an algebraic expression for each problem.

121. *Triangle.* The lengths of the sides of a triangular flower bed are s feet, $s + 2$ feet, and $s + 4$ feet. What is its perimeter? $3s + 6$ ft

122. *Parallelogram.* The lengths of the sides of a lot in the shape of a parallelogram are w feet and $w + 50$ feet. What is its perimeter? Is it possible to find the area from this information? $4w + 100$ ft, no

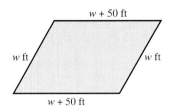

$w + 50$ ft
w ft w ft
$w + 50$ ft

FIGURE FOR EXERCISE 122

123. *Parthenon.* To obtain a pleasing rectangular shape, the ancient Greeks constructed buildings with a length that was about $\frac{1}{6}$ longer than the width. If the width of the

Parthenon is x meters and its length is $x + \frac{1}{6}x$ meters, then what is the perimeter? What is the area?
$\frac{13}{3}x$ m, $\frac{7}{6}x^2$ m²

124. *Square.* If the length of each side of a square sign is x inches, then what are the perimeter and area of the square? $4x$ in., x^2 in.²

FIGURE FOR EXERCISE 124

COLLABORATIVE ACTIVITIES

OOOP! Order of Operations Game

In this game you will be reviewing the established order of operations for real numbers. You may evaluate any compound expression while playing this game. Use a new piece of paper for each game.

1. Before Play Begins

You will need three to five players on a team. Assign each player a role or operation: **E**—Exponents, **M**—Multiply/Divide (may have two players working as a team, one who multiplies and one who divides); **A**—Add/Subtract (may have two players working as a team, one who adds and one who subtracts).

2. Determining a Player's Turn at Play

All players working together analyze the expression and decide which part to complete. If there are parentheses, then the players decide whether what is inside the parentheses needs simplification. **E** *performs his or her operation before* **M,** *and* **M** *performs his or her operation before* **A.** Each player's turn ends when he or she encounters an operation that precedes his or hers or when he or she reaches the end of the expression.

Grouping: Three to five students per group
Topic: Order of operations, learning to work in groups

3. Each Player's Task

> **E—Exponents:** Working left to right, **E** evaluates exponential expressions in order.

> **M—Multiply/Divide:** Working left to right, **M** performs multiplications and divisions in order. Multiplication and division may be done by two players working as a team. If these tasks are split, then the two players perform their assigned operations in order, taking turns as needed.

> **A—Add/Subtract:** Working left to right, **A** performs additions and subtractions in order. Addition and subtraction may be done by two players working as a team. If these tasks are split, then the two players perform their assigned operations in order, taking turns as needed.

4. Recording Results

Results are recorded on one sheet of paper. As each player finishes his or her operation, she or he passes the paper to the player with the next task. Each player rewrites the new form of the expression on the next line of the page and initials his or her work with **E, M,** or **A.**

W R A P - U P

SUMMARY

Sets		**Examples**
Set-builder notation	Notation for describing a set using variables.	$\{x \mid x$ is a natural number smaller than 4$\}$ $D = \{3, 4\}$
Membership	The symbol $\in$ means "is an element of."	$1 \in C, 4 \notin C$
Union	$A \cup B = \{x \mid x \in A$ or $x \in B\}$	$C \cup D = \{1, 2, 3, 4\}$
Intersection	$A \cap B = \{x \mid x \in A$ and $x \in B\}$	$C \cap D = \{3\}$
Subset	A is a subset of B if every element of A is also an element of B. $\varnothing \subseteq A$ for any set A.	$\{1, 2\} \subseteq C$ $\varnothing \subseteq C, \varnothing \subseteq D$

Real Numbers		**Examples**
Rational numbers	$Q = \left\{ \dfrac{a}{b} \mid a$ and b are integers with $b \neq 0 \right\}$	$\dfrac{3}{2}, 5, -6, 0, 0.25252525 \ldots$
Irrational numbers	$I = \{x \mid x$ is a real number that is not rational$\}$	$\sqrt{2}, \sqrt{3}, \pi, 0.1515515551 \ldots$
Real numbers	$R = \{x \mid x$ is the coordinate of a point on the number line$\}$	$R = Q \cup I$

Operations with Real Numbers		**Examples**
Absolute value	$|a| = \begin{cases} a & \text{if } a \text{ is positive or zero} \\ -a & \text{if } a \text{ is negative} \end{cases}$	$|6| = 6, |0| = 0$ $|-6| = 6$
Addition and subtraction	To find the sum of two numbers with the same sign, add their absolute values. The sum has the same sign as the original numbers.	$-2 + (-7) = -9$
	To find the sum of two numbers with unlike signs, subtract their absolute values. The sum is positive if the number with the larger absolute value is positive. The sum is negative if the number with the larger absolute value is negative.	$-6 + 9 = 3$ $-9 + 6 = -3$
	Subtraction: $a - b = a + (-b)$ (Change the sign and add.)	$4 - 7 = 4 + (-7) = -3$ $5 - (-3) = 5 + 3 = 8$

Multiplication and division	To find the product or quotient of two numbers, multiply or divide their absolute values: Same signs $\leftrightarrow$ positive result Opposite signs $\leftrightarrow$ negative result	$(-4)(-2) = 8, (-4)(2) = -8$ $-8 \div (-2) = 4, -8 \div 2 = -4$
Exponential expressions	In the expression a^n, a is the base and n is the exponent.	$2^3 = 2 \cdot 2 \cdot 2 = 8$
Square roots	If $a^2 = b$, then a is a square root of b. If $a \geq 0$ and $a^2 = b$, then $\sqrt{b} = a$.	Both 3 and -3 are square roots of 9. Because $3 \geq 0$, $\sqrt{9} = 3$.
Order of operations	In an expression without parentheses or absolute value: 1. Evaluate exponential expressions. 2. Perform multiplication and division. 3. Perform addition and subtraction. With parentheses or absolute value: First evaluate within each set of parentheses or absolute value, using the above order.	$7 + 2^3 = 15$ $3 + 4 \cdot 6 = 27$ $5 + 4 \cdot 3^2 = 41$ $(2 + 4)(5 - 9) = -24$ $3 + 4\,\lvert\,2 - 3\,\rvert = 7$

Properties of the Real Numbers		**Examples**
	For any real numbers a, b, and c:	
Commutative property of		
addition	$a + b = b + a$	$3 + 7 = 7 + 3$
multiplication	$ab = ba$	$4 \cdot 3 = 3 \cdot 4$
Associative property of		
addition	$(a + b) + c = a + (b + c)$	$(1 + 3) + 5 = 1 + (3 + 5)$
multiplication	$(ab)c = a(bc)$	$(3 \cdot 5)7 = 3(5 \cdot 7)$
Distributive property	$a(b + c) = ab + ac$	$3(4 + x) = 12 + 3x$ $5x - 10 = 5(x - 2)$
Additive identity property	$a + 0 = 0 + a = a$	$6 + 0 = 0 + 6 = 6$
Multiplicative identity property	$1 \cdot a = a \cdot 1 = a$	$1 \cdot 6 = 6 \cdot 1 = 6$
Additive inverse property	$a + (-a) = -a + a = 0$	$8 + (-8) = 0, -8 + 8 = 0$
Multiplicative inverse property	$a \cdot \dfrac{1}{a} = \dfrac{1}{a} \cdot a = 1$ for $a \neq 0$	$8 \cdot \dfrac{1}{8} = 1, -2\left(-\dfrac{1}{2}\right) = 1$
Multiplication property of zero	$0 \cdot a = a \cdot 0 = 0$	$9 \cdot 0 = 0$ $(0)(-4) = 0$

Algebraic Concepts		**Examples**
Algebraic expressions	Any meaningful combination of numbers, variables, and operations	$x^2 + y^2$, $-5abc$
Term	An expression containing a number or the product of a number and one or more variables raised to powers	$3x^2$, $-7x^2y$, 8
Like terms	Terms with identical variable parts	$4bc - 8bc = -4bc$

ENRICHING YOUR MATHEMATICAL WORD POWER

For each mathematical term, choose the correct meaning.

1. **term**
   a. an expression containing a number or the product of a number and one or more variables raised to powers
   b. the amount of time spent in this course
   c. a word that describes a number
   d. a variable   a

2. **like terms**
   a. terms that are identical
   b. the terms of a sum
   c. terms that have the same variables with the same exponents
   d. terms with the same variables   c

3. **variable**
   a. a letter that is used to represent some numbers
   b. the letter x
   c. an equation with a letter in it
   d. not the same   a

4. **additive inverse**
   a. the number -1
   b. the number 0
   c. the opposite of addition
   d. opposite   d

5. **order of operations**
   a. the order in which operations are to be performed in the absence of grouping symbols
   b. the order in which the operations were invented
   c. the order in which operations are written
   d. a list of operations in alphabetical order   a

6. **absolute value**
   a. a definite value
   b. a positive number
   c. the distance from 0 on the number line
   d. the opposite of a number   c

7. **natural numbers**
   a. the counting numbers
   b. numbers that are not irrational
   c. the nonnegative numbers
   d. numbers that we find in nature   a

8. **rational numbers**
   a. the numbers 1, 2, 3, and so on
   b. the integers
   c. numbers that make sense
   d. numbers of the form $\frac{a}{b}$ where a and b are integers with $b \neq 0$ d

9. **irrational numbers**
   a. cube roots
   b. numbers that cannot be expressed as a ratio of integers
   c. numbers that do not make sense
   d. integers   b

10. **additive identity**
    a. the number 0
    b. the number 1
    c. the opposite of a number
    d. when two sums are identical   a

11. **multiplicative identity**
    a. the number 0
    b. the number 1
    c. the reciprocal
    d. when two products are identical   b

12. **dividend**
    a. a in $\dfrac{a}{b}$ b. b in $\dfrac{a}{b}$
    c. the result of $\dfrac{a}{b}$ d. what a bank pays on deposits   a

13. divisor

a: a in $\dfrac{a}{b}$ b: b in $\dfrac{a}{b}$

c: the result of $\dfrac{a}{b}$ d: two visors   b

14. quotient

a: a in $\dfrac{a}{b}$ b: b in $\dfrac{a}{b}$

c: $\dfrac{a}{b}$ d: the divisor plus the remainder   c

REVIEW EXERCISES

1.1 *Let* $A = \{1, 2, 3\}$, $B = \{3, 4, 5\}$, $C = \{1, 2, 3, 4, 5\}$, $D = \{3\}$, *and* $E = \{4, 5\}$. *Determine whether each statement is true or false.*

1. $A \cap B = D$ True

2. $A \cap B = E$ False

3. $A \cup B = E$ False

4. $A \cup B = C$ True

5. $B \cup C = C$ True

6. $A \cap C = B$ False

7. $A \cap \varnothing = A$ False

8. $A \cup \varnothing = \varnothing$ False

9. $(A \cap B) \cup E = B$ True

10. $(C \cap B) \cap A = D$ True

11. $B \subseteq C$ True

12. $A \subseteq E$ False

13. $A = B$ False

14. $B = C$ False

15. $3 \in D$ True

16. $5 \notin A$ True

17. $0 \in E$ False

18. $D \subseteq \varnothing$ False

19. $\varnothing \subseteq E$ True

20. $1 \in A$ True

study tip

Note how the review exercises are arranged according to the sections in this chapter. If you are having trouble with a certain type of problem, refer back to the appropriate section for examples and explanations.

1.2 *Which elements of the set*

$$\left\{-\sqrt{2},\ -1,\ 0,\ 1,\ 1.732,\ \sqrt{3},\ \pi,\ \frac{22}{7},\ 31\right\}$$

are members of the following sets?

21. Whole numbers $\{0, 1, 31\}$

22. Natural numbers $\{1, 31\}$

23. Integers $\{-1, 0, 1, 31\}$

24. Rational numbers $\left\{-1, 0, 1, 1.732, \frac{22}{7}, 31\right\}$

25. Irrational numbers $\{-\sqrt{2}, \sqrt{3}, \pi\}$

26. Real numbers   All

True or false? Explain.

27. The set of whole numbers is a subset of the set of natural numbers.   False

28. Zero is not a real number.   False

29. The set of natural numbers larger than -3 is $\{-2, -1, 0, 1, 2, 3, \ldots\}$.   False

30. The set of rational numbers is finite.   False

31. The ratio of the circumference to the diameter of any circle is exactly 3.14.   False

32. Every terminating decimal number is an integer.   False

33. Every repeating decimal number is a rational number.   True

34. The number $\sqrt{9}$ is an irrational number.   False

35. Zero is both rational and irrational.   False

36. The irrational numbers have no decimal representation.   False

1.3 *Evaluate.*

37. $-4 + 9$ 5

38. $-3 + (-5)$ -8

39. $25 - 37$ -12

40. $-6 - 10$ -16

41. $(-4)(6)$ -24

42. $(-7)(-6)$ 42

43. $(-8) \div (-4)$ 2

44. $40 \div (-8)$ -5

45. $-\dfrac{1}{4} + \dfrac{1}{12}$ $-\dfrac{1}{6}$

46. $\dfrac{1}{3} - \left(-\dfrac{1}{12}\right)$ $\dfrac{5}{12}$

47. $\dfrac{-20}{-2}$ 10

48. $\dfrac{30}{-6}$ -5

49. $-0.04 + 10$ 9.96

50. $-0.05 + (-3)$ -3.05

51. $-6 - (-2)$ -4

52. $-0.2 - (-0.04)$ -0.16

53. $-0.5 + 0.5$ 0

54. $-0.04 \div 0.2$ -0.2

55. $3.2 \div (-0.8)$ -4

56. $(0.2)(-0.9)$ -0.18

57. $0 \div (-0.3545)$ 0

58. $(-6)(-0.5)$ 3

59. $\dfrac{1}{4}(-12)$ -3

60. $-7 - (-9)$ 2

1.4 *Evaluate each expression.*

61. $4 + 7(5)$ 39

62. $(4 + 7)5$ 55

63. $(4 + 7)^2$ 121

64. $4 + 7^2$ 53

65. $5 + 3 \cdot |6 - 4 \cdot 3|$ 23

66. $6 - (7 - 8)$ 7

67. $(6 - 8) - (5 - 9)$ 2

68. $5 - 6 - 8 - 10$ -19

69. $3 - 5(6 - 2 \cdot 5)$ 23

70. $4^2 - 9 + 3^2$ 16

71. $5^2 - (6 + 5)^2$ -96

72. $|3 - 4 \cdot 2| - |5 - 8|$ 2

73. $\sqrt{3^2 + 4^2}$ 5

74. $\sqrt{13^2 - 5^2}$ 12

75. $\dfrac{-4 - 5}{7 - (-2)}$ -1

76. $\dfrac{5 - 9}{2 - 4}$ 2

77. $1 - (0.8)(0.3)$ 0.76

78. $5 - (0.2)(0.1)$ 4.98

79. $(-3)^2 - (4)(-1)(-2)$ 1

80. $3^2 - 4(1)(-3)$ 21

81. $3 - 2|4 - 7|$ -3

82. $3|4 - 6| + 1$ 7

83. $\dfrac{-5 - 3}{-4 + 2}$ 4

84. $\dfrac{-6 + 3}{3 - 9}$ $\dfrac{1}{2}$

Let $a = -2$, $b = 3$, and $c = -1$. Find the value of each algebraic expression.

85. $\sqrt{b^2 - 4ac}$ 1

86. $\sqrt{a^2 + 4b}$ 4

87. $(c - b)(c + b)$ -8

88. $(a + b)(a - b)$ -5

89. $a^2 + 2ab + b^2$ 1

90. $a^2 - 2ab + b^2$ 25

91. $a^3 - b^3$ -35

92. $a^3 + b^3$ 19

93. $\dfrac{b + c}{a + b}$ 2

94. $\dfrac{b - c}{2b - a}$ $\dfrac{1}{2}$

95. $|a - b|$ 5

96. $|b - a|$ 5

97. $(a + b)c$ -1

98. $ac + bc$ -1

1.5 *Name the property that justifies each equation.*

99. $a + x = x + a$ Commutative property of addition

100. $0 \cdot 5 = 0$ Multiplication property of zero

101. $3(x - 1) = 3x - 3$ Distributive property

102. $10 + (-10) = 0$ Additive inverse property

103. $5(2x) = (5 \cdot 2)x$ Associative property of multiplication

104. $w + y = y + w$ Commutative property of addition

105. $1 \cdot y = y$ Multiplicative identity property

106. $4 \cdot \dfrac{1}{4} = 1$ Multiplicative inverse property

107. $5(0.2) = 1$ Multiplicative inverse property

108. $3 \cdot 1 = 3$ Multiplicative identity property

109. $12 \cdot 0 = 0$ Multiplication property of zero

110. $x + 1 = 1 + x$ Commutative property of addition

111. $18 + 0 = 18$ Additive identity property

112. $2w + 2m = 2(w + m)$ Distributive property

113. $-5 + 5 = 0$ Additive inverse property

114. $2 + (3 + 4) = (2 + 3) + 4$
Associative property of addition

Use the distributive property to rewrite each sum or difference as a product, and each product as a sum of difference, whichever is appropriate.

115. $3x - 3a$ $3(x - a)$

116. $5x - 5y$ $5(x - y)$

117. $3(w + 1)$ $3w + 3$

118. $2(m + 14)$ $2m + 28$

119. $7x + 7$ $7(x + 1)$

120. $3w + 3$ $3(w + 1)$

121. $5(x - 5)$ $5x - 25$

122. $13(b - 3)$ $13b - 39$

123. $-3(2x - 5)$ $-6x + 15$

124. $-2(5 - 4x)$ $-10 + 8x$

125. $p - pt$ $p(1 - t)$

126. $ab + b$ $b(a + 1)$

1.6 *Simplify each expression.*

127. $3a + 7 + 4a - 5$ $7a + 2$

128. $2m + 6 + m - 2$ $3m + 4$

129. $5(t - 4) -3(2t - 6)$ $-t - 2$

130. $2(x - 3) + 2(3 - x)$ 0

131. $-(a - 2) + 2 - a$ $-2a + 4$

132. $-(w - y) + 3(y - w)$ $-4w + 4y$

133. $5 - 3(x - 2) + 7(x + 4) - 6$ $4x + 33$

134. $7 - 2(x - 7) + 7 - x$ $-3x + 28$

135. $0.2(x + 0.1) - (x + 0.5)$ $-0.8x - 0.48$

136. $0.1(x - 0.2) - (x + 0.1)$ $-0.9x - 0.12$

137. $0.05(x + 3) - 0.1(x + 20)$ $-0.05x - 1.85$

138. $0.02(x + 100) - 0.2(x + 50)$ $-0.18x - 8$

139. $\dfrac{1}{2}(x + 4) - \dfrac{1}{4}(x - 8)$ $\dfrac{1}{4}x + 4$

140. $\dfrac{1}{2}(2x - 1) + \dfrac{1}{4}(x + 1)$ $\dfrac{5}{4}x - \dfrac{1}{4}$

141. $\dfrac{-9x^2 - 6x + 3}{3}$ $-3x^2 - 2x + 1$

142. $\dfrac{4x - 2}{-2} + \dfrac{4x + 2}{2}$ 2

MISCELLANEOUS

Evaluate the following expressions for $w = 24$, $x = -6$, $y = 6$, and $z = 4$. Name the property or properties used.

143. $32z(x + y)$
0, additive inverse, multiplication property of zero

144. $(wz)\dfrac{1}{w}$ 4, multiplicative inverse, multiplicative identity

145. $768z + 768y$ 7,680, distributive

146. $28z + 28y$ 280, distributive

147. $(12z + x) + y$
48, associative property of addition, additive inverse

148. $(42 + x) + y$
42, associative property of addition, additive inverse, additive identity

149. $752x + 752y$ 0, distributive, additive inverse

150. $37y + 37x$
0, distributive, additive inverse, multiplication property of zero

151. $(47y)\dfrac{z}{w}$
47, associative property of multiplication, multiplicative inverse

152. $3w + 3y$ 90, distributive

153. $(xw)\dfrac{1}{y}$
-24, commutative property of multiplication, associative property of multiplication

154. $(xz)\dfrac{1}{x}$
4, commutative property of multiplication, associative property of multiplication, multiplicative inverse

155. $5(x + y)(z + w)$
0, additive inverse, multiplication property of zero

156. $(4x + 7y)(w + xz)$ 0, multiplication property of zero

Solve each problem.

157. *Teamwork.* Istvan can attach one shingle using a nail gun in 0.2 minute while Robin takes 0.5 minute to attach one shingle using a hammer. How many shingles per minute will they be attaching if they work together?
Seven shingles per minute

158. *Carpeting costs.* Write an algebraic expression for the cost of carpeting a rectangular room that is x yards by $x + 2$ yards if carpeting costs \$20 per square yard?
$20x^2 + 4x$ dollars

159. *Inflationary spiral.* If car prices increase 5% annually, then in n years a car that currently costs P dollars will cost $P(1.05)^n$ dollars.

   a) Use this algebraic expression to predict the price of a new 2010 Camaro Z28 Coupe, if the price of the 1998 model was \$20,470 (Edmund's New Car Prices, www.edmunds.com).

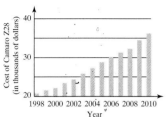 

FIGURE FOR EXERCISE 159

b) Use the accompanying graph to determine the first year in which the price of this car will be over \$30,000.
\$36,761, 2006

160. *Lots of water.* The volume of water in a round swimming pool with radius r feet and depth h feet is $7.5\pi r^2 h$ gallons. Find the volume of water in a pool that has diameter 24 feet and depth 3 feet.
10,179 gallons

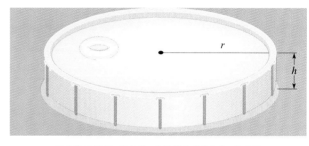

FIGURE FOR EXERCISE 160

CHAPTER 1 TEST

Let $A = \{2, 4, 6, 8, 10\}$, $B = \{3, 4, 5, 6, 7\}$, and $C = \{6, 7, 8, 9, 10\}$. List the elements in each of the following sets.

  1. $A \cup B$ $\{2, 3, 4, 5, 6, 7, 8, 10\}$ **2.** $B \cap C$ $\{6, 7\}$
  3. $A \cap (B \cup C)$ $\{4, 6, 8, 10\}$

Which elements of $\left\{-4, -\sqrt{3}, -\dfrac{1}{2}, 0, 1.65, \sqrt{5}, \pi, 8\right\}$ are members of the following sets?

  4. Whole numbers $\{0, 8\}$ **5.** Integers $\{-4, 0, 8\}$

  6. Rational numbers $\left\{-4, -\dfrac{1}{2}, 0, 1.65, 8\right\}$

  7. Irrational numbers $\{-\sqrt{3}, \sqrt{5}, \pi\}$

Evaluate each expression.

  8. $6 + 3(-5)$ -9
  9. $\sqrt{(-2)^2 - 4(3)(-5)}$ 8
  10. $-5 + 6 - 12$ -11
  11. $0.02 - 2$ -1.98
  12. $\dfrac{-3 - (-7)}{3 - 5}$ -2

13. $\dfrac{-6 - 2}{4 - 2}$ -4

14. $\left(\dfrac{2}{3} - 1\right)\left(\dfrac{1}{3} - \dfrac{1}{2}\right)$ $\dfrac{1}{18}$

15. $-\dfrac{4}{7} - \dfrac{1}{2}\left(24 - \dfrac{8}{7}\right)$ -12

16. $|\, 3 - 5(2)\,|$ 7
17. $5 - 2\,|\,6 - 10\,|$ -3
18. $(452 + 695)[2(-4) + 8]$ 0
19. $478(8) + 478(2)$ $4,780$
20. $-8 \cdot 3 - 4(6 - 9 \cdot 2^3)$ 240
21. $-3 + 5 - 6 \cdot 3 - 4 + 8 - 9 \cdot 2$ -30

Evaluate each expression for $a = -3$, $b = -4$, and $c = 2$.

22. $b^2 - 4ac$ 40 **23.** $\dfrac{a^2 - b^2}{b - a}$ 7 **24.** $\dfrac{ab - 6c}{b^2 - c^2}$ 0

Identify the property that justifies each equation.

25. $2(5 + 7) = 10 + 14$ Distributive property
26. $57 \cdot 4 = 4 \cdot 57$ Commutative property of multiplication
27. $2 + (6 + x) = (2 + 6) + x$
   Associative property of addition
28. $-6 + 6 = 0$ Additive inverse property
29. $1 \cdot (-6) = (-6) \cdot 1$
   Commutative property of multiplication
30. $0 \cdot 28 = 0$ Multiplication property of zero

Simplify each expression.

31. $3(m - 5) - 4(-2m - 3)$ $11m - 3$

32. $x + 3 - 0.05(x + 2)$ $0.95x + 2.9$

33. $\frac{1}{2}(x - 4) + \frac{1}{4}(x + 3)$ $\frac{3}{4}x - \frac{5}{4}$

34. $-3(x^2 - 2y) - 2(3y - 4x^2)$ $5x^2$

35. $\frac{-6x^2 - 4x + 2}{-2}$ $3x^2 + 2x - 1$

36. $\frac{-8xy}{2} - \frac{6xy}{2}$ $-7xy$

Use the distributive property to rewrite each expression as a product.

37. $5x - 40$ $5(x - 8)$ **38.** $7t - 7$ $7(t - 1)$

Solve each problem.

39. If Celeste and her crew of loggers can cut and load one tree in 0.0625 hour, then how many trees per hour can they cut and load?   16 trees per hour

40. The rectangular table for table tennis is x feet long and $x - 4$ feet wide. Write algebraic expressions for the perimeter and the area of the table. Find the actual perimeter and area using $x = 9$.
   $4x - 8, x^2 - 4x$, 28 feet, 45 square feet

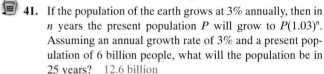 **41.** If the population of the earth grows at 3% annually, then in n years the present population P will grow to $P(1.03)^n$. Assuming an annual growth rate of 3% and a present population of 6 billion people, what will the population be in 25 years?   12.6 billion

Linear Equations and Inequalities in One Variable

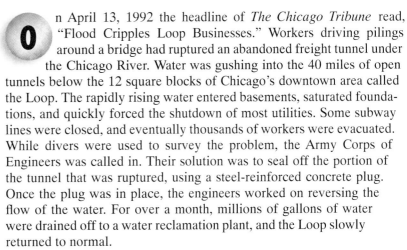

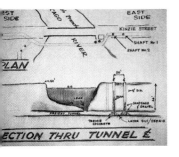

O n April 13, 1992 the headline of *The Chicago Tribune* read, "Flood Cripples Loop Businesses." Workers driving pilings around a bridge had ruptured an abandoned freight tunnel under the Chicago River. Water was gushing into the 40 miles of open tunnels below the 12 square blocks of Chicago's downtown area called the Loop. The rapidly rising water entered basements, saturated foundations, and quickly forced the shutdown of most utilities. Some subway lines were closed, and eventually thousands of workers were evacuated. While divers were used to survey the problem, the Army Corps of Engineers was called in. Their solution was to seal off the portion of the tunnel that was ruptured, using a steel-reinforced concrete plug. Once the plug was in place, the engineers worked on reversing the flow of the water. For over a month, millions of gallons of water were drained off to a water reclamation plant, and the Loop slowly returned to normal.

In this chapter we will study algebraic equations and formulas. In Exercises 85 and 86 of Section 2.2 you will see how the engineers used very simple algebraic formulas to calculate the amount of force the water would have on the plug and how much force the concrete plug would withstand. The story of the Chicago flood is just one example of how algebraic formulas are used daily in practical situations.

2.1 LINEAR EQUATIONS IN ONE VARIABLE

The applications of algebra often lead to equations. The skills that you learned in Chapter 1, such as combining like terms and performing operations with algebraic expressions, will now be used to solve equations.

Basic Ideas and Definitions

An **equation** is a sentence that expresses the equality of two algebraic expressions. Consider the equation

$$2x + 1 = 7.$$

Because $2(3) + 1 = 7$ is true, we say that 3 **satisfies** the equation. No other number in place of x will make the statement $2x + 1 = 7$ true. However, an equation might be satisfied by more than one number. For example, both 3 and -3 satisfy $x^2 = 9$. Any number that satisfies an equation is called a **solution** or **root** to the equation.

> **Solution Set**
>
> The set of all solutions to an equation is called the **solution set** to the equation.

The solution set to $2x + 1 = 7$ is $\{3\}$. To determine whether a number is in the solution set to an equation, we simply replace the variable by the number and see whether the equation is correct.

EXAMPLE 1

Satisfying an equation

Determine whether each equation is satisfied by the number following the equation.

a) $3x + 7 = -8,\quad -5$ **b)** $2(x - 1) = 2x + 3,\quad 4$

Solution

a) Replace x by -5 and evaluate each side of the equation.

$$3x + 7 = -8$$
$$3(-5) + 7 = -8$$
$$-15 + 7 = -8$$
$$-8 = -8 \quad \text{Correct}$$

Because -5 satisfies the equation, -5 is in the solution set to the equation.

b) Replace x by 4 and evaluate each side of the equation.

$$2(x - 1) = 2x + 3$$
$$2(4 - 1) = 2(4) + 3 \quad \text{Replace } x \text{ by 4.}$$
$$2(3) = 8 + 3$$
$$6 = 11 \quad \text{Incorrect}$$

The two sides of the equation have different values when $x = 4$. So 4 is *not* in the solution set to the equation.

Solving Equations

To **solve** an equation means to find its solution set. It is easy to determine whether a given number is in the solution set of an equation as in Example 1, but that example does not provide a method for *solving* equations. The most basic method for solving equations involves the **properties of equality.**

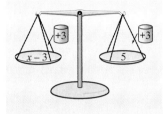
Properties of Equality

Addition Property of Equality
Adding the same number to both sides of an equation does not change the solution set to the equation. In symbols, if $a = b$, then $a + c = b + c$.

Multiplication Property of Equality
Multiplying both sides of an equation by the same nonzero number does not change the solution set to the equation. In symbols, if $a = b$ and $c \neq 0$, then $ca = cb$.

Because subtraction is defined in terms of addition, the addition property of equality also allows us to subtract the same number from both sides. For example, subtracting 3 from both sides is equivalent to adding -3 to both sides. Because division is defined in terms of multiplication, the multiplication property of equality also allows us to divide both sides by the same nonzero number. For example, dividing both sides by 2 is equivalent to multiplying both sides by $\frac{1}{2}$.

Equations that have the same solution set are called **equivalent equations.** In the next example we use the properties of equality to solve an equation by writing an equivalent equation with x isolated on one side of the equation.

E X A M P L E   2

Using the properties of equality
Solve the equation $6 - 3x = 8 - 2x$.

Solution

We want to obtain an equivalent equation with only a single x on the left-hand side and a number on the other side.

$$6 - 3x = 8 - 2x$$
$$6 - 3x - 6 = 8 - 2x - 6 \qquad \text{Subtract 6 from each side.}$$
$$-3x = 2 - 2x \qquad \text{Simplify.}$$
$$-3x + 2x = 2 - 2x + 2x \qquad \text{Add } 2x \text{ to each side.}$$
$$-x = 2 \qquad \text{Combine like terms.}$$
$$-1 \cdot (-x) = -1 \cdot 2 \qquad \text{Multiply each side by } -1.$$
$$x = -2$$

Replacing x by -2 in the original equation gives us

$$6 - 3(-2) = 8 - 2(-2),$$

which is correct. So the solution set to the original equation is $\{-2\}$.

The addition property of equality allows us to add $2x$ to each side of the equation in Example 2 because $2x$ represents a real number.

CAUTION   If you add an expression to each side that does not always represent a real number, then the equations might not be equivalent. For example,

$$x = 0 \qquad \text{and} \qquad x + \frac{1}{x} = 0 + \frac{1}{x}$$

are *not* equivalent because 0 satisfies the first equation but not the second one. (The expression $\frac{1}{x}$ is not defined if x is 0.)

To solve some equations, we must simplify the equation before using the properties of equality.

E X A M P L E 3

Simplifying the equation first

Solve the equation $2(x - 4) + 5x = 34$.

Solution

Before using the properties of equality, we simplify the expression on the left-hand side of the equation:

$$
\begin{aligned}
2(x - 4) + 5x &= 34 \\
2x - 8 + 5x &= 34 && \text{Distributive property} \\
7x - 8 &= 34 && \text{Combine like terms.} \\
7x - 8 + 8 &= 34 + 8 && \text{Add 8 to each side.} \\
7x &= 42 && \text{Simplify.} \\
\frac{7x}{7} &= \frac{42}{7} && \text{Divide each side by 7 to get} \\
&&& \text{a single } x \text{ on the left side.} \\
x &= 6
\end{aligned}
$$

To check, we replace x by 6 in the original equation and simplify:

$$
\begin{aligned}
2(6 - 4) + 5 \cdot 6 &= 34 \\
2(2) + 30 &= 34 \\
34 &= 34
\end{aligned}
$$

The solution set to the equation is $\{6\}$.

When an equation involves fractions, we can simplify it by multiplying each side by a number that is evenly divisible by all of the denominators. The smallest such number is called the **least common denominator (LCD).** Multiplying each side of the equation by the LCD will eliminate all of the fractions.

E X A M P L E 4

An equation with fractions

Find the solution set for the equation

$$\frac{x}{2} - \frac{1}{3} = \frac{x}{3} + \frac{5}{6}.$$

Solution

To solve this equation, we multiply each side by 6, the LCD for 2, 3, and 6:

$$6\left(\frac{x}{2} - \frac{1}{3}\right) = 6\left(\frac{x}{3} + \frac{5}{6}\right) \qquad \text{Multiply each side by 6.}$$

$$\frac{6x}{2} - \frac{6}{3} = \frac{6x}{3} + \frac{30}{6} \qquad \text{Distributive property}$$

$$3x - 2 = 2x + 5 \qquad \text{Divide to eliminate the fractions.}$$

$$3x - 2 - 2x = 2x + 5 - 2x \qquad \text{Subtract } 2x \text{ from each side.}$$

$$x - 2 = 5 \qquad \text{Combine like terms.}$$

$$x - 2 + 2 = 5 + 2 \qquad \text{Add 2 to each side.}$$

$$x = 7 \qquad \text{Combine like terms.}$$

Check 7 in the original equation. The solution set is $\{7\}$.

Equations that involve decimal numbers can be solved like equations involving fractions. If we multiply a decimal number by 10, 100, or 1000, the decimal point is moved one, two, or three places to the right, respectively. If the decimal points are all moved far enough to the right, the decimal numbers will be replaced by whole numbers. The next example shows how to use the multiplication property of equality to eliminate decimal numbers in an equation.

helpful hint

Solving equations is like playing football. In football you run a play and then pick up the injured players, regroup, and get ready for the next play. In equations you apply a property of equality and then remove parentheses, do arithmetic, simplify, and get ready to apply another property of equality.

EXAMPLE 5

An equation with decimals

Solve the equation $x - 0.1x = 0.75x + 4.5$.

Solution

Because the number with the most decimal places in this equation is 0.75 (75 hundredths), multiplying each side by 100 will eliminate all decimals.

$$100(x - 0.1x) = 100(0.75x + 4.5) \qquad \text{Multiply each side by 100.}$$

$$100x - 10x = 75x + 450 \qquad \text{Distributive property}$$

$$90x = 75x + 450 \qquad \text{Combine like terms.}$$

$$90x - 75x = 75x + 450 - 75x \qquad \text{Subtract } 75x \text{ from each side.}$$

$$15x = 450 \qquad \text{Combine like terms.}$$

$$\frac{15x}{15} = \frac{450}{15} \qquad \text{Divide each side by 15.}$$

$$x = 30$$

Check that 30 satisfies the original equation. The solution set is $\{30\}$.

calculator

close-up

To check 30 in Example 5 you can calculate the value of each side of the equation as shown here. Another way to check is to display the whole equation and then press ENTER. (Look in the TEST menu for the "=" symbol.) The calculator returns a 1 if the equation is correct or a 0 if the equation is incorrect.

```
30-.1*30
                27
.75*30+4.5
                27
30-.1*30=.75*30+
4.5
                1
```

Types of Equations

We often think of an equation such as $3x + 4x = 7x$ as an "addition fact" because the equation is satisfied by all real numbers. However, some equations that we think of as facts are not satisfied by all real numbers. For example, $\frac{x}{x} = 1$ is satisfied by every real number except 0 because $\frac{0}{0}$ is undefined. The equation $x + 1 = x + 1$ is satisfied by all real numbers because both sides are identical. All of these equations are called *identities*.

The equation $2x + 1 = 7$ is true only on condition that we choose $x = 3$. For this reason, it is called a *conditional equation*. The equations in Examples 2 through 5 are conditional equations.

Some equations are false no matter what value is used to replace the variable. For example, no number satisfies $x = x + 1$. The solution set to this *inconsistent* equation is the empty set, $\varnothing$.

Identity, Conditional Equation, Inconsistent Equation

An **identity** is an equation that is satisfied by every number for which both sides are defined.

A **conditional equation** is an equation that is satisfied by at least one number but is not an identity.

An **inconsistent equation** is an equation whose solution set is the empty set.

It is easy to classify $2x = 2x$ as an identity and $x = x + 2$ as an inconsistent equation, but some equations must be simplified before they can be classified.

E X A M P L E   6

An inconsistent equation and an identity

Solve each equation.

a) $8 - 3(x - 5) + 7 = 3 - (x - 5) - 2(x - 11)$

b) $5 - 3(x - 6) = 4(x - 9) - 7x$

Solution

a) First simplify each side of the equation:

$$8 - 3(x - 5) + 7 = 3 - (x - 5) - 2(x - 11)$$
$$8 - 3x + 15 + 7 = 3 - x + 5 - 2x + 22 \quad \text{Distributive property}$$
$$30 - 3x = 30 - 3x \quad \text{Combine like terms.}$$

This last equation is satisfied by any value of x because the two sides are identical. Because the last equation is equivalent to the original equation, the original equation is satisfied by any value of x and is an identity. The solution set is R, the set of all real numbers.

b) First simplify each side of the equation.

$$5 - 3(x - 6) = 4(x - 9) - 7x$$
$$5 - 3x + 18 = 4x - 36 - 7x \quad \text{Distributive property}$$
$$23 - 3x = -36 - 3x \quad \text{Combine like terms.}$$
$$23 - 3x + 3x = -36 - 3x + 3x \quad \text{Add } 3x \text{ to each side.}$$
$$23 = -36 \quad \text{Combine like terms.}$$

The equation $23 = -36$ is false for any choice of x. Because these equations are all equivalent, the original equation is also false for any choice of x. The solution set to this inconsistent equation is the empty set, $\varnothing$.

> **helpful hint**
>
> Removing parentheses with the distributive property and combining like terms was discussed in Chapter 1. If you are having trouble with these equations, your problem might be in the preceding chapter.

Strategy for Solving Linear Equations

The most basic equations of algebra are linear equations. In Chapter 3 we will see a connection between linear equations in one variable and straight lines.

Linear Equation in One Variable

A **linear equation in one variable** x is an equation of the form $ax + b = 0$, where a and b are real numbers, with $a \neq 0$.

The equations in Examples 2 through 5 are called linear equations in one variable, or simply linear equations, because they could all be rewritten in the form $ax + b = 0$. At first glance the equations in Example 6 appear to be linear equations. However, they cannot be written in the form $ax + b = 0$, with $a \neq 0$, so they are not linear equations. A linear equation has exactly one solution. The strategy that we use for solving linear equations is summarized in the following box.

Strategy for Solving a Linear Equation

1. If fractions are present, multiply each side by the LCD to eliminate them.
2. Use the distributive property to remove parentheses.
3. Combine any like terms.
4. Use the addition property of equality to get all variables on one side and numbers on the other side.
5. Use the multiplication property of equality to get a single variable on one side.
6. Check by replacing the variable in the original equation with your solution.

Note that not all equations require all of the steps.

EXAMPLE 7

Using the equation-solving strategy

Solve the equation $\dfrac{y}{2} - \dfrac{y-4}{5} = \dfrac{23}{10}$.

Solution

We first multiply each side of the equation by 10, the LCD for 2, 5, and 10. However, we do not have to write down that step. We can simply use the distributive property to multiply each term of the equation by 10.

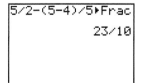
$$\frac{y}{2} - \frac{y-4}{5} = \frac{23}{10}$$

$$\overset{5}{\cancel{10}}\left(\frac{y}{2}\right) - \overset{2}{\cancel{10}}\left(\frac{y-4}{\cancel{5}}\right) = \cancel{10}\left(\frac{23}{\cancel{10}}\right) \qquad \text{Multiply each side by 10.}$$

$$5y - 2(y - 4) = 23 \qquad \text{Divide each denominator into 10 to eliminate fractions.}$$

$$5y - 2y + 8 = 23 \qquad \text{Be careful to change all signs:}\ -2(y-4) = -2y + 8$$

$$3y + 8 = 23 \qquad \text{Combine like terms.}$$

$$3y + 8 - 8 = 23 - 8 \qquad \text{Subtract 8 from each side.}$$

$$3y = 15 \qquad \text{Simplify.}$$

$$\frac{3y}{3} = \frac{15}{3} \qquad \text{Divide each side by 3.}$$

$$y = 5$$

Check that 5 satisfies the original equation. The solution set is $\{5\}$.

Techniques

Writing down every step when solving an equation is not always necessary. Solving an equation is often part of a larger problem, and anything that we can do to make the process more efficient will make solving the entire problem faster and easier. For example, we can combine some steps.

Combining Steps	*Writing Every Step*

$$\begin{array}{l} 4x - 5 = 23 \\ \quad 4x = 28 \quad \text{\small Add 5 to each side.} \\ \quad\quad x = 7 \quad \text{\small Divide each side by 4.} \end{array}$$

$$\begin{array}{c} 4x - 5 = 23 \\ 4x - 5 + 5 = 23 + 5 \\ 4x = 28 \\ \dfrac{4x}{4} = \dfrac{28}{4} \\ x = 7 \end{array}$$

The same steps are used in each of the solutions. However, when 5 is added to each side in the solution on the left, only the result is written. When each side is divided by 4, only the result is written.

The equation $-x = -5$ says that the additive inverse of x is -5. Since the additive inverse of 5 is -5, we conclude that x is 5. So instead of multiplying each side of $-x = -5$ by -1, we solve the equation as follows:

$$\begin{array}{ll} -x = -5 \\ \quad x = 5 & \text{\small Additive inverse property} \end{array}$$

Sometimes it is simpler to isolate x on the right-hand side of the equation:

$$\begin{array}{ll} 3x + 1 = 4x - 5 \\ \quad\quad 6 = x & \text{\small Subtract } 3x \text{ from each side and add 5 to each side.} \end{array}$$

You can rewrite $6 = x$ as $x = 6$ or leave it as is. Either way, 6 is the solution.

For some equations with fractions it is more efficient to multiply by a multiplicative inverse instead of multiplying by the LCD:

$$-\frac{2}{3}x = \frac{1}{2}$$

$$-\frac{3}{2}\left(-\frac{2}{3}x\right) = -\frac{3}{2}\left(\frac{1}{2}\right) \quad \text{\small Multiply each side by } -\frac{3}{2}, \text{ the reciprocal of } -\frac{2}{3}.$$

$$x = -\frac{3}{4}$$

The techniques shown here should not be attempted until you have become proficient at solving equations by writing out every step. The more efficient techniques shown here are not a requirement of algebra, but they can be a labor-saving tool that will be useful when we solve more complicated problems.

E X A M P L E 8 **Efficient solutions**

Solve each equation.

a) $3x + 4 = 0$

b) $2 - (x + 5) = -2(3x - 1) + 6x$

Solution

a) Combine steps to solve the equation efficiently.

$$3x + 4 = 0$$
$$3x = -4 \quad \text{Subtract 4 from each side.}$$
$$x = -\frac{4}{3} \quad \text{Divide each side by 3.}$$

Check $-\frac{4}{3}$ in the original equation:

$$3\left(-\frac{4}{3}\right) + 4 = -4 + 4 = 0$$

The solution set is $\left\{-\frac{4}{3}\right\}$.

b) $2 - (x + 5) = -2(3x - 1) + 6x$

$$-x - 3 = 2 \quad \text{Simplify each side.}$$
$$-x = 5 \quad \text{Add 3 to each side.}$$
$$x = -5 \quad \text{Additive inverse property}$$

Check that -5 satisfies the original equation. The solution set is $\{-5\}$.

Applications

In the next example we show how a linear equation can occur in an application.

EXAMPLE 9

Completing high school

In 1940 only 24% of persons 25 years and over had completed 4 years of high school or more, but that percentage has been growing steadily as shown in Fig. 2.1 (Census Bureau, www.census.gov). The expression $1.05n + 24$ gives the percentage of persons 25 and over who have completed 4 years of high school in the year $1940 + n$.

a) What was the percentage in 1998?

b) When will the percentage reach 94%?

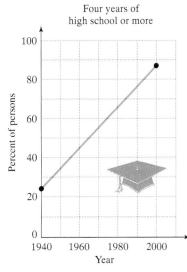

FIGURE 2.1

Solution

a) Since 1998 is 58 years after 1940, $n = 58$ and

$$1.05(58) + 24 = 84.9.$$

So in 1998 approximately 85% of persons 25 and over had completed 4 years of high school or more.

b) To find when the percentage will reach 94%, solve the following equation:

$$1.05n + 24 = 94$$
$$1.05n = 70$$
$$n = \frac{70}{1.05} \approx 67$$

So in 2007 (1940 + 67) the percentage will reach 94%.

WARM-UPS

True or false? Explain your answer.

1. The equation $-2x + 3 = 8$ is equivalent to $-2x = 11$.   False
2. The equation $x - (x - 3) = 5x$ is equivalent to $3 = 5x$.   True
3. To solve $\frac{3}{4}x = 12$, we should multiply each side by $\frac{3}{4}$.   False
4. The equation $-x = -6$ is equivalent to $x = 6$.   True
5. To eliminate fractions, we multiply each side of an equation by the LCD.   True
6. The solution set to $3x + 5 = 7$ is $\left\{\frac{2}{3}\right\}$.   True
7. The equation $2(3x + 4) = 6x + 12$ is an inconsistent equation.   True
8. The equation $4(x + 3) = x + 3$ is a conditional equation.   True
9. The equation $x - 0.2x = 0.8x$ is an identity.   True
10. The equation $3x - 5 = 7$ is a linear equation.   True

2.1   EXERCISES

Reading and Writing   *After reading this section, write out the answers to these questions. Use complete sentences.*

1. What is an equation?
   An equation is a sentence that expresses the equality of two algebraic expressions.

2. How do you know if a number satisfies an equation?
   A number satisfies the equation if the equation is true when the variable is replaced by the number.

3. What are equivalent equations?
   Equivalent equations are equations that have the same solution set.

4. What is a linear equation in one variable?
   A linear equation in one variable is an equation of the form $ax + b = 0$.

5. What is the usual first step in solving an equation that involves fractions?
   If the equation involves fractions, then multiply each side by the LCD.

6. What is an identity?
   An identity is an equation that is satisfied by all values of the variable for which both sides are defined.

7. What is a conditional equation?
   A conditional equation is an equation that has at least one solution but is not an identity.

8. What is an inconsistent equation?
   An inconsistent equation is an equation that has no solution.

Determine whether each equation is satisfied by the given number. See Example 1.

9. $3x + 7 = -5$, -4 Yes
10. $-3x - 5 = 13$, -6 Yes
11. $\frac{1}{2}x - 4 = \frac{1}{3}x - 2$,   12   Yes
12. $\frac{y - 7}{2} - \frac{1}{3} = \frac{y - 7}{3}$,   9   Yes

13. $0.2(x - 50) = 20 - 0.05x,$ 200   No

14. $0.12x - (4 - x) = 1.02x + 1,$ 50   Yes

15. $0.1x - 30 = 16 - 0.06x,$ 80   No

16. $0.08x + 3.2 = 0.1x + 2.8,$ 20   Yes

Solve each linear equation. Show your work and check your answer. See Examples 2 and 3.

17. $-72 - x = 15$
$\{-87\}$

18. $51 - x = -9$
$\{60\}$

19. $-3x - 19 = 5 - 2x$
$\{-24\}$

20. $-5x + 4 = -9 - 4x$
$\{13\}$

21. $2x - 3 = 0$
$\left\{\dfrac{3}{2}\right\}$

22. $5x + 7 = 0$
$\left\{-\dfrac{7}{5}\right\}$

23. $-2x + 5 = 7$
$\{-1\}$

24. $-3x - 4 = 11$
$\{-5\}$

25. $-12x - 15 = 21$
$\{-3\}$

26. $-13x + 7 = -19$
$\{2\}$

27. $26 = 4x + 16$
$\left\{\dfrac{5}{2}\right\}$

28. $14 = -5x - 21$
$\{-7\}$

29. $-3(x - 16) = 12 - x$
$\{18\}$

30. $-2(x + 17) = 13 - x$
$\{-47\}$

31. $2(x + 9) - x = 36$
$\{18\}$

32. $3(x - 13) - x = 9$
$\{24\}$

33. $2 + 3(x - 1) = x - 1$
$\{0\}$

34. $x + 9 = 1 - 4(x - 2)$
$\{0\}$

Solve each equation. See Example 4.

35. $-\dfrac{3}{7}x = 4$ $\left\{-\dfrac{28}{3}\right\}$

36. $\dfrac{5}{6}x = -2$ $\left\{-\dfrac{12}{5}\right\}$

37. $-\dfrac{5}{7}x - 1 = 3$
$\left\{-\dfrac{28}{5}\right\}$

38. $4 - \dfrac{3}{5}x = -6$
$\left\{\dfrac{50}{3}\right\}$

39. $\dfrac{x}{3} + \dfrac{1}{2} = \dfrac{7}{6}$
$\{2\}$

40. $\dfrac{1}{4} + \dfrac{1}{5} = \dfrac{x}{2}$
$\left\{\dfrac{9}{10}\right\}$

41. $\dfrac{2}{3}x + 5 = -\dfrac{1}{3}x + 17$
$\{12\}$

42. $\dfrac{1}{4}x - 6 = -\dfrac{3}{4}x + 14$
$\{20\}$

43. $\dfrac{1}{2}x + \dfrac{1}{4} = \dfrac{1}{4}(x - 6)$
$\{-7\}$

44. $\dfrac{1}{3}(x - 2) = \dfrac{2}{3}x - \dfrac{13}{3}$
$\{11\}$

45. $8 - \dfrac{x - 2}{2} = \dfrac{x}{4}$
$\{12\}$

46. $\dfrac{x}{3} - \dfrac{x - 5}{5} = 3$
$\{15\}$

47. $\dfrac{y - 3}{3} - \dfrac{y - 2}{2} = -1$
$\{6\}$

48. $\dfrac{x - 2}{2} - \dfrac{x - 3}{4} = \dfrac{7}{4}$
$\{8\}$

Solve each equation. See Example 5.

49. $x - 0.2x = 72$ $\{90\}$

50. $x - 0.1x = 63$ $\{70\}$

51. $0.03(x + 200) + 0.05x = 86$ $\{1000\}$

52. $0.02(x - 100) + 0.06x = 62$ $\{800\}$

53. $0.1x + 0.05(x - 300) = 105$ $\{800\}$

54. $0.2x - 0.05(x - 100) = 35$ $\{200\}$

Solve each equation. Identify each as a conditional equation, an inconsistent equation, or an identity. See Example 6.

55. $2(x + 1) = 2(x + 3)$ $\varnothing$, inconsistent

56. $2x + 3x = 6x$ $\{0\}$, conditional

57. $x + x = 2x$ R, identity

58. $4x - 3x = x$ R, identity

59. $x + x = 2$ $\{1\}$, conditional

60. $4x - 3x = 5$ $\{5\}$, conditional

61. $\dfrac{4x}{4} = x$ R, identity

62. $5x \div 5 = x$ R, identity

63. $x \cdot x = x^2$ R, identity

64. $\dfrac{2x}{2x} = 1$ $\{x \mid x \neq 0\}$, identity

65. $2(x + 3) - 7 = 5(5 - x) + 7(x + 1)$ $\varnothing$, inconsistent

66. $2(x + 4) - 8 = 2x + 1$ $\varnothing$, inconsistent

67. $2\left(\dfrac{1}{2}x + \dfrac{3}{2}\right) - \dfrac{7}{2} = \dfrac{3}{2}(x + 1) - \left(\dfrac{1}{2}x + 2\right)$ R, identity

68. $2\left(\dfrac{1}{4}x + 1\right) - 2 = \dfrac{1}{2}x$ R, identity

69. $2(0.5x + 1.5) - 3.5 = 3(0.5x + 0.5)$ $\{-4\}$, conditional

70. $2(0.25x + 1) - 2 = 0.75x - 1.75$ $\{7\}$, conditional

Solve each equation. See Example 7.

71. $4 - 6(2x - 3) + 1 = 3 + 2(5 - x)$ $\{1\}$

72. $3x - 5(6 - 2x) = 4(x - 8) + 3$ $\left\{\dfrac{1}{9}\right\}$

73. $\dfrac{1}{2}\left(y - \dfrac{1}{6}\right) + \dfrac{2}{3} = \dfrac{5}{6} + \dfrac{1}{3}\left(\dfrac{1}{2} - 3y\right)$ $\left\{\dfrac{5}{18}\right\}$

74. $\dfrac{3}{4} - \dfrac{1}{3}\left(\dfrac{1}{2}y - 2\right) = 3\left(y - \dfrac{1}{4}\right)$ $\left\{\dfrac{13}{19}\right\}$

75. $8 - \dfrac{2}{3}(60x - 900) = \dfrac{1}{2}(400x + 6)$ $\left\{\dfrac{121}{48}\right\}$

76. $\dfrac{40x - 5}{2} + \dfrac{5}{2} = \dfrac{33 - 2x}{3} - 11$ $\{0\}$

77. $\dfrac{a - 3}{4} - \dfrac{2a - 5}{2} = \dfrac{a + 1}{3} - \dfrac{1}{6}$ $\left\{\dfrac{19}{13}\right\}$

78. $\dfrac{1}{2}\left(\dfrac{b}{3} - \dfrac{4b}{5}\right) + \dfrac{1}{6} = \dfrac{1}{3} - \dfrac{b - 1}{2}$ $\left\{\dfrac{5}{2}\right\}$

79. $1.3 - 0.2(6 - 3x) = 0.1(0.2x + 3)$ $\left\{\dfrac{10}{29}\right\}$

80. $0.01(500 - 30x) = 5.4x + 200$ $\left\{-\dfrac{650}{19}\right\}$

Solve each equation. Practice combining some steps. Look for more efficient ways to solve each equation. See Example 8.

81. $3x - 9 = 0$ $\{3\}$

82. $5x + 1 = 0$ $\left\{-\dfrac{1}{5}\right\}$

83. $7 - z = -9$ $\{16\}$

84. $-3 - z = 3$ $\{-6\}$

85. $\dfrac{2}{3}x = \dfrac{1}{2}$ $\left\{\dfrac{3}{4}\right\}$

86. $\dfrac{3}{2}x = -\dfrac{9}{5}$ $\left\{-\dfrac{6}{5}\right\}$

87. $-\dfrac{3}{5}y = 9$ $\{-15\}$

88. $-\dfrac{2}{7}w = 4$ $\{-14\}$

89. $3y + 5 = 4y - 1$ $\{6\}$

90. $2y - 7 = 3y + 1$ $\{-8\}$

91. $5x + 10(x + 2) = 110$ $\{6\}$

92. $1 - 3(x - 2) = 4(x - 1) - 3$ $\{2\}$

Solve each equation.

93. $\dfrac{P + 7}{3} - \dfrac{P - 2}{5} = \dfrac{7}{3} - \dfrac{P}{15}$ $\{-2\}$

94. $\dfrac{w - 3}{8} - \dfrac{5 - w}{4} = \dfrac{4w - 1}{8} - 1$ $\{-4\}$

 95. $x - 0.06x = 50,000$ $\{53,191.49\}$

96. $x - 0.05x = 800$ $\{842.11\}$

97. $2.365x + 3.694 = 14.8095$ $\{4.7\}$

98. $-3.48x + 6.981 = 4.329x - 6.851$ $\{1.7713\}$

Solve each problem. See Example 9.

99. *Public school enrollment.* The expression
$$0.45x + 39.05$$
can be used to approximate in millions the total enrollment in public elementary and secondary schools in the year $1985 + x$ (National Center for Education Statistics, www.nces.ed.gov).
a) What was the public school enrollment in 1992?
b) In which year will enrollment reach 50 million students?
c) Judging from the accompanying graph, is enrollment increasing or decreasing?
   a) 42.2 million   b) 2010   c) increasing

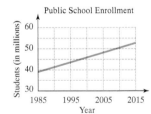 

FIGURE FOR EXERCISE 99

100. *Teacher's average salary.* The expression
$$553.7x + 27,966$$
can be used to approximate the average annual salary in dollars of public school teachers in the year $1985 + x$ (National Center for Education Statistics, www.nces.ed.gov).

a) What was the average teacher's salary in 1993?
b) In which year will the average salary reach $40,000?
   a) $32,396   b) 2007

101. *Solid waste recovery.* In 1960 the United States generated 87.1 million tons of municipal solid waste and recovered (or recycled) only 4.3% of it (U.S. Department of Energy, www.doe.gov). The amount of solid waste generated in the United States in the year $1960 + n$ is given by
$$w = 3.14n + 87.1,$$
whereas the amount recovered is given by
$$w = 0.576n + 3.78,$$
where w is in millions of tons.
a) Use the accompanying graph to estimate the first year in which the United States generated over 100 million tons of municipal solid waste.
b) Find the year in which 13% of the municipal solid waste will be recovered by solving
$$0.576n + 3.78 = 0.13(3.14n + 87.1).$$
   a) 1964   b) 2005

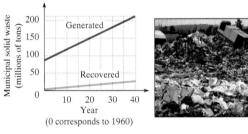

FIGURE FOR EXERCISE 101

 102. *Recycling progress.* Find the year in which 14% of the municipal solid waste will be recovered by solving
$$0.576n + 3.78 = 0.14(3.14n + 87.1).$$
See the previous exercise.   2022

GETTING MORE INVOLVED

103. *Exploration.* If you solved the equations in the two previous exercises, then you found the years in which recovery of municipal solid waste will reach 13% and 14%.
a) Write equations corresponding to recovery rates of 15%, 16%, 17%, 18%, and 19%.
b) Solve your equations to find the years in which those recovery rates will be achieved.
c) Use your results to judge whether we are making real progress in recovery of municipal solid waste.

104. *Writing.* Explain how to eliminate the decimals in an equation that involves numbers with decimal points. Would you use the same technique when using a calculator?

105. *Discussion.* Explain why the multiplication property of equality does not allow us to multiply each side of an equation by zero.

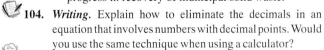

2.2 FORMULAS

A real-life problem may involve many variable quantities that are related to each other. The relationship between these variables may be expressed by a formula. In this section we will combine our knowledge of evaluating expressions from Chapter 1 with the equation-solving skills of Section 2.1 to work with formulas.

Solving for a Variable

A **formula** or **literal equation** is an equation involving two or more variables. For example, the formula

$$A = LW$$

expresses the known relationship among the length L, width W, and area A of a rectangle. The formula

$$C = \frac{5}{9}(F - 32)$$

expresses the relationship between the Fahrenheit and Celsius measurements of temperature. The Celsius temperature is determined by the Fahrenheit temperature. For example, if the Fahrenheit temperature F is 95°, we can use the formula to find the Celsius temperature C as follows:

$$C = \frac{5}{9}(95 - 32) = \frac{5}{9}(63) = 35$$

A temperature of 95°F is equivalent to 35°C.

Sometimes it is necessary to solve a formula for one variable in terms of the others without substituting numbers for the variables. We will use the same steps in solving for a particular variable as we did in solving linear equations.

E X A M P L E   1

Solving for a variable

Solve the formula $C = \frac{5}{9}(F - 32)$ for F.

Solution

To solve the formula for F, we isolate F on one side of the equation. We can eliminate both the 9 and the 5 from the right-hand side of the equation by multiplying by $\frac{9}{5}$, the reciprocal of $\frac{5}{9}$:

$$C = \frac{5}{9}(F - 32)$$

$$\frac{9}{5}C = \frac{9}{5} \cdot \frac{5}{9}(F - 32) \quad \text{Multiply each side by } \frac{9}{5}.$$

$$\frac{9}{5}C = F - 32 \qquad\qquad \frac{9}{5} \cdot \frac{5}{9} = 1$$

$$\frac{9}{5}C + 32 = F \qquad\qquad \text{Add 32 to each side.}$$

So the formula $F = \frac{9}{5}C + 32$ expresses F in terms of C. With this formula, we can use the value of C to determine the corresponding value of F. ■

Note that both $F = \frac{9}{5}C + 32$ and $C = \frac{5}{9}(F - 32)$ express the relationship between C and F. The formula $F = \frac{9}{5}C + 32$ gives F in terms of C and $C = \frac{5}{9}(F - 32)$ gives C in terms of F. If we substitute 35 for C in $F = \frac{9}{5}C + 32$, we get

$$F = \frac{9}{5}(35) + 32 = 63 + 32 = 95.$$

In Chapter 3 we will study linear equations involving x and y. We often need to solve such equations for one of the variables.

E X A M P L E   2

Equations involving x and y

Write y in terms of x, if $3x - 2y = 6$.

Solution

We can isolate y on the left-hand side:

$$3x - 2y = 6$$
$$-2y = -3x + 6 \qquad \text{Subtract } 3x \text{ from each side.}$$
$$\frac{-2y}{-2} = \frac{-3x + 6}{-2} \qquad \text{Divide each side by } -2.$$
$$y = \frac{3}{2}x - 3 \qquad \text{This equation expresses } y \text{ in terms of } x.$$

The formula $A = P + Prt$ is used to find the amount A after t years for an investment of P dollars at simple interest rate r. Note that the variable P occurs twice in the formula. To solve the formula for P, we use the distributive property as shown in the next example.

E X A M P L E   3

Specified variable occurring twice

Solve $A = P + Prt$ for P.

Solution

We can use the distributive property to write the sum $P + Prt$ as a product of P and $1 + rt$:

$$A = P + Prt$$
$$A = P(1 + rt) \qquad \text{Distributive property}$$
$$\frac{A}{1 + rt} = \frac{P(1 + rt)}{1 + rt} \qquad \text{Divide each side by } 1 + rt.$$
$$\frac{A}{1 + rt} = P$$

The formula $P = \dfrac{A}{1 + rt}$ expresses P in terms of A, r, and t. Note that parentheses are not needed around the expression $1 + rt$ in the denominator because the fraction bar acts as a grouping symbol.

   C A U T I O N   If you write $A = P + Prt$ as $P = A - Prt$, then you have not solved the formula for P. When a formula is solved for a specified variable, that variable must be isolated on one side, and it must not occur on the other side.

When the variable for which we are solving occurs on opposite sides of the equation, we must move all terms involving that variable to the same side and then use the distributive property to write the expression as a product.

EXAMPLE 4

Specified variable occurring on both sides

Suppose $3a + 7 = -5ab + b$. Solve for a.

Solution

Get all terms involving a onto one side and all other terms onto the other side:

$$3a + 7 = -5ab + b$$

$$3a + 5ab + 7 = b \qquad \text{Add } 5ab \text{ to each side.}$$

$$3a + 5ab = b - 7 \qquad \text{Subtract 7 from each side.}$$

$$a(3 + 5b) = b - 7 \qquad \text{Use the distributive property to write the left-hand side as a product.}$$

$$\frac{a(3 + 5b)}{3 + 5b} = \frac{b - 7}{3 + 5b} \qquad \text{Divide each side by } 3 + 5b.$$

$$a = \frac{b - 7}{3 + 5b}$$

> **helpful hint**
>
> If you do the steps in Example 4 in a different way, you might end up with
>
> $$a = \frac{7 - b}{-3 - 5b}.$$
>
> This answer is correct because a is isolated. However, we usually prefer to see fewer negative signs. So we multiply this numerator and denominator by -1 and get the answer in Example 4.

When solving an equation in one variable that contains many decimal numbers, we usually use a calculator for the arithmetic. However, if you use a calculator at every step and round off the result of every computation, the final answer can differ greatly from the correct answer. The next example shows how to avoid this problem. The numbers are treated as if they were variables and no arithmetic is performed until all of the numbers are on one side of the equation. This technique is similar to solving an equation for a specified variable.

EXAMPLE 5

Doing computations last

Solve $3.24x - 6.78 = 6.31(x + 23.45)$.

Solution

Use the distributive property on the right-hand side, but simply write $(6.31)(23.45)$ rather than the result obtained on a calculator.

$$3.24x - 6.78 = 6.31x + (6.31)(23.45) \qquad \text{Distributive property}$$

$$3.24x - 6.31x = (6.31)(23.45) + 6.78 \qquad \text{Get all } x\text{-terms on the left.}$$

$$(3.24 - 6.31)x = (6.31)(23.45) + 6.78 \qquad \text{Distributive property}$$

$$x = \frac{(6.31)(23.45) + 6.78}{3.24 - 6.31} \qquad \text{Divide each side by } (3.24 - 6.31).$$

$$\approx -50.407 \qquad \text{Round to three decimal places.}$$

Check -50.407 in the original equation. When you check an approximate answer, you will get approximately the same value for each side of the equation. The solution set is $\{-50.407\}$.

> **calculator**
>
> **close-up**
>
> A graphing calculator allows you to enter the entire expression in Example 5 and to evaluate it in one step. The ANS key holds the last value calculated. If we use ANS for x in the original equation, the calculator returns a 1, indicating that the equation is satisfied.
>
>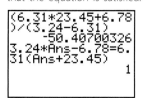
>
> ```
> (6.31*23.45+6.78
> )/(3.24-6.31)
>        -50.40700326
> 3.24*Ans-6.78=6.
> 31(Ans+23.45)
>                  1
> ```

Finding the Value of a Variable

In many situations we know the values of all of the variables in a formula except one. We can use the formula to determine the unknown value. A list of common formulas and their meanings is given at the back of the text. This list may be helpful for doing the exercises at the end of this section.

E X A M P L E   6

Finding the value of a variable

Use the formula $-2x + 3y = 9$ to find y given that $x = -3$.

Solution

To find y, we first write y in terms of x:

$$-2x + 3y = 9 \qquad \text{Original equation}$$
$$3y = 2x + 9 \qquad \text{Add } 2x \text{ to each side.}$$
$$y = \frac{2}{3}x + 3 \qquad \text{Divide each side by 3.}$$

helpful hint

It doesn't matter what form to use when solving for y here. If you use

$$y = \frac{2x + 9}{3}$$

and let $x = -3$, you get $y = 1$.

Now replace x by -3:

$$y = \frac{2}{3}(-3) + 3$$
$$y = 1$$

E X A M P L E   7

Finding the interest rate

The simple interest on a loan is \$50, the principal is \$500, and the time is 2 years. What is the simple interest rate?

Solution

The formula $I = Prt$ expresses the interest I in terms of the principal P, rate r, and time t. To find the rate, we first solve the formula for r, then insert the values of P, I, and t:

$$Prt = I$$
$$\frac{Prt}{Pt} = \frac{I}{Pt} \qquad \text{Divide each side by } Pt.$$
$$r = \frac{I}{Pt} \qquad \text{This formula expresses the rate in terms of } I, P, \text{ and } t.$$
$$r = \frac{50}{500(2)} \qquad \text{Substitute values for } I, P, \text{ and } t.$$
$$r = 0.05 = 5\% \qquad \text{A rate is usually written as a percent.}$$

study tip

Note how the exercises are keyed to the examples. This serves two purposes. If you have missed class and are studying on your own, you should study an example and then immediately try to work the corresponding exercises. If you have seen an explanation in class, then you can start the exercises and refer back to the examples as necessary.

In Example 7 we solved the formula for r and then inserted the values of the other variables. If we had to find the interest rate for many different loans, this method would be the most direct. But we could also have inserted the values of I, P, and t into the original formula and then solved for r. Examples 8 and 9 illustrate this second approach.

Geometric Formulas

Appendix A contains some geometric formulas that will be useful in problems that involve geometric shapes. In geometry it is common to use variables with subscripts. A subscript is a slightly lowered number following the variable. For example, the areas of two triangles might be referred to as A_1 and A_2. (We read A_1 as "A sub one" or simply "A one.") You will see subscripts in the next example.

E X A M P L E   8

Area of a trapezoid

The wildlife sanctuary shown in Fig. 2.2 on the next page has a trapezoidal shape with an area of 30 square kilometers. If one base, b_1, of the trapezoid is 10 kilometers and its height is 5 kilometers, find the length of the other base, b_2.

Solution

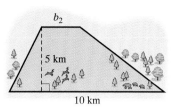

FIGURE 2.2

In any geometric problem it is helpful to have a diagram, as in Fig. 2.2. Substitute the given values into the formula for the area of a trapezoid, found in Appendix A, and then solve for b_2:

$$A = \frac{1}{2}h(b_1 + b_2)$$ The area depends on h, b_1, and b_2.

$$30 = \frac{1}{2} \cdot 5(10 + b_2)$$ Substitute given values into the formula for the area of a trapezoid.

$$60 = 5(10 + b_2)$$ Multiply each side by 2.

$$12 = 10 + b_2$$ Divide each side by 5.

$$2 = b_2$$ Subtract 10 from each side.

The length of the base b_2 is 2 kilometers.

E X A M P L E 9

Volume of a rectangular solid

Millie has just completed pouring 14 cubic yards of concrete to construct a rectangular driveway. If the concrete is 4 inches thick and the driveway is 18 feet wide, then how long is her driveway?

Solution

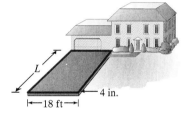

FIGURE 2.3

First draw a diagram as in Fig. 2.3. The driveway is a rectangular solid. The formula for the volume of a rectangular solid is $V = LWH$ (from Appendix A). Before we insert the values of the variables into the formula, we must convert all of them to the same unit of measurement. We will convert feet and inches to yards:

$$4 \text{ inches} = 4 \text{ in.} \cdot \frac{1 \text{ yd}}{36 \text{ in.}} = \frac{1}{9} \text{ yard}$$

$$18 \text{ feet} = 18 \text{ ft} \cdot \frac{1 \text{ yd}}{3 \text{ ft}} = 6 \text{ yards}$$

Now replace W, H, and V by the appropriate values:

$$V = LWH$$ The volume is determined by the length, width, and height.

$$14 = L \cdot 6 \cdot \frac{1}{9}$$

$$\frac{9}{6} \cdot 14 = L$$ Multiply each side by $\frac{9}{6}$.

$$21 = L$$

The length of the driveway is 21 yards, or 63 feet.

WARM-UPS

True or false? Explain your answer.

1. The formula $A = P + Prt$ solved for P is $P = A - Prt$.   False
2. In solving $A = P + Prt$ for P, we do not need the distributive property.   False
3. Solving $I = Prt$ for t gives us $t = I - Pr$.   False
4. If $a = \frac{bh}{2}$, $b = 5$, and $h = 6$, then $a = 15$.   True
5. The perimeter of a rectangle is found by multiplying its length and width.   False

WARM-UPS

(continued)

6. The volume of a rectangular box is the product of its length, width, and height.   True
7. The area of a trapezoid with parallel sides b_1 and b_2 is $\frac{1}{2}(b_1 + b_2)$.   False
8. If $x - y = 5$, then $y = x - 5$ expresses y in terms of x.   True
9. If $x = -3$ and $y = -2x - 4$, then $y = 2$.   True
10. The area of a rectangle is the total distance around the outside edge.   False

2.2 EXERCISES

Reading and Writing   *After reading this section, write out the answers to these questions. Use complete sentences.*

1. What is a formula?
   A formula is an equation involving two or more variables.
2. What is a formula used for?
   Formulas are used to express relationships between variables.
3. What does it mean to solve a formula for a particular variable?
   Solving for a variable means to rewrite the formula by isolating the indicated variable.
4. How do you solve for a variable that occurs twice in a formula?
   If a variable occurs twice, then we usually use the distributive property to isolate it.
5. How do you find the value of a variable in a formula?
   To find the value of a variable, solve for that variable, then replace all other variables with the given numbers.
6. What formula expresses the volume of a rectangular solid in terms of its length, width, and height?
   The formula for the volume of a rectangular solid is $V = LWH$.

Solve each formula for the specified variable. See Example 1.

7. $I = Prt$ for t
   $t = \dfrac{I}{Pr}$

8. $d = rt$ for r
   $r = \dfrac{d}{t}$

9. $F = \dfrac{9}{5}C + 32$ for C
   $C = \dfrac{5}{9}(F - 32)$

10. $A = \dfrac{1}{2}bh$ for h
    $h = \dfrac{2A}{b}$

11. $A = LW$ for W
    $W = \dfrac{A}{L}$

12. $C = 2\pi r$ for r
    $r = \dfrac{C}{2\pi}$

13. $A = \dfrac{1}{2}(b_1 + b_2)$ for b_1
    $b_1 = 2A - b_2$

14. $A = \dfrac{1}{2}(b_1 + b_2)$ for b_2
    $b_2 = 2A - b_1$

Write y in terms of x. See Example 2.

15. $2x + 3y = 9$
    $y = -\dfrac{2}{3}x + 3$

16. $4y + 5x = 8$
    $y = -\dfrac{5}{4}x + 2$

17. $x - y = 4$
    $y = x - 4$

18. $y - x = 6$
    $y = x + 6$

19. $\dfrac{1}{2}x - \dfrac{1}{3}y = 2$
    $y = \dfrac{3}{2}x - 6$

20. $\dfrac{1}{3}x - \dfrac{1}{4}y = 1$
    $y = \dfrac{4}{3}x - 4$

21. $y - 2 = \dfrac{1}{2}(x - 3)$
    $y = \dfrac{1}{2}x + \dfrac{1}{2}$

22. $y - 3 = \dfrac{1}{3}(x - 4)$
    $y = \dfrac{1}{3}x + \dfrac{5}{3}$

Solve for the specified variable. See Examples 3 and 4.

23. $A = P + Prt$ for t
    $t = \dfrac{A - P}{Pr}$

24. $A = P + Prt$ for r
    $r = \dfrac{A - P}{Pt}$

25. $ab + a = 1$ for a
    $a = \dfrac{1}{b + 1}$

26. $y - wy = m$ for y
    $y = \dfrac{m}{1 - w}$

27. $xy + 5 = y - 7$ for y
    $y = \dfrac{12}{1 - x}$

28. $xy + 5 = x + 7$ for x
    $x = \dfrac{2}{y - 1}$

29. $xy^2 + xz^2 = xw^2 - 6$ for x
    $x = \dfrac{6}{w^2 - y^2 - z^2}$

30. $xz^2 + xw^2 = xy^2 + 5$ for x
    $x = \dfrac{5}{z^2 + w^2 - y^2}$

31. $\dfrac{1}{R} = \dfrac{1}{R_1} + \dfrac{1}{R_2}$ for R_1
    $R_1 = \dfrac{RR_2}{R_2 - R}$

32. $\dfrac{1}{a} + \dfrac{1}{b} = \dfrac{1}{2}$ for a
    $a = \dfrac{2b}{b - 2}$

   Solve each equation. Use a calculator only on the last step. Round answers to three decimal places and use your calculator to check your answer. See Example 5.

33. $3.35x - 54.6 = 44.3 - 4.58x$ 12.472
34. $-4.487x - 33.41 = 55.83 - 22.49x$ 4.957
35. $4.59x - 66.7 = 3.2(x - 5.67)$ 34.932
36. $457(36x - 99) = 34(28x - 239)$ 2.395

37. $\dfrac{x}{19} - \dfrac{3}{23} = \dfrac{4}{31} - \dfrac{3x}{7}$ 0.539

38. $\dfrac{1}{8} - \dfrac{5}{7}\left(x - \dfrac{5}{22}\right) = \dfrac{4x}{9} + \dfrac{1}{12}$ 0.176

Find y given that x = 3. See Example 6.

39. $2x - 3y = 5$ $\dfrac{1}{3}$

40. $-3x - 4y = 4$ $-\dfrac{13}{4}$

41. $-4x + 2y = 1$ $\dfrac{13}{2}$

42. $x - y = 7$ -4

43. $y = -2x + 5$ -1

44. $y = -3x - 6$ -15

45. $-x + 2y = 5$ 4

46. $-x - 3y = 6$ -3

47. $y - 1.046 = 2.63(x - 5.09)$ -4.4507

48. $y - 2.895 = -1.07(x - 2.89)$ 2.7773

Find x in each formula given that y = 2, z = −3, and w = 4. See Example 6.

49. $wxy = 5$ $\dfrac{5}{8}$

50. $wxz = 4$ $-\dfrac{1}{3}$

51. $x + xz = 7$ $-\dfrac{7}{2}$

52. $xw - x = 3$ 1

53. $w(x - z) = y(x - 4)$ -10

54. $z(x - y) = y(x + 5)$ $-\dfrac{4}{5}$

55. $w = \dfrac{1}{2}xz$ $-\dfrac{8}{3}$

56. $y = \dfrac{1}{2}wx$ 1

57. $\dfrac{1}{w} + \dfrac{1}{x} = \dfrac{1}{y}$ 4

58. $\dfrac{1}{w} + \dfrac{1}{y} = \dfrac{1}{x}$ $\dfrac{4}{3}$

Find the geometric formula in each case.

59. The area of a circle in terms of its radius
$A = \pi r^2$

60. The circumference of a circle in terms of its diameter
$C = \pi d$

61. The radius of a circle in terms of its circumference
$r = \dfrac{C}{2\pi}$

62. The diameter of a circle in terms of its circumference
$d = \dfrac{C}{\pi}$

63. The width of a rectangle in terms of its length and perimeter
$W = \dfrac{P - 2L}{2}$

64. The length of a rectangle in terms of its width and area
$L = \dfrac{A}{W}$

Solve each problem. Draw a diagram for each geometric problem. See Examples 7–9.

65. *Simple interest.* If the simple interest on $1,000 for 2 years is $300, then what is the rate?
15%

66. *Finding time.* If the simple interest on $2,000 at 18% is $180, then what is the time?
One-half year

67. *Rectangular floor.* The area of a rectangular floor is 23 square yards. The width is 4 yards. Find the length.
5.75 yards

68. *Rectangular garden.* The area of a rectangular garden is 55 square meters. The length is 7 meters. Find the width.
$\dfrac{55}{7}$ meters

69. *Ice sculpture.* The volume of a rectangular block of ice is 36 cubic feet. The bottom is 2 feet by 2.5 feet. Find the height of the block.   7.2 feet

70. *Cardboard box.* A shipping box has a volume of 2.5 cubic meters. The box measures 1 meter high by 1.25 meters wide. How long is the box?   2 meters

71. *Fish tank.* The volume of a rectangular aquarium is 900 gallons. The bottom is 4 feet by 6 feet. Find the height of the tank. (*Hint:* There are 7.5 gallons per cubic foot.) See figure for Exercise 71.   5 feet

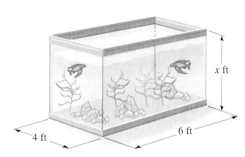

4 ft   6 ft   x ft

FIGURE FOR EXERCISE 71

72. *Reflecting pool.* A rectangular reflecting pool with a horizontal bottom holds 60,000 gallons of water. If the pool is 40 feet by 100 feet, how deep is the water?   2 feet

73. *Area of a triangle.* The area of a triangle is 30 square feet. If the base is 4 feet, then what is the height?   15 feet

74. *Larger triangle.* The area of a triangle is 40 square meters. If the height is 10 meters, then what is the length of the base?   8 meters

75. *Second base.* The area of a trapezoid is 300 square inches. If the height is 20 inches and the lower base is 16 inches, then what is the length of the upper base?   14 inches

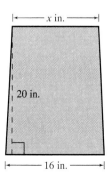

x in.

20 in.

16 in.

FIGURE FOR EXERCISE 75

76. *Height of a trapezoid.* The area of a trapezoid is 200 square centimeters. The bases are 16 centimeters and 24 centimeters. Find the height.   10 centimeters

77. *Fencing.* If it takes 600 feet of fence to enclose a rectangular lot that is 132 feet wide, then how deep is the lot?   168 feet

78. *Football.* The perimeter of a football field in the NFL, excluding the end zones, is $306\frac{2}{3}$ yards. How wide is the field?   160 feet

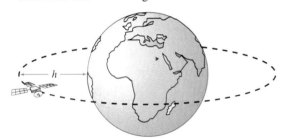

FIGURE FOR EXERCISE 78

79. *Radius of a circle.* If the circumference of a circle is 3π meters, then what is the radius?   1.5 meters

80. *Diameter of a circle.* If the circumference of a circle is 12π inches, then what is the diameter?   12 inches

 81. *Radius of the earth.* If the circumference of the earth is 25,000 miles, then what is the radius?   3,979 miles

 82. *Altitude of a satellite.* If a satellite travels 26,000 miles in each circular orbit of the earth, then how high above the earth is the satellite orbiting? See Exercise 81.   159 miles

 83. *Height of a can.* If the volume of a can is 30 cubic inches and the diameter of the top is 2 inches, then what is the height of the can?   9.55 inches

 84. *Height of a cylinder.* If the volume of a cylinder is 6.3 cubic meters and the diameter of the lid is 1.2 meters, then what is the height of the cylinder?   5.57 meters

85. *Great Chicago flood.* The great Chicago flood of April 1992 occurred when an old freight tunnel connecting buildings in the Loop ruptured. As shown in the figure, engineers plugged the tunnel with concrete on each side of the hole. They used the formula $F = WDA$ to find the force F of the water on the plug. In this formula the weight of water W is 62 pounds per cubic foot (lb/ft^3), the average depth D of the tunnel below the surface of the river is 32 ft, and the cross-sectional area A of the tunnel is 48 ft^2. Find the force on the plug.   95,232 pounds

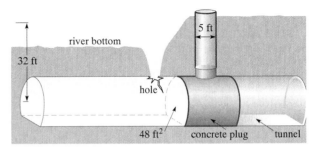

FIGURE FOR EXERCISE 85

86. *Will it hold?* To plug the tunnel described in the previous exercise, engineers drilled a 5-foot-diameter shaft down to the tunnel. The concrete plug was made so that it extended up into the shaft. For the plug to remain in place, the shear strength of the concrete in the shaft would have to be greater than the force of the water. The amount of force F that it would take for the water to shear the concrete in the shaft is given by $F = s\pi r^2$, where s is the shear strength of concrete and r is the radius of the shaft in inches. If the shear strength of concrete is 38 lb/in.^{2}, then what force of water would shear the concrete in the shaft? Use the result from Exercise 85 to determine whether the concrete would be strong enough to hold back the water.
107,442 pounds, strong enough

87. *Estimating armaments.* During World War II the Allies captured some German tanks on which the smallest serial number was S and the biggest was B. Assuming the entire production of tanks was numbered 1 through N, the Allies used the formula $N = B + S - 1$ to estimate the number of tanks in the German army (*New Scientist,* May 1998).
a) Find N if $B = 2{,}003$ and $S = 455$.
2,457
b) If this formula was used to estimate $N = 1{,}452$ and the largest serial number was 1,033, what was the smallest serial number?
420

88. *Cigarette usage.* The percentage of Americans aged 18 to 25 who use cigarettes has been decreasing at an approximately constant rate since 1974 (National Institute on Drug Abuse, www.nida.nih.gov). The formula

$$P = 47.9 - 0.949n$$

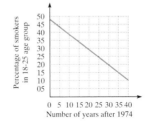

FIGURE FOR EXERCISE 88

can be used to estimate the percentage of smokers in this age group *n* years after 1974.

a) Use the formula to find the percentage of smokers in this age group in 1998.

b) Use the accompanying graph to estimate the year in which smoking will be eliminated from this group.

c) Use the formula to find the year in which smoking will be eliminated from this group.

 a) 25.1% b) 2024 c) 2024

89. *Distance between streets.* Harold Johnson lives on a four-sided, 50,000-square-foot lot that is bounded on two sides by parallel streets. The city has assessed him $1,000 for curb repair, $2 for each foot of property bordering on these two streets. How far apart are the streets?
 200 feet

FIGURE FOR EXERCISES 89–91

90. *Assessed for repairs.* Harold's sister, Maude, lives next door on a triangular lot of 25,000 square feet that also extends from street to street but has frontage only on one street. What will her assessment be? (See Exercise 89.) $500

91. *Juniper's lot.* Harold's other sister, Juniper, lives on the other side of him on a lot of 60,000 square feet in the shape of a parallelogram. What will her assessment be? (See Exercise 89.) $1,200

92. *Mother's driveway.* Harold's mother, who lives across the street, is pouring a concrete driveway, 12 feet wide and 4 inches thick, from the street straight to her house. This is too much work for Harold to do in one day, so his mother has agreed to buy 4 cubic yards of concrete each Saturday for three consecutive Saturdays. How far is it from the street to her house? 81 feet

GETTING MORE INVOLVED

93. *Exploration.* Electric companies often point out the low cost of electricity in performing common household tasks.

a) Find the cost of a kilowatt-hour of electricity in your area.

b) Write a formula for finding the cost of electricity for a household appliance to perform a certain task and to explain what each variable represents.

c) Use your formula to find the cost in your area for baking a $1\frac{1}{2}$-pound loaf of bread for 5 hours in a 750-watt Welbilt breadmaker.

2.3 APPLICATIONS

We often use algebra to solve problems by translating them into algebraic equations. Sometimes we can use formulas such as those in Appendix A. More often we have to set up a new equation that represents or **models** the problem. We begin with translating verbal expressions into algebraic expressions.

Writing Algebraic Expressions

Consider the three consecutive integers 5, 6, and 7. Note that each integer is 1 larger than the previous integer. To represent three *unknown* consecutive integers, we let

$$x = \text{the first integer,}$$
$$x + 1 = \text{the second integer,}$$

and
$$x + 2 = \text{the third integer.}$$

Consider the three consecutive odd integers 7, 9, and 11. Note that each odd integer is 2 larger than the previous odd integer. To represent three *unknown* consecutive odd integers, we let

$$x = \text{the first odd integer,}$$
$$x + 2 = \text{the second odd integer,}$$

and
$$x + 4 = \text{the third odd integer.}$$

Note that consecutive even integers as well as consecutive odd integers differ by 2. So the same expressions are used in either case.

How would we represent two numbers that have a sum of 8? If one of the numbers is 2, the other is certainly 6, or $8 - 2$. So if x is one of the numbers, then $8 - x$ is the other number. The expressions x and $8 - x$ have a sum of 8 for any value of x.

E X A M P L E   1

Writing algebraic expressions

Write algebraic expressions to represent each verbal expression.

a) Two numbers that differ by 12

b) Two consecutive even integers

c) Two investments that total $5,000

d) The length of a rectangle if the width is x meters and the perimeter is 10 meters

Solution

a) The expressions x and $x + 12$ differ by 12. Note that we could also use x and $x - 12$ for two numbers that differ by 12.

b) The expressions x and $x + 2$ represent two consecutive even integers.

c) If x represents the amount of one investment, then $5,000 - x$ represents the amount of the other investment.

d) Because the perimeter is 10 meters and $P = 2L + 2W = 2(L + W)$, the sum of the length and width is 5 meters. Because the width is x, the length is $5 - x$. ■

M A T H   A T   W O R K $x^2 + (x + 1)^2 = 5^2$

COFFEE STORE MANAGER

Mark Cromett, General Manager of the Charles Street Starbucks Coffee Store, arrives at work early in the morning to make sure each customer receives a perfect cup of coffee. Coffee beans from Central and South America, East Africa, and the Pacific are ground daily. Careful calibrations for the grinding are done by weighing each specific type of coffee. Even humidity frequently becomes part of the equation on how coffee is prepared. Besides the geographical area where the beans originate, customers have many choices for coffee. Selections are made among full city, espresso, Italian, and French roasts. Even seven decaffeinated coffees are available. But even with all of the choices, customers sometimes prefer their own special blend, requesting a mixture of different types of beans. Mr. Cromett is glad to brew or grind any special blend for a customer, and his charge depends on the different prices of the coffees mixed together. In Exercise 49 of this section you will determine the price of a specially blended coffee.

Many verbal phrases occur repeatedly in applications. The following list of some frequently occurring verbal phrases and their translations into algebraic expressions will help you to translate words into algebra.

Translating Words into Algebra

	Verbal Phrase	**Algebraic Expression**
Addition:	The sum of a number and 8	$x + 8$
	Five is added to a number	$x + 5$
	Two more than a number	$x + 2$
	A number increased by 3	$x + 3$
Subtraction:	Four is subtracted from a number	$x - 4$
	Three less than a number	$x - 3$
	The difference between 7 and a number	$7 - x$
	Some number decreased by 2	$x - 2$
	A number less 5	$x - 5$
Multiplication:	The product of 5 and a number	$5x$
	Seven times a number	$7x$
	Twice a number	$2x$
	One-half of a number	$\frac{1}{2}x \left(\text{or } \frac{x}{2}\right)$
Division:	The ratio of a number to 6	$\frac{x}{6}$
	The quotient of 5 and a number	$\frac{5}{x}$
	Three divided by some number	$\frac{3}{x}$

More than one operation can be combined in a single expression. For example, 7 less than twice a number is written as $2x - 7$.

Solving Problems

We will now see how algebraic expressions can be used to form an equation. If the equation correctly models a problem, then we may be able to solve the equation to get the solution to the problem. Some problems in this section could be solved without using algebra. However, the purpose of this section is to gain experience in setting up equations and using algebra to solve problems. We will show a complete solution to each problem so that you can gain the experience needed to solve more complex problems. We begin with a simple number problem.

E X A M P L E 2   **A number problem**

The sum of three consecutive integers is 228. Find the integers.

Solution

We first represent the unknown quantities with variables. The unknown quantities are the three consecutive integers. Let

$$x = \text{the first integer,}$$
$$x + 1 = \text{the second integer,}$$

and
$$x + 2 = \text{the third integer.}$$

helpful **hint**

Making a guess can be a good way to become familiar with the problem. For example, let's guess that the answers to Example 2 are 50, 51, and 52. Since $50 + 51 + 52 = 153$, these are not the correct numbers. But now we realize that we should use x, $x + 1$, and $x + 2$ and that the equation should be

$x + x + 1 + x + 2 = 228$.

Since the sum of these three expressions for the consecutive integers is 228, we can write the following equation and solve it:

$$x + (x + 1) + (x + 2) = 228 \qquad \text{The sum of the integers is 228.}$$
$$3x + 3 = 228$$
$$3x = 225$$
$$x = 75$$
$$x + 1 = 76 \qquad \text{Identify the other unknown quantities.}$$
$$x + 2 = 77$$

To verify that these values are the correct integers, we compute

$$75 + 76 + 77 = 228.$$

The three consecutive integers that have a sum of 228 are 75, 76, and 77.

General Strategy for Problem Solving

The steps to follow in providing a complete solution to a verbal problem can be stated as follows.

study **tip**

Don't simply work exercises to get answers. Keep reminding yourself of what you are actually doing. Keep trying to obtain the big picture. How does this section relate to what we did in the previous section? Where are we going next? When is the picture complete?

> **Strategy for Solving Word Problems**
>
> 1. Read the problem until you understand the problem. Making a guess and checking it will help you to understand the problem.
> 2. If possible, draw a diagram to illustrate the problem.
> 3. Choose a variable and write down what it represents.
> 4. Represent any other unknowns in terms of that variable.
> 5. Write an equation that models the situation.
> 6. Solve the equation.
> 7. Be sure that your solution answers the question posed in the original problem.
> 8. Check your answer by using it to solve the original problem (not the equation).

We will now see how this strategy can be applied to various types of problems.

Geometric Problems

Any problem that involves a geometric figure may be referred to as a **geometric problem.** For geometric problems the equation is often a geometric formula.

EXAMPLE 3

Finding the length and width of a rectangle

The length of a rectangular piece of property is 1 foot more than twice the width. If the perimeter is 302 feet, find the length and width.

Solution

First draw a diagram as in Fig. 2.4. Because the length is 1 foot more than twice the width, we let

$$x = \text{the width}$$

and

$$2x + 1 = \text{the length.}$$

x

$2x + 1$

FIGURE 2.4

The perimeter of a rectangle is modeled by the equation $2L + 2W = P$:

$$2L + 2W = P$$
$$2(2x + 1) + 2(x) = 302 \quad \text{Replace } L \text{ by } 2x + 1 \text{ and } W \text{ by } x.$$
$$4x + 2 + 2x = 302 \quad \text{Remove the parentheses.}$$
$$6x = 300$$
$$x = 50$$
$$2x + 1 = 101 \quad \text{Because } 2(50) + 1 = 101$$

Because $P = 2(101) + 2(50) = 302$ and 101 is 1 more than twice 50, we can be sure that the answer is correct. So the length is 101 feet, and the width is 50 feet.

Investment Problems

Investment problems involve sums of money invested at various interest rates. In this chapter we consider simple interest only.

E X A M P L E   4

Investing at two rates

Greg Smith invested some money in a certificate of deposit (CD) with an annual yield of 9%. He invested twice as much money in a mutual fund with an annual yield of 12%. His interest from the two investments at the end of the year was $396. How much money was invested at each rate?

Solution

Recall the formula $I = Prt$. In this problem the time t is 1 year, so $I = Pr$. If we let x represent the amount invested at the 9% rate, then $2x$ is the amount invested at 12%. The interest on these investments is the principal times the rate, or $0.09x$ and $0.12(2x)$. It is often helpful to make a table for the unknown quantities.

	Principal	Rate	Interest
Certificate of deposit	x	9%	$0.09x$
Mutual fund	$2x$	12%	$0.12(2x)$

The fact that the total interest from the investments was $396 is expressed in the following equation:

$$0.09x + 0.12(2x) = 396$$
$$0.09x + 0.24x = 396 \quad \text{We could multiply each side by 100}$$
$$0.33x = 396 \quad \text{to eliminate the decimals.}$$
$$x = \frac{396}{0.33}$$
$$x = 1200$$
$$2x = 2400$$

To check this answer, we find that $0.09(\$1200) = \$108$ and $0.12(\$2400) = \$288$. Now $\$108 + \$288 = \$396$. So Greg invested $1200 at 9% and $2400 at 12%.

Mixture Problems

Mixture problems involve solutions containing various percentages of a particular ingredient.

EXAMPLE 5   **Mixing milk**

How many gallons of milk containing 5% butterfat must be mixed with 90 gallons of 1% milk to obtain 2% milk?

Solution

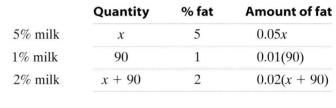

If x represents the number of gallons of 5% milk, then $0.05x$ represents the amount of fat in that milk. If we mix x gallons of 5% milk with 90 gallons of 1% milk, we will have $x + 90$ gallons of 2% milk. See Fig. 2.5. We can make a table to classify all of the unknown quantities.

helpful hint

To become familiar with the problem, let's guess that 100 gallons of 5% milk should be mixed with 90 gallons of 1% milk. The total amount of fat would be 0.05(100) + 0.01(90) or 5.9 gallons of fat. But 2% of 190 is 3.8 gallons of fat. Since the amounts of fat should be equal, our guess is incorrect.

	Quantity	% fat	Amount of fat
5% milk	x	5	$0.05x$
1% milk	90	1	$0.01(90)$
2% milk	$x + 90$	2	$0.02(x + 90)$

In mixture problems we always write an equation that accounts for one of the ingredients in the process. In this case we write an equation to express the fact that the total amount of fat from the first two types of milk is the same as the amount of fat in the mixture.

$$0.05x + 0.01(90) = 0.02(x + 90)$$
$$0.05x + 0.9 = 0.02x + 1.8 \quad \text{Remove parentheses.}$$
$$0.03x = 0.9 \quad \text{Note that we chose to work with the decimals}$$
$$x = 30 \quad \text{rather than eliminate them.}$$

FIGURE 2.5

We should use 30 gallons of 5% milk. There are 1.5 gallons of fat in the 30 gallons of 5% milk. The 1% milk will contribute 0.9 gallon of fat, and there will be 2.4 gallons of fat in 120 gallons of 2% milk. Because $1.5 + 0.9 = 2.4$, we have the correct solution. ■

EXAMPLE 6   **Blending gasoline**

A dealer has 10,000 gallons of unleaded gasoline. He wants to add just enough ethanol to make the fuel a 10% ethanol mixture. How many gallons of ethanol should be added?

Solution

Let x represent the number of gallons of 100% pure ethanol that should be added. (The original gasoline has no ethanol in it.) We can classify all of this information in a table.

	Amount	% ethanol	Amount of ethanol
Gasoline	10,000	0	0
Ethanol	x	100	x
Mixture	$x + 10,000$	10	$0.1(x + 10,000)$

We can write $0 + x = 0.1(x + 10,000)$ to express the fact that the amount of ethanol in the gasoline plus the amount of ethanol added is equal to the amount of ethanol in the final mixture. This model accounts for all of the ethanol in the process.

$$0 + x = 0.1(x + 10,000)$$
$$x = 0.1x + 1000 \quad \text{Remove parentheses.}$$
$$0.9x = 1000 \quad \text{Subtract } 0.1x \text{ from each side of the equation.}$$
$$x = \frac{1000}{0.9} \quad \text{Divide each side by 0.9.}$$
$$x = 1,111.1 \text{ gallons}$$

The amount of ethanol has been rounded to the nearest tenth of a gallon, so we cannot expect checking to be exact. If we combine 10,000 gallons of gasoline with 1,111.1 gallons of ethanol, we obtain 11,111.1 gallons of fuel. Notice that the amount of ethanol is 10% of the total mixture, 10% of $(10,000 + x)$. ■

Uniform Motion Problems

Problems that involve motion at a constant rate are referred to as **uniform motion problems.**

E X A M P L E   7

Driving Miss Jennifer

Jennifer drove her car for 3 hours in a dust storm. When the skies cleared, she increased her speed by 30 miles per hour and drove for 4 more hours, completing her 295-mile trip. How fast did she travel during the dust storm?

Solution

If x was Jennifer's speed during the dust storm, then her speed under clear skies was $x + 30$. For problems involving motion we use the formula $D = RT$ (distance equals rate times time). It is again helpful to make a table to classify the given information.

	Rate	Time	Distance
Dust storm	x	3	$3x$
Clear skies	$x + 30$	4	$4(x + 30)$

The following equation indicates that the total distance traveled was 295 miles:

$$3x + 4(x + 30) = 295$$
$$3x + 4x + 120 = 295 \quad \text{Remove parentheses.}$$
$$7x = 175$$
$$x = 25 \text{ miles per hour}$$

Check this answer in the original problem. Jennifer traveled 25 miles per hour (mph) during the storm. ■

Commission Problems

When property is sold, the percentage of the selling price that the selling agent receives is the **commission.**

E X A M P L E   8

Selling price of a house

Sonia is selling her house through a real estate agent whose commission is 6% of the selling price. What should be the selling price so that Sonia can get $84,600?

Solution

Let x be the selling price. The commission is 6% of x (not 6% of $84,600). Sonia receives the selling price less the sales commission.

$$\text{Selling price} - \text{commission} = \text{Sonia's share}$$
$$x - 0.06x = 84,600$$
$$0.94x = 84,600$$
$$x = \frac{84,600}{0.94}$$
$$= \$90,000$$

> **helpful hint**
>
> To become familiar with the problem, let's guess that the selling price is $100,000. The commission is 6% of the selling price: 0.06(100,000) or $6,000. So Sonia receives $94,000, which is incorrect.

The commission is 0.06($90,000), or $5,400. Sonia's share is $90,000 $-$ $5,400, or $84,600. The house should sell for $90,000. ■

W A R M - U P S

True or false? Explain your answer.

1. The recommended first step in solving a word problem is to write the equation.   False
2. When solving word problems, always write what the variable stands for.   True
3. Any solution to your equation must solve the word problem.   False
4. To represent two consecutive odd integers, we use x and $x + 1$.   False
5. We can represent two numbers that have a sum of 6 by x and $6 - x$.   True
6. Two numbers that differ by 7 can be represented by x and $x + 7$.   True
7. If $5x$ feet is 2 feet more than $3(x + 20)$ feet, then $5x + 2 = 3(x + 20)$.   False
8. If x is the selling price and the commission is 8% of the selling price, then the commission is $0.08x$.   True
9. If you need $80,000 for your house and the agent gets 10% of the selling price, then the agent gets $8,000, and the house sells for $88,000.   False
10. When we mix a 10% acid solution with a 14% acid solution, we can obtain a solution that is 24% acid.   False

---

2.3   EXERCISES

Reading and Writing   *After reading this section, write out the answers to these questions. Use complete sentences.*

1. How do you algebraically represent three unknown consecutive integers?

   Three unknown consecutive integers are represented by x, $x + 1$, and $x + 2$.

2. What is the difference between representing three unknown consecutive even or odd integers?

   In either case we use x, $x + 2$, and $x + 4$, but for odd integers x represents an odd integer and for even integers x represents an even integer.

3. What formula expresses the perimeter of a rectangle in terms of length and width?

The formula $P = 2L + 2W$ expresses the perimeter in terms of length and width.

4. What verbal phrases are used to indicate the operation of addition?

Addition can be indicated by the words, "sum," "more than," or "plus."

5. What is the commission when a real estate agent sells property?

The commission is a percentage of the selling price.

6. What is uniform motion?

Uniform motion is motion at a constant rate.

Find algebraic expressions for each of the following. See Example 1.

7. Two consecutive even integers $x, x + 2$

8. Two consecutive odd integers $x, x + 2$

9. Two numbers with a sum of 10 $x, 10 - x$

10. Two numbers with a difference of 3 $x, x + 3$

11. Eighty-five percent of the selling price $0.85x$

12. The product of a number and 3 $3x$

13. The distance traveled in 3 hours at x miles per hour $3x$ miles

14. The time it takes to travel 100 miles at $x + 5$ miles per hour $\dfrac{100}{x + 5}$ hours

15. The perimeter of a rectangle if the width is x feet and the length is 5 feet longer than the width $4x + 10$

16. The width of a rectangle if the length is x meters and the perimeter is 20 meters $10 - x$ meters

Show a complete solution for each number problem. See Example 2.

17. The sum of three consecutive integers is 84. Find the integers.   27, 28, 29

18. Find three consecutive integers whose sum is 171.
56, 57, 58

19. Find three consecutive even integers whose sum is 252.
82, 84, 86

20. Find three consecutive even integers whose sum is 84.
26, 28, 30

21. Two consecutive odd integers have a sum of 128. What are the integers?   63, 65

22. Four consecutive odd integers have a sum of 56. What are the integers?   11, 13, 15, 17

Show a complete solution to each geometric problem. See Example 3.

23. *Length and width.* If the perimeter of a rectangle is 278 meters and the length is 1 meter longer than twice the width, then what are the length and width?
Width 46 meters, length 93 meters

24. *Dimensions of a frame.* A frame maker made a large picture frame using 10 feet of frame molding. If the length of

the finished frame was 2 feet more than the width, then what were the dimensions of the frame?
Width 1.5 feet, length 3.5 feet

25. *Perimeter of a lot.* Having finished fencing the perimeter of a triangular piece of land, Lance observed that the second side was just 10 feet short of being twice as long as the first side, and the third side was exactly 50 feet longer than the first side. If he used 684 feet of fencing, what are the lengths of the three sides?
161 feet, 312 feet, 211 feet

26. *Isosceles triangle.* A flag in the shape of an isosceles triangle has a base that is 3.5 inches shorter than either of the equal sides. If the perimeter of the triangle is 49 inches, what is the length of the equal sides?
17.5 inches

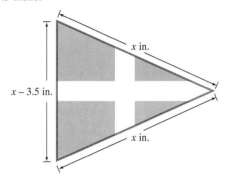

FIGURE FOR EXERCISE 26

27. *Hog heaven.* Farmer Hodges has 50 feet of fencing to make a rectangular hog pen beside a very large barn. He needs to fence only three sides because the barn will form the fourth side. Studies have shown that under those conditions the side parallel to the barn should be 5 feet longer than twice the width. If Farmer Hodges uses all of the fencing, what should the dimensions be?
Width 11.25 feet, length 27.5 feet

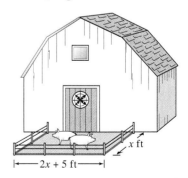

FIGURE FOR EXERCISE 27

28. *Doorway dimensions.* A carpenter made a doorway that is 1 foot taller than twice the width. If she used three pieces of door edge molding with a total length of 17 feet, then what are the approximate dimensions of the doorway?
3 feet wide, 7 feet high

Show a complete solution to each investment problem. See Example 4.

29. Investing money. Mr. and Mrs. Jackson invested some money at 6% simple interest and some money at 10% simple interest. In the second investment they put $1,000 more than they put in the first. If the income from both investments for one year was $340, then how much did they invest at each rate? $1,500 at 6%, $2,500 at 10%

30. Sibling rivalry. Samantha lent her brother some money at 9% simple interest and her sister one-half as much money at 16% simple interest. If she received a total of 34 cents in interest, then how much did she lend to each one?
Brother $2, sister $1

31. Investing inheritance. Norman invested one-half of his inheritance in a CD that had a 10% annual yield. He lent one-quarter of his inheritance to his brother-in-law at 12% simple interest. His income from these two investments was $6,400 for one year. How much was the inheritance?
$80,000

32. Insurance settlement. Gary invested one-third of his insurance settlement in a CD that yielded 12%. He also invested one-third in Tara's computer business. Tara paid Gary 15% on this investment. If Gary's total income from these investments was $10,800 for one year, then what was the amount of his insurance settlement? $120,000

Show a complete solution to each mixture problem. See Examples 5 and 6.

33. Acid solutions. How many gallons of 5% acid solution should be mixed with 20 gallons of a 10% acid solution to obtain an 8% acid solution? $\frac{40}{3}$ gallons

34. Alcohol solutions. How many liters of a 10% alcohol solution should be mixed with 12 liters of a 20% alcohol solution to obtain a 14% alcohol solution?   18 liters

35. Increasing acidity. A gallon of Del Monte White Vinegar is labeled 5% acidity. How many fluid ounces of pure acid must be added to get 6% acidity?   1.36 ounces

36. Chlorine bleach. A gallon of Clorox bleach is labeled "5.25% sodium hypochlorite by weight." If a gallon of bleach weighs 8.3 pounds, then how many ounces of sodium hypochlorite must be added so that the bleach will be 6% sodium hypochlorite?   1.0596 ounces

Show a complete solution to each uniform motion problem. See Example 7.

37. Driving in a fog. Carlo drove for 3 hours in a fog, then increased his speed by 30 miles per hour (mph) and drove 6 more hours. If his total trip was 540 miles, then what was his speed in the fog?   40 mph

38. Walk, don't run. Louise walked for 2 hours then ran for $1\frac{1}{2}$ hours. If she runs twice as fast as she walks and the total trip was 20 miles, then how fast does she run?   8 mph

39. Commuting to work. A commuter bus takes 2 hours to get downtown; an express bus, averaging 25 mph faster, takes 45 minutes to cover the same route. What is the average speed for the commuter bus?   15 mph

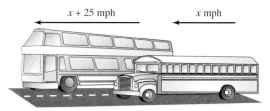

x + 25 mph   *x* mph

FIGURE FOR EXERCISE 39

40. Passengers versus freight. A freight train takes $1\frac{1}{4}$ hours to get to the city; a passenger train averaging 40 mph faster takes only 45 minutes to cover the same distance. What is the average speed of the passenger train?   100 mph

Show a complete solution to each problem. See Example 8.

41. Listing a house. Karl wants to get $80,000 for his house. The real estate agent charges 8% of the selling price for selling the house. What should the selling price be?
$86,957

42. Hot tamales. Martha sells hot tamales at a sidewalk stand. Her total receipts including the 5% sales tax were $915.60. What amount of sales tax did she collect? $43.60

43. Mustang Sally. Sally bought a used Mustang. The selling price plus the 7% state sales tax was $9,041.50. What was the selling price? $8,450

44. Choosing a selling price. Roy is selling his car through a broker. Roy wants to get $3,000 for himself, but the broker gets a commission of 10% of the selling price. What should the selling price be? $3,333.33

Show a complete solution to each problem.

45. Tennis. The distance from the baseline to the service line on a tennis court is 3 feet longer than the distance from the service line to the net. If the distance from the baseline to the net is 39 feet, then what is the distance from the service line to the net?   18 feet

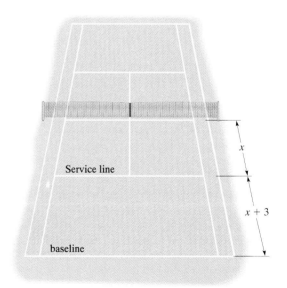

Service line

baseline

x

x + 3

FIGURE FOR EXERCISE 45

46. *Mixed doubles.* The doubles court in tennis is one-third wider than the singles court. If the doubles court is 36 feet wide, then what is the width of the singles court?
27 feet

47. *First Super Bowl.* In the first Super Bowl game in the Los Angeles Coliseum in 1967, the Green Bay Packers outscored the Kansas City Chiefs by 25 points. If 45 points were scored in that game, then what was the final score?
Packers 35, Chiefs 10

48. *Toy sales.* In 1998 Toys "R" Us and Wal-Mart together held 36% of the toy market share (*Fortune*, June 1, 1998, www.fortune.com). If the market share for Toys "R" Us was 4 percentage points higher than the market share for Wal-Mart, then what was the market share for each company?
16% Wal-Mart, 20% Toys "R" Us

49. *Blending coffee.* Mark blends $\frac{3}{4}$ of a pound of premium Brazilian coffee with $1\frac{1}{2}$ pounds of standard Colombian coffee. If the Brazilian coffee sells for $10 per pound and the Colombian coffee sells for $8 per pound, then what should the price per pound be for the blended coffee?
$8.67 per pound

FIGURE FOR EXERCISE 49

50. *'Tis the seasoning.* Cheryl's Famous Pumpkin Pie Seasoning consists of a blend of cinnamon, nutmeg, and cloves. When Cheryl mixes up a batch, she uses 200 ounces of cinnamon, 100 ounces of nutmeg, and 100 ounces of cloves. If cinnamon sells for $1.80 per ounce, nutmeg sells for $1.60 per ounce, and cloves sell for $1.40 per ounce, what should be the price per ounce of the mixture?
$1.65 per ounce

51. *Health food mix.* Dried bananas sell for $0.80 per quarter-pound, and dried apricots sell for $1.00 per quarter-pound. How many pounds of apricots should be mixed with 10 pounds of bananas to get a mixture that sells for $0.95 per quarter-pound?   30 pounds

52. *Mixed nuts.* Cashews sell for $1.20 per quarter-pound, and Brazil nuts sell for $1.50 per quarter-pound. How many pounds of cashews should be mixed with 20 pounds of Brazil nuts to get a mix that sells for $1.30 per quarter-pound?   40 pounds

53. *Antifreeze mixture.* A mechanic finds that a car with a 20-quart radiator has a mixture containing 30% antifreeze in it. How much of this mixture would he have to drain out and replace with pure antifreeze to get a 50% antifreeze mixture? $\frac{40}{7}$ quarts

54. *Increasing the percentage.* A mechanic has found that a car with a 16-quart radiator has a 40% antifreeze mixture in the radiator. She has on hand a 70% antifreeze solution. How much of the 40% solution would she have to replace with the 70% solution to get the solution in the radiator up to 50%? $\frac{16}{3}$ quarts

55. *Fortune 500 profits.* The total profit for Fortune 500 companies was $324 billion in 1997 (*Fortune*, July 2, 1998). This figure represents an increase of 7.8% from the previous year. What was the total profit in 1996?
$300.6 billion

56. *Decreasing fertility.* The fertility rate in developing countries has decreased 48% from 1960 to 1998 (U.N. Population Division, www.un.org). The fertility rate was 3.1 children per woman in 1998. What was the fertility rate in developing countries in 1960?
5.96 children per woman

57. *Dividing the estate.* Uncle Albert's estate is to be divided among his three nephews. The will specifies that Daniel receive one-half of the amount that Brian receives and that Raymond receive $1,000 less than one-third of the amount that Brian receives. If the estate amounts to $25,400, then how much does each inherit?
Brian $14,400, Daniel $7,200, Raymond $3,800

58. *Mary's assets.* Mary Hall's will specifies that her lawyer is to liquidate her assets and divide the proceeds among her three sisters. Lena's share is to be one-half of Lisa's, and Lisa's share is to be one-half of Lauren's. If the lawyer has agreed to a fee that is equal to 10% of the largest share and the proceeds amount to $164,428, then how much does each person get?
Lauren $88,880, Lisa $44,440, Lena $22,220, lawyer $8,888

59. *Missing integers.* If the larger of two consecutive integers is subtracted from twice the smaller integer, then the result is 21. Find the integers.   22, 23

60. *Really odd integers.* If the smaller of two consecutive odd integers is subtracted from twice the larger one, then the result is 13. Find the integers.   9, 11

61. *Highway miles.* Berenice and Jarrett drive a rig for Continental Freightways. In one day Berenice averaged 50 mph and Jarrett averaged 56 mph, but Berenice drove for two more hours than Jarrett. If together they covered 683 miles, then for how many hours did Berenice drive?   7.5 hours

62. *Spring break.* Fernell and Dabney shared the driving to Florida for spring break. Fernell averaged 50 mph, and Dabney averaged 64 mph. If Fernell drove for 3 hours longer than Dabney but covered 18 miles less than Dabney, then for how many hours did Fernell drive?   15 hours

63. *Stacy's square.* Stacy has 70 meters of fencing and plans to make a square pen. In one side she is going to leave an opening that is one-half the length of the side. If she uses all 70 meters of fencing, how large can the square be?
20 meters by 20 meters

64. ***Shawn's shed.*** Shawn is building a tool shed with a square foundation and has enough siding to cover 32 linear feet of walls. If he leaves a 4-foot space for a door, then what size foundation would use up all of his siding?

9 feet by 9 feet

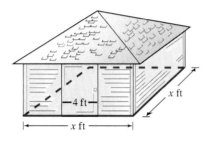

FIGURE FOR EXERCISE 64

65. ***Splitting investments.*** Joan had $3,000 to invest. She invested part of it in an investment paying 8% and the remainder in an investment paying 10%. If the total income on these investments was $290, then how much did she invest at each rate?

$500 at 8%, $2,500 at 10%

66. ***Financial independence.*** Dorothy had $8,000 to invest. She invested part of it in an investment paying 6% and the rest in an investment paying 9%. If the total income from these investments was $690, then how much did she invest at each rate?

$1,000 at 6%, $7,000 at 9%

67. ***Alcohol solutions.*** Amy has two solutions available in the laboratory, one with 5% alcohol and the other with 10% alcohol. How much of each should she mix together to obtain 5 gallons of an 8% solution?

2 gallons of 5% solution, 3 gallons of 10% solution

68. ***Alcohol and water.*** Joy has a solution containing 12% alcohol. How much of this solution and how much water must she use to get 6 liters of a solution containing 10% alcohol?

5 liters of 12% alcohol, 1 liter of water

69. ***Chance meeting.*** In 6 years Todd will be twice as old as Darla was when they met 6 years ago. If their ages total 78 years, then how old are they now?

Todd 46, Darla 32

70. ***Centennial Plumbing Company.*** The three Hoffman brothers advertise that together they have a century of plumbing experience. Bart has twice the experience of Al, and in 3 years Carl will have twice the experience that Al had a year ago. How many years of experience does each of them have?

Al 21, Bart 42, Carl 37

 **2.4**   **INEQUALITIES**

So far, we have been working with equations in this chapter. Equations express the equality of two algebraic expressions. But we are often concerned with two algebraic expressions that are not equal, one expression being greater than or less than the other. In this section we will begin our study of inequalities.

Basic Ideas

Statements that express the inequality of algebraic expressions are called **inequalities.** The symbols that we use to express inequality are given below with their meanings.

Inequality Symbols

Symbol	Meaning
$<$	Is less than
$\leq$	Is less than or equal to
$>$	Is greater than
$\geq$	Is greater than or equal to

It is clear that 5 is less than 10, but how do we compare -5 and -10? If we think of negative numbers as debts, we would say that -10 is the larger debt. However, in algebra the size of a number is determined only by its position on the number line. For two numbers a and b we say that *a is less than b* if and only if a is to the *left* of b on the number line. To compare -5 and -10, we locate each point on the number line in Fig. 2.6. Because -10 is to the left of -5 on the number line, we say that -10 is less than -5. In symbols,

$$-10 < -5.$$

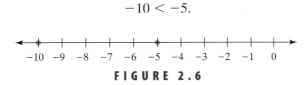

FIGURE 2.6

We say that a is greater than b if and only if a is to the *right* of b on the number line. Thus we can also write

$$-5 > -10.$$

The statement $a \leq b$ is true if a is less than b or if a is equal to b. The statement $a \geq b$ is true if a is greater than b or if a equals b. For example, the statement $3 \leq 5$ is true, and so is the statement $5 \leq 5$.

E X A M P L E   1

Inequalities
Determine whether each statement is true or false.

a) $-5 < 3$ **b)** $-9 > -6$

c) $-3 \leq 2$ **d)** $4 \geq 4$

Solution

a) The statement $-5 < 3$ is true because -5 is to the left of 3 on the number line. In fact, any negative number is less than any positive number.

b) The statement $-9 > -6$ is false because -9 lies to the left of -6.

c) The statement $-3 \leq 2$ is true because -3 is less than 2.

d) The statement $4 \geq 4$ is true because $4 = 4$ is true. ◼

Interval Notation and Graphs

If an inequality involves a variable, then which real numbers can be used in place of the variable to obtain a correct statement? The set of all such numbers is the **solution set** to the inequality. For example, $x < 3$ is correct if x is replaced by any number that lies to the left of 3 on the number line:

$$1.5 < 3, \qquad 0 < 3, \qquad \text{and} \qquad -2 < 3$$

The set of real numbers to the left of 3 is written in set notation as $\{x \mid x < 3\}$, in **interval notation** as $(-\infty, 3)$, and graphed in Fig. 2.7:

FIGURE 2.7

Note that $-\infty$ (negative infinity) is not a number, but it indicates that there is no end to the real numbers less than 3. The parenthesis used next to the 3 in the interval notation and on the graph means that 3 is not included in the solution set to $x < 3$.

An inequality such as $x \geq 1$ is satisfied by 1 and any real number that lies to the right of 1 on the number line. The solution set to $x \geq 1$ is written in set notation as $\{x \mid x \geq 1\}$, in interval notation as $[1, \infty)$, and graphed in Fig. 2.8:

FIGURE 2.8

The bracket used next to the 1 in the interval notation and on the graph means that 1 is in the solution set to $x \geq 1$.

The solution set to an inequality can be stated symbolically with set notation and interval notation, or visually with a graph. Interval notation is popular because it is simpler to write than set notation. The interval notation and graph for each of the four basic inequalities is summarized as follows.

Basic Interval Notation (k any real number)

Inequality	Solution Set with Interval Notation	Graph
$x > k$	(k, ∞)	
$x \geq k$	$[k, \infty)$	
$x < k$	$(-\infty, k)$	
$x \leq k$	$(-\infty, k]$	

E X A M P L E   2   **Interval notation and graphs**

Write the solution set to each inequality in interval notation and graph it.

a) $x > -5$ **b)** $x \leq 2$

Solution

a) The solution set to the inequality $x > -5$ is $\{x \mid x > -5\}$. The solution set is the interval of all numbers to the right of -5 on the number line. This set is written in interval notation as $(-5, \infty)$, and it is graphed in Fig. 2.9.

FIGURE 2.9

b) The solution set to $x \leq 2$ is $\{x \mid x \leq 2\}$. This set includes 2 and all real numbers to the left of 2. Because 2 is included, we use a bracket at 2. The interval notation for this set is $(-\infty, 2]$. The graph is shown in Fig. 2.10.

FIGURE 2.10

Solving Linear Inequalities

In Section 2.1 we defined a linear equation as an equation of the form $ax + b = 0$. If we replace the equality symbol in a linear equation with an inequality symbol, we have a linear inequality.

Linear Inequality

A **linear inequality** in one variable x is any inequality of the form $ax + b < 0$, where a and b are real numbers, with $a \neq 0$. In place of $<$ we may also use $\leq$, $>$, or $\geq$.

study tip

What's on the final exam? Chances are that if your instructor thinks a question is important enough for a test or quiz, that question is also important enough for the final exam. So keep all tests and quizzes, and make sure that you have corrected any mistakes on them. To study for the final exam, write the old questions/problems on note cards, one to a card. Shuffle the note cards and see if you can answer the questions or solve the problems in a random order.

Inequalities that can be *rewritten* in the form of a linear inequality are also called linear inequalities.

Before we solve linear inequalities, let's examine the results of performing various operations on each side of an inequality. If we start with the inequality $2 < 6$ and add 2 to each side, we get the true statement $4 < 8$. Examine the results in the following table.

Perform these operations on each side of $2 < 6$:

	Add 2	**Subtract 2**	**Multiply by 2**	**Divide by 2**
Resulting inequality	$4 < 8$	$0 < 4$	$4 < 12$	$1 < 3$

All of the resulting inequalities are correct. However, if we perform operations on each side of $2 < 6$ using -2, the situation is not as simple. For example, $-2 \cdot 2 = -4$ and $-2 \cdot 6 = -12$, but -4 is greater than -12. To get a correct inequality when each side is multiplied or divided by -2, we must reverse the inequality symbol, as shown in the following table.

Perform these operations on each side of $2 < 6$:

	Add -2	**Subtract -2**	**Multiply by -2**	**Divide by -2**
Resulting inequality	$0 < 4$	$4 < 8$	$-4 > -12$	$-1 > -3$

Inequality reverses

These examples illustrate the properties that we use for solving inequalities.

Properties of Inequality

Addition Property of Inequality
If the same number is added to both sides of an inequality, then the solution set to the inequality is unchanged.

Multiplication Property of Inequality
If both sides of an inequality are multiplied by the same *positive number,* then the solution set to the inequality is unchanged.
If both sides of an inequality are multiplied by the same *negative number* and *the inequality symbol is reversed,* then the solution set to the inequality is unchanged.

Because subtraction is defined in terms of addition, the addition property of inequality also allows us to subtract the same number from both sides. Because division is defined in terms of multiplication, the multiplication property of inequality also allows us to divide both sides by the same nonzero number *as long as we reverse the inequality symbol when dividing by a negative number.*

Equivalent inequalities are inequalities with the same solution set. We find the solution to a linear inequality by using the properties to convert it into an equivalent inequality with an obvious solution set, just as we do when solving equations.

E X A M P L E   3   **Solving inequalities**

Solve each inequality. State and graph the solution set.

a) $2x - 7 < -1$ **b)** $5 - 3x < 11$

Solution

a) We proceed exactly as we do when solving equations:

$$2x - 7 < -1 \quad \text{Original inequality}$$
$$2x < 6 \quad \text{Add 7 to each side.}$$
$$x < 3 \quad \text{Divide each side by 2.}$$

FIGURE 2.11

The solution set is written in set notation as $\{x \mid x < 3\}$ and in interval notation as $(-\infty, 3)$. The graph is shown in Fig. 2.11.

b) We divide by a negative number to solve this inequality.

$$5 - 3x < 11 \quad \text{Original equation}$$
$$-3x < 6 \quad \text{Subtract 5 from each side.}$$
$$x > -2 \quad \text{Divide each side by } -3 \text{ and reverse the inequality symbol.}$$

FIGURE 2.12

The solution set is written in set notation as $\{x \mid x > -2\}$ and in interval notation as $(-2, \infty)$. The graph is shown in Fig. 2.12.   ■

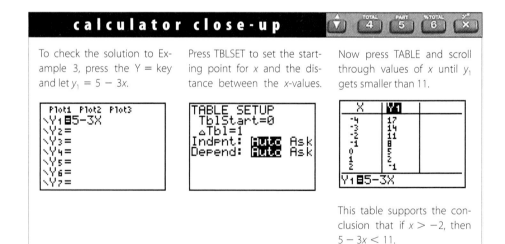

calculator close-up

To check the solution to Example 3, press the Y = key and let $y_1 = 5 - 3x$.

Press TBLSET to set the starting point for x and the distance between the x-values.

Now press TABLE and scroll through values of x until y_1 gets smaller than 11.

```
Plot1 Plot2 Plot3
\Y1■5-3X
\Y2=
\Y3=
\Y4=
\Y5=
\Y6=
\Y7=
```

```
TABLE SETUP
 TblStart=0
 ΔTbl=1
Indpnt: Auto Ask
Depend: Auto Ask
```

```
  X  │ Y1
─────┼─────
 -4  │ 17
 -3  │ 14
 -2  │ 11
 -1  │ 8
  0  │ 5
  1  │ 2
  2  │ -1
─────┴─────
Y1■5-3X
```

This table supports the conclusion that if $x > -2$, then $5 - 3x < 11$.

E X A M P L E   4   **Solving inequalities**

Solve $\dfrac{8 + 3x}{-5} \geq -4$. State and graph the solution set.

Solution

$$\frac{8 + 3x}{-5} \geq -4 \qquad \text{Original inequality}$$

$$-5\left(\frac{8 + 3x}{-5}\right) \leq -5(-4) \qquad \text{Multiply each side by } -5 \text{ and reverse the inequality symbol.}$$

$$8 + 3x \leq 20 \qquad \text{Simplify.}$$

$$3x \leq 12 \qquad \text{Subtract 8 from each side.}$$

$$x \leq 4 \qquad \text{Divide each side by 3.}$$

The solution set is $(-\infty, 4]$, and its graph is shown in Fig. 2.13.

FIGURE 2.13

E X A M P L E 5

An inequality with fractions

Solve $\frac{1}{2}x - \frac{2}{3} \leq x + \frac{4}{3}$. State and graph the solution set.

Solution

First multiply each side of the inequality by 6, the LCD:

$$\frac{1}{2}x - \frac{2}{3} \leq x + \frac{4}{3} \qquad \text{Original inequality}$$

$$6\left(\frac{1}{2}x - \frac{2}{3}\right) \leq 6\left(x + \frac{4}{3}\right) \qquad \text{Multiplying by positive 6 does not reverse the inequality.}$$

$$3x - 4 \leq 6x + 8 \qquad \text{Distributive property}$$

$$3x \leq 6x + 12 \qquad \text{Add 4 to each side.}$$

$$-3x \leq 12 \qquad \text{Subtract } 6x \text{ from each side.}$$

$$x \geq -4 \qquad \text{Divide each side by } -3 \text{ and reverse the inequality.}$$

The solution set is the interval $[-4, \infty)$. Its graph is shown in Fig. 2.14.

FIGURE 2.14

> **helpful hint**
>
> Notice that we use the same strategy for solving inequalities as we do for solving equations. But we must remember to reverse the inequality symbol when we multiply or divide by a negative number. For inequalities it is usually best to isolate the variable on the left-hand side.

Applications

Inequalities have applications just as equations do. To use inequalities, we must be able to translate a verbal problem into an algebraic inequality. Inequality can be expressed verbally in a variety of ways.

E X A M P L E 6

Writing inequalities

Identify the variable and write an inequality that describes the situation.

a) Chris paid more than $200 for a suit.

b) A candidate for president must be at least 35 years old.

c) The capacity of an elevator is at most 1,500 pounds.

d) The company must hire no fewer than 10 programmers.

Solution

a) If c is the cost of the suit in dollars, then $c > 200$.

b) If a is the age of the candidate in years, then $a \geq 35$.

c) If x is the capacity of the elevator in pounds, then $x \leq 1{,}500$.

d) If n represents the number of programmers and n is not less than 10, then $n \geq 10$. ■

In Example 6(d) we knew that n was not less than 10. So there were exactly two other possibilities: n was greater than 10 or equal to 10. The fact that there are only three possible ways to position two real numbers on a number line is called the **trichotomy property.**

> **Trichotomy Property**
>
> For any two real numbers a and b, exactly one of the following is true:
> $$a < b, \qquad a = b, \qquad \text{or} \qquad a > b$$

We follow the same steps to solve problems involving inequalities as we do to solve problems involving equations.

E X A M P L E   7

Price range

Lois plans to spend less than $500 on an electric dryer, including the 9% sales tax and a $64 setup charge. In what range is the selling price of the dryer that she can afford?

Solution

If we let x represent the selling price in dollars for the dryer, then the amount of sales tax is $0.09x$. Because her total cost must be less than $500, we can write the following inequality:

$$x + 0.09x + 64 < 500$$
$$1.09x < 436 \qquad \text{Subtract 64 from each side.}$$
$$x < \frac{436}{1.09} \qquad \text{Divide each side by 1.09.}$$
$$x < 400$$

The selling price of the dryer must be less than $400. ■

study tip

When studying for an exam, start by working the exercises in the Chapter Review. If you find exercises that you cannot do, then go back to the section where the appropriate concepts were introduced. Study the appropriate examples in the section and work some problems. Then go back to the Chapter Review and continue.

Note that if we had written the equation $x + 0.09x + 64 = 500$ for the last example, we would have gotten $x = 400$. We could then have concluded that the selling price must be less than $400. This would certainly solve the problem, but it would not illustrate the use of inequalities. The original problem describes an inequality, and we should solve it as an inequality.

E X A M P L E   8

Paying off the mortgage

Tessie owns a piece of land on which she owes $12,760 to a bank. She wants to sell the land for enough money to at least pay off the mortgage. The real estate agent gets 6% of the selling price, and her city has a $400 real estate transfer tax paid by the seller. What should the range of the selling price be for Tessie to get at least enough money to pay off her mortgage?

Solution

If x is the selling price in dollars, then the commission is $0.06x$. We can write an inequality expressing the fact that the selling price minus the real estate commission minus the $400 tax must be at least $12,760:

$$x - 0.06x - 400 \geq 12{,}760$$
$$0.94x - 400 \geq 12{,}760 \quad \text{\small $1 - 0.06 = 0.94$}$$
$$0.94x \geq 13{,}160 \quad \text{\small Add 400 to each side.}$$
$$x \geq \frac{13{,}160}{0.94} \quad \text{\small Divide each side by 0.94.}$$
$$x \geq 14{,}000$$

The selling price must be at least $14,000 for Tessie to pay off the mortgage.

WARM-UPS

True or false? Explain your answer.

1. $0 < 0$ False   2. $-300 > -2$ False   3. $-60 \leq -60$ True
4. The inequality $6 < x$ is equivalent to $x < 6$.   False
5. The inequality $-2x < 10$ is equivalent to $x < -5$.   False
6. The solution set to $3x \geq -12$ is $(-\infty, -4]$.   False
7. The solution set to $-x > 4$ is $(-\infty, -4)$.   True
8. If x is no larger than 8, then $x \leq 8$.   True
9. If m is any real number, then exactly one of the following is true: $m < 0$, $m = 0$, or $m > 0$.   True
10. The number -2 is a member of the solution set to the inequality $3 - 4x \leq 11$.   True

2.4   EXERCISES

Reading and Writing   After reading this section, write out the answers to these questions. Use complete sentences.

1. What is an inequality?
   An inequality is a sentence that expresses inequality between two algebraic expressions.

2. What symbols are used to express inequality?
   To express inequality we use the symbols $<$, $\leq$, $>$, and $\geq$.

3. What does it mean when we say that a is less than b?
   If a is less than b, then a lies to the left of b on the number line.

4. What is a linear inequality?
   A linear inequality is an inequality of the form $ax + b > 0$ or with any of the other inequality symbols used in place of $>$.

5. How does solving linear inequalities differ from solving linear equations?
   When you multiply or divide by a negative number, the inequality symbol is reversed.

6. What verbal phrases are used to indicate an inequality?
   We can verbally indicate inequality with words like "less than," "at least," "greater than," and "at most."

Determine whether each inequality is true or false. See Example 1.

7. $-3 < -9$
   False

8. $-8 > -7$
   False

9. $0 \leq 8$
   True

10. $-6 \geq -8$
   True

11. $(-3)20 > (-3)40$
    True

12. $(-1)(-3) < (-1)(5)$
    False

13. $9 - (-3) \leq 12$
    True

14. $(-4)(-5) + 2 \geq 21$
    True

Determine whether each inequality is satisfied by the given number.

15. $2x - 4 < 8, -3$
    Yes

16. $5 - 3x > -1, 6$
    No

17. $2x - 3 \leq 3x - 9, 5$
    No

18. $6 - 3x \geq 10 - 2x, -4$
    Yes

19. $5 - x < 4 - 2x, -1$
    No

20. $3x - 7 \geq 3x - 10, 9$
    Yes

Write the solution set in interval notation and graph it. See Example 2.

21. $x \leq -1$ $(-\infty, -1]$

22. $x \geq -7$ $[-7, \infty)$

23. $x > 20$ $(20, \infty)$

24. $x < 30$ $(-\infty, 30)$

25. $3 \leq x$ $[3, \infty)$

26. $-2 > x$ $(-\infty, -2)$

27. $x < 2.3$ $(-\infty, 2.3)$

28. $x \leq 4.5$ $(-\infty, 4.5]$

Rewrite each set in interval notation.

29. $\{x \mid x > 1\}$
    $(1, \infty)$

30. $\{x \mid x < 3\}$
    $(-\infty, 3)$

31. $\{x \mid x \leq -3\}$
    $(-\infty, -3]$

32. $\{x \mid x \geq -2\}$
    $[-2, \infty)$

33. $\{x \mid x < 5\}$
    $(-\infty, 5)$

34. $\{x \mid x > -7\}$
    $(-7, \infty)$

35. $\{x \mid x \geq -4\}$
    $[-4, \infty)$

36. $\{x \mid x \leq -9\}$
    $(-\infty, -9]$

Fill in the blank with an inequality symbol so that the two statements are equivalent.

37. $x + 5 > 12$
    $x \underset{>}{\_\_} 7$

38. $2x - 3 \leq -4$
    $2x \underset{\leq}{\_\_} -1$

39. $-x < 6$
    $x \underset{>}{\_\_} -6$

40. $-5 \geq -x$
    $5 \underset{\leq}{\_\_} x$

41. $-2x \geq 8$
    $x \underset{\leq}{\_\_} -4$

42. $-5x > -10$
    $x \underset{<}{\_\_} 2$

43. $4 < x$
    $x \underset{>}{\_\_} 4$

44. $-9 \leq -x$
    $x \underset{\leq}{\_\_} 9$

Solve each of the following inequalities. Express the solution set in interval notation and graph it. See Examples 3–5.

45. $7x > -14$ $(-2, \infty)$

46. $4x \leq -8$ $(-\infty, -2]$

47. $-3x \leq 12$ $[-4, \infty)$

48. $-2x > -6$ $(-\infty, 3)$

49. $2x - 3 > 7$ $(5, \infty)$

50. $3x - 2 < 6$ $\left(-\infty, \dfrac{8}{3}\right)$

51. $3 - 5x \leq 18$ $[-3, \infty)$

52. $5 - 4x \geq 19$ $\left(-\infty, -\dfrac{7}{2}\right)$

53. $\dfrac{x - 3}{-5} < -2$ $(13, \infty)$

54. $\dfrac{2x - 3}{4} > 6$ $\left(\dfrac{27}{2}, \infty\right)$

55. $\dfrac{5 - 3x}{4} \leq 2$ $[-1, \infty)$

56. $\dfrac{7 - 5x}{-2} \geq -1$ $[1, \infty)$

57. $3 - \dfrac{1}{4}x \geq 2$ $(-\infty, 4]$

58. $5 - \dfrac{1}{3}x > 2$ $(-\infty, 9)$

59. $\dfrac{1}{4}x - \dfrac{1}{2} < \dfrac{1}{2}x - \dfrac{2}{3}$ $\left(\dfrac{2}{3}, \infty\right)$

60. $\dfrac{1}{3}x - \dfrac{1}{6} < \dfrac{1}{6}x - \dfrac{1}{2}$
    $(-\infty, -2)$

61. $\dfrac{y - 3}{2} > \dfrac{1}{2} - \dfrac{y - 5}{4}$
    $\left(\dfrac{13}{3}, \infty\right)$

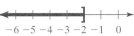
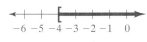

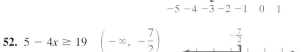
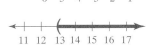

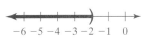
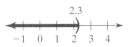

62. $\dfrac{y-1}{3} - \dfrac{y+1}{5} > 1$ $\quad \left(\dfrac{23}{2}, \infty\right)$

Solve each inequality and graph the solution set.

63. $2x + 3 > 2(x - 4)$ $\quad (-\infty, \infty)$

64. $-2(5x - 1) \le -5(5 + 2x)$ $\quad \varnothing$

65. $-4(2x - 5) \le 2(6 - 4x)$ $\quad \varnothing$

66. $-3(2x - 1) \le 2(5 - 3x)$ $\quad (-\infty, \infty)$

67. $-\dfrac{1}{2}(x - 6) < \dfrac{1}{2}x + 2$ $\quad (1, \infty)$

68. $-3\left(\dfrac{1}{2}x - \dfrac{1}{4}\right) > \dfrac{x}{2} - \dfrac{1}{4}$ $\quad \left(-\infty, \dfrac{1}{2}\right)$

69. $-\dfrac{1}{2}(2x - 3) + \dfrac{1}{3}(4 - 6x) \ge \dfrac{1}{4}(7 - 2x) - 3$ $\quad \left(-\infty, \dfrac{49}{30}\right)$

70. $\dfrac{3}{5}(x - 3) - \dfrac{1}{4}(7 - 5x) < \dfrac{2}{3}(3 - x) - 5$ $\quad \left(-\infty, \dfrac{33}{151}\right)$

71. $4.273 + 2.8x \le 10.985$ $\quad (-\infty, 2.397]$

72. $1.064 < 5.94 - 3.2x$ $\quad (-\infty, 1.52375)$

73. $3.25x - 27.39 > 4.06 + 5.1x$ $\quad (-\infty, -17)$

74. $4.86(3.2x - 1.7) > 5.19 - x$ $\quad (0.8127, \infty)$

Identify the variable and write an inequality that describes each situation. See Example 6.

75. Tony is taller than 6 feet.
$x = $ Tony's height, $x > 6$ feet

76. Glenda is under 60 years old.
$a = $ Glenda's age, $a < 60$ years

77. Wilma makes less than $80,000 per year.
$s = $ Wilma's salary, $s < \$80,000$

78. Bubba weighs over 80 pounds.
$w = $ Bubba's weight, $w > 80$ pounds

79. The maximum speed for the Concorde is 1,450 miles per hour (mph).
$v = $ speed of the Concorde, $v \le 1,450$ mph

80. The minimum speed on the freeway is 45 mph.
$s = $ minimum speed, $s \ge 45$ mph

81. Julie can afford at most $400 per month.
$a = $ amount Julie can afford, $a \le \$400$

82. Fred must have at least a 3.2 grade point average.
$a = $ Fred's grade point average, $a \ge 3.2$

83. Burt is no taller than 5 feet.
$b = $ Burt's height, $b \le 5$ feet

84. Ernie cannot run faster than 10 mph.
$r = $ Ernie's speed, $r \le 10$ mph

85. Tina makes no more than $8.20 per hour.
$t = $ Tina's hourly wage, $t \le \$8.20$

86. Rita will not take less than $12,000 for the car.
$s = $ selling price, $s \ge \$12,000$

Solve each problem by using an inequality. See Examples 7 and 8.

87. *Car shopping.* Jennifer is shopping for a new car. In addition to the price of the car, there is an 8% sales tax and a $172 title and license fee. If Jennifer decides that she will spend less than $10,000 total, then what is the price range for the car?
$x = $ price of car, $x < \$9,100$

88. *Sewing machines.* Charles wants to buy a sewing machine in a city with a 10% sales tax. He has at most $700 to spend. In what price range should he look?
$x = $ price of sewing machine, $x \le \$636.36$

89. *Truck shopping.* Linda and Bob are shopping for a new truck in a city with a 9% sales tax. There is also an $80 title and license fee to pay. They want to get a good truck and plan to spend at least $10,000. What is the price range for the truck?
$x = $ price of truck, $x \ge \$9,100.92$

90. *Curly's contribution.* Larry, Curly, and Moe are going to buy their mother a color television set. Larry has a better job than Curly and agrees to contribute twice as much as Curly. Moe is unemployed and can spare only $50. If the kind of television Mama wants costs at least $600, then what is the price range for Curly's contribution?
$x = $ Curly's contribution, $x \ge \$183.33$

91. *Bachelor's degrees.* The graph on the next page shows the number of bachelor's degrees awarded in the United States each year since 1985 (National Center for Education Statistics, www.nces.ed.gov).
a) Has the number of bachelor's degrees been increasing or decreasing since 1985? Increasing
b) The formula $B = 16.45n + 980.20$ can be used to approximate the number of degrees awarded in thousands in the year $1985 + n$. What is the first year in which the number of bachelor's degrees will exceed 1.3 million? 2005

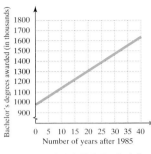

FIGURE FOR EXERCISE 91

92. *Master's degrees.* In 1985, 15.9% of all degrees awarded in U.S. higher education were master's degrees (National Center for Education Statistics). If the formulas $M = 7.79n + 287.87$ and $T = 30.95n + 1,808.22$ give the number of master's degrees and the total number of higher education degrees awarded in thousands, respectively, in the year $1985 + n$, then what is the first year in which more than 20% of all degrees awarded will be master's degrees?
2031

93. *Weighted average.* Professor Jorgenson gives only a midterm exam and a final exam. The semester average is computed by taking $\frac{1}{3}$ of the midterm exam score plus $\frac{2}{3}$ of the final exam score. The grade is determined from the semester average by using the grading scale given in the table. If Stanley scored only 56 on the midterm, then for what range of scores on the final exam would he get a C or better in the course?
$x =$ final exam score, $x \geq 77$

Grading	Scale
90–100	A
80–89	B
70–79	C
60–69	D

TABLE FOR EXERCISES 93 AND 94

94. *C or better.* Professor Brown counts her midterm as $\frac{2}{3}$ of the grade and her final as $\frac{1}{3}$ of the grade. Wilbert scored only 56 on the midterm. If Professor Brown also uses the grading scale given in the table, then what range of scores on the final exam would give Wilbert a C or better in the course?
$x =$ final exam score, $x \geq 98$

95. *Designer jeans.* A pair of ordinary jeans at A-Mart costs $50 less than a pair of designer jeans at Enrico's. In fact, you can buy four pairs of A-Mart jeans for less than one pair of Enrico's jeans. What is the price range for a pair of A-Mart jeans?
$x =$ the price of A-Mart jeans, $x < \$16.67$

96. *United Express.* Al and Rita both drive parcel delivery trucks for United Express. Al averages 20 mph less than Rita. In fact, Al is so slow that in 5 hours he covered fewer miles than Rita did in 3 hours. What are the possible values for Al's rate of speed?
$x =$ Al's rate, $x < 30$ mph

GETTING MORE INVOLVED

97. *Discussion.* If 3 is added to every number in $(4, \infty)$, the resulting set is $(7, \infty)$. In each of the following cases, write the resulting set of numbers in interval notation. Explain your results.
   a) The number -6 is subtracted from every number in $[2, \infty)$.
   b) Every number in $(-\infty, -3)$ is multiplied by 2.
   c) Every number in $(8, \infty)$ is divided by 4.
   d) Every number in $(6, \infty)$ is multiplied by -2.
   e) Every number in $(-\infty, -10)$ is divided by -5.
       a) $[8, \infty)$ **b)** $(-\infty, -6)$ **c)** $(2, \infty)$
       d) $(-\infty, -12)$ **e)** $(2, \infty)$

98. *Writing.* Explain why saying that x is *at least* 9 is equivalent to saying that x is *greater than or equal to* 9. Explain why saying that x is *at most* 5 is equivalent to saying that x is *less than or equal to* 5.

 **2.5**   **COMPOUND INEQUALITIES**

In this section we will use the ideas of union and intersection from Chapter 1 along with our knowledge of inequalities from Section 2.4 to work with compound inequalities.

In this
        section

- Basics
- Graphing the Solution Set
- Applications

Basics

The inequalities that we studied in Section 2.4 are referred to as **simple inequalities.** If we join two simple inequalities with the connective "and" or the connective "or," we get a **compound inequality.** A compound inequality using the connective "and" is true if and only if *both* simple inequalities are true.

E X A M P L E   1

Compound inequalities using the connective "and"

Determine whether each compound inequality is true.

a) $3 > 2$ and $3 < 5$

b) $6 > 2$ and $6 < 5$

Solution

a) The compound inequality is true because $3 > 2$ is true and $3 < 5$ is true.

b) The compound inequality is false because $6 < 5$ is false.

A compound inequality using the connective "or" is true if one or the other or both of the simple inequalities are true. It is false only if both simple inequalities are false.

E X A M P L E   2

Compound inequalities using the connective "or"

Determine whether each compound inequality is true.

a) $2 < 3$ or $2 > 7$

b) $4 < 3$ or $4 \geq 7$

Solution

a) The compound inequality is true because $2 < 3$ is true.

b) The compound inequality is false because both $4 < 3$ and $4 \geq 7$ are false.

> **helpful hint**
>
> There is a big difference between "and" and "or." To get money from an automatic teller you must have a bank card *and* know a secret number (PIN). There would be a lot of problems if you could get money by having a bank card *or* knowing a PIN.

If a compound inequality involves a variable, then we are interested in the solution set to the inequality. The solution set to an "and" inequality consists of all numbers that satisfy both simple inequalities, whereas the solution set to an "or" inequality consists of all numbers that satisfy at least one of the simple inequalities.

E X A M P L E   3

Solutions of compound inequalities

Determine whether 5 satisfies each compound inequality.

a) $x < 6$ and $x < 9$

b) $2x - 9 \leq 5$ or $-4x \geq -12$

Solution

a) Because $5 < 6$ and $5 < 9$ are both true, 5 satisfies the compound inequality.

b) Because $2 \cdot 5 - 9 \leq 5$ is true, it does not matter that $-4 \cdot 5 \geq -12$ is false. So 5 satisfies the compound inequality.

Graphing the Solution Set

The solution set to a compound inequality such as

$$x > 2 \qquad \text{and} \qquad x < 5$$

consists of all numbers that are in the solution sets to both simple inequalities. So the solution set to this compound inequality is the intersection of those two solution sets. In symbols,

$$\{x \mid x > 2 \text{ and } x < 5\} = \{x \mid x > 2\} \cap \{x \mid x < 5\}.$$

E X A M P L E   4

Graphing compound inequalities

Graph the solution set to the compound inequality $x > 2$ and $x < 5$.

Solution

We first sketch the graph of $x > 2$ and then the graph of $x < 5$, as shown in the top two number lines in Fig. 2.15. The intersection of these two solution sets is the portion of the number line that is shaded on both graphs, just the part between 2 and 5, not including the endpoints. The graph of $\{x \mid x > 2 \text{ and } x < 5\}$ is shown at the bottom of Fig. 2.15. We write this set in interval notation as $(2, 5)$.

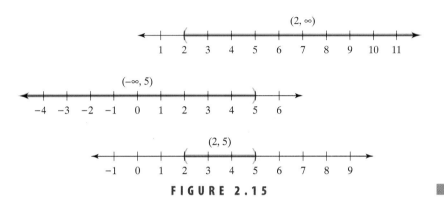

FIGURE 2.15

The solution set to a compound inequality such as

$$x > 4 \qquad \text{or} \qquad x < -1$$

consists of all numbers that satisfy one or the other or both of the simple inequalities. So the solution set to the compound inequality is the union of the solution sets to the simple inequalities. In symbols,

$$\{x \mid x > 4 \text{ or } x < -1\} = \{x \mid x > 4\} \cup \{x \mid x < -1\}.$$

E X A M P L E   5

Graphing compound inequalities

Graph the solution set to the compound inequality $x > 4$ or $x < -1$.

Solution

To find the union of the solution sets to the simple inequalities, we sketch their graphs as shown at the top of Fig. 2.16. We graph the union of these two sets by putting both shaded regions together on the same line as shown in the bottom graph in Fig. 2.16. This set is written in interval notation as $(-\infty, -1) \cup (4, \infty)$.

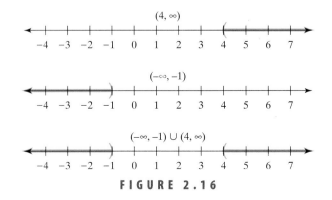

FIGURE 2.16

C A U T I O N   When graphing the intersection of two simple inequalities, do not draw too much. For the intersection, graph only numbers that satisfy *both* inequalities. Omit numbers that satisfy one but not the other inequality. Graphing a union is usually easier because we can simply draw both solution sets on the same number line.

It is not always necessary to graph the solution set to each simple inequality before graphing the solution set to the compound inequality. We can save time and work if we learn to think of the two preliminary graphs but draw only the final one.

E X A M P L E   6

Overlapping intervals

Sketch the graph and write the solution set in interval notation to each compound inequality.

a) $x < 3$ and $x < 5$ **b)** $x > 4$ or $x > 0$

Solution

a) To graph $x < 3$ and $x < 5$, we shade only the numbers that are both less than 3 and less than 5. So numbers between 3 and 5 are not shaded in Fig. 2.17. The compound inequality $x < 3$ and $x < 5$ is equivalent to the simple inequality $x < 3$. The solution set can be written as $(-\infty, 3)$.

FIGURE 2.17

b) To graph $x > 4$ or $x > 0$, we shade both regions on the same number line as shown in Fig. 2.18. The compound inequality $x > 4$ or $x > 0$ is equivalent to the simple inequality $x > 0$. The solution set is $(0, \infty)$.

FIGURE 2.18

The next example shows a compound inequality that has no solution and one that is satisfied by every real number.

E X A M P L E   7

All or nothing

Sketch the graph and write the solution set in interval notation to each compound inequality.

a) $x < 2$ and $x > 6$ **b)** $x < 3$ or $x > 1$

Solution

a) A number satisfies $x < 2$ and $x > 6$ if it is both less than 2 *and* greater than 6. There are no such numbers. The solution set is the empty set, $\varnothing$.

b) To graph $x < 3$ or $x > 1$, we shade both regions on the same number line as shown in Fig. 2.19. Since the two regions cover the entire line, the solution set is the set of all real numbers $(-\infty, \infty)$.

FIGURE 2.19

If we start with a more complicated compound inequality, we first simplify each part of the compound inequality and then find the union or intersection.

EXAMPLE 8

Intersection

Solve $x + 2 > 3$ and $x - 6 < 7$. Graph the solution set.

Solution

First simplify each simple inequality:

$$x + 2 - 2 > 3 - 2 \qquad \text{and} \qquad x - 6 + 6 < 7 + 6$$
$$x > 1 \qquad\qquad \text{and} \qquad\qquad x < 13$$

The intersection of these two solution sets is the set of numbers between (but not including) 1 and 13. Its graph is shown in Fig. 2.20. The solution set is written in interval notation as (1, 13).

FIGURE 2.20

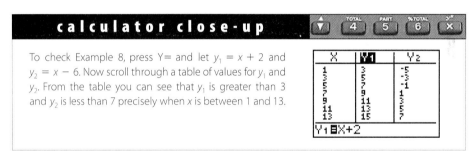

calculator close-up

To check Example 8, press Y= and let $y_1 = x + 2$ and $y_2 = x - 6$. Now scroll through a table of values for y_1 and y_2. From the table you can see that y_1 is greater than 3 and y_2 is less than 7 precisely when x is between 1 and 13.

EXAMPLE 9

calculator

close-up

To check Example 9, press Y= and let $y_1 = 5 - 7x$ and $y_2 = 3x - 2$. Now scroll through a table of values for y_1 and y_2. From the table you can see that either $y_1 \geq 12$ or $y_2 < 7$ is true for $x < 3$. Note also that for $x \geq 3$ both $y_1 \geq 12$ and $y_2 < 7$ are incorrect. The table supports the conclusion of Example 9.

Union

Graph the solution set to the inequality

$$5 - 7x \geq 12 \qquad \text{or} \qquad 3x - 2 < 7.$$

Solution

First solve each of the simple inequalities:

$$5 - 7x - 5 \geq 12 - 5 \qquad \text{or} \qquad 3x - 2 + 2 < 7 + 2$$
$$-7x \geq 7 \qquad\quad \text{or} \qquad\qquad 3x < 9$$
$$x \leq -1 \qquad\quad \text{or} \qquad\qquad x < 3$$

The union of the two solution intervals is $(-\infty, 3)$. The graph is shown in Fig. 2.21.

FIGURE 2.21

An inequality may be read from left to right or from right to left. Consider the inequality $1 < x$. If we read it in the usual way, we say, "1 is less than x." The meaning is clearer if we read the variable first. Reading from right to left, we say, "x is greater than 1."

Another notation is commonly used for the compound inequality

$$x > 1 \qquad \text{and} \qquad x < 13.$$

This compound inequality can also be written as

$$1 < x < 13.$$

Reading from left to right, we read $1 < x < 13$ as "1 is less than x is less than 13." The meaning of this inequality is clearer if we read the variable first and read the first inequality symbol from right to left. Reading the variable first, $1 < x < 13$ is read as "x is greater than 1 and less than 13." So x is between 1 and 13, and reading x first makes it clear.

CAUTION   We write $a < x < b$ only if $a < b$, and we write $a > x > b$ only if $a > b$. Similar rules hold for $\leq$ and $\geq$. So $4 < x < 9$ and $-6 \geq x \geq -8$ are correct uses of this notation, but $5 < x < 2$ is not correct. Also, the inequalities should *not* point in opposite directions as in $5 < x > 7$.

E X A M P L E   1 0   **Another notation**

Solve the inequality and graph the solution set:

$$-2 \leq 2x - 3 < 7$$

Solution

This inequality could be written as the compound inequality

$$2x - 3 \geq -2 \qquad \text{and} \qquad 2x - 3 < 7.$$

However, there is no need to rewrite the inequality because we can solve it in its original form.

$$-2 + 3 \leq 2x - 3 + 3 < 7 + 3 \qquad \text{Add 3 to each part.}$$

$$1 \leq 2x \leq 10$$

$$\frac{1}{2} \leq \frac{2x}{2} < \frac{10}{2} \qquad \text{Divide each part by 2.}$$

$$\frac{1}{2} \leq x < 5$$

The solution set is $\left[\frac{1}{2}, 5\right)$, and its graph is shown in Fig. 2.22.

F I G U R E   2 . 2 2

E X A M P L E   1 1   **Solving a compound inequality**

Solve the inequality $-1 < 3 - 2x < 9$ and graph the solution set.

Solution

$$-1 - 3 < 3 - 2x - 3 < 9 - 3 \qquad \text{Subtract 3 from each part of the inequality.}$$

$$-4 < -2x < 6$$

$$2 > x > -3 \qquad \text{Divide each part by } -2 \text{ and reverse both inequality symbols.}$$

$$-3 < x < 2 \qquad \text{Rewrite the inequality with the smallest number on the left.}$$

The solution set is $(-3, 2)$, and its graph is shown in Fig. 2.23.

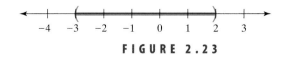

F I G U R E   2 . 2 3

c a l c u l a t o r

c l o s e - u p

Do not use a table on your calculator as a method for solving an inequality. Use a table to check your algebraic solution and you will get a better understanding of inequalities.

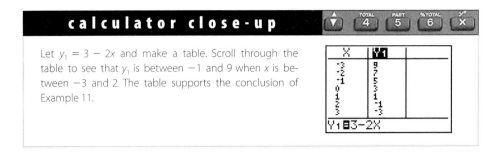

Let $y_1 = 3 - 2x$ and make a table. Scroll through the table to see that y_1 is between -1 and 9 when x is between -3 and 2. The table supports the conclusion of Example 11.

Applications

When final exams are approaching, students are often interested in finding the final exam score that would give them a certain grade for a course.

E X A M P L E   1 2

Final exam scores

Fiana made a score of 76 on her midterm exam. For her to get a B in the course, the average of her midterm exam and final exam must be between 80 and 89 inclusive. What possible scores on the final exam would give Fiana a B in the course?

Solution

Let x represent her final exam score. Between 80 and 89 inclusive means that an average between 80 and 89 as well as an average of exactly 80 or 89 will get a B. So the average of the two scores must be greater than or equal to 80 and less than or equal to 89.

$$80 \leq \frac{x + 76}{2} \leq 89$$

$$160 \leq x + 76 \leq 178 \quad \text{Multiply by 2.}$$

$$160 - 76 \leq x \leq 178 - 76 \quad \text{Subtract 76.}$$

$$84 \leq x \leq 102$$

If Fiana scores between 84 and 102 inclusive, she will get a B in the course.

> **helpful hint**
>
> When you use two inequality symbols as in Example 12, they must both point in the same direction. In fact, we usually have them both point to the left so that the numbers increase in size from left to right.

WARM-UPS

True or false? Explain your answer.

1. $3 < 5$ and $3 \leq 10$ True
2. $3 < 5$ or $3 < 10$ True
3. $3 > 5$ and $3 < 10$ False
4. $3 \geq 5$ or $3 \leq 10$ True
5. $4 < 8$ and $4 > 2$ True
6. $4 < 8$ or $4 > 2$ True
7. $-3 < 0 < -2$ False
8. $(3, \infty) \cap (8, \infty) = (8, \infty)$ True
9. $(3, \infty) \cup [8, \infty) = [8, \infty)$ False
10. $(-2, \infty) \cap (-\infty, 9) = (-2, 9)$ True

2.5   EXERCISES

Reading and Writing   *After reading this section, write out the answers to these questions. Use complete sentences.*

1. What is a compound inequality?

   A compound inequality consists of two inequalities joined with the words "and" or "or."

2. When is a compound inequality using "and" true?

   A compound inequality using "and" is true only when both simple inequalities are true.

3. When is a compound inequality using "or" true?

   A compound inequality using "or" is true when either one or the other or both inequalities is true.

4. How do we solve compound inequalities?

   Solve each simple inequality and then either union or intersect the solution sets.

5. What is the meaning of $a < b < c$?

   The inequality $a < b < c$ means that $a < b$ and $b < c$.

6. What is the meaning of $5 < x > 7$?

   The inequality $5 < x > 7$ has no meaning. All inequality symbols must point in the same direction in this notation.

Determine whether each compound inequality is true. See Examples 1 and 2.

7. $-6 < 5$ and $-6 > -3$ No

8. $3 < 5$ or $0 < -3$ Yes

9. $4 \leq 4$ and $-4 \leq 0$ Yes

10. $1 < 5$ and $1 > -3$ Yes

11. $6 < 5$ or $-4 > -3$ No

12. $4 \leq -4$ or $0 \leq 0$ Yes

Determine whether -4 satisfies each compound inequality. See Example 3.

13. $x < 5$ and $x > -3$ No

14. $x < 5$ or $x > -3$ Yes

15. $x - 3 \geq -7$ or $x + 1 > 1$ Yes

16. $2x \leq -8$ and $5x \leq 0$ Yes

17. $2x - 1 < -7$ or $-2x > 18$ Yes

18. $-3x > 0$ and $3x - 4 < 11$ Yes

Graph the solution set to each compound inequality. See Examples 4–7.

19. $x > -1$ and $x < 4$

20. $x \leq 3$ and $x \leq 0$

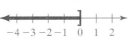

21. $x \geq 2$ or $x \geq 5$

22. $x < -1$ or $x < 3$

23. $x \leq 6$ or $x > -2$

24. $x > -2$ and $x \leq 4$

25. $x \leq 6$ and $x > 9$ $\varnothing$

26. $x < 7$ or $x > 0$

27. $x \leq 6$ or $x > 9$

28. $x \geq 4$ and $x \leq -4$ $\varnothing$

29. $x \geq 6$ and $x \leq 1$ $\varnothing$

30. $x > 3$ or $x < -3$

Solve each compound inequality. Write the solution set using interval notation and graph it. See Examples 8 and 9.

31. $x - 3 > 7$ or $3 - x > 2$

   $(-\infty, 1) \cup (10, \infty)$

32. $x - 5 > 6$ or $2 - x > 4$

   $(-\infty, -2) \cup (11, \infty)$

33. $3 < x$ and $1 + x > 10$

   $(9, \infty)$

34. $-0.3x < 9$ and $0.2x > 2$

   $(10, \infty)$

35. $\frac{1}{2}x > 5$ or $-\frac{1}{3}x < 2$

   $(-6, \infty)$

36. $5 < x$ or $3 - \frac{1}{2}x < 7$

   $(-8, \infty)$

37. $2x - 3 \leq 5$ and $x - 1 > 0$

   $(1, 4]$

38. $\frac{3}{4}x < 9$ and $-\frac{1}{3}x \leq -15$ $\varnothing$

39. $\frac{1}{2}x - \frac{1}{3} \geq -\frac{1}{6}$ or $\frac{2}{7}x \leq \frac{1}{10}$

   $(-\infty, \infty)$

40. $\frac{1}{4}x - \frac{1}{3} > -\frac{1}{5}$ and $\frac{1}{2}x < 2$

   $\left(\frac{8}{15}, 4\right)$

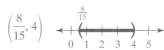

41. $0.5x < 2$ and $-0.6x < -3$ $\varnothing$

42. $0.3x < 0.6$ or $0.05x > -4$

   $(-\infty, \infty)$

Solve each compound inequality. Write the solution set in interval notation and graph it. See Examples 10 and 11.

43. $5 < 2x - 3 < 11$ $(4, 7)$

44. $-2 < 3x + 1 < 10$ $(-1, 3)$

45. $-1 < 5 - 3x \leq 14$ $[-3, 2)$

46. $-1 \leq 3 - 2x < 11$ $[-4, 2)$

47. $-3 < \dfrac{3m + 1}{2} \leq 5$ $\left[-\dfrac{7}{3}, 3\right)$

48. $0 \leq \dfrac{3 - 2x}{2} < 5$ $\left(-\dfrac{7}{2}, \dfrac{3}{2}\right]$

49. $-2 < \dfrac{1 - 3x}{-2} < 7$ $(-1, 5)$

50. $-3 < \dfrac{2x - 1}{3} < 7$ $(-4, 11)$

51. $3 \leq 3 - 5(x - 3) \leq 8$ $[2, 3]$

52. $2 \leq 4 - \dfrac{1}{2}(x - 8) \leq 10$

$[-4, 12]$

Write each union or intersection of intervals as a single interval if possible.

53. $(2, \infty) \cup (4, \infty)$
$(2, \infty)$

54. $(-3, \infty) \cup (-6, \infty)$
$(-6, \infty)$

55. $(-\infty, 5) \cap (-\infty, 9)$
$(-\infty, 5)$

56. $(-\infty, -2) \cap (-\infty, 1)$
$(-\infty, -2)$

57. $(-\infty, 4] \cap [2, \infty)$
$[2, 4]$

58. $(-\infty, 8) \cap [3, \infty)$
$[3, 8]$

59. $(-\infty, 5) \cup [-3, \infty)$
$(-\infty, \infty)$

60. $(-\infty, -2] \cup (2, \infty)$
$(-\infty, -2] \cup (2, \infty)$

61. $(3, \infty) \cap (-\infty, 3]$
$\varnothing$

62. $[-4, \infty) \cap (-\infty, -6]$
$\varnothing$

63. $(3, 5) \cap [4, 8)$
$[4, 5)$

64. $[-2, 4] \cap (0, 9]$
$(0, 4]$

65. $[1, 4) \cup (2, 6]$
$[1, 6]$

66. $[1, 3) \cup (0, 5)$
$(0, 5)$

Write either a simple or a compound inequality that has the given graph as its solution set.

67.

$x > 2$

68.

$x \leq 5$

69.

$x < 3$

70.

$x < -4$ or $x > 3$

71.

$x > 2$ or $x \leq -1$

72.

$-1 < x < 2$

73.

$-2 \leq x < 3$

74.

$x < 2$

75.

$x \geq -3$

76.

$x \leq 0$ or $x > 1$

Solve each compound inequality and write the solution set using interval notation.

77. $2 < x < 7$ and $2x > 10$ $(5, 7)$

78. $3 < 5 - x < 8$ or $-3x < 0$ $(-3, \infty)$

79. $-1 < 3x + 2 \leq 5$ or $\dfrac{3}{2}x - 6 > 9$ $(-1, 1] \cup (10, \infty)$

80. $0 < 5 - 2x \leq 10$ and $-6 < 4 - x < 0$ $\varnothing$

81. $-3 < \dfrac{x - 1}{2} < 5$ and $-1 < \dfrac{1 - x}{2} < 2$ $(-3, 3)$

82. $-3 < \dfrac{3x - 1}{5} < \dfrac{1}{2}$ and $\dfrac{1}{3} < \dfrac{3 - 2x}{6} < \dfrac{9}{2}$ $\left(-\dfrac{14}{3}, \dfrac{1}{2}\right)$

Solve each problem by using a compound inequality. See Example 12.

83. *Aiming for a C.* Professor Johnson gives only a midterm exam and a final exam. The semester average is computed by taking $\frac{1}{3}$ of the midterm exam score plus $\frac{2}{3}$ of the final exam score. To get a C, Beth must have a semester average between 70 and 79 inclusive. If Beth scored only 64 on the midterm, then for what range of scores on the final exam would Beth get a C?
$x = $ final exam score, $73 \leq x \leq 86.5$

84. Two tests only. Professor Davis counts his midterm as $\frac{2}{3}$ of the grade, and his final as $\frac{1}{3}$ of the grade. Jason scored only 64 on the midterm. What range of scores on the final exam would put Jason's average between 70 and 79 inclusive?
x = final exam score, $82 \le x \le 109$

85. Keep on truckin'. Abdul is shopping for a new truck in a city with an 8% sales tax. There is also an $84 title and license fee to pay. He wants to get a good truck and plans to spend at least $12,000 but no more than $15,000. What is the price range for the truck?
x = price of truck, $\$11{,}033 \le x \le \$13{,}811$

86. Selling-price range. Renee wants to sell her car through a broker who charges a commission of 10% of the selling price. The book value of the car is $14,900, but Renee still owes $13,104 on it. Although the car is in only fair condition and will not sell for more than the book value, Renee must get enough to at least pay off the loan. What is the range of the selling price?
x = selling price, $\$14{,}560 \le x \le \$14{,}900$

87. Hazardous to her health. Trying to break her smoking habit, Jane calculates that she smokes only three full cigarettes a day, one after each meal. The rest of the time she smokes on the run and smokes only half of the cigarette. She estimates that she smokes the equivalent of 5 to 12 cigarettes per day. How many times a day does she light up on the run?
x = number of cigarettes on the run, $4 \le x \le 18$

88. Possible width. The length of a rectangle is 20 meters longer than the width. The perimeter must be between 80 and 100 meters. What are the possible values for the width of the rectangle? w = width, $10 < w < 15$

89. Higher education. The formulas

$$B = 16.45n + 980.20$$

and $$M = 7.79n + 287.87$$

can be used to approximate the number of bachelor's and master's degrees in thousands, respectively, awarded in the year 1985 + n (National Center for Education Statistics, www.nces.ed.gov).
a) How many bachelor's degrees were awarded in 1995?
b) In what year will the number of bachelor's degrees that are awarded reach 1.26 million?
c) What is the first year in which both B is greater than 1.3 million and M is greater than 0.5 million?

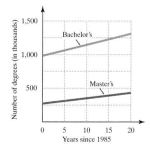

FIGURE FOR EXERCISE 89

d) What is the first year in which either B is greater than 1.3 million or M is greater than 0.5 million?
   a) 1,144,700   **b)** 2002   **c)** 2013   **d)** 2005

90. Senior citizens. The number of senior citizens (65 years old and over) in the United States in millions in the year 1970 + n can be estimated by using the formula

$$S = 0.48n + 19.71$$

(U.S. Bureau of the Census, www.census.gov). The percentage of senior citizens living below the poverty level in the year 1970 + n can be estimated by using the formula

$$p = -0.72n + 24.2.$$

a) How many senior citizens were there in 1998?
b) In what year did the percentage of seniors living below the poverty level reach 2.6%?
c) What is the first year in which we can expect both the number of seniors to be greater than 36 million and fewer than 2.6% living below the poverty level?
   a) 33.15 million
   b) 2000
   c) 2004

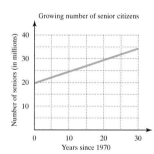

FIGURE FOR EXERCISE 90

91. Discussion. If $-x$ is between a and b, then what can you say about x? $-b < x < -a$ provided $a < b$

92. Discussion. For which of the inequalities is the notation used correctly?
   a) $-2 \le x < 3$ **b)** $-4 \ge x < 7$ **c)** $-1 \le x > 0$
   d) $6 < x \le -8$ **e)** $5 \ge x \ge -9$
   Notation is used correctly in (a) and (e).

93. Discussion. In each case, write the resulting set of numbers in interval notation. Explain your answers.
   a) Every number in $(3, 8)$ is multiplied by 4. $(12, 32)$
   b) Every number in $[-2, 4)$ is multiplied by -5. $(-20, 10]$
   c) Three is added to every number in $(-3, 6)$. $(0, 9)$
   d) Every number in $[3, 9]$ is divided by -3. $[-3, -1]$

94. Discussion. Write the solution set using interval notation for each of the following inequalities in terms of s and t. State any restrictions on s and t. For what values of s and t is the solution set empty?
   a) $x > s$ and $x < t$ (s, t) if $s < t$, no solution if $t < s$
   b) $x > s$ and $x > t$ (s, ∞) if $s > t$, (t, ∞) if $t > s$

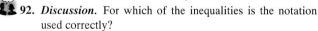

2.6   ABSOLUTE VALUE EQUATIONS AND INEQUALITIES

In Chapter 1 we learned that absolute value measures the distance of a number from 0 on the number line. In this section we will learn to solve equations and inequalities involving absolute value.

Absolute Value Equations

Solving equations involving absolute value requires some techniques that are different from those studied in previous sections. For example, the solution set to the equation

$$|x| = 5$$

is $\{-5, 5\}$ because both 5 and -5 are five units from 0 on the number line, as shown in Fig. 2.24. So $|x| = 5$ is equivalent to the compound equation

$$x = 5 \text{ or } x = -5.$$

5 units   5 units

$$\begin{array}{ccccccccccccc} -6 & -5 & -4 & -3 & -2 & -1 & 0 & 1 & 2 & 3 & 4 & 5 & 6 \end{array}$$

FIGURE 2.24

The equation $|x| = 0$ is equivalent to the equation $x = 0$ because 0 is the only number whose distance from 0 is zero. The solution set to $|x| = 0$ is $\{0\}$.

The equation $|x| = -7$ is inconsistent because absolute value measures distance, and distance is never negative. So the solution set is empty. These ideas are summarized as follows.

helpful hint

Some students grow up believing that the only way to solve an equation is to "do the same thing to each side." Then along comes absolute value equations. For an absolute value equation we write an equivalent compound equation that is not obtained by "doing the same thing to each side."

Basic Absolute Value Equations

Absolute Value Equation	Equivalent Equation	Solution Set
$\|x\| = k \; (k > 0)$	$x = k$ or $x = -k$	$\{k, -k\}$
$\|x\| = 0$	$x = 0$	$\{0\}$
$\|x\| = k \; (k < 0)$		$\varnothing$

We can use these ideas to solve more complicated absolute value equations.

EXAMPLE 1   **Absolute value equal to a positive number**

Solve each equation.

a) $|x - 7| = 2$ **b)** $|3x - 5| = 7$

Solution

a) First rewrite $|x - 7| = 2$ without absolute value:

$$\begin{array}{ccc} x - 7 = 2 & \text{or} & x - 7 = -2 \quad \text{Equivalent equation} \\ x = 9 & \text{or} & x = 5 \end{array}$$

The solution set is $\{5, 9\}$. The distance from 5 to 7 or from 9 to 7 is 2 units.

b) First rewrite $|3x - 5| = 7$ without absolute value:

$$3x - 5 = 7 \qquad \text{or} \qquad 3x - 5 = -7 \quad \text{Equivalent equation}$$
$$3x = 12 \qquad \text{or} \qquad 3x = -2$$
$$x = 4 \qquad \text{or} \qquad x = -\frac{2}{3}$$

The solution set is $\left\{-\frac{2}{3}, 4\right\}$.

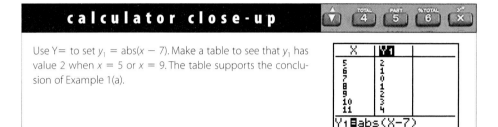

calculator close-up

Use Y= to set $y_1 = \text{abs}(x - 7)$. Make a table to see that y_1 has value 2 when $x = 5$ or $x = 9$. The table supports the conclusion of Example 1(a).

E X A M P L E   2

Absolute value equal to zero
Solve $|2(x - 6) + 7| = 0$.

Solution

Since 0 is the only number whose absolute value is 0, the expression within the absolute value bars must be 0.

$$2(x - 6) + 7 = 0 \quad \text{Equivalent equation}$$
$$2x - 12 + 7 = 0$$
$$2x - 5 = 0$$
$$2x = 5$$
$$x = \frac{5}{2}$$

The solution set is $\left\{\frac{5}{2}\right\}$.

E X A M P L E   3

Absolute value equal to a negative number
Solve $-5|3x - 7| + 4 = 14$.

Solution

First subtract 4 from each side to isolate the absolute value expression:

$$-5|3x - 7| + 4 = 14 \quad \text{Original equation}$$
$$-5|3x - 7| = 10 \quad \text{Subtract 4 from each side.}$$
$$|3x - 7| = -2 \quad \text{Divide each side by } -5.$$

There is no solution because no number has a negative absolute value.

The equation in the next example has an absolute value on both sides.

EXAMPLE 4

Absolute value on both sides

Solve $|2x - 1| = |x + 3|$.

Solution

Two quantities have the same absolute value only if they are equal or opposites. So we can write an equivalent compound equation:

$$
\begin{array}{lcl}
2x - 1 = x + 3 & \text{or} & 2x - 1 = -(x + 3) \\
x - 1 = 3 & \text{or} & 2x - 1 = -x - 3 \\
x = 4 & \text{or} & 3x = -2 \\
x = 4 & \text{or} & x = -\dfrac{2}{3}
\end{array}
$$

Check that both 4 and $-\dfrac{2}{3}$ satisfy the original equation. The solution set is $\left\{-\dfrac{2}{3}, 4\right\}$.

Absolute Value Inequalities

Since absolute value measures distance from 0 on the number line, $|x| > 5$ indicates that x is more than five units from 0. Any number on the number line to the right of 5 or to the left of -5 is more than five units from 0. So $|x| > 5$ is equivalent to

$$x > 5 \qquad \text{or} \qquad x < -5.$$

The solution set to this inequality is the union of the solution sets to the two simple inequalities. The solution set is $(-\infty, -5) \cup (5, \infty)$. The graph of $|x| > 5$ is shown in Fig. 2.25.

FIGURE 2.25

The inequality $|x| \leq 3$ indicates that x is less than or equal to three units from 0. Any number between -3 and 3 inclusive satisfies that condition. So $|x| \leq 3$ is equivalent to

$$-3 \leq x \leq 3.$$

The graph of $|x| \leq 3$ is shown in Fig. 2.26. These examples illustrate the basic types of absolute value inequalities.

FIGURE 2.26

Basic Absolute Value Inequalities ($k > 0$)

Absolute Value Inequality	Equivalent Inequality	Solution Set	Graph of Solution Set
$\lvert x \rvert > k$	$x > k$ or $x < -k$	$(-\infty, -k) \cup (k, \infty)$	
$\lvert x \rvert \geq k$	$x \geq k$ or $x \leq -k$	$(-\infty, -k] \cup [k, \infty)$	
$\lvert x \rvert < k$	$-k < x < k$	$(-k, k)$	
$\lvert x \rvert \leq k$	$-k \leq x \leq k$	$[-k, k]$	

We can solve more complicated inequalities in the same manner as simple ones.

EXAMPLE 5

Absolute value inequality

Solve $\lvert x - 9 \rvert < 2$ and graph the solution set.

Solution

Because $\lvert x \rvert < k$ is equivalent to $-k < x < k$, we can rewrite $\lvert x - 9 \rvert < 2$ as follows:

$$-2 < x - 9 < 2$$
$$-2 + 9 < x - 9 + 9 < 2 + 9 \qquad \text{Add 9 to each part of the inequality.}$$
$$7 < x < 11$$

The graph of the solution set (7, 11) is shown in Fig. 2.27. Note that the graph consists of all real numbers that are within two units of 9.

FIGURE 2.27

EXAMPLE 6

Absolute value inequality

Solve $\lvert 3x + 5 \rvert > 2$ and graph the solution set.

Solution

$$3x + 5 > 2 \qquad \text{or} \qquad 3x + 5 < -2 \qquad \text{Equivalent compound inequality}$$
$$3x > -3 \qquad \text{or} \qquad 3x < -7$$
$$x > -1 \qquad \text{or} \qquad x < -\frac{7}{3}$$

The solution set is $\left(-\infty, -\frac{7}{3}\right) \cup (-1, \infty)$, and its graph is shown in Fig. 2.28.

FIGURE 2.28

E X A M P L E   7

Use Y= to   set y_1 = abs(5 − 3x). The table supports the conclusion that $y \le 6$ when x is between $-\frac{1}{3}$ and $\frac{11}{3}$ even though $-\frac{1}{3}$ and $\frac{11}{3}$ do not appear in the table. For more accuracy, make a table in which the change in x is $\frac{1}{3}$.

X	Y1	
-1	8	
0	5	
1	2	
2	1	
3	4	
4	7	
5	10	

Y1■abs(5−3X)

Absolute value inequality

Solve $|\,5 - 3x\,| \le 6$ and graph the solution set.

Solution

$$-6 \le 5 - 3x \le 6 \quad \text{Equivalent inequality}$$

$$-11 \le -3x \le 1 \quad \text{Subtract 5 from each part.}$$

$$\frac{11}{3} \ge x \ge -\frac{1}{3} \quad \text{Divide by } -3 \text{ and reverse each inequality symbol.}$$

$$-\frac{1}{3} \le x \le \frac{11}{3} \quad \text{Write } -\frac{1}{3} \text{ on the left because it is smaller than } \frac{11}{3}.$$

The solution set is $\left[-\dfrac{1}{3}, \dfrac{11}{3}\right]$ and its graph is shown in Fig. 2.29.

FIGURE 2.29

There are a few absolute value inequalities that do not fit the preceding categories. They are easy to solve using the definition of absolute value.

E X A M P L E   8

Special case

Solve $3 + |\,7 - 2x\,| \ge 3$.

Solution

Subtract 3 from each side to isolate the absolute value expression.

$$|\,7 - 2x\,| \ge 0$$

Because the absolute value of any real number is greater than or equal to 0, the solution set is R, the set of all real numbers.

E X A M P L E   9

An impossible case

Solve $|\,5x - 12\,| < -2$.

Solution

We write an equivalent inequality only when the value of k is positive. With -2 on the right-hand side, we do not write an equivalent inequality. Since the absolute value of any quantity is greater than or equal to 0, no value for x can make this absolute value less than -2. The solution set is $\varnothing$, the empty set.

Applications

A simple example will show how absolute value inequalities can be used in applications.

E X A M P L E   1 0

Controlling water temperature

The water temperature in a certain manufacturing process must be kept at 143°F. The computer is programmed to shut down the process if the water temperature is more than $7°$ away from what it is supposed to be. For what temperature readings is the process shut down?

Solution

If we let x represent the water temperature, then $x - 143$ represents the difference between the actual temperature and the desired temperature. The quantity $x - 143$ could be positive or negative. The process is shut down if the absolute value of $x - 143$ is greater than 7.

$$|x - 143| > 7$$
$$x - 143 > 7 \quad \text{or} \quad x - 143 < -7$$
$$x > 150 \quad \text{or} \quad x < 136$$

The process is shut down for temperatures greater than $150°F$ or less than $136°F$.

WARM-UPS

True or false? Explain your answer.

1. The equation $|x| = 2$ is equivalent to $x = 2$ or $x = -2$.   True
2. All absolute value equations have two solutions.   False
3. The equation $|2x - 3| = 7$ is equivalent to $2x - 3 = 7$ or $2x + 3 = 7$.   False
4. The inequality $|x| > 5$ is equivalent to $x > 5$ or $x < -5$.   True
5. The equation $|x| = -5$ is equivalent to $x = 5$ or $x = -5$.   False
6. There is only one solution to the equation $|3 - x| = 0$.   True
7. We should write the inequality $x > 3$ or $x < -3$ as $3 < x < -3$.   False
8. The inequality $|x| < 7$ is equivalent to $-7 \leq x \leq 7$.   False
9. The equation $|x| + 2 = 5$ is equivalent to $|x| = 3$.   True
10. The inequality $|x| < -2$ is equivalent to $x < 2$ and $x > -2$.   False

---

2.6   EXERCISES

Reading and Writing   *After reading this section, write out the answers to these questions. Use complete sentences.*

1. What does absolute value measure?
   Absolute value of a number is the number's distance from 0 on the number line.

2. Why does $|x| = 0$ have only one solution?
   Only 0 is 0 units from 0 on the number line.

3. Why does $|x| = 4$ have two solutions?
   Since both 4 and -4 are four units from 0, $|x| = 4$ has two solutions.

4. Why is $|x| = -3$ inconsistent?
   Since $|x| \geq 0$ for every real number x, $|x| = -3$ is impossible.

5. Why do all real numbers satisfy $|x| \geq 0$?
   Since the distance from 0 for every number on the number line is greater than or equal to 0, $|x| \geq 0$.

6. Why do no real numbers satisfy $|x| < -3$?
   Since $|x| \geq 0$ for all x, $|x| < -3$ is impossible.

Solve each absolute value equation. See Examples 1–3.

7. $|a| = 5$
   $\{-5, 5\}$

8. $|x| = 2$
   $\{-2, 2\}$

9. $|x - 3| = 1$
   $\{2, 4\}$

10. $|x - 5| = 2$
    $\{3, 7\}$

11. $|3 - x| = 6$
    $\{-3, 9\}$

12. $|7 - x| = 6$
    $\{1, 13\}$

13. $|3x - 4| = 12$
    $\left\{-\dfrac{8}{3}, \dfrac{16}{3}\right\}$

14. $|5x + 2| = -3$
    $\varnothing$

15. $\left|\dfrac{2}{3}x - 8\right| = 0$
    $\{12\}$

16. $\left|3 - \dfrac{3}{4}x\right| = \dfrac{1}{4}$
    $\left\{\dfrac{11}{3}, \dfrac{13}{3}\right\}$

17. $|6 - 0.2x| = 10$ $\{-20, 80\}$

18. $|5 - 0.1x| = 0$ $\{50\}$

19. $|7(x - 6)| = -3$ $\varnothing$

20. $|2(a + 3)| = 15$ $\{-10.5, 4.5\}$

21. $|2(x - 4) + 3| = 5$ $\{0, 5\}$

22. $|3(x - 2) + 7| = 6$ $\left\{\dfrac{5}{3}, -\dfrac{7}{3}\right\}$

23. $|7.3x - 5.26| = 4.215$ $\{0.143, 1.298\}$

24. $|5.74 - 2.17x| = 10.28$ $\{-2.092, 7.382\}$

Solve each absolute value equation. See Examples 3 and 4.

25. $3 + |x| = 5$ $\{-2, 2\}$

26. $|x| - 10 = -3$ $\{-7, 7\}$

27. $2 - |x + 3| = -6$ $\{-11, 5\}$

28. $4 - 3|x - 2| = -8$ $\{-2, 6\}$

29. $5 - \dfrac{|3 - 2x|}{3} = 4$ $\{0, 3\}$

30. $3 - \dfrac{1}{2}\left|\dfrac{1}{2}x - 4\right| = 2$ $\{4, 12\}$

31. $|x - 5| = |2x + 1|$ $\left\{-6, \dfrac{4}{3}\right\}$

32. $|w - 6| = |3 - 2w|$ $\{-3, 3\}$

33. $\left|\dfrac{5}{2} - x\right| = \left|2 - \dfrac{x}{2}\right|$ $\{1, 3\}$

34. $\left|x - \dfrac{1}{4}\right| = \left|\dfrac{1}{2}x - \dfrac{3}{4}\right|$ $\left\{-1, \dfrac{2}{3}\right\}$

35. $|x - 3| = |3 - x|$ $(-\infty, \infty)$

36. $|a - 6| = |6 - a|$ $(-\infty, \infty)$

Write an absolute value inequality whose solution set is shown by the graph. See Examples 5–7.

37.
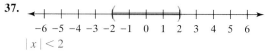

$-6\ -5\ -4\ -3\ -2\ -1\ \ 0\ \ 1\ \ 2\ \ 3\ \ 4\ \ 5\ \ 6$

$|x| < 2$

38.

$-6\ -5\ -4\ -3\ -2\ -1\ \ 0\ \ 1\ \ 2\ \ 3\ \ 4\ \ 5\ \ 6$

$|x| \le 5$

39.

$-6\ -5\ -4\ -3\ -2\ -1\ \ 0\ \ 1\ \ 2\ \ 3\ \ 4\ \ 5\ \ 6$

$|x| > 3$

40.

$-8\ -7\ -6\ -5\ -4\ -3\ -2\ -1\ \ 0\ \ 1\ \ 2\ \ 3\ \ 4\ \ 5\ \ 6\ \ 7\ \ 8$

$|x| \ge 6$

41.

$-6\ -5\ -4\ -3\ -2\ -1\ \ 0\ \ 1\ \ 2\ \ 3\ \ 4\ \ 5\ \ 6$

$|x| \le 1$

42.

$-6\ -5\ -4\ -3\ -2\ -1\ \ 0\ \ 1\ \ 2\ \ 3\ \ 4\ \ 5\ \ 6$

$|x| < 1$

43.

$-6\ -5\ -4\ -3\ -2\ -1\ \ 0\ \ 1\ \ 2\ \ 3\ \ 4\ \ 5\ \ 6$

$|x| \ge 2$

44.

$-6\ -5\ -4\ -3\ -2\ -1\ \ 0\ \ 1\ \ 2\ \ 3\ \ 4\ \ 5\ \ 6$

$|x| > 4$

Determine whether each absolute value inequality is equivalent to the inequality following it. See Examples 5–7.

45. $|x| < 3, x < 3$ No

46. $|x| > 3, x > 3$ No

47. $|x - 3| > 1, x - 3 > 1$ or $x - 3 < -1$ Yes

48. $|x - 3| \le 1, -1 \le x - 3 \le 1$ Yes

49. $|x - 3| \ge 1, x - 3 \ge 1$ or $x - 3 \le 1$ No

50. $|x - 3| > 0, x - 3 > 0$ No

51. $|4 - x| < 1, 4 - x < 1$ and $-(4 - x) < 1$ Yes

52. $|4 - x| > 1, 4 - x > 1$ or $-(4 - x) > 1$ Yes

Solve each absolute value inequality and graph the solution set. See Examples 5–7.

53. $|x| > 6$

$(-\infty, -6) \cup (6, \infty)$ $-8\ -6\ -4\ -2\ \ 0\ \ 2\ \ 4\ \ 6\ \ 8$

54. $|w| > 3$

$(-\infty, -3) \cup (3, \infty)$ $-4\ -3\ -2\ -1\ \ 0\ \ 1\ \ 2\ \ 3\ \ 4$

55. $|2a| < 6$

$(-3, 3)$ $-3\ -2\ -1\ \ 0\ \ 1\ \ 2\ \ 3$

56. $|3x| < 21$

$(-7, 7)$ $-7\ -5\ -3\ -1\ \ 1\ \ 3\ \ 5\ \ 7$

57. $|x - 2| \ge 3$

$(-\infty, -1] \cup [5, \infty)$ $-3\ -2\ -1\ 0\ 1\ 2\ 3\ 4\ 5\ 6\ 7$

58. $|x - 5| \ge 1$

$(-\infty, 4] \cup [6, \infty)$ $2\ \ 3\ \ 4\ \ 5\ \ 6\ \ 7\ \ 8$

59. $\dfrac{1}{5}|2x - 4| < 1$

$\left(-\dfrac{1}{2}, \dfrac{9}{2}\right)$ $-1\ \ 0\ \ 1\ \ 2\ \ 3\ \ 4\ \ 5$

60. $\dfrac{1}{3}|2x - 1| < 1$

$(-1, 2)$ $-2\ -1\ \ 0\ \ 1\ \ 2\ \ 3$

61. $-2|5 - x| \ge -14$

$[-2, 12]$ $-2\ \ 0\ \ 2\ \ 4\ \ 6\ \ 8\ \ 10\ \ 12$

62. $-3|6 - x| \ge -3$

$[5, 7]$ $3\ \ 4\ \ 5\ \ 6\ \ 7\ \ 8\ \ 9$

63. $2|3 - 2x| - 6 \ge 18$

$\left(-\infty, -\dfrac{9}{2}\right] \cup \left[\dfrac{15}{2}, \infty\right)$ $-4\ -2\ \ 0\ \ 2\ \ 4\ \ 6\ \ 8$

64. $2|5 - 2x| - 15 \ge 5$

$\left(-\infty, -\dfrac{5}{2}\right] \cup \left[\dfrac{15}{2}, \infty\right)$ $-4\ -2\ \ 0\ \ 2\ \ 4\ \ 6\ \ 8$

Solve each absolute value inequality and graph the solution set. See Examples 8 and 9.

65. $|x - 2| > 0$

$(-\infty, 2) \cup (2, \infty)$

66. $|6 - x| \geq 0$ $(-\infty, \infty)$

67. $|x - 5| \geq 0$ $(-\infty, \infty)$

68. $|3x - 7| \geq -3$ $(-\infty, \infty)$

69. $-2|3x - 7| > 6$ $\varnothing$

70. $-3|7x - 42| > 18$ $\varnothing$

71. $|2x + 3| + 6 > 0$ $(-\infty, \infty)$

72. $|5 - x| + 5 > 5$

$(-\infty, 5) \cup (5, \infty)$

Solve each inequality. Write the solution set using interval notation.

73. $1 < |x + 2|$ $(-\infty, -3) \cup (-1, \infty)$

74. $5 \geq |x - 4|$ $[-1, 9]$

75. $5 > |x| + 1$ $(-4, 4)$

76. $4 \leq |x| - 6$ $(-\infty, -10] \cup [10, \infty)$

77. $3 - 5|x| > -2$ $(-1, 1)$

78. $1 - 2|x| < -7$ $(-\infty, -4) \cup (4, \infty)$

79. $|5.67x - 3.124| < 1.68$ $(0.255, 0.847)$

80. $|4.67 - 3.2x| \geq 1.43$ $(-\infty, 1.0125] \cup [1.90625, \infty)$

81. $|2x - 1| < 3$ and $2x - 3 > 2$ $\varnothing$

82. $|5 - 3x| \geq 3$ and $5 - 2x > 3$ $\left(-\infty, \dfrac{2}{3}\right]$

83. $|x - 2| < 3$ and $|x - 7| < 3$ $(4, 5)$

84. $|x - 5| < 4$ and $|x - 6| > 2$ $(1, 4) \cup (8, 9)$

Solve each problem by using an absolute value equation or inequality. See Example 10.

85. *Famous battles.* In the Hundred Years' War, Henry V defeated a French army in the battle of Agincourt and Joan of Arc defeated an English army in the battle of Orleans (*The Doubleday Almanac*). Suppose you know only that these two famous battles were 14 years apart and that the battle of Agincourt occurred in 1415. Use an absolute value equation to find the possibilities for the year in which the battle of Orleans occurred.

1401 or 1429

86. *World records.* In July 1985 Steve Cram of Great Britain set a world record of 3 minutes 29.67 seconds for the 1,500-meter race and a world record of 3 minutes 46.31 seconds for the 1-mile race (*The Doubleday Almanac*). Suppose you know only that these two events occurred 11 days apart and that the 1,500-meter record was set on July 16. Use an absolute value equation to find the possible dates for the 1-mile record run.

July 5 or July 27

87. *Weight difference.* Research at a major university has shown that identical twins generally differ by less than 6 pounds in body weight. If Kim weighs 127 pounds, then

in what range is the weight of her identical twin sister Kathy?

Between 121 and 133 pounds

88. *Intelligence quotient.* Jude's IQ score is more than 15 points away from Sherry's. If Sherry scored 110, then in what range is Jude's score?

Greater than 125 or less than 95

89. *Unidentified flying objects.* The formula

$$S = -16t^2 + v_0 t + s_0$$

gives height in feet above the earth at time t seconds for an object projected into the air with an initial velocity of v_0 feet per second (ft/sec) from an initial height of s_0 feet. Two balls are tossed into the air simultaneously, one from the ground at 50 ft/sec and one from a height of 10 feet at 40 ft/sec. See the accompanying graph.

a) Use the graph to estimate the time at which the balls are at the same height.

b) Find the time from part (a) algebraically.

c) For what values of t will their heights above the ground differ by less than 5 feet (while they are both in the air)?

a) 1 second   **b)** 1 second   **c)** $0.5 < t < 1.5$

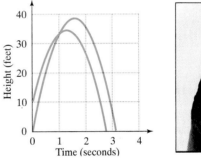 

FIGURE FOR EXERCISE 89

90. *Playing catch.* A circus clown at the top of a 60-foot platform is playing catch with another clown on the ground.

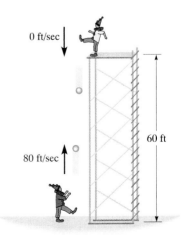

FIGURE FOR EXERCISE 90

The clown on the platform drops a ball at the same time as the one on the ground tosses a ball upward at 80 ft/sec. For what length of time is the distance between the balls less than or equal to 10 feet? (*Hint:* Use the formula given in Exercise 89. The initial velocity of a ball that is dropped is 0 ft/sec.)   0.25 second

GETTING MORE INVOLVED

91. *Discussion.* For which real numbers m and n is each equation satisfied?

   a) $|m - n| = |n - m|$

     $(-\infty, \infty)$

   b) $|mn| = |m| \cdot |n|$

     $(-\infty, \infty)$

   c) $\left|\dfrac{m}{n}\right| = \dfrac{|m|}{|n|}$

     all reals except $n = 0$

92. *Exploration.* **a)** Evaluate $|m + n|$ and $|m| + |n|$ for

   i) $m = 3$ and $n = 5$

   ii) $m = -3$ and $n = 5$

   iii) $m = 3$ and $n = -5$

   iv) $m = -3$ and $n = -5$

  b) What can you conclude about the relationship between $|m + n|$ and $|m| + |n|$?

    $|m + n| \le |m| + |n|$

COLLABORATIVE ACTIVITIES

Everyday Algebra

Grouping: Two to four students per group
Topic: Use of algebra in common occurrences

Every day, people use algebra without even knowing it. Any time you solve for an unknown quantity, you are using algebra. Here is an example of a simple problem that you could solve without even thinking of algebra.

• While shopping Joe notices a store brand that is available at a lower price than the name brand. If the name brand product costs $3.79 and Joe has a coupon for $0.50, what would the store brand price need to be for Joe to save money with the coupon?

This is a problem you could solve mentally by subtracting the two quantities and finding that if the store brand costs $3.29 or more, Joe would save money with the coupon. If the store brand is less than $3.29, then Joe would save money by buying it instead. The beauty of algebra is not apparent in cases like this because you already know what operation to perform to find the unknown quantity. Algebra becomes useful when it is not clear what to do.

We will consider another situation in which we want to find the best price of an item. For this situation we will use the following formulas.

The formula for markup of an item is

$$P = C + rC,$$

where P is the price of the item, C is the wholesale cost of the item, and r is the percent of markup.

The formula for discounting an item is

$$S = P - dP,$$

where S is the discounted price and d is the percent discount.

• Lane belongs to a wholesale buying club. She can order items through the club with a markup of 8% above the wholesale cost. She also can buy the same items in a store where she can get a 10% discount off the shelf price. Lane wants to know what the store markup on any particular item must be for it to be cheaper to order through the club.

Form groups of two to four people. Assign a role to each person: **Recorder, Moderator, Messenger,** or **Quality Manager** (roles may be combined if there are fewer than four people in a group). In your groups:

1. Decide how to rewrite Lane's problem using algebra. Decide what the unknown quantities are and assign them variable names.

2. Write the equations or inequalities on your paper using the variables you defined above.

3. Solve the problem and state the group's decision on what Lane should do.

Extension: Pick a similar problem from your own lives to use algebra to solve.

WRAP-UP

CHAPTER 2

SUMMARY

Equations		Examples
Solution set	The set of all numbers that satisfy an equation (or inequality)	$x + 2 = 6$ has solution set $\{4\}$.
Equivalent equations	Equations with the same solution set	$2x + 1 = 5$ $2x = 4$
Properties of equality	We may perform the same operation $(+, -, \cdot, \div)$ with the same real number on each side of an equation without changing the solution set (excluding multiplication and division by 0).	$x = 4$ $x + 1 = 5$ $x - 1 = 3$ $2x = 8$ $\dfrac{x}{2} = 2$
Identity	An equation that is satisfied by every number for which both sides are defined	$x + x = 2x$
Conditional equation	An equation whose solution set contains at least one real number but is not an identity	$5x - 10 = 0$
Inconsistent equation	An equation whose solution set is $\varnothing$	$x = x + 1$
Linear equation in one variable	An equation of the form $ax + b = 0$ with $a \neq 0$	$3x + 8 = 0$ $5x - 1 = 2x - 9$
Strategy for solving a linear equation	1. If fractions are present, multiply each side by the LCD to eliminate the fractions. 2. Use the distributive property to remove parentheses. 3. Combine any like terms. 4. Use the addition property of equality to get all variables on one side and numbers on the other side. 5. Use the multiplication property of equality to get a single variable on one side. 6. Check your work by replacing the variable in the original equation with your solution.	
Strategy for solving word problems	1. Read the problem until you understand the problem. 2. If possible, draw a diagram to illustrate the problem. 3. Choose a variable and write down what it represents. 4. Represent any other unknowns in terms of that variable.	

5. Write an equation that models the situation.
6. Solve the equation.
7. Be sure that your solution answers the question posed in the original problem.
8. Check your answer by using it to solve the original problem (not the equation).

Inequalities		**Examples**
Linear inequality in one variable	Any inequality of the form $ax + b < 0$ with $a \neq 0$ In place of $<$ we can use $\leq$, $>$ or $\geq$.	$2x + 9 < 0$ $x - 2 \geq 7$ $-3x - 1 \geq 2x + 5$
Properties of inequality	We may perform the same operation $(+, -, \cdot, \div)$ on each side of an inequality just as we do in solving equations, with one exception: When multiplying or dividing by a negative number, the inequality symbol is reversed.	$-3x > 6$ $x < -2$
Trichotomy property	For any two real numbers a and b, exactly one of the following statements is true: $a < b$, $a = b$, or $a > b$	If w is not greater than 7, then $w \leq 7$.
Compound inequality	Two simple inequalities connected with the word "and" or "or" *And* corresponds to *intersection*. *Or* corresponds to *union*.	$x > 1$ and $x < 5$ $x > 3$ or $x < 1$

Absolute value

	Absolute Value Equation	**Equivalent Equation**	**Solution Set**
Basic absolute value equations	$\lvert x \rvert = k\ (k > 0)$	$x = k$ or $x = -k$	$\{k, -k\}$
	$\lvert x \rvert = 0$	$x = 0$	$\{0\}$
	$\lvert x \rvert = k\ (k < 0)$		$\varnothing$

	Absolute Value Inequality	**Equivalent Inequality**	**Solution Set**	**Graph of Solution Set**
Basic absolute value inequalities $(k > 0)$	$\lvert x \rvert > k$	$x > k$ or $x < -k$	$(-\infty, -k) \cup (k, \infty)$	
	$\lvert x \rvert \geq k$	$x \geq k$ or $x \leq -k$	$(-\infty, -k] \cup [k, \infty)$	
	$\lvert x \rvert < k$	$-k < x < k$	$(-k, k)$	
	$\lvert x \rvert \leq k$	$-k \leq x \leq k$	$[-k, k]$	

ENRICHING YOUR MATHEMATICAL WORD POWER

For each mathematical term, choose the correct meaning.

1. equation
   a. an expression
   b. an inequality
   c. a sentence that expresses the equality of two algebraic expressions
   d. an algebraic sentence   c

2. linear equation
   a. an equation in which the terms are in line
   b. an equation of the form $ax + b = 0$ where $a \neq 0$
   c. the equation of a line
   d. an equation of the form $a^2 + b^2 = c^2$ b

3. identity
   a. an equation that is satisfied by all real numbers
   b. an equation that is satisfied by every real number
   c. an equation that is identical
   d. an equation that is satisfied by every real number for which both sides are defined   d

4. conditional equation
   a. an equation that has at least one real solution
   b. an equation that is correct
   c. an equation that is satisfied by at least one real number but is not an identity
   d. an equation that we are not sure how to solve   c

5. inconsistent equation
   a. an equation that is wrong
   b. an equation that is only sometimes consistent
   c. an equation that has no solution
   d. an equation with two variables   c

6. equivalent equations
   a. equations that are identical
   b. equations that are correct
   c. equations that are equal
   d. equations that have the same solution   d

7. formula
   a. a form
   b. a type of race car

   c. a process
   d. an equation involving two or more variables   d

8. literal equation
   a. a formula
   b. an equation with words
   c. a false equation
   d. a fact   a

9. uniform motion
   a. movement of an army
   b. movement in a straight line
   c. consistent motion
   d. motion at a constant rate   d

10. least common denominator
   a. the smallest divisor of all denominators
   b. the denominator that appears the least
   c. the smallest identical denominator
   d. the least common multiple of the denominators   d

11. equivalent inequalities
   a. the inequality reverses when dividing by a negative number
   b. $a < b$ and $b < c$
   c. $a < b$ and $a \leq b$
   d. inequalities that have the same solution set   d

12. inequality
   a. an equation that is not correct
   b. two different numbers
   c. a statement that expresses the inequality of two algebraic expressions
   d. a larger number   c

13. compound inequality
   a. an inequality that is complicated
   b. an inequality that reverses when divided by a negative number
   c. an inequality of negative numbers
   d. two simple inequalities joined with "and" or "or"   d

REVIEW EXERCISES

2.1 *Solve each equation.*

1. $2x - 7 = 9$ $\{8\}$

2. $5x - 7 = 38$ $\{9\}$

3. $11 = 5 - 4x$ $\left\{-\dfrac{3}{2}\right\}$

4. $-8 = 7 - 3x$ $\{5\}$

5. $x - 6 - (x - 6) = 0$ R

6. $x - 6 - 2(x - 3) = 0$ $\{0\}$

7. $2(x - 3) - 5 = 5 - (3 - 2x)$ $\varnothing$

8. $2(x - 4) + 5 = -(3 - 2x)$ R

> **study tip**
>
> Note how the review exercises are arranged according to the sections in this chapter. If you have trouble with a certain type of problem, refer back to the appropriate section for examples and explanations.

9. $\frac{3}{17}x = 0$ $\{0\}$

10. $-\frac{3}{8}x = \frac{1}{2}$ $\left\{-\frac{4}{3}\right\}$

11. $\frac{1}{4}x - \frac{1}{5} = \frac{1}{5}x + \frac{4}{5}$ $\{20\}$

12. $\frac{1}{2}x - 1 = \frac{1}{3}x$ $\{6\}$

13. $\frac{t}{2} - \frac{t-2}{3} = \frac{3}{2}$ $\{5\}$

14. $\frac{y+1}{4} - \frac{y-1}{6} = y + 5$ $\{-5\}$

15. $1 - 0.4(x - 4) + 0.6(x - 7) = -0.6$ $\{5\}$

16. $0.04x - 0.06(x - 8) = 0.1x$ $\{4\}$

2.2 *Solve each equation for x.*

17. $ax + b = 0$

$x = \dfrac{-b}{a}$

18. $mx + c = d$

$x = \dfrac{d-c}{m}$

19. $ax + 2 = cx$

$x = \dfrac{2}{c-a}$

20. $mx = 3 - x$

$x = \dfrac{3}{m+1}$

21. $mwx = P$

$x = \dfrac{P}{mw}$

22. $xyz = 2$

$x = \dfrac{2}{yz}$

23. $\dfrac{1}{x} + \dfrac{1}{2} = w$

$x = \dfrac{2}{2w-1}$

24. $\dfrac{1}{x} + \dfrac{1}{a} = 2$

$x = \dfrac{a}{2a-1}$

Write y in terms of x.

25. $3x - 2y = -6$

$y = \dfrac{3}{2}x + 3$

26. $4x - 3y + 9 = 0$

$y = \dfrac{4}{3}x + 3$

27. $y - 2 = -\dfrac{1}{3}(x - 6)$

$y = -\dfrac{1}{3}x + 4$

28. $y + 6 = \dfrac{1}{2}(x - 4)$

$y = \dfrac{1}{2}x - 8$

29. $\dfrac{1}{2}x - \dfrac{1}{4}y = 5$

$y = 2x - 20$

30. $-\dfrac{x}{3} + \dfrac{y}{2} = \dfrac{5}{8}$

$y = \dfrac{2}{3}x + \dfrac{5}{4}$

2.3 *Solve each problem.*

31. *Legal pad.* If the perimeter of a legal-size note pad is 45 inches and the pad is 5.5 inches longer than it is wide, then what are its length and width?
Length 8.5 inches, width 14 inches

32. *Area of a trapezoid.* The height of a trapezoid is 5 feet, and the upper base is 2 feet shorter than the lower base. If the area of the trapezoid is 45 square feet, then how long is the lower base?   10 feet

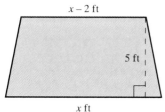

$x - 2$ ft

5 ft

x ft

FIGURE FOR EXERCISE 32

33. *Saving for retirement.* Roy makes $8,000 more per year than his wife. Roy saves 10% of his income for retirement, and his wife saves 8%. Together they save $5,660 per year. How much does each make?
Wife $27,000, Roy $35,000

34. *Charitable contributions.* Duane makes $1,000 less per year than his wife. Duane gives 5% of his income to charity, and his wife gives 10% of her income to charity. Together they contribute $2,500 to charity. How much does each make?   Duane $16,000, wife $17,000

35. *Dealer discounts.* Sarah is buying a car for $7,600. The dealer gave her a 20% discount off the list price. What was the list price? $9,500

36. *Gold sale.* At 25% off, a jeweler is selling a gold chain for $465. What was the list price? $620

37. *Nickels and dimes.* Rebecca has 15 coins consisting of dimes and nickels. The total value of the coins is $0.95. How many of each does she have?
11 nickels, 4 dimes

38. *Nickels, dimes, and quarters.* Camille has 19 coins consisting of nickels, dimes, and quarters. The value of the coins is $1.60. If she has six times as many nickels as quarters, then how many of each does she have?
2 quarters, 12 nickels, 5 dimes

39. *Tour de desert.* On a recent bicycle trip across the desert Barbara rode for 5 hours. Her bicycle then developed mechanical difficulties, and she walked the bicycle for 3 hours to the nearest town. Altogether, she covered 85 miles. If she rides 9 miles per hour (mph) faster than she walks, then how far did she walk?   15 miles

40. *Motor city.* Delmas flew to Detroit in 90 minutes and drove his new car back home in 6 hours. If he drove 150 mph slower than he flew, then how fast did he fly?
200 mph

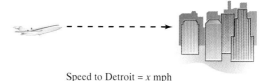

Speed to Detroit = x mph

Speed from Detroit = $x - 150$ mph

FIGURE FOR EXERCISE 40

2.4 *Solve each inequality. State the solution set using interval notation and graph it.*

41. $3 - 4x < 15$ $(-3, \infty)$

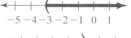

42. $5 - 6x > 35$ $(-\infty, -5)$

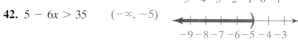

43. $2(x - 3) > -6$ $(0, \infty)$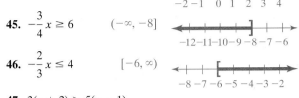

44. $4(5 - x) < 20$ $(0, \infty)$

45. $-\dfrac{3}{4}x \geq 6$ $(-\infty, -8]$

46. $-\dfrac{2}{3}x \leq 4$ $[-6, \infty)$

47. $3(x + 2) > 5(x - 1)$
$\left(-\infty, \dfrac{11}{2}\right)$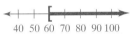

48. $4 - 2(x - 3) < 0$ $(5, \infty)$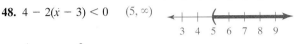

49. $\dfrac{1}{2}x + 7 \leq \dfrac{3}{4}x - 5$ $[48, \infty)$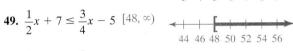

50. $\dfrac{5}{6}x - 3 \geq \dfrac{2}{3}x + 7$ $[60, \infty)$

2.5 *Solve each compound inequality. State the solution set using interval notation and graph it.*

51. $x + 2 > 3$ or $x - 6 < -10$
$(-\infty, -4) \cup (1, \infty)$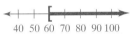

52. $x - 2 > 5$ or $x - 2 < -1$
$(-\infty, 1) \cup (7, \infty)$

53. $x > 0$ and $x - 6 < 3$
$(0, 9)$

54. $x \leq 0$ and $x + 6 > 3$
$(-3, 0]$

55. $6 - x < 3$ or $-x < 0$
$(0, \infty)$

56. $-x > 0$ or $x + 2 < 7$
$(-\infty, 5)$

57. $2x < 8$ and $2(x - 3) < 6$
$(-\infty, 4)$

58. $\dfrac{1}{3}x > 2$ and $\dfrac{1}{4}x > 2$
$(8, \infty)$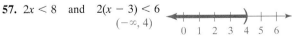

59. $x - 6 > 2$ and $6 - x > 0$ $\varnothing$

60. $-\dfrac{1}{2}x < 6$ or $\dfrac{2}{3}x < 4$ $(-\infty, \infty)$

61. $0.5x > 10$ or $0.1x < 3$ $(-\infty, \infty)$

62. $0.02x > 4$ and $0.2x < 3$ $\varnothing$

63. $-2 \leq \dfrac{2x - 3}{10} \leq 1$
$\left[-\dfrac{17}{2}, \dfrac{13}{2}\right]$

64. $-3 < \dfrac{4 - 3x}{5} < 2$
$\left(-2, \dfrac{19}{3}\right)$

Write each union or intersection of intervals as a single interval.

65. $[1, 4) \cup (2, \infty)$ $[1, \infty)$ **66.** $(2, 5) \cup (-1, \infty)$ $(-1, \infty)$

67. $(3, 6) \cap [2, 8]$ $(3, 6)$ **68.** $[-1, 3] \cap [0, 8]$ $[0, 3]$

69. $(-\infty, 5) \cup [5, \infty)$ $(-\infty, \infty)$

70. $(-\infty, 1) \cup (0, \infty)$ $(-\infty, \infty)$

71. $(-3, -1] \cap [-2, 5]$ $[-2, -1]$

72. $[-2, 4] \cap (4, 7]$ $\varnothing$

2.6 *Solve each absolute value equation or inequality and graph the solution set.*

73. $|2x| \geq 8$

74. $|5x - 1| \leq 14$

75. $|x| = -5$ $\varnothing$

76. $|x| + 2 = 16$

77. $\left|1 - \dfrac{x}{5}\right| > \dfrac{9}{5}$

78. $\left|1 - \dfrac{1}{6}x\right| < \dfrac{1}{2}$

79. $|x - 3| < -3$ $\varnothing$ **80.** $|x - 7| \leq -4$ $\varnothing$

81. $\left|\dfrac{x}{2}\right| - 5 = -1$

82. $\left|\dfrac{x}{2} - 5\right| = -1$ $\varnothing$

83. $|x + 4| \geq -1$ R **84.** $|6x - 1| \geq 0$ R

85. $1 - \dfrac{3}{2}|x - 2| < -\dfrac{1}{2}$

86. $1 > \dfrac{1}{2}|6 - x| - \dfrac{3}{4}$

MISCELLANEOUS

Solve each problem by using equations or inequalities.

87. *Rockbuster video.* Stephen plans to open a video rental store in Edmonton. Industry statistics show that 45% of the rental price goes for overhead. If the maximum that anyone will pay to rent a tape is $5 and Stephen wants a profit of at least $1.65 per tape, then in what range should the rental price be? x = rental price, $3 \le x \le 5$

88. ***Working girl.*** Regina makes $6.80 per hour working in the snack bar. To keep her grant, she may not earn more than $51 per week. What is the range of the number of hours per week that she may work?   [0, 7.5]

89. ***Skeletal remains.*** Forensic scientists use the formula $h = 60.089 + 2.238F$ to predict the height h (in centimeters) for a male whose femur measures F centimeters. (See the accompanying figure.) In what range is the length of the femur for males between 150 centimeter and 180 centimeter in height? Round to the nearest tenth of a centimeter. (40.2, 53.6)

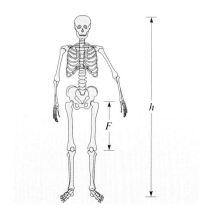

FIGURE FOR EXERCISE 89

90. ***Female femurs.*** Forensic scientists use the formula $h = 61.412 + 2.317F$ to predict the height h in centimeters for a female whose femur measures F centimeters.
   a) Use the accompanying graph to estimate the femur length for a female with height of 160 centimeters.
   b) In what range is the length of the femur for females who are over 170 centimeters tall?
      a) 43 centimeter   b) $F > 47.0$

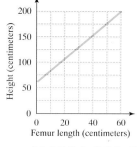 

FIGURE FOR EXERCISE 90

91. ***Car trouble.*** Dane's car was found abandoned at mile marker 86 on the interstate. If Dane was picked up by the police on the interstate exactly 5 miles away, then at what mile marker was he picked up?   81 or 91

92. ***Comparing scores.*** Scott scored 72 points on the midterm, and Katie's score was more than 16 points away from Scott's. What was Katie's score?
Greater than 88 or less than 56

93. ***Year-end bonus.*** A law firm has agreed to distribute 20% of its profits to its employees as a year-end bonus. To the firm's accountant, the bonus is an expense that must be used to determine the profit. That is, bonus = 20% × (profit before bonus − bonus). Given that the profit before the bonus is $300,000, find the amount of the bonus using the accountant's method. How does this answer compare to 20% of $300,000, which is what the employees want?
$50,000 accountant, $60,000 employees

94. ***Higher rate.*** Suppose that the employees in Exercise 93 got the bonus that they wanted. To the accountant, what percent of the profits was given in bonuses?   25%

95. ***Dairy cattle.*** Thirty percent of the dairy cattle in Washington County are Holsteins, whereas 60% of the dairy cattle in neighboring Cade County are Holsteins. In the combined two-county area, 50% of the 3,600 dairy cattle are Holsteins. How many dairy cattle are in each county?
Washington County 1,200, Cade County 2,400

96. ***Profitable business.*** United Home Improvement (UHI) makes 20% profit on its good grade of vinyl siding, 30% profit on its better grade, and 60% profit on its best grade. So far this year, UHI has $40,000 in sales of good siding and $50,000 in sales of better siding. The company goal is to have at least an overall profit of 50% of total sales. What would the sales figures for the best grade of siding have to be to reach this goal?   At least $220,000

For each graph in Exercises 97–114, write an equation or inequality that has the solution set shown by the graph. Use absolute value when possible.

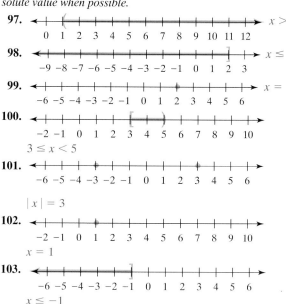

97. $x > 1$

98. $x \le 2$

99. $x = 2$

100. $3 \le x < 5$

101. $|x| = 3$

102. $x = 1$

103. $x \le -1$

104. $|x| > 2$

105.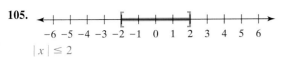

$-6\ -5\ -4\ -3\ -2\ -1\ \ 0\ \ 1\ \ 2\ \ 3\ \ 4\ \ 5\ \ 6$

$|x| \leq 2$

106.

$-6\ -5\ -4\ -3\ -2\ -1\ \ 0\ \ 1\ \ 2\ \ 3\ \ 4\ \ 5\ \ 6$

$|x| = 5$

107.

$-1\ \ 0\ \ 1\ \ 2\ \ 3\ \ 4\ \ 5\ \ 6\ \ 7\ \ 8\ \ 9\ \ 10\ \ 11$

$x \leq 2$ or $x \geq 7$

108.

$-6\ -5\ -4\ -3\ -2\ -1\ \ 0\ \ 1\ \ 2\ \ 3\ \ 4\ \ 5\ \ 6$

$|x| \leq 1$

109.

$-6\ -5\ -4\ -3\ -2\ -1\ \ 0\ \ 1\ \ 2\ \ 3\ \ 4\ \ 5\ \ 6$

$|x| > 3$

110.

$-5\ -4\ -3\ -2\ -1\ \ 0\ \ 1\ \ 2\ \ 3\ \ 4\ \ 5\ \ 6\ \ 7$

$x > 3$ or $x < -1$

111.

$0\ \ 1\ \ 2\ \ 3\ \ 4\ \ 5\ \ 6\ \ 7\ \ 8\ \ 9\ \ 10\ \ 11\ \ 12$

$5 < x < 7$

112.

$-6\ -5\ -4\ -3\ -2\ -1\ \ 0\ \ 1\ \ 2\ \ 3\ \ 4\ \ 5\ \ 6$

$|x| > 4$

113.

$-6\ -5\ -4\ -3\ -2\ -1\ \ 0\ \ 1\ \ 2\ \ 3\ \ 4\ \ 5\ \ 6$

$|x| > 0$

114.

$-6\ -5\ -4\ -3\ -2\ -1\ \ 0\ \ 1\ \ 2\ \ 3\ \ 4\ \ 5\ \ 6$

$-6 \leq x < 6$

CHAPTER 2 TEST

Solve each equation.

1. $-10x - 5 + 4x = -4x + 3$ $\{-4\}$

2. $\dfrac{y}{2} - \dfrac{y-3}{3} = \dfrac{y+6}{6}$ R

3. $|w| + 3 = 9$ $\{-6, 6\}$

4. $|3 - 2(5 - x)| = 3$ $\{2, 5\}$

Write y in terms of x.

5. $2x - 5y = 20$

$y = \dfrac{2}{5}x - 4$

6. $y = 3xy + 5$

$y = \dfrac{5}{1 - 3x}$

Solve each inequality. State the solution set using interval notation and graph the solution set.

7. $|m - 6| \leq 2$ $[4, 8]$

$3\ \ 4\ \ 5\ \ 6\ \ 7\ \ 8\ \ 9$

8. $2|x - 3| - 5 > 15$

$(-\infty, -7) \cup (13, \infty)$

$-15\ -10\ -5\ \ 0\ \ 5\ \ 10\ \ 15$

(-7 and 13 marked)

9. $2 - 3(w - 1) < -2w$

$(5, \infty)$

$3\ \ 4\ \ 5\ \ 6\ \ 7\ \ 8\ \ 9$

10. $2 < \dfrac{5 - 2x}{3} < 7$

$\left(-8, -\dfrac{1}{2}\right)$

$-8\ -7\ -6\ -5\ -4\ -3\ -2\ -1\ \ 0$ ($-\dfrac{1}{2}$ marked)

11. $3x - 2 < 7$ and $-3x \leq 15$

$[-5, 3)$

$-5\ -4\ -3\ -2\ -1\ \ 0\ \ 1\ \ 2\ \ 3$

12. $\dfrac{2}{3}y < 4$ or $y - 3 < 12$

$(-\infty, 15)$

$11\ \ 12\ \ 13\ \ 14\ \ 15\ \ 16\ \ 17$

Solve each equation or inequality.

13. $|2x - 7| = 3$ $\{2, 5\}$

14. $x - 4 > 1$ or $x < 12$ $(-\infty, \infty)$

15. $3x < 0$ and $x - 5 > 2$ $\varnothing$

16. $|2x - 5| \leq 0$ $\{2.5\}$

17. $|x - 3| < 0$ $\varnothing$

18. $x + 3x = 4x$ R

19. $2(x + 7) = 2x + 9$ $\varnothing$

20. $|x - 6| > -6$ R

21. $x - 0.04(x - 10) = 96.4$ $\{100\}$

Write a complete solution to each problem.

22. The perimeter of a rectangle is 84 meters. If the width is 16 meters less than the length, then what is the width of the rectangle?   13 meters

23. If the area of a triangle is 21 square inches and the base is 3 inches, then what is the height?   14 inches

24. Joan bought a gold chain marked 30% off. If she paid $210, then what was the original price? $300

25. How many liters of an 11% alcohol solution should be mixed with 60 liters of a 5% alcohol solution to obtain a mixture that is 7% alcohol?   30 liters

26. Al and Brenda do the same job, but their annual salaries differ by more than $3,000. Assume, Al makes $28,000 per year and write an absolute value inequality to describe this situation. What are the possibilities for Brenda's salary? $|x - 28,000| > 3,000$, Brenda makes more than $31,000 or less than $25,000.

Simplify each expression.

1. $5x + 6x$
    $11x$

2. $5x \cdot 6x$
    $30x^2$

3. $\dfrac{6x + 2}{2}$
    $3x + 1$

4. $5 - 4(2 - x)$
    $4x - 3$

5. $(30 - 1)(30 + 1)$
    899

6. $(30 + 1)^2$
    961

7. $(30 - 1)^2$
    841

8. $(2 + 3)^2$
    25

9. $2^2 + 3^2$
    13

10. $(8 - 3)(3 - 8)$
    -25

11. $(-1)(3 - 8)$
    5

12. -2^2
    -4

13. $3x + 8 - 5(x - 1)$
    $-2x + 13$

14. $(-6)^2 - 4(-3)2$
    60

15. $3^2 \cdot 2^3$
    72

16. $4(-6) - (-5)(3)$
    -9

17. $-3x \cdot x \cdot x$
    $-3x^3$

18. $(-1)(-1)(-1)(-1)(-1)(-1)$
    1

Solve each equation.

19. $5x + 6x = 8x$
    $\{0\}$

20. $5x + 6x = 11x$
    R

21. $5x + 6x = 0$
    $\{0\}$

22. $5x + 6 = 11x$
    $\{1\}$

23. $3x + 1 = 0$
    $\left\{-\dfrac{1}{3}\right\}$

24. $5 - 4(2 - x) = 1$
    $\{1\}$

25. $3x + 6 = 3(x + 2)$
    R

26. $x - 0.01x = 990$
    $\{1000\}$

27. $|5x + 6| = 11$
    $\left\{-\dfrac{17}{5}, 1\right\}$

Solve the problem.

28. *Cost analysis.* Diller Electronics can rent a copy machine for 5 years from American Business Supply for $75 per month plus 6 cents per copy. The same copier can be purchased for $8,000, but then it costs only 2 cents per copy for supplies and maintenance. The purchased copier has no value after 5 years.

  a) Use the accompanying graph to estimate the number of copies for 5 years for which the cost of renting would equal the cost of buying.

  b) Write a formula for the 5-year cost under each plan.

  c) Algebraically find the number of copies for which the 5-year costs would be equal.

  d) If Diller makes 120,000 copies in 5 years, which plan is cheaper and by how much?

  e) For what range of copies do the two plans differ by less than $500?

  a) 87,500
  b) $C_r = 4{,}500 + 0.06x$, $C_b = 8{,}000 + 0.02x$
  c) 87,500
  d) Buying is $1,300 cheaper
  e) (75,000, 100,000)

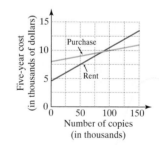

FIGURE FOR EXERCISE 28

CHAPTER 3

Graphs and Functions in the Cartesian Coordinate System

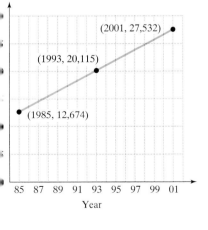

(2001, 27,532)

(1993, 20,115)

(1985, 12,674)

85 87 89 91 93 95 97 99 01

Year

The first self-propelled automobile to carry passengers was built in 1801 by the British inventor Richard Trevithick. By 1911 about 600,000 automobiles were operated in the United States alone. Some were powered by steam and some by electricity, but most were powered by gasoline. In 1913, to meet the ever growing demand, Henry Ford increased production by introducing a moving assembly line to carry automobile parts. Today the United States is a nation of cars. Over 11 million automobiles are produced here annually, and total car registrations number over 114 million.

Henry Ford's early model T sold for $850. In 1994 the average price of a new car was $18,000, with some selling for more than $50,000. Prices for new cars rise every year, but, unfortunately for the buyer, the moment a new automobile is bought, its value begins to decrease. Much of the behavior of automobile prices can be modeled by linear equations. In Exercises 53 and 54 of Section 3.1 you will use linear equations to find increasing new car prices and depreciating used car prices.

3.1 GRAPHING LINES IN THE COORDINATE PLANE

In Chapter 1 we graphed numbers on a number line. In Chapter 2 we used number lines to illustrate the solution sets to inequalities. In this section we graph pairs of numbers in a coordinate system made from two number lines to illustrate solution sets.

Ordered Pairs

An equation in two variables, such as $y = 2x + 3$, is satisfied only if we find a value for x and a value for y that make it true. For example, if $x = 4$ and $y = 11$, then the equation becomes $11 = 2 \cdot 4 + 3$, which is a true statement. We write $x = 4$ and $y = 11$ as the **ordered pair** $(4, 11)$. The order of the numbers in an ordered pair is important. For example, $(5, 13)$ also satisfies $y = 2x + 3$ because using 5 for x and 13 for y gives $13 = 2 \cdot 5 + 3$. However, $(13, 5)$ does not satisfy this equation because $5 \neq 2 \cdot 13 + 3$. In an ordered pair the value for x, the **x-coordinate,** is always written first and the value for y, the **y-coordinate,** is second: (x, y).

EXAMPLE 1   **Writing ordered pairs**

Complete the following ordered pairs so that each ordered pair satisfies the equation $4x + y = 5$.

a) $(-2, \quad)$ **b)** $(\quad , 3)$

Solution

a) Replace x with -2 in $4x + y = 5$ because the x-coordinate is -2:

$$4(-2) + y = 5$$
$$-8 + y = 5$$
$$y = 13$$

So the y-coordinate is 13 and the ordered pair is $(-2, 13)$.

b) Replace y with 3 in $4x + y = 5$ because the y-coordinate is 3:

$$4x + 3 = 5$$
$$4x = 2$$
$$x = \frac{1}{2}$$

So the x-coordinate is $\frac{1}{2}$ and the ordered pair is $\left(\frac{1}{2}, 3\right)$. ■

Plotting Points

To graph ordered pairs of real numbers, we need a new coordinate system. The **rectangular** or **Cartesian coordinate system** consists of a horizontal number line, the **x-axis,** and a vertical number line, the **y-axis,** as shown in Fig. 3.1. The intersection of the axes is the **origin.** The axes divide the coordinate plane, or the **xy-plane,** into four regions called **quadrants.** The quadrants are numbered as shown in Fig. 3.1, and they do not include any points on the axes.

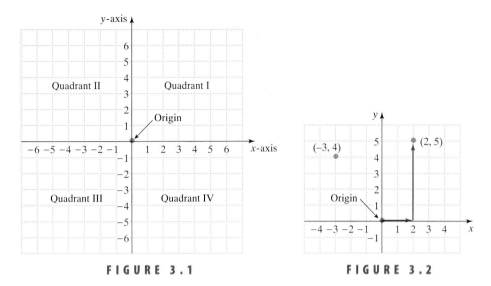

FIGURE 3.1   **FIGURE 3.2**

Just as every real number corresponds to a point on the number line, every pair of real numbers corresponds to a point in the rectangular coordinate system. For example, the pair $(2, 5)$ corresponds to the point that lies two units to the right of the origin and five units up. See Fig. 3.2. To locate $(-3, 4)$, start at the origin and move three units to the left and four units upward, as in Fig. 3.2. Locating a point in the rectangular coordinate system is referred to as **plotting** or **graphing** the point.

EXAMPLE 2

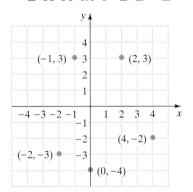

FIGURE 3.3

Graphing ordered pairs

Plot the points $(2, 3)$, $(-2, -3)$, $(-1, 3)$, $(0, -4)$, and $(4, -2)$.

Solution

To plot $(2, 3)$, start at the origin, move two units to the right, then up three units. To graph $(-2, -3)$, start at the origin, move two units to the left, then down three units. All five points are shown in Fig. 3.3. ∎

Graphing a Linear Equation

The solution set to an equation in two variables consists of all ordered pairs that satisfy the equation. For example, the solution set to $y = 2x + 3$ can be written in set notation as $\{(x, y) \mid y = 2x + 3\}$. However, this set notation does not shed any light on the solution set to $y = 2x + 3$. We can get a better understanding of the solution set with a visual image or **graph** of the solution set.

EXAMPLE 3

Graphing a linear equation

Graph the solution set to $y = 2x + 3$.

Solution

If we arbitrarily choose $x = -4$, then y is determined by the equation $y = 2x + 3$:

$$y = 2(-4) + 3 = -5$$

So the ordered pair $(-4, -5)$ satisfies the equation. In this manner we can make the following table of ordered pairs:

x	-4	-3	-2	-1	0	1
$y = 2x + 3$	-5	-3	-1	1	3	5

helpful hint

The graph of a linear equation is a straight line that exists in our minds. The straight line in our minds has no thickness, is perfectly straight, and extends infinitely. All attempts to draw it on paper fall short. The best we can do is to use a sharp pencil to keep it thin, a ruler to make it straight, and arrows to indicate that it is infinite.

Plot these ordered pairs as shown in Fig. 3.4. Of course, there are infinitely many ordered pairs that satisfy $y = 2x + 3$, but they all lie along the line in Fig. 3.4. The arrows on the ends of the line indicate that it extends without bound in both directions. The line in Fig. 3.4 is a graph of the solution set to $y = 2x + 3$. ■

In an equation such as $y = 2x + 3$ the value of y depends on the value of x. So x is the **independent variable** and y is the **dependent variable.** Because the graph of $y = 2x + 3$ is a line, the equation is a **linear equation.**

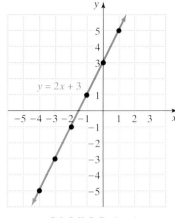

FIGURE 3.4

E X A M P L E   4

Graphing a linear equation

Graph $y + 2x = 1$. Plot at least four points.

Solution

If we write y in terms of x, we get $y = -2x + 1$. Now arbitrarily select four values for x and calculate the corresponding values for y:

x	-1	0	1	2
$y = -2x + 1$	3	1	-1	-3

Plot these points and draw a line through them as shown in Fig. 3.5. ■

If the coefficient of a variable in a linear equation is 0, then that variable is usually omitted from the equation. For example, the equation $y = 0 \cdot x + 2$ is written as $y = 2$. Because x is multiplied by 0, any value of x can be used as long as y is 2. Because the y-coordinates are all the same, the graph is a horizontal line.

FIGURE 3.5

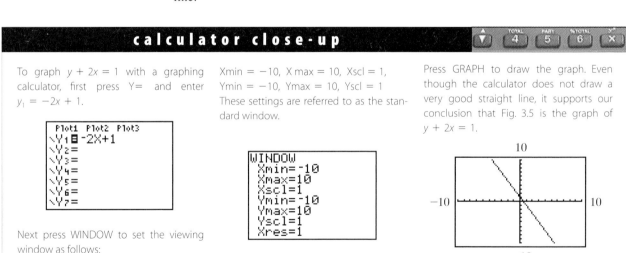

E X A M P L E   5

Graphing a horizontal line

Graph $y = 2$. Plot at least four points.

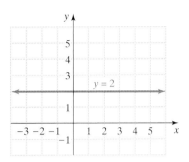

F I G U R E   3 . 6

Solution

The following table gives four points that satisfy $y = 2$, or $y = 0 \cdot x + 2$. Note that it is easy to determine y in this case because y is always 2.

x	-2	-1	0	1
$y = 0 \cdot x + 2$	2	2	2	2

The horizontal line through these points is shown in Fig. 3.6.

If the coefficient of y is 0 in a linear equation, then the graph is a vertical line.

E X A M P L E   6

Graphing a vertical line

Graph $x = 4$. Plot at least four points.

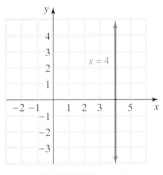

F I G U R E   3 . 7

Solution

We can think of the equation $x = 4$ as $x = 4 + 0 \cdot y$. Because y is multiplied by 0, the equation is satisfied by any ordered pair with an x-coordinate of 4.

$x = 4 + 0 \cdot y$	4	4	4	4
y	-2	-1	0	1

The vertical line through these points is shown in Fig. 3.7.

Using Intercepts for Graphing

The **x-intercept** of a line is the point where the line crosses the x-axis. The x-intercept has a y-coordinate of 0. Similarly, the **y-intercept** of a line is the point where the line crosses the y-axis. The y-intercept has an x-coordinate of 0. If a line has distinct x- and y-intercepts, then these intercepts can be used as two points that determine the location of the line. (Horizontal lines, vertical lines, and lines through the origin do not have two distinct intercepts.)

E X A M P L E   7

Using intercepts to graph

Use the intercepts to graph the line $3x - 4y = 6$.

helpful hint

You can find the intercepts for $3x - 4y = 6$ using the *cover-up method*. Cover up $-4y$ with your pencil, then solve $3x = 6$ mentally to get $x = 2$ and an x-intercept of $(2, 0)$. Now cover up $3x$ and solve $-4y = 6$ to get $y = -3/2$ and a y-intercept of $(0, -3/2)$.

Solution

Let $x = 0$ in $3x - 4y = 6$ to find the y-intercept:

$$3(0) - 4y = 6$$
$$-4y = 6$$
$$y = -\frac{3}{2}$$

Let $y = 0$ in $3x - 4y = 6$ to find the x-intercept:

$$3x - 4(0) = 6$$
$$3x = 6$$
$$x = 2$$

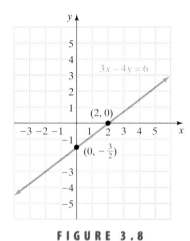

FIGURE 3.8

The y-intercept is $\left(0, -\frac{3}{2}\right)$, and the x-intercept is $(2, 0)$. The line through the intercepts is shown in Fig. 3.8. To check, find another point that satisfies the equation. The point $(-2, -3)$ satisfies the equation and is on the line in Fig. 3.8.   ■

CAUTION   Even though two points determine the location of a line, finding at least three points will help you to avoid errors.

Applications

In applications we often use variables such as C for cost, R for revenue, and n for the number of items so that it is easier to remember what the variables represent. In this case we rename the axes. Which axis is labeled with which variable is somewhat arbitrary. However, when one variable depends on, or is determined by another, the dependent variable is usually on the vertical axis and the independent variable is on the horizontal axis.

E X A M P L E   8

Graphing a linear equation in an application

The cost per week C (in dollars) of producing n pairs of shoes for the Reebop Shoe Company is given by the linear equation $C = 2n + 8000$. Graph the equation for n between 0 and 800 inclusive $(0 \leq n \leq 800)$.

Solution

Make a table of values for n and C as follows:

n	0	200	400	600	800
$C = 2n + 8000$	8000	8400	8800	9200	9600

Graph the line as shown in Fig. 3.9. Because $C = 2n + 8000$ expresses C in terms of n, C is the dependent variable and the vertical axis is labeled C. To accommodate the large numbers, we let each unit on the n-axis represent 200 pairs and each unit on the C-axis represent $1,000.   ■

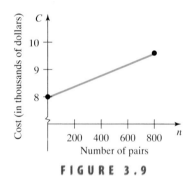

FIGURE 3.9

Note how the axes in Fig. 3.9 are labeled. The n-axis starts at 0 and each unit represents 200 pairs. Because all of the costs were between $8,000 and $9,600 we omitted the tick marks for 1 through 7 and put a wave in the C-axis to indicate that some numbers are missing. Omitting the numbers from 1 through 7 makes the difference between the $8,000 and $9,600 costs look greater.

WARM-UPS

True or false? Explain your answer.

1. The point $(2, 5)$ satisfies the equation $3y - 2x = -4$.   False

2. The vertical axis is usually called the x-axis.   False

3. The point $(0, 0)$ is in quadrant I.   False

4. The point $(0, 1)$ is on the y-axis.   True

5. The graph of $x = 7$ is a vertical line.   True

6. The graph of $8 - y = 0$ is a horizontal line.   True

7. The y-intercept for the line $y = 2x - 3$ is $(0, -3)$.   True

8. If $C = 3n + 4$, then $C = 10$ when $n = 2$.   True

9. If $P = 3x$ and $P = 12$, then $x = 36$.   False

10. The vertical axis should be A when graphing $A = \pi r^2$.   True

3.1 EXERCISES

Reading and Writing After reading this section, write out the answers to these questions. Use complete sentences.

1. What is the point called at the intersection of the x- and y-axis?

    The origin is the point where the x-axis and y-axis intersect.

2. What is an ordered pair?

    An ordered pair is a pair of real numbers in which there is a first number and a second one.

3. What are the x- and y-intercepts?

    Intercepts are points where a graph crosses the axes.

4. What type of equation has a graph that is a horizontal line?

    The graph of an equation of the type $y = k$ where k is a fixed number is a horizontal line.

5. What type of equation has a graph that is a vertical line?

    The graph of an equation of the type $x = k$ where k is a fixed number is a vertical line.

6. Which variable usually goes on the vertical axis?

    The dependent variable usually goes on the vertical axis.

Complete the given ordered pairs so that each ordered pair satisfies the given equation. See Example 1.

7. $(2, \ ), (\ , -3)$, $y = -3x + 6$ $(2, 0), (3, -3)$

8. $(-1, \ ), (\ , 4)$, $y = \frac{1}{2}x + 2$ $\left(-1, \frac{3}{2}\right), (4, 4)$

9. $(-4, \ ), (\ , 6)$, $\frac{1}{2}x - \frac{1}{3}y = 9$ $(-4, -33), (22, 6)$

10. $(3, \ ), (\ , -1)$, $2x - 3y = 5$ $\left(3, \frac{1}{3}\right), (1, -1)$

Plot the following points in a rectangular coordinate system. For each point, name the quadrant in which it lies or the axis on which it lies. See Example 2.

11. $(2, 5)$ I

12. $(-5, 1)$ II

13. $\left(-3, -\frac{1}{2}\right)$ III

14. $(-2, -6)$ III

15. $(0, 4)$ y-axis

16. $(0, 2)$ y-axis

17. $(\pi, -1)$ IV

18. $\left(\frac{4}{3}, 0\right)$ x-axis

19. $(-4, 3)$ II

20. $(0, -3)$ y-axis

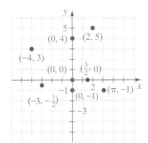

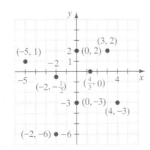

21. $\left(\frac{3}{2}, 0\right)$ x-axis

22. $(3, 2)$ I

23. $(0, -1)$ y-axis

24. $(4, -3)$ IV

Graph each linear equation. Plot four points for each line. See Examples 3–6.

25. $y = x + 1$

26. $y = x - 1$

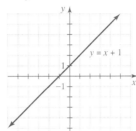

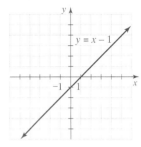

27. $y = -2x + 3$

28. $y = 2x - 3$

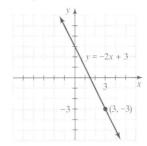

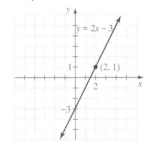

29. $y = x$

30. $y = -x$

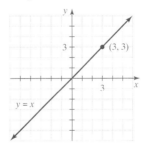

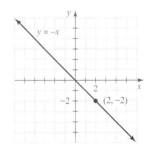

31. $y = 3$

32. $y = -2$

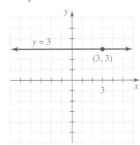

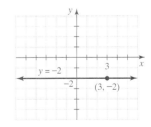

33. $y = 1 - x$

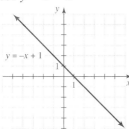

34. $y = 2 - x$

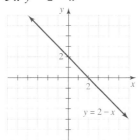

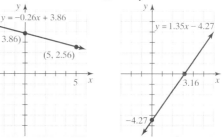

43. $y = -0.26x + 3.86$

44. $y = 1.35x - 4.27$

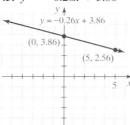

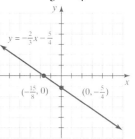

35. $x = 2$

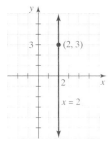

36. $x = -3$

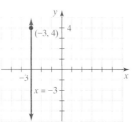

Find the x- and y-intercepts for each line and use them to graph the line. See Example 7.

45. $4x - 3y = 12$

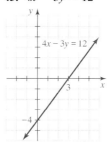

46. $2x + 5y = 20$

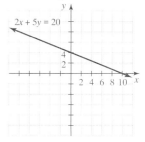

37. $y = \dfrac{1}{2}x - 1$

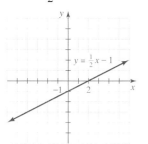

38. $y = \dfrac{1}{3}x - 2$

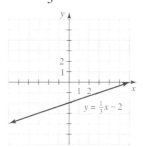

47. $x - y + 5 = 0$

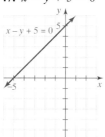

48. $x + y + 7 = 0$

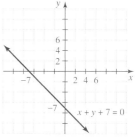

39. $x - 4 = 0$

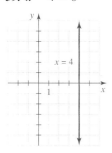

40. $y + 3 = 0$

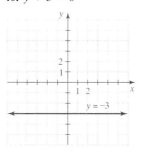

49. $2x + 3y = 5$

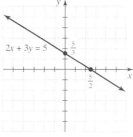

50. $3x - 4y = 7$

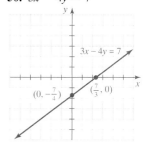

41. $3x + y = 5$

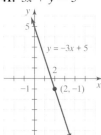

42. $x + 2y = 4$

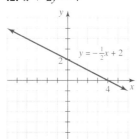

51. $y = \dfrac{3}{5}x + \dfrac{2}{3}$

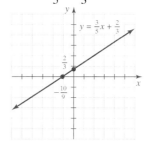

52. $y = -\dfrac{2}{3}x - \dfrac{5}{4}$

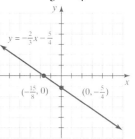

Solve the following. See Example 8.

53. Camaro inflation. The rising list price P (in dollars) for a new Camaro Z28 Coupe can be modeled by the equation $P = 19,663 + 269n$, where n is the number of years since 1995 (Edmund's New Car Prices, www.edmunds.com).
a) What will be the list price for a new Z28 in 2005?
b) What is the annual increase in list price?
c) Sketch a graph of this equation for $0 \le n \le 10$.
   a) $22,353   b) $269

c)

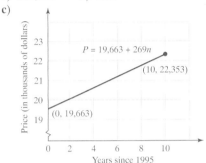

54. Camaro Z28 depreciation. The 1998 average retail price P (in dollars) for an n-year-old Camaro Z28 Coupe can be modeled by the equation $P = 18,675 - 1,960n$, where $1 \le n \le 4$ (Edmund's Used Car Prices, www.edmunds.com).
a) What was the average retail price of a 4-year-old Z28 in 1998?
b) How much does this model depreciate each year?
c) Sketch a graph of this equation for $1 \le n \le 4$.
   a) $10,835   b) $1960

c)

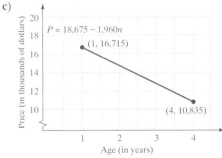

55. Rental cost. For a one-day car rental the X-press Car Company charges C dollars, where C is determined by the formula $C = 0.26m + 42$ and m is the number of miles driven. What is the charge for a car driven 400 miles? Sketch a graph of the equation for m ranging from 0 to 1000.
$146

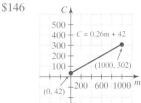

56. Measuring risk. The Friendly Bob Loan Company gives each applicant a rating, t, from 0 to 10 according to the applicant's ability to repay, a higher rating indicating higher risk. The interest rate, r, is then determined by the formula $r = 0.02t + 0.15$. If your rating were 8, then what would be your interest rate? Sketch the graph of the equation for t ranging from 0 to 10.
31%

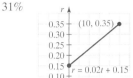

57. Little Chicago Pizza. The equation $C = 0.50t + 8.95$ gives the customer's cost in dollars for a pan pizza, where t is the number of toppings.
a) Find the cost of a five-topping pizza.
b) Find t if $C = 14.45$ and interpret your result.
   a) $11.45 is the cost of a five-topping pizza.
   b) 11 is the number of toppings on a $14.45 pizza.

58. Long distance charges. The formula $L = 0.10n + 4.95$ gives the monthly bill in dollars for AT&T's one rate plan, where n is the number of minutes of long distance used during the month. Answer each question and interpret the result.
a) Find the bill for 120 minutes. $16.95
b) Find n if $L = 23.45$.   185
c) What are the coordinates for the L-intercept shown on the accompanying graph? $(0, 4.95)$
d) What are the coordinates of the n-intercept? $(-49.5, 0)$

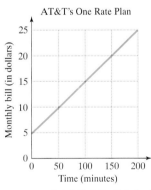

FIGURE FOR EXERCISE 58

59. Cost, Revenue, and Profit. Hillary sells roses at a busy Los Angeles intersection. The formulas

$$C = 0.55x + 50,$$
$$R = 1.50x,$$

and

$$P = 0.95x - 50$$

give her weekly cost, revenue, and profit in terms of x, where x is the number of roses that she sells in one week.
a) Find C, R, and P if $x = 850$. Interpret your results.
b) Find x if $P = 995$ and interpret your result.

c) Find $R - C$ if $x = 1100$ and interpret your result.

   a) Her weekly cost, revenue, and profit are $517.50, $1,275, and $757.50.

   b) 1,100. She had a profit of $995 on selling 1,100 roses.

   c) 995. The difference between revenue and cost is $995, which is her profit.

60. *Velocity of a pop up.* A pop up off the bat of Mark McGwire goes straight into the air at 88 feet per second (ft/sec). The formula $v = -32t + 88$ gives the velocity of the ball in feet per second, t seconds after the ball is hit.

a) Find the velocity for $t = 2$ and $t = 3$ seconds. What does a negative velocity mean?

b) For what value of t is $v = 0$? Where is the ball at this time?

c) What are the two intercepts on the accompanying graph? Interpret this answer.

d) If the ball takes the same time going up as it does coming down, then what is its velocity as it hits the ground?

   a) 24 ft/sec, -8 ft/sec, going down

   b) 2.75 seconds, at maximum height

   c) (0, 88) indicates that at $t = 0$ second the velocity was 88 ft/sec, (2.75, 0) indicates that at $t = 2.75$ seconds the velocity was 0 ft/sec

   d) -88 ft/sec

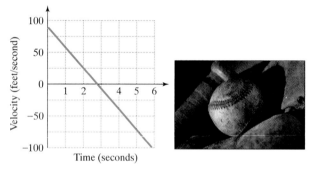

FIGURE FOR EXERCISE 60

GETTING MORE INVOLVED

61. *Midpoint formula.* The point M midway between (x_1, y_1) and (x_2, y_2) can be found using the formula

$$M = \left(\frac{x_1 + x_2}{2}, \frac{y_1 + y_2}{2}\right).$$

Find the midpoint for each of the following pairs of points and plot the points.

a) (1, 2) and (5, 6)   (3, 4)

b) (6, -5) and (-2, -1)   (2, -3)

c) $\left(\frac{1}{2}, \frac{1}{4}\right)$ and $\left(\frac{1}{3}, \frac{1}{2}\right)$ $\left(\frac{5}{12}, \frac{3}{8}\right)$

62. *Intersecting medians.* Using graph paper, draw a triangle with vertices (0, 6), (6, 2), and (2, -2).

a) Find the midpoint of each side of your triangle.

b) A *median* is a line segment connecting a vertex of a triangle with the midpoint of the opposite side. Draw the three medians for your triangle.

c) According to a theorem in geometry, the three medians of any triangle intersect at a single point. Estimate the point of intersection of the three medians in your triangle.

   a) (3, 4), (4, 0), (1, 2)   **c)** $\left(\frac{8}{3}, 2\right)$

GRAPHING CALCULATOR EXERCISES

Graph each linear equation on a graphing calculator. Choose a viewing window that shows both intercepts. Answers may vary.

63. $y = -3x + 20$

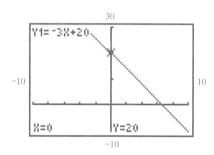

64. $y = 5 - 50x$

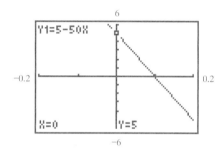

65. $y = 300x - 2$

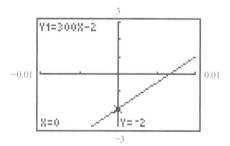

66. $y = 5x - 800$

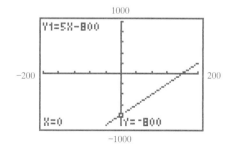

67. $x + 2y = 600$

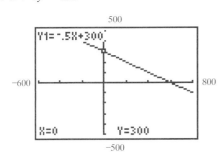

68. $3x - 2y = 1500$

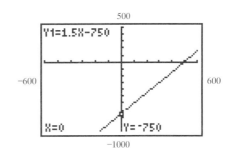

3.2   SLOPE OF A LINE

In Section 3.1 we saw some equations whose graphs were straight lines. In this section we look at graphs of straight lines in more detail and study the concept of slope of a line.

Slope

If a highway has a 6% grade, then in 100 feet (measured horizontally) the road rises 6 feet (measured vertically). See Fig. 3.10. The ratio of 6 to 100 is 6%. If a roof rises 9 feet in a horizontal distance (or run) of 12 feet, then the roof has a 9–12 pitch. A roof with a 9–12 pitch is steeper than a roof with a 6–12 pitch. The grade of a road and the pitch of a roof are measurements of steepness. In each case the measurement is a ratio of rise (vertical change) to run (horizontal change).

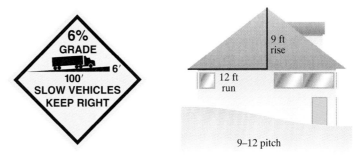

FIGURE 3.10

helpful hint

Since the amount of run is arbitrary, we can choose the run to be 1. In this case

$$\text{slope} = \frac{\text{rise}}{1} = \text{rise}.$$

So the slope is the amount of change in y for a change of 1 in the x-coordinate. This is why rates like 50 miles per hour (mph), 8 hours per day, and two people per car are all slopes.

We measure the steepness of a line in the same way that we measure steepness of a road or a roof. The slope of a line is the ratio of the change in y-coordinate, or the **rise,** to the change in x-coordinate, or the **run,** between two points on the line.

Slope

$$\text{Slope} = \frac{\text{change in } y\text{-coordinate}}{\text{change in } x\text{-coordinate}} = \frac{\text{rise}}{\text{run}}$$

Consider the line in Fig. 3.11(a) on the next page. In going from $(0, 1)$ to $(1, 3)$, there is a change of $+1$ in the x-coordinate and a change of $+2$ in the y-coordinate,

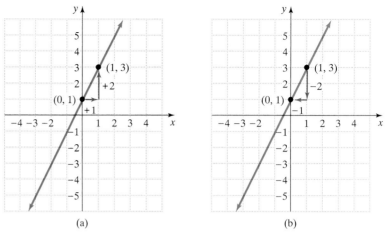

FIGURE 3.11

or a run of 1 and a rise of 2. So the slope is $\frac{2}{1}$ or 2. If we move from $(1, 3)$ to $(0, 1)$ as in Fig. 3.11(b) the rise is -2 and the run is -1. So the slope is $\frac{-2}{-1}$ or 2. If we start at either point and move to the other point, we get the same slope.

E X A M P L E   1   **Finding the slope from a graph**

Find the slope of each line by going from point A to point B.

a)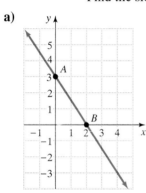

b)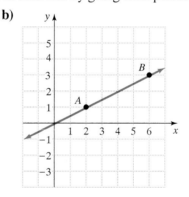

c)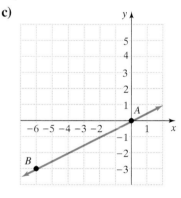

Solution

a) A is located at $(0, 3)$ and B at $(2, 0)$. In going from A to B, the change in y is -3 and the change in x is 2. So

$$\text{slope} = -\frac{3}{2}.$$

b) In going from $A(2, 1)$ to $B(6, 3)$, we must rise 2 and run 4. So

$$\text{slope} = \frac{2}{4} = \frac{1}{2}.$$

c) In going from $A(0, 0)$ to $B(-6, -3)$, we find that the rise is -3 and the run is -6. So

$$\text{slope} = \frac{-3}{-6} = \frac{1}{2}.$$

Note that in Example 1(c) we found the slope of the line of Example 1(b) by using two different points. The slope is the ratio of the lengths of the two legs of a right triangle whose hypotenuse is on the line. See Fig. 3.12. As long as one leg is vertical and the other leg is horizontal, all such triangles for a given line have the same shape: They are similar triangles. Because ratios of corresponding sides in similar triangles are equal, the slope has the same value no matter which two points of the line are used to find it.

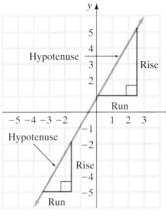

FIGURE 3.12   **FIGURE 3.13**

Using Coordinates to Find Slope

We can obtain the rise and run from a graph, or we can get them without a graph by subtracting the y-coordinates to get the rise and the x-coordinates to get the run for two points on the line. See Fig. 3.13.

Slope Using Coordinates

The slope m of the line containing the points (x_1, y_1) and (x_2, y_2) is given by

$$m = \frac{y_2 - y_1}{x_2 - x_1}, \qquad \text{provided that } x_2 - x_1 \neq 0.$$

E X A M P L E   2

Finding slope from coordinates

Find the slope of each line.

a) The line through $(2, 5)$ and $(6, 3)$

b) The line through $(-2, 3)$ and $(-5, -1)$

c) The line through $(-6, 4)$ and the origin

Solution

a) Let $(x_1, y_1) = (2, 5)$ and $(x_2, y_2) = (6, 3)$. The assignment of (x_1, y_1) and (x_2, y_2) is arbitrary.

$$m = \frac{y_2 - y_1}{x_2 - x_1} = \frac{3 - 5}{6 - 2} = \frac{-2}{4} = -\frac{1}{2}$$

b) Let $(x_1, y_1) = (-5, -1)$ and $(x_2, y_2) = (-2, 3)$:

$$m = \frac{y_2 - y_1}{x_2 - x_1} = \frac{3 - (-1)}{-2 - (-5)} = \frac{4}{3}$$

c) Let $(x_1, y_1) = (0, 0)$ and $(x_2, y_2) = (-6, 4)$:

$$m = \frac{4 - 0}{-6 - 0} = \frac{4}{-6} = -\frac{2}{3}$$

CAUTION Do not reverse the order of subtraction from numerator to denominator when finding the slope. If you divide $y_2 - y_1$ by $x_1 - x_2$, you will get the wrong sign for the slope.

EXAMPLE 3

Slope for horizontal and vertical lines
Find the slope of each line.

a)

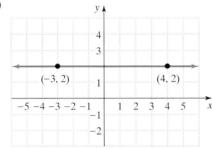

b)

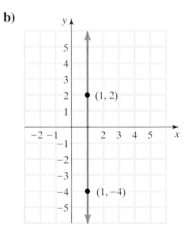

Think about what slope means to skiers. No one skis on cliffs or even refers to them as slopes.

Zero slope

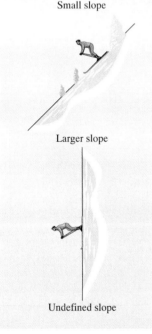

Small slope

Larger slope

Undefined slope

Solution

a) Using $(-3, 2)$ and $(4, 2)$ to find the slope of the horizontal line, we get

$$m = \frac{2 - 2}{-3 - 4}$$

$$= \frac{0}{-7} = 0.$$

b) Using $(1, -4)$ and $(1, 2)$ to find the slope of the vertical line, we get $x_2 - x_1 = 0$. Because the definition of slope using coordinates says that $x_2 - x_1$ must be nonzero, the slope is undefined for this line.

Since the y-coordinates are equal for any two points on a horizontal line, $y_2 - y_1 = 0$ and the slope is 0. Since the x-coordinates are equal for any two points on a vertical line, $x_2 - x_1 = 0$ and the slope is undefined.

Horizontal and Vertical Lines

The slope of any horizontal line is 0.
Slope is undefined for any vertical line.

CAUTION Do not say that a vertical line has no slope because "no slope" could be confused with 0 slope, the slope of a horizontal line.

As you move the tip of your pencil from left to right along a line with positive slope, the y-coordinates are increasing. As you move the tip of your pencil from

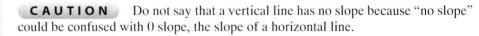

left to right along a line with negative slope, the y-coordinates are decreasing. See Fig. 3.14.

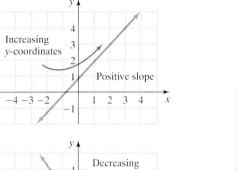

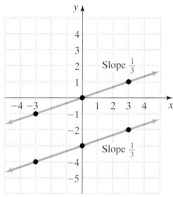

FIGURE 3.14   FIGURE 3.15

Parallel Lines

Consider the two lines shown in Fig. 3.15. Each of these lines has a slope of $\frac{1}{3}$, and these lines are parallel. In general, we have the following fact.

Parallel Lines

Nonvertical parallel lines have equal slopes.

Of course, any two vertical lines are parallel, but we cannot say that they have equal slopes because slope is not defined for vertical lines.

E X A M P L E   4

Parallel lines

Line l goes through the origin and is parallel to the line through $(-2, 3)$ and $(4, -5)$. Find the slope of line l.

Solution

The line through $(-2, 3)$ and $(4, -5)$ has slope

$$m = \frac{-5 - 3}{4 - (-2)} = \frac{-8}{6} = -\frac{4}{3}.$$

Because line l is parallel to a line with slope $-\frac{4}{3}$, the slope of line l is $-\frac{4}{3}$ also. ■

FIGURE 3.16

Perpendicular Lines

The lines shown in Fig. 3.16 have slopes 2 and $-\frac{1}{2}$. These two lines appear to be perpendicular to each other. It can be shown that *a line is perpendicular to another line if its slope is the negative of the reciprocal of the slope of the other.*

Perpendicular Lines

Two lines with slopes m_1 and m_2 are perpendicular if and only if

$$m_1 = -\frac{1}{m_2}.$$

Of course, any vertical line and any horizontal line are perpendicular, but we cannot give a relationship between their slopes because slope is undefined for vertical lines.

E X A M P L E   5   **Perpendicular lines**

Line l contains the point $(1, 6)$ and is perpendicular to the line through $(-4, 1)$ and $(3, -2)$. Find the slope of line l.

Solution

The line through $(-4, 1)$ and $(3, -2)$ has slope

$$m = \frac{1 - (-2)}{-4 - 3} = \frac{3}{-7} = -\frac{3}{7}.$$

Because line l is perpendicular to a line with slope $-\frac{3}{7}$, the slope of line l is $\frac{7}{3}$.   ■

Applications of Slope

When a geometric figure is located in a coordinate system, we can use slope to determine whether it has any parallel or perpendicular sides.

E X A M P L E   6   **Using slope with geometric figures**

Determine whether $(-3, 2)$, $(-2, -1)$, $(4, 1)$, and $(3, 4)$ are the vertices of a rectangle.

Solution

Figure 3.17 shows the quadrilateral determined by these points. If a parallelogram has at least one right angle, then it is a rectangle. Calculate the slope of each side.

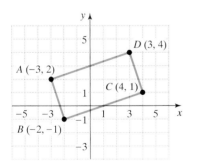

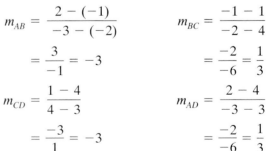

$$m_{AB} = \frac{2 - (-1)}{-3 - (-2)} \qquad m_{BC} = \frac{-1 - 1}{-2 - 4}$$

$$= \frac{3}{-1} = -3 \qquad = \frac{-2}{-6} = \frac{1}{3}$$

$$m_{CD} = \frac{1 - 4}{4 - 3} \qquad m_{AD} = \frac{2 - 4}{-3 - 3}$$

$$= \frac{-3}{1} = -3 \qquad = \frac{-2}{-6} = \frac{1}{3}$$

FIGURE 3.17

Because the opposite sides have the same slope, they are parallel, and the figure is a parallelogram. Because $\frac{1}{3}$ is the opposite of the reciprocal of -3, the intersecting sides are perpendicular. Therefore the figure is a rectangle.   ■

The slope of a line is a rate. The slope tells us how much the dependent variable changes for a change of 1 in the independent variable. For example, if the horizontal axis is hours and the vertical axis is miles, then the slope is miles per hour (mph).

If the horizontal axis is days and the vertical axis is dollars, then the slope is dollars per day.

E X A M P L E   7

Slope as a rate

Worldwide carbon dioxide (CO_2) emissions have increased from 14 billion tons in 1970 to 24 billion tons in 1995 (World Resources Institute, www.wri.org).

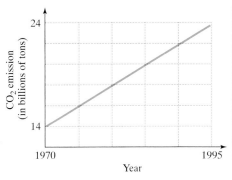

FIGURE FOR EXAMPLE 7

a) Find and interpret the slope of the line in the accompanying figure.

b) Predict the amount of worldwide CO_2 emissions in 2005.

Solution

a) Find the slope of the line through (1970, 14) and (1995, 24):

$$m = \frac{24 - 14}{1995 - 1970} = 0.4$$

The slope of the line is 0.4 billion tons per year.

b) If the (CO_2) emissions keep increasing at 0.4 billion tons per year, then in 10 years the level will go up 10(0.4) or 4 billion tons. So in 2005 CO_2 emissions will be 28 billion tons.

WARM-UPS

True or false? Explain your answer.

1. Slope is a measurement of the steepness of a line.   True
2. Slope is run divided by rise.   False
3. The line through (4, 5) and (−3, 5) has undefined slope.   False
4. The line through (−2, 6) and (−2, −5) has undefined slope.   True
5. Slope cannot be negative.   False
6. The slope of the line through (0, −2) and (5, 0) is $-\frac{2}{5}$.   False
7. The line through (4, 4) and (5, 5) has slope $\frac{5}{4}$.   False
8. If a line contains points in quadrants I and III, then its slope is positive.   True
9. Lines with slope $\frac{2}{3}$ and $-\frac{2}{3}$ are perpendicular to each other.   False
10. Any two parallel lines have equal slopes.   False

3.2 EXERCISES

Reading and Writing After reading this section, write out the answers to these questions. Use complete sentences.

1. What does slope measure?

Slope measures the steepness of a line.

2. What is the rise and what is the run?

The rise is the change in *y*-coordinates and run is the change in *x*-coordinates.

3. Why does a horizontal line have zero slope?

A horizontal line has zero slope because it has no rise.

4. Why is slope undefined for vertical lines?

Slope is undefined for vertical lines because the run is zero and division by zero is undefined.

5. What is the relationship between the slopes of perpendicular lines?

If m_1 and m_2 are the slopes of perpendicular lines, then $m_1 = \frac{-1}{m_2}$.

6. What is the relationship between the slopes of parallel lines?

If m_1 and m_2 are the slopes of parallel lines, then $m_1 = m_2$.

Determine the slope of each line. See Example 1.

7.

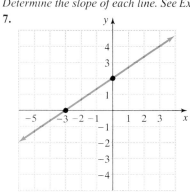

$\frac{2}{3}$

8.

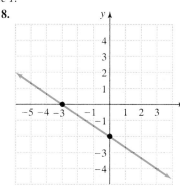

$-\frac{2}{3}$

9.

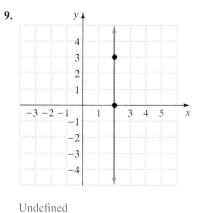

Undefined

10.

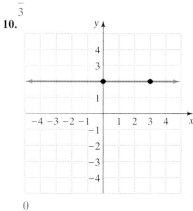

0

11.

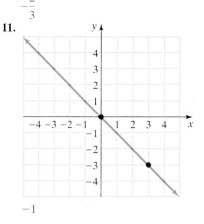

-1

12.

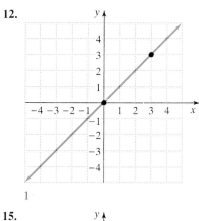

1

13.

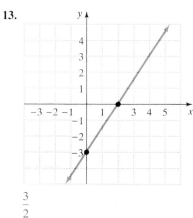

$\frac{3}{2}$

14.

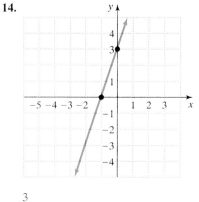

3

15.

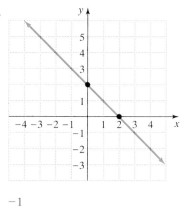

-1

16.

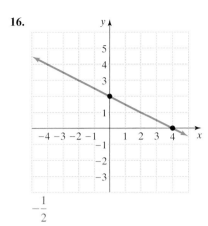

$$-\frac{1}{2}$$

Find the slope of the line that contains each of the following pairs of points. See Examples 2 and 3.

17. $(2, 6), (5, 1)$ $-\dfrac{5}{3}$ **18.** $(3, 4), (6, 10)$ 2

19. $(-3, -1), (4, 3)$ $\dfrac{4}{7}$ **20.** $(-2, -3), (1, 3)$ 2

21. $(-2, 2), (-1, 7)$ 5 **22.** $(-3, 5), (1, -6)$ $-\dfrac{11}{4}$

23. $(3, -5), (0, 0)$ $-\dfrac{5}{3}$ **24.** $(0, 0), (-2, -1)$ $\dfrac{1}{2}$

25. $(0, 3), (5, 0)$ $-\dfrac{3}{5}$

26. $(3, 0), (0, 10)$ $-\dfrac{10}{3}$

27. $\left(\dfrac{3}{4}, -1\right), \left(-\dfrac{1}{2}, -\dfrac{1}{2}\right)$ $-\dfrac{2}{5}$

28. $\left(\dfrac{1}{2}, 2\right), \left(\dfrac{1}{4}, \dfrac{1}{2}\right)$ 6

29. $(6, 212), (7, 209)$ -3

30. $(1988, 306), (1990, 315)$ $\dfrac{9}{2}$

31. $(4, 7), (-12, 7)$ 0

32. $(5, -3), (9, -3)$ 0

33. $(2, 6), (2, -6)$ Undefined

34. $(-3, 2), (-3, 0)$ Undefined

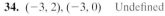 **35.** $(24.3, 11.9), (3.57, 8.4)$ 0.169

36. $(-2.7, 19.3), (5.46, -3.28)$ -2.767

37. $\left(\dfrac{\pi}{4}, 1\right), \left(\dfrac{\pi}{2}, 0\right)$ -1.273

38. $\left(\dfrac{\pi}{3}, -1\right), \left(\dfrac{\pi}{6}, 0\right)$ -1.910

In each case, make a sketch and find the slope of line l. See Examples 4 and 5.

39. Line *l* contains the point $(3, 4)$ and is perpendicular to the line through $(-5, 1)$ and $(3, -2)$. $\dfrac{8}{3}$

40. Line *l* goes through $(-3, -5)$ and is perpendicular to the line through $(-2, 6)$ and $(5, 3)$. $\dfrac{7}{3}$

41. Line *l* goes through $(2, 5)$ and is parallel to the line through $(-3, -2)$ and $(4, 1)$. $\dfrac{3}{7}$

42. Line *l* goes through the origin and is parallel to the line through $(-3, -5)$ and $(4, -1)$. $\dfrac{4}{7}$

43. Line *l* is perpendicular to a line with slope $\dfrac{4}{5}$. Both lines contain the origin. $-\dfrac{5}{4}$

44. Line *l* is perpendicular to a line with slope -5. Both lines contain the origin. $\dfrac{1}{5}$

Solve each geometric figure problem. See Example 6.

45. If the opposite sides of a quadrilateral are parallel, then it is a parallelogram. Use slope to determine whether the points $(-6, 1), (-2, -1), (0, 3),$ and $(4, 1)$ are the vertices of a parallelogram.   Yes

46. Use slope to determine whether the points $(-7, 0), (-1, 6), (-1, -2),$ and $(6, 5)$ are the vertices of a parallelogram. See Exercise 45.   No

47. A trapezoid is a quadrilateral with one pair of parallel sides. Use slope to determine whether the points $(-3, 2), (-1, -1), (3, 6),$ and $(6, 4)$ are the vertices of a trapezoid.   No

48. A parallelogram with at least one right angle is a rectangle. Determine whether the points $(-4, 4), (-1, -2), (0, 6),$ and $(3, 0)$ are the vertices of a rectangle.   Yes

49. If a triangle has one right angle, then it is a right triangle. Use slope to determine whether the points $(-3, 3), (-1, 6),$ and $(0, 0)$ are the vertices of a right triangle.   No

50. Use slope to determine whether the points $(0, -1), (2, 5),$ and $(5, 4)$ are the vertices of a right triangle. See Exercise 49.   Yes

Solve each problem. See Example 7.

51. ***Pricing the Crown Victoria.*** The list price of a new Ford Crown Victoria four-door sedan was \$20,115 in 1993 and \$21,135 in 1998 (Edmund's New Car Prices, www.edmunds.com).
  a) Find the slope of the line shown in the figure.   204
  b) Use the graph to predict the price in 2005.   \$22,500
  c) Use the slope to predict the price of a new Crown Victoria in 2005.   \$22,563

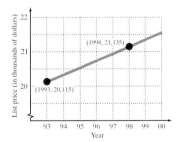

FIGURE FOR EXERCISE 51

52. ***Depreciating Monte Carlo.*** In 1998 the average retail price of a one-year-old Chevrolet Monte Carlo was \$13,595, whereas the average retail price of a 3-year-old Monte Carlo was \$11,095 (*Edmund's Used Car Prices*).

a) Use the graph on the next page to estimate the average retail price of a 2-year-old car in 1998. $12,000

b) Find the slope of the line shown in the figure.   −1250

c) Use the slope to predict the price of a 2-year-old car. $12,345

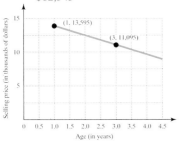

FIGURE FOR EXERCISE 52

MISCELLANEOUS

53. The points (3,) and (, −7) are on the line that passes through (2, 1) and has slope 4. Find the missing coordinates of the points.   (3, 5), (0, −7)

54. If a line passes through (5, 2) and has slope $\frac{2}{3}$, then what is the value of y on this line when $x = 8, x = 11$, and $x = 12$? $4, 6, 6\frac{2}{3}$

55. Find k so that the line through (2, k) and (−3, −5) has slope $\frac{1}{2}$. $-\frac{5}{2}$

56. Find k so that the line through (k, 3) and (−2, 0) has slope 3. −3 or 1

57. What is the slope of a line that is perpendicular to a line with slope 0.247? −4.049

58. What is the slope of a line that is perpendicular to the line through (3.27, −1.46) and (−5.48, 3.61)? 1.726

GETTING MORE INVOLVED

59. *Writing.* What is the difference between zero slope and undefined slope?

A horizontal line has a zero slope and a vertical line has undefined slope.

60. *Writing.* Is it possible for a line to be in only one quadrant? Two quadrants? Write a rule for determining whether a line has positive, negative, zero, or undefined slope from knowing in which quadrants the line is found.

Every line goes through at least two quadrants. A nonhorizontal, nonvertical line that misses quadrant II or IV or both has a positive slope. A nonhorizontal, nonvertical line that misses quadrant I or III or both has a negative slope.

61. *Exploration.* A rhombus is a quadrilateral with four equal sides. Draw a rhombus with vertices (−3, −1), (0, 3), (2, −1), and (5, 3). Find the slopes of the diagonals of the rhombus. What can you conclude about the diagonals of this rhombus?

$-2, \frac{1}{2}$, perpendicular

62. *Exploration.* Draw a square with vertices (−5, 3), (−3, −3), (1, 5), and (3, −1). Find the slopes of the diagonals of the square. What can you conclude about the diagonals of this square?

$2, -\frac{1}{2}$, perpendicular

GRAPHING CALCULATOR EXERCISES

63. Graph $y = 1x$, $y = 2x$, $y = 3x$, and $y = 4x$ together in the standard viewing window. These equations are all of the form $y = mx$. What effect does increasing m have on the graph of the equation? What are the slopes of these four lines?

Increasing m makes the graph increase faster. The slopes of these lines are 1, 2, 3, and 4.

64. Graph $y = -1x$, $y = -2x$, $y = -3x$, and $y = -4x$ together in the standard viewing window. These equations are all of the form $y = mx$. What effect does decreasing m have on the graph of the equation? What are the slopes of these four lines?

Decreasing m makes the graph decrease faster. The slopes of these lines are −1, −2, −3, and −4.

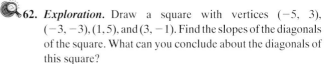

In this **section**

- Point-Slope Form
- Slope-Intercept Form
- Standard Form
- Using Slope-Intercept Form for Graphing
- Linear Functions

3.3   THREE FORMS FOR THE EQUATION OF A LINE

In Section 3.1 you learned how to graph a straight line corresponding to a linear equation. The line contains all of the points that satisfy the equation. In this section we start with a line or a description of a line and write an equation corresponding to the line.

Point-Slope Form

Figure 3.18 shows the line that has slope $\frac{2}{3}$ and contains the point (3, 5). In Section 3.2 you learned that the slope is the same no matter which two points of the line are used to calculate it. So if we find the slope m for this line using an arbitrary point of the line, say (x, y), and the specific point $(3, 5)$, we get

$$m = \frac{y - 5}{x - 3}.$$

Because the slope of this line is $\frac{2}{3}$, we can write

$$\frac{y - 5}{x - 3} = \frac{2}{3}.$$

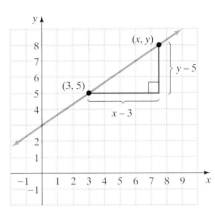

FIGURE 3.18

Multiplying each side by $x - 3$, we get

$$y - 5 = \frac{2}{3}(x - 3).$$

Because (x, y) was an arbitrary point on the line, this equation is satisfied by every point on the line.

If we use (x_1, y_1) as the specific point and (x, y) as an arbitrary point on a line with slope m, we can write

$$\frac{y - y_1}{x - x_1} = m.$$

Multiplying each side of this equation by $x - x_1$ gives us the **point-slope form** of the equation of the line.

> ### Point-Slope Form
>
> The equation of the line through (x_1, y_1) with slope m in point-slope form is
>
> $$y - y_1 = m(x - x_1).$$

E X A M P L E   1

Writing an equation for a line given a point and the slope

Find an equation for the line through $(-2, 5)$ with slope -3 and solve it for y.

Solution

Use $x_1 = -2$, $y_1 = 5$, and $m = -3$ in the point-slope form:

$$y - 5 = -3[x - (-2)]$$

Now solve the equation for y:

$$y - 5 = -3[x + 2]$$
$$y - 5 = -3x - 6$$
$$y = -3x - 1$$

■

If you know two points on a line, then you can graph the line (two points determine a line). In the next example we will see that two points of a line also determine an equation for the line.

calculator

close-up

Graph $y = -3x - 1$ and check that the line goes through $(-2, 5)$ by using the TRACE feature.

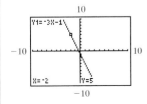

E X A M P L E   2

Writing an equation for a line given two points on the line

Find an equation for the line through $(3, -2)$ and $(-1, 1)$ and solve it for y.

Solution

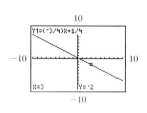

We are not given the slope, but we can find it because the points $(3, -2)$ and $(-1, 1)$ are on the line:

$$m = \frac{1 - (-2)}{-1 - 3}$$

$$= \frac{3}{-4} = -\frac{3}{4}$$

Now use this slope and one of the points, say $(3, -2)$, to write the equation in point-slope form:

$$y - (-2) = -\frac{3}{4}(x - 3) \quad \text{Point-slope form}$$

$$y + 2 = -\frac{3}{4}x + \frac{9}{4} \quad \text{Distributive property}$$

$$y = -\frac{3}{4}x + \frac{1}{4} \quad \text{Solve for } y: \frac{9}{4} - 2 = \frac{9}{4} - \frac{8}{4} = \frac{1}{4}.$$

Note that we would get the same equation if we had used slope $-\frac{3}{4}$ and the other point $(-1, 1)$. Try it.

For the next example, recall that if a line has slope m, then the slope of any line perpendicular to it is $-\frac{1}{m}$, provided that $m \neq 0$.

E X A M P L E   3

An equation of a line perpendicular to another line

Line l goes through $(2, 0)$ and is perpendicular to the line through $(5, -1)$ and $(-1, 3)$. Find the equation of line l and then solve it for y.

Solution

First find the slope of the line through $(5, -1)$ and $(-1, 3)$:

$$m = \frac{3 - (-1)}{-1 - 5} = \frac{4}{-6} = -\frac{2}{3}$$

Because line l is perpendicular to this line, line l has slope $\frac{3}{2}$. Now use $(2, 0)$ and the slope $\frac{3}{2}$ in the point-slope formula to get the equation of line l:

$$y - 0 = \frac{3}{2}(x - 2)$$

$$y = \frac{3}{2}x - 3 \quad \text{Distributive property}$$

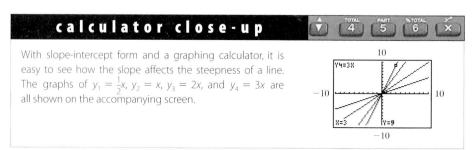

With slope-intercept form and a graphing calculator, it is easy to see how the slope affects the steepness of a line. The graphs of $y_1 = \frac{1}{2}x$, $y_2 = x$, $y_3 = 2x$, and $y_4 = 3x$ are all shown on the accompanying screen.

Slope-Intercept Form

The line $y = -3x - 1$ in Example 1 has slope -3. To find the y-intercept of this line, let $x = 0$ in $y = -3x - 1$: $y = -3(0) - 1 = -1$. The y-intercept is $(0, -1)$. Its y-coordinate appears in the equation:

$$y = -3x - 1$$

Slope y-intercept $(0, -1)$

Because the slope and y-intercept can be read from the equation when it is solved for y, this form of the equation of the line is called slope-intercept form.

> ### Slope-Intercept Form
>
> The equation of a line in **slope-intercept form** is
>
> $$y = mx + b,$$
>
> where m is the slope and $(0, b)$ is the y-intercept.

E X A M P L E   4

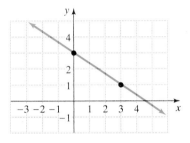

FIGURE 3.19

Writing an equation given its slope and y-intercept

Write the slope-intercept form of the equation of the line shown in Fig. 3.19.

Solution

From Fig. 3.19 we see that the y-intercept is $(0, 3)$. If we start at the y-intercept and move down 2 and 3 to the right, we get to another point on the line. So the slope is $-\frac{2}{3}$. The equation of this line in slope-intercept form is

$$y = -\frac{2}{3}x + 3.$$

Standard Form

If x students paid \$5 each and y adults paid \$7 each to attend a play for which the ticket sales totaled \$1900, then we can write the equation $5x + 7y = 1900$. This form of a linear equation is common in applications. It is called standard form.

> ### Standard Form
>
> The equation of a line in **standard form** is
>
> $$Ax + By = C,$$
>
> where A, B, and C are real numbers with A and B not both zero.

The numbers A, B, and C in standard form can be any real numbers, but it is a common practice to write standard form using only integers and a positive coefficient for x.

E X A M P L E   5

Changing to standard form

Write the equation $y = \frac{1}{2}x - \frac{3}{4}$ in standard form using only integers and a positive coefficient for x.

pause**Solution**

Use the properties of equality to get the equation in the form $Ax + By = C$:

$$y = \frac{1}{2}x - \frac{3}{4} \qquad \text{Original equation}$$

$$-\frac{1}{2}x + y = -\frac{3}{4} \qquad \text{Subtract } \frac{1}{2}x \text{ from each side.}$$

$$4\left(-\frac{1}{2}x + y\right) = 4\left(-\frac{3}{4}\right) \qquad \begin{array}{l}\text{Multiply each side by 4 to}\\ \text{get integral coefficients.}\end{array}$$

$$-2x + 4y = -3 \qquad \text{Distributive property}$$

$$2x - 4y = 3 \qquad \begin{array}{l}\text{Multiply by } -1 \text{ to make the}\\ \text{coefficient of } x \text{ positive.}\end{array}$$

To find the slope and y-intercept of a line written in standard form, we convert the equation to slope-intercept form.

helpful hint

Solve $Ax + By = C$ for y, to get

$$y = \frac{-A}{B}x + \frac{C}{B}.$$

So the slope of $Ax + By = C$ is $-\frac{A}{B}$. This fact can be used in checking standard form. The slope of $2x - 4y = 3$ in Example 5 is $\frac{-2}{-4}$ or $\frac{1}{2}$, which is the slope of the original equation.

E X A M P L E   6   **Changing to slope-intercept form**

Find the slope and y-intercept of the line $3x - 2y = 5$.

Solution

Solve for y to get slope-intercept form:

$$3x - 2y = 5 \qquad \text{Original equation}$$

$$-2y = -3x + 5 \qquad \text{Subtract } 3x \text{ from each side.}$$

$$y = \frac{3}{2}x - \frac{5}{2} \qquad \text{Divide each side by } -2.$$

The slope is $\frac{3}{2}$, and the y-intercept is $\left(0, -\frac{5}{2}\right)$.

helpful hint

Note that every term in a linear equation in two variables is either a constant or a multiple of a variable. That is why equations in one variable of the form $ax + b = 0$ were called linear equations in Chapter 2.

You learned in Section 3.1 that the graph of the equation $x = 4$ is a vertical line. Because slope is undefined for vertical lines, the equation of this line cannot be written in slope-intercept form or point-slope form. Only nonvertical lines can be written in those forms. However, a vertical line can be written in standard form. For example,

$$x = 4$$

can be written as

$$1 \cdot x + 0 \cdot y = 4.$$

Every line has an equation in standard form.

E X A M P L E   7   **Finding the equation of a line**

Write an equation in standard form with integral coefficients for the line l through $(2, 5)$ that is perpendicular to the line $2x + 3y = 1$.

Solution

First solve the equation $2x + 3y = 1$ for y to find its slope:

$$2x + 3y = 1$$

$$3y = -2x + 1$$

$$y = -\frac{2}{3}x + \frac{1}{3} \qquad \text{The slope is } -\frac{2}{3}.$$

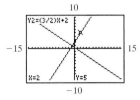

Graph $y_1 = \left(-\frac{2}{3}\right)x + \frac{1}{3}$ and $y_2 = \left(\frac{3}{2}\right)x + 2$ to check that y_2 is perpendicular to y_1 and that y_2 goes through (2, 5). The lines will look perpendicular only if the same unit length is used on both axes.

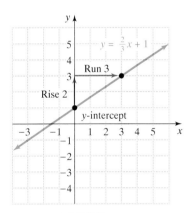

Some calculators have a feature that adjusts the window to get the same unit length on both axes.

The slope of line l is the opposite of the reciprocal of $-\frac{2}{3}$. So line l has slope $\frac{3}{2}$ and goes through (2, 5). Now use the point-slope form to write the equation:

$$y - 5 = \frac{3}{2}(x - 2) \quad \text{Point-slope form}$$

$$y - 5 = \frac{3}{2}x - 3 \quad \text{Distributive property}$$

$$y = \frac{3}{2}x + 2$$

$$-\frac{3}{2}x + y = 2$$

$$3x - 2y = -4 \quad \text{Multiply each side by } -2.$$

So $3x - 2y = -4$ is the standard form of the equation of the line through (2, 5) that is perpendicular to $2x + 3y = 1$.   ■

Using Slope-Intercept Form for Graphing

In the slope-intercept form, a point on the line (the y-intercept) and the slope are readily available. To graph a line, we can start at the y-intercept and count off the rise and run to get a second point on the line.

E X A M P L E   8   **Using slope and y-intercept to graph**
Graph the line $2x - 3y = -3$.

Solution
First write the equation in slope-intercept form:

$$2x - 3y = -3$$
$$-3y = -2x - 3$$
$$y = \frac{2}{3}x + 1$$

The slope is $\frac{2}{3}$, and the y-intercept is (0, 1). Start at (0, 1) on the y-axis, then rise 2 and run 3 to locate a second point on the line. Because there is only one line containing any two given points, these two points determine the line. See Fig. 3.20.   ■

The three methods that we used for graphing linear equations are summarized as follows.

FIGURE 3.20

Methods for Graphing a Linear Equation
1. Arbitrarily select some points that satisfy the equation, and draw a line through them.
2. Find the x- and y-intercepts (provided that they are not the origin), and draw a line through them.
3. Start at the y-intercept and use the slope to locate a second point, then draw a line through the two points.

If the y-coordinate of the y-intercept is an integer and the slope is a rational number, then it is usually the easiest to use the y-intercept and slope.

Linear Functions

The linear equation $y = mx + b$ with $m \neq 0$ is a formula that shows how to determine a value of y from a value of x. We say that y is a linear function of x. Functions in general will be discussed in Section 3.5. In the next example we use the point-slope formula to write Fahrenheit temperature as a linear function of Celsius temperature.

E X A M P L E   9   **Writing a linear function given two points**

Fahrenheit temperature F is a linear function of Celsius temperature C. Water freezes at 0°C or 32°F and boils at 100°C or 212°F. Find the linear equation that expresses F as a linear function of C.

Solution

We want the equation of the line that contains the points (0, 32) and (100, 212) as shown in Fig. 3.21. Use C as the independent variable (x) and F as the dependent variable (y). The slope of the line is

$$m = \frac{F_2 - F_1}{C_2 - C_1} = \frac{212 - 32}{100 - 0} = \frac{180}{100} = \frac{9}{5}.$$

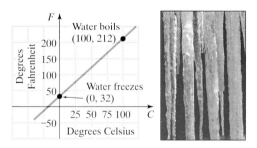

F I G U R E   3 . 2 1

Using a slope of $\frac{9}{5}$ and the point (100, 212) in the point-slope formula, we get

$$F - 212 = \frac{9}{5}(C - 100).$$

We can solve this equation for F to get the familiar formula relating Celsius and Fahrenheit temperature:

$$F = \frac{9}{5}C + 32$$

Because we knew the intercept (0, 32), we could have used it and the slope $\frac{9}{5}$ in slope-intercept form to write $F = \frac{9}{5}C + 32$.

True or false? Explain your answer.

1. There is exactly one line through a given point with a given slope.   True
2. The line $y - a = m(x - b)$ goes through (a, b) and has slope m.   False
3. The equation of the line through (a, b) with slope m is $y = mx + b$.   False

WARM-UPS

(*continued*)

4. The *x*-coordinate of the *y*-intercept of a nonvertical line is 0.   True

5. The *y*-coordinate of the *x*-intercept of a nonhorizontal line is 0.   True

6. Every line in the *xy*-plane has an equation in slope-intercept form.   False

7. The line $2y + 3x = 7$ has slope $-\frac{3}{2}$.   True

8. The line $y = 3x - 1$ is perpendicular to the line $y = \frac{1}{3}x - 1$.   False

9. The line $2y = 3x + 5$ has a *y*-intercept of $(0, 5)$.   False

10. Every line in the *xy*-plane has an equation in standard form.   True

3.3   EXERCISES

Reading and Writing   After reading this section, write out the answers to these questions. Use complete sentences.

1. What is point-slope form?
   Point-slope form is $y - y_1 = m(x - x_1)$, where *m* is the slope and (x_1, y_1) is a point on the line.

2. What is slope-intercept form?
   Slope-intercept form is $y = mx + b$, where *m* is the slope and $(0, b)$ is the *y*-intercept.

3. What two bits of information must you have to write the equation of a line from a description of the line?
   To write an equation of a line, we need the slope and a point on the line.

4. What is standard form?
   Standard form is $Ax + Bx = C$, where *A*, *B*, and *C* are real numbers with *A* and *B* not both zero.

5. How do you find the slope of a line when its equation is given in standard form?
   To find the slope from standard form, solve the equation for *y* to get the form $y = mx + b$, where *m* is the slope.

6. How do you graph a line when its equation is given in slope-intercept form?
   To graph a line knowing the slope and *y*-intercept, start at the *y*-intercept and count off the rise and run to locate a second point. Then draw a line through the *y*-intercept and your second point.

Find the equation of line l in each case and solve it for y. See Examples 1–3.

7. Line *l* goes through $(2, -3)$ and has slope 2.
   $y = 2x - 7$

8. Line *l* goes through $(-2, 5)$ and has slope 6.
   $y = 6x + 17$

9. Line *l* goes through $(-2, 3)$ and has slope $-\frac{1}{2}$.
   $y = -\frac{1}{2}x + 2$

10. Line *l* goes through $(3, 5)$ and has slope $\frac{2}{3}$.
    $y = \frac{2}{3}x + 3$

11. Line *l* goes through $(2, 3)$ and $(-5, 6)$.
    $y = -\frac{3}{7}x + \frac{27}{7}$

12. Line *l* goes through $(-2, 1)$ and $(3, -4)$.
    $y = -x - 1$

13. Line *l* goes through $(3, 4)$ and is perpendicular to the line through $(-3, 1)$ and $(5, -1)$.
    $y = 4x - 8$

14. Line *l* goes through $(0, 0)$ and is perpendicular to the line through $(0, 6)$ and $(-5, 0)$.
    $y = -\frac{5}{6}x$

15. Line *l* goes through $(0, 0)$ and is parallel to the line through $(9, -3)$ and $(-3, 6)$.
    $y = -\frac{3}{4}x$

16. Line *l* goes through $(2, -4)$ and is parallel to the line through $(6, 2)$ and $(-2, 6)$.
    $y = -\frac{1}{2}x - 3$

In Exercises 17–26, write an equation in slope-intercept form (if possible) for each of the lines shown. See Example 4.

17.

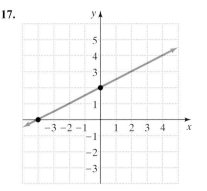

$y = \frac{1}{2}x + 2$

18.

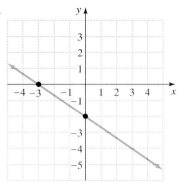

$$y = -\frac{2}{3}x - 2$$

19.

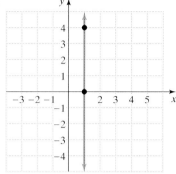

$$x = 1$$

20.

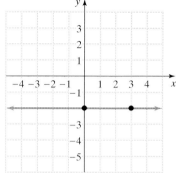

$$y = -2$$

21.

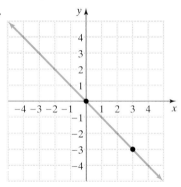

$$y = -x$$

22.

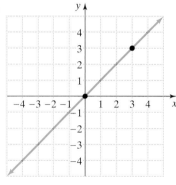

$$y = x$$

23.

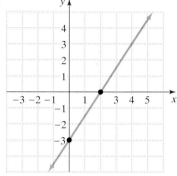

$$y = \frac{3}{2}x - 3$$

24.

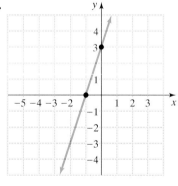

$$y = 3x + 3$$

25.

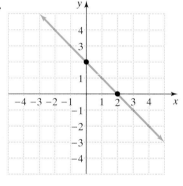

$$y = -x + 2$$

26.

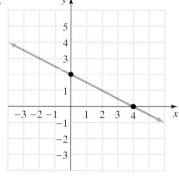

$$y = -\frac{1}{2}x + 2$$

Write each equation in standard form using only integers and a positive coefficient for x. See Example 5.

27. $y = \frac{1}{3}x - 2$ $x - 3y = 6$

28. $y = \frac{1}{2}x + 7$ $x - 2y = -14$

29. $y - 5 = \frac{1}{2}(x + 3)$ $x - 2y = -13$

30. $y - 1 = \frac{1}{4}(x - 6)$ $x - 4y = 2$

31. $y + \frac{1}{2} = \frac{1}{3}(x - 4)$ $2x - 6y = 11$

32. $y + \frac{1}{3} = \frac{1}{4}(x - 3)$ $3x - 12y = 13$

33. $0.05x + 0.06y - 8.9 = 0$ **34.** $0.03x - 0.07y = 2$
$\quad\ 5x + 6y = 890$ $\qquad\qquad\qquad 3x - 7y = 200$

Write each equation in slope-intercept form, and identify the slope and y-intercept. See Example 6.

35. $2x + 5y = 1$ $y = -\frac{2}{5}x + \frac{1}{5}, \ -\frac{2}{5}, \left(0, \frac{1}{5}\right)$

36. $3x - 3y = 2$ $y = x - \dfrac{2}{3}, 1, \left(0, -\dfrac{2}{3}\right)$

37. $3x - y - 2 = 0$ $y = 3x - 2, 3, (0, -2)$

38. $5 - x - 2y = 0$ $y = -\dfrac{1}{2}x + \dfrac{5}{2}, -\dfrac{1}{2}, \left(0, \dfrac{5}{2}\right)$

39. $y + 3 = 5$ $y = 2, 0, (0, 2)$

40. $y - 9 = 0$ $y = 9, 0, (0, 9)$

41. $y - 2 = 3(x - 1)$ $y = 3x - 1, 3, (0, -1)$

42. $y + 4 = -2(x - 5)$ $y = -2x + 6, -2, (0, 6)$

43. $\dfrac{y - 5}{x + 4} = \dfrac{3}{2}$ $y = \dfrac{3}{2}x + 11, \dfrac{3}{2}, (0, 11)$

44. $\dfrac{y - 6}{x - 2} = -\dfrac{3}{5}$ $y = -\dfrac{3}{5}x + \dfrac{36}{5}, -\dfrac{3}{5}, \left(0, \dfrac{36}{5}\right)$

45. $y - \dfrac{1}{2} = \dfrac{1}{3}\left(x + \dfrac{1}{4}\right)$ $y = \dfrac{1}{3}x + \dfrac{7}{12}, \dfrac{1}{3}, \left(0, \dfrac{7}{12}\right)$

46. $y - \dfrac{1}{3} = -\dfrac{1}{2}\left(x - \dfrac{1}{4}\right)$ $y = -\dfrac{1}{2}x + \dfrac{11}{24}, -\dfrac{1}{2}, \left(0, \dfrac{11}{24}\right)$

47. $y - 6000 = 0.01(x + 5700)$
$y = 0.01x + 6057, 0.01, (0, 6057)$

48. $y - 5000 = 0.05(x - 1990)$
$y = 0.05x + 4900.5, 0.05, (0, 4900.5)$

Find the equation of line l in each case and then write it in standard form with integral coefficients. See Example 7.

49. Line l has slope $\dfrac{1}{2}$ and goes through $(0, 5)$.
$x - 2y = -10$

50. Line l has slope 5 and goes through $\left(0, \dfrac{1}{2}\right)$.
$10x - 2y = -1$

51. Line l has x-intercept $(2, 0)$ and y-intercept $(0, 4)$.
$2x + y = 4$

52. Line l has y-intercept $(0, 5)$ and x-intercept $(4, 0)$.
$5x + 4y = 20$

53. Line l goes through $(-2, 1)$ and is parallel to $y = 2x + 6$.
$2x - y = -5$

54. Line l goes through $(1, -3)$ and is parallel to $y = -3x - 5$.
$3x + y = 0$

55. Line l is parallel to $2x + 4y = 1$ and goes through $(-3, 5)$.
$x + 2y = 7$

56. Line l is parallel to $3x - 5y = -7$ and goes through $(2, 4)$.
$3x - 5y = -14$

57. Line l goes through $(1, 1)$ and is perpendicular to $y = \dfrac{1}{2}x - 3$. $2x + y = 3$

58. Line l goes through $(-1, -2)$ and is perpendicular to $y = -3x + 7$. $x - 3y = 5$

59. Line l goes through $(-2, 3)$ and is perpendicular to $x + 3y = 4$. $3x - y = -9$

60. Line l is perpendicular to $2y + 5 - 3x = 0$ and goes through $(2, 7)$. $2x + 3y = 25$

61. Line l goes through $(2, 5)$ and is parallel to the x-axis.
$y = 5$

62. Line l goes through $(-1, 6)$ and is parallel to the y-axis.
$x = -1$

Graph each line. Use the slope and y-intercept when possible. See Example 8.

63. $y = \dfrac{1}{2}x$

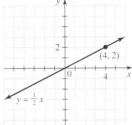

64. $y = -\dfrac{2}{3}x$

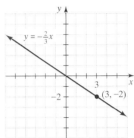

65. $y = 2x - 3$

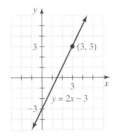

66. $y = -x + 1$

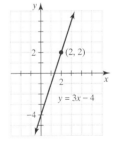

67. $y = -\dfrac{2}{3}x + 2$

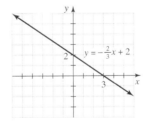

68. $y = 3x - 4$

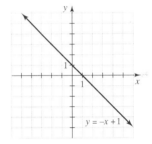

69. $3y + x = 0$

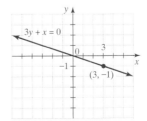

70. $4y - x = 0$

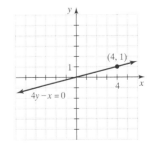

71. $y - x + 3 = 0$

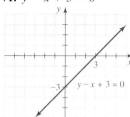

72. $x - y = 4$

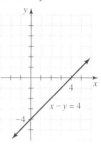

73. $3x - 2y = 6$

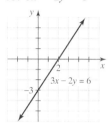

74. $3x + 5y = 10$

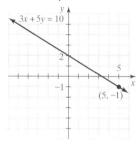

75. $y + 2 = 0$

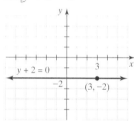

76. $y - 3 = 0$

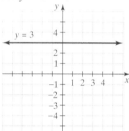

77. $x + 3 = 0$

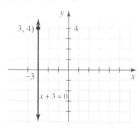

78. $x - 5 = 0$

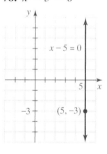

Determine whether each pair of lines is parallel or perpendicular.

79. $y = 3x - 8$, $x + 3y = 7$ Perpendicular

80. $y = \dfrac{1}{2}x - 4$, $\dfrac{1}{2}x + \dfrac{1}{4}y = 1$ Perpendicular

81. $2x - 4y = 9$, $\dfrac{1}{3}x = \dfrac{2}{3}y - 8$ Parallel

82. $\dfrac{1}{4}x - \dfrac{1}{6}y = \dfrac{1}{3}$, $\dfrac{1}{3}y = \dfrac{1}{2}x - 2$ Parallel

83. $x - 6 = 9$, $y - 4 = 12$ Perpendicular

84. $9 - x = 3$, $\dfrac{1}{2}x = 8$ Parallel

Solve each problem. See Example 9.

85. *Heating water.* Suppose the temperature, t, of a cup of water is a linear function of the number of seconds, s, that it is in the microwave. If the temperature at $s = 0$ second is $t = 60°F$ and the temperature at $s = 120$ seconds is $200°F$, find the linear equation that expresses t as a function of s. What should the temperature be after 30 seconds? (*Hint:* Write the equation of the line containing the points $(0, 60)$ and $(120, 200)$ in the form $t = ms + b$.) Draw a graph of this linear function. $t = \dfrac{7}{6}s + 60, 95°F$

86. *Making circuit boards.* The accountant at Apollo Manufacturing has determined that the cost, C, per week in dollars for making circuit boards is a linear function of the number, n, of circuit boards produced in a week. If $C = \$1500$ when $n = 1000$, and $C = \$2000$ when $n = 2000$, find the linear equation that expresses C in terms of n. What is the cost if Apollo produces only one circuit board in a week? Draw a graph of this linear function. $C = \dfrac{1}{2}n + 1000, \$1000.50$

87. *Carbon dioxide emission.* Worldwide emission of carbon dioxide (CO_2) increased from 14 billion tons in 1970 to 24 billion tons in 1995 (World Resources Institute, www.wri.org).
a) Find the equation of the line through $(1970, 14)$ and $(1995, 24)$. $y = 0.4x - 774$
b) Use the equation to predict the worldwide emission of CO_2 in 2005.   28 billion tons

88. *World energy use.* Worldwide energy use in all forms increased from the equivalent of 3.5 billion tons of oil in 1970 to the equivalent of 6 billion tons of oil in 1995 (World Resources Institute, www.wri.org).
a) Find the equation of the line through $(70, 3.5)$ and $(95, 6)$. $y = 0.1x - 3.5$
b) Use the equation to predict the worldwide energy use in 2005.   7 billion tons

89. *Depth and flow.* On May 1, 1998 the depth of the water in the Tangipahoa River at Robert, Louisiana was 8.24 feet and the flow was 1015.5 cubic feet per second (ft^3/sec). On May 8 the depth was 7.26 feet and the flow was 717.1 cubic feet per second (*U.S. Geological Survey, Water Resources Data for Louisiana, 1998*). The flow w is a linear function of the depth d.

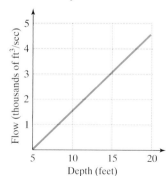

FIGURE FOR EXERCISE 89

a) Write the equation of the line through (8.24, 1015.5) and (7.26, 717.1) and express *w* as a linear function of *d*.

b) What is the flow when the depth is 7.81 feet?

c) Is the flow increasing or decreasing as the depth increases?

    a) $w = 304.5d - 1493.5$ b) $884.6\ \text{ft}^3/\text{sec}$

    c) increasing

90. *Buying stock.* On July 2, 1998 a mutual fund manager spent \$5,031,250 on *x* shares of Ford Motor Stock at \$58.25 per share and *y* shares of General Motors stock at \$47.50 per share.

a) Write a linear equation that models this situation.

b) If 35,000 shares of Ford were purchased, then how many shares of GM were purchased?

c) What are the intercepts of the graph of the linear equation? Interpret the intercepts.

d) As the number of shares of Ford increases, does the number of shares of GM increase or decrease?

    a) $58.25x + 47.50y = 5,031,250$ b) 63,000

    c) (0, 105,921.1), (86,373.4, 0), The intercepts give the number of shares if all of the money was spent on only one type of stock.   d) decrease

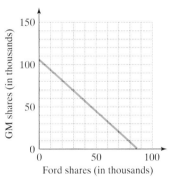

FIGURE FOR EXERCISE 90

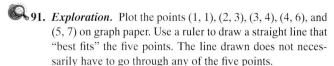

GETTING MORE INVOLVED

91. *Exploration.* Plot the points (1, 1), (2, 3), (3, 4), (4, 6), and (5, 7) on graph paper. Use a ruler to draw a straight line that "best fits" the five points. The line drawn does not necessarily have to go through any of the five points.

a) Estimate the slope and *y*-intercept for the line drawn and write an equation for the line in slope-intercept form.

b) For each *x*-coordinate from 1 through 5, find the difference between the given *y*-coordinate and the *y*-coordinate on your line.

c) To determine how well you have done, square each difference that you found in part (b) and then find the sum of those squares. Compare your sum with your classmates' sums. The person with the smallest sum has done the best job of fitting a line to the five given points.

GRAPHING CALCULATOR EXERCISES

92. Graph the equation $y = 0.5x - 1$ using the standard viewing window. Adjust the range of *y*-values so that the line goes from the lower left corner of your viewing window to the upper right corner.

93. Graph $y = x - 3000$, using a viewing window that shows both the *x*-intercept and the *y*-intercept.

94. Graph $y = 2x - 400$ and $y = -0.5x + 1$ on the same screen, using the viewing window $-500 \le x \le 500$ and $-1000 \le y \le 1000$. Should these lines be perpendicular? Explain.

The lines are perpendicular and will appear so in a window in which the length of one unit on the *x*-axis is equal to the length of one unit on the *y*-axis.

95. The lines $y = 2x - 3$ and $y = 1.9x + 2$ are not parallel. Find a viewing window in which the lines intersect. Estimate the point of intersection.

The lines intersect at (50, 97).

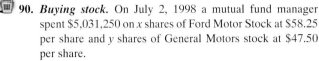

3.4

LINEAR INEQUALITIES AND THEIR GRAPHS

In the first three sections of this chapter you studied linear equations. We now turn our attention to linear inequalities.

In this section

- Definition
- Graphing Linear Inequalities
- The Test Point Method
- Graphing Compound Inequalities
- Applications

Definition

A linear inequality is a linear equation with the equal sign replaced by an inequality symbol.

> ### Linear Inequality
>
> If *A*, *B*, and *C* are real numbers with *A* and *B* not both zero, then
> $$Ax + By \le C$$
> is called a **linear inequality.** In place of $\le$, we can also use $\ge$, $<$, or $>$.

Graphing Linear Inequalities

Consider the inequality $-x + y > 1$. If we solve the inequality for y, we get

$$y > x + 1.$$

Which points in the xy-plane satisfy this inequality? We want the points where the y-coordinate is larger than the x-coordinate plus 1. If we locate a point on the line $y = x + 1$, say $(2, 3)$, then the y-coordinate is equal to the x-coordinate plus 1. If we move upward from that point, to say $(2, 4)$, the y-coordinate is larger than the x-coordinate plus 1. Because this argument can be made at every point on the line, all points above the line satisfy $y > x + 1$. Likewise, points below the line satisfy $y < x + 1$. The solution sets, or graphs, for the inequality $y > x + 1$ and the inequality $y < x + 1$ are the shaded regions shown in Figs. 3.22(a) and 3.22(b). In each case the line $y = x + 1$ is dashed to indicate that points on the line do not satisfy the inequality and so are not in the solution set. If the inequality symbol is $\leq$ or $\geq$, then points on the boundary line also satisfy the inequality, and the line is drawn solid.

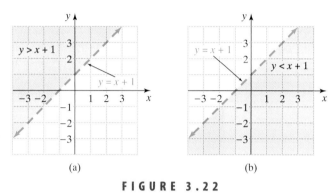

(a)   (b)

FIGURE 3.22

Every nonvertical line divides the xy-plane into two regions. One region is above the line, and the other is below the line. A vertical line also divides the plane into two regions, but one is on the left side of the line and the other is on the right side of the line. An inequality involving only x has a vertical boundary line, and its graph is one of those regions.

Graphing a Linear Inequality

1. Solve the inequality for y, then graph $y = mx + b$.

$y > mx + b$ is satisfied above the line.

$y = mx + b$ is satisfied on the line itself.

$y < mx + b$ is satisfied below the line.

2. If the inequality involves x and not y, then graph the vertical line $x = k$.

$x > k$ is satisfied to the right of the line.

$x = k$ is satisfied on the line itself.

$x < k$ is satisfied to the left of the line.

E X A M P L E   1   **Graphing linear inequalities**

Graph each inequality.

a) $y < \dfrac{1}{2}x - 1$ **b)** $y \geq -2x + 1$ **c)** $3x - 2y < 6$

Solution

a) The set of points satisfying this inequality is the region below the line $y = \frac{1}{2}x - 1$. To show this region, we first graph the boundary line $y = \frac{1}{2}x - 1$. The slope of the line is $\frac{1}{2}$, and the y-intercept is $(0, -1)$. Start at $(0, -1)$ on the y-axis, then rise 1 and run 2 to get a second point of the line. We draw the line dashed because points on the line do not satisfy this inequality. The solution set to the inequality is the shaded region shown in Fig. 3.23.

b) Because the inequality symbol is $\geq$, every point on or above the line satisfies this inequality. To show that the line $y = -2x + 1$ is included, we make it a solid line. See Fig. 3.24.

c) First solve for y:

$$3x - 2y < 6$$
$$-2y < -3x + 6$$
$$y > \frac{3}{2}x - 3 \quad \text{Divide by } -2 \text{ and reverse the inequality.}$$

To graph this inequality, use a dashed line for the boundary $y = \frac{3}{2}x - 3$ and shade the region above the line. See Fig. 3.25 for the graph.

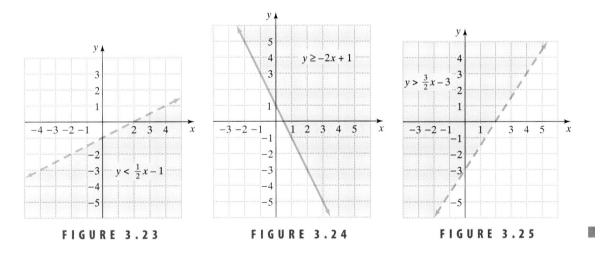

FIGURE 3.23   **FIGURE 3.24**   **FIGURE 3.25**

CAUTION   In Example 1(c) we solved the inequality for y before graphing the line. We did that because $<$ corresponds to the region below the line and $>$ corresponds to the region above the line only when the inequality is solved for y.

E X A M P L E   2

Inequalities with horizontal and vertical boundaries

Graph the inequalities.

a) $y \leq 5$ **b)** $x > 4$

Solution

a) The line $y = 5$ is the horizontal line with y-intercept $(0, 5)$. Draw a solid horizontal line and shade below it as in Fig. 3.26 on the next page.

b) The points that satisfy $x > 4$ lie to the right of the vertical line $x = 4$. The solution set is shown in Fig. 3.27 on the next page.

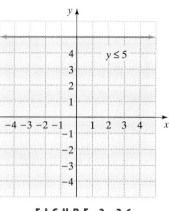

FIGURE 3.26

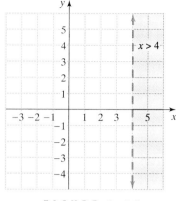

FIGURE 3.27

The Test Point Method

The graph of any line $Ax + By = C$ separates the xy-plane into two regions. Every point on one side of the line satisfies the inequality $Ax + By < C$, and every point on the other side satisfies the inequality $Ax + By > C$. We can use these facts to graph an inequality by the **test point method:**

1. Graph the corresponding equation.

2. Choose any point *not* on the line.

3. Test to see whether the point satisfies the inequality.

If the point satisfies the inequality, then the solution set is the region containing the test point. If not, then the solution set is the other region. With this method, it is not necessary to solve the inequality for y.

E X A M P L E   3

Using the test point method

Graph the inequality $3x - 4y > 7$.

Solution

First graph the equation $3x - 4y = 7$ using the x-intercept and the y-intercept. If $x = 0$, then $y = -\frac{7}{4}$. If $y = 0$, then $x = \frac{7}{3}$. Use the x-intercept $\left(\frac{7}{3}, 0\right)$ and the y-intercept $\left(0, -\frac{7}{4}\right)$ to graph the line as shown in Fig. 3.28(a). Select a point on one

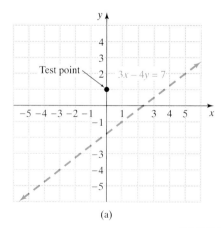

(a)

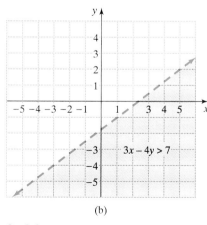

(b)

FIGURE 3.28

side of the line, say $(0, 1)$, to test in the inequality. Because

$$3(0) - 4(1) > 7$$

is false, the region on the other side of the line satisfies the inequality. The graph of $3x - 4y > 7$ is shown in Fig. 3.28(b). ■

Graphing Compound Inequalities

We can write compound inequalities with two variables just as we do for one variable. For example,

$$y > x - 3 \qquad \text{and} \qquad y < -\frac{1}{2}x + 2$$

is a compound inequality. Because the inequalities are connected by the word *and,* a point is in the solution set to the compound inequality if and only if it is in the solution sets to *both* of the individual inequalities. So the graph of this compound inequality is the intersection of the solution sets to the individual inequalities.

E X A M P L E   4

Graphing a compound inequality with *and*

Graph the compound inequality $y > x - 3$ and $y < -\frac{1}{2}x + 2$.

Solution

We first graph the equations $y = x - 3$ and $y = -\frac{1}{2}x + 2$. These lines divide the plane into four regions as shown in Fig. 3.29(a). Now test one point of each region to determine which region satisfies the compound inequality. Test the points $(3, 3)$, $(0, 0)$, $(4, -5)$, and $(5, 0)$:

$$3 > 3 - 3 \qquad \text{and} \qquad 3 < -\frac{1}{2} \cdot 3 + 2 \qquad \text{Second inequality is incorrect.}$$

$$0 > 0 - 3 \qquad \text{and} \qquad 0 < -\frac{1}{2} \cdot 0 + 2 \qquad \text{Both inequalities are correct.}$$

$$-5 > 4 - 3 \qquad \text{and} \qquad -5 < -\frac{1}{2} \cdot 4 + 2 \qquad \text{First inequality is incorrect.}$$

$$0 > 5 - 3 \qquad \text{and} \qquad 5 < -\frac{1}{2} \cdot 0 + 2 \qquad \text{Both inequalities are incorrect.}$$

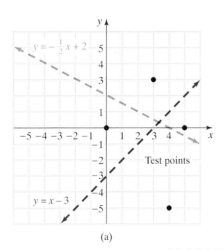

(a)

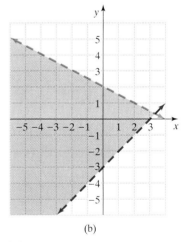

(b)

FIGURE 3.29

The only point that satisfies both inequalities is $(0, 0)$. So the solution set to the compound inequality consists of all points in the region containing $(0, 0)$. The graph of the compound inequality is shown in Fig. 3.29(b). ■

Compound inequalities are also formed by connecting individual inequalities with the word *or*. A point satisfies a compound inequality connected by *or* if and only if it satisfies one or the other or both of the individual inequalities. The graph is the union of the graphs of the individual inequalities.

E X A M P L E   5

Graphing a compound inequality with *or*

Graph the compound inequality

$$2x - 3y \le -6 \qquad \text{or} \qquad x + 2y \ge 4.$$

Solution

First graph the lines $2x - 3y = -6$ and $x + 2y = 4$. If we graph the lines using x- and y-intercepts, then we do not have to solve the equations for y. The lines are shown in Fig. 3.30(a). The graph of the compound inequality is the set of all points that satisfy either one inequality or the other (or both). Test the points $(0, 0)$, $(3, 2)$, $(0, 5)$, and $(-3, 2)$. You should verify that only $(0, 0)$ fails to satisfy at least one of the inequalities. So only the region containing the origin is left unshaded. The graph of the compound inequality is shown in Fig. 3.30(b).

> **helpful hint**
>
> When graphing a compound inequality connected with "or," shade the region that satisfies the first inequality and then shade the region that satisfies the second inequality. If the inequalities are connected with "and," then you must be careful not to shade too much.

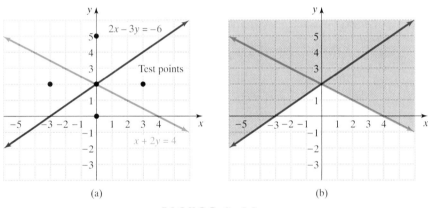

FIGURE 3.30

In the next example we graph absolute value inequalities by writing equivalent compound inequalities.

E X A M P L E   6

Graphing absolute value inequalities

Graph each absolute value inequality.

a) $|y - 2x| \le 3$ **b)** $|x - y| > 1$

Solution

a) The inequality $|y - 2x| \le 3$ is equivalent to $-3 \le y - 2x \le 3$, which is equivalent to the compound inequality

$$y - 2x \le 3 \qquad \text{and} \qquad y - 2x \ge -3.$$

First graph the lines $y - 2x = 3$ and $y - 2x = -3$ as shown in Fig. 3.31(a) on the next page. These lines divide the plane into three regions. Test a point from each region in the original inequality, say $(-5, 0)$, $(0, 1)$, and $(5, 0)$:

$$|0 - 2(-5)| \le 3 \qquad |1 - 2 \cdot 0| \le 3 \qquad |0 - 2 \cdot 5| \le 3$$

$$10 \le 3 \qquad\qquad 1 \le 3 \qquad\qquad 10 \le 3$$

helpful hint

Remember that absolute value of a quantity is its distance from 0 (Section 2.6). If $|w| < 3$, then w is less than 3 units from 0:

$$-3 < w < 3$$

If $|w| > 1$, then w is more than 1 unit away from 0:

$$w > 1 \quad \text{or} \quad w < -1$$

In Example 6 we are using an expression in place of w.

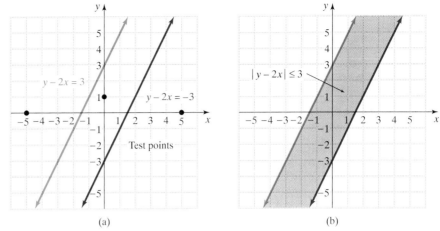

FIGURE 3.31

Only $(0, 1)$ satisfies the original inequality. So the region satisfying the absolute value inequality is the shaded region containing $(0, 1)$ as shown in Fig. 3.31(b). The boundary lines are solid because of the $\leq$ symbol.

b) The inequality $|x - y| > 1$ is equivalent to

$$x - y > 1 \quad \text{or} \quad x - y < -1.$$

First graph the lines $x - y = 1$ and $x - y = -1$ as shown in Fig. 3.32(a). Test a point from each region in the original inequality, say $(-4, 0)$, $(0, 0)$, and $(4, 0)$:

$$|-4 - 0| > 1 \qquad |0 - 0| > 1 \qquad |4 - 0| > 1$$
$$4 > 1 \qquad\qquad 0 > 1 \qquad\qquad 4 > 1$$

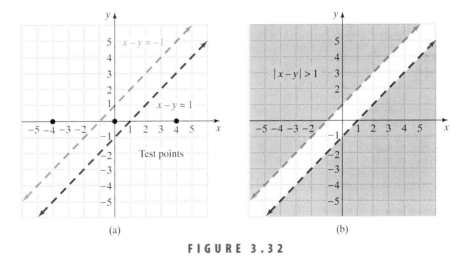

FIGURE 3.32

Because $(-4, 0)$ and $(4, 0)$ satisfy the inequality, we shade those regions as shown in Fig. 3.32(b). The boundary lines are dashed because of the $>$ symbol.

Applications

In real situations x and y often represent quantities or amounts, which cannot be negative. In this case our graphs are restricted to the first quadrant, where x and y are both nonnegative.

EXAMPLE 7   **Inequalities in business**

The manager of a furniture store can spend a maximum of $3000 on advertising per week. It costs $50 to run a 30-second ad on an AM radio station and $75 to run the ad on an FM station. Graph the region that shows the possible numbers of AM and FM ads that can be purchased and identify some possibilities.

Solution

If x represents the number of AM ads and y represents the number of FM ads, then x and y must satisfy the inequality $50x + 75y \leq 3000$. Because the number of ads cannot be negative, we also have $x \geq 0$ and $y \geq 0$. So we graph only points in the

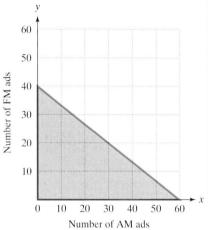

FIGURE 3.33

M A T H   A T   W O R K $x^2 + (x+1)^2 = 5^2$

"We will return after these messages." We often hear these words on television and radio just before several minutes of commercials. Carolanne Johnson, Account Executive and Media Salesperson for WBOQ, a classical radio station, is involved in every step of creating such advertisements.

MEDIA
SALESPERSON

The first step is finding clients that are consistent with the station's image. Ms. Johnson generates her own leads from a number of sources, such as print ads and billboards. The next steps are sitting down with the client, gathering information about the product or service, assessing the competition, and finally determining how much of the client's advertising budget should be spent on radio. Typically, this can be 2% to 4% of the total budget.

Radio ads usually run for 60 seconds, but reminder ads can be as short as 30 seconds. Some of the radio spots are time-sensitive and run 40 to 60 times a month for a specific month. Other clients are concerned with image building and may sponsor one particular broadcast every day for the whole year.

Ms. Johnson is concerned that the clients receive an adequate return on their investment. She is constantly reviewing the budget and making sure that the commercials present what the client wishes to project.

Example 7 and Exercise 73 of this section give problems that involve allocation of advertising dollars.

first quadrant that satisfy $50x + 75y \leq 3000$. The line $50x + 75y = 3000$ goes through (0, 40) and (60, 0). The inequality is satisfied below this line. The region showing the possible numbers of AM ads and FM ads is shown in Fig. 3.33. We shade the entire region in Fig. 3.33, but only points in the shaded region in which both coordinates are whole numbers actually satisfy the given condition. For example, 40 AM ads and 10 FM ads could be purchased. Other possibilities are 30 AM ads and 20 FM ads, or 10 AM ads and 10 FM ads.

WARM-UPS

True or false? Explain your answer.

1. The point (2, −3) satisfies the inequality $y > -3x + 2$.   True
2. The graph of $3x - y > 2$ is the region above the line $3x - y = 2$.   False
3. The graph of $3x + y < 5$ is the region below the line $y = -3x + 5$.   True
4. The graph of $x < -3$ is the region to the left of the vertical line $x = 3$.   False
5. The graph of $y > x + 3$ and $y < 2x - 6$ is the intersection of two regions.   True
6. The graph of $y \leq 2x - 3$ or $y \geq 3x + 5$ is the union of two regions.   True
7. The ordered pair (2, −5) satisfies $y > -3x + 5$ and $y < 2x - 3$.   False
8. The ordered pair (−3, 2) satisfies $y \leq 3x - 6$ or $y \leq x + 5$.   True
9. The inequality $|2x - y| \leq 4$ is equivalent to $2x - y \leq 4$ and $2x + y \leq 4$.   False
10. The inequality $|x - y| > 3$ is equivalent to $x - y > 3$ or $x - y < -3$.   True

3.4   EXERCISES

Reading and Writing   *After reading this section, write out the answers to these questions. Use complete sentences.*

1. What is a linear inequality?
   A linear inequality is an inequality of the form $Ax + By \leq C$ (or using $<$, $>$, or $\geq$), where A, B, and C are real numbers and A and B are not both zero.

2. How do we usually illustrate the solution set to a linear inequality in two variables.
   The solution set to a linear inequality in two variables is usually illustrated with a graph.

3. How do you know whether the line should be solid or dashed when graphing a linear inequality?
   If the inequality includes equality, then the line should be solid.

4. How do you know which side of the line to shade when graphing a linear inequality?
   We shade the side on which the inequality is satisfied.

5. What is the test point method used for?
   The test point method is used to determine which side of the boundary line to shade.

6. How do you graph a compound inequality?
   To graph a compound inequality, we find either the union or intersection of the regions determined by each simple inequality.

Graph each linear inequality. See Examples 1 and 2.

7. $y < x + 2$

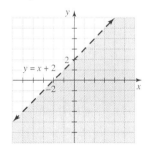

8. $y < x - 1$

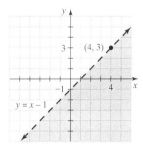

9. $y \leq -2x + 1$

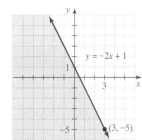

10. $y \geq -3x + 4$

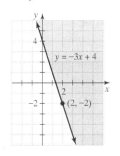

11. $x + y > 3$

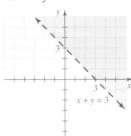

12. $x + y \leq -1$

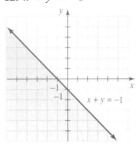

13. $2x + 3y < 9$

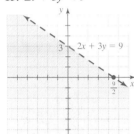

14. $-3x + 2y > 6$

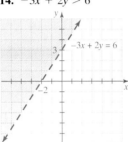

15. $3x - 4y \leq 8$

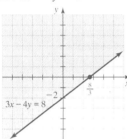

16. $4x - 5y > 10$

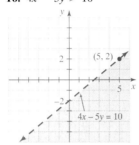

17. $x - y > 0$

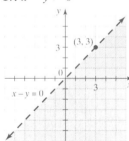

18. $2x - y < 0$

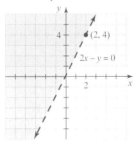

19. $x \geq 1$

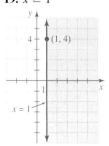

20. $x < 0$

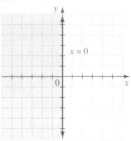

21. $y < 3$

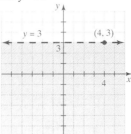

22. $y > -1$

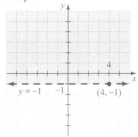

Graph each linear inequality by using a test point. See Example 3.

23. $2x - 3y < 5$

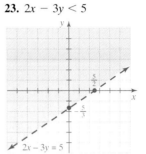

24. $5x - 4y > 3$

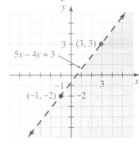

25. $x + y + 3 \geq 0$

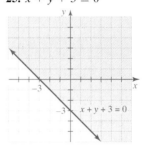

26. $x - y - 6 \leq 0$

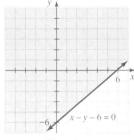

27. $\dfrac{1}{2}x + \dfrac{1}{3}y < 1$

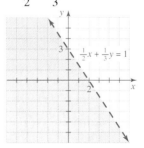

28. $2 - \dfrac{2}{5}y > \dfrac{1}{2}x$

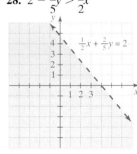

Graph each compound inequality. See Examples 4 and 5.

29. $y > x$ and $y > -2x + 3$

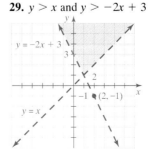

30. $y < x$ and $y < -3x + 2$

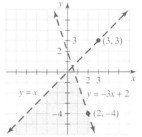

31. $y < x + 3$ or
$\quad\ \ y > -x + 2$

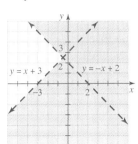

32. $y \geq x - 5$ or
$\quad\ \ y \leq -2x + 1$

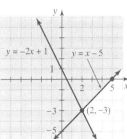

39. $y \geq x$ and $x \leq 2$

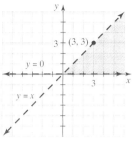

40. $y < x$ and $y > 0$

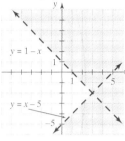

33. $x + y \leq 5$ and
$\quad\ \ x - y \leq 3$

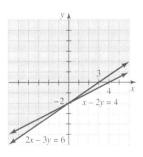

34. $2x - y < 3$ and
$\quad\ \ 3x - y > 0$

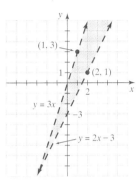

41. $2x < y + 3$ or
$\quad\ \ y > 2 - x$

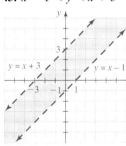

42. $3 - x < y + 2$ or
$\quad\ \ x > y + 5$

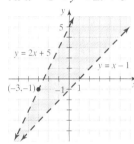

43. $x - 1 < y < x + 3$

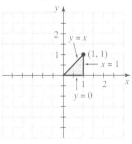

44. $x - 1 < y < 2x + 5$

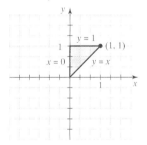

35. $x - 2y \leq 4$ or
$\quad\ \ 2x - 3y \leq 6$

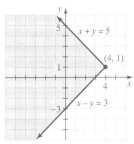

36. $4x - 3y \leq 3$ or
$\quad\ \ 2x + y \geq 2$

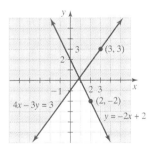

45. $0 \leq y \leq x$ and $x \leq 1$

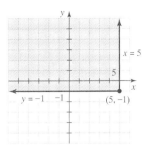

46. $x \leq y \leq 1$ and $x \geq 0$

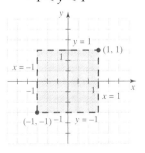

37. $y > 2$ and $x < 3$

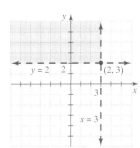

38. $x \leq 5$ and $y \geq -1$

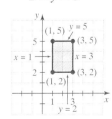

47. $1 \leq x \leq 3$ and
$\quad\ \ 2 \leq y \leq 5$

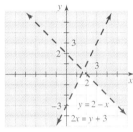

48. $-1 < x < 1$ and
$\quad\ \ -1 < y < 1$

Graph the absolute value inequalities. See Example 6.

49. $|x + y| < 2$

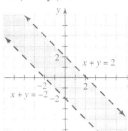

50. $|2x + y| < 1$

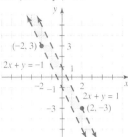

51. $|2x + y| \geq 1$

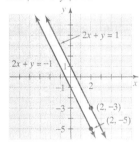

52. $|x + 2y| \geq 6$

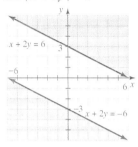

53. $|x - y - 3| > 5$

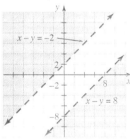

54. $|x - 2y + 4| > 2$

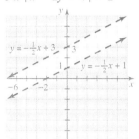

55. $|x - 2y| \leq 4$

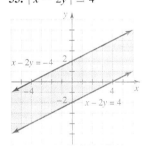

56. $|x - 3y| \leq 6$

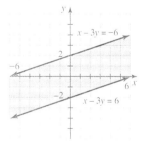

57. $|x| > 2$

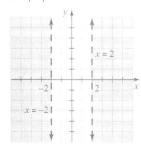

58. $|x| \leq 3$

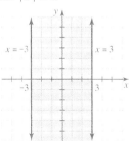

59. $|y| < 1$

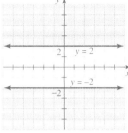

60. $|y| \geq 2$

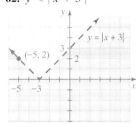

61. $y < |x|$

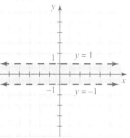

62. $y < |x + 3|$

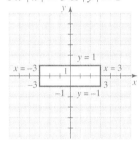

63. $|x| < 2$ and $|y| < 3$

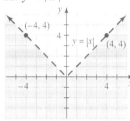

64. $|x| \geq 3$ or $|y| \geq 1$

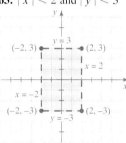

65. $|x - 3| < 1$ and $|y - 2| < 1$

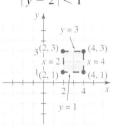

66. $|x - 2| \geq 3$ or $|y - 5| \geq 2$

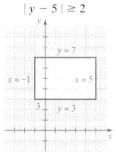

Solve each problem. See Example 7.

67. Budget planning. The Highway Patrol can spend a maximum of \$120,000 on new vehicles this year. They can get a fully equipped compact car for \$15,000 or a fully equipped full-size car for \$20,000. Graph the region that shows the number of cars of each type that could be purchased.

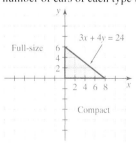

68. *Allocating resources.* A furniture maker has a shop that can employ 12 workers for 40 hours per week at its maximum capacity. The shop makes tables and chairs. It takes 16 hours of labor to make a table and 8 hours of labor to make a chair. Graph the region that shows the possibilities for the number of tables and chairs that could be made in one week.

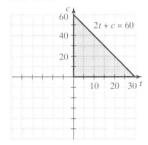

69. *More restrictions.* In Exercise 67, add the condition that the number of full-size cars must be greater than or equal to the number of compact cars. Graph the region showing the possibilities for the number of cars of each type that could be purchased.

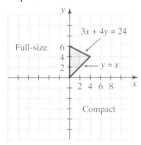

70. *Chairs per table.* In Exercise 68, add the condition that the number of chairs must be at least four times the number of tables and at most six times the number of tables. Graph the region showing the possibilities for the number of tables and chairs that could be made in one week.

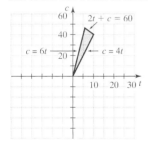

71. *Building fitness.* To achieve cardiovascular fitness, you should exercise so that your target heart rate is between 70% and 85% of its maximum rate. Your target heart rate h depends on your age a. For building fitness, you should have $h \le 187 - 0.85a$ and $h \ge 154 - 0.70a$ (NordicTrack brochure). Graph this compound inequality

for $20 \le a \le 75$ to see the heart rate target zone for building fitness.

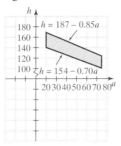

72. *Waist-to-hip ratio.* A study by Dr. Aaron R. Folsom concluded that waist-to-hip ratios are a better predictor of 5-year survival than more traditional height-to-weight ratios. Dr. Folsom concluded that for good health the waist size of a woman aged 50 to 69 should be less than or equal to 80% of her hip size, $w \le 0.80h$. Make a graph showing possible waist and hip sizes for good health for women in this age group for which hip size is no more than 50 inches.

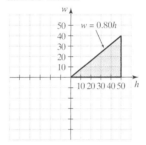

73. *Advertising dollars.* A restaurant manager can spend at most $9000 on advertising per month and has two choices for advertising. The manager can purchase an ad in the *Daily Chronicle* (a 7-day-per-week newspaper) for $300 per day or a 30-second ad on WBTU television for $1000 each time the ad is aired. Graph the region that shows the possible number of days that an ad can be run in the newspaper and the possible number of times that an ad can be aired on television.

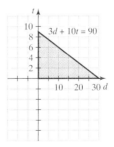

74. *Shipping restrictions.* The graph on the next page shows all of the possibilities for the number of refrigerators and the number of TVs that will fit into an 18-wheeler.

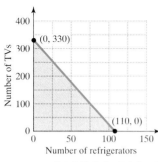

FIGURE FOR EXERCISE 74

a) Write an inequality to describe this region.

b) Will the truck hold 71 refrigerators and 118 TVs?

c) Will the truck hold 51 refrigerators and 176 TVs?

a) $3r + t \leq 330$ b) no c) yes

GETTING MORE INVOLVED

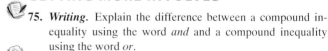

75. *Writing.* Explain the difference between a compound inequality using the word *and* and a compound inequality using the word *or*.

76. *Discussion.* Explain how to write an absolute value inequality as a compound inequality.

3.5 **RELATIONS AND FUNCTIONS**

Earlier in this chapter we used the phrase "is a function of" to describe a special relationship between variables. The area of a circle is a function of its radius because the area *is determined by* the radius using $A = \pi r^2$. The area of rectangle is not a function of its length because the area *is not determined by* the length alone. So "is a function of" means "is determined by." In this section we will learn that *function* is also a noun. A function is a special kind of set. By studying functions as sets, we can make the concept of functions more precise.

Definitions

Apple Imagewriter ribbons are sold in boxes of six in the K-LOG Catalog and are priced as follows.

Number of boxes	1	2–3	4
Cost per ribbon	$4.85	$4.60	$4.35

We can write this data as a set of ordered pairs in which the first coordinate is the number of boxes purchased and the second is the cost per ribbon in dollars:

$$\{(1, 4.85), (2, 4.60), (3, 4.60), (4, 4.35)\}$$

Because the number of boxes determines the cost per ribbon, we say that *the cost is a function of the number of boxes purchased.*

Suppose the following table appeared in the K-LOG Catalog:

Number of boxes	1	2	2	4
Cost per ribbon	$4.85	$4.60	$4.45	$4.35

Something is wrong with this table. The cost per ribbon when you buy two boxes is not clear because the ordered pairs (2, 4.60) and (2, 4.45) have the same first coordinate and different second coordinates. In this case the cost per ribbon is *not* a function of the number of boxes.

These examples illustrate the definition of function.

The key word here is "determines." According to the dictionary, determine means to settle conclusively. If the second coordinate of an ordered pair is inconclusive, then the set of ordered pairs is not a function.

Function—A Set of Ordered Pairs

A **function** is a set of ordered pairs in which no two ordered pairs have the same first coordinate and different second coordinates.

Any set of ordered pairs is called a **relation.** So every function is a relation, but a relation in which two ordered pairs have the same first coordinate and different second coordinates is not a function.

EXAMPLE 1

Functions defined from a list of ordered pairs

Determine whether each relation is a function.

a) {(1, 2), (1, 5), (3, 7)}   **b)** {(4, 5), (3, 5), (2, 6), (1, 7)}

Solution

a) This relation is not a function because (1, 2) and (1, 5) have the same first coordinates but different second coordinates.

b) This relation is a function. Note that the same second coordinate with different first coordinates is permitted in a function.   ■

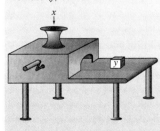

helpful hint

For an example of a function in real life consider the function that pairs up each universal product code at your grocery store with the price of the item.

If variables are used to represent the numbers of the ordered pairs, then the variable for the first coordinate is the **independent variable** and the variable for the second coordinate is the **dependent variable.** When we use variables x and y, we always assume x is the independent variable and y is the dependent variable. If we state that B is a function of A, then A is the independent variable (first coordinate) and B is the dependent variable (second coordinate), and we write (A, B).

EXAMPLE 2

Functions defined with set notation

Determine whether each relation is a function.

a) $\{(x, y) \mid y = 5x^2 - 7x + 2\}$ **b)** $\{(s, t) \mid t^2 = s\}$ **c)** $\{(u, v) \mid 2u + 3v = 6\}$

Solution

a) This relation is a function, or y is a function of x, because for any x there is only one value for y determined by the equation $y = 5x^2 - 7x + 2$.

b) Because the dependent variable is squared in $t^2 = s$, a positive value for s corresponds to both a positive and a negative value for t. For example, if $s = 4$, then $t^2 = 4$, or $t = \pm 2$. The ordered pairs $(4, -2)$ and $(4, 2)$ belong to this set. So this relation is not a function; that is, t is not a function of s.

c) If we solve $2u + 3v = 6$ for v, we get $v = -\frac{2}{3}u + 2$. Because each value of u determines only one value for v, this relation is a function. In this case, v is a *linear function* of u.   ■

We often omit the set notation and refer to an equation as a relation or a function. There are many well-known formulas that express the value of one variable as a function of another variable. For example, $A = \pi r^2$ gives the area of a circle as a function of its radius. The formula $F = \frac{9}{5}C + 32$ expresses the Fahrenheit temperature as a linear function of the Celsius temperature.

helpful hint

Some people like to think of a function as a machine. The first coordinate (x) is put into the machine, the handle is turned, and the second coordinate (y) comes out.

CAUTION   Any equation with two variables is a relation because it determines a set of ordered pairs. However, an equation is a function only if the set of ordered pairs that satisfy the equation is a function.

EXAMPLE 3

Functions defined by equations

Determine whether each relation defines y as a function of x.

a) $x = y^4$ **b)** $y = x^2$ **c)** $x^2 + y^2 = 16$

Solution

a) Because the dependent variable appears to the fourth power in $x = y^4$, a positive value for x corresponds to both a positive and a negative value for y. For example, both $(1, -1)$ and $(1, 1)$ satisfy $x = y^4$, and these ordered pairs have the same first coordinate and different second coordinates. So $x = y^4$ does not define y as a function of x. The equation $x = y^4$ is not a function.

b) For each value of x the equation $y = x^2$ determines only one value for y. So $y = x^2$ defines y as a function of x. We also say that $y = x^2$ is a function. Note that $(1, 1)$ and $(-1, 1)$ satisfy $y = x^2$, but ordered pairs with the same second coordinate and different first coordinates are allowed in a function.

c) The even power of the dependent variable in $x^2 + y^2 = 16$ indicates that we can use a number or its opposite for y and get the same value for y^2. For example, if $y = \pm 4$, then both $(0, -4)$ and $(0, 4)$ satisfy $x^2 + y^2 = 16$. So the equation is not a function. The equation does not define y as a function of x. ■

Domain and Range

The **domain** of a relation (or function) is the set of first coordinates, and the **range** is the set of second coordinates of the ordered pairs. If the ordered pairs of a relation are listed, then the domain and range can be read from the list. Relations are often defined by equations with no domain stated. *If the domain is not stated, we agree that the domain consists of all real numbers that, when substituted for the independent variable, produce real numbers for the dependent variable.*

E X A M P L E   4

Identifying domain and range

State the domain and range of each relation.

a) $\{(2, 5), (2, 7), (4, 3)\}$
b) $y = \sqrt{x}$

Solution

a) The domain is the set of numbers used as first coordinates, $\{2, 4\}$. The range is the set of second coordinates, $\{3, 5, 7\}$.

b) Because $\sqrt{x}$ is a real number only for $x \geq 0$, the domain is $[0, \infty)$, the nonnegative real numbers. The range is the set of numbers that result from taking the principal square root of every nonnegative real number. Thus the range is also the set of nonnegative real numbers, $[0, \infty)$. ■

For some relations it is easier to determine the domain than the range. In Section 3.6 we will see that the graph of a relation can be helpful for finding the range.

The Rule Definition of Function

There is another definition of function that is equivalent to the ordered-pair definition of function that we have been using:

Function—A Rule

A **function** is a rule that assigns to each element of one set (the domain) exactly one element of another set (the range).

If a set of ordered pairs is a function, then the ordered pairs give us a rule for assigning each element of one set with exactly one element of another set. Conversely, if we have a rule for assigning elements, then we could write all of the assignments as ordered pairs that satisfy the ordered-pair definition of function. So a function by one definition is also a function by the other definition.

Function Notation or *f*-notation

If concert tickets are $30 each, we write the cost of n tickets as $C = 30n$. We could also write $C(n) = 30n$, where $C(n)$ is read as "C of n." If $n = 5$, then $C(5) = \$150$. A third way to express this function is to write $C = f(n)$, which is read "C equals f of n or C is a function of n." With this notation, f is not thought of as a variable, but rather as a name for the function that pairs a value for n with a value for C. This notation is called **function notation** or **f-notation.** If we are discussing two functions, we can name them with different letters. For example, if $f(x) = x^2 + x - 9$ and $g(x) = 3x - 1$, we can refer to the functions as f and g without mentioning the formulas. To find $f(4)$, we write

$$f(4) = 4^2 + 4 - 9 = 11.$$

CAUTION The notation $f(4)$ does not mean f times 4.

EXAMPLE 5

Using *f*-notation

Let $f(x) = 3x - 2$ and $g(x) = x^2 - x$. Find the following:

a) $f(-5)$ **b)** $g(-3)$ **c)** x, if $f(x) = 0$

Solution

a) Replace x by -5 in the equation defining the function f:

$$f(x) = 3x - 2$$
$$f(-5) = 3(-5) - 2$$
$$= -17$$

So $f(-5) = -17$.

b) Replace x by -3 in the equation defining the function g:

$$g(x) = x^2 - x$$
$$g(-3) = (-3)^2 - (-3)$$
$$= 12$$

So $g(-3) = 12$.

c) Because $f(x) = 3x - 2$ and we are given that $f(x) = 0$, we can conclude that $3x - 2 = 0$. To find x, solve this equation:

$$3x - 2 = 0$$
$$3x = 2$$
$$x = \frac{2}{3}$$

The letter used for the independent variable in $f(x) = x^2 + 2$ is not important. The notation $f(t) = t^2 + 2$ or even

$$f(\text{first coordinate}) = (\text{first coordinate})^2 + 2$$

could be used to define the same set of ordered pairs. The second coordinate of an ordered pair of this function is determined by squaring the first coordinate and adding 2.

c a l c u l a t o r

c l o s e - u p

A graphing calculator has a built-in *f*-notation. Define $y_1 = 3x - 2$ and $y_2 = x^2 - x$ using Y =.

```
Plot1 Plot2 Plot3
\Y1■3X-2
\Y2■X²-X
\Y3=
\Y4=
\Y5=
\Y6=
\Y7=
```

Now use the variables feature (VARS) to find $y_1(-5)$ and $y_2(-3)$.

```
Y1(-5)
              -17
Y2(-3)
               12
```

The first coordinate of an ordered pair can be an expression rather than a real number. For example, if the first coordinate is the expression $a + 4$ and the function is $f(x) = 3x + 2$, then

$$\begin{aligned} f(a + 4) &= 3(a + 4) + 2 && \text{Replace } x \text{ by } a + 4. \\ &= 3a + 12 + 2 && \text{Distributive property} \\ &= 3a + 14 \end{aligned}$$

The first coordinate can even be an expression involving x. For example, if the first coordinate is $x + 5$ and the function is $f(x) = 3x + 2$, then

$$\begin{aligned} f(x + 5) &= 3(x + 5) + 2 && \text{Replace } x \text{ by } x + 5. \\ &= 3x + 17 \end{aligned}$$

Note that $x + 5$ replaces x and $x + 5$ is multiplied by 3, not just x.

EXAMPLE 6 **Using an expression for the independent variable**

Let $f(x) = 3x - 2$ and $g(x) = \frac{1}{x}$. Find the following:

a) $f(a - 3)$ **b)** $g(x + h)$

Solution

a) Replace x by $a - 3$ in the equation defining the function f:

$$\begin{aligned} f(x) &= 3x - 2 \\ f(a - 3) &= 3(a - 3) - 2 \\ &= 3a - 11 \end{aligned}$$

So $f(a - 3) = 3a - 11$.

b) Replace x by $x + h$ in the equation defining the function g:

$$g(x) = \frac{1}{x}$$

$$g(x + h) = \frac{1}{x + h}$$

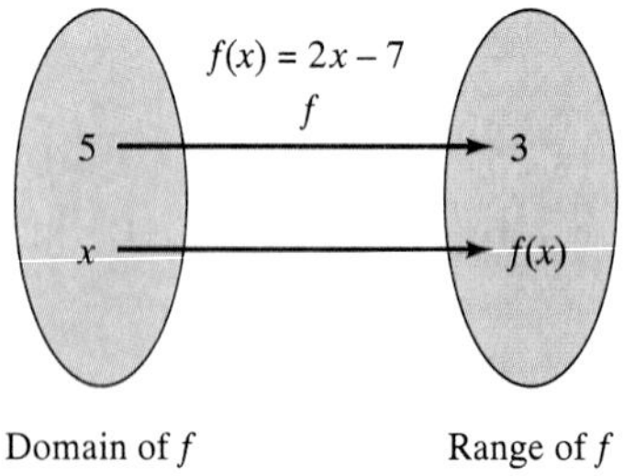

f(x) = 2x – 7
f
5
3
x
f(x)
Domain of f
Range of f

FIGURE 3.34

A function is a rule for pairing numbers in one set (the domain) with numbers in another set (the range). Figure 3.34 shows the domain and range of a function. The arrow represents the function and illustrates the pairing of values in the domain with values in the range.

Average Rate of Change (Optional)

The U.S. Bureau of Labor Statistics reports on the number of people (in millions) in the civilian labor force of the United States. The size of the labor force is a function of the year. Two ordered pairs of this function are (1985, 88.4) and (1989, 97.3). To see how the function is changing in time, we can find the slope of the line through these two points:

$$\frac{97.3 - 88.4}{1989 - 1985} \approx 2.2 \text{ million people per year}$$

During this 4-year period the civilian labor force increased on the average by 2.2 million people per year. Note that 2.2 million is not necessarily the increase for any particular year, but it is the average rate at which the labor force was changing during those 4 years.

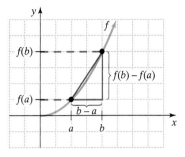

FIGURE 3.35

In general, the average rate of change of a function over an interval in its domain is defined as follows.

Average Rate of Change of a Function

If a and b are in the domain of the function f and $a < b$, then the **average rate of change** of f over the interval $[a, b]$ is

$$\frac{f(b) - f(a)}{b - a}.$$

Note that the average rate of change of f over $[a, b]$ is the slope of the line through $(a, f(a))$ and $(b, f(b))$, as shown in Fig. 3.35.

E X A M P L E   7   Average rate of change of a function (Optional)

If a stunt man jumps from a bridge that is 144 feet above the water, then his height above the water in feet at time t in seconds is given by the function $h(t) = -16t^2 + 144$. Find the average rate of change of his height above the water over the time interval $[1, 3]$. See Fig. 3.36.

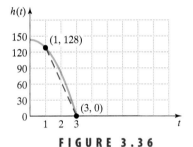

FIGURE 3.36

Solution

Because $h(1) = 128$ and $h(3) = 0$, we have

$$\frac{h(3) - h(1)}{3 - 1} = \frac{0 - 128}{2}$$

$$= -64 \text{ ft/sec.}$$

So the average rate of change of the height is -64 ft/sec on the interval $[1, 3]$. ■

WARM-UPS

True or false? Explain your answer.

1. Any set of ordered pairs is a function.   False
2. The circumference of a circle is a function of the diameter.   True
3. The set $\{(1, 2), (3, 2), (5, 2)\}$ is a function.   True
4. Every relation is a function.   False
5. The set $\{(1, 5), (3, 6), (1, 7)\}$ is a function.   False
6. The domain of $f(x) = \sqrt{x}$ is the set of positive real numbers.   False
7. The range of $g(x) = |x|$ is the set of nonnegative real numbers.   True
8. The set $\{(x, y) \mid x = 4y\}$ is a function.   True
9. The set $\{(u, v) \mid u = v^2\}$ is a function.   False
10. If $h(x) = x^2 - 3$, then $h(-2) = 1$.   True

3.5   EXERCISES

Reading and Writing   *After reading this section, write out the answers to these questions. Use complete sentences.*

1. What is a function?
   A function is a set of ordered pairs in which no two have the same first coordinate and different second coordinates.

2. What is a relation?
   A relation is any set of ordered pairs.

3. What is the domain of a relation?
   The domain of a relation is the set of all possible first coordinates.

4. What is the range of a relation?

The range of relation is the set of all possible second coordinates.

5. What is *f*-notation?

The *f*-notation is the notation in which we use $f(x)$ rather than y as the dependent variable.

6. What is the average rate of change of a function?

The average rate of change of the function $f(x)$ on the interval $[a, b]$ is $(f(b) - f(a))/(b - a)$.

Determine whether each relation is a function. See Example 1.

7. $\{(2, 4), (3, 4), (4, 5)\}$ **8.** $\{(2, -5), (2, 5), (3, 10)\}$

Yes   No

9. $\{(-2, 4), (-2, 6), (3, 6)\}$ **10.** $\{(3, 6), (6, 3)\}$

No   Yes

11. $\{(\pi, -1), (\pi, 1)\}$ No

12. $\{(-0.3, -0.3), (-0.2, 0), (-0.3, 1)\}$ No

13. $\left\{\left(\dfrac{1}{2}, \dfrac{1}{2}\right)\right\}$ Yes

14. $\left\{\left(\dfrac{1}{3}, 7\right), \left(-\dfrac{1}{3}, 7\right), \left(\dfrac{1}{6}, 7\right)\right\}$ Yes

Determine whether each relation is a function. See Example 2.

15. $\{(x, y) \mid y = (x - 1)^2\}$ Yes

16. $\{(x, y) \mid y = x^2 - 12x + 1\}$ Yes

17. $\{(x, y) \mid x = |y|\}$ No   **18.** $\{(x, y) \mid x = y^2 + 2\}$ No

19. $\{(s, t) \mid t = s\}$ Yes   **20.** $\left\{(u, v) \mid v = \dfrac{1}{u}\right\}$ Yes

21. $\{(x, y) \mid x = 5y + 2\}$ Yes   **22.** $\{(x, y) \mid x = 3y\}$ Yes

Determine whether each relation defines y as a function of x. See Example 3.

23. $x = 2y^2$ **24.** $y = 3x^4$ **25.** $y = 3x - 4$

No   Yes   Yes

26. $y = 2x + 9$ **27.** $x^2 + y^2 = 1$ **28.** $x^4 - y^4 = 0$

Yes   No   No

29. $y = \sqrt{x}$ **30.** $x = \sqrt{y}$ **31.** $x = 2|y|$

Yes   Yes   No

32. $y = |x - 1|$ Yes

Determine the domain and range of each relation. See Example 4.

33. $\{(2, 3), (2, 5), (2, 7)\}$ **34.** $\{(3, 1), (5, 1), (4, 1)\}$

$\{2\}, \{3, 5, 7\}$ $\{3, 4, 5\}, \{1\}$

35. $y = |x|$ **36.** $y = 2x + 1$

R, $[0, \infty)$ R, R

37. $x = \sqrt{y}$ **38.** $y = \sqrt{x} + 1$

$[0, \infty), [0, \infty)$ $[0, \infty), [1, \infty)$

Let $f(x) = 3x - 2$, $g(x) = x^2 - 3x + 2$, and $h(x) = |x + 2|$. Find the following. See Example 5.

39. $f(4)$ 10   **40.** $f(100)$ 298   **41.** $g(-2)$ 12

42. $g(6)$ 20   **43.** $h(-3)$ 1   **44.** $h(-19)$ 17

45. x, if $f(x) = 5$ $\dfrac{7}{3}$ **46.** x, if $f(x) = 49$ 17

47. x, if $h(x) = 3$ -5 or 1   **48.** x, if $h(x) = 7$ 5 or -9

Let $f(x) = 4x - 1$ and $g(x) = \dfrac{1}{x + 2}$. Find the following. See Example 6.

49. $f(a)$ **50.** $f(a + 1)$ **51.** $f(x + 2)$

$4a - 1$ $4a + 3$ $4x + 7$

52. $f(x + h)$ **53.** $g(x + 3)$ **54.** $g(x - 2)$

$4x + 4h - 1$ $\dfrac{1}{x + 5}$ $\dfrac{1}{x}$

55. $g(x + h)$ **56.** $g(a - 2)$

$\dfrac{1}{x + h + 2}$ $\dfrac{1}{a}$

Find the average rate of change of the given function over the given interval. See Example 7.

57. $f(x) = x^2$, $[1, 3]$ 4   **58.** $g(x) = 2x - 5$, $[3, 9]$ 2

59. $h(x) = 2x^2 - 3x$, $[4, 8]$ 21   **60.** $f(x) = \sqrt{x}$, $[4, 9]$ $\dfrac{1}{5}$

61. $g(x) = \dfrac{1}{x}$, $[2, 4]$ $-\dfrac{1}{8}$ **62.** $h(x) = \dfrac{2}{x^2}$, $[1, 2]$ $-\dfrac{3}{2}$

 Let $f(x) = \sqrt{x + 2}$ and $g(x) = 3x^2 - 8x + 2$. Use a calculator to find the following. Round answers to three decimal places.

63. $f(3.46)$ 2.337   **64.** $g(-1.37)$ 18.591

65. $g(-3.5)$ 66.75   **66.** $f(-1.2)$ 0.894

Let $f(x) = 3x - 2$ and $g(x) = 3 - 5x$. Find and simplify each expression.

67. $f(a - 5) - f(a)$ -15 **68.** $f(x + h) - f(x)$ $3h$

69. $g(a + 2) - g(a)$ -10 **70.** $g(x + h) - g(x)$ $-5h$

71. $\dfrac{f(x + h) - f(x)}{h}$ 3   **72.** $\dfrac{f(n + 3) - f(n)}{3}$ 3

Solve each problem.

73. *Area of a square.* Express the area of a square, A, as a function of its side, s. $A = s^2$

74. *Perimeter of a square.* Express the perimeter of a square, P, as a function of its side, s. $P = 4s$

75. *Cost of fabric.* If a certain fabric is priced at \$3.98 per yard, express the cost of a purchase, C, as a function of the number of yards purchased, y. $C = 3.98y$

76. *Earned income.* If Mildred earns \$14.50 per hour, express her total pay, P, as a function of the number of hours worked, h. $P = 14.5h$

77. *Cost of pizza.* A pizza parlor in Victoria, B.C. charges \$14.95 for a pizza plus \$0.50 for each topping. Express the cost of a pizza, C, as a function of the number of toppings, n. $C = 0.50n + 14.95$

78. *Dealing in gravel.* A gravel dealer charges \$120 for a minimum load of 9 cubic yards and \$10 more for each additional cubic yard. Express the total charge, C, as a function of the number of yards sold, n, where $n \geq 9$. $C = 10n + 30$

79. *Staying fit.* Suppose the heart rate of a certain individual is a linear function of the number of minutes she spends on a treadmill. A heart rate of 78 was measured after 2 minutes, and a heart rate of 86 was measured after 4 minutes. Write the heart rate, h, as a linear function of the number of minutes, t, on the treadmill, for $0 \leq t \leq 8$. $h = 4t + 70$

80. Printing costs. To determine the cost of printing a book, a printer uses a linear function of the number of pages. If the cost is $8.60 for a 400-page book and $12.20 for a 580 page book, then what is the linear function that is used?

$C = 0.02p + 0.60$

81. Depreciating Beretta. A Chevrolet Beretta that sold new for $11,640 in 1990 sold used for $3,590 in 1998 (Edmund's Used Car Prices, www.edmunds.com). Find the average rate of change of the value of this car over the time interval [1990, 1998].

−$1,006.25 per year

82. More depreciation. A Porsche 928S that sold new for $69,680 in 1988 sold used for $16,550 in 1998 (Edmund's Used Car Prices, www.edmunds.com). Find the average rate of change of the value of this car over the time interval [1988, 1998].

−$5,313 per year

83. Fast cat. The 1999 Mercury Cougar had a base price of $17,095 and could go from 0 to 60 mph in 8.0 seconds (*Fortune*, June 26, 1998, www.pathfinder.com). Find the average rate of change of the velocity of this car over the time interval [0, 8.0].

7.5 mph per second

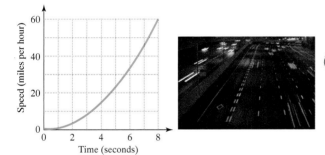

FIGURE FOR EXERCISE 83

84. Total construction. For May 1996 the rate for total value of construction put in place was $580 billion per year. For May 1998 the rate was $634 billion per year (U.S. Census Bureau, www.census.gov). Find the average rate of change of the rate for this 24-month period, [0, 24].

$2.25 billion per year per month

FIGURE FOR EXERCISE 84

GETTING MORE INVOLVED

85. Exploration. In Example 7 we found that the stunt man traveled 128 feet in 2 seconds for an average velocity of −64 ft/sec over the interval [1, 3]. His velocity actually starts out at 0 ft/sec and keeps increasing as he falls. Find the average rate of change of his height, or his average velocity, over the time intervals [2.8, 3], [2.9, 3], and [2.99, 3].

−92.8 ft/sec, −94.4 ft/sec, −95.84 ft/sec

86. Exploration. In Exercise 85 we found the stunt man's average velocity for some time intervals right before he hits the water. His velocity at the moment he hits the water is his *instantaneous velocity* at $t = 3$. Find the average velocity over the time intervals [2.999, 3] and [2.9999, 3]. What do you think his instantaneous velocity is at time $t = 3$? Explain your answer.

−95.984 ft/sec, −95.9984, −96 ft/sec

3.6 GRAPHS OF FUNCTIONS

Earlier in this chapter we used graphs to visualize the solution sets to linear equations. In this section we will use graphs to visualize various types of functions.

Linear and Constant Functions

Linear Function

A **linear function** is a function of the form

$$f(x) = mx + b,$$

where m and b are real numbers with $m \neq 0$.

The graph of the linear function $f(x) = mx + b$ is exactly the same as the graph of the linear equation $y = mx + b$. If $m = 0$, then we get $f(x) = b$, which is called a **constant function.** If $m = 1$ and $b = 0$, then we get the function $f(x) = x$, which is called the **identity function.** When we graph a function given in f-notation, we usually label the vertical axis as $f(x)$ rather than y.

E X A M P L E   1

Graphing a constant function

Graph $f(x) = 3$ and state the domain and range.

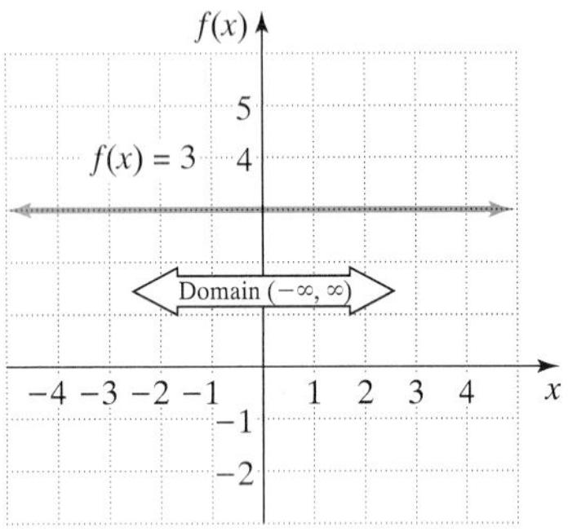

Solution

The graph of $f(x) = 3$ is the same as the graph of $y = 3$, which is the horizontal line in Fig. 3.37. Since any real number can be used for x in $f(x) = 3$ and since the line in Fig. 3.37 extends without bounds to the left and right, the domain is the set of all real numbers, $(-\infty, \infty)$. Since the only y-coordinate for $f(x) = 3$ is 3, the range is $\{3\}$.

FIGURE 3.37

The domain and range of a function can be determined from the formula or the graph. However, the graph is usually very helpful for understanding domain and range.

E X A M P L E   2

Graphing a linear function

Graph the function $f(x) = 3x - 4$ and state the domain and range.

Solution

The y-intercept is $(0, -4)$ and the slope of the line is 3. We can use the y-intercept and the slope to draw the graph in Fig. 3.38. Since any real number can be used for x in $f(x) = 3x - 4$, and since the line in Fig. 3.38 extends without bounds to the left and right, the domain is the set of all real numbers, $(-\infty, \infty)$. Since the graph extends without bounds upward and downward, the range is the set of all real numbers, $(-\infty, \infty)$.

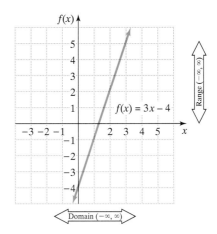

FIGURE 3.38

Absolute Value Functions

The equation $y = |x|$ defines a function because every value of x determines a unique value of y. We call this function the absolute value function.

Absolute Value Function

The **absolute value function** is the function defined by

$$f(x) = |x|.$$

To graph the absolute value function, we simply plot enough ordered pairs of the function to see what the graph looks like.

E X A M P L E   3

The absolute value function

Graph $f(x) = |x|$ and state the domain and range.

Solution

To graph this function, we find points that satisfy the equation $f(x) = |x|$.

x	-2	-1	0	1	2
$f(x) = |x|$	2	1	0	1	2

helpful hint

The most important feature of an absolute value function is its V-shape. If we had plotted only points in the first quadrant, we would not have seen the V-shape. So for an absolute value function we always plot enough points to see the V-shape.

Plotting these points, we see that they lie along the V-shaped graph shown in Fig. 3.39. Since any real number can be used for x in $f(x) = |x|$ and since the graph extends without bounds to the left and right, the domain is $(-\infty, \infty)$. Because the graph does not go below the x-axis and because $|x|$ is never negative, the range is the set of nonnegative real numbers, $[0, \infty)$.

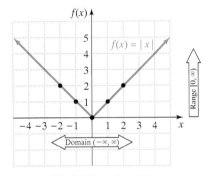

F I G U R E   3 . 3 9

Many functions involving absolute value have graphs that are V-shaped, as in Fig. 3.39. To graph functions involving absolute value, we must choose points that determine the correct shape and location of the V-shaped graph.

E X A M P L E   4

Other functions involving absolute value

Graph each function and state the domain and range.

a) $f(x) = |x| - 2$ **b)** $g(x) = |2x - 6|$

Solution

a) Choose values for x and find $f(x)$.

x	-2	-1	0	1	2
$f(x) = |x| - 2$	0	-1	-2	-1	0

Plot these points and draw a V-shaped graph through them as shown in Fig. 3.40 on the next page. The domain is $(-\infty, \infty)$, and the range is $[-2, \infty)$.

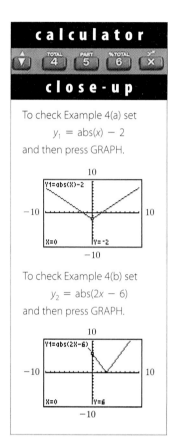

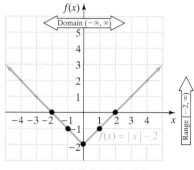

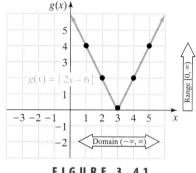

FIGURE 3.40   **FIGURE 3.41**

b) Make a table of values for x and $g(x)$.

x	1	2	3	4	5
$g(x) = \mid 2x - 6 \mid$	4	2	0	2	4

Draw the graph as shown in Fig. 3.41. The domain is $(-\infty, \infty)$, and the range is $[0, \infty)$.

Quadratic Functions

Quadratic Function

A **quadratic function** is a function of the form
$$f(x) = ax^2 + bx + c,$$
where a, b, and c are real numbers, with $a \neq 0$.

Without the term ax^2 this function would be a linear function, which is why we specify $a \neq 0$.

E X A M P L E   5

The simplest quadratic function

Graph the function $h(x) = x^2$ and state the domain and range.

Solution

Make a table of values for x and $h(x)$:

x	-2	-1	0	1	2
$h(x) = x^2$	4	1	0	1	4

See Fig. 3.42 for the graph. The domain is $(-\infty, \infty)$. From the graph we see that the smallest y-coordinate of the function is 0. So the range is $[0, \infty)$.

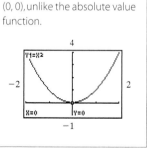

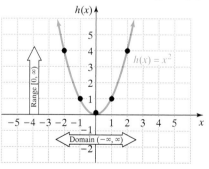

FIGURE 3.42

The graph of any quadratic function is called a **parabola.** All parabolas are similar in shape to the one in Fig. 3.42. This parabola **opens upward.** The lowest point on a parabola that opens upward is the **vertex.** The y-coordinate of the vertex is the **minimum value** of the function. So for $h(x) = x^2$ the vertex is $(0, 0)$ and the minimum value of the function is 0. The parabola in the next example opens downward.

E X A M P L E   6

A quadratic function

Graph the function $g(x) = 4 - x^2$ and state the domain and range.

Solution

We plot enough points to get the correct shape of the graph.

x	-2	-1	0	1	2
$g(x) = 4 - x^2$	0	3	4	3	0

See Fig. 3.43 for the graph. The domain is $(-\infty, \infty)$. From the graph we see that the largest y-coordinate is 4. So the range is $(-\infty, 4]$.

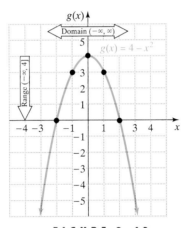

F I G U R E   3.43

The parabola in Fig. 3.43 **opens downward.** The highest point on a parabola that opens downward is the **vertex** and the y-coordinate of the vertex is the **maximum value** of the function. For $g(x) = 4 - x^2$ the vertex is $(0, 4)$ and the maximum function value is 4. Parabolas will be discussed further when we study conic sections later in this text.

Square-Root Functions

Functions involving square roots typically have graphs that look like half a parabola.

> **Square-Root Function**
>
> The **square-root function** is the function defined by
> $$f(x) = \sqrt{x}.$$

E X A M P L E   7

Square-root functions

Graph each equation and state the domain and range.

a) $y = \sqrt{x}$ \qquad\qquad\qquad **b)** $y = \sqrt{x} + 3$

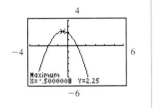

Solution

a) The graph of the equation $y = \sqrt{x}$ and the graph of the function $f(x) = \sqrt{x}$ are the same. Because $\sqrt{x}$ is a real number only if $x \geq 0$, the domain of this function is the set of nonnegative real numbers. The following ordered pairs are on the graph:

x	0	1	4	9
$y = \sqrt{x}$	0	1	2	3

The graph goes through these ordered pairs as shown in Fig. 3.44. Note that x is chosen from the nonnegative numbers. The domain is $[0, \infty)$ and the range is $[0, \infty)$.

b) Note that $\sqrt{x + 3}$ is a real number only if $x + 3 \geq 0$, or $x \geq -3$. So we make a table of ordered pairs in which $x \geq -3$:

x	-3	-2	1	6
$y = \sqrt{x + 3}$	0	1	2	3

The graph goes through these ordered pairs as shown in Fig. 3.45. The domain is $[-3, \infty)$ and the range is $[0, \infty)$.

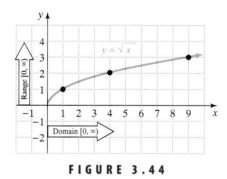

FIGURE 3.44

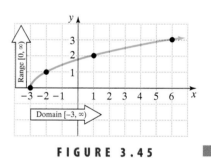

FIGURE 3.45

Graphs of Relations

In the next example we graph a relation by simply finding enough points to see the shape of the graph.

E X A M P L E   8

The graph of a relation

Graph $x = y^2$ and state the domain and range.

Solution

The set of ordered pairs that satisfy $x = y^2$ is not a function. The equation $x = y^2$ does not express y as a function of x. However, we can still make a table of ordered pairs and sketch the graph. Because the equation $x = y^2$ expresses x in terms of y, it is easier to choose the y-coordinate first and then find the x-coordinate.

$x = y^2$	4	1	0	1	4
y	-2	-1	0	1	2

Figure 3.46 on the next page shows the graph. The domain is $[0, \infty)$ and the range is $(-\infty, \infty)$.

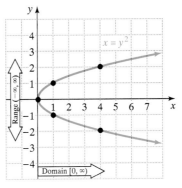

FIGURE 3.46

Vertical-Line Test

The relation $x = y^2$ from Example 8 does not define y as a function of x. A relation is not a function if there are two ordered pairs with the same first coordinate and different second coordinates. Whether a relation has such ordered pairs can be determined by a simple visual test called the **vertical-line test.**

Vertical-Line Test

If it is possible to draw a vertical line that crosses the graph of a relation two or more times, then the graph is not the graph of a function.

If there is a vertical line that crosses a graph twice (or more), then we have two points (or more) with the same x-coordinate and different y-coordinates, and so the graph is not the graph of a function. If you mentally consider every possible vertical line and none of them crosses the graph more than once, then you can conclude that the graph is the graph of a function.

EXAMPLE 9

Using the vertical-line test

Which of the following graphs are graphs of functions?

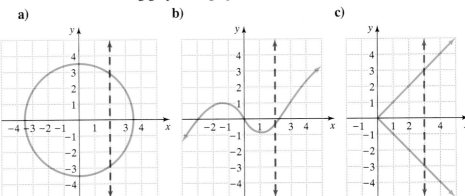

Solution

Neither (a) nor (c) is the graph of a function, since we can draw vertical lines that cross these graphs twice. The graph (b) is the graph of a function, since no vertical line crosses it twice.

Applications

We can determine much about a function by examining its graph. In the next example we find the maximum altitude reached by a projectile by examining a graph. The maximum and minimum values of quadratic functions will be discussed in more detail when we study conic sections later in this text.

E X A M P L E   1 0   **Projectile motion**

A flare is fired upward from the surface of the earth with an initial velocity of 96 feet per second. The altitude (in feet) of the flare t seconds after it is fired is given by the function $h(t) = -16t^2 + 96t$.

a) Graph the function for $0 \le t \le 6$.

b) For what value of t does the flare reach its maximum altitude?

c) What is the maximum altitude of the flare?

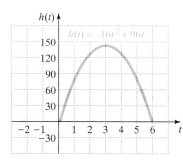

FIGURE 3.47

Solution

a) Make a table of values for t between 0 and 6:

t	0	1	2	3	4	5	6
$h(t) = -16t^2 + 96t$	0	80	128	144	128	80	0

Now sketch the graph as shown in Fig. 3.47.

b) From the graph it appears that the flare reaches its maximum altitude when $t = 3$ seconds.

c) Since $h(3) = 144$, the maximum altitude reached by the flare is 144 feet.   ◼

WARM-UPS

True or false? Explain your answer.

1. The graph of a function is a picture of all ordered pairs of the function.   True

2. The graph of every linear function is a straight line.   True

3. The absolute value function has a V-shaped graph.   True

4. The domain of $f(x) = \frac{1}{x}$ is $(-\infty, \infty)$.   False

5. The graph of a quadratic function is a parabola.   True

6. The range of any quadratic function is $(-\infty, \infty)$.   False

7. The y-axis and the $f(x)$-axis are the same.   True

8. The vertical-line test is used to determine whether a relation is a function.   True

9. The domain of $f(x) = \sqrt{x - 1}$ is $(1, \infty)$.   False

10. The domain of any quadratic function is $(-\infty, \infty)$.   True

3.6   EXERCISES

Reading and Writing   *After reading this section, write out the answers to these questions. Use complete sentences.*

1. What is a linear function?
   A linear function is a function of the form $f(x) = mx + b$, where m and b are real numbers with $m \neq 0$.

2. What is a constant function?
   A constant function is a function of the form $f(x) = k$, where k is a real number.

3. What is the graph of a constant function?
   The graph of a constant function is a horizontal line.

4. What shape is the graph of an absolute value function?
The absolute value function has a V-shaped graph.

5. What is the graph of quadratic function called?
The graph of a quadratic function is a parabola.

6. How can you tell at a glance if a graph is the graph of a function?
If there is a vertical line that crosses a graph more than once, the graph is not the graph of a function.

Graph each function and state its domain and range. See Examples 1 and 2.

7. $h(x) = -2$
R, $[-2]$

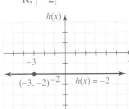

8. $f(x) = 4$
R, $[4]$

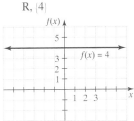

9. $f(x) = 2x - 1$
R, R

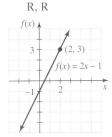

10. $g(x) = x + 2$
R, R

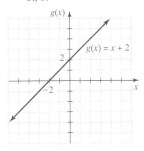

11. $g(x) = \frac{1}{2}x + 2$
R, R

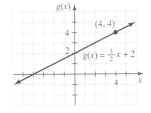

12. $h(x) = \frac{2}{3}x - 4$
R, R

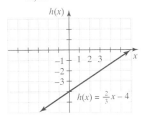

13. $y = -\frac{2}{3}x + 3$
R, R

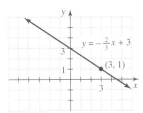

14. $y = -\frac{3}{4}x + 4$
R, R

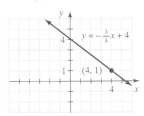

15. $y = -0.3x + 6.5$ R, R

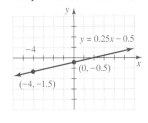

16. $y = 0.25x - 0.5$ R, R

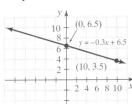

Graph each absolute value function and state its domain and range. See Examples 3 and 4.

17. $f(x) = |x| + 1$
$(-\infty, \infty), [1, \infty)$

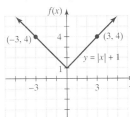

18. $g(x) = |x| - 3$
$(-\infty, \infty), [-3, \infty)$

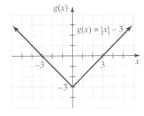

19. $h(x) = |x + 1|$
$(-\infty, \infty), [0, \infty)$

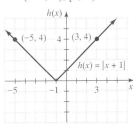

20. $f(x) = |x - 2|$
$(-\infty, \infty), [0, \infty)$

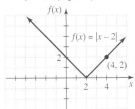

21. $g(x) = |3x|$
$(-\infty, \infty), [0, \infty)$

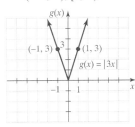

22. $h(x) = |-2x|$
$(-\infty, \infty), [0, \infty)$

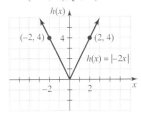

23. $f(x) = |2x - 1|$
$(-\infty, \infty), [0, \infty)$

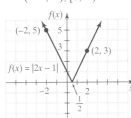

24. $y = |2x - 3|$
$(-\infty, \infty), [0, \infty)$

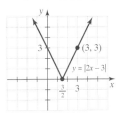

25. $f(x) = |x - 2| + 1$
$(-\infty, \infty), [1, \infty)$

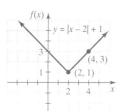

26. $y = |x - 1| + 2$
$(-\infty, \infty), [2, \infty)$

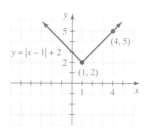

33. $y = -x^2 + 2x + 1$
$(-\infty, \infty), (-\infty, 2]$

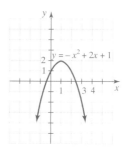

34. $y = x^2 + 4x + 1$
$(-\infty, \infty), [-3, \infty)$

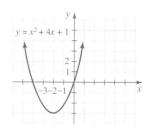

Graph each quadratic function and state its domain and range. See Examples 5 and 6.

27. $g(x) = x^2 + 2$
$(-\infty, \infty), [2, \infty)$

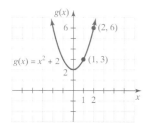

28. $f(x) = x^2 - 4$
$(-\infty, \infty), [-4, \infty)$

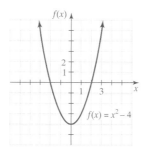

35. $f(x) = x^2 + 4x + 3$
$(-\infty, \infty), [-1, \infty)$

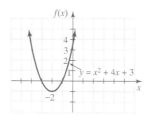

36. $f(x) = -x^2 + 4x - 3$
$(-\infty, \infty), (-\infty, 1]$

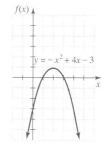

29. $f(x) = 2x^2$
$(-\infty, \infty), [0, \infty)$

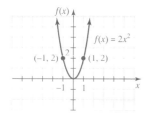

30. $h(x) = -3x^2$
$(-\infty, \infty), (-\infty, 0]$

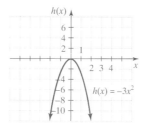

Graph each square-root function and state its domain and range. See Example 7.

37. $g(x) = 2\sqrt{x}$
$[0, \infty), [0, \infty)$

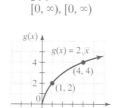

38. $g(x) = \sqrt{x} - 1$
$[0, \infty), [-1, \infty)$

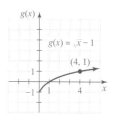

31. $y = 6 - x^2$
$(-\infty, \infty), (-\infty, 6]$

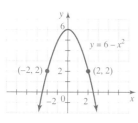

32. $y = -2x^2 + 3$
$(-\infty, \infty), (-\infty, 3]$

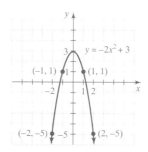

39. $f(x) = \sqrt{x - 1}$
$[1, \infty), [0, \infty)$

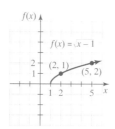

40. $f(x) = \sqrt{x + 1}$
$[-1, \infty), [0, \infty)$

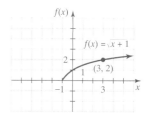

41. $h(x) = -\sqrt{x}$
$[0, \infty), (-\infty, 0]$

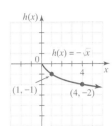

42. $h(x) = -\sqrt{x-1}$
$[1, \infty), (-\infty, 0]$

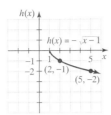

49. $x = 5$
$\{5\}, (-\infty, \infty)$

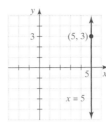

50. $x = -3$
$\{-3\}, (-\infty, \infty)$

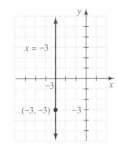

43. $y = \sqrt{x} + 2$
$[0, \infty), [2, \infty)$

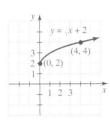

44. $y = 2\sqrt{x} + 1$
$[0, \infty), [1, \infty)$

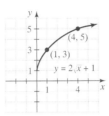

51. $x + 9 = y^2$
$[-9, \infty), (-\infty, \infty)$

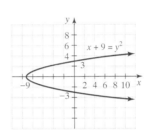

52. $x + 3 = |y|$
$[-3, \infty), (-\infty, \infty)$

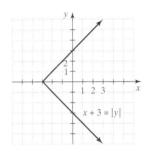

Graph each relation and state its domain and range. See Example 8.

45. $x = |y|$
$[0, \infty), (-\infty, \infty)$

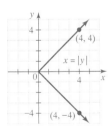

46. $x = -|y|$
$(-\infty, 0], (-\infty, \infty)$

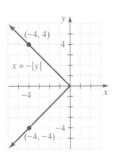

53. $x = \sqrt{y}$
$[0, \infty), [0, \infty)$

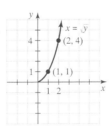

54. $x = -\sqrt{y}$
$(-\infty, 0], [0, \infty)$

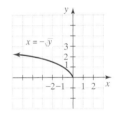

47. $x = -y^2$
$(-\infty, 0], (-\infty, \infty)$

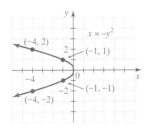

48. $x = 1 - y^2$
$(-\infty, 1], (-\infty, \infty)$

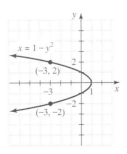

55. $x = (y - 1)^2$
$[0, \infty), (-\infty, \infty)$

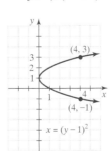

56. $x = (y + 2)^2$
$[0, \infty), (-\infty, \infty)$

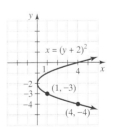

Use the vertical-line test to determine which of the graphs are graphs of functions. See Example 9.

57.

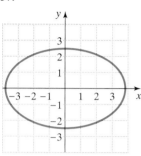

No

58.

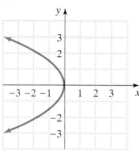

No

59.

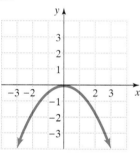

Yes

60.

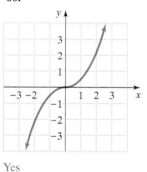

Yes

61.

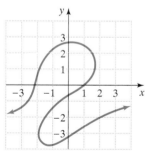

No

62.

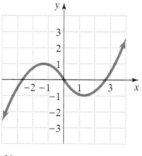

Yes

Graph each function and state the domain and range.

63. $f(x) = 1 - |x|$
$(-\infty, \infty), (-\infty, 1]$

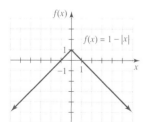

64. $h(x) = \sqrt{x - 3}$
$[3, \infty), [0, \infty)$

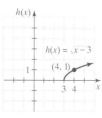

65. $y = (x - 3)^2 - 1$
$(-\infty, \infty), [-1, \infty)$

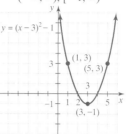

66. $y = x^2 - 2x - 3$
$(-\infty, \infty), [-4, \infty)$

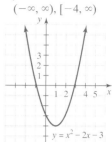

67. $y = |x + 3| + 1$
$(-\infty, \infty), [1, \infty)$

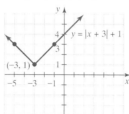

68. $f(x) = -2x + 4$
$(-\infty, \infty), (-\infty, \infty)$

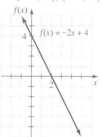

69. $y = \sqrt{x} - 3$
$[0, \infty), [-3, \infty)$

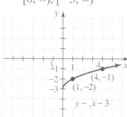

70. $y = 2 |x|$
$(-\infty, \infty), [0, \infty)$

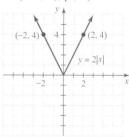

71. $y = 3x - 5$
$(-\infty, \infty), (-\infty, \infty)$

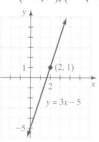

72. $g(x) = (x + 2)^2$
$(-\infty, \infty), [0, \infty)$

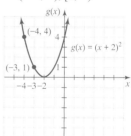

73. $y = -x^2 + 4x - 4$
$(-\infty, \infty), (-\infty, 0]$

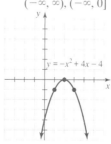

74. $y = -2 |x - 1| + 4$
$(-\infty, \infty), (-\infty, 4]$

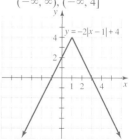

Solve each problem. See Example 10.

75. *Maximum height.* If an object is projected upward with an initial velocity of 64 feet per second, then its height is a function of time, given by $h(t) = -16t^2 + 64t$. Graph this function for $0 \le t \le 4$. What is the maximum height reached by this object?   64 feet

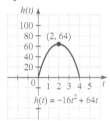

76. *Maximum height.* If a soccer ball is kicked straight up with an initial velocity of 32 feet per second, then its height above the earth is a function of time given by $h(t) = -16t^2 + 32t$. Graph this function for $0 \le t \le 2$. What is the maximum height reached by this ball?   16 feet

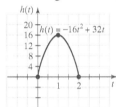

77. *Air pollution.* The amount of nitrogen dioxide (NO_2) present in the air in the city of Springfield on a certain day in July is shown in the accompanying figure.

a) Use the graph to estimate the time of day when the NO_2 level was at its maximum.   2 P.M.

b) The equation for the graph is

$$A(t) = -2t^2 + 32t + 42,$$

where t is the number of hours after 6:00 A.M. Use this function to determine the maximum level of NO_2 on that day.   170 parts per million

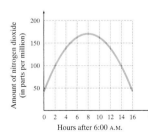

FIGURE FOR EXERCISE 77

78. *Prisoner cost.* The sheriff has determined that the cost in dollars per day per prisoner for housing prisoners in the Macon County jail is a function of the number of

prisoners n, where

$$C(n) = 0.015n^2 - 4.5n + 400.$$

a) Graph this function for $0 \le n \le 300$.
b) Use the graph to estimate the number of prisoners for which the cost is at a minimum.
c) What is the minimum cost per day per prisoner?
  b) 150
  c) $62.50

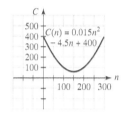

Solve each problem.

79. *Thinning eggshells.* Acid rain is causing the thinning of song thrush shells in Britain (*New Scientist*, April 1998). Scientists use the function

$$S(t) = -0.000133t + 0.24$$

to model the shell thickness in the year $1850 + t$, where $S(t)$ is in milligrams per square millimeters (mg/mm^2).

a) If this trend continues, then what will be the thickness of song thrush shells in 2050?
b) In what year was the shell thickness 0.23 mg/mm²?
  a) 0.2134 mg/mm²
  b) 1925

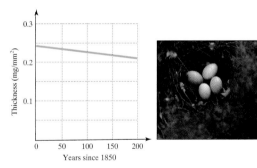

FIGURE FOR EXERCISE 79

80. *Radius of a circle.* The function $r(A) = \sqrt{\dfrac{A}{\pi}}$ gives the radius of a circle as a function of its area A. Find the radius of a circle that has area 500 square centimeters (cm^2). Round your answer to the nearest tenth of a centimeter (cm). Find the area of a circle that has a radius of 30 feet. Round the answer to the nearest tenth of a square foot.   12.6 cm, 2,827.4 ft²

GRAPHING CALCULATOR EXERCISES

81. Graph the function $f(x) = \sqrt{x^2}$ and explain what this graph illustrates.

The graph of $f(x) = \sqrt{x^2}$ is the same as the graph of $f(x) = |x|$.

82. Graph the function $f(x) = \frac{1}{x}$ and state the domain and range. $(-\infty, 0) \cup (0, \infty), (-\infty, 0) \cup (0, \infty)$

83. Graph $y = x^2$, $y = \frac{1}{2}x^2$, and $y = 2x^2$ on the same coordinate system. What can you say about the graph of $y = kx^2$?

For large values of k the graph gets narrower and for smaller values of k the graph gets broader.

84. Graph $y = x^2$, $y = x^2 + 2$, and $y = x^2 - 3$ on the same screen. What can you say about the position of $y = x^2 + k$ relative to $y = x^2$.

The graph of $y = x^2 + k$ moves upward for $k > 0$ and downward for $k < 0$.

85. Graph $y = x^2$, $y = (x + 5)^2$, and $y = (x - 2)^2$ on the same screen. What can you say about the position of $y = (x - k)^2$ relative to $y = x^2$.

The graph of $y = (x - k)^2$ moves to the right for $k > 0$ and to the left for $k < 0$.

86. You can graph the relation $x = y^2$ by graphing the two functions $y = \sqrt{x}$ and $y = -\sqrt{x}$. Try it and explain why this works.

The equation $x = y^2$ is equivalent to $y = \sqrt{x}$ or $y = -\sqrt{x}$.

87. Graph $y = (x - 3)^2$, $y = |x - 3|$, and $y = \sqrt{x - 3}$ on the same coordinate system. How does the graph of $y = f(x - k)$ compare to the graph of $y = f(x)$?

The graph of $y = f(x - k)$ lies to the right of the graph of $y = f(x)$ when $k > 0$.

COLLABORATIVE ACTIVITIES

Parallel and Perpendicular Explorations

Grouping: Two students per group
Topic: Parallel and perpendicular lines

Part I: One person in your pair will do the graphing (**G**—Grapher) while the other one is answering the questions (**A**—Answerer). Discuss your answers so that you both know how to find them. Use either graph paper or a graphing calculator so that your graphs will be accurate.

1. **G:** Graph the line through the points $(2, 3)$ and $(-1, 4)$. We will call this line S. **A:** Find the equation for line S.

2. **G:** Plot points three units up (positive y-direction) from each of the points given on line S and draw a line through the new points. We will call this line P. **A:** What are the new points? Find the equation for line P.

3. **G:** On the same graph, plot points five units to the left (negative x-direction) from each of the points given on line S and draw a line through the new points. We will call this line L. **A:** What are the new points? Find the equation for line L.

4. **G, A:** Looking only at your graph, state how these three lines are the same and how they are different. Make sure your equations are in slope-intercept form. State how the equations are the same and how they are different.

5. **G, A:** Find another line parallel to line S. Graph it and state its equation.

Part II: Switch roles and get a new sheet of graph paper or clear the graph on your calculator.

6. **G:** Graph the line through the points $(3, 1)$ and $(-1, -2)$. We will call this line T. **A:** Find the equation for line T.

7. **G:** Plot one point four units down (negative y-direction) and three units to the right (positive x-direction) of the point $(3, 1)$. Draw the line connecting $(3, 1)$ and your new point. Call this line Q. **A:** What is this point? Find the equation for line Q. Explain how the direction and number of units you moved from $(3, 1)$ relate to the slope of your new equation.

8. **G:** Plot the point $(-4, 2)$ and draw a line from the point $(-1, -2)$ to this new point. Call this line M. **A:** Find the equation for line M.

9. **G, A:** Looking at the graph, state how the three lines are related. State how their equations compare.

10. **G, A:** Starting from any point on line Q, how many units left or right (negative or positive x-direction) and how many units up or down (positive or negative y-direction) would you need to move to get to a point that would be on a line perpendicular to line Q? Find a line perpendicular to line Q. Graph it and state its equation. How does it compare to line T?

WRAP-UP

C H A P T E R   3

SUMMARY

Rectangular Coordinate System		**Examples**

Rectangular Coordinate System

x-intercept   The point where a nonhorizontal line intersects the *x*-axis

y-intercept   The point where a nonvertical line intersects the *y*-axis

Examples

For the line $2x + y = 6$, the *x*-intercept is $(3, 0)$ and the *y*-intercept is $(0, 6)$.

Slope

Examples

Slope of a line $\text{Slope} = \dfrac{\text{change in } y\text{-coordinate}}{\text{change in } x\text{-coordinate}}$

$= \dfrac{\text{rise}}{\text{run}}$

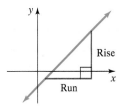

Slope using coordinates   Slope of line through (x_1, y_1) and (x_2, y_2) is $m = \dfrac{y_2 - y_1}{x_2 - x_1},$ provided that $x_2 - x_1 \neq 0$.

If $(x_1, y_1) = (4, -2)$ and $(x_2, y_2) = (3, -6)$, then

$$m = \frac{-6 - (-2)}{3 - 4} = 4.$$

Types of slope

Perpendicular lines   The slope of one line is the negative of the reciprocal of the slope of the other line.

The lines $y = -\dfrac{1}{3}x + 5$ and $y = 3x - 9$ are perpendicular.

Parallel lines   Nonvertical parallel lines have equal slopes.

The lines $y = 2x - 3$ and $y = 2x + 7$ are parallel.

Forms of Linear Equations

Examples

Point-slope form $y - y_1 = m(x - x_1)$
(x_1, y_1) is a point on the line, and *m* is the slope.

Line through $(5, -3)$ with slope 2: $y + 3 = 2(x - 5)$

Slope-intercept form $y = mx + b$
m is the slope, $(0, b)$ is the *y*-intercept, and *y* is a linear function of *x*.
We also write $f(x) = mx + b$.

Line through $(0, -3)$ with slope 2: $y = 2x - 3$
$f(x) = 2x - 3, f(4) = 5$

Standard form	$Ax + By = C$ A and B are not both 0.	$3x - 2y = 12$
Vertical line	$x = k$, where k is any real number. Slope is undefined for vertical lines.	$x = 5$
Horizontal line	$y = k$, where k is any real number. Slope is zero for horizontal lines.	$y = -2$

Graphing Linear Equations

Examples

Point-plotting	Arbitrarily select some points that satisfy the equation, and draw a line through them.	For $y = 2x + 1$, draw a line through $(0, 1)$, $(1, 3)$, and $(2, 5)$.
Intercepts	Find the x- and y-intercepts (provided that they are not the origin), and draw a line through them.	For $x + y = 4$ the intercepts are $(0, 4)$ and $(4, 0)$.
y-intercept and slope	Start at the y-intercept and use the slope to locate a second point, then draw a line through the two points.	For $y = 3x - 2$ start at $(0, -2)$, rise 3 and run 1 to get to $(1, 1)$. Draw a line through the two points.

Linear Inequalities

Examples

Linear inequality	$Ax + By \leq C$, where A and B are not both zero. The symbols $<$, $>$, and $\geq$ are also used.	$2x - 3y \leq 7$ $x - y > 6$
Graphing linear inequalities	Solve for y, then graph the line $y = mx + b$. $y > mx + b$ is the region above the line. $y < mx + b$ is the region below the line.	Graph of $y = x + 2$ is a line. $y > x + 2$ above $y = x + 2$. $y < x + 2$ below $y = x + 2$.
	For inequalities without y, graph $x = k$. $x > k$ is the region to the right of $x = k$. $x < k$ is the region to the left of $x = k$.	The graph of $x > 5$ is to the right of the vertical line $x = 5$, and the graph of $x < 5$ is to left of $x = 5$.
Test points	A linear inequality may also be graphed by graphing the corresponding line and then testing a point to determine which region satisfies the inequality.	
Compound inequalities	Simple inequalities connected by *and* or *or.* For *and,* find the intersection of the regions. For *or,* find the union of the regions.	$x < 3$ and $y > x - 4$ $x + y \geq 1$ or $y \leq x - 1$

Relations and Functions

Examples

Relation	Any set of ordered pairs of real numbers	$\{(1, 2), (1, 3)\}$
Function	A relation in which no two ordered pairs have the same first coordinate and different second coordinates.	$\{(1, 2), (3, 5), (4, 5)\}$ $\{(x, y) \mid y = x^2\}$
	If y is a function of x, then y is uniquely determined by x. A function may be defined by a table, a listing of ordered pairs in set notation, or an equation.	

Domain	The set of first coordinates of the ordered pairs	Function: $y = x^2$ Domain: $(-\infty, \infty)$
Range	The set of second coordinates of the ordered pairs.	Range: $[0, \infty)$
f-notation	If y is a function of x, we write $y = f(x)$.	$y = 2x + 3$ $f(x) = 2x + 3$
Average rate of change of f over $[a, b]$	$\dfrac{f(b) - f(a)}{b - a}$, where a and b are in the domain of f $\dfrac{f(b) - f(a)}{b - a}$ is the slope of the line through $(a, f(a))$ and $(b, f(b))$.	Average rate of change of $f(x) = x^2$ over $[1, 3]$ is $\dfrac{3^2 - 1^2}{3 - 1} = 4.$

Types of Functions		**Examples**
Linear function	$y = mx + b$ or $f(x) = mx + b$ for $m \neq 0$ Domain $(-\infty, \infty)$, range $(-\infty, \infty)$ If $m = 0$, $y = b$ is a constant function. Domain $(-\infty, \infty)$, range $\{b\}$	$f(x) = 2x - 3$
Absolute value function	$y = |x|$ or $f(x) = |x|$ Domain $(-\infty, \infty)$, range $[0, \infty)$	$f(x) = |x + 5|$
Quadratic function	$f(x) = ax^2 + bx + c$ for $a \neq 0$	$f(x) = x^2 - 4x + 3$
Square-root function	$f(x) = \sqrt{x}$ Domain $[0, \infty)$, range $[0, \infty)$	$f(x) = \sqrt{x - 4}$
Vertical-line test	If a graph can be crossed more than once by a vertical line, then it is not the graph of a function.	

ENRICHING YOUR MATHEMATICAL WORD POWER

For each mathematical term, choose the correct meaning.

1. **graph of an equation**
   a. the Cartesian coordinate system
   b. two number lines that intersect at a right angle
   c. the x-axis and y-axis
   d. an illustration in the coordinate plane that shows all ordered pairs that satisfy an equation   d

2. **origin**
   a. the point of intersection of the x- and y-axes
   b. the beginning of algebra
   c. the number 0
   d. the x-axis   a

3. **x-coordinate**
   a. the first number in an ordered pair
   b. the second number in an ordered pair

   c. a point on the x-axis
   d. a point where a graph crosses the x-axis   a

4. **y-intercept**
   a. the second number in an ordered pair
   b. a point at which a graph intersects the y-axis
   c. any point on the y-axis
   d. the point where the y-axis intersects the x-axis   b

5. **coordinate plane**
   a. a matching plane
   b. when the x-axis is coordinated with the y-axis
   c. a plane with a rectangular coordinate system
   d. a coordinated system for graphs   c

6. **independent variable**
   a. the first coordinate of an ordered pair
   b. the second coordinate of an ordered pair

c. the x-axis
d. the y-axis   a

7. dependent variable
a. the first coordinate of an ordered pair
b. the second coordinate of an ordered pair
c. the x-axis
d. the y-axis   b

8. slope
a. the change in x divided by the change in y
b. a measure of the steepness of a line
c. the run divided by the rise
d. the slope of a line   b

9. slope-intercept form
a. $y = mx + b$
b. rise over run
c. the point at which a line crosses the y-axis
d. $y - y_1 = m(x - x_1)$ a

10. point-slope form
a. $Ax + By = C$
b. rise over run
c. $y - y_1 = m(x - x_1)$
d. the slope of a line at a single point   c

11. standard form
a. $y = mx + b$
b. $Ax + By = C$, where A and B are not both 0
c. $y - y_1 = m(x - x_1)$
d. the most common form   b

12. linear inequality in two variables
a. when two lines are not equal
b. line segments that are unequal in length
c. an inequality of the form $Ax + By \geq C$ or with another symbol of inequality
d. an inequality of the form $Ax^2 + By^2 < C^2$ c

13. function
a. a set of ordered pairs of real numbers
b. a set of ordered pairs of real numbers in which no two have the same first coordinates and different second coordinates
c. a set of ordered pairs of real numbers in which no two have the same second coordinates and different first coordinates
d. an equation   b

14. relation
a. a set of ordered pairs of real numbers
b. a set of ordered pairs of real numbers in which no two have the same first coordinates and different second coordinates
c. cousins and second cousins
d. a fraction   a

15. domain
a. the range
b. the set of second coordinates of a relation
c. the independent variable
d. the set of first coordinates of a relation   d

16. f-notation
a. a notation where $f(x)$ is used as the independent variable
b. a notation where $f(x)$ is used as the dependent variable
c. the notation of algebra
d. the notation of exponents   b

17. average rate of change
a. average speed
b. the total divided by the number of scores
c. the average y-coordinate
d. the ratio of the change in y to the change in x d

REVIEW EXERCISES

3.1 *For each point, name the quadrant in which it lies or the axis on which it lies.*

1. $(-3, -2)$ III   **2.** $(0, \pi)$ y-axis   **3.** $(\pi, 0)$ x-axis

4. $(-5, 4)$ II   **5.** $(0, -1)$ y-axis   **6.** $\left(\dfrac{\pi}{2}, 1\right)$ I

7. $(\sqrt{2}, -3)$ IV   **8.** $(6, -3)$ IV

| study | tip |

Note how the review exercises are arranged according to the sections in this chapter. If you are having trouble with a certain type of problem, refer back to the appropriate section for examples and explanations.

Complete the given ordered pairs so that each ordered pair satisfies the given equation.

9. $(0, \ \ ), (\ \ , 0), (4, \ \ ), (\ \ , -3), y = -3x + 2$
$(0, 2), \left(\dfrac{2}{3}, 0\right), (4, -10), \left(\dfrac{5}{3}, -3\right)$

10. $(0, \ \ ), (\ \ , 0), (-6, \ \ ), (\ \ , 5), 2x + 3y = 5$
$\left(0, \dfrac{5}{3}\right), \left(\dfrac{5}{2}, 0\right), \left(-6, \dfrac{17}{3}\right), (-5, 5)$

3.2 *Find the slope of the line through each pair of points.*

11. $(-5, 6), (-2, 9)$ 1   **12.** $(-2, 7), (3, -4)$ $-\dfrac{11}{5}$

13. $(4, 1), (-3, -2)$ $\dfrac{3}{7}$ **14.** $(6, 0), (0, -3)$ $\dfrac{1}{2}$

Solve each problem.

15. What is the slope of any line that is parallel to the line through $(-3, -4)$ and $(5, -1)$? $\dfrac{3}{8}$

16. What is the slope of the line through $(4, 6)$ that is parallel to the line through $(-2, 1)$ and $(7, 1)$? 0

17. What is the slope of any line that is perpendicular to the line through $(-3, 5)$ and $(4, -6)$? $\dfrac{7}{11}$

18. What is the slope of the line through $(1, 2)$ that is perpendicular to the line through $(5, 4)$ and $(5, -2)$? 0

3.3 *Find the slope and y-intercept for each line.*

19. $y = -3x + 4$ $-3, (0, 4)$

20. $2y - 3x + 1 = 0$ $\dfrac{3}{2}, \left(0, -\dfrac{1}{2}\right)$

21. $y - 3 = \dfrac{2}{3}(x - 1)$ $\dfrac{2}{3}, \left(0, \dfrac{7}{3}\right)$

22. $y - 3 = 5$ $0, (0, 8)$

Write each equation in standard form with integral coefficients.

23. $y = \dfrac{2}{3}x - 4$ $2x - 3y = 12$

24. $y = -0.05x + 0.26$ $5x + 100y = 26$

25. $y - 1 = \dfrac{1}{2}(x + 3)$ $x - 2y = -5$

26. $\dfrac{1}{2}x - \dfrac{1}{3}y = \dfrac{1}{4}$ $6x - 4y = 3$

Write the equation of the line containing the given point and having the given slope. Rewrite each equation in standard form with integral coefficients.

27. $(1, -3), m = \dfrac{1}{2}$ $x - 2y = 7$

28. $(0, 2), m = 3$ $3x - y = -2$

29. $(-2, 6), m = -\dfrac{3}{4}$ $3x + 4y = 18$

30. $\left(2, \dfrac{1}{2}\right), m = \dfrac{1}{4}$ $x - 4y = 0$

31. $(3, 5), m = 0$ $y = 5$

32. $(0, 0), m = -1$ $x + y = 0$

Graph each equation.

33. $y = 2x - 3$

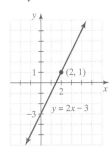

34. $y = \dfrac{2}{3}x + 1$

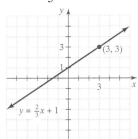

35. $3x - 2y = -6$

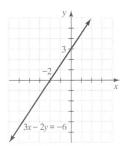

36. $4x + 5y = 10$

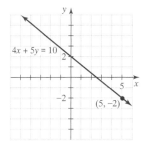

37. $y - 3 = 10$

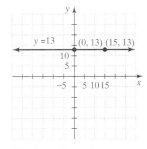

38. $2x = 8$

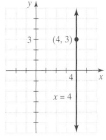

39. $5x - 3y = 7$

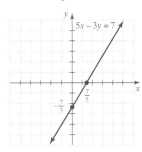

40. $3x + 4y = -1$

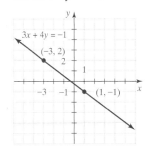

41. $5x + 4y = 100$

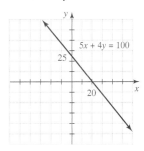

42. $2x - y = 120$

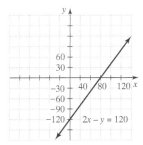

43. $x - 80y = 400$

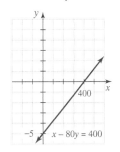

44. $75x + y = 300$

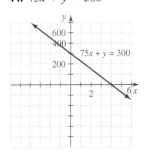

3.4 *Graph each linear inequality.*

45. $y > 3x - 2$

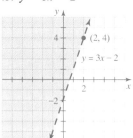

46. $y \leq 2x + 3$

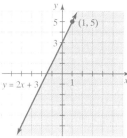

55. $5x - 2y < 9$

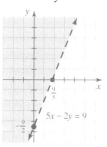

56. $3x + 4y \leq -1$

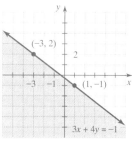

47. $x - y \leq 5$

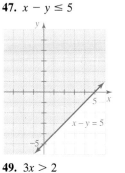

48. $2x + y > 1$

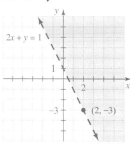

Graph each compound or absolute value inequality.

57. $y > 3$ and $y - x < 5$

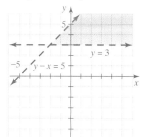

58. $x + y \leq 1$ or $y \leq 4$

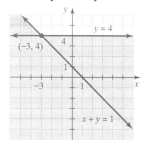

49. $3x > 2$

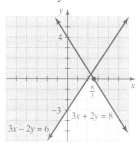

50. $x + 2 \leq 0$

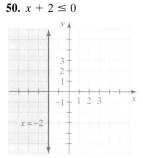

59. $3x + 2y \geq 8$ or
$3x - 2y \leq 6$

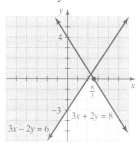

60 $x + 8y > 8$ and
$x - 2y < 10$

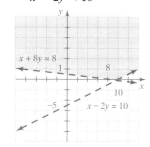

51. $4y \leq 0$

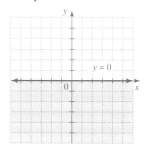

52. $4y - 4 > 0$

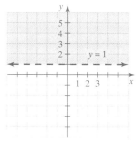

61. $|x + 2y| < 10$

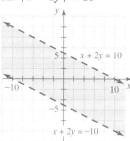

62. $|x - 3y| \geq 9$

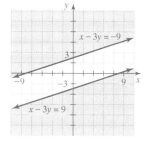

53. $4x - 2y \geq 6$

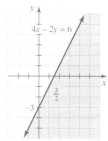

54. $-5x - 3y > 6$

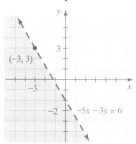

63. $|x| \leq 5$

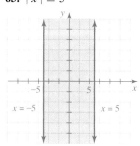

64. $|y| > 6$

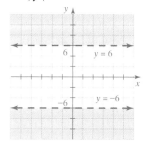

65. $|y - x| > 2$

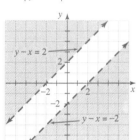

66. $|x - y| \le 1$

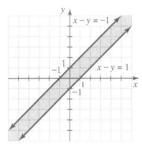

3.5 *Determine whether each relation is a function.*

67. $\{(5, 7), (5, 10)\}$ No

68. $\{(10, 7), (6, 7)\}$ Yes

69. $\{(x, y) \mid y = x^2\}$ Yes

70. $x^2 = 1 + y^2$ No

71. $\{(x, y) \mid x = y^4\}$ No

72. $y = \sqrt{x - 1}$ Yes

Determine the domain and range of each relation.

73. $\{(1, 3), (2, 3)\}$ $\{1, 2\}, \{3\}$

74. $\{(u, v) \mid v = -u^2\}$ $(-\infty, \infty), (-\infty, 0]$

75. $f(t) = 2t - 7$ $(-\infty, \infty), (-\infty, \infty)$

76. $x = \sqrt{y}$ $[0, \infty), [0, \infty)$

77. $y = |x|$ $(-\infty, \infty), [0, \infty)$

78. $y = \sqrt{x - 1}$ $[0, \infty), [-1, \infty)$

Let $f(x) = 2x - 5$ and $g(x) = x^2 + x - 6$. Find the following.

79. $f(0)$ -5

80. $f(-3)$ -11

81. $g(0)$ -6

82. $g(-2)$ -4

83. $g\left(\dfrac{1}{2}\right)$ $-\dfrac{21}{4}$

84. $g\left(-\dfrac{1}{2}\right)$ $-\dfrac{25}{4}$

85. $f(a)$ $2a - 5$

86. $f(x + 3)$ $2x + 1$

87. $f(a - 1)$ $2a - 7$

88. $f(x + h)$ $2x + 2h - 5$

89. a, if $f(a) = 1$ 3

90. x, if $f(x) = 0$ $\dfrac{5}{2}$

Find the average rate of change of the given function on the given interval.

91. $f(x) = 20x^2, [2, 8]$ 200

92. $g(x) = \sqrt{x - 8}, [9, 89]$ $\dfrac{1}{10}$

93. $h(x) = \dfrac{1}{x + 2}, [2, 4]$ $-\dfrac{1}{24}$

94. $j(x) = x^2 - x, [1, 3]$ 3

3.6 *Graph each function and state the domain and range.*

95. $f(x) = 3x - 4$

$(-\infty, \infty), (-\infty, \infty)$

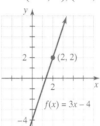

96. $y = 0.3x$

$(-\infty, \infty), (-\infty, \infty)$

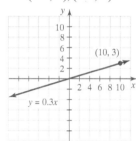

97. $h(x) = |x| - 2$

$(-\infty, \infty), [-2, \infty)$

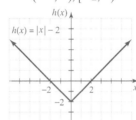

98. $y = |x - 2|$

$(-\infty, \infty), [0, \infty)$

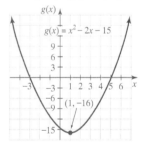

99. $y = x^2 - 2x + 1$

$(-\infty, \infty), [0, \infty)$

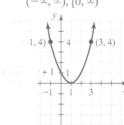

100. $g(x) = x^2 - 2x - 15$

$(-\infty, \infty), [-16, \infty)$

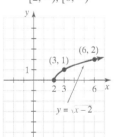

10 $k(x) = \sqrt{x + 2}$

$\infty), [2, \infty)$

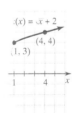

102. $y = \sqrt{x - 2}$

$[2, \infty), [0, \infty)$

103. $y = 30 - x^2$
$(-\infty, \infty), (-\infty, 30]$

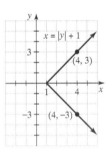

104. $y = 4 - x^2$
$(-\infty, \infty), (-\infty, 4]$

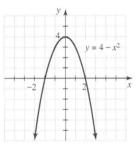

Graph each relation and state its domain and range.

105. $x = 2$
$\{2\}, (-\infty, \infty)$

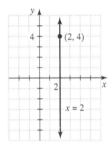

106. $x = y^2 - 1$
$[-1, \infty), (-\infty, \infty)$

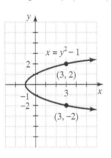

107. $x = |y| + 1$
$[1, \infty), (-\infty, \infty)$

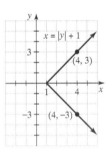

108. $x = \sqrt{y - 1}$
$[0, \infty), [1, \infty)$

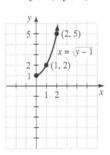

MISCELLANEOUS

Write an equation in standard form with integral coefficients for each line described.

109. The line that crosses the *x*-axis at $(2, 0)$ and the *y*-axis at $(0, -6)$ $3x - y = 6$

110. The line with an *x*-intercept of $(4, 0)$ and slope $-\frac{1}{2}$
$x + 2y = 4$

111. The line through $(-1, 4)$ with slope $-\frac{1}{2}$ $x + 2y = 7$

112. The line through $(2, -3)$ with slope 0 $y = -3$

113. The line through $(2, -6)$ and $(2, 5)$ $x = 2$

114. The line through $(-3, 6)$ and $(4, 2)$ $4x + 7y = 30$

115. The line through $(0, 0)$ perpendicular to $x = 5$ $y = 0$

116. The line through $(2, -3)$ perpendicular to $y = -3x + 5$
$x - 3y = 11$

117. The line through $(-1, 4)$ parallel to $y = 2x + 1$
$2x - y = -6$

118. The line through $(2, 1)$ perpendicular to $y = 10$ $x = 2$

For Exercises 119–122, write an equation in standard form with integral coefficients for each line.

119.

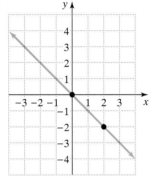

$3x - y = -6$

120.

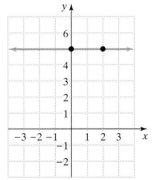

$x + y = 0$

121.

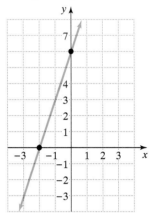

$y = 5$

122.

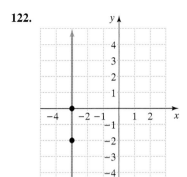

$x = -3$

Use slope to solve each geometric problem.

123. Show that the points $(-5, -5)$, $(-3, -1)$, $(6, 2)$, and $(4, -2)$ are the vertices of a parallelogram.

124. Show that the points $(-5, -5)$, $(4, -2)$, and $(3, 1)$ are the vertices of a right triangle.

125. Show that the points $(-2, 2)$, $(0, 0)$, $(2, 6)$, and $(4, 4)$ are the vertices of a rectangle.

126. Determine whether the points $(2, 1)$, $(4, 7)$, and $(-3, -14)$ lie on a straight line.   Yes

Solve the following problems.

127. *Maximum heart rate.* The maximum heart rate during exercise for a 20 year old is 200 beats per minute, and the maximum heart rate for a 70 year old is 150 (NordicTrack brochure) as shown in the accompanying figure.
  a) Write the maximum heart rate h as a linear function of age a.
  b) What is the maximum heart rate for a 40 year old?
  c) Does your maximum heart rate increase or decrease as you get older?
  a) $h = 220 - a$
  b) 180
  c) decreases

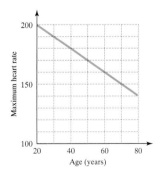

FIGURE FOR EXERCISE 127

128. *Resting heart rate.* A subject is given 3 milligrams (mg) of an experimental drug, and a resting heart rate of 82 is recorded. Another subject is given 5 mg of the same drug, and a resting heart rate of 89 is recorded. If we assume the heart rate, h, is a linear function of the dosage, d, find the linear equation expressing h in terms of d. If a subject is given 10 mg of the drug, what would be the expected heart rate? $h = 3.5d + 71.5$, 106.5

129. *Rental costs.* The charge, C, in dollars, for renting an air hammer from the Tools Is Us Rental Company is determined from the formula $C = 26 + 17d$, where d is the number of days in the rental period. Graph this function for d from 1 to 30. If the air hammer is worth $1,080, then in how many days would the rental charge equal the value of the air hammer?   62 days

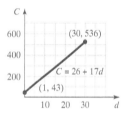

130. *Waist-to-hip ratio.* Dr. Aaron R. Folsom, from the University of Minnesota School of Public Health, has concluded that for a man aged 50 to 69 to be in good health, his waist size w should be less than or equal to 95% of his hip size h as shown in the accompanying figure.
  a) Write an inequality that describes the region shown in the figure.
  b) Is a man in this group with a 36-inch waist and 37-inch hips in good health?
  c) If a man in this group has a waist of 38 inches, then what is his minimum hip size for good health?
  a) $w \le 0.95h$
  b) No
  c) 40 inches

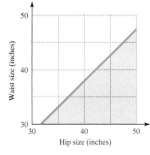

FIGURE FOR EXERCISE 130

CHAPTER 3 TEST

Complete each ordered pair so that it satisfies the given equation.

1. $(0, \ ), (\ , 0), (\ , -8), 2x + y = 5$

$(0, 5), \left(\frac{5}{2}, 0\right), \left(\frac{13}{2}, -8\right)$

Solve each problem.

2. Find the slope of the line through $(-3, 7)$ and $(2, 1)$. $-\frac{6}{5}$

3. Determine the slope and y-intercept for the line $8x - 5y = -10$. $\frac{8}{5}, (0, 2)$

4. Show that $(-1, -2), (0, 0), (6, 2),$ and $(5, 0)$ are the vertices of a parallelogram.

5. Suppose the value, V, in dollars, of a boat is a linear function of its age, a, in years. If a boat was valued at $22,000 brand new and it is worth $16,000 when it is 3 years old, find the linear equation that expresses V in terms of a.
$V = -2000a + 22,000$

For each line described below, write its equation in standard form with integral coefficients.

6. The line with y-intercept $(0, 3)$ and slope $-\frac{1}{2}$ $x + 2y = 6$

7. The line through $(-3, 5)$ with slope -4 $4x + y = -7$

8. The line through $(2, 3)$ that is perpendicular to $3x - 5y = 7$
$5x + 3y = 19$

9. The line shown in the graph:

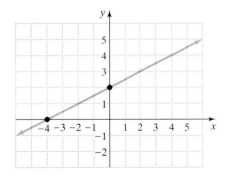

FIGURE FOR EXERCISE 9

$x - 2y = -4$

Sketch the graph of each inequality.

10. $y > -\frac{1}{2}x + 3$

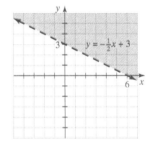

11. $x > 2$ and $x + y > 0$

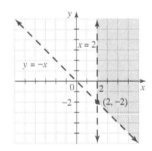

12. $|2x + y| \geq 3$

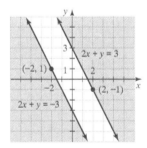

Sketch the graph of each function or relation.

13. $f(x) = -\frac{2}{3}x + 1$

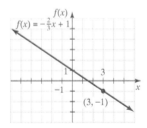

14. $y = |x| - 4$

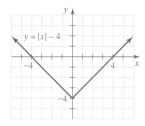

15. $g(x) = x^2 + 2x - 8$

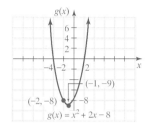

16. $x = y^2$

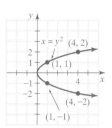

17. $h(x) = 2\sqrt{x} - 1$

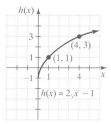

Solve each problem.

18. Determine whether $\{(0, 5), (9, 5), (4, 5)\}$ is a function.

Yes

19. If $f(x) = -2x + 5$, find $f(-3)$ and $f(a + 1)$.

$11, \ -2a + 3$

20. Find the average rate of change of the function $f(x) = 3x^2$ over the interval $[1, 4]$.

15

21. Find the domain and range of the function $f(x) = |x|$.

$(-\infty, \infty), [0, \infty)$

22. A mail order firm charges its customers a shipping and handling fee of $3.00 plus $0.50 per pound for each order shipped. Express the shipping and handling fee as a function of the weight of the order, n.

$S = 0.50n + 3$

23. If a ball is tossed into the air from a height of 6 feet with a velocity of 32 feet per second, then its altitude at time t (in seconds) can be described by the quadratic function $A(t) = -16t^2 + 32t + 6$. For what value of t is the altitude at its maximum?

1 second

Graph Paper

Use these grids for graphing. Make as many copies of this page as you need.

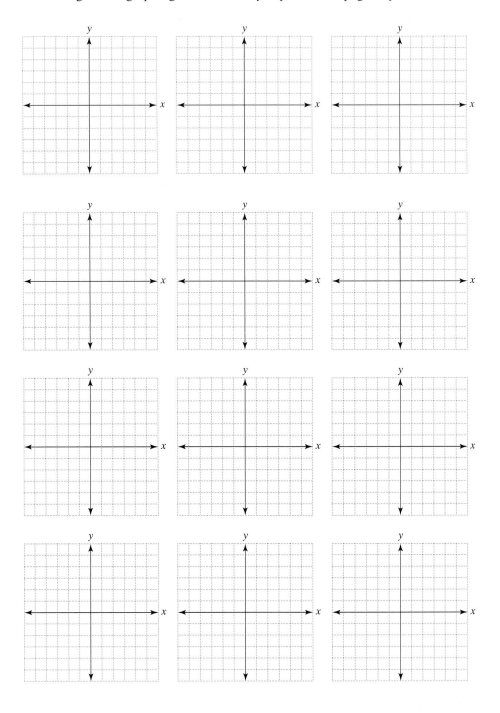

Evaluate each expression.

1. $2^3 \cdot 4^2$ 128

2. $2^7 - 2^6$ 64

3. $3^2 - 4(5)(-2)$ 49

4. $3 - 2\,|\,5 - 7 \cdot 3\,|$ -29

5. $\dfrac{2 - (-3)}{5 - 6}$ -5

6. $\dfrac{-3 - 7}{-1 - (-3)}$ -5

Simplify each expression.

7. $3t \cdot 4t$ $12t^2$

8. $3t + 4t$ $7t$

9. $\dfrac{4x + 8}{4}$ $x + 2$

10. $\dfrac{-8y}{-4} - \dfrac{10y}{-2}$ $7y$

11. $3(x - 4) - 4(5 - x)$ $7x - 32$

12. $-2(3x^2 - x) + 3(2x - 5x^2)$ $-21x^2 + 8x$

Solve each equation.

13. $15(b - 27) = 0$ $\{27\}$

14. $0.05a - 0.04(a - 50) = 4$ $\{200\}$

15. $|\,3v - 7\,| = 0$ $\left\{\dfrac{7}{3}\right\}$

16. $|\,3u - 7\,| = 3$ $\left\{\dfrac{4}{3}, \dfrac{10}{3}\right\}$

17. $|\,3x - 7\,| = -77$ $\varnothing$

18. $|\,3x - 7\,| + 1 = 8$ $\left\{0, \dfrac{14}{3}\right\}$

Graph the solution set to each inequality or compound inequality in one variable on the number line.

19. $2x - 1 > 7$

20. $5 - 3x \le -1$

21. $x - 5 \le 4$ and $3x - 1 < 8$

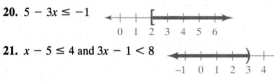

22. $2x \le -6$ or $5 - 2x < -7$

23. $|\,x - 3\,| < 2$

24. $|\,1 - 2x\,| \ge 7$

Graph the solution set to each linear inequality or compound inequality in a rectangular coordinate system.

25. $y < 2x - 1$

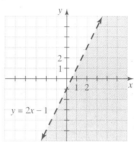

26. $3x - y \le 2$

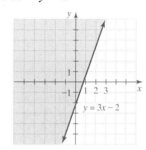

27. $y > x$ and $y < 5 - 3x$

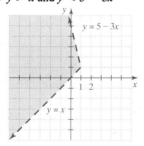

28. $y \le 2$ or $x \ge -3$

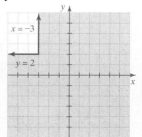

Solve the following problem.

29. *Social Security.* A person retiring in the year 2005 who earned a lifetime average annual salary of $25,000 will

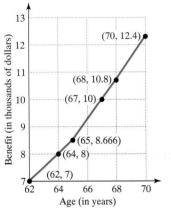

(70, 12.4)

(68, 10.8)

(67, 10)

(65, 8.666)

(64, 8)

(62, 7)

Benefit (in thousands of dollars)

Age (in years)

FIGURE FOR EXERCISE 29

receive a benefit based on age (Social Security Administration, www.ssa.gov). For ages 62 through 64 the benefit in dollars will be $b = 7000 + 500(a - 62)$, for ages 65 through 67 the benefit will be $b = 10{,}000 + 667(a - 67)$, and for ages 68 through 70 the benefit will be $b = 10{,}000 + 800(a - 67)$.

a) Write each benefit formula in slope-intercept form.
b) What will be the annual Social Security benefit for a person who retires in 2005 at age 64?
c) If a person retires in 2005 and gets an $11,600 benefit, then what is the age of that person in 2005?
d) Find the slope of each line segment in the accompanying figure and interpret your results.

    a) $b = 500a - 24{,}000$, $b = 667a - 34{,}689$,
        $b = 800a - 43{,}600$
    b) $8,000
    c) 69
    d) The slopes 500, 667, and 800 indicate the additional amount per year received beyond the basic amount in each category.

Systems of Linear Equations

n his letter to M. Leroy in 1789 Benjamin Franklin said, "in this world nothing is certain but death and taxes." Since that time taxes have become not only inevitable, but also intricate and complex.

Each year the U.S. Congress revises parts of the Federal Income Tax Code. To help clarify these revisions, the Internal Revenue Service issues frequent revenue rulings. In addition, there are seven tax courts that further interpret changes and revisions, sometimes in entirely different ways. Is it any wonder that tax preparation has become complicated and few individuals actually prepare their own taxes? Both corporate and individual tax preparation is a growing business, and there are over 500,000 tax counselors helping more than 60 million taxpayers to file their returns correctly.

Everyone knows that doing taxes involves a lot of arithmetic, but not everyone knows that computing taxes can also involve algebra. In fact, to find state and federal taxes for certain corporations, you must solve a system of equations. You will see an example of using algebra to find amounts of income taxes in Exercises 53 and 54 of Section 4.1.

4.1 SOLVING SYSTEMS BY GRAPHING AND SUBSTITUTION

In Chapter 3 we studied linear equations in two variables, but we have usually considered only one equation at a time. In this chapter we will see problems that involve more than one equation. Any collection of two or more equations is called a **system** of equations. If the equations of a system involve two variables, then the set of ordered pairs that satisfy all of the equations is the **solution set of the system.** In this section we solve systems of linear equations in two variables and use systems to solve problems.

Solving a System by Graphing

Because the graph of each linear equation is a line, points that satisfy both equations lie on both lines. For some systems these points can be found by graphing.

E X A M P L E   1

A system with only one solution

Solve the system by graphing:

$$y = x + 2$$
$$x + y = 4$$

Solution

First write the equations in slope-intercept form:

$$y = x + 2$$
$$y = -x + 4$$

Use the y-intercept and the slope to graph each line. The graph of the system is shown in Fig. 4.1. From the graph it appears that these lines intersect at $(1, 3)$. To be certain, we can check that $(1, 3)$ satisfies both equations. Let $x = 1$ and $y = 3$ in $y = x + 2$ to get

$$3 = 1 + 2.$$

Let $x = 1$ and $y = 3$ in $x + y = 4$ to get

$$1 + 3 = 4.$$

Because $(1, 3)$ satisfies both equations, the solution set to the system is $\{(1, 3)\}$.

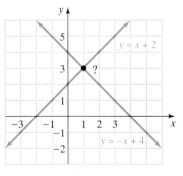

F I G U R E   4 . 1

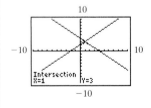

The graphs of the equations in the next example are parallel lines, and there is no point of intersection.

E X A M P L E   2

A system with no solution

Solve the system by graphing:

$$2x - 3y = 6$$
$$3y - 2x = 3$$

Solution

First write each equation in slope-intercept form:

$$2x - 3y = 6 \qquad\qquad 3y - 2x = 3$$
$$-3y = -2x + 6 \qquad\qquad 3y = 2x + 3$$
$$y = \frac{2}{3}x - 2 \qquad\qquad y = \frac{2}{3}x + 1$$

The graph of the system is shown in Fig. 4.2. Because the two lines in Fig. 4.2 are parallel, there is no ordered pair that satisfies both equations. The solution set to the system is the empty set, $\varnothing$.

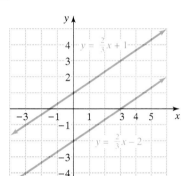

FIGURE 4.2

The equations in the next example are two equations that look different for the same straight line.

E X A M P L E   3

A system with infinitely many solutions

Solve the system by graphing:

$$2(y + 2) = x$$
$$x - 2y = 4$$

Solution

Write each equation in slope-intercept form:

$$2(y + 2) = x \qquad\qquad x - 2y = 4$$
$$2y + 4 = x \qquad\qquad -2y = -x + 4$$
$$y = \frac{1}{2}x - 2 \qquad\qquad y = \frac{1}{2}x - 2$$

Because the equations have the same slope-intercept form, the original equations are equivalent. Their graphs are the same straight line as shown in Fig. 4.3. Every point on the line satisfies both equations of the system. There are infinitely many points in the solution set. The solution set is $\{(x, y) \mid x - 2y = 4\}$.

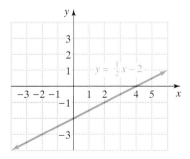

FIGURE 4.3

Independent, Inconsistent, and Dependent Equations

Our first three examples illustrate the three possible ways in which two lines can be positioned in a plane. In Example 1 the lines intersect in a single point. In this case we say that the equations are **independent** or the system is independent. If the two lines are parallel, as in Example 2, then there is no solution to the system, and the equations are **inconsistent** or the system is inconsistent. If the two equations of a system are equivalent, as in Example 3, the equations are **dependent** or the system is dependent. Figure 4.4 shows the types of graphs that correspond to independent, inconsistent, and dependent systems.

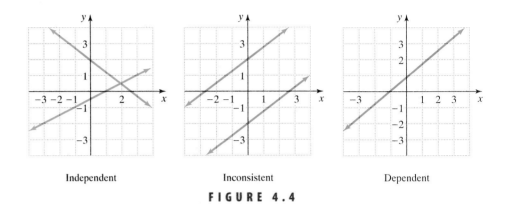

Independent   Inconsistent   Dependent

FIGURE 4.4

Solving by Substitution

Solving a system by graphing is certainly limited by the accuracy of the graph. If the lines intersect at a point whose coordinates are not integers, then it is difficult to determine those coordinates from the graph. The method of solving a system by **substitution** does not depend on a graph and is totally accurate. For substitution we replace a variable in one equation with an equivalent expression obtained from the other equation. Our intention in this substitution step is to eliminate a variable and to give us an equation involving only one variable.

E X A M P L E   4   **An independent system solved by substitution**
Solve the system by substitution:

$$2x + 3y = 8$$
$$y + 2x = 6$$

Solution

We can easily solve $y + 2x = 6$ for y to get $y = -2x + 6$. Now replace y in the first equation by $-2x + 6$:

$$2x + 3y = 8$$
$$2x + 3(-2x + 6) = 8 \quad \text{Substitute } -2x + 6 \text{ for } y.$$
$$2x - 6x + 18 = 8$$
$$-4x = -10$$
$$x = \frac{5}{2}$$

calculator

close-up

To check Example 4, graph
$$y_1 = (8 - 2x)/3$$
and
$$y_2 = -2x + 6.$$
From the CALC menu, choose intersect to have the calculator locate the point of intersection of the two lines.

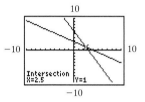

To find y, we let $x = \frac{5}{2}$ in the equation $y = -2x + 6$:

$$y = -2\left(\frac{5}{2}\right) + 6 = -5 + 6 = 1$$

The next step is to check $x = \frac{5}{2}$ and $y = 1$ in each equation. If $x = \frac{5}{2}$ and $y = 1$ in $2x + 3y = 8$, we get

$$2\left(\frac{5}{2}\right) + 3(1) = 8.$$

If $x = \frac{5}{2}$ and $y = 1$ in $y + 2x = 6$, we get

$$1 + 2\left(\frac{5}{2}\right) = 6.$$

Because both of these equations are true, the solution set to the system is $\left\{\left(\frac{5}{2}, 1\right)\right\}$. The equations of this system are independent. ■

EXAMPLE 5

An inconsistent system solved by substitution
Solve by substitution:

$$x - 2y = 3$$
$$2x - 4y = 7$$

helpful hint

The purpose of Example 5 is to show what happens when you try to solve an inconsistent system by substitution. If we had first written the equations in slope-intercept form, we would have known that the lines are parallel and the solution set is the empty set.

Solution

Solve the first equation for x to get $x = 2y + 3$. Substitute $2y + 3$ for x in the second equation:

$$2x - 4y = 7$$
$$2(2y + 3) - 4y = 7$$
$$4y + 6 - 4y = 7$$
$$6 = 7$$

Because $6 = 7$ is incorrect no matter what values are chosen for x and y, there is no solution to this system of equations. The equations are inconsistent. To check, we write each equation in slope-intercept form:

$$x - 2y = 3 \qquad\qquad 2x - 4y = 7$$
$$-2y = -x + 3 \qquad\qquad -4y = -2x + 7$$
$$y = \frac{1}{2}x - \frac{3}{2} \qquad\qquad y = \frac{1}{2}x - \frac{7}{4}$$

The graphs of these equations are parallel lines with different y-intercepts. The solution set to the system is the empty set, $\varnothing$. ■

EXAMPLE 6

A dependent system solved by substitution
Solve by substitution:

$$2x + 3y = 5 + x + 4y$$
$$y = x - 5$$

Solution

Substitute $y = x - 5$ into the first equation:

$$2x + 3(x - 5) = 5 + x + 4(x - 5)$$
$$2x + 3x - 15 = 5 + x + 4x - 20$$
$$5x - 15 = 5x - 15$$

Because the last equation is an identity, any ordered pair that satisfies $y = x - 5$ will also satisfy $2x + 3y = 5 + x + 4y$. The equations of this system are dependent. The solution set to the system is the set of all points that satisfy $y = x - 5$. We write the solution set in set notation as

$$\{(x, y) \mid y = x - 5\}.$$

We can verify this result by writing $2x + 3y = 5 + x + 4y$ in slope-intercept form:

$$2x + 3y = 5 + x + 4y$$
$$3y = -x + 5 + 4y$$
$$-y = -x + 5$$
$$y = x - 5$$

Because this slope-intercept form is identical to the slope-intercept form of the other equation, they are two equations that look different for the same straight line.

If a system is dependent, then an identity will result after the substitution. If the system is inconsistent, then an inconsistent equation will result after the substitution. The strategy for solving an independent system by substitution can be summarized as follows.

The Substitution Method

1. Solve one of the equations for one variable in terms of the other.
2. Substitute into the other equation to get an equation in one variable.
3. Solve for the remaining variable (if possible).
4. Insert the value just found into one of the original equations to find the value of the other variable.
5. Check the two values in both equations.

Applications

Many of the problems that we solved in previous chapters involved more than one unknown quantity. To solve them, we wrote expressions for all of the unknowns in terms of one variable. Now we can solve problems involving two unknowns by using two variables and writing a system of equations.

E X A M P L E   7   **Perimeter of a rectangle**

The length of a rectangular swimming pool is twice the width. If the perimeter is 120 feet, then what are the length and width?

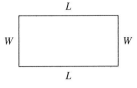

L

W W

L

FIGURE 4.5

Solution

Draw a diagram as shown in Fig. 4.5. If L represents the length and W represents the width, then we can write the following system.

$$L = 2W$$
$$2L + 2W = 120$$

Since $L = 2W$, we can replace L in $2L + 2W = 120$ with $2W$:

$$2(2W) + 2W = 120$$
$$4W + 2W = 120$$
$$6W = 120$$
$$W = 20$$

So the width is 20 feet and the length is 2(20) or 40 feet.   ■

E X A M P L E   8

Tale of two investments

Belinda had \$20,000 to invest. She invested part of it at 10% and the remainder at 12%. If her income from the two investments was \$2160, then how much did she invest at each rate?

helpful hint

In Chapter 2 we would have done Example 8 with one variable by letting x represent the amount invested at 10% and $20,000 - x$ represent the amount invested at 12%.

Solution

Let x be the amount invested at 10% and y be the amount invested at 12%. We can summarize all of the given information in a table:

	Amount	Rate	Interest
First investment	x	10%	$0.10x$
Second investment	y	12%	$0.12y$

We can write one equation about the amounts invested and another about the interest from the investments:

$$x + y = 20,000 \quad \text{Total amount invested}$$
$$0.10x + 0.12y = 2160 \quad \text{Total interest}$$

Solve the first equation for x to get $x = 20,000 - y$. Substitute $20,000 - y$ for x in the second equation:

$$0.10x + 0.12y = 2160$$
$$0.10(20,000 - y) + 0.12y = 2160 \quad \text{Replace } x \text{ by } 20,000 - y.$$
$$2000 - 0.10y + 0.12y = 2160 \quad \text{Solve for } y.$$
$$0.02y = 160$$
$$y = 8,000$$
$$x = 12,000 \quad \text{Because } x = 20,000 - y$$

calculator

close-up

To check Example 8, graph
$$y_1 = 20,000 - x$$
and
$$y_2 = (2160 - 0.1x)/0.12.$$
The viewing window needs to be large enough to contain the point of intersection. Use the intersection feature to find the point of intersection.

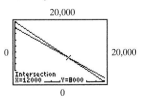

To check this answer, find 10% of \$12,000 and 12% of \$8,000:

$$0.10(12,000) = 1,200$$
$$0.12(8,000) = 960$$

Because \$1,200 + \$960 = \$2,160 and \$8,000 + \$12,000 = \$20,000, we can be certain that Belinda invested \$12,000 at 10% and \$8,000 at 12%.   ■

M A T H   A T   W O R K $x^2 + (x+1)^2 = 5^2$

Accountants work with both people and numbers. When Maria L. Manning, an auditor at Deloitte & Touche LLP, is preparing for an audit, she studies the numbers on the balance sheet and income statement, comparing the current fiscal year to the prior one. The purpose of an independent audit is to give investors a realistic view of a company's finances. To determine that a company's financial statements are fairly stated in accordance with the General Accepted Accounting Procedures (GAAP), she first interviews the comptroller to get the story behind the numbers. Typical questions she could ask are: How productive was your year? Are there any new products? What was behind the big stories in the newspapers? Then Ms. Manning and members of the audit team test the financial statement in detail, closely examining accounts relating to unusual losses or profits.

ACCOUNTANT

Ms. Manning is responsible for both manufacturing and mutual fund companies. At a manufacturing company, accounts receivable and inventory are two key components of an audit. For example, to test inventory, Ms. Manning visits a company's warehouse and physically counts all the items for sale to verify a company's assets. For a mutual fund company the audit team pays close attention to current events, for they indirectly affect the financial industry.

In Exercises 55 and 56 of this section you will work problems that involve one aspect of cost accounting: calculating the amount of taxes and bonuses paid by a company.

WARM-UPS

True or false? Explain your answer.

1. The ordered pair (1, 2) is in the solution set to the equation $2x + y = 4$. True

2. The ordered pair (1, 2) satisfies $2x + y = 4$ and $3x - y = 6$.   False

3. The ordered pair (2, 3) satisfies $4x - y = 5$ and $4x - y = -5$.   False

4. If two distinct straight lines in the coordinate plane are not parallel, then they intersect in exactly one point.   True

5. The substitution method is used to eliminate a variable.   True

6. No ordered pair satisfies $y = 3x - 5$ and $y = 3x + 1$.   True

7. The equations $y = 3x - 6$ and $y = 2x + 4$ are independent.   True

8. The equations $y = 2x + 7$ and $y = 2x + 8$ are inconsistent.   True

9. The graphs of dependent equations are the same.   True

10. The graphs of independent linear equations intersect at exactly one point. True

4.1 EXERCISES

Reading and Writing *After reading this section, write out the answers to these questions. Use complete sentences.*

1. How do we solve a system of linear equations by graphing?
   The intersection point of the graphs is the solution to an independent system.

2. How can you determine whether a system has no solution by graphing?
   The lines do not intersect if the system has no solution.

3. What is the major disadvantage to solving a system by graphing?
   The graphing method can be very inaccurate.

4. How do we solve systems by substitution?
   For substitution we eliminate a variable by substituting one equation into the other.

5. How can you identify an inconsistent system when solving by substitution?
   If the equation you get after substituting turns out to be incorrect, such as $0 = 9$, then the system has no solution.

6. How can you identify a dependent system when solving by substitution?
   If the substitution results in an identity, then the system is dependent.

Solve each system by graphing. See Examples 1–3.

7. $y = 2x$
   $y = -x + 3$
   $\{(1, 2)\}$

8. $y = x - 3$
   $y = -x + 1$
   $\{(2, -1)\}$

9. $y = 2x - 1$
   $2y = x - 2$
   $\{(0, -1)\}$

10. $y = 2x + 1$
    $x + y = -2$
    $\{(-1, -1)\}$

11. $y = x - 3$
    $x - 2y = 4$
    $\{(2, -1)\}$

12. $y = -3x$
    $x + y = 2$
    $\{(-1, 3)\}$

13. $2y - 2x = 2$
    $2y - 2x = 6$
    $\varnothing$

14. $3y - 3x = 9$
    $x - y = 1$
    $\varnothing$

15. $y = -\dfrac{1}{2}x + 4$
    $x + 2y = 8$
    $\{(x, y) \mid x + 2y = 8\}$

16. $2x - 3y = 6$
    $y = \dfrac{2}{3}x - 2$
    $\{(x, y) \mid 2x - 3y = 6\}$

The graphs of the following systems are given in (a) through (d). Match each system with the correct graph.

17. $5x + 4y = 7$
    $x - 3y = 9$
    c

18. $3x - 5y = -9$
    $5x - 6y = -8$
    d

19. $4x - 5y = -2$
    $3y - x = -3$
    b

20. $4x + 5y = -2$
    $4y - x = 11$
    a

a)

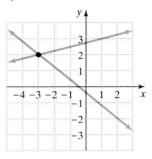

b)

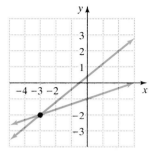

c)

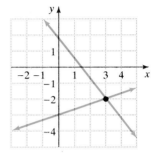

d)

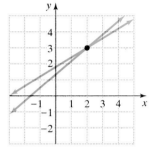

Solve each system by the substitution method. Determine whether the equations are independent, dependent, or inconsistent. See Examples 4–6.

21. $y = x - 5$
    $2x - 5y = 1$
    $\{(8, 3)\}$, independent

22. $y = x + 4$
    $3y - 5x = 6$
    $\{(3, 7)\}$, independent

23. $x = 2y - 7$
    $3x + 2y = -5$
    $\{(-3, 2)\}$, independent

24. $x = y + 3$
    $3x - 2y = 4$
    $\{(-2, -5)\}$, independent

25. $x - y = 5$
    $2x = 2y + 14$
    $\varnothing$, inconsistent

26. $2x - y = 3$
    $2y = 4x - 6$
    $\{(x, y) \mid 2x - y = 3\}$,
    dependent

27. $y = 2x - 5$
    $y + 1 = 2(x - 2)$
    $\{(x, y) \mid y = 2x - 5\}$,
    dependent

28. $3x - 6y = 5$
    $2y = 4x - 6$
    $\left\{\left(\dfrac{13}{9}, -\dfrac{1}{9}\right)\right\}$,
    independent

29. $2x + y = 9$
    $2x - 5y = 15$
    $\{(5, -1)\}$, independent

30. $3y - x = 0$
    $x - 4y = -2$
    $\{(6, 2)\}$, independent

31. $x - y = 0$
    $2x + 3y = 35$
    $\{(7, 7)\}$, independent

32. $2y = x + 6$
    $-3x + 2y = -2$
    $\{(4, 5)\}$, independent

33. $x + y = 40$
    $0.1x + 0.08y = 3.5$
    $\{(15, 25)\}$, independent

34. $x - y = 10$
    $0.2x + 0.05y = 7$
    $\{(30, 20)\}$, independent

35. $y = 2x - 30$
$\frac{1}{5}x - \frac{1}{2}y = -1$
$\{(20, 10)\}$, independent

36. $3x - 5y = 4$
$y = \frac{3}{4}x - 2$
$\{(8, 4)\}$, independent

37. $x + y = 4$
$x - y = 5$
$\left\{\left(\frac{9}{2}, -\frac{1}{2}\right)\right\}$, independent

38. $y = 2x - 3$
$y = 3x - 3$
$\{(0, -3)\}$, independent

39. $2x - y = 4$
$2x - y = 3$
$\varnothing$, inconsistent

40. $y = 3(x - 4)$
$3x - y = 12$
$\{(x, y) \mid 3x - y = 12\}$, dependent

41. $3(y - 1) = 2(x - 3)$
$3y - 2x = -3$
$\{(x, y) \mid 3y - 2x = -3\}$, dependent

42. $y = 3x$
$y = 3x + 1$
$\varnothing$, inconsistent

43. $x - y = -0.375$
$1.5x - 3y = -2.25$
$\{(0.75, 1.125)\}$, independent

44. $y - 2x = 1.875$
$2.5y - 3.5x = 11.8125$
$\{(4.75, 11.375)\}$, independent

In Exercises 45–58, write a system of two equations in two unknowns for each problem. Solve each system by substitution. See Examples 7 and 8.

45. *Perimeter of a rectangle.* The length of a rectangular swimming pool is 15 feet longer than the width. If the perimeter is 82 feet, then what are the length and width?
Width 13 feet, length 28 feet

46. *Household income.* Alkena and Hsu together earn $84,326 per year. If Alkena earns $12,468 more per year than Hsu, then how much does each of them earn per year?
Hsu $35,929, Alkena $48,397

47. *Different interest rates.* Mrs. Brighton invested $30,000 and received a total of $2,300 in interest. If she invested part of the money at 10% and the remainder at 5%, then how much did she invest at each rate?
$14,000 at 5%, $16,000 at 10%

48. *Different growth rates.* The combined population of Marysville and Springfield was 25,000 in 1990. By 1995 the population of Marysville had increased by 10%, while Springfield had increased by 9%. If the total population increased by 2,380 people, then what was the population of each city in 1990?
Marysville 13,000, Springfield 12,000

49. *Finding numbers.* The sum of two numbers is 2, and their difference is 26. Find the numbers.
-12 and 14

50. *Finding more numbers.* The sum of two numbers is -16, and their difference is 8. Find the numbers.
-4 and -12

51. *Toasters and vacations.* During one week a land developer gave away Florida vacation coupons or toasters to 100 potential customers who listened to a sales presentation. It costs the developer $6 for a toaster and $24 for a Florida

vacation coupon. If his bill for prizes that week was $708, then how many of each prize did he give away?
94 toasters, 6 vacation coupons

52. *Ticket sales.* Tickets for a concert were sold to adults for $3 and to students for $2. If the total receipts were $824 and twice as many adult tickets as student tickets were sold, then how many of each were sold?
103 student tickets, 206 adult tickets

53. *Corporate taxes.* According to Bruce Harrell, CPA, the amount of federal income tax for a class C corporation is deductible on the Louisiana state tax return, and the amount of state income tax for a class C corporation is deductible on the federal tax return. So for a state tax rate of 5% and a federal tax rate of 30%, we have

state tax = 0.05(taxable income − federal tax)

and

federal tax = 0.30(taxable income − state tax).

Find the amounts of state and federal income tax for a class C corporation that has a taxable income of $100,000.
State tax $3,553, federal tax $28,934

54. *More taxes.* Use information given in Exercise 53 to find the amounts of state and federal income tax for a class C corporation that has a taxable income of $300,000. Use a state tax rate of 6% and a federal tax rate of 40%.
State tax $11,066, federal tax $115,574

55. *Cost accounting.* The problems presented in this exercise and the next are encountered in cost accounting. A company has agreed to distribute 20% of its net income N to its employees as a bonus; $B = 0.20N$. If the company has income of $120,000 before the bonus, the bonus B is deducted from the $120,000 as an expense to determine net income; $N = 120,000 - B$. Solve the system of two equations in N and B to find the amount of the bonus.
$20,000

56. *Bonus and taxes.* A company has an income of $100,000 before paying taxes and a bonus. The bonus B is to be 20% of the income after deducting income taxes T but before deducting the bonus. So

$$B = 0.20(100{,}000 - T).$$

Because the bonus is a deductible expense, the amount of income tax T at a 40% rate is 40% of the income after

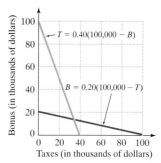

y-axis: Bonus (in thousands of dollars): 100, 80, 60, 40, 20, 0
$T = 0.40(100{,}000 - B)$
$B = 0.20(100{,}000 - T)$
x-axis: 0, 20, 40, 60, 80, 100 — Taxes (in thousands of dollars)

FIGURE FOR EXERCISE 56

deducting the bonus. So

$$T = 0.40(100,000 - B).$$

a) Use the accompanying graph to estimate the values of T and B that satisfy both equations. (35,000, 15,000)
b) Solve the system algebraically to find the bonus and the amount of tax. Bonus $13,043, taxes $34,783

57. *Textbook case.* The accompanying graph shows the cost of producing textbooks and the revenue from the sale of those textbooks.

a) What is the cost of producing 10,000 textbooks?
b) What is the revenue when 10,000 textbooks are sold?
c) For what number of textbooks is the cost equal to the revenue?
d) The cost of producing zero textbooks is called the *fixed cost.* Find the fixed cost.
   a) $500,000   b) $300,000   c) 20,000   d) $400,000

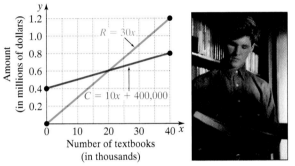

FIGURE FOR EXERCISE 57

58. *Free market.* The function $S = 5000 + 200x$ and $D = 9500 - 100x$ express the supply S and the demand D, respectively, for a popular compact disk brand as a function of its price x (in dollars).

a) Graph the functions on the same coordinate system.
b) What happens to the supply as the price increases?
c) What happens to the demand as the price increases?

d) The price at which supply and demand are equal is called the *equilibrium price.* What is the equilibrium price?

a)

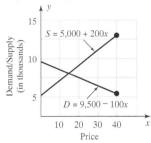

b) The supply increases.   c) The demand decreases.
d) Equilibrium price is $15.

GETTING MORE INVOLVED

 59. *Discussion.* Which of the following equations is not equivalent to $2x - 3y = 6$?

a) $3y - 2x = 6$ b) $y = \dfrac{2}{3}x - 2$

c) $x = \dfrac{3}{2}y + 3$ d) $2(x - 5) = 3y - 4$ a

60. *Discussion.* Which of the following equations is inconsistent with the equation $3x + 4y = 8$?

a) $y = \dfrac{3}{4}x + 2$ b) $6x + 8y = 16$

c) $y = -\dfrac{3}{4}x + 8$ d) $3x - 4y = 8$ c

GRAPHING CALCULATOR EXERCISES

61. Solve each system by graphing each pair of equations on a graphing calculator and using the trace feature or intersect feature to estimate the point of intersection. Find the coordinates of the intersection to the nearest tenth.

a) $y = 3.5x - 7.2$
   $y = -2.3x + 9.1$
   (2.8, 2.6)

b) $2.3x - 4.1y = 3.3$
   $3.4x + 9.2y = 1.3$
   (1.0, -0.2)

4.2 THE ADDITION METHOD

In Section 4.1 you used substitution to eliminate a variable in a system of equations. In this section we see another method for eliminating a variable in a system of equations.

The Addition Method

In the **addition method** we eliminate a variable by adding the equations.

E X A M P L E   1   **An independent system solved by addition**

Solve the system by the addition method:

$$3x - 5y = -9$$
$$4x + 5y = 23$$

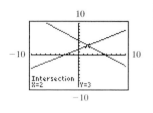
Solution

The addition property of equality allows us to add the same number to each side of an equation. We can also use the addition property of equality to add the two left sides and add the two right sides:

$$3x - 5y = -9$$
$$4x + 5y = 23$$
$$7x \quad\quad = 14 \quad\text{Add.}$$
$$x = 2$$

The y-term was eliminated when we added the equations because the coefficients of the y-terms were opposites. Now use $x = 2$ in one of the original equations to find y. It does not matter which original equation we use. In this example we will use both equations to see that we get the same y in either case.

$$3x - 5y = -9 \qquad\qquad 4x + 5y = 23$$
$$3(2) - 5y = -9 \quad\text{Replace } x \text{ by 2.} \quad 4(2) + 5y = 23$$
$$6 - 5y = -9 \quad\text{Solve for } y. \qquad 8 + 5y = 23$$
$$-5y = -15 \qquad\qquad\qquad 5y = 15$$
$$y = 3 \qquad\qquad\qquad\qquad y = 3$$

Because $3(2) - 5(3) = -9$ and $4(2) + 5(3) = 23$ are both true, $(2, 3)$ satisfies both equations. The solution set is $\{(2, 3)\}$.

Actually the addition method can be used to eliminate any variable whose coefficients are opposites. If neither variable has coefficients that are opposites, then we use the multiplication property of equality to change the coefficients of the variables, as shown in Examples 2 and 3.

E X A M P L E   2

Using multiplication and addition

Solve the system by the addition method:

$$2x - 3y = -13$$
$$5x - 12y = -46$$

Solution

If we multiply both sides of the first equation by -4, the coefficients of y will be 12 and -12, and y will be eliminated by addition.

$$(-4)(2x - 3y) = (-4)(-13) \quad\text{Multiply each side by } -4.$$
$$5x - 12y = -46$$

$$-8x + 12y = 52$$
$$\underline{5x - 12y = -46} \qquad\text{Add.}$$
$$-3x \quad\quad = 6$$
$$x = -2$$

Replace x by -2 in one of the original equations to find y:

$$2x - 3y = -13$$
$$2(-2) - 3y = -13$$
$$-4 - 3y = -13$$
$$-3y = -9$$
$$y = 3$$

Because $2(-2) - 3(3) = -13$ and $5(-2) - 12(3) = -46$ are both true, the solution set is $\{(-2, 3)\}$. ◼

E X A M P L E   3

Multiplying both equations before adding

Solve the system by the addition method:

$$-2x + 3y = 6$$
$$3x - 5y = -11$$

Solution

To eliminate x, we multiply the first equation by 3 and the second by 2:

$$3(-2x + 3y) = 3(6) \qquad \text{Multiply each side by 3.}$$
$$2(3x - 5y) = 2(-11) \qquad \text{Multiply each side by 2.}$$

$$\begin{array}{r} -6x + 9y = 18 \\ \underline{6x - 10y = -22} \qquad \text{Add.} \\ -y = -4 \\ y = 4 \end{array}$$

Note that we could have eliminated y by multiplying by 5 and 3. Now insert $y = 4$ into one of the original equations to find x:

$$-2x + 3(4) = 6 \qquad \text{Let } y = 4 \text{ in } -2x + 3y = 6.$$
$$-2x + 12 = 6$$
$$-2x = -6$$
$$x = 3$$

Check that $(3, 4)$ satisfies both equations. The solution set is $\{(3, 4)\}$. ◼

We can always use the addition method as long as the equations in a system are in the same form.

E X A M P L E   4

To check Example 4, graph
$$y_1 = (5x + 7)/-4$$
and
$$y_2 = (-5x + 12)/4.$$
Since the lines appear to be parallel, the graph supports the conclusion that the system is inconsistent.

Using the addition method for an inconsistent system

Solve the system:

$$-4y = 5x + 7$$
$$4y = -5x + 12$$

Solution

If these equations are added, both variables are eliminated:

$$\begin{array}{r} -4y = 5x + 7 \\ \underline{4y = -5x + 12} \\ 0 = 19 \end{array}$$

Because this equation is inconsistent, the original equations are inconsistent. The solution set to the system is the empty set, $\varnothing$. ◼

Equations Involving Fractions or Decimals

When a system of equations involves fractions or decimals, we can use the multiplication property of equality to eliminate the fractions or decimals.

E X A M P L E   5

A system with fractions

Solve the system:

$$\frac{1}{2}x - \frac{2}{3}y = 7$$

$$\frac{2}{3}x - \frac{3}{4}y = 11$$

Solution

Multiply the first equation by 6 and the second equation by 12:

$$6\left(\frac{1}{2}x - \frac{2}{3}y\right) = 6(7) \qquad \rightarrow \qquad 3x - 4y = 42$$

$$12\left(\frac{2}{3}x - \frac{3}{4}y\right) = 12(11) \qquad \rightarrow \qquad 8x - 9y = 132$$

To eliminate x, multiply the first equation by -8 and the second by 3:

$$-8(3x - 4y) = -8(42) \qquad \rightarrow \qquad -24x + 32y = -336$$

$$3(8x - 9y) = 3(132) \qquad \rightarrow \qquad \underline{24x - 27y = 396}$$

$$5y = 60$$

$$y = 12$$

Substitute $y = 12$ into the first of the original equations:

$$\frac{1}{2}x - \frac{2}{3}(12) = 7$$

$$\frac{1}{2}x - 8 = 7$$

$$\frac{1}{2}x = 15$$

$$x = 30$$

Check $(30, 12)$ in the original system. The solution set is $\{(30, 12)\}$.

The strategy for solving a system by addition is summarized as follows.

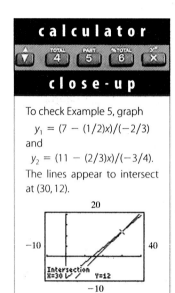

The Addition Method

1. Write both equations in the same form (usually $Ax + By = C$).
2. Multiply one or both of the equations by appropriate numbers (if necessary) so that one of the variables will be eliminated by addition.
3. Add the equations to get an equation in one variable.
4. Solve the equation in one variable.
5. Substitute the value obtained for one variable into one of the original equations to obtain the value of the other variable.
6. Check the two values in both of the original equations.

Applications

Any system of two linear equations in two variables can be solved by either the addition method or substitution. In applications we use whichever method appears to be the simpler for the problem at hand.

EXAMPLE 6

Fajitas and burritos

At the Cactus Cafe the total price for four fajita dinners and three burrito dinners is $48, and the total price for three fajita dinners and two burrito dinners is $34. What is the price of each type of dinner?

Solution

Let x represent the price (in dollars) of a fajita dinner, and let y represent the price (in dollars) of a burrito dinner. We can write two equations to describe the given information:

$$4x + 3y = 48$$
$$3x + 2y = 34$$

Because 12 is the least common multiple of 4 and 3 (the coefficients of x), we multiply the first equation by -3 and the second by 4:

$$-3(4x + 3y) = -3(48) \quad \text{Multiply each side by } -3.$$
$$4(3x + 2y) = 4(34) \quad \text{Multiply each side by } 4.$$

$$-12x - 9y = -144$$
$$\underline{12x + 8y = 136} \quad \text{Add.}$$
$$-y = -8$$
$$y = 8$$

To find x, use $y = 8$ in the first equation $4x + 3y = 48$:

$$4x + 3(8) = 48$$
$$4x + 24 = 48$$
$$4x = 24$$
$$x = 6$$

So the fajita dinners are $6 each, and the burrito dinners are $8 each. Check this solution in the original problem. ∎

EXAMPLE 7

Mixing cooking oil

Canola oil is 7% saturated fat, and corn oil is 14% saturated fat. Crisco sells a blend, Crisco Canola and Corn Oil, which is 11% saturated fat. How many gallons of each type of oil must be mixed to get 280 gallons of this blend?

Solution

Let x represent the number of gallons of canola oil, and let y represent the number of gallons of corn oil. Make a table to summarize all facts:

	Amount (gallons)	% fat	Amount of fat (gallons)
Canola oil	x	7	$0.07x$
Corn oil	y	14	$0.14y$
Canola and Corn Oil	280	11	0.11(280) or 30.8

We can write two equations to express the following facts: (1) the total amount of oil is 280 gallons and (2) the total amount of fat is 30.8 gallons. Then we can use multiplication and addition to solve the system.

$$(1) \qquad x + y = 280 \qquad \text{Multiply by } -0.07. \quad -0.07x - 0.07y = -19.6$$
$$(2) \quad 0.07x + 0.14y = 30.80 \qquad \qquad \underline{\quad 0.07x + 0.14y = \quad 30.8}$$
$$0.07y = \quad 11.2$$
$$y = \frac{11.2}{0.07} = 160$$

If $y = 160$ and $x + y = 280$, then $x = 120$. Check that $0.07(120) + 0.14(160) = 30.8$. So it takes 120 gallons of canola oil and 160 gallons of corn oil to make 280 gallons of Crisco Canola and Corn Oil. ∎

WARM-UPS

True or false? Explain your answer.

Exercises 1–6 refer to the following systems.

a) $3x - y = 9$ **b)** $4x - 2y = 20$ **c)** $x - y = 6$
     $2x + y = 6$ $-2x + y = -10$ $x - y = 7$

1. To solve system (a) by addition, we simply add the equations. True
2. To solve system (a) by addition, we can multiply the first equation by 2 and the second by 3 and then add. False
3. To solve system (b) by addition, we can multiply the second equation by 2 and then add. True
4. Both $(0, -10)$ and $(5, 0)$ are in the solution set to system (b). True
5. The solution set to system (b) is the set of all real numbers. False
6. System (c) has no solution. True
7. Both the addition method and substitution method are used to eliminate a variable from a system of two linear equations in two variables. True
8. For the addition method, both equations must be in standard form. False
9. To eliminate fractions in an equation, we multiply each side by the least common denominator of all fractions involved. True
10. We can eliminate either variable by using the addition method. True

4.2   EXERCISES

Reading and Writing   *After reading this section, write out the answers to these questions. Use complete sentences.*

1. What method is presented in this section for solving a system of linear equations?
   In this section we learned the addition method.

2. What are we trying to accomplish by adding the equations?
   We try to eliminate a variable by adding the equations.

3. What must we sometimes do before we add the equations?
   In some cases we multiply one or both of the equations on each side to change the coefficients of the variable that we are trying to eliminate.

4. How can you recognize an inconsistent system when solving by addition?
   If a false equation, such as $3 = 4$, results from addition of the equations, then the equations are inconsistent.

5. How can you recognize a dependent system when solving by addition?
   If an identity, such as $0 = 0$, results from addition of the equations, then the equations are dependent.

6. For which systems is the addition method easier to use than substitution?
   Addition is usually easier to use when the equations are in the same form.

Solve each system by the addition method. See Examples 1–3.

7. $x + y = 7$
    $x - y = 9$
    $\{(8, -1)\}$

8. $3x - 4y = 11$
    $-3x + 2y = -7$
    $\{(1, -2)\}$

9. $x - y = 12$
    $2x + y = 3$
    $\{(5, -7)\}$

10. $x - 2y = -1$
    $-x + 5y = 4$
    $\{(1, 1)\}$

11. $2x - y = -5$
    $3x + 2y = 3$
    $\{(-1, 3)\}$

12. $3x + 5y = -11$
    $x - 2y = 11$
    $\{(3, -4)\}$

13. $2x - 5y = 13$
    $3x + 4y = -15$
    $\{(-1, -3)\}$

14. $3x + 4y = -5$
    $5x + 6y = -7$
    $\{(1, -2)\}$

15. $2x = 3y + 11$
    $7x - 4y = 6$
    $\{(-2, -5)\}$

16. $2x = 2 - y$
    $3x + y = -1$
    $\{(-3, 8)\}$

17. $x + y = 48$
    $12x + 14y = 628$
    $\{(22, 26)\}$

18. $x + y = 13$
    $22x + 36y = 356$
    $\{(8, 5)\}$

Solve each system by the addition method. Determine whether the equations are independent, dependent, or inconsistent. See Example 4.

19. $3x - 4y = 9$
    $-3x + 4y = 12$
    $\varnothing$, inconsistent

20. $x - y = 3$
    $-6x + 6y = 17$
    $\varnothing$, inconsistent

21. $5x - y = 1$
    $10x - 2y = 2$
    $\{(x, y) \mid 5x - y = 1\}$, dependent

22. $4x + 3y = 2$
    $-12x - 9y = -6$
    $\{(x, y) \mid 4x + 3y = 2\}$, dependent

23. $2x - y = 5$
    $2x + y = 5$
    $\left\{\left(\dfrac{5}{2}, 0\right)\right\}$, independent

24. $-3x + 2y = 8$
    $3x + 2y = 8$
    $\{(0, 4)\}$, independent

Solve each system by the addition method. See Example 5.

25. $\dfrac{1}{4}x + \dfrac{1}{3}y = 5$
    $x - y = 6$
    $\{(12, 6)\}$

26. $\dfrac{3x}{2} - \dfrac{2y}{3} = 10$
    $\dfrac{1}{2}x + \dfrac{1}{2}y = -1$
    $\{(4, -6)\}$

27. $\dfrac{x}{4} - \dfrac{y}{3} = -4$
    $\dfrac{x}{8} + \dfrac{y}{6} = 0$
    $\{(-8, 6)\}$

28. $\dfrac{x}{3} - \dfrac{y}{2} = -\dfrac{5}{6}$
    $\dfrac{x}{5} - \dfrac{y}{3} = -\dfrac{3}{5}$
    $\{(2, 3)\}$

29. $\dfrac{1}{8}x + \dfrac{1}{4}y = 5$
    $\dfrac{1}{16}x + \dfrac{1}{2}y = 7$
    $\{(16, 12)\}$

30. $\dfrac{3}{7}x + \dfrac{5}{9}y = 27$
    $\dfrac{1}{9}x + \dfrac{2}{7}y = 7$
    $\{(63, 0)\}$

31. $0.05x + 0.10y = 1.30$
    $x + y = 19$
    $\{(12, 7)\}$

32. $0.1x + 0.06y = 9$
    $0.09x + 0.5y = 52.7$
    $\{(30, 100)\}$

33. $x + y = 1200$
    $0.12x + 0.09y = 120$
    $\{(400, 800)\}$

34. $x - y = 100$
    $0.20x + 0.06y = 150$
    $\{(600, 500)\}$

35. $1.5x - 2y = -0.25$
    $3x + 1.5y = 6.375$
    $\{(1.5, 1.25)\}$

36. $3x - 2.5y = 7.125$
    $2.5x - 3y = 7.3125$
    $\{(1.125, -1.5)\}$

Write a system of two equations in two unknowns for each problem. Solve each system by the method of your choice. See Example 6.

37. *Coffee and doughnuts.* On Monday, Archie paid $2.54 for three doughnuts and two coffees. On Tuesday he paid $2.46 for two doughnuts and three coffees. On Wednesday he was tired of paying the tab and went out for coffee by himself. What was his bill for one doughnut and one coffee?
$1.00

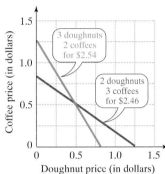

FIGURE FOR EXERCISE 37

38. *Books and magazines.* At Gwen's garage sale, all books were one price, and all magazines were another price. Harriet bought four books and three magazines for $1.45, and June bought two books and five magazines for $1.25. What was the price of a book and what was the price of a magazine?
Books $0.25 each, magazines $0.15 each

39. *Boys and girls.* One-half of the boys and one-third of the girls of Freemont High attended the homecoming game, whereas one-third of the boys and one-half of the girls attended the homecoming dance. If there were 570 students at the game and 580 at the dance, then how many students are there at Freemont High?
1,380 students

40. *Girls and boys.* There are 385 surfers in Surf City. Two-thirds of the boys are surfers and one-twelfth of the girls are surfers. If there are two girls for every boy, then how many boys and how many girls are there in Surf City?
462 boys, 924 girls

41. *Nickels and dimes.* Winborne has 35 coins consisting of dimes and nickels. If the value of his coins is $3.30, then how many of each type does he have?
31 dimes, 4 nickels

42. *Pennies and nickels.* Wendy has 52 coins consisting of nickels and pennies. If the value of the coins is $1.20, then how many of each type does she have?

17 nickels, 35 pennies

43. *Blending fudge.* The Chocolate Factory in Vancouver blends its double-dark-chocolate fudge, which is 35% fat, with its peanut butter fudge, which is 25% fat, to obtain double-dark-peanut fudge, which is 29% fat.

a) Use the accompanying graph to estimate the number of pounds of each type that must be mixed to obtain 50 pounds of double-dark-peanut fudge.

b) Write a system of equations and solve it algebraically to find the exact amount of each type that should be used to obtain 50 pounds of double-dark-peanut fudge.

   a) (20, 30)

   b) 20 pounds chocolate, 30 pounds peanut butter

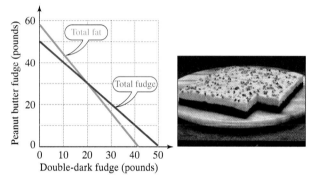

FIGURE FOR EXERCISE 43

44. *Low-fat yogurt.* Ziggy's Famous Yogurt blends regular yogurt that is 3% fat with its no-fat yogurt to obtain low-fat yogurt that is 1% fat. How many pounds of regular yogurt and how many pounds of no-fat yogurt should be mixed to obtain 60 pounds of low-fat yogurt?

20 pounds regular, 40 pounds no-fat

45. *Keystone state.* Judy averaged 42 miles per hour (mph) driving from Allentown to Harrisburg and 51 mph driving

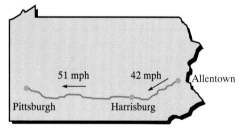

FIGURE FOR EXERCISE 45

from Harrisburg to Pittsburgh. If she drove a total of 288 miles in 6 hours, then how long did it take her to drive from Harrisburg to Pittsburgh?

4 hours

46. *Empire state.* Spike averaged 45 mph driving from Rochester to Syracuse and 49 mph driving from Syracuse to Albany. If he drove a total of 237 miles in 5 hours, then how far is it from Syracuse to Albany?

147 miles

47. *Probability of rain.* If Valerie Voss states that the probability of rain tomorrow is four times the probability that it doesn't rain, then what is the probability of rain tomorrow? (*Hint:* The probability that it rains plus the probability that it doesn't rain is 1.)

80%

48. *Super Bowl contender.* A Las Vegas odds-maker believes that the probability that San Francisco plays in the next Super Bowl is nine times the probability that they do not play in the next Super Bowl. What is the odds-maker's probability that San Francisco plays in the next Super Bowl?

90%

49. *Rectangular lot.* The width of a rectangular lot is 75% of its length. If the perimeter is 700 meters, then what are the length and width?

Width 150 meters, length 200 meters

50. *Fence painting.* Darren and Douglas must paint the 792-foot fence that encircles their family home. Because Darren is older, he has agreed to paint 20% more than Douglas. How much of the fence will each boy paint?

Darren 432 feet, Douglas 360 feet

GETTING MORE INVOLVED

51. *Discussion.* Explain how you decide whether it is easier to solve a system by substitution or addition.

52. *Exploration.* **a)** Write a linear equation in two variables that is satisfied by $(-3, 5)$.

b) Write another linear equation in two variables that is satisfied by $(-3, 5)$.

c) Are your equations independent or dependent?

d) Explain how to select the second equation so that it will be independent of the first.

53. *Exploration.* **a)** Make up a system of two linear equations in two variables such that both $(-1, 2)$ and $(4, 5)$ are in the solution set.

b) Are your equations independent or dependent?

c) Is it possible to find an independent system that is satisfied by both ordered pairs? Explain.

SYSTEMS OF LINEAR EQUATIONS IN THREE VARIABLES

The techniques that you learned in Section 4.2 can be extended to systems of equations in more than two variables. In this section we use elimination of variables to solve systems of equations in three variables.

Definition

The equation $5x - 4y = 7$ is called a linear equation in two variables because its graph is a straight line. The equation $2x + 3y - 4z = 12$ is similar in form, and so it is a linear equation in three variables. An equation in three variables is graphed in a three-dimensional coordinate system. The graph of a linear equation in three variables is a plane, not a line. We will not graph equations in three variables in this text, but we can solve systems without graphing. In general, we make the following definition.

> **Linear Equation in Three Variables**
>
> If A, B, C, and D are real numbers, with A, B, and C not all zero, then
> $$Ax + By + Cz = D$$
> is called a **linear equation in three variables.**

study tip

Everyone knows that you must practice to be successful with musical instruments, foreign languages, and sports. Success in algebra also requires regular practice. Thus budget your time so that you have a regular practice period for algebra.

Solving a System by Elimination

A solution to an equation in three variables is an **ordered triple** such as $(-2, 1, 5)$, where the first coordinate is the value of x, the second coordinate is the value of y, and the third coordinate is the value of z. There are infinitely many solutions to a linear equation in three variables.

The solution to a system of equations in three variables is the set of all ordered triples that satisfy all of the equations of the system. The techniques for solving a system of linear equations in three variables are similar to those used on systems of linear equations in two variables. We eliminate variables by either substitution or addition.

EXAMPLE 1

A linear system with a single solution

Solve the system:

$$(1) \quad x + y - z = -1$$
$$(2) \quad 2x - 2y + 3z = 8$$
$$(3) \quad 2x - y + 2z = 9$$

Solution

We can eliminate z from Eqs. (1) and (2) by multiplying Eq. (1) by 3 and adding it to Eq. (2):

$$\begin{array}{ll} 3x + 3y - 3z = -3 & \text{Eq. (1) multiplied by 3} \\ 2x - 2y + 3z = 8 & \text{Eq. (2)} \\ \hline (4) \quad 5x + y \phantom{{}- 3z} = 5 & \end{array}$$

helpful hint

Note that we could have chosen to eliminate x, y, or z first in Example 1. You should solve this same system by eliminating x first and then by eliminating y first. To eliminate x first, multiply the first equation by -2 and add it with the second and third equations.

calculator

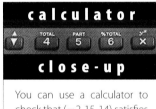

close-up

You can use a calculator to check that $(-2, 15, 14)$ satisfies all three equations of the original system.

```
-2+15-14
              -1
2*-2-2*15+3*14
               8
2*-2-15+2*14
               9
```

Now we must eliminate the same variable, z, from another pair of equations. Eliminate z from (1) and (3):

$$
\begin{aligned}
2x + 2y - 2z &= -2 \quad \text{Eq. (1) multiplied by 2} \\
2x - y + 2z &= 9 \quad \text{Eq. (3)} \\
\hline
(5) \quad 4x + y \phantom{{}+ 2z} &= 7
\end{aligned}
$$

Equations (4) and (5) give us a system with two variables. We now solve this system. Eliminate y by multiplying Eq. (5) by -1 and adding the equations:

$$
\begin{aligned}
5x + y &= 5 \quad \text{Eq. (4)} \\
-4x - y &= -7 \quad \text{Eq. (5) multiplied by } -1 \\
\hline
x \phantom{{}+ y} &= -2
\end{aligned}
$$

Now that we have x, we can replace x by -2 in Eq. (5) to find y:

$$
\begin{aligned}
4x + y &= 7 \\
4(-2) + y &= 7 \\
-8 + y &= 7 \\
y &= 15
\end{aligned}
$$

Now replace x by -2 and y by 15 in Eq. (1) to find z:

$$
\begin{aligned}
x + y - z &= -1 \\
-2 + 15 - z &= -1 \\
13 - z &= -1 \\
-z &= -14 \\
z &= 14
\end{aligned}
$$

Check that $(-2, 15, 14)$ satisfies all three of the original equations. The solution set is $\{(-2, 15, 14)\}$. ∎

The strategy that we follow for solving a system of three linear equations in three variables is stated as follows.

Solving a System in Three Variables

1. Use substitution or addition to eliminate any one of the variables from a pair of equations of the system. Look for the easiest variable to eliminate.
2. Eliminate the same variable from another pair of equations of the system.
3. Solve the resulting system of two equations in two unknowns.
4. After you have found the values of two of the variables, substitute into one of the original equations to find the value of the third variable.
5. Check the three values in all of the original equations.

In the next example we use a combination of addition and substitution.

E X A M P L E   2   **Using addition and substitution**

Solve the system:

$$
\begin{aligned}
(1) \quad x + y \phantom{{}+ 2x} &= 4 \\
(2) \quad 2x \phantom{{}+ y} - 3z &= 14 \\
(3) \quad 2y + z &= 2
\end{aligned}
$$

Solution

From Eq. (1) we get $y = 4 - x$. If we substitute $y = 4 - x$ into Eq. (3), then Eqs. (2) and (3) will be equations involving x and z only.

$$(3) \qquad 2y + z = 2$$
$$2(4 - x) + z = 2 \quad \text{Replace } y \text{ by } 4 - x.$$
$$8 - 2x + z = 2 \quad \text{Simplify.}$$
$$(4) \qquad -2x + z = -6$$

Now solve the system consisting of Eqs. (2) and (4) by addition:

$$\begin{array}{rcl} 2x - 3z &=& 14 \quad \text{Eq. (2)} \\ -2x + z &=& -6 \quad \text{Eq. (4)} \\ \hline -2z &=& 8 \\ z &=& -4 \end{array}$$

Use Eq. (3) to find y:

$$2y + z = 2 \quad \text{Eq. (3)}$$
$$2y + (-4) = 2 \quad \text{Let } z = -4.$$
$$2y = 6$$
$$y = 3$$

Use Eq. (1) to find x:

$$x + y = 4 \quad \text{Eq. (1)}$$
$$x + 3 = 4 \quad \text{Let } y = 3.$$
$$x = 1$$

Check that $(1, 3, -4)$ satisfies all three of the original equations. The solution set is $\{(1, 3, -4)\}$. ∎

CAUTION In solving a system in three variables it is essential to keep your work organized and neat. Writing short notes that explain your steps (as was done in the examples) will allow you to go back and check your work.

Graphs of Equations in Three Variables

The graph of any equation in three variables can be drawn on a three-dimensional coordinate system. The graph of a linear equation in three variables is a plane. To solve a system of three linear equations in three variables by graphing, we would have to draw the three planes and then identify the points that lie on all three of them. This method would be difficult even when the points have simple coordinates. So we will not attempt to solve these systems by graphing.

By considering how three planes might intersect, we can better understand the different types of solutions to a system of three equations in three variables. Figure 4.6, on the next page, shows some of the possibilities for the positioning of three planes in three-dimensional space. In most of the problems that we will solve the planes intersect at a single point as in Fig. 4.6(a). The solution set consists of one ordered triple. However, the system may include two equations corresponding to parallel planes that have no intersection. In this case the equations are said to be **inconsistent.** If the system has at least two inconsistent equations, then the solution set is the empty set [see Figs. 4.6(b) and 4.6(c)].

There are two ways in which the intersection of three planes can consist of infinitely many points. The intersection could be a line or a plane. To get a line, we can have either three different planes intersecting along a line, as in Fig. 4.6(d) or two equations for the same plane, with the third plane intersecting that plane. If all three equations are equations of the same plane, we get that plane for the intersection. We will not solve systems corresponding to all of the possible configurations described. The following examples illustrate two of these cases.

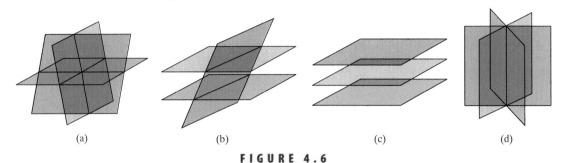

(a)   (b)   (c)   (d)

FIGURE 4.6

E X A M P L E 3

An inconsistent system of three linear equations

Solve the system:

$$\begin{aligned}(1) \qquad x + y - z &= 5\\(2) \qquad 3x - 2y + z &= 8\\(3) \qquad 2x + 2y - 2z &= 7\end{aligned}$$

Solution

We can eliminate the variable z from Eqs. (1) and (2) by adding them:

$$\begin{aligned}(1) \qquad x + y - z &= 5\\(2) \qquad \underline{3x - 2y + z = 8}\\4x - y \quad\;\; &= 13\end{aligned}$$

To eliminate z from Eqs. (1) and (3), multiply Eq. (1) by -2 and add the resulting equation to Eq. (3):

$$\begin{aligned}-2x - 2y + 2z &= -10 \qquad \text{Eq. (1) multiplied by } -2\\\underline{2x + 2y - 2z = 7} \qquad & \text{Eq. (3)}\\0 &= -3\end{aligned}$$

Because the last equation is false, there are two inconsistent equations in the system. Therefore the solution set is the empty set. ◼

E X A M P L E 4

A dependent system of three equations

Solve the system:

$$\begin{aligned}(1) \qquad 2x - 3y - z &= 4\\(2) \qquad -6x + 9y + 3z &= -12\\(3) \qquad 4x - 6y - 2z &= 8\end{aligned}$$

Solution

We will first eliminate x from Eqs. (1) and (2). Multiply Eq. (1) by 3 and add the resulting equation to Eq. (2):

$$\begin{aligned}6x - 9y - 3z &= 12 \qquad \text{Eq. (1) multiplied by 3}\\\underline{-6x + 9y + 3z = -12} \qquad & \text{Eq. (2)}\\0 &= 0\end{aligned}$$

helpful hint

If you recognize that multiplying Eq. (1) by -3 will produce Eq. (2), and multiplying Eq. (1) by 2 will produce Eq. (3), then you can conclude that all three equations are equivalent and there is no need to add the equations.

The last statement is an identity. The identity occurred because Eq. (2) is a multiple of Eq. (1). In fact, Eq. (3) is also a multiple of Eq. (1). These equations are dependent. They are all equations for the same plane. The solution set is the set of all points on that plane,

$$\{(x, y, z) \mid 2x - 3y - z = 4\}.$$

Applications

Problems involving three unknown quantities can often be solved by using a system of three equations in three variables.

E X A M P L E   5

Finding three unknown rents

Theresa took in a total of $1,240 last week from the rental of three condominiums. She had to pay 10% of the rent from the one-bedroom condo for repairs, 20% of the rent from the two-bedroom condo for repairs, and 30% of the rent from the three-bedroom condo for repairs. If the three-bedroom condo rents for twice as much as the one-bedroom condo and her total repair bill was $276, then what is the rent for each condo?

Solution

Let x, y, and z represent the rent on the one-bedroom, two-bedroom, and three-bedroom condos, respectively. We can write one equation for the total rent, another equation for the total repairs, and a third equation expressing the fact that the rent for the three-bedroom condo is twice that for the one-bedroom condo:

$$x + y + z = 1240$$
$$0.1x + 0.2y + 0.3z = 276$$
$$z = 2x$$

Substitute $z = 2x$ into both of the other equations to eliminate z:

$$x + y + 2x = 1240$$
$$0.1x + 0.2y + 0.3(2x) = 276$$

$$3x + \quad y = 1240$$
$$0.7x + 0.2y = 276$$

$$-2(3x + y) = -2(1240) \quad \text{Multiply each side by } -2.$$
$$10(0.7x + 0.2y) = 10(276) \quad \text{Multiply each side by 10.}$$

$$-6x - 2y = -2480$$
$$\underline{7x + 2y = 2760} \quad \text{Add.}$$
$$x \quad\quad = 280$$

$$z = 2(280) = 560 \quad \text{Because } z = 2x$$
$$280 + y + 560 = 1240 \quad \text{Because } x + y + z = 1240$$
$$y = 400$$

Check that (280, 400, 560) satisfies all three of the original equations. The condos rent for $280, $400, and $560 per week.

WARM-UPS

True or false? Explain your answer.

1. The point $(1, -2, 3)$ is in the solution set to the equation $x + y - z = 4$. False

2. The point $(4, 1, 1)$ is the only solution to the equation $x + y - z = 4$. False

3. The ordered triple $(1, -1, 2)$ satisfies $x + y + z = 2$, $x - y - z = 0$, and $2x + y - z = -1$. True

4. Substitution cannot be used on three equations in three variables. False

5. Two distinct planes are either parallel or intersect in a single point. False

6. The equations $x - y + 2z = 6$ and $x - y + 2z = 4$ are inconsistent. True

7. The equations $3x + 2y - 6z = 4$ and $-6x - 4y + 12z = -8$ are dependent. True

8. The graph of $y = 2x - 3z + 4$ is a straight line. False

9. The value of x nickels, y dimes, and z quarters is $0.05x + 0.10y + 0.25z$ cents. False

10. If $x = -2$, $z = 3$, and $x + y + z = 6$, then $y = 7$. False

4.3   EXERCISES

Reading and Writing   *After reading this section, write out the answers to these questions. Use complete sentences.*

1. What is a linear equation in three variables?

   A linear equation in three variables is an equation of the form $Ax + By + Cz = D$ where A, B, and C cannot all be zero.

2. What is an ordered triple?

   An ordered triple is a collection of three numbers [written as (a, b, c)] in which the order of the numbers is important.

3. What is a solution to a system of linear equations in three variables?

   A solution to a system of linear equations in three variables is an ordered triple that satisfies all of the equations in the system.

4. How do we solve systems of linear equations in three variables?

   We solve systems in three variables by using addition or substitution to eliminate variables.

5. What does the graph of a linear equation in three variables look like?

   The graph of a linear equation in three variables is a plane in a three-dimensional coordinate system.

6. How are the planes positioned when a system of linear equations in three variables is inconsistent?

   For an inconsistent system at least two of the planes are parallel.

Solve each system of equations. See Examples 1 and 2.

7. $x + y + z = 2$
    $x + 2y - z = 6$
    $2x + y - z = 5$
    $\{(1, 2, -1)\}$

8. $2x - y + 3z = 14$
    $x + y - 2z = -5$
    $3x + y - z = 2$
    $\{(2, -1, 3)\}$

9. $x - 2y + 4z = 3$
    $x + 3y - 2z = 6$
    $x - 4y + 3z = -5$
    $\{(1, 3, 2)\}$

10. $2x + 3y + z = 13$
    $-3x + 2y + z = -4$
    $4x - 4y + z = 5$
    $\{(3, 2, 1)\}$

11. $2x - y + z = 10$
    $3x - 2y - 2z = 7$
    $x - 3y - 2z = 10$
    $\{(1, -5, 3)\}$

12. $x - 3y + 2z = -11$
    $2x - 4y + 3z = -15$
    $3x - 5y - 4z = 5$
    $\{(1, 2, -3)\}$

13. $2x - 3y + z = -9$
    $-2x + y - 3z = 7$
    $x - y + 2z = -5$
    $\{(-1, 2, -1)\}$

14. $3x - 4y + z = 19$
    $2x + 4y + z = 0$
    $x - 2y + 5z = 17$
    $\{(3, -2, 2)\}$

15. $2x - 5y + 2z = 16$
    $3x + 2y - 3z = -19$
    $4x - 3y + 4z = 18$
    $\{(-1, -2, 4)\}$

16. $-2x + 3y - 4z = 3$
    $3x - 5y + 2z = 4$
    $-4x + 2y - 3z = 0$
    $\{(1, -1, -2)\}$

17. $x + y = 4$
    $y - z = -2$
    $x + y + z = 9$
    $\{(1, 3, 5)\}$

18. $x + y - z = 0$
    $x - y = -2$
    $y + z = 10$
    $\{(2, 4, 6)\}$

19. $x + y \quad\quad = 7$
$\quad\quad y - z = -1$
$x \quad\quad + 3z = 18$
$\{(3, 4, 5)\}$

20. $2x - y \quad\quad = -8$
$\quad\quad y + 3z = 22$
$x \quad\quad - z = -8$
$\{(-2, 4, 6)\}$

Solve each system. See Examples 3 and 4.

21. $x - y + 2z = 3$
$2x + y - z = 5$
$3x - 3y + 6z = 4 \quad \varnothing$

22. $2x - 4y + 6z = 12$
$6x - 12y + 18z = 36$
$-x + 2y - 3z = -6$
$\{(x, y, z) \mid -x + 2y - 3z = -6\}$

23. $3x - y + z = 5$
$9x - 3y + 3z = 15$
$-12x + 4y - 4z = -20$
$\{(x, y, z) \mid 3x - y + z = 5\}$

24. $4x - 2y - 2z = 5$
$2x - y - z = 7$
$-4x + 2y + 2z = 6 \quad \varnothing$

25. $x - y \quad\quad = 3$
$\quad\quad y + z = 8$
$2x \quad\quad + 2z = 7 \quad \varnothing$

26. $2x - y \quad\quad = 6$
$\quad\quad 2y + z = -4$
$8x \quad\quad + 2z = 3 \quad \varnothing$

27. $0.10x + 0.08y - 0.04z = 3$
$5x + 4y - 2z = 150$
$0.3x + 0.24y - 0.12z = 9$
$\{(x, y, z) \mid 5x + 4y - 2z = 150\}$

28. $0.06x - 0.04y + z = 6$
$3x - 2y + 50z = 300$
$0.03x - 0.02y + 0.5z = 3$
$\{(x, y, z) \mid 3x - 2y + 50z = 300\}$

 Use a calculator to solve each system.

29. $3x + 2y - 0.4z = 0.1$
$3.7x - 0.2y + 0.05z = 0.41$
$-2x + 3.8y - 2.1z = -3.26 \quad \{(0.1, 0.3, 2)\}$

30. $3x - 0.4y + 9z = 1.668$
$0.3x + 5y - 8z = -0.972$
$5x - 4y - 8z = 1.8 \quad \{(0.36, -0.12, 0.06)\}$

Solve each problem by using a system of three equations in three unknowns. See Example 5.

31. *Diversification.* Ann invested a total of $12,000 in stocks, bonds, and a mutual fund. She received a 10% return on her stock investment, an 8% return on her bond investment, and a 12% return on her mutual fund. Her total return was $1,230. If the total investment in stocks and bonds equaled her mutual fund investment, then how much did she invest in each?
$1,500 stocks, $4,500 bonds, $6,000 mutual fund

32. *Paranoia.* Fearful of a bank failure, Norman split his life savings of $60,000 among three banks. He received 5%, 6%, and 7% on the three deposits. In the account earning 7% interest, he deposited twice as much as in the account earning 5% interest. If his total earnings were $3,760, then how much did he deposit in each account?
$16,000 at 5%, $12,000 at 6%, $32,000 at 7%

33. *Big tipper.* On Monday Headley paid $1.70 for two cups of coffee and one doughnut, including the tip. On Tuesday he paid $1.65 for two doughnuts and a cup of coffee, including the tip. On Wednesday he paid $1.30 for one coffee and one doughnut, including the tip. If he always tips the same amount, then what is the amount of each item?
Coffee $0.40, doughnut $0.35, tip $0.55

34. *Weighing in.* Anna, Bob, and Chris will not disclose their weights but agree to be weighed in pairs. Anna and Bob together weigh 226 pounds. Bob and Chris together weigh 210 pounds. Anna and Chris together weigh 200 pounds. How much does each student weigh?
Anna 108 pounds, Bob 118 pounds, Chris 92 pounds

Anna & Bob   Bob & Chris   Anna & Chris

FIGURE FOR EXERCISE 34

35. *Lunch-box special.* Salvador's Fruit Mart sells variety packs. The small pack contains three bananas, two apples, and one orange for $1.80. The medium pack contains four bananas, three apples, and three oranges for $3.05. The family size contains six bananas, five apples, and four oranges for $4.65. What price should Salvador charge for his lunch-box special that consists of one banana, one apple, and one orange? $0.95

36. *Three generations.* Edwin, his father, and his grandfather have an average age of 53. One-half of his grandfather's age, plus one-third of his father's age, plus one-fourth of Edwin's age is 65. If 4 years ago, Edwin's grandfather was four times as old as Edwin, then how old are they all now?
Edwin 24, father 51, grandfather 84

37. *Error in the scale.* Alex is using a scale that is known to have a constant error. A can of soup and a can of tuna are placed on this scale, and it reads 24 ounces. Now four identical cans of soup and three identical cans of tuna are placed on an accurate scale, and a weight of 80 ounces is recorded. If two cans of tuna weigh 18 ounces on the bad scale, then what is the amount of error in the scale and what is the correct weight of each type of can?
Soup 14 ounces, tuna 8 ounces, error 2 ounces

38. *Three-digit number.* The sum of the digits of a three-digit number is 11. If the digits are reversed, the new number is 46 more than five times the old number. If the hundreds digit plus twice the tens digit is equal to the units digit, then what is the number?   137

39. *Working overtime.* To make ends meet, Ms. Farnsby works three jobs. Her total income last year was $48,000. Her income from teaching was just $6,000 more than her income from house painting. Royalties from her textbook sales were one-seventh of the total money she received from teaching and house painting. How much did she make from each source last year?

$24,000 teaching, $18,000 painting, $6,000 royalties

40. *Pocket change.* Harry has $2.25 in nickels, dimes, and quarters. If he had twice as many nickels, half as many dimes, and the same number of quarters, he would have $2.50. If he has 27 coins altogether, then how many of each does he have?   15 nickels, 10 dimes, 2 quarters

GETTING MORE INVOLVED

 41. *Exploration.* Draw diagrams showing the possible ways to position three planes in three-dimensional space.

 42. *Discussion.* Make up a system of three linear equations in three variables for which the solution set is $\{(0, 0, 0)\}$. A system with this solution set is called a *homogeneous* system. Why do you think it is given that name?

 43. *Cooperative learning.* Working in groups, do parts (a)–(d) below. Then write a report on your findings.
   a) Find values of a, b, and c so that the graph of $y = ax^2 + bx + c$ goes through the points $(-1, -2)$, $(1, 0)$, and $(2, 7)$.
   b) Arbitrarily select three ordered pairs and find the equation of the parabola that goes through the three points.
   c) Could more than one parabola pass through three given points? Give reasons for your answer.
   d) Explain how to pick three points for which no parabola passes through all of them.

4.4   SOLVING LINEAR SYSTEMS USING MATRICES

You solved linear systems in two variables by substitution and addition in Sections 4.1 and 4.2. Those methods are done differently on each system. In this section you will learn the Gaussian elimination method, which is related to the addition method. The Gaussian elimination method is performed in the same way on every system. We first need to introduce some new terminology.

In this section

- Matrices
- The Augmented Matrix
- The Gaussian Elimination Method
- Inconsistent and Dependent Equations

Matrices

A **matrix** is a rectangular array of numbers. The **rows** of a matrix run horizontally, and the **columns** of a matrix run vertically. A matrix with m rows and n columns has **order** $m \times n$ (read "m by n"). Each number in a matrix is called an **element** or **entry** of the matrix.

E X A M P L E   1

Order of a matrix

Determine the order of each matrix.

a) $\begin{bmatrix} -1 & 2 \\ 5 & \sqrt{2} \\ 0 & 3 \end{bmatrix}$ **b)** $\begin{bmatrix} 2 & 3 \\ -1 & 5 \end{bmatrix}$ **c)** $\begin{bmatrix} 1 & 2 & 3 \\ 4 & 5 & 6 \\ -1 & 0 & 2 \end{bmatrix}$ **d)** $\begin{bmatrix} 1 & 3 & 6 \end{bmatrix}$

Solution

Because matrix (a) has 3 rows and 2 columns, its order is 3×2. Matrix (b) is a 2×2 matrix, matrix (c) is a 3×3 matrix, and matrix (d) is a 1×3 matrix.   ■

The Augmented Matrix

The solution to a system of linear equations such as

$$x - 2y = -5$$
$$3x + y = 6$$

study tip

As soon as possible after class, find a quiet place and work on your homework. The longer you wait, the harder it is to remember what happened in class.

depends on the coefficients of x and y and the constants on the right-hand side of the equation. The matrix of coefficients for this system is the 2×2 matrix

$$\begin{bmatrix} 1 & -2 \\ 3 & 1 \end{bmatrix}.$$

If we insert the constants from the right-hand side of the system into the matrix of coefficients, we get the 2×3 matrix

$$\left[\begin{array}{cc|c} 1 & -2 & -5 \\ 3 & 1 & 6 \end{array}\right].$$

We use a vertical line between the coefficients and the constants to represent the equal signs. This matrix is the **augmented matrix** of the system. Two systems of linear equations are **equivalent** if they have the same solution set. Two augmented matrices are **equivalent** if the systems they represent are equivalent.

E X A M P L E 2

Writing the augmented matrix

Write the augmented matrix for each system of equations.

a) $3x - 5y = 7$
$x + y = 4$

b) $x + y - z = 5$
$2x + z = 3$
$2x - y + 4z = 0$

c) $x + y = 1$
$y + z = 6$
$z = -5$

Solution

a) $\left[\begin{array}{cc|c} 3 & -5 & 7 \\ 1 & 1 & 4 \end{array}\right]$

b) $\left[\begin{array}{ccc|c} 1 & 1 & -1 & 5 \\ 2 & 0 & 1 & 3 \\ 2 & -1 & 4 & 0 \end{array}\right]$

c) $\left[\begin{array}{ccc|c} 1 & 1 & 0 & 1 \\ 0 & 1 & 1 & 6 \\ 0 & 0 & 1 & -5 \end{array}\right]$ ■

E X A M P L E 3

Writing the system

Write the system of equations represented by each augmented matrix.

a) $\left[\begin{array}{cc|c} 1 & 4 & -2 \\ 1 & -1 & 3 \end{array}\right]$

b) $\left[\begin{array}{cc|c} 1 & 0 & 5 \\ 0 & 1 & 1 \end{array}\right]$

c) $\left[\begin{array}{ccc|c} 2 & 3 & 4 & 6 \\ -1 & 0 & 5 & -2 \\ 1 & -2 & 3 & 1 \end{array}\right]$

Solution

a) Use the first two numbers in each row as the coefficients of x and y and the last number as the constant to get the following system:

$$x + 4y = -2$$
$$x - y = 3$$

b) Use the first two numbers in each row as the coefficients of x and y and the last number as the constant to get the following system:

$$x = 5$$
$$y = 1$$

c) Use the first three numbers in each row as the coefficients of x, y, and z and the last number as the constant to get the following system:

$$2x + 3y + 4z = 6$$
$$-x + 5z = -2$$
$$x - 2y + 3z = 1$$ ■

The Gaussian Elimination Method

When we solve a single equation, we write simpler and simpler equivalent equations to get an equation whose solution is obvious. In the **Gaussian elimination method** we write simpler and simpler equivalent augmented matrices until we get an augmented matrix (like the one in Example 3(b)) in which the solution to the corresponding system is obvious.

Because each row of an augmented matrix represents an equation, we can perform the **row operations** on the augmented matrix. These row operations, which follow, correspond to the usual operations with equations used in the addition method.

Row Operations

The following row operations on an augmented matrix give an equivalent augmented matrix:

1. Interchange two rows of the matrix.
2. Multiply every element in a row by a nonzero real number.
3. Add to a row a multiple of another row.

In the Gaussian elimination method our goal is to use row operations to obtain an augmented matrix that has ones on the **diagonal** in its matrix of coefficients and zeros elsewhere:

$$\begin{bmatrix} 1 & 0 & | & a \\ 0 & 1 & | & b \end{bmatrix}$$

The system corresponding to this augmented matrix is $x = a$ and $y = b$. So the solution set to the system is $\{(a, b)\}$.

E X A M P L E   4

Gaussian elimination with two equations in two variables

Use the Gaussian elimination method to solve the system:

$$x - 3y = 11$$
$$2x + y = 1$$

Solution

Start with the augmented matrix:

$$\begin{bmatrix} 1 & -3 & | & 11 \\ 2 & 1 & | & 1 \end{bmatrix}$$

Multiply row 1 by -2 and add the result to row 2 (in symbols, $-2R_1 + R_2 \rightarrow R_2$). Because $-2R_1 = [-2, 6, -22]$ and $R_2 = [2, 1, 1]$, $-2R_1 + R_2 = [0, 7, -21]$. Note that the coefficient of x in the second equation is now 0. We get the following matrix:

$$\begin{bmatrix} 1 & -3 & | & 11 \\ 0 & 7 & | & -21 \end{bmatrix} \quad -2R_1 + R_2 \rightarrow R_2$$

Multiply each element of row 2 by $\frac{1}{7}$ (in symbols, $\frac{1}{7}R_2 \rightarrow R_2$):

$$\begin{bmatrix} 1 & -3 & | & 11 \\ 0 & 1 & | & -3 \end{bmatrix} \quad \frac{1}{7}R_2 \rightarrow R_2$$

Multiply row 2 by 3 and add the result to row 1. Because $3R_2 = [0, 3, -9]$ and $R_1 = [1, -3, 11]$, $3R_2 + R_1 = [1, 0, 2]$. Note that the coefficient of y in the first equation is now 0. We get the following matrix:

$$\left[\begin{array}{cc|c} 1 & 0 & 2 \\ 0 & 1 & -3 \end{array}\right] \quad 3R_2 + R_1 \rightarrow R_1$$

This augmented matrix represents the system $x = 2$ and $y = -3$. So the solution set to the system is $\{(2, -3)\}$. Check in the original system. ■

In the next example we use the row operations on the augmented matrix of a system of three linear equations in three variables.

E X A M P L E   5   **Gaussian elimination with three equations in three variables**

Use the Gaussian elimination method to solve the following system:

$$2x - y + z = -3$$
$$x + y - z = 6$$
$$3x - y - z = 4$$

Solution

Start with the augmented matrix and interchange the first and second rows to get a 1 in the upper left position in the matrix:

$$\left[\begin{array}{ccc|c} 2 & -1 & 1 & -3 \\ 1 & 1 & -1 & 6 \\ 3 & -1 & -1 & 4 \end{array}\right] \quad \text{The augmented matrix}$$

$$\left[\begin{array}{ccc|c} 1 & 1 & -1 & 6 \\ 2 & -1 & 1 & -3 \\ 3 & -1 & -1 & 4 \end{array}\right] \quad R_1 \leftrightarrow R_2$$

helpful hint

It is not necessary to perform the row operations in exactly the same order as is shown in Example 5. As long as you use the legitimate row operations and get to the final form, you will get the solution to the system. Of course, you must double check your arithmetic at every step if you want to be successful at Gaussian elimination.

Now multiply the first row by -2 and add the result onto the second row. Multiply the first row by -3 and add the result onto the third row. These two steps eliminate the variable x from the second and third rows:

$$\left[\begin{array}{ccc|c} 1 & 1 & -1 & 6 \\ 0 & -3 & 3 & -15 \\ 0 & -4 & 2 & -14 \end{array}\right] \quad \begin{array}{l} -2R_1 + R_2 \rightarrow R_2 \\ -3R_1 + R_3 \rightarrow R_3 \end{array}$$

Multiply the second row by $-\frac{1}{3}$ to get 1 in the second position on the diagonal:

$$\left[\begin{array}{ccc|c} 1 & 1 & -1 & 6 \\ 0 & 1 & -1 & 5 \\ 0 & -4 & 2 & -14 \end{array}\right] \quad -\frac{1}{3}R_2 \rightarrow R_2$$

Use the second row to eliminate the variable y from the first and third rows:

$$\left[\begin{array}{ccc|c} 1 & 0 & 0 & 1 \\ 0 & 1 & -1 & 5 \\ 0 & 0 & -2 & 6 \end{array}\right] \quad \begin{array}{l} -1R_2 + R_1 \rightarrow R_1 \\ 4R_2 + R_3 \rightarrow R_3 \end{array}$$

Multiply the third row by $-\frac{1}{2}$ to get a 1 in the third position on the diagonal:

$$\left[\begin{array}{ccc|c} 1 & 0 & 0 & 1 \\ 0 & 1 & -1 & 5 \\ 0 & 0 & 1 & -3 \end{array}\right] \quad -\tfrac{1}{2}R_3 \rightarrow R_3$$

Use the third row to eliminate the variable z from the second row:

$$\left[\begin{array}{ccc|c} 1 & 0 & 0 & 1 \\ 0 & 1 & 0 & 2 \\ 0 & 0 & 1 & -3 \end{array}\right] \quad R_3 + R_2 \rightarrow R_2$$

This last augmented matrix represents the system $x = 1$, $y = 2$, and $z = -3$. So the solution set to the system is $\{(1, 2, -3)\}$. ∎

Inconsistent and Dependent Equations

Inconsistent and dependent equations are easily recognized in using the Gaussian elimination method.

E X A M P L E 6

Gaussian elimination with an inconsistent system

Solve the system:

$$\begin{aligned} x - y &= 1 \\ -3x + 3y &= 4 \end{aligned}$$

helpful hint

The point of Example 6 is to recognize an inconsistent system with Gaussian elimination. We could also observe that -3 times the first equation yields

$$-3x + 3y = -3,$$

which is inconsistent with

$$-3x + 3y = 4.$$

Solution

Start with the augmented matrix:

$$\left[\begin{array}{cc|c} 1 & -1 & 1 \\ -3 & 3 & 4 \end{array}\right]$$

Multiply row 1 by 3 and add the result to row 2. We get the following matrix:

$$\left[\begin{array}{cc|c} 1 & -1 & 1 \\ 0 & 0 & 7 \end{array}\right] \quad 3R_1 + R_2 \rightarrow R_2$$

The second row of the augmented matrix corresponds to the equation $0 = 7$. So the equations are inconsistent, and there is no solution to the system. ∎

E X A M P L E 7

Gaussian elimination with a dependent system

Solve the system:

$$\begin{aligned} 3x + y &= 1 \\ 6x + 2y &= 2 \end{aligned}$$

Solution

Start with the augmented matrix:

$$\left[\begin{array}{cc|c} 3 & 1 & 1 \\ 6 & 2 & 2 \end{array}\right]$$

Multiply row 1 by -2 and add the result to row 2. We get the following matrix:

$$\left[\begin{array}{cc|c} 3 & 1 & 1 \\ 0 & 0 & 0 \end{array}\right] \quad -2R_1 + R_2 \rightarrow R_2$$

In the second row of the augmented matrix we have the equation $0 = 0$. So the equations are dependent. Every ordered pair that satisfies the first equation satisfies both equations. The solution set is $\{(x, y) \mid 3x + y = 1\}$. ■

The Gaussian elimination method may be applied to a system of n linear equations in n unknowns, where $n \geq 2$. However, it is a rather tedious method to perform when n is greater than 2, especially when fractions are involved. Computers are programmed to work with matrices, and the Gaussian elimination method is a popular method for computers.

WARM-UPS

True or false? Explain your answer.

Statements 1–7 refer to the following matrices:

a) $\begin{bmatrix} 1 & 3 & | & 5 \\ -1 & -3 & | & 2 \end{bmatrix}$

b) $\begin{bmatrix} 1 & 3 & | & 5 \\ 0 & 0 & | & 7 \end{bmatrix}$

c) $\begin{bmatrix} -1 & 2 & | & -3 \\ 2 & -4 & | & 3 \end{bmatrix}$

d) $\begin{bmatrix} 1 & 3 & | & 5 \\ 0 & 0 & | & 0 \end{bmatrix}$

1. The augmented matrix for $x + 3y = 5$ and $-x - 3y = 2$ is matrix (a). True

2. The augmented matrix for $2y - x = -3$ and $2x - 4y = 3$ is matrix (c). True

3. Matrix (a) is equivalent to matrix (b). True

4. Matrix (c) is equivalent to matrix (d). False

5. The system corresponding to matrix (b) is inconsistent. True

6. The system corresponding to matrix (c) is dependent. False

7. The system corresponding to matrix (d) is independent. False

8. The augmented matrix for a system of two linear equations in two unknowns is a 2×2 matrix. False

9. The notation $2R_1 + R_3 \rightarrow R_3$ means to replace R_3 by $2R_1 + R_3$. True

10. The notation $R_1 \leftrightarrow R_2$ means to replace R_2 by R_1. False

4.4 EXERCISES

Reading and Writing *After reading this section, write out the answers to these questions. Use complete sentences.*

1. What is a matrix?

A marix is a rectangular array of numbers.

2. What is the difference between a row and a column of a matrix?

A row runs horizontally and a column runs vertically.

3. What is the order of a matrix?

The order of a matrix is the number of rows and columns.

4. What is an element of a matrix?

An element of a matrix is a number that occupies a position in the matrix.

5. What is an augmented matrix?

An augmented matrix is a matrix where the entries in the first column are the coefficients of x, the entries in the second column are the coefficients of y, and the entries in the third column are the constants from a system of two linear equations in two unknowns.

6. What is the goal of Gaussian elimination?

The goal of Gaussian elimination is to get ones on the diagonal.

Determine the order of each matrix. See Example 1.

7. $\begin{bmatrix} 5 & 0 \\ -2 & 3 \end{bmatrix}$
2×2

8. $\begin{bmatrix} 1 & 3 & 6 \\ -7 & 0 & 2 \end{bmatrix}$
2×3

9. $\begin{bmatrix} a & c \\ 0 & d \\ 3 & w \end{bmatrix}$
3×2

10. $\begin{bmatrix} 0 & a & b \\ 5 & 7 & -8 \\ a & b & 2 \end{bmatrix}$ **11.** $\begin{bmatrix} -\sqrt{3} \\ \pi \\ \dfrac{1}{2} \end{bmatrix}$ **12.** $\begin{bmatrix} 3 & 0 & 4 \end{bmatrix}$

    3×3 3×1 1×3

Write the augmented matrix for each system of equations. See Example 2.

13. $\begin{aligned} 2x - 3y &= 9 \\ -3x + y &= -1 \end{aligned}$ **14.** $\begin{aligned} x - y &= 4 \\ 2x + y &= 3 \end{aligned}$

$\begin{bmatrix} 2 & -3 & 9 \\ -3 & 1 & -1 \end{bmatrix}$ $\begin{bmatrix} 1 & -1 & 4 \\ 2 & 1 & 3 \end{bmatrix}$

15. $\begin{aligned} x - y + z &= 1 \\ x + y - 2z &= 3 \\ y - 3z &= 4 \end{aligned}$ **16.** $\begin{aligned} x + y \quad\;\; &= 2 \\ y - 3z &= 5 \\ -3x \quad\;\; + 2z &= 8 \end{aligned}$

$\begin{bmatrix} 1 & -1 & 1 & 1 \\ 1 & 1 & -2 & 3 \\ 0 & 1 & -3 & 4 \end{bmatrix}$ $\begin{bmatrix} 1 & 1 & 0 & 2 \\ 0 & 1 & -3 & 5 \\ -3 & 0 & 2 & 8 \end{bmatrix}$

Write the system of equations represented by each augmented matrix. See Example 3.

17. $\begin{bmatrix} 5 & 1 & | & -1 \\ 2 & -3 & | & 0 \end{bmatrix}$ **18.** $\begin{bmatrix} 1 & 0 & | & 4 \\ 0 & 1 & | & -3 \end{bmatrix}$

$\begin{aligned} 5x + y &= -1 \\ 2x - 3y &= 0 \end{aligned}$ $\begin{aligned} x &= 4 \\ y &= -3 \end{aligned}$

19. $\begin{bmatrix} 1 & 0 & 0 & | & 6 \\ -1 & 0 & 1 & | & -3 \\ 1 & 1 & 0 & | & 1 \end{bmatrix}$ **20.** $\begin{bmatrix} 1 & 0 & 4 & | & 3 \\ 0 & 2 & 1 & | & -1 \\ 1 & 1 & 1 & | & 1 \end{bmatrix}$

$\begin{aligned} x \quad\;\;\; &= 6 \\ -x + z &= -3 \\ x + y &= 1 \end{aligned}$ $\begin{aligned} x + \quad\;\; 4z &= 3 \\ 2y + z &= -1 \\ x + y + z &= 1 \end{aligned}$

Determine the row operation that was used to convert each given augmented matrix into the equivalent augmented matrix that follows it. See Example 4.

21. $\begin{bmatrix} 3 & 2 & | & 12 \\ 1 & -1 & | & -1 \end{bmatrix}, \begin{bmatrix} 1 & -1 & | & -1 \\ 3 & 2 & | & 12 \end{bmatrix}$ $R_1 \leftrightarrow R_2$

22. $\begin{bmatrix} 1 & -1 & | & -1 \\ 3 & 2 & | & 12 \end{bmatrix}, \begin{bmatrix} 1 & -1 & | & -1 \\ 0 & 5 & | & 15 \end{bmatrix}$ $-3R_1 + R_2 \rightarrow R_2$

23. $\begin{bmatrix} 1 & -1 & | & -1 \\ 0 & 5 & | & 15 \end{bmatrix}, \begin{bmatrix} 1 & -1 & | & -1 \\ 0 & 1 & | & 3 \end{bmatrix}$ $\dfrac{1}{5}R_2 \rightarrow R_2$

24. $\begin{bmatrix} 1 & -1 & | & -1 \\ 0 & 1 & | & 3 \end{bmatrix}, \begin{bmatrix} 1 & 0 & | & 2 \\ 0 & 1 & | & 3 \end{bmatrix}$ $R_2 + R_1 \rightarrow R_1$

Solve each system using the Gaussian elimination method. See Examples 4–7.

25. $\begin{aligned} x + y &= 3 \\ -3x + y &= -1 \end{aligned}$ **26.** $\begin{aligned} x - y &= -1 \\ 2x - y &= 2 \end{aligned}$

$\{(1, 2)\}$ $\{(3, 4)\}$

27. $\begin{aligned} 2x - y &= 3 \\ x + y &= 9 \end{aligned}$ **28.** $\begin{aligned} 3x - 4y &= -1 \\ x - y &= 0 \end{aligned}$

$\{(4, 5)\}$ $\{(1, 1)\}$

29. $\begin{aligned} 3x - y &= 4 \\ 2x + y &= 1 \end{aligned}$ **30.** $\begin{aligned} 2x - y &= -3 \\ 3x + y &= -2 \end{aligned}$

$\{(1, -1)\}$ $\{(-1, 1)\}$

31. $\begin{aligned} 6x - 7y &= 0 \\ 2x + y &= 20 \end{aligned}$ **32.** $\begin{aligned} 2x + y &= 11 \\ 2x - y &= 1 \end{aligned}$

$\{(7, 6)\}$ $\{(3, 5)\}$

33. $\begin{aligned} 2x - 3y &= 4 \\ -2x + 3y &= 5 \end{aligned}$ **34.** $\begin{aligned} x - 3y &= 8 \\ 2x - 6y &= 1 \end{aligned}$

$\varnothing$ $\varnothing$

35. $\begin{aligned} x + 2y &= 1 \\ 3x + 6y &= 3 \end{aligned}$ **36.** $\begin{aligned} 2x - 3y &= 1 \\ -6x + 9y &= -3 \end{aligned}$

$\{(x, y) \mid x + 2y = 1\}$ $\{(x, y) \mid 2x - 3y = 1\}$

37. $\begin{aligned} x + y + z &= 6 \\ x - y + z &= 2 \\ 2y - z &= 1 \end{aligned}$ **38.** $\begin{aligned} x - y - z &= 0 \\ -x - y + z &= -4 \\ -x + y - z &= -2 \end{aligned}$

$\{(1, 2, 3)\}$ $\{(3, 2, 1)\}$

39. $\begin{aligned} 2x + y + z &= 4 \\ x + y - z &= 1 \\ x - y + 2z &= 2 \end{aligned}$ **40.** $\begin{aligned} 3x - y \quad\;\; &= 1 \\ x + y + z &= 4 \\ x \quad\;\; + 2z &= 3 \end{aligned}$

$\{(1, 1, 1)\}$ $\{(1, 2, 1)\}$

41. $\begin{aligned} 2x - y + z &= 0 \\ x + y - 3z &= 3 \\ x - y + z &= -1 \end{aligned}$ **42.** $\begin{aligned} x - y - z &= 0 \\ -x - y + 2z &= -1 \\ -x + y - 2z &= -3 \end{aligned}$

$\{(1, 2, 0)\}$ $\{(2, 1, 1)\}$

43. $\begin{aligned} -x + 3y + z &= 0 \\ x - y - 4z &= -3 \\ x + y + 2z &= 3 \end{aligned}$ **44.** $\begin{aligned} -x \quad\;\; + z &= -2 \\ 2x - y \quad\;\; &= 5 \\ y + 3z &= 9 \end{aligned}$

$\{(1, 0, 1)\}$ $\{(4, 3, 2)\}$

45. $\begin{aligned} x - y + z &= 1 \\ 2x - 2y + 2z &= 2 \\ -3x + 3y - 3z &= -3 \end{aligned}$ **46.** $\begin{aligned} 4x - 2y + 2z &= 2 \\ 2x - y + z &= 1 \\ -2x + y - z &= -1 \end{aligned}$

$\{(x, y, z) \mid x - y + z = 1\}$ $\{(x, y, z) \mid 2x - y + z = 1\}$

47. $\begin{aligned} x + y - z &= 2 \\ 2x - y + z &= 1 \\ 3x + 3y - 3z &= 8 \end{aligned}$ **48.** $\begin{aligned} x + y + z &= 5 \\ x - y - z &= 8 \\ -x + y + z &= 2 \end{aligned}$

$\varnothing$ $\varnothing$

GETTING MORE INVOLVED

49. *Cooperative learning.* Write a step-by-step procedure for solving any system of two linear equations in two variables by the Gaussian elimination method. Have a classmate evaluate your procedure by using it to solve a system.

50. *Cooperative learning.* Repeat Exercise 49 for a system of three linear equations in three variables.

4.5   CRAMER'S RULE FOR SYSTEMS IN TWO VARIABLES

The Gaussian elimination method of Section 4.4 can be performed the same way on every system. Another method that is applied the same way for every system is Cramer's rule, which we study in this section. Before you learn Cramer's rule, we need to introduce a new number associated with a matrix, called a *determinant*.

Determinants

The determinant of a square matrix is a real number corresponding to the matrix. For a 2×2 matrix the determinant is defined as follows.

> **Determinant of a 2 × 2 Matrix**
>
> The **determinant** of the matrix $\begin{bmatrix} a & b \\ c & d \end{bmatrix}$ is defined to be the real number $ad - bc$. We write
> $$\begin{vmatrix} a & b \\ c & d \end{vmatrix} = ad - bc.$$

Note that the symbol for determinant is a pair of vertical lines similar to the absolute value symbol, while a matrix is enclosed in brackets.

EXAMPLE 1

Using the definition of determinant

Find the determinant of each matrix.

a) $\begin{bmatrix} 1 & 3 \\ -2 & 5 \end{bmatrix}$ b) $\begin{bmatrix} 2 & 4 \\ 6 & 12 \end{bmatrix}$

Solution

a) $\begin{vmatrix} 1 & 3 \\ -2 & 5 \end{vmatrix} = 1 \cdot 5 - 3(-2)$ b) $\begin{vmatrix} 2 & 4 \\ 6 & 12 \end{vmatrix} = 2 \cdot 12 - 4 \cdot 6$

$\qquad\qquad = 5 + 6 \qquad\qquad\qquad\qquad\qquad = 24 - 24$

$\qquad\qquad = 11 \qquad\qquad\qquad\qquad\qquad\qquad = 0$

Cramer's Rule

To understand Cramer's rule, we first solve a general system of two linear equations in two variables. Consider the system

$$(1) \qquad a_1 x + b_1 y = c_1$$
$$(2) \qquad a_2 x + b_2 y = c_2$$

where a_1, b_1, c_1, a_2, b_2, and c_2 represent real numbers. To eliminate y, we multiply Eq. (1) by b_2 and Eq. (2) by $-b_1$:

$$a_1 b_2 x + b_1 b_2 y = c_1 b_2 \qquad \text{Eq. (1) multiplied by } b_2$$
$$\underline{-a_2 b_1 x - b_1 b_2 y = -c_2 b_1} \qquad \text{Eq. (2) multiplied by } -b_1$$
$$a_1 b_2 x - a_2 b_1 x \qquad\quad = c_1 b_2 - c_2 b_1 \qquad \text{Add.}$$
$$(a_1 b_2 - a_2 b_1)x = c_1 b_2 - c_2 b_1$$
$$x = \frac{c_1 b_2 - c_2 b_1}{a_1 b_2 - a_2 b_1} \qquad \text{Provided that } a_1 b_2 - a_2 b_1 \neq 0$$

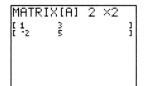

calculator

close-up

With a graphing calculator you can define matrix A using MATRX EDIT.

```
MATRIX[A]  2 ×2
[ 1    3        ]
[ -2   5        ]
```

Then use the determinant function (det) found in MATRX MATH and the A from MATRX NAMES to find its determinant.

```
det([A])
              11
```

Using similar steps to eliminate x from the system, we get

$$y = \frac{a_1 c_2 - a_2 c_1}{a_1 b_2 - a_2 b_1},$$

provided that $a_1 b_2 - a_2 b_1 \neq 0$. These formulas for x and y can be written by using determinants. In the determinant form they are known as **Cramer's rule.**

Cramer's Rule

The solution to the system

$$a_1 x + b_1 y = c_1$$
$$a_2 x + b_2 y = c_2$$

is given by $x = \frac{D_x}{D}$ and $y = \frac{D_y}{D}$, where

$$D = \begin{vmatrix} a_1 & b_1 \\ a_2 & b_2 \end{vmatrix}, \qquad D_x = \begin{vmatrix} c_1 & b_1 \\ c_2 & b_2 \end{vmatrix}, \qquad \text{and} \qquad D_y = \begin{vmatrix} a_1 & c_1 \\ a_2 & c_2 \end{vmatrix},$$

provided that $D \neq 0$.

Note that D is the determinant made up of the original coefficients of x and y. D is used in the denominator for both x and y. D_x is obtained by replacing the first (or x) column of D by the constants c_1 and c_2. D_y is found by replacing the second (or y) column of D by the constants c_1 and c_2.

E X A M P L E   2

An independent system

Use Cramer's rule to solve the system:

$$3x - 2y = 4$$
$$2x + \ y = -3$$

Solution

First find the determinants D, D_x, and D_y:

$$D = \begin{vmatrix} 3 & -2 \\ 2 & 1 \end{vmatrix} = 3 - (-4) = 7$$

$$D_x = \begin{vmatrix} 4 & -2 \\ -3 & 1 \end{vmatrix} = 4 - 6 = -2, \qquad D_y = \begin{vmatrix} 3 & 4 \\ 2 & -3 \end{vmatrix} = -9 - 8 = -17$$

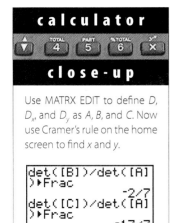

By Cramer's rule, we have

$$x = \frac{D_x}{D} = -\frac{2}{7} \qquad \text{and} \qquad y = \frac{D_y}{D} = -\frac{17}{7}.$$

Check in the original equations. The solution set is $\left\{ \left(-\frac{2}{7}, -\frac{17}{7} \right) \right\}$. ■

C A U T I O N   Cramer's rule works *only* when the determinant D is *not* equal to zero. Cramer's rule solves only those systems that have a single point in their solution set. If $D = 0$, we use elimination to determine whether the solution set is empty or contains all points of a line.

E X A M P L E   3

An inconsistent system

Solve the system:

$$2x + 3y = 9$$
$$-2x - 3y = 5$$

Solution

Cramer's rule does not work because

$$D = \begin{vmatrix} 2 & 3 \\ -2 & -3 \end{vmatrix} = -6 - (-6) = 0.$$

Because Cramer's rule fails to solve the system, we apply the addition method:

$$2x + 3y = 9$$
$$\underline{-2x - 3y = 5}$$
$$0 = 14$$

Because this last statement is false, the solution set is empty. The original equations are inconsistent.

E X A M P L E   4

A dependent system

Solve the system:

$$(1) \qquad 3x - 5y = 7$$
$$(2) \qquad 6x - 10y = 14$$

Solution

Cramer's rule does not apply because

$$D = \begin{vmatrix} 3 & -5 \\ 6 & -10 \end{vmatrix} = -30 - (-30) = 0.$$

Multiply Eq. (1) by -2 and add it to Eq. (2):

$$-6x + 10y = -14 \qquad \text{Eq. (1) times } -2$$
$$\underline{6x - 10y = 14} \qquad \text{Eq. (2)}$$
$$0 = 0 \qquad \text{Add.}$$

Because the last statement is an identity, the equations are dependent. The solution set is $\{(x, y) \mid 3x - 5y = 7\}$.

To apply Cramer's rule, we need the equations in standard form.

E X A M P L E   5

Standard form for Cramer's rule

Use Cramer's rule to solve the system:

$$(1) \qquad 2x - 3(y + 1) = -3$$
$$(2) \qquad 2y = 3x - 5$$

Solution

First write Eq. (1) in standard form, $Ax + By = C$:

$$2x - 3(y + 1) = -3$$
$$2x - 3y - 3 = -3$$
$$2x - 3y = 0$$

In standard form, Eq. (2) is $-3x + 2y = -5$. Now solve the system:

$$2x - 3y = 0$$
$$-3x + 2y = -5$$

Find D, D_x, and D_y:

$$D = \begin{vmatrix} 2 & -3 \\ -3 & 2 \end{vmatrix} = 4 - 9 = -5$$

$$D_x = \begin{vmatrix} 0 & -3 \\ -5 & 2 \end{vmatrix} = 0 - 15 = -15, \qquad D_y = \begin{vmatrix} 2 & 0 \\ -3 & -5 \end{vmatrix} = -10 - 0 = -10$$

Using Cramer's rule, we get

$$x = \frac{D_x}{D} = \frac{-15}{-5} = 3 \qquad \text{and} \qquad y = \frac{D_y}{D} = \frac{-10}{-5} = 2.$$

Because $(3, 2)$ satisfies both of the original equations, the solution set is $\{(3, 2)\}$.

WARM-UPS

True or false? Explain your answer.

1. $\begin{vmatrix} -1 & 2 \\ 3 & -5 \end{vmatrix} = -1$ True   2. $\begin{vmatrix} 2 & 4 \\ -4 & 8 \end{vmatrix} = 0$ False

3. Cramer's rule solves any system of two linear equations in two variables.   False

4. The determinant of a 2×2 matrix is a real number.   True

5. If $D = 0$, then there might be no solution to the system.   True

6. Cramer's rule is used to solve systems of linear equations only.   True

7. If the graphs of a pair of linear equations intersect at exactly one point, then this point can be found by using Cramer's rule.   True

8. If x and y represent the digits of a two-digit number, then the value of the number is either $10x + y$ or $10y + x$.   True

9. If x represents the length of a side of a square and y represents the length of a side of a separate equilateral triangle, then $4x + y$ represents the total of their perimeters.   False

10. If a and b are the digits of a two-digit number whose value is $10a + b$, then $b + 10a$ is the value of the number when the digits are reversed.   False

4.5   EXERCISES

Reading and Writing   *After reading this section, write out the answers to these questions. Use complete sentences.*

1. What is a determinant?

A determinant is a real number associated with a square matrix.

2. What is Cramer's rule used for?

Cramer's rule can be used to solve systems of linear equations.

3. Which systems can be solved using Cramer's rule?

Cramer's rule works on systems that have exactly one solution.

4. What is the determinant of the matrix of coefficients for inconsistent and dependent systems?

For inconsistent and dependent systems the determinant of the matrix of coefficients is zero.

Find the value of each determinant. See Example 1.

5. $\begin{vmatrix} 2 & 5 \\ 3 & 7 \end{vmatrix}$
-1

6. $\begin{vmatrix} -1 & 0 \\ 1 & 1 \end{vmatrix}$
-1

7. $\begin{vmatrix} 0 & 3 \\ 1 & 5 \end{vmatrix}$
-3

8. $\begin{vmatrix} 2 & 4 \\ 6 & 12 \end{vmatrix}$
0

9. $\begin{vmatrix} -3 & -2 \\ -4 & 2 \end{vmatrix}$
-14

10. $\begin{vmatrix} -2 & 2 \\ -3 & -5 \end{vmatrix}$
16

11. $\begin{vmatrix} 0.05 & 0.06 \\ 10 & 20 \end{vmatrix}$
0.4

12. $\begin{vmatrix} 0.02 & -0.5 \\ 30 & 50 \end{vmatrix}$
16

Solve each system using Cramer's rule. See Example 2.

13. $2x - y = 5$
$3x + 2y = -3$
$\{(1, -3)\}$

14. $3x + y = -1$
$x + 2y = 8$
$\{(-2, 5)\}$

15. $3x - 5y = -2$
$2x + 3y = 5$
$\{(1, 1)\}$

16. $x - y = 1$
$3x - 2y = 0$
$\{(-2, -3)\}$

17. $4x - 3y = 5$
$2x + 5y = 7$
$\left\{\left(\dfrac{23}{13}, \dfrac{9}{13}\right)\right\}$

18. $2x - y = 2$
$3x - 2y = 1$
$\{(3, 4)\}$

19. $0.5x + 0.2y = 8$
$0.4x - 0.6y = -5$
$\{(10, 15)\}$

20. $0.6x + 0.5y = 18$
$0.5x - 0.25y = 7$
$\{(20, 12)\}$

21. $\dfrac{1}{2}x + \dfrac{1}{4}y = 5$
$\dfrac{1}{3}x - \dfrac{1}{2}y = -1$
$\left\{\left(\dfrac{27}{4}, \dfrac{13}{2}\right)\right\}$

22. $\dfrac{1}{2}x + \dfrac{2}{3}y = 4$
$\dfrac{3}{4}x + \dfrac{1}{3}y = -2$
$\{(-8, 12)\}$

Solve each system. Use Cramer's rule when possible. See Examples 3 and 4.

23. $2x - 3y = 5$
$4x - 6y = 8$
$\varnothing$

24. $-x + 3y = 6$
$3x - 9y = -18$
$\{(x, y) \mid -x + 3y = 6\}$

25. $x - y = 4$
$x + 2y = 6$
$\left\{\left(\dfrac{14}{3}, \dfrac{2}{3}\right)\right\}$

26. $2x - y = 7$
$3x + 2y = -7$
$\{(1, -5)\}$

27. $4x - y = 6$
$-8x + 2y = -12$
$\{(x, y) \mid 4x - y = 6\}$

28. $-x + 2y = 3$
$3x - 6y = 10$
$\varnothing$

Solve each system using Cramer's rule. See Example 5.

29. $y = 3x - 12$
$3(x + 1) - 11 = y + 4$
$\{(x, y) \mid y = 3x - 12\}$

30. $y = x - 7$
$y = -2x + 5$
$\{(4, -3)\}$

31. $x - 6 = y + 1$
$y = -3x + 1$
$\{(2, -5)\}$

32. $1 - x = 6 + y$
$5 - 2x = 6 - y$
$\{(-2, -3)\}$

33. $y = 0.05x$
$x + y = 504$
$\{(480, 24)\}$

34. $0.05x + 0.10y = 6$
$y = 98 - x$
$\{(76, 22)\}$

Solve each problem by using two equations in two variables and Cramer's rule.

35. *Peas and beets.* One serving of canned peas contains 3 grams of protein and 11 grams of carbohydrates. One serving of canned beets contains 1 gram of protein and 8 grams of carbohydrates. A dietitian wants to determine the number of servings of each that would provide 38 grams of protein and 187 grams of carbohydrates.

   a) Use the accompanying graph to estimate the number of servings of each.

   b) Use Cramer's rule to find the number of servings of each.

   a) (9, 11)   **b)** 9 servings peas, 11 servings beets

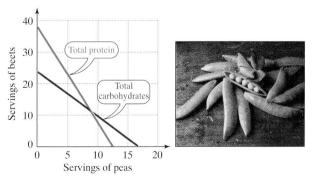

FIGURE FOR EXERCISE 35

36. *Protein and carbohydrates.* One serving of Cornies breakfast cereal contains 2 grams of protein and 25 grams of carbohydrates. One serving of Oaties breakfast cereal contains 4 grams of protein and 20 grams of carbohydrates. How many servings of each would provide exactly 24 grams of protein and 210 grams of carbohydrates?

Cornies 6 servings, Oaties 3 servings

37. *Two-digit number.* The sum of the digits of a two-digit number is 10. If the digits are reversed, the new number is 1 less than twice the original number. Find the original number.   37

38. *Another two-digit number.* The units digit of a two-digit number is twice the tens digit. If the digits are reversed, the new number is 36 more than the original number. Find the original number.   48

39. *Milk and a magazine.* Althia bought a gallon of milk and a magazine for a total of $4.65, excluding tax. Including the tax, the bill was $4.95. If there is a 5% sales tax on milk and an 8% sales tax on magazines, then what was the price of each item?
Milk $2.40, magazine $2.25

40. *Washing machines and refrigerators.* A truck carrying 3,600 cubic feet of cargo consisting of washing machines and refrigerators was hijacked. The washing machines are worth $300 each and are shipped in 36-cubic-foot cartons. The refrigerators are worth $900 each and are shipped in 45-cubic-foot cartons. If the total value of the cargo was $51,000, then how many of each were there on the truck?
50 washing machines, 40 refrigerators

41. *Singles and doubles.* Windy's Hamburger Palace sells singles and doubles. Toward the end of the evening, Windy himself noticed that he had on hand only 32 patties and 34 slices of tomatoes. If a single takes 1 patty and 2 slices, and a double takes 2 patties and 1 slice, then how many more singles and doubles must Windy sell to use up all of his patties and tomato slices?
12 singles, 10 doubles

42. *Valuable wrenches.* Carmen has a total of 28 wrenches, all of which are either box wrenches or open-end wrenches. For insurance purposes she values the box wrenches at $3.00 each and the open-end wrenches at $2.50 each. If the value of her wrench collection is $78, then how many of each type does she have?
16 box wrenches, 12 open-end wrenches

43. *Gary and Harry.* Gary is 5 years older than Harry. Twenty-nine years ago, Gary was twice as old as Harry. How old are they now?
Gary 39, Harry 34

44. *Acute angles.* One acute angle of a right triangle is 3° more than twice the other acute angle. What are the sizes of the acute angles?
29° and 61°

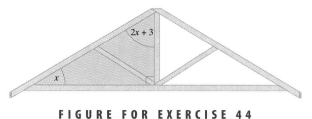

FIGURE FOR EXERCISE 44

45. *Equal perimeters.* A rope of length 80 feet is to be cut into two pieces. One piece will be used to form a square, and the other will be used to form an equilateral triangle. If the figures are to have equal perimeters, then what should be the length of a side of each?
Square 10 feet, triangle $\frac{40}{3}$ feet

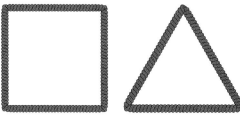

FIGURE FOR EXERCISE 45

46. *Coffee and doughnuts.* For a cup of coffee and a doughnut, Thurrel spent $2.25, including a tip. Later he spent $4.00 for two coffees and three doughnuts, including a tip. If he always tips $1.00, then what is the price of a cup of coffee? $0.75

47. *Chlorine mixture.* A 10% chlorine solution is to be mixed with a 25% chlorine solution to obtain 30 gallons of 20% solution. How many gallons of each must be used?
10 gallons of 10% solution, 20 gallons of 25% solution

48. *Safe drivers.* Emily and Camille started from the same city and drove in opposite directions on the freeway. After 3 hours they were 354 miles apart. If they had gone in the same direction, they would have been only 18 miles apart. How fast did each woman drive?
Emily 62 mph, Camille 56 mph

GETTING MORE INVOLVED

49. *Writing.* Explain what to do when you are trying to use Cramer's rule and $D = 0$.

50. *Exploration.* For what value of a does the system
$$ax - y = 3$$
$$x + 2y = 1$$
have a single solution?

51. *Exploration.* Can Cramer's rule be used to solve the following system? Explain.
$$2x^2 - y = 3$$
$$3x^2 + 2y = 22$$

GRAPHING CALCULATOR EXERCISES

52. Use the determinant feature on your graphing calculator to find the determinants in Exercises 5–12 of this section.

53. Solve the systems in Exercises 13–22 of this section by using your graphing calculator to find the necessary determinants.

4.6   CRAMER'S RULE FOR SYSTEMS IN THREE VARIABLES

The solution of linear systems involving three variables using determinants is very similar to the solution of linear systems in two variables using determinants. However, you first must learn to find the determinant of a 3 × 3 matrix.

Minors

To each element of a 3 × 3 matrix there corresponds a 2 × 2 matrix that is obtained by deleting the row and column of that element. The determinant of the 2 × 2 matrix is called the **minor** of that element.

EXAMPLE 1   **Finding minors**

Find the minors for the elements 2, 3, and −6 of the 3 × 3 matrix

$$\begin{bmatrix} 2 & -1 & -8 \\ 0 & -2 & 3 \\ 4 & -6 & 7 \end{bmatrix}.$$

Solution

To find the minor for 2, delete the first row and first column of the matrix:

$$\begin{bmatrix} 2 & -1 & -8 \\ 0 & -2 & 3 \\ 4 & -6 & 7 \end{bmatrix}$$

Now find the determinant of $\begin{bmatrix} -2 & 3 \\ -6 & 7 \end{bmatrix}$:

$$\begin{vmatrix} -2 & 3 \\ -6 & 7 \end{vmatrix} = (-2)(7) - (-6)(3) = 4$$

The minor for 2 is 4. To find the minor for 3, delete the second row and third column of the matrix:

$$\begin{bmatrix} 2 & -1 & -8 \\ 0 & -2 & 3 \\ 4 & -6 & 7 \end{bmatrix}$$

Now find the determinant of $\begin{bmatrix} 2 & -1 \\ 4 & -6 \end{bmatrix}$:

$$\begin{vmatrix} 2 & -1 \\ 4 & -6 \end{vmatrix} = (2)(-6) - (4)(-1) = -8$$

The minor for 3 is −8. To find the minor for −6, delete the third row and the second column of the matrix:

$$\begin{bmatrix} 2 & -1 & -8 \\ 0 & -2 & 3 \\ 4 & -6 & 7 \end{bmatrix}$$

Now find the determinant of $\begin{bmatrix} 2 & -8 \\ 0 & 3 \end{bmatrix}$:

$$\begin{vmatrix} 2 & -8 \\ 0 & 3 \end{vmatrix} = (2)(3) - (0)(-8) = 6$$

The minor for -6 is 6. ∎

Evaluating a 3 × 3 Determinant

The determinant of a 3×3 matrix is defined in terms of the determinants of minors.

Determinant of a 3 × 3 Matrix

The determinant of a 3×3 matrix is defined as follows:

$$\begin{vmatrix} a_1 & b_1 & c_1 \\ a_2 & b_2 & c_2 \\ a_3 & b_3 & c_3 \end{vmatrix} = a_1 \cdot \begin{vmatrix} b_2 & c_2 \\ b_3 & c_3 \end{vmatrix} - a_2 \cdot \begin{vmatrix} b_1 & c_1 \\ b_3 & c_3 \end{vmatrix} + a_3 \cdot \begin{vmatrix} b_1 & c_1 \\ b_2 & c_2 \end{vmatrix}$$

Note that the determinants following a_1, a_2, and a_3 are the minors for a_1, a_2, and a_3, respectively. Writing the determinant of a 3×3 matrix in terms of minors is called **expansion by minors.** In the definition we expanded by minors about the first column. Later we will see how to expand by minors using any row or column and get the same value for the determinant.

study tip

Remember that everything we do in solving problems is based on principles (which are also called rules, theorems, and definitions). These principles justify the steps we take. Be sure that you understand the reasons. If you just memorize procedures without understanding, you will soon forget the procedures.

E X A M P L E   2

Determinant of a 3 × 3 matrix

Find the determinant of the matrix by expansion by minors about the first column.

$$\begin{bmatrix} 1 & 3 & -5 \\ -2 & 4 & 6 \\ 0 & -7 & 9 \end{bmatrix}$$

Solution

$$\begin{vmatrix} 1 & 3 & -5 \\ -2 & 4 & 6 \\ 0 & -7 & 9 \end{vmatrix} = 1 \cdot \begin{vmatrix} 4 & 6 \\ -7 & 9 \end{vmatrix} - (-2) \cdot \begin{vmatrix} 3 & -5 \\ -7 & 9 \end{vmatrix} + 0 \cdot \begin{vmatrix} 3 & -5 \\ 4 & 6 \end{vmatrix}$$

$$= 1 \cdot [36 - (-42)] + 2 \cdot (27 - 35) + 0 \cdot [18 - (-20)]$$

$$= 1 \cdot 78 + 2 \cdot (-8) + 0$$

$$= 78 - 16$$

$$= 62$$

In the next example we evaluate a determinant using expansion by minors about the second row. In expanding about any row or column, the signs of the coefficients of the minors alternate according to the **sign array** that follows:

$$\begin{bmatrix} + & - & + \\ - & + & - \\ + & - & + \end{bmatrix}$$

The sign array is easily remembered by observing that there is a "+" sign in the upper left position and then alternating signs for all of the remaining positions.

EXAMPLE 3

Determinant of a 3 × 3 matrix

Evaluate the determinant of the matrix by expanding by minors about the second row.

$$\begin{bmatrix} 1 & 3 & -5 \\ -2 & 4 & 6 \\ 0 & -7 & 9 \end{bmatrix}$$

Solution

For expansion using the second row we prefix the signs "$- + -$" from the second row of the sign array to the corresponding numbers in the second row of the matrix, -2, 4, and 6. Note that the signs from the sign array are used in addition to any signs that occur on the numbers in the second row.

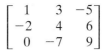

From the sign array, second row

$$\begin{vmatrix} 1 & 3 & -5 \\ -2 & 4 & 6 \\ 0 & -7 & 9 \end{vmatrix} = -(-2) \cdot \begin{vmatrix} 3 & -5 \\ -7 & 9 \end{vmatrix} + 4 \cdot \begin{vmatrix} 1 & -5 \\ 0 & 9 \end{vmatrix} - 6 \cdot \begin{vmatrix} 1 & 3 \\ 0 & -7 \end{vmatrix}$$

$$= 2(27 - 35) + 4(9 - 0) - 6(-7 - 0)$$
$$= 2(-8) + 4(9) - 6(-7)$$
$$= -16 + 36 + 42$$
$$= 62$$

Note that 62 is the same value that was obtained for this determinant in Example 2.

It can be shown that expanding by minors using any row or column prefixed by the corresponding signs from the sign array yields the same value for the determinant. Because we can use any row or column to evaluate a determinant of a 3 × 3 matrix, we can choose a row or column that makes the work easier. We can shorten the work considerably by picking a row or column with zeros in it.

EXAMPLE 4

Choosing the simplest row or column

Find the determinant of the matrix

$$\begin{bmatrix} 3 & -5 & 0 \\ 4 & -6 & 0 \\ 7 & 9 & 2 \end{bmatrix}$$

Solution

We choose to expand by minors about the third column of the matrix because the third column contains two zeros. Prefix the third-column entries 0, 0, 2 by the signs "$+ - +$" from the third column of the sign array:

$$\begin{vmatrix} 3 & -5 & 0 \\ 4 & -6 & 0 \\ 7 & 9 & 2 \end{vmatrix} = 0 \cdot \begin{vmatrix} 4 & -6 \\ 7 & 9 \end{vmatrix} - 0 \cdot \begin{vmatrix} 3 & -5 \\ 7 & 9 \end{vmatrix} + 2 \cdot \begin{vmatrix} 3 & -5 \\ 4 & -6 \end{vmatrix}$$

$$= 0 - 0 + 2[-18 - (-20)]$$
$$= 4$$

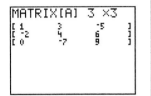

A calculator is very useful for finding the determinant of a 3 × 3 matrix. Define A using MATRX EDIT.

Now use the determinant function from MATRX MATH and the A from MATRX NAMES to find the determinant.

Cramer's Rule

A system of three linear equations in three variables can be solved by using determinants in a manner similar to that of the previous section. This rule is also called Cramer's rule.

Cramer's Rule for Three Equations in Three Unknowns

The solution to the system

$$a_1x + b_1y + c_1z = d_1$$
$$a_2x + b_2y + c_2z = d_2$$
$$a_3x + b_3y + c_3z = d_3$$

is given by $x = \dfrac{D_x}{D}$, $y = \dfrac{D_y}{D}$, and $z = \dfrac{D_z}{D}$, where

$$D = \begin{vmatrix} a_1 & b_1 & c_1 \\ a_2 & b_2 & c_2 \\ a_3 & b_3 & c_3 \end{vmatrix}, \qquad D_x = \begin{vmatrix} d_1 & b_1 & c_1 \\ d_2 & b_2 & c_2 \\ d_3 & b_3 & c_3 \end{vmatrix},$$

$$D_y = \begin{vmatrix} a_1 & d_1 & c_1 \\ a_2 & d_2 & c_2 \\ a_3 & d_3 & c_3 \end{vmatrix}, \qquad D_z = \begin{vmatrix} a_1 & b_1 & d_1 \\ a_2 & b_2 & d_2 \\ a_3 & b_3 & d_3 \end{vmatrix},$$

provided that $D \neq 0$.

Note that D_x, D_y, and D_z are obtained from D by replacing the x-, y-, or z-column with the constants d_1, d_2, and d_3.

E X A M P L E   5   **Solving an independent system**

Use Cramer's rule to solve the system:

$$x + y + z = 4$$
$$x - y = -3$$
$$x + 2y - z = 0$$

Solution

We first calculate D, D_x, D_y, and D_z. To calculate D, expand by minors about the third column because the third column has a zero in it:

$$D = \begin{vmatrix} 1 & 1 & 1 \\ 1 & -1 & 0 \\ 1 & 2 & -1 \end{vmatrix} = 1 \cdot \begin{vmatrix} 1 & -1 \\ 1 & 2 \end{vmatrix} - 0 \cdot \begin{vmatrix} 1 & 1 \\ 1 & 2 \end{vmatrix} + (-1) \cdot \begin{vmatrix} 1 & 1 \\ 1 & -1 \end{vmatrix}$$

$$= 1 \cdot [2 - (-1)] - 0 + (-1)[-1 - 1]$$
$$= 3 - 0 + 2$$
$$= 5$$

For D_x, expand by minors about the first column:

$$D_x = \begin{vmatrix} 4 & 1 & 1 \\ -3 & -1 & 0 \\ 0 & 2 & -1 \end{vmatrix} = 4 \cdot \begin{vmatrix} -1 & 0 \\ 2 & -1 \end{vmatrix} - (-3) \cdot \begin{vmatrix} 1 & 1 \\ 2 & -1 \end{vmatrix} + 0 \cdot \begin{vmatrix} 1 & 1 \\ -1 & 0 \end{vmatrix}$$

$$= 4 \cdot (1 - 0) + 3 \cdot (-1 - 2) + 0$$

$$= 4 - 9 + 0 = -5$$

For D_y, expand by minors about the third row:

$$D_y = \begin{vmatrix} 1 & 4 & 1 \\ 1 & -3 & 0 \\ 1 & 0 & -1 \end{vmatrix} = 1 \cdot \begin{vmatrix} 4 & 1 \\ -3 & 0 \end{vmatrix} - 0 \cdot \begin{vmatrix} 1 & 1 \\ 1 & 0 \end{vmatrix} + (-1) \cdot \begin{vmatrix} 1 & 4 \\ 1 & -3 \end{vmatrix}$$

$$= 1 \cdot 3 - 0 + (-1)(-7) = 10$$

To get D_z, expand by minors about the third row:

$$D_z = \begin{vmatrix} 1 & 1 & 4 \\ 1 & -1 & -3 \\ 1 & 2 & 0 \end{vmatrix} = 1 \cdot \begin{vmatrix} 1 & 4 \\ -1 & -3 \end{vmatrix} - 2 \cdot \begin{vmatrix} 1 & 4 \\ 1 & -3 \end{vmatrix} + 0 \cdot \begin{vmatrix} 1 & 1 \\ 1 & -1 \end{vmatrix}$$

$$= 1 \cdot 1 - 2(-7) + 0 = 15$$

Now, by Cramer's rule,

$$x = \frac{D_x}{D} = \frac{-5}{5} = -1, \qquad y = \frac{D_y}{D} = \frac{10}{5} = 2, \qquad \text{and} \qquad z = \frac{D_z}{D} = \frac{15}{5} = 3.$$

Check $(-1, 2, 3)$ in the original equations. The solution set is $\{(-1, 2, 3)\}$. ■

If $D = 0$, Cramer's rule does not apply. Cramer's rule provides the solution only to a system of three equations with three variables that has a single point in the solution set. If $D = 0$, then the solution set either is empty or consists of infinitely many points, and we use elimination of variables to find the solution.

E X A M P L E   6

Solving a dependent system

Solve the system:

$$\begin{aligned} (1) & \qquad x + y - z = 2 \\ (2) & \qquad 2x + 2y - 2z = 4 \\ (3) & \qquad -3x - 3y + 3z = -6 \end{aligned}$$

Solution

Calculate D by expanding about the first column:

$$D = \begin{vmatrix} 1 & 1 & -1 \\ 2 & 2 & -2 \\ -3 & -3 & 3 \end{vmatrix} = 1 \cdot \begin{vmatrix} 2 & -2 \\ -3 & 3 \end{vmatrix} - 2 \cdot \begin{vmatrix} 1 & -1 \\ -3 & 3 \end{vmatrix} + (-3) \cdot \begin{vmatrix} 1 & -1 \\ 2 & -2 \end{vmatrix}$$

$$= 1 \cdot 0 - 2 \cdot 0 + (-3) \cdot 0 = 0$$

Because $D = 0$, Cramer's rule does not apply to this system. If we multiply Eq. (1) by 2, we get Eq. (2). If we multiply Eq. (1) by -3, we get Eq. (3). Thus all three equations are equivalent, and they are dependent. The solution set to the system is $\{(x, y, z) \,|\, x + y - z = 2\}$. ■

True or false? Explain your answer.

1. A minor is the determinant of a 2×2 matrix.   True

2. The minor for an element is found by deleting the element from the matrix. False

3. The determinant of a 3×3 matrix is found by using minors.   True

4. Expansion by minors converts a 3×3 matrix into a 4×4 matrix.   False

5. Using Cramer's rule, we use $\frac{D}{D_x}$ to get the value of x.   False

6. Expansion by minors about any row or any column gives the same value for the determinant of a 3×3 matrix.   True

7. The sign array is used in evaluating the determinant of a 3×3 matrix. True

8. It is easier to find the determinant of a 3×3 matrix with several zero elements than one with no zero elements.   True

9. If $D = 0$, then x, y, and z are all zero.   False

10. Cramer's rule solves nonlinear systems of three equations in three unknowns.   False

4.6   EXERCISES

Reading and Writing   *After reading this section, write out the answers to these questions. Use complete sentences.*

1. What is a minor?

   A minor for an element in a 3×3 matrix is the determinant of a 2×2 matrix.

2. How do you find the minor for an element of a 3×3 matrix?

   A minor for an element is obtained by deleting the row and column of the element and finding the determinant of the 2×2 matrix that remains.

3. What is the purpose of the sign array?

   The sign array tells what signs to use in the expansion by minors.

4. Which systems can be solved by Cramer's rule for three equations in three unknowns?

   Cramer's rule solves only those systems that have a unique solution.

Find the indicated minors using the following matrix. See Example 1.

$$\begin{bmatrix} 3 & -2 & 5 \\ 4 & -3 & 7 \\ 0 & 1 & -6 \end{bmatrix}$$

5. Minor for 3   11

6. Minor for -2 -24

7. Minor for 5   4

8. Minor for -3 -18

9. Minor for 7   3

10. Minor for 0   1

11. Minor for 1   1

12. Minor for -6 -1

Find the determinant of each 3×3 matrix by using expansion by minors about the first column. See Example 2.

13. $\begin{bmatrix} 1 & 1 & 2 \\ 2 & 3 & 1 \\ 3 & 1 & 5 \end{bmatrix}$ -7

14. $\begin{bmatrix} 2 & 1 & 3 \\ 1 & 1 & 2 \\ 3 & 4 & 6 \end{bmatrix}$ -1

15. $\begin{bmatrix} 2 & 1 & 0 \\ 1 & 0 & 1 \\ 3 & 1 & 2 \end{bmatrix}$ -1

16. $\begin{bmatrix} 1 & 0 & 2 \\ 2 & 1 & 3 \\ 4 & 3 & 0 \end{bmatrix}$ -5

17. $\begin{bmatrix} -2 & 1 & 2 \\ -3 & 3 & 1 \\ -5 & 4 & 0 \end{bmatrix}$ 9

18. $\begin{bmatrix} -2 & 1 & 3 \\ -1 & 4 & 2 \\ 2 & 1 & 1 \end{bmatrix}$ -26

19. $\begin{bmatrix} 1 & 1 & 5 \\ 0 & 3 & 2 \\ 0 & 2 & 3 \end{bmatrix}$ 5

20. $\begin{bmatrix} 1 & 0 & 6 \\ 0 & 1 & 4 \\ 0 & 0 & 9 \end{bmatrix}$ 9

Evaluate the determinant of each 3×3 matrix using expansion by minors about the row or column of your choice. See Examples 3 and 4.

21. $\begin{bmatrix} 3 & 1 & 5 \\ 2 & 0 & 6 \\ 4 & 0 & 1 \end{bmatrix}$ 22

22. $\begin{bmatrix} 2 & 1 & 2 \\ 1 & 2 & 5 \\ 3 & 0 & 0 \end{bmatrix}$ 3

23. $\begin{bmatrix} -2 & 1 & 3 \\ 0 & 1 & -1 \\ 2 & -4 & -3 \end{bmatrix}$ 6

24. $\begin{bmatrix} -2 & 0 & 1 \\ -3 & 2 & -5 \\ 4 & -2 & 6 \end{bmatrix}$ -6

25. $\begin{bmatrix} -2 & -3 & 0 \\ 4 & -1 & 0 \\ 0 & 3 & 5 \end{bmatrix}$ 70

26. $\begin{bmatrix} -2 & 6 & 3 \\ 0 & 4 & 0 \\ -1 & -4 & 5 \end{bmatrix}$ -28

27. $\begin{bmatrix} 2 & 1 & 1 \\ 0 & 0 & 5 \\ 5 & 0 & 4 \end{bmatrix}$ 25

28. $\begin{bmatrix} 2 & 3 & 0 \\ 6 & 4 & 1 \\ 1 & 2 & 0 \end{bmatrix}$ -1

Use Cramer's rule to solve each system. See Example 5.

29. $x + y + z = 6$
$x - y + z = 2$
$2x + y + z = 7$
$\{(1, 2, 3)\}$

30. $x + y + z = 2$
$x - y - 2z = -3$
$2x - y + z = 7$
$\{(1, -2, 3)\}$

31. $x - 3y + 2z = 0$
$x + y + z = 2$
$x - y + z = 0$
$\{(-1, 1, 2)\}$

32. $3x + 2y + 2z = 0$
$x - y + z = 1$
$x + y - z = 3$
$\{(2, -1, -2)\}$

33. $x + y = -1$
$2y - z = 3$
$x + y + z = 0$
$\{(-3, 2, 1)\}$

34. $x - y = 8$
$x - 2z = 0$
$x + y - z = 1$
$\{(6, -2, 3)\}$

35. $x + y - z = 0$
$2x + 2y + z = 6$
$x - 3y = 0$
$\left\{\left(\dfrac{3}{2}, \dfrac{1}{2}, 2\right)\right\}$

36. $x + y + z = 1$
$5x - y = 0$
$3x + y + 2z = 0$
$\left\{\left(\dfrac{1}{2}, \dfrac{5}{2}, -2\right)\right\}$

37. $x + y + z = 0$
$2y + 2z = 0$
$3x - y = -1$
$\{(0, 1, -1)\}$

38. $x + z = 0$
$x - 3y = 1$
$4y - 3z = 3$
$\{(1, 0, -1)\}$

Solve each system. Use Cramer's rule if possible. See Example 6.

39. $2x - y + z = 1$
$-6x + 3y - 3z = -3$
$4x - 2y + 2z = 2$
$\{(x, y, z) \mid 2x - y + z = 1\}$

40. $x - y + z = 4$
$2x - 2y + 2z = 3$
$4x - y - z = 1$
$\varnothing$

41. $x + y = 1$
$y + 2z = 3$
$x + 2y + 2z = 5$
$\varnothing$

42. $x - y + z = 5$
$2x - 2y + 2z = 10$
$3x - 3y + 3z = 15$
$\{(x, y, z) \mid x - y + z = 5\}$

43. $x + y = 4$
$y + z = -3$
$x + z = -5$
$\{(1, 3, -6)\}$

44. $x + y = 0$
$x - z = -1$
$y + z = 3$
$\varnothing$

Write a system of three equations in three variables for each word problem. Use Cramer's rule to solve each system.

45. *Weighing dogs.* Cassandra wants to determine the weights of her two dogs, Mimi and Mitzi. However, neither dog will sit on the scale by herself. Cassandra, Mimi, and Mitzi altogether weigh 175 pounds. Cassandra and Mimi together weigh 143 pounds. Cassandra and Mitzi together weigh 139 pounds. How much does each weigh individually?
Mimi 36 pounds, Mitzi 32 pounds, Cassandra 107 pounds

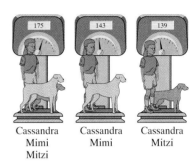

Cassandra   Cassandra   Cassandra
Mimi   Mimi   Mitzi
Mitzi

F I G U R E   F O R   E X E R C I S E   4 5

46. *Nickels, dimes, and quarters.* Bernard has 41 coins consisting of nickels, dimes, and quarters, and they are worth a total of $4.00. If the number of dimes plus the number of quarters is one more than the number of nickels, then how many of each does he have?
20 nickels, 15 dimes, 6 quarters

47. *Finding three angles.* If the two acute angles of a right triangle differ by 12°, then what are the measures of the three angles of this triangle?
39°, 51°, 90°

48. *Two acute and one obtuse.* The obtuse angle of a triangle is twice as large as the sum of the two acute angles. If the smallest angle is only one-eighth as large as the sum of the other two, then what is the measure of each angle?
20°, 40°, 120°

GETTING MORE INVOLVED

49. *Writing.* For what values of a, b, c, and d is the determinant of the matrix

$$\begin{bmatrix} a & b & 0 \\ c & d & 0 \\ b & a & 0 \end{bmatrix}$$

equal to zero? Explain your answer.

50. *Exploration.* The determinant of a 4×4 matrix is found by expanding by 3×3 minors and using a 4×4 sign array. Find the determinant of a 4×4 matrix of your choice. Use the definition of the determinant of a 3×3 matrix as a guide.

51. Use the determinant feature of your graphing calculator to find the determinant of each matrix in Exercises 13–20 of this section.

52. Solve the systems in Exercises 29–38 by using Cramer's rule and your graphing calculator to find the determinants.

53. The solution to an independent system of four linear equations in four variables can be found by using determinants of 4 × 4 matrices in the same manner as Cramer's rule for three variables. Write Cramer's rule for four variables. Make up a system of four linear equations in four variables and solve it using your new rule and a graphing calculator to evaluate the determinants.

 4.7   LINEAR PROGRAMMING

In this section

- Graphing the Constraints
- Maximizing or Minimizing a Linear Function

In this section we graph the solution set to a system of several linear inequalities, much as we graphed compound linear inequalities in Chapter 3. We then use the solution set as the domain of a function for which we are seeking the maximum or minimum value. The method that we use is called **linear programming,** and it can be applied to problems such as finding maximum profit or minimum cost.

Graphing the Constraints

In linear programming we have two variables that must satisfy several linear inequalities. These inequalities are called the **constraints** because they restrict the variables to only certain values. A graph in the coordinate plane is used to indicate the points that satisfy all of the constraints.

EXAMPLE 1

Graphing the constraints

Graph the solution set to the system of inequalities and identify each vertex of the region:

$$x \geq 0, \quad y \geq 0$$
$$3x + 2y \leq 12$$
$$x + 2y \leq 8$$

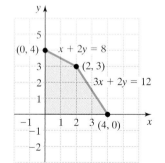

FIGURE 4.7

Solution

The points on or to the right of the y-axis satisfy $x \geq 0$. The points on or above the x-axis satisfy $y \geq 0$. The points on or below the line $3x + 2y = 12$ satisfy $3x + 2y \leq 12$. The points on or below the line $x + 2y = 8$ satisfy $x + 2y \leq 8$. Graph each straight line and shade the region that satisfies all four inequalities as shown in Fig. 4.7. Three of the vertices are easily identified as $(0, 0)$, $(0, 4)$, and $(4, 0)$. The fourth vertex is found by solving the system $3x + 2y = 12$ and $x + 2y = 8$. The fourth vertex is $(2, 3)$. ■

In linear programming the constraints usually come from physical limitations in some problem. In the next example we write the constraints and then graph the points in the coordinate plane that satisfy all of the constraints.

EXAMPLE 2

Writing the constraints

Jules is in the business of constructing dog houses. A small dog house requires 8 square feet (ft^2) of plywood and 6 ft^2 of insulation. A large dog house requires 16 ft^2 of plywood and 3 ft^2 of insulation. Jules has available only 48 ft^2 of plywood and 18 ft^2 of insulation. Write the constraints on the number of small and large dog houses that he can build with the available supplies and graph the solution set to the system of constraints.

Solution

Let x represent the number of small dog houses and y represent the number of large dog houses. We have two natural constraints $x \geq 0$ and $y \geq 0$ since he cannot build a negative number of dog houses. Since the total plywood available for use is at most 48 ft^2, $8x + 16y \leq 48$. Since the total insulation available is at most 18 ft^2, $6x + 3y \leq 18$. Simplify the inequalities to get the following constraints:

$$x \geq 0, \quad y \geq 0$$
$$x + 2y \leq 6$$
$$2x + y \leq 6$$

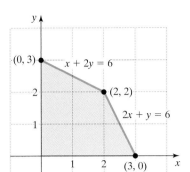

FIGURE 4.8

The graph of the solution set to the system of inequalities is shown in Fig. 4.8. ■

Maximizing or Minimizing a Linear Function

In Example 2 any ordered pair within the region is a possible solution to the number of dog houses of each type that could be built. If a small dog house sells for $15 and a large dog house sells for $20, then the total revenue in dollars from x small and y large dog houses is $R = 15x + 20y$. Since the revenue is a function of x and y, we write $R(x, y) = 15x + 20y$. The function R is a linear function of x and y. The domain of R is the region graphed in Fig. 4.8.

Linear Function of Two Variables

A function of the form $f(x, y) = Ax + By + C$, where A, B, and C are real numbers, is called a **linear function of two variables.**

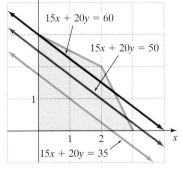

FIGURE 4.9

Naturally, we are interested in the maximum revenue subject to the constraints on x and y. To investigate some possible revenues, replace R in $R = 15x + 20y$ with, say, 35, 50, and 60. The graphs of the parallel lines $15x + 20y = 35$, $15x + 20y = 50$, and $15x + 20y = 60$ are shown in Fig. 4.9. The revenue at any point on the line $15x + 20y = 35$ is $35. We get a larger revenue on a higher revenue line (and lower revenue on a lower line). The maximum revenue possible will be on the highest revenue line that still intersects the region. Because the sides of the region are straight-line segments, the intersection of the highest (or lowest) revenue line with the region must include a vertex of the region. This is the fundamental principle behind linear programming.

The Principle of Linear Programming

The maximum or minimum value of a linear function subject to linear constraints occurs at a vertex of the region determined by the constraints.

E X A M P L E   3

Maximizing a linear function with linear constraints

A small dog house requires 8 ft^2 of plywood and 6 ft^2 of insulation. A large dog house requires 16 ft^2 of plywood and 3 ft^2 of insulation. Only 48 ft^2 of plywood and 18 ft^2 of insulation are available. If a small dog house sells for $15 and a large dog house sells for $20, then how many dog houses of each type should be built to maximize the revenue and to satisfy the constraints?

Solution

Let x be the number of small dog houses and y be the number of large dog houses. We wrote and graphed the constraints for this problem in Example 2, so we will not repeat that here. The graph in Fig. 4.8 has four vertices: (0, 0), (0, 3), (3, 0), and (2, 2). The revenue function is $R(x, y) = 15x + 20y$. Since the maximum value of this function must occur at a vertex, we evaluate the function at each vertex:

$$R(0, 0) = 15(0) + 20(0) = \$0$$
$$R(0, 3) = 15(0) + 20(3) = \$60$$
$$R(3, 0) = 15(3) + 20(0) = \$45$$
$$R(2, 2) = 15(2) + 20(2) = \$70$$

From this list we can see that the maximum revenue is \$70 when two small and two large dog houses are built. We also see that the minimum revenue is \$0 when no dog houses of either type are built.

We can summarize the procedure for solving linear programming problems with the following strategy.

Strategy for Linear Programming

Use the following steps to find the maximum or minimum value of a linear function subject to linear constraints.

1. Graph the region that satisfies all of the constraints.
2. Determine the coordinates of each vertex of the region.
3. Evaluate the function at each vertex of the region.
4. Identify which vertex gives the maximum or minimum value of the function.

In the next example we solve another linear programming problem.

E X A M P L E   4   **Minimizing a linear function with linear constraints**

One serving of food A contains 2 grams of protein and 6 grams of carbohydrates. One serving of food B contains 4 grams of protein and 3 grams of carbohydrates. A dietitian wants a meal that contains at least 12 grams of protein and at least 18 grams of carbohydrates. If the cost of food A is 9 cents per serving and the cost of food B is 20 cents per serving, then how many servings of each food would minimize the cost and satisfy the constraints?

Solution

Let x equal the number of servings of food A and y equal the number of servings of food B. If the meal is to contain at least 12 grams of protein, then $2x + 4y \geq 12$. If the meal is to contain at least 18 grams of carbohydrates, then $6x + 3y \geq 18$. Simplify each inequality and use the two natural constraints to get the following system:

$$x \geq 0, \quad y \geq 0$$
$$x + 2y \geq 6$$
$$2x + \quad y \geq 6$$

The graph of the constraints is shown in Fig. 4.10. The vertices are $(0, 6)$, $(6, 0)$, and $(2, 2)$. The cost in cents for x servings of A and y servings of B is $C(x, y) = 9x + 20y$. Evaluate the cost at each vertex:

$$C(0, 6) = 9(0) + 20(6) = 120 \text{ cents}$$
$$C(6, 0) = 9(6) + 20(0) = 54 \text{ cents}$$
$$C(2, 2) = 9(2) + 20(2) = 58 \text{ cents}$$

The minimum cost of 54 cents is attained by using six servings of food A and no servings of food B.

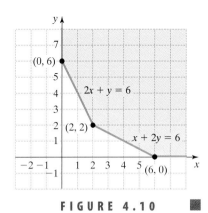

FIGURE 4.10

WARM-UPS

True or false? Explain your answer.

1. The graph of $x \geq 0$ in the coordinate plane consists of the points on or above the x-axis.   False

2. The graph of $y \geq 0$ in the coordinate plane consists of the points on or to the right of the y-axis.   False

3. The graph of $x + y \leq 6$ consists of the points below the line $x + y = 6$.   False

4. The graph of $2x + 3y = 30$ has x-intercept $(0, 10)$ and y-intercept $(15, 0)$.   False

5. The graph of a system of inequalities is a union of their individual solution sets.   False

6. In linear programming, constraints are inequalities that restrict the possible values that the variables can assume.   True

7. The function $F(x, y) = Ax^2 + By^2 + C$ is a linear function of x and y.   False

8. The value of $R(x, y) = 3x + 5y$ at the point $(2, 4)$ is 26.   True

9. If $C(x, y) = 12x + 10y$, then $C(0, 5) = 62$.   False

10. In solving a linear programming problem, we must determine the vertices of the region defined by the constraints.   True

---

4.7   EXERCISES

Reading and Writing   *After reading this section, write out the answers to these questions. Use complete sentences.*

1. What is a constraint?
   A constraint is an inequality that restricts the values of the variables.

2. What is linear programming?
   Linear programming is the process used to maximize or minimize a linear function subject to linear constraints.

3. Where do the constraints come from in a linear programming problem?
   Constraints may be limitations on the amount of available supplies, money, or other resources.

4. What is a linear function of two variables?
   A linear function of two variables is a function of the form $f(x, y) = Ax + By + C$.

5. Where does the maximum or minimum value of a linear function subject to linear constraints occur?

The maximum or minimum of a linear function subject to linear constraints occurs at a vertex of the region determined by the constraints.

6. What is the strategy for solving a linear programming problem?

Write the constraints, graph the region that they determine, locate each vertex, then evaluate the function at each vertex, and identify the maximum or minimum.

Graph the solution set to each system of inequalities and identify each vertex of the region. See Example 1.

7. $x \geq 0, y \geq 0$
$x + y \leq 5$

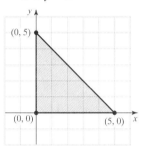

8. $x \geq 0, y \geq 0$
$y \leq 5, y \geq x$

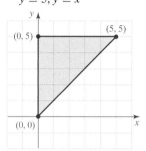

9. $x \geq 0, y \geq 0$
$2x + y \leq 4$
$x + y \leq 3$

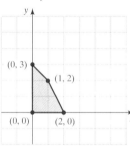

10. $x \geq 0, y \geq 0$
$x + y \leq 4$
$x + 2y \leq 6$

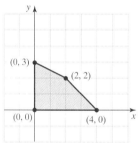

11. $x \geq 0, y \geq 0$
$2x + y \geq 3$
$x + y \geq 2$

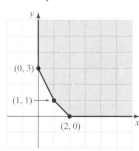

12. $x \geq 0, y \geq 0$
$3x + 2y \geq 12$
$2x + y \geq 7$

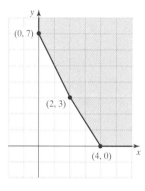

13. $x \geq 0, y \geq 0$
$x + 3y \leq 15$
$2x + y \leq 10$

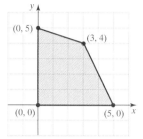

14. $x \geq 0, y \geq 0$
$2x + 3y \leq 15$
$x + y \leq 7$

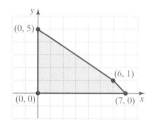

15. $x \geq 0, y \geq 0$
$x + y \geq 4$
$3x + y \geq 6$

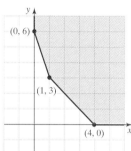

16. $x \geq 0, y \geq 0$
$x + 3y \geq 6$
$2x + y \geq 7$

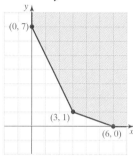

Solve each problem. See Examples 2–4.

17. *Phase I advertising.* The publicity director for Mercy Hospital is planning to bolster the hospital's image by running a TV ad and a radio ad. Due to budgetary and other constraints, the number of times that she can run the TV ad, x, and the number of times that she can run the radio ad, y, must be in the region shown in the figure on the next page. The function

$$A = 9000x + 4000y$$

gives the total number of people reached by the ads.

a) Find the total number of people reached by the ads at each vertex of the region.

b) What mix of TV and radio ads maximizes the number of people reached?

   a) 0, 320,000, 510,000, 450,000

   b) 30 TV ads and 60 radio ads

18. *Phase II advertising.* Suppose the radio station in Exercise 17 starts playing country music and the function for the total number of people changes to

$$A = 9000x + 2000y.$$

a) Find A at each vertex of the region using this function.

b) What mix of TV and radio ads maximizes the number of people reached?

   a) 0, 160,000, 390,000, 450,000

   b) 50 TV ads and 0 radio ads

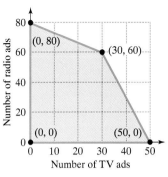

FIGURE FOR EXERCISES 17 AND 18

19. At Burger Heaven a double contains 2 meat patties and 6 pickles, whereas a triple contains 3 meat patties and 3 pickles. Near closing time one day, only 24 meat patties and 48 pickles are available. If a double burger sells for $1.20 and a triple burger sells for $1.50, then how many of each should be made to maximize the total revenue?

6 doubles, 4 triples

20. Sam and Doris manufacture rocking chairs and porch swings in the Ozarks. Each rocker requires 3 hours of work from Sam and 2 hours from Doris. Each swing requires 2 hours of work from Sam and 2 hours from Doris. Sam cannot work more than 48 hours per week, and Doris cannot work more than 40 hours per week. If a rocker sells for $160 and a swing sells for $100, then how many of each should be made per week to maximize the revenue?

16 chairs, 0 swings

21. If a double burger sells for $1.00 and a triple burger sells for $2.00, then how many of each should be made to maximize the total revenue subject to the constraints of Exercise 19?

0 doubles, 8 triples

22. If a rocker sells for $120 and a swing sells for $100, then how many of each should be made to maximize the total revenue subject to the constraints of Exercise 20?

8 chairs, 12 swings

23. One cup of Doggie Dinner contains 20 grams of protein and 40 grams of carbohydrates. One cup of Puppy Power contains 30 grams of protein and 20 grams of carbohydrates. Susan wants her dog to get at least 200 grams of protein and 180 grams of carbohydrates per day. If Doggie Dinner costs 16 cents per cup and Puppy Power costs 20 cents per cup, then how many cups of each would satisfy the constraints and minimize the total cost?

1.75 cups Doggie Dinner, 5.5 cups Puppy Power

24. Mammoth Muffler employs supervisors and helpers. According to the union contract, a supervisor does 2 brake jobs and 3 mufflers per day, whereas a helper does 6 brake jobs and 3 mufflers per day. The home office requires enough staff for at least 24 brake jobs and for at least 18 mufflers per day. If a supervisor makes $90 per day and a helper makes $100 per day, then how many of each should be employed to satisfy the constraints and to minimize the daily labor cost?

3 supervisors, 3 helpers

25. Suppose in Exercise 23 Doggie Dinner costs 4 cents per cup and Puppy Power costs 10 cents per cup. How many cups of each would satisfy the constraints and minimize the total cost?

10 cups Doggie Dinner, 0 cups Puppy Power

26. Suppose in Exercise 24 the supervisor makes $110 per day and the helper makes $100 per day. How many of each should be employed to satisfy the constraints and to minimize the daily labor cost?

0 supervisors, 6 helpers

27. Anita has at most $24,000 to invest in her brother-in-law's laundromat and her nephew's car wash. Her brother-in-law has high blood pressure and heart disease but he will pay 18%, whereas her nephew is healthier but will pay only 12%. So the amount she will invest in the car wash will be at least twice the amount that she will invest in the laundromat but not more than three times as much. How much should she invest in each to maximize her total income from the two investments?

Laundromat $8,000, car wash $16,000

28. Herbert assembles computers in his shop. The parts for each economy model are shipped to him in a carton with a volume of 2 cubic feet (ft^3) and the parts for each deluxe model are shipped to him in a carton with a volume of 3 ft^3. After assembly, each economy model is shipped out in a carton with a volume of 4 ft^3, and each deluxe model is shipped out in a carton with a volume of 4 ft^3. The truck that delivers the parts has a maximum capacity of 180 ft^3, and the truck that takes out the completed computers has a maximum capacity of 280 ft^3. He can receive only one shipment of parts and send out one shipment of computers per week. If his profit on an economy model is $60 and his profit on a deluxe model is $100, then how many of each should he order per week to maximize his profit?

0 economy models, 60 deluxe models

COLLABORATIVE ACTIVITIES

Types of Systems

Assign roles in your groups. Set up the system of equations for each scenario below and attempt to solve it. Analyze your results.

Part I: Talia's Tallitot. Talia plans to make tallitot (prayer shawls) and matching kipot (prayer hats) from a bolt of material. The fabric she plans to use is 250 yards long. She will need 2 yards of material for each tallit (prayer shawl) and $\frac{1}{6}$ of a yard for each kipah (prayer hat). She also plans to put tzitzit (fringes) on the tallit and will need 1 yard of cord for each tallit. The kipot will not have any fringes. She has 100 yards of cord to use for the tzitzit.

1. Write two equations to use to solve for how many kipot and tallitot she can make from the supplies on hand. Let t equal the number of tallitot made and k equal the number of kipot made.

2. Try to solve this system of equations. If you get a solution, state what it is. If not, describe what your result means.

Part II: Karif's Kerchiefs. Karif plans to make red and green kerchiefs in two sizes, medium and large. He has 60 yards of each color of material. He will need $\frac{1}{4}$ of a yard of the appropriate colored fabric for each medium kerchief and $\frac{1}{2}$ of a yard for each large kerchief. He also plans to trim each kerchief with the alternate color of ribbon. He has 240 yards of each color of ribbon. He will need 1 yard of ribbon for each medium kerchief and 2 yards for each large kerchief.

Grouping: Two to four students per group
Topic: Types of solutions to systems of equations

3. Let m equal the number of medium kerchiefs and l equal the number of large kerchiefs. Set up two equations to use to solve for the number of each color of kerchiefs Karif can make from the supplies on hand.

4. Try to solve this system of equations. If you get a solution, state what it is. If not, describe what your result means.

Part III: Maria's Mantillas. Maria plans to make two different kinds of mantillas (veils), some out of lace and some out of a sheer cotton fabric. She also plans to make two sizes of mantillas, one size for women and one size for young girls. She has 150 yards of the lace fabric and 200 yards of the cotton fabric. She will need 2 yards of either fabric for the smaller mantillas and $3\frac{1}{2}$ yards of either fabric for the larger mantillas.

5. Let w equal the number of mantillas for the women and y equal the number of mantillas for the young girls. Set up two equations to use to solve for the number of each mantilla Maria can make from each type of fabric.

6. Try to solve this system of equations. If you get a solution, state what it is. If not, describe what your result means.

Extension: Explain what you would need to do for the systems of equations you could not solve to make them solvable.

WRAP-UP   CHAPTER 4

SUMMARY

Systems of Linear Equations

Methods for solving systems in two variables	Graphing: Sketch the graphs to see the solution.	**Examples** The graphs of $y = x - 1$ and $x + y = 3$ intersect at (2, 1).
	Substitution: Solve one equation for one variable in terms of the other, then substitute into the other equation.	Substitution: $x + (x - 1) = 3$
	Addition: Multiply each equation as necessary to eliminate a variable upon addition of the equations.	$-x + y = -1$ $\underline{x + y = 3}$ $2y = 2$

Types of linear systems in two variables	Independent: One point in solution set The lines intersect at one point.	$y = x - 5$ $y = 2x + 3$
	Inconsistent: Empty solution set The lines are parallel.	$2x + y = 1$ $2x + y = 5$
	Dependent: Infinite solution set The lines are the same.	$2x + 3y = 4$ $4x + 6y = 8$
Linear equation in three variables	$Ax + By + Cz = D$ In a three-dimensional coordinate system the graph is a plane.	$2x - y + 3z = 5$
Linear systems in three variables	Use substitution or addition to eliminate variables in the system. The solution set may be a single point, the empty set, or an infinite set of points.	$x + y - z = 3$ $2x - 3y + z = 2$ $x - y - 4z = 14$

Matrices and Determinants

Examples

Matrix	A rectangular array of real numbers An $n \times m$ matrix has n rows and m columns.	$\begin{bmatrix} 1 & -3 \\ 2 & 5 \end{bmatrix}, \begin{bmatrix} 1 & 0 & 1 \\ 2 & 1 & 4 \end{bmatrix}$
Augmented matrix	The matrix of coefficients and constants from a system of linear equations	$x - 3y = -7$ $2x + 5y = 19$ Augmented matrix: $\begin{bmatrix} 1 & -3 & -7 \\ 2 & 5 & 19 \end{bmatrix}$
Gaussian elimination method	Use the row operations to get ones on the diagonal and zeros elsewhere for the coefficients in the augmented matrix.	$\begin{bmatrix} 1 & 0 & 2 \\ 0 & 1 & 3 \end{bmatrix}$ $x = 2$ and $y = 3$
Determinant	A real number corresponding to a square matrix	
Determinant of a 2×2 matrix	$\begin{vmatrix} a_1 & b_1 \\ a_2 & b_2 \end{vmatrix} = a_1 b_2 - a_2 b_1$	$\begin{vmatrix} 1 & -3 \\ 2 & 5 \end{vmatrix} = 5 - (-6)$ $= 11$
Determinant of a 3×3 matrix	Expand by minors about any row or column, using signs from the sign array. $\begin{vmatrix} a_1 & b_1 & c_1 \\ a_2 & b_2 & c_2 \\ a_3 & b_3 & c_3 \end{vmatrix} = a_1 \cdot \begin{vmatrix} b_2 & c_2 \\ b_3 & c_3 \end{vmatrix} - a_2 \cdot \begin{vmatrix} b_1 & c_1 \\ b_3 & c_3 \end{vmatrix} + a_3 \cdot \begin{vmatrix} b_1 & c_1 \\ b_2 & c_2 \end{vmatrix}$	Sign array: $\begin{bmatrix} + & - & + \\ - & + & - \\ + & - & + \end{bmatrix}$

Cramer's Rules

Two linear equations in two variables	The solution to the system $$a_1 x + b_1 y = c_1$$ $$a_2 x + b_2 y = c_2$$ is given by $x = \dfrac{D_x}{D}$ and $y = \dfrac{D_y}{D}$, where $$D = \begin{vmatrix} a_1 & b_1 \\ a_2 & b_2 \end{vmatrix}, \qquad D_x = \begin{vmatrix} c_1 & b_1 \\ c_2 & b_2 \end{vmatrix}, \qquad \text{and} \qquad D_y = \begin{vmatrix} a_1 & c_1 \\ a_2 & c_2 \end{vmatrix}$$ provided that $D \neq 0$.

Three linear equations in three variables	The solution to the system

$$a_1x + b_1y + c_1z = d_1$$
$$a_2x + b_2y + c_2z = d_2$$
$$a_3x + b_3y + c_3z = d_3$$

is given by $x = \dfrac{D_x}{D}$, $y = \dfrac{D_y}{D}$, and $z = \dfrac{D_z}{D}$, where

$$D = \begin{vmatrix} a_1 & b_1 & c_1 \\ a_2 & b_2 & c_2 \\ a_3 & b_3 & c_3 \end{vmatrix}, \qquad D_x = \begin{vmatrix} d_1 & b_1 & c_1 \\ d_2 & b_2 & c_2 \\ d_3 & b_3 & c_3 \end{vmatrix},$$

$$D_y = \begin{vmatrix} a_1 & d_1 & c_1 \\ a_2 & d_2 & c_2 \\ a_3 & d_3 & c_3 \end{vmatrix}, \qquad D_z = \begin{vmatrix} a_1 & b_1 & d_1 \\ a_2 & b_2 & d_2 \\ a_3 & b_3 & d_3 \end{vmatrix},$$

provided that $D \neq 0$.

Linear Programming

Use the following steps to find the maximum or minimum value of a linear function subject to linear constraints.
1. Graph the region that satisfies all of the constraints.
2. Determine the coordinates of each vertex of the region.
3. Evaluate the function at each vertex of the region.
4. Identify which vertex gives the maximum or minimum value of the function.

ENRICHING YOUR MATHEMATICAL WORD POWER

For each mathematical term, choose the correct meaning.

1. **system of equations**
   a. a systematic method for classifying equations
   b. a method for solving an equation
   c. two or more equations
   d. the properties of equality   c

2. **independent linear system**
   a. a system with exactly one solution
   b. an equation that is satisfied by every real number
   c. equations that are identical
   d. a system of lines   a

3. **inconsistent system**
   a. a system with no solution
   b. a system of inconsistent equations
   c. a system that is incorrect
   d. a system that we are not sure how to solve   a

4. **dependent system**
   a. a system that is independent
   b. a system that depends on a variable
   c. a system that has no solution
   d. a system for which the graphs coincide   d

5. **substitution method**
   a. replacing the variables by the correct answer
   b. a method of eliminating a variable by substituting one equation into the other
   c. the replacement method
   d. any method of solving a system   b

6. **addition method**
   a. adding the same number to each side of an equation
   b. adding fractions
   c. eliminating a variable by adding two equations
   d. the sum of a number and its additive inverse is zero
     c

7. **linear equation in three variables**
   a. $Ax + By + Cz = D$ with A, B, and C not all zero
   b. $Ax + By = C$ with A and B not both zero
   c. the equation of a line
   d. $A/x + B/y = C$ with A and B not both zero   a

8. **matrix**
   a. a television screen
   b. a maze
   c. a rectangular array of numbers
   d. coordinates in four dimensions   c

9. **augmented matrix**
   a. a matrix with a power booster
   b. a matrix with no solution
   c. a square matrix
   d. a matrix containing the coefficients and constants of a system of equations   d

10. **order**
    a. the length of a matrix
    b. the number of rows and columns in a matrix
    c. the highest power of a matrix
    d. the lowest power of a matrix   b

11. **determinant**
    a. a number corresponding to a square matrix
    b. a number that is determined by any matrix
    c. the first entry of a matrix
    d. a number that determines whether a matrix has a solution   a

12. **sign array**
    a. the signs of the entries of a matrix
    b. the sign of the determinant
    c. the signs of the answers
    d. a matrix of $+$ and $-$ signs used in computing a determinant   d

REVIEW EXERCISES

4.1 *Solve by graphing. Indicate whether each system is independent, inconsistent, or dependent.*

1. $y = 2x - 1$
$x + y = 2$
$\{(1, 1)\}$, independent

2. $y = -x$
$y = -x + 3$
$\varnothing$, inconsistent

3. $y = 3x - 4$
$y = -2x + 1$
$\{(1, -1)\}$, independent

4. $x + y = 5$
$x - y = -1$
$\{(2, 3)\}$, independent

Solve each system by the substitution method. Indicate whether each system is independent, inconsistent, or dependent.

5. $y = 3x + 11$
$2x + 3y = 0$
$\{(-3, 2)\}$, independent

6. $x - y = 3$
$3x - 2y = 3$
$\{(-3, -6)\}$, independent

7. $x = y + 5$
$2x - 2y = 12$
$\varnothing$, inconsistent

8. $2x - y = 3$
$6x - 9 = 3y$
$\{(x, y) \mid 2x - y = 3\}$, dependent

4.2 *Solve each system by the addition method. Indicate whether each system is independent, inconsistent, or dependent.*

9. $5x - 3y = -20$
$3x + 2y = 7$
$\{(-1, 5)\}$, independent

10. $-3x + y = 3$
$2x - 3y = 5$
$\{(-2, -3)\}$, independent

11. $2(y - 5) + 4 = 3(x - 6)$
$3x - 2y = 12$
$\{(x, y) \mid 3x - 2y = 12\}$, dependent

12. $3x - 4(y - 5) = x + 2$
$2y - x = 7$
$\varnothing$, inconsistent

4.3 *Solve each system by elimination of variables.*

13. $2x - y - z = 3$
$3x + y + 2z = 4$
$4x + 2y - z = -4$
$\{(1, -3, 2)\}$

14. $2x + 3y - 2z = -11$
$3x - 2y + 3z = 7$
$x - 4y + 4z = 14$
$\{(-2, 1, 5)\}$

15. $x - 3y + z = 5$
$2x - 4y - z = 7$
$2x - 6y + 2z = 6$
$\varnothing$

16. $x - y + z = 1$
$2x - 2y + 2z = 2$
$-3x + 3y - 3z = -3$
$\{(x, y, z) \mid x - y + z = 1\}$

4.4 *Solve each system by using the Gaussian elimination method.*

17. $2x + y = 0$
$x - 3y = 14$
$\{(2, -4)\}$

18. $2x - y = 8$
$3x + 2y = -2$
$\{(2, -4)\}$

19. $x + y - z = 0$
$x - y + 2z = 4$
$2x + y - z = 1$
$\{(1, 1, 2)\}$

20. $2x - y + 2z = 9$
$x + 3y = 5$
$3x + z = 9$
$\{(2, 1, 3)\}$

4.5 *Evaluate each determinant.*

21. $\begin{vmatrix} 1 & 3 \\ 0 & 2 \end{vmatrix}$

2

22. $\begin{vmatrix} -1 & 2 \\ -3 & 5 \end{vmatrix}$

1

23. $\begin{vmatrix} 0.01 & 0.02 \\ 50 & 80 \end{vmatrix}$

-0.2

24. $\begin{vmatrix} \dfrac{1}{2} & \dfrac{1}{3} \\ \dfrac{1}{4} & \dfrac{1}{5} \end{vmatrix}$

$\dfrac{1}{60}$

> **study tip**
>
> Note how the review exercises are arranged according to the sections in this chapter. If you are having trouble with a certain type of problem, refer back to the appropriate section for examples and explanations.

Solve each system. Use Cramer's rule if possible.

25. $2x - y = 0$
$3x + y = -5$
$\{(-1, -2)\}$

26. $3x - 2y = 14$
$2x + 3y = -8$
$\{(2, -4)\}$

27. $y = 2x - 3$
$3x - 2y = 4$
$\{(2, 1)\}$

28. $2x - 5y = -1$
$10y - 4x = 2$
$\{(x, y) \mid 2x - 5y = -1\}$

29. $3x - y = -1$
$2y - 6x = 5$
$\varnothing$

30. $y = 2x - 5$
$y = 3x - 3y$
$\{(4, 3)\}$

4.6 *Evaluate each determinant.*

31. $\begin{vmatrix} 2 & 3 & 1 \\ -1 & 2 & 4 \\ 6 & 1 & 1 \end{vmatrix}$
58

32. $\begin{vmatrix} 1 & -1 & 0 \\ -2 & 0 & 0 \\ 3 & 1 & 5 \end{vmatrix}$
-10

33. $\begin{vmatrix} 2 & 3 & -2 \\ 2 & 0 & 4 \\ -1 & 0 & 3 \end{vmatrix}$
-30

34. $\begin{vmatrix} 3 & -1 & 4 \\ 2 & -1 & 1 \\ -2 & 0 & 1 \end{vmatrix}$
-7

Solve each system. Use Cramer's rule if possible.

35. $x + y = 3$
$x + y + z = 0$
$x - y - z = 2$
$\{(1, 2, -3)\}$

36. $2x - 4y + 2z = 6$
$x - y - z = 1$
$x - 2y + z = 4$
$\varnothing$

37. $2x - y + z = 0$
$4x + 6y - 2z = 0$
$x - 2y - z = -9$
$\{(-1, 2, 4)\}$

38. $3x - 3y - 3z = 3$
$2x - 2y - 2z = 2$
$x - y - z = 1$
$\{(x, y, z) \mid x - y - z = 1\}$

4.7 *Graph each system of inequalities and identify each vertex of the region.*

39. $x \geq 0, y \geq 0$
$x + 2y \leq 6$
$x + y \leq 5$
$(0, 0), (0, 3), (4, 1), (5, 0)$

40. $x \geq 0, y \geq 0$
$3x + 2y \geq 12$
$x + 2y \geq 8$
$(0, 6), (2, 3), (8, 0)$

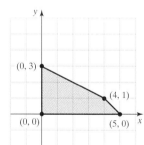

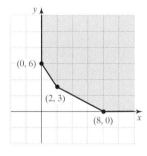

Solve each problem by linear programming.

41. Find the maximum value of the function $R(x, y) = 6x + 9y$ subject to the following constraints:

$$x \geq 0, y \geq 0$$
$$2x + y \leq 6$$
$$x + 2y \leq 6 \quad 30$$

42. Find the minimum value of the function $C(x, y) = 9x + 10y$ subject to the following constraints:

$$x \geq 0, y \geq 0$$
$$x + y \geq 4$$
$$3x + y \geq 6 \quad 36$$

MISCELLANEOUS

Use a system of equations in two or three variables to solve each word problem. Solve by the method of your choice.

43. *Two-digit number.* The sum of the digits in a two-digit number is 15. When the digits are reversed, the new number is 9 more than the original number. What is the original number?   78

44. *Two-digit number.* The sum of the digits in a two-digit number is 8. When the digits are reversed, the new number is 18 less than the original number. What is the original number?   53

45. *Traveling by boat.* Alonzo can travel from his camp downstream to the mouth of the river in 30 minutes. If it takes him 45 minutes to come back, then how long would it take him to go that same distance in the lake with no current?   36 minutes

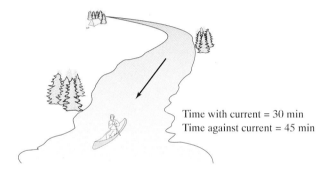

Time with current = 30 min
Time against current = 45 min

FIGURE FOR EXERCISE 45

46. *Driving and dating.* In 4 years Gasper will be old enough to drive. His parents said that he must have a driver's license for 2 years before he can date. Three years ago, Gasper's age was only one-half of the age necessary to date. How old must Gasper be to drive, and how old is he now?   16 years old to drive, 12 years old at present

47. *Three solutions.* A chemist has three solutions of acid that must be mixed to obtain 20 liters of a solution that is 38% acid. Solution A is 30% acid, solution B is 20% acid, and solution C is 60% acid. Because of another chemical in these solutions, the chemist must keep the ratio of solution C to solution A at 2 to 1. How many liters of each should she mix together?
4 liters of 30% solution A, 8 liters of 20% solution B, 8 liters of 60% solution C

48. *Mixing investments.* Darlene invested a total of $20,000. The part that she invested in Dell Computer stock returned 70% and the part that she invested in U.S. Treasury bonds returned 5%. Her total return on these two investments was $9,580.
   a) Use the graph on the next page to estimate the amount that she put into each investment.   (13,000, 7,000)
   b) Solve a system of equations to find the exact amount that she put into each investment.
   $13,200 in Dell, $6,800 in bonds

FIGURE FOR EXERCISE 48

49. ***Beets and beans.*** One serving of canned beets contains 1 gram of protein and 6 grams of carbohydrates. One serving of canned red beans contains 6 grams of protein and 20 grams of carbohydrates. How many servings of each would it take to get exactly 21 grams of protein and 78 grams of carbohydrates?
Three servings of each

CHAPTER 4 TEST

Solve the system by graphing.

1. $x + y = 4$
$y = 2x + 1$
$\{(1, 3)\}$

Solve each system by substitution.

2. $y = 2x - 8$
$4x + 3y = 1$
$\left\{ \left(\dfrac{5}{2}, -3 \right) \right\}$

3. $y = x - 5$
$3x - 4(y - 2) = 28 - x$
$\{(x, y) \mid y = x - 5\}$

Solve each system by the addition method.

4. $3x + 2y = 3$
$4x - 3y = -13$
$\{(-1, 3)\}$

5. $3x - y = 5$
$-6x + 2y = 1$
$\varnothing$

Determine whether each system is independent, inconsistent, or dependent.

6. $y = 3x - 5$
$y = 3x + 2$
Inconsistent

7. $2x + 2y = 8$
$x + y = 4$
Dependent

8. $y = 2x - 3$
$y = 5x - 14$
Independent

Solve the following system by elimination of variables.

9. $x + y - z = 2$
$2x - y + 3z = -5$
$x - 3y + z = 4$
$\{(1, -2, -3)\}$

Solve by the Gaussian elimination method.

10. $3x - y = 1$
$x + 2y = 12$

$\{(2, 5)\}$

11. $x - y - z = 1$
$-x - y + 2z = -2$
$-x - 3y + z = -5$
$\{(3, 1, 1)\}$

Evaluate each determinant.

12. $\begin{vmatrix} 2 & 3 \\ 4 & -3 \end{vmatrix}$

-18

13. $\begin{vmatrix} 1 & -2 & -1 \\ 2 & 3 & 1 \\ 1 & 1 & 0 \end{vmatrix}$

-2

Solve each system by using Cramer's rule.

14. $2x - y = -4$
$3x + y = -1$
$\{(-1, 2)\}$

15. $x + y = 0$
$x - y + 2z = 6$
$2x + y - z = 1$
$\{(2, -2, 1)\}$

For each problem, write a system of equations in two or three variables. Use the method of your choice to solve each system.

16. One night the manager of the Sea Breeze Motel rented 5 singles and 12 doubles for a total of $390. The next night he rented 9 singles and 10 doubles for a total of $412. What is the rental charge for each type of room?
Singles $18, doubles $25

17. Jill, Karen, and Betsy studied a total of 93 hours last week. Jill's and Karen's study time totaled only one-half as much as Betsy's. If Jill studied 3 hours more than Karen, then how many hours did each one of the girls spend studying?
Jill 17 hours, Karen 14 hours, Betsy 62 hours

Solve the following problem by linear programming.

18. Find the maximum value of the function

$$P(x, y) = 8x + 10y$$

subject to the following constraints:

$$x \geq 0, y \geq 0$$
$$2x + 3y \leq 12$$
$$x + y \leq 5 \quad 44$$

Simplify each expression.

1. -3^4 -81

2. $\frac{1}{3}(3) + 6$ 7

3. $(-5)^2 - 4(-2)(6)$ 73

4. $6 - (0.2)(0.3)$ 5.94

5. $5(t - 3) - 6(t - 2)$ $-t - 3$

6. $0.1(x - 1) - (x - 1)$ $-0.9x + 0.9$

7. $\frac{-9x^2 - 6x + 3}{-3}$ $3x^2 + 2x - 1$

8. $\frac{4y - 6}{2} - \frac{3y - 9}{3}$ y

Solve each equation for y.

9. $3x - 5y = 7$ $y = \frac{3}{5}x - \frac{7}{5}$

10. $Cx - Dy = W$ $y = \frac{C}{D}x - \frac{W}{D}$

11. $Cy = Wy - K$ $y = \frac{K}{W - C}$

12. $A = \frac{1}{2}b(w - y)$ $y = \frac{bw - 2A}{b}$

Solve each system.

13. $y = x - 5$
  $2x + 3y = 5$
  $\{(4, -1)\}$

14. $0.05x + 0.06y = 67$
  $x + y = 1200$
  $\{(500, 700)\}$

15. $3x - 15y = -51$
  $x + 17 = 5y$
  $\{(x, y) \mid x + 17 = 5y\}$

16. $0.07a + 0.3b = 6.70$
  $7a + 30b = 67$
  $\varnothing$

Find the equation of each line.

17. The line through $(0, 55)$ and $(-99, 0)$
  $y = \frac{5}{9}x + 55$

18. The line through $(2, -3)$ and $(-4, 8)$
  $y = -\frac{11}{6}x + \frac{2}{3}$

19. The line through $(-4, 6)$ that is parallel to $y = 5x$
  $y = 5x + 26$

20. The line through $(4, 7)$ that is perpendicular to
  $y = -2x + 1$
  $y = \frac{1}{2}x + 5$

21. The line through $(3, 5)$ that is parallel to the *x*-axis
  $y = 5$

22. The line through $(-7, 0)$ that is perpendicular to the *x*-axis
  $x = -7$

Solve.

23. *Comparing copiers.* A self-employed consultant has prepared the accompanying graph to compare the total cost of purchasing and using two different copy machines.

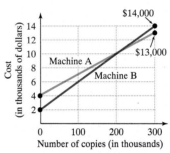

FIGURE FOR EXERCISE 23

a) Which machine has the larger purchase price?

b) What is the per copy cost for operating each machine, not including the purchase price?

c) Find the slope of each line and interpret your findings.

d) Find the equation of each line.

e) Find the number of copies for which the total cost is the same for both machines.

  a) Machine A

  b) Machine B $0.04 per copy, machine A $0.03 per copy

  c) The slopes 0.04 and 0.03 are the per copy cost for each machine.

  d) B: $y = 0.04x + 2000$, A: $y = 0.03x + 4000$

  e) 200,000

Exponents and Polynomials

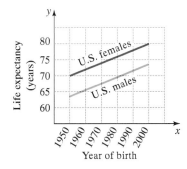

One statistic that can be used to measure the general health of a nation or group within a nation is life expectancy. This data is considered more accurate than many other statistics because it is easy to determine the precise number of years in a person's lifetime.

According to the National Center for Health Statistics, an American born in 1996 has a life expectancy of 76.1 years. However, an American male born in 1996 has a life expectancy of only 73.0 years, whereas a female can expect 79.0 years. A male who makes it to 65 can expect to live 15.7 more years, whereas a female who makes it to 65 can expect 18.9 more years. In the next few years, thanks in part to advances in health care and science, longevity is expected to increase significantly worldwide. In fact, the World Health Organization predicts that by 2025 no country will have a life expectancy of less than 50 years.

In this chapter we study algebraic expressions involving exponents. In Exercises 95 and 96 of Section 5.2 you will see how formulas involving exponents can be used to find the life expectancies of men and women.

5.1 INTEGRAL EXPONENTS AND SCIENTIFIC NOTATION

In Chapter 1 we defined positive integral exponents and learned to evaluate expressions involving exponents. In this section we will extend the definition of exponents to include all integers and to learn some rules for working with integral exponents. In Chapter 7 we will see that any rational number can be used as an exponent.

Positive and Negative Exponents

Positive integral exponents provide a convenient way to write repeated multiplication or very large numbers. For example,

$$2 \cdot 2 \cdot 2 = 2^3, \qquad y \cdot y \cdot y \cdot y = y^4, \qquad \text{and} \qquad 1{,}000{,}000{,}000 = 10^9.$$

We refer to 2^3 as "2 cubed," "2 raised to the third power," or "a power of 2."

> **Positive Integral Exponents**
>
> If a is a nonzero real number and n is a positive integer, then
>
> $$a^n = a \cdot \underbrace{a \cdot a \cdot \ldots \cdot a}_{n \text{ factors of } a}.$$

In the **exponential expression** a^n, the **base** is a, and the **exponent** is n.

We use 2^{-3} to represent the reciprocal of 2^3. Because $2^3 = 8$, we have $2^{-3} = \frac{1}{8}$. In general, a^{-n} is defined as the reciprocal of a^n.

> **Negative Integral Exponents**
>
> If a is a nonzero real number and n is a positive integer, then
>
> $$a^{-n} = \frac{1}{a^n}. \qquad \text{(If } n \text{ is positive, } -n \text{ is negative.)}$$

To evaluate 2^{-3}, you can first cube 2 to get 8 and then find the reciprocal to get $\frac{1}{8}$, or you can first find the reciprocal of 2 $\left(\text{which is } \frac{1}{2}\right)$ and then cube $\frac{1}{2}$ to get $\frac{1}{8}$. So

$$2^{-3} = \left(\frac{1}{2}\right)^3.$$

The power and the reciprocal can be found in either order. If the exponent is -1, we simply find the reciprocal. For example,

$$2^{-1} = \frac{1}{2}, \qquad \left(\frac{2}{3}\right)^{-1} = \frac{3}{2}, \qquad \text{and} \qquad \left(-\frac{1}{4}\right)^{-1} = -4.$$

Because 2^3 and 2^{-3} are reciprocals of each other, we have

$$2^{-3} = \frac{1}{2^3} \qquad \text{and} \qquad \frac{1}{2^{-3}} = 2^3.$$

These examples illustrate the following rules.

helpful hint

A negative exponent does not cause an expression to have a negative value. The negative exponent "causes" the reciprocal:

$$2^{-3} = \frac{1}{8},$$

$$(-3)^{-4} = \frac{1}{81},$$

$$(-4)^{-3} = -\frac{1}{64}.$$

Rules for Negative Exponents

If a is a nonzero real number and n is a positive integer, then

$$a^{-n} = \left(\frac{1}{a}\right)^n, \qquad a^{-1} = \frac{1}{a}, \qquad \text{and} \qquad \frac{1}{a^{-n}} = a^n.$$

EXAMPLE 1

Negative exponents

Evaluate each expression.

a) 3^{-2} **b)** $(-3)^{-2}$ **c)** -3^{-2} **d)** $\left(\frac{3}{4}\right)^{-3}$ **e)** $\frac{1}{5^{-3}}$

Solution

a) $3^{-2} = \frac{1}{3^2} = \frac{1}{9}$ Definition of negative exponent

b) $(-3)^{-2} = \frac{1}{(-3)^2} = \frac{1}{9}$ Definition of negative exponent

c) $-3^{-2} = -\frac{1}{3^2} = -\frac{1}{9}$ Evaluate 3^{-2}, then take the opposite.

d) $\left(\frac{3}{4}\right)^{-3} = \left(\frac{4}{3}\right)^3$ The reciprocal of $\frac{3}{4}$ is $\frac{4}{3}$.

$\qquad = \frac{64}{27}$ The cube of $\frac{4}{3}$ is $\frac{64}{27}$.

e) $\frac{1}{5^{-3}} = 5^3 = 125$ The reciprocal of 5^{-3} is 5^3.

CAUTION   In Chapter 1 we agreed to evaluate -3^2 by squaring 3 first and then taking the opposite. So $-3^2 = -9$, whereas $(-3)^2 = 9$. The same agreement also holds for negative exponents. That is why the answer to Example 1(c) is negative.

Product Rule

We can simplify an expression such as $2^3 \cdot 2^5$ using the definition of exponents.

$$2^3 \cdot 2^5 = \overbrace{(2 \cdot 2 \cdot 2)}^{\text{three 2's}}\overbrace{(2 \cdot 2 \cdot 2 \cdot 2 \cdot 2)}^{\text{five 2's}} = 2^8$$
$$\underbrace{}_{\text{eight 2's}}$$

Notice that the exponent 8 is the sum of the exponents 3 and 5. This example illustrates the **product rule for exponents.**

Product Rule for Exponents

If m and n are integers and $a \neq 0$, then

$$a^m \cdot a^n = a^{m+n}.$$

EXAMPLE 2

Using the product rule

Simplify each expression. Write answers with positive exponents and assume all variables represent nonzero real numbers.

a) $3^4 \cdot 3^6$ **b)** $4x^{-3} \cdot 5x$ **c)** $-2y^{-3}(-5y^{-4})$

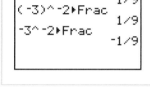

calculator

close-up

You can evaluate expressions with negative exponents using a graphing calculator. Use the fraction feature to get fractional answers.

```
3^-2▶Frac
              1/9
(-3)^-2▶Frac
              1/9
-3^-2▶Frac
             -1/9
```

calculator

close-up

A graphing calculator cannot prove that the product rule is correct, but it can provide numerical support for the product rule.

```
2^3*2^5
              256
2^8
              256
```

Solution

a) $3^4 \cdot 3^6 = 3^{4+6} = 3^{10}$ Product rule

b) $4x^{-3} \cdot 5x = 4 \cdot 5 \cdot x^{-3} \cdot x^1$

$\quad\quad\quad = 20x^{-2}$ Product rule: $x^{-3} \cdot x^1 = x^{-3+1} = x^{-2}$

$\quad\quad\quad = \dfrac{20}{x^2}$ Definition of negative exponent

c) $-2y^{-3}(-5y^{-4}) = (-2)(-5)y^{-3}y^{-4}$

$\quad\quad\quad\quad\quad = 10y^{-7}$ Product rule: $-3 + (-4) = -7$

$\quad\quad\quad\quad\quad = \dfrac{10}{y^7}$ Definition of negative exponent

C A U T I O N   The product rule cannot be applied to $2^3 \cdot 3^2$ because the bases are not identical. Even when the bases are identical, we do not multiply the bases. For example, $2^5 \cdot 2^4 \neq 4^9$. Using the rule correctly, we get $2^5 \cdot 2^4 = 2^9$.

Zero Exponent

We have used positive and negative integral exponents, but we have not yet seen the integer 0 used as an exponent. Note that the product rule was stated to hold for *any* integers m and n. If we use the product rule on $2^3 \cdot 2^{-3}$, we get

$$2^3 \cdot 2^{-3} = 2^0.$$

Because $2^3 = 8$ and $2^{-3} = \dfrac{1}{8}$, we must have $2^3 \cdot 2^{-3} = 1$. So for consistency we define 2^0 and the zero power of any nonzero number to be 1.

> **Zero Exponent**
>
> If a is any nonzero real number, then $a^0 = 1$.

E X A M P L E   3   **Using zero as an exponent**

Simplify each expression. Write answers with positive exponents and assume all variables represent nonzero real numbers.

a) -3^0 b) $\left(\dfrac{1}{4} - \dfrac{3}{2}\right)^0$ c) $-2a^5b^{-6} \cdot 3a^{-5}b^2$

Solution

a) To evaluate -3^0, we find 3^0 and then take the opposite. So $-3^0 = -1$.

b) $\left(\dfrac{1}{4} - \dfrac{3}{2}\right)^0 = 1$ Definition of zero exponent

c) $-2a^5b^{-6} \cdot 3a^{-5}b^2 = -6a^5 \cdot a^{-5} \cdot b^{-6} \cdot b^2$

$\quad\quad\quad\quad\quad\quad\quad = -6a^0b^{-4}$ Product rule

$\quad\quad\quad\quad\quad\quad\quad = -\dfrac{6}{b^4}$ Definitions of negative and zero exponent

Changing the Sign of an Exponent

Because a^{-n} and a^n are reciprocals of each other, we know that

$$a^{-n} = \dfrac{1}{a^n} \quad\quad \text{and} \quad\quad \dfrac{1}{a^{-n}} = a^n.$$

So a negative exponent in the numerator or denominator can be changed to positive by relocating the exponential expression. In the next example we use these facts to remove negative exponents from exponential expressions.

E X A M P L E   4

Simplifying expressions with negative exponents

Write each expression without negative exponents and simplify. All variables represent nonzero real numbers.

a) $\dfrac{5a^{-3}}{a^2 \cdot 2^{-2}}$ b) $\dfrac{-2x^{-3}}{y^{-2}z^3}$

Solution

a) $\dfrac{5a^{-3}}{a^2 \cdot 2^{-2}} = 5 \cdot a^{-3} \cdot \dfrac{1}{a^2} \cdot \dfrac{1}{2^{-2}}$ Rewrite division as multiplication.

$\qquad\qquad = 5 \cdot \dfrac{1}{a^3} \cdot \dfrac{1}{a^2} \cdot 2^2$ Change the signs of the negative exponents.

$\qquad\qquad = \dfrac{20}{a^5}$ Product rule: $a^3 \cdot a^2 = a^5$

Note that in $5a^{-3}$ the negative exponent applies only to a.

b) $\dfrac{-2x^{-3}}{y^{-2}z^3} = -2 \cdot x^{-3} \cdot \dfrac{1}{y^{-2}} \cdot \dfrac{1}{z^3}$ Rewrite as multiplication.

$\qquad\qquad = -2 \cdot \dfrac{1}{x^3} \cdot y^2 \cdot \dfrac{1}{z^3}$ Definition of negative exponent

$\qquad\qquad = \dfrac{-2y^2}{x^3z^3}$ Simplify.

In Example 4 we showed more steps than are necessary. For instance, in part (b) we could simply write

$$\frac{-2x^{-3}}{y^{-2}z^3} = \frac{-2y^2}{x^3z^3}.$$

Exponential expressions (that are factors) can be moved from numerator to denominator (or vice versa) as long as we change the sign of the exponent.

C A U T I O N   If an exponential expression is *not* a factor, you *cannot* move it from numerator to denominator (or vice versa). For example,

$$\frac{2^{-1} + 1^{-1}}{1^{-1}} \neq \frac{1}{2 + 1}.$$

Because $2^{-1} = \frac{1}{2}$ and $1^{-1} = 1$, we get

$$\frac{2^{-1} + 1^{-1}}{1^{-1}} = \frac{\frac{1}{2} + 1}{1} = \frac{\frac{3}{2}}{1} = \frac{3}{2} \qquad \text{not} \qquad \frac{1}{2 + 1} = \frac{1}{3}.$$

Quotient Rule

We can use arithmetic to simplify the quotient of two exponential expressions. For example,

$$\frac{2^5}{2^3} = \frac{\cancel{2} \cdot \cancel{2} \cdot \cancel{2} \cdot 2 \cdot 2}{\cancel{2} \cdot \cancel{2} \cdot \cancel{2}} = 2^2.$$

calculator

close-up

A graphing calculator cannot prove that the quotient rule is correct, but it can provide numerical support for the quotient rule.

```
2^15/2^5
             1024
2^10
             1024
```

There are five 2's in the numerator and three 2's in the denominator. After dividing, two 2's remain. The exponent in 2^2 can be obtained by subtracting the exponents 3 and 5. This example illustrates the **quotient rule for exponents.**

Quotient Rule for Exponents

If m and n are any integers and $a \neq 0$, then

$$\frac{a^m}{a^n} = a^{m-n}.$$

E X A M P L E   5

Using the quotient rule

Simplify each expression. Write answers with positive exponents only. All variables represent nonzero real numbers.

a) $\dfrac{2^9}{2^4}$ b) $\dfrac{m^5}{m^{-3}}$ c) $\dfrac{y^{-4}}{y^{-2}}$

Solution

a) $\dfrac{2^9}{2^4} = 2^{9-4} = 2^5$

b) $\dfrac{m^5}{m^{-3}} = m^{5-(-3)} = m^8$

c) $\dfrac{y^{-4}}{y^{-2}} = y^{-4-(-2)} = y^{-2} = \dfrac{1}{y^2}$

The next example further illustrates the rules of exponents. Remember that the bases must be identical for the quotient rule or the product rule.

E X A M P L E   6

Using the product and quotient rules

Use the rules of exponents to simplify each expression. Write answers with positive exponents only. All variables represent nonzero real numbers.

a) $\dfrac{2x^{-7}}{x^{-7}}$ b) $\dfrac{w(2w^{-4})}{3w^{-2}}$ c) $\dfrac{x^{-1}x^{-3}y^5}{x^{-2}y^2}$

Solution

a) $\dfrac{2x^{-7}}{x^{-7}} = 2x^0$ Quotient rule: $-7 - (-7) = 0$

$\qquad\quad = 2$ Definition of zero exponent

b) $\dfrac{w(2w^{-4})}{3w^{-2}} = \dfrac{2w^{-3}}{3w^{-2}}$ Product rule: $w^1 \cdot w^{-4} = w^{-3}$

$\qquad\quad = \dfrac{2w^{-1}}{3}$ Quotient rule: $-3 - (-2) = -1$

$\qquad\quad = \dfrac{2}{3w}$ Definition of negative exponent

c) $\dfrac{x^{-1}x^{-3}y^5}{x^{-2}y^2} = \dfrac{x^{-4}y^5}{x^{-2}y^2} = x^{-2}y^3 = \dfrac{y^3}{x^2}$

Scientific Notation

Many of the numbers that are encountered in science are either very large or very small. For example, the distance from the earth to the sun is 93,000,000 miles, and a hydrogen atom weighs 0.0000000000000000000000017 gram. Scientific notation provides a convenient way of writing very large and very small numbers. In scientific notation the distance from the earth to the sun is 9.3×10^7 miles and a hydrogen atom weighs 1.7×10^{-24} gram. In scientific notation the times symbol, $\times$, is used to indicate multiplication. Converting a number from scientific notation to standard notation is simply a matter of multiplication.

E X A M P L E   7

Scientific notation to standard notation

Write each number using standard notation.

a) 7.62×10^5 **b)** 6.35×10^{-4}

Solution

a) Multiplying a number by 10^5 moves the decimal point five places to the right:
$$7.62 \times 10^5 = 762000. = 762{,}000$$

b) Multiplying a number by 10^{-4} or 0.0001 moves the decimal point four places to the left:
$$6.35 \times 10^{-4} = 0.000635 = 0.000635$$ ■

The procedure for converting a number from scientific notation to standard notation is summarized as follows.

c a l c u l a t o r

c l o s e - u p

In normal mode, display a number in scientific notation and press ENTER to convert to standard notation. You can use a power of 10 or the EE key to get the E for the built-in scientific notation.

```
7.62ᴇ5
              762000
7.62*10^5
              762000
```

> **Strategy for Converting from Scientific Notation to Standard Notation**
>
> 1. Determine the number of places to move the decimal point by examining the exponent on the 10.
> 2. Move to the right for a positive exponent and to the left for a negative exponent.

A positive number in scientific notation is written as a product of a number between 1 and 10, and a power of 10. Numbers in scientific notation are written with only one digit to the left of the decimal point. A number larger than 10 is written with a positive power of 10, and a positive number smaller than 1 is written with a negative power of 10. Numbers between 1 and 10 are usually not written in scientific notation. To convert to scientific notation, we reverse the strategy for converting from scientific notation.

> **Strategy for Converting from Standard Notation to Scientific Notation**
>
> 1. Count the number of places (n) that the decimal point must be moved so that it will follow the first nonzero digit of the number.
> 2. If the original number was larger than 10, use 10^n.
> 3. If the original number was smaller than 1, use 10^{-n}.

E X A M P L E   8

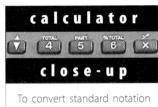

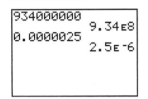

Standard notation to scientific notation

Convert each number to scientific notation.

a) 934,000,000 **b)** 0.0000025

Solution

a) In 934,000,000 the decimal point must be moved eight places to the left to get it to follow 9, the first nonzero digit.

$$934,000,000 = 9.34 \times 10^8 \quad \text{Use 8 because } 934,000,000 > 10.$$

b) The decimal point in 0.0000025 must be moved six places to the right to get the 2 to the left of the decimal point.

$$0.0000025 = 2.5 \times 10^{-6} \quad \text{Use } -6 \text{ because } 0.0000025 < 1.$$ ■

We can perform computations with numbers in scientific notation by using the rules of exponents on the powers of 10.

E X A M P L E   9

Using scientific notation in computations

Evaluate $\frac{(10,000)(0.000025)}{0.000005}$ by first converting each number to scientific notation.

Solution

$$\frac{(10,000)(0.000025)}{0.000005} = \frac{(1 \times 10^4)(2.5 \times 10^{-5})}{5 \times 10^{-6}}$$

$$= \frac{2.5}{5} \cdot \frac{10^4 \cdot 10^{-5}}{10^{-6}} \quad \text{Commutative and associative properties}$$

$$= 0.5 \times 10^5$$

$$= 5 \times 10^{-1} \times 10^5 \quad \text{Write 0.5 in scientific notation.}$$

$$= 5 \times 10^4$$ ■

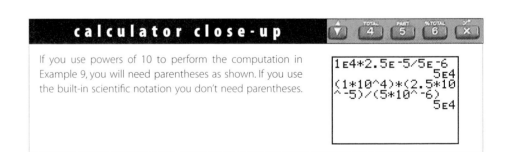

calculator close-up

If you use powers of 10 to perform the computation in Example 9, you will need parentheses as shown. If you use the built-in scientific notation you don't need parentheses.

E X A M P L E   10

Counting hydrogen atoms

If the weight of hydrogen is 1.7×10^{-24} gram per atom, then how many hydrogen atoms are there in one kilogram of hydrogen?

Solution

There are 1000 or 1×10^3 grams in one kilogram. So to find the number of hydrogen atoms in one kilogram of hydrogen, we divide 1×10^3 by 1.7×10^{-24}:

$$\frac{1 \times 10^3 \text{ g/kg}}{1.7 \times 10^{-24} \text{ g/atom}} \approx 5.9 \times 10^{26} \text{ atom per kilogram (atom/kg)}$$

To divide by grams per atom, we invert and multiply: $\frac{g}{kg} \cdot \frac{atom}{g} = \frac{atom}{kg}$. Keeping track of the units as we did here helps us to be sure that we performed the correct operation. So there are approximately 5.9×10^{26} hydrogen atoms in one kilogram of hydrogen.   ◼

WARM-UPS

True or false? Explain your answer.

1. $3^5 \cdot 3^4 = 3^9$ True

2. $3 \cdot 3 \cdot 3^{-1} = \frac{1}{27}$ False

3. $10^{-3} = 0.0001$·   False

4. $1^{-1} = 1$ True

5. $\frac{2^5}{2^{-2}} = 2^7$ True

6. $2^3 \cdot 5^2 = 10^5$ False

7. $-2^{-2} = -\frac{1}{4}$ True

8. $46.7 \times 10^5 = 4.67 \times 10^6$ True

9. $0.512 \times 10^{-3} = 5.12 \times 10^{-4}$ True

10. $\frac{8 \times 10^{30}}{2 \times 10^{-5}} = 4 \times 10^{25}$ False

---

5.1   EXERCISES

Reading and Writing   After reading this section, write out the answers to these questions. Use complete sentences.

1. What is an exponential expression?

An exponential expression is an expression of the form a^n.

2. What is the meaning of a negative exponent?

The expression $a^{-n} = 1/a^n$ for any positive integer n and $a \neq 0$.

3. What is the product rule?

The product rule says that $a^m a^n = a^{m+n}$.

4. What is the quotient rule?

The quotient rule says that $a^m/a^n = a^{m-n}$.

5. How do you convert a number from scientific notation to standard notation?

To convert a number in scientific notation to standard notation, move the decimal point n places to the left if the exponent on 10 is $-n$ or move the decimal point n places to the right if the exponent on 10 is n, assuming n is a positive integer.

6. How do you convert a number from standard notation to scientific notation?

To convert from standard notation to scientific notation, count the number of decimal places, n, required to move the decimal point so that there is one nonzero digit to the left of the decimal point. Use 10^n if you moved the decimal point to the left and use 10^{-n} if you moved the decimal point to the right.

For all exercises in this section, assume that the variables represent nonzero real numbers and use only positive exponents in your answers.

Evaluate each expression. See Example 1.

7. 4^2 16

8. 3^3 27

9. 4^{-2} $\frac{1}{16}$

10. 3^{-3} $\frac{1}{27}$

11. $\left(\frac{1}{4}\right)^{-2}$ 16

12. $\left(\frac{3}{5}\right)^{-1}$ $\frac{5}{3}$

13. $\left(-\dfrac{3}{4}\right)^{-2}$ $\quad\dfrac{16}{9}$

14. $10^2 \cdot \left(-\dfrac{1}{10}\right)^{-2}$ $\quad 10{,}000$

15. $-2^{-1} \cdot (-6)^2$ $\quad -18$

16. $-2^{-3} \cdot (-4)^2$ $\quad -2$

17. $3 \cdot 10^{-3}$ $\quad 0.003$

18. $5 \cdot 10^4$ $\quad 50{,}000$

19. $\dfrac{1}{2^{-4}}$ $\quad 16$

20. $\dfrac{2}{2^{-3}}$ $\quad 16$

Simplify. See Examples 2 and 3.

21. $2^5 \cdot 2^{12}$ $\quad 2^{17}$

22. $2x^2 \cdot 3x^3$ $\quad 6x^5$

23. $2y^3 \cdot y^2 \cdot y^{-7}$ $\quad\dfrac{2}{y^2}$

24. $2 \cdot 10^6 \cdot 5 \cdot 10^{-7}$ $\quad 1$

25. $5x^{-3} \cdot 6x^{-4}$ $\quad\dfrac{30}{x^7}$

26. $4a^{-13} \cdot a^{-4}$ $\quad\dfrac{4}{a^{17}}$

27. $-8^0 - 8^0$ $\quad -2$

28. $(-8 - 8)^0$ $\quad 1$

29. $(x - y)^0$ $\quad 1$

30. $(x^2 - y^2)^0$ $\quad 1$

31. $2w^{-3}(w^7 \cdot w^{-4})$ $\quad 2$

32. $5y^2z(y^{-3}z^{-1})$ $\quad\dfrac{5}{y}$

Write each expression without negative exponents and simplify. See Example 4.

33. $\dfrac{2}{4^{-2}}$ $\quad 32$

34. $\dfrac{5}{10^{-3}}$ $\quad 5000$

35. $\dfrac{3^{-1}}{10^{-2}}$ $\quad\dfrac{100}{3}$

36. $\dfrac{2y^{-2}}{3^{-1}}$ $\quad\dfrac{6}{y^2}$

37. $\dfrac{2x^{-3}(4x)}{5y^{-2}}$ $\quad\dfrac{8y^2}{5x^2}$

38. $\dfrac{5^{-2}xy^{-3}}{3x^{-2}}$ $\quad\dfrac{x^3}{75y^3}$

39. $\dfrac{4^{-2}x^3x^{-6}}{3x^{-3}x^2}$ $\quad\dfrac{1}{48x^2}$

40. $\dfrac{3y^{-4}y^{-6}}{2^{-3}y^2y^{-7}}$ $\quad\dfrac{24}{y^5}$

Simplify each expression. See Examples 5 and 6.

41. $\dfrac{x^5}{x^3}$ $\quad x^2$

42. $\dfrac{a^8}{a^3}$ $\quad a^5$

43. $\dfrac{3^6}{3^{-2}}$ $\quad 3^8$

44. $\dfrac{6^2}{6^{-5}}$ $\quad 6^7$

45. $\dfrac{4a^{-5}}{12a^{-2}}$ $\quad\dfrac{1}{3a^3}$

46. $\dfrac{-3a^{-3}}{-21a^{-4}}$ $\quad\dfrac{a}{7}$

47. $\dfrac{-6w^{-5}}{2w^3}$ $\quad\dfrac{-3}{w^8}$

48. $\dfrac{10x^{-6}}{-2x^2}$ $\quad\dfrac{-5}{x^8}$

49. $\dfrac{3^3w^{-2}w^5}{3^{-5}w^{-3}}$ $\quad 3^8w^6$

50. $\dfrac{2^{-3}w^5}{2^5w^3w^{-7}}$ $\quad\dfrac{w^9}{2^8}$

51. $\dfrac{3x^{-6} \cdot x^2y^{-1}}{6x^{-5}y^{-2}}$ $\quad\dfrac{xy}{2}$

52. $\dfrac{2r^{-3}t^{-1}}{10r^5r^2 \cdot t^{-3}}$ $\quad\dfrac{1}{5r^8}$

Use the rules of exponents to simplify each expression.

53. $3^{-1}\left(\dfrac{1}{3}\right)^{-3}$ $\quad 9$

54. $2^{-2}\left(\dfrac{1}{4}\right)^{-3}$ $\quad 16$

55. $-2^4 + \left(\dfrac{1}{2}\right)^{-1}$ $\quad -14$

56. $-3^4 - (-3)^4$ $\quad -162$

57. $-(-2)^{-3} \cdot 2^{-1}$ $\quad\dfrac{1}{16}$

58. $-(-3)^{-1} \cdot 9^{-1}$ $\quad\dfrac{1}{27}$

59. $(1 + 2^{-1})^{-2}$ $\quad\dfrac{4}{9}$

60. $(2^{-1} + 2^{-1})^{-2}$ $\quad 1$

61. $-5a^2 \cdot 3a^2$ $\quad -15a^4$

62. $2x^2 \cdot 5x^{-5}$ $\quad\dfrac{10}{x^3}$

63. $5a \cdot a^2 + 3a^{-2} \cdot a^5$ $\quad 8a^3$

64. $2x^2 \cdot 5y^{-5}$ $\quad\dfrac{10x^2}{y^5}$

65. $\dfrac{-3a^5(-2a^{-1})}{6a^3}$ $\quad a$

66. $\dfrac{6a(-ab^{-2})}{-2a^2b^{-3}}$ $\quad 3b$

67. $\dfrac{(-3x^3y^2)(-2xy^{-3})}{-9x^2y^{-5}}$ $\quad\dfrac{-2x^2y^4}{3}$

68. $\dfrac{(-2x^{-5}y)(-3xy^6)}{-6x^{-6}y^2}$ $\quad -x^2y^5$

69. $\dfrac{2^{-1} + 3^{-1}}{2^{-1}}$ $\quad\dfrac{5}{3}$

70. $\dfrac{3^{-1} + 4^{-1}}{12^{-1}}$ $\quad 7$

71. $\dfrac{(2 + 3)^{-1}}{2^{-1} + 3^{-1}}$ $\quad\dfrac{6}{25}$

72. $\dfrac{(3 - 4)^{-1}}{3 \cdot 2^{-1}}$ $\quad -\dfrac{2}{3}$

For each equation, find the integer that can be used as the exponent to make the equation correct.

73. $8 = 2^?$ $\quad 3$

74. $27 = 3^?$ $\quad 3$

75. $\dfrac{1}{4} = 2^?$ $\quad -2$

76. $\dfrac{1}{125} = 5^?$ $\quad -3$

77. $16 = \left(\dfrac{1}{2}\right)^?$ $\quad -4$

78. $81 = \left(\dfrac{1}{3}\right)^?$ $\quad -4$

79. $10^? = 0.001$ $\quad -3$

80. $10^? = 10{,}000$ $\quad 4$

Write each number in standard notation. See Example 7.

81. 4.86×10^8
$486{,}000{,}000$

82. 3.80×10^2
380

83. 2.37×10^{-6}
0.00000237

84. 1.62×10^{-3}
0.00162

85. 4×10^6
$4{,}000{,}000$

86. 496×10^3
$496{,}000$

87. 5×10^{-6}
0.000005

88. 48×10^{-3}
0.048

Write each number in scientific notation. See Example 8.

89. $320{,}000$
3.2×10^5

90. $43{,}298{,}000$
4.3298×10^7

91. 0.00000071
7.1×10^{-7}

92. 0.00000894
8.94×10^{-6}

93. 0.00007
7×10^{-5}

94. $8{,}295{,}100$
8.2951×10^6

95. 235×10^5
2.35×10^7

96. 0.43×10^{-9}
4.3×10^{-10}

Evaluate each expression using scientific notation without a calculator. See Example 9.

97. $\dfrac{(5{,}000{,}000)(0.0003)}{2000}$
7.5×10^{-1}

98. $\dfrac{(6000)(0.00004)}{(30{,}000)(0.002)}$
4×10^{-3}

99. $\dfrac{6 \times 10^{40}}{2 \times 10^{18}}$

3×10^{22}

100. $\dfrac{4.6 \times 10^{12}}{2.3 \times 10^{5}}$

2×10^{7}

101. $\dfrac{(-4 \times 10^{5})(6 \times 10^{-9})}{2 \times 10^{-16}}$

-1.2×10^{13}

102. $\dfrac{(4.8 \times 10^{-3})(5 \times 10^{-8})}{(1.2 \times 10^{-6})(2 \times 10^{12})}$

1×10^{-16}

 Evaluate each expression using a calculator. Write answers in scientific notation. Round the decimal part to three decimal places. See Example 10.

103. $(4.3 \times 10^{9})(3.67 \times 10^{-5})$ 1.578×10^{5}

104. $(2.34 \times 10^{6})(8.7 \times 10^{5})$ 2.036×10^{12}

105. $(4.37 \times 10^{-6}) + (8.75 \times 10^{-5})$ 9.187×10^{-5}

106. $(6.72 \times 10^{5}) + (8.98 \times 10^{6})$ 9.652×10^{6}

107. $(9.27 \times 10^{80})(6.43 \times 10^{76})$ 5.961×10^{157}

108. $(1.35 \times 10^{66})(2.7 \times 10^{74})$ 3.645×10^{140}

109. $\dfrac{(5.6 \times 10^{14})^{2}(3.2 \times 10^{-6})}{(6.4 \times 10^{-3})^{3}}$ 3.828×10^{30}

110. $\dfrac{(3.51 \times 10^{-6})^{3}(4000)^{5}}{2\pi}$ 7.048

Solve each problem.

111. Distance to the sun. The distance from the earth to the sun is 93 million miles. Express this distance in feet using scientific notation (1 mile = 5,280 feet).
4.910×10^{11} feet

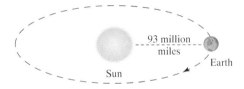

FIGURE FOR EXERCISE 111

112. Traveling time. The speed of light is 9.83569×10^{8} feet per second. How long does it take light to get from the sun to the earth? (See Exercise 111.)
8.3 minutes

113. Space travel. How long does it take a spacecraft traveling 1.2×10^{5} kilometers per second to travel 4.6×10^{12} kilometers?
3.83×10^{7} seconds

114. Diameter of a dot. If the circumference of a very small circle is 2.35×10^{-8} meter, then what is the diameter of the circle?
7.480×10^{-9} meters

115. Solid waste per person. In 1960 the 1.80863×10^{8} people in the United States generated 8.71×10^{7} tons of municipal solid waste (Environmental Protection Agency, www.epa.gov). How many pounds per person per day were generated in 1960?
2.6 lb/person/day

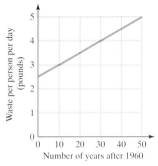

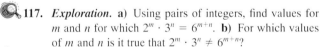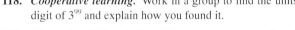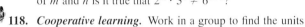

FIGURE FOR EXERCISES 115 AND 116

116. An increasing problem. According to the EPA, in 1998 the 2.70058×10^{8} people in the United States generated 4.324×10^{11} pounds of solid municipal waste.

a) How many pounds per person per day were generated in 1998?
4.4 lb/person/day in 1998

b) Use the graph to predict the number of pounds per person per day that will be generated in the year 2010.
5.0 lb/person/day in 2010

GETTING MORE INVOLVED

117. Exploration. **a)** Using pairs of integers, find values for m and n for which $2^{m} \cdot 3^{n} = 6^{m+n}$. **b)** For which values of m and n is it true that $2^{m} \cdot 3^{n} \neq 6^{m+n}$?

118. Cooperative learning. Work in a group to find the units digit of 3^{99} and explain how you found it.

119. Discussion. What is the difference between $-a^{n}$ and $(-a)^{n}$, where n is an integer? For which values of a and n do they have the same value, and for which values of a and n do they have different values?

120. Exploration. If $a + b = a$, then what can you conclude about b? Use scientific notation on your calculator to find $5 \times 10^{20} + 3 \times 10^{6}$. Explain why your calculator displays the answer that it gets.

5.2 THE POWER RULES

In Section 5.1 you learned some of the basic rules for working with exponents. All of the rules of exponents are designed to make it easier to work with exponential expressions. In this section we will extend our list of rules to include three new ones.

Raising an Exponential Expression to a Power

An expression such as $(x^3)^2$ consists of the exponential expression x^3 raised to the power 2. We can use known rules to simplify this expression.

$$(x^3)^2 = x^3 \cdot x^3 \quad \text{Exponent 2 indicates two factors of } x^3.$$
$$= x^6 \quad \text{Product rule: } 3 + 3 = 6$$

Note that the exponent 6 is the *product* of the exponents 2 and 3. This example illustrates the **power of a power rule.**

> **Power of a Power Rule**
>
> If m and n are any integers and $a \neq 0$, then
>
> $$(a^m)^n = a^{mn}.$$

EXAMPLE 1

Using the power of a power rule

Use the rules of exponents to simplify each expression. Write the answer with positive exponents only. Assume all variables represent nonzero real numbers.

a) $(2^3)^5$

b) $(x^2)^{-6}$

c) $3(y^{-3})^{-2}y^{-5}$

d) $\dfrac{(x^2)^{-1}}{(x^{-3})^3}$

Solution

a) $(2^3)^5 = 2^{15}$ Power of a power rule

b) $(x^2)^{-6} = x^{-12}$ Power of a power rule

$$= \frac{1}{x^{12}} \quad \text{Definition of a negative exponent}$$

c) $3(y^{-3})^{-2}y^{-5} = 3y^6 y^{-5}$ Power of a power rule

$$= 3y \quad \text{Product rule}$$

d) $\dfrac{(x^2)^{-1}}{(x^{-3})^3} = \dfrac{x^{-2}}{x^{-9}}$ Power of a power rule

$$= x^7 \quad \text{Quotient rule}$$

Raising a Product to a Power

Consider how we would simplify a product raised to a positive power and a product raised to a negative power using known rules.

$$(2x)^3 = \overbrace{2x \cdot 2x \cdot 2x}^{3 \text{ factors of } 2x} = 2^3 \cdot x^3 = 8x^3$$

$$(ay)^{-3} = \frac{1}{(ay)^3} = \frac{1}{(ay)(ay)(ay)} = \frac{1}{a^3y^3} = a^{-3}y^{-3}$$

In each of these cases the original exponent is applied to each factor of the product. These examples illustrate the **power of a product rule.**

Power of a Product Rule

If a and b are nonzero real numbers and n is any integer, then

$$(ab)^n = a^n \cdot b^n.$$

E X A M P L E   2

Using the power of a product rule

Simplify. Assume the variables represent nonzero real numbers. Write the answers with positive exponents only.

a) $(-3x)^4$ **b)** $(-2x^2)^3$ **c)** $(3x^{-2}y^3)^{-2}$

Solution

a) $(-3x)^4 = (-3)^4 x^4$ Power of a product rule

        $= 81x^4$

b) $(-2x^2)^3 = (-2)^3(x^2)^3$ Power of a product rule

         $= -8x^6$ Power of a power rule

c) $(3x^{-2}y^3)^{-2} = (3)^{-2}(x^{-2})^{-2}(y^3)^{-2} = \frac{1}{9}x^4 y^{-6} = \frac{x^4}{9y^6}$

Raising a Quotient to a Power

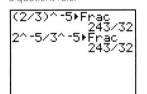

Now consider an example of applying known rules to a power of a quotient:

$$\left(\frac{x}{5}\right)^3 = \frac{x}{5} \cdot \frac{x}{5} \cdot \frac{x}{5} = \frac{x^3}{5^3}$$

We get a similar result with a negative power:

$$\left(\frac{x}{5}\right)^{-3} = \left(\frac{5}{x}\right)^3 = \frac{5}{x} \cdot \frac{5}{x} \cdot \frac{5}{x} = \frac{5^3}{x^3} = \frac{x^{-3}}{5^{-3}}$$

In each of these cases the original exponent applies to both the numerator and denominator. These examples illustrate the **power of a quotient rule.**

Power of a Quotient Rule

If a and b are nonzero real numbers and n is any integer, then

$$\left(\frac{a}{b}\right)^n = \frac{a^n}{b^n}.$$

E X A M P L E   3

Using the power of a quotient rule

Use the rules of exponents to simplify each expression. Write your answers with positive exponents only. Assume the variables are nonzero real numbers.

a) $\left(\dfrac{x}{2}\right)^3$
b) $\left(-\dfrac{2x^3}{3y^2}\right)^3$
c) $\left(\dfrac{x^{-2}}{2^3}\right)^{-1}$
d) $\left(-\dfrac{3}{4x^3}\right)^{-2}$

Solution

a) $\left(\dfrac{x}{2}\right)^3 = \dfrac{x^3}{2^3}$ Power of a quotient rule

$= \dfrac{x^3}{8}$

b) $\left(-\dfrac{2x^3}{3y^2}\right)^3 = \dfrac{(-2)^3 x^9}{3^3 y^6}$ Because $(x^3)^3 = x^9$ and $(y^2)^3 = y^6$

$= \dfrac{-8x^9}{27y^6} = -\dfrac{8x^9}{27y^6}$

c) $\left(\dfrac{x^{-2}}{2^3}\right)^{-1} = \dfrac{x^2}{2^{-3}} = 8x^2$

d) $\left(-\dfrac{3}{4x^3}\right)^{-2} = \dfrac{(-3)^{-2}}{4^{-2}x^{-6}} = \dfrac{4^2 x^6}{(-3)^2} = \dfrac{16x^6}{9}$ ∎

A fraction to a negative power can be simplified by using the power of a quotient rule as in Example 3. Another method is to find the reciprocal of the fraction first, then use the power of a quotient rule as shown in the next example.

E X A M P L E   4

Negative powers of fractions

Simplify. Assume the variables are nonzero real numbers and write the answers with positive exponents only.

a) $\left(\dfrac{3}{4}\right)^{-3}$
b) $\left(\dfrac{x^2}{5}\right)^{-2}$
c) $\left(-\dfrac{2y^3}{3}\right)^{-2}$

Solution

a) $\left(\dfrac{3}{4}\right)^{-3} = \left(\dfrac{4}{3}\right)^3$ The reciprocal of $\dfrac{3}{4}$ is $\dfrac{4}{3}$.

$= \dfrac{4^3}{3^3}$ Power of a quotient rule

$= \dfrac{64}{27}$

b) $\left(\dfrac{x^2}{5}\right)^{-2} = \left(\dfrac{5}{x^2}\right)^2 = \dfrac{5^2}{(x^2)^2} = \dfrac{25}{x^4}$
c) $\left(-\dfrac{2y^3}{3}\right)^{-2} = \left(-\dfrac{3}{2y^3}\right)^2 = \dfrac{9}{4y^6}$ ∎

Variable Exponents

So far, we have used the rules of exponents only on expressions with integral exponents. However, we can use the rules to simplify expressions having variable exponents that represent integers.

E X A M P L E   5

Expressions with variables as exponents

Simplify. Assume the variables represent integers.

a) $3^{4y} \cdot 3^{5y}$
b) $(5^{2x})^{3x}$
c) $\left(\dfrac{2^n}{3^m}\right)^{5n}$

Solution

a) $3^{4y} \cdot 3^{5y} = 3^{9y}$ Product rule: $4y + 5y = 9y$

b) $(5^{2x})^{3x} = 5^{6x^2}$ Power of a power rule: $2x \cdot 3x = 6x^2$

c) $\left(\dfrac{2^n}{3^m}\right)^{5n} = \dfrac{(2^n)^{5n}}{(3^m)^{5n}}$ Power of a quotient rule

           $= \dfrac{2^{5n^2}}{3^{5mn}}$ Power of a power rule

Summary of the Rules

The definitions and rules that were introduced in the last two sections are summarized in the following box.

Rules for Integral Exponents

For these rules m and n are integers and a and b are nonzero real numbers.

1. $a^{-n} = \dfrac{1}{a^n}$ Definition of negative exponent

2. $a^{-n} = \left(\dfrac{1}{a}\right)^n$, $a^{-1} = \dfrac{1}{a}$, and $\dfrac{1}{a^{-n}} = a^n$ Negative exponent rules

3. $a^0 = 1$ Definition of zero exponent

4. $a^m a^n = a^{m+n}$ Product rule

5. $\dfrac{a^m}{a^n} = a^{m-n}$ Quotient rule

6. $(a^m)^n = a^{mn}$ Power of a power rule

7. $(ab)^n = a^n b^n$ Power of a product rule

8. $\left(\dfrac{a}{b}\right)^n = \dfrac{a^n}{b^n}$ Power of a quotient rule

---

Applications

Both positive and negative exponents occur in formulas used in investment situations. The amount of money invested is the **principal,** and the value of the principal after a certain time period is the **amount.** Interest rates are annual percentage rates.

> **Amount Formula**
>
> The amount A of an investment of P dollars with interest rate r compounded annually for n years is given by the formula
> $$A = P(1 + r)^n.$$

E X A M P L E   6

Finding the amount

According to Fidelity Investments of Boston, U.S. common stocks have returned an average of 10% annually since 1926. If your great-grandfather had invested $100 in the stock market in 1926 and obtained the average increase each year, then how much would the investment be worth in the year 2006 after 80 years of growth?

close-up

Did we forget to include the rule $(a + b)^n = a^n + b^n$? You can easily check with a calculator that this rule is not correct.

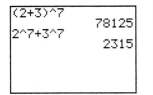

```
(2+3)^7
              78125
2^7+3^7
               2315
```

\helpful\hint

In this section we use the amount formula for interest compounded annually only. But you probably have money in a bank where interest is compounded daily. In this case r represents the daily rate (APR/365) and n is the number of days that the money is on deposit.

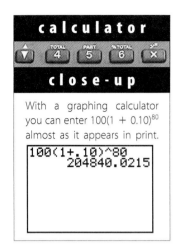

calculator

close-up

With a graphing calculator you can enter $100(1 + 0.10)^{80}$ almost as it appears in print.

```
100(1+.10)^80
        204840.0215
```

Solution

Use $n = 80, P = \$100$, and $r = 0.10$ in the amount formula:

$$A = P(1 + r)^n$$
$$A = 100(1 + 0.10)^{80}$$
$$= 100(1.1)^{80}$$
$$= 204{,}840.02$$

So $100 invested in 1926 would have amounted to $204,840.02 in 2006. ■

When we are interested in the principal that must be invested today to grow to a certain amount, the principal is called the **present value** of the investment. We can find a formula for present value by solving the amount formula for P:

$$A = P(1 + r)^n$$
$$P = \frac{A}{(1 + r)^n} \qquad \text{Divide each side by } (1 + r)^n.$$
$$P = A(1 + r)^{-n} \qquad \text{Definition of a negative exponent}$$

Present Value Formula

The present value P that will amount to A dollars after n years with interest compounded annually at annual interest rate r is given by

$$P = A(1 + r)^{-n}.$$

E X A M P L E   7

Finding the present value

If your great-grandfather wanted you to have $1,000,000 in 2006, then how much could he have invested in the stock market in 1926 to achieve this goal? Assume he could get the average annual return of 10% (from Example 6) for 80 years.

Solution

Use $r = 0.10, n = 80$, and $A = 1{,}000{,}000$ in the present value formula:

$$P = A(1 + r)^{-n}$$
$$P = 1{,}000{,}000(1 + 0.10)^{-80}$$
$$P = 1{,}000{,}000(1.1)^{-80} \qquad \text{Use a calculator with an exponent key.}$$
$$P = 488.19$$

A deposit of $488.19 in 1926 would have grown to $1,000,000 in 80 years at a rate of 10% compounded annually. ■

WARM-UPS

True or false? Explain your answer. Assume all variables represent nonzero real numbers.

1. $(2^2)^3 = 2^5$ False
2. $(2^{-3})^{-1} = 8$ True
3. $(x^{-3})^3 = x^{-9}$ True

4. $(2^3)^3 = 2^{27}$ False
5. $(2x)^3 = 6x^3$ False
6. $(-3y^3)^2 = 9y^9$ False

7. $\left(\frac{2}{3}\right)^{-1} = \frac{3}{2}$ True
8. $\left(\frac{2}{3}\right)^3 = \frac{8}{27}$ True
9. $\left(\frac{x^2}{2}\right)^3 = \frac{x^6}{8}$ True

10. $\left(\frac{2}{x}\right)^{-2} = \frac{x^2}{4}$ True

5.2 EXERCISES

Reading and Writing *After reading this section, write out the answers to these questions. Use complete sentences.*

1. What is the power of a power rule?
The power of a power rule says that $(a^m)^n = a^{mn}$.

2. What is the power of a product rule?
The power of a product rule says that $(ab)^m = a^m b^m$.

3. What is the power of a quotient rule?
The power of a quotient rule says that $(a/b)^m = a^m/b^m$.

4. What is principal?
Principal is the amount of money invested initially.

5. What formula is used for computing the amount of an investment for which interest is compounded annually?
To compute the amount A when interest is compounded annually, use $A = P(1 + i)^n$, where P is the principal, i is the annual interest rate, and n is the number of years.

6. What formula is used for computing the present value of an amount in the future with interest compounded annually?
To compute the present value P for the amount A in n years at annual interest rate i, use $P = A(1 + i)^{-n}$.

For all exercises in this section, assume the variables represent nonzero real numbers and use positive exponents only in your answers.

Use the rules of exponents to simplify each expression. See Example 1.

7. $(2^2)^3$ 64

8. $(3^2)^2$ 81

9. $(y^2)^5$ y^{10}

10. $(x^6)^2$ x^{12}

11. $(x^2)^{-4}$ $\dfrac{1}{x^8}$

12. $(x^{-2})^7$ $\dfrac{1}{x^{14}}$

13. $(m^{-3})^{-6}$ m^{18}

14. $(a^{-3})^{-3}$ a^9

15. $(x^{-2})^3(x^{-3})^{-2}$ 1

16. $(m^{-3})^{-1}(m^2)^{-4}$ $\dfrac{1}{m^5}$

17. $\dfrac{(x^3)^{-4}}{(x^2)^{-5}}$ $\dfrac{1}{x^2}$

18. $\dfrac{(a^2)^{-3}}{(a^{-2})^4}$ a^2

Simplify. See Example 2.

19. $(-9y)^2$ $81y^2$

20. $(-2a)^3$ $-8a^3$

21. $(-5w^3)^2$ $25w^6$

22. $(-2w^{-5})^3$ $\dfrac{-8}{w^{15}}$

23. $(x^3y^{-2})^3$ $\dfrac{x^9}{y^6}$

24. $(a^2b^{-3})^2$ $\dfrac{a^4}{b^6}$

25. $(3ab^{-1})^{-2}$ $\dfrac{b^2}{9a^2}$

26. $(2x^{-1}y^2)^{-3}$ $\dfrac{x^3}{8y^6}$

27. $\dfrac{2xy^{-2}}{(3x^2y)^{-1}}$ $\dfrac{6x^3}{y}$

28. $\dfrac{3ab^{-1}}{(5ab^2)^{-1}}$ $15a^2b$

29. $\dfrac{(2ab)^{-2}}{2ab^2}$ $\dfrac{1}{8a^3b^4}$

30. $\dfrac{(3xy)^{-3}}{3xy^3}$ $\dfrac{1}{81x^4y^6}$

Simplify. See Example 3.

31. $\left(\dfrac{w}{2}\right)^3$ $\dfrac{w^3}{8}$

32. $\left(\dfrac{m}{5}\right)^2$ $\dfrac{m^2}{25}$

33. $\left(-\dfrac{3a}{4}\right)^3$ $-\dfrac{27a^3}{64}$

34. $\left(-\dfrac{2}{3b}\right)^4$ $\dfrac{16}{81b^4}$

35. $\left(\dfrac{2x^{-1}}{y}\right)^{-2}$ $\dfrac{x^2y^2}{4}$

36. $\left(\dfrac{2a^2b}{3}\right)^{-3}$ $\dfrac{27}{8a^6b^3}$

37. $\left(\dfrac{-3x^3}{y}\right)^{-2}$ $\dfrac{y^2}{9x^6}$

38. $\left(\dfrac{-2y^2}{x}\right)^{-3}$ $-\dfrac{x^3}{8y^6}$

Simplify. See Example 4

39. $\left(\dfrac{2}{5}\right)^{-2}$ $\dfrac{25}{4}$

40. $\left(\dfrac{3}{4}\right)^{-2}$ $\dfrac{16}{9}$

41. $\left(-\dfrac{1}{2}\right)^{-2}$ 4

42. $\left(-\dfrac{2}{3}\right)^{-2}$ $\dfrac{9}{4}$

43. $\left(-\dfrac{2x}{3}\right)^{-3}$ $-\dfrac{27}{8x^3}$

44. $\left(-\dfrac{ab}{c}\right)^{-1}$ $-\dfrac{c}{ab}$

45. $\left(\dfrac{2x^2}{3y}\right)^{-3}$ $\dfrac{27y^3}{8x^6}$

46. $\left(\dfrac{ab^{-3}}{a^2b}\right)^{-2}$ a^2b^8

Simplify each expression. Assume that the variables represent integers. See Example 5.

47. $5^{2t} \cdot 5^{4t}$ 5^{6t}

48. $3^{2n-3} \cdot 3^{4-2n}$ 3

49. $(2^{-3w})^{-2w}$ 2^{6w^2}

50. $6^{8x} \cdot (6^{2x})^{-3}$ 6^{2x}

51. $\dfrac{7^{2m+6}}{7^{m+3}}$ 7^{m+3}

52. $\dfrac{4^{-3p}}{4^{-4p}}$ 4^p

53. $8^{2a-1} \cdot (8^{a+4})^3$ 8^{5a+11}

54. $(5^{4-3y})^3(5^{y-2})^2$ 5^{8-7y}

Use the rules of exponents to simplify each expression. If possible, write down only the answer.

55. $3x^4 \cdot 2x^5$ $6x^9$

56. $(3x^4)^2$ $9x^8$

57. $(-2x^2)^3$ $-8x^6$

58. $3x^2 \cdot 2x^{-4}$ $\dfrac{6}{x^2}$

59. $\dfrac{3x^{-2}y^{-1}}{z^{-1}}$ $\dfrac{3z}{x^2y}$

60. $\dfrac{2^{-1}x^2}{y^{-2}}$ $\dfrac{x^2y^2}{2}$

61. $\left(\dfrac{-2}{3}\right)^{-1}$ $-\dfrac{3}{2}$

62. $\left(\dfrac{-1}{5}\right)^{-1}$ -5

63. $\left(\dfrac{2x^3}{3}\right)^2$ $\dfrac{4x^6}{9}$

64. $\left(\dfrac{-2y^4}{x}\right)^3$ $-\dfrac{8y^{12}}{x^3}$

65. $(-2x^{-2})^{-1}$ $-\dfrac{x^2}{2}$

66. $(-3x^{-2})^3$ $-\dfrac{27}{x^6}$

Use the rules of exponents to simplify each expression.

67. $\left(\dfrac{2x^2y}{xy^2}\right)^{-3}$ $\dfrac{y^3}{8x^3}$

68. $\left(\dfrac{2x^3y^2}{3xy^3}\right)^{-1}$ $\dfrac{3y}{2x^2}$

69. $\dfrac{(5a^{-1}b^2)^3}{(5ab^{-2})^4}$ $\dfrac{b^{14}}{5a^7}$

70. $\dfrac{(2m^2n^{-3})^4}{mn^5}$ $\dfrac{16m^7}{n^{17}}$

71. $\dfrac{(2x^{-2}y)^{-3}}{(2xy^{-1})^2}(2x^2y^{-7})$

$\dfrac{x^6}{16y^8}$

72. $\dfrac{(3x^{-1}y^3)^{-2}}{(3xy^{-1})^3}(9x^{-9}y^5)$

$\dfrac{y^2}{27x^{10}}$

73. $\left(\dfrac{6a^{-2}b^3}{2c^4}\right)^{-2}(3a^{-1}b^2)^3$

$3ac^8$

74. $(7xz^2)^{-4}\left(\dfrac{7xy^{-1}}{z}\right)^3$

$\dfrac{1}{7xy^3z^{11}}$

Write each expression as 2 raised to a power. Assume that the variables represent integers.

75. $32 \cdot 64$
2^{11}

76. 8^{20}
2^{60}

77. $81 \cdot 6^{-4}$
2^{-4}

78. $10^{-6} \cdot 20^6$
2^6

79. 4^{3n}
2^{6n}

80. $6^{n-5} \cdot 3^{5-n}$
2^{n-5}

81. $\left(\dfrac{1}{16^{-m}}\right)^m$
2^{4m^2}

82. $\dfrac{32^n}{128^{3n}}$
2^{-16n}

 Use a calculator to evaluate each expression. Round approximate answers to three decimal places.

83. $\dfrac{1}{5^{-2}}$ 25

84. $\dfrac{(2.5)^{-3}}{(2.5)^{-5}}$ 6.25

85. $2^{-1} + 2^{-2}$ 0.75

86. $\left(\dfrac{2}{3}\right)^{-1} + 2^{-1}$ 2

87. $(0.036)^{-2} + (4.29)^3$
850.559

88. $3(4.71)^2 - 5(0.471)^{-3}$
18.700

89. $\dfrac{(5.73)^{-1} + (4.29)^{-1}}{(3.762)^{-1}}$
1.533

90. $[5.29 + (0.374)^{-1}]^3$
505.080

Solve each problem. See Examples 6 and 7.

91. *Deeper in debt.* Melissa borrowed $40,000 at 12% compounded annually and made no payments for 3 years. How much did she owe the bank at the end of the 3 years? (Use the compound interest formula.) $56,197.12

92. *Comparing stocks and bonds.* According to Fidelity Investments of Boston, throughout the 1990s annual returns on common stocks averaged 19%, whereas annual returns on bonds averaged 9%.
   a) If you had invested $10,000 in bonds in 1990 and achieved the average return, then what would your investment be worth after 10 years in 2000? $23,673.64

b) How much more would your $10,000 investment be worth in 2000 if you had invested in stocks? $33,273.20

93. *Saving for college.* Mr. Watkins wants to have $10,000 in a savings account when his little Wanda is ready for college. How much must he deposit today in an account paying 7% compounded annually to have $10,000 in 18 years? $2,958.64

94. *Saving for retirement.* In the 1990s returns on Treasury Bills fell to an average of 4.5% per year (Fidelity Investments). Wilma wants to have $2,000,000 when she retires in 45 years. If she assumes an average annual return of 4.5%, then how much must she invest now in Treasury Bills to achieve her goal? $275,928.73

 95. *Life expectancy of white males.* Strange as it may seem, your life expectancy increases as you get older. The function

$$L = 72.2(1.002)^a$$

can be used to model life expectancy L for U.S. white males with present age a (National Center for Health Statistics, www.cdc.gov/nchswww).
   a) To what age can a 20-year-old white male expect to live?   75.1 years
   b) To what age can a 60-year-old white male expect to live? (See also Chapter Review Exercises 153 and 154.)   81.4 years

 96. *Life expectancy of white females.* Life expectancy improved more for females than for males during the 1940s and 1950s due to a dramatic decrease in maternal mortality rates. The function

$$L = 78.5(1.001)^a$$

can be used to model life expectancy L for U.S. white females with present age a.
   a) To what age can a 20-year-old white female expect to live?   80.1 years
   b) Bob, 30, and Ashley, 26, are an average white couple. How many years can Ashley expect to live as a widow?   7.9 years
   c) Why do the life expectancy curves intersect in the accompanying figure?
At 80 both males and females can expect about 5 more years.

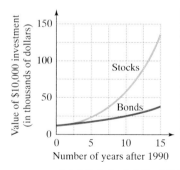

FIGURE FOR EXERCISE 92

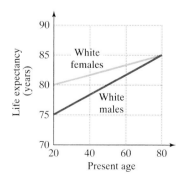

FIGURE FOR EXERCISES 95 AND 96

GETTING MORE INVOLVED

97. *Discussion*. For which values of a and b is it true that $(ab)^{-1} = a^{-1}b^{-1}$? Find a pair of nonzero values for a and b for which $(a + b)^{-1} \neq a^{-1} + b^{-1}$.

98. *Writing*. Explain how to evaluate $\left(-\frac{2}{3}\right)^{-3}$ in three different ways.

99. *Discussion*. Which of the following expressions has a value different from the others? Explain.

a) -1^{-1} b) -3^{0} c) $-2^{-1} - 2^{-1}$

d) $(-1)^{-2}$ e) $(-1)^{-3}$ d

100. *True or False?* Explain your answer.

a) The square of a product is the product of the squares.

b) The square of a sum is the sum of the squares.

   a) True   b) False

GRAPHING CALCULATOR EXERCISES

101. At 12% compounded annually the value of an investment of $10,000 after x years is given by

$$y = 10,000(1.12)^x.$$

a) Graph $y = 10,000(1.12)^x$ and the function $y = 20,000$ on a graphing calculator. Use a viewing window that shows the intersection of the two graphs.

b) Use the intersect feature of your calculator to find the point of intersection.

c) The x-coordinate of the point of intersection is the number of years that it will take for the $10,000 investment to double. What is that number of years?

   b) $(6.116, 20,000)$

   c) 6.116 years

102. The function $y = 72.2(1.002)^x$ gives the life expectancy y of a U.S. white male with present age x. (See Exercise 95.)

a) Graph $y = 72.2(1.002)^x$ and $y = 86$ on a graphing calculator. Use a viewing window that shows the intersection of the two graphs.

b) Use the intersect feature of your calculator to find the point of intersection.

c) What does the x-coordinate of the point of intersection tell you?

   b) $(87.54, 86)$

   c) At 87.54 years of age you can expect to live until 86. The model fails here.

5.3   ADDITION, SUBTRACTION, AND MULTIPLICATION OF POLYNOMIALS

A polynomial is a particular type of algebraic expression that serves as a fundamental building block in algebra. We used polynomials in Chapters 1 and 2, but we did not identify them as polynomials. In this section you will learn to recognize polynomials and to add, subtract, and multiply them.

In this section

• Polynomials

• Evaluating Polynomials

• Addition and Subtraction of Polynomials

• Multiplication of Polynomials

Polynomials

The expression $3x^3 - 15x^2 + 7x - 2$ is an example of a polynomial in one variable. Because this expression could be written as

$$3x^3 + (-15x^2) + 7x + (-2),$$

we say that this polynomial is a sum of four terms:

$$3x^3, \quad -15x^2, \quad 7x, \quad \text{and} \quad -2.$$

A **term** of a polynomial is a single number or the product of a number and one or more variables raised to whole number powers. The number preceding the variable in each term is called the **coefficient** of that variable. In $3x^3 - 15x^2 + 7x - 2$ the coefficient of x^3 is 3, the coefficient of x^2 is -15, and the coefficient of x is 7. In algebra a number is frequently referred to as a **constant,** and so the last term -2 is called the **constant term.** A **polynomial** is defined as a single term or a sum of a finite number of terms.

EXAMPLE 1

Identifying polynomials

Determine whether each algebraic expression is a polynomial.

a) -3 **b)** $3x + 2^{-1}$ **c)** $3x^{-2} + 4y^2$

d) $\dfrac{1}{x} + \dfrac{1}{x^2}$ **e)** $x^{49} - 8x^2 + 11x - 2$

Solution

a) The number -3 is a polynomial of one term, a constant term.

b) Since $3x + 2^{-1}$ can be written as $3x + \frac{1}{2}$, it is a polynomial of two terms.

c) The expression $3x^{-2} + 4y^2$ is not a polynomial because x has a negative exponent.

d) If this expression is rewritten as $x^{-1} + x^{-2}$, then it fails to be a polynomial because of the negative exponents. So a polynomial does not have variables in denominators, and

$$\frac{1}{x} + \frac{1}{x^2}$$

is not a polynomial.

e) The expression $x^{49} - 8x^2 + 11x - 2$ is a polynomial.   ◼

For simplicity we usually write polynomials in one variable with the exponents in decreasing order from left to right. Thus we would write

$$3x^3 - 15x^2 + 7x - 2 \qquad \text{rather than} \qquad -15x^2 - 2 + 7x + 3x^3.$$

When a polynomial is written in decreasing order, the coefficient of the first term is called the **leading coefficient.**

Certain polynomials have special names depending on the number of terms. A **monomial** is a polynomial that has one term, a **binomial** is a polynomial that has two terms, and a **trinomial** is a polynomial that has three terms. The **degree** of a polynomial in one variable is the highest power of the variable in the polynomial. The number 0 is considered to be a monomial without degree because $0 = 0x^n$, where n could be any number.

EXAMPLE 2

Identifying coefficients and degree

State the degree of each polynomial and the coefficient of x^2. Determine whether the polynomial is monomial, binomial, or trinomial.

a) $\dfrac{x^2}{3} - 5x^3 + 7$ **b)** $x^{48} - x^2$ **c)** 6

Solution

a) The degree of this trinomial is 3, and the coefficient of x^2 is $\frac{1}{3}$.

b) The degree of this binomial is 48, and the coefficient of x^2 is -1.

c) Because $6 = 6x^0$, the number 6 is a monomial with degree 0. Because x^2 does not appear in this polynomial, the coefficient of x^2 is 0.   ◼

study tip

Effective time management will allow adequate time for school, social life, and free time. However, at times you will have to sacrifice to do well.

Although we are mainly concerned here with polynomials in one variable, we will also encounter polynomials in more than one variable, such as

$$4x^2 - 5xy + 6y^2, \qquad x^2 + y^2 + z^2, \qquad \text{and} \qquad ab^2 - c^3.$$

In a term containing more than one variable, the coefficient of a variable consists of all other numbers and variables in the term. For example, the coefficient of x in $-5xy$ is $-5y$, and the coefficient of y is $-5x$.

Evaluating Polynomial Functions

The formula $D = -16t^2 + v_0t + s_0$ is used to model the effect of gravity on an object tossed straight upward with initial velocity v_0 feet per second from an initial height of s_0 feet. For example, if a ball is tossed into the air at 64 feet per second from a height of 4 feet, then $D = -16t^2 + 64t + 4$ gives the ball's distance above the ground in feet, t seconds after it is tossed. Because D is determined by t, we say that D is a function of t. The values of t range from $t = 0$ when the ball is tossed to the time when it hits the ground. To emphasize that the value of D depends on t, we can use the function notation introduced in Chapter 3 and write

$$D(t) = -16t^2 + 64t + 4.$$

We read $D(t)$ as "D of t." The expression $D(t)$ is the value of the polynomial at time t. To find the value when $t = 2$, replace t by 2:

$$D(2) = -16 \cdot 2^2 + 64 \cdot 2 + 4$$
$$= -16 \cdot 4 + 128 + 4$$
$$= 68$$

The statement $D(2) = 68$ means that the ball is 68 feet above the ground 2 seconds after the ball was tossed upward. Note that $D(2)$ does not mean D times 2.

E X A M P L E   3

Finding the value of a polynomial

Suppose $Q(x) = 2x^3 - 3x^2 - 7x - 6$. Find $Q(3)$ and $Q(-1)$.

Solution

To find $Q(3)$, replace x by 3 in $Q(x) = 2x^3 - 3x^2 - 7x - 6$:

$$Q(3) = 2 \cdot 3^3 - 3 \cdot 3^2 - 7 \cdot 3 - 6$$
$$= 54 - 27 - 21 - 6$$
$$= 0$$

To find $Q(-1)$, replace x by -1 in $Q(x) = 2x^3 - 3x^2 - 7x - 6$:

$$Q(-1) = 2(-1)^3 - 3(-1)^2 - 7(-1) - 6$$
$$= -2 - 3 + 7 - 6$$
$$= -4$$

So $Q(3) = 0$ and $Q(-1) = -4$.

Addition and Subtraction of Polynomials

When evaluating a polynomial, we get a real number. So the operations that we perform with real numbers can be performed with polynomials. Actually, we have been adding and subtracting polynomials since Chapter 1. To add two polynomials, we simply add the like terms.

E X A M P L E   4

Adding polynomials

Find the sums.

a) $(x^2 - 5x - 7) + (7x^2 - 4x + 10)$

b) $(3x^3 - 5x^2 - 7) + (4x^2 - 2x + 3)$

Solution

a) $(x^2 - 5x - 7) + (7x^2 - 4x + 10) = 8x^2 - 9x + 3$ Combine like terms.

b) For illustration we will write this addition vertically:

$$\begin{array}{r} 3x^3 - 5x^2 - 7 \\ \underline{4x^2 - 2x + 3} \quad \text{Line up like terms.} \\ 3x^3 - x^2 - 2x - 4 \quad \text{Add.} \end{array}$$

When we subtract polynomials, we subtract like terms. Because $a - b = a + (-b)$, we often perform subtraction by changing signs and adding.

E X A M P L E   5

Subtracting polynomials

Find the differences.

a) $(x^2 - 7x - 2) - (5x^2 + 6x - 4)$

b) $(6y^3z - 5yz + 7) - (4y^2z - 3yz - 9)$

Solution

a) We find the first difference horizontally:

$(x^2 - 7x - 2) - (5x^2 + 6x - 4) = x^2 - 7x - 2 - 5x^2 - 6x + 4$ Change signs.

$ = -4x^2 - 13x + 2$ Combine like terms.

b) For illustration we write $(6y^3z - 5yz + 7) - (4y^2z - 3yz - 9)$ vertically:

$$\begin{array}{r} 6y^3z - 5yz + 7 \\ \underline{-4y^2z + 3yz + 9} \quad \text{Change signs.} \\ 6y^3z - 4y^2z - 2yz + 16 \quad \text{Add.} \end{array}$$

It is certainly not necessary to write out all of the steps shown in Examples 4 and 5, but we must use the following rule.

Addition and Subtraction of Polynomials

To add two polynomials, add the like terms.
To subtract two polynomials, subtract the like terms.

Multiplication of Polynomials

We learned how to multiply monomials when we learned the product rule in Section 5.1. For example,

$$-2x^3 \cdot 4x^2 = -8x^5.$$

To multiply a monomial and a polynomial of two or more terms, we apply the distributive property. For example,

$$3x(x^3 - 5) = 3x^4 - 15x.$$

E X A M P L E   6

Multiplying a monomial and a polynomial

Find the products.

a) $2ab^2 \cdot 3a^2b$ **b)** $(-1)(5 - x)$ **c)** $(x^3 - 5x + 2)(-3x)$

Solution

a) $2ab^2 \cdot 3a^2b = 6a^3b^3$

b) $(-1)(5 - x) = -5 + x = x - 5$

c) Each term of $x^3 - 5x + 2$ is multiplied by $-3x$:

$$(x^3 - 5x + 2)(-3x) = -3x^4 + 15x^2 - 6x$$ ■

Note what happened to the binomial in Example 6(b) when we multiplied it by -1. If we multiply any difference by -1, we get the same type of result:

$$-1(a - b) = -a + b = b - a.$$

Because multiplying by -1 is the same as taking the opposite, we can write this equation as

$$-(a - b) = b - a.$$

This equation says that $a - b$ and $b - a$ are opposites or additive inverses of each other. Note that the opposite of $a + b$ is $-a - b$, not $a - b$.

To multiply a binomial and a trinomial, we can use the distributive property or set it up like multiplication of whole numbers.

E X A M P L E   7

Multiplying a binomial and a trinomial

Find the product $(x + 2)(x^2 + 3x - 5)$.

Solution

We can find this product by applying the distributive property twice. First we multiply the binomial and each of the trinomial:

$$(x + 2)(x^2 + 3x - 5) = (x + 2)x^2 + (x + 2)3x + (x + 2)(-5)$$ Distributive property

$$= x^3 + 2x^2 + 3x^2 + 6x - 5x - 10$$ Distributive property

$$= x^3 + 5x^2 + x - 10$$ Combine like terms.

We could have found this product vertically:

$$
\begin{array}{r}
x^2 + 3x - 5 \\
x + 2 \\
\hline
2x^2 + 6x - 10 \\
x^3 + 3x^2 - 5x \\
\hline
x^3 + 5x^2 + x - 10
\end{array}
$$

$2(x^2 + 3x - 5) = 2x^2 + 6x - 10$

$x(x^2 + 3x - 5) = x^3 + 3x^2 - 5x$

Add.   ■

Multiplication of Polynomials

To multiply polynomials, multiply each term of the first polynomial by each term of the second polynomial and then combine like terms.

In the next example we multiply binomials.

E X A M P L E   8   **Multiplying binomials**

Find the products.

a) $(x + y)(z + 4)$ **b)** $(x - 3)(2x + 5)$

Solution

a) $(x + y)(z + 4) = (x + y)z + (x + y)4$ Distributive property

$\qquad\qquad\qquad = xz + yz + 4x + 4y$ Distributive property

Notice that this product does not have any like terms to combine.

b) Multiply:

$$
\begin{array}{r}
x - 3 \\
2x + 5 \\
\hline
5x - 15 \\
2x^2 - 6x \\
\hline
2x^2 - x - 15
\end{array}
$$

WARM-UPS

True or false? Explain your answers.

1. The expression $3x^{-2} - 5x + 2$ is a trinomial.   False

2. In the polynomial $3x^2 - 5x + 3$ the coefficient of x is 5.   False

3. The degree of the polynomial $x^2 + 3x - 5x^3 + 4$ is 2.   False

4. If $C(x) = x^2 - 3$, then $C(5) = 22$.   True

5. If $P(t) = 30t + 10$, then $P(0) = 40$.   False

6. $(2x^2 - 3x + 5) + (x^2 + 5x - 7) = 3x^2 + 2x - 2$ for any value of x.   True

7. $(x^2 - 5x) - (x^2 - 3x) = -8x$ for any value of x.   False

8. $-2x(3x - 4x^2) = 8x^3 - 6x^2$ for any value of x.   True

9. $-(x - 7) = 7 - x$ for any value of x.   True

10. The opposite of $y + 5$ is $y - 5$ for any value of y.   False

---

5.3   EXERCISES

Reading and Writing   After reading this section, write out the answers to these questions. Use complete sentences.

1. What is a term of a polynomial?
A term of a polynomial is a single number or the product of a number and one or more variables raised to whole number powers.

2. What is a coefficient?
The number preceding the variable in each term is the coefficient of that variable.

3. What is a constant?
A constant is simply a number.

4. What is a polynomial?
A polynomial is a single term or a finite sum of terms.

5. What is the degree of a polynomial?
The degree of a polynomial in one variable is the highest power of the variable in the polynomial.

6. What property is used when multiplying a binomial and a trinomial?
To multiply a binomial and a trinomial, we use the distributive property.

Determine whether each algebraic expression is a polynomial. See Example 1.

7. $3x$ Yes

8. -9 Yes

9. $x^{-1} + 4$ No

10. $3x^{-3} + 4x - 1$ No

11. $x^2 - 3x + 1$ Yes

12. $\dfrac{x^3}{3} - \dfrac{3x^2}{2} + 4$ Yes

13. $\dfrac{1}{x} + y - 3$ No

14. $x^{50} - \dfrac{9}{y^2}$ No

State the degree of each polynomial and the coefficient of x^3. Determine whether each polynomial is a monomial, binomial, or trinomial. See Example 2.

15. $x^4 - 8x^3$ 4, -8, binomial

16. $15 - x^3$ $3, -1$, binomial

17. -8 $0, 0$, monomial

18. 17 $0, 0$, monomial

19. $\dfrac{x^7}{15}$ $7, 0$, monomial

20. $5x^4$ $4, 0$, monomial

21. $x^3 + 3x^4 - 5x^6$ $6, 1$, trinomial

22. $\dfrac{x^3}{2} + \dfrac{5x}{2} - 7$ $3, \frac{1}{2}$, trinomial

For each given polynomial, find the indicated value of the polynomial. See Example 3.

23. $P(x) = x^4 - 1$, $P(3)$ 80

24. $P(x) = x^2 - x - 2$, $P(-1)$ 0

25. $M(x) = -3x^2 + 4x - 9$, $M(-2)$ -29

26. $C(w) = 3w^2 - w$, $C(0)$ 0

27. $R(x) = x^5 - x^4 + x^3 - x^2 + x - 1$, $R(1)$ 0

28. $T(a) = a^7 + a^6$, $T(-1)$ 0

Perform the indicated operations. See Examples 4 and 5.

29. $(2a - 3) + (a + 5)$ $3a + 2$

30. $(2w - 6) + (w + 5)$ $3w - 1$

31. $(7xy + 30) - (2xy + 5)$ $5xy + 25$

32. $(5ab + 7) - (3ab + 6)$ $2ab + 1$

33. $(x^2 - 3x) + (-x^2 + 5x - 9)$ $2x - 9$

34. $(2y^2 - 3y - 8) + (y^2 + 4y - 1)$ $3y^2 + y - 9$

35. $(2x^3 - 4x - 3) - (x^2 - 2x + 5)$ $2x^3 - x^2 - 2x - 8$

36. $(2x - 5) - (x^2 - 3x + 2)$ $-x^2 + 5x - 7$

Perform the indicated operations vertically. See Examples 4 and 5.

37. Add
$$\begin{array}{r} x^3 + 3x^2 - 5x - 2 \\ -x^3 + 8x^2 + 3x - 7 \\ \hline 11x^2 - 2x - 9 \end{array}$$

38. Add
$$\begin{array}{r} x^2 - 3x + 7 \\ -2x^2 - 5x + 2 \\ \hline -x^2 - 8x + 9 \end{array}$$

39. Subtract
$$\begin{array}{r} 5x + 2 \\ 4x - 3 \\ \hline x + 5 \end{array}$$

40. Subtract
$$\begin{array}{r} 4x + 3 \\ 2x - 6 \\ \hline 2x + 9 \end{array}$$

41. Subtract
$$\begin{array}{r} -x^2 + 3x - 5 \\ 5x^2 - 2x - 7 \\ \hline -6x^2 + 5x + 2 \end{array}$$

42. Subtract
$$\begin{array}{r} -3x^2 + 5x - 2 \\ x^2 - 5x - 6 \\ \hline -4x^2 + 10x + 4 \end{array}$$

43. Add
$$\begin{array}{r} x - y \\ x + y \\ \hline 2x \end{array}$$

44. Add
$$\begin{array}{r} -w + 4 \\ 2w - 3 \\ \hline w + 1 \end{array}$$

Find each product. See Examples 6–8.

45. $-3x^2 \cdot 5x^4$ $-15x^6$

46. $(-ab^5)(-2a^2b)$ $2a^3b^6$

47. $-1(3x - 2)$ $-3x + 2$

48. $-1(-x^2 + 3x - 9)$ $x^2 - 3x + 9$

49. $5x^2y^3(3x^2y - 4x)$ $15x^4y^4 - 20x^3y^3$

50. $3y^4z(8y^2z^2 - 3yz + 2y)$ $24y^6z^3 - 9y^5z^2 + 6y^5z$

51. $(x - 2)(x + 2)$ $x^2 - 4$

52. $(x - 1)(x + 1)$ $x^2 - 1$

53. $(x^2 + x + 2)(2x - 3)$ $2x^3 - x^2 + x - 6$

54. $(x^2 - 3x + 2)(x - 4)$ $x^3 - 7x^2 + 14x - 8$

Find each product vertically. See Examples 6–8.

55. Multiply
$$\begin{array}{r} 2x - 3 \\ -5x \\ \hline -10x^2 + 15x \end{array}$$

56. Multiply
$$\begin{array}{r} 3a^3 - 5a^2 + 7 \\ -2a \\ \hline -6a^4 + 10a^3 - 14a \end{array}$$

57. Multiply
$$\begin{array}{r} x + 5 \\ x + 5 \\ \hline x^2 + 10x + 25 \end{array}$$

58. Multiply
$$\begin{array}{r} a + b \\ a - b \\ \hline a^2 - b^2 \end{array}$$

59. Multiply
$$\begin{array}{r} x + 6 \\ 2x - 3 \\ \hline 2x^2 + 9x - 18 \end{array}$$

60. Multiply
$$\begin{array}{r} 3x^2 + 2 \\ 2x^2 - 5 \\ \hline 6x^4 - 11x^2 - 10 \end{array}$$

61. Multiply
$$\begin{array}{r} x^2 + xy + y^2 \\ x - y \\ \hline x^3 - y^3 \end{array}$$

62. Multiply
$$\begin{array}{r} a^2 - ab + b^2 \\ a + b \\ \hline a^3 + b^3 \end{array}$$

Perform the indicated operations.

63. $(x - 7) + (2x - 3) + (5 - x)$ $2x - 5$

64. $(5x - 3) + (x^3 + 3x - 2) + (-2x - 3)$ $x^3 + 6x - 8$

65. $(a^2 - 5a + 3) + (3a^2 - 6a - 7)$ $4a^2 - 11a - 4$

66. $(w^2 - 3w + 2) + (2w - 3 + w^2)$ $2w^2 - w - 1$

67. $(w^2 - 7w - 2) - (w - 3w^2 + 5)$ $4w^2 - 8w - 7$

68. $(a^3 - 3a) - (1 - a - 2a^2)$ $a^3 + 2a^2 - 2a - 1$

69. $(x - 2)(x^2 + 2x + 4)$ $x^3 - 8$

70. $(a - 3)(a^2 + 3a + 9)$ $a^3 - 27$

71. $(x - w)(z + 2w)$ $xz - wz + 2xw - 2w^2$

72. $(w^2 - a)(t^2 + 3)$ $w^2t^2 - at^2 + 3w^2 - 3a$

73. $(2xy - 1)(3xy + 5)$ $6x^2y^2 + 7xy - 5$

74. $(3ab - 4)(ab + 8)$ $3a^2b^2 + 20ab - 32$

  Perform the following operations using a calculator.

75. $(2.31x - 5.4)(6.25x + 1.8)$
$14.4375x^2 - 29.592x - 9.72$

76. $(x - 0.28)(x^2 - 34.6x + 21.2)$
$x^3 - 34.88x^2 + 30.888x - 5.936$

77. $(3.759x^2 - 4.71x + 2.85) + (11.61x^2 + 6.59x - 3.716)$
$15.369x^2 + 1.88x - 0.866$

78. $(43.19x^3 - 3.7x^2 - 5.42x + 3.1) - (62.7x^3 - 7.36x - 12.3)$
$-19.51x^3 - 3.7x^2 + 1.94x + 15.4$

Perform the indicated operations.

79. $\left(\dfrac{1}{2}x + 2\right) + \left(\dfrac{1}{4}x - \dfrac{1}{2}\right)$ $\dfrac{3}{4}x + \dfrac{3}{2}$

80. $\left(\dfrac{1}{3}x + 1\right) + \left(\dfrac{1}{3}x - \dfrac{3}{2}\right)$ $\dfrac{2}{3}x - \dfrac{1}{2}$

81. $\left(\frac{1}{2}x^2 + \frac{1}{3}x - \frac{1}{5}\right) - \left(x^2 - \frac{2}{3}x - \frac{1}{5}\right)$ $\quad -\frac{1}{2}x^2 + x$

82. $\left(\frac{2}{3}x^2 - \frac{1}{3}x + \frac{1}{6}\right) - \left(-\frac{1}{3}x^2 + x + 1\right)$ $\quad x^2 - \frac{4}{3}x - \frac{5}{6}$

83. $[x^2 - 3 - (x^2 + 5x - 4)] - [x - 3(x^2 - 5x)]$
$3x^2 - 21x + 1$

84. $[x^3 - 4x(x^2 - 3x + 2) - 5x] + [x^2 - 5(4 - x^2) + 3]$
$-3x^3 + 18x^2 - 13x - 17$

85. $[5x - 4(x - 3)][3x - 7(x + 2)]$ $\quad -4x^2 - 62x - 168$

86. $[x^2 - (5x - 2)][x^2 + (5x - 2)]$ $\quad x^4 - 25x^2 + 20x - 4$

87. $[x^2 - (m + 2)][x^2 + (m + 2)]$ $\quad x^4 - m^2 - 4m - 4$

88. $[3x^2 - (x - 2)][3x^2 + (x + 2)]$ $\quad 9x^4 + 11x^2 + 4$

89. $2x(5x - 4) - 3x[5x^2 - 3x(x - 7)]$
$-6x^3 - 53x^2 - 8x$

90. $-3x(x - 2) - 5[2x - 4(x + 6)]$
$-3x^2 + 16x + 120$

Perform the indicated operations. A variable used in an expo-
nent represents an integer; a variable used as a base represents
a nonzero real number.

91. $(a^{2m} + 3a^m - 3) + (-5a^{2m} - 7a^m + 8)$
$-4a^{2m} - 4a^m + 5$

92. $(b^{3z} - 6) - (4b^{3z} - b^{2z} - 7)$ $\quad -3b^{3z} + b^{2z} + 1$

93. $(x^n - 1)(x^n + 3)$ $\quad x^{2n} + 2x^n - 3$

94. $(2y^t - 3)(4y^t + 7)$ $\quad 8y^{2t} + 2y^t - 21$

95. $z^{3w} - z^{2w}(z^{1-w} - 4z^w)$ $\quad 5z^{3w} - z^{1+w}$

96. $(w^p - 1)(w^{2p} + w^p + 1)$ $\quad w^{3p} - 1$

97. $(x^{2r} + y)(x^{4r} - x^{2r}y + y^2)$ $\quad x^{6r} + y^3$

98. $(2x^a - z)(2x^a + z)$ $\quad 4x^{2a} - z^2$

Solve each problem. See Exercises 99–104.

99. *Cost of gravel.* The cost in dollars of x cubic yards of
gravel is given by the formula

$$C(x) = 20x + 15.$$

Find $C(3)$, the cost of 3 cubic yards of gravel.
$75

100. *Annual bonus.* Sheila's annual bonus in dollars for
selling n life insurance policies is given by the formula

$$B(n) = 0.1n^2 + 3n + 50.$$

Find $B(20)$, her bonus for selling 20 policies.
$150

101. *Marginal cost.* A company uses the formula $C(n) = 50n - 0.01n^4$ to find the daily cost in dollars of manufacturing n aluminum windows. The *marginal cost* of the
nth window is the additional cost incurred for manufacturing that window. For example, the marginal cost of the
third window is $C(3) - C(2)$. Find the marginal cost for
manufacturing the third window. What is the marginal
cost for manufacturing the tenth window?
$49.35, $15.61

102. *Marginal profit.* A company uses the formula $P(n) = 4n + 0.9n^3$ to estimate its daily profit in dollars for producing n automatic garage door openers. The *marginal
profit* of the nth opener is the amount of additional profit
made for that opener. For example, the marginal profit for
the fourth opener is $P(4) - P(3)$. Find the marginal profit
for the fourth opener. What is the marginal profit for the
tenth opener? Use the bar graph to explain why the
marginal profit increases as production goes up.
$37.30, $247.90. There is a greater difference in height
between adjacent bars as the number of openers increases.

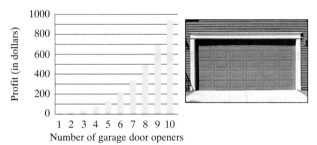

FIGURE FOR EXERCISE 102

103. *Male and female life expectancy.* Since 1950 the life
expectancies of U.S. males and females born in year y
can be modeled by the formulas

$$M(y) = 0.16252y - 251.91$$

and

$$F(y) = 0.18268y - 284.98,$$

respectively (National Center for Health Statistics,
www.cdc.gov/nchswww).
a) How much greater was the life expectancy of a female
born in 1950 than a male born in 1950?
b) Are the lines in the accompanying figure parallel?
c) In what year will female life expectancy be 8 years
greater than male life expectancy?
   a) 6.2 years **b)** no **c)** 2037

104. *More life expectancy.* Use the functions from the previous exercise for these questions.
a) A male born in 1975 does not want his future wife to
outlive him. What should be the year of birth for his

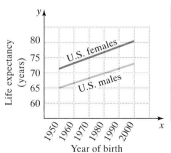

FIGURE FOR EXERCISES 103 AND 104

wife so that they both can be expected to die in the same year?

b) Find $\dfrac{M(y) + F(y)}{2}$ to get a formula for the life expectancy of a person born in year y.

   a) 1969   b) $0.1726y - 268.445$

GETTING MORE INVOLVED

 105. *Discussion.* Is it possible for a binomial to have degree 4? If so, give an example.

  106. *Discussion.* Give an example of two fourth-degree trinomials whose sum is a third-degree binomial.

 107. *Cooperative learning.* Work in a group to find the product $(a + b)(c + d)$. How many terms does it have? Find the product $(a + b)(c + d)(e + f)$. How many terms does it have? How many terms are there in a product of four binomials in which there are no like terms to combine? How many terms are there in a product of n binomials in which there are no like terms?

5.4   MULTIPLYING BINOMIALS

In this / **section**

- The FOIL Method
- The Square of a Binomial
- Product of a Sum and a Difference

In Section 5.3 you learned to multiply polynomials. In this section you will learn rules to make multiplication of binomials simpler.

The FOIL Method

Consider how we find the product of two binomials $x + 3$ and $x + 5$ using the distributive property twice:

$$
\begin{aligned}
(x + 3)(x + 5) &= (x + 3)x + (x + 3)5 & \text{Distributive property} \\
&= x^2 + 3x + 5x + 15 & \text{Distributive property} \\
&= x^2 + 8x + 15 & \text{Combine like terms.}
\end{aligned}
$$

There are four terms in the product. The term x^2 is the product of the first term of each binomial. The term $5x$ is the product of the two outer terms, 5 and x. The term $3x$ is the product of the two inner terms, 3 and x. The term 15 is the product of the last two terms in each binomial, 3 and 5. It may be helpful to connect the terms multiplied by lines.

helpful hint

The product of two binomials always has four terms before combining like terms. The product of two trinomials always has nine terms before combining like terms. How many terms are there in the product of a binomial and trinomial?

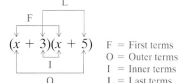

$(x + 3)(x + 5)$

F = First terms
O = Outer terms
I = Inner terms
L = Last terms

So instead of writing out all of the steps in using the distributive property, we can get the result by finding the products of the first, outer, inner, and last terms. This method is called the **FOIL method.**

For example, let's apply FOIL to the product $(x - 3)(x + 4)$:

$$(x - 3)(x + 4) = x^2 + 4x - 3x - 12 = x^2 + x - 12$$

If the outer and inner products are like terms, you can save a step by writing down only their sum.

EXAMPLE 1

Multiplying binomials

Use FOIL to find the products of the binomials.

a) $(2x - 3)(3x + 4)$ **b)** $(2x^3 + 5)(2x^3 - 5)$

c) $(m + w)(2m - w)$ **d)** $(a + b)(a - 3)$

Solution

a) $(2x - 3)(3x + 4) = \overset{\text{F}}{6x^2} + \overset{\text{O}}{8x} - \overset{\text{I}}{9x} - \overset{\text{L}}{12} = 6x^2 - x - 12$

b) $(2x^3 + 5)(2x^3 - 5) = 4x^6 - 10x^3 + 10x^3 - 25 = 4x^6 - 25$

c) $(m + w)(2m - w) = 2m^2 - mw + 2mw - w^2 = 2m^2 + mw - w^2$

d) $(a + b)(a - 3) = a^2 - 3a + ab - 3b$ There are no like terms.

The Square of a Binomial

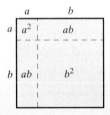
To find $(a + b)^2$, the square of a sum, we can use FOIL on $(a + b)(a + b)$:

$$(a + b)(a + b) = a^2 + ab + ab + b^2$$
$$= a^2 + 2ab + b^2$$

You can use the result $a^2 + 2ab + b^2$ that we obtained from FOIL to quickly find the square of *any* sum. *To square a sum, we square the first term* (a^2), *add twice the product of the two terms* ($2ab$), *then add the square of the last term* (b^2).

Rule for the Square of a Sum

$$(a + b)^2 = a^2 + 2ab + b^2$$

In general, the square of a sum $(a + b)^2$ is not equal to the sum of the squares $a^2 + b^2$. The square of a sum has the middle term $2ab$.

EXAMPLE 2

Squaring a binomial

Square each sum, using the new rule.

a) $(x + 5)^2$ **b)** $(2w + 3)^2$ **c)** $(2y^4 + 3)^2$

Solution

a) $(x + 5)^2 = \underset{\substack{\uparrow \\ \text{Square} \\ \text{of} \\ \text{first}}}{x^2} + \underset{\substack{\uparrow \\ \text{Twice} \\ \text{the} \\ \text{product}}}{2(x)(5)} + \underset{\substack{\uparrow \\ \text{Square} \\ \text{of} \\ \text{last}}}{5^2} = x^2 + 10x + 25$

b) $(2w + 3)^2 = (2w)^2 + 2(2w)(3) + 3^2 = 4w^2 + 12w + 9$

c) $(2y^4 + 3)^2 = (2y^4)^2 + 2(2y^4)(3) + 3^2 = 4y^8 + 12y^4 + 9$

CAUTION   Squaring $x + 5$ correctly, as in Example 2(a), gives us the identity

$$(x + 5)^2 = x^2 + 10x + 25,$$

which is satisfied by any x. If you forget the middle term and write $(x + 5)^2 = x^2 + 25$, then you have an equation that is satisfied only if $x = 0$.

To find $(a - b)^2$, the square of a difference, we can use FOIL:

$$(a - b)(a - b) = a^2 - ab - ab + b^2$$
$$= a^2 - 2ab + b^2$$

As in squaring a sum, it is simply better to remember the result of using FOIL. *To square a difference, square the first term, subtract twice the product of the two terms, and add the square of the last term.*

Rule for the Square of a Difference

$$(a - b)^2 = a^2 - 2ab + b^2$$

E X A M P L E   3

Squaring a binomial
Square each difference, using the new rule.

a) $(x - 6)^2$ **b)** $(3w - 5y)^2$ **c)** $(-4 - st)^2$ **d)** $(3 - 5a^3)^2$

Solution

a) $(x - 6)^2 = x^2 - 2(x)(6) + 6^2$ For the middle term, subtract twice the product: $2(x)(6)$.

$$= x^2 - 12x + 36$$

b) $(3w - 5y)^2 = (3w)^2 - 2(3w)(5y) + (5y)^2$
$$= 9w^2 - 30wy + 25y^2$$

c) $(-4 - st)^2 = (-4)^2 - 2(-4)(st) + (st)^2$
$$= 16 + 8st + s^2t^2$$

d) $(3 - 5a^3)^2 = 3^2 - 2(3)(5a^3) + (5a^3)^2$
$$= 9 - 30a^3 + 25a^6$$

> **helpful** **hint**
>
> Many students keep using FOIL to find the square of a sum or a difference. However, you will be greatly rewarded if you learn the new rules for squaring a sum or a difference.

Product of a Sum and a Difference

If we multiply the sum $a + b$ and the difference $a - b$ by using FOIL, we get

$$(a + b)(a - b) = a^2 - ab + ab - b^2$$
$$= a^2 - b^2.$$

The inner and outer products add up to zero, canceling each other out. So *the product of a sum and a difference is the difference of two squares,* as shown in the following rule.

> **helpful** **hint**
>
> You can use
> $(a + b)(a - b) = a^2 - b^2$
> to perform mental arithmetic tricks such as
> $59 \cdot 61 = 3600 - 1 = 3599.$
> What is $49 \cdot 51$? $28 \cdot 32$?

Rule for the Product of a Sum and a Difference

$$(a + b)(a - b) = a^2 - b^2$$

E X A M P L E   4

Finding the product of a sum and a difference
Find the products.

a) $(x + 3)(x - 3)$ **b)** $(a^3 + 8)(a^3 - 8)$ **c)** $(3x^2 - y^3)(3x^2 + y^3)$

Solution

a) $(x + 3)(x - 3) = x^2 - 9$

b) $(a^3 + 8)(a^3 - 8) = a^6 - 64$

c) $(3x^2 - y^3)(3x^2 + y^3) = 9x^4 - y^6$

The square of a sum, the square of a difference, and the product of a sum and a difference are referred to as **special products.** Although the special products can be found by using the distributive property or FOIL, they occur so frequently in algebra that it is essential to learn the new rules. In the next example we use the special product rules to multiply two trinomials and to square a trinomial.

E X A M P L E   5   **Using special product rules to multiply trinomials**

Find the products.

a) $[(x + y) + 3][(x + y) - 3]$

b) $[(m - n) + 5]^2$

Solution

a) Use the rule $(a + b)(a - b) = a^2 - b^2$ with $a = x + y$ and $b = 3$:

$$[(x + y) + 3][(x + y) - 3] = (x + y)^2 - 3^2$$
$$= x^2 + 2xy + y^2 - 9$$

b) Use the rule $(a + b)^2 = a^2 + 2ab + b^2$ with $a = m - n$ and $b = 5$:

$$[(m - n) + 5]^2 = (m - n)^2 + 2(m - n)5 + 5^2$$
$$= m^2 - 2mn + n^2 + 10m - 10n + 25$$

W A R M - U P S

True or false? Explain your answer.

1. $(x + 2)(x + 5) = x^2 + 7x + 10$ for any value of x.   True
2. $(2x - 3)(3x + 5) = 6x^2 + x - 15$ for any value of x.   True
3. $(2 + 3)^2 = 2^2 + 3^2$ False
4. $(x + 7)^2 = x^2 + 14x + 49$ for any value of x.   True
5. $(8 - 3)^2 = 64 - 9$ False
6. The product of a sum and a difference of the same two terms is equal to the difference of two squares.   True
7. $(60 - 1)(60 + 1) = 3600 - 1$ True
8. $(x - y)^2 = x^2 - 2xy + y^2$ for any values of x and y.   True
9. $(x - 3)^2 = x^2 - 3x + 9$ for any value of x.   False
10. The expression $3x \cdot 5x$ is a product of two binomials.   False

---

5.4   E X E R C I S E S

Reading and Writing   After reading this section, write out the answers to these questions. Use complete sentences.

1. What property is used to multiply two binomials?
   The distributive property is used in multiplying binomials.

2. What does FOIL stand for?
   FOIL stands for first, outer, inner, last.

3. What is the purpose of the FOIL method?
   The purpose of FOIL is to provide a fast way to find the product of two binomials.

4. How do you square a sum?
   The square of a sum is the square of the first term plus twice the product of the two terms plus the square of the last term.

5. How do you square a difference?
   The square of a difference is the square of the first term minus twice the product of the two terms plus the square of the last term.

6. How do you find the product of a sum and a difference?
   The product of a sum and a difference of the same two terms is the square of the first term minus the square of the second term.

Find each product. When possible, write down only the answer. See Example 1.

7. $(x - 2)(x + 4)$ $x^2 + 2x - 8$
8. $(x - 3)(x + 5)$ $x^2 + 2x - 15$

9. $(2x + 1)(x + 3)$ $2x^2 + 7x + 3$

10. $(2y + 3)(y + 2)$ $2y^2 + 7y + 6$

11. $(-2a - 3)(-a + 5)$ $2a^2 - 7a - 15$

12. $(-3x - 5)(-x + 6)$ $3x^2 - 13x - 30$

13. $(2x^2 - 7)(2x^2 + 7)$ $4x^4 - 49$

14. $(3y^3 + 8)(3y^3 - 8)$ $9y^6 - 64$

15. $(2x^3 - 1)(x^3 + 4)$ $2x^6 + 7x^3 - 4$

16. $(3t^2 - 4)(2t^2 + 3)$ $6t^4 + t^2 - 12$

17. $(6z + w)(w - z)$ $w^2 + 5wz - 6z^2$

18. $(4y + w)(w - 2y)$ $w^2 + 2wy - 8y^2$

19. $(3k - 2t)(4t + 3k)$ $9k^2 + 6kt - 8t^2$

20. $(7a - 2x)(x + a)$ $7a^2 + 5ax - 2x^2$

21. $(x - 3)(y + w)$ $xy - 3y + xw - 3w$

22. $(z - 1)(y + 2)$ $yz - y + 2z - 2$

Find the square of each sum or difference. When possible, write down only the answer. See Examples 2 and 3.

23. $(m + 3)^2$ $m^2 + 6m + 9$

24. $(a + 2)^2$ $a^2 + 4a + 4$

25. $(a - 4)^2$ $a^2 - 8a + 16$

26. $(b - 3)^2$ $b^2 - 6b + 9$

27. $(2w + 1)^2$ $4w^2 + 4w + 1$

28. $(3m + 4)^2$ $9m^2 + 24m + 16$

29. $(3t - 5u)^2$ $9t^2 - 30tu + 25u^2$

30. $(3w - 2x)^2$ $9w^2 - 12wx + 4x^2$

31. $(-x - 1)^2$ $x^2 + 2x + 1$

32. $(-d - 5)^2$ $d^2 + 10d + 25$

33. $(a - 3y^3)^2$ $a^2 - 6ay^3 + 9y^6$

34. $(3m - 5n^3)^2$ $9m^2 - 30mn^3 + 25n^6$

35. $\left(\dfrac{1}{2}x - 1\right)^2$ $\dfrac{1}{4}x^2 - x + 1$

36. $\left(\dfrac{2}{3}x - \dfrac{1}{2}\right)^2$ $\dfrac{4}{9}x^2 - \dfrac{2}{3}x + \dfrac{1}{4}$

Find each product. See Example 4.

37. $(w - 9)(w + 9)$ $w^2 - 81$

38. $(m - 4)(m + 4)$ $m^2 - 16$

39. $(w^3 + y)(w^3 - y)$ $w^6 - y^2$

40. $(a^3 - x)(a^3 + x)$ $a^6 - x^2$

41. $(2x - 7)(2x + 7)$ $4x^2 - 49$

42. $(5x + 3)(5x - 3)$ $25x^2 - 9$

43. $(3x^2 - 2)(3x^2 + 2)$ $9x^4 - 4$

44. $(4y^2 + 1)(4y^2 - 1)$ $16y^4 - 1$

Use the special product rules to find each product. See Example 5.

45. $[(m + t) + 5][(m + t) - 5]$ $m^2 + 2mt + t^2 - 25$

46. $[(2x + 3) - y][(2x + 3) + y]$ $4x^2 + 12x + 9 - y^2$

47. $[y - (r + 5)][y + (r + 5)]$ $y^2 - r^2 - 10r - 25$

48. $[x + (3 - k)][x - (3 - k)]$ $x^2 - 9 + 6k - k^2$

49. $[(2y - t) + 3]^2$ $4y^2 - 4yt + t^2 + 12y - 6t + 9$

50. $[(u - 3v) - 4]^2$ $u^2 - 6uv + 9v^2 - 8u + 24v + 16$

51. $[3h + (k - 1)]^2$ $9h^2 + 6hk - 6h + k^2 - 2k + 1$

52. $[2p - (3q + 6)]^2$ $4p^2 - 12pq - 24p + 9q^2 + 36q + 36$

Perform the operations and simplify.

53. $(x - 6)(x + 9)$ $x^2 + 3x - 54$

54. $(2x^2 - 3)(3x^2 + 4)$ $6x^4 - x^2 - 12$

55. $(5 - x)(5 + x)$ $25 - x^2$

56. $(4 - ab)(4 + ab)$ $16 - a^2b^2$

57. $(3x - 4a)(2x + 5a)$ $6x^2 + 7ax - 20a^2$

58. $(x^5 + 2)(x^5 - 2)$ $x^{10} - 4$

59. $(2t - 3)(t + w)$ $2t^2 + 2tw - 3t - 3w$

60. $(5x - 9)(ax + b)$ $5ax^2 - 9ax + 5bx - 9b$

61. $(3x^2 + 2y^3)^2$ $9x^4 + 12x^2y^3 + 4y^6$

62. $(5a^4 - 2b)^2$ $25a^8 - 20a^4b + 4b^2$

63. $(2y + 2)(3y - 5)$ $6y^2 - 4y - 10$

64. $(3b - 3)(2b + 3)$ $6b^2 + 3b - 9$

65. $(50 + 2)(50 - 2)$ 2496

66. $(100 - 1)(100 + 1)$ 9999

67. $(3 + 7x)^2$ $49x^2 + 42x + 9$

68. $(1 - pq)^2$ $1 - 2pq + p^2q^2$

69. $4y\left(3y + \dfrac{1}{2}\right)^2$ $36y^3 + 12y^2 + y$

70. $25y\left(2y - \dfrac{1}{5}\right)^2$ $100y^3 - 20y^2 + y$

71. $(a + h)^2 - a^2$ $2ah + h^2$

72. $\dfrac{(x + h)^2 - x^2}{h}$ $2x + h$

73. $(x + 2)(x + 2)^2$ $x^3 + 6x^2 + 12x + 8$

74. $(a + 1)^2(a + 1)^2$ $a^4 + 4a^3 + 6a^2 + 4a + 1$

75. $(y - 3)^3$ $y^3 - 9y^2 + 27y - 27$

76. $(2b + 1)^4$ $16b^4 + 32b^3 + 24b^2 + 8b + 1$

 Use a calculator to help you perform the following operations.

77. $(3.2x - 4.5)(5.1x + 3.9)$ $16.32x^2 - 10.47x - 17.55$

78. $(5.3x - 9.2)^2$ $28.09x^2 - 97.52x + 84.64$

79. $(3.6y + 4.4)^2$ $12.96y^2 + 31.68y + 19.36$

80. $(3.3a - 7.9b)(3.3a + 7.9b)$ $10.89a^2 - 62.41b^2$

Find the products. Assume all variables are nonzero and variables used in exponents represent integers.

81. $(x^m + 2)(x^{2m} + 3)$ $x^{3m} + 2x^{2m} + 3x^m + 6$

82. $(a^n - b)(a^n + b)$ $a^{2n} - b^2$

83. $a^{n+1}(a^{2n} + a^n - 3)$ $a^{3n+1} + a^{2n+1} - 3a^{n+1}$

84. $x^{3b}(x^{-3b} + 3x^{-b} + 5)$ $1 + 3x^{2b} + 5x^{3b}$

85. $(a^m + a^n)^2$ $a^{2m} + 2a^{m+n} + a^{2n}$

86. $(x^w - x^t)^2$ $x^{2w} - 2x^{w+t} + x^{2t}$

87. $(5y^m + 8z^k)(3y^{2m} + 4z^{3-k})$
$15y^{3m} + 24y^{2m}z^k + 20y^m z^{3-k} + 32z^3$

88. $(4x^{a-1} + 3y^{b+5})(x^{2a-3} - 2y^{4-b})$
$4x^{3a-4} + 3y^{b+5}x^{2a-3} - 8x^{a-1}y^{4-b} - 6y^9$

Solve each problem.

89. Area of a room. Suppose the length of a rectangular room is $x + 3$ meters and the width is $x + 1$ meters. Find a trinomial that can be used to represent the area of the room.
$x^2 + 4x + 3$

90. House plans. Barbie and Ken planned to build a square house with area x^2 square feet. Then they revised the plan so that one side was lengthened by 20 feet and the other side was shortened by 6 feet. Find a trinomial that can be used to represent the area of the revised house.
$x^2 + 14x - 120$ square feet

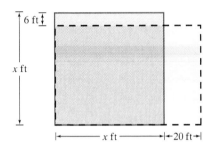

FIGURE FOR EXERCISE 90

91. Available habitat. A wild animal will generally stay more than x kilometers from the edge of a forest preserve. So the available habitat for the animal excludes an area of uniform width x on the edge of the rectangular forest preserve shown in the figure. The value of x depends on the animal. Find a trinomial in x that gives the area of the available habitat in square kilometers (km^2) for the forest preserve shown. What is the available habitat in this forest preserve for a bobcat for which $x = 0.4$ kilometers.
$4x^2 - 36x + 80$, 66.24 km^2

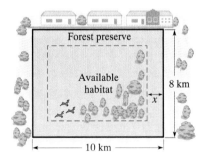

FIGURE FOR EXERCISE 91

92. Cubic coating. A cubic metal box x inches on each side is designed for transporting frozen specimens. The box is surrounded on all sides by a 2-inch-thick layer of styrofoam insulation. Find a polynomial that represents the total volume of the cube and styrofoam.
$x^3 + 12x^2 + 48x + 64$ $in.^3$

93. Overflow pan. An air conditioning contractor makes an overflow pan for a condenser by cutting squares with side of length x feet from the corners of a 4 foot by 6 foot piece of galvanized sheet metal as shown in the figure. The sides

are then folded up, and the corners are sealed. Find a polynomial that gives the volume of the pan in cubic feet (ft^3). What is the volume of the pan if $x = 4$ inches?
$4x^3 - 20x^2 + 24x$, 5.9 ft^3

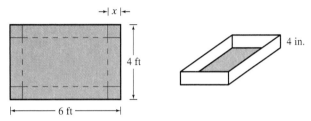

FIGURE FOR EXERCISE 93

94. Energy efficient. A manufacturer of mobile homes makes a custom model that is x feet long, 12 feet wide, and 8 feet high (all inside dimensions). The insulation is 3 inches thick in the walls, 6 inches thick in the floor, and 8 inches thick in the ceiling. Given that the insulation costs the manufacturer 25 cents per cubic foot and doors and windows take up 80 square feet of wall space, find a polynomial in x that gives the cost in dollars for insulation in this model. State any assumptions that you are making to solve this problem.
$4.5x + 7$ dollars

GETTING MORE INVOLVED

95. Exploration. **a)** Find $(a + b)^3$ by multiplying $(a + b)^2$ by $a + b$.
b) Next find $(a + b)^4$ and $(a + b)^5$.
c) How many terms are in each of these powers of $a + b$ after combining like terms?
d) Make a general statement about the number of terms in $(a + b)^n$.

96. Cooperative learning. Make a four-column table with columns for a, b, $(a + b)^2$, and $a^2 - b^2$. Work with a group to fill in the table with five pairs of numbers for a and b for which $(a + b)^2 \neq a^2 + b^2$. For what values of a and b does $(a + b)^2 = a^2 + b^2$?

97. Discussion. The area of the large square shown in the figure is $(a + b)^2$. Find the area of each of the four smaller regions in the figure, and then find the sum of those areas. What conclusion can you draw from these areas about $(a + b)^2$?

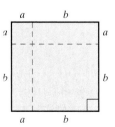

FIGURE FOR EXERCISE 97

5.5   DIVISION OF POLYNOMIALS

We began our study of polynomials in Section 5.3 by learning how to add, subtract, and multiply polynomials. In this section we will study division of polynomials.

Dividing a Polynomial by a Monomial

You learned how to divide monomials in Section 5.1. For example,

$$6x^3 \div (3x) = \frac{6x^3}{3x} = 2x^2.$$

We check by multiplying. Because $2x^2 \cdot 3x = 6x^3$, this answer is correct. Recall that $a \div b = c$ if and only if $c \cdot b = a$. We call a the **dividend,** b the **divisor,** and c the **quotient.** We may also refer to $a \div b$ and $\frac{a}{b}$ as quotients.

We can use the distributive property to find that

$$3x(2x^2 + 5x - 4) = 6x^3 + 15x^2 - 12x.$$

So if we divide $6x^3 + 15x^2 - 12x$ by the monomial $3x$, we must get $2x^2 + 5x - 4$. We can perform this division by dividing $3x$ into *each term* of $6x^3 + 15x^2 - 12x$:

$$\frac{6x^3 + 15x^2 - 12x}{3x} = \frac{6x^3}{3x} + \frac{15x^2}{3x} - \frac{12x}{3x}$$
$$= 2x^2 + 5x - 4$$

In this case the divisor is $3x$, the dividend is $6x^3 + 15x^2 - 12x$, and the quotient is $2x^2 + 5x - 4$.

EXAMPLE 1

Dividing polynomials

Find the quotient.

a) $-12x^5 \div (2x^3)$ **b)** $(-20x^6 + 8x^4 - 4x^2) \div (4x^2)$

Solution

a) When dividing x^5 by x^3, we subtract the exponents:

$$-12x^5 \div (2x^3) = \frac{-12x^5}{2x^3} = -6x^2$$

The quotient is $-6x^2$. Check:

$$-6x^2 \cdot 2x^3 = -12x^5$$

b) Divide each term of $-20x^6 + 8x^4 - 4x^2$ by $4x^2$:

$$\frac{-20x^6 + 8x^4 - 4x^2}{4x^2} = \frac{-20x^6}{4x^2} + \frac{8x^4}{4x^2} - \frac{4x^2}{4x^2}$$
$$= -5x^4 + 2x^2 - 1$$

The quotient is $-5x^4 + 2x^2 - 1$. Check:

$$4x^2(-5x^4 + 2x^2 - 1) = -20x^6 + 8x^4 - 4x^2$$

helpful **hint**

Recall that the order of operations gives multiplication and division an equal ranking and says to do them in order from left to right. So without parentheses,

$$-12x^5 \div 2x^3$$

actually means

$$\frac{-12x^5}{2} \cdot x^3.$$

Dividing a Polynomial by a Binomial

We can multiply $x - 2$ and $x + 5$ to get

$$(x - 2)(x + 5) = x^2 + 3x - 10.$$

So if we divide $x^2 + 3x - 10$ by the factor $x - 2$, we should get the other factor $x + 5$. This division is not done like division by a monomial; it is done like long division of whole numbers. We get the first term of the quotient by dividing the first term of $x - 2$ into the first term of $x^2 + 3x - 10$. Divide x^2 by x to get x.

$$
\begin{array}{r}
x \\
x - 2 \overline{)x^2 + 3x - 10} \\
\underline{x^2 - 2x} \\
5x
\end{array}
$$

$x^2 \div x = x$

Multiply: $x(x - 2) = x^2 - 2x$.

Subtract: $3x - (-2x) = 5x$.

Now bring down -10. We get the second term of the quotient (below) by dividing the first term of $x - 2$ into the first term of $5x - 10$. Divide $5x$ by x to get 5.

$$
\begin{array}{r}
x + 5 \\
x - 2 \overline{)x^2 + 3x - 10} \\
\underline{x^2 - 2x} \\
5x - 10 \\
\underline{5x - 10} \\
0
\end{array}
$$

$5x \div x = 5$

Multiply: $5(x - 2) = 5x - 10$.

Subtract: $-10 - (-10) = 0$.

So the quotient is $x + 5$ and the **remainder** is 0. If the remainder is not 0, then

$$\textbf{dividend} = \textbf{(divisor)(quotient)} + \textbf{(remainder).}$$

If we divide each side of this equation by the divisor, we get

$$\frac{\textbf{dividend}}{\textbf{divisor}} = \textbf{quotient} + \frac{\textbf{remainder}}{\textbf{divisor}}.$$

When dividing polynomials, we must write the terms of the divisor and the dividend in descending order of the exponents. If any terms are missing, as in the next example, we insert terms with a coefficient of 0 as placeholders. When dividing polynomials, we stop the process when the degree of the remainder is smaller than the degree of the divisor.

E X A M P L E   2

Dividing polynomials

Find the quotient and remainder for $(3x^4 - 2 - 5x) \div (x^2 - 3x)$.

Solution

Rearrange $3x^4 - 2 - 5x$ as $3x^4 - 5x - 2$ and insert the terms $0x^3$ and $0x^2$:

$$
\begin{array}{r}
3x^2 + 9x + 27 \\
x^2 - 3x \overline{)3x^4 + 0x^3 + 0x^2 - 5x - 2} \\
\underline{3x^4 - 9x^3} \\
9x^3 + 0x^2 \\
\underline{9x^3 - 27x^2} \\
27x^2 - 5x \\
\underline{27x^2 - 81x} \\
76x - 2
\end{array}
$$

$0x^3 - (-9x^3) = 9x^3$

The quotient is $3x^2 + 9x + 27$, and the remainder is $76x - 2$. Note that the degree of the remainder is 1, and the degree of the divisor is 2. To check, verify that
$$(x^2 - 3x)(3x^2 + 9x + 27) + 76x - 2 = 3x^4 - 5x - 2. \qquad ■$$

EXAMPLE 3

Rewriting a ratio of two polynomials

Write $\dfrac{4x^3 - x - 9}{2x - 3}$ in the form

$$\text{quotient} + \frac{\text{remainder}}{\text{divisor}}.$$

Solution

Divide $4x^3 - x - 9$ by $2x - 3$. Insert $0 \cdot x^2$ for the missing term.

$$
\require{enclose}
\begin{array}{r}
2x^2 + 3x + 4 \\
2x - 3 \enclose{longdiv}{4x^3 + 0x^2 - x - 9} \\
\underline{4x^3 - 6x^2} \\
6x^2 - x \\
\underline{6x^2 - 9x} \\
8x - 9 \\
\underline{8x - 12} \\
3
\end{array}
$$

$4x^3 \div (2x) = 2x^2$

$0x^2 - (-6x^2) = 6x^2$

$-x - (-9x) = 8x$

$-9 - (-12) = 3$

Since the quotient is $2x^2 + 3x + 4$ and the remainder is 3, we have
$$\frac{4x^3 - x - 9}{2x - 3} = 2x^2 + 3x + 4 + \frac{3}{2x - 3}.$$

To check the answer, we must verify that
$$(2x - 3)(2x^2 + 3x + 4) + 3 = 4x^3 - x - 9. \qquad ■$$

Synthetic Division

When dividing a polynomial by a binomial of the form $x - c$, we can use **synthetic division** to speed up the process. For synthetic division we write only the essential parts of ordinary division. For example, to divide $x^3 - 5x^2 + 4x - 3$ by $x - 2$, we write only the coefficients of the dividend 1, -5, 4, and -3 in order of descending exponents. From the divisor $x - 2$ we use 2 and start with the following arrangement:

$$
\begin{array}{r|rrrr}
2 & 1 & -5 & 4 & -3
\end{array}
$$
$\qquad (1 \cdot x^3 - 5x^2 + 4x - 3) \div (x - 2)$

Next we bring the first coefficient, 1, straight down:

$$
\begin{array}{r|rrrr}
2 & 1 & -5 & 4 & -3 \\
& \downarrow & \text{Bring down} \\
\hline
& 1
\end{array}
$$

We then multiply the 1 by the 2 from the divisor, place the answer under the -5, and then add that column. Using 2 for $x - 2$ allows us to add the column rather than subtract as in ordinary division:

$$
\begin{array}{r|rrrr}
2 & 1 & -5 & 4 & -3 \\
& & 2 & & \text{Add} \\
\hline
& 1 & -3
\end{array}
$$
Multiply

We then repeat the multiply-and-add step for each of the remaining columns:

$$
\begin{array}{r|rrrr}
2 & 1 & -5 & 4 & -3 \\
& & 2 & -6 & -4 \\
\hline
& 1 & -3 & -2 & -7
\end{array}
$$

Multiply $\leftarrow$ Remainder

Quotient

From the bottom row we can read the quotient and remainder. Since the degree of the quotient is one less than the degree of the dividend, the quotient is $1x^2 - 3x - 2$. The remainder is -7.

The strategy for getting the quotient $Q(x)$ and remainder R by synthetic division can be stated as follows.

Strategy for Using Synthetic Division

1. List the coefficients of the polynomial (the dividend).
2. Be sure to include zeros for any missing terms in the dividend.
3. For dividing by $x - c$, place c to the left.
4. Bring the first coefficient down.
5. Multiply by c and add for each column.
6. Read $Q(x)$ and R from the bottom row.

CAUTION Synthetic division is used only for dividing a polynomial by the binomial $x - c$, where c is a constant. If the binomial is $x - 7$, then $c = 7$. For the binomial $x + 7$ we have $x + 7 = x - (-7)$ and $c = -7$.

EXAMPLE 4

Using synthetic division

Find the quotient and remainder when $2x^4 - 5x^2 + 6x - 9$ is divided by $x + 2$.

Solution

Since $x + 2 = x - (-2)$, we use -2 for the divisor. Because x^3 is missing in the dividend, use a zero for the coefficient of x^3:

$$
\begin{array}{r|rrrrr}
-2 & 2 & 0 & -5 & 6 & -9 \\
& & -4 & 8 & -6 & 0 \\
\hline
& 2 & -4 & 3 & 0 & -9
\end{array}
$$

$\leftarrow 2x^4 + 0 \cdot x^3 - 5x^2 + 6x - 9$ Add

Multiply $\leftarrow$ Quotient and remainder

Because the degree of the dividend is 4, the degree of the quotient is 3. The quotient is $2x^3 - 4x^2 + 3x$, and the remainder is -9. We can also express the results of this division in the form quotient $+ \frac{\text{remainder}}{\text{divisor}}$:

$$
\frac{2x^4 - 5x^2 + 6x - 9}{x + 2} = 2x^3 - 4x^2 + 3x + \frac{-9}{x + 2}
$$

Division and Factoring

To **factor** a polynomial means to write it as a product of two or more simpler polynomials. If we divide two polynomials and get 0 remainder, then we can write

$$\text{dividend} = (\text{divisor})(\text{quotient})$$

and we have factored the dividend. *The dividend factors as the divisor times the quotient if and only if the remainder is* 0. We can use division to help us discover factors of polynomials. To use this idea, however, we must know a factor or a possible factor to use as the divisor.

E X A M P L E 5

Using synthetic division to determine factors

Is $x - 1$ a factor of $6x^3 - 5x^2 - 4x + 3$?

Solution

We can use synthetic division to divide $6x^3 - 5x^2 - 4x + 3$ by $x - 1$:

$$
\begin{array}{r|rrrr}
1 & 6 & -5 & -4 & 3 \\
  & \downarrow & 6 & 1 & -3 \\
\hline
  & 6 & 1 & -3 & 0
\end{array}
$$

Because the remainder is 0, $x - 1$ is a factor, and

$$6x^3 - 5x^2 - 4x + 3 = (x - 1)(6x^2 + x - 3).$$

E X A M P L E 6

Using division to determine factors

Is $a - b$ a factor of $a^3 - b^3$?

Solution

Divide $a^3 - b^3$ by $a - b$. Insert zeros for the missing a^2b- and ab^2-terms.

$$
\begin{array}{r}
a^2 + ab + b^2 \\
a - b \overline{)a^3 + 0 + 0 - b^3} \\
\underline{a^3 - a^2b} \\
a^2b + 0 \\
\underline{a^2b - ab^2} \\
ab^2 - b^3 \\
\underline{ab^2 - b^3} \\
0
\end{array}
$$

Because the remainder is 0, $a - b$ is a factor, and

$$a^3 - b^3 = (a - b)(a^2 + ab + b^2).$$

W A R M - U P S

True or false? Explain your answer.

1. If $a \div b = c$, then c is the dividend.   False

2. The quotient times the dividend plus the remainder equals the divisor.   False

3. $(x + 2)(x + 3) + 1 = x^2 + 5x + 7$ is true for any value of x.   True

4. The quotient of $(x^2 + 5x + 7) \div (x + 3)$ is $x + 2$.   True

5. If $x^2 + 5x + 7$ is divided by $x + 2$, the remainder is 1.   True

6. To divide $x^3 - 4x + 1$ by $x - 3$, we use -3 in synthetic division.   False

7. We can use synthetic division to divide $x^3 - 4x^2 - 6$ by $x^2 - 5$.   False

8. If $3x^5 - 4x^2 - 3$ is divided by $x + 2$, the quotient has degree 4.   True

9. If the remainder is zero, then the divisor is a factor of the dividend.   True

10. If the remainder is zero, then the quotient is a factor of the dividend.   True

Reading and Writing   *After reading this section, write out the answers to these questions. Use complete sentences.*

1. What are the dividend, divisor, and quotient?

If $a \div b = c$, then the divisor is b, the dividend is a, and the quotient is c.

2. In what form should polynomials be written for long division?

For long division polynomials should be written with the exponents in descending order.

3. What do you do about missing terms when dividing polynomials?

If the term x^n is missing in the dividend, insert the term $0 \cdot x^n$ for the missing term.

4. When do you stop the long division process for dividing polynomials?

Stop the long division process when the remainder has a smaller degree than the divisor.

5. What is synthetic division used for?

Synthetic division is used only for dividing by a binomial of the form $x - c$.

6. What is the relationship between division of polynomials and factoring polynomials?

The remainder is zero if and only if the dividend is a factor of the divisor.

Find the quotient. See Example 1.

7. $36x^7 \div (3x^3)$ $12x^4$

8. $-30x^3 \div (-5x)$ $6x^2$

9. $16x^2 \div (-8x^2)$ -2

10. $-22a^3 \div (11a^2)$ $-2a$

11. $(6b - 9) \div 3$ $2b - 3$

12. $(8x^2 - 6x) \div 2$ $4x^2 - 3x$

13. $(3x^2 + 6x) \div (3x)$ $x + 2$

14. $(5x^3 - 10x^2 + 20x) \div (5x)$ $x^2 - 2x + 4$

15. $(10x^4 - 8x^3 + 6x^2) \div (-2x^2)$ $-5x^2 + 4x - 3$

16. $(-9x^3 + 6x^2 - 12x) \div (-3x)$ $3x^2 - 2x + 4$

17. $(7x^3 - 4x^2) \div (2x)$ $\dfrac{7}{2}x^2 - 2x$

18. $(6x^3 - 5x^2) \div (4x^2)$ $\dfrac{3}{2}x - \dfrac{5}{4}$

Find the quotient and remainder as in Example 2. Check by using the formula

$$\text{dividend} = (\text{divisor})(\text{quotient}) + \text{remainder}.$$

19. $(x^2 + 8x + 13) \div (x + 3)$ $x + 5, -2$

20. $(x^2 + 5x + 7) \div (x + 3)$ $x + 2, 1$

21. $(x^2 - 2x) \div (x + 2)$ $x - 4, 8$

22. $(3x) \div (x - 1)$ $3, 3$

23. $(x^3 + 8) \div (x + 2)$ $x^2 - 2x + 4, 0$

24. $(y^3 - 1) \div (y - 1)$ $y^2 + y + 1, 0$

25. $(a^3 + 4a - 5) \div (a - 2)$ $a^2 + 2a + 8, 11$

26. $(w^3 + w^2 - 3) \div (w - 2)$ $w^2 + 3w + 6, 9$

27. $(x^3 - x^2 + x - 3) \div (x + 1)$ $x^2 - 2x + 3, -6$

28. $(a^3 - a^2 + a - 4) \div (a + 2)$ $a^2 - 3a + 7, -18$

29. $(x^4 - x + x^3 - 1) \div (x - 2)$ $x^3 + 3x^2 + 6x + 11, 21$

30. $(3x^4 + 6 - x^2 + 3x) \div (x + 2)$
$3x^3 - 6x^2 + 11x - 19, 44$

31. $(5x^2 - 3x^4 + x - 2) \div (x^2 - 2)$ $-3x^2 - 1, x - 4$

32. $(x^4 - 2 + x^3) \div (x^2 + 3)$ $x^2 + x - 3, -3x + 7$

33. $(x^2 - 4x + 2) \div (2x - 3)$ $\dfrac{1}{2}x - \dfrac{5}{4}, -\dfrac{7}{4}$

34. $(x^2 - 5x + 1) \div (3x + 6)$ $\dfrac{1}{3}x - \dfrac{7}{3}, 15$

35. $(2x^2 - x + 6) \div (3x - 2)$ $\dfrac{2}{3}x + \dfrac{1}{9}, \dfrac{56}{9}$

36. $(3x^2 + 4x - 1) \div (2x + 1)$ $\dfrac{3}{2}x + \dfrac{5}{4}, -\dfrac{9}{4}$

Write each expression in the form

$$\text{quotient} + \frac{\text{remainder}}{\text{divisor}}.$$

See Example 3.

37. $\dfrac{2x}{x - 5}$

$2 + \dfrac{10}{x - 5}$

38. $\dfrac{x}{x - 1}$

$1 + \dfrac{1}{x - 1}$

39. $\dfrac{x^2}{x + 1}$

$x - 1 + \dfrac{1}{x + 1}$

40. $\dfrac{x^2 + 9}{x + 3}$

$x - 3 + \dfrac{18}{x + 3}$

41. $\dfrac{x^3}{x + 2}$

$x^2 - 2x + 4 + \dfrac{-8}{x + 2}$

42. $\dfrac{x^3 - 1}{x - 2}$

$x^2 + 2x + 4 + \dfrac{7}{x - 2}$

43. $\dfrac{x^3 + 2x}{x^2}$

$x + \dfrac{2}{x}$

44. $\dfrac{2x^2 + 3}{2x}$

$x + \dfrac{3}{2x}$

45. $\dfrac{x^2 - 4x + 9}{x + 2}$

$x - 6 + \dfrac{21}{x + 2}$

46. $\dfrac{x^2 - 5x - 10}{x - 3}$

$x - 2 + \dfrac{-16}{x - 3}$

47. $\dfrac{3x^3 - 4x^2 + 7}{x - 1}$

$3x^2 - x - 1 + \dfrac{6}{x - 1}$

48. $\dfrac{-2x^3 + x^2 - 3}{x + 2}$

$-2x^2 + 5x - 10 + \dfrac{17}{x + 2}$

Use synthetic division to find the quotient and remainder when the first polynomial is divided by the second. See Example 4.

49. $x^3 - 5x^2 + 6x - 3$, $x - 2$
$x^2 - 3x, -3$

50. $x^3 + 6x^2 - 3x - 5$, $x - 3$
$x^2 + 9x + 24, 67$

51. $2x^2 - 4x + 5$, $x + 1$ $2x - 6, 11$

52. $3x^2 - 7x + 4$, $x + 2$ $3x - 13, 30$

53. $3x^4 - 15x^2 + 7x - 9$, $x - 3$
$3x^3 + 9x^2 + 12x + 43, 120$

54. $-2x^4 + 3x^2 - 5$, $x - 2$
$-2x^3 - 4x^2 - 5x - 10, -25$

55. $x^5 - 1$, $x - 1$
$x^4 + x^3 + x^2 + x + 1, 0$

56. $x^6 - 1$, $x + 1$
$x^5 - x^4 + x^3 - x^2 + x - 1, 0$

57. $x^3 - 5x + 6$, $x + 2$
$x^2 - 2x - 1, 8$

58. $x^3 - 3x - 7$, $x - 4$
$x^2 + 4x + 13, 45$

 59. $2.3x^2 - 0.14x + 0.6$, $x - 0.32$
$2.3x + 0.596, 0.79072$

 60. $1.6x^2 - 3.5x + 4.7$, $x + 1.8$
$1.6x - 6.38, 16.184$

For each pair of polynomials, determine whether the first polynomial is a factor of the second. Use synthetic division when possible. See Examples 5 and 6.

61. $x + 4$, $x^3 + x^2 - 11x + 8$ No

62. $x + 4$, $x^3 + x^2 + x + 48$ No

63. $x - 4$, $x^3 - 13x - 12$ Yes

64. $x - 1$, $x^3 + 3x^2 - 5x$ No

65. $2x - 3$, $2x^3 - 3x^2 - 4x + 6$ Yes

66. $3x - 5$, $6x^2 - 7x - 6$ No

67. $3w + 1$, $27w^3 + 1$ Yes

68. $2w + 3$, $8w^3 + 27$ Yes

69. $a - 5$, $a^3 - 125$ Yes

70. $a - 2$, $a^6 - 64$ Yes

71. $x^2 - 2$, $x^4 + 3x^3 - 6x - 4$ Yes

72. $x^2 - 3$, $x^4 + 2x^3 - 4x^2 - 6x + 3$ Yes

Factor each polynomial given that the binomial following each polynomial is a factor of the polynomial.

73. $x^2 - 6x + 8$, $x - 4$ $(x - 4)(x - 2)$

74. $x^2 + 3x - 40$, $x + 8$ $(x + 8)(x - 5)$

75. $w^3 - 27$, $w - 3$ $(w - 3)(w^2 + 3w + 9)$

76. $w^3 + 125$, $w + 5$ $(w + 5)(w^2 - 5w + 25)$

77. $x^3 - 4x^2 + 6x - 4$, $x - 2$ $(x^2 - 2x + 2)(x - 2)$

78. $2x^3 + 5x + 7$, $x + 1$ $(2x^2 - 2x + 7)(x + 1)$

79. $z^2 + 6z + 9$, $z + 3$ $(z + 3)(z + 3)$

80. $4a^2 - 20a + 25$, $2a - 5$ $(2a - 5)(2a - 5)$

81. $6y^2 + 5y + 1$, $2y + 1$ $(2y + 1)(3y + 1)$

82. $12y^2 - y - 6$, $4y - 3$ $(4y - 3)(3y + 2)$

Solve each problem.

83. *Average cost.* The total cost in dollars for manufacturing x professional racing bicycles in one week is given by the polynomial function

$$C(x) = 0.03x^2 + 300x.$$

The average cost per bicycle is given by

$$AC(x) = \frac{C(x)}{x}.$$

a) Find a formula for $AC(x)$. $AC(x) = 0.03x + 300$

b) Is $AC(x)$ a constant function?   No

c) Why does the average cost look constant in the accompanying figure?
Because $AC(x)$ is very close to 300 for x less than 15, the graph looks horizontal.

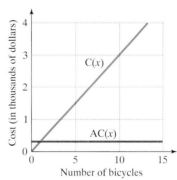

FIGURE FOR EXERCISE 83

84. *Average profit.* The weekly profit in dollars for manufacturing x bicycles is given by the polynomial $P(x) = 100x + 2x^2$. The average profit per bicycle is given by $AP(x) = \frac{P(x)}{x}$. Find $AP(x)$. Find the average profit per bicycle when 12 bicycles are manufactured.
$AP(x) = 100 + 2x$, $124

85. *Area of a poster.* The area of a rectangular poster advertising a Pearl Jam concert is $x^2 - 1$ square feet. If the length is $x + 1$ feet, then what is the width? $x - 1$ feet

86. *Volume of a box.* The volume of a shipping crate is $h^3 + 5h^2 + 6h$. If the height is h and the length is $h + 2$, then what is the width? $h + 3$

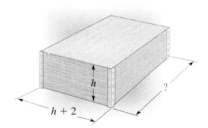

FIGURE FOR EXERCISE 86

87. *Volume of a pyramid.* Ancient Egyptian pyramid builders knew that the volume of the truncated pyramid shown in the figure on the next page is given by

$$V = \frac{H(a^3 - b^3)}{3(a - b)},$$

where a^2 is the area of the square base, b^2 is the area of the square top, and H is the distance from the base to the top. Find the volume of a truncated pyramid that has a base of 900 square meters, a top of 400 square meters, and a height H of 10 meters.

6,333.3 cubic meters

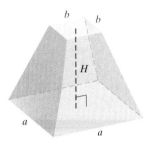

FIGURE FOR EXERCISE 87

88. *Egyptian pyramid formula.* Rewrite the formula of the previous exercise so that the denominator contains the number 3 only.

$$V = \frac{H(a^2 + ab + b^2)}{3}$$

GETTING MORE INVOLVED

 89. *Discussion.* On a test a student divided $3x^3 - 5x^2 - 3x + 7$ by $x - 3$ and got a quotient of $3x^2 + 4x$ and remainder $9x + 7$. Verify that the divisor times the quotient plus the remainder is equal to the dividend. Why was the student's answer incorrect?

 90. *Exploration.* Use synthetic division to find the quotient when $x^5 - 1$ is divided by $x - 1$ and the quotient when $x^6 - 1$ is divided by $x - 1$. Observe the pattern in the first two quotients and then write the quotient for $x^9 - 1$ divided by $x - 1$ without dividing.

5.6 FACTORING POLYNOMIALS

In this section

- Factoring Out the Greatest Common Factor (GCF)
- Factoring Out the Opposite of the GCF
- Factoring the Difference of Two Squares
- Factoring Perfect Square Trinomials
- Factoring a Difference or Sum of Two Cubes
- Factoring a Polynomial Completely
- Factoring by Substitution

In Section 5.5 you learned that a polynomial could be factored by using division: If we know one factor of a polynomial, then we can use it as a divisor to obtain the other factor, the quotient. However, this technique is not very practical because the division process can be somewhat tedious, and it is not easy to obtain a factor to use as the divisor. In this section and the next two sections we will develop better techniques for factoring polynomials. These techniques will be used for solving equations and problems in the last section of this chapter.

Factoring Out the Greatest Common Factor (GCF)

A natural number larger than 1 that has no factors other than itself and 1 is called a **prime number.** The numbers

$$2, 3, 5, 7, 11, 13, 17, 19, 23$$

are the first nine prime numbers. There are infinitely many prime numbers.

To factor a natural number **completely** means to write it as a product of prime numbers. In factoring 12 we might write $12 = 4 \cdot 3$. However, 12 is not factored completely as $4 \cdot 3$ because 4 is not a prime. To factor 12 completely, we write $12 = 2 \cdot 2 \cdot 3$ (or $2^2 \cdot 3$).

We use the distributive property to multiply a monomial and a binomial:

$$6x(2x - 1) = 12x^2 - 6x$$

If we start with $12x^2 - 6x$, we can use the distributive property to get

$$12x^2 - 6x = 6x(2x - 1).$$

We have **factored out** $6x$, which is a common factor of $12x^2$ and $-6x$. We could have factored out just 3 to get

$$12x^2 - 6x = 3(4x^2 - 2x),$$

but this would not be factoring out the *greatest* common factor. The **greatest common factor** (GCF) is a monomial that includes every number or variable that is a factor of all of the terms of the polynomial.

We can use the following strategy for finding the greatest common factor of a group of terms.

Strategy for Finding the Greatest Common Factor (GCF)

1. Factor each term completely.
2. Write a product using each factor that is common to all of the terms.
3. On each of these factors, use an exponent equal to the smallest exponent that appears on that factor in any of the terms.

E X A M P L E   1

The greatest common factor

Find the greatest common factor (GCF) for each group of terms.

a) $8x^2y, 20xy^3$ **b)** $30a^2, 45a^3b^2, 75a^4b$

Solution

a) First factor each term completely:

$$8x^2y = 2^3x^2y$$
$$20xy^3 = 2^2 \cdot 5xy^3$$

The factors common to both terms are $2, x$, and y. In the GCF we use the smallest exponent that appears on each factor in either of the terms. So the GCF is 2^2xy or $4xy$.

b) First factor each term completely:

$$30a^2 = 2 \cdot 3 \cdot 5a^2$$
$$45a^3b^2 = 3^2 \cdot 5a^3b^2$$
$$75a^4b = 3 \cdot 5^2a^4b$$

The GCF is $3 \cdot 5a^2$ or $15a^2$.

To factor out the GCF from a polynomial, find the GCF for the terms, then use the distributive property to factor it out.

E X A M P L E   2

Factoring out the greatest common factor

Factor each polynomial by factoring out the GCF.

a) $5x^4 - 10x^3 + 15x^2$ **b)** $8xy^2 + 20x^2y$ **c)** $60x^5 + 24x^3 + 36x^2$

Solution

a) First factor each term completely:

$$5x^4 = 5x^4, \qquad 10x^3 = 2 \cdot 5x^3, \qquad 15x^2 = 3 \cdot 5x^2.$$

The GCF of the three terms is $5x^2$. Now factor $5x^2$ out of each term:

$$5x^4 - 10x^3 + 15x^2 = 5x^2(x^2 - 2x + 3)$$

b) The GCF for $8xy^2$ and $20x^2y$ is $4xy$:

$$8xy^2 + 20x^2y = 4xy(2y + 5x)$$

c) First factor each coefficient in $60x^5 + 24x^3 + 36x^2$:

$$60 = 2^2 \cdot 3 \cdot 5, \qquad 24 = 2^3 \cdot 3, \qquad 36 = 2^2 \cdot 3^2.$$

The GCF of the three terms is $2^2 \cdot 3x^2$ or $12x^2$:

$$60x^5 + 24x^3 + 36x^2 = 12x^2(5x^3 + 2x + 3)$$

In the next example the common factor in each term is a binomial.

E X A M P L E   3   **Factoring out a binomial**

Factor.

a) $(x + 3)w + (x + 3)a$ b) $x(x - 9) - 4(x - 9)$

Solution

a) We treat $x + 3$ like a common monomial when factoring:

$$(x + 3)w + (x + 3)a = (x + 3)(w + a)$$

b) Factor out the common binomial $x - 9$:

$$x(x - 9) - 4(x - 9) = (x - 4)(x - 9)$$

Factoring Out the Opposite of the GCF

The GCF, the greatest common factor, for $-6x^2 - 4x$ is $2x$, but we can factor out either $2x$ or its opposite, $-2x$:

$$-6x^2 - 4x = 2x(-3x - 2)$$
$$= -2x(3x + 2)$$

In Example 8 of this section it will be necessary to factor out the opposite of the GCF.

E X A M P L E   4   **Factoring out the opposite of the GCF**

Factor out the GCF, then factor out the opposite of the GCF.

a) $5x - 5y$ b) $-x^2 - 3$ c) $-x^3 + 3x^2 - 5x$

Solution

a) $5x - 5y = 5(x - y)$ Factor out 5.

$\qquad\quad = -5(-x + y)$ Factor out -5.

b) $-x^2 - 3 = 1(-x^2 - 3)$ The GCF is 1.

$\qquad\quad = -1(x^2 + 3)$ Factor out -1.

c) $-x^3 + 3x^2 - 5x = x(-x^2 + 3x - 5)$ Factor out x.

$\qquad\qquad\qquad = -x(x^2 - 3x + 5)$ Factor out $-x$.

Factoring the Difference of Two Squares

A first-degree polynomial in one variable, such as $3x - 5$, is called a linear polynomial. (The equation $3x - 5 = 0$ is a linear equation.)

Linear Polynomial

If a and b are real numbers with $a \neq 0$, then $ax + b$ is called a **linear polynomial.**

A second-degree polynomial such as $x^2 + 5x - 6$ is called a quadratic polynomial.

Quadratic Polynomial

If a, b, and c are real numbers with $a \neq 0$, then $ax^2 + bx + c$ is called a **quadratic polynomial.**

helpful hint

The prefix "quad" means four. So why is a polynomial of three terms called quadratic? Perhaps it is because a quadratic polynomial can often be factored into a product of two binomials.

One of the main goals of this chapter is to write a quadratic polynomial (when possible) as a product of linear factors.

Consider the quadratic polynomial $x^2 - 25$. We recognize that $x^2 - 25$ is a difference of two squares, $x^2 - 5^2$. We recall that the product of a sum and a difference is a difference of two squares: $(a + b)(a - b) = a^2 - b^2$. If we reverse this special product rule, we get a rule for factoring the difference of two squares.

Factoring the Difference of Two Squares

$$a^2 - b^2 = (a + b)(a - b)$$

The difference of two squares factors as the product of a sum and a difference. To factor $x^2 - 25$, we replace a by x and b by 5 to get

$$x^2 - 25 = (x + 5)(x - 5).$$

This equation expresses a quadratic polynomial as a product of two linear factors.

E X A M P L E   5

Factoring the difference of two squares

Factor each polynomial.

a) $y^2 - 36$ **b)** $9x^2 - 1$ **c)** $4x^2 - y^2$

Solution

Each of these binomials is a difference of two squares. Each binomial factors into a product of a sum and a difference.

a) $y^2 - 36 = (y + 6)(y - 6)$ We could also write $(y - 6)(y + 6)$ because the factors can be written in any order.

b) $9x^2 - 1 = (3x + 1)(3x - 1)$

c) $4x^2 - y^2 = (2x + y)(2x - y)$ ■

Factoring Perfect Square Trinomials

The trinomial that results from squaring a binomial is called a **perfect square trinomial.** We can reverse the rules from Section 5.4 for the square of a sum or a difference to get rules for factoring.

Factoring Perfect Square Trinomials

$$a^2 + 2ab + b^2 = (a + b)^2$$
$$a^2 - 2ab + b^2 = (a - b)^2$$

Consider the polynomial $x^2 + 6x + 9$. If we recognize that

$$x^2 + 6x + 9 = x^2 + 2 \cdot x \cdot 3 + 3^2,$$

then we can see that it is a perfect square trinomial. It fits the rule if $a = x$ and $b = 3$:

$$x^2 + 6x + 9 = (x + 3)^2$$

Perfect square trinomials can be identified by using the following strategy.

> **Strategy for Identifying Perfect Square Trinomials**
>
> A trinomial is a perfect square trinomial if
>
> **1.** the first and last terms are of the form a^2 and b^2,
>
> **2.** the middle term is 2 or -2 times the product of a and b.

We use this strategy in the next example.

E X A M P L E   6   **Factoring perfect square trinomials**

Factor each polynomial.

a) $x^2 - 8x + 16$ **b)** $a^2 + 14a + 49$ **c)** $4x^2 + 12x + 9$

Solution

a) Because the first term is x^2, the last is 4^2, and $-2(x)(4)$ is equal to the middle term $-8x$, the trinomial $x^2 - 8x + 16$ is a perfect square trinomial:
$$x^2 - 8x + 16 = (x - 4)^2$$

b) Because $49 = 7^2$ and $14a = 2(a)(7)$, we have a perfect square trinomial:
$$a^2 + 14a + 49 = (a + 7)^2$$

c) Because $4x^2 = (2x)^2$, $9 = 3^2$, and the middle term $12x$ is equal to $2(2x)(3)$, the trinomial $4x^2 + 12x + 9$ is a perfect square trinomial:
$$4x^2 + 12x + 9 = (2x + 3)^2 \qquad ■$$

Factoring a Difference or a Sum of Two Cubes

In Example 6 of Section 5.5 we divided $a^3 - b^3$ by $a - b$ to get the quotient $a^2 + ab + b^2$ and no remainder. So $a - b$ is a factor of $a^3 - b^3$, a difference of two cubes. If you divide $a^3 + b^3$ by $a + b$, you will get the quotient $a^2 - ab + b^2$ and no remainder. Try it. So $a + b$ is a factor of $a^3 + b^3$, a sum of two cubes. These results give us two more factoring rules.

Factoring a Difference or a Sum of Two Cubes

$$a^3 - b^3 = (a - b)(a^2 + ab + b^2)$$
$$a^3 + b^3 = (a + b)(a^2 - ab + b^2)$$

E X A M P L E   7   **Factoring a difference or a sum of two cubes**

Factor each polynomial.

a) $x^3 - 8$ **b)** $y^3 + 1$ **c)** $8z^3 - 27$

Solution

a) Because $8 = 2^3$, we can use the formula for factoring the difference of two cubes. In the formula $a^3 - b^3 = (a - b)(a^2 + ab + b^2)$, let $a = x$ and $b = 2$:
$$x^3 - 8 = (x - 2)(x^2 + 2x + 4)$$

b) $y^3 + 1 = y^3 + 1^3$ Recognize a sum of two cubes.

$\qquad\qquad = (y + 1)(y^2 - y + 1)$ Let $a = y$ and $b = 1$ in the formula for the sum of two cubes.

c) $8z^3 - 27 = (2z)^3 - 3^3$ Recognize a difference of two cubes.

$\qquad\qquad = (2z - 3)(4z^2 + 6z + 9)$ Let $a = 2z$ and $b = 3$ in the formula for a difference of two cubes.

Factoring a Polynomial Completely

Polynomials that cannot be factored are called **prime polynomials.** Because binomials such as $x + 5$, $a - 6$, and $3x + 1$ cannot be factored, they are prime polynomials. A polynomial is **factored completely** when it is written as a product of prime polynomials. To factor completely, always factor out the GCF (or its opposite) first. Then continue to factor until all of the factors are prime.

E X A M P L E 8

Factoring completely

Factor each polynomial completely.

a) $5x^2 - 20$ **b)** $3a^3 - 30a^2 + 75a$ **c)** $-2b^4 + 16b$

Solution

a) $5x^2 - 20 = 5(x^2 - 4)$ Greatest common factor

$\qquad\qquad\quad = 5(x - 2)(x + 2)$ Difference of two squares

b) $3a^3 - 30a^2 + 75a = 3a(a^2 - 10a + 25)$ Greatest common factor

$\qquad\qquad\qquad\qquad = 3a(a - 5)^2$ Perfect square trinomial

c) $-2b^4 + 16b = -2b(b^3 - 8)$ Factor out $-2b$ to make the next step easier.

$\qquad\qquad\quad = -2b(b - 2)(b^2 + 2b + 4)$ Difference of two cubes

Factoring by Substitution

So far, the polynomials that we have factored, without common factors, have all been of degree 2 or 3. Some polynomials of higher degree can be factored by substituting a single variable for a variable with a higher power. After factoring, we replace the single variable by the higher-power variable. This method is called **substitution.**

E X A M P L E 9

Factoring by substitution

Factor each polynomial.

a) $x^4 - 9$ **b)** $y^8 - 14y^4 + 49$

Solution

a) We recognize $x^4 - 9$ as a difference of two squares in which $x^4 = (x^2)^2$ and $9 = 3^2$. If we let $w = x^2$, then $w^2 = x^4$. So we can replace x^4 by w^2 and factor:

$\qquad x^4 - 9 = w^2 - 9$ Replace x^4 by w^2.

$\qquad\qquad\quad = (w + 3)(w - 3)$ Difference of two squares

$\qquad\qquad\quad = (x^2 + 3)(x^2 - 3)$ Replace w by x^2.

b) We recognize $y^8 - 14y^4 + 49$ as a perfect square trinomial in which $y^8 = (y^4)^2$ and $49 = 7^2$. We let $w = y^4$ and $w^2 = y^8$:

$\qquad y^8 - 14y^4 + 49 = w^2 - 14w + 49$ Replace y^4 by w and y^8 by w^2.

$\qquad\qquad\qquad\qquad = (w - 7)^2$ Perfect square trinomial

$\qquad\qquad\qquad\qquad = (y^4 - 7)^2$ Replace w by y^4.

helpful hint

It is not actually necessary to perform the substitution step. If you can recognize that

$x^4 - 9 = (x^2 - 3)(x^2 + 3)$

then skip the substitution.

CAUTION The polynomials that we factor by substitution must contain just the right powers of the variable. We can factor $y^8 - 14y^4 + 49$ because $(y^4)^2 = y^8$, but we cannot factor $y^7 - 14y^4 + 49$ by substitution.

In the next example we use substitution to factor polynomials that have variables as exponents.

E X A M P L E   1 0 **Polynomials with variable exponents**

Factor completely. The variables used in the exponents represent positive integers.

a) $x^{2m} - y^2$ 　　　　　　　　　　 **b)** $z^{2n+1} - 6z^{n+1} + 9z$

Solution

a) Notice that $x^{2m} = (x^m)^2$. So if we let $w = x^m$, then $w^2 = x^{2m}$:

$$
\begin{aligned}
x^{2m} - y^2 &= w^2 - y^2 && \text{Substitution} \\
&= (w + y)(w - y) && \text{Difference of two squares} \\
&= (x^m + y)(x^m - y) && \text{Replace } w \text{ by } x^m.
\end{aligned}
$$

b) First factor out the common factor z:

$$
\begin{aligned}
z^{2n+1} - 6z^{n+1} + 9z &= z(z^{2n} - 6z^n + 9) \\
&= z(a^2 - 6a + 9) && \text{Let } a = z^n. \\
&= z(a - 3)^2 && \text{Perfect square trinomial} \\
&= z(z^n - 3)^2 && \text{Replace } a \text{ by } z^n.
\end{aligned}
$$

WARM-UPS

True or false? Explain your answer.

1. For the polynomial $3x^2y - 6xy^2$ we can factor out either $3xy$ or $-3xy$.　　True

2. The greatest common factor for the polynomial $8a^3 - 15b^2$ is 1.　　True

3. $2x - 4 = -2(2 - x)$ for any value of x.　　True

4. $x^2 - 16 = (x - 4)(x + 4)$ for any value of x.　　True

5. The polynomial $x^2 + 6x + 36$ is a perfect square trinomial.　　False

6. The polynomial $y^2 + 16$ is a perfect square trinomial.　　False

7. $9x^2 + 21x + 49 = (3x + 7)^2$ for any value of x.　　False

8. The polynomial $x + 1$ is a factor of $x^3 + 1$.　　True

9. $x^3 - 27 = (x - 3)(x^2 + 6x + 9)$ for any value of x.　　False

10. $x^3 - 8 = (x - 2)^3$ for any value of x.　　False

5.6   EXERCISES

Reading and Writing   After reading this section, write out the answers to these questions. Use complete sentences.

1. What is a prime number?

A prime number is a natural number greater than 1 that has no factors other than itself and 1.

2. When is a natural number factored completely?

A natural number is factored completely when it is expressed as a product of prime numbers.

3. What is the greatest common factor for the terms of a polynomial?

The greatest common factor for the terms of a polynomial is a monomial that includes every number or variable that is a factor of all of the terms of the polynomial.

4. What are the two ways to factor out the greatest common factor?

The greatest common factor can be factored out with a positive coefficient or a negative coefficient.

5. What is a linear polynomial?

A linear polynomial is a polynomial of the form $ax + b$ with $a \neq 0$.

6. What is a quadratic polynomial?

A quadratic polynomial is a polynomial of the form $ax^2 + bx + c$ with $a \neq 0$.

7. What is a prime polynomial?

A prime polynomial is a polynomial that cannot be factored.

8. When is a polynomial factored completely?

A polynomial is factored completely when it is expressed as a product of prime polynomials.

Find the greatest common factor for each group of terms. See Example 1.

9. $48, 36x$ 12

10. $42a, 28a^2$ $14a$

11. $9wx, 21wy, 15xy$ 3

12. $70x^2, 84x, 42x^3$ $14x$

13. $24x^2y, 42xy^2, 66xy^3$ $6xy$

14. $60a^2b^5, 140a^9b^2, 40a^3b^6$ $20a^2b^2$

Factor out the greatest common factor in each expression. See Examples 2 and 3.

15. $x^3 - 5x$ $x(x^2 - 5)$

16. $10x^2 - 20y^3$ $10(x^2 - 2y^3)$

17. $48wx + 36wy$ $12w(4x + 3y)$

18. $42wz + 28wa$ $14w(3z + 2a)$

19. $2x^3 - 4x^2 + 6x$ $2x(x^2 - 2x + 3)$

20. $6x^3 - 12x^2 + 18x$ $6x(x^2 - 2x + 3)$

21. $36a^3b^6 - 24a^4b^2 + 60a^5b^3$ $12a^3b^2(3b^4 - 2a + 5a^2b)$

22. $44x^8y^6z - 110x^6y^9z^2$ $22x^6y^6z(2x^2 - 5y^3z)$

23. $(x - 6)a + (x - 6)b$ $(x - 6)(a + b)$

24. $(y - 4)3 + (y - 4)b$ $(y - 4)(3 + b)$

25. $(y - 1)^2y + (y - 1)^2z$ $(y - 1)^2(y + z)$

26. $(w - 2)^2 \cdot w + (w - 2)^2 \cdot 3$ $(w - 2)^2(w + 3)$

Factor out the greatest common factor, then factor out the opposite of the greatest common factor. See Example 4.

27. $2x - 2y$ $2(x - y), -2(-x + y)$

28. $-3x + 6$ $3(-x + 2), -3(x - 2)$

29. $6x^2 - 3x$ $3x(2x - 1), -3x(-2x + 1)$

30. $10x^2 + 5x$ $5x(2x + 1), -5x(-2x - 1)$

31. $-w^3 + 3w^2$ $w^2(-w + 3), -w^2(w - 3)$

32. $-2w^4 + 6w^3$ $2w^3(-w + 3), -2w^3(w - 3)$

33. $-a^3 + a^2 - 7a$ $a(-a^2 + a - 7), -a(a^2 - a + 7)$

34. $-2a^4 - 4a^3 + 6a^2$
$2a^2(-a^2 - 2a + 3), -2a^2(a^2 + 2a - 3)$

Factor each polynomial. See Example 5.

35. $x^2 - 100$ $(x - 10)(x + 10)$

36. $81 - y^2$ $(9 - y)(9 + y)$

37. $4y^2 - 49$ $(2y - 7)(2y + 7)$

38. $16b^2 - 1$ $(4b - 1)(4b + 1)$

39. $9x^2 - 25a^2$ $(3x - 5a)(3x + 5a)$

40. $121a^2 - b^2$ $(11a - b)(11a + b)$

41. $144w^2z^2 - 1$ $(12wz - 1)(12wz + 1)$

42. $x^2y^2 - 9c^2$ $(xy - 3c)(xy + 3c)$

Factor each polynomial. See Example 6.

43. $x^2 - 20x + 100$ $(x - 10)^2$

44. $y^2 + 10y + 25$ $(y + 5)^2$

45. $4m^2 - 4m + 1$ $(2m - 1)^2$

46. $9t^2 + 30t + 25$ $(3t + 5)^2$

47. $w^2 - 2wt + t^2$ $(w - t)^2$

48. $4r^2 + 20rt + 25t^2$ $(2r + 5t)^2$

Factor. See Example 7.

49. $a^3 - 1$ $(a - 1)(a^2 + a + 1)$

50. $w^3 + 1$ $(w + 1)(w^2 - w + 1)$

51. $w^3 + 27$ $(w + 3)(w^2 - 3w + 9)$

52. $x^3 - 64$ $(x - 4)(x^2 + 4x + 16)$

53. $8x^3 - 1$ $(2x - 1)(4x^2 + 2x + 1)$

54. $27x^3 + 1$ $(3x + 1)(9x^2 - 3x + 1)$

55. $a^3 + 8$ $(a + 2)(a^2 - 2a + 4)$

56. $m^3 - 8$ $(m - 2)(m^2 + 2m + 4)$

Factor each polynomial completely. See Example 8.

57. $2x^2 - 8$ $2(x + 2)(x - 2)$

58. $3x^3 - 27x$ $3x(x - 3)(x + 3)$

59. $x^3 + 10x^2 + 25x$ $x(x + 5)^2$

60. $5a^4m - 45a^2m$ $5a^2m(a - 3)(a + 3)$

61. $4x^2 + 4x + 1$ $(2x + 1)^2$

62. $ax^2 - 8ax + 16a$ $a(x - 4)^2$

63. $(x + 3)x + (x + 3)7$ $(x + 3)(x + 7)$

64. $(x - 2)x - (x - 2)5$ $(x - 2)(x - 5)$

65. $6y^2 + 3y$ $3y(2y + 1)$

66. $4y^2 - y$ $y(4y - 1)$

67. $4x^2 - 20x + 25$ $(2x - 5)^2$

68. $a^3x^3 - 6a^2x^2 + 9ax$ $ax(ax - 3)^2$

69. $2m^4 - 2mn^3$ $2m(m - n)(m^2 + mn + n^2)$

70. $5x^3y^2 - y^5$ $y^2(5x^3 - y^3)$

71. $(2x - 3)x - (2x - 3)2$ $(2x - 3)(x - 2)$

72. $(2x + 1)x + (2x + 1)3$ $(2x + 1)(x + 3)$

73. $9a^3 - aw^2$ $a(3a + w)(3a - w)$

74. $2bn^2 - 4b^2n + 2b^3$ $2b(n - b)^2$

75. $-5a^2 + 30a - 45$ $-5(a - 3)^2$

76. $-2x^2 + 50$ $-2(x - 5)(x + 5)$

77. $16 - 54x^3$ $2(2 - 3x)(4 + 6x + 9x^2)$

78. $27x^2y - 64x^2y^4$ $x^2y(3 - 4y)(9 + 12y + 16y^2)$

79. $-3y^3 - 18y^2 - 27y$ $-3y(y + 3)^2$

80. $-2m^2n - 8mn - 8n$ $-2n(m + 2)^2$

81. $-7a^2b^2 + 7$ $-7(ab + 1)(ab - 1)$

82. $-17a^2 - 17a$ $-17a(a + 1)$

Factor each polynomial completely. See Example 9.

83. $x^{10} - 9$ $(x^5 - 3)(x^5 + 3)$

84. $y^8 - 4$ $(y^4 - 2)(y^4 + 2)$

85. $z^{12} - 6z^6 + 9$ $(z^6 - 3)^2$

86. $a^6 + 10a^3 + 25$ $(a^3 + 5)^2$

87. $2x^7 + 8x^4 + 8x$ $2x(x^3 + 2)^2$

88. $x^{13} - 6x^7 + 9x$ $x(x^6 - 3)^2$

89. $4x^5 + 4x^3 + x$ $x(2x^2 + 1)^2$

90. $18x^6 + 24x^3 + 8$ $2(3x^3 + 2)^2$

91. $x^6 - 8$ $(x^2 - 2)(x^4 + 2x^2 + 4)$

92. $y^6 - 27$ $(y^2 - 3)(y^4 + 3y^2 + 9)$

93. $2x^9 + 16$ $2(x^3 + 2)(x^6 - 2x^3 + 4)$

94. $x^{13} + x$ $x(x^4 + 1)(x^8 - x^4 + 1)$

Factor each polynomial completely. The variables used as exponents represent positive integers. See Example 10.

95. $a^{2n} - 1$ $(a^n - 1)(a^n + 1)$

96. $b^{4n} - 9$ $(b^{2n} - 3)(b^{2n} + 3)$

97. $a^{2r} + 6a^r + 9$ $(a^r + 3)^2$

98. $u^{6n} - 4u^{3n} + 4$ $(u^{3n} - 2)^2$

99. $x^{3m} - 8$ $(x^m - 2)(x^{2m} + 2x^m + 4)$

100. $y^{3n} + 1$ $(y^n + 1)(y^{2n} - y^n + 1)$

101. $a^{3m} - b^3$ $(a^m - b)(a^{2m} + a^m b + b^2)$

102. $r^{3m} + 8t^3$ $(r^m + 2t)(r^{2m} - 2r^m t + 4t^2)$

103. $k^{2w+1} - 10k^{w+1} + 25k$ $k(k^w - 5)^2$

104. $4a^{2t+1} + 4a^{t+1} + a$ $a(2a^t + 1)^2$

105. $uv^{6k} - 2u^2v^{4k} + u^3v^{2k}$ $uv^{2k}(v^{2k} - u)^2$

106. $u^{3m}v^n + 2u^{2m}v^{2n} + u^m v^{3n}$ $u^m v^n (u^m + v^n)^2$

Replace k in each trinomial by a number that makes the trinomial a perfect square trinomial.

107. $x^2 + 6x + k$ 9 **108.** $y^2 - 8y + k$ 16

109. $4a^2 - ka + 25$ 20 **110.** $9u^2 + kuv + 49v^2$ 42

111. $km^2 - 24m + 9$ 16 **112.** $kz^2 + 40z + 16$ 25

113. $81y^2 - 180y + k$ 100 **114.** $36a^2 + 60a + k$ 25

Solve each problem.

115. *Volume of a bird cage.* A company makes rectangular shaped bird cages with height b inches and square bottoms. The volume of these cages is given by the function

$$V = b^3 - 6b^2 + 9b.$$

a) What is the length of a side of the square bottom?

b) Use the function to find the volume of a cage with a height of 18 inches.

c) Use the accompanying graph to estimate the height of a cage for which the volume is 20,000 cubic inches.
    a) $b - 3$ **b)** 4,050 cubic inches (in.^{3})
    c) 30 inches

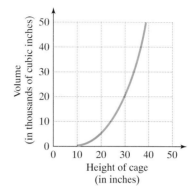

FIGURE FOR EXERCISE 115

116. *Pyramid power.* A powerful crystal pyramid has a square base and a volume of $3y^3 + 12y^2 + 12y$ cubic centimeters. If its height is y centimeters, then what polynomial represents the length of a side of the square base? $\left(\text{The volume of a pyramid with a square base of area } a^2 \text{ and height } h \text{ is given by } V = \frac{ha^2}{3}.\right)$
$3y + 6$ centimeters

FIGURE FOR EXERCISE 116

GETTING MORE INVOLVED

117. *Cooperative learning.* List the perfect square trinomials corresponding to $(x + 1)^2$, $(x + 2)^2$, $(x + 3)^2$, ..., $(x + 12)^2$. Use your list to quiz a classmate. Read a perfect square trinomial at random from your list and ask your classmate to write its factored form. Repeat until both of you have mastered these 12 perfect square trinomials.

5.7 FACTORING $ax^2 + bx + c$

In Section 5.5 you learned to factor certain special polynomials. In this section you will learn to factor general quadratic polynomials. We first factor $ax^2 + bx + c$ with $a = 1$, and then we consider the case $a \neq 1$.

Factoring Trinomials with Leading Coefficient 1

Let's look closely at an example of finding the product of two binomials using the distributive property:

$$(x + 3)(x + 4) = (x + 3)x + (x + 3)4 \quad \text{Distributive property}$$
$$= x^2 + 3x + 4x + 12 \quad \text{Distributive property}$$
$$= x^2 + 7x + 12$$

To factor $x^2 + 7x + 12$, we need to reverse these steps. First observe that the coefficient 7 is the sum of two numbers that have a product of 12. The only numbers that have a product of 12 and a sum of 7 are 3 and 4. So write $7x$ as $3x + 4x$:

$$x^2 + 7x + 12 = x^2 + 3x + 4x + 12$$

Now factor the common factor x out of the first two terms and the common factor 4 out of the last two terms. This method is called **factoring by grouping.**

$$
\begin{array}{c}
\overbrace{\text{Factor out } x} \quad \overbrace{\text{Factor out } 4} \\
x^2 + 7x + 12 = x^2 + 3x + 4x + 12 \quad \text{Rewrite } 7x \text{ as } 3x + 4x. \\
= (x + 3)x + (x + 3)4 \quad \text{Factor out common factors.} \\
= (x + 3)(x + 4) \quad \text{Factor out the common factor } x + 3.
\end{array}
$$

EXAMPLE 1

Factoring a trinomial by grouping

Factor each trinomial by grouping.

a) $x^2 + 9x + 18$ **b)** $x^2 - 2x - 24$

Solution

a) We need to find two integers with a product of 18 and a sum of 9. For a product of 18 we could use 1 and 18, 2 and 9, or 3 and 6. Only 3 and 6 have a sum of 9. So we replace $9x$ with $3x + 6x$ and factor by grouping:

$$x^2 + 9x + 18 = x^2 + 3x + 6x + 18 \quad \text{Replace } 9x \text{ by } 3x + 6x.$$
$$= (x + 3)x + (x + 3)6 \quad \text{Factor out common factors.}$$
$$= (x + 3)(x + 6) \quad \text{Check by using FOIL.}$$

b) We need to find two integers with a product of -24 and a sum of -2. For a product of 24 we have 1 and 24, 2 and 12, 3 and 8, or 4 and 6. To get a product of -24 and a sum of -2, we must use 4 and -6:

$$x^2 - 2x - 24 = x^2 - 6x + 4x - 24 \quad \text{Replace } -2x \text{ with } -6x + 4x.$$
$$= (x - 6)x + (x - 6)4 \quad \text{Factor out common factors.}$$
$$= (x - 6)(x + 4) \quad \text{Check by using FOIL.} \quad ■$$

The method shown in Example 1 can be shortened greatly. Once we discover that 3 and 6 have a product of 18 and a sum of 9, we can simply write

$$x^2 + 9x + 18 = (x + 3)(x + 6).$$

Once we discover that 4 and -6 have a product of -24 and a sum of -2, we can simply write

$$x^2 - 2x - 24 = (x - 6)(x + 4).$$

In the next example we use this shortcut.

E X A M P L E   2   **Factoring $ax^2 + bx + c$ with $a = 1$**

Factor each quadratic polynomial.

a) $x^2 + 4x + 3$ **b)** $x^2 + 3x - 10$ **c)** $a^2 - 5a + 6$

Solution

a) Two integers with a product of 3 and a sum of 4 are 1 and 3:
$$x^2 + 4x + 3 = (x + 1)(x + 3)$$
Check by using FOIL.

b) Two integers with a product of -10 and a sum of 3 are 5 and -2:
$$x^2 + 3x - 10 = (x + 5)(x - 2)$$
Check by using FOIL.

c) Two integers with a product of 6 and a sum of -5 are -3 and -2:
$$a^2 - 5a + 6 = (a - 3)(a - 2)$$
Check by using FOIL.

Factoring Trinomials with Leading Coefficient Not 1

If the leading coefficient of a quadratic trinomial is not 1, we can again use grouping to factor the trinomial. However, the procedure is slightly different.

Consider the trinomial $2x^2 + 11x + 12$, for which $a = 2$, $b = 11$, and $c = 12$. First find ac, the product of the leading coefficient and the constant term. In this case $ac = 2 \cdot 12 = 24$. Now find two integers with a product of 24 and a sum of 11. The pairs of integers with a product of 24 are 1 and 24, 2 and 12, 3 and 8, and 4 and 6. Only 3 and 8 have a product of 24 and a sum of 11. Now replace $11x$ by $3x + 8x$ and factor by grouping:

$$\begin{aligned} 2x^2 + 11x + 12 &= 2x^2 + 3x + 8x + 12 \\ &= (2x + 3)x + (2x + 3)4 \\ &= (2x + 3)(x + 4) \end{aligned}$$

This strategy for factoring a quadratic trinomial, known as the *ac* **method,** is summarized in the following box. The *ac* method works also when $a = 1$.

Strategy for Factoring $ax^2 + bx + c$ by the *ac* Method

To factor the trinomial $ax^2 + bx + c$

1. find two integers that have a product equal to ac and a sum equal to b,
2. replace bx by two terms using the two new integers as coefficients,
3. then factor the resulting four-term polynomial by grouping.

E X A M P L E   3

Factoring $ax^2 + bx + c$ with $a \neq 1$

Factor each trinomial.

a) $2x^2 + 9x + 4$ **b)** $2x^2 + 5x - 12$

Solution

a) Because $2 \cdot 4 = 8$, we need two numbers with a product of 8 and a sum of 9. The numbers are 1 and 8. Replace $9x$ by $x + 8x$ and factor by grouping:

$$2x^2 + 9x + 4 = 2x^2 + x + 8x + 4$$
$$= (2x + 1)x + (2x + 1)4$$
$$= (2x + 1)(x + 4) \quad \text{Check by FOIL.}$$

Note that if you start with $2x^2 + 8x + x + 4$, and factor by grouping, you get the same result.

b) Because $2(-12) = -24$, we need two numbers with a product of -24 and a sum of 5. The pairs of numbers with a product of 24 are 1 and 24, 2 and 12, 3 and 8, and 4 and 6. To get a product of -24, one of the numbers must be negative and the other positive. To get a sum of positive 5, we need -3 and 8:

$$2x^2 + 5x - 12 = 2x^2 - 3x + 8x - 12$$
$$= (2x - 3)x + (2x - 3)4$$
$$= (2x - 3)(x + 4) \quad \text{Check by FOIL.}$$

Trial and Error

After we have gained some experience at factoring by grouping, we can often find the factors without going through the steps of grouping. Consider the polynomial

$$2x^2 - 7x + 6.$$

The factors of $2x^2$ can only be $2x$ and x. The factors of 6 could be 2 and 3 or 1 and 6. We can list all of the possibilities that give the correct first and last terms without putting in the signs:

$$(2x \quad 2)(x \quad 3) \qquad (2x \quad 6)(x \quad 1)$$
$$(2x \quad 3)(x \quad 2) \qquad (2x \quad 1)(x \quad 6)$$

Before actually trying these out, we make an important observation. If $(2x \quad 2)$ or $(2x \quad 6)$ were one of the factors, then there would be a common factor 2 in the original trinomial, but there is not. *If the original trinomial has no common factor, there can be no common factor in either of its linear factors.* Since 6 is positive and the middle term is $-7x$, both of the missing signs must be negative. So the only possibilities are $(2x - 1)(x - 6)$ and $(2x - 3)(x - 2)$. The middle term of the first product is $-13x$, and the middle term of the second product is $-7x$. So we have found the factors:

$$2x^2 - 7x + 6 = (2x - 3)(x - 2)$$

Even though there may be many possibilities in some factoring problems, often we find the correct factors without writing down every possibility. We can use a bit of guesswork in factoring trinomials. *Try* whichever possibility you think might work. *Check* it by multiplying. If it is not right, then *try again.* That is why this method is called **trial and error.**

E X A M P L E   4

Trial and error

Factor each quadratic trinomial using trial and error.

a) $2x^2 + 5x - 3$ **b)** $3x^2 - 11x + 6$

Solution

a) Because $2x^2$ factors only as $2x \cdot x$ and 3 factors only as $1 \cdot 3$, there are only two possible ways to factor this trinomial to get the correct first and last terms:

$$(2x \quad 1)(x \quad 3) \qquad \text{and} \qquad (2x \quad 3)(x \quad 1)$$

Because the last term of the trinomial is negative, one of the missing signs must be $+$, and the other must be $-$. Now we try the various possibilities until we get the correct middle term:

$$(2x + 1)(x - 3) = 2x^2 - 5x - 3$$
$$(2x + 3)(x - 1) = 2x^2 + x - 3$$
$$(2x - 1)(x + 3) = 2x^2 + 5x - 3$$

Since the last product has the correct middle term, the trinomial is factored as

$$2x^2 + 5x - 3 = (2x - 1)(x + 3).$$

b) There are four possible ways to factor $3x^2 - 11x + 6$:

$$(3x \quad 1)(x \quad 6) \qquad (3x \quad 2)(x \quad 3)$$
$$(3x \quad 6)(x \quad 1) \qquad (3x \quad 3)(x \quad 2)$$

Because the last term is positive and the middle term is negative, both signs must be negative. Now try possible factors until we get the correct middle term:

$$(3x - 1)(x - 6) = 3x^2 - 19x + 6$$
$$(3x - 2)(x - 3) = 3x^2 - 11x + 6$$

The trinomial is factored correctly as

$$3x^2 - 11x + 6 = (3x - 2)(x - 3).$$ ■

Higher Degrees and Variable Exponents

It is not necessary always to use substitution to factor polynomials with higher degrees or variable exponents as we did in Section 5.6. In the next example we use trial and error to factor two polynomials of higher degree and one with variable exponents. Remember that if there is a common factor to all terms, factor it out first.

E X A M P L E   5

Higher-degree and variable exponent trinomials

Factor each polynomial completely. Variables used as exponents represent positive integers.

a) $x^8 - 2x^4 - 15$ **b)** $-18y^7 + 21y^4 + 15y$ **c)** $2u^{2m} - 5u^m - 3$

Solution

a) To factor by trial and error, notice that $x^8 = x^4 \cdot x^4$. Now 15 is $3 \cdot 5$ or $1 \cdot 15$. Using 1 and 15 will not give the required -2 for the coefficient of the middle term. So choose 3 and -5 to get the -2 in the middle term:

$$x^8 - 2x^4 - 15 = (x^4 - 5)(x^4 + 3)$$

b) $-18y^7 + 21y^4 + 15y = -3y(6y^6 - 7y^3 - 5)$ Factor out the common factor $-3y$ first.

$$= -3y(2y^3 + 1)(3y^3 - 5)$$ Factor the trinomial by trial and error.

c) Notice that $2u^{2m} = 2u^m \cdot u^m$ and $3 = 3 \cdot 1$. Using trial and error, we get

$$2u^{2m} - 5u^m - 3 = (2u^m + 1)(u^m - 3).$$ ■

WARM-UPS

True or false? Answer true if the polynomial is factored correctly and false otherwise.

1. $x^2 + 9x + 18 = (x + 3)(x + 6)$ True
2. $y^2 + 2y - 35 = (y + 5)(y - 7)$ False
3. $x^2 + 4 = (x + 2)(x + 2)$ False
4. $x^2 - 5x - 6 = (x - 3)(x - 2)$ False
5. $x^2 - 4x - 12 = (x - 6)(x + 2)$ True
6. $x^2 + 15x + 36 = (x + 4)(x + 9)$ False
7. $3x^2 + 4x - 15 = (3x + 5)(x - 3)$ False
8. $4x^2 + 4x - 3 = (4x - 1)(x + 3)$ False
9. $4x^2 - 4x - 3 = (2x + 1)(2x - 3)$ True
10. $4x^2 + 8x + 3 = (2x + 1)(2x + 3)$ True

5.7   EXERCISES

Reading and Writing   *After reading this section, write out the answers to these questions. Use complete sentences.*

1. How do we factor trinomials that have a leading coefficient of 1?
   To factor $x^2 + bx + c$, find two integers whose sum is b and whose product is c.

2. How do we factor trinomials in which the leading coefficient is not 1?
   To factor $ax^2 + bx + c$, find two integers whose product is ac and whose sum is b and then use grouping.

3. What is trial-and-error factoring?
   Trial and error means simply to write down possible factors and then to use FOIL to check until you get the correct factors.

4. What should you always first look for when factoring a polynomial?
   When factoring a polynomial, first factor out the greatest common factor.

Factor each polynomial. See Examples 1 and 2.

5. $x^2 + 4x + 3$
   $(x + 1)(x + 3)$
6. $y^2 + 5y + 6$
   $(y + 2)(y + 3)$
7. $a^2 + 15a + 50$
   $(a + 10)(a + 5)$
8. $t^2 + 11t + 24$
   $(t + 8)(t + 3)$
9. $y^2 - 5y - 14$
   $(y - 7)(y + 2)$
10. $x^2 - 3x - 18$
    $(x - 6)(x + 3)$
11. $x^2 - 6x + 8$
    $(x - 2)(x - 4)$
12. $y^2 - 13y + 30$
    $(y - 10)(y - 3)$
13. $a^2 - 12a + 27$
    $(a - 9)(a - 3)$
14. $x^2 - x - 30$
    $(x - 6)(x + 5)$
15. $a^2 + 7a - 30$
    $(a + 10)(a - 3)$
16. $w^2 + 29w - 30$
    $(w + 30)(w - 1)$

Factor each polynomial using the ac method. See Example 3.

17. $6w^2 + 5w + 1$
    $(3w + 1)(2w + 1)$
18. $4x^2 + 11x + 6$
    $(4x + 3)(x + 2)$
19. $2x^2 - 5x - 3$
    $(2x + 1)(x - 3)$
20. $2a^2 + 3a - 2$
    $(2a - 1)(a + 2)$
21. $4x^2 + 16x + 15$
    $(2x + 5)(2x + 3)$
22. $6y^2 + 17y + 12$
    $(2y + 3)(3y + 4)$
23. $6x^2 - 5x + 1$
    $(2x - 1)(3x - 1)$
24. $6m^2 - m - 12$
    $(3m + 4)(2m - 3)$
25. $12y^2 + y - 1$
    $(3y + 1)(4y - 1)$
26. $12x^2 + 5x - 2$
    $(4x - 1)(3x + 2)$
27. $6a^2 + a - 5$
    $(6a - 5)(a + 1)$
28. $30b^2 - b - 3$
    $(10b + 3)(3b - 1)$

Factor each polynomial using trial and error. See Example 4.

29. $2x^2 + 15x - 8$
    $(2x - 1)(x + 8)$
30. $3a^2 + 20a + 12$
    $(3a + 2)(a + 6)$
31. $3b^2 - 16b - 35$
    $(3b + 5)(b - 7)$
32. $2y^2 - 17y + 21$
    $(2y - 3)(y - 7)$
33. $6w^2 - 35w + 36$
    $(3w - 4)(2w - 9)$
34. $15x^2 - x - 6$
    $(3x - 2)(5x + 3)$
35. $4x^2 - 5x + 1$
    $(4x - 1)(x - 1)$
36. $4x^2 + 7x + 3$
    $(4x + 3)(x + 1)$
37. $5m^2 + 13m - 6$
    $(5m - 2)(m + 3)$
38. $5t^2 - 9t - 2$
    $(5t + 1)(t - 2)$
39. $6y^2 - 7y - 20$
    $(3y + 4)(2y - 5)$
40. $7u^2 + 11u - 6$
    $(7u - 3)(u + 2)$

Factor each polynomial completely. See Example 5. The variables used in exponents represent positive integers.

41. $x^6 - 2x^3 - 35$
    $(x^3 + 5)(x^3 - 7)$
42. $x^4 + 7x^2 - 30$
    $(x^2 + 10)(x^2 - 3)$
43. $a^{20} - 20a^{10} + 100$
    $(a^{10} - 10)^2$
44. $b^{16} + 22b^8 + 121$
    $(b^8 + 11)^2$

45. $-12a^5 - 10a^3 - 2a$
$-2a(3a^2 + 1)(2a^2 + 1)$

46. $-4b^7 + 4b^4 + 3b$
$-b(2b^3 - 3)(2b^3 + 1)$

47. $x^{2a} + 2x^a - 15$
$(x^a + 5)(x^a - 3)$

48. $y^{2b} + y^b - 20$
$(y^b + 5)(y^b - 4)$

49. $x^{2a} - y^{2b}$
$(x^a - y^b)(x^a + y^b)$

50. $w^{4m} - a^2$
$(w^{2m} - a)(w^{2m} + a)$

51. $x^8 - x^4 - 6$
$(x^4 - 3)(x^4 + 2)$

52. $m^{10} - 5m^5 - 6$
$(m^5 - 6)(m^5 + 1)$

53. $x^{a+2} - x^a$
$x^a(x - 1)(x + 1)$

54. $y^{2a+1} - y$
$y(y^a - 1)(y^a + 1)$

55. $x^{2a} + 6x^a + 9$
$(x^a + 3)^2$

56. $x^{2a} - 2x^a y^b + y^{2b}$
$(x^a - y^b)^2$

57. $4y^3z^6 + 5z^3y^2 - 6y$
$y(4yz^3 - 3)(yz^3 + 2)$

58. $2u^6v^6 + 5u^4v^3 - 12u^2$
$u^2(2u^2v^3 - 3)(u^2v^3 + 4)$

Factor each polynomial completely.

59. $2x^2 + 20x + 50$
$2(x + 5)^2$

60. $3a^2 + 6a + 3$
$3(a + 1)^2$

61. $a^3 - 36a$
$a(a - 6)(a + 6)$

62. $x^3 - 5x^2 - 6x$
$x(x - 6)(x + 1)$

63. $5a^2 + 25a - 30$
$5(a + 6)(a - 1)$

64. $2a^2 - 2a - 84$
$2(a - 7)(a + 6)$

65. $2x^2 - 128y^2$
$2(x + 8y)(x - 8y)$

66. $a^3 - 6a^2 + 9a$
$a(a - 3)^2$

67. $-3x^2 + 3x + 36$
$-3(x + 3)(x - 4)$

68. $xy^2 - 3xy - 70x$
$x(y - 10)(y + 7)$

69. $m^5 + 20m^4 + 100m^3$
$m^3(m + 10)^2$

70. $4a^2 - 16a + 16$
$4(a - 2)^2$

71. $6x^2 + 23x + 20$
$(3x + 4)(2x + 5)$

72. $2y^2 - 13y + 6$
$(2y - 1)(y - 6)$

73. $y^2 - 12y + 36$
$(y - 6)^2$

74. $m^2 - 2m + 1$
$(m - 1)^2$

75. $9m^2 - 25n^2$
$(3m - 5n)(3m + 5n)$

76. $m^2n^2 - 2mn^3 + n^4$
$n^2(m - n)^2$

77. $5a^2 + 20a - 60$
$5(a + 6)(a - 2)$

78. $-3y^2 + 9y + 30$
$-3(y - 5)(y + 2)$

79. $-2w^2 + 18w + 20$
$-2(w - 10)(w + 1)$

80. $x^2z + 2xyz + y^2z$
$z(x + y)^2$

81. $w^2x^2 - 100x^2$
$x^2(w + 10)(w - 10)$

82. $9x^2 + 30x + 25$
$(3x + 5)^2$

83. $9x^2 - 1$
$(3x + 1)(3x - 1)$

84. $6w^2 - 19w - 36$
$(3w + 4)(2w - 9)$

85. $8x^2 - 2x - 15$
$(4x + 5)(2x - 3)$

86. $4w^2 + 12w + 9$
$(2w + 3)^2$

87. $4x^2 - 20x + 25$
$(2x - 5)^2$

88. $9m^2 - 121$
$(3m - 11)(3m + 11)$

89. $9a^2 + 60a + 100$
$(3a + 10)^2$

90. $10w^2 + 20w - 80$
$10(w + 4)(w - 2)$

91. $4a^2 + 24a + 32$
$4(a + 2)(a + 4)$

92. $20x^2 - 23x + 6$
$(5x - 2)(4x - 3)$

93. $3m^4 - 24m$
$3m(m - 2)(m^2 + 2m + 4)$

94. $6w^3z + 6z$
$6z(w + 1)(w^2 - w + 1)$

GETTING MORE INVOLVED

 95. *Discussion.* Which of the following is not a perfect square trinomial? Explain.

a) $4a^6 - 6a^3b^4 + 9b^8$ **b)** $1000x^2 + 200ax + a^2$
c) $900y^4 - 60y^2 + 1$ **d)** $36 - 36z^7 + 9z^{14}$
    a and b

96. *Discussion.* Which of the following is not a difference of two squares? Explain.

a) $16a^8y^4 - 25c^{12}$ **b)** $a^9 - b^4$
c) $t^{90} - 1$ **d)** $x^2 - 196$
    b

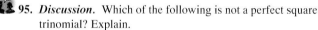

5.8 FACTORING STRATEGY

In previous sections we established the general idea of factoring and many special cases. In this section we will see that a polynomial can have as many factors as its degree, and we will factor higher-degree polynomials completely. We will also see a general strategy for factoring polynomials.

Prime Polynomials

A polynomial that cannot be factored is a prime polynomial. Binomials with no common factors, such as $2x + 1$ and $a - 3$, are prime polynomials. To determine whether a polynomial such as $x^2 + 1$ is a prime polynomial, we must try all possibilities for factoring it. If $x^2 + 1$ could be factored as a product of two binomials, the only possibilities that would give a first term of x^2 and a last term of 1 are $(x + 1)(x + 1)$ and $(x - 1)(x - 1)$. However,

$$(x + 1)(x + 1) = x^2 + 2x + 1 \quad \text{and} \quad (x - 1)(x - 1) = x^2 - 2x + 1.$$

Both products have an x-term. Of course, $(x + 1)(x - 1)$ has no x-term, but
$$(x + 1)(x - 1) = x^2 - 1.$$
Because none of these possibilities results in $x^2 + 1$, the polynomial $x^2 + 1$ is a prime polynomial. Note that $x^2 + 1$ is a sum of two squares. A sum of two squares of the form $a^2 + b^2$ is always a prime polynomial.

E X A M P L E   1

Prime polynomials

Determine whether the polynomial $x^2 + 3x + 4$ is a prime polynomial.

Solution

To factor $x^2 + 3x + 4$, we must find two integers with a product of 4 and a sum of 3. The only pairs of positive integers with a product of 4 are 1 and 4, and 2 and 2. Because the product is positive 4, both numbers must be negative or both positive. Under these conditions it is impossible to get a sum of positive 3. The polynomial is prime.

Factoring Polynomials Completely

So far, a typical polynomial has been a product of two factors, with possibly a common factor removed first. However, it is possible that the factors can still be factored again. A polynomial in a single variable may have as many factors as its degree. We have factored a polynomial completely when all of the factors are prime polynomials.

E X A M P L E   2

Factoring higher-degree polynomials completely

Factor $x^4 + x^2 - 2$ completely.

Solution

Two numbers with a product of -2 and a sum of 1 are 2 and -1:
$$x^4 + x^2 - 2 = (x^2 + 2)(x^2 - 1)$$
$$= (x^2 + 2)(x - 1)(x + 1) \quad \text{Difference of two squares}$$
Since $x^2 + 2$, $x - 1$, and $x + 1$ are prime, the polynomial is factored completely.

In the next example we factor a sixth-degree polynomial.

E X A M P L E   3

Factoring completely

Factor $3x^6 - 3$ completely.

Solution

To factor $3x^6 - 3$, we must first factor out the common factor 3 and then recognize that x^6 is a perfect square: $x^6 = (x^3)^2$:

$$3x^6 - 3 = 3(x^6 - 1) \qquad \text{Factor out the common factor.}$$
$$= 3((x^3)^2 - 1) \qquad \text{Write } x^6 \text{ as a perfect square.}$$
$$= 3(x^3 - 1)(x^3 + 1) \qquad \text{Difference of two squares}$$
$$= 3(x - 1)(x^2 + x + 1)(x + 1)(x^2 - x + 1) \qquad \text{Difference of two cubes and sum of two cubes}$$

Since $x^2 + x + 1$ and $x^2 - x + 1$ are prime, the polynomial is factored completely.

In Example 3 we recognized $x^6 - 1$ as a difference of two squares. However, $x^6 - 1$ is also a difference of two cubes, and we can factor it using the rule for the difference of two cubes:

$$x^6 - 1 = (x^2)^3 - 1 = (x^2 - 1)(x^4 + x^2 + 1)$$

Now we can factor $x^2 - 1$, but it is difficult to see how to factor $x^4 + x^2 + 1$. (It is not prime.) Although x^6 can be thought of as a perfect square or a perfect cube, in this case thinking of it as a perfect square is better.

In the next example we use substitution to simplify the polynomial before factoring. This fourth-degree polynomial has four factors.

E X A M P L E   4   **Using substitution to simplify**

Factor $(w^2 - 1)^2 - 11(w^2 - 1) + 24$ completely.

Solution

Let $a = w^2 - 1$ to simplify the polynomial:

$$
\begin{aligned}
(w^2 - 1)^2 - 11(w^2 - 1) + 24 &= a^2 - 11a + 24 && \text{Replace } w^2 - 1 \text{ by } a. \\
&= (a - 8)(a - 3) \\
&= (w^2 - 1 - 8)(w^2 - 1 - 3) && \text{Replace } a \text{ by } w^2 - 1. \\
&= (w^2 - 9)(w^2 - 4) \\
&= (w + 3)(w - 3)(w + 2)(w - 2)
\end{aligned}
$$

Factoring Polynomials with Four Terms

In Section 5.6 we rewrote a trinomial as a polynomial with four terms and then used factoring by grouping. Factoring by grouping can also be used on other types of polynomials with four terms.

E X A M P L E   5   **Polynomials with four terms**

Use grouping to factor each polynomial completely.

a) $x^3 + x^2 + 4x + 4$ **b)** $3x^3 - x^2 - 27x + 9$ **c)** $ax - bw + bx - aw$

Solution

a) Note that the first two terms of $x^3 + x^2 + 4x + 4$ have a common factor of x^2, and the last two terms have a common factor of 4.

$$
\begin{aligned}
x^3 + x^2 + 4x + 4 &= x^2(x + 1) + 4(x + 1) && \text{Factor by grouping.} \\
&= (x^2 + 4)(x + 1) && \text{Factor out } x + 1.
\end{aligned}
$$

Since $x^2 + 4$ is a sum of two squares, it is prime and the polynomial is factored completely.

b) We can factor x^2 out of the first two terms of $3x^3 - x^2 - 27x + 9$ and 9 or -9 from the last two terms. We choose -9 to get the factor $3x - 1$ in each case.

$$
\begin{aligned}
3x^3 - x^2 - 27x + 9 &= x^2(3x - 1) - 9(3x - 1) && \text{Factor by grouping.} \\
&= (x^2 - 9)(3x - 1) && \text{Factor out } 3x - 1. \\
&= (x - 3)(x + 3)(3x - 1) && \text{Difference of two squares}
\end{aligned}
$$

This third-degree polynomial has three factors.

c) First rearrange the terms so that the first two and the last two have common factors:

$$ax - bw + bx - aw = ax + bx - aw - bw \qquad \text{Rearrange the terms.}$$
$$= x(a + b) - w(a + b) \qquad \text{Common factors}$$
$$= (x - w)(a + b) \qquad \text{Factor out } a + b.$$

Summary

A strategy for factoring polynomials is given in the following box.

Strategy for Factoring Polynomials

1. If there are any common factors, factor them out first.

2. When factoring a binomial, look for the special cases: difference of two squares, difference of two cubes, and sum of two cubes. Remember that a sum of two squares $a^2 + b^2$ is prime.

3. When factoring a trinomial, check to see whether it is a perfect square trinomial.

4. When factoring a trinomial that is not a perfect square, use grouping or trial and error.

5. When factoring a polynomial of high degree, use substitution to get a polynomial of degree 2 or 3, or use trial and error.

6. If the polynomial has four terms, try factoring by grouping.

E X A M P L E   6

Using the factoring strategy

Factor each polynomial completely.

a) $3w^3 - 3w^2 - 18w$ **b)** $10x^2 + 160$

c) $16a^2b - 80ab + 100b$ **d)** $aw + mw + az + mz$

Solution

a) The greatest common factor (GCF) for the three terms is $3w$:
$$3w^3 - 3w^2 - 18w = 3w(w^2 - w - 6) \qquad \text{Factor out } 3w.$$
$$= 3w(w - 3)(w + 2) \qquad \text{Factor completely.}$$

b) The GCF in $10x^2 + 160$ is 10:
$$10x^2 + 160 = 10(x^2 + 16)$$

Because $x^2 + 16$ is prime, the polynomial is factored completely.

c) The GCF in $16a^2b - 80ab + 100b$ is $4b$:
$$16a^2b - 80ab + 100b = 4b(4a^2 - 20a + 25)$$
$$= 4b(2a - 5)^2$$

d) The polynomial has four terms, and we can factor it by grouping:
$$aw + mw + az + mz = w(a + m) + z(a + m)$$
$$= (w + z)(a + m)$$

h e l p f u l   h i n t

When factoring integers, we write $4 = 2 \cdot 2$. However, when factoring polynomials we usually do not factor any of the integers that appear. So we say that $4b(2a - 5)^2$ is factored completely.

True or false? Explain your answer.

1. $x^2 - 9 = (x - 3)^2$ for any value of x.   False
2. The polynomial $4x^2 + 12x + 9$ is a perfect square trinomial.   True
3. The sum of two squares $a^2 + b^2$ is prime.   True
4. The polynomial $x^4 - 16$ is factored completely as $(x^2 - 4)(x^2 + 4)$.   False
5. $y^3 - 27 = (y + 3)(y^2 + 3y - 9)$ for any value of y.   False
6. The polynomial $y^6 - 1$ is a difference of two squares.   True
7. The polynomial $2x^2 + 2x - 12$ is factored completely as $(2x - 4)(x + 3)$.   False
8. The polynomial $x^2 - 4x - 4$ is a prime polynomial.   True
9. The polynomial $a^6 - 1$ is the difference of two cubes.   True
10. The polynomial $x^2 + 3x - ax + 3a$ can be factored by grouping.   False

5.8   EXERCISES

Reading and Writing *After reading this section, write out the answers to these questions. Use complete sentences.*

1. What should you do first when factoring a polynomial?
   Always factor out the greatest common factor first.
2. If you are factoring a binomial, then what should you look for?
   In factoring a binomial, look for a difference of two squares and a sum or difference of two cubes.
3. When factoring a trinomial what should you look for?
   In factoring a trinomial, look for the perfect square trinomials.
4. What should you look for when factoring a four-term polynomial?
   In factoring a four-term polynomial, try to factor by grouping.

Determine whether each polynomial is a prime polynomial. See Example 1.

5. $y^2 + 100$ Prime
6. $3x^2 + 27$ Not prime
7. $-9w^2 - 9$ Not prime
8. $25y^2 + 36$ Prime
9. $x^2 - 2x - 3$ Not prime
10. $x^2 - 2x + 3$ Prime
11. $x^2 + 2x + 3$ Prime
12. $x^2 + 4x + 3$ Not prime
13. $x^2 - 4x - 3$ Prime
14. $x^2 + 4x - 3$ Prime
15. $6x^2 + 3x - 4$ Prime
16. $4x^2 - 5x - 3$ Prime

Factor each polynomial completely. See Examples 2–4.

17. $a^4 - 10a^2 + 25$ $(a^2 - 5)^2$
18. $9y^4 + 12y^2 + 4$ $(3y^2 + 2)^2$
19. $x^4 - 6x^2 + 8$ $(x^2 - 2)(x - 2)(x + 2)$
20. $x^6 + 2x^3 - 3$ $(x^3 + 3)(x - 1)(x^2 + x + 1)$

21. $(3x - 5)^2 - 1$ $3(x - 2)(3x - 4)$
22. $(2x + 1)^2 - 4$ $(2x - 1)(2x + 3)$
23. $2y^6 - 128$ $2(y - 2)(y^2 + 2y + 4)(y + 2)(y^2 - 2y + 4)$
24. $6 - 6y^6$ $6(1 - y)(1 + y + y^2)(1 + y)(1 - y + y^2)$
25. $32a^4 - 18$ $2(4a^2 + 3)(4a^2 - 3)$
26. $2a^4 - 32$ $2(a - 2)(a + 2)(a^2 + 4)$
27. $x^4 - (x - 6)^2$ $(x + 3)(x - 2)(x^2 - x + 6)$
28. $y^4 - (2y + 1)^2$ $(y^2 - 2y - 1)(y + 1)^2$
29. $(m + 2)^2 + 2(m + 2) - 3$ $(m + 5)(m + 1)$
30. $(2w - 3)^2 - 2(2w - 3) - 15$ $4w(w - 4)$
31. $3(y - 1)^2 + 11(y - 1) - 20$ $(3y - 7)(y + 4)$
32. $2(w + 2)^2 + 5(w + 2) - 3$ $(2w + 3)(w + 5)$
33. $(y^2 - 3)^2 - 4(y^2 - 3) - 12$
    $(y - 3)(y + 3)(y - 1)(y + 1)$
34. $(m^2 - 8)^2 - 4(m^2 - 8) - 32$
    $(m - 4)(m + 4)(m - 2)(m + 2)$

Use grouping to factor each polynomial completely. See Example 5.

35. $ax + ay + bx + by$ $(a + b)(x + y)$
36. $7x + 7z + kx + kz$ $(7 + k)(x + z)$
37. $x^3 + x^2 - 9x - 9$ $(x - 3)(x + 3)(x + 1)$
38. $x^3 + x^2 - 25x - 25$ $(x - 5)(x + 5)(x + 1)$
39. $aw - bw - 3a + 3b$ $(w - 3)(a - b)$
40. $wx - wy + 2x - 2y$ $(w + 2)(x - y)$
41. $a^4 + 3a^3 + 27a + 81$ $(a + 3)^2(a^2 - 3a + 9)$
42. $ac - bc - 5a + 5b$ $(c - 5)(a - b)$

43. $y^4 - 5y^3 + 8y - 40$ $(y - 5)(y + 2)(y^2 - 2y + 4)$

44. $x^3 + ax - 3a - 3x^2$ $(x - 3)(x^2 + a)$

45. $ady + d - w - awy$ $(d - w)(ay + 1)$

46. $xy + by + ax + ab$ $(y + a)(x + b)$

47. $x^2y - a + ax^2 - y$ $(a + y)(x - 1)(x + 1)$

48. $a^2d - b^2c + a^2c - b^2d$ $(c + d)(a - b)(a + b)$

49. $y^4 + b + by^3 + y$ $(y + b)(y + 1)(y^2 - y + 1)$

50. $ab + mw^2 + bm + aw^2$ $(b + w^2)(a + m)$

Use the factoring strategy to factor each polynomial completely. See Example 6.

51. $9x^2 - 24x + 16$ $(3x - 4)^2$

52. $-3x^2 + 18x + 48$ $-3(x - 8)(x + 2)$

53. $12x^2 - 13x + 3$ $(3x - 1)(4x - 3)$

54. $2x^2 - 3x - 6$ Prime

55. $3a^4 + 81a$ $3a(a + 3)(a^2 - 3a + 9)$

56. $-a^3 + 25a$ $-a(a - 5)(a + 5)$

57. $32 + 2x^2$ $2(x^2 + 16)$

58. $x^3 + 4x^2 + 4x$ $x(x + 2)^2$

59. $6x^2 - 5x + 12$ Prime

60. $x^4 + 2x^3 - x - 2$ $(x - 1)(x^2 + x + 1)(x + 2)$

61. $(x + y)^2 - 1$ $(x + y - 1)(x + y + 1)$

62. $x^3 + 9x$ $x(x^2 + 9)$

63. $a^3b - ab^3$ $ab(a - b)(a + b)$

64. $2m^3 - 250$ $2(m - 5)(m^2 + 5m + 25)$

65. $x^4 + 2x^3 - 8x - 16$ $(x + 2)(x - 2)(x^2 + 2x + 4)$

66. $(x + 5)^2 - 4$ $(x + 3)(x + 7)$

67. $m^2n + 2mn + n^3$ $n(m + n)^2$

68. $a^2b - 6ab + 9b$ $b(a - 3)^2$

69. $2m + wn + 2n + wm$ $(2 + w)(m + n)$

70. $aw - 5b + bw - 5a$ $(w - 5)(a + b)$

71. $4w^2 + 4w - 3$ $(2w + 3)(2w - 1)$

72. $4w^2 + 8w - 63$ Prime

73. $t^4 + 4t^2 - 21$ $(t^2 + 7)(t^2 - 3)$

74. $m^4 + 5m^2 + 4$ $(m^2 + 1)(m^2 + 4)$

75. $-a^3 - 7a^2 + 30a$ $-a(a + 10)(a - 3)$

76. $2y^4 + 3y^3 - 20y^2$ $y^2(2y - 5)(y + 4)$

77. $(y + 5)^2 - 2(y + 5) - 3$ $(y + 2)(y + 6)$

78. $(2t - 1)^2 + 7(2t - 1) + 10$ $2(2t + 1)(t + 2)$

79. $-2w^4 + 1250$ $-2(w - 5)(w + 5)(w^2 + 25)$

80. $5a^5 - 5a$ $5a(a - 1)(a + 1)(a^2 + 1)$

81. $8a^3 + 8a$ $8a(a^2 + 1)$

82. $awx + ax$ $ax(w + 1)$

83. $(w + 5)^2 - 9$ $(w + 2)(w + 8)$

84. $(a - 6)^2 - 1$ $(a - 7)(a - 5)$

85. $4aw^2 - 12aw + 9a$ $a(2w - 3)^2$

86. $9an^3 + 15an^2 - 14an$ $an(3n - 2)(3n + 7)$

87. $x^2 - 6x + 9$ $(x - 3)^2$

88. $x^3 + 12x^2 + 36x$ $x(x + 6)^2$

89. $3x^4 - 75x^2$ $3x^2(x - 5)(x + 5)$

90. $3x^2 + 9x + 12$ $3(x^2 + 3x + 4)$

91. $m^3n - n$ $n(m - 1)(m^2 + m + 1)$

92. $m^4 + 16m^2$ $m^2(m^2 + 16)$

93. $12x^2 + 2x - 30$ $2(3x + 5)(2x - 3)$

94. $90x^2 + 3x - 60$ $3(5x - 4)(6x + 5)$

95. $2a^3 - 32$ $2(a^3 - 16)$

96. $12x^2 - 28x + 15$ $(2x - 3)(6x - 5)$

Factor completely. Assume variables used as exponents represent positive integers.

97. $a^{3m} - 1$ $(a^m - 1)(a^{2m} + a^m + 1)$

98. $x^{6a} + 8$ $(x^{2a} + 2)(x^{4a} - 2x^{2a} + 4)$

99. $a^{3w} - b^{6n}$ $(a^w - b^{2n})(a^{2w} + a^w b^{2n} + b^{4n})$

100. $x^{2n} - 9$ $(x^n - 3)(x^n + 3)$

101. $t^{4n} - 16$ $(t^n - 2)(t^n + 2)(t^{2n} + 4)$

102. $a^{3n+2} + a^2$ $a^2(a^n + 1)(a^{2n} - a^n + 1)$

103. $a^{2n+1} - 2a^{n+1} - 15a$ $a(a^n - 5)(a^n + 3)$

104. $x^{3m} + x^{2m} - 6x^m$ $x^m(x^m + 3)(x^m - 2)$

105. $a^{2n} - 3a^n + a^n b - 3b$ $(a^n - 3)(a^n + b)$

106. $x^m z + 5z + x^{m+1} + 5x$ $(z + x)(x^m + 5)$

GETTING MORE INVOLVED

 107. *Cooperative learning.* Write down 10 trinomials of the form $ax^2 + bx + c$ "at random" using integers for a, b, and c. What percent of your 10 trinomials are prime? Would you say that prime trinomials are the exception or the rule? Compare your results with those of your classmates.

 108. *Writing.* The polynomial

$$x^5 + x^4 - 9x^3 - 13x^2 + 8x + 12$$

is a product of five factors of the form $x \pm n$, where n is a natural number smaller than 4. Factor this polynomial completely and explain your procedure.

5.9 SOLVING EQUATIONS BY FACTORING

The techniques of factoring can be used to solve equations involving polynomials that cannot be solved by the other methods that you have learned. After you learn to solve equations by factoring, we will use this technique to solve some new applied problems in this section and in Chapters 6 and 8.

The Zero Factor Property

The equation $ab = 0$ indicates that the product of two unknown numbers is 0. But the product of two real numbers is zero only when one or the other of the numbers is 0. So even though we do not know exactly the values of a and b from $ab = 0$, we do know that $a = 0$ or $b = 0$. This idea is called the **zero factor property.**

helpful hint

Note that the zero factor property is our second example of getting an equivalent equation without "doing the same thing to each side." What was the first?

Zero Factor Property

The equation $ab = 0$ is equivalent to the compound equation

$$a = 0 \quad \text{or} \quad b = 0.$$

The next example shows how to use the zero factor property to solve an equation in one variable.

E X A M P L E   1

Using the zero factor property

Solve $x^2 + x - 12 = 0$.

Solution

We factor the left-hand side of the equation to get a product of two factors that are equal to 0. Then we write an equivalent equation using the zero factor property.

$$x^2 + x - 12 = 0$$
$$(x + 4)(x - 3) = 0 \quad \text{Factor the left-hand side.}$$
$$x + 4 = 0 \quad \text{or} \quad x - 3 = 0 \quad \text{Zero factor property}$$
$$x = -4 \quad \text{or} \quad x = 3 \quad \text{Solve each part of the compound equation.}$$

Check that both -4 and 3 satisfy $x^2 + x - 12 = 0$. If $x = -4$, we get

$$(-4)^2 + (-4) - 12 = 16 - 4 - 12 = 0.$$

If $x = 3$, we get

$$(3)^2 + 3 - 12 = 9 + 3 - 12 = 0.$$

So the solution set is $\{-4, 3\}$. ■

The zero factor property is used only in solving polynomial equations that have zero on one side and a polynomial that can be factored on the other side. The polynomials that we factored most often were the quadratic polynomials. The equations that we will solve most often using the zero factor property will be quadratic equations.

Quadratic Equation

If a, b, and c are real numbers, with $a \neq 0$, then the equation

$$ax^2 + bx + c = 0$$

is called a **quadratic equation.**

M A T H   A T   W O R K $x^2 + (x+1)^2 = 5^2$

Seamas Mercado, professional bodyboarder and 1988 National Champion, charges the waves off Hawaii, Tahiti, Indonesia, Mexico, and California. In choosing a board for competition and for the maneuvers he wants to perform, Mercado factors in his height and weight as well as the size, power, and temperature of the waves he will be riding. In colder water a softer, more flexible board is used; in warmer water a stiffer board is chosen. When waves crash on shore, the ride usually lasts 3 to 5 seconds, and a shorter board with a narrow tail is chosen for greater control. When waves break along a sand bar or reef, the ride can sometimes last as long as 2 minutes, and a straighter board with more surface area is chosen so that the board will move faster and allow the rider to pull off more maneuvers. Basic maneuvers include bottom turns, aerials, forward and reverse 360's, and el rollos.

**B O D Y B O A R D
D E S I G N E R**

As one of the top 10 bodyboarders in the world, Mercado helps to design the boards he uses. "Performance levels are greatly increased with fine-tuned equipment and techniques," he says. In Exercise 63 of Section 5.9 you will find the dimensions of a given bodyboard.

In Chapter 8 we will study quadratic equations further and solve quadratic equations that cannot be solved by factoring. Keep the following strategy in mind when solving equations by factoring.

> **Strategy for Solving Equations by Factoring**
>
> **1.** Write the equation with 0 on the right-hand side.
> **2.** Factor the left-hand side.
> **3.** Use the zero factor property to get simpler equations. (Set each factor equal to 0.)
> **4.** Solve the simpler equations.
> **5.** Check the answers in the original equation.

E X A M P L E   2

Solving a quadratic equation by factoring

Solve each equation.

a) $10x^2 = 5x$

b) $3x - 6x^2 = -9$

Solution

a) Use the steps in the strategy for solving equations by factoring:

$$10x^2 = 5x \qquad \text{Original equation}$$
$$10x^2 - 5x = 0 \qquad \text{Rewrite with zero on the right-hand side.}$$
$$5x(2x - 1) = 0 \qquad \text{Factor the left-hand side.}$$
$$5x = 0 \quad \text{or} \quad 2x - 1 = 0 \qquad \text{Zero factor property}$$
$$x = 0 \quad \text{or} \quad x = \frac{1}{2} \qquad \text{Solve for } x.$$

The solution set is $\left\{0, \frac{1}{2}\right\}$. Check each solution in the original equation.

b) First rewrite the equation with 0 on the right-hand side and the left-hand side in order of descending exponents:

$$3x - 6x^2 = -9 \quad \text{Original equation}$$
$$-6x^2 + 3x + 9 = 0 \quad \text{Add 9 to each side.}$$
$$2x^2 - x - 3 = 0 \quad \text{Divide each side by } -3.$$
$$(2x - 3)(x + 1) = 0 \quad \text{Factor.}$$
$$2x - 3 = 0 \quad \text{or} \quad x + 1 = 0 \quad \text{Zero factor property}$$
$$x = \frac{3}{2} \quad \text{or} \quad x = -1 \quad \text{Solve for } x.$$

The solution set is $\left\{-1, \frac{3}{2}\right\}$. Check each solution in the original equation.

CAUTION   If we divide each side of $10x^2 = 5x$ by $5x$, we get $2x = 1$, or $x = \frac{1}{2}$. We do not get $x = 0$. By dividing by $5x$ we have lost one of the factors and one of the solutions.

In the next example there are more than two factors, but we can still write an equivalent equation by setting each factor equal to 0.

EXAMPLE 3

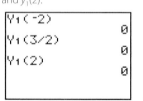
Solving a cubic equation by factoring

Solve $2x^3 - 3x^2 - 8x + 12 = 0$.

Solution

First notice that the first two terms have the common factor x^2 and the last two terms have the common factor -4.

$$x^2(2x - 3) - 4(2x - 3) = 0 \quad \text{Factor by grouping.}$$
$$(x^2 - 4)(2x - 3) = 0 \quad \text{Factor out } 2x - 3.$$
$$(x - 2)(x + 2)(2x - 3) = 0 \quad \text{Factor completely.}$$
$$x - 2 = 0 \quad \text{or} \quad x + 2 = 0 \quad \text{or} \quad 2x - 3 = 0 \quad \text{Set each factor equal to 0.}$$
$$x = 2 \quad \text{or} \quad x = -2 \quad \text{or} \quad x = \frac{3}{2}$$

The solution set is $\left\{-2, \frac{3}{2}, 2\right\}$. Check each solution in the original equation.

The equation in the next example involves absolute value.

EXAMPLE 4

Solving an absolute value equation by factoring

Solve $|x^2 - 2x - 16| = 8$.

Solution

First write an equivalent compound equation without absolute value:

$$x^2 - 2x - 16 = 8 \quad \text{or} \quad x^2 - 2x - 16 = -8$$
$$x^2 - 2x - 24 = 0 \quad \text{or} \quad x^2 - 2x - 8 = 0$$
$$(x - 6)(x + 4) = 0 \quad \text{or} \quad (x - 4)(x + 2) = 0$$
$$x - 6 = 0 \quad \text{or} \quad x + 4 = 0 \quad \text{or} \quad x - 4 = 0 \quad \text{or} \quad x + 2 = 0$$
$$x = 6 \quad \text{or} \quad x = -4 \quad \text{or} \quad x = 4 \quad \text{or} \quad x = -2$$

The solution set is $\{-2, -4, 4, 6\}$. Check each solution.

Applications

Many applied problems can be solved by using equations such as those we have been solving.

EXAMPLE 5

Area of a room

Ronald's living room is 2 feet longer than it is wide, and its area is 168 square feet. What are the dimensions of the room?

Solution

Let x be the width and $x + 2$ be the length. See Figure 5.1. Because the area of a rectangle is the length times the width, we can write the equation

$$x(x + 2) = 168.$$

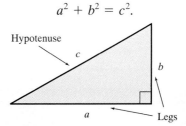

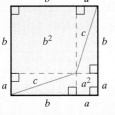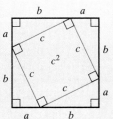

helpful hint

To prove the Pythagorean theorem, draw two squares with sides of length $a + b$, and partition them as shown.

Erase the four triangles in each picture. Since we started with equal areas, we must have equal areas after erasing the triangles:

$$a^2 + b^2 = c^2$$

FIGURE 5.1

We solve the equation by factoring:

$$x^2 + 2x - 168 = 0$$
$$(x - 12)(x + 14) = 0$$

$$x - 12 = 0 \quad \text{or} \quad x + 14 = 0$$
$$x = 12 \quad \text{or} \quad x = -14$$

Because the width of a room is a positive number, we disregard the solution $x = -14$. We use $x = 12$ and get a width of 12 feet and a length of 14 feet. Check this answer by multiplying 12 and 14 to get 168. ∎

Applications involving quadratic equations often require a theorem called the **Pythagorean theorem.** This theorem states that *in any right triangle the sum of the squares of the lengths of the legs is equal to the length of the hypotenuse squared.*

The Pythagorean Theorem

The triangle shown is a right triangle if and only if

$$a^2 + b^2 = c^2.$$

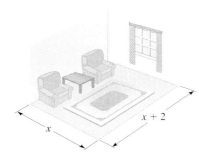

We use the Pythagorean theorem in the next example.

E X A M P L E   6   **Using the Pythagorean theorem**

Shirley used 14 meters of fencing to enclose a rectangular region. To be sure that the region was a rectangle, she measured the diagonals and found that they were 5 meters each. (If the opposite sides of a quadrilateral are equal and the diagonals are equal, then the quadrilateral is a rectangle.) What are the length and width of the rectangle?

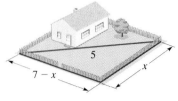

F I G U R E   5 . 2

Solution

The perimeter of a rectangle is twice the length plus twice the width, $P = 2L + 2W$. Because the perimeter is 14 meters, the sum of one length and one width is 7 meters. If we let x represent the width, then $7 - x$ is the length. We use the Pythagorean theorem to get a relationship among the length, width, and diagonal. See Figure 5.2.

$$x^2 + (7 - x)^2 = 5^2 \quad \text{Pythagorean theorem}$$
$$x^2 + 49 - 14x + x^2 = 25 \quad \text{Simplify.}$$
$$2x^2 - 14x + 24 = 0 \quad \text{Simplify.}$$
$$x^2 - 7x + 12 = 0 \quad \text{Divide each side by 2.}$$
$$(x - 3)(x - 4) = 0 \quad \text{Factor the left-hand side.}$$
$$x - 3 = 0 \quad \text{or} \quad x - 4 = 0 \quad \text{Zero factor property}$$
$$x = 3 \quad \text{or} \quad x = 4$$
$$7 - x = 4 \quad \text{or} \quad 7 - x = 3$$

Solving the equation gives two possible rectangles: a 3 by 4 rectangle or a 4 by 3 rectangle. However, those are identical rectangles. The rectangle is 3 meters by 4 meters. ∎

WARM-UPS

True or false? Explain your answer.

1. The equation $(x - 1)(x + 3) = 12$ is equivalent to $x - 1 = 3$ or $x + 3 = 4$. False

2. Equations solved by factoring may have two solutions.   True

3. The equation $c \cdot d = 0$ is equivalent to $c = 0$ or $d = 0$.   True

4. The equation $|x^2 + 4| = 5$ is equivalent to the compound equation $x^2 + 4 = 5$ or $x^2 - 4 = 5$.   False

5. The solution set to the equation $(2x - 1)(3x + 4) = 0$ is $\left\{\frac{1}{2}, -\frac{4}{3}\right\}$.   True

6. The Pythagorean theorem states that the sum of the squares of any two sides of any triangle is equal to the square of the third side.   False

7. If the perimeter of a rectangular room is 38 feet, then the sum of the length and width is 19 feet.   True

8. Two numbers that have a sum of 8 can be represented by x and $8 - x$.   True

9. The solution set to the equation $x(x - 1)(x - 2) = 0$ is $\{1, 2\}$.   False

10. The solution set to the equation $3(x + 2)(x - 5) = 0$ is $\{3, -2, 5\}$. False

5.9 EXERCISES

Reading and Writing   *After reading this section, write out the answers to these questions. Use complete sentences.*

1. What is the zero factor property?
   The zero factor property says that if $a \cdot b = 0$ then either $a = 0$ or $b = 0$.

2. What is a quadratic equation?
   A quadratic equation is an equation of the form $ax^2 + bx + c = 0$ with $a \neq 0$.

3. Where is the hypotenuse in a right triangle?
   The hypotenuse of a right triangle is the side opposite the right angle.

4. Where are the legs in a right triangle?
   The legs of a right triangle are the sides that form the right angle.

5. What is the Pythagorean theorem?
   The Pythagorean theorem says that a triangle is a right triangle if and only if the sum of the squares of the legs is equal to the square of the hypotenuse.

6. Where is the diagonal of a rectangle?
   The diagonal of a rectangle is the line segment that joins two opposite vertices.

Solve each equation. See Examples 1–3.

7. $(x - 5)(x + 4) = 0$ $\{-4, 5\}$
8. $(a - 6)(a + 5) = 0$ $\{-5, 6\}$
9. $(2x - 5)(3x + 4) = 0$ $\left\{\dfrac{5}{2}, -\dfrac{4}{3}\right\}$
10. $(3k + 8)(4k - 3) = 0$ $\left\{-\dfrac{8}{3}, \dfrac{3}{4}\right\}$
11. $w^2 + 5w - 14 = 0$ $\{-7, 2\}$
12. $t^2 - 6t - 27 = 0$ $\{-3, 9\}$
13. $m^2 - 7m = 0$ $\{0, 7\}$
14. $h^2 - 5h = 0$ $\{0, 5\}$
15. $a^2 - a = 20$ $\{-4, 5\}$
16. $p^2 - p = 42$ $\{-6, 7\}$
17. $3x^2 - 3x - 36 = 0$ $\{-3, 4\}$
18. $-2x^2 - 16x - 24 = 0$ $\{-6, -2\}$
19. $z^2 + \dfrac{3}{2}z = 10$ $\left\{-4, \dfrac{5}{2}\right\}$
20. $m^2 + \dfrac{11}{3}m = -2$ $\left\{-3, -\dfrac{2}{3}\right\}$
21. $x^3 - 4x = 0$ $\{-2, 0, 2\}$
22. $16x - x^3 = 0$ $\{-4, 0, 4\}$
23. $w^3 + 4w^2 - 25w - 100 = 0$ $\{-5, -4, 5\}$
24. $a^3 + 2a^2 - 16a - 32 = 0$ $\{-4, -2, 4\}$
25. $n^3 - 2n^2 - n + 2 = 0$ $\{-1, 1, 2\}$
26. $w^3 - w^2 - 25w + 25 = 0$ $\{-5, 1, 5\}$

Solve each equation. See Example 4.

27. $|x^2 - 5| = 4$ $\{-3, -1, 1, 3\}$

28. $|x^2 - 17| = 8$ $\{-5, -3, 3, 5\}$
29. $|x^2 + 2x - 36| = 12$ $\{-8, -6, 4, 6\}$
30. $|x^2 + 2x - 19| = 16$ $\{-7, -3, 1, 5\}$
31. $|x^2 + 4x + 2| = 2$ $\{-4, -2, 0\}$
32. $|x^2 + 8x + 8| = 8$ $\{-8, -4, 0\}$
33. $|x^2 + 6x + 1| = 8$ $\{-7, -3, 1\}$
34. $|x^2 - x - 21| = 9$ $\{-5, -3, 4, 6\}$

Solve each equation.

35. $2x^2 - x = 6$ $\left\{-\dfrac{3}{2}, 2\right\}$
36. $3x^2 + 14x = 5$ $\left\{-5, \dfrac{1}{3}\right\}$
37. $|x^2 + 5x| = 6$ $\{-6, -3, -2, 1\}$
38. $|x^2 + 6x - 4| = 12$ $\{-8, -4, -2, 2\}$
39. $x^2 + 5x = 6$ $\{-6, 1\}$
40. $x + 5x = 6$ $\{1\}$
41. $(x + 2)(x + 1) = 12$ $\{-5, 2\}$
42. $(x + 2)(x + 3) = 20$ $\{-7, 2\}$
43. $y^3 + 9y^2 + 20y = 0$ $\{-5, -4, 0\}$
44. $m^3 - 2m^2 - 3m = 0$ $\{-1, 0, 3\}$
45. $5a^3 = 45a$ $\{-3, 0, 3\}$
46. $5x^3 = 125x$ $\{-5, 0, 5\}$
47. $(2x - 1)(x^2 - 9) = 0$ $\left\{-3, \dfrac{1}{2}, 3\right\}$
48. $(x - 1)(x + 3)(x - 9) = 0$ $\{-3, 1, 9\}$
49. $4x^2 - 12x + 9 = 0$ $\left\{\dfrac{3}{2}\right\}$
50. $16x^2 + 8x + 1 = 0$ $\left\{-\dfrac{1}{4}\right\}$

Solve each equation for y. Assume a and b are positive numbers.

51. $y^2 + by = 0$ $\{0, -b\}$
52. $y^2 + ay + by + ab = 0$ $\{-a, -b\}$
53. $a^2y^2 - b^2 = 0$ $\left\{-\dfrac{b}{a}, \dfrac{b}{a}\right\}$
54. $9y^2 + 6ay + a^2 = 0$ $\left\{-\dfrac{a}{3}\right\}$
55. $4y^2 + 4by + b^2 = 0$ $\left\{-\dfrac{b}{2}\right\}$
56. $y^2 - b^2 = 0$ $\{-b, b\}$
57. $ay^2 + 3y - ay = 3$ $\left\{-\dfrac{3}{a}, 1\right\}$
58. $a^2y^2 + 2aby + b^2 = 0$ $\left\{-\dfrac{b}{a}\right\}$

Solve each problem. See Examples 5 and 6.

59. **Color print.** The length of a new "super size" color print is 2 inches more than the width. If the area is 24 square inches, what are the length and width?
   Width 4 inches, length 6 inches

60. *Tennis court dimensions.* In singles competition, each player plays on a rectangular area of 117 square yards. Given that the length of that area is 4 yards greater than its width, find the length and width.
Width 9 yards, length 13 yards

61. *Missing numbers.* The sum of two numbers is 13 and their product is 36. Find the numbers. 4 and 9

62. *More missing numbers.* The sum of two numbers is 6.5, and their product is 9. Find the numbers. 2 and 4.5

63. *Bodyboarding.* The Seamas Channel pro bodyboard shown in the figure has a length that is 21 inches greater than its width. Any rider weighing up to 200 pounds can use it because its surface area is 946 square inches. Find the length and width. Length 43 inches, width 22 inches

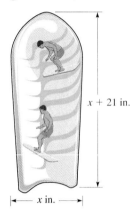

FIGURE FOR EXERCISE 63

64. *New dimensions in gardening.* Mary Gold has a rectangular flower bed that measures 4 feet by 6 feet. If she wants to increase the length and width by the same amount to have a flower bed of 48 square feet, then what will be the new dimensions? 6 feet by 8 feet

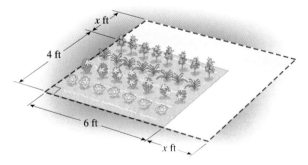

FIGURE FOR EXERCISE 64

65. *Shooting arrows.* An archer shoots an arrow straight upward at 64 feet per second. The height of the arrow $h(t)$ (in feet) at time t seconds is given by the function

$$h(t) = -16t^2 + 64t.$$

a) Use the accompanying graph to estimate the amount of time that the arrow is in the air. 4 seconds

b) Algebraically find the amount of time that the arrow is in the air. 4 seconds

c) Use the accompanying graph to estimate the maximum height reached by the arrow. 64 feet

d) At what time does the arrow reach its maximum height? 2 seconds

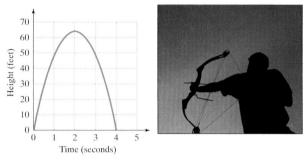

FIGURE FOR EXERCISE 65

66. *Time until impact.* If an object is dropped from a height of s_0 feet, then its altitude after t seconds is given by the formula $S = -16t^2 + s_0$. If a pack of emergency supplies is dropped from an airplane at a height of 1600 feet, then how long does it take for it to reach the ground? 10 seconds

67. *Yolanda's closet.* The length of Yolanda's closet is 2 feet longer than twice its width. If the diagonal measures 13 feet, then what are the length and width?
Width 5 feet, length 12 feet

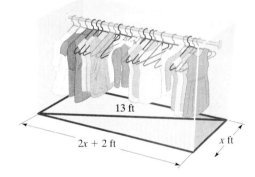

FIGURE FOR EXERCISE 67

68. *Ski jump.* The base of a ski ramp forms a right triangle. One leg of the triangle is 2 meters longer than the other. If the hypotenuse is 10 meters, then what are the lengths of the legs? 6 feet and 8 feet

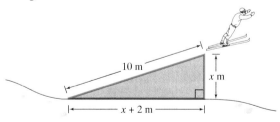

FIGURE FOR EXERCISE 68

69. *Trimming a gate.* A total of 34 feet of 1×4 lumber is used around the perimeter of the gate shown in the figure on the next page. If the diagonal brace is 13 feet long, then what are the length and width of the gate?
Width 5 feet, length 12 feet

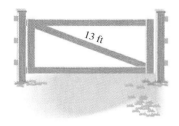

FIGURE FOR EXERCISE 69

70. *Perimeter of a rectangle.* The perimeter of a rectangle is 28 inches, and the diagonal measures 10 inches. What are the length and width of the rectangle?

Length 8 inches, width 6 inches

71. *Consecutive integers.* The sum of the squares of two consecutive integers is 25. Find the integers.

3 and 4, or -4 and -3

72. *Pete's garden.* Each row in Pete's garden is 3 feet wide. If the rows run north and south, he can have two more rows than if they run east and west. If the area of Pete's garden is 135 square feet, then what are the length and width?

Length 15 feet, width 9 feet

73. *House plans.* In the plans for their dream house the Baileys have a master bedroom that is 240 square feet in area. If they increase the width by 3 feet, they must decrease the length by 4 feet to keep the original area. What are the original dimensions of the bedroom?

Length 20 feet, width 12 feet

74. *Arranging the rows.* Mr. Converse has 112 students in his algebra class with an equal number in each row. If he arranges the desks so that he has one fewer rows, he will have two more students in each row. How many rows did he have originally?   8

GETTING MORE INVOLVED

 75. *Writing.* If you divide each side of $x^2 = x$ by x, you get $x = 1$. If you subtract x from each side of $x^2 = x$, you get $x^2 - x = 0$, which has two solutions. Which method is correct? Explain.

 76. *Cooperative learning.* Work with a group to examine the following solution to $x^2 - 2x = -1$:

$$x(x - 2) = -1$$
$$x = -1 \quad \text{or} \quad x - 2 = -1$$
$$x = -1 \quad \text{or} \quad x = 1$$

Is this method correct? Explain.

77. *Cooperative learning.* Work with a group to examine the following steps in the solution to $5x^2 - 5 = 0$

$$5(x^2 - 1) = 0$$
$$5(x - 1)(x + 1) = 0$$
$$x - 1 = 0 \quad \text{or} \quad x + 1 = 0$$
$$x = 1 \quad \text{or} \quad x = -1$$

What happened to the 5? Explain.

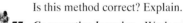

COLLABORATIVE ACTIVITIES

Magic Tricks

Jim and Sadar are talking one day after class.

Sadar: Jim, I have a trick for you. Think of a number between 1 and 10. I will ask you to do some things to this number. Then at the end tell me your result, and I will tell you your number.

Jim: Oh, yeah you probably rig it so the result is my number.

Sadar: Come on Jim, give it a try and see.

Jim: Okay, okay, I thought of a number.

Sadar: Good, now write it down, and don't let me see your paper. Now add x. Got that? Now multiply everything by 2.

Jim: Hey, I didn't know you were going to make me think! This is algebra!

Sadar: I know, now just do it. Okay, now square the polynomial. Got that? Now subtract $4x^2$.

Jim: How did you know I had a $4x^2$? I told you this was rigged!

Sadar: Of course it's rigged, or it wouldn't work. Do you want to finish or not?

Jim: Yeah, I guess so. Go ahead, what do I do next?

Sadar: Divide by 4. Okay, now subtract the x-term.

Jim: Just any old x-term? Got any particular coefficient in mind?

Grouping: Two students per group
Topic: Practice with exponent rules, multiplying polynomials

Sadar: Now stop teasing me. I know you only have one x-term left, so subtract it.

Jim: Ha, ha, I *could* give you a hint about the coefficient, but that wouldn't be fair, would it?

Sadar: Well you could, and then I could tell you your number, or you could just tell me the number you have left after subtracting.

Jim: Okay, the number I had left at the end was 25. Let's see if you can tell me what the coefficient of the x-term I subtracted is.

Sadar: Aha, then the number you chose at the beginning was 5, and the coefficient was 10!

Jim: Hey, you're right! How did you do that?

In your group, follow Sadar's instructions and determine why she knew Jim's number. Make up another set of instructions to use as a magic trick. Be sure to use variables and some of the exponent rules or rules for multiplying polynomials that you learned in this chapter. Exchange instructions with another group and see whether you can figure out how their trick works.

W R A P - U P

SUMMARY

Definitions		**Examples**
Definition of negative integral exponents	If a is a nonzero real number and n is a positive integer, then $$a^{-n} = \frac{1}{a^n}.$$	$2^{-3} = \frac{1}{2^3} = 8$
Definition of zero exponent	If a is any nonzero real number, then $a^0 = 1$. The expression 0^0 is undefined.	$3^0 = 1$

Rules of Exponents		**Examples**

If a and b are nonzero real numbers and m and n are integers, then the following rules hold.

Negative exponent rules	$a^{-n} = \left(\frac{1}{a}\right)^n, \quad a^{-1} = \frac{1}{a}, \quad \text{and} \quad \frac{1}{a^{-n}} = a^n$ Find the power and reciprocal in either order.	$5^{-1} = \frac{1}{5}, \frac{1}{5^{-3}} = 5^3$ $\left(\frac{2}{3}\right)^{-2} = \left(\frac{3}{2}\right)^2$
Product rule	$a^m \cdot a^n = a^{m+n}$	$3^5 \cdot 3^7 = 3^{12}, 2^{-3} \cdot 2^{10} = 2^7$
Quotient rule	$\dfrac{a^m}{a^n} = a^{m-n}$	$\dfrac{x^8}{x^5} = x^3, \dfrac{5^4}{5^{-7}} = 5^{11}$
Power of a power rule	$(a^m)^n = a^{mn}$	$(5^2)^3 = 5^6$
Power of a product rule	$(ab)^n = a^n b^n$	$(2x)^3 = 8x^3$ $(2x^3)^4 = 16x^{12}$
Power of a quotient rule	$\left(\dfrac{a}{b}\right)^n = \dfrac{a^n}{b^n}$	$\left(\dfrac{x}{3}\right)^2 = \dfrac{x^2}{9}$

Scientific Notation		**Examples**
Converting from scientific notation	1. Determine the number of places to move the decimal point by examining the exponent on the 10.	$4 \times 10^3 = 4000$
	2. Move to the right for a positive exponent and to the left for a negative exponent.	$3 \times 10^{-4} = 0.0003$

| Converting to scientific notation | 1. Count the number of places (n) that the decimal point must be moved so that it will follow the first nonzero digit of the number.
2. If the original number was larger than 10, use 10^n.
3. If the original number was smaller than 1, use 10^{-n}. | $67{,}000 = 6.7 \times 10^4$

$0.009 = 9 \times 10^{-3}$ |

Polynomials

		Examples
Term of a polynomial	The product of a number (coefficient) and one or more variables raised to whole number powers	$3x^4,\ -2xy^2,\ 5$
Polynomial	A single term or a finite sum of terms	$x^5 - 3x^2 + 7$
Adding or subtracting polynomials	Add or subtract the like terms.	$(x + 3) + (x - 7) = 2x - 4$ $(x^2 - 2x) - (3x^2 - x) = -2x^2 - x$
Multiplying two polynomials	Multiply each term of the first polynomial by each term of the second polynomial, then combine like terms.	$(x^2 + 2x + 3)(x + 1)$ $= (x^2 + 2x + 3)x + (x^2 + 2x + 3)1$ $= x^3 + 2x^2 + 3x + x^2 + 2x + 3$ $= x^3 + 3x^2 + 5x + 3$

Dividing polynomials	Ordinary division or long division dividend = (quotient)(divisor) + remainder $\dfrac{\text{dividend}}{\text{divisor}} = \text{quotient} + \dfrac{\text{remainder}}{\text{divisor}}$	$$\begin{array}{r} x - 7 \\ x + 2\overline{)x^2 - 5x - 14} \\ \underline{x^2 + 2x} \\ -7x - 14 \\ \underline{-7x - 14} \\ 0 \end{array}$$

Synthetic division	A condensed version of long division, used only for dividing by a polynomial of the form $x - c$ If the remainder is 0, then the dividend factors as dividend = (quotient)(divisor).	$$\begin{array}{r} -2\ \big|\ \begin{array}{rrr} 1 & -5 & -14 \\ & -2 & 14 \end{array} \\ \hline \begin{array}{rrr} 1 & -7 & 0 \end{array} \end{array}$$ $x^2 - 5x - 14 = (x - 7)(x + 2)$

Shortcuts for Multiplying Two Binomials

		Examples
FOIL	The product of two binomials can be found quickly by multiplying their **F**irst, **O**uter, **I**nner, and **L**ast terms.	$(x + 2)(x + 3) = x^2 + 5x + 6$
Square of a sum	$(a + b)^2 = a^2 + 2ab + b^2$	$(x + 5)^2 = x^2 + 10x + 25$
Square of a difference	$(a - b)^2 = a^2 - 2ab + b^2$	$(m - 3)^2 = m^2 - 6m + 9$
Product of a sum and a difference	$(a + b)(a - b) = a^2 - b^2$	$(x + 3)(x - 3) = x^2 - 9$

Factoring		**Examples**
Factoring a polynomial	Write a polynomial as a product of two or more polynomials. A polynomial is factored completely if it is a product of prime polynomials.	$3x^2 - 3 = 3(x^2 - 1)$ $\quad\quad = 3(x + 1)(x - 1)$
Common factors	Factor out the greatest common factor (GCF).	$2x^3 - 6x = 2x(x^2 - 3)$
Difference of two squares	$a^2 - b^2 = (a + b)(a - b)$ (The sum of two squares $a^2 + b^2$ is prime.)	$m^2 - 25 = (m + 5)(m - 5)$ $m^2 + 25$ is prime.
Perfect square trinomials	$a^2 + 2ab + b^2 = (a + b)^2$ $a^2 - 2ab + b^2 = (a - b)^2$	$x^2 + 10x + 25 = (x + 5)^2$ $x^2 - 6x + 9 = (x - 3)^2$
Difference of two cubes	$a^3 - b^3 = (a - b)(a^2 + ab + b^2)$	$x^3 - 8 = (x - 2)(x^2 + 2x + 4)$
Sum of two cubes	$a^3 + b^3 = (a + b)(a^2 - ab + b^2)$	$x^3 + 27 = (x + 3)(x^2 - 3x + 9)$
Grouping	Factor out common factors from groups of terms.	$3x + 3w + bx + bw$ $= 3(x + w) + b(x + w)$ $= (3 + b)(x + w)$
Factoring $ax^2 + bx + c$	By the *ac* method: 1. Find two numbers that have a product equal to *ac* and a sum equal to *b*. 2. Replace *bx* by two terms using the two new numbers as coefficients. 3. Factor the resulting four-term polynomial by grouping.	$2x^2 + 7x + 3$ $ac = 6, b = 7, 1 \cdot 6 = 6, 1 + 6 = 7$ $2x^2 + 7x + 3$ $= 2x^2 + x + 6x + 3$ $= (2x + 1)x + (2x + 1)3$ $= (2x + 1)(x + 3)$
	By trial and error: Try possibilities by considering factors of the first term and factors of the last term. Check them by FOIL.	$12x^2 + 19x - 18$ $= (3x - 2)(4x + 9)$
Substitution	Use substitution on higher-degree polynomials to reduce the degree to 2 or 3.	$x^4 - 3x^2 - 18$ Let $a = x^2$. $a^2 - 3a - 18$

Solving Equations by Factoring		**Examples**
Strategy	1. Write the equation with 0 on the right-hand side. 2. Factor the left-hand side. 3. Set each factor equal to 0. 4. Solve the simpler equations. 5. Check the answers in the original equation.	$x^2 - 3x - 18 = 0$ $(x - 6)(x + 3) = 0$ $x - 6 = 0$ or $x + 3 = 0$ $\quad\quad x = 6$ or $\quad\quad x = -3$ $6^2 - 3(6) - 18 = 0$ $(-3)^2 - 3(-3) - 18 = 0$

ENRICHING YOUR MATHEMATICAL WORD POWER

For each mathematical term, choose the correct meaning.

1. **polynomial**
   a. four or more terms
   b. many numbers
   c. a sum of four or more numbers
   d. a single term or a finite sum of terms   d

2. **degree of a polynomial**
   a. the number of terms in a polynomial
   b. the highest degree of any of the terms of a polynomial
   c. the value of a polynomial when $x = 0$
   d. the largest coefficient of any of the terms of a polynomial   b

3. **leading coefficient**
   a. the first coefficient
   b. the largest coefficient
   c. the coefficient of the first term when a polynomial is written with decreasing exponents
   d. the most important coefficient   c

4. **monomial**
   a. a single polynomial
   b. one number
   c. an equation that has only one solution
   d. a polynomial that has one term   d

5. **FOIL**
   a. a method for adding polynomials
   b. first, outer, inner, last
   c. an equation with no solution
   d. a polynomial with five terms   b

6. **binomial**
   a. a polynomial with two terms
   b. any two numbers
   c. the two coordinates in an ordered pair
   d. an equation with two variables   a

7. **scientific notation**
   a. the notation of rational exponents
   b. the notation of algebra
   c. a notation for expressing large or small numbers with powers of 10
   d. radical notation   c

8. **trinomial**
   a. a polynomial with three terms
   b. an ordered triple of real numbers
   c. a sum of three numbers
   d. a product of three numbers   a

9. **synthetic division**
   a. division of nonreal numbers
   b. division by zero
   c. multiplication that looks like division
   d. a quick method for dividing by $x - c$ d

10. **factor**
    a. to write an expression as a product
    b. to multiply
    c. what two numbers have in common
    d. to FOIL   a

11. **prime number**
    a. a polynomial that cannot be factored
    b. a number with no divisors
    c. an integer between 1 and 10
    d. an integer larger than 1 that has no integral factors other than itself and 1   d

12. **greatest common factor**
    a. the least common multiple
    b. the least common denominator
    c. the largest integer that is a factor of two or more integers
    d. the largest number in a product   c

13. **prime polynomial**
    a. a polynomial that has no factors
    b. a product of prime numbers
    c. a first-degree polynomial
    d. a monomial   a

14. **factor completely**
    a. to factor by grouping
    b. to factor out a prime number
    c. to write as a product of primes
    d. to factor by trial-and-error   c

15. **sum of two cubes**
    a. $(a + b)^3$
    b. $a^3 + b^3$
    c. $a^3 - b^3$
    d. $a^3 b^3$ b

16. **quadratic equation**
    a. $ax + b = 0$, where $a \neq 0$
    b. $ax + b = cx + d$
    c. $ax^2 + bx + c = 0$, where $a \neq 0$
    d. any equation with four terms   c

17. **zero factor property**
    a. If $ab = 0$, then $a = 0$ or $b = 0$
    b. $a \cdot 0 = 0$ for any a
    c. $a = a + 0$ for any real number a
    d. $a + (-a) = 0$ for any real number a a

18. **difference of two squares**
    a. $a^3 - b^3$
    b. $2a - 2b$
    c. $a^2 - b^2$
    d. $(a - b)^2$ c

REVIEW EXERCISES

5.1 *Simplify each expression. Assume all variables represent nonzero real numbers. Write your answers with positive exponents.*

1. $2 \cdot 2 \cdot 2^{-1}$ 2

2. $5^{-1} \cdot 5$ 1

3. $2^2 \cdot 3^2$ 36

4. $3^2 \cdot 5^2$ 225

5. $(-3)^{-3}$ $-\dfrac{1}{27}$

6. $(-2)^{-2}$ $\dfrac{1}{4}$

7. $-(-1)^{-3}$ 1

8. $3^4 \cdot 3^7$ 3^{11}

9. $2x^3 \cdot 4x^{-6}$ $\dfrac{8}{x^3}$

10. $-3a^{-3} \cdot 4a^{-4}$ $-\dfrac{12}{a^7}$

11. $\dfrac{y^{-5}}{y^{-3}}$ $\dfrac{1}{y^2}$

12. $\dfrac{w^3}{w^{-3}}$ w^6

13. $\dfrac{a^5 \cdot a^{-2}}{a^{-4}}$ a^7

14. $\dfrac{2m^3 \cdot m^6}{2m^{-2}}$ m^{11}

15. $\dfrac{6x^{-2}}{3x^2}$ $\dfrac{2}{x^4}$

16. $\dfrac{-5y^2 x^{-3}}{5y^{-2} x^7}$ $-\dfrac{y^4}{x^{10}}$

Write each number in standard notation.

17. 8.36×10^6 $8{,}360{,}000$

18. 3.4×10^7 $34{,}000{,}000$

19. 5.7×10^{-4} 0.00057

20. 4×10^{-3} 0.004

Write each number in scientific notation.

21. $8{,}070{,}000$ 8.07×10^6

22. $90{,}000$ 9×10^4

23. 0.000709 7.09×10^{-4}

24. 0.0000005 5×10^{-7}

Perform each computation without a calculator. Write the answer in scientific notation.

25. $\dfrac{(4{,}000{,}000{,}000)(0.0000006)}{(0.000012)(2{,}000{,}000)}$ 1×10^2

26. $\dfrac{(1.2 \times 10^{32})(2 \times 10^{-5})}{4 \times 10^{-7}}$ 6×10^{33}

5.2 *Simplify each expression. Assume all variables represent nonzero real numbers. Write your answers with positive exponents.*

27. $(a^{-3})^{-2} \cdot a^{-7}$ $\dfrac{1}{a}$

28. $(-3x^{-2}y)^{-4}$ $\dfrac{x^8}{81y^4}$

29. $(m^2 n^3)^{-2}(m^{-3}n^2)^4$ $\dfrac{n^2}{m^{16}}$

30. $(w^{-3}xy)^{-1}(wx^{-3}y)^2$ $\dfrac{w^5 y}{x^7}$

31. $\left(\dfrac{2}{3}\right)^{-4}$ $\dfrac{81}{16}$

32. $\left(\dfrac{a^4}{3}\right)^{-2}$ $\dfrac{9}{a^8}$

33. $\left(\dfrac{1}{2} + \dfrac{1}{3}\right)^2$ $\dfrac{25}{36}$

34. $\left(\dfrac{1}{2} - \dfrac{1}{3}\right)^{-2}$ 36

35. $\left(-\dfrac{3a}{4b^{-1}}\right)^{-1}$ $-\dfrac{4}{3ab}$

36. $\left(-\dfrac{4x^5}{5y^{-3}}\right)^{-1}$ $-\dfrac{5}{4x^5 y^3}$

37. $\dfrac{(a^{-3}b)^4}{(ab^2)^{-5}}$ $\dfrac{b^{14}}{a^7}$

38. $\dfrac{(2x^3)^3}{(3x^2)^{-2}}$ $72x^{13}$

Simplify each expression. Assume that the variables represent integers.

39. $5^{2w} \cdot 5^{4w} \cdot 5^{-1}$ 5^{6w-1}

40. $3^y (3^{2y})^3$ 3^{7y}

41. $\left(\dfrac{7^{3a}}{7^8}\right)^5$ 7^{15a-40}

42. $\left(\dfrac{2^{6-k}}{2^{2-3k}}\right)^3$ 2^{12+6k}

5.3 *Perform the indicated operations.*

43. $(2w - 3) + (6w + 5)$ $8w + 2$

44. $(3a - 2xy) + (5xy - 7a)$ $3xy - 4a$

45. $(x^2 - 3x - 4) - (x^2 + 3x - 7)$ $-6x + 3$

46. $(7 - 2x - x^2) - (x^2 - 5x + 6)$ $-2x^2 + 3x + 1$

47. $(x^2 - 2x + 4)(x - 2)$ $x^3 - 4x^2 + 8x - 8$

48. $(x + 5)(x^2 - 2x + 10)$ $x^3 + 3x^2 + 50$

49. $xy + 7z - 5(xy - 3z)$ $-4xy + 22z$

50. $7 - 4(x - 3)$ $-4x + 19$

51. $m^2(5m^3 - m + 2)$ $5m^5 - m^3 + 2m^2$

52. $(a + 2)^3$ $a^3 + 6a^2 + 12a + 8$

5.4 *Perform the following computations mentally. Write down only the answers.*

53. $(x - 3)(x + 7)$ $x^2 + 4x - 21$

54. $(k - 5)(k + 4)$ $k^2 - k - 20$

55. $(z - 5y)(z + 5y)$ $z^2 - 25y^2$

56. $(m - 3)(m + 3)$ $m^2 - 9$

57. $(m + 8)^2$ $m^2 + 16m + 64$

58. $(b + 2a)^2$ $b^2 + 4ab + 4a^2$

59. $(w - 6x)(w - 4x)$ $w^2 - 10xw + 24x^2$

60. $(2w - 3)(w + 6)$ $2w^2 + 9w - 18$

61. $(k - 3)^2$ $k^2 - 6k + 9$

62. $(n - 5)^2$ $n^2 - 10n + 25$

63. $(m^2 - 5)(m^2 + 5)$ $m^4 - 25$

64. $(3k^2 - 5t)(2k^2 + 6t)$ $6k^4 + 8k^2 t - 30t^2$

5.5 *Find the quotient and remainder.*

65. $(x^3 + x^2 - 11x + 10) \div (x - 2)$ $x^2 + 3x - 5, 0$

66. $(2x^3 + 5x^2 + 9) \div (x + 3)$ $2x^2 - x + 3, 0$

67. $(m^4 - 1) \div (m + 1)$ $m^3 - m^2 + m - 1, 0$

68. $(x^4 - 1) \div (x - 1)$ $x^3 + x^2 + x + 1, 0$

69. $(a^9 - 8) \div (a^3 - 2)$ $a^6 + 2a^3 + 4, 0$

70. $(a^2 - b^2) \div (a - b)$ $a + b, 0$

71. $(3m^3 + 6m^2 - 18m) \div (3m)$ $m^2 + 2m - 6, 0$

72. $(w - 3) \div (3 - w)$ $-1, 0$

Rewrite each expression in the form

$$quotient + \frac{remainder}{divisor}.$$

Use synthetic division.

73. $\dfrac{x^2 - 5}{x - 1}$ $x + 1 + \dfrac{-4}{x - 1}$

74. $\dfrac{x^2 + 3x + 2}{x + 3}$ $x + \dfrac{2}{x + 3}$

75. $\dfrac{3x}{x - 2}$ $3 + \dfrac{6}{x - 2}$

76. $\dfrac{4x}{x - 5}$ $4 + \dfrac{20}{x - 5}$

Determine whether the first polynomial is a factor of the second. Use synthetic division when possible.

77. $x + 2$, $x^3 - 2x^2 + 3x + 22$ Yes

78. $x - 2$, $x^3 + x - 10$ Yes

79. $x - 5$, $x^3 - x - 120$ Yes

80. $x + 3$, $x^3 + 2x + 15$ No

81. $x - 1$, $x^3 + x^2 - 3$ No

82. $x - 1$, $x^3 + 1$ No

83. $x^2 + 2$, $x^4 + x^3 + 5x^2 + 2x + 6$ Yes

84. $x^2 + 1$, $x^4 - 1$ Yes

5.6 *Complete the factoring by filling in the parentheses.*

85. $3x - 6 = 3(\quad)$ $3(x - 2)$

86. $7x^2 - x = x(\quad)$ $x(7x - 1)$

87. $4a - 20 = -4(\quad)$ $-4(-a + 5)$

88. $w^2 - w = -w(\quad)$ $-w(-w + 1)$

89. $3w - w^2 = -w(\quad)$ $-w(w - 3)$

90. $3x - 6 = (\quad)(2 - x)$ $(-3)(2 - x)$

Factor each polynomial.

91. $y^2 - 81$ $(y - 9)(y + 9)$

92. $r^2t^2 - 9v^2$ $(rt - 3v)(rt + 3v)$

93. $4x^2 + 28x + 49$ $(2x + 7)^2$

94. $y^2 - 20y + 100$ $(y - 10)^2$

95. $t^2 - 18t + 81$ $(t - 9)^2$

96. $4w^2 + 4ws + s^2$ $(2w + s)^2$

97. $t^3 - 125$ $(t - 5)(t^2 + 5t + 25)$

98. $8y^3 + 1$ $(2y + 1)(4y^2 - 2y + 1)$

5.7 *Factor each polynomial.*

99. $x^2 - 7x - 30$ $(x - 10)(x + 3)$

100. $y^2 + 4y - 32$ $(y + 8)(y - 4)$

101. $w^2 - 3w - 28$ $(w - 7)(w + 4)$

102. $6t^2 - 5t + 1$ $(2t - 1)(3t - 1)$

103. $2m^2 + 5m - 7$ $(2m + 7)(m - 1)$

104. $12x^2 - 17x + 6$ $(4x - 3)(3x - 2)$

105. $m^7 - 3m^4 - 10m$ $m(m^3 - 5)(m^3 + 2)$

106. $6w^5 - 7w^3 - 5w$ $w(3w^2 - 5)(2w^2 + 1)$

5.8 *Factor each polynomial completely.*

107. $5x^3 + 40$ $5(x + 2)(x^2 - 2x + 4)$

108. $w^3 - 6w^2 + 9w$ $w(w - 3)^2$

109. $9x^2 + 9x + 2$ $(3x + 2)(3x + 1)$

110. $ax^3 + a$ $a(x + 1)(x^2 - x + 1)$

111. $x^3 + x^2 - x - 1$ $(x - 1)(x + 1)^2$

112. $16x^2 - 4x - 2$ $2(4x + 1)(2x - 1)$

113. $-x^2y + 16y$ $-y(x - 4)(x + 4)$

114. $-5m^2 + 5$ $-5(m - 1)(m + 1)$

115. $-a^3b^2 + 2a^2b^2 - ab^2$ $-ab^2(a - 1)^2$

116. $-2w^2 - 16w - 32$ $-2(w + 4)^2$

117. $x^3 - x^2 + 9x - 9$ $(x - 1)(x^2 + 9)$

118. $w^4 + 2w^2 - 3$ $(w^2 + 3)(w - 1)(w + 1)$

119. $x^4 - x^2 - 12$ $(x - 2)(x + 2)(x^2 + 3)$

120. $8x^3 - 1$ $(2x - 1)(4x^2 + 2x + 1)$

121. $a^6 - a^3$ $a^3(a - 1)(a^2 + a + 1)$

122. $a^2 - ab + 2a - 2b$ $(a + 2)(a - b)$

123. $-8m^2 - 24m - 18$ $-2(2m + 3)^2$

124. $-3x^2 - 9x + 30$ $-3(x + 5)(x - 2)$

125. $(2x - 3)^2 - 16$ $(2x - 7)(2x + 1)$

126. $(m - 6)^2 - (m - 6) - 12$ $(m - 10)(m - 3)$

127. $x^6 + 7x^3 - 8$ $(x + 2)(x^2 - 2x + 4)(x - 1)(x^2 + x + 1)$

128. $32a^5 - 2a$ $2a(2a - 1)(2a + 1)(4a^2 + 1)$

129. $(a^2 - 9)^2 - 5(a^2 - 9) + 6$ $(a^2 - 11)(a^2 - 12)$

130. $x^3 - 9x + x^2 - 9$ $(x + 1)(x - 3)(x + 3)$

Factor each polynomial completely. Variables used as exponents represent positive integers.

131. $x^{2k} - 49$ $(x^k - 7)(x^k + 7)$

132. $x^{6k} - 1$ $(x^k - 1)(x^{2k} + x^k + 1)(x^k + 1)(x^{2k} - x^k + 1)$

133. $m^{2a} - 2m^a - 3$ $(m^a - 3)(m^a + 1)$

134. $2y^{2n} - 7y^n + 6$ $(2y^n - 3)(y^n - 2)$

135. $9z^{2k} - 12z^k + 4$ $(3z^k - 2)^2$

136. $25z^{6m} + 20z^{3m} + 4$ $(5z^{3m} + 2)^2$

137. $y^{2a} - by^a + cy^a - bc$ $(y^a - b)(y^a + c)$

138. $x^3y^b - xy^b + 2x^3 - 2x$ $x(y^b + 2)(x - 1)(x + 1)$

5.9 *Solve each equation.*

139. $x^3 - 5x^2 = 0$ $\{0, 5\}$

140. $2m^2 + 10m + 12 = 0$ $\{-3, -2\}$

141. $(a - 2)(a - 3) = 6$ $\{0, 5\}$

142. $(w - 2)(w + 3) = 50$ $\{-8, 7\}$

143. $2m^2 - 9m - 5 = 0$ $\{-1/2, 5\}$

144. $m^3 + 4m^2 - 9m - 36 = 0$ $\{-4, -3, 3\}$

145. $w^3 + 5w^2 - w - 5 = 0$ $\{-5, -1, 1\}$

146. $12x^2 + 5x - 3 = 0$ $\left\{-\dfrac{3}{4}, \dfrac{1}{3}\right\}$

147. $|x^2 - 5| = 4$ $\{-3, -1, 1, 3\}$

148. $|x^2 - 3x - 7| = 3$ $\{-2, -1, 4, 5\}$

MISCELLANEOUS

Solve each problem.

149. *Roadrunner and the coyote.* The roadrunner has just taken a position atop a giant saguaro cactus. While positioning a 10-foot Acme ladder against the cactus, Wile E. Coyote notices a warning label on the ladder. For safety, Acme recommends that the distance from the ground to the top of the ladder, measured vertically along the cactus, must be 2 feet longer than the distance between the bottom of the ladder and the cactus. How far from the cactus should he place the bottom of this ladder?   6 feet

150. *Three consecutive integers.* Find three consecutive integers such that the sum of their squares is 50.
$-5, -4, -3$, or $3, 4, 5$

151. *Perimeter of a square.* If the area of a square field is $9a^2 + 6a + 1$ square kilometers, then what is its perimeter? $12a + 4$ kilometers

152. *Landscape design.* Rico planted red tulips in a square flower bed with an area of x^2 square feet (ft²). He plans to surround the tulips with daffodils in a uniform border with a width of 3 feet. Write a polynomial that represents the area planted in daffodils. $12x + 36$ ft²

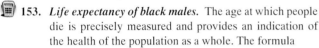 **153.** *Life expectancy of black males.* The age at which people die is precisely measured and provides an indication of the health of the population as a whole. The formula

$$L = 64.3(1.0033)^a$$

can be used to model life expectancy L for U.S. black males with present age a (National Center for Health Statistics, www.cdc.gov/nchswww).

a) To what age can a 20-year-old black male expect to live? 68.7 years

b) How many more years is a 20-year-old white male expected to live than a 20-year-old black male? (See Section 5.2 Exercise 95.) 6.5 years

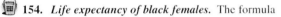 **154.** *Life expectancy of black females.* The formula

$$L = 72.9(1.002)^a$$

can be used to model life expectancy for U.S. black females with present age a. How long can a 20-year-old black female expect to live? 75.9 years

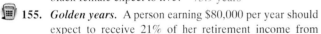 **155.** *Golden years.* A person earning $80,000 per year should expect to receive 21% of her retirement income from Social Security and the rest from personal savings. To calculate the amount of regular savings, we use the formula

$$S = R \cdot \frac{(1 + i)^n - 1}{i},$$

where S is the amount at the end of n years of n investments of R dollars each year earning interest rate i compounded annually.

a) Use the accompanying graph to estimate the interest rate needed to get an investment of $1 per year for 20 years to amount to $100. 15%

b) Use the formula to determine the annual savings for 20 years that would amount to $500,000 at 7% compounded annually. $12,196.46

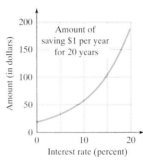

FIGURE FOR EXERCISE 155

156. *Costly education.* The average cost for tuition, room, and board for one year at a private college was $16,222 for 1994–1995 (National Center for Educational Statistics, www.nces.ed.gov). Use the formula in the previous exercise to find the annual savings for 18 years that would amount to $16,222 with an annual return of 8%. $433.16

CHAPTER 5 TEST

Simplify each expression. Assume all variables represent nonzero real numbers. Exponents in your answers should be positive exponents.

1. 3^{-2} $\dfrac{1}{9}$

2. $\dfrac{1}{6^{-2}}$ 36

3. $\left(\dfrac{1}{2}\right)^{-3}$ 8

4. $3x^4 \cdot 4x^3$ $12x^7$

5. $\dfrac{8y^9}{2y^{-3}}$ $4y^{12}$

6. $(4a^2b)^3$ $64a^6b^3$

7. $\left(\dfrac{x^2}{3}\right)^{-3}$ $\dfrac{27}{x^6}$

8. $\dfrac{(2^{-1}a^2b)^{-3}}{4a^{-9}}$ $\dfrac{2a^3}{b^3}$

Convert to standard notation.

9. 3.24×10^9 3,240,000,000

10. 8.673×10^{-4} 0.0008673

Perform each computation by converting each number to scientific notation. Give the answer in scientific notation.

11. $\dfrac{(80,000)(0.0006)}{2,000,000}$ 2.4×10^{-5}

12. $\dfrac{(0.00006)^2(500)}{(30,000)^2(0.01)}$ 2×10^{-13}

Perform the indicated operations.

13. $(3x^3 - x^2 + 6) + (4x^2 - 2x - 3)$ $3x^3 + 3x^2 - 2x + 3$

14. $(x^2 - 6x - 7) - (3x^2 + 2x - 4)$ $-2x^2 - 8x - 3$

15. $(x^2 - 3x + 7)(x - 2)$ $x^3 - 5x^2 + 13x - 14$

16. $(x^3 + 7x^2 + 7x - 15) \div (x + 3)$ $x^2 + 4x - 5$

17. $(x - 2)^3$ $x^3 - 6x^2 + 12x - 8$

18. $(x - 3) \div (3 - x)$ -1

Find the products.

19. $(x - 7)(2x + 3)$
$2x^2 - 11x - 21$

20. $(x - 6)^2$
$x^2 - 12x + 36$

21. $(2x + 5)^2$
$4x^2 + 20x + 25$

22. $(3y^2 - 5)(3y^2 + 5)$
$9y^4 - 25$

Rewrite each expression in the form

$$quotient + \frac{remainder}{divisor}.$$

Use synthetic division.

23. $\dfrac{5x}{x + 3}$ $5 + \dfrac{-15}{x + 3}$

24. $\dfrac{x^2 + 3x - 6}{x - 2}$ $x + 5 + \dfrac{4}{x - 2}$

Factor completely.

25. $a^2 - 2a - 24$ $(a - 6)(a + 4)$

26. $4x^2 + 28x + 49$ $(2x + 7)^2$

27. $3m^3 - 24$ $3(m - 2)(m^2 + 2m + 4)$

28. $2x^2y - 32y$ $2y(x - 4)(x + 4)$

29. $2xa + 3a - 10x - 15$ $(a - 5)(2x + 3)$

30. $x^4 + 3x^2 - 4$ $(x - 1)(x + 1)(x^2 + 4)$

Solve each equation.

31. $2m^2 + 7m - 15 = 0$ $\left\{-5, \dfrac{3}{2}\right\}$

32. $x^3 - 4x = 0$ $\{-2, 0, 2\}$

33. $|x^2 + x - 9| = 3$ $\{-4, -3, 2, 3\}$

Write a complete solution for each problem.

34. A portable television is advertised as having a 10-inch diagonal measure screen. If the width of the screen is 2 inches more than the height, then what are the dimensions of the screen?   Width 8 inches, height 6 inches

35. The infant mortality rate for the United States, the number of deaths per 100,000 live births, has decreased dramatically since 1950. The formula

$$d = (1.8 \times 10^{28})(1.032)^{-y}$$

gives the infant mortality rate d as a function of the year y (National Center for Health Statistics, www.cdc.gov/nchswww). Find the infant mortality rates in 1950, 1990, and 2000.   38.0, 10.8, 7.9

Simplify each expression.

1. 4^2 16

2. $4(-2)$ -8

3. 4^{-2} $\dfrac{1}{16}$

4. $2^3 \cdot 4^{-1}$ 2

5. $2^{-1} + 2^{-1}$ 1

6. $2^{-1} \cdot 3^{-1}$ $\dfrac{1}{6}$

7. $3^{-1} - 2^{-2}$ $\dfrac{1}{12}$

8. $3^2 - 4(5)(-2)$ 49

9. $2^7 - 2^6$ 64

10. $0.08(32) + 0.08(68)$ 8

11. $3 - 2\,|\,5 - 7 \cdot 3\,|$ -29

12. $5^{-1} + 6^{-1}$ $\dfrac{11}{30}$

Solve each equation.

13. $0.05a - 0.04(a - 50) = 4$ $\{200\}$

14. $15b - 27 = 0$ $\left\{\dfrac{9}{5}\right\}$

15. $2c^2 + 15c - 27 = 0$ $\left\{-9, \dfrac{3}{2}\right\}$

16. $2t^2 + 15t = 0$ $\left\{-\dfrac{15}{2}, 0\right\}$

17. $|15u - 27| = 3$ $\left\{2, \dfrac{8}{5}\right\}$

18. $|15v - 27| = 0$ $\left\{\dfrac{9}{5}\right\}$

19. $|15x - 27| = -78$ 0

20. $|x^2 + x - 4| = 2$ $\{-3, -2, 1, 2\}$

21. $(2x - 1)(x + 5) = 0$ $\left\{-5, \dfrac{1}{2}\right\}$

22. $|3x - 1| + 6 = 9$ $\left\{-\dfrac{2}{3}, \dfrac{4}{3}\right\}$

23. $(1.5 \times 10^{-4})w - 5 \times 10^5 = 7 \times 10^6$ 5×10^{10}

24. $(3 \times 10^7)(y - 5 \times 10^3) = 6 \times 10^{12}$ 2.05×10^5

Solve each problem.

25. *Negative income tax.* In a negative income tax proposal, the function

$$D = 0.75E + 5000$$

is used to determine the disposable income D (the amount available for spending) for an earned income E (the amount earned). If $E > D$, then the difference is paid in federal taxes. If $D > E$, then the difference is paid to the wage earner by Uncle Sam.

a) Find the amount of tax paid by a person who earns $100,000. $20,000

b) Find the amount received from Uncle Sam by a person who earns $10,000. $2,500

c) The accompanying graph shows the lines $D = 0.75E + 5000$ and $D = E$. Find the intersection of these lines. $20,000

d) How much tax does a person pay whose earned income is at the intersection found in part (c)?   0

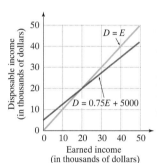

FIGURE FOR EXERCISE 25

CHAPTER 6

Rational Expressions

nformation is everywhere—in the newspapers and magazines we read, the televisions we watch, and the computers we use. And now people are talking about the Information Superhighway, which will deliver vast amounts of information directly to consumers' homes. In the future the combination of telephone, television, and computer will give us on-the-spot health care recommendations, video conferences, home shopping, and perhaps even electronic voting and driver's license renewal, to name just a few. There is even talk of 500 television channels!

Some experts are concerned that the consumer will give up privacy for this technology. Others worry about regulation, access, and content of the enormous international computer network.

Whatever the future of this technology, few people understand how all their electronic devices work. However, this vast array of electronics rests on physical principles, which are described by mathematical formulas. In Exercises 49 and 50 of Section 6.6 we will see that the formula governing resistance for receivers connected in parallel involves rational expressions, which are the subject of this chapter.

6.1 PROPERTIES OF RATIONAL EXPRESSIONS

A ratio of two integers is called a rational number; a ratio of two polynomials is called a rational expression. Rational expressions are as fundamental to algebra as rational numbers are to arithmetic. In this section we look carefully at some of the properties of rational numbers and see how they extend to rational expressions.

Definition of Rational Expressions

A **rational expression** is the ratio of two polynomials with the denominator not equal to zero. For example,

$$\frac{2}{3}, \qquad 3a + 5, \qquad \frac{x - 3}{2x^2 - 2}, \qquad \frac{y + 2}{5y}, \qquad \text{and} \qquad \frac{x - 2}{x + 1}$$

are rational expressions. The rational number $\frac{2}{3}$ is a rational expression because 2 and 3 are monomials and $\frac{2}{3}$ is a ratio of two monomials. If the denominator of a rational expression is 1, it is usually omitted, as in the expression $3a + 5$.

Domain

If the domain consists of all real numbers except -5, some people write $R - \{-5\}$ for the domain. Even though there are several ways to indicate the domain, you should keep practicing interval notation because it is used in algebra, trigonometry, and calculus.

The **domain** of a rational expression is the set of all real numbers that can be used in place of the variable. Because the denominator of a rational expression cannot be zero, the domain of a rational expression consists of the set of real numbers except those that cause the denominator to be zero. The domain of

$$\frac{x}{x + 5}$$

is the set of all real numbers excluding -5. In set-builder notation this set is written as

$$\{x \mid x \neq -5\},$$

and in interval notation it is written as

$$(-\infty, -5) \cup (-5, \infty).$$

EXAMPLE 1   **Domain**

Find the domain of each rational expression.

a) $\dfrac{x - 2}{x + 9}$

b) $\dfrac{y + 2}{5y}$

c) $\dfrac{x - 3}{2x^2 - 2}$

Solution

a) The denominator is zero if $x + 9 = 0$ or $x = -9$. The domain is $\{x \mid x \neq -9\}$ or

$$(-\infty, -9) \cup (-9, \infty).$$

b) The denominator is zero if $5y = 0$ or $y = 0$. The domain is $\{y \mid y \neq 0\}$ or

$$(-\infty, 0) \cup (0, \infty).$$

c) The denominator is zero if $2x^2 - 2 = 0$. Solve this equation.

$$2x^2 - 2 = 0$$
$$2(x^2 - 1) = 0 \quad \text{Factor out 2.}$$
$$2(x + 1)(x - 1) = 0 \quad \text{Factor completely.}$$
$$x + 1 = 0 \quad \text{or} \quad x - 1 = 0 \quad \text{Zero factor property}$$
$$x = -1 \quad \text{or} \quad x = 1$$

The domain is the set of all real numbers except -1 and 1. This set is written as $\{x \mid x \neq -1 \text{ and } x \neq 1\}$, or in interval notation as

$$(-\infty, -1) \cup (-1, 1) \cup (1, \infty). \qquad \blacksquare$$

CAUTION   The numbers that you find when you set the denominator equal to zero and solve for x are *not* in the domain of the rational expression. The solutions to that equation are excluded from the domain.

Reducing to Lowest Terms

Each rational number can be written in infinitely many equivalent forms. For example,

$$\frac{2}{3} = \frac{4}{6} = \frac{6}{9} = \frac{8}{12} = \frac{10}{15} = \cdots .$$

Each equivalent form of $\frac{2}{3}$ is obtained from $\frac{2}{3}$ by multiplying both numerator and denominator by the same nonzero number. For example,

$$\frac{2}{3} = \frac{2}{3} \cdot 1 = \frac{2}{3} \cdot \frac{2}{2} = \frac{4}{6} \quad \text{and} \quad \frac{2}{3} = \frac{2}{3} \cdot \frac{3}{3} = \frac{6}{9}.$$

Note that we are actually multiplying $\frac{2}{3}$ by equivalent forms of 1, the multiplicative identity.

If we start with $\frac{4}{6}$ and convert it into $\frac{2}{3}$, we are simplifying by *reducing* $\frac{4}{6}$ to its *lowest terms*. We can reduce as follows:

$$\frac{4}{6} = \frac{\cancel{2} \cdot 2}{\cancel{2} \cdot 3} = \frac{2}{3}.$$

A rational number is expressed in its lowest terms when the numerator and denominator have no common factors other than 1. In reducing $\frac{4}{6}$, we divide the numerator and denominator by the common factor 2, or "divide out" the common factor 2. We can multiply or divide both numerator and denominator of a rational number by the same nonzero number without changing the value of the rational number. This fact is called the **basic principle of rational numbers.**

Basic Principle of Rational Numbers

If $\frac{a}{b}$ is a rational number and c is a nonzero real number, then

$$\frac{a}{b} = \frac{ac}{bc}.$$

CAUTION Although it is true that

$$\frac{5}{6} = \frac{2+3}{2+4},$$

we cannot divide out the 2's in this expression because the 2's are not factors. We can divide out only common *factors* when reducing fractions.

Just as a rational number has infinitely many equivalent forms, a rational expression also has infinitely many equivalent forms. To reduce rational expressions to its lowest terms, we follow exactly the same procedure as we do for rational numbers: *Factor the numerator and denominator completely, then divide out all common factors.*

E X A M P L E 2

Reducing

Reduce each rational expression to its lowest terms.

a) $\dfrac{18}{42}$

b) $\dfrac{-2a^7b}{a^2b^3}$

Solution

a) Factor 18 as $2 \cdot 3^2$ and 42 as $2 \cdot 3 \cdot 7$:

$$\frac{18}{42} = \frac{2 \cdot 3^2}{2 \cdot 3 \cdot 7} \quad \text{Factor.}$$

$$= \frac{3}{7} \qquad \text{Divide out the common factors.}$$

b) Because this expression is already factored, we use the quotient rule for exponents to reduce:

$$\frac{-2a^7b}{a^2b^3} = \frac{-2a^5}{b^2}$$

In the next example we use the techniques for factoring polynomials that we learned in Chapter 5.

> **helpful hint**
>
> A negative sign in a fraction can be placed in three locations:
>
> $$\frac{-1}{2} = \frac{1}{-2} = -\frac{1}{2}$$
>
> The same goes for rational expressions:
>
> $$\frac{-3x^2}{5y} = \frac{3x^2}{-5y} = -\frac{3x^2}{5y}$$

E X A M P L E 3

Reducing

Reduce each rational expression to its lowest terms.

a) $\dfrac{2x^2 - 18}{x^2 + x - 6}$

b) $\dfrac{w - 2}{2 - w}$

c) $\dfrac{2a^3 - 16}{16 - 4a^2}$

Solution

a) $\dfrac{2x^2 - 18}{x^2 + x - 6} = \dfrac{2(x^2 - 9)}{(x - 2)(x + 3)}$ Factor.

$= \dfrac{2(x - 3)(x + 3)}{(x - 2)(x + 3)}$ Factor completely.

$= \dfrac{2x - 6}{x - 2}$ Divide out the common factors.

b) Factor out -1 from the numerator to get a common factor:

$$\frac{w - 2}{2 - w} = \frac{-1(2 - w)}{(2 - w)} = -1$$

c) $\dfrac{2a^3 - 16}{16 - 4a^2} = \dfrac{2(a^3 - 8)}{-4(a^2 - 4)}$ Factoring out -4 will give the common factor $a - 2$.

$$= \frac{2(a - 2)(a^2 + 2a + 4)}{-2 \cdot 2(a - 2)(a + 2)}$$ Difference of two cubes, difference of two squares

$$= -\frac{a^2 + 2a + 4}{2a + 4}$$ Divide out common factors.

The rational expressions in Example 3(a) are equivalent because they have the same value for any replacement of the variables, provided that the replacement is in the domain of both expressions. In other words, the equation

$$\frac{2x^2 - 18}{x^2 + x - 6} = \frac{2x - 6}{x - 2}$$

is an identity. It is true for any value of x except 2 and -3.

The main points to remember for reducing rational expressions are summarized as follows.

> **Strategy for Reducing Rational Expressions**
>
> 1. All reducing is done by dividing out common factors.
> 2. Factor the numerator and denominator completely to see the common factors.
> 3. Use the quotient rule to reduce a ratio of two monomials involving exponents.
> 4. We may have to factor out a common factor with a negative sign to get identical factors in the numerator and denominator.

Building Up the Denominator

In Section 6.3 we will see that only rational expressions with identical denominators can be added or subtracted. Fractions without identical denominators can be converted to equivalent fractions with a common denominator by reversing the procedure for reducing fractions to its lowest terms. This procedure is called **building up the denominator.**

Consider converting the fraction $\frac{1}{3}$ into an equivalent fraction with a denominator of 51. Any fraction that is equivalent to $\frac{1}{3}$ can be obtained by multiplying the numerator and denominator of $\frac{1}{3}$ by the same nonzero number. Because $51 = 3 \cdot 17$, we multiply the numerator and denominator of $\frac{1}{3}$ by 17 to get an equivalent fraction with a denominator of 51:

$$\frac{1}{3} = \frac{1}{3} \cdot 1 = \frac{1}{3} \cdot \frac{17}{17} = \frac{17}{51}$$

E X A M P L E   4

Building up the denominator

Convert each rational expression into an equivalent rational expression that has the indicated denominator.

a) $\dfrac{2}{7}, \dfrac{?}{42}$

b) $\dfrac{5}{3a^2b}, \dfrac{?}{9a^3b^4}$

Solution

a) Factor 42 as $42 = 2 \cdot 3 \cdot 7$, then multiply the numerator and denominator of $\dfrac{2}{7}$ by the missing factors, 2 and 3:

$$\frac{2}{7} = \frac{2 \cdot 2 \cdot 3}{7 \cdot 2 \cdot 3} = \frac{12}{42}$$

b) Because $9a^3b^4 = 3ab^3 \cdot 3a^2b$, we multiply the numerator and denominator by $3ab^3$:

$$\frac{5}{3a^2b} = \frac{5 \cdot 3ab^3}{3a^2b \cdot 3ab^3}$$
$$= \frac{15ab^3}{9a^3b^4}$$

When building up a denominator to match a more complicated denominator, we factor both denominators completely to see which factors are missing from the simpler denominator. Then we multiply the numerator and denominator of the simpler expression by the missing factors.

E X A M P L E   5

Building up the denominator

Convert each rational expression into an equivalent rational expression that has the indicated denominator.

a) $\dfrac{5}{2a - 2b}, \dfrac{?}{6b - 6a}$

b) $\dfrac{x + 2}{x + 3}, \dfrac{?}{x^2 + 7x + 12}$

Solution

a) Factor both $2a - 2b$ and $6b - 6a$ to see which factor is missing in $2a - 2b$. Note that we factor out -6 from $6b - 6a$ to get the factor $a - b$:

$$2a - 2b = 2(a - b)$$
$$6b - 6a = -6(a - b) = -3 \cdot 2(a - b)$$

Now multiply the numerator and denominator by the missing factor, -3:

$$\frac{5}{2a - 2b} = \frac{5(-3)}{(2a - 2b)(-3)} = \frac{-15}{6b - 6a}$$

b) Because $x^2 + 7x + 12 = (x + 3)(x + 4)$, multiply the numerator and denominator by $x + 4$:

$$\frac{x + 2}{x + 3} = \frac{(x + 2)(x + 4)}{(x + 3)(x + 4)} = \frac{x^2 + 6x + 8}{x^2 + 7x + 12}$$

Rational Functions

A rational expression can be used to determine the value of a variable. For example, if

$$y = \frac{3x - 1}{x^2 - 4},$$

then we say that y is a **rational function** of x. We can also use function notation as shown in the next example.

EXAMPLE 6

Evaluating a rational function

Find $R(3)$, $R(-1)$, and $R(2)$ for the rational function

$$R(x) = \frac{3x - 1}{x^2 - 4}.$$

calculator

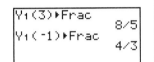

close-up

To check, use Y= to enter
$y_1 = (3x - 1)/(x^2 - 4)$.
Then use the variables feature
(VARS) to find $y_1(3)$ and $y_1(-1)$.

```
Y₁(3)▶Frac
              8/5
Y₁(-1)▶Frac
              4/3
```

Solution

To find $R(3)$, replace x by 3 in the formula:

$$R(3) = \frac{3 \cdot 3 - 1}{3^2 - 4} = \frac{8}{5}$$

To find $R(-1)$, replace x by -1 in the formula:

$$R(-1) = \frac{3(-1) - 1}{(-1)^2 - 4}$$

$$= \frac{-4}{-3} = \frac{4}{3}$$

We cannot find $R(2)$ because 2 is not in the domain of the rational expression. ■

Applications

A rational expression can occur in finding an average cost. The average cost of making a product is the total cost divided by the number of products made.

EXAMPLE 7

Average cost function

Mercedes Benz spent $700 million to develop its new 1999 M class SUV, which will sell for around $40,000 (Motor Trend, July 1998, www.motortrend.com). If the cost of manufacturing the SUV is $30,000 each, then what rational function gives the average cost of developing and manufacturing x vehicles? Compare the average cost per vehicle for manufacturing levels of 10,000 vehicles and 100,000 vehicles.

Solution

The polynomial $30,000x + 700,000,000$ gives the cost in dollars of developing and manufacturing x vehicles. The average cost per vehicle is given by the rational function

$$AC(x) = \frac{30,000x + 700,000,000}{x}.$$

If $x = 10,000$, then

$$AC(10,000) = \frac{30,000(10,000) + 700,000,000}{10,000} = 100,000.$$

If $x = 100,000$, then

$$AC(100,000) = \frac{30,000(100,000) + 700,000,000}{100,000} = 37,000.$$

The average cost per vehicle when 10,000 vehicles are made is $100,000, whereas the average cost per vehicle when 100,000 vehicles are made is $37,000. ■

WARM-UPS

True or false? Explain.

1. A rational number is a rational expression.   True
2. The expression $\frac{2+x}{x-1}$ is a rational expression.   True
3. The domain of the rational expression $\frac{3}{x-2}$ is $\{2\}$.   False
4. The domain of $\frac{2x+5}{(x-9)(2x+1)}$ is $\left\{x \mid x \neq 9 \text{ and } x \neq -\frac{1}{2}\right\}$.   True
5. The domain of $\frac{x-1}{x+2}$ is $(-\infty, -2) \cup (-2, 1) \cup (1, \infty)$.   False
6. The rational expression $\frac{5x+2}{15}$ reduces to $\frac{x+2}{3}$.   False
7. Multiplying the numerator and denominator of $\frac{x}{x-1}$ by x yields $\frac{x^2}{x^2-1}$.
   False
8. The expression $\frac{2}{3-x}$ is equivalent to $\frac{-2}{x-3}$.   True
9. The equation $\frac{4x^3}{6x} = \frac{2x^2}{3}$ is an identity.   True
10. The expression $\frac{x^2-y^2}{x-y}$ reduced to its lowest terms is $x - y$.   False

6.1   EXERCISES

Reading and Writing   *After reading this section, write out the answers to these questions. Use complete sentences.*

1. What is a rational expression?
   A rational expression is a ratio of two polynomials with the denominator not equal to zero.

2. What is the domain of a rational expression?
   The domain of a rational expression is all real numbers except those that cause the denominator to be zero.

3. What is the basic principle of rational numbers?
   The basic principle of rational numbers says that $(ab)/(ac) = b/c$, provided a and c are not zero.

4. How do we reduce a rational expression to lowest terms?
   To reduce a rational expression, factor the numerator and denominator completely and then divide out the common factors.

5. How do you build up the denominator of a rational expression?
   We build up the denominator by multiplying the numerator and denominator by the same expression.

6. What is average cost?
   Average cost is total cost divided by the number of items.

Find the domain of each rational expression. See Example 1.

7. $\frac{3x}{x-1}$ $\{x \mid x \neq 1\}$

8. $\frac{x}{x+5}$ $\{x \mid x \neq -5\}$

9. $\frac{2z-5}{7z}$ $\{z \mid z \neq 0\}$

10. $\frac{z-12}{4z}$ $\{z \mid z \neq 0\}$

11. $\frac{5y-1}{y^2-4}$ $\{y \mid y \neq -2 \text{ and } y \neq 2\}$

12. $\frac{2y-1}{y^2-9}$ $\{y \mid y \neq -3 \text{ and } y \neq 3\}$

13. $\frac{2a-3}{a^2+5a+6}$ $\{a \mid a \neq -2 \text{ and } a \neq -3\}$

14. $\frac{3b+1}{2b^2-7b-4}$ $\left\{b \mid b \neq -\frac{1}{2} \text{ and } b \neq 4\right\}$

15. $\frac{x-1}{x^2+4x}$ $\{x \mid x \neq -4 \text{ and } x \neq 0\}$

16. $\frac{2x}{3x^2+9x}$ $\{x \mid x \neq -3 \text{ and } x \neq 0\}$

17. $\frac{x+1}{x^3+x^2-6x}$ $\{x \mid x \neq -3 \text{ and } x \neq 0 \text{ and } x \neq 2\}$

18. $\frac{x^2-3x-4}{2x^5-2x}$ $\{x \mid x \neq -1 \text{ and } x \neq 0 \text{ and } x \neq 1\}$

Reduce each rational expression to its lowest terms. See Examples 2 and 3.

19. $\frac{6}{57}$
    $\frac{2}{19}$

20. $\frac{14}{91}$
    $\frac{2}{13}$

21. $\frac{42}{210}$
    $\frac{1}{5}$

22. $\frac{242}{154}$
    $\frac{11}{7}$

23. $\frac{2x+2}{4}$
    $\frac{x+1}{2}$

24. $\frac{3a+3}{3}$
    $a+1$

25. $\dfrac{3x - 6y}{10y - 5x}$

$-\dfrac{3}{5}$

26. $\dfrac{5b - 10a}{2a - b}$

-5

27. $\dfrac{ab^2}{a^3b}$

$\dfrac{b}{a^2}$

28. $\dfrac{36y^3z^8}{54y^2z^9}$

$\dfrac{2y}{3z}$

29. $\dfrac{-2w^2x^3y}{6wx^5y^2}$

$\dfrac{-w}{3x^2y}$

30. $\dfrac{6a^3b^{12}c^5}{-8ab^4c^9}$

$\dfrac{-3a^2b^8}{4c^4}$

31. $\dfrac{a^3b^2}{a^3 + a^4}$

$\dfrac{b^2}{1 + a}$

32. $\dfrac{b^8 - ab^5}{ab^5}$

$\dfrac{b^3 - a}{a}$

33. $\dfrac{a - b}{2b - 2a}$

$-\dfrac{1}{2}$

34. $\dfrac{2m - 2n}{4n - 4m}$

$-\dfrac{1}{2}$

35. $\dfrac{3x + 6}{3x}$

$\dfrac{x + 2}{x}$

36. $\dfrac{7x - 14}{7x}$

$\dfrac{x - 2}{x}$

37. $\dfrac{a^3 - b^3}{a - b}$

$a^2 + ab + b^2$

38. $\dfrac{27x^3 + y^3}{6x + 2y}$

$\dfrac{9x^2 - 3xy + y^2}{2}$

39. $\dfrac{4x^2 - 4}{4x^2 + 4}$

$\dfrac{x^2 - 1}{x^2 + 1}$

40. $\dfrac{2a^2 - 2b^2}{2a^2 + 2b^2}$

$\dfrac{a^2 - b^2}{a^2 + b^2}$

41. $\dfrac{2x^2 + 2x - 12}{4x^2 - 36}$

$\dfrac{x - 2}{2x - 6}$

42. $\dfrac{2x^2 + 10x + 12}{2x^2 - 8}$

$\dfrac{x + 3}{x - 2}$

43. $\dfrac{x^3 + 7x^2 - 4x}{x^3 - 16x}$

$\dfrac{x^2 + 7x - 4}{x^2 - 16}$

44. $\dfrac{2x^4 - 32}{4x - 8}$

$\dfrac{(x + 2)(x^2 + 4)}{2}$

45. $\dfrac{ab + 3a - by - 3y}{a^2 - y^2}$

$\dfrac{b + 3}{a + y}$

46. $\dfrac{2x^2 - 5x - 3}{2x^2 + 11x + 5}$

$\dfrac{x - 3}{x + 5}$

Convert each rational expression into an equivalent rational expression that has the indicated denominator. See Examples 4 and 5.

47. $\dfrac{1}{5}, \dfrac{?}{50}$

$\dfrac{10}{50}$

48. $\dfrac{2}{3}, \dfrac{?}{9}$

$\dfrac{6}{9}$

49. $\dfrac{1}{x}, \dfrac{?}{3x^2}$

$\dfrac{3x}{3x^2}$

50. $\dfrac{3}{ab^2}, \dfrac{?}{a^3b^5}$

$\dfrac{3a^2b^3}{a^3b^5}$

51. $\dfrac{5}{x - 1}, \dfrac{?}{x^2 - 2x + 1}$

$\dfrac{5x - 5}{x^2 - 2x + 1}$

52. $\dfrac{x}{x - 3}, \dfrac{?}{x^2 - 9}$

$\dfrac{x^2 + 3x}{x^2 - 9}$

53. $\dfrac{1}{2x + 2}, \dfrac{?}{-6x - 6}$

$\dfrac{-3}{-6x - 6}$

54. $\dfrac{-2}{-3x + 4}, \dfrac{?}{15x - 20}$

$\dfrac{10}{15x - 20}$

55. $5, \dfrac{?}{a}$

$\dfrac{5a}{a}$

56. $3, \dfrac{?}{a + 1}$

$\dfrac{3a + 3}{a + 1}$

57. $\dfrac{x + 2}{x + 3}, \dfrac{?}{x^2 + 2x - 3}$

$\dfrac{x^2 + x - 2}{x^2 + 2x - 3}$

58. $\dfrac{x}{x - 5}, \dfrac{?}{x^2 - x - 20}$

$\dfrac{x^2 + 4x}{x^2 - x - 20}$

59. $\dfrac{7}{x - 1}, \dfrac{?}{1 - x}$

$\dfrac{-7}{1 - x}$

60. $\dfrac{1}{a - b}, \dfrac{?}{2b - 2a}$

$\dfrac{-2}{2b - 2a}$

61. $\dfrac{3}{x + 2}, \dfrac{?}{x^3 + 8}$

$\dfrac{3x^2 - 6x + 12}{x^3 + 8}$

62. $\dfrac{x}{x - 2}, \dfrac{?}{x^3 - 8}$

$\dfrac{x^3 + 2x^2 + 4x}{x^3 - 8}$

63. $\dfrac{x + 2}{3x - 1}, \dfrac{?}{6x^2 + 13x - 5}$

$\dfrac{2x^2 + 9x + 10}{6x^2 + 13x - 5}$

64. $\dfrac{a}{2a + 1}, \dfrac{?}{4a^2 - 16a - 9}$

$\dfrac{2a^2 - 9a}{4a^2 - 16a - 9}$

Find the indicated value for each given rational expression. See Example 6.

65. $R(x) = \dfrac{3x - 5}{x + 4}, R(3)$ $\dfrac{4}{7}$

66. $T(x) = \dfrac{5 - x}{x - 5}, T(-9)$ -1

67. $H(y) = \dfrac{y^2 - 5}{3y - 4}, H(-2)$ $\dfrac{1}{10}$

68. $G(a) = \dfrac{3 - 5a}{2a + 7}, G(5)$ $-\dfrac{22}{17}$

69. $W(b) = \dfrac{4b^3 - 1}{b^2 - b - 6}, W(-2)$ Undefined

70. $N(x) = \dfrac{x + 3}{x^3 - 2x^2 - 2x - 3}, N(3)$ Undefined

In place of each question mark in Exercises 71–90, put an expression that will make the rational expressions equivalent.

71. $\dfrac{1}{3} = \dfrac{?}{21}$

$\dfrac{7}{21}$

72. $4 = \dfrac{?}{3}$

$\dfrac{12}{3}$

73. $5 = \dfrac{10}{?}$

$\dfrac{10}{2}$

74. $\dfrac{3}{4} = \dfrac{12}{?}$

$\dfrac{12}{16}$

75. $\dfrac{3}{a} = \dfrac{?}{a^2}$

$\dfrac{3a}{a^2}$

76. $\dfrac{5}{y} = \dfrac{10}{?}$

$\dfrac{10}{2y}$

77. $\dfrac{2}{a-b} = \dfrac{?}{b-a}$

$\dfrac{-2}{b-a}$

78. $\dfrac{3}{x-4} = \dfrac{?}{4-x}$

$\dfrac{-3}{4-x}$

79. $\dfrac{2}{x-1} = \dfrac{?}{x^2-1}$

$\dfrac{2x+2}{x^2-1}$

80. $\dfrac{5}{x+3} = \dfrac{?}{x^2-9}$

$\dfrac{5x-15}{x^2-9}$

81. $\dfrac{2}{w-3} = \dfrac{-2}{?}$

$\dfrac{-2}{3-w}$

82. $\dfrac{-2}{5-x} = \dfrac{2}{?}$

$\dfrac{2}{x-5}$

83. $\dfrac{2x+4}{6} = \dfrac{?}{3}$

$\dfrac{x+2}{3}$

84. $\dfrac{2x-3}{4x-6} = \dfrac{1}{?}$

$\dfrac{1}{2}$

85. $\dfrac{x+4}{x^2-16} = \dfrac{1}{?}$

$\dfrac{1}{x-4}$

86. $\dfrac{2x+2}{2x} = \dfrac{x+1}{?}$

$\dfrac{x+1}{x}$

87. $\dfrac{3a+3}{3a} = \dfrac{?}{a}$

$\dfrac{a+1}{a}$

88. $\dfrac{x-3}{x^2-9} = \dfrac{1}{?}$

$\dfrac{1}{x+3}$

89. $\dfrac{1}{x-1} = \dfrac{?}{x^3-1}$

$\dfrac{x^2+x+1}{x^3-1}$

90. $\dfrac{x^2+2x+4}{x+2} = \dfrac{?}{x^2-4}$

$\dfrac{x^3-8}{x^2-4}$

Reduce each rational expression to its lowest terms. Variables used in exponents represent integers.

91. $\dfrac{x^{2a}-4}{x^a+2}$

$x^a - 2$

92. $\dfrac{x^{2b}+3x^b-18}{x^{2b}-36}$

$\dfrac{x^b-3}{x^b-6}$

93. $\dfrac{x^a+m+wx^a+wm}{x^{2a}-m^2}$

$\dfrac{1+w}{x^a-m}$

94. $\dfrac{x^{3a}-8}{x^{2a}+2x^a+4}$

$x^a - 2$

95. $\dfrac{x^{3b+1}-x}{x^{2b+1}-x}$

$\dfrac{x^{2b}+x^b+1}{x^b+1}$

96. $\dfrac{2x^{2a+1}+3x^{a+1}+x}{4x^{2a+1}-x}$

$\dfrac{x^a+1}{2x^a-1}$

Solve each problem. See Example 7.

97. *Driving speed.* If Jeremy drives 500 miles in $2x$ hours, then what rational expression represents his speed in miles per hour (mph)?

$\dfrac{250}{x}$ mph

98. *Filing suit.* If Marsha files 48 suits in $2x+2$ work days, then what rational expression represents the rate (in suits per day) at which she is filing suits?

$\dfrac{24}{x+1}$ suits per day

99. *Wedding bells.* Wheeler Printing Co. charges $45 plus $0.50 per invitation to print wedding invitations.
a) Write a rational function that gives the average cost in dollars per invitation for printing n invitations.
b) How much less does it cost per invitation to print 300 invitations rather than 200 invitations?
c) As the number of invitations increases, does the average cost per invitation increase or decrease?
d) As the number of invitations increases, does the total cost of the invitations increase or decrease?

a) $A(n) = \dfrac{0.50n+45}{n}$ dollars

b) 7.5 cents   c) decreases   d) increases

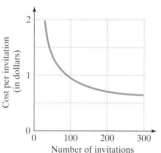

FIGURE FOR EXERCISE 99

100. *Rose Bowl bound.* A travel agent offers a Rose Bowl package including hotel, tickets, and transportation. It costs the travel agent $50,000 plus $300 per person to charter the airplane. Find a rational function that gives the average cost in dollars per person for the charter flight. How much lower is the average cost per person when 200 people go compared to 100 people?

$A(n) = \dfrac{50{,}000+300n}{n}$ dollars, $250 per person

 101. *Solid waste recovery.* The amount of municipal solid waste generated in the United States in the year $1960+n$ is given by the polynomial

$$3.43n + 87.24,$$

whereas the amount recycled is given by the polynomial

$$0.053n^2 - 0.64n + 6.71,$$

where the amounts are in millions of tons (U.S. Environmental Protection Agency, www.epa.gov).

a) Write a rational function $p(n)$ that gives the fraction of solid waste that is recovered in the year $1960 + n$.

b) Find $p(0)$, $p(30)$, and $p(50)$.

a) $p(n) = \dfrac{0.053n^2 - 0.64n + 6.71}{3.43n + 87.24}$

b) 7.7%, 18.5%, 41.4%

102. Higher education. The total number of degrees awarded in U.S. higher education in the year $1990 + n$ is given in thousands by the polynomial $41.7n + 1429$, whereas the number of bachelor's degrees awarded is given in thousands by the polynomial $25.2n + 1069$ (National Center for Education Statistics, www.nces.ed.gov).

a) Write a rational function $p(n)$ that gives the percentage of bachelor's degrees among the total number of degrees conferred for the year $1990 + n$.

b) What percentage of the degrees awarded in 2010 will be bachelor's degrees?

a) $p(n) = \dfrac{25.2n + 1069}{41.7n + 1429}$ b) 69.5%

103. Exploration. Use a calculator to find $R(2)$, $R(30)$, $R(500)$, $R(9,000)$, and $R(80,000)$ for the rational expression

$$R(x) = \frac{x - 3}{2x + 1}.$$

Round answers to four decimal places. What can you conclude about the value of $R(x)$ as x gets larger and larger without bound?

The value of $R(x)$ gets closer and closer to $\frac{1}{2}$.

104. Exploration. Use a calculator to find $H(1,000)$, $H(100,000)$, $H(1,000,000)$, and $H(10,000,000)$ for the rational expression

$$H(x) = \frac{7x - 50}{3x + 91}.$$

Round answers to four decimal places. What can you conclude about the value of $H(x)$ as x gets larger and larger without bound?

The value of $H(x)$ gets closer and closer to $\frac{7}{3}$.

6.2   MULTIPLICATION AND DIVISION

In Chapter 5 you learned to add, subtract, multiply, and divide polynomials. In this chapter you will learn to perform the same operations with rational expressions. We begin in this section with multiplication and division.

In this

section

- Multiplying Rational Expressions
- Dividing $a - b$ by $b - a$
- Dividing Rational Expressions

Multiplying Rational Expressions

We multiply two rational numbers by multiplying their numerators and multiplying their denominators. For example,

$$\frac{6}{7} \cdot \frac{14}{15} = \frac{84}{105} = \frac{\cancel{21} \cdot 4}{\cancel{21} \cdot 5} = \frac{4}{5}.$$

Instead of reducing the rational number after multiplying, it is often easier to reduce before multiplying. We first factor all terms, then divide out the common factors, then multiply:

$$\frac{6}{7} \cdot \frac{14}{15} = \frac{2 \cdot \cancel{3} \cdot 2 \cdot \cancel{7}}{\cancel{7} \cdot \cancel{3} \cdot 5} = \frac{4}{5}.$$

When we multiply rational numbers, we use the following definition.

Multiplication of Rational Numbers

If $\dfrac{a}{b}$ and $\dfrac{c}{d}$ are rational numbers, then $\dfrac{a}{b} \cdot \dfrac{c}{d} = \dfrac{ac}{bd}$.

We multiply rational expressions in the same way that we multiply rational numbers: Factor all polynomials, divide out the common factors, then multiply the remaining factors.

E X A M P L E   1

Multiplying rational expressions

Find each product of rational expressions.

a) $\dfrac{3a^8b^3}{6b} \cdot \dfrac{10a}{a^2b^6}$

b) $\dfrac{x^2 + 7x + 12}{x^2 + 3x} \cdot \dfrac{x^2}{x^2 - 16}$

Solution

a) First factor the coefficients in each numerator and denominator:

$$\dfrac{3a^8b^3}{6b} \cdot \dfrac{10a}{a^2b^6} = \dfrac{3a^8b^3}{2 \cdot 3b} \cdot \dfrac{2 \cdot 5a}{a^2b^6} \qquad \text{Factor.}$$

$$= \dfrac{5a^9b^3}{a^2b^7} \qquad \text{Divide out the common factors.}$$

$$= \dfrac{5a^7}{b^4} \qquad \text{Quotient rule}$$

b) $\dfrac{x^2 + 7x + 12}{x^2 + 3x} \cdot \dfrac{x^2}{x^2 - 16} = \dfrac{(x + 3)(x + 4)}{x(x + 3)} \cdot \dfrac{x \cdot x}{(x - 4)(x + 4)} = \dfrac{x}{x - 4}$

CAUTION Do not attempt to divide out the x in $\dfrac{x}{x-4}$. This expression cannot be reduced because x is not a factor of *both* terms in the denominator. Compare this expression to the following:

$$\dfrac{3x}{x - xy} = \dfrac{x \cdot 3}{x(1 - y)} = \dfrac{3}{1 - y}$$

In Example 2(a) we will multiply a rational expression and a polynomial. For Example 2(b) we will use the rule for factoring the difference of two cubes.

E X A M P L E   2

Multiplying rational expressions

Find each product.

a) $(a^2 - 1) \cdot \dfrac{6}{2a^2 + 4a + 2}$

b) $\dfrac{a^3 - b^3}{b - a} \cdot \dfrac{6}{2a^2 + 2ab + 2b^2}$

Solution

a) First factor the polynomials completely:

$$(a^2 - 1) \cdot \dfrac{6}{2a^2 + 4a + 2} = \dfrac{(a + 1)(a - 1)}{1} \cdot \dfrac{2 \cdot 3}{2(a + 1)(a + 1)}$$

$$= \dfrac{3(a - 1)}{a + 1} \qquad \text{Divide out the common factors.}$$

$$= \dfrac{3a - 3}{a + 1} \qquad \text{Multiply.}$$

b) Note that $a - b$ is a factor of $a^3 - b^3$ and $b - a$ occurs in the denominator. We can factor $b - a$ as $-1(a - b)$ to get a common factor:

$$\dfrac{a^3 - b^3}{b - a} \cdot \dfrac{6}{2a^2 + 2ab + 2b^2} = \dfrac{(a - b)(a^2 + ab + b^2)}{-1(a - b)} \cdot \dfrac{2 \cdot 3}{2(a^2 + ab + b^2)}$$

$$= \dfrac{3}{-1}$$

$$= -3$$

Dividing $a - b$ by $b - a$

In Example 2(b) we factored $b - a$ as $-1(a - b)$ to get the common factor $a - b$. Instead of factoring out -1, we could have used the fact that $(a - b) \div (b - a) = -1$, as shown in the next example.

EXAMPLE 3

Dividing $a - b$ by $b - a$

Find the product:

$$\frac{m - 4}{3} \cdot \frac{6}{4 - m}$$

Solution

Instead of factoring out -1 from $m - 4$, we divide $m - 4$ by $4 - m$ to get -1:

$$\frac{m - 4}{3} \cdot \frac{6}{4 - m} = \frac{\overset{-1}{\cancel{m - 4}}}{\cancel{3}} \cdot \frac{\overset{2}{\cancel{6}}}{\cancel{4 - m}} \qquad \text{Note that } (m - 4) \div (4 - m) = -1.$$

$$= -2$$

Dividing Rational Expressions

We divide rational numbers by multiplying by the reciprocal or multiplicative inverse of the divisor. For example,

$$\frac{3}{4} \div \frac{15}{2} = \frac{3}{4} \cdot \frac{2}{15} = \frac{\cancel{3}}{2 \cdot \cancel{2}} \cdot \frac{\cancel{2} \cdot 1}{\cancel{3} \cdot 5} = \frac{1}{10}.$$

When we divide rational numbers, we use the following definition.

Division of Rational Numbers

If $\frac{a}{b}$ and $\frac{c}{d}$ are rational numbers with $\frac{c}{d} \neq 0$, then

$$\frac{a}{b} \div \frac{c}{d} = \frac{a}{b} \cdot \frac{d}{c}.$$

We use the same method to divide rational expressions: We invert the divisor and multiply.

EXAMPLE 4

Dividing rational expressions

Find each quotient.

a) $\dfrac{10}{3x} \div \dfrac{6}{5x}$

b) $\dfrac{5a^2b^8}{c^3} \div (4ab^3c)$

Solution

a) The reciprocal of the divisor $\frac{6}{5x}$ is $\frac{5x}{6}$.

$$\frac{10}{3x} \div \frac{6}{5x} = \frac{10}{3x} \cdot \frac{5x}{6} \qquad \text{Invert and multiply.}$$

$$= \frac{2 \cdot 5}{3\cancel{x}} \cdot \frac{5\cancel{x}}{2 \cdot 3} = \frac{25}{9}$$

b) The reciprocal of $4ab^3c$ is $\frac{1}{4ab^3c}$.

$$\frac{5a^2b^8}{c^3} \div (4ab^3c) = \frac{5a^2b^8}{c^3} \cdot \frac{1}{4ab^3c} = \frac{5ab^5}{4c^4} \quad \text{Quotient rule}$$

In the next example we factor the polynomials in the rational expressions.

E X A M P L E   5   **Dividing rational expressions**

Find the quotient:

$$\frac{25 - x^2}{x^2 + x} \div \frac{x - 5}{x^2 - 1}$$

Solution

$$\frac{25 - x^2}{x^2 + x} \div \frac{x - 5}{x^2 - 1} = \frac{25 - x^2}{x^2 + x} \cdot \frac{x^2 - 1}{x - 5} \qquad \text{Invert and multiply.}$$

$$= \frac{\overset{-1}{\cancel{(5 - x)}}(5 + x)}{x\cancel{(x + 1)}} \cdot \frac{\cancel{(x + 1)}(x - 1)}{\cancel{x - 5}} \qquad (5 - x) \div (x - 5) = -1.$$

$$= \frac{-1(5 + x)(x - 1)}{x} \qquad \begin{array}{l}\text{Divide out the common}\\\text{factors.}\end{array}$$

$$= \frac{-x^2 - 4x + 5}{x} \qquad \begin{array}{l}\text{Multiply the factors}\\\text{in the numerator.}\end{array}$$

C A U T I O N   When dividing rational expressions, you can factor the polynomials at any time, but do not reduce until after you have inverted the divisor.

In the next example division is indicated by a fraction bar.

E X A M P L E   6   **Dividing rational expressions**

Perform the operations indicated.

a) $\dfrac{\dfrac{a + b}{3}}{\dfrac{1}{2}}$ **b)** $\dfrac{\dfrac{x^2 - 4}{2}}{\dfrac{x - 2}{3}}$ **c)** $\dfrac{\dfrac{m^2 + 1}{5}}{3}$

Solution

a) $\dfrac{\dfrac{a + b}{3}}{\dfrac{1}{2}} = \dfrac{a + b}{3} \div \dfrac{1}{2}$

$$= \frac{a + b}{3} \cdot \frac{2}{1} \qquad \text{Invert the divisor.}$$

$$= \frac{2a + 2b}{3} \qquad \text{Multiply.}$$

b) $\dfrac{\dfrac{x^2 - 4}{2}}{\dfrac{x - 2}{3}} = \dfrac{x^2 - 4}{2} \cdot \dfrac{3}{x - 2}$ Invert and multiply.

$= \dfrac{(x - 2)(x + 2)}{2} \cdot \dfrac{3}{x - 2}$ Factor.

$= \dfrac{3x + 6}{2}$ Reduce.

c) $\dfrac{\dfrac{m^2 + 1}{5}}{3} = \dfrac{m^2 + 1}{5} \cdot \dfrac{1}{3} = \dfrac{m^2 + 1}{15}$ Multiply by $\frac{1}{3}$, the reciprocal of 3.

WARM-UPS

True or false? Explain.

1. We can multiply only fractions that have identical denominators.   False
2. $\frac{2}{7} \cdot \frac{3}{7} = \frac{6}{7}$ False
3. To divide rational expressions, invert the divisor and multiply.   True
4. $a \div b = \frac{1}{a} \cdot b$ for any nonzero a and b.   False
5. $\frac{1}{2x} \cdot 8x^2 = 4x$ for any nonzero real number x.   True
6. One-half of one-third is one-sixth.   True
7. One-third divided by one-half is two-thirds.   True
8. The quotient of $w - z$ divided by $z - w$ is -1, provided that $z - w \neq 0$.   True
9. $\frac{x}{3} \div 2 = \frac{x}{6}$ for any real number x.   True
10. $\frac{a}{b} \div \frac{b}{a} = 1$ for any nonzero real numbers a and b.   False

---

6.2   EXERCISES

Reading and Writing   *After reading this section, write out the answers to these questions. Use complete sentences.*

1. How do you multiply rational numbers?
   To multiply rational numbers, multiply the numerators and the denominators.

2. What is the procedure for multiplying rational expressions?
   To multiply rational expressions, multiply the numerators and the denominators.

3. What is the relationship between $a - b$ and $b - a$?
   The expressions $a - b$ and $b - a$ are opposites.

4. How do we divide rational numbers?
   To divide rational numbers, invert the divisor and multiply.

In Exercises 5–20, perform the indicated operations. See Examples 1–3.

5. $\dfrac{12}{42} \cdot \dfrac{35}{22}$ $\dfrac{5}{11}$

6. $\dfrac{3}{8} \cdot \dfrac{20}{21}$ $\dfrac{5}{14}$

7. $\dfrac{3a}{10b} \cdot \dfrac{5b^2}{6}$ $\dfrac{ab}{4}$

8. $\dfrac{3x}{7y} \cdot \dfrac{14y^2}{9x}$ $\dfrac{2y}{3}$

9. $\dfrac{3x - 3}{6} \cdot \dfrac{x}{x^2 - x}$ $\dfrac{1}{2}$

10. $\dfrac{-2x - 4}{2} \cdot \dfrac{6}{3x + 6}$ -2

11. $\dfrac{10x + 5}{5x^2 + 5} \cdot \dfrac{2x^2 + x - 1}{4x^2 - 1}$ $\dfrac{x + 1}{x^2 + 1}$

12. $\dfrac{x^3 + x}{5} \cdot \dfrac{5x - 5}{x^3 - x}$ $\dfrac{x^2 + 1}{x + 1}$

13. $\dfrac{ax + aw + bx + bw}{x^2 - w^2} \cdot \dfrac{x - w}{a^2 - b^2}$ $\dfrac{1}{a - b}$

14. $\dfrac{3a - 3y}{3a - 3y - ab + by} \cdot \dfrac{b^2 - 9}{6b + 18}$ $\dfrac{1}{2}$

15. $\dfrac{a^2 - 2a + 4}{a^3 + 8} \cdot \dfrac{(a + 2)^3}{2a + 4}$ $\dfrac{a + 2}{2}$

16. $\dfrac{w^3 - 1}{(w-1)^2} \cdot \dfrac{w^2 - 1}{w^2 + w + 1}$ $w + 1$

17. $\dfrac{x - 9}{12y} \cdot \dfrac{8y}{9 - x}$ $-\dfrac{2}{3}$

18. $\dfrac{19x^2}{12y - 1} \cdot \dfrac{1 - 12y}{3x}$ $-\dfrac{19x}{3}$

19. $(a^2 - 4) \cdot \dfrac{7}{2 - a}$ $-7a - 14$

20. $\dfrac{10x - 4x^2}{4x^2 - 20x + 25} \cdot (2x^3 - 5x^2)$ $-2x^3$

Perform the indicated operations. See Examples 4 and 5.

21. $\dfrac{15}{17} \div \dfrac{10}{17}$ $\dfrac{3}{2}$

22. $\dfrac{3}{4} \div \dfrac{1}{8}$ 6

23. $\dfrac{36x}{5y} \div \dfrac{20x}{35y}$ $\dfrac{63}{5}$

24. $\dfrac{18a^3b^4}{c^9} \div \dfrac{12ab^6}{7c^2}$ $\dfrac{21a^2}{2b^2c^7}$

25. $\dfrac{24a^5b^2}{5c^3} \div (4a^5bc^5)$ $\dfrac{6b}{5c^8}$

26. $\dfrac{60x^9y^2}{z} \div (48x^4y^3)$ $\dfrac{5x^5}{4yz}$

27. $(w + 1) \div \dfrac{w^2 - 1}{w}$ $\dfrac{w}{w - 1}$

28. $(a - 3) \div \dfrac{9 - a^2}{4}$ $\dfrac{-4}{a + 3}$

29. $\dfrac{x - y}{5} \div \dfrac{x^2 - 2xy + y^2}{10}$ $\dfrac{2}{x - y}$

30. $\dfrac{x^2 + 6x + 9}{18} \div \dfrac{(x + 3)^2}{36}$ 2

31. $\dfrac{4x - 2}{x^2 - 5x} \div \dfrac{2x^2 + 9x - 5}{x^2 - 25}$ $\dfrac{2}{x}$

32. $\dfrac{2x^2 - 5x - 12}{6 + 4x} \div \dfrac{x^2 - 16}{2}$ $\dfrac{1}{x + 4}$

Perform the indicated operations. See Example 6.

33. $\dfrac{\dfrac{x - y}{3}}{\dfrac{1}{6}}$ $2x - 2y$

34. $\dfrac{\dfrac{2a - b}{10}}{\dfrac{1}{5}}$ $\dfrac{2a - b}{2}$

35. $\dfrac{\dfrac{x^2 - 25}{3}}{\dfrac{x - 5}{6}}$ $2x + 10$

36. $\dfrac{\dfrac{3x^2 + 3}{5}}{\dfrac{3x + 3}{5}}$ $x^2 + 1$ $x + 1$

37. $\dfrac{\dfrac{a - b}{2}}{3}$ $\dfrac{a - b}{6}$

38. $\dfrac{\dfrac{a + b}{5}}{10}$ $\dfrac{2}{a + b}$

39. $\dfrac{\dfrac{a^2 - b^2}{a + b}}{3}$ $3a - 3b$

40. $\dfrac{\dfrac{x^2 + 5x + 6}{x + 2}}{x + 3}$ $x^2 + 6x + 9$

Perform the indicated operations. When possible write down only the answer.

41. $\dfrac{5x}{2} \div 3$ $\dfrac{5x}{6}$

42. $\dfrac{x}{a} \div 2$ $\dfrac{x}{2a}$

43. $\dfrac{3}{4} \div \dfrac{1}{4}$ 3

44. $\dfrac{1}{4} \div \dfrac{1}{2}$ $\dfrac{1}{2}$

45. One-half of $\dfrac{1}{6}$ $\dfrac{1}{12}$

46. One-half of $\dfrac{b}{a}$ $\dfrac{b}{2a}$

47. One-half of $\dfrac{4x}{3}$ $\dfrac{2x}{3}$

48. One-third of $\dfrac{6x}{y}$ $\dfrac{2x}{y}$

49. $(a - b) \div (b - a)$ -1

50. $(a - b) \div (-1)$ $b - a$

51. $\dfrac{x - y}{3} \cdot \dfrac{6}{y - x}$ -2

52. $\dfrac{5x - 5y}{x} \cdot \dfrac{1}{x - y}$ $\dfrac{5}{x}$

53. $\dfrac{2a + 2b}{a} \cdot \dfrac{1}{2}$ $\dfrac{a + b}{a}$

54. $\dfrac{x - y}{y - x} \cdot \dfrac{1}{2}$ $-\dfrac{1}{2}$

55. $\dfrac{a + b}{\dfrac{1}{2}}$ $2a + 2b$

56. $\dfrac{x + 3}{\dfrac{1}{3}}$ $3x + 9$

57. $\dfrac{\dfrac{3x}{5}}{y}$ $\dfrac{3x}{5y}$

58. $\dfrac{\dfrac{b^2 - 4a}{2}}{a}$ $\dfrac{b^2 - 4a}{2a}$

59. $\dfrac{\dfrac{3a}{5b}}{2}$ $\dfrac{3a}{10b}$

60. $\dfrac{\dfrac{6x}{a}}{x}$ $\dfrac{6}{a}$

Perform the indicated operations.

61. $\dfrac{3x^2 + 13x - 10}{x} \cdot \dfrac{x^3}{9x^2 - 4} \cdot \dfrac{7x - 35}{x^2 - 25}$ $\dfrac{7x^2}{3x + 2}$

62. $\dfrac{x^2 + 5x + 6}{x} \cdot \dfrac{x^2}{3x + 6} \cdot \dfrac{9}{x^2 - 4}$ $\dfrac{3x^2 + 9x}{x^2 - 4}$

63. $\dfrac{(a^2b^3c)^2}{(-2ab^2c)^3} \cdot \dfrac{(a^3b^2c)^3}{(abc)^4}$ $-\dfrac{a^6b^2}{8c^2}$

64. $\dfrac{(-wy^2)^3}{3w^2y} \cdot \dfrac{(2wy)^2}{4wy^3}$ $-\dfrac{w^2y^4}{3}$

65. $\dfrac{(2mn)^3}{6mn^2} \div \dfrac{2m^2n^3}{(m^2n)^4}$ $\dfrac{2m^8n^2}{3}$

66. $\dfrac{(rt)^3}{rt^4} \div \dfrac{(rt^2)^3}{r^2t^3}$ $\dfrac{r}{t^4}$

67. $\dfrac{2x^2 + 7x - 15}{4x^2 - 100} \cdot \dfrac{2x^2 - 9x - 5}{4x^2 - 1}$ $\dfrac{2x - 3}{8x - 4}$

68. $\dfrac{x^3 + 1}{x^2 - 1} \cdot \dfrac{3x - 3}{x^3 - x^2 + x}$ $\dfrac{3}{x}$

69. $\dfrac{k^2 + 2km + m^2}{k^2 - 2km + m^2} \cdot \dfrac{m^2 + 3m - mk - 3k}{m^2 + mk + 3m + 3k}$ $\dfrac{k + m}{m - k}$

70. $\dfrac{a^2 + 2ab + b^2}{ac + bc - ad - bd} \div \dfrac{ac + ad - bc - bd}{c^2 - d^2}$ $\dfrac{a + b}{a - b}$

Perform the indicated operations. Variables in exponents represent integers.

71. $\dfrac{x^a}{y^2} \cdot \dfrac{y^{b+2}}{x^{2a}}$

$\dfrac{y^b}{x^a}$

72. $\dfrac{x^{3a+1}}{y^{2b-3}} \cdot \dfrac{y^{3b+4}}{x^{2a-1}}$

$x^{a+2}y^{b+7}$

73. $\dfrac{x^{2a} + x^a - 6}{x^{2a} + 6x^a + 9} \div \dfrac{x^{2a} - 4}{x^{2a} + 2x^a - 3}$

$\dfrac{x^a - 1}{x^a + 2}$

74. $\dfrac{w^{2b} + 2w^b - 8}{w^{2b} + 3w^b - 4} \div \dfrac{w^{2b} - w^b - 2}{w^{2b} - 1}$

1

75. $\dfrac{m^k v^k + 3v^k - 2m^k - 6}{m^{2k} - 9} \cdot \dfrac{m^{2k} - 2m^k - 3}{v^k m^k - 2m^k + 2v^k - 4}$

$\dfrac{m^k + 1}{m^k + 2}$

76. $\dfrac{m^{3k} - 1}{m^{3k} + 1} \cdot \dfrac{m^{2k+1} - m^{k+1} + m}{m^{3k} + m^{2k} + m^k}$

$\dfrac{m^k - 1}{m^{2k-1} + m^{k-1}}$

Solve each problem.

77. *School enrollment.* In 2005, $\frac{1}{50}$ of the children enrolled in U.S. schools will be enrolled in private secondary schools (National Center for Education Statistics, www.nces.ed.gov). Use the accompanying figure to determine the percentage of secondary school children who will be in private schools in 2005.
7.1%

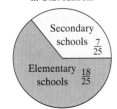

2005 distribution of students in U.S. schools

Secondary schools $\frac{7}{25}$

Elementary schools $\frac{18}{25}$

FIGURE FOR EXERCISE 77

78. *The golden state.* In 2000, $\frac{3}{25}$ of the U.S. population will be living in California (U.S. Census Bureau). Use the accompanying figure to determine the percentage of the population of the western region living in California in 2000.
54%

Distribution of U.S. Population in 2000

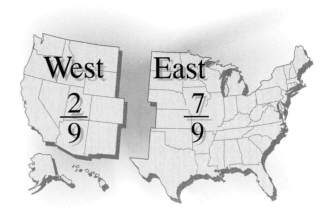

West $\dfrac{2}{9}$ East $\dfrac{7}{9}$

FIGURE FOR EXERCISE 78

79. *Distance traveled.* Bonita drove 100 miles in x hours. Assuming she continued to drive at the same speed, write a rational expression for the distance that she traveled in the next $\frac{3}{4}$ of an hour.

$\frac{75}{x}$ miles

80. *Increasing speed.* Before lunch Avonda drove 200 miles at x miles per hour (mph). After lunch she drove 250 miles in the same amount of time. Write a rational expression for her speed after lunch.

$\frac{5x}{4}$ mph

GETTING MORE INVOLVED

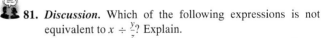

81. *Discussion.* Which of the following expressions is not equivalent to $x \div \frac{y}{z}$? Explain.

a) $x \cdot \dfrac{z}{y}$ **b)** $\dfrac{x}{y} \cdot z$ **c)** $zx \cdot \dfrac{1}{y}$

d) $\dfrac{x}{\frac{y}{z}}$ **e)** $\dfrac{xy}{z}$

e

82. *Discussion.* Which of the following equations is not an identity? Explain.

a) $\dfrac{x^2 - 1}{2} \cdot \dfrac{2}{x - 1} = x + 1$

b) $\dfrac{x - 1}{x^2 - 1} = x + 1$

c) $x^2 - 1 = (x - 1)(x + 1)$

d) $\dfrac{1}{x^2 - 1} \div \dfrac{1}{x + 1} = \dfrac{1}{x - 1}$

b

6.3 ADDITION AND SUBTRACTION

We can multiply or divide any rational expressions, but we add or subtract only rational expressions with identical denominators. So when the denominators are not the same, we must find equivalent forms of the expressions that have identical denominators. In this section we will review the idea of the least common denominator and will learn to use it for addition and subtraction of rational expressions.

Adding and Subtracting with Identical Denominators

It is easy to add or subtract fractions with identical denominators. For example,

$$\frac{1}{7} + \frac{3}{7} = \frac{4}{7} \quad \text{and} \quad \frac{3}{5} - \frac{2}{5} = \frac{1}{5}.$$

In general, we have the following definition.

Addition and Subtraction of Rational Numbers

If $b \neq 0$, then

$$\frac{a}{b} + \frac{c}{b} = \frac{a + c}{b} \quad \text{and} \quad \frac{a}{b} - \frac{c}{b} = \frac{a - c}{b}.$$

Rational expressions with identical denominators are added or subtracted in the same manner as fractions.

EXAMPLE 1

Identical denominators

Perform the indicated operations.

a) $\dfrac{3}{2x} + \dfrac{5}{2x}$

b) $\dfrac{5x - 3}{x - 1} + \dfrac{5 - 7x}{x - 1}$

c) $\dfrac{x^2 + 4x + 7}{x^2 - 1} - \dfrac{x^2 - 2x + 1}{x^2 - 1}$

Solution

a) $\dfrac{3}{2x} + \dfrac{5}{2x} = \dfrac{8}{2x}$ Add the numerators.

$= \dfrac{4}{x}$ Reduce.

b) $\dfrac{5x - 3}{x - 1} + \dfrac{5 - 7x}{x - 1} = \dfrac{5x - 3 + 5 - 7x}{x - 1}$ Add the numerators.

$= \dfrac{-2x + 2}{x - 1}$ Combine like terms.

$= \dfrac{-2(x - 1)}{x - 1}$ Factor.

$= -2$ Reduce to its lowest terms.

helpful hint

You can remind yourself of the difference between addition and multiplication of fractions with a simple example: If you and your spouse each own 1/7 of Microsoft, then together you own 2/7 of Microsoft. If you own 1/7 of Microsoft, and give 1/7 of your stock to your child, then your child owns 1/49 of Microsoft.

c) The polynomials in the numerators are treated as if they were in parentheses:

$$\frac{x^2 + 4x + 7}{x^2 - 1} - \frac{x^2 - 2x + 1}{x^2 - 1} = \frac{x^2 + 4x + 7 - (x^2 - 2x + 1)}{x^2 - 1}$$

$$= \frac{x^2 + 4x + 7 - x^2 + 2x - 1}{x^2 - 1}$$

$$= \frac{6x + 6}{x^2 - 1}$$

$$= \frac{6(x + 1)}{(x + 1)(x - 1)} = \frac{6}{x - 1}$$

Least Common Denominator

To add fractions with denominators that are not identical, we use the basic principle of rational numbers to build up the denominators to the **least common denominator (LCD).** For example,

$$\frac{1}{4} + \frac{1}{6} = \frac{1 \cdot 3}{4 \cdot 3} + \frac{1 \cdot 2}{6 \cdot 2} = \frac{3}{12} + \frac{2}{12} = \frac{5}{12}.$$

The LCD 12 is the **least common multiple (LCM)** of the numbers 4 and 6.

Finding the LCM for a pair of large numbers such as 24 and 126 will help you to understand the procedure for finding the LCM for any polynomials. First factor the numbers completely:

$$24 = 2^3 \cdot 3$$
$$126 = 2 \cdot 3^2 \cdot 7$$

Any number that is a multiple of both 24 and 126 must have all of the factors of 24 and all of the factors of 126 in its factored form. So in the LCM we use the factors 2, 3, and 7, and for each factor we use the highest power that appears on that factor. The highest power of 2 is 3, the highest power of 3 is 2, and the highest power of 7 is 1. So the LCM is $2^3 \cdot 3^2 \cdot 7$. If we write this product without exponents, we can see clearly that it is a multiple of both 24 and 126:

$$\underbrace{2 \cdot 2 \cdot 2 \cdot 3 \cdot 3 \cdot 7}_{\substack{24 \\ \overbrace{}^{126}}} = 504 \qquad \begin{array}{l} 504 = 126 \cdot 4 \\ 504 = 24 \cdot 21 \end{array}$$

The strategy for finding the LCM for a group of polynomials can be stated as follows.

Strategy for Finding the LCM for Polynomials

1. Factor each polynomial completely. Use exponents to express repeated factors.
2. Write the product of all of the different factors that appear in the polynomials.
3. For each factor, use the highest power of that factor in any of the polynomials.

E X A M P L E   2   **Finding the LCM**

Find the least common multiple for each group of polynomials.

a) $4x^2y$, $6y$ **b)** a^2bc, ab^3c^2, a^3bc **c)** $x^2 + 5x + 6$, $x^2 + 6x + 9$

Solution

a) Factor $4x^2y$ and $6y$ as follows:
$$4x^2y = 2^2 \cdot x^2y, \qquad 6y = 2 \cdot 3y$$

To get the LCM, we use 2, 3, x, and y the maximum number of times that each appears in either of the expressions. The LCM is $2^2 \cdot 3 \cdot x^2y$, or $12x^2y$.

b) The expressions a^2bc, ab^3c^2, and a^3bc are already factored. To get the LCM, we use a, b, and c the maximum number of times that each appears in any of the expressions. The LCM is $a^3b^3c^2$.

c) Factor $x^2 + 5x + 6$ and $x^2 + 6x + 9$ completely:
$$x^2 + 5x + 6 = (x + 2)(x + 3), \qquad x^2 + 6x + 9 = (x + 3)^2$$

The LCM is $(x + 2)(x + 3)^2$.

Adding and Subtracting with Different Denominators

To add or subtract rational expressions with different denominators, we must build up each rational expression to equivalent forms with identical denominators, as we did in Section 6.1. Of course, it is most efficient to use the LCD as in the following examples.

E X A M P L E   3   **Different denominators**

Perform the indicated operations.

a) $\dfrac{3}{a^2b} + \dfrac{5}{ab^3}$ **b)** $\dfrac{x + 1}{6} - \dfrac{2x - 3}{4}$

Solution

a) The LCD for a^2b and ab^3 is a^2b^3. To build up each denominator to a^2b^3, multiply the numerator and denominator of the first expression by b^2, and multiply the numerator and denominator of the second expression by a:

$$\frac{3}{a^2b} + \frac{5}{ab^3} = \frac{3(b^2)}{a^2b(b^2)} + \frac{5(a)}{ab^3(a)} \qquad \text{Build up each denominator to the LCD.}$$

$$= \frac{3b^2}{a^2b^3} + \frac{5a}{a^2b^3}$$

$$= \frac{3b^2 + 5a}{a^2b^3} \qquad \text{Add the numerators.}$$

b)
$$\frac{x + 1}{6} - \frac{2x - 3}{4} = \frac{(x + 1)(2)}{6(2)} - \frac{(2x - 3)(3)}{4(3)} \qquad \begin{array}{l}\text{Build up each denominator} \\ \text{to the LCD 12.}\end{array}$$

$$= \frac{2x + 2}{12} - \frac{6x - 9}{12} \qquad \text{Distributive property}$$

$$= \frac{2x + 2 - (6x - 9)}{12} \qquad \begin{array}{l}\text{Subtract the numerators.} \\ \text{Note that } 6x - 9 \text{ is put in} \\ \text{parentheses.}\end{array}$$

$$= \frac{2x + 2 - 6x + 9}{12} \qquad \text{Remove the parentheses.}$$

$$= \frac{-4x + 11}{12} \qquad \text{Combine like terms.}$$

CAUTION   Before you add or subtract rational expressions, they must be written with identical denominators. For multiplication and division it is not necessary to have identical denominators.

In the next example we must first factor polynomials to find the LCD.

E X A M P L E   4

Different denominators

Perform the indicated operations.

a) $\dfrac{1}{x^2 - 1} + \dfrac{2}{x^2 + x}$ **b)** $\dfrac{5}{a - 2} - \dfrac{3}{2 - a}$

Solution

a) Because $x^2 - 1 = (x + 1)(x - 1)$ and $x^2 + x = x(x + 1)$, the LCD is $x(x - 1)(x + 1)$. The first denominator is missing the factor x, and the second denominator is missing the factor $x - 1$.

> **helpful hint**
>
> It is not actually necessary to identify the LCD. Once the denominators are factored, simply look at each denominator and ask, "What factor does the other denominator have that is missing from this one?" Then use the missing factor to build up the denominator and you will obtain the LCD.

$$\dfrac{1}{x^2 - 1} + \dfrac{2}{x^2 + x} = \underbrace{\dfrac{1}{(x - 1)(x + 1)}}_{\text{Missing } x} + \underbrace{\dfrac{2}{x(x + 1)}}_{\text{Missing } x - 1} \qquad \begin{array}{l}\text{The LCD is}\\ x(x - 1)(x + 1).\end{array}$$

$$= \dfrac{1(x)}{(x - 1)(x + 1)(x)} + \dfrac{2(x - 1)}{x(x + 1)(x - 1)} \qquad \begin{array}{l}\text{Build up the}\\\text{denominators to}\\\text{the LCD.}\end{array}$$

$$= \dfrac{x}{x(x - 1)(x + 1)} + \dfrac{2x - 2}{x(x - 1)(x + 1)}$$

$$= \dfrac{3x - 2}{x(x - 1)(x + 1)} \qquad \begin{array}{l}\text{Add the}\\\text{numerators.}\end{array}$$

For this type of answer we usually leave the denominator in factored form. That way, if we need to work with the expression further, we do not have to factor the denominator again.

b) Because $-1(2 - a) = a - 2$, we can convert the denominator $2 - a$ to $a - 2$.

$$\dfrac{5}{a - 2} - \dfrac{3}{2 - a} = \dfrac{5}{a - 2} - \dfrac{3(-1)}{(2 - a)(-1)}$$

$$= \dfrac{5}{a - 2} - \dfrac{-3}{a - 2} \qquad \text{The LCD is } a - 2.$$

$$= \dfrac{5 - (-3)}{a - 2} \qquad \text{Subtract the numerators.}$$

$$= \dfrac{8}{a - 2} \qquad \text{Simplify.}$$

Note that if we had changed the denominator of the first expression to $2 - a$, we would have gotten the answer

$$\dfrac{-8}{2 - a},$$

but this rational expression is equivalent to the first answer.

Shortcuts

Consider the following addition:

$$\frac{a}{b} + \frac{c}{d} = \frac{a(d)}{b(d)} + \frac{c(b)}{d(b)} = \frac{ad + bc}{bd} \qquad \text{The LCD is } bd.$$

We can use this result as a rule for adding simple fractions in which the LCD is the product of the denominators. A similar rule works for subtraction.

> **Adding or Subtracting Simple Fractions**
>
> If $b \neq 0$ and $d \neq 0$, then
>
> $$\frac{a}{b} + \frac{c}{d} = \frac{ad + bc}{bd} \qquad \text{and} \qquad \frac{a}{b} - \frac{c}{d} = \frac{ad - bc}{bd}.$$

E X A M P L E   5   **Adding and subtracting simple fractions**

Use the rules for adding and subtracting simple fractions to find the sums and differences.

a) $\dfrac{1}{2} + \dfrac{1}{3}$

b) $\dfrac{1}{a} - \dfrac{1}{x}$

c) $\dfrac{a}{5} + \dfrac{a}{3}$

d) $x - \dfrac{2}{3}$

Solution

a) For the numerator, compute $ad + bc = 1 \cdot 3 + 2 \cdot 1 = 5$. Use $2 \cdot 3$ or 6 for the denominator:

$$\frac{1}{2} + \frac{1}{3} = \frac{5}{6}$$

b) $\dfrac{1}{a} - \dfrac{1}{x} = \dfrac{1 \cdot x - 1 \cdot a}{ax} = \dfrac{x - a}{ax}$

c) $\dfrac{a}{5} + \dfrac{a}{3} = \dfrac{3a + 5a}{15} = \dfrac{8a}{15}$

d) $x - \dfrac{2}{3} = \dfrac{x}{1} - \dfrac{2}{3} = \dfrac{3x - 2}{3}$

C A U T I O N   The rules for adding or subtracting simple fractions can be applied to any rational expressions, but they work best when the LCD is the product of the two denominators. Always make sure that the answer is in its lowest terms. If the product of the two denominators is too large, these rules are not helpful because then reducing can be difficult.

Applications

Rational expressions occur often in expressing rates. For example, if you can process one application in 2 hours, then you are working at the rate of $\frac{1}{2}$ of an application per hour. If you can complete one task in x hours, then you are working at the rate of $\frac{1}{x}$ task per hour.

EXAMPLE 6

Work rates

Susan takes an average of x hours to process a mortgage application, whereas Betty's average is 1 hour longer. Write a rational expression for the number of applications that they can process in 40 hours.

Solution

The number of applications processed by Susan is the product of her rate and her time:

$$\frac{1}{x} \frac{\text{application}}{\text{hr}} \cdot 40 \text{ hr} = \frac{40}{x} \text{ applications}$$

The number of applications processed by Betty is the product of her rate and her time:

$$\frac{1}{x+1} \frac{\text{application}}{\text{hr}} \cdot 40 \text{ hr} = \frac{40}{x+1} \text{ applications}$$

Find the sum of the rational expressions:

$$\frac{40}{x} + \frac{40}{x+1} = \frac{40x + 40 + 40x}{x(x+1)} = \frac{80x + 40}{x(x+1)}$$

So together in 40 hours they process $\frac{80x + 40}{x(x+1)}$ applications.

WARM-UPS

True or false? Explain.

1. The LCM of 6 and 10 is 60.   False

2. The LCM of $6a^2b$ and $8ab^3$ is $24ab$.   False

3. The LCM of $x^2 - 1$ and $x - 1$ is $x^2 - 1$.   True

4. The LCD for the rational expressions $\frac{5}{x}$ and $\frac{x-3}{x+1}$ is $x + 1$.   False

5. $\frac{1}{2} + \frac{2}{3} = \frac{3}{5}$ False

6. $5 + \frac{1}{x} = \frac{6}{x}$ for any nonzero real number x.   False

7. $\frac{7}{a} + 3 = \frac{7 + 3a}{a}$ for any $a \neq 0$.   True

8. $\frac{c}{3} - \frac{d}{5} = \frac{5c - 3d}{15}$ for any real numbers c and d.   True

9. $\frac{2}{3} + \frac{3}{4} = \frac{17}{12}$ True

10. If Jamal uses x reams of paper in one day, then he uses $\frac{1}{x}$ ream per day. False

6.3 EXERCISES

Reading and Writing *After reading this section, write out the answers to these questions. Use complete sentences.*

1. How do you add rational numbers?

The sum of a/b and c/b is $(a + c)/b$.

2. What is the least common denominator (LCD)?

The LCD is the least common multiple of the denominators.

3. What is the least common multiple?

The least common multiple (LCM) of some numbers is the smallest number that is a multiple of all of the numbers.

4. How do we find the LCM for a group of polynomials?

To find the LCM for some polynomials, first factor completely, then use the LCM of the coefficients and every other factor with the highest exponent that occurs on that factor in any of the polynomials.

5. How do we add or subtract rational expressions with different denominators?

To add rational expressions with different denominators, you must build up the expressions to equivalent expressions with the same denominator.

6. For which operations with rational expressions is it not necessary to have identical denominators?

You do not need identical denominators for division and multiplication.

Perform the indicated operations. Reduce answers to their lowest terms. See Example 1.

7. $\dfrac{3x}{2} + \dfrac{5x}{2}$ $4x$

8. $\dfrac{5x^2}{3} + \dfrac{4x^2}{3}$ $3x^2$

9. $\dfrac{x - 3}{2x} - \dfrac{3x - 5}{2x}$ $\dfrac{-x + 1}{x}$

10. $\dfrac{9 - 4y}{3y} - \dfrac{6 - y}{3y}$ $\dfrac{1 - y}{y}$

11. $\dfrac{3x - 4}{2x - 4} + \dfrac{2x - 6}{2x - 4}$ $\dfrac{5}{2}$

12. $\dfrac{a^3}{a + b} + \dfrac{b^3}{a + b}$ $a^2 - ab + b^2$

13. $\dfrac{x^2 + 4x - 6}{x^2 - 9} - \dfrac{x^2 + 2x - 12}{x^2 - 9}$ $\dfrac{2}{x - 3}$

14. $\dfrac{x^2 + 3x - 3}{x - 4} - \dfrac{x^2 + 4x - 7}{x - 4}$ -1

Find the least common multiple for each group of polynomials. See Example 2.

15. $24, 20$ 120

16. $12, 18, 22$ 396

17. $10x^3y, 15x$ $30x^3y$

18. $12a^3b^2, 18ab^5$ $36a^3b^5$

19. a^3b, ab^4c, ab^5c^2 $a^3b^5c^2$

20. x^2yz, xy^2z^3, xy^6 $x^2y^6z^3$

21. $x, x + 2, x - 2$ $x(x + 2)(x - 2)$

22. $y, y - 5, y + 2$ $y(y - 5)(y + 2)$

23. $4a + 8, 6a + 12$ $12a + 24$

24. $4a - 6, 2a^2 - 3a$ $4a^2 - 6a$

25. $x^2 - 1, x^2 + 2x + 1$ $(x - 1)(x + 1)^2$

26. $y^2 - 2y - 15, y^2 + 6y + 9$ $(y - 5)(y + 3)^2$

27. $x^2 - 4x, x^2 - 16, x^2 + 6x + 8$ $x(x - 4)(x + 4)(x + 2)$

28. $z^2 - 25, 5z - 25, 5z + 25$ $5z^2 - 125$

Perform the indicated operations. Reduce answers to its lowest terms. See Examples 3 and 4.

29. $\dfrac{1}{28} + \dfrac{3}{35}$ $\dfrac{17}{140}$

30. $\dfrac{7}{48} - \dfrac{5}{36}$ $\dfrac{1}{144}$

31. $\dfrac{7}{24} - \dfrac{4}{15}$ $\dfrac{1}{40}$

32. $\dfrac{7}{52} + \dfrac{3}{40}$ $\dfrac{109}{520}$

33. $\dfrac{3}{wz^2} + \dfrac{5}{w^2z}$ $\dfrac{3w + 5z}{w^2z^2}$

34. $\dfrac{2}{a^2b} - \dfrac{3}{ab^2}$ $\dfrac{2b - 3a}{a^2b^2}$

35. $\dfrac{2x - 3}{8} - \dfrac{x - 2}{6}$ $\dfrac{2x - 1}{24}$

36. $\dfrac{a - 5}{10} + \dfrac{3 - 2a}{15}$ $\dfrac{-a - 9}{30}$

37. $\dfrac{x}{2a} + \dfrac{3x}{5a}$ $\dfrac{11x}{10a}$

38. $\dfrac{x}{6y} - \dfrac{3x}{8y}$ $\dfrac{-5x}{24y}$

39. $\dfrac{9}{4y} - x$ $\dfrac{9 - 4xy}{4y}$

40. $\dfrac{b^2}{4a} - c$ $\dfrac{b^2 - 4ac}{4a}$

41. $\dfrac{5}{a + 2} - \dfrac{7}{a}$ $\dfrac{-2a - 14}{a(a + 2)}$

42. $\dfrac{2}{x + 1} - \dfrac{3}{x}$ $\dfrac{-x - 3}{x(x + 1)}$

43. $\dfrac{1}{a - b} + \dfrac{2}{a + b}$ $\dfrac{3a - b}{(a - b)(a + b)}$

44. $\dfrac{5}{x + 2} + \dfrac{3}{x - 2}$ $\dfrac{8x - 4}{(x + 2)(x - 2)}$

45. $\dfrac{x}{x^2 - 9} + \dfrac{3}{x - 3}$ $\dfrac{4x + 9}{(x + 3)(x - 3)}$

46. $\dfrac{x}{x^2 - 25} + \dfrac{5}{x - 5}$ $\dfrac{6x + 25}{(x - 5)(x + 5)}$

47. $\dfrac{1}{a - b} + \dfrac{1}{b - a}$ 0

48. $\dfrac{3}{x - 5} + \dfrac{7}{5 - x}$ $\dfrac{-4}{x - 5}$

49. $\dfrac{5}{2x - 4} - \dfrac{3}{2 - x}$ $\dfrac{11}{2x - 4}$

50. $\dfrac{4}{3x - 9} - \dfrac{7}{3 - x}$ $\dfrac{25}{3x - 9}$

51. $\dfrac{5}{x^2 + x - 2} - \dfrac{6}{x^2 + 2x - 3}$ $\dfrac{-x + 3}{(x - 1)(x + 2)(x + 3)}$

52. $\dfrac{2}{x^2 - 4} - \dfrac{5}{x^2 - 3x - 10}$ $\dfrac{-3x}{(x - 2)(x + 2)(x - 5)}$

53. $\dfrac{x}{x^2 - 9} + \dfrac{6}{x^2 + 4x + 3}$ $\dfrac{x^2 + 7x - 18}{(x + 1)(x + 3)(x - 3)}$

54. $\dfrac{2x - 1}{x^2 - x - 12} + \dfrac{x + 5}{x^2 + 5x + 6}$ $\dfrac{3x^2 + 4x - 22}{(x + 2)(x + 3)(x - 4)}$

55. $\dfrac{1}{x} + \dfrac{2}{x - 1} - \dfrac{3}{x + 2}$ $\dfrac{8x - 2}{x(x - 1)(x + 2)}$

56. $\dfrac{2}{a} - \dfrac{3}{a + 1} + \dfrac{5}{a - 1}$ $\dfrac{4a^2 + 8a - 2}{a(a + 1)(a - 1)}$

Perform the following operations. Write down only the answer. See Example 5.

57. $\dfrac{1}{3} + \dfrac{1}{4}$ $\dfrac{7}{12}$

58. $\dfrac{3}{5} + \dfrac{1}{4}$ $\dfrac{17}{20}$

59. $\dfrac{1}{8} - \dfrac{3}{5}$ $-\dfrac{19}{40}$

60. $\dfrac{a}{2} + \dfrac{5}{3}$ $\dfrac{3a + 10}{6}$

61. $\dfrac{x}{3} + \dfrac{x}{2}$ $\dfrac{5x}{6}$

62. $\dfrac{y}{4} - \dfrac{y}{3}$ $-\dfrac{y}{12}$

63. $\dfrac{a}{b} - \dfrac{2}{3}$ $\dfrac{3a - 2b}{3b}$

64. $\dfrac{3}{x} + \dfrac{1}{9}$ $\dfrac{x + 27}{9x}$

65. $a + \dfrac{2}{3}$ $\dfrac{3a + 2}{3}$

66. $\dfrac{m}{3} + y$ $\dfrac{m + 3y}{3}$

67. $\dfrac{3}{a} + 1$ $\dfrac{a + 3}{a}$

68. $\dfrac{1}{x} + 1$ $\dfrac{x + 1}{x}$

69. $\dfrac{3 + x}{x} - 1$ $\dfrac{3}{x}$

70. $\dfrac{a + 2}{a} + 3$ $\dfrac{4a + 2}{a}$

71. $\dfrac{2}{3} + \dfrac{1}{4x}$ $\dfrac{8x + 3}{12x}$

72. $\dfrac{1}{5} + \dfrac{1}{5x}$ $\dfrac{x + 1}{5x}$

Perform the indicated operations.

73. $\dfrac{w^2 - 3w + 6}{w - 5} + \dfrac{9 - w^2}{w - 5}$ -3

74. $\dfrac{2z^2 - 3z + 6}{z^2 - 1} - \dfrac{z^2 - 5z + 9}{z^2 - 1}$ $\dfrac{z + 3}{z + 1}$

75. $\dfrac{1}{x + 2} - \dfrac{2}{x + 3}$ $\dfrac{-x - 1}{(x + 2)(x + 3)}$

76. $\dfrac{2}{a + 3} - \dfrac{3}{a + 1}$ $\dfrac{-a - 7}{(a + 3)(a + 1)}$

77. $\dfrac{1}{a^3 - 1} - \dfrac{1}{a^3 + 1}$ $\dfrac{2}{(a^3 - 1)(a^3 + 1)}$

78. $\dfrac{1}{x^3 - 8} + \dfrac{1}{x^3 + 8}$ $\dfrac{2x^3}{(x^3 + 8)(x^3 - 8)}$

79. $\dfrac{(a^2 b^3)^4}{(ab^4)^3} \cdot \dfrac{(ab)^3}{(a^4 b)^2}$ b

80. $\dfrac{(ab)^2}{(a + b)^2} \cdot \dfrac{(a + b)^3}{(ab)^3}$ $\dfrac{a + b}{ab}$

81. $\dfrac{x^2 - 3x}{x^3 - 1} + \dfrac{4}{x - 1}$ $\dfrac{5x^2 + x + 4}{(x - 1)(x^2 + x + 1)}$

82. $\dfrac{4a}{a^2 + 2a + 1} - \dfrac{3}{a + 1}$ $\dfrac{a - 3}{(a + 1)^2}$

83. $\dfrac{x^2 + 25}{x^2 - 25} \cdot \dfrac{x^2 + 10x + 25}{x^2 + 5x}$ $\dfrac{x^2 + 25}{x(x - 5)}$

84. $\dfrac{a^2 + a - 2}{a^3 - 1} \cdot \dfrac{a^2 + a + 1}{a^2 - 4}$ $\dfrac{1}{a - 2}$

85. $\dfrac{w^2 - 3}{3w^3 + 81} - \dfrac{2}{6w + 18} - \dfrac{w - 4}{w^2 - 3w + 9}$
$\dfrac{-w^2 + 2w + 8}{(w + 3)(w^2 - 3w + 9)}$

86. $\dfrac{a - 3}{a^3 + 8} - \dfrac{2}{a + 2} - \dfrac{a - 3}{a^2 - 2a + 4}$ $\dfrac{-3a^2 + 6a - 5}{a^3 + 8}$

87. $\dfrac{a^2 - 6a + 9}{a^3 - 8} \div \dfrac{a^2 - a - 6}{a^2 - 4}$ $\dfrac{a - 3}{a^2 + 2a + 4}$

88. $\dfrac{1}{z^2 + 4} \div \dfrac{z^3 - 8}{z^4 - 16}$ $\dfrac{z + 2}{z^2 + 2z + 4}$

89. $\dfrac{w^2 + 3}{w^3 - 8} - \dfrac{2w}{w^2 - 4}$ $\dfrac{-w^3 - 2w^2 - 5w + 6}{(w - 2)(w + 2)(w^2 + 2w + 4)}$

90. $\dfrac{x + 5}{x^3 + 27} - \dfrac{x - 1}{x^2 - 9}$ $\dfrac{-x^3 + 5x^2 - 10x - 6}{(x^2 - 9)(x^2 - 3x + 9)}$

91. $\dfrac{1}{x^3 - 1} - \dfrac{1}{x^2 - 1} + \dfrac{1}{x - 1}$ $\dfrac{x^3 + x^2 + 2x + 1}{(x - 1)(x + 1)(x^2 + x + 1)}$

92. $\dfrac{x - 4}{x^3 - 1} + \dfrac{x - 2}{x^2 - 1}$ $\dfrac{x^3 - 4x - 6}{(x^3 - 1)(x + 1)}$

In Exercises 93–98, solve each problem. See Example 6.

93. *Processing.* Joe takes x hours on the average to process a claim, whereas Ellen averages $x + 1$ hours to process a claim. Write a rational expression for the number of claims that they will process while working together for an 8-hour shift.
$\dfrac{16x + 8}{x^2 + x}$

94. *Roofing.* Bill attaches one bundle of shingles in an average of x minutes using a hammer, whereas Julio can attach one bundle in an average of $x - 6$ minutes using a pneumatic stapler. Write a rational expression for the number of bundles that they can attach while working together for 10 hours.

$$\frac{1200x - 3600}{x^2 - 6x}$$

FIGURE FOR EXERCISE 94

95. *Selling.* George sells one magazine subscription every 20 minutes, whereas Theresa sells one every x minutes. Write a rational expression for the number of magazine subscriptions that they will sell when working together for one hour.

$$\frac{3x + 60}{x}$$

96. *Painting.* Harry can paint his house by himself in 6 days. His wife Judy can paint the house by herself in x days. Write a rational expression for the portion of the house that they paint when working together for 2 days.

$$\frac{x + 6}{3x}$$

97. *Driving.* Joan drove for 100 miles at one speed and then increased her speed by 5 miles per hour and drove 200 additional miles. Write a rational expression for her total travel time.

$$\frac{300x + 500}{x^2 + 5x} \text{ hours}$$

98. *Running.* Willard jogged for 3 miles at one speed and then doubled his speed for an additional mile. Write a rational expression for his total running time.

$$\frac{7}{2x} \text{ hours}$$

GETTING MORE INVOLVED

 99. *Discussion.* Explain why fractions must have common denominators for addition but not for multiplication.

 100. *Discussion.* Find each "infinite sum" and explain your answer.

a) $\dfrac{3}{10} + \dfrac{3}{10^2} + \dfrac{3}{10^3} + \dfrac{3}{10^4} + \cdots$

b) $\dfrac{9}{10} + \dfrac{9}{10^2} + \dfrac{9}{10^3} + \dfrac{9}{10^4} + \cdots$

 6.4 **COMPLEX FRACTIONS**

In this section we will use the techniques of Section 6.3 to simplify complex fractions. As their name suggests, complex fractions are rather messy-looking expressions.

- Simplifying Complex Fractions
- Simplifying Expressions with Negative Exponents
- Applications

Simplifying Complex Fractions

A **complex fraction** is a fraction that has rational expressions in the numerator, the denominator, or both. For example,

$$\frac{\dfrac{1}{2} + \dfrac{1}{3}}{\dfrac{1}{4} + \dfrac{1}{5}}, \qquad \frac{3 - \dfrac{2}{x}}{\dfrac{1}{x^2} - \dfrac{1}{4}}, \qquad \text{and} \qquad \frac{\dfrac{x + 2}{x^2 - 9}}{\dfrac{x}{x^2 - 6x + 9} + \dfrac{4}{x - 3}}$$

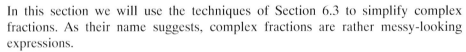

are complex fractions. In the next example we show two methods for simplifying a complex fraction.

EXAMPLE 1

A complex fraction without variables

Simplify $\dfrac{\frac{1}{2} + \frac{1}{3}}{\frac{1}{4} + \frac{1}{5}}$.

Solution

Method A For this method we perform the computations of the numerator and denominator separately and then divide:

$$\frac{\frac{1}{2} + \frac{1}{3}}{\frac{1}{4} + \frac{1}{5}} = \frac{\frac{5}{6}}{\frac{9}{20}} = \frac{5}{6} \div \frac{9}{20} = \frac{5}{6} \cdot \frac{20}{9} = \frac{5 \cdot 2 \cdot 10}{2 \cdot 3 \cdot 9} = \frac{50}{27}$$

Method B For this method we find the LCD for all of the fractions in the complex fraction. Then we multiply the numerator and denominator of the complex fraction by the LCD. The LCD for the denominators 2, 3, 4, and 5 is 60. So we multiply the numerator and denominator of the complex fraction by 60:

$$\frac{\frac{1}{2} + \frac{1}{3}}{\frac{1}{4} + \frac{1}{5}} = \frac{\left(\frac{1}{2} + \frac{1}{3}\right)60}{\left(\frac{1}{4} + \frac{1}{5}\right)60} = \frac{30 + 20}{15 + 12} = \frac{50}{27} \qquad \frac{1}{2} \cdot 60 = 30, \frac{1}{3} \cdot 60 = 20$$
$$\frac{1}{4} \cdot 60 = 15, \frac{1}{5} \cdot 60 = 12$$

In most cases Method B of Example 1 is the faster method for simplifying complex fractions, and we will continue to use it.

EXAMPLE 2

A complex fraction with variables

Simplify $\dfrac{3 - \frac{2}{x}}{\frac{1}{x^2} - \frac{1}{4}}$.

Solution

The LCD of x, x^2, and 4 is $4x^2$. Multiply the numerator and denominator by $4x^2$:

$$\frac{3 - \dfrac{2}{x}}{\dfrac{1}{x^2} - \dfrac{1}{4}} = \frac{\left(3 - \dfrac{2}{x}\right)(4x^2)}{\left(\dfrac{1}{x^2} - \dfrac{1}{4}\right)(4x^2)}$$

$$= \frac{3(4x^2) - \dfrac{2}{x}(4x^2)}{\dfrac{1}{x^2}(4x^2) - \dfrac{1}{4}(4x^2)} \qquad \text{Distributive property}$$

$$= \frac{12x^2 - 8x}{4 - x^2}$$

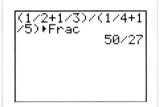

```
(1/2+1/3)/(1/4+1
/5)▶Frac
              50/27
```

helpful hint

When students see addition or subtraction in a complex fraction, they often convert all of the fractions to the same denominator. This is not wrong, but it is not necessary. Simply multiplying ever fraction by the LCD eliminates the denominators of the original fractions.

EXAMPLE 3

More complicated denominators

Simplify $\dfrac{\dfrac{x+2}{x^2-9}}{\dfrac{x}{x^2-6x+9}+\dfrac{4}{x-3}}$.

Solution

Because $x^2 - 9 = (x - 3)(x + 3)$ and $x^2 - 6x + 9 = (x - 3)^2$, the LCD is $(x - 3)^2(x + 3)$. Multiply the numerator and denominator by the LCD:

$$\frac{\dfrac{x+2}{x^2-9}}{\dfrac{x}{x^2-6x+9}+\dfrac{4}{x-3}} = \frac{\dfrac{x+2}{(x-3)(x+3)}(x-3)^2(x+3)}{\dfrac{x}{(x-3)^2}(x-3)^2(x+3)+\dfrac{4}{x-3}(x-3)^2(x+3)}$$

$$= \frac{(x+2)(x-3)}{x(x+3)+4(x-3)(x+3)} \qquad \text{Simplify.}$$

$$= \frac{(x+2)(x-3)}{(x+3)[x+4(x-3)]} \qquad \text{Factor out } x + 3.$$

$$= \frac{(x+2)(x-3)}{(x+3)(5x-12)}$$

Simplifying Expressions with Negative Exponents

Consider the expression

$$\frac{3a^{-1} - 2^{-1}}{1 - b^{-1}}.$$

Using the definition of negative exponents, we can rewrite this expression as a complex fraction:

$$\frac{3a^{-1} - 2^{-1}}{1 - b^{-1}} = \frac{\dfrac{3}{a} - \dfrac{1}{2}}{1 - \dfrac{1}{b}}$$

The LCD for the complex fraction is $2ab$. Note that $2ab$ could also be obtained from the bases of the expressions with the negative exponents. To simplify the complex fraction, we could use Method B as we have been doing. However, it is not necessary to rewrite the original expression as a complex fraction. The next example shows how to use Method B with the original expression.

EXAMPLE 4

A complex fraction with negative exponents

Simplify the complex fraction $\frac{3a^{-1} - 2^{-1}}{1 - b^{-1}}$.

Solution

Multiply the numerator and denominator by $2ab$, the LCD of the fractions. Remember that $a^{-1} \cdot a = a^0 = 1$.

$$\frac{3a^{-1} - 2^{-1}}{1 - b^{-1}} = \frac{(3a^{-1} - 2^{-1})2ab}{(1 - b^{-1})2ab}$$

$$= \frac{3a^{-1}(2ab) - 2^{-1}(2ab)}{1(2ab) - b^{-1}(2ab)} \qquad \text{Distributive property}$$

$$= \frac{6b - ab}{2ab - 2a}$$

EXAMPLE 5

A complex fraction with negative exponents

Simplify the complex fraction $\dfrac{a^{-1} + b^{-2}}{ab^{-2} + ba^{-3}}$.

Solution

If we rewrote a^{-1}, b^{-2}, b^{-2}, and a^{-3}, then the denominators would be a, b^2, b^2, and a^3. So the LCD is a^3b^2. If we multiply the numerator and denominator by a^3b^2, the negative exponents will be eliminated:

$$\frac{a^{-1} + b^{-2}}{ab^{-2} + ba^{-3}} = \frac{(a^{-1} + b^{-2})a^3b^2}{(ab^{-2} + ba^{-3})a^3b^2}$$

$$= \frac{a^{-1} \cdot a^3b^2 + b^{-2} \cdot a^3b^2}{ab^{-2} \cdot a^3b^2 + ba^{-3} \cdot a^3b^2} \quad \text{Distributive property}$$

$$= \frac{a^2b^2 + a^3}{a^4 + b^3} \qquad \begin{matrix} b^{-2}b^2 = b^0 = 1 \\ a^{-3}a^3 = a^0 = 1 \end{matrix}$$

Note that the positive exponents of a^3b^2 are just large enough to eliminate all of the negative exponents when we multiply. ∎

The next example is not exactly a complex fraction, but we can use the same technique as in the previous example.

EXAMPLE 6

More negative exponents

Eliminate negative exponents and simplify $p + p^{-1}q^{-2}$.

Solution

If we multiply the numerator and denominator by pq^2, we will eliminate the negative exponents:

$$p + p^{-1}q^{-2} = \frac{(p + p^{-1}q^{-2})}{1} \cdot \frac{pq^2}{pq^2}$$

$$= \frac{p^2q^2 + 1}{pq^2} \quad \begin{matrix} p \cdot pq^2 = p^2q^2 \\ p^{-1}q^{-2} \cdot pq^2 = 1 \end{matrix}$$

∎

Applications

The next example illustrates how complex fractions can occur in a problem.

EXAMPLE 7

An application of complex fractions

Eastside Elementary has the same number of students as Westside Elementary. One-half of the students at Eastside ride buses to school, and two-thirds of the students at Westside ride buses to school. One-sixth of the students at Eastside are female, and one-third of the students at Westside are female. If all of the female students ride the buses, then what percentage of the students who ride the buses are female?

Solution

To find the required percentage, we must divide the number of females who ride the buses by the total number of students who ride the buses. Let

$$x = \text{the number of students at Eastside.}$$

Because the number of students at Westside is also x, we have

$$\frac{1}{2}x + \frac{2}{3}x = \text{the total number of students who ride the buses}$$

and

$$\frac{1}{6}x + \frac{1}{3}x = \text{the total number of female students.}$$

Because all of the female students ride the buses, we can express the percentage of riders who are female by the following rational expression:

$$\frac{\dfrac{1}{6}x + \dfrac{1}{3}x}{\dfrac{1}{2}x + \dfrac{2}{3}x}$$

Multiply the numerator and denominator by 6, the LCD for 2, 3, and 6:

$$\frac{\left(\dfrac{1}{6}x + \dfrac{1}{3}x\right)6}{\left(\dfrac{1}{2}x + \dfrac{2}{3}x\right)6} = \frac{x + 2x}{3x + 4x} = \frac{3x}{7x} = \frac{3}{7} \approx 0.43 = 43\%$$

So 43% of the students who ride the buses are female.

WARM-UPS

True or false? Explain.

1. The LCM for 2, x, 6, and x^2 is $6x^3$.   False
2. The LCM for $a - b$, $2b - 2a$, and 6 is $6a - 6b$.   True
3. The LCD is the LCM of the denominators.   True
4. $\dfrac{\dfrac{1}{2} + \dfrac{1}{3}}{1 + \dfrac{1}{2}} = \dfrac{5}{6} \div \dfrac{3}{2}$ True
5. $2^{-1} + 3^{-1} = (2 + 3)^{-1}$ False
6. $(2^{-1} + 3^{-1})^{-1} = 2 + 3$ False
7. $2 + 3^{-1} = 5^{-1}$ False
8. $x + 2^{-1} = \frac{x}{2}$ for any real number x.   False
9. To simplify $\frac{a^{-1} - b^{-1}}{a - b}$, multiply the numerator and denominator by ab.   True
10. To simplify $\frac{ab^{-2} + a^{-5}b^2}{a^{-3}b - a^5b^{-1}}$, multiply the numerator and denominator by a^5b^2.   True

6.4   EXERCISES

Reading and Writing   *After reading this section, write out the answers to these questions. Use complete sentences.*

1. What is a complex fraction?

   A complex fraction is a fraction that contains fractions in the numerator, denominator, or both.

2. What are the two methods for simplifying complex fractions?

   One method is to perform the operations in the numerator and then in the denominator, and then divide the results. The other method is to multiply the numerator and the denominator by the LCD for all of the fractions.

Simplify each complex fraction. Use either method. See Example 1.

3. $\dfrac{\dfrac{1}{2}-\dfrac{1}{3}}{\dfrac{1}{4}-\dfrac{1}{5}}$ $\quad\dfrac{10}{3}$

4. $\dfrac{\dfrac{1}{3}+\dfrac{1}{4}}{\dfrac{1}{5}+\dfrac{1}{6}}$ $\quad\dfrac{35}{22}$

5. $\dfrac{\dfrac{2}{3}+\dfrac{5}{6}-\dfrac{1}{2}}{\dfrac{1}{8}-\dfrac{1}{3}+\dfrac{1}{12}}$ $\quad-8$

6. $\dfrac{\dfrac{2}{5}-\dfrac{x}{9}-\dfrac{1}{3}}{\dfrac{1}{3}+\dfrac{x}{5}+\dfrac{2}{15}}$ $\quad\dfrac{3-5x}{9x+21}$

Simplify the complex fractions. Use Method B. See Example 2.

7. $\dfrac{\dfrac{a+b}{b}}{\dfrac{a-b}{ab}}$ $\quad\dfrac{a^2+ab}{a-b}$

8. $\dfrac{\dfrac{m-n}{m^2}}{\dfrac{m-3}{mn^3}}$ $\quad\dfrac{mn^3-n^4}{m^2-3m}$

9. $\dfrac{a+\dfrac{3}{b}}{\dfrac{b}{a}+\dfrac{1}{b}}$ $\quad\dfrac{a^2b+3a}{a+b^2}$

10. $\dfrac{m-\dfrac{2}{n}}{\dfrac{1}{m}-\dfrac{3}{n}}$ $\quad\dfrac{m^2n-2m}{n-3m}$

11. $\dfrac{\dfrac{x-3y}{xy}}{\dfrac{1}{x}+\dfrac{1}{y}}$ $\quad\dfrac{x-3y}{x+y}$

12. $\dfrac{\dfrac{2}{w}+\dfrac{3}{t}}{\dfrac{w-t}{4wt}}$ $\quad\dfrac{8t+12w}{w-t}$

13. $\dfrac{3-\dfrac{m-2}{6}}{\dfrac{4}{9}+\dfrac{2}{m}}$ $\quad\dfrac{60m-3m^2}{8m+36}$

14. $\dfrac{6-\dfrac{2-z}{z}}{\dfrac{1}{3z}-\dfrac{1}{6}}$ $\quad\dfrac{42z-12}{2-z}$

15. $\dfrac{\dfrac{a^2-b^2}{a^2b^3}}{\dfrac{a+b}{a^3b}}$ $\quad\dfrac{a^2-ab}{b^2}$

16. $\dfrac{\dfrac{4x^2-1}{x^2y}}{\dfrac{4x-2}{xy^2}}$ $\quad\dfrac{2xy+y}{2x}$

17. $\dfrac{\dfrac{1}{x^2y^2}+\dfrac{1}{xy^3}}{\dfrac{1}{x^3y}-\dfrac{1}{xy}}$ $\quad\dfrac{xy+x^2}{y^2-x^2y^2}$

18. $\dfrac{\dfrac{1}{2a^3b}-\dfrac{1}{ab^4}}{\dfrac{1}{6a^2b^2}+\dfrac{1}{3a^4b}}$ $\quad\dfrac{3ab^3-6a^3}{a^2b^2+2b^3}$

Simplify each complex fraction. See Examples 1–3.

19. $\dfrac{x+\dfrac{4}{x+4}}{x-\dfrac{4x+4}{x+4}}$ $\quad\dfrac{x+2}{x-2}$

20. $\dfrac{x-\dfrac{x+6}{x+2}}{x-\dfrac{4x+15}{x+2}}$ $\quad\dfrac{x-2}{x-5}$

21. $\dfrac{1-\dfrac{1}{y-1}}{3+\dfrac{1}{y+1}}$ $\quad\dfrac{y^2-y-2}{(y-1)(3y+4)}$

22. $\dfrac{2-\dfrac{3}{a-2}}{4-\dfrac{1}{a+2}}$ $\quad\dfrac{2a^2-3a-14}{4a^2-a-14}$

23. $\dfrac{\dfrac{2}{3-x}-4}{\dfrac{1}{x-3}-1}$ $\quad\dfrac{4x-10}{x-4}$

24. $\dfrac{\dfrac{x}{x-5}-2}{\dfrac{2x}{5-x}-1}$ $\quad\dfrac{x-10}{3x-5}$

25. $\dfrac{\dfrac{w+2}{w-1}-\dfrac{w-3}{w}}{\dfrac{w+4}{w}+\dfrac{w-2}{w-1}}$ $\quad\dfrac{6w-3}{2w^2+w-4}$

26. $\dfrac{\dfrac{x-1}{x+2}-\dfrac{x-2}{x+3}}{\dfrac{x-3}{x+3}+\dfrac{x+1}{x+2}}$ $\quad\dfrac{2x+1}{2x^2+3x-3}$

27. $\dfrac{\dfrac{1}{a-b}-\dfrac{3}{a+b}}{\dfrac{2}{b-a}+\dfrac{4}{b+a}}$ $\quad\dfrac{2b-a}{a-3b}$

28. $\dfrac{\dfrac{3}{2+x}-\dfrac{4}{2-x}}{\dfrac{1}{x+2}-\dfrac{3}{x-2}}$ $\quad\dfrac{7x+2}{-2x-8}$

29. $\dfrac{3-\dfrac{4}{a-1}}{5-\dfrac{3}{1-a}}$ $\quad\dfrac{3a-7}{5a-2}$

30. $\dfrac{\dfrac{x}{3}-\dfrac{x-1}{9-x}}{\dfrac{x}{6}-\dfrac{2-x}{x-9}}$ $\quad\dfrac{2x^2-12x-6}{x^2-3x-12}$

31. $\dfrac{\dfrac{2}{m-3}+\dfrac{4}{m}}{\dfrac{3}{m-2}+\dfrac{1}{m}}$ $\quad\dfrac{3m^2-12m+12}{(m-3)(2m-1)}$

32. $\dfrac{\dfrac{1}{y+2}-\dfrac{4}{3y}}{\dfrac{3}{y}-\dfrac{2}{y+3}}$ $\quad\dfrac{-y^2-11y-24}{3y^2+33y+54}$

33. $\dfrac{\dfrac{3}{x^2-1}-\dfrac{x-2}{x^3-1}}{\dfrac{3}{x^2+x+1}+\dfrac{x-3}{x^3-1}}$ $\quad\dfrac{2x^2+4x+5}{4x^2-2x-6}$

34. $\dfrac{\dfrac{2}{a^3+8}-\dfrac{3}{a^2-2a+4}}{\dfrac{4}{a^2-4}+\dfrac{a-3}{a^3+8}}$ $\quad\dfrac{-3a^2+2a+8}{5a^2-13a+22}$

Simplify. See Examples 4–6.

35. $\dfrac{w^{-1}+y^{-1}}{z^{-1}+y^{-1}}$ $\quad\dfrac{yz+wz}{wy+wz}$

36. $\dfrac{a^{-1}-b^{-1}}{a^{-1}+b^{-1}}$ $\quad\dfrac{b-a}{b+a}$

37. $\dfrac{1-x^{-1}}{1-x^{-2}}$ $\quad\dfrac{x}{x+1}$

38. $\dfrac{4-a^{-2}}{2-a^{-1}}$ $\quad\dfrac{2a+1}{a}$

39. $\dfrac{a^{-2}+b^{-2}}{a^{-1}b}$ $\quad\dfrac{a^2+b^2}{ab^3}$

40. $\dfrac{m^{-3}+n^{-3}}{mn^{-2}}$ $\quad\dfrac{n^3+m^3}{m^4n}$

41. $\dfrac{1 - a^{-1}}{\dfrac{a - 1}{a}}$

42. $\dfrac{m^{-1} - a^{-1}}{\dfrac{a - m}{am}}$

43. $\dfrac{\dfrac{x^{-1} + x^{-2}}{x + x^{-2}}}{\dfrac{1}{x^2 - x + 1}}$

44. $\dfrac{\dfrac{x - x^{-2}}{1 - x^{-2}}}{\dfrac{x^2 + x + 1}{x + 1}}$

45. $\dfrac{\dfrac{2m^{-1} - 3m^{-2}}{m^{-2}}}{2m - 3}$

46. $\dfrac{\dfrac{4x^{-3} - 6x^{-5}}{2x^{-5}}}{2x^2 - 3}$

47. $\dfrac{\dfrac{a^{-1} - b^{-1}}{a - b}}{-\dfrac{1}{ab}}$

48. $\dfrac{\dfrac{a^2 - b^2}{a^{-2} - b^{-2}}}{-a^2b^2}$

49. $\dfrac{\dfrac{x^3 - y^3}{x^{-3} - y^{-3}}}{-x^3y^3}$

50. $\dfrac{\dfrac{(a - b)^2}{a^{-2} - b^{-2}}}{\dfrac{a^2b^3 - a^3b^2}{a + b}}$

51. $\dfrac{\dfrac{1 - 8x^{-3}}{x^{-1} + 2x^{-2} + 4x^{-3}}}{x - 2}$

52. $\dfrac{\dfrac{a + 27a^{-2}}{1 - 3a^{-1} + 9a^{-2}}}{a + 3}$

53. $\dfrac{(x^{-1} + y^{-1})^{-1}}{\dfrac{xy}{x + y}}$

54. $\dfrac{(a^{-1} - b^{-1})^{-2}}{\dfrac{a^2b^2}{b^2 - 2ab + a^2}}$

 Use a calculator to evaluate each complex fraction. Round answers to four decimal places. If your calculator does fractions, then also find the fractional answer.

55. $\dfrac{\dfrac{5}{3} - \dfrac{4}{5}}{\dfrac{1}{3} - \dfrac{5}{6}}$

$-1.7333, -\dfrac{26}{15}$

56. $\dfrac{\dfrac{1}{12} + \dfrac{1}{2} - \dfrac{3}{4}}{\dfrac{3}{5} + \dfrac{5}{6}}$

$-0.1163, -\dfrac{5}{43}$

57. $\dfrac{4^{-1} - 9^{-1}}{2^{-1} + 3^{-1}}$

$0.1667, \dfrac{1}{6}$

58. $\dfrac{2^{-1} + 3^{-1} - 6^{-1}}{3^{-1} - 5^{-1} + 4^{-1}}$

$1.7391, \dfrac{40}{23}$

Solve each problem. See Example 7.

59. *Racial balance.* Clarksville has three elementary schools. Northside has one-half as many students as Central, and Southside has two-thirds as many students as Central. One-third of the students at Northside are African-American, three-fourths of the students at Central are African-American, and one-sixth of the students at Southside are African-American. What percent of the city's elementary students are African-American?
47.4%

60. *Explosive situation.* All of the employees at Acme Explosives are in either development, manufacturing, or sales.

One-fifth of the employees in development are women, one-third of the employees in manufacturing are women, and one-half of the employees in sales are women. Use the accompanying figure to determine the percentage of workers at Acme who are women. What percent of the women at Acme are in sales?
38.3%, 65.2%

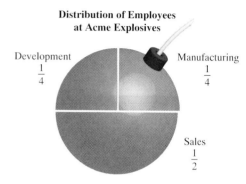

Distribution of Employees at Acme Explosives

Development $\dfrac{1}{4}$ Manufacturing $\dfrac{1}{4}$ Sales $\dfrac{1}{2}$

FIGURE FOR EXERCISE 60

61. *Average speed.* Mary drove from Clarksville to Leesville at 45 miles per hour (mph). At Leesville she discovered that she had forgotten her purse. She immediately returned to Clarksville at 55 mph. What was her average speed for the entire trip? (The answer is *not* 50 mph.)
49.5 mph

 62. *Average price.* On her way to New York, Jenny spent the same amount for gasoline each of the three times that she filled up. She paid 99.9 cents per gallon the first time, 109.9 cents per gallon the second time, and 119.9 cents per gallon the third time. What was the average price per gallon to the nearest tenth of a cent for the gasoline that she bought?
109.3 cents per gallon

FIGURE FOR EXERCISE 62

GETTING MORE INVOLVED

63. *Cooperative learning.* Write a step-by-step strategy for simplifying complex fractions with negative exponents. Have a classmate use your strategy to simplify some complex fractions from Exercises 35–54.

64. *Discussion.* **a)** Find the exact value of each expression.

i)
$$\cfrac{1}{1 + \cfrac{1}{1 + \cfrac{1}{1 + \cfrac{1}{2}}}}$$

ii)
$$\cfrac{1}{1 + \cfrac{1}{1 + \cfrac{1}{1 + \cfrac{1}{3}}}}$$

$$\frac{5}{8}, \frac{11}{18}$$

b) Explain why in each case the exact value must be less than 1.

The denominator is larger than the numerator in the first fraction.

65. *Cooperative Learning.* Work with a group to simplify the complex fraction. For what values of x is the complex fraction undefined?

$$\cfrac{1}{1 + \cfrac{1}{1 + \cfrac{1}{1 + \cfrac{1}{x}}}}$$

$$\frac{2x + 1}{3x + 2}, x \neq 0, -1, -\frac{1}{2}, -\frac{2}{3}$$

6.5   SOLVING EQUATIONS INVOLVING RATIONAL EXPRESSIONS

Many problems in algebra are modeled by equations involving rational expressions. In this section you will learn how to solve equations that have rational expressions, and in Section 6.6 we will solve problems using these equations.

In this section

- Multiplying by the LCD
- Extraneous Roots
- Proportions

Multiplying by the LCD

To solve equations having rational expressions, we multiply each side of the equation by the LCD of the rational expressions.

EXAMPLE 1   An equation with rational expressions

Solve $\frac{1}{x} + \frac{1}{4} = \frac{1}{6}$.

Solution

The LCD for the denominators 4, 6, and x is $12x$:

$$12x\left(\frac{1}{x} + \frac{1}{4}\right) = 12x\left(\frac{1}{6}\right) \quad \text{Multiply each side by } 12x.$$

$$12x \cdot \frac{1}{x} + \overset{3}{12x} \cdot \frac{1}{4} = \overset{2}{12x} \cdot \frac{1}{6} \quad \text{Distributive property}$$

$$12 + 3x = 2x \quad \text{Divide out the common factors.}$$

$$12 + x = 0$$

$$x = -12$$

Check -12 in the original equation. The solution set is $\{-12\}$.

> **helpful hint**
>
> Note that it is not necessary to convert each fraction into an equivalent fraction with a common denominator here. Since we can multiply both sides of an equation by any expression we choose, we choose to multiply by the LCD. This tactic eliminates the fractions in one step and that is good.

EXAMPLE 2

An equation with rational expressions

Solve $\frac{1}{x} + \frac{2}{3x} = \frac{1}{5}$.

Solution

Multiply each side by $15x$, the LCD for x, $3x$, and 5:

$$15x\left(\frac{1}{x} + \frac{2}{3x}\right) = 15x\left(\frac{1}{5}\right)$$

$$15x \cdot \frac{1}{x} + 15x \cdot \frac{2}{3x} = 3x$$

$$15 + 10 = 3x$$

$$25 = 3x$$

$$\frac{25}{3} = x$$

Check $\frac{25}{3}$ in the original equation. The solution set is $\left\{\frac{25}{3}\right\}$. ∎

> **CAUTION**   To solve an equation with rational expressions, we do *not* convert the rational expressions to ones with a common denominator. Instead, we multiply each side by the LCD to *eliminate* the denominators.

EXAMPLE 3

An equation with two solutions

Solve $\frac{200}{x} + \frac{300}{x + 20} = 10$.

Solution

$$x(x + 20)\left(\frac{200}{x} + \frac{300}{x + 20}\right) = x(x + 20)10 \qquad \text{Multiply each side by } x(x + 20).$$

$$x(x + 20)\frac{200}{x} + x(x + 20)\frac{300}{x + 20} = x(x + 20)10 \qquad \text{Distributive property}$$

$$(x + 20)200 + x(300) = (x^2 + 20x)10 \qquad \text{Simplify.}$$

$$200x + 4000 + 300x = 10x^2 + 200x$$

$$4000 + 300x = 10x^2 \qquad \text{Combine like terms.}$$

$$400 + 30x = x^2 \qquad \text{Divide each side by 10.}$$

$$0 = x^2 - 30x - 400$$

$$0 = (x - 40)(x + 10) \qquad \text{Factor.}$$

$$x - 40 = 0 \quad \text{or} \quad x + 10 = 0 \qquad \text{Set each factor equal to 0.}$$

$$x = 40 \quad \text{or} \qquad x = -10$$

Check these values in the original equation. The solution set is $\{-10, 40\}$. ∎

Extraneous Roots

Because equations involving rational expressions have variables in denominators, a root to the equation might cause a 0 to appear in a denominator. In this case the root does not satisfy the original equation, and so it is called an **extraneous root.**

EXAMPLE 4   **An equation with an extraneous root**

Solve $\dfrac{3}{x} + \dfrac{6}{x-2} = \dfrac{12}{x^2 - 2x}$.

Solution

Because $x^2 - 2x = x(x-2)$, the LCD for x, $x - 2$, and $x^2 - 2x$ is $x(x-2)$.

$$x(x-2)\frac{3}{x} + x(x-2)\frac{6}{x-2} = x(x-2)\frac{12}{x(x-2)} \qquad \text{Multiply each side by } x(x-2).$$

$$3(x-2) + 6x = 12$$
$$3x - 6 + 6x = 12$$
$$9x - 6 = 12$$
$$9x = 18$$
$$x = 2$$

Neither 0 nor 2 could be a solution because replacing x by either 0 or 2 would cause 0 to appear in a denominator in the original equation. So 2 is an extraneous root and the solution set is the empty set, $\varnothing$. ■

EXAMPLE 5   **An equation with an extraneous root**

Solve $x + 2 + \dfrac{x}{x-2} = \dfrac{2}{x-2}$.

Solution

Because the LCD is $x - 2$, we multiply each side by $x - 2$:

$$(x-2)(x+2) + (x-2)\frac{x}{x-2} = (x-2)\frac{2}{x-2}$$

$$x^2 - 4 + x = 2$$
$$x^2 + x - 6 = 0$$
$$(x+3)(x-2) = 0$$
$$x + 3 = 0 \quad \text{or} \quad x - 2 = 0$$
$$x = -3 \quad \text{or} \quad x = 2$$

Replacing x by 2 in the original equation would cause 0 to appear in a denominator. So 2 is an extraneous root. Check that the original equation is satisfied if $x = -3$. The solution set is $\{-3\}$. ■

Proportions

An equation that expresses the equality of two rational expressions is called a **proportion.** The equation

$$\frac{a}{b} = \frac{c}{d}$$

is a proportion. The terms in the position of b and c are called the **means.** The terms in the position of a and d are called the **extremes.** If we multiply this proportion by the LCD, bd, we get

$$bd \cdot \frac{a}{b} = bd \cdot \frac{c}{d}$$

or

$$ad = bc.$$

The equation $ad = bc$ says that *the product of the extremes is equal to the product of the means*. When solving a proportion, we can omit multiplication by the LCD and just remember the result, $ad = bc$, as the **extremes-means property.**

Extremes-Means Property

If $\frac{a}{b} = \frac{c}{d}$, then $ad = bc$.

The extremes-means property makes it easier to solve proportions.

E X A M P L E   6

A proportion with one solution

Solve $\frac{20}{x} = \frac{30}{x + 20}$.

Solution

Rather than multiplying by the LCD, we use the extremes-means property to eliminate the denominators:

$$\frac{20}{x} = \frac{30}{x + 20}$$

$$20(x + 20) = 30x \qquad \text{Extremes-means property}$$

$$20x + 400 = 30x$$

$$400 = 10x$$

$$40 = x$$

Check 40 in the original equation. The solution set is $\{40\}$.

E X A M P L E   7

A proportion with two solutions

Solve $\frac{2}{x} = \frac{x + 3}{5}$.

Solution

Use the extremes-means property to write an equivalent equation:

$$x(x + 3) = 2 \cdot 5 \quad \text{Extremes-means property}$$

$$x^2 + 3x = 10$$

$$x^2 + 3x - 10 = 0$$

$$(x + 5)(x - 2) = 0 \qquad \text{Factor.}$$

$$x + 5 = 0 \quad \text{or} \quad x - 2 = 0 \qquad \text{Zero factor property}$$

$$x = -5 \quad \text{or} \qquad x = 2$$

Both -5 and 2 satisfy the original equation. The solution set is $\{-5, 2\}$.

CAUTION Use the extremes-means property only when solving a *proportion*. It cannot be used on an equation such as

$$\frac{3}{x} = \frac{2}{x + 1} + 5.$$

Cargo has been lost, or the hull of a ship has been damaged. What is the amount of money that should be paid to the insured party? Lisa M. Paccione, Ocean Marine Claim Representative for the St. Paul Insurance Company, investigates, evaluates, resolves, and pays these types of claims. Ms. Paccione does this by gathering data, occasionally doing a visual inspection, interviewing witnesses, and negotiating with attorneys.

**M A R I N E
I N S U R A N C E
A G E N T**

   Decisions about losses are based on the insured party's individual policy as well as traditional marine practices and maritime law. When consignees suffer a cargo loss, they not only are compensated for the actual amount of the damaged goods, but also receive an additional "advance" in the settlement. Customarily, the advance is 10% over the value of the goods. The amount that St. Paul pays the insured party for a valid claim is computed by using a proportion. In Exercises 59 and 60 of this section you will solve problems involving this proportion.

E X A M P L E   8

Ratios and proportions

The ratio of men to women at a football game was 4 to 3. If there were 12,000 more men than women in attendance, then how many men and how many women were in attendance?

Solution

Let x represent the number of men in attendance and $x - 12,000$ represent the number of women in attendance. Because the ratio of men to women was 4 to 3, we can write the following proportion:

$$\frac{4}{3} = \frac{x}{x - 12,000}$$

$$4x - 48,000 = 3x$$

$$x = 48,000$$

So there were 48,000 men and 36,000 women at the game.

W A R M - U P S

True or false? Explain.

1. In solving an equation involving rational expressions, multiply each side by the LCD for all of the denominators.   True

2. To solve $\frac{1}{x} + \frac{1}{2x} = \frac{1}{3}$, first change each rational expression to an equivalent rational expression with a denominator of $6x$.   False

3. Extraneous roots are not real numbers.   False

4. To solve $\frac{1}{x - 2} + 3 = \frac{1}{x + 2}$, multiply each side by $x^2 - 4$.   True

5. The solution set to $\frac{x}{3x + 4} - \frac{6}{2x + 1} = \frac{7}{5}$ is $\left\{-\frac{4}{3}, -\frac{1}{2}\right\}$.   False

(continued)

6. The solution set to $\frac{3}{x} = \frac{2}{5}$ is $\left\{\frac{15}{2}\right\}$.   True

7. We should use the extremes-means property to solve $\frac{x-2}{x+3} + 1 = \frac{1}{x}$.   False

8. The equation $x^2 = x$ is equivalent to the equation $x = 0$.   False

9. The solution set to $(2x - 3)(3x + 4) = 0$ is $\left\{\frac{3}{2}, \frac{4}{3}\right\}$.   False

10. The equation $\frac{2}{x+1} = \frac{x-1}{4}$ is equivalent to $x^2 - 1 = 8$.   True

6.5 EXERCISES

Reading and Writing   *After reading this section, write out the answers to these questions. Use complete sentences.*

1. What is the usual first step in solving an equation involving rational expressions?

The first step is to multiply each side of the equation by the LCD.

2. How can an equation involving rational expressions have an extraneous root?

A solution to the equation can cause 0 to appear in a denominator.

3. What is a proportion?

A proportion is an equation expressing equality of two rational expressions.

4. What are the means?

In $a/b = c/d$ the means are b and c.

5. What are the extremes?

In $a/b = c/d$ the extremes are a and d.

6. What is the extremes-means property?

The extremes-means property says that if $a/b = c/d$ then $bc = ad$.

Find the solution set to each equation. See Examples 1–5.

7. $\frac{1}{x} + \frac{1}{6} = \frac{1}{8}$ $\{-24\}$

8. $\frac{3}{x} + \frac{1}{5} = \frac{1}{2}$ $\{10\}$

9. $\frac{2}{3x} + \frac{1}{15x} = \frac{1}{2}$ $\left\{\frac{22}{15}\right\}$

10. $\frac{5}{6x} - \frac{1}{8x} = \frac{17}{24}$ $\{1\}$

11. $\frac{3}{x-2} + \frac{5}{x} = \frac{10}{x}$ $\{5\}$

12. $\frac{5}{x-1} + \frac{1}{2x} = \frac{1}{x}$ $\left\{-\frac{1}{9}\right\}$

13. $\frac{x}{x-2} + \frac{3}{x} = 2$ $\{1, 6\}$

14. $\frac{x}{x-5} + \frac{5}{x} = \frac{11}{6}$ $\{2, 15\}$

15. $\frac{100}{x} = \frac{150}{x+5} - 1$ $\{20, 25\}$

16. $\frac{30}{x} = \frac{50}{x+10} + \frac{1}{2}$ $\{-60, 10\}$

17. $\frac{3x-5}{x-1} = 2 - \frac{2x}{x-1}$ $\varnothing$

18. $\frac{x-3}{x+2} = 3 - \frac{1-2x}{x+2}$ $\varnothing$

19. $x + 1 + \frac{2x-5}{x-5} = \frac{x}{x-5}$ $\{-2\}$

20. $\frac{x-3}{2} - \frac{1}{x-3} = \frac{8-3x}{x-3}$ $\{-3\}$

21. $\frac{2}{x+2} + \frac{x}{x-3} + \frac{1}{x^2-x-6} = 0$ $\{-5, 1\}$

22. $\frac{x-4}{x^2+2x-15} = 2 - \frac{2}{x-3}$ $\left\{-\frac{9}{2}, 4\right\}$

Find the solution set to each equation. See Examples 6 and 7.

23. $\frac{2}{x} = \frac{3}{4}$ $\left\{\frac{8}{3}\right\}$

24. $\frac{5}{x} = \frac{7}{9}$ $\left\{\frac{45}{7}\right\}$

25. $\frac{a}{3} = \frac{-1}{4}$ $\left\{-\frac{3}{4}\right\}$

26. $\frac{b}{5} = \frac{-3}{7}$ $\left\{-\frac{15}{7}\right\}$

27. $-\frac{5}{7} = \frac{2}{x}$ $\left\{-\frac{14}{5}\right\}$

28. $-\frac{3}{8} = \frac{5}{x}$ $\left\{-\frac{40}{3}\right\}$

29. $\frac{10}{x} = \frac{20}{x+20}$ $\{20\}$

30. $\frac{x}{5} = \frac{x+2}{3}$ $\{-5\}$

31. $\frac{2}{x+1} = \frac{x-1}{4}$ $\{-3, 3\}$

32. $\frac{3}{x-2} = \frac{x+2}{7}$ $\{-5, 5\}$

33. $\frac{x}{6} = \frac{5}{x-1}$ $\{-5, 6\}$

34. $\frac{x+5}{2} = \frac{3}{x}$ $\{-6, 1\}$

35. $\frac{x}{x-3} = \frac{x+2}{x}$ $\{-6\}$

36. $\frac{x+1}{x-5} = \frac{x+2}{x-4}$ $\varnothing$

37. $\frac{x-2}{x-3} = \frac{x+5}{x+2}$ $\left\{\frac{11}{2}\right\}$

38. $\frac{x}{x+5} = \frac{x}{x-2}$ $\{0\}$

Solve each equation.

39. $\frac{a}{9} = \frac{4}{a}$ $\{-6, 6\}$

40. $\frac{y}{3} = \frac{27}{y}$ $\{-9, 9\}$

41. $\frac{1}{2x-4} + \frac{1}{x-2} = \frac{1}{4}$ $\{8\}$

42. $\frac{7}{3x-9} - \frac{1}{x-3} = \frac{4}{9}$ $\{6\}$

43. $\frac{x-2}{4} = \frac{x-2}{x}$ $\{2, 4\}$

44. $\frac{y+5}{2} = \frac{y+5}{y}$ $\{-5, 2\}$

45. $\frac{5}{2x+4} - \frac{1}{x-1} = \frac{3}{x+2}$ $\{-1\}$

46. $\dfrac{5}{2w + 6} - \dfrac{1}{w - 1} = \dfrac{1}{w + 3}$ $\{9\}$

47. $\dfrac{5}{x - 3} = \dfrac{x}{x - 3}$ $\{5\}$ **48.** $\dfrac{6}{a + 2} = \dfrac{a}{a + 2}$ $\{6\}$

49. $\dfrac{w}{6} = \dfrac{3}{2w}$ $\{-3, 3\}$ **50.** $\dfrac{2m}{5} = \dfrac{10}{m}$ $\{-5, 5\}$

51. $\dfrac{5}{4x - 2} - \dfrac{1}{1 - 2x} = \dfrac{7}{3x + 6}$ $\{8\}$

52. $\dfrac{5}{x + 1} - \dfrac{1}{1 - x} = \dfrac{1}{x^2 - 1}$ $\left\{\dfrac{5}{6}\right\}$

53. $\dfrac{5}{x} = \dfrac{2}{5}$ $\left\{\dfrac{25}{2}\right\}$

54. $\dfrac{-3}{2x} = \dfrac{1}{-5}$ $\left\{\dfrac{15}{2}\right\}$

55. $\dfrac{5}{x^2 - 9} + \dfrac{2}{x + 3} = \dfrac{1}{x - 3}$ $\{4\}$

56. $\dfrac{1}{x - 2} - \dfrac{2}{x + 3} = \dfrac{11}{x^2 + x - 6}$ $\{-4\}$

57. $\dfrac{9}{x^3 - 1} - \dfrac{1}{x - 1} = \dfrac{2}{x^2 + x + 1}$ $\{-5, 2\}$

58. $\dfrac{x + 4}{x^3 + 8} + \dfrac{x + 2}{x^2 - 2x + 4} = \dfrac{11}{2x + 4}$ $\left\{\dfrac{14}{9}, 2\right\}$

In Exercises 59–68, solve each problem. See Example 8.

59. *Maritime losses.* The amount paid to an insured party by the American Insurance Company is computed by using the proportion

$$\frac{\text{value shipped}}{\text{amount of loss}} = \frac{\text{amount of declared premium}}{\text{amount insured party gets paid}}.$$

If the value shipped was $300,000, the amount of loss was $250,000, and the amount of declared premium was $200,000, then what amount is paid to the insured party? $166,666.67

60. *Maritime losses.* Suppose the value shipped was $400,000, the amount of loss was $300,000, and the amount that the insured party got paid was $150,000. Use the proportion of Exercise 59 to find the amount of declared premium. $200,000

61. *Capture-recapture method.* To estimate the size of the grizzly bear population in a national park, rangers tagged and released 12 bears. Later it was observed that in 23 sightings of grizzly bears, only two had been tagged. Assuming the proportion of tagged bears in the later sightings is the same as the proportion of tagged bears in the population, estimate the number of bears in the population.   138

62. *Please rewind.* In a sample of 24 returned videotapes, it was found that only 3 were rewound as requested. If 872 videos are returned in a day, then how many of them would you expect to find that are not rewound?   763

63. *Pleasing painting.* The ancient Greeks often used the ratio of length to width for a rectangle as 7 to 6 to give the rectangle a pleasing shape. If the length of a pleasantly shaped

Greek painting is 22 centimeters (cm) longer than its width, then what are its length and width?
Width 132 cm, length 154 cm

64. *Pickups and cars.* The ratio of pickups to cars sold at a dealership is 2 to 3. If the dealership sold 142 more cars than pickups in 1999, then how many of each did it sell?
426 cars, 284 pickups

65. *Cleaning up the river.* Pollution in the Tickfaw River has been blamed primarily on pesticide runoff from area farms. The formula

$$C = \frac{4{,}000{,}000p}{100 - p}$$

has been used to model the cost in dollars for removing $p\%$ of the pollution in the river. If the state gets a $1 million federal grant for cleaning up the river, then what percentage of the pollution can be removed? Use the bar graph to estimate the percentage that can be cleaned up with a $100 million grant.   20%, 96%

FIGURE FOR EXERCISE 65

66. *Campaigning for governor.* A campaign manager for a gubernatorial candidate estimates that the cost in dollars for an advertising campaign that will get his candidate $p\%$ of the votes is given by

$$C = \frac{1{,}000{,}000 + 2{,}000{,}000p}{100 - p}.$$

If the candidate can spend only $2 million for advertising, then what percentage of the votes can she expect to receive? Use the bar graph to estimate the percentage of votes expected if $4 million is spent.   49.75%, 66%

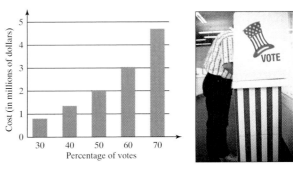

FIGURE FOR EXERCISE 66

67. *Wealth-building portfolio.* Misty decided to invest her annual bonus in a wealth-building portfolio as shown in the figure *(Fidelity Investments, Boston).*
  a) If the amount that she invested in stocks was $20,000 greater than her investment in bonds, then how much did she invest in bonds?
  b) What was the amount of her annual bonus?
   a) $17,142.86   **b)** $57,142.86

Designing a retirement portfolio

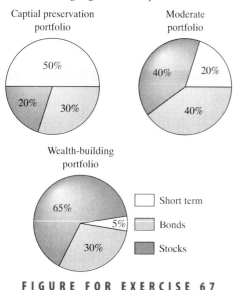

FIGURE FOR EXERCISE 67

68. *Estimating weapons.* When intelligence agents obtain enemy weapons marked with serial numbers, they use the formula $N = (1 + 1/C)B - 1$ to estimate the total number of such weapons N that the enemy has produced *(New Scientist,* May 1998). B is the biggest serial number obtained and C is the number of weapons obtained. It is assumed the weapons are numbered 1 through N.
  a) Find N if agents obtain five nerve gas containers numbered 45, 143, 258, 301, and 465.
  b) Find C if agents estimate that the enemy has 255 tanks from a group of captured tanks on which the biggest serial number is 224.
   a) 557   **b)** 7

GETTING MORE INVOLVED

 69. *Writing.* In this chapter the LCD is used to add rational expressions and to solve equations. Explain the difference between using the LCD to solve the equation

$$\frac{3}{x-2} + \frac{7}{x+2} = 2$$

and using the LCD to find the sum

$$\frac{3}{x-2} + \frac{7}{x+2}.$$

70. *Discussion.* For each equation, find the values for x that *cannot* be solutions to the equation. Do not solve the equations.
  a) $\dfrac{1}{x} + \dfrac{1}{x-1} = \dfrac{1}{2}$ **b)** $\dfrac{x}{x-1} = \dfrac{1}{2}$

  c) $\dfrac{1}{x^2+1} = \dfrac{1}{x+1}$
   a) 0, 1   **b)** 1   **c)** -1

6.6 APPLICATIONS

In this section we will use the techniques of Section 6.5 to rewrite formulas involving rational expressions and to solve some problems.

Formulas

Rewriting formulas having rational expressions is similar to solving equations having rational expressions. Generally, the first step is to multiply each side by the LCD for the rational expressions.

EXAMPLE 1   **Solving a formula**

In Chapter 3, we wrote the equation of a line by starting with an equation involving a rational expression:

$$\frac{y - y_1}{x - x_1} = m$$

Solve the equation for y.

helpful hint

When this equation was written in the form
$$y - y_1 = m(x - x_1)$$
in Chapter 3, we called it the point-slope formula for the equation of a line.

Solution

$$(x - x_1)\frac{y - y_1}{x - x_1} = (x - x_1)m \quad \text{Multiply each side by the denominator } x - x_1.$$

$$y - y_1 = (x - x_1)m \quad \text{Reduce.}$$

$$y = (x - x_1)m + y_1$$

In the next example we solve for a variable that occurs twice in the original formula. Remember that when a formula is solved for a certain variable, that variable must appear only once in the final formula.

EXAMPLE 2

Solving for a variable

The formula

$$\frac{P}{P_W} = \frac{2L}{2L + d}$$

is used in physics to find the relative density of a substance. Since P_W has subscript W, we treat P and P_W as two different variables. Solve the formula for L.

Solution

$$\frac{P}{P_W} = \frac{2L}{2L + d}$$

$$P(2L + d) = P_W(2L) \qquad \text{The extremes-means property}$$

$$2PL + Pd = 2LP_W \qquad \text{Simplify.}$$

$$Pd = 2LP_W - 2PL \qquad \text{Get all terms involving } L \text{ onto the same side.}$$

$$Pd = (2P_W - 2P)L \qquad \text{Factor out } L.$$

$$\frac{Pd}{2P_W - 2P} = L$$

In the next example we find the value of one variable when given the values of the remaining variables.

study tip

As you study from the text, think about the material. Ask yourself questions. If you were the professor, what questions would you ask on the test?

EXAMPLE 3

Evaluating a formula

Find x if $x_1 = 2, y_1 = -3, y = -1, m = \dfrac{1}{2}$, and

$$\frac{y - y_1}{x - x_1} = m.$$

Solution

Substitute all of the values into the formula and solve for x:

$$\frac{-1 - (-3)}{x - 2} = \frac{1}{2} \qquad \text{Substitute.}$$

$$\frac{2}{x - 2} = \frac{1}{2}$$

$$x - 2 = 4 \qquad \text{Extremes-means property}$$

$$x = 6 \qquad \text{Check in the original formula.}$$

Uniform Motion Problems

The uniform motion problems here are similar to those of Chapter 2, but in this chapter the equations have rational expressions.

E X A M P L E   4

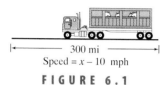

300 mi
Speed = x mph

300 mi
Speed = $x - 10$ mph

F I G U R E   6 . 1

Uniform motion

Michele drove her empty rig 300 miles to Salina to pick up a load of cattle. When her rig was fully loaded, her average speed was 10 miles per hour less than when the rig was empty. If the return trip took her 1 hour longer, then what was her average speed with the rig empty? (See Fig. 6.1.)

Solution

Let x be Michele's average speed empty and let $x - 10$ be her average speed full. Because the time can be determined from the distance and the rate, $T = \dfrac{D}{R}$, we can make the following table.

	Rate	Time	Distance
Empty	$x \dfrac{\text{mi}}{\text{hr}}$	$\dfrac{300}{x}$ hr	300 mi
Full	$x - 10 \dfrac{\text{mi}}{\text{hr}}$	$\dfrac{300}{x - 10}$ hr	300 mi

We now write an equation expressing the fact that her time empty was 1 hour less than her time full:

$$\frac{300}{x} = \frac{300}{x - 10} - 1$$

$$x(x - 10)\frac{300}{x} = x(x - 10)\frac{300}{x - 10} - x(x - 10)1 \qquad \text{Multiply each side by } x(x - 10).$$

$$300x - 3000 = 300x - x^2 + 10x \qquad \text{Reduce.}$$

$$-3000 = -x^2 + 10x$$

$$x^2 - 10x - 3000 = 0 \qquad \text{Get 0 on one side.}$$

$$(x + 50)(x - 60) = 0 \qquad \text{Factor.}$$

$$x + 50 = 0 \qquad \text{or} \qquad x - 60 = 0 \qquad \text{Zero factor property}$$

$$x = -50 \qquad \text{or} \qquad x = 60$$

The equation is satisfied if $x = -50$, but because -50 is negative, it cannot be the speed of the truck. Michele's average speed empty was 60 miles per hour (mph). Checking this answer, we find that if she traveled 300 miles at 60 mph, it would take her 5 hours. If she traveled 300 miles at 50 mph with the loaded rig, it would take her 6 hours. Because Michele's time with the empty rig was 1 hour less than her time with the loaded rig, 60 mph is the correct answer. ■

Work Problems

Problems involving different rates for completing a task are referred to as **work problems.** We did not solve work problems earlier because they usually require equations with rational expressions. Work problems are similar to uniform motion problems in which $RT = D$. The product of a person's time and rate is the amount of work completed. For example, if your puppy gains 1 pound every 3 days, then he is growing at the rate of $\frac{1}{3}$ pound per day. If he grows at the rate of $\frac{1}{3}$ pound per day for a period of 30 days, then he gains 10 pounds.

EXAMPLE 5

Working together

Linda can mow a certain lawn with her riding lawn mower in 4 hours. When Linda uses the riding mower and Rebecca operates the push mower, it takes them 3 hours to mow the lawn. How long would it take Rebecca to mow the lawn by herself using the push mower?

Solution

If x is the number of hours it takes for Rebecca to complete the lawn alone, then her rate is $\frac{1}{x}$ of the lawn per hour. Because Linda can mow the entire lawn in 4 hours, her rate is $\frac{1}{4}$ of the lawn per hour. In the 3 hours that they work together, Rebecca completes $\frac{3}{x}$ of the lawn while Linda completes $\frac{3}{4}$ of the lawn. We can classify all of the necessary information in a table that looks a lot like the one we used in Example 4.

	Rate	Time	Amount of Work
Linda	$\dfrac{1\ \text{lawn}}{4\ \text{hr}}$	3 hr	$\dfrac{3}{4}$ lawn
Rebecca	$\dfrac{1\ \text{lawn}}{x\ \text{hr}}$	3 hr	$\dfrac{3}{x}$ lawn

> **helpful hint**
>
> The secret to work problems is remembering that the individual amounts of work or the individual rates can be added when people work together. If your painting rate is $\frac{1}{10}$ of the house per day and your helper's rate is $\frac{1}{5}$ of the house per day, then your rate together will be $\frac{3}{10}$ of the house per day.

Because the lawn is finished in 3 hours, the two portions of the lawn (in the work column) mowed by each girl have a sum of 1:

$$\frac{3}{4} + \frac{3}{x} = 1$$

$$4x \cdot \frac{3}{4} + 4x \cdot \frac{3}{x} = 4x \cdot 1 \quad \text{Multiply each side by } 4x.$$

$$3x + 12 = 4x$$

$$12 = x$$

If $x = 12$, then in the 3 hours that they work together, Rebecca does $\frac{3}{12}$ or $\frac{1}{4}$ of the job while Linda does $\frac{3}{4}$ of the job. So it would take Rebecca 12 hours to mow the lawn by herself using the push mower. ◼

Miscellaneous Problems

EXAMPLE 6

Hamburger and steak

Patrick bought 50 pounds of meat consisting of hamburger and steak. Steak costs twice as much per pound as hamburger. If he bought $30 worth of hamburger and $90 worth of steak, then how many pounds of each did he buy?

Solution

Let x be the number of pounds of hamburger and $50 - x$ be the number of pounds of steak. Because Patrick got x pounds of hamburger for $30, he paid $\frac{30}{x}$ dollars per pound for the hamburger. We can classify all of the given information in a table.

	Price per pound	Amount	Total price
Hamburger	$\dfrac{30}{x}\ \dfrac{\text{dollars}}{\text{lb}}$	x lb	30 dollars
Steak	$\dfrac{90}{50-x}\ \dfrac{\text{dollars}}{\text{lb}}$	$50 - x$ lb	90 dollars

Because the price per pound of steak is twice that of hamburger, we can write the following equation:

$$2\left(\frac{30}{x}\right) = \frac{90}{50 - x}$$

$$\frac{60}{x} = \frac{90}{50 - x}$$

$$90x = 3000 - 60x \quad \text{The extremes-means property}$$

$$150x = 3000$$

$$x = 20$$

$$50 - x = 30$$

Patrick purchased 20 pounds of hamburger and 30 pounds of steak. Check this answer. ◼

WARM-UPS

True or false? Explain.

1. The formula $w = \frac{1-t}{t}$, solved for t, is $t = \frac{1-t}{w}$. False
2. To solve $\frac{1}{p} + \frac{1}{q} = \frac{1}{s}$ for s, multiply each side by pqs. True
3. If 50 pounds of steak cost x dollars, then the price is $\frac{50}{x}$ dollars per pound. False
4. If Claudia drives x miles in 3 hours, then her rate is $\frac{x}{3}$ miles per hour. True
5. If Takenori mows his entire lawn in $x + 2$ hours, then he mows $\frac{1}{x+2}$ of the lawn per hour. True
6. If Kareem drives 200 nails in 12 hours, then he is driving $\frac{200}{12}$ nails per hour. True
7. If x hours is 1 hour less than y hours, then $x - 1 = y$. False
8. If $A = \frac{mv^2}{B}$ and m and B are nonzero, then $v^2 = \frac{AB}{m}$. True
9. If a and y are nonzero and $a = \frac{x}{y}$, then $y = ax$. False
10. If x hours is 3 hours more than y hours, then $x + 3 = y$. False

6.6   EXERCISES

Solve each equation for y. See Example 1.

1. $\dfrac{y-3}{x-2} = 5$ $y = 5x - 7$

2. $\dfrac{y-4}{x-7} = -6$ $y = -6x + 46$

3. $\dfrac{y+1}{x-6} = -\dfrac{1}{3}$ $y = -\dfrac{1}{3}x + 1$

4. $\dfrac{y+7}{x-2} = \dfrac{-2}{3}$ $y = -\dfrac{2}{3}x - \dfrac{17}{3}$

5. $\dfrac{y-a}{x-b} = m$ $y = mx - bm + a$

6. $\dfrac{y-h}{x-k} = a$ $y = ax - ak + h$

7. $\dfrac{y-2}{x+5}=-\dfrac{7}{3}$ $y=-\dfrac{7}{3}x-\dfrac{29}{3}$

8. $\dfrac{y-3}{x+1}=-\dfrac{9}{4}$ $y=-\dfrac{9}{4}x+\dfrac{3}{4}$

Solve each formula for the indicated variable. See Example 2.

9. $M=\dfrac{F}{f}$ for f

$f=\dfrac{F}{M}$

10. $P=\dfrac{A}{1+rt}$ for A

$A=P(1+rt)$

11. $A=\dfrac{\pi}{4}\cdot D^2$ for D^2

$D^2=\dfrac{4A}{\pi}$

12. $V=\pi r^2 h$ for r^2

$r^2=\dfrac{V}{\pi h}$

13. $F=k\dfrac{m_1 m_2}{r^2}$ for m_1

$m_1=\dfrac{Fr^2}{km_2}$

14. $F=\dfrac{mv^2}{r}$ for v^2

$v^2=\dfrac{rF}{m}$

15. $\dfrac{1}{p}+\dfrac{1}{q}=\dfrac{1}{f}$ for q

$q=\dfrac{pf}{p-f}$

16. $\dfrac{1}{R}=\dfrac{1}{R_1}+\dfrac{1}{R_2}$ for R_1

$R_1=\dfrac{RR_2}{R_2-R}$

17. $e^2=1-\dfrac{b^2}{a^2}$ for a^2

$a^2=\dfrac{b^2}{1-e^2}$

18. $e^2=1-\dfrac{b^2}{a^2}$ for b^2

$b^2=a^2-a^2e^2$

19. $\dfrac{P_1V_1}{T_1}=\dfrac{P_2V_2}{T_2}$ for T_1

$T_1=\dfrac{P_1V_1T_2}{P_2V_2}$

20. $\dfrac{P_1V_1}{T_1}=\dfrac{P_2V_2}{T_2}$ for P_2

$P_2=\dfrac{P_1V_1T_2}{T_1V_2}$

21. $V=\dfrac{4}{3}\pi r^2 h$ for h

$h=\dfrac{3V}{4\pi r^2}$

22. $h=\dfrac{S-2\pi r^2}{2\pi r}$ for S

$S=2\pi rh+2\pi r^2$

Use the formula from the indicated exercise to find the value of the indicated variable. See Example 3. For calculator problems, round answers to three decimal places.

23. If $M=10$ and $F=5$ in Exercise 9, find f. $\dfrac{1}{2}$

24. If $A=550$, $P=500$, and $t=2$ in Exercise 10, find r.   0.05

25. If $A=6\pi$ in Exercise 11, find D^2.   24

26. If $V=12\pi$ and $r=3$ in Exercise 12, find h. $\dfrac{4}{3}$

27. If $F=32$, $r=4$, $m_1=6$, and $m_2=8$ in Exercise 13, find k. $\dfrac{32}{3}$

28. If $F=10$, $m=8$, and $v=6$ in Exercise 14, find r. $\dfrac{144}{5}$

29. If $f=2.3$ and $q=1.7$ in Exercise 15, find p. -6.517

30. If $R=1.29$ and $R_1=0.045$ in Exercise 16, find R_2. -0.046

31. If $e=0.62$ and $b=3.5$ in Exercise 17, find a^2.   19.899

32. If $a=3.61$ and $e=2.4$ in Exercise 18, find b^2. -62.033

33. If $V=25.6$ and $h=3.2$ in Exercise 21, find r^2.   1.910

34. If $h=3.6$ and $r=2.45$ in Exercise 22, find S.   93.133

Solve each problem. See Examples 4–6.

35. *Walking and riding.* Karen can ride her bike from home to school in the same amount of time as she can walk from home to the post office. She rides 10 miles per hour (mph) faster than she walks. The distance from her home to school is 7 miles, and the distance from her home to the post office is 2 miles. How fast does Karen walk?   4 mph

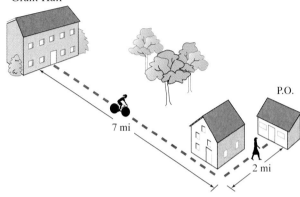

Grant Hall   P.O.   7 mi   2 mi

FIGURE FOR EXERCISE 35

36. *Fast driving.* Beverly can drive 600 miles in the same time as it takes Susan to drive 500 miles. If Beverly drives 10 mph faster than Susan, then how fast does Beverly drive?   60 mph

37. *Faster driving.* Patrick drives 40 miles to work, and Guy drives 60 miles to work. Guy claims that he drives at the same speed as Patrick, but it takes him only 12 minutes longer to get to work. If this is true, then how long does it take each of them to get to work? What are their speeds?   Patrick 24 minutes; Guy 36 minutes, 100 mph

38. *Route drivers.* David and Keith are route drivers for a fast-photo company. David's route is 80 miles, and Keith's is 100 miles. Keith averages 10 mph more than David and finishes his route 10 minutes before David. What is David's speed?   30 mph

39. *Physically fit.* Every morning, Yong Yi runs 5 miles, then walks one mile. He runs 6 mph faster than he walks. If his total time yesterday was 45 minutes, then how fast did he run?   10 mph

40. *Row, row, row your boat.* Norma can row her boat 12 miles in the same time as it takes Marietta to cover 36 miles in her motorboat. If Marietta's boat travels 15 mph faster than Norma's boat, then how fast is Norma rowing her boat?   7.5 mph

41. *Pumping out the pool.* A large pump can drain an 80,000-gallon pool in 3 hours. With a smaller pump also operating, the job takes only 2 hours. How long would it take the smaller pump to drain the pool by itself?   6 hours

42. *Trimming hedges.* Lourdes can trim the hedges around her property in 8 hours by using an electric hedge trimmer. Rafael can do the same job in 15 hours by using a manual trimmer. How long would it take them to trim the hedges working together?

$\dfrac{120}{23}$ hours

43. *Filling the tub.* It takes 10 minutes to fill Alisha's bathtub and 12 minutes to drain the water out. How long would it take to fill it with the drain accidentally left open?

60 minutes

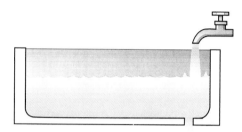

FIGURE FOR EXERCISE 43

44. *Eating machine.* Charles can empty the cookie jar in $1\frac{1}{2}$ hours. It takes his mother 2 hours to bake enough cookies to fill it. If the cookie jar is full when Charles comes home from school, and his mother continues baking and restocking the cookie jar, then how long will it take him to empty the cookie jar?

6 hours

45. *Filing the invoices.* It takes Gina 90 minutes to file the monthly invoices. If Hilda files twice as fast as Gina does, how long will it take them working together?

30 minutes

46. *Painting alone.* Julie can paint a fence by herself in 12 hours. With Betsy's help, it takes only 5 hours. How long would it take Betsy by herself?

$\dfrac{60}{7}$ hours

47. *Buying fruit.* Molly bought $5.28 worth of oranges and $8.80 worth of apples. She bought 2 more pounds of oranges than apples. If apples cost twice as much per pound as oranges, then how many pounds of each did she buy?

10 pounds apples, 12 pounds oranges

48. *Raising rabbits.* Luke raises rabbits and raccoons to sell for meat. The price of raccoon meat is three times the price of rabbit meat. One day Luke sold 160 pounds of meat, $72 worth of each type. What is the price per pound of each type of meat?

Rabbit $0.60 per pound, raccoon $1.80 per pound

49. *Total resistance.* If two receivers with resistances R_1 and R_2 are connected in parallel, then the formula

$$\frac{1}{R} = \frac{1}{R_1} + \frac{1}{R_2}$$

relates the total resistance for the circuit R with R_1 and R_2. Given that R_1 is 3 ohms and R is 2 ohms, find R_2.

6 ohms

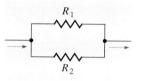

FIGURE FOR EXERCISE 49

50. *More resistance.* Use the formula from Exercise 49 to find R_1 and R_2 given that the total resistance is 1.2 ohms and R_1 is 1 ohm larger than R_2.

$R_2 = 2$ ohms, $R_1 = 3$ ohms

51. *Las Vegas vacation.* Brenda of Horizon Travel has arranged for a group of gamblers to share the $24,000 cost of a charter flight to Las Vegas. If Brenda can get 40 more people to share the cost, then the cost per person will decrease by $100.

a) How many people were in the original group?

b) Write the cost per person as a function of the number of people sharing the cost.

a) 80   **b)** $C(n) = \dfrac{24,000}{n}$

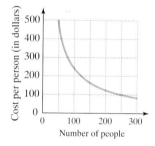

FIGURE FOR EXERCISE 51

52. *White-water rafting.* Adventures, Inc. has a $1,500 group rate for an overnight rafting trip on the Colorado River. For the last trip five people failed to show, causing the price per person to increase by $25. How many were originally scheduled for the trip?

20

 53. *Doggie bag.* Muffy can eat a 25-pound bag of dog food in 28 days, whereas Missy eats a 25-pound bag in 23 days. How many days would it take them together to finish a 50-pound bag of dog food.

25.255 days

 54. *Rodent food.* A pest control specialist has found that 6 rats can eat an entire box of sugar-coated breakfast cereal in 13.6 minutes, and it takes a dozen mice 34.7 minutes to devour the same size box of cereal. How long would it take all 18 rodents, in a cooperative manner, to finish off a box of cereal?

9.7706 minutes

COLLABORATIVE ACTIVITIES

Beorg's Business

In manufacturing or other businesses in which time is money and tasks are easily shared, problems involving work appear. An owner or manager who wants to know how to bid a job often develops a table of times needed to complete the job as determined by how much work is required and who could be assigned to the job.

Beorg owns a kaleidoscope-manufacturing company with two employees, Scott and Salina. It takes Scott one hour to make one kaleidoscope, and it takes Salina $\frac{1}{2}$ hour to make one kaleidoscope. Beorg wants to know how long it would take to complete a certain number of kaleidoscopes. Using the information given and answering the questions below, fill in the following table for Beorg.

Name of Employee	Time for one kaleidoscope	Time for 20 kaleidoscopes
Scott	1 hr	
Salina	$\frac{1}{2}$ hr	
Scott & Salina		
Sammy		
Scott & Sammy	$\frac{3}{4}$ hr	
Salina & Sammy		
Scott, Salina, & Sammy		

Grouping: Four students per group
Topic: Applications of work problems

1. How long will it take Scott and Salina working together to make one kaleidoscope?

2. Beorg hires a third person, Sammy, and has him and Scott make one kaleidoscope. Working together, it takes them $\frac{3}{4}$ hour to make one kaleidoscope. How long would it take Sammy by himself to make one kaleidoscope?

3. How long would it take Salina and Sammy working together to make one kaleidoscope? How long would it take for all three working together?

   Now Beorg wants to finish his time table. He would like to have 20 kaleidoscopes completed each day.

4. Finish the preceding table, and find the best combination or combinations of employees to use to have 20 kaleidoscopes at the end of an 8-hour day.

Extension: Is Sammy in the combination or combinations you found in the last question? Is it worth having Sammy work? Remember that when someone is starting a new job, he or she may work more slowly until he or she learns how to do the job more efficiently. Find out how fast Sammy would need to work for production to double (40 kaleidoscopes in an 8-hour day).

WRAP-UP   CHAPTER 6

SUMMARY

Rational Expressions

Examples

Rational expression	The ratio of two polynomials with the denominator not equal to zero	$\dfrac{x^2 - 1}{2x - 3}$
Domain of a rational expression	The set of all possible numbers that can be used as replacements for the variable	$D = \left\{ x \mid x \neq \dfrac{3}{2} \right\}$

Operations with Rational Expressions		**Examples**
Basic principle of rational numbers	If $\frac{a}{b}$ is a rational number and c is a nonzero real number, then $$\frac{a}{b} = \frac{ac}{bc}.$$	Used for reducing: $$\frac{14}{16} = \frac{2 \cdot 7}{2 \cdot 8} = \frac{7}{8}$$ Used for building: $$\frac{2}{x} = \frac{2 \cdot 3}{x \cdot 3} = \frac{6}{3x}$$
Multiplication of rational numbers	If $\frac{a}{b}$ and $\frac{c}{d}$ are rational numbers, then $$\frac{a}{b} \cdot \frac{c}{d} = \frac{ac}{bd}.$$	$$\frac{3}{x} \cdot \frac{6}{x^2} = \frac{18}{x^3}$$
Division of rational numbers	If $\frac{a}{b}$ and $\frac{c}{d}$ are rational numbers with $\frac{c}{d} \neq 0$, then $$\frac{a}{b} \div \frac{c}{d} = \frac{a}{b} \cdot \frac{d}{c}. \quad \text{(Invert and multiply.)}$$	$$\frac{a}{x} \div \frac{5}{4x} = \frac{a}{x} \cdot \frac{4x}{5} = \frac{4a}{5}$$
Least common multiple	The LCM is the product of all of the different factors that appear in the polynomials. The exponent on each factor is the highest power that occurs on that factor in any of the polynomials.	$4a^3b, 6ab^2$ $$\text{LCM} = 12a^3b^2$$
Least common denominator	The LCD for a group of denominators is the LCM of the denominators.	$$\frac{1}{4a^3b} + \frac{1}{6ab^2}$$ $$\text{LCD} = 12a^3b^2$$
Addition and subtraction	If $b \neq 0$, then $$\frac{a}{b} + \frac{c}{b} = \frac{a + c}{b} \quad \text{and} \quad \frac{a}{b} - \frac{c}{b} = \frac{a - c}{b}.$$ If the denominators are not identical, we must build up each fraction to an equivalent fraction with the LCD as denominator.	$$\frac{2x}{x-3} + \frac{7x}{x-3} = \frac{9x}{x-3}$$ $$\frac{1}{2} + \frac{1}{x} = \frac{x}{2x} + \frac{2}{2x} = \frac{x+2}{2x}$$
Rules for adding and subtracting simple fractions	If $b \neq 0$ and $d \neq 0$, then $$\frac{a}{b} + \frac{c}{d} = \frac{ad + bc}{bd}$$ and $$\frac{a}{b} - \frac{c}{d} = \frac{ad - bc}{bd}.$$	$$\frac{1}{2} + \frac{1}{3} = \frac{5}{6}$$ $$\frac{2}{5} - \frac{3}{7} = \frac{-1}{35}$$

| Simplifying complex fractions | Multiply the numerator and denominator by the LCD. | $$\dfrac{\left(\dfrac{1}{2} + \dfrac{1}{x}\right)6x}{\left(\dfrac{1}{x} - \dfrac{1}{3}\right)6x} = \dfrac{3x + 6}{6 - 2x}$$ |

Equations with Rational Expressions

Examples

| Solving equations with rational expressions | Multiply each side by the LCD to eliminate all denominators. | $$\dfrac{1}{x} - \dfrac{1}{3} = \dfrac{1}{2x} - \dfrac{1}{6}$$ $$6x\left(\dfrac{1}{x} - \dfrac{1}{3}\right) = 6x\left(\dfrac{1}{2x} - \dfrac{1}{6}\right)$$ $$6 - 2x = 3 - x$$ |
| Solving proportions by the extremes-means property | If $\dfrac{a}{b} = \dfrac{c}{d}$, then $ad = bc$. | $$\dfrac{2}{x - 3} = \dfrac{5}{6}$$ $$12 = 5x - 15$$ |

ENRICHING YOUR MATHEMATICAL WORD POWER

For each mathematical term, choose the correct meaning.

1. **rational expression**
   a. a fraction
   b. a ratio of two polynomials with the denominator not equal to zero
   c. an expression involving fractions
   d. a fraction in which the numerator and denominator contain fractions   b

2. **domain of a rational expression**
   a. all real numbers
   b. the denominator of the rational expression
   c. the set of all real numbers that cannot be used in place of the variable
   d. the set of all real numbers that can be used in place of the variable   d

3. **lowest terms**
   a. the numerator is smaller than the denominator
   b. no common factors
   c. the best interest rate
   d. when the numerator is 1   b

4. **reducing**
   a. less than
   b. losing weight
   c. making equivalent
   d. dividing out common factors   d

5. **equivalent fractions**
   a. identical fractions
   b. fractions that represent the same number
   c. fractions with the same denominator
   d. fractions with the same numerator   b

6. **complex fraction**
   a. a fraction having rational expressions in the numerator, denominator, or both
   b. a fraction with a large denominator
   c. the sum of two fractions
   d. a fraction with a variable in the denominator   a

7. **building up the denominator**
   a. the opposite of reducing a fraction
   b. finding the least common denominator
   c. adding the same number to the numerator and denominator
   d. writing a fraction larger   a

8. **least common denominator**
   a. the largest number that is a multiple of all denominators
   b. the sum of the denominators
   c. the product of the denominators
   d. the smallest number that is a multiple of all denominators   d

9. **extraneous root**
   a. a number that appears to be a solution to an equation but does not satisfy the equation
   b. an extra solution to an equation
   c. the second solution
   d. a nonreal solution   a

10. **ratio of *a* to *b***
    a. b/a
    b. a/b
    c. $a/(a + b)$
    d. ab b

11. **proportion**
   a. a ratio
   b. two ratios
   c. the product of the means equals the product of the extremes
   d. a statement expressing the equality of two rational expressions   d

12. **extremes**
   a. a and d in $a/b = c/d$
   b. b and c in $a/b = c/d$
   c. the extremes-means property
   d. if $a/b = c/d$, then $ad = bc$ a

13. **means**
   a. the average of a, b, c, and d
   b. a and d in $a/b = c/d$
   c. b and c in $a/b = c/d$
   d. if $a/b = c/d$, then $(a + b)/2 = (c + d)/2$ c

14. **extremes-means property**
   a. $ab = ba$ for any real numbers a and b
   b. $(a - b)^2 = (b - a)^2$ for any real numbers a and b
   c. if $a/b = c/d$, then $ab = cd$
   d. if $a/b = c/d$, then $ad = bc$ d

REVIEW EXERCISES

6.1 *State the domain of each rational expression.*

1. $\dfrac{5 - x}{3x - 3}$
   $\{x \mid x \neq 1\}$

2. $\dfrac{x - 4}{x^2 - 25}$
   $\{x \mid x \neq -5 \text{ and } x \neq 5\}$

3. $\dfrac{x}{x^2 - x - 2}$
   $\{x \mid x \neq -1 \text{ and } x \neq 2\}$

4. $\dfrac{1}{x^3 - x^2}$
   $\{x \mid x \neq 0 \text{ and } x \neq 1\}$

Reduce each rational expression to its lowest terms.

5. $\dfrac{a^3bc^3}{a^5b^2c}$
   $\dfrac{c^2}{a^2b}$

6. $\dfrac{x^4 - 1}{3x^2 - 3}$
   $\dfrac{x^2 + 1}{3}$

7. $\dfrac{68x^3}{51xy}$
   $\dfrac{4x^2}{3y}$

8. $\dfrac{5x^2 - 15x + 10}{5x - 10}$
   $x - 1$

6.2 *Perform the indicated operations.*

9. $\dfrac{a^3b^2}{b^3a} \cdot \dfrac{ab - b^2}{ab - a^2}$ $-a$

10. $\dfrac{x^3 - 1}{3x} \cdot \dfrac{6x^2}{x - 1}$ $2x^3 + 2x^2 + 2x$

11. $\dfrac{w - 4}{3w} \div \dfrac{2w - 8}{9w}$ $\dfrac{3}{2}$

12. $\dfrac{x^3 - xy^2}{y} \div \dfrac{x^3 + 2x^2y + xy^2}{3y}$ $\dfrac{3x - 3y}{x + y}$

6.3 *Find the least common multiple for each group of polynomials.*

13. $6x,\ 3x - 6,\ x^2 - 2x$
   $6x(x - 2)$

14. $x^3 - 8,\ x^2 - 4,\ 2x + 8$
   $2(x - 2)(x + 2)(x + 4)(x^2 + 2x + 4)$

15. $6ab^3,\ 4a^5b^2$
   $12a^5b^3$

16. $4x^2 - 9,\ 4x^2 + 12x + 9$
   $(2x - 3)(2x + 3)^2$

Perform the indicated operations.

17. $\dfrac{3}{2x - 6} + \dfrac{1}{x^2 - 9}$
   $\dfrac{3x + 11}{2(x - 3)(x + 3)}$

18. $\dfrac{3}{x - 3} - \dfrac{5}{x + 4}$
   $\dfrac{-2x + 27}{(x - 3)(x + 4)}$

19. $\dfrac{w}{ab^2} - \dfrac{5}{a^2b}$
   $\dfrac{aw - 5b}{a^2b^2}$

20. $\dfrac{x}{x - 1} + \dfrac{3x}{x^2 - 1}$
   $\dfrac{x^2 + 4x}{(x - 1)(x + 1)}$

6.4 *Simplify the complex fractions.*

21. $\dfrac{\dfrac{3}{2x} - \dfrac{4}{5x}}{\dfrac{1}{3} - \dfrac{2}{x}}$
   $\dfrac{21}{10x - 60}$

22. $\dfrac{\dfrac{5}{x - 2} - \dfrac{4}{4 - x^2}}{\dfrac{3}{x + 2} - \dfrac{1}{2 - x}}$
   $\dfrac{5x + 14}{4x - 4}$

23. $\dfrac{\dfrac{1}{y - 2} - 3}{\dfrac{5}{y - 2} + 4}$
   $\dfrac{7 - 3y}{4y - 3}$

24. $\dfrac{\dfrac{a}{b^2} - \dfrac{b}{a^3}}{\dfrac{a}{b} + \dfrac{b}{a^2}}$
   $\dfrac{a^4 - b^3}{a^4b + ab^3}$

25. $\dfrac{a^{-2} - b^{-3}}{a^{-1}b^{-2}}$
   $\dfrac{b^3 - a^2}{ab}$

26. $p^{-1} + pq^{-2}$
   $\dfrac{q^2 + p^2}{pq^2}$

6.5 *Solve each equation.*

27. $\dfrac{-3}{8} = \dfrac{2}{x}$ $\left\{-\dfrac{16}{3}\right\}$

28. $\dfrac{2}{x} + \dfrac{5}{2x} = 1$ $\left\{\dfrac{9}{2}\right\}$

29. $\dfrac{15}{a^2 - 25} + \dfrac{1}{a - 5} = \dfrac{6}{a + 5}$ {10}

30. $2 + \dfrac{3}{x - 5} = \dfrac{x - 1}{x - 5}$ {6}

6.6 *Solve each formula for the indicated variable.*

31. $\dfrac{y - b}{m} = x$ for y

$y = mx + b$

32. $\dfrac{2A}{h} = b_1 + b_2$ for A

$A = \dfrac{h}{2}(b_1 + b_2)$

33. $F = \dfrac{mv^2}{r}$ for m

$m = \dfrac{Fr}{v^2}$

34. $P = \dfrac{A}{1 + rt}$ for r

$r = \dfrac{A - P}{Pt}$

35. $A = \dfrac{2}{3}\pi rh$ for r

$r = \dfrac{3A}{2\pi h}$

36. $\dfrac{a}{w^2} = \dfrac{2}{b}$ for b

$b = \dfrac{2w^2}{a}$

37. $\dfrac{y + 3}{x - 7} = 2$ for y

$y = 2x - 17$

38. $\dfrac{y - 5}{x + 4} = \dfrac{-1}{2}$ for y

$y = -\dfrac{1}{2}x + 3$

MISCELLANEOUS

Either perform the indicated operation or solve the equation, whichever is appropriate.

39. $\dfrac{1}{x} + \dfrac{1}{3x}$ $\dfrac{4}{3x}$

40. $\dfrac{1}{y} + \dfrac{3}{2y} = 5$ $\left\{\dfrac{1}{2}\right\}$

41. $\dfrac{5}{3xy} + \dfrac{7}{6x}$ $\dfrac{10 + 7y}{6xy}$

42. $\dfrac{2}{x - 2} - \dfrac{3}{x}$ $\dfrac{-x + 6}{x(x - 2)}$

43. $\dfrac{5}{a - 5} - \dfrac{3}{-a - 5}$ $\dfrac{8a + 10}{(a - 5)(a + 5)}$

44. $\dfrac{2}{x - 2} - \dfrac{3}{x} = \dfrac{-1}{5x}$ {7}

45. $\dfrac{1}{x - 2} - \dfrac{1}{x + 2} = \dfrac{1}{15}$ {−8, 8}

46. $\dfrac{2}{x - 3} \cdot \dfrac{6x - 18}{30}$ $\dfrac{2}{5}$

47. $\dfrac{-3}{x + 2} \cdot \dfrac{5x + 10}{10}$ $-\dfrac{3}{2}$

48. $\dfrac{x}{10} = \dfrac{10}{x}$ {−10, 10}

49. $\dfrac{x}{-3} = \dfrac{-27}{x}$ {−9, 9}

50. $\dfrac{x^2 - 4}{x} \div \dfrac{x^3 - 8}{x}$ $\dfrac{x + 2}{x^2 + 2x + 4}$

51. $\dfrac{wx + wm + 3x + 3m}{w^2 - 9} \div \dfrac{x^2 - m^2}{w - 3}$ $\dfrac{1}{x - m}$

52. $\dfrac{-5}{7} = \dfrac{3}{x}$ $\left\{-\dfrac{21}{5}\right\}$

53. $\dfrac{5}{a^2 - 25} + \dfrac{3}{a^2 - 4a - 5}$ $\dfrac{8a + 20}{(a - 5)(a + 5)(a + 1)}$

54. $\dfrac{3}{w^2 - 1} + \dfrac{2}{2w + 2}$ $\dfrac{w + 2}{(w + 1)(w - 1)}$

55. $\dfrac{-7}{2a^2 - 18} - \dfrac{4}{a^2 + 5a + 6}$ $\dfrac{-15a + 10}{2(a + 2)(a + 3)(a - 3)}$

56. $\dfrac{-5}{3a^2 - 12} - \dfrac{1}{a^2 - 3a + 2}$ $\dfrac{-8a - 1}{3(a - 2)(a + 2)(a - 1)}$

57. $\dfrac{7}{a^2 - 1} + \dfrac{2}{1 - a} = \dfrac{1}{a + 1}$ {2}

58. $2 + \dfrac{4}{x - 1} = \dfrac{3x + 1}{x - 1}$ $\varnothing$

59. $\dfrac{2x}{x - 3} + \dfrac{3}{x - 2} = \dfrac{6}{(x - 2)(x - 3)}$ $\left\{-\dfrac{5}{2}\right\}$

60. $\dfrac{a - 3}{a + 3} \div \dfrac{9 - a^2}{3}$ $\dfrac{-3}{(a + 3)^2}$

61. $\dfrac{x - 2}{6} \div \dfrac{2 - x}{2}$ $-\dfrac{1}{3}$

62. $\dfrac{x}{x + 4} - \dfrac{2}{x + 1} = \dfrac{-2}{(x + 1)(x + 4)}$ {−2, 3}

63. $\dfrac{x - 3}{x^2 + 3x + 2} \cdot \dfrac{x^2 - 4}{3x - 9}$ $\dfrac{x - 2}{3x + 3}$

64. $\dfrac{x^2 - 1}{x^2 + 2x + 1} \cdot \dfrac{x^3 + 1}{2x - 2}$ $\dfrac{x^2 - x + 1}{2}$

65. $\dfrac{a + 4}{a^3 - 8} - \dfrac{3}{2 - a}$ $\dfrac{3a^2 + 7a + 16}{a^3 - 8}$

66. $\dfrac{x + 2}{5} = \dfrac{3}{x}$ {−5, 3}

67. $\dfrac{x^3 - 9x}{1 - x^2} \div \dfrac{x^3 + 6x^2 + 9x}{x - 1}$ $\dfrac{3 - x}{(x + 1)(x + 3)}$

68. $\dfrac{x + 3}{2x + 3} = \dfrac{x - 3}{x - 1}$ {−1, 6}

69. $\dfrac{a^2 + 3a + 3w + aw}{a^2 + 6a + 8} \cdot \dfrac{a^2 - aw - 2w + 2a}{a^2 + 3a - 3w - aw}$ $\dfrac{a + w}{a + 4}$

70. $\dfrac{3}{4 - 2y} + \dfrac{6}{y^2 - 4} + \dfrac{3}{2 + y}$ $\dfrac{3}{2y + 4}$

71. $\dfrac{5}{x} - \dfrac{4}{x + 2} = \dfrac{1}{5} + \dfrac{1}{5x}$ {−6, 8}

72. $\dfrac{1}{x} + \dfrac{1}{x-5} = \dfrac{2x+1}{x^2-25} + \dfrac{9}{x^2+5x}$ {4}

Replace each question mark by an expression that makes the equation an identity.

73. $\dfrac{6}{x} = \dfrac{?}{3x}$ $\dfrac{18}{3x}$

74. $\dfrac{?}{a} = \dfrac{8}{4a}$ $\dfrac{2}{a}$

75. $\dfrac{3}{a-b} = \dfrac{?}{b-a}$ $\dfrac{-3}{b-a}$

76. $\dfrac{-2}{a-x} = \dfrac{2}{?}$ $\dfrac{2}{x-a}$

77. $4 = \dfrac{?}{x}$ $\dfrac{4x}{x}$

78. $5a = \dfrac{?}{b}$ $\dfrac{5ab}{b}$

79. $5x \div \dfrac{1}{2} = ?$ $10x$

80. $3a \div \dfrac{1}{a} = ?$ $3a^2$

81. $4a \div ? = 12a$ $\dfrac{1}{3}$

82. $14x \div ? = 28x^2$ $\dfrac{1}{2x}$

83. $\dfrac{a-3}{a^2-9} = \dfrac{1}{?}$ $\dfrac{1}{a+3}$

84. $\dfrac{?}{x^2-4} = \dfrac{1}{x-2}$ $\dfrac{x+2}{x^2-4}$

85. $\dfrac{1}{2} - \dfrac{1}{5} = ?$ $\dfrac{3}{10}$

86. $\dfrac{1}{4} - \dfrac{1}{5} = ?$ $\dfrac{1}{20}$

87. $\dfrac{a}{3} + \dfrac{a}{2} = ?$ $\dfrac{5a}{6}$

88. $\dfrac{x}{5} + \dfrac{x}{3} = ?$ $\dfrac{8x}{15}$

89. $\dfrac{1}{a} - \dfrac{1}{b} = ?$ $\dfrac{b-a}{ab}$

90. $\dfrac{3}{w} - \dfrac{2}{b} = ?$ $\dfrac{3b-2w}{bw}$

91. $\dfrac{a}{3} - 1 = ?$ $\dfrac{a-3}{3}$

92. $\dfrac{x}{y} - 1 = ?$ $\dfrac{x-y}{y}$

93. $2 + \dfrac{1}{a} = ?$ $\dfrac{2a+1}{a}$

94. $3 - \dfrac{1}{x} = ?$ $\dfrac{3x-1}{x}$

95. $\dfrac{a}{5} - ? = \dfrac{a-5}{5}$ 1

96. $? - \dfrac{1}{y} = \dfrac{y^2-1}{y}$ y

97. $(x-1) \div (-1) = ?$ $1 - x$

98. $(a-3) \div (3-a) = ?$ -1

99. $(m-2) \div (2-m) = ?$ -1

100. $(x-2) \div (2x-4) = ?$ $\dfrac{1}{2}$

101. $\dfrac{\dfrac{b}{3a}}{2} = ?$ $\dfrac{b}{6a}$

102. $\dfrac{a-3}{\dfrac{1}{2}} = ?$ $2a - 6$

103. $\dfrac{a-6}{5} \cdot \dfrac{10}{6-a} = ?$ -2

104. $\dfrac{3}{w-2} + \dfrac{2}{2-w} = ?$ $\dfrac{1}{w-2}$

Perform the indicated operations. Variables in exponents represent integers.

105. $\dfrac{2}{x^a-2} + \dfrac{1}{x^a+3}$ $\dfrac{3x^a+4}{(x^a-2)(x^a+3)}$

106. $\dfrac{x^k+1}{x^{k+1}-x} - \dfrac{x^{k-1}}{x^k-1}$ $\dfrac{1}{x(x^k-1)}$

107. $\dfrac{x^{2k}-9}{3x^{k+3}+9x^3} \cdot \dfrac{6x^{k+5}}{x^{k+2}-3x^2}$ $2x^k$

108. $\dfrac{y^{2a}-y^a-12}{y^{2a}-4y^a} \div \dfrac{y^{3a}+3y^{2a}}{y^{6a}}$ y^{3a}

Solve each problem.

109. *Male and female AIDS cases.* The ratio of reported male AIDS cases to reported female AIDS cases for the years 1985 through 1996 was 5.98 to 1 (Center for Disease Control, www.cdc.gov). See the accompanying figure. If there were 373,486 more reported male AIDS cases than female AIDS cases, then how many reported male AIDS cases were there?
448,483

110. *AIDS in children.* The ratio of reported AIDS cases for females to the reported AIDS cases for children was 10.8 to 1 for the years 1985 through 1996.
  a) If there were 68,041 fewer AIDS cases reported in children than in females, then how many AIDS cases were reported in children during those years?
  b) What was the total number of AIDS cases reported during the years 1985 through 1996?
    a) 6,943   **b)** 530,410

Distribution of AIDS cases: 1985–1996

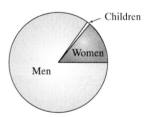

FIGURE FOR EXERCISES 109 AND 110

111. *Just passing through.* Nikita drove 310 miles on his way to Louisville in the same amount of time that he drove 360 miles after passing through Louisville. If his average speed after passing Louisville was 10 miles per hour (mph) more than his average speed on his way to Louisville, then for how many hours did he drive?
10 hours

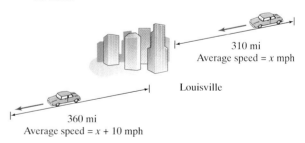

FIGURE FOR EXERCISE 111

112. *Pushing a barge.* A tug can push a barge 144 miles down the Mississippi River in the same time that it takes to push the barge 84 miles in the Gulf of Mexico. If the tug's speed

FIGURE FOR EXERCISE 112

is 5 mph greater going down the river, then what is its speed in the Gulf of Mexico?

7 mph

113. **Quilting bee.** Debbie can make a hand-sewn quilt in 2,000 hours, and Rosalina can make an identical quilt in 1,000 hours. If Cheryl works just as fast as Rosalina, then how long will it take all three of them working together to make one quilt?

400 hours

114. **Blood out of a turnip.** A small pump can pump all of the blood out of an average turnip in 30 minutes. A larger pump can pump all of the blood from the same turnip in 20 minutes. If both pumps are hooked to the turnip, then how long would it take to get all of the blood out?

12 minutes

CHAPTER 6 TEST

State the domain of each rational expression.

1. $\dfrac{5}{4 - 3x}$ 2. $\dfrac{2x - 1}{x^2 - 9}$ 3. $\dfrac{17}{x^2 + 9}$

$\left\{x \mid x \neq \dfrac{4}{3}\right\}$ $\{x \mid x \neq 3 \text{ and } x \neq -3\}$ $(-\infty, \infty)$

Reduce to its lowest terms.

4. $\dfrac{12a^9 b^8}{(2a^2 b^3)^3}$ $\dfrac{3a^3}{2b}$ 5. $\dfrac{y^2 - x^2}{2x^2 - 4xy + 2y^2}$ $\dfrac{-x - y}{2x - 2y}$

Perform the indicated operations. Write answers in lowest terms.

6. $\dfrac{5}{12} - \dfrac{4}{9}$ $-\dfrac{1}{36}$ 7. $\dfrac{3}{y} + 7y$ $\dfrac{7y^2 + 3}{y}$

8. $\dfrac{4}{a - 9} - \dfrac{1}{9 - a}$ $\dfrac{5}{a - 9}$ 9. $\dfrac{1}{6ab^2} + \dfrac{1}{8a^2 b}$ $\dfrac{4a + 3b}{24a^2 b^2}$

10. $\dfrac{3a^3 b}{20ab} \cdot \dfrac{2a^2 b}{9ab^3}$ $\dfrac{a^3}{30b^2}$

11. $\dfrac{a - b}{7} \div \dfrac{b^2 - a^2}{21}$ $-\dfrac{3}{a + b}$

12. $\dfrac{x - 3}{x - 1} \div (x^2 - 2x - 3)$ $\dfrac{1}{x^2 - 1}$

13. $\dfrac{2}{x^2 - 4} - \dfrac{6}{x^2 - 3x - 10}$ $\dfrac{-4x + 2}{(x + 2)(x - 2)(x - 5)}$

14. $\dfrac{m^3 - 1}{(m - 1)^2} \cdot \dfrac{m^2 - 1}{3m^2 + 3m + 3}$ $\dfrac{m + 1}{3}$

Find the solution set to each equation.

15. $\dfrac{3}{x} = \dfrac{7}{4}$ 16. $\dfrac{x}{x - 2} - \dfrac{5}{x} = \dfrac{3}{4}$ 17. $\dfrac{3m}{2} = \dfrac{6}{m}$

$\left\{\dfrac{12}{7}\right\}$ $\{4, 10\}$ $\{-2, 2\}$

Solve each formula for the indicated variable.

18. $W = \dfrac{a^2}{t}$ for t 19. $\dfrac{1}{a} + \dfrac{1}{b} = \dfrac{1}{2}$ for b

$t = \dfrac{a^2}{W}$ $b = \dfrac{2a}{a - 2}$

Simplify.

20. $\dfrac{\dfrac{1}{x} + \dfrac{1}{3x}}{\dfrac{3}{4x} - \dfrac{1}{2}}$ $\dfrac{16}{9 - 6x}$

21. $\dfrac{m^{-2} - w^{-2}}{m^{-2}w^{-1} + m^{-1}w^{-2}}$ $w - m$

22. $\dfrac{\dfrac{a^2 b^3}{4a}}{\dfrac{ab^3}{6a^2}}$ $\dfrac{3a^2}{2}$

Solve each problem.

23. When Jane's wading pool was new, it could be filled in 6 minutes with water from the hose. Now that the pool has several leaks, it takes only 8 minutes for all of the water to leak out of a full pool. How long does it take to fill the leaky pool?

24 minutes

24. Milton and Bonnie are hiking the Appalachian Trail together. Milton averages 4 miles per hour (mph), and Bonnie averages 3 mph. If they start out together in the morning, but Milton gets to camp 2 hours and 30 minutes ahead of Bonnie, then how many miles did they hike that day?

30 miles

25. A group of sailors plans to share equally the cost and use of a $72,000 boat. If they can get three more sailors to join their group, then the cost per person will be reduced by $2,000. How many sailors are in the original group?

9

Find the solution set to each equation.

1. $\dfrac{3}{x} = \dfrac{4}{5}$ $\left\{\dfrac{15}{4}\right\}$

2. $\dfrac{2}{x} = \dfrac{x}{8}$ $\{-4, 4\}$

3. $\dfrac{x}{3} = \dfrac{4}{5}$ $\left\{\dfrac{12}{5}\right\}$

4. $\dfrac{3}{x} = \dfrac{x+3}{6}$ $\{-6, 3\}$

5. $\dfrac{1}{x} = 4$ $\left\{\dfrac{1}{4}\right\}$

6. $\dfrac{2}{3}x = 4$ $\{6\}$

7. $2x + 3 = 4$ $\left\{\dfrac{1}{2}\right\}$

8. $2x + 3 = 4x$ $\left\{\dfrac{3}{2}\right\}$

9. $\dfrac{2a}{3} = \dfrac{6}{a}$ $\{-3, 3\}$

10. $\dfrac{12}{x} - \dfrac{14}{x+1} = \dfrac{1}{2}$ $\{-8, 3\}$

11. $|6x - 3| = 1$ $\left\{\dfrac{1}{3}, \dfrac{2}{3}\right\}$

12. $\dfrac{x}{2x+9} = \dfrac{3}{x}$ $\{-3, 9\}$

13. $4(6x - 3)(2x + 9) = 0$ $\left\{-\dfrac{9}{2}, \dfrac{1}{2}\right\}$

14. $\dfrac{x-1}{x+2} - \dfrac{1}{5(x+2)} = 1$ $\varnothing$

Solve each equation for y. Assume A, B, and C are constants for which all expressions are defined.

15. $Ax + By = C$

$y = \dfrac{C - Ax}{B}$

16. $\dfrac{y-3}{x+5} = -\dfrac{1}{3}$

$y = -\dfrac{1}{3}x + \dfrac{4}{3}$

17. $Ay = By + C$

$y = \dfrac{C}{A - B}$

18. $\dfrac{A}{y} = \dfrac{y}{A}$

$y = A \quad \text{or} \quad y = -A$

19. $\dfrac{A}{y} - \dfrac{1}{2} = \dfrac{B}{y}$

$y = 2A - 2B$

20. $\dfrac{A}{y} - \dfrac{1}{2} = \dfrac{B}{C}$

$y = \dfrac{2AC}{2B + C}$

21. $3x - 4y = 6$

$y = \dfrac{3}{4}x - \dfrac{3}{2}$

22. $y^2 - 2y - Ay + 2A = 0$

$y = A \quad \text{or} \quad y = 2$

23. $A = \dfrac{1}{2}B(C + y)$

$y = \dfrac{2A - BC}{B}$

24. $y^2 + Cy = BC + By$

$y = B \quad \text{or} \quad y = -C$

Simplify each expression.

25. $3x^5 \cdot 4x^8$

$12x^{13}$

26. $3x^2(x^3 + 5x^6)$

$3x^5 + 15x^8$

27. $(5x^6)^2$ $25x^{12}$

28. $(3a^3b^2)^3$ $27a^9b^6$

29. $\dfrac{12a^9b^4}{-3a^3b^{-2}}$ $-4a^6b^6$

30. $\left(\dfrac{x^{-2}}{2}\right)^5$ $\dfrac{1}{32x^{10}}$

31. $\left(\dfrac{2x^{-4}}{3y^5}\right)^{-3}$ $\dfrac{27x^{12}y^{15}}{8}$

32. $(-2a^{-1}b^3c)^{-2}$ $\dfrac{a^2}{4b^6c^2}$

33. $\dfrac{a^{-1} + b^3}{a^{-2} + b^{-1}}$ $\dfrac{ab + a^2b^4}{b + a^2}$

34. $\dfrac{(a + b)^{-1}}{(a + b)^{-2}}$ $a + b$

Solve.

35. ***Basic energy requirement.*** Clinical dietitians must design diets that meet patients' energy requirements and are suitable for the condition of their health (*Snapshots of Applications in Mathematics*). The basic energy requirement B (in calories) for a male is a function of three variables,

$$B = 655 + 9.56W + 1.85H - 4.68A,$$

where W is the patient's weight in kilograms, H is the height in centimeters, and A is the age in years.

a) Find the basic energy requirement for Chicago Bulls' center Luc Longley. Longley is 30 years old, has a height of 7 ft 2 in., and weight of 292 pounds (www.nba.com). (1 in. ≈ 2.54 cm, 1 kg ≈ 2.2 lb.)

b) The accompanying graph shows the basic energy requirement for a 7 ft 2 in. male at age 30 as a function of his weight. As the weight increases, does the basic energy requirement increase or decrease?

c) What is the equation for the line in the accompanying figure?

d) Write the basic energy requirement for Luc Longley as a function of his age and graph this function for $20 \le A \le 70$. Assume his size stays fixed.

a) 2,188 calories   **b)** increases

c) $B = 9.56W + 918.7$ **d)** $B = 2328 - 4.68A$

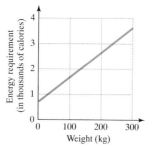

FIGURE FOR EXERCISE 35

Rational Exponents and Radicals

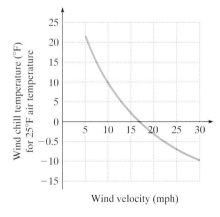

J ust how cold is it in Fargo, North Dakota, in winter? According to local meteorologists, the mercury hit a low of –33°F on January 18, 1994. But air temperature alone is not always a reliable indicator of how cold you feel. On the same date the average wind velocity was 13.8 miles per hour. This dramatically affected how cold people felt when they stepped outside. High winds along with cold temperatures make exposed skin feel colder because the wind significantly speeds up the loss of body heat. Meteorologists use the terms "wind chill factor," "wind chill index," and "wind chill temperature" to take into account both air temperature and wind velocity.

Through experimentation in Antarctica, Paul A. Siple developed a formula in the 1940s that measures the wind chill from the velocity of the wind and the air temperature. His complex formula involving the square root of the velocity of the wind is still used today to calculate wind chill temperatures. Siple's formula is unlike most scientific formulas in that it is not based on theory. Siple experimented with various formulas involving wind velocity and temperature until he found a formula that seemed to predict how cold the air felt. His formula is stated and used in Exercises 105 and 106 of Section 7.2.

7.1 RATIONAL EXPONENTS

In Sections 5.1 and 5.2 you studied integral exponents. In this section we will extend the idea of exponents to include rational numbers as exponents.

Roots

If a square has a side of length 7 feet, then it has an area of 7^2, or 49 square feet. If a square has an area of 36 square feet, then the length of its side is 6 feet. If we know the length of a side, then we square it to find the area. If we know the area, then we must undo the process of squaring to find the length of a side. Undoing the process of squaring is called **taking the square root.**

Because $3^2 = 9$, $2^3 = 8$, and $(-2)^4 = 16$, we say that 3 is a square root of 9, 2 is the cube root of 8, and -2 is a fourth root of 16. In general, undoing an nth power is referred to as **taking an nth root.**

> **nth Roots**
>
> The number b is an **nth root** of a if
> $$b^n = a.$$

Both 3 and -3 are **square roots** of 9 because $3^2 = 9$ and $(-3)^2 = 9$. Because $2^4 = 16$ and $(-2)^4 = 16$, there are two real fourth roots of 16: 2 and -2. If n is a positive even integer and a is any positive real number, then there are two real nth roots of a. We call these roots **even roots.** The positive even root of a positive number is called the **principal root.** The principal square root of 9 is 3, and the principal fourth root of 16 is 2. When n is even, the exponent $1/n$ is used to indicate the principal nth root. The principal nth root can also be indicated by the radical symbol $\sqrt[n]{}$, which will be discussed in Section 7.2.

> **Exponent $1/n$ When n Is Even**
>
> If n is a positive even integer and a is a positive real number, then $a^{1/n}$ denotes the positive real nth root of a and is called the **principal nth root of a.**

CAUTION An exponent of n indicates nth power, an exponent of $-n$ indicates the reciprocal of the nth power, and an exponent of $1/n$ indicates nth root.

We will see later that choosing $1/n$ to indicate nth root fits in nicely with the rules of exponents that we have already studied.

EXAMPLE 1 **Finding even roots**

Evaluate each expression.

a) $4^{1/2}$ b) $16^{1/4}$ c) $-81^{1/4}$ d) $\left(\dfrac{1}{4}\right)^{1/2}$

Solution

a) Because $2^2 = 4$, we have $4^{1/2} = 2$. Note that $4^{1/2} \neq -2$.

b) Because $2^4 = 16$, we have $16^{1/4} = 2$. Note that $16^{1/4} \neq -2$.

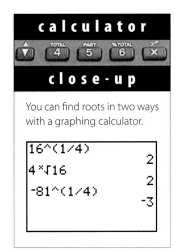

c) Following the accepted order of operations from Chapter 1, we find the root first and then take the opposite of it. Because $3^4 = 81$, we have $81^{1/4} = 3$ and $-81^{1/4} = -3$.

d) Because $\left(\frac{1}{2}\right)^2 = \frac{1}{4}$, we have $\left(\frac{1}{4}\right)^{1/2} = \frac{1}{2}$. ◼

Note that $2^3 = 8$ but $(-2)^3 = -8$. The **cube root** of 8 is 2, and the cube root of -8 is -2. If n is a positive odd integer and a is any real number, then there is only one real nth root of a. We call this root an **odd root.**

Exponent $1/n$ When n Is Odd

If n is a positive odd integer and a is any real number, then $a^{1/n}$ denotes the real nth root of a.

E X A M P L E   2

Finding odd roots

Evaluate each expression.

a) $8^{1/3}$ **b)** $(-27)^{1/3}$ **c)** $-32^{1/5}$

Solution

a) Because $2^3 = 8$, we have $8^{1/3} = 2$.

b) Because $(-3)^3 = -27$, we have $(-27)^{1/3} = -3$.

c) Because $2^5 = 32$, we have $-32^{1/5} = -2$. ◼

We do not allow 0 as the base when we use negative exponents because division by zero is undefined. However, positive powers of zero are defined, and so are roots of zero; for example, $0^4 = 0$, and so $0^{1/4} = 0$.

nth Root of Zero

If n is a positive integer, then $0^{1/n} = 0$.

CAUTION   An expression such as $(-9)^{1/2}$ is not included in the definition of roots because there is no real number whose square is -9. The definition of roots does not include an even root of any negative number because no even power of a real number is negative. However, in Section 7.6 we will define expressions such as $(-9)^{1/2}$ when we study complex numbers.

The expression $3^{1/2}$ represents the unique positive real number whose square is 3. Because there is no rational number that has a square equal to 3, the number $3^{1/2}$ is an irrational number. If we use a calculator, we find that $3^{1/2}$ is approximately equal to the rational number 1.732. Because the square root of 3 is not a rational number, the simplest representation for the exact value of the square root of 3 is $3^{1/2}$.

Positive Rational Exponents

When a rational number is used as an exponent, the denominator indicates "root" and the numerator indicates "power." For example, $8^{2/3}$ means $(8^{1/3})^2$. We take the cube root of 8 to get 2, then square 2 to get 4. Thus

$$8^{2/3} = 4.$$

In general, $a^{m/n}$ is the mth power of the nth root of a.

Rational Exponents

If m and n are positive integers, then
$$a^{m/n} = (a^{1/n})^m,$$
provided that $a^{1/n}$ is defined.

E X A M P L E   3   **Evaluating expressions with positive rational exponents**

Evaluate each expression.

a) $16^{3/4}$ **b)** $27^{4/3}$ **c)** $(-8)^{2/3}$ **d)** $-4^{3/2}$

Solution

a) To evaluate $16^{3/4}$, take the fourth root of 16 to get 2 and then cube 2 to get 8. These steps are written as follows:
$$16^{3/4} = (16^{1/4})^3 = 2^3 = 8$$

b) To evaluate $27^{4/3}$, take the cube root of 27 to get 3, and then raise 3 to the fourth power to get 81:
$$27^{4/3} = (27^{1/3})^4 = 3^4 = 81$$

c) $(-8)^{2/3} = [(-8)^{1/3}]^2 = (-2)^2 = 4$

d) $-4^{3/2} = -(4^{1/2})^3 = -2^3 = -8$ ∎

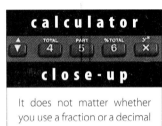

close-up

It does not matter whether you use a fraction or a decimal as the exponent as long as the decimal is the exact value of the fraction.

```
16^(3/4)
            8
16^.75
            8
27^(4/3)
           81
```

CAUTION   We can evaluate $-4^{3/2}$ because the negative sign is saved until last. But $(-4)^{3/2}$ is not a real number because $(-4)^{1/2}$ is not a real number.

By the definition of rational exponents we find the root and then the power. For example,
$$8^{2/3} = (8^{1/3})^2 = 2^2 = 4.$$

However, we get the same result if we find the power and then the root:
$$8^{2/3} = (8^2)^{1/3} = 64^{1/3} = 4$$

In general, $a^{m/n}$ is also the nth root of the mth power of a.

Evaluating $a^{m/n}$ in Either Order

If m and n are positive integers, then
$$a^{m/n} = (a^{1/n})^m = (a^m)^{1/n},$$
provided that $a^{1/n}$ is defined.

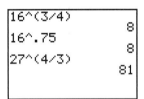

close-up

You will get the same result whether you find the root first or the power first.

```
(8^2)^(1/3)
            4
(8^(1/3))^2
            4
```

The fact that the power can be found before the root will be used later in this chapter. It is not useful in simply evaluating expressions because if the power is evaluated first, we might have to find the root of a very large number. For example, finding the power first in Example 3(b), we get
$$27^{4/3} = (27^4)^{1/3} = 531,441^{1/3} = 81.$$

Negative Rational Exponents

Negative integral exponents were defined by using reciprocals, and so are negative rational exponents. For example, $8^{-2/3}$ is the reciprocal of $8^{2/3}$. So

$$8^{-2/3} = \frac{1}{8^{2/3}} = \frac{1}{4}.$$

Negative Rational Exponents

If m and n are positive integers, then

$$a^{-m/n} = \frac{1}{a^{m/n}},$$

provided that $a^{1/n}$ is defined and nonzero.

Three operations are involved in evaluating the expression $a^{-m/n}$. The operations (root, power, reciprocal) can be performed in any order, but the simplest way to evaluate $a^{-m/n}$ is usually the following order.

Evaluating $a^{-m/n}$

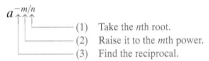

$a^{-m/n}$
- (1)   Take the nth root.
- (2)   Raise it to the mth power.
- (3)   Find the reciprocal.

E X A M P L E   4

Evaluating expressions with negative rational exponents

Evaluate each expression.

a) $4^{-3/2}$ **b)** $(-27)^{-1/3}$ **c)** $(-16)^{-3/4}$

Solution

a) The square root of 4 is 2. The cube of 2 is 8. The reciprocal of 8 is $\frac{1}{8}$. So

$$4^{-3/2} = \frac{1}{8}.$$

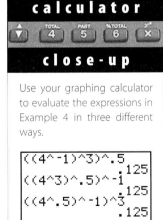
b) The cube root of -27 is -3. The first power of -3 is -3. The reciprocal of -3 is $-\frac{1}{3}$. So

$$(-27)^{-1/3} = -\frac{1}{3}.$$

c) The expression $(-16)^{-3/4}$ is not a real number because it involves the fourth root (an even root) of a negative number. ■

Using the Rules of Exponents

All of the rules for exponents hold for rational exponents as well as integral exponents. Of course, we cannot apply the rules of exponents to expressions that are not real numbers.

Rules for Rational Exponents

The following rules hold for any nonzero real numbers a and b and rational numbers r and s for which the expressions represent real numbers.

1. $a^r a^s = a^{r+s}$		Product rule
2. $\dfrac{a^r}{a^s} = a^{r-s}$		Quotient rule
3. $(a^r)^s = a^{rs}$		Power of a power rule
4. $(ab)^r = a^r b^r$		Power of a product rule
5. $\left(\dfrac{a}{b}\right)^r = \dfrac{a^r}{b^r}$		Power of a quotient rule

We can use the product rule to add rational exponents. For example,

$$16^{1/4} \cdot 16^{1/4} = 16^{2/4}.$$

The fourth root of 16 is 2, and 2 squared is 4. So $16^{2/4} = 4$. Because we also have $16^{1/2} = 4$, we see that a rational exponent can be reduced to its lowest terms. If an exponent can be reduced, it is usually simpler to reduce the exponent before we evaluate the expression. We can simplify $16^{1/4} \cdot 16^{1/4}$ as follows:

$$16^{1/4} \cdot 16^{1/4} = 16^{2/4} = 16^{1/2} = 4$$

E X A M P L E   5   **Using the product and quotient rules with rational exponents**

Simplify each expression.

a) $27^{1/6} \cdot 27^{1/2}$ **b)** $\dfrac{5^{3/4}}{5^{1/4}}$

Solution

a) $27^{1/6} \cdot 27^{1/2} = 27^{1/6+1/2}$ Product rule for exponents

$$= 27^{2/3}$$

$$= 9$$

b) $\dfrac{5^{3/4}}{5^{1/4}} = 5^{3/4-1/4} = 5^{2/4} = 5^{1/2}$ We used the quotient rule to subtract the exponents.

E X A M P L E   6   **Using the power rules with rational exponents**

Simplify each expression.

a) $3^{1/2} \cdot 12^{1/2}$ **b)** $(3^{10})^{1/2}$ **c)** $\left(\dfrac{2^6}{3^9}\right)^{-1/3}$

Solution

a) Because the bases 3 and 12 are different, we cannot use the product rule to add the exponents. Instead, we use the power of a product rule to place the $1/2$ power outside the parentheses:

$$3^{1/2} \cdot 12^{1/2} = (3 \cdot 12)^{1/2} = 36^{1/2} = 6$$

b) Use the power of a power rule to multiply the exponents:

$$(3^{10})^{1/2} = 3^5$$

c) $\left(\dfrac{2^6}{3^9}\right)^{-1/3} = \dfrac{(2^6)^{-1/3}}{(3^9)^{-1/3}}$ Power of a quotient rule

$\qquad\qquad = \dfrac{2^{-2}}{3^{-3}}$ Power of a power rule

$\qquad\qquad = \dfrac{3^3}{2^2}$ Definition of negative exponent

$\qquad\qquad = \dfrac{27}{4}$ ■

Simplifying Expressions Involving Variables

When simplifying expressions involving rational exponents and variables, we must be careful to write equivalent expressions. For example, in the equation

$$(x^2)^{1/2} = x$$

it looks as if we are correctly applying the power of a power rule. However, this statement is false if x is negative because the $1/2$ power on the left-hand side indicates the positive square root of x^2. For example, if $x = -3$, we get

$$[(-3)^2]^{1/2} = 9^{1/2} = 3,$$

which is not equal to -3. To write a simpler equivalent expression for $(x^2)^{1/2}$, we use absolute value as follows.

Square Root of x^2

$$(x^2)^{1/2} = |x| \text{ for any real number } x.$$

Note that the equation $(x^2)^{1/2} = |x|$ is an identity. It is also necessary to use absolute value when writing identities for other even roots of expressions involving variables.

E X A M P L E   7

Using absolute value symbols with exponents

Simplify each expression. Assume the variables represent any real numbers and use absolute value symbols as necessary.

a) $(x^8y^4)^{1/4}$
 b) $\left(\dfrac{x^9}{8}\right)^{1/3}$

Solution

a) Apply the power of a product rule to get the equation $(x^8y^4)^{1/4} = x^2y$. The left-hand side is nonnegative for any choices of x and y, but the right-hand side is negative when y is negative. So for any real values of x and y we have

$$(x^8y^4)^{1/4} = x^2|y|.$$

b) Using the power of a quotient rule, we get

$$\left(\dfrac{x^9}{8}\right)^{1/3} = \dfrac{x^3}{2}.$$

This equation is valid for every real number x, so no absolute value signs are used. ■

Because there are no real even roots of negative numbers, the expressions

$$a^{1/2}, \quad x^{-3/4}, \quad \text{and} \quad y^{1/6}$$

are not real numbers if the variables have negative values. To simplify matters, we sometimes assume the variables represent only positive numbers when we are working with expressions involving variables with rational exponents. That way we do not have to be concerned with undefined expressions and absolute value.

E X A M P L E   8

Expressions involving variables with rational exponents

Use the rules of exponents to simplify the following. Write your answers with positive exponents. Assume all variables represent *positive* real numbers.

a) $x^{2/3}x^{4/3}$

b) $\dfrac{a^{1/2}}{a^{1/4}}$

c) $(x^{1/2}y^{-3})^{1/2}$

d) $\left(\dfrac{x^2}{y^{1/3}}\right)^{-1/2}$

Solution

a) $x^{2/3}x^{4/3} = x^{6/3}$ Use the product rule to add the exponents.

           $= x^2$ Reduce the exponent.

b) $\dfrac{a^{1/2}}{a^{1/4}} = a^{1/2-1/4}$ Use the quotient rule to subtract the exponents.

         $= a^{1/4}$ Simplify.

c) $(x^{1/2}y^{-3})^{1/2} = (x^{1/2})^{1/2}(y^{-3})^{1/2}$ Power of a product rule

         $= x^{1/4}y^{-3/2}$ Power of a power rule

         $= \dfrac{x^{1/4}}{y^{3/2}}$ Definition of negative exponent

d) Because this expression is a negative power of a quotient, we can first find the reciprocal of the quotient, then apply the power of a power rule:

$$\left(\frac{x^2}{y^{1/3}}\right)^{-1/2} = \left(\frac{y^{1/3}}{x^2}\right)^{1/2} = \frac{y^{1/6}}{x} \quad \frac{1}{3}\cdot\frac{1}{2} = \frac{1}{6}$$

True or false? Explain.

1. $4^{-1/2} = \dfrac{1}{2}$ True

2. $16^{1/2} = 8$ False

3. $(3^{2/3})^3 = 9$ True

4. $8^{-2/3} = -4$ False

5. $2^{1/2} \cdot 2^{1/2} = 2$ True

6. $\left(\dfrac{1}{4}\right)^{1/2} = \dfrac{1}{2}$ True

7. $\dfrac{3}{3^{1/2}} = 3^{1/2}$ True

8. $(2^9)^{1/2} = 2^3$ False

9. $3^{1/3} \cdot 6^{1/3} = 18^{2/3}$ False

10. $2^{3/4} \cdot 2^{1/4} = 4$ False

7.1   EXERCISES

Reading and Writing   After reading this section, write out the answers to these questions. Use complete sentences.

1. What does it mean to say that a is an nth root of b?
   If $a^n = b$, then a is an nth root of b.

2. What is the difference between an even root and an odd root?
   If $a^n = b$, then a is an even root of b, provided n is even, and a is an odd root of b, provided n is odd.

3. What is the principal root?
   The principal root is the positive even root of a positive number.

4. How do we symbolically indicate an nth root?
   The nth root of b is written as $b^{1/n}$.

5. What is the nth root of zero?
   The nth root of 0 is 0.

6. What three operations are indicated by a negative rational exponent?
   The expression $a^{-m/n}$ represents the reciprocal of the nth root of the mth power of a.

Simplify. Some of these expressions are not real numbers. See Examples 1 and 2.

7. $100^{1/2}$ 10
8. $169^{1/2}$ 13
9. $81^{1/4}$ 3
10. $64^{1/6}$ 2
11. $-9^{1/2}$ -3
12. $-4^{1/2}$ -2
13. $\left(\dfrac{1}{64}\right)^{1/6}$ $\dfrac{1}{2}$
14. $\left(-\dfrac{1}{8}\right)^{1/3}$ $-\dfrac{1}{2}$
15. $(-25)^{1/2}$
    Not a real number
16. $(-16)^{1/4}$
    Not a real number
17. $1000^{1/3}$ 10
18. $27^{1/3}$ 3
19. $(-64)^{1/3}$ -4
20. $(-32)^{1/5}$ -2
21. $-1^{1/5}$ -1
22. $-125^{1/3}$ -5

Evaluate each expression. Some of these expressions are not real numbers. See Examples 3 and 4.

23. $32^{3/5}$ 8
24. $25^{3/2}$ 125
25. $(-27)^{2/3}$ 9
26. $(-32)^{3/5}$ -8
27. $-25^{3/2}$ -125
28. $-100^{3/2}$ -1000
29. $(-25)^{3/2}$
    Not a real number
30. $(-64)^{5/6}$
    Not a real number
31. $4^{-1/2}$ $\dfrac{1}{2}$
32. $9^{-1/2}$ $\dfrac{1}{3}$
33. $8^{-4/3}$ $\dfrac{1}{16}$
34. $4^{-3/2}$ $\dfrac{1}{8}$
35. $(-32)^{-3/5}$ $-\dfrac{1}{8}$
36. $(-27)^{-4/3}$ $\dfrac{1}{81}$
37. $(-9)^{-1/2}$
    Not a real number
38. $(-4)^{-1/2}$
    Not a real number

Use the rules of exponents to simplify each expression. See Examples 5 and 6.

39. $3^{1/3}3^{1/4}$ $3^{7/12}$
40. $2^{1/2}\,2^{1/3}$ $2^{5/6}$
41. $3^{1/3}3^{-1/3}$ 1
42. $5^{1/4}5^{-1/4}$ 1

43. $\dfrac{8^{1/3}}{8^{2/3}}$ $\dfrac{1}{2}$
44. $\dfrac{27^{-2/3}}{27^{-1/3}}$ $\dfrac{1}{3}$
45. $4^{3/4} \div 4^{1/4}$ 2
46. $9^{1/4} \div 9^{3/4}$ $\dfrac{1}{3}$
47. $18^{1/2}2^{1/2}$ 6
48. $8^{1/2}2^{1/2}$ 4
49. $(2^6)^{1/3}$ 4
50. $(3^{10})^{1/5}$ 9
51. $(3^8)^{1/2}$ 81
52. $(3^{-6})^{1/3}$ $\dfrac{1}{9}$
53. $(2^{-4})^{1/2}$ $\dfrac{1}{4}$
54. $(5^4)^{1/2}$ 25
55. $\left(\dfrac{3^4}{2^6}\right)^{1/2}$ $\dfrac{9}{8}$
56. $\left(\dfrac{5^4}{3^6}\right)^{1/2}$ $\dfrac{25}{27}$

Simplify each expression. Assume the variables represent any real numbers and use absolute value as necessary. See Example 7.

57. $(x^4)^{1/4}$ $|x|$
58. $(y^6)^{1/6}$ $|y|$
59. $(a^8)^{1/2}$ a^4
60. $(b^{10})^{1/2}$ $|b^5|$
61. $(y^3)^{1/3}$ y
62. $(w^9)^{1/3}$ w^3
63. $(9x^6y^2)^{1/2}$ $|3x^3y|$
64. $(16a^8b^4)^{1/4}$ $|2a^2b|$
65. $\left(\dfrac{81x^{12}}{y^{20}}\right)^{1/4}$ $\left|\dfrac{3x^3}{y^5}\right|$
66. $\left(\dfrac{144a^8}{9y^{18}}\right)^{1/2}$ $\dfrac{4a^4}{|y^9|}$

Simplify. Assume all variables represent positive numbers. Write answers with positive exponents only. See Example 8.

67. $x^{1/2}x^{1/4}$ $x^{3/4}$
68. $y^{1/3}y^{1/3}$ $y^{2/3}$
69. $(x^{1/2}y)(x^{-3/4}y^{1/2})$ $\dfrac{y^{3/2}}{x^{1/4}}$
70. $(a^{1/2}b^{-1/3})(ab)$ $a^{3/2}b^{2/3}$
71. $\dfrac{w^{1/3}}{w^3}$ $\dfrac{1}{w^{8/3}}$
72. $\dfrac{a^{1/2}}{a^2}$ $\dfrac{1}{a^{3/2}}$
73. $\dfrac{x^{1/2}y}{x}$ $\dfrac{y}{x^{1/2}}$
74. $\dfrac{x^{1/3}y^{-1/2}}{xy^{-1}}$ $\dfrac{y^{1/2}}{x^{2/3}}$
75. $(144x^{16})^{1/2}$ $12x^8$
76. $(125a^8)^{1/3}$ $5a^{8/3}$
77. $(4x^{-1/2}yz^{1/2})^{-1/2}$ $\dfrac{x^{1/4}}{2y^{1/2}z^{1/4}}$
78. $(9x^8y^{-10}z^{12})^{1/2}$ $\dfrac{3x^4z^6}{y^5}$
79. $\left(\dfrac{a^{-1/2}}{b^{-1/4}}\right)^{-4}$ $\dfrac{a^2}{b}$
80. $\left(\dfrac{2a^{1/2}}{b^{1/3}}\right)^6$ $\dfrac{64a^3}{b^2}$

Simplify each expression. Write your answers with positive exponents. Assume that all variables represent positive real numbers.

81. $(9^2)^{1/2}$ 9
82. $(4^{16})^{1/2}$ 4^8
83. $-16^{-3/4}$ $-\dfrac{1}{8}$
84. $-25^{-3/2}$ $-\dfrac{1}{125}$
85. $125^{-4/3}$ $\dfrac{1}{625}$
86. $27^{-2/3}$ $\dfrac{1}{9}$
87. $2^{1/2}2^{-1/4}$ $2^{1/4}$
88. $9^{-1}9^{1/2}$ $\dfrac{1}{3}$
89. $3^{0.26}3^{0.74}$ 3
90. $2^{1.5}2^{0.5}$ 4
91. $3^{1/4}27^{1/4}$ 3
92. $3^{2/3}9^{2/3}$ 9
93. $\left(-\dfrac{8}{27}\right)^{2/3}$ $\dfrac{4}{9}$
94. $\left(-\dfrac{8}{27}\right)^{-1/3}$ $-\dfrac{3}{2}$

95. $\left(-\dfrac{1}{16}\right)^{-3/4}$

Not a real number

96. $\left(\dfrac{9}{16}\right)^{-1/2}$

$\dfrac{4}{3}$

97. $(9x^9)^{1/2}$

$3x^{9/2}$

98. $(-27x^9)^{1/3}$

$-3x^3$

99. $(3a^{-2/3})^{-3}$

$\dfrac{a^2}{27}$

100. $(5x^{-1/2})^{-2}$

$\dfrac{x}{25}$

101. $(a^{1/2}b)^{1/2}(ab^{1/2})$

$a^{5/4}b$

102. $(m^{1/4}n^{1/2})^2(m^2n^3)^{1/2}$

$m^{3/2}n^{5/2}$

103. $(km^{1/2})^3(k^3m^5)^{1/2}$

$k^{9/2}m^4$

104. $(tv^{1/3})^2(t^2v^{-3})^{-1/2}$

$tv^{13/6}$

Use a scientific calculator with a power key (x^y) to find the decimal value of each expression. Round answers to four decimal places.

105. $2^{1/3}$ 1.2599

106. $5^{1/2}$ 2.2361

107. $-2^{1/2}$ -1.4142

108. $(-3)^{1/3}$ -1.4422

109. $1024^{1/10}$ 2

110. $7776^{0.2}$ 6

111. $8^{0.33}$ 1.9862

112. $289^{0.5}$ 17

113. $\left(\dfrac{64}{15{,}625}\right)^{-1/6}$ 2.5

114. $\left(\dfrac{32}{243}\right)^{-3/5}$ 3.375

Simplify each expression. Assume a and b are positive real numbers and m and n are rational numbers.

115. $a^{m/2} \cdot a^{m/4}$ $a^{3m/4}$

116. $b^{n/2} \cdot b^{-n/3}$ $b^{n/6}$

117. $\dfrac{a^{-m/5}}{a^{-m/3}}$ $a^{2m/15}$

118. $\dfrac{b^{-n/4}}{b^{-n/3}}$ $b^{n/12}$

119. $(a^{-1/m}b^{-1/n})^{-mn}$ $a^n b^m$

120. $(a^{-m/2}b^{-n/3})^{-6}$ $a^{3m}b^{2n}$

121. $\left(\dfrac{a^{-3m}b^{-6n}}{a^{9m}}\right)^{-1/3}$ $a^{4m}b^{2n}$

122. $\left(\dfrac{a^{-3/m}b^{6/n}}{a^{-6/m}b^{9/n}}\right)^{-1/3}$ $\dfrac{b^{1/n}}{a^{1/m}}$

In Exercises 123–130, solve each problem.

123. *Diagonal of a box.* The length of the diagonal of a box can be found from the formula

$$D = (L^2 + W^2 + H^2)^{1/2},$$

where L, W, and H represent the length, width, and height of the box, respectively. If the box is 12 inches long, 4 inches wide, and 3 inches high, then what is the length of the diagonal?   13 inches

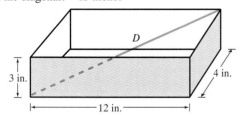

FIGURE FOR EXERCISE 123

124. *Radius of a sphere.* The radius of a sphere is a function of its volume, given by the formula

$$r = \left(\dfrac{0.75V}{\pi}\right)^{1/3}.$$

Find the radius of a spherical tank that has a volume of $\dfrac{32\pi}{3}$ cubic meters.   2 meters

FIGURE FOR EXERCISE 124

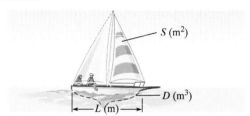 **125.** *Maximum sail area.* According to the new International America's Cup Class Rules, the maximum sail area in square meters for a yacht in the America's Cup race is given by

$$S = (13.0368 + 7.84D^{1/3} - 0.8L)^2,$$

where D is the displacement in cubic meters (m^3), and L is the length in meters (m). (*Scientific American,* May 1992). Find the maximum sail area for a boat that has a displacement of 18.42 m^3 and a length of 21.45 m.   274.96 m^2

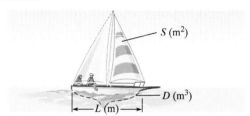

FIGURE FOR EXERCISE 125

126. *Orbits of the planets.* According to Kepler's third law of planetary motion, the average radius R of the orbit of a planet around the sun is determined by $R = T^{2/3}$, where T is the number of years for one orbit and R is measured in astronomical units or AUs (Windows to the Universe, www.windows.umich.edu).

a) It takes Mars 1.881 years to make one orbit of the sun. What is the average radius (in AUs) of the orbit of Mars?

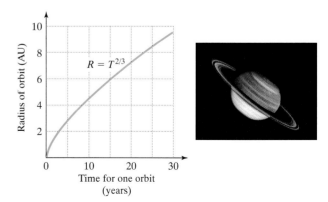

FIGURE FOR EXERCISE 126

b) The average radius of the orbit of Saturn is 9.05 AU. Use the accompanying graph to estimate the number of years it takes Saturn to make one orbit of the sun.

  a) 1.52 AU   **b)** 27 years

127. *Best stock fund.* The average annual return for an investment is given by the formula

$$r = \left(\frac{S}{P}\right)^{1/n} - 1,$$

where P is the initial investment and S is the amount it is worth after n years. The top mutual fund for 1997 in the 3-year category was Fidelity Select-Energy Services (Money Guide to Mutual Funds, 1998), in which an investment of \$10,000 grew to \$31,895.06 from 1994 to 1997. Find the 3-year average annual return for this fund. 47.2%

128. *Best bond fund.* The top bond fund for 1997 in the 5-year category was GT Global High Income B. An investment of \$10,000 in 1992 grew to \$21,830.95 in 1997. Use the formula from the previous exercise to find the 5-year average annual return for this fund. 16.9%

129. *Overdue loan payment.* In 1777 a wealthy Pennsylvania merchant, Jacob DeHaven, lent \$450,000 to the Continental Congress to rescue the troops at Valley Forge. The loan was not repaid. In 1990 DeHaven's descendants filed suit for \$141.6 billion (*New York Times,* May 27, 1990). What average annual rate of return were they using to calculate the value of the debt after 213 years? (See Exercise 127.) 6.1%

130. *California growin'.* The population of California grew from 19.9 million in 1970 to 32.5 million in 2000 (U.S. Census Bureau, www.census.gov). Find the average annual rate of growth for that time period. (Use the formula from Exercise 127 with P being the initial population and S being the population n years later.) 1.65%

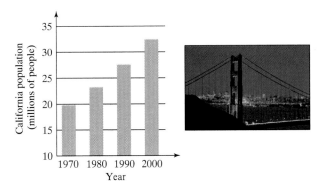

FIGURE FOR EXERCISE 130

GETTING MORE INVOLVED

131. *Discussion.* If we use the product rule to simplify $(-1)^{1/2} \cdot (-1)^{1/2}$, we get

$$(-1)^{1/2} \cdot (-1)^{1/2} = (-1)^1 = -1.$$

If we use the power of a product rule, we get

$$(-1)^{1/2} \cdot (-1)^{1/2} = (-1 \cdot -1)^{1/2} = 1^{1/2} = 1.$$

Which of these computations is incorrect? Explain your answer.

132. *Discussion.* Determine whether each equation is an identity. Explain.

  a) $(w^2 x^2)^{1/2} = |w| \cdot |x|$

  b) $(w^2 x^2)^{1/2} = |wx|$

  c) $(w^2 x^2)^{1/2} = w|x|$

7.2   RADICALS

In Section 7.1, we used the exponent $1/n$ to represent the nth root. The symbol $\sqrt[n]{\ }$ has exactly the same meaning. In this section we will learn to use the rules of exponents with this new notation.

Radical Notation

Any expression involving an nth root can be written using radical notation. The symbol $\sqrt{\ }$ is called the **radical symbol.**

> **Radicals**
>
> If n is a positive integer and a is a real number for which $a^{1/n}$ is defined, then the expression $\sqrt[n]{a}$ is called a **radical,** and
>
> $$\sqrt[n]{a} = a^{1/n}.$$

The number a is called the **radicand.** The number n is called the **index** of the radical. Radicals of index 2 and 3 are referred to as "square roots" and "cube roots," respectively, whereas the expression $\sqrt[n]{a}$ is "the nth root of a." The index 2 is usually omitted. We write $\sqrt{a}$ rather than $\sqrt[2]{a}$. Remember that $\sqrt[n]{a}$ is simply another notation for $a^{1/n}$. So $\sqrt[n]{a}$ is the *positive* nth root of a when n is even and a is positive, whereas $\sqrt[n]{a}$ is the real nth root of a when n is odd and a is any real number. The expressions

$$\sqrt{-4}, \quad \sqrt[4]{-16}, \quad \text{and} \quad \sqrt[6]{-2}$$

are not real numbers because there are no even roots of negative numbers in the real number system.

E X A M P L E   1

Changing notations

Write each exponential expression using radical notation and each radical expression using exponential notation. Assume the variables represent positive numbers. Do not simplify.

a) $\sqrt{36}$ **b)** $\sqrt[3]{-8}$ **c)** $\sqrt[3]{y^6}$ **d)** $x^{3/4}$ **e)** $x^{-5/2}$

Solution

a) $\sqrt{36} = 36^{1/2}$ **b)** $\sqrt[3]{-8} = (-8)^{1/3}$ **c)** $\sqrt[3]{y^6} = y^{6/3}$

d) $x^{3/4} = \sqrt[4]{x^3} = \left(\sqrt[4]{x}\right)^3$ We can write either the root or the power first.

e) $x^{-5/2} = \sqrt{x^{-5}} = \left(\sqrt{x}\right)^{-5}$ ◼

E X A M P L E   2

Simplifying radical expressions

Simplify.

a) $\sqrt{49}$ **b)** $\sqrt[3]{-125}$ **c)** $\sqrt[4]{81}$ **d)** $\sqrt{x^{20}}$ **e)** $\sqrt[3]{2^{21}}$

Solution

a) $\sqrt{49} = 7$ because $7^2 = 49$. **b)** $\sqrt[3]{-125} = -5$ because $(-5)^3 = -125$.

c) $\sqrt[4]{81} = 3$ because $3^4 = 81$. **d)** $\sqrt{x^{20}} = x^{20/2} = x^{10}$

e) $\sqrt[3]{2^{21}} = 2^{21/3} = 2^7 = 128$ ◼

Product Rule for Radicals

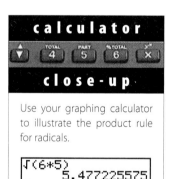

calculator

close-up

Use your graphing calculator to illustrate the product rule for radicals.

```
√(6*5)
       5.477225575
√(6)*√(5)
       5.477225575
```

Recall that the power of a product rule is valid for rational exponents as well as integers. For example, the power of a product rule allows us to write

$$(4y)^{1/2} = 4^{1/2} \cdot y^{1/2} \quad \text{and} \quad (8 \cdot 7)^{1/3} = 8^{1/3} \cdot 7^{1/3}.$$

These equations can be written using radical notation as

$$\sqrt{4y} = \sqrt{4} \cdot \sqrt{y} \quad \text{and} \quad \sqrt[3]{8 \cdot 7} = \sqrt[3]{8} \cdot \sqrt[3]{7}.$$

The power of a product rule (for the power $1/n$) can be stated using radical notation. In this form the rule is called the **product rule for radicals.**

Product Rule for Radicals

The nth root of a product is equal to the product of the nth roots. In symbols,

$$\sqrt[n]{ab} = \sqrt[n]{a} \cdot \sqrt[n]{b},$$

provided that all of the expressions represent real numbers.

The numbers 1, 4, 9, 16, 25, 49, 64, and so on are called **perfect squares** because they are the squares of the positive integers. If the radicand of a square root has a perfect square (other than 1) as a factor, the product rule can be used to simplify the radical expression. For example, the radicand of $\sqrt{50}$ has 25 as a factor, so we can use the product rule to factor $\sqrt{50}$ into a product of two square roots:

$$\sqrt{50} = \sqrt{25} \cdot \sqrt{2} = 5\sqrt{2}.$$

When simplifying a cube root, we check the radicand for factors that are perfect cubes: 8, 27, 64, 125, and so on. In general, when simplifying an nth root, we look for a perfect nth power as a factor of the radicand.

EXAMPLE 3

Using the product rule to simplify radicals

Simplify each expression. Assume all variables represent positive numbers.

a) $\sqrt{4y}$　　　　　b) $\sqrt{18}$　　　　　c) $\sqrt[3]{56}$　　　　　d) $\sqrt[4]{32}$

Solution

a) The radicand $4y$ has the perfect square 4 as a factor. So

$$\sqrt{4y} = \sqrt{4} \cdot \sqrt{y}$$
$$= 2\sqrt{y}.$$

helpful hint

We could get by without the rules for radicals. If we converted every radical expression to an exponential expression, then we could apply the rules for exponents. However, it is simpler to learn a few rules for radicals.

b) The radicand 18 has a factor of 9. So

$$\sqrt{18} = \sqrt{9} \cdot \sqrt{2}$$
$$= 3\sqrt{2}.$$

c) The radicand 56 in this cube root has the perfect cube 8 as a factor. So

$$\sqrt[3]{56} = \sqrt[3]{8} \cdot \sqrt[3]{7}$$
$$= 2\sqrt[3]{7}.$$

d) The radicand in this fourth root has the perfect fourth power 16 as a factor. So

$$\sqrt[4]{32} = \sqrt[4]{16} \cdot \sqrt[4]{2}$$
$$= 2\sqrt[4]{2}. \qquad ■$$

Quotient Rule for Radicals

The power of a quotient rule is also valid for integral and rational exponents. This rule allows us to write

$$\left(\frac{y}{9}\right)^{1/2} = \frac{y^{1/2}}{9^{1/2}} = \frac{y^{1/2}}{3} \quad \text{and} \quad \left(\frac{3}{32}\right)^{1/5} = \frac{3^{1/5}}{32^{1/5}} = \frac{3^{1/5}}{2}.$$

These equations can be written using radical notation as

$$\sqrt{\frac{y}{9}} = \frac{\sqrt{y}}{\sqrt{9}} = \frac{\sqrt{y}}{3} \quad \text{and} \quad \sqrt[5]{\frac{3}{32}} = \frac{\sqrt[5]{3}}{\sqrt[5]{32}} = \frac{\sqrt[5]{3}}{2}.$$

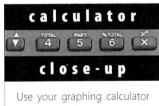

calculator

close-up

Use your graphing calculator to illustrate the quotient rule for radicals.

```
√(2/3)
        .8164965809
√(2)/√(3)
        .8164965809
```

The power of a quotient rule (for the power $1/n$) can be stated using radical notation. When written with radicals, it is called the **quotient rule for radicals.**

Quotient Rule for Radicals

The nth root of a quotient is equal to the quotient of the nth roots. In symbols,

$$\sqrt[n]{\frac{a}{b}} = \frac{\sqrt[n]{a}}{\sqrt[n]{b}},$$

provided that all of the expressions represent real numbers and $b \neq 0$.

The quotient rule is used to simplify radicals by rewriting the root of a quotient as the quotient of the roots.

E X A M P L E   4

Using the quotient rule to simplify radicals

Simplify each expression.

a) $\sqrt{\dfrac{4}{25}}$

b) $\sqrt{\dfrac{8}{9}}$

c) $\sqrt[3]{-\dfrac{8}{27}}$

d) $\sqrt[3]{\dfrac{24y}{125}}$

Solution

a) $\sqrt{\dfrac{4}{25}} = \dfrac{\sqrt{4}}{\sqrt{25}} = \dfrac{2}{5}$

b) $\sqrt{\dfrac{8}{9}} = \dfrac{\sqrt{8}}{\sqrt{9}} = \dfrac{\sqrt{4}\cdot\sqrt{2}}{3} = \dfrac{2\sqrt{2}}{3}$

c) $\sqrt[3]{-\dfrac{8}{27}} = \dfrac{\sqrt[3]{-8}}{\sqrt[3]{27}} = \dfrac{-2}{3} = -\dfrac{2}{3}$

d) $\sqrt[3]{\dfrac{24y}{125}} = \dfrac{\sqrt[3]{8}\cdot\sqrt[3]{3y}}{\sqrt[3]{125}} = \dfrac{2\sqrt[3]{3y}}{5}$

Rationalizing the Denominator

Square roots such as $\sqrt{2}$, $\sqrt{3}$, and $\sqrt{5}$ are irrational numbers. If roots of this type appear in the denominator of a fraction, it is customary to rewrite the fraction with

M A T H   A T   W O R K $x^2 + (x+1)^2 = 5^2$

Ernie Godshalk, avid sailor and owner of the sloop *Golden Eye,* has competed in races all over the world. He has learned that success in a race depends on the winds, a good crew, lots of skill, and knowing what makes his boat go fast.

Because of its shape and hull design, the maximum speed (in knots) of a sailboat can be calculated by finding the value 1.3 times the square root of her waterline (in feet). Thus Godshalk knows the maximum possible speed of the *Golden Eye.* But decisions made while she is under sail make the differ-

YACHTSMAN

ence between attaining her maximum speed and only approaching it. For example, Godshalk knows that the pressure on the sails is what makes the boat move at a certain speed. It is especially important to sail where the wind is the strongest. This means that choosing the correct tack, as well as "going where the wind is" can make the difference between coming in first and just finishing the race. In Exercise 109 of this section you will calculate the maximum speed of the *Golden Eye.*

a rational number in the denominator, or **rationalize** it. We rationalize a denominator by multiplying both the numerator and denominator by another radical that makes the denominator rational.

You can find products of radicals in two ways. By definition, $\sqrt{2}$ is the positive number that you multiply by itself to get 2. So

$$\sqrt{2} \cdot \sqrt{2} = 2.$$

By the product rule, $\sqrt{2} \cdot \sqrt{2} = \sqrt{4} = 2$. Note that $\sqrt[3]{2} \cdot \sqrt[3]{2} = \sqrt[3]{4}$ by the product rule, but $\sqrt[3]{4} \neq 2$. By definition of a cube root,

$$\sqrt[3]{2} \cdot \sqrt[3]{2} \cdot \sqrt[3]{2} = 2.$$

E X A M P L E   5

Rationalizing the denominator

Rewrite each expression with a rational denominator.

a) $\dfrac{\sqrt{3}}{\sqrt{5}}$

b) $\dfrac{3}{\sqrt[3]{2}}$

Solution

a) Because $\sqrt{5} \cdot \sqrt{5} = 5$, multiplying both the numerator and denominator by $\sqrt{5}$ will rationalize the denominator:

$$\frac{\sqrt{3}}{\sqrt{5}} = \frac{\sqrt{3}}{\sqrt{5}} \cdot \frac{\sqrt{5}}{\sqrt{5}} = \frac{\sqrt{15}}{5} \qquad \text{By the product rule, } \sqrt{3} \cdot \sqrt{5} = \sqrt{15}.$$

b) We must build up the denominator to be the cube root of a perfect cube. So we multiply by $\sqrt[3]{4}$ to get $\sqrt[3]{4} \cdot \sqrt[3]{2} = \sqrt[3]{8}$:

$$\frac{3}{\sqrt[3]{2}} = \frac{3}{\sqrt[3]{2}} \cdot \frac{\sqrt[3]{4}}{\sqrt[3]{4}} = \frac{3\sqrt[3]{4}}{\sqrt[3]{8}} = \frac{3\sqrt[3]{4}}{2}$$

C A U T I O N   To rationalize a denominator with a single square root, you simply multiply by that square root. If the denominator has a cube root, you build the denominator to a cube root of a perfect cube, as in Example 5(b). For a fourth root you build to a fourth root of a perfect fourth power, and so on.

Simplifying Radicals

When simplifying any expression, we try to make it look "simpler." When simplifying a radical expression, we have three specific conditions to satisfy.

Simplified Radical Form for Radicals of Index n

A radical expression of index n is in **simplified radical form** if it has

1. *no* perfect nth powers as factors of the radicand,
2. *no* fractions inside the radical, and
3. *no* radicals in the denominator.

The radical expressions in the next example do not satisfy the three conditions for simplified radical form. To rewrite an expression in simplified form, we use the product rule, the quotient rule, and rationalizing the denominator.

EXAMPLE 6   **Writing radical expressions in simplified radical form**

Simplify.

a) $\dfrac{\sqrt{10}}{\sqrt{6}}$ b) $\sqrt[3]{\dfrac{5}{9}}$ c) $\sqrt[4]{\dfrac{1}{3}}$

Solution

a) To rationalize the denominator, multiply the numerator and denominator by $\sqrt{6}$:

$$\dfrac{\sqrt{10}}{\sqrt{6}} = \dfrac{\sqrt{10}}{\sqrt{6}} \cdot \dfrac{\sqrt{6}}{\sqrt{6}}$$ Rationalize the denominator.

$$= \dfrac{\sqrt{60}}{6}$$

$$= \dfrac{\sqrt{4}\sqrt{15}}{6}$$ Remove the perfect square from $\sqrt{60}$.

$$= \dfrac{2\sqrt{15}}{6}$$

$$= \dfrac{\sqrt{15}}{3}$$ Reduce $\dfrac{2}{6}$ to $\dfrac{1}{3}$. Note that $\sqrt{15} \div 3 \neq \sqrt{5}$.

b) To rationalize the denominator, build up the denominator to a cube root of a perfect cube. Because $\sqrt[3]{9} \cdot \sqrt[3]{3} = \sqrt[3]{27} = 3$, we multiply by $\sqrt[3]{3}$:

$$\sqrt[3]{\dfrac{5}{9}} = \dfrac{\sqrt[3]{5}}{\sqrt[3]{9}}$$ Quotient rule for radicals

$$= \dfrac{\sqrt[3]{5}}{\sqrt[3]{9}} \cdot \dfrac{\sqrt[3]{3}}{\sqrt[3]{3}}$$ Rationalize the denominator.

$$= \dfrac{\sqrt[3]{15}}{\sqrt[3]{27}}$$

$$= \dfrac{\sqrt[3]{15}}{3}$$

c) To rationalize the denominator, observe that $\sqrt[4]{3} \cdot \sqrt[4]{27} = \sqrt[4]{81} = 3$.

$$\sqrt[4]{\dfrac{1}{3}} = \dfrac{\sqrt[4]{1}}{\sqrt[4]{3}}$$ Quotient rule for radicals

$$= \dfrac{1}{\sqrt[4]{3}} \cdot \dfrac{\sqrt[4]{27}}{\sqrt[4]{27}}$$ Rationalize the denominator.

$$= \dfrac{\sqrt[4]{27}}{\sqrt[4]{81}} = \dfrac{\sqrt[4]{27}}{3}$$ Do not omit the index of the radical in any step.

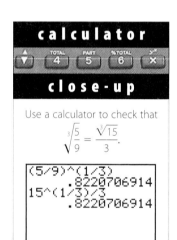

calculator

close-up

Use a calculator to check that
$$\sqrt[3]{\dfrac{5}{9}} = \dfrac{\sqrt[3]{15}}{3}.$$

```
(5/9)^(1/3)
        .822070 6914
15^(1/3)/3
        .822070 6914
```

Simplifying Radicals Involving Variables

To simplify radicals involving variables, we must recognize exponential expressions that are perfect squares, perfect cubes, and so on. The expressions

$$x^2, \quad w^4, \quad y^8, \quad z^{14}, \quad \text{and} \quad x^{50}$$

are **perfect squares** because they are squares of variables with integral powers. *Any even power of a variable is a perfect square.* If we assume the variables represent positive numbers, we can write

$$\sqrt{x^2} = x, \quad \sqrt{w^4} = w^2, \quad \sqrt{y^8} = y^4, \quad \sqrt{z^{14}} = z^7, \quad \text{and} \quad \sqrt{x^{50}} = x^{25}.$$

helpful hint

If you use exponential notation, then it is clear why the square root takes half of the exponent:

$$\sqrt{z^{14}} = z^{14/2} = z^7$$

Note that when we find the square root, the result has one-half of the original exponent.

E X A M P L E   7

Radicals with variables

Simplify each expression. Assume all variables represent positive real numbers.

a) $\sqrt{12x^6}$ b) $\sqrt{98x^5y^9}$

Solution

a) Use the product rule to place all perfect squares under the first radical symbol and the remaining factors under the second:

$$\sqrt{12x^6} = \sqrt{4x^6 \cdot 3} \qquad \text{Factor out the perfect squares.}$$
$$= \sqrt{4x^6} \cdot \sqrt{3} \qquad \text{Product rule for radicals}$$
$$= 2x^3\sqrt{3}$$

b) $\sqrt{98x^5y^9} = \sqrt{49x^4y^8} \cdot \sqrt{2xy} \qquad \text{Product rule for radicals}$
$$= 7x^2y^4\sqrt{2xy}$$

In the next example we start with a square root of a quotient.

E X A M P L E   8

Radicals with variables

Simplify each expression. Assume all variables represent positive real numbers.

a) $\sqrt{\dfrac{a}{b}}$ b) $\sqrt{\dfrac{x^3}{y^5}}$

Solution

a) $\sqrt{\dfrac{a}{b}} = \dfrac{\sqrt{a}}{\sqrt{b}}$ Quotient rule for radicals

$$= \dfrac{\sqrt{a} \cdot \sqrt{b}}{\sqrt{b} \cdot \sqrt{b}} \qquad \text{Rationalize the denominator.}$$

$$= \dfrac{\sqrt{ab}}{b}$$

b) $\sqrt{\dfrac{x^3}{y^5}} = \dfrac{\sqrt{x^3}}{\sqrt{y^5}}$ Quotient rule for radicals

$$= \dfrac{\sqrt{x^2} \cdot \sqrt{x}}{\sqrt{y^4} \cdot \sqrt{y}} \qquad \text{Product rule for radicals}$$

$$= \dfrac{x\sqrt{x}}{y^2\sqrt{y}} \qquad \text{Simplify.}$$

$$= \dfrac{x\sqrt{x} \cdot \sqrt{y}}{y^2\sqrt{y} \cdot \sqrt{y}} \qquad \text{Rationalize the denominator.}$$

$$= \dfrac{x\sqrt{xy}}{y^2 \cdot y} = \dfrac{x\sqrt{xy}}{y^3}$$

Any variable with an exponent that is a multiple of 3 is a **perfect cube.** For example,

$$a^3, \quad b^6, \quad c^{15}, \quad \text{and} \quad w^{39}$$

are perfect cubes. Each of these expressions is the cube of a variable with an integral exponent. For any values of the variables we can write

$$\sqrt[3]{a^3} = a, \qquad \sqrt[3]{b^6} = b^2, \qquad \sqrt[3]{c^{15}} = c^5, \qquad \text{and} \qquad \sqrt[3]{w^{39}} = w^{13}.$$

Note that when we find the cube root, the result has one-third of the original exponent.

If the exponent on a variable is a multiple of 4, we have a perfect fourth power; if the exponent is a multiple of 5, we have a perfect fifth power; and so on. In the next example we simplify radicals with an index higher than 2.

EXAMPLE 9

Simplifying higher-index radicals with variables

Simplify. Assume the variables represent positive numbers.

a) $\sqrt[3]{40x^8}$ 　　　　　**b)** $\sqrt[4]{x^{12}y^5}$ 　　　　　**c)** $\sqrt[3]{\dfrac{x}{y}}$

Solution

a) Use the product rule to place the largest perfect cube factors under the first radical and the remaining factors under the second:

$$\sqrt[3]{40x^8} = \sqrt[3]{8x^6} \cdot \sqrt[3]{5x^2} = 2x^2\sqrt[3]{5x^2}$$

b) Place the largest perfect fourth power factors under the first radical and the remaining factors under the second:

$$\sqrt[4]{x^{12}y^5} = \sqrt[4]{x^{12}y^4} \cdot \sqrt[4]{y} = x^3y\sqrt[4]{y}$$

c) Multiply by $\sqrt[3]{y^2}$ to rationalize the denominator:

$$\sqrt[3]{\frac{x}{y}} = \frac{\sqrt[3]{x}}{\sqrt[3]{y}} = \frac{\sqrt[3]{x}}{\sqrt[3]{y}} \cdot \frac{\sqrt[3]{y^2}}{\sqrt[3]{y^2}} = \frac{\sqrt[3]{xy^2}}{\sqrt[3]{y^3}} = \frac{\sqrt[3]{xy^2}}{y}$$

WARM-UPS

True or false? Explain.

1. $2^{1/2} = \sqrt{2}$　True 　　　　　**2.** $3^{1/3} = \sqrt{3}$　False

3. $2^{2/3} = \sqrt[3]{4}$　True 　　　　　**4.** $\sqrt{81} = \sqrt{9}$　False

5. $\sqrt{417^2} = 417$　True 　　　　　**6.** $\sqrt[3]{a^{27}} = a^3$　False

7. $\dfrac{\sqrt{2}}{2} = \dfrac{1}{\sqrt{2}}$　True 　　　　　**8.** $\dfrac{\sqrt{10}}{2} = \sqrt{5}$　False

9. $\sqrt{2^{-4}} = \dfrac{1}{4}$　True 　　　　　**10.** $\dfrac{\sqrt{6}}{\sqrt{3}} = \sqrt{2}$　True

7.2 　 EXERCISES

Reading and Writing 　 *After reading this section, write out the answers to these questions. Use complete sentences.*

1. What is a radical?

　The expression $\sqrt[n]{a}$ is called a radical.

2. What are the two ways to indicate an *n*th root of a number?

　The expressions $\sqrt[n]{a}$ and $a^{1/n}$ both represent the *n*th root of *a*.

3. What is the product rule for radicals?

　The product rule for radicals says that $\sqrt[n]{a} \cdot \sqrt[n]{b} = \sqrt[n]{ab}$.

4. How do we use the product rule to simplify a radical?

　The product rule can be used to factor out a perfect square from the radicand as in $\sqrt{18} = \sqrt{9}\sqrt{2} = 3\sqrt{2}$.

5. What is the quotient rule for radicals?

　The quotient rule for radicals says that $\sqrt[n]{a}/\sqrt[n]{b} = \sqrt[n]{a/b}$.

6. What is simplified form for a radical?

The simplified form for a radical expression has no perfect nth powers as factors of the radicand, no fractions inside the radical, and no radicals in the denominator.

All variables in Exercises 7–96 represent positive numbers.

Write the radical expressions in exponential notation and the exponential expressions in radical notation. Do not simplify. See Example 1.

7. $\sqrt{27}$
$27^{1/2}$

8. $\sqrt[3]{-27}$
$(-27)^{1/3}$

9. $\sqrt{x^5}$
$x^{5/2}$

10. $\sqrt{a^3}$
$a^{3/2}$

11. $\sqrt[3]{a^{12}}$
$a^{12/13}$

12. $\sqrt[3]{w^{-27}}$
$w^{-27/3}$

13. $-5^{1/3}$
$-\sqrt[3]{5}$

14. $-7^{1/2}$
$-\sqrt{7}$

15. $2^{2/5}$
$\sqrt[5]{2^2}$

16. $3^{2/3}$
$\sqrt[3]{3^2}$

17. $x^{-2/3}$
$\sqrt[3]{x^{-2}}$

18. $x^{-2/5}$
$\sqrt[5]{x^{-2}}$

Simplify each radical expression. See Example 2.

19. $\sqrt{121}$
11

20. $\sqrt{64}$
8

21. $\sqrt[3]{-1000}$
-10

22. $\sqrt[5]{-1}$
-1

23. $\sqrt[4]{-81}$
Not a real number

24. $\sqrt[4]{16}$
2

25. $\sqrt{a^{16}}$
a^8

26. $\sqrt{b^{36}}$
b^{18}

27. $\sqrt{4^{16}}$
4^8

28. $\sqrt{w^4}$
w^2

29. $\sqrt[5]{w^{30}}$
w^6

30. $\sqrt[5]{a^{20}}$
a^4

Simplify each expression. See Example 3.

31. $\sqrt{9w}$
$3\sqrt{w}$

32. $\sqrt{36m}$
$6\sqrt{m}$

33. $\sqrt{20}$
$2\sqrt{5}$

34. $\sqrt{50}$
$5\sqrt{2}$

35. $\sqrt{45w}$
$3\sqrt{5w}$

36. $\sqrt{48t}$
$4\sqrt{3t}$

37. $\sqrt{288}$
$12\sqrt{2}$

38. $\sqrt{242}$
$11\sqrt{2}$

39. $\sqrt[3]{54}$
$3\sqrt[3]{2}$

40. $\sqrt[3]{-48}$
$-2\sqrt[3]{6}$

41. $\sqrt[4]{32a}$
$2\sqrt[4]{2a}$

42. $\sqrt[4]{80xy}$
$2\sqrt[4]{5xy}$

Simplify each expression. See Example 4.

43. $\sqrt{\dfrac{9}{100}}$ $\dfrac{3}{10}$

44. $\sqrt{\dfrac{25}{4}}$ $\dfrac{5}{2}$

45. $\sqrt{\dfrac{50}{9}}$ $\dfrac{5\sqrt{2}}{3}$

46. $\sqrt{\dfrac{18}{25}}$ $\dfrac{3\sqrt{2}}{5}$

47. $\sqrt[3]{-\dfrac{125}{8}}$ $-\dfrac{5}{2}$

48. $\sqrt[4]{\dfrac{16}{81}}$ $\dfrac{2}{3}$

49. $\sqrt[3]{\dfrac{16x}{27}}$ $\dfrac{2\sqrt[3]{2x}}{3}$

50. $\sqrt[3]{\dfrac{-81a}{1000}}$ $-\dfrac{3\sqrt[3]{3a}}{10}$

Rewrite each expression with a rational denominator. See Example 5.

51. $\dfrac{2}{\sqrt{5}}$ $\dfrac{2\sqrt{5}}{5}$

52. $\dfrac{5}{\sqrt{3}}$ $\dfrac{5\sqrt{3}}{3}$

53. $\dfrac{\sqrt{3}}{\sqrt{7}}$ $\dfrac{\sqrt{21}}{7}$

54. $\dfrac{\sqrt{6}}{\sqrt{5}}$ $\dfrac{\sqrt{30}}{5}$

55. $\dfrac{1}{\sqrt[3]{4}}$ $\dfrac{\sqrt[3]{2}}{2}$

56. $\dfrac{7}{\sqrt[3]{3}}$ $\dfrac{7\sqrt[3]{9}}{3}$

57. $\dfrac{\sqrt[3]{6}}{\sqrt[3]{5}}$ $\dfrac{\sqrt[3]{150}}{5}$

58. $\dfrac{\sqrt[4]{2}}{\sqrt[4]{27}}$ $\dfrac{\sqrt[4]{6}}{3}$

Write each radical expression in simplified radical form. See Example 6.

59. $\dfrac{\sqrt{5}}{\sqrt{12}}$ $\dfrac{\sqrt{15}}{6}$

60. $\dfrac{\sqrt{7}}{\sqrt{18}}$ $\dfrac{\sqrt{14}}{6}$

61. $\dfrac{\sqrt{3}}{\sqrt{12}}$ $\dfrac{1}{2}$

62. $\dfrac{\sqrt{2}}{\sqrt{18}}$ $\dfrac{1}{3}$

63. $\sqrt{\dfrac{1}{2}}$ $\dfrac{\sqrt{2}}{2}$

64. $\sqrt{\dfrac{3}{8}}$ $\dfrac{\sqrt{6}}{4}$

65. $\sqrt[3]{\dfrac{7}{4}}$ $\dfrac{\sqrt[3]{14}}{2}$

66. $\sqrt[4]{\dfrac{1}{5}}$ $\dfrac{\sqrt[4]{125}}{5}$

Simplify. See Examples 7 and 8.

67. $\sqrt{12x^8}$
$2x^4\sqrt{3}$

68. $\sqrt{72x^{10}}$
$6x^5\sqrt{2}$

69. $\sqrt{60a^9b^3}$
$2a^4b\sqrt{15ab}$

70. $\sqrt{63w^{15}z^7}$
$3w^7z^3\sqrt{7wz}$

71. $\sqrt{\dfrac{x}{y}}$
$\dfrac{\sqrt{xy}}{y}$

72. $\sqrt{\dfrac{x^2}{a}}$
$\dfrac{x\sqrt{a}}{a}$

73. $\sqrt{\dfrac{a^3}{b^7}}$
$\dfrac{a\sqrt{ab}}{b^4}$

74. $\sqrt[2]{\dfrac{w^5}{y^8}}$
$\dfrac{w^2\sqrt{w}}{y^4}$

Simplify. See Example 9.

75. $\sqrt[3]{16x^{13}}$ $2x^4\sqrt[3]{2x}$

76. $\sqrt[3]{24x^{17}}$ $2x^5\sqrt[3]{3x^2}$

77. $\sqrt[4]{x^9y^6}$ $x^2y\sqrt[4]{xy^2}$

78. $\sqrt[4]{w^{14}y^7}$ $w^3y\sqrt[4]{w^2y^3}$

79. $\sqrt[5]{64x^{22}}$ $2x^4\sqrt[5]{2x^2}$

80. $\sqrt[5]{x^{12}y^5z^3}$ $x^2y\sqrt[5]{x^2z^3}$

81. $\sqrt[3]{\dfrac{a}{b}}$ $\dfrac{\sqrt[3]{ab^2}}{b}$

82. $\sqrt[3]{\dfrac{a}{w^2}}$ $\dfrac{\sqrt[3]{aw}}{w}$

Simplify.

83. $\sqrt[4]{3^{12}}$ 27

84. $\sqrt[3]{2^{-9}}$ $\dfrac{1}{8}$

85. $\sqrt{10^{-2}}$ $\dfrac{1}{10}$

86. $\sqrt{-10^{-4}}$ Not a real number

87. $\sqrt{\dfrac{8x}{49}}$ $\dfrac{2\sqrt{2x}}{7}$

88. $\sqrt{\dfrac{12b}{121}}$ $\dfrac{2\sqrt{3b}}{11}$

89. $\sqrt[4]{\dfrac{32a}{81}}$ $\dfrac{2\sqrt[4]{2a}}{3}$

90. $\sqrt[4]{\dfrac{162y}{625}}$ $\dfrac{3\sqrt[4]{2y}}{5}$

91. $\sqrt[3]{-27x^9y^8}$ $-3x^3y^2\sqrt[3]{y^2}$

92. $\sqrt[4]{32y^8z^{11}}$ $2y^2z^2\sqrt[4]{2z^3}$

93. $\dfrac{\sqrt{ab^3}}{\sqrt{a^3b^2}}$ $\dfrac{\sqrt{b}}{a}$

94. $\dfrac{\sqrt{m^3n^5}}{\sqrt{m^5n}}$ $\dfrac{n^2}{m}$

95. $\dfrac{\sqrt[3]{a^2b}}{\sqrt[3]{4ab^2}\sqrt[3]{3ab^5}}$ $\dfrac{\sqrt[3]{18}}{6b^2}$

96. $\dfrac{\sqrt[3]{5xy^2}}{\sqrt[3]{18x^2y}}$ $\dfrac{\sqrt[3]{60x^2y}}{6x}$

Use a calculator to find a decimal approximation to each radical expression. Round to three decimal places.

97. $\dfrac{5}{\sqrt{3}}$ 2.887

98. $\sqrt{\dfrac{2}{27}}$ 0.272

99. $\sqrt[3]{\dfrac{1}{3}}$ 0.693

100. $\sqrt[3]{56}$ 3.826

101. $\dfrac{\sqrt[3]{9}}{\sqrt[3]{4}}$ 1.310

102. $\dfrac{\sqrt[4]{25}}{\sqrt{5}}$ 1

103. $\dfrac{\sqrt[6]{16}}{\sqrt[3]{4}}$ 1

104. $\sqrt[5]{2.48832}$ 1.2

In Exercises 105–112, solve each problem.

105. *Factoring in the wind.* Through experimentation in Antarctica, Paul Siple developed the formula

$$W = 91.4 - \dfrac{(10.5 + 6.7\sqrt{v} - 0.45v)(457 - 5t)}{110}$$

to calculate the wind chill temperature W (in degrees Fahrenheit) from the wind velocity v [in miles per hour (mph)] and the air temperature t (in degrees Fahrenheit). Find the wind chill temperature when the air temperature is 25°F and the wind velocity is 20 mph. Use the accompanying graph to estimate the wind chill temperature when the air temperature is 25°F and the wind velocity is 30 mph. $-4°\text{F}, -10°\text{F}$

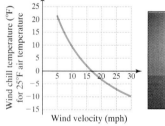

FIGURE FOR EXERCISE 105

106. *Comparing wind chills.* Use the formula from Exercise 105 to determine who will feel colder: a person in Minneapolis at 10°F with a 15-mph wind or a person in Chicago at 20°F with a 25-mph wind. Minneapolis

107. *Diving time.* The time t (in seconds) that it takes for a cliff diver to reach the water is a function of the height h (in feet) from which he dives:

$$t = \sqrt{\dfrac{h}{16}}$$

a) Use the properties of radicals to simplify this formula.

b) Find the exact time (according to the formula) that it takes for a diver to hit the water when diving from a height of 40 feet.

c) Use the accompanying graph to estimate the height if a diver takes 2.5 seconds to reach the water?
a) $t = \dfrac{\sqrt{h}}{4}$ b) $\dfrac{\sqrt{10}}{2}$ sec   c) 100 ft

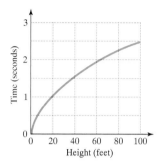

FIGURE FOR EXERCISE 107

108. *Sky diving.* The formula in the previous exercise accounts for the effect of gravity only on a falling object. According to that formula, how long would it take a sky diver to reach the earth when jumping from 17,000 feet? (A sky diver can actually get about twice as much falling time by spreading out and using the air to slow the fall.) 32.6 seconds

109. *Maximum sailing speed.* To find the maximum possible speed in knots (nautical miles per hour) for a sailboat, sailors use the formula $M = 1.3\sqrt{w}$, where w is the length of the waterline in feet. If the waterline for the sloop *Golden Eye* is 20 feet, then what is the maximum speed of the *Golden Eye*?   5.8 knots

110. *America's Cup.* Since 1988 basic yacht dimensions for the America's Cup competition have satisfied the inequality

$$L + 1.25\sqrt{S} - 9.8\sqrt[3]{D} \le 16.296,$$

where L is the boat's length in meters (m), S is the sail area in square meters, and D is the displacement in cubic meters (*Scientific American,* May 1992). A team of naval architects is planning to build a boat with a displacement of 21.44 cubic meters (m^3), a sail area of 320.13 square meters (m^2), and a length of 21.22 m. Does this boat satisfy the inequality? If the length and displacement of this boat cannot be changed, then how many square meters of sail area must be removed so that the boat satisfies the inequality?   No, 1.84 m^2

111. *Landing a Piper Cheyenne.* Aircraft design engineers determine the proper landing speed V [in feet per second (ft/sec)] for an airplane from the formula

$$V = \sqrt{\dfrac{841L}{CS}},$$

where L is the gross weight of the aircraft in pounds (lb), C is the coefficient of lift, and S is the wing surface area in square feet. According to Piper Aircraft of Vero Beach, Florida, the Piper Cheyenne has a gross weight of

8,700 lb, a coefficient of lift of 2.81, and a wing surface area of 200 ft². Find the proper landing speed for this plane. What is the landing speed in miles per hour (mph)? 114.1 ft/sec, 77.8 mph

 112. *Landing speed and weight.* Because the gross weight of the Piper Cheyenne depends on how much fuel and cargo are on board, the proper landing speed (from Exercise 111) is not always the same. The formula $V = \sqrt{1.496L}$ gives the landing speed as a function of the gross weight only.
  a) Find the landing speed if the gross weight is 7,000 lb.
  b) What gross weight corresponds to a landing speed of 115 ft/sec?
    a) 102.3 ft/sec   **b)** 8,840 lb

GETTING MORE INVOLVED

113. *Cooperative learning.* Work in a group to determine whether each equation is an identity. Explain your answers.

a) $\sqrt{x^2} = |x|$ **b)** $\sqrt[3]{x^3} = |x|$
c) $\sqrt{x^4} = x^2$ **d)** $\sqrt[4]{x^4} = |x|$
For which values of n is $\sqrt[n]{x^n} = x$ an identity?

 114. *Cooperative learning.* Work in a group to determine whether each inequality is correct.

a) $\sqrt{0.9} > 0.9$ **b)** $\sqrt{1.01} > 1.01$
c) $\sqrt[3]{0.99} > 0.99$ **d)** $\sqrt[3]{1.001} > 1.001$
For which values of x and n is $\sqrt[n]{x} > x$?

 115. *Discussion.* If your test scores are 80 and 100, then the arithmetic mean of your scores is 90. The geometric mean of the scores is a number h such that

$$\frac{80}{h} = \frac{h}{100}.$$

Are you better off with the arithmetic mean or the geometric mean?

7.3   OPERATIONS WITH RADICALS

In this section we will use the ideas of Section 7.2 in performing arithmetic operations with radical expressions.

Adding and Subtracting Radicals

To find the sum of $\sqrt{2}$ and $\sqrt{3}$, we can use a calculator to get $\sqrt{2} \approx 1.414$ and $\sqrt{3} \approx 1.732$. (The symbol $\approx$ means "is approximately equal to.") We can then add the decimal numbers and get

$$\sqrt{2} + \sqrt{3} \approx 1.414 + 1.732 = 3.146.$$

We cannot write an exact decimal form for $\sqrt{2} + \sqrt{3}$; the number 3.146 is an approximation of $\sqrt{2} + \sqrt{3}$. To represent the exact value of $\sqrt{2} + \sqrt{3}$, we just use the form $\sqrt{2} + \sqrt{3}$. This form cannot be simplified any further. However, a sum of like radicals can be simplified. **Like radicals** are radicals that have the same index and the same radicand.

To simplify the sum $3\sqrt{2} + 5\sqrt{2}$, we can use the fact that $3x + 5x = 8x$ is true for any value of x. Substituting $\sqrt{2}$ for x gives us $3\sqrt{2} + 5\sqrt{2} = 8\sqrt{2}$. So like radicals can be combined just as like terms are combined.

E X A M P L E   1   **Adding and subtracting like radicals**
Simplify the following expressions. Assume the variables represent positive numbers.

a) $3\sqrt{5} + 4\sqrt{5}$ **b)** $\sqrt[4]{w} - 6\sqrt[4]{w}$
c) $\sqrt{3} + \sqrt{5} - 4\sqrt{3} + 6\sqrt{5}$ **d)** $3\sqrt[3]{6x} + 2\sqrt[3]{x} + \sqrt[3]{6x} + \sqrt[3]{x}$

Solution

a) $3\sqrt{5} + 4\sqrt{5} = 7\sqrt{5}$ **b)** $\sqrt[4]{w} - 6\sqrt[4]{w} = -5\sqrt[4]{w}$
c) $\sqrt{3} + \sqrt{5} - 4\sqrt{3} + 6\sqrt{5} = -3\sqrt{3} + 7\sqrt{5}$ Only like radicals are combined.
d) $3\sqrt[3]{6x} + 2\sqrt[3]{x} + \sqrt[3]{6x} + \sqrt[3]{x} = 4\sqrt[3]{6x} + 3\sqrt[3]{x}$ ■

We may have to simplify the radicals to determine which ones can be combined.

EXAMPLE 2   Simplifying radicals before combining

Perform the indicated operations. Assume the variables represent positive numbers.

a) $\sqrt{8} + \sqrt{18}$

b) $\sqrt{\dfrac{1}{5}} + \sqrt{20}$

c) $\sqrt{2x^3} - \sqrt{4x^2} + 5\sqrt{18x^3}$

d) $\sqrt[3]{16x^4y^3} - \sqrt[3]{54x^4y^3}$

Solution

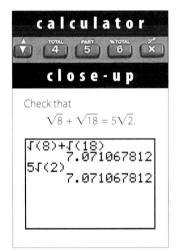

a) $\sqrt{8} + \sqrt{18} = \sqrt{4} \cdot \sqrt{2} + \sqrt{9} \cdot \sqrt{2}$

$\qquad\qquad\qquad = 2\sqrt{2} + 3\sqrt{2}$ Simplify each radical.

$\qquad\qquad\qquad = 5\sqrt{2}$ Add like radicals.

$\qquad$ Note that $\sqrt{8} + \sqrt{18} \neq \sqrt{26}$.

b) $\sqrt{\dfrac{1}{5}} + \sqrt{20} = \dfrac{\sqrt{5}}{5} + 2\sqrt{5}$ Because $\sqrt{\dfrac{1}{5}} = \dfrac{1}{\sqrt{5}} \cdot \dfrac{\sqrt{5}}{\sqrt{5}} = \dfrac{\sqrt{5}}{5}$ and $\sqrt{20} = 2\sqrt{5}$

$\qquad\qquad\qquad = \dfrac{\sqrt{5}}{5} + \dfrac{10\sqrt{5}}{5}$ Use the LCD of 5.

$\qquad\qquad\qquad = \dfrac{11\sqrt{5}}{5}$

c) $\sqrt{2x^3} - \sqrt{4x^2} + 5\sqrt{18x^3} = \sqrt{x^2} \cdot \sqrt{2x} - 2x + 5 \cdot \sqrt{9x^2} \cdot \sqrt{2x}$

$\qquad\qquad\qquad = x\sqrt{2x} - 2x + 15x\sqrt{2x}$ Simplify each radical.

$\qquad\qquad\qquad = 16x\sqrt{2x} - 2x$ Add like radicals only.

d) $\sqrt[3]{16x^4y^3} - \sqrt[3]{54x^4y^3} = \sqrt[3]{8x^3y^3} \cdot \sqrt[3]{2x} - \sqrt[3]{27x^3y^3} \cdot \sqrt[3]{2x}$

$\qquad\qquad\qquad = 2xy\sqrt[3]{2x} - 3xy\sqrt[3]{2x}$ Simplify each radical.

$\qquad\qquad\qquad = -xy\sqrt[3]{2x}$

Multiplying Radicals

We have already multiplied radicals, in Section 7.2 when we rationalized denominators. The product rule for radicals, $\sqrt[n]{a} \cdot \sqrt[n]{b} = \sqrt[n]{ab}$, allows multiplication of radicals with the same index, such as

$$\sqrt{5} \cdot \sqrt{3} = \sqrt{15}, \qquad \sqrt[3]{2} \cdot \sqrt[3]{5} = \sqrt[3]{10}, \qquad \text{and} \qquad \sqrt[5]{x^2} \cdot \sqrt[5]{x} = \sqrt[5]{x^3}.$$

CAUTION   The product rule does not allow multiplication of radicals that have different indices. We cannot use the product rule to multiply $\sqrt{2}$ and $\sqrt[3]{5}$.

EXAMPLE 3   Multiplying radicals with the same index

Multiply and simplify the following expressions. Assume the variables represent positive numbers.

a) $5\sqrt{6} \cdot 4\sqrt{3}$

b) $\sqrt{3a^2} \cdot \sqrt{6a}$

c) $\sqrt[3]{4} \cdot \sqrt[3]{4}$

d) $\sqrt[4]{\dfrac{x^3}{2}} \cdot \sqrt[4]{\dfrac{x^2}{4}}$

Solution

a) $5\sqrt{6} \cdot 4\sqrt{3} = 5 \cdot 4 \cdot \sqrt{6} \cdot \sqrt{3}$

$= 20\sqrt{18}$ Product rule for radicals

$= 20 \cdot 3\sqrt{2}$ $\sqrt{18} = \sqrt{9} \cdot \sqrt{2} = 3\sqrt{2}$

$= 60\sqrt{2}$

b) $\sqrt{3a^2} \cdot \sqrt{6a} = \sqrt{18a^3}$ Product rule for radicals

$= \sqrt{9a^2} \cdot \sqrt{2a}$

$= 3a\sqrt{2a}$ Simplify.

c) $\sqrt[3]{4} \cdot \sqrt[3]{4} = \sqrt[3]{16}$

$= \sqrt[3]{8} \cdot \sqrt[3]{2}$ Simplify.

$= 2\sqrt[3]{2}$

d) $\sqrt[4]{\dfrac{x^3}{2}} \cdot \sqrt[4]{\dfrac{x^2}{4}} = \sqrt[4]{\dfrac{x^5}{8}}$ Product rule for radicals

$= \dfrac{\sqrt[4]{x^4} \cdot \sqrt[4]{x}}{\sqrt[4]{8}}$ Simplify

$= \dfrac{x\sqrt[4]{x}}{\sqrt[4]{8}}$

$= \dfrac{x\sqrt[4]{x} \cdot \sqrt[4]{2}}{\sqrt[4]{8} \cdot \sqrt[4]{2}}$ Rationalize the denominator.

$= \dfrac{x\sqrt[4]{2x}}{2}$ $\sqrt[4]{8} \cdot \sqrt[4]{2} = \sqrt[4]{16} = 2$

> **helpful** **hint**
>
> Students often write
>
> $\sqrt{15} \cdot \sqrt{15} = \sqrt{225} = 15.$
>
> Although this is correct, you should get used to the idea that
>
> $\sqrt{15} \cdot \sqrt{15} = 15.$
>
> Because of the definition of a square root, $\sqrt{a} \cdot \sqrt{a} = a$ for any positive number a.

We find a product such as $3\sqrt{2}(4\sqrt{2} - \sqrt{3})$ by using the distributive property as we do when multiplying a monomial and a binomial. A product such as $(2\sqrt{3} + \sqrt{5})(3\sqrt{3} - 2\sqrt{5})$ can be found by using FOIL as we do for the product of two binomials.

E X A M P L E   4

Multiplying radicals

Multiply and simplify.

a) $3\sqrt{2}\left(4\sqrt{2} - \sqrt{3}\right)$ 　　　　　**b)** $\sqrt[3]{a}\left(\sqrt[3]{a} - \sqrt[3]{a^2}\right)$

c) $(2\sqrt{3} + \sqrt{5})(3\sqrt{3} - 2\sqrt{5})$ 　　**d)** $\left(3 + \sqrt{x - 9}\right)^2$

Solution

a) $3\sqrt{2}\left(4\sqrt{2} - \sqrt{3}\right) = 3\sqrt{2} \cdot 4\sqrt{2} - 3\sqrt{2} \cdot \sqrt{3}$ Distributive property

$= 12 \cdot 2 - 3\sqrt{6}$ Because $\sqrt{2} \cdot \sqrt{2} = 2$ and $\sqrt{2} \cdot \sqrt{3} = \sqrt{6}$

$= 24 - 3\sqrt{6}$

b) $\sqrt[3]{a}\left(\sqrt[3]{a} - \sqrt[3]{a^2}\right) = \sqrt[3]{a^2} - \sqrt[3]{a^3}$ Distributive property

$= \sqrt[3]{a^2} - a$

c) $(2\sqrt{3} + \sqrt{5})(3\sqrt{3} - 2\sqrt{5})$

$$= \overbrace{2\sqrt{3} \cdot 3\sqrt{3}}^{\text{F}} - \overbrace{2\sqrt{3} \cdot 2\sqrt{5}}^{\text{O}} + \overbrace{\sqrt{5} \cdot 3\sqrt{3}}^{\text{I}} - \overbrace{\sqrt{5} \cdot 2\sqrt{5}}^{\text{L}}$$

$= 18 - 4\sqrt{15} + 3\sqrt{15} - 10$

$= 8 - \sqrt{15}$ Combine like radicals.

d) To square a sum, we use $(a + b)^2 = a^2 + 2ab + b^2$:

$$\left(3 + \sqrt{x - 9}\right)^2 = 3^2 + 2 \cdot 3\sqrt{x - 9} + \left(\sqrt{x - 9}\right)^2$$
$$= 9 + 6\sqrt{x - 9} + x - 9$$
$$= x + 6\sqrt{x - 9}$$

In the next example we multiply radicals that have different indices.

EXAMPLE 5

Multiplying radicals with different indices

Write each product as a single radical expression.

a) $\sqrt[3]{2} \cdot \sqrt[4]{2}$

b) $\sqrt[3]{2} \cdot \sqrt{3}$

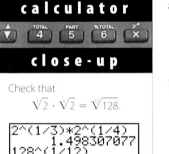

calculator

close-up

Check that
$$\sqrt[3]{2} \cdot \sqrt[4]{2} = \sqrt[12]{128}.$$

```
2^(1/3)*2^(1/4)
         1.498307077
128^(1/12)
         1.498307077
```

Solution

a) $\sqrt[3]{2} \cdot \sqrt[4]{2} = 2^{1/3} \cdot 2^{1/4}$ Write in exponential notation.

$\qquad\qquad = 2^{7/12}$ Product rule for exponents: $\frac{1}{3} + \frac{1}{4} = \frac{7}{12}$

$\qquad\qquad = \sqrt[12]{2^7}$ Write in radical notation.

$\qquad\qquad = \sqrt[12]{128}$

b) $\sqrt[3]{2} \cdot \sqrt{3} = 2^{1/3} \cdot 3^{1/2}$ Write in exponential notation.

$\qquad\qquad = 2^{2/6} \cdot 3^{3/6}$ Write the exponents with the LCD of 6.

$\qquad\qquad = \sqrt[6]{2^2} \cdot \sqrt[6]{3^3}$ Write in radical notation.

$\qquad\qquad = \sqrt[6]{2^2 \cdot 3^3}$ Product rule for radicals

$\qquad\qquad = \sqrt[6]{108}$ $2^2 \cdot 3^3 = 4 \cdot 27 = 108$

CAUTION   Because the bases in $2^{1/3} \cdot 2^{1/4}$ are identical, we can add the exponents [Example 5(a)]. Because the bases in $2^{2/6} \cdot 3^{3/6}$ are not the same, we cannot add the exponents [Example 5(b)]. Instead, we write each factor as a sixth root and use the product rule for radicals.

Conjugates

helpful hint

The word "conjugate" is used in many contexts in mathematics. According to the dictionary, conjugate means joined together, especially as in a pair.

Recall the special product rule $(a + b)(a - b) = a^2 - b^2$. The product of the sum $4 + \sqrt{3}$ and the difference $4 - \sqrt{3}$ can be found by using this rule:

$$\left(4 + \sqrt{3}\right)\left(4 - \sqrt{3}\right) = 4^2 - \left(\sqrt{3}\right)^2 = 16 - 3 = 13$$

The product of the irrational number $4 + \sqrt{3}$ and the irrational number $4 - \sqrt{3}$ is the rational number 13. For this reason the expressions $4 + \sqrt{3}$ and $4 - \sqrt{3}$ are called **conjugates** of one another. We will use conjugates in Section 7.4 to rationalize some denominators.

EXAMPLE 6

Multiplying conjugates

Find the products. Assume the variables represent positive real numbers.

a) $\left(2 + 3\sqrt{5}\right)\left(2 - 3\sqrt{5}\right)$

b) $\left(\sqrt{3} - \sqrt{2}\right)\left(\sqrt{3} + \sqrt{2}\right)$

c) $\left(\sqrt{2x} - \sqrt{y}\right)\left(\sqrt{2x} + \sqrt{y}\right)$

Solution

a) $(2 + 3\sqrt{5})(2 - 3\sqrt{5}) = 2^2 - (3\sqrt{5})^2$ $(a + b)(a - b) = a^2 - b^2$

$$= 4 - 45 \qquad (3\sqrt{5})^2 = 9 \cdot 5 = 45$$

$$= -41$$

b) $(\sqrt{3} - \sqrt{2})(\sqrt{3} + \sqrt{2}) = 3 - 2$

$$= 1$$

c) $(\sqrt{2x} - \sqrt{y})(\sqrt{2x} + \sqrt{y}) = 2x - y$

WARM-UPS

True or false? Explain.

1. $\sqrt{3} + \sqrt{3} = \sqrt{6}$ False
2. $\sqrt{8} + \sqrt{2} = 3\sqrt{2}$ True
3. $2\sqrt{3} \cdot 3\sqrt{3} = 6\sqrt{3}$ False
4. $\sqrt[3]{2} \cdot \sqrt[3]{2} = 2$ False
5. $2\sqrt{5} \cdot 3\sqrt{2} = 6\sqrt{10}$ True
6. $2\sqrt{5} + 3\sqrt{5} = 5\sqrt{10}$ False
7. $\sqrt{2}(\sqrt{3} - \sqrt{2}) = \sqrt{6} - 2$ True
8. $\sqrt{12} = 2\sqrt{6}$ False
9. $(\sqrt{2} + \sqrt{3})^2 = 2 + 3$ False
10. $(\sqrt{3} - \sqrt{2})(\sqrt{3} + \sqrt{2}) = 1$ True

7.3   EXERCISES

Reading and Writing   *After reading this section, write out the answers to these questions. Use complete sentences.*

1. What are like radicals?

 Like radicals are radicals with the same index and the same radicand.

2. How do we combine like radicals?

 Like radicals are combined using the distributive property just as we combine like terms.

3. Does the product rule allow multiplication of unlike radicals?

 In the product rule the radicals must have the same index but do not have to have the same radicand.

4. How do we multiply radicals of different indices?

 To multiply radicals of different indices, we convert them to equivalent radicals with the same index.

All variables in the following exercises represent positive numbers.

Simplify the sums and differences. Give exact answers. See Example 1.

5. $\sqrt{3} - 2\sqrt{3}$ $-\sqrt{3}$
6. $\sqrt{5} - 3\sqrt{5}$ $-2\sqrt{5}$
7. $5\sqrt{7x} + 4\sqrt{7x}$ $9\sqrt{7x}$
8. $3\sqrt{6a} + 7\sqrt{6a}$ $10\sqrt{6a}$
9. $2\sqrt[3]{2} + 3\sqrt[3]{2}$ $5\sqrt[3]{2}$
10. $\sqrt[3]{4} + 4\sqrt[3]{4}$ $5\sqrt[3]{4}$
11. $\sqrt{3} - \sqrt{5} + 3\sqrt{3} - \sqrt{5}$ $4\sqrt{3} - 2\sqrt{5}$
12. $\sqrt{2} - 5\sqrt{3} - 7\sqrt{2} + 9\sqrt{3}$ $-6\sqrt{2} + 4\sqrt{3}$
13. $\sqrt[3]{2} + \sqrt[3]{x} - \sqrt[3]{2} + 4\sqrt[3]{x}$ $5\sqrt[3]{x}$
14. $\sqrt[3]{5y} - 4\sqrt[3]{5y} + \sqrt[3]{x} + \sqrt[3]{x}$ $-3\sqrt[3]{5y} + 2\sqrt[3]{x}$
15. $\sqrt[3]{x} - \sqrt{2x} + \sqrt[3]{x}$ $2\sqrt[3]{x} - \sqrt{2x}$
16. $\sqrt[3]{ab} + \sqrt{a} + 5\sqrt{a} + \sqrt[3]{ab}$ $2\sqrt[3]{ab} + 6\sqrt{a}$

Simplify each expression. Give exact answers. See Example 2.

17. $\sqrt{8} + \sqrt{28}$ $2\sqrt{2} + 2\sqrt{7}$
18. $\sqrt{12} + \sqrt{24}$ $2\sqrt{3} + 2\sqrt{6}$
19. $\sqrt{2} - \sqrt{8}$ $-\sqrt{2}$
20. $\sqrt{20} - \sqrt{125}$ $-3\sqrt{5}$
21. $\dfrac{\sqrt{2}}{2} + \sqrt{2}$ $\dfrac{3\sqrt{2}}{2}$
22. $\dfrac{\sqrt{3}}{3} - \sqrt{3}$ $-\dfrac{2\sqrt{3}}{3}$
23. $\sqrt{80} + \sqrt{\dfrac{1}{5}}$ $\dfrac{21\sqrt{5}}{5}$

24. $\sqrt{32} + \sqrt{\dfrac{1}{2}}$ $\dfrac{9\sqrt{2}}{2}$

25. $\sqrt{45x^3} - \sqrt{18x^2} + \sqrt{50x^2} - \sqrt{20x^3}$ $x\sqrt{5x} + 2x\sqrt{2}$

26. $\sqrt{12x^5} - \sqrt{18x} - \sqrt{300x^5} + \sqrt{98x}$ $4\sqrt{2x} - 8x^2\sqrt{3x}$

27. $\sqrt[3]{24} + \sqrt[3]{81}$ $5\sqrt[3]{3}$

28. $\sqrt[3]{24} + \sqrt[3]{375}$ $7\sqrt[3]{3}$

29. $\sqrt[4]{48} - \sqrt[4]{243}$ $-\sqrt[4]{3}$

30. $\sqrt[5]{64} + \sqrt[5]{2}$ $3\sqrt[5]{2}$

31. $\sqrt[3]{54t^4y^3} - \sqrt[3]{16t^4y^3}$ $ty\sqrt[3]{2t}$

32. $\sqrt[3]{2000w^2z^5} - \sqrt[3]{16w^2z^5}$ $8z\sqrt[3]{2w^2z^2}$

Simplify the products. Give exact answers. See Examples 3 and 4.

33. $\sqrt{3} \cdot \sqrt{5}$ $\sqrt{15}$

34. $\sqrt{5} \cdot \sqrt{7}$ $\sqrt{35}$

35. $2\sqrt{5} \cdot 3\sqrt{10}$ $30\sqrt{2}$

36. $(3\sqrt{2})(-4\sqrt{10})$ $-24\sqrt{5}$

37. $2\sqrt{7a} \cdot 3\sqrt{2a}$ $6a\sqrt{14}$

38. $2\sqrt{5c} \cdot 5\sqrt{5}$ $50\sqrt{c}$

39. $\sqrt[4]{9} \cdot \sqrt[4]{27}$ $3\sqrt[4]{3}$

40. $\sqrt[3]{5} \cdot \sqrt[3]{100}$ $5\sqrt[3]{4}$

41. $(2\sqrt{3})^2$ 12

42. $(-4\sqrt{2})^2$ 32

43. $\sqrt[3]{\dfrac{4x^2}{3}} \cdot \sqrt[3]{\dfrac{2x^2}{3}}$ $\dfrac{2x\sqrt[3]{3x}}{3}$

44. $\sqrt[4]{\dfrac{4x^2}{5}} \cdot \sqrt[4]{\dfrac{4x^3}{25}}$ $\dfrac{2x\sqrt[4]{5x}}{5}$

45. $2\sqrt{3}(\sqrt{6} + 3\sqrt{3})$ $6\sqrt{2} + 18$

46. $2\sqrt{5}(\sqrt{3} + 3\sqrt{5})$ $2\sqrt{15} + 30$

47. $\sqrt{5}(\sqrt{10} - 2)$ $5\sqrt{2} - 2\sqrt{5}$

48. $\sqrt{6}(\sqrt{15} - 1)$ $3\sqrt{10} - \sqrt{6}$

49. $\sqrt[3]{3t}(\sqrt[3]{9t} - \sqrt[3]{t^2})$ $3\sqrt[3]{t^2} - t\sqrt[3]{3}$

50. $\sqrt[3]{2}(\sqrt[3]{12x} - \sqrt[3]{2x})$ $2\sqrt[3]{3x} - \sqrt[3]{4x}$

51. $(\sqrt{3} + 2)(\sqrt{3} - 5)$ $-7 - 3\sqrt{3}$

52. $(\sqrt{5} + 2)(\sqrt{5} - 6)$ $-7 - 4\sqrt{5}$

53. $(\sqrt{11} - 3)(\sqrt{11} + 3)$ 2

54. $(\sqrt{2} + 5)(\sqrt{2} + 5)$ $27 + 10\sqrt{2}$

55. $(2\sqrt{5} - 7)(2\sqrt{5} + 4)$ $-8 - 6\sqrt{5}$

56. $(2\sqrt{6} - 3)(2\sqrt{6} + 4)$ $12 + 2\sqrt{6}$

57. $(2\sqrt{3} - \sqrt{6})(\sqrt{3} + 2\sqrt{6})$ $-6 + 9\sqrt{2}$

58. $(3\sqrt{3} - \sqrt{2})(\sqrt{2} + \sqrt{3})$ $2\sqrt{6} + 7$

Write each product as a single radical expression. See Example 5.

59. $\sqrt[3]{3} \cdot \sqrt{3}$ $\sqrt[6]{3^5}$

60. $\sqrt{3} \cdot \sqrt[4]{3}$ $\sqrt[4]{27}$

61. $\sqrt[3]{5} \cdot \sqrt[4]{5}$ $\sqrt[12]{5^7}$

62. $\sqrt[3]{2} \cdot \sqrt[5]{2}$ $\sqrt[15]{2^8}$

63. $\sqrt[3]{2} \cdot \sqrt{5}$ $\sqrt[6]{500}$

64. $\sqrt{6} \cdot \sqrt[3]{2}$ $\sqrt[6]{864}$

65. $\sqrt[3]{2} \cdot \sqrt[4]{3}$ $\sqrt[12]{432}$

66. $\sqrt[3]{3} \cdot \sqrt[4]{2}$ $\sqrt[12]{648}$

Find the product of each pair of conjugates. See Example 6.

67. $(\sqrt{3} - 2)(\sqrt{3} + 2)$ -1

68. $(7 - \sqrt{3})(7 + \sqrt{3})$ 46

69. $(\sqrt{5} + \sqrt{2})(\sqrt{5} - \sqrt{2})$ 3

70. $(\sqrt{6} + \sqrt{5})(\sqrt{6} - \sqrt{5})$ 1

71. $(2\sqrt{5} + 1)(2\sqrt{5} - 1)$ 19

72. $(3\sqrt{2} - 4)(3\sqrt{2} + 4)$ 2

73. $(3\sqrt{2} + \sqrt{5})(3\sqrt{2} - \sqrt{5})$ 13

74. $(2\sqrt{3} - \sqrt{7})(2\sqrt{3} + \sqrt{7})$ 5

75. $(5 - 3\sqrt{x})(5 + 3\sqrt{x})$ $25 - 9x$

76. $(4\sqrt{y} + 3\sqrt{z})(4\sqrt{y} - 3\sqrt{z})$ $16y - 9z$

Simplify each expression.

77. $\sqrt{300} + \sqrt{3}$ $11\sqrt{3}$

78. $\sqrt{50} + \sqrt{2}$ $6\sqrt{2}$

79. $2\sqrt{5} \cdot 5\sqrt{6}$ $10\sqrt{30}$

80. $3\sqrt{6} \cdot 5\sqrt{10}$ $30\sqrt{15}$

81. $(3 + 2\sqrt{7})(\sqrt{7} - 2)$ $8 - \sqrt{7}$

82. $(2 + \sqrt{7})(\sqrt{7} - 2)$ 3

83. $4\sqrt{w} \cdot 4\sqrt{w}$ $16w$

84. $3\sqrt{m} \cdot 5\sqrt{m}$ $15m$

85. $\sqrt{3x^3} \cdot \sqrt{6x^2}$ $3x^2\sqrt{2x}$

86. $\sqrt{2t^5} \cdot \sqrt{10t^4}$ $2t^4\sqrt{5t}$

87. $\dfrac{1}{\sqrt{2}} - \dfrac{1}{\sqrt{8}} + \dfrac{1}{\sqrt{18}}$ $\dfrac{5\sqrt{2}}{12}$

88. $\dfrac{1}{\sqrt{3}} + \sqrt{\dfrac{1}{3}} - \sqrt{3}$ $\dfrac{-\sqrt{3}}{3}$

89. $(2\sqrt{5} + \sqrt{2})(3\sqrt{5} - \sqrt{2})$ $28 + \sqrt{10}$

90. $(3\sqrt{2} - \sqrt{3})(2\sqrt{2} + 3\sqrt{3})$ $3 + 7\sqrt{6}$

91. $\dfrac{\sqrt{2}}{3} + \dfrac{\sqrt{2}}{5}$ $\dfrac{8\sqrt{2}}{15}$

92. $\dfrac{\sqrt{2}}{4} + \dfrac{\sqrt{3}}{5}$ $\dfrac{5\sqrt{2} + 4\sqrt{3}}{20}$

93. $(5 + 2\sqrt{2})(5 - 2\sqrt{2})$ 17

94. $(3 - 2\sqrt{7})(3 + 2\sqrt{7})$ -19

95. $(3 + \sqrt{x})^2$ $9 + 6\sqrt{x} + x$

96. $(1 - \sqrt{x})^2$ $1 - 2\sqrt{x} + x$

97. $(5\sqrt{x} - 3)^2$ $25x - 30\sqrt{x} + 9$

98. $(3\sqrt{a} + 2)^2$ $9a + 12\sqrt{a} + 4$

99. $(1 + \sqrt{x + 2})^2$ $x + 3 + 2\sqrt{x + 2}$

100. $(\sqrt{x - 1} + 1)^2$ $x + 2\sqrt{x - 1}$

101. $\sqrt{4w} - \sqrt{9w}$ $-\sqrt{w}$

102. $10\sqrt{m} - \sqrt{16m}$ $6\sqrt{m}$

103. $2\sqrt{a^3} + 3\sqrt{a^3} - 2a\sqrt{4a}$ $a\sqrt{a}$

104. $5\sqrt{w^2y} - 7\sqrt{w^2y} + 6\sqrt{w^2y}$ $4w\sqrt{y}$

105. $\sqrt{50a} + \sqrt{18a} - \sqrt{2a}$ $7\sqrt{2a}$

106. $\sqrt{200z} + \sqrt{128z} - \sqrt{8z}$ $16\sqrt{2z}$

107. $\sqrt{x^5} + 2x\sqrt{x^3}$ $3x^2\sqrt{x}$

108. $\sqrt{8x^3} + \sqrt{50x^3} - x\sqrt{2x}$ $6x\sqrt{2x}$

109. $(\sqrt{a} + a^3)(\sqrt{a} - a^3)$ $a - a^6$

110. $(\sqrt{wz} - 2y^4)(\sqrt{wz} + 2y^4)$ $wz - 4y^8$

111. $\sqrt[3]{-16x^4} + 5x\sqrt[3]{54x}$ $13x\sqrt[3]{2x}$

112. $\sqrt[3]{3x^5y^7} - \sqrt[3]{24x^5y^7}$ $-xy^2\sqrt[3]{3x^2y}$

113. $\sqrt[3]{\dfrac{y^7}{4x}}$ $\dfrac{y^2\sqrt[3]{2x^2y}}{2x}$ **114.** $\sqrt[4]{\dfrac{16}{9z^3}}$ $\dfrac{2\sqrt[4]{9z}}{3z}$

115. $\sqrt[3]{\dfrac{x}{5}} \cdot \sqrt[3]{\dfrac{x^5}{5}}$ $\dfrac{x^2\sqrt[3]{5}}{5}$ **116.** $\sqrt[4]{a^3}(\sqrt[4]{a} - \sqrt[4]{a^5})$ $a - a^2$

117. $\sqrt[3]{2x} \cdot \sqrt[6]{2x}$ $\sqrt[6]{32x^5}$ **118.** $\sqrt[3]{2m} \cdot \sqrt[4]{2n}$ $\sqrt[12]{128m^4n^3}$

In Exercises 119–122, solve each problem.

119. ***Area of a rectangle.*** Find the exact area of a rectangle that has a length of $\sqrt{6}$ feet and a width of $\sqrt{3}$ feet.
$3\sqrt{2}$ square feet (ft^2)

120. ***Volume of a cube.*** Find the exact volume of a cube with sides of length $\sqrt{3}$ meters.
$3\sqrt{3}$ cubic meters (m^3)

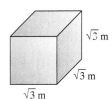

$\sqrt{3}$ m

$\sqrt{3}$ m

$\sqrt{3}$ m

FIGURE FOR EXERCISE 120

121. ***Area of a trapezoid.*** Find the exact area of a trapezoid with a height of $\sqrt{6}$ feet and bases of $\sqrt{3}$ feet and $\sqrt{12}$ feet.
$\dfrac{9\sqrt{2}}{2}$ ft^2

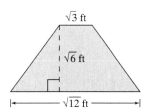

$\sqrt{3}$ ft

$\sqrt{6}$ ft

$\sqrt{12}$ ft

FIGURE FOR EXERCISE 121

122. ***Area of a triangle.*** Find the exact area of a triangle with a base of $\sqrt{30}$ meters and a height of $\sqrt{6}$ meters.
$3\sqrt{5}$ square meters (m^2)

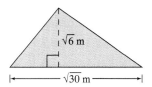

$\sqrt{6}$ m

$\sqrt{30}$ m

FIGURE FOR EXERCISE 122

GETTING MORE INVOLVED

123. ***Discussion.*** Is $\sqrt{a} + \sqrt{b} = \sqrt{a + b}$ for all values of a and b?
No

124. ***Discussion.*** Which of the following equations are identities? Explain your answers.
a) $\sqrt{9x} = 3\sqrt{x}$
b) $\sqrt{9 + x} = 3 + \sqrt{x}$
c) $\sqrt{x - 4} = \sqrt{x} - 2$
d) $\sqrt{\dfrac{x}{4}} = \dfrac{\sqrt{x}}{2}$
a and d

125. ***Exploration.*** Because 3 is the square of $\sqrt{3}$, a binomial such as $y^2 - 3$ is a difference of two squares.
a) Factor $y^2 - 3$ and $2a^2 - 7$ using radicals.
b) Use factoring with radicals to solve the equations $x^2 - 8 = 0$ and $3y^2 - 11 = 0$.
c) Assuming a and b are positive real numbers, solve the equations $x^2 - a = 0$ and $ax^2 - b = 0$.

 7.4   **MORE OPERATIONS WITH RADICALS**

<table>
<tr><td>

In this

section

</td><td>

In this section you will continue studying operations with radicals. We learn to rationalize some denominators that are different from those rationalized in Section 7.2.

</td></tr>
</table>

- Dividing Radicals
- Rationalizing the Denominator
- Powers of Radical Expressions

Dividing Radicals

In Section 7.3 you learned how to add, subtract, and multiply radical expressions. To divide two radical expressions, simply write the quotient as a ratio and then simplify, as we did in Section 7.2. In general, we have

$$\sqrt[n]{a} \div \sqrt[n]{b} = \frac{\sqrt[n]{a}}{\sqrt[n]{b}} = \sqrt[n]{\frac{a}{b}},$$

provided that all expressions represent real numbers. Note that the quotient rule is applied only to radicals that have the same index.

E X A M P L E   1

Dividing radicals with the same index

Divide and simplify. Assume the variables represent positive numbers.

a) $\sqrt{10} \div \sqrt{5}$

b) $(3\sqrt{2}) \div (2\sqrt{3})$

c) $\sqrt[3]{10x^2} \div \sqrt[3]{5x}$

Solution

a) $\sqrt{10} \div \sqrt{5} = \dfrac{\sqrt{10}}{\sqrt{5}}$ $a \div b = \dfrac{a}{b}$, provided that $b \neq 0$.

$\qquad\qquad\quad = \sqrt{\dfrac{10}{5}}$ Quotient rule for radicals

$\qquad\qquad\quad = \sqrt{2}$ Reduce.

b) $(3\sqrt{2}) \div (2\sqrt{3}) = \dfrac{3\sqrt{2}}{2\sqrt{3}}$

$\qquad\qquad\qquad\quad = \dfrac{3\sqrt{2}}{2\sqrt{3}} \cdot \dfrac{\sqrt{3}}{\sqrt{3}}$ Rationalize the denominator.

$\qquad\qquad\qquad\quad = \dfrac{3\sqrt{6}}{2 \cdot 3}$

$\qquad\qquad\qquad\quad = \dfrac{\sqrt{6}}{2}$ Note that $\sqrt{6} \div 2 \neq \sqrt{3}$.

c) $\sqrt[3]{10x^2} \div \sqrt[3]{5x} = \dfrac{\sqrt[3]{10x^2}}{\sqrt[3]{5x}}$

$\qquad\qquad\qquad\quad = \sqrt[3]{\dfrac{10x^2}{5x}}$ Quotient rule for radicals

$\qquad\qquad\qquad\quad = \sqrt[3]{2x}$ Reduce.

Note that in Example 1(a) we applied the quotient rule to get $\sqrt{10} \div \sqrt{5} = \sqrt{2}$. In Example 1(b) we did not use the quotient rule because 2 is not evenly divisible by 3. Instead, we rationalized the denominator to get the result in simplified form.

In Chapter 8 it will be necessary to simplify expressions of the type found in the next example.

E X A M P L E   2

Simplifying radical expressions

Simplify.

a) $\dfrac{4 - \sqrt{12}}{4}$

b) $\dfrac{-6 + \sqrt{20}}{-2}$

Solution

a) First write $\sqrt{12}$ in simplified form. Then simplify the expression.

$$\frac{4 - \sqrt{12}}{4} = \frac{4 - 2\sqrt{3}}{4} \qquad \text{Simplify } \sqrt{12}.$$

$$= \frac{\cancel{2}(2 - \sqrt{3})}{\cancel{2} \cdot 2} \qquad \text{Factor.}$$

$$= \frac{2 - \sqrt{3}}{2} \qquad \text{Divide out the common factor.}$$

b) $\dfrac{-6 + \sqrt{20}}{-2} = \dfrac{-6 + 2\sqrt{5}}{-2}$

$$= \frac{-\cancel{2}(3 - \sqrt{5})}{-\cancel{2}}$$

$$= 3 - \sqrt{5}$$

■

CAUTION   To simplify the expressions in Example 2, you must simplify the radical, factor the numerator, and then divide out the common factors. You cannot simply "cancel" the 4's in $\frac{4 - \sqrt{12}}{4}$ or the 2's in $\frac{2 - \sqrt{3}}{2}$ because they are not common factors.

Rationalizing the Denominator

In Section 7.2 you learned that a simplified expression involving radicals does not have radicals in the denominator. If an expression such as $4 - \sqrt{3}$ appears in a denominator, we can multiply both the numerator and denominator by its conjugate $4 + \sqrt{3}$ to get a rational number in the denominator.

E X A M P L E   3

Rationalizing the denominator using conjugates

Write in simplified form.

a) $\dfrac{2 + \sqrt{3}}{4 - \sqrt{3}}$

b) $\dfrac{\sqrt{5}}{\sqrt{6} + \sqrt{2}}$

Solution

a) $\dfrac{2 + \sqrt{3}}{4 - \sqrt{3}} = \dfrac{(2 + \sqrt{3})(4 + \sqrt{3})}{(4 - \sqrt{3})(4 + \sqrt{3})}$ Multiply the numerator and denominator by $4 + \sqrt{3}$.

$$= \frac{8 + 6\sqrt{3} + 3}{13} \qquad (4 - \sqrt{3})(4 + \sqrt{3}) = 16 - 3 = 13$$

$$= \frac{11 + 6\sqrt{3}}{13} \qquad \text{Simplify.}$$

b) $\dfrac{\sqrt{5}}{\sqrt{6} + \sqrt{2}} = \dfrac{\sqrt{5}(\sqrt{6} - \sqrt{2})}{(\sqrt{6} + \sqrt{2})(\sqrt{6} - \sqrt{2})}$ Multiply the numerator and denominator by $\sqrt{6} - \sqrt{2}$.

$$= \frac{\sqrt{30} - \sqrt{10}}{4} \qquad (\sqrt{6} + \sqrt{2})(\sqrt{6} - \sqrt{2}) = 6 - 2 = 4$$

■

Powers of Radical Expressions

We can use the power of a product rule and the power of a power rule to simplify a radical expression raised to a power. In the next example we also use the fact that a root and a power can be found in either order.

E X A M P L E   4   **Finding powers of rational expressions**

Simplify. Assume the variables represent positive numbers.

a) $(5\sqrt{2})^3$

b) $(2\sqrt{x^3})^4$

c) $(3w\sqrt[3]{2w})^3$

d) $(2t\sqrt[4]{3t})^3$

Solution

a) $(5\sqrt{2})^3 = 5^3(\sqrt{2})^3$ Power of a product rule

           $= 125\sqrt{8}$ $(\sqrt{2})^3 = \sqrt{2^3} = \sqrt{8}$

           $= 125 \cdot 2\sqrt{2}$ $\sqrt{8} = \sqrt{4}\sqrt{2} = 2\sqrt{2}$

           $= 250\sqrt{2}$

b) $(2\sqrt{x^3})^4 = 2^4(\sqrt{x^3})^4$

           $= 16\sqrt{x^{12}}$

           $= 16x^6$

c) $(3w\sqrt[3]{2w})^3 = 3^3w^3(\sqrt[3]{2w})^3$

             $= 27w^3(2w)$

             $= 54w^4$

d) $(2t\sqrt[4]{3t})^3 = 2^3t^3(\sqrt[4]{3t})^3 = 8t^3\sqrt[4]{27t^3}$

WARM-UPS

True or false? Explain.

1. $\dfrac{\sqrt{6}}{\sqrt{2}} = \sqrt{3}$

    True

2. $\dfrac{2}{\sqrt{2}} = \sqrt{2}$

    True

3. $\dfrac{4 - \sqrt{10}}{2} = 2 - \sqrt{10}$

    False

4. $\dfrac{1}{\sqrt{3}} = \dfrac{\sqrt{3}}{3}$

    True

5. $\dfrac{8\sqrt{7}}{2\sqrt{7}} = 4\sqrt{7}$

    False

6. $\dfrac{2(2 + \sqrt{3})}{(2 - \sqrt{3})(2 + \sqrt{3})} = 4 + 2\sqrt{3}$

    True

7. $\dfrac{\sqrt{12}}{3} = \sqrt{4}$

    False

8. $\dfrac{\sqrt{20}}{\sqrt{5}} = 2$

    True

9. $(2\sqrt{4})^2 = 16$

    True

10. $(3\sqrt{5})^3 = 27\sqrt{125}$

    True

7.4   EXERCISES

All variables in the following exercises represent positive numbers.

Divide and simplify. See Example 1.

1. $\sqrt{15} \div \sqrt{5}$
$\sqrt{3}$

2. $\sqrt{14} \div \sqrt{7}$
$\sqrt{2}$

3. $\sqrt{3} \div \sqrt{5}$
$\dfrac{\sqrt{15}}{5}$

4. $\sqrt{5} \div \sqrt{7}$
$\dfrac{\sqrt{35}}{7}$

5. $(3\sqrt{3}) \div (5\sqrt{6})$
$\dfrac{3\sqrt{2}}{10}$

6. $(2\sqrt{2}) \div (4\sqrt{10})$
$\dfrac{\sqrt{5}}{10}$

7. $(2\sqrt{3}) \div (3\sqrt{6})$
$\dfrac{\sqrt{2}}{3}$

8. $(5\sqrt{12}) \div (4\sqrt{6})$
$\dfrac{5\sqrt{2}}{4}$

9. $\sqrt[3]{20} \div \sqrt[3]{2}$
$\sqrt[3]{10}$

10. $\sqrt[4]{48} \div \sqrt[4]{3}$
2

11. $\sqrt[3]{8x^7} \div \sqrt[3]{2x}$
$x^2\sqrt[3]{4}$

12. $\sqrt[4]{4a^{10}} \div \sqrt[4]{2a^2}$
$a^2\sqrt[4]{2}$

Simplify. See Example 2.

13. $\dfrac{6 + \sqrt{45}}{3}$
$2 + \sqrt{5}$

14. $\dfrac{10 + \sqrt{50}}{5}$
$2 + \sqrt{2}$

15. $\dfrac{-2 + \sqrt{12}}{-2}$
$1 - \sqrt{3}$

16. $\dfrac{-6 + \sqrt{72}}{-6}$
$1 - \sqrt{2}$

Simplify each expression by rationalizing the denominator. See Example 3.

17. $\dfrac{1 + \sqrt{2}}{\sqrt{3} - 1}$
$\dfrac{1 + \sqrt{6} + \sqrt{2} + \sqrt{3}}{2}$

18. $\dfrac{2 - \sqrt{3}}{\sqrt{2} + \sqrt{6}}$
$\dfrac{3\sqrt{6} - 5\sqrt{2}}{4}$

19. $\dfrac{\sqrt{2}}{\sqrt{6} + \sqrt{3}}$
$\dfrac{2\sqrt{3} - \sqrt{6}}{3}$

20. $\dfrac{5}{\sqrt{7} - \sqrt{5}}$
$\dfrac{5\sqrt{7} + 5\sqrt{5}}{2}$

21. $\dfrac{2\sqrt{3}}{3\sqrt{2} - \sqrt{5}}$
$\dfrac{6\sqrt{6} + 2\sqrt{15}}{13}$

22. $\dfrac{3\sqrt{5}}{5\sqrt{2} + \sqrt{6}}$
$\dfrac{15\sqrt{10} - 3\sqrt{30}}{44}$

23. $\dfrac{1 + 3\sqrt{2}}{2\sqrt{6} + 3\sqrt{10}}$
$\dfrac{18\sqrt{5} + 3\sqrt{10} - 2\sqrt{6} - 12\sqrt{3}}{66}$

24. $\dfrac{3\sqrt{3} + 1}{4 - 5\sqrt{3}}$
$-\dfrac{17\sqrt{3} + 49}{59}$

Simplify. See Example 4.

25. $(2\sqrt{2})^5$ $128\sqrt{2}$

26. $(3\sqrt{3})^4$ 729

27. $(\sqrt{x})^5$ $x^2\sqrt{x}$

28. $(2\sqrt{y})^3$ $8y\sqrt{y}$

29. $(-3\sqrt{x^3})^3$ $-27x^4\sqrt{x}$

30. $(-2\sqrt{x^3})^4$ $16x^6$

31. $(2x\sqrt[3]{x^2})^3$ $8x^5$

32. $(2y\sqrt[3]{4y})^3$ $32y^4$

33. $(-2\sqrt[3]{5})^2$ $4\sqrt[3]{25}$

34. $(-3\sqrt[3]{4})^2$ $18\sqrt[3]{2}$

35. $(\sqrt[3]{x^2})^6$ x^4

36. $(2\sqrt[4]{y^3})^3$ $8y^2\sqrt[4]{y}$

In Exercises 37–74, simplify.

37. $\dfrac{\sqrt{3}}{\sqrt{2}} + \dfrac{2}{\sqrt{2}}$
$\dfrac{\sqrt{6} + 2\sqrt{2}}{2}$

38. $\dfrac{2}{\sqrt{7}} + \dfrac{5}{\sqrt{7}}$
$\sqrt{7}$

39. $\dfrac{\sqrt{3}}{\sqrt{2}} + \dfrac{3\sqrt{6}}{2}$
$2\sqrt{6}$

40. $\dfrac{\sqrt{3}}{2\sqrt{2}} + \dfrac{\sqrt{5}}{3\sqrt{2}}$
$\dfrac{3\sqrt{6} + 2\sqrt{10}}{12}$

41. $\dfrac{\sqrt{6}}{2} \cdot \dfrac{1}{\sqrt{3}}$
$\dfrac{\sqrt{2}}{2}$

42. $\dfrac{\sqrt{6}}{\sqrt{7}} \cdot \dfrac{\sqrt{14}}{\sqrt{3}}$
2

43. $(2\sqrt{w}) \div (3\sqrt{w})$
$\dfrac{2}{3}$

44. $2 \div (3\sqrt{a})$
$\dfrac{2\sqrt{a}}{3a}$

45. $\dfrac{8 - \sqrt{32}}{20}$
$\dfrac{2 - \sqrt{2}}{5}$

46. $\dfrac{4 - \sqrt{28}}{6}$
$\dfrac{2 - \sqrt{7}}{3}$

47. $\dfrac{5 + \sqrt{75}}{10}$
$\dfrac{1 + \sqrt{3}}{2}$

48. $\dfrac{3 + \sqrt{18}}{6}$
$\dfrac{1 + \sqrt{2}}{2}$

49. $\sqrt{a}(\sqrt{a} - 3)$
$a - 3\sqrt{a}$

50. $3\sqrt{m}(2\sqrt{m} - 6)$
$6m - 18\sqrt{m}$

51. $4\sqrt{a}(a + \sqrt{a})$
$4a\sqrt{a} + 4a$

52. $\sqrt{3ab}(\sqrt{3a} + \sqrt{3})$
$3a\sqrt{b} + 3\sqrt{ab}$

53. $(2\sqrt{3m})^2$
$12m$

54. $(-3\sqrt{4y})^2$
$36y$

55. $(-2\sqrt{xy^2z})^2$
$4xy^2z$

56. $(5a\sqrt{ab})^2$
$25a^3b$

57. $\sqrt[3]{m}(\sqrt[3]{m^2} - \sqrt[3]{m^5})$
$m - m^2$

58. $\sqrt[4]{w}(\sqrt[4]{w^3} - \sqrt[4]{w^7})$
$w - w^2$

59. $\sqrt[3]{8x^4} + \sqrt[3]{27x^4}$
$5x\sqrt[3]{x}$

60. $\sqrt[3]{16a^4} + a\sqrt[3]{2a}$
$3a\sqrt[3]{2a}$

61. $(2m\sqrt[4]{2m^2})^3$
$8m^4\sqrt[4]{8m^2}$

62. $(-2t\sqrt[6]{2t^2})^5$
$-32t^6\sqrt[6]{32t^4}$

63. $\dfrac{4}{2 + \sqrt{8}}$
$2\sqrt{2} - 2$

64. $\dfrac{6}{3 - \sqrt{18}}$
$-2 - 2\sqrt{2}$

65. $\dfrac{5}{\sqrt{2} - 1} + \dfrac{3}{\sqrt{2} + 1}$
$2 + 8\sqrt{2}$

66. $\dfrac{\sqrt{3}}{\sqrt{6} - 1} - \dfrac{\sqrt{3}}{\sqrt{6} + 1}$
$\dfrac{2\sqrt{3}}{5}$

67. $\dfrac{1}{\sqrt{2}} + \dfrac{1}{\sqrt{3}}$

$\dfrac{3\sqrt{2} + 2\sqrt{3}}{6}$

68. $\dfrac{4}{2\sqrt{3}} + \dfrac{1}{\sqrt{5}}$

$\dfrac{10\sqrt{3} + 3\sqrt{5}}{15}$

69. $\dfrac{3}{\sqrt{2} - 1} + \dfrac{4}{\sqrt{2} + 1}$

$7\sqrt{2} - 1$

70. $\dfrac{3}{\sqrt{5} - \sqrt{3}} - \dfrac{2}{\sqrt{5} + \sqrt{3}}$

$\dfrac{\sqrt{5} + 5\sqrt{3}}{2}$

71. $\dfrac{\sqrt{x}}{\sqrt{x} + 2} + \dfrac{3\sqrt{x}}{\sqrt{x} - 2}$

$\dfrac{4x + 4\sqrt{x}}{x - 4}$

72. $\dfrac{\sqrt{5}}{3 - \sqrt{y}} - \dfrac{\sqrt{5y}}{3 + \sqrt{y}}$

$\dfrac{3\sqrt{5} + y\sqrt{5} - 2\sqrt{5y}}{9 - y}$

73. $\dfrac{1}{\sqrt{x}} + \dfrac{1}{1 - \sqrt{x}}$

$\dfrac{x + \sqrt{x}}{x - x^2}$

74. $\dfrac{\sqrt{x}}{\sqrt{x} - 3} + \dfrac{5}{\sqrt{x}}$

$\dfrac{x^2 + 8x\sqrt{x} - 45\sqrt{x}}{x^2 - 9x}$

Replace the question mark by an expression that makes the equation correct. Equations involving variables are to be identities.

75. $\dfrac{\sqrt{2}}{\sqrt{3}} = \dfrac{\sqrt{6}}{?}$

$\dfrac{\sqrt{6}}{3}$

76. $\dfrac{2}{?} = \sqrt{2}$

$\dfrac{2}{\sqrt{2}}$

77. $\dfrac{1}{\sqrt{2} - 1} = \dfrac{\sqrt{2} + 1}{?}$

$\dfrac{\sqrt{2} + 1}{1}$

78. $\dfrac{\sqrt{6}}{\sqrt{6} + 2} = \dfrac{?}{2}$

$\dfrac{6 - 2\sqrt{6}}{2}$

79. $\dfrac{1}{\sqrt{x} - 1} = \dfrac{?}{x - 1}$

$\dfrac{\sqrt{x} + 1}{x - 1}$

80. $\dfrac{5}{3 - \sqrt{x}} = \dfrac{?}{9 - x}$

$\dfrac{15 + 5\sqrt{x}}{9 - x}$

81. $\dfrac{3}{\sqrt{2} + x} = \dfrac{?}{2 - x^2}$

$\dfrac{3\sqrt{2} - 3x}{2 - x^2}$

82. $\dfrac{4}{2\sqrt{3} + a} = \dfrac{?}{12 - a^2}$

$\dfrac{8\sqrt{3} - 4a}{12 - a^2}$

Use a calculator to find a decimal approximation for each radical expression. Round your answers to three decimal places.

83. $\sqrt{3} + \sqrt{5}$
3.968

84. $\sqrt{5} + \sqrt{7}$
4.882

85. $2\sqrt{3} + 5\sqrt{3}$
12.124

86. $7\sqrt{3}$
12.124

87. $(2\sqrt{3})(3\sqrt{2})$
14.697

88. $6\sqrt{6}$
14.697

89. $\sqrt{5}(\sqrt{5} + \sqrt{3})$
8.873

90. $5 + \sqrt{15}$
8.873

91. $\dfrac{-1 + \sqrt{6}}{2}$
0.725

92. $\dfrac{-1 - \sqrt{6}}{2}$
-1.725

93. $\dfrac{4 - \sqrt{10}}{-2}$
-0.419

94. $\dfrac{4 + \sqrt{10}}{-2}$
-3.581

GETTING MORE INVOLVED

95. Exploration. A polynomial is prime if it cannot be factored by using integers, but many prime polynomials can be factored if we use radicals.

a) Find the product $(x - \sqrt[3]{2})(x^2 + \sqrt[3]{2}x + \sqrt[3]{4})$.

b) Factor $x^3 + 5$ using radicals.

c) Find the product
$$(\sqrt[3]{5} - \sqrt[3]{2})(\sqrt[3]{25} + \sqrt[3]{10} + \sqrt[3]{4}).$$

d) Use radicals to factor $a + b$ as a sum of two cubes and $a - b$ as a difference of two cubes.

  a) $x^3 - 2$ b) $(x + \sqrt[3]{5})(x^2 - \sqrt[3]{5}x + \sqrt[3]{25})$ c) 3
  d) $(\sqrt[3]{a} + \sqrt[3]{b})(\sqrt[3]{a^2} - \sqrt[3]{ab} + \sqrt[3]{b^2})$
     $(\sqrt[3]{a} - \sqrt[3]{b})(\sqrt[3]{a^2} + \sqrt[3]{ab} + \sqrt[3]{b^2})$

96. Discussion. Which one of the following expressions is not equivalent to the others?

a) $(\sqrt[3]{x})^4$ b) $\sqrt[4]{x^3}$ c) $\sqrt[3]{x^4}$

d) $x^{4/3}$ e) $(x^{1/3})^4$

b

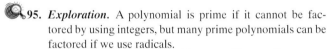

In this

section

- The Odd-Root Property
- The Even-Root Property
- Raising Each Side to a Power
- Equations Involving Rational Exponents
- Summary of Methods
- The Distance Formula

7.5 SOLVING EQUATIONS WITH RADICALS AND EXPONENTS

One of our goals in algebra is to keep increasing our knowledge of solving equations because the solutions to equations can give us the answers to various applied questions. In this section we will apply our knowledge of radicals and exponents to solving some new types of equations.

The Odd-Root Property

Because $(-2)^3 = -8$ and $2^3 = 8$, the equation $x^3 = 8$ is equivalent to $x = 2$. The equation $x^3 = -8$ is equivalent to $x = -2$. Because there is only one real odd root of each real number, there is a simple rule for writing an equivalent equation in this situation.

Odd-Root Property

If n is an odd positive integer,

$$x^n = k \quad \text{is equivalent to} \quad x = \sqrt[n]{k}$$

for any real number k.

E X A M P L E   1

Using the odd-root property

Solve each equation.

a) $x^3 = 27$

b) $x^5 + 32 = 0$

c) $(x - 2)^3 = 24$

Solution

a) $x^3 = 27$

$\quad x = \sqrt[3]{27}$ Odd-root property

$\quad x = 3$

Check 3 in the original equation. The solution set is $\{3\}$.

b) $x^5 + 32 = 0$

$\quad x^5 = -32$ Isolate the variable.

$\quad x = \sqrt[5]{-32}$ Odd-root property

$\quad x = -2$

Check -2 in the original equation. The solution set is $\{-2\}$.

c) $(x - 2)^3 = 24$

$\quad x - 2 = \sqrt[3]{24}$ Odd-root property

$\quad x = 2 + 2\sqrt[3]{3}$ $\sqrt[3]{24} = \sqrt[3]{8} \cdot \sqrt[3]{3} = 2\sqrt[3]{3}$

Check. The solution set is $\{2 + 2\sqrt[3]{3}\}$.   ■

The Even-Root Property

In solving the equation $x^2 = 4$, you might be tempted to write $x = 2$ as an equivalent equation. But $x = 2$ is not equivalent to $x^2 = 4$ because $2^2 = 4$ and $(-2)^2 = 4$. So the solution set to $x^2 = 4$ is $\{-2, 2\}$. The equation $x^2 = 4$ is equivalent to the compound sentence $x = 2$ or $x = -2$, which we can abbreviate as $x = \pm 2$. The equation $x = \pm 2$ is read "x equals positive or negative 2."

Equations involving other even powers are handled like the squares. Because $2^4 = 16$ and $(-2)^4 = 16$, the equation $x^4 = 16$ is equivalent to $x = \pm 2$. So $x^4 = 16$ has two real solutions. Note that $x^4 = -16$ has no real solutions. The equation $x^6 = 5$ is equivalent to $x = \pm\sqrt[6]{5}$. We can now state a general rule.

Even-Root Property

Suppose n is a positive even integer.

If $k > 0$, then $x^n = k$ is equivalent to $x = \pm\sqrt[n]{k}$.

If $k = 0$, then $x^n = k$ is equivalent to $x = 0$.

If $k < 0$, then $x^n = k$ has no real solution.

E X A M P L E   2

Using the even-root property

Solve each equation.

a) $x^2 = 10$ b) $w^8 = 0$ c) $x^4 = -4$

Solution

a) $x^2 = 10$

    $x = \pm\sqrt{10}$ Even-root property

    The solution set is $\{-\sqrt{10}, \sqrt{10}\}$, or $\{\pm\sqrt{10}\}$.

b) $w^8 = 0$

    $w = 0$ Even-root property

    The solution set is $\{0\}$.

c) By the even-root property, $x^4 = -4$ has no real solution. (The fourth power of any real number is nonnegative.)

    In the next example the even-root property is used to solve some equations that are a bit more complicated than those of Example 2.

E X A M P L E   3

Using the even-root property

Solve each equation.

a) $(x - 3)^2 = 4$ b) $2(x - 5)^2 - 7 = 0$ c) $x^4 - 1 = 80$

Solution

a) $(x - 3)^2 = 4$

    $x - 3 = 2$ or $x - 3 = -2$ Even-root property

       $x = 5$ or $x = 1$ Add 3 to each side.

    The solution set is $\{1, 5\}$.

b) $2(x - 5)^2 - 7 = 0$

    $2(x - 5)^2 = 7$ Add 7 to each side.

    $(x - 5)^2 = \dfrac{7}{2}$ Divide each side by 2.

    $x - 5 = \sqrt{\dfrac{7}{2}}$ or $x - 5 = -\sqrt{\dfrac{7}{2}}$ Even-root property

    $x = 5 + \dfrac{\sqrt{14}}{2}$ or $x = 5 - \dfrac{\sqrt{14}}{2}$ $\sqrt{\dfrac{7}{2}} = \dfrac{\sqrt{7} \cdot \sqrt{2}}{\sqrt{2} \cdot \sqrt{2}} = \dfrac{\sqrt{14}}{2}$

    $x = \dfrac{10 + \sqrt{14}}{2}$ or $x = \dfrac{10 - \sqrt{14}}{2}$

    The solution set is $\left\{\dfrac{10 + \sqrt{14}}{2}, \dfrac{10 - \sqrt{14}}{2}\right\}$.

c) $x^4 - 1 = 80$

    $x^4 = 81$

    $x = \pm\sqrt[4]{81} = \pm 3$

    The solution set is $\{-3, 3\}$.

> **study tip**
>
> Review, review, review! Don't wait until the end of a chapter to review. Do a little review every time you study for this course.

    In Chapter 5 we solved quadratic equations by factoring. The quadratic equations that we encounter in this chapter can be solved by using the even-root property as in parts (a) and (b) of Example 3. In Chapter 8 you will learn general methods for solving any quadratic equation.

Raising Each Side to a Power

If we start with the equation $x = 3$ and square both sides, we get $x^2 = 9$. The solution set to $x^2 = 9$ is $\{-3, 3\}$; the solution set to the original equation is $\{3\}$. Squaring both sides of an equation might produce a *nonequivalent* equation that has more solutions than the original equation. We call these additional solutions **extraneous solutions.** However, any solution of the original must be among the solutions to the new equation.

> **CAUTION**   When you solve an equation by raising each side to a power, you must check your answers. Raising each side to an odd power will always give an equivalent equation; raising each side to an even power might not.

EXAMPLE   4

Raising each side to a power to eliminate radicals

Solve each equation.

a) $\sqrt{2x - 3} - 5 = 0$ b) $\sqrt[3]{3x + 5} = \sqrt[3]{x - 1}$ c) $\sqrt{3x + 18} = x$

Solution

a) Eliminate the square root by raising each side to the power 2:

$$\sqrt{2x - 3} - 5 = 0 \qquad \text{Original equation}$$
$$\sqrt{2x - 3} = 5 \qquad \text{Isolate the radical.}$$
$$(\sqrt{2x - 3})^2 = 5^2 \qquad \text{Square both sides.}$$
$$2x - 3 = 25$$
$$2x = 28$$
$$x = 14$$

Check by evaluating $x = 14$ in the original equation:

$$\sqrt{2(14) - 3} - 5 = 0$$
$$\sqrt{28 - 3} - 5 = 0$$
$$\sqrt{25} - 5 = 0$$
$$0 = 0$$

The solution set is $\{14\}$.

b) $\sqrt[3]{3x + 5} = \sqrt[3]{x - 1}$ Original equation
   $(\sqrt[3]{3x + 5})^3 = (\sqrt[3]{x - 1})^3$ Cube each side.
   $3x + 5 = x - 1$
   $2x = -6$
   $x = -3$

Check $x = -3$ in the original equation:

$$\sqrt[3]{3(-3) + 5} = \sqrt[3]{-3 - 1}$$
$$\sqrt[3]{-4} = \sqrt[3]{-4}$$

Note that $\sqrt[3]{-4}$ is a real number. The solution set is $\{-3\}$. In this example we checked for arithmetic mistakes. There was no possibility of extraneous solutions here because we raised each side to an odd power.

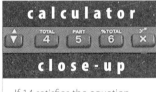

close-up

If 14 satisfies the equation
$\sqrt{2x - 3} - 5 = 0$,
then $(14, 0)$ is an x-intercept for the graph of
$y = \sqrt{2x - 3} - 5$.
So the calculator graph shown here provides visual support for the conclusion that 14 is the only solution to the equation.

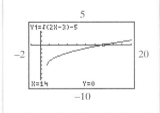

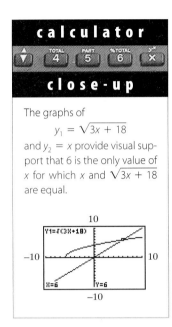

c)

$$\sqrt{3x + 18} = x$$ Original equation

$$\left(\sqrt{3x + 18}\right)^2 = x^2$$ Square both sides.

$$3x + 18 = x^2$$ Simplify.

$$-x^2 + 3x + 18 = 0$$ Subtract x^2 from each side to get zero on one side.

$$x^2 - 3x - 18 = 0$$ Multiply each side by -1 for easier factoring.

$$(x - 6)(x + 3) = 0$$ Factor.

$$x - 6 = 0 \quad \text{or} \quad x + 3 = 0$$ Zero factor property

$$x = 6 \quad \text{or} \quad x = -3$$

Because we squared both sides, we must check for extraneous solutions. If $x = -3$ in the original equation $\sqrt{3x + 18} = x$, we get

$$\sqrt{3(-3) + 18} = -3$$
$$\sqrt{9} = -3$$
$$3 = -3,$$

which is not correct. If $x = 6$ in the original equation, we get

$$\sqrt{3(6) + 18} = 6,$$

which is correct. The solution set is $\{6\}$.   ■

In the next example the radicals are not eliminated after squaring both sides of the equation. In this case we must square both sides a second time. Note that we square the side with two terms the same way we square a binomial.

E X A M P L E   5

Squaring both sides twice
Solve $\sqrt{5x - 1} - \sqrt{x + 2} = 1$.

Solution

It is easier to square both sides if the two radicals are not on the same side, so we first rewrite the equation:

$$\sqrt{5x - 1} - \sqrt{x + 2} = 1$$ Original equation

$$\sqrt{5x - 1} = 1 + \sqrt{x + 2}$$ Add $\sqrt{x + 2}$ to each side.

$$\left(\sqrt{5x - 1}\right)^2 = \left(1 + \sqrt{x + 2}\right)^2$$ Square both sides.

$$5x - 1 = 1 + 2\sqrt{x + 2} + x + 2$$ Square the right side like a binomial.

$$5x - 1 = 3 + x + 2\sqrt{x + 2}$$ Combine like terms on the right side.

$$4x - 4 = 2\sqrt{x + 2}$$ Isolate the square root.

$$2x - 2 = \sqrt{x + 2}$$ Divide each side by 2.

$$(2x - 2)^2 = \left(\sqrt{x + 2}\right)^2$$ Square both sides.

$$4x^2 - 8x + 4 = x + 2$$ Square the binomial on the left side.

$$4x^2 - 9x + 2 = 0$$

$$(4x - 1)(x - 2) = 0$$

$$4x - 1 = 0 \quad \text{or} \quad x - 2 = 0$$

$$x = \frac{1}{4} \quad \text{or} \quad x = 2$$

Check to see whether $\sqrt{5x - 1} - \sqrt{x + 2} = 1$ for $x = \frac{1}{4}$ and for $x = 2$:

$$\sqrt{5 \cdot \frac{1}{4} - 1} - \sqrt{\frac{1}{4} + 2} = \sqrt{\frac{1}{4}} - \sqrt{\frac{9}{4}} = \frac{1}{2} - \frac{3}{2} = -1$$

$$\sqrt{5 \cdot 2 - 1} - \sqrt{2 + 2} = \sqrt{9} - \sqrt{4} = 3 - 2 = 1$$

Because $\frac{1}{4}$ does not satisfy the original equation, the solution set is $\{2\}$.

Equations Involving Rational Exponents

Equations involving rational exponents can be solved by combining the methods that you just learned for eliminating radicals and integral exponents. For equations involving rational exponents, always eliminate the root first and the power second.

E X A M P L E   6

Eliminating the root, then the power

Solve each equation.

a) $x^{2/3} = 4$

b) $(w - 1)^{-2/5} = 4$

helpful hint

Note how we eliminate the root first by raising each side to an integer power, and then apply the even-root property to get two solutions in Example 6(a). A common mistake is to raise each side to the 3/2 power and get $x = 4^{3/2} = 8$. If you do not use the even-root property you can easily miss the solution -8.

Solution

a) Because the exponent $2/3$ indicates a cube root, raise each side to the power 3:

$$x^{2/3} = 4 \qquad \text{Original equation}$$
$$(x^{2/3})^3 = 4^3 \qquad \text{Cube each side.}$$
$$x^2 = 64 \qquad \text{Multiply the exponents: } \frac{2}{3} \cdot 3 = 2.$$
$$x = 8 \quad \text{or} \quad x = -8 \qquad \text{Even-root property}$$

All of the equations are equivalent. Check 8 and -8 in the original equation. The solution set is $\{-8, 8\}$.

b)
$$(w - 1)^{-2/5} = 4 \qquad \text{Original equation}$$
$$[(w - 1)^{-2/5}]^{-5} = 4^{-5} \qquad \text{Raise each side to the power } -5 \text{ to eliminate the negative exponent.}$$
$$(w - 1)^2 = \frac{1}{1024} \qquad \text{Multiply the exponents: } -\frac{2}{5}(-5) = 2.$$
$$w - 1 = \pm\sqrt{\frac{1}{1024}} \qquad \text{Even-root property}$$
$$w - 1 = \frac{1}{32} \quad \text{or} \quad w - 1 = -\frac{1}{32}$$
$$w = \frac{33}{32} \quad \text{or} \quad w = \frac{31}{32}$$

calculator

close-up

Check that 31/32 and 33/32 satisfy the original equation.

```
(31/32-1)^(-2/5)
                4
(33/32-1)^(-2/5)
                4
```

Check the values in the original equation. The solution set is $\left\{\frac{31}{32}, \frac{33}{32}\right\}$.

An equation with a rational exponent might not have a real solution because all even powers of real numbers are nonnegative.

E X A M P L E   7

An equation with no solution

Solve $(2t - 3)^{-2/3} = -1$.

Solution

Raise each side to the power -3 to eliminate the root and the negative sign in the exponent:

$$(2t - 3)^{-2/3} = -1 \qquad \text{Original equation}$$
$$[(2t - 3)^{-2/3}]^{-3} = (-1)^{-3} \qquad \text{Raise each side to the } -3 \text{ power.}$$
$$(2t - 3)^2 = -1 \qquad \text{Multiply the exponents: } -\frac{2}{3}(-3) = 2.$$

By the even-root property this equation has no real solution. The square of every real number is nonnegative. ∎

Summary of Methods

The three most important rules for solving equations with exponents and radicals are restated here.

> ### Strategy for Solving Equations with Exponents and Radicals
>
> **1.** In raising each side of an equation to an even power, we can create an equation that gives extraneous solutions. We must check all possible solutions in the original equation.
>
> **2.** When applying the even-root property, remember that there is a positive and a negative even root for any positive real number.
>
> **3.** For equations with rational exponents, raise each side to a positive or negative integral power first, then apply the even- or odd-root property. (Positive fraction—raise to a positive power; negative fraction—raise to a negative power.)

The Distance Formula

Consider the points (x_1, y_1) and (x_2, y_2) as shown in Fig. 7.1. The distance between these points is the length of the hypotenuse of a right triangle as shown in the figure. The length of side a is $y_2 - y_1$ and the length of side b is $x_2 - x_1$. Using the Pythagorean theorem, we can write

$$d^2 = (x_2 - x_1)^2 + (y_2 - y_1)^2.$$

If we apply the even-root property and omit the negative square root (because the distance is positive), we can express this formula as follows.

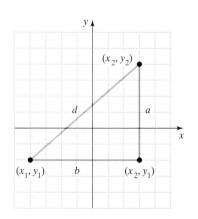

FIGURE 7.1

Distance Formula

The distance d between (x_1, y_1) and (x_2, y_2) is given by the formula

$$d = \sqrt{(x_2 - x_1)^2 + (y_2 - y_1)^2}.$$

EXAMPLE 8 **Using the distance formula**

Find the length of the line segment with endpoints $(-8, -10)$ and $(6, -4)$.

Solution

Let $(x_1, y_1) = (-8, -10)$ and $(x_2, y_2) = (6, -4)$. Now substitute the appropriate values into the distance formula:

$$\begin{aligned} d &= \sqrt{[6 - (-8)]^2 + [-4 - (-10)]^2} \\ &= \sqrt{(14)^2 + (6)^2} \\ &= \sqrt{196 + 36} \\ &= \sqrt{232} \\ &= \sqrt{4 \cdot 58} \\ &= 2\sqrt{58} \quad \text{Simplified form} \end{aligned}$$

The exact length of the segment is $2\sqrt{58}$.

In the next example we find the distance between two points without the distance formula. Although we could solve the problem using a coordinate system and the distance formula, that is not necessary.

EXAMPLE 9 **Diagonal of a baseball diamond**

A baseball diamond is actually a square, 90 feet on each side. What is the distance from third base to first base?

Solution

First make a sketch as in Fig. 7.2. The distance x from third base to first base is the length of the diagonal of the square shown in Fig. 7.2. The Pythagorean theorem can be applied to the right triangle formed from the diagonal and two sides of the square. The sum of the squares of the sides is equal to the diagonal squared:

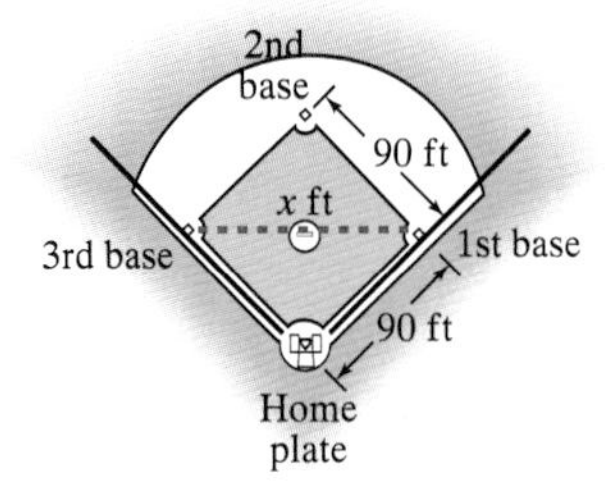

$$\begin{aligned} x^2 &= 90^2 + 90^2 \\ x^2 &= 8100 + 8100 \\ x^2 &= 16{,}200 \\ x &= \pm\sqrt{16{,}200} = \pm 90\sqrt{2} \end{aligned}$$

FIGURE 7.2

The length of the diagonal of a square must be positive, so we disregard the negative solution. Checking the answer in the original equation verifies that the *exact* length of the diagonal is $90\sqrt{2}$ feet.

WARM-UPS

True or false? Explain.

1. The equations $x^2 = 4$ and $x = 2$ are equivalent.   False
2. The equation $x^2 = -25$ has no real solution.   True
3. There is no solution to the equation $x^2 = 0$.   False
4. The equation $x^3 = 8$ is equivalent to $x = \pm 2$.   False
5. The equation $-\sqrt{x} = 16$ has no real solution.   True
6. To solve $\sqrt{x - 3} = \sqrt{2x + 5}$, first apply the even-root property.   False

(*continued*)

7. Extraneous solutions are solutions that cannot be found.   False

8. Squaring both sides of $\sqrt{x} = -7$ yields an equation with an extraneous solution.   True

9. The equations $x^2 - 6 = 0$ and $x = \pm\sqrt{6}$ are equivalent.   True

10. Cubing each side of an equation will not produce an extraneous solution.   True

7.5   EXERCISES

Reading and Writing   After reading this section, write out the answers to these questions. Use complete sentences.

1. What is the odd-root property?

The odd-root property says that if n is an odd positive integer, then $x^n = k$ is equivalent to $x = \sqrt[n]{k}$ for any real number k.

2. What is the even-root property?

The even-root property says that if n is a positive even integer, then $x^n = k$ is equivalent to $x = \pm\sqrt[n]{k}$ for $k > 0$, $x = 0$ for $k = 0$, and has no solution for $k < 0$.

3. What is an extraneous solution?

An extraneous solution is a solution that appears when solving an equation but does not satisfy the original equation.

4. Why can raising each side to a power produce an extraneous solution?

Raising each side to an even power can produce an extraneous root because the even powers of both negative and positive numbers are positive. For example, if $\sqrt{x} = -2$, then squaring each side produces an extraneous root.

Solve each equation. See Example 1.

5. $x^3 = -1000$
$\{-10\}$

6. $y^3 = 125$
$\{5\}$

7. $32m^5 - 1 = 0$
$\left\{\dfrac{1}{2}\right\}$

8. $243a^5 + 1 = 0$
$\left\{-\dfrac{1}{3}\right\}$

9. $(y - 3)^3 = -8$
$\{1\}$

10. $(x - 1)^3 = -1$
$\{0\}$

11. $\dfrac{1}{2}x^3 + 4 = 0$
$\{-2\}$

12. $3(x - 9)^7 = 0$
$\{9\}$

Solve each equation. See Examples 2 and 3.

13. $x^2 = 25$
$\{-5, 5\}$

14. $x^2 = 36$
$\{-6, 6\}$

15. $x^2 - 20 = 0$
$\{-2\sqrt{5}, 2\sqrt{5}\}$

16. $a^2 - 40 = 0$
$\{-2\sqrt{10}, 2\sqrt{10}\}$

17. $x^2 = -9$
$\varnothing$

18. $w^2 + 49 = 0$
$\varnothing$

19. $(x - 3)^2 = 16$
$\{-1, 7\}$

20. $(a - 2)^2 = 25$
$\{-3, 7\}$

21. $(x + 1)^2 - 8 = 0$
$\{-1 - 2\sqrt{2}, -1 + 2\sqrt{2}\}$

22. $(w + 3)^2 - 12 = 0$
$\{-3 - 2\sqrt{3}, -3 + 2\sqrt{3}\}$

23. $\dfrac{1}{2}x^2 = 5$
$\{-\sqrt{10}, \sqrt{10}\}$

24. $\dfrac{1}{3}x^2 = 6$
$\{\pm 3\sqrt{2}\}$

25. $(y - 3)^4 = 0$
$\{3\}$

26. $(2x - 3)^6 = 0$
$\left\{\dfrac{3}{2}\right\}$

27. $2x^6 = 128$
$\{-2, 2\}$

28. $3y^4 = 48$
$\{-2, 2\}$

Solve each equation and check for extraneous solutions. See Example 4.

29. $\sqrt{x - 3} - 7 = 0$
$\{52\}$

30. $\sqrt{a - 1} - 6 = 0$
$\{37\}$

31. $2\sqrt{w + 4} = 5$
$\left\{\dfrac{9}{4}\right\}$

32. $3\sqrt{w + 1} = 6$
$\{3\}$

33. $\sqrt[3]{2x + 3} = \sqrt[3]{x + 12}$
$\{9\}$

34. $\sqrt[3]{a + 3} = \sqrt[3]{2a - 7}$
$\{10\}$

35. $\sqrt{2t + 4} = \sqrt{t - 1}$
$\varnothing$

36. $\sqrt{w - 3} = \sqrt{4w + 15}$
$\varnothing$

37. $\sqrt{4x^2 + x - 3} = 2x$
$\{3\}$

38. $\sqrt{x^2 - 5x + 2} = x$
$\left\{\dfrac{2}{5}\right\}$

39. $\sqrt{x^2 + 2x - 6} = 3$
$\{-5, 3\}$

40. $\sqrt{x^2 - x - 4} = 4$
$\{-4, 5\}$

41. $\sqrt{2x^2 - 1} = x$
$\{1\}$

42. $\sqrt{2x^2 - 3x - 10} = x$
$\{5\}$

43. $\sqrt{2x^2 + 5x + 6} = x$
$\varnothing$

44. $\sqrt{5x^2 - 9} = 2x$
$\{3\}$

Solve each equation and check for extraneous solutions. See Example 5.

45. $\sqrt{x + 3} - \sqrt{x - 2} = 1$ $\{6\}$

46. $\sqrt{2x + 1} - \sqrt{x} = 1$ $\{0, 4\}$
47. $\sqrt{2x + 2} - \sqrt{x - 3} = 2$ $\{7\}$
48. $\sqrt{3x} - \sqrt{x - 2} = 4$ $\{27\}$
49. $\sqrt{4 - x} - \sqrt{x + 6} = 2$ $\{-5\}$
50. $\sqrt{6 - x} - \sqrt{x - 2} = 2$ $\{2\}$

Solve each equation. See Examples 6 and 7.

51. $x^{2/3} = 3$
$\{-3\sqrt{3}, 3\sqrt{3}\}$

52. $a^{2/3} = 2$
$\{-2\sqrt{2}, 2\sqrt{2}\}$

53. $y^{-2/3} = 9$
$\left[-\dfrac{1}{27}, \dfrac{1}{27}\right]$

54. $w^{-2/3} = 4$
$\left[-\dfrac{1}{8}, \dfrac{1}{8}\right]$

55. $w^{1/3} = 8$
$\{512\}$

56. $a^{1/3} = 27$
$\{19{,}683\}$

57. $t^{-1/2} = 9$
$\left[\dfrac{1}{81}\right]$

58. $w^{-1/4} = \dfrac{1}{2}$
$\{16\}$

59. $(3a - 1)^{-2/5} = 1$
$\left[0, \dfrac{2}{3}\right]$

60. $(r - 1)^{-2/3} = 1$
$\{0, 2\}$

61. $(t - 1)^{-2/3} = 2$
$\left[\dfrac{4 - \sqrt{2}}{4}, \dfrac{4 + \sqrt{2}}{4}\right]$

62. $(w + 3)^{-1/3} = \dfrac{1}{3}$
$\{24\}$

63. $(x - 3)^{2/3} = -4$
$\varnothing$

64. $(x + 2)^{3/2} = -1$
$\varnothing$

Find the distance between each given pair of points. See Example 8.

65. $(6, 5), (4, 2)$
$\sqrt{13}$

66. $(7, 3), (5, 1)$
$2\sqrt{2}$

67. $(3, 5), (1, -3)$
$2\sqrt{17}$

68. $(6, 2), (3, -5)$
$\sqrt{58}$

69. $(4, -2), (-3, -6)$
$\sqrt{65}$

70. $(-2, 3), (1, -4)$
$\sqrt{58}$

Solve each equation.

71. $2x^2 + 3 = 7$ $\{-\sqrt{2}, \sqrt{2}\}$
72. $3x^2 - 5 = 16$ $\{-\sqrt{7}, \sqrt{7}\}$
73. $\sqrt[3]{2w + 3} = \sqrt[3]{w - 2}$ $\{-5\}$
74. $\sqrt[3]{2 - w} = \sqrt[3]{2w - 28}$ $\{10\}$

75. $9x^2 - 1 = 0$ $\left[-\dfrac{1}{3}, \dfrac{1}{3}\right]$

76. $4x^2 - 1 = 0$ $\left[-\dfrac{1}{2}, \dfrac{1}{2}\right]$

77. $(w + 1)^{2/3} = -3$ $\varnothing$
78. $(x - 2)^{3/4} = 2$ $\{2 + 2\sqrt[3]{2}\}$
79. $(a + 1)^{1/3} = -2$ $\{-9\}$
80. $(a - 1)^{1/3} = -3$ $\{-26\}$

81. $(4y - 5)^7 = 0$ $\left[\dfrac{5}{4}\right]$

82. $(5x)^9 = 0$ $\{0\}$

83. $\sqrt{x^2 + 5x} = 6$ $\{-9, 4\}$
84. $\sqrt{x^2 - 8x} = -3$ $\varnothing$

85. $\sqrt{4x^2} = x + 2$ $\left[-\dfrac{2}{3}, 2\right]$

86. $\sqrt{9x^2} = x + 6$ $\left[-\dfrac{3}{2}, 3\right]$

87. $(t + 2)^4 = 32$ $\{-2 - 2\sqrt[4]{2}, -2 + 2\sqrt[4]{2}\}$
88. $(w + 1)^4 = 48$ $\{-1 - 2\sqrt[4]{3}, -1 + 2\sqrt[4]{3}\}$
89. $\sqrt{x^2 - 3x} = x$ $\{0\}$
90. $\sqrt[4]{4x^4 - 48} = -x$ $\{-2\}$

91. $x^{-3} = 8$ $\left[\dfrac{1}{2}\right]$

92. $x^{-2} = 4$ $\left[\pm\dfrac{1}{2}\right]$

93. $a^{-2} = 3$ $\left[-\dfrac{\sqrt{3}}{3}, \dfrac{\sqrt{3}}{3}\right]$

94. $w^{-2} = 18$ $\left[-\dfrac{\sqrt{2}}{6}, \dfrac{\sqrt{2}}{6}\right]$

Solve each problem by writing an equation and solving it. Find the exact answer and simplify it using the rules for radicals. See Example 9.

95. *Side of a square.* Find the length of the side of a square whose diagonal is 8 feet.
$4\sqrt{2}$ feet

96. *Diagonal of a patio.* Find the length of the diagonal of a square patio with an area of 40 square meters.
$4\sqrt{5}$ meters

97. *Side of a sign.* Find the length of the side of a square sign whose area is 50 square feet.
$5\sqrt{2}$ feet

98. *Side of a cube.* Find the length of the side of a cubic box whose volume is 80 cubic feet.
$2\sqrt[3]{10}$ feet

99. *Diagonal of a rectangle.* If the sides of a rectangle are 30 feet and 40 feet in length, find the length of the diagonal of the rectangle.
50 feet

100. *Diagonal of a sign.* What is the length of the diagonal of a rectangular billboard whose sides are 5 meters and 12 meters?
13 meters

101. *Sailboat stability.* To be considered safe for ocean sailing, the capsize screening value C should be less than 2 (*Sail*, May 1997). For a boat with a beam (or width) b in feet and displacement d in pounds, C is determined by the formula

$$C = 4d^{-1/3}b.$$

a) Find the capsize screening value for the Tartan 4100, which has a displacement of 23,245 pounds and a beam of 13.5 feet.
1.89

b) Solve this formula for d.

$$d = \frac{64b^3}{C^3}$$

c) The accompanying graph shows C as a function of d for the Tartan 4100 ($b = 13.5$). For what displacement is the Tartan 4100 safe for ocean sailing?
$d > 19{,}683$ pounds

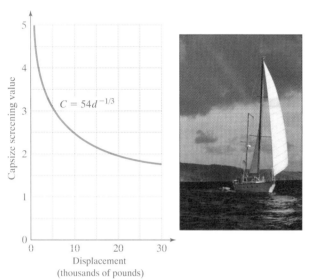

FIGURE FOR EXERCISE 101

102. *Sailboat speed.* The sail area-displacement ratio S provides a measure of the sail power available to drive a boat. For a boat with a displacement of d pounds and a sail area of A square feet

$$S = 16Ad^{-2/3}.$$

a) Find S for the Tartan 4100, which has a sail area of 810 square feet and a displacement of 23,245 pounds.
15.9

b) Solve the formula for d.

$$d = \left(\frac{16A}{S}\right)^{3/2}$$

103. *Diagonal of a side.* Find the length of the diagonal of a side of a cubic packing crate whose volume is 2 cubic meters.
$\sqrt[6]{32}$ meters

104. *Volume of a cube.* Find the volume of a cube on which the diagonal of a side measures 2 feet.
$2\sqrt{2}$ cubic feet (ft^3)

105. *Length of a road.* An architect designs a public park in the shape of a trapezoid. Find the length of the diagonal road marked a in the figure.
$\sqrt{73}$ kilometers (km)

106. *Length of a boundary.* Find the length of the border of the park marked b in the trapezoid shown in the figure.
$\sqrt{13}$ km

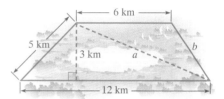

FIGURE FOR EXERCISES 105 AND 106

107. *Average annual return.* The formula

$$r = \left(\frac{S}{P}\right)^{1/n} - 1$$

was used to find the average annual return on an investment in Exercise 127 in Section 7.1. Solve the formula for S (the amount). Solve it for P (the original principal).
$S = P(1 + r)^n, P = S(1 + r)^{-n}$

108. *Surface area of a cube.* The formula $A = 6V^{2/3}$ gives the surface area of a cube in terms of its volume V. What is the volume of a cube with surface area 12 square feet?
$2\sqrt{2}$ ft^3

 109. *Kepler's third law.* According to Kepler's third law of planetary motion, the ratio $\frac{T^2}{R^3}$ has the same value for every planet in our solar system. R is the average radius of the orbit of the planet measured in astronomical units (AU), and T is the number of years it takes for one complete orbit of the sun. Jupiter orbits the sun in 11.86 years with an average radius of 5.2 AU, whereas Saturn orbits the sun in 29.46 years. Find the average radius of the orbit of Saturn. (One AU is the distance from the earth to the sun.)
9.5 AU

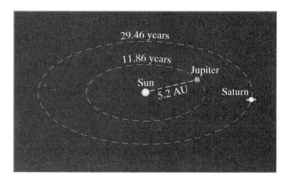

FIGURE FOR EXERCISE 109

 110. *Orbit of Venus.* If the average radius of the orbit of Venus is 0.723 AU, then how many years does it take for

Venus to complete one orbit of the sun? Use the information in Exercise 109.

0.61 year

Use a calculator to find approximate solutions to the following equations. Round your answers to three decimal places.

111. $x^2 = 3.24$ $\{-1.8, 1.8\}$

112. $(x + 4)^3 = 7.51$ $\{-2.042\}$

113. $\sqrt{x - 2} = 1.73$ $\{4.993\}$

114. $\sqrt[3]{x - 5} = 3.7$ $\{55.653\}$

115. $x^{2/3} = 8.86$ $\{-26.372, 26.372\}$

116. $(x - 1)^{-3/4} = 7.065$
$\{1.074\}$

GETTING MORE INVOLVED

117. *Cooperative learning.* Work in a small group to write a formula that gives the side of a cube in terms of the volume of the cube and explain the formula to the other groups.

118. *Cooperative learning.* Work in a small group to write a formula that gives the side of a square in terms of the diagonal of the square and explain the formula to the other groups.

7.6   COMPLEX NUMBERS

In Chapter 1 we discussed the real numbers and the various subsets of the real numbers. In this section we define a set of numbers that has the real numbers as a subset.

In this

section

- Definition
- Addition, Subtraction, and Multiplication
- Division of Complex Numbers
- Square Roots of Negative Numbers
- Imaginary Solutions to Equations

Definition

The equation $2x = 1$ has no solution in the set of integers, but in the set of rational numbers, $2x = 1$ has a solution. The situation is similar for the equation $x^2 = -4$. It has no solution in the set of real numbers because the square of every real number is nonnegative. However, in the set of complex numbers $x^2 = -4$ has two solutions. The complex numbers were developed so that equations such as $x^2 = -4$ would have solutions.

The complex numbers are based on the symbol $\sqrt{-1}$. In the real number system this symbol has no meaning. In the set of complex numbers this symbol is given meaning. We call it i. We make the definition that

$$i = \sqrt{-1} \qquad \text{and} \qquad i^2 = -1.$$

Complex Numbers

The set of **complex numbers** is the set of all numbers of the form

$$a + bi,$$

where a and b are real numbers, $i = \sqrt{-1}$, and $i^2 = -1$.

In the complex number $a + bi$, a is called the **real part** and b is called the **imaginary part.** If $b \neq 0$, the number $a + bi$ is called an **imaginary number.**

In dealing with complex numbers, we treat $a + bi$ as if it were a binomial, with i being a variable. Thus we would write $2 + (-3)i$ as $2 - 3i$. We agree that $2 + i3$, $3i + 2$, and $i3 + 2$ are just different ways of writing $2 + 3i$ (the standard form). Some examples of complex numbers are

$$-3 - 5i, \quad \frac{2}{3} - \frac{3}{4}i, \quad 1 + i\sqrt{2}, \quad 9 + 0i, \quad \text{and} \quad 0 + 7i.$$

study tip

Make sure that you know what your instructor expects from you. You can determine what your instructor feels is important by looking at the examples that your instructor works in class and the homework assignments. When in doubt, ask your instructor what you will be responsible for and write down the answer.

For simplicity we write only $7i$ for $0 + 7i$. The complex number $9 + 0i$ is the real number 9, and $0 + 0i$ is the real number 0. Any complex number with $b = 0$ is a real number. For any real number a,

$$a + 0i = a.$$

The set of real numbers is a subset of the set of complex numbers. See Fig. 7.3.

Complex numbers	
Real numbers	Imaginary numbers
$3, \pi, \frac{5}{2}, 0, -9, \sqrt{2}$	$i, 2 + 3i, \sqrt{-5}, -3 - 8i$

FIGURE 7.3

Addition, Subtraction, and Multiplication

Addition and subtraction of complex numbers are performed as if the complex numbers were algebraic expressions with i being a variable.

E X A M P L E   1

Addition and subtraction of complex numbers

Find the sums and differences.

a) $(2 + 3i) + (6 + i)$ **b)** $(-2 + 3i) + (-2 - 5i)$

c) $(3 + 5i) - (1 + 2i)$ **d)** $(-2 - 3i) - (1 - i)$

Solution

a) $(2 + 3i) + (6 + i) = 8 + 4i$

b) $(-2 + 3i) + (-2 - 5i) = -4 - 2i$

c) $(3 + 5i) - (1 + 2i) = 2 + 3i$

d) $(-2 - 3i) - (1 - i) = -3 - 2i$

We can give a symbolic definition of addition and subtraction as follows.

Addition and Subtraction of Complex Numbers

The sum and difference of $a + bi$ and $c + di$ are defined as follows:

$$(a + bi) + (c + di) = (a + c) + (b + d)i$$
$$(a + bi) - (c + di) = (a - c) + (b - d)i$$

Complex numbers are multiplied as if they were algebraic expressions. Whenever i^2 appears, we replace it by -1.

E X A M P L E   2

Products of complex numbers

Find each product.

a) $2i(1 + i)$ **b)** $(2 + 3i)(4 + 5i)$

c) $(3 + i)(3 - i)$

Solution

a) $2i(1 + i) = 2i + 2i^2$ Distributive property

$\qquad\qquad\quad = 2i + 2(-1)$ $i^2 = -1$

$\qquad\qquad\quad = -2 + 2i$

b) Use the FOIL method to find the product:

$$(2 + 3i)(4 + 5i) = 8 + 10i + 12i + 15i^2$$

$$= 8 + 22i + 15(-1)$$ Replace i^2 by -1.

$$= 8 + 22i - 15$$

$$= -7 + 22i$$

c) This product is the product of a sum and a difference.

$$(3 + i)(3 - i) = 9 - 3i + 3i - i^2$$

$$= 9 - (-1)$$ $i^2 = -1$

$$= 10$$

We can find powers of i using the fact that $i^2 = -1$. For example,

$$i^3 = i^2 \cdot i = -1 \cdot i = -i.$$

The value of i^4 is found from the value of i^3:

$$i^4 = i^3 \cdot i = -i \cdot i = -i^2 = 1.$$

In the next example we find more powers of imaginary numbers.

E X A M P L E   3

Powers of imaginary numbers

Write each expression in the form $a + bi$.

a) $(2i)^2$ **b)** $(-2i)^2$ **c)** i^6

Solution

a) $(2i)^2 = 2^2 \cdot i^2 = 4(-1) = -4$

b) $(-2i)^2 = (-2)^2 \cdot i^2 = 4i^2 = 4(-1) = -4$

c) $i^6 = i^2 \cdot i^4 = -1 \cdot 1 = -1$

For completeness we give the following symbolic definition of multiplication of complex numbers. However, it is simpler to find products as we did in Examples 2 and 3 than to use this definition.

Multiplication of Complex Numbers

The complex numbers $a + bi$ and $c + di$ are multiplied as follows:

$$(a + bi)(c + di) = (ac - bd) + (ad + bc)i$$

Division of Complex Numbers

To divide a complex number by a real number, divide each term by the real number, just as we would divide a binomial by a number. For example,

$$\frac{4 + 6i}{2} = \frac{2(2 + 3i)}{2}$$

$$= 2 + 3i.$$

helpful hint

Here is that word "conjugate" again. It is generally used to refer to two things that go together in some way.

To understand division by a complex number, we first look at imaginary numbers that have a real product. The product of the two imaginary numbers in Example 2(c) is a real number:

$$(3 + i)(3 - i) = 10$$

We say that $3 + i$ and $3 - i$ are complex conjugates of each other.

Complex Conjugates

The complex numbers $a + bi$ and $a - bi$ are called **complex conjugates** of one another. Their product is the real number $a^2 + b^2$.

E X A M P L E 4

Products of conjugates

Find the product of the given complex number and its conjugate.

a) $2 + 3i$ **b)** $5 - 4i$

Solution

a) The conjugate of $2 + 3i$ is $2 - 3i$.

$$(2 + 3i)(2 - 3i) = 4 - 9i^2$$
$$= 4 + 9$$
$$= 13$$

b) The conjugate of $5 - 4i$ is $5 + 4i$.

$$(5 - 4i)(5 + 4i) = 25 + 16$$
$$= 41$$

We use the idea of complex conjugates to divide complex numbers. The process is similar to rationalizing the denominator. Multiply the numerator and denominator of the quotient by the complex conjugate of the denominator.

E X A M P L E 5

Dividing complex numbers

Find each quotient. Write the answer in the form $a + bi$.

a) $\dfrac{5}{3 - 4i}$ **b)** $\dfrac{3 - i}{2 + i}$ **c)** $\dfrac{3 + 2i}{i}$

Solution

a) Multiply the numerator and denominator by $3 + 4i$, the conjugate of $3 - 4i$:

$$\frac{5}{3 - 4i} = \frac{5(3 + 4i)}{(3 - 4i)(3 + 4i)}$$

$$= \frac{15 + 20i}{9 - 16i^2}$$

$$= \frac{15 + 20i}{25} \qquad 9 - 16i^2 = 9 - 16(-1) = 25$$

$$= \frac{15}{25} + \frac{20}{25}i$$

$$= \frac{3}{5} + \frac{4}{5}i$$

b) Multiply the numerator and denominator by $2 - i$, the conjugate of $2 + i$:

$$\frac{3 - i}{2 + i} = \frac{(3 - i)(2 - i)}{(2 + i)(2 - i)}$$

$$= \frac{6 - 5i + i^2}{4 - i^2}$$

$$= \frac{6 - 5i - 1}{4 - (-1)}$$

$$= \frac{5 - 5i}{5}$$

$$= 1 - i$$

c) Multiply the numerator and denominator by $-i$, the conjugate of i:

$$\frac{3 + 2i}{i} = \frac{(3 + 2i)(-i)}{i(-i)}$$

$$= \frac{-3i - 2i^2}{-i^2}$$

$$= \frac{-3i + 2}{1}$$

$$= 2 - 3i$$

The symbolic definition of division of complex numbers follows.

Division of Complex Numbers

We divide the complex number $a + bi$ by the complex number $c + di$ as follows:

$$\frac{a + bi}{c + di} = \frac{(a + bi)(c - di)}{(c + di)(c - di)}$$

Square Roots of Negative Numbers

In Examples 3(a) and 3(b) we saw that both

$$(2i)^2 = -4 \quad \text{and} \quad (-2i)^2 = -4.$$

Because the square of each of these complex numbers is -4, both $2i$ and $-2i$ are square roots of -4. We write $\sqrt{-4} = 2i$. In the complex number system the square root of any negative number is an imaginary number.

Square Root of a Negative Number

For any positive real number b,

$$\sqrt{-b} = i\sqrt{b}.$$

For example, $\sqrt{-9} = i\sqrt{9} = 3i$ and $\sqrt{-7} = i\sqrt{7}$. Note that the expression $\sqrt{7}i$ could easily be mistaken for the expression $\sqrt{7i}$, where i is under the radical. For this reason, when the coefficient of i is a radical, we write i preceding the radical.

E X A M P L E   6

Square roots of negative numbers

Write each expression in the form $a + bi$, where a and b are real numbers.

a) $3 + \sqrt{-9}$ **b)** $\sqrt{-12} + \sqrt{-27}$ **c)** $\dfrac{-1 - \sqrt{-18}}{3}$

Solution

a) $3 + \sqrt{-9} = 3 + i\sqrt{9}$
$$= 3 + 3i$$

b) $\sqrt{-12} + \sqrt{-27} = i\sqrt{12} + i\sqrt{27}$
$$= 2i\sqrt{3} + 3i\sqrt{3} \qquad \begin{array}{l} \sqrt{12} = \sqrt{4}\sqrt{3} = 2\sqrt{3} \\ \sqrt{27} = \sqrt{9}\sqrt{3} = 3\sqrt{3} \end{array}$$
$$= 5i\sqrt{3}$$

c) $\dfrac{-1 - \sqrt{-18}}{3} = \dfrac{-1 - i\sqrt{18}}{3}$

$$= \dfrac{-1 - 3i\sqrt{2}}{3}$$

$$= -\dfrac{1}{3} - i\sqrt{2}$$

Imaginary Solutions to Equations

In the complex number system the even-root property can be restated so that $x^2 = k$ is equivalent to $x = \pm\sqrt{k}$ for any $k \neq 0$. So an equation such as $x^2 = -9$ that has no real solutions has two imaginary solutions in the complex numbers.

E X A M P L E   7

Complex solutions to equations

Find the complex solutions to each equation.

a) $x^2 = -9$ **b)** $3x^2 + 2 = 0$

Solution

a) First apply the even-root property:
$$x^2 = -9$$
$$x = \pm\sqrt{-9} \qquad \text{Even-root property}$$
$$= \pm i\sqrt{9}$$
$$= \pm 3i$$

Check these solutions in the original equation:
$$(3i)^2 = 9i^2 = 9(-1) = -9$$
$$(-3i)^2 = 9i^2 = -9$$

The solution set is $\{\pm 3i\}$.

b) First solve the equation for x^2:
$$3x^2 + 2 = 0$$
$$x^2 = -\dfrac{2}{3}$$
$$x = \pm\sqrt{-\dfrac{2}{3}}$$
$$= \pm i\sqrt{\dfrac{2}{3}} = \pm i\dfrac{\sqrt{6}}{3}$$

Check these solutions in the original equation. The solution set is $\left\{ \pm i\dfrac{\sqrt{6}}{3} \right\}$.

The basic facts about complex numbers are listed in the following box.

> **Complex Numbers**
>
> 1. Definition of i: $i = \sqrt{-1}$, and $i^2 = -1$.
> 2. A complex number has the form $a + bi$, where a and b are real numbers.
> 3. The complex number $a + 0i$ is the real number a.
> 4. If b is a positive real number, then $\sqrt{-b} = i\sqrt{b}$.
> 5. The numbers $a + bi$ and $a - bi$ are called complex conjugates of each other. Their product is the real number $a^2 + b^2$.
> 6. Add, subtract, and multiply complex numbers as if they were algebraic expressions with i being the variable, and replace i^2 by -1.
> 7. Divide complex numbers by multiplying the numerator and denominator by the conjugate of the denominator.
> 8. In the complex number system $x^2 = k$ for any real number k is equivalent to $x = \pm\sqrt{k}$.

WARM–UPS

True or false? Explain.

1. The set of real numbers is a subset of the set of complex numbers. True
2. $2 - \sqrt{-6} = 2 - 6i$ False   3. $\sqrt{-9} = \pm3i$ False
4. The solution set to the equation $x^2 = -9$ is $\{\pm3i\}$. True
5. $2 - 3i - (4 - 2i) = -2 - i$ True
6. $i^4 = 1$ True   7. $(2 - i)(2 + i) = 5$ True
8. $i^3 = i$ False   9. $i^{48} = 1$ True
10. The equation $x^2 = k$ has two complex solutions for any real number k. False

7.6 EXERCISES

Reading and Writing *After reading this section, write out the answers to these questions. Use complete sentences.*

1. What are complex numbers?
   A complex number is a number of the form $a + bi$, where a and b are real numbers.

2. What is an imaginary number?
   An imaginary number is a complex number in which $b \neq 0$.

3. What is the relationship among the real numbers, the imaginary numbers, and the complex numbers?
   The union of the real numbers and the imaginary numbers is the set of complex numbers.

4. How do we add, subtract, and multiply complex numbers?
   Addition, subtraction, and multiplication of complex numbers are done as if the complex numbers were binomials with i being a variable. When i^2 occurs, we replace it with -1.

5. What is the conjugate of a complex number?
   The conjugate of $a + bi$ is $a - bi$.

6. How do we divide complex numbers?
   To divide complex numbers, write the quotient as a fraction and multiply the numerator and denominator by the conjugate of the denominator.

Find the indicated sums and differences of complex numbers. See Example 1.

7. $(2 + 3i) + (-4 + 5i)$
   $-2 + 8i$

8. $(-1 + 6i) + (5 - 4i)$
   $4 + 2i$

9. $(2 - 3i) - (6 - 7i)$
   $-4 + 4i$

10. $(2 - 3i) - (6 - 2i)$
    $-4 - i$

11. $(-1 + i) + (-1 - i)$
    -2

12. $(-5 + i) + (-5 - i)$
    -10

13. $(-2 - 3i) - (6 - i)$
    $-8 - 2i$

14. $(-6 + 4i) - (2 - i)$
    $-8 + 5i$

Find the indicated products of complex numbers. See Example 2.

15. $3(2 + 5i)$
$6 + 15i$

16. $4(1 - 3i)$
$4 - 12i$

17. $2i(i - 5)$
$-2 - 10i$

18. $3i(2 - 6i)$
$18 + 6i$

19. $-4i(3 - i)$
$-4 - 12i$

20. $-5i(2 + 3i)$
$15 - 10i$

21. $(2 + 3i)(4 + 6i)$
$-10 + 24i$

22. $(2 + i)(3 + 4i)$
$2 + 11i$

23. $(-1 + i)(2 - i)$
$-1 + 3i$

24. $(3 - 2i)(2 - 5i)$
$-4 - 19i$

25. $(-1 - 2i)(2 + i)$
$-5i$

26. $(1 - 3i)(1 + 3i)$
10

27. $(5 - 2i)(5 + 2i)$
29

28. $(4 + 3i)(4 + 3i)$
$7 + 24i$

29. $(1 - i)(1 + i)$
2

30. $(2 + 6i)(2 - 6i)$
40

31. $(4 + 2i)(4 - 2i)$
20

32. $(4 - i)(4 + i)$
17

Find the indicated powers of complex numbers. See Example 3.

33. $(3i)^2$ -9 **34.** $(5i)^2$ -25 **35.** $(-5i)^2$ -25

36. $(-9i)^2$ -81 **37.** $(2i)^4$ 16 **38.** $(-2i)^3$ $8i$

39. i^9 i **40.** i^{12} 1

Find the product of the given complex number and its conjugate. See Example 4.

41. $3 + 5i$ 34 **42.** $3 + i$ 10 **43.** $1 - 2i$ 5

44. $4 - 6i$ 52 **45.** $-2 + i$ 5 **46.** $-3 - 2i$ 13

47. $2 - i\sqrt{3}$ 7 **48.** $\sqrt{5} - 4i$ 21

Find each quotient. See Example 5.

49. $\dfrac{3}{4 + i}$ $\dfrac{12}{17} - \dfrac{3}{17}i$

50. $\dfrac{6}{7 - 2i}$ $\dfrac{42}{53} + \dfrac{12}{53}i$

51. $\dfrac{2 + i}{3 - 2i}$ $\dfrac{4}{13} + \dfrac{7}{13}i$

52. $\dfrac{3 + 5i}{2 - i}$ $\dfrac{1}{5} + \dfrac{13}{5}i$

53. $\dfrac{4 + 3i}{i}$ $3 - 4i$

54. $\dfrac{5 - 6i}{3i}$ $-2 - \dfrac{5}{3}i$

55. $\dfrac{2 + 6i}{2}$ $1 + 3i$

56. $\dfrac{9 - 3i}{-6}$ $-\dfrac{3}{2} + \dfrac{1}{2}i$

Write each expression in the form a + bi, where a and b are real numbers. See Example 6.

57. $2 + \sqrt{-4}$
$2 + 2i$

58. $3 + \sqrt{-9}$
$3 + 3i$

59. $2\sqrt{-9} + 5$
$5 + 6i$

60. $3\sqrt{-16} + 2$
$2 + 12i$

61. $7 - \sqrt{-6}$
$7 - i\sqrt{6}$

62. $\sqrt{-5} + 3$
$3 + i\sqrt{5}$

63. $\sqrt{-8} + \sqrt{-18}$
$5i\sqrt{2}$

64. $2\sqrt{-20} - \sqrt{-45}$
$i\sqrt{5}$

65. $\dfrac{2 + \sqrt{-12}}{2}$
$1 + i\sqrt{3}$

66. $\dfrac{-6 - \sqrt{-18}}{3}$
$-2 - i\sqrt{2}$

67. $\dfrac{-4 - \sqrt{-24}}{4}$
$-1 - \dfrac{1}{2}i\sqrt{6}$

68. $\dfrac{8 + \sqrt{-20}}{-4}$
$-2 - \dfrac{1}{2}i\sqrt{5}$

Find the complex solutions to each equation. See Example 7.

69. $x^2 = -36$ $\{\pm 6i\}$ **70.** $x^2 + 4 = 0$ $\{\pm 2i\}$

71. $x^2 = -12$ $\{\pm 2i\sqrt{3}\}$ **72.** $x^2 = -25$ $\{\pm 5i\}$

73. $2x^2 + 5 = 0$ $\left\{\pm\dfrac{i\sqrt{10}}{2}\right\}$ **74.** $3x^2 + 4 = 0$ $\left\{\pm\dfrac{2i\sqrt{3}}{3}\right\}$

75. $3x^2 + 6 = 0$ $\{\pm i\sqrt{2}\}$ **76.** $x^2 + 1 = 0$ $\{\pm i\}$

Write each expression in the form a + bi, where a and b are real numbers.

77. $(2 - 3i)(3 + 4i)$
$18 - i$

78. $(2 - 3i)(2 + 3i)$
13

79. $(2 - 3i) + (3 + 4i)$
$5 + i$

80. $(3 - 5i) - (2 - 7i)$
$1 + 2i$

81. $\dfrac{2 - 3i}{3 + 4i}$ $-\dfrac{6}{25} - \dfrac{17}{25}i$

82. $\dfrac{-3i}{3 - 6i}$ $\dfrac{2}{5} - \dfrac{1}{5}i$

83. $i(2 - 3i)$ $3 + 2i$ **84.** $-3i(4i - 1)$ $12 + 3i$

85. $(-3i)^2$ -9 **86.** $(-2i)^6$ -64

87. $\sqrt{-12} + \sqrt{-3}$ $3i\sqrt{3}$ **88.** $\sqrt{-49} - \sqrt{-25}$ $2i$

89. $(2 - 3i)^2$ $-5 - 12i$ **90.** $(5 + 3i)^2$ $16 + 30i$

91. $\dfrac{-4 + \sqrt{-32}}{2}$

$-2 + 2i\sqrt{2}$

92. $\dfrac{-2 - \sqrt{-27}}{-6}$

$\dfrac{1}{3} + \dfrac{1}{2}i\sqrt{3}$

GETTING MORE INVOLVED

93. *Writing.* Explain why $2 - i$ is a solution to
$$x^2 - 4x + 5 = 0.$$

94. *Cooperative learning.* Work with a group to verify that $-1 + i\sqrt{3}$ and $-1 - i\sqrt{3}$ satisfy the equation
$$x^3 - 8 = 0.$$
In the complex number system there are three cube roots of 8. What are they?

95. *Discussion.* What is wrong with using the product rule for radicals to get
$$\sqrt{-4} \cdot \sqrt{-4} = \sqrt{(-4)(-4)} = \sqrt{16} = 4?$$
What is the correct product?

COLLABORATIVE ACTIVITIES

Laws of Falling Bodies

Jaki Sena is a private investigator working on a case. Her client is accused of killing a woman who had lived in the same apartment building as he did. She pulls over the file of notes from across her desk and looks through it yet again.

"Okay, so my client said he was on the roof of the building just before 4:00 A.M., dropping blocks over into the alleyway. He dropped the last block just as the clock tower began striking the hour. I wonder how high the building is," Jaki says to herself.

During her initial investigation of the crime scene Jaki had asked the building manager whether he knew the height of the five-story building. He had been very condescending and had told her nothing. Another tenant of the building, some sort of math instructor at the city college, had told her about an experiment he had done with shadows and a climbing rope. He had given her a drawing.

Jaki looked through the file until she found the following drawing.

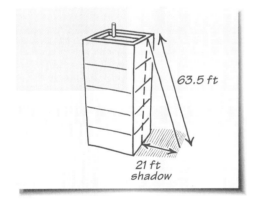

Grouping: Two to four students per group
Topic: Applications of formulas with square roots

She pulled out a blank piece of paper, and using a calculator, she soon had a number.

"Now I can find out how long it took the block to fall!" Jaki exclaimed. "Okay, now the witness who saw my client on the top of the building also swears he saw the woman walking past the steps next door as the clock tower started to chime 4:00," Jaki said out loud as she got up from her desk and looked out the window. "I have no idea how fast the woman was walking. That may be a hard one to answer. But I do know the steps begin about $9\frac{1}{2}$ feet from where the block fell." Jaki turned back to the file of notes on her desk. She soon found another piece of paper with measurements on it.

"And I do know the woman was 5 feet tall!" Jaki exclaimed, taking out her calculator again.

1. Working in your groups, find the height of the building from the information on Jaki's paper. Round your answer to the nearest tenth of a foot.

2. Find out how many seconds it would take the block to fall from the top of the building to hit the woman on the head. Round your answer to the nearest tenth of a second. (*Hint:* You may need to look in other chapters of the book to find the formulas you need here.)

3. How fast would the woman have to walk to have been hit on the head by the block? Round your answer to the nearest tenth of a foot per second.

4. Could Jaki's client have "done it?" Give a reason for your answer.

WRAP-UP   CHAPTER 7

SUMMARY

Roots and Rational Exponents

Examples

| Definition of *n*th roots | The number *a* is an *n*th root of *b* if $a^n = b$. | $3^4 = 81$, $(-3)^4 = 81$ Both 3 and -3 are fourth roots of 81. |

| Definition of $1/n$ as an exponent | If n is a positive *even* integer and a is positive, then $a^{1/n}$ denotes the positive real nth root of a. If n is a positive *odd* integer and a is any real number, then $a^{1/n}$ is the real nth root of a. | $81^{1/4} = 3$ $(-8)^{1/3} = -2$ $8^{1/3} = 2$ |

Definition of nth root of zero	If n is a positive integer, then $0^{1/n} = 0$.	$0^{1/6} = 0$
Definition of rational exponents	If m and n are positive integers, then $a^{m/n} = (a^{1/n})^m$, provided that $a^{1/n}$ is real. We can find the root first or the power first.	$8^{2/3} = (8^{1/3})^2$ $= 2^2 = 4$ $(-16)^{3/4}$ is not real.
Definition of negative rational exponents	If m and n are positive integers, then $$a^{-m/n} = \frac{1}{a^{m/n}},$$ provided that $a^{1/n}$ is defined and nonzero.	Root $8^{-2/3} = \dfrac{1}{4}$ Power Reciprocal

Rules for Rational Exponents

Examples

If a and b are nonzero real numbers and r and s are rational numbers, then the following rules hold, provided all expressions represent real numbers.

Product rule	$a^r \cdot a^s = a^{r+s}$	$3^{1/4} \cdot 3^{1/2} = 3^{3/4}$
Quotient rule	$\dfrac{a^r}{a^s} = a^{r-s}$	$\dfrac{x^{3/4}}{x^{1/4}} = x^{1/2}$
Power of a power rule	$(a^r)^s = a^{rs}$	$(2^{1/2})^{-1/2} = 2^{-1/4}$ $(x^{3/4})^4 = x^3$
Power of a product rule	$(ab)^r = a^r b^r$	$(a^2 b^6)^{1/2} = ab^3$
Power of a quotient rule	$\left(\dfrac{a}{b}\right)^r = \dfrac{a^r}{b^r}$	$\left(\dfrac{8}{x^6}\right)^{2/3} = \dfrac{4}{x^4}$

Radicals

Examples

Definition of radicals	If n is a positive integer and a is a number for which $a^{1/n}$ is real, then $\sqrt[n]{a} = a^{1/n}$. If $n = 2$, omit the 2 and write $\sqrt[2]{a}$ as $\sqrt{a}$.	$8^{1/3} = \sqrt[3]{8}$ $4^{1/2} = \sqrt{4}$
Product rule for radicals	Provided that all roots are real, $$\sqrt[n]{ab} = \sqrt[n]{a} \cdot \sqrt[n]{b}.$$	$\sqrt{2} \cdot \sqrt{3} = \sqrt{6}$ $\sqrt{4x} = 2\sqrt{x}$
Quotient rule for radicals	Provided that all roots are real and $b \neq 0$, $$\sqrt[n]{\dfrac{a}{b}} = \dfrac{\sqrt[n]{a}}{\sqrt[n]{b}}.$$	$\sqrt{\dfrac{5}{9}} = \dfrac{\sqrt{5}}{3}$ $\sqrt{10} \div \sqrt{5} = \sqrt{2}$

Simplified radical form for radicals of index n	A simplified radical of index n has 1. *no* perfect nth powers as factors of the radicand, 2. *no* fractions inside the radical, and 3. *no* radicals in the denominator.	$\sqrt{20} = \sqrt{4 \cdot 5} = 2\sqrt{5}$ $\sqrt{\dfrac{3}{2}} = \dfrac{\sqrt{3}}{\sqrt{2}}$ $\dfrac{\sqrt{3}}{\sqrt{2}} = \dfrac{\sqrt{3}}{\sqrt{2}} \cdot \dfrac{\sqrt{2}}{\sqrt{2}} = \dfrac{\sqrt{6}}{2}$

Equations   **Examples**

Equations with radicals and exponents	1. In raising each side of an equation to an even power, we can create an equation that gives extraneous solutions. We must check. 2. When applying the even-root property, remember that there is a positive and a negative root. 3. For equations with rational exponents, raise each side to a positive or a negative power first, then apply the even- or odd-root property.	$\sqrt{x} = -3$ $x = 9$ $x^2 = 36$ $x = \pm 6$ $x^{-2/3} = 4$ $(x^{-2/3})^{-3} = 4^{-3}$ $x^2 = \dfrac{1}{64}$ $x = \pm\dfrac{1}{8}$
Distance formula	The distance between (x_1, y_1) and (x_2, y_2) is $\sqrt{(x_2 - x_1)^2 + (y_2 - y_1)^2}$.	Distance between $(1, -2)$ and $(3, -4)$ is $\sqrt{2^2 + (-2)^2}$ or $2\sqrt{2}$.

Complex Numbers   **Examples**

Complex numbers	Numbers of form $a + bi$, where a and b are real numbers: $i = \sqrt{-1}$, $i^2 = -1$	$2 + 3i$ $-6i$ $\sqrt{2} + i$
Complex conjugates	Complex numbers of the form $a + bi$ and $a - bi$: Their product is the real number $a^2 + b^2$.	$(2 + 3i)(2 - 3i) = 2^2 + 3^2$ $= 13$
Complex number operations	Add, subtract, and multiply as algebraic expressions with i being the variable. Simplify using $i^2 = -1$. Divide complex numbers by multiplying numerator and denominator by the conjugate of the denominator.	$(2 + 5i) + (4 - 2i) = 6 + 3i$ $(2 + 5i) - (4 - 2i) = -2 + 7i$ $(2 + 5i)(4 - 2i) = 18 + 16i$ $(2 + 5i) \div (4 - 2i)$ $= \dfrac{(2 + 5i)(4 + 2i)}{(4 - 2i)(4 + 2i)}$ $= \dfrac{-2 + 24i}{20} = -\dfrac{1}{10} + \dfrac{6}{5}i$
Square root of a negative number	For any positive real number b, $\sqrt{-b} = i\sqrt{b}$.	$\sqrt{-9} = i\sqrt{9} = 3i$
Imaginary solutions to equations	In the complex number system, $x^2 = k$ for any real k is equivalent to $x = \pm\sqrt{k}$.	$x^2 = -25$ $x = \pm\sqrt{-25} = \pm 5i$

ENRICHING YOUR MATHEMATICAL WORD POWER

For each mathematical term, choose the correct meaning.

1. **nth root of a**
   a. a square root
   b. the root of a^n
   c. a number b such that $a^n = b$
   d. a number b such that $b^n = a$ d

2. **square of a**
   a. a number b such that $b^2 = a$
   b. a^2
   c. $|a|$
   d. $\sqrt{a}$ b

3. **cube root of a**
   a. a^3
   b. a number b such that $b^3 = a$
   c. $a/3$
   d. a number b such that $b = a^3$ b

4. **principal root**
   a. the main root
   b. the positive even root of a positive number
   c. the positive odd root of a negative number
   d. the negative odd root of a negative number   b

5. **odd root of a**
   a. the number b such that $b^n = a$, where a is an odd number
   b. the opposite of the even root of a
   c. the nth root of a
   d. the number b such that $b^n = a$, where n is an odd number   d

6. **index of a radical**
   a. the number n in $n\sqrt{a}$
   b. the number n in $\sqrt[n]{a}$
   c. the number n in a^n
   d. the number n in $\sqrt{a^n}$ b

7. **like radicals**
   a. radicals with the same index
   b. radicals with the same radicand
   c. radicals with the same radicand and the same index
   d. radicals with even indices   c

8. **integral exponent**
   a. an exponent that is an integer
   b. a positive exponent
   c. a rational exponent
   d. a fractional exponent   a

9. **rational exponent**
   a. an exponent that produces a rational number
   b. an integral exponent
   c. an exponent that is a real number
   d. an exponent that is a rational number   d

10. **radicand**
    a. the expression $\sqrt[n]{a}$
    b. the expression $\sqrt{a}$
    c. the number a in $\sqrt[n]{a}$
    d. the number n in $\sqrt[n]{a}$ c

11. **complex numbers**
    a. $a + bi$, where a and b are real
    b. irrational numbers
    c. imaginary numbers
    d. $\sqrt{-1}$ a

12. **imaginary unit**
    a. 1
    b. -1
    c. i
    d. $\sqrt{1}$ c

13. **imaginary number**
    a. $a + bi$, where a and b are real
    b. i
    c. a complex number
    d. a complex number in which $b \neq 0$ d

14. **complex conjugates**
    a. i and $\sqrt{-1}$
    b. $a + bi$ and $a - bi$
    c. $(a + b)(a - b)$
    d. i and -1 b

REVIEW EXERCISES

7.1 *Simplify the expressions involving rational exponents. Assume all variables represent positive real numbers. Write your answers with positive exponents.*

1. $(-27)^{-2/3}$ $\dfrac{1}{9}$

2. $-25^{3/2}$ -125

3. $(2^6)^{1/3}$ 4

4. $(5^2)^{1/2}$ 5

5. $100^{-3/2}$ $\dfrac{1}{1000}$

6. $1000^{-2/3}$ $\dfrac{1}{100}$

7. $\dfrac{3x^{-1/2}}{3^{-2}x^{-1}}$ $27x^{1/2}$

8. $\dfrac{(x^2y^{-3}z)^{1/2}}{x^{1/2}yz^{-1/2}}$ $\dfrac{x^{1/2}z}{y^{5/2}}$

9. $(a^{1/2}b)^3(ab^{1/4})^2$ $a^{7/2}b^{7/2}$

10. $(t^{-1/2})^{-2}(t^{-2}v^2)$ $\dfrac{v^2}{t}$

11. $(x^{1/2}y^{1/4})(x^{1/4}y)$ $x^{3/4}y^{5/4}$

12. $(a^{1/3}b^{1/6})^2(a^{1/3}b^{2/3})$ ab

7.2 *Simplify each radical expression. Assume all variables represent positive real numbers.*

13. $\sqrt{72x^5}$ $6x^2\sqrt{2x}$

14. $\sqrt{90y^9z^4}$ $3y^4z^2\sqrt{10y}$

15. $\sqrt[3]{72x^5}$ $2x\sqrt[3]{9x^2}$

16. $\sqrt[3]{81a^8b^9}$ $3a^2b^3\sqrt[3]{3a^2}$

17. $\sqrt{2^6}$ 8

18. $\sqrt{2^7}$ $8\sqrt{2}$

19. $\sqrt{\dfrac{2}{5}}$ $\dfrac{\sqrt{10}}{5}$

20. $\sqrt{\dfrac{1}{6}}$ $\dfrac{\sqrt{6}}{6}$

21. $\sqrt[3]{\dfrac{2}{3}}$ $\dfrac{\sqrt[3]{18}}{3}$

22. $\sqrt[3]{\dfrac{1}{9}}$ $\dfrac{\sqrt[3]{3}}{3}$

23. $\dfrac{2}{\sqrt{3x}}$ $\dfrac{2\sqrt{3x}}{3x}$

24. $\dfrac{3}{\sqrt{2y}}$ $\dfrac{3\sqrt{2y}}{2y}$

25. $\dfrac{\sqrt{10y^3}}{\sqrt{6}}$ $\dfrac{y\sqrt{15y}}{3}$

26. $\dfrac{\sqrt{5x^5}}{\sqrt{8}}$ $\dfrac{x^2\sqrt{10x}}{4}$

27. $\dfrac{3}{\sqrt[3]{2a}}$ $\dfrac{3\sqrt[3]{4a^2}}{2a}$

28. $\dfrac{a}{\sqrt[3]{a^2}}$ $\sqrt[3]{a}$

29. $\dfrac{5}{\sqrt[4]{3x^2}}$ $\dfrac{5\sqrt[4]{27x^2}}{3x}$

30. $\dfrac{b}{\sqrt[4]{a^2b^3}}$ $\dfrac{\sqrt[4]{a^2b}}{a}$

31. $\sqrt[4]{48x^5y^{12}}$ $2xy^3\sqrt[4]{3x}$

32. $\sqrt[5]{32x^{10}y^{12}}$ $2x^2y^2\sqrt[5]{y^2}$

7.3 *Perform the operations and simplify. Assume the variables represent positive real numbers.*

33. $\sqrt{13} \cdot \sqrt{13}$ 13

34. $\sqrt[3]{14} \cdot \sqrt[3]{14} \cdot \sqrt[3]{14}$ 14

35. $\sqrt{27} + \sqrt{45} - \sqrt{75}$ $3\sqrt{5} - 2\sqrt{3}$

36. $\sqrt{12} - \sqrt{50} + \sqrt{72}$ $2\sqrt{3} + \sqrt{2}$

37. $\sqrt{\dfrac{1}{3}} + \sqrt{27}$ $\dfrac{10\sqrt{3}}{3}$

38. $\sqrt{\dfrac{1}{2}} - \sqrt{\dfrac{1}{8}}$ $\dfrac{\sqrt{2}}{4}$

39. $3\sqrt{2}\left(5\sqrt{2} - 7\sqrt{3}\right)$ $30 - 21\sqrt{6}$

40. $-2\sqrt{a}\left(\sqrt{a} - \sqrt{ab^6}\right)$ $-2a + 2ab^3$

41. $(2 - \sqrt{3})(3 + \sqrt{2})$ $6 - 3\sqrt{3} + 2\sqrt{2} - \sqrt{6}$

42. $(2\sqrt{x} - \sqrt{y})(\sqrt{x} + \sqrt{y})$ $2x + \sqrt{xy} - y$

43. $\sqrt[3]{40} - \sqrt[3]{5}$ $\sqrt[3]{5}$

44. $\sqrt[3]{54x^4} - x\sqrt[3]{16x}$ $5x\sqrt[3]{2x}$

7.4 *Perform the operations and simplify.*

45. $5 \div \sqrt{2}$ $\dfrac{5\sqrt{2}}{2}$

46. $\left(10\sqrt{6}\right) \div \left(2\sqrt{2}\right)$ $5\sqrt{3}$

47. $\left(\sqrt{3}\right)^4$ 9

48. $\left(-2\sqrt{x}\right)^9$ $-512x^4\sqrt{x}$

49. $\dfrac{2 - \sqrt{8}}{2}$ $1 - \sqrt{2}$

50. $\dfrac{-3 - \sqrt{18}}{-6}$ $\dfrac{1 + \sqrt{2}}{2}$

51. $\dfrac{\sqrt{6}}{1 - \sqrt{3}}$ $\dfrac{-\sqrt{6} - 3\sqrt{2}}{2}$

52. $\dfrac{\sqrt{15}}{2 + \sqrt{5}}$ $-2\sqrt{15} + 5\sqrt{3}$

53. $\dfrac{2\sqrt{3}}{3\sqrt{6} - \sqrt{12}}$ $\dfrac{3\sqrt{2} + 2}{7}$

54. $\dfrac{-\sqrt{xy}}{3\sqrt{x} + \sqrt{xy}}$ $\dfrac{-3\sqrt{y} + y}{9 - y}$

55. $\left(2w\sqrt[3]{2w^2}\right)^6$ $256w^{10}$

56. $\left(m\sqrt[4]{m^3}\right)^8$ m^{14}

7.5 *Find all real solutions to each equation.*

57. $x^2 = 16$ $\{-4, 4\}$

58. $w^2 = 100$ $\{-10, 10\}$

59. $(a - 5)^2 = 4$ $\{3, 7\}$

60. $(m - 7)^2 = 25$ $\{2, 12\}$

61. $(a + 1)^2 = 5$ $\{-1 - \sqrt{5}, -1 + \sqrt{5}\}$

62. $(x + 5)^2 = 3$ $\{-5 - \sqrt{3}, -5 + \sqrt{3}\}$

63. $(m + 1)^2 = -8$ $\varnothing$

64. $(w + 4)^2 = 16$ $\{-8, 0\}$

65. $\sqrt{m - 1} = 3$ $\{10\}$

66. $3\sqrt{x + 5} = 12$ $\{11\}$

67. $\sqrt[3]{2x + 9} = 3$ $\{9\}$

68. $\sqrt[4]{2x - 1} = 2$ $\left\{\dfrac{17}{2}\right\}$

69. $w^{2/3} = 4$ $\{-8, 8\}$

70. $m^{-4/3} = 16$ $\left\{-\dfrac{1}{8}, \dfrac{1}{8}\right\}$

71. $(m + 1)^{1/3} = 5$ $\{124\}$

72. $(w - 3)^{-2/3} = 4$ $\left\{\dfrac{23}{8}, \dfrac{25}{8}\right\}$

73. $\sqrt{x-3} = \sqrt{x+2} - 1$ $\{7\}$

74. $\sqrt{x^2+3x+6} = 4$ $\{-5, 2\}$

75. $\sqrt{5x-x^2} = \sqrt{6}$ $\{2, 3\}$

76. $\sqrt{x+4} - 2\sqrt{x-1} = -1$ $\{5\}$

77. $\sqrt{x+7} - 2\sqrt{x} = -2$ $\{9\}$

78. $\sqrt{x} - \sqrt{x-1} = 1$ $\{1\}$

79. $2\sqrt{x} - \sqrt{x-3} = 3$ $\{4\}$

80. $1 + \sqrt{x+7} = \sqrt{2x+7}$ $\{9\}$

7.6 *Perform the indicated operations. Write answers in the form a + bi.*

81. $(2-3i)(-5+5i)$ $5 + 25i$

82. $(2+i)(5-2i)$ $12 + i$

83. $(2+i) + (5-4i)$ $7 - 3i$

84. $(2+i) + (3-6i)$ $5 - 5i$

85. $(1-i) - (2-3i)$ $-1 + 2i$

86. $(3-2i) - (1-i)$ $2 - i$

87. $\dfrac{6+3i}{3}$ $2 + i$

88. $\dfrac{8+12i}{4}$ $2 + 3i$

89. $\dfrac{4-\sqrt{-12}}{2}$ $2 - i\sqrt{3}$

90. $\dfrac{6+\sqrt{-18}}{3}$ $2 + i\sqrt{2}$

91. $\dfrac{2-3i}{4+i}$ $\dfrac{5}{17} - \dfrac{14}{17}i$

92. $\dfrac{3+i}{2-3i}$ $\dfrac{3}{13} + \dfrac{11}{13}i$

Find the imaginary solutions to each equation.

93. $x^2 + 100 = 0$ $\{\pm 10i\}$

94. $25a^2 + 3 = 0$ $\left\{\pm\dfrac{i\sqrt{3}}{5}\right\}$

95. $2b^2 + 9 = 0$ $\left\{\pm\dfrac{3i\sqrt{2}}{2}\right\}$

96. $3y^2 + 8 = 0$ $\left\{\pm\dfrac{2i\sqrt{6}}{3}\right\}$

MISCELLANEOUS

Determine whether each equation is true or false and explain your answer. An equation involving variables should be marked true only if it is an identity. Do not use a calculator.

97. $2^3 \cdot 3^2 = 6^5$ False

98. $16^{1/4} = 4^{1/2}$ True

99. $(\sqrt{2})^3 = 2\sqrt{2}$ True

100. $\sqrt[3]{9} = 3$ False

101. $8^{200} \cdot 8^{200} = 64^{200}$ True

102. $\sqrt{295} \cdot \sqrt{295} = 295$ True

103. $4^{1/2} = \sqrt{2}$ False

104. $\sqrt{a^2} = |a|$ True

105. $5^2 \cdot 5^2 = 25^4$ False

106. $\sqrt{6} \div \sqrt{2} = \sqrt{3}$ True

107. $\sqrt{w^{10}} = w^5$ False

108. $\sqrt{a^{16}} = a^4$ False

109. $\sqrt{x^6} = x^3$ False

110. $\sqrt[6]{16} = \sqrt[3]{4}$ True

111. $\sqrt{x^8} = x^4$ True

112. $\sqrt[9]{2^6} = 2^{2/3}$ True

113. $\sqrt{16} = 2$ False

114. $2^{1/2} \cdot 2^{1/4} = 2^{3/4}$ True

115. $2^{600} = 4^{300}$ True

116. $\sqrt{2} \cdot \sqrt[4]{2} = \sqrt[6]{2}$ False

117. $\dfrac{2+\sqrt{6}}{2} = 1 + \sqrt{6}$ False

118. $\dfrac{4+2\sqrt{3}}{2} = 2 + \sqrt{3}$ True

119. $\sqrt{\dfrac{4}{6}} = \dfrac{2}{3}$ False

120. $8^{200} \cdot 8^{200} = 8^{400}$ True

121. $81^{2/4} = 81^{1/2}$ True

122. $(-64)^{2/6} = (-64)^{1/3}$ False

123. $(a^4 b^2)^{1/2} = |a^2 b|$ True

124. $\left(\dfrac{a^2}{b^6}\right)^{1/2} = \dfrac{|a|}{b^3}$ False

In Exercises 125–138, solve each problem.

125. Find the distance between $(-4, 6)$ and $(2, -8)$.
$2\sqrt{58}$

126. Find the distance between $(-3, -5)$ and $(5, -7)$.
$2\sqrt{17}$

127. *Falling objects.* If we neglect air resistance, the number of feet s that an object falls from rest during t seconds is given by the equation $s = 16t^2$. How long would it take the landing gear of an airplane to reach the earth if it fell off the airplane at 12,000 feet?
$5\sqrt{30}$ seconds

128. *Timber.* Anne is pulling on a 60-foot rope attached to the top of a 48-foot tree while Walter is cutting the tree at its base. How far from the base of the tree is Anne standing?

36 feet

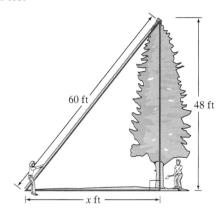

FIGURE FOR EXERCISE 128

129. *Guy wire.* If a guy wire of length 40 feet is attached to an antenna at a height of 30 feet, then how far from the base of the antenna is the wire attached to the ground?

$10\sqrt{7}$ feet

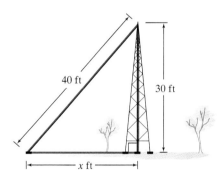

FIGURE FOR EXERCISE 129

130. *Touchdown.* Suppose at the kickoff of a football game, the receiver catches the football at the left side of the goal line and runs for a touchdown diagonally across the field. How many yards would he run? (A football field is 100 yards long and 160 feet wide.)

$113\frac{1}{3}$ yards

131. *Long guy wires.* The manufacturer of an antenna recommends that guy wires from the top of the antenna to the ground be attached to the ground at a distance from the base equal to the height of the antenna. How long would the guy wires be for a 200-foot antenna?

$200\sqrt{2}$ feet

132. *Height of a post.* Betty observed that the lamp post in front of her house casts a shadow of length 8 feet when the angle of inclination of the sun is 60 degrees. How tall is the lamp post? (In a 30-60-90 right triangle, the side opposite 30 is one-half the length of the hypotenuse.)

$8\sqrt{3}$ feet

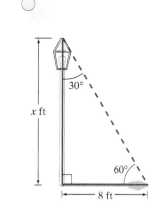

FIGURE FOR EXERCISE 132

133. *Manufacturing a box.* A cubic box has a volume of 40 cubic feet. The amount of recycled cardboard that it takes to make the six-sided box is 10% larger than the surface area of the box. Find the exact amount of recycled cardboard used in manufacturing the box.

$26.4\sqrt[3]{25}$ ft^2

134. *Shipping parts.* A cubic box with a volume of 32 cubic feet is to be used to ship some machine parts. All of the parts are small except for a long, straight steel connecting rod. What is the maximum length of a connecting rod that will fit into this box?

$2\sqrt[6]{432}$ ft

135. *Rising costs of health care.* Total annual expenditures on health care in the United States grew from $700 billion in 1990 to $1,035 billion in 1996 (Statistical Abstract of the United States, www.census.gov). Find the average annual rate of growth r for that period by solving

$$1035 = 700(1 + r)^6.$$

6.7%

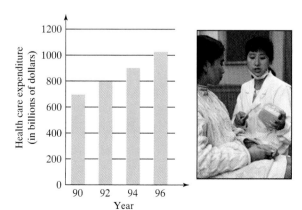

FIGURE FOR EXERCISE 135

136. *Population growth rate.* The formula $P = P_0(1 + r)^n$ gives the population P at the end of an n-year time period, where P_0 is the initial population and r is the average annual growth rate. The U.S. population grew from 248.7 million in 1990 to 270.1 million in 1998 (U.S. Census Bureau). Find the average annual rate of growth for the U.S. population for that period.

1.04%

137. *Landing speed.* Aircraft engineers determine the proper landing speed V (in feet per second) for an airplane from the formula

$$V = \sqrt{\frac{841L}{CS}},$$

where L is the gross weight of the aircraft in pounds, C is the coefficient of lift, and S is the wing surface area in square feet. Rewrite the formula so that the expression on the right-hand side is in simplified radical form.

$$V = \frac{29\sqrt{LCS}}{CS}$$

138. *Spillway capacity.* Civil engineers use the formula

$$Q = 3.32LH^{3/2}$$

to find the maximum discharge that the dam (a broad-crested weir) shown in the figure can pass before the water breaches its abutments (*Standard Handbook for Civil Engineers*, 1968). In the formula Q is the discharge in cubic feet per second, L is the length of the spillway in feet, and H is the depth of the spillway. Find Q given that $L = 60$ feet and $H = 5$ feet. Find H given that $Q = 3,000$ cubic feet per second and $L = 70$ feet.

2,227 ft³/sec, 5.5 ft

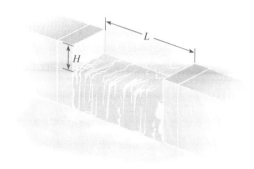

FIGURE FOR EXERCISE 138

CHAPTER 7 TEST

Simplify each expression. Assume all variables represent positive numbers.

1. $8^{2/3}$ 4

2. $4^{-3/2}$ $\dfrac{1}{8}$

3. $\sqrt{21} \div \sqrt{7}$ $\sqrt{3}$

4. $2\sqrt{5} \cdot 3\sqrt{5}$ 30

5. $\sqrt{20} + \sqrt{5}$ $3\sqrt{5}$

6. $\sqrt{5} + \dfrac{1}{\sqrt{5}}$ $\dfrac{6\sqrt{5}}{5}$

7. $2^{1/2} \cdot 2^{1/2}$ 2

8. $\sqrt{72}$ $6\sqrt{2}$

9. $\sqrt{\dfrac{5}{12}}$ $\dfrac{\sqrt{15}}{6}$

10. $\dfrac{6 + \sqrt{18}}{6}$ $\dfrac{2 + \sqrt{2}}{2}$

11. $(2\sqrt{3} + 1)(\sqrt{3} - 2)$ $4 - 3\sqrt{3}$

12. $\sqrt[4]{32a^5y^8}$ $2ay^2\sqrt[4]{2a}$

13. $\dfrac{1}{\sqrt[3]{2x^2}}$ $\dfrac{\sqrt[3]{4x}}{2x}$

14. $\sqrt{\dfrac{8a^9}{b^3}}$ $\dfrac{2a^4\sqrt{2ab}}{b^2}$

15. $\sqrt[3]{-27x^9}$ $-3x^3$

16. $\sqrt{20m^3}$ $2m\sqrt{5m}$

17. $x^{1/2} \cdot x^{1/4}$ $x^{3/4}$

18. $(9y^4x^{1/2})^{1/2}$ $3y^2x^{1/4}$

19. $\sqrt[3]{40x^7}$ $2x^2\sqrt[3]{5x}$

20. $(4 + \sqrt{3})^2$ $19 + 8\sqrt{3}$

Rationalize the denominator and simplify.

21. $\dfrac{2}{5 - \sqrt{3}}$ $\dfrac{5 + \sqrt{3}}{11}$

22. $\dfrac{\sqrt{6}}{4\sqrt{3} + \sqrt{2}}$ $\dfrac{6\sqrt{2} - \sqrt{3}}{23}$

Write each expression in the form $a + bi$.

23. $(3 - 2i)(4 + 5i)$ $22 + 7i$

24. $i^4 - i^5$ $1 - i$

25. $\dfrac{3 - i}{1 + 2i}$ $\dfrac{1}{5} - \dfrac{7}{5}i$

26. $\dfrac{-6 + \sqrt{-12}}{8}$ $-\dfrac{3}{4} + \dfrac{1}{4}i\sqrt{3}$

Find all real or imaginary solutions to each equation.

27. $(x - 2)^2 = 49$ $\{-5, 9\}$

28. $2\sqrt{x + 4} = 3$ $\left\{-\dfrac{7}{4}\right\}$

29. $w^{2/3} = 4$ $\{-8, 8\}$

30. $9y^2 + 16 = 0$ $\left\{\pm\dfrac{4}{3}i\right\}$

31. $\sqrt{2x^2 + x - 12} = x$ $\{3\}$

32. $\sqrt{x - 1} + \sqrt{x + 4} = 5$ $\{5\}$

Show a complete solution to each problem.

33. Find the distance between $(-1, 4)$ and $(1, 6)$. $2\sqrt{2}$

34. Find the exact length of the side of a square whose diagonal is 3 feet. $\dfrac{3\sqrt{2}}{2}$ feet

35. Two positive numbers differ by 11, and their square roots differ by 1. Find the numbers.
25 and 36

36. If the perimeter of a rectangle is 20 feet and the diagonal is $2\sqrt{13}$ feet, then what are the length and width?
Length 6 ft, width 4 ft

37. The average radius R of the orbit of a planet around the sun is determined by $R = T^{2/3}$, where T is the number of years for one orbit and R is measured in astronomical units (AU). If it takes Pluto 248.530 years to make one orbit of the sun, then what is the average radius of the orbit of Pluto? If the average radius of the orbit of Neptune is 30.08 AU, then how many years does it take Neptune to complete one orbit of the sun?
39.53 AU, 164.97 years

Find all real solutions to each equation or inequality. For the inequalities, also sketch the graph of the solution set.

1. $3(x - 2) + 5 = 7 - 4(x + 3)$ $\left\{-\dfrac{4}{7}\right\}$

2. $\sqrt{6x + 7} = 4$ $\left\{\dfrac{3}{2}\right\}$

3. $|2x + 5| > 1$ $(-\infty, -3) \cup (-2, \infty)$

$$\xleftarrow{\hspace{1cm}}\overset{-5\ -4\ -3\ -2\ -1\ \ 0}{)\ \ (\hspace{1cm}}\rightarrow$$

4. $8x^3 - 27 = 0$ $\left\{\dfrac{3}{2}\right\}$

5. $2x - 3 > 3x - 4$ $(-\infty, 1)$

$$\xleftarrow{\hspace{1cm}}\overset{-3\ -2\ -1\ \ 0\ \ 1\ \ 2\ \ 3}{)\hspace{1cm}}\rightarrow$$

6. $\sqrt{2x - 3} - \sqrt{3x + 4} = 0$ $\varnothing$

7. $\dfrac{w}{3} + \dfrac{w - 4}{2} = \dfrac{11}{2}$ $\{9\}$

8. $2(x + 7) - 4 = x - (10 - x)$ $\varnothing$

9. $(x + 7)^2 = 25$ $\{-12, -2\}$

10. $a^{-1/2} = 4$ $\left\{\dfrac{1}{16}\right\}$

11. $x - 3 > 2$ or $x < 2x + 6$ $(-6, \infty)$

$$\xleftarrow{\hspace{1cm}}\overset{-8\ -7\ -6\ -5\ -4\ -3\ -2}{(\hspace{1cm}}\rightarrow$$

12. $a^{-2/3} = 16$
$\left\{-\dfrac{1}{64}, \dfrac{1}{64}\right\}$

13. $3x^2 - 1 = 0$
$\left\{-\dfrac{\sqrt{3}}{3}, \dfrac{\sqrt{3}}{3}\right\}$

14. $5 - 2(x - 2) = 3x - 5(x - 2) - 1$ R

15. $|3x - 4| < 5$
$\left(-\dfrac{1}{3}, 3\right)$

$$\xleftarrow{\hspace{1cm}}\overset{-\frac{1}{3}}{(}\overset{-2\ -1\ \ 0\ \ 1\ \ 2\ \ 3\ \ 4}{\hspace{0.3cm})\hspace{1cm}}\rightarrow$$

16. $3x - 1 = 0$ $\left\{\dfrac{1}{3}\right\}$

17. $\sqrt{y - 1} = 9$ $\{82\}$

18. $|5(x - 2) + 1| = 3$ $\left\{\dfrac{6}{5}, \dfrac{12}{5}\right\}$

19. $0.06x - 0.04(x - 20) = 2.8$ $\{100\}$

20. $|3x - 1| > -2$ R

21. $\dfrac{3\sqrt{2}}{x} = \dfrac{\sqrt{3}}{4\sqrt{5}}$ $\{4\sqrt{30}\}$

22. $\dfrac{\sqrt{x} - 4}{x} = \dfrac{1}{\sqrt{x} + 5}$ $\{400\}$

23. $\dfrac{3\sqrt{2} + 4}{\sqrt{2}} = \dfrac{x\sqrt{18}}{3\sqrt{2} + 2}$ $\left\{\dfrac{13 + 9\sqrt{2}}{3}\right\}$

24. $\dfrac{x}{2\sqrt{5} - \sqrt{2}} = \dfrac{2\sqrt{5} + \sqrt{2}}{x}$ $\{-3\sqrt{2}, 3\sqrt{2}\}$

25. $\dfrac{\sqrt{2x} - 5}{x} = \dfrac{-3}{\sqrt{2x} + 5}$ $\{5\}$

26. $\dfrac{\sqrt{6} + 2}{x} = \dfrac{2}{\sqrt{6} + 4}$ $\{7 + 3\sqrt{6}\}$

27. $\dfrac{x - 1}{\sqrt{6}} = \dfrac{\sqrt{6}}{x}$ $\{-2, 3\}$

28. $\dfrac{x + 3}{\sqrt{10}} = \dfrac{\sqrt{10}}{x}$ $\{-5, 2\}$

29. $\dfrac{1}{x} - \dfrac{1}{x - 1} = -\dfrac{1}{6}$ $\{-2, 3\}$

30. $\dfrac{1}{x^2 - 2x} + \dfrac{1}{x} = \dfrac{2}{3}$ $\left\{\dfrac{1}{2}, 3\right\}$

The expression $\dfrac{-b + \sqrt{b^2 - 4ac}}{2a}$ will be used in Chapter 8 to solve quadratic equations. Evaluate it for the given values of a, b, and c.

31. $a = 1, b = 2, c = -15$ 3

32. $a = 1, b = 8, c = 12$ -2

33. $a = 2, b = 5, c = -3$ $\dfrac{1}{2}$

34. $a = 6, b = 7, c = -3$ $\dfrac{1}{3}$

Solve each problem.

35. **Popping corn.** The results of an experiment by D.D. Metzger to determine the relationship between the moisture content of popcorn and the volume of popped corn are shown in the figure (*Cereal Chemistry,* 1989). The formula $v = -94.8 + 21.4x - 0.761x^2$ can be used to model the data shown in the graph. In the formula v is the number of cubic centimeters (cm^3) of popped corn that result from popping 1 gram of corn with moisture content $x\%$ in a hot-air popper. Use the formula to find the volume that results when 1 gram of corn with a moisture content of 11% is popped.
48.5 cm^3

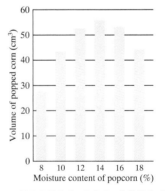

FIGURE FOR EXERCISES 35 AND 36

36. *Maximizing the volume of popped corn.* Use the graph to estimate the moisture content that will produce the maximum amount of popped corn. What is the maximum possible volume for popping 1 gram of corn in a hot-air popper?
14%, 56 cm^3

CHAPTER

8

Quadratic Equations and Inequalities

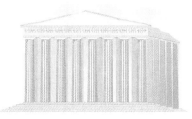

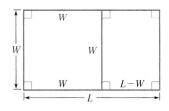

Is it possible to measure beauty? For thousands of years artists and philosophers have been challenged to answer this question. The seventeenth-century philosopher John Locke said, "Beauty consists of a certain composition of color and figure causing delight in the beholder." Over the centuries many architects, sculptors, and painters have searched for beauty in their work by exploring numerical patterns in various art forms.

Today many artists and architects still use the concepts of beauty given to us by the ancient Greeks. One principle, called the Golden Rectangle, concerns the most pleasing proportions of a rectangle. The Golden Rectangle appears in nature as well as in many cultures. Examples of it can be seen in Leonardo da Vinci's *Proportions of the Human Figure* as well as in Indonesian temples and Chinese pagodas. Perhaps one of the best-known examples of the Golden Rectangle is in the façade and floor plan of the Parthenon, built in Athens in the fifth century B.C. In Exercise 63 of Section 8.3 we will see that the principle of the Golden Rectangle is based on a proportion that we can solve using the quadratic formula.

8.1 FACTORING AND COMPLETING THE SQUARE

Factoring and the even-root property were used to solve quadratic equations in Chapters 5, 6, and 7. In this section we first review those methods. Then you will learn the method of completing the square, which can be used to solve any quadratic equation.

Review of Factoring

A quadratic equation is a second-degree polynomial equation of the form

$$ax^2 + bx + c = 0,$$

where a, b, and c are real numbers with $a \neq 0$. If the second-degree polynomial on the left-hand side can be factored, then we can solve the equation by breaking it into two first-degree polynomial equations (linear equations) using the following strategy.

> **Strategy for Solving Quadratic Equations by Factoring**
>
> 1. Write the equation with 0 on the right-hand side.
> 2. Factor the left-hand side.
> 3. Use the zero factor property to set each factor equal to zero.
> 4. Solve the simpler equations.
> 5. Check the answers in the original equation.

EXAMPLE 1 **Solving a quadratic equation by factoring**

Solve $3x^2 - 4x = 15$ by factoring.

Solution

Subtract 15 from each side to get 0 on the right-hand side:

$$3x^2 - 4x - 15 = 0$$
$$(3x + 5)(x - 3) = 0 \quad \text{Factor the left-hand side.}$$
$$3x + 5 = 0 \quad \text{or} \quad x - 3 = 0 \quad \text{Zero factor property}$$
$$3x = -5 \quad \text{or} \quad x = 3$$
$$x = -\frac{5}{3}$$

The solution set is $\left\{ -\frac{5}{3}, 3 \right\}$. Check the solutions in the original equation.

Review of the Even-Root Property

In Chapter 7 we solved quadratic equations by using the even-root property.

E X A M P L E   2

Solving a quadratic equation by the even-root property

Solve $(a - 1)^2 = 9$.

Solution

By the even-root property $x^2 = k$ is equivalent to $x = \pm\sqrt{k}$.

$$(a - 1)^2 = 9$$
$$a - 1 = \pm\sqrt{9} \quad \text{Even-root property}$$

$$a - 1 = 3 \quad \text{or} \quad a - 1 = -3$$
$$a = 4 \quad \text{or} \quad a = -2$$

Check these solutions in the original equation. The solution set is $\{-2, 4\}$. ■

Completing the Square

We cannot solve every quadratic by factoring because not all quadratic polynomials can be factored. However, we can write any quadratic equation in the form of Example 2 and then apply the even-root property to solve it. This method is called **completing the square.**

The essential part of completing the square is to recognize a perfect square trinomial when given its first two terms. For example, if we are given $x^2 + 6x$, how do we recognize that these are the first two terms of the perfect square trinomial $x^2 + 6x + 9$? To answer this question, recall that $x^2 + 6x + 9$ is a perfect square trinomial because it is the square of the binomial $x + 3$:

$$(x + 3)^2 = x^2 + 2 \cdot 3x + 3^2 = x^2 + 6x + 9$$

Notice that the 6 comes from multiplying 3 by 2 and the 9 comes from squaring the 3. So to find the missing 9 in $x^2 + 6x$, divide 6 by 2 to get 3, then square 3 to get 9. This procedure can be used to find the last term in any perfect square trinomial in which the coefficient of x^2 is 1.

Rule for Finding the Last Term

The last term of a perfect square trinomial is the square of one-half of the coefficient of the middle term. In symbols, the perfect square trinomial whose first two terms are $x^2 + bx$ is $x^2 + bx + \left(\dfrac{b}{2}\right)^2$.

E X A M P L E   3

Finding the last term

Find the perfect square trinomial whose first two terms are given.

a) $x^2 + 8x$ **b)** $x^2 - 5x$ **c)** $x^2 + \dfrac{4}{7}x$ **d)** $x^2 - \dfrac{3}{2}x$

Solution

a) One-half of 8 is 4, and 4 squared is 16. So the perfect square trinomial is

$$x^2 + 8x + 16.$$

b) One-half of -5 is $-\dfrac{5}{2}$, and $-\dfrac{5}{2}$ squared is $\dfrac{25}{4}$. So the perfect square trinomial is $x^2 - 5x + \dfrac{25}{4}.$

c) One-half of $\frac{4}{7}$ is $\frac{2}{7}$, and $\frac{2}{7}$ squared is $\frac{4}{49}$. So the perfect square trinomial is

$$x^2 + \frac{4}{7}x + \frac{4}{49}.$$

d) One-half of $-\frac{3}{2}$ is $-\frac{3}{4}$, and $\left(-\frac{3}{4}\right)^2 = \frac{9}{16}$. So the perfect square trinomial is

$$x^2 - \frac{3}{2}x + \frac{9}{16}.$$

     Another essential step in completing the square is to write the perfect square trinomial as the square of a binomial. Recall that

$$a^2 + 2ab + b^2 = (a + b)^2$$

and

$$a^2 - 2ab + b^2 = (a - b)^2.$$

E X A M P L E 4

Factoring perfect square trinomials

Factor each trinomial.

a) $x^2 + 12x + 36$ **b)** $y^2 - 7y + \dfrac{49}{4}$ **c)** $z^2 - \dfrac{4}{3}z + \dfrac{4}{9}$

Solution

a) The trinomial $x^2 + 12x + 36$ is of the form $a^2 + 2ab + b^2$ with $a = x$ and $b = 6$. So

$$x^2 + 12x + 36 = (x + 6)^2.$$

Check by squaring $x + 6$.

b) The trinomial $y^2 - 7y + \frac{49}{4}$ is of the form $a^2 - 2ab + b^2$ with $a = y$ and $b = \frac{7}{2}$. So

$$y^2 - 7y + \frac{49}{4} = \left(y - \frac{7}{2}\right)^2.$$

Check by squaring $y - \frac{7}{2}$.

c) The trinomial $z^2 - \frac{4}{3}z + \frac{4}{9}$ is of the form $a^2 - 2ab + b^2$ with $a = z$ and $b = -\frac{2}{3}$. So

$$z^2 - \frac{4}{3}z + \frac{4}{9} = \left(z - \frac{2}{3}\right)^2.$$

    In the next example we use the skills that we practiced in Examples 2, 3, and 4 to solve the quadratic equation $ax^2 + bx + c = 0$ with $a = 1$ by the method of completing the square.

E X A M P L E 5

Completing the square with $a = 1$

Solve $x^2 + 6x + 5 = 0$ by completing the square.

calculator

close-up

The solutions to

$$x^2 + 6x + 5 = 0$$

correspond to the x-intercepts for the graph of the function

$$y = x^2 + 6x + 5.$$

So we can check our solutions by graphing the function and using the TRACE feature as shown here.

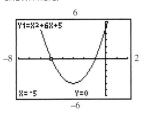

Solution

The perfect square trinomial whose first two terms are $x^2 + 6x$ is

$$x^2 + 6x + 9.$$

So we move 5 to the right-hand side of the equation, then add 9 to each side:

$$x^2 + 6x = -5 \qquad \text{Subtract 5 from each side.}$$

$$x^2 + 6x + 9 = -5 + 9 \qquad \begin{array}{l}\text{Add 9 to each side to get} \\ \text{a perfect square trinomial.}\end{array}$$

$$(x + 3)^2 = 4 \qquad \text{Factor the left-hand side.}$$

$$x + 3 = \pm\sqrt{4} \qquad \text{Even-root property}$$

$$x + 3 = 2 \qquad \text{or} \qquad x + 3 = -2$$

$$x = -1 \qquad \text{or} \qquad x = -5$$

Check in the original equation:

$$(-1)^2 + 6(-1) + 5 = 0$$

and

$$(-5)^2 + 6(-5) + 5 = 0$$

The solution set is $\{-1, -5\}$.

CAUTION All of the perfect square trinomials that we have used so far had a leading coefficient of 1. If $a \neq 1$, then we must divide each side of the equation by a to get an equation with a leading coefficient of 1.

The strategy for solving a quadratic equation by completing the square is stated in the following box.

Strategy for Solving Quadratic Equations by Completing the Square

1. The coefficient of x^2 must be 1.
2. Get only the x^2 and the x terms on the left-hand side.
3. Add to each side the square of $\frac{1}{2}$ the coefficient of x.
4. Factor the left-hand side as the square of a binomial.
5. Apply the even-root property.
6. Solve for x.
7. Simplify.

In our procedure for completing the square the coefficient of x^2 must be 1. We can solve $ax^2 + bx + c = 0$ with $a \neq 1$ by completing the square if we first divide each side of the equation by a.

EXAMPLE 6

Completing the square with $a \neq 1$

Solve $2x^2 + 3x - 2 = 0$ by completing the square.

Solution

For completing the square, the coefficient of x^2 must be 1. So we first divide each side of the equation by 2:

$$\frac{2x^2 + 3x - 2}{2} = \frac{0}{2} \qquad \text{Divide each side by 2.}$$

$$x^2 + \frac{3}{2}x - 1 \quad = 0 \qquad \text{Simplify.}$$

$$x^2 + \frac{3}{2}x \quad = 1 \qquad \begin{array}{l}\text{Get only } x^2 \text{ and } x \text{ terms on the}\\ \text{left-hand side.}\end{array}$$

$$x^2 + \frac{3}{2}x + \frac{9}{16} = 1 + \frac{9}{16} \qquad \text{One-half of } \tfrac{3}{2} \text{ is } \tfrac{3}{4}, \text{ and } \left(\tfrac{3}{4}\right)^2 = \tfrac{9}{16}.$$

$$\left(x + \frac{3}{4}\right)^2 = \frac{25}{16} \qquad \text{Factor the left-hand side.}$$

$$x + \frac{3}{4} = \pm\sqrt{\frac{25}{16}} \qquad \text{Even-root property}$$

$$x + \frac{3}{4} = \frac{5}{4} \qquad \text{or} \qquad x + \frac{3}{4} = -\frac{5}{4}$$

$$x = \frac{2}{4} = \frac{1}{2} \qquad \text{or} \qquad x = -\frac{8}{4} = -2$$

Check these values in the original equation. The solution set is $\left\{-2, \frac{1}{2}\right\}$.

In Examples 5 and 6 the solutions were rational numbers, and the equations could have been solved by factoring. In the next example the solutions are irrational numbers, and factoring will not work.

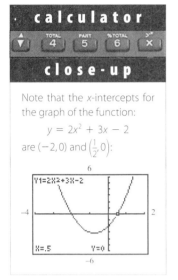

EXAMPLE 7

A quadratic equation with irrational solutions

Solve $x^2 - 3x - 6 = 0$ by completing the square.

Solution

Because $a = 1$, we first get the x^2 and x terms on the left-hand side:

$$x^2 - 3x - 6 = 0$$

$$x^2 - 3x \quad = 6 \qquad \text{Add 6 to each side.}$$

$$x^2 - 3x + \frac{9}{4} = 6 + \frac{9}{4} \qquad \text{One-half of } -3 \text{ is } -\tfrac{3}{2}, \text{ and } \left(-\tfrac{3}{2}\right)^2 = \tfrac{9}{4}.$$

$$\left(x - \frac{3}{2}\right)^2 = \frac{33}{4} \qquad 6 + \tfrac{9}{4} = \tfrac{24}{4} + \tfrac{9}{4} = \tfrac{33}{4}$$

$$x - \frac{3}{2} = \pm\sqrt{\frac{33}{4}} \qquad \text{Even-root property}$$

$$x = \frac{3}{2} \pm \frac{\sqrt{33}}{2} \qquad \text{Add } \tfrac{3}{2} \text{ to each side.}$$

$$x = \frac{3 \pm \sqrt{33}}{2}$$

The solution set is $\left\{\dfrac{3 + \sqrt{33}}{2}, \dfrac{3 - \sqrt{33}}{2}\right\}$.

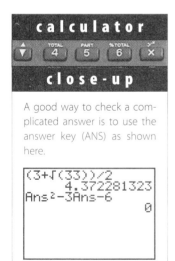

Miscellaneous Equations

The next two examples show equations that are not originally in the form of quadratic equations. However, after simplifying these equations, we get quadratic equations. Even though completing the square can be used on any quadratic equation, factoring and the square root property are usually easier and we can use them when applicable. In the next examples we will use the most appropriate method.

E X A M P L E   8

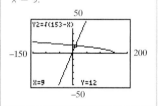

calculator

close-up

You can provide graphical support for the solution to Example 8 by graphing

$$y_1 = x + 3$$

and

$$y_2 = \sqrt{153 - x}.$$

It appears that the only point of intersection occurs when $x = 9$.

An equation containing a radical

Solve $x + 3 = \sqrt{153 - x}$.

Solution

Square both sides of the equation to eliminate the radical:

$$x + 3 = \sqrt{153 - x} \qquad \text{The original equation}$$
$$(x + 3)^2 = \left(\sqrt{153 - x}\right)^2 \qquad \text{Square each side.}$$
$$x^2 + 6x + 9 = 153 - x \qquad \text{Simplify.}$$
$$x^2 + 7x - 144 = 0$$
$$(x - 9)(x + 16) = 0 \qquad \text{Factor.}$$
$$x - 9 = 0 \quad \text{or} \quad x + 16 = 0 \qquad \text{Zero factor property}$$
$$x = 9 \quad \text{or} \quad x = -16$$

Because we squared each side of the original equation, we must check for extraneous roots. Let $x = 9$ in the original equation:

$$9 + 3 = \sqrt{153 - 9}$$
$$12 = \sqrt{144} \quad \text{Correct}$$

Let $x = -16$ in the original equation:

$$-16 + 3 = \sqrt{153 - (-16)}$$
$$-13 = \sqrt{169} \quad \text{Incorrect because } \sqrt{169} = 13$$

Because -16 is an extraneous root, the solution set is $\{9\}$.

E X A M P L E   9

An equation containing rational expressions

Solve $\dfrac{1}{x} + \dfrac{3}{x - 2} = \dfrac{5}{8}$.

Solution

The least common denominator (LCD) for x, $x - 2$, and 8 is $8x(x - 2)$.

$$\frac{1}{x} + \frac{3}{x - 2} = \frac{5}{8}$$
$$8x(x - 2)\frac{1}{x} + 8x(x - 2)\frac{3}{x - 2} = 8x(x - 2)\frac{5}{8} \qquad \text{Multiply each side by the LCD.}$$
$$8x - 16 + 24x = 5x^2 - 10x$$
$$32x - 16 = 5x^2 - 10x$$
$$-5x^2 + 42x - 16 = 0$$
$$5x^2 - 42x + 16 = 0 \qquad \text{Multiply each side by } -1 \text{ for easier factoring.}$$
$$(5x - 2)(x - 8) = 0 \qquad \text{Factor.}$$
$$5x - 2 = 0 \quad \text{or} \quad x - 8 = 0$$
$$x = \frac{2}{5} \quad \text{or} \quad x = 8$$

Check these values in the original equation. The solution set is $\left\{\frac{2}{5}, 8\right\}$.

Imaginary Solutions

In Chapter 7 we found imaginary solutions to quadratic equations using the even-root property. We can get imaginary solutions also by completing the square.

E X A M P L E   1 0   **An equation with imaginary solutions**

Find the complex solutions to $x^2 - 4x + 12 = 0$.

calculator

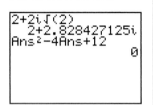

close-up

The answer key (ANS) can be used to check imaginary answers as shown here.

```
2+2i√(2)
      2+2.828427125i
Ans²-4Ans+12
                    0
```

Solution

Because the quadratic polynomial cannot be factored, we solve the equation by completing the square.

$$x^2 - 4x + 12 = 0 \qquad \text{The original equation}$$
$$x^2 - 4x = -12 \qquad \text{Subtract 12 from each side.}$$
$$x^2 - 4x + 4 = -12 + 4 \qquad \text{One-half of } -4 \text{ is } -2, \text{ and } (-2)^2 = 4.$$
$$(x - 2)^2 = -8$$
$$x - 2 = \pm\sqrt{-8} \qquad \text{Even-root property}$$
$$x = 2 \pm i\sqrt{8}$$
$$= 2 \pm 2i\sqrt{2}$$

Check these values in the original equation. The solution set is $\{2 \pm 2i\sqrt{2}\}$. ◼

WARM-UPS

True or false? Explain.

1. Completing the square means drawing the fourth side.   False
2. The equation $(x - 3)^2 = 12$ is equivalent to $x - 3 = 2\sqrt{3}$.   False
3. Every quadratic equation can be solved by factoring.   False
4. The trinomial $x^2 + \frac{4}{3}x + \frac{16}{9}$ is a perfect square trinomial.   False
5. Every quadratic equation can be solved by completing the square.   True
6. To complete the square for $2x^2 + 6x = 4$, add 9 to each side.   False
7. $(2x - 3)(3x + 5) = 0$ is equivalent to $x = \frac{3}{2}$ or $x = \frac{5}{3}$.   False
8. In completing the square for $x^2 - 3x = 4$, add $\frac{9}{4}$ to each side.   True
9. The equation $x^2 = -8$ is equivalent to $x = \pm 2\sqrt{2}$.   False
10. All quadratic equations have two distinct complex solutions.   False

8.1   EXERCISES

Reading and Writing   *After reading this section, write out the answers to these questions. Use complete sentences.*

1. What are the three methods discussed in this section for solving a quadratic equation?
    In this section quadratic equations are solved by factoring, the even-root property, and completing the square.

2. Which quadratic equations can be solved by the even-root property?
    If $b = 0$ in $ax^2 + bx + c = 0$, then the equation can be solved by the even-root property.

3. How do you find the last term for a perfect square trinomial when completing the square?
    The last term is the square of one-half the coefficient of the middle term.

4. How do you complete the square when the leading coefficient is not 1?
    If the leading coefficient is not 1, then the first step is to divide each side by the leading coefficient.

Solve by factoring. See Example 1.

5. $x^2 - x - 6 = 0$ $\{-2, 3\}$ 6. $x^2 + 6x + 8 = 0$ $\{-4, -2\}$

7. $a^2 + 2a = 15$
$\{-5, 3\}$

8. $w^2 - 2w = 15$
$\{-3, 5\}$

9. $2x^2 - x - 3 = 0$
$\left\{-1, \dfrac{3}{2}\right\}$

10. $6x^2 - x - 15 = 0$
$\left\{-\dfrac{3}{2}, \dfrac{5}{3}\right\}$

11. $y^2 + 14y + 49 = 0$
$\{-7\}$

12. $a^2 - 6a + 9 = 0$
$\{3\}$

13. $a^2 - 16 = 0$
$\{-4, 4\}$

14. $4w^2 - 25 = 0$
$\left\{-\dfrac{5}{2}, \dfrac{5}{2}\right\}$

Use the even-root property to solve each equation. See Example 2.

15. $x^2 = 81$
$\{-9, 9\}$

16. $x^2 = \dfrac{9}{4}$
$\left\{-\dfrac{3}{2}, \dfrac{3}{2}\right\}$

17. $x^2 = \dfrac{16}{9}$
$\left\{-\dfrac{4}{3}, \dfrac{4}{3}\right\}$

18. $a^2 = 32$
$\{-4\sqrt{2}, 4\sqrt{2}\}$

19. $(x - 3)^2 = 16$
$\{-1, 7\}$

20. $(x + 5)^2 = 4$
$\{-7, -3\}$

21. $(z + 1)^2 = 5$
$\{-1 - \sqrt{5}, -1 + \sqrt{5}\}$

22. $(a - 2)^2 = 8$
$\{2 - 2\sqrt{2}, 2 + 2\sqrt{2}\}$

23. $\left(w - \dfrac{3}{2}\right)^2 = \dfrac{7}{4}$
$\left\{\dfrac{3 - \sqrt{7}}{2}, \dfrac{3 + \sqrt{7}}{2}\right\}$

24. $\left(w + \dfrac{2}{3}\right)^2 = \dfrac{5}{9}$
$\left\{\dfrac{-2 - \sqrt{5}}{3}, \dfrac{-2 + \sqrt{5}}{3}\right\}$

Find the perfect square trinomial whose first two terms are given. See Example 3.

25. $x^2 + 2x$
$x^2 + 2x + 1$

26. $m^2 + 14m$
$m^2 + 14m + 49$

27. $x^2 - 3x$
$x^2 - 3x + \dfrac{9}{4}$

28. $w^2 - 5w$
$w^2 - 5w + \dfrac{25}{4}$

29. $y^2 + \dfrac{1}{4}y$
$y^2 + \dfrac{1}{4}y + \dfrac{1}{64}$

30. $z^2 + \dfrac{3}{2}z$
$z^2 + \dfrac{3}{2}z + \dfrac{9}{16}$

31. $x^2 + \dfrac{2}{3}x$
$x^2 + \dfrac{2}{3}x + \dfrac{1}{9}$

32. $p^2 + \dfrac{6}{5}p$
$p^2 + \dfrac{6}{5}p + \dfrac{9}{25}$

Factor each perfect square trinomial. See Example 4.

33. $x^2 + 8x + 16$
$(x + 4)^2$

34. $x^2 - 10x + 25$
$(x - 5)^2$

35. $y^2 - 5y + \dfrac{25}{4}$
$\left(y - \dfrac{5}{2}\right)^2$

36. $w^2 + w + \dfrac{1}{4}$
$\left(w + \dfrac{1}{2}\right)^2$

37. $z^2 - \dfrac{4}{7}z + \dfrac{4}{49}$
$\left(z - \dfrac{2}{7}\right)^2$

38. $m^2 - \dfrac{6}{5}m + \dfrac{9}{25}$
$\left(m - \dfrac{3}{5}\right)^2$

39. $t^2 + \dfrac{3}{5}t + \dfrac{9}{100}$
$\left(t + \dfrac{3}{10}\right)^2$

40. $h^2 + \dfrac{3}{2}h + \dfrac{9}{16}$
$\left(h + \dfrac{3}{4}\right)^2$

Solve by completing the square. See Examples 5–7. Use your calculator to check.

41. $x^2 - 2x - 15 = 0$ $\{-3, 5\}$

42. $x^2 - 6x - 7 = 0$ $\{-1, 7\}$

43. $x^2 + 8x = 20$ $\{-10, 2\}$

44. $x^2 + 10x = -9$ $\{-1, -9\}$

45. $2x^2 - 4x = 70$ $\{-5, 7\}$

46. $3x^2 - 6x = 24$ $\{-2, 4\}$

47. $w^2 - w - 20 = 0$ $\{-4, 5\}$

48. $y^2 - 3y - 10 = 0$ $\{-2, 5\}$

49. $q^2 + 5q = 14$ $\{-7, 2\}$

50. $z^2 + z = 2$ $\{-2, 1\}$

51. $2h^2 - h - 3 = 0$ $\left\{-1, \dfrac{3}{2}\right\}$

52. $2m^2 - m - 15 = 0$ $\left\{-\dfrac{5}{2}, 3\right\}$

53. $x^2 + 4x = 6$ $\{-2 - \sqrt{10}, -2 + \sqrt{10}\}$

54. $x^2 + 6x - 8 = 0$ $\{-3 - \sqrt{17}, -3 + \sqrt{17}\}$

55. $x^2 + 8x - 4 = 0$ $\{-4 - 2\sqrt{5}, -4 + 2\sqrt{5}\}$

56. $x^2 + 10x - 3 = 0$ $\{-5 - 2\sqrt{7}, -5 + 2\sqrt{7}\}$

57. $2x^2 + 3x - 4 = 0$ $\left\{\dfrac{-3 - \sqrt{41}}{4}, \dfrac{-3 + \sqrt{41}}{4}\right\}$

58. $2x^2 + 5x - 1 = 0$ $\left\{\dfrac{-5 - \sqrt{33}}{4}, \dfrac{-5 + \sqrt{33}}{4}\right\}$

Solve each equation by an appropriate method. See Examples 8 and 9.

59. $\sqrt{2x + 1} = x - 1$
$\{4\}$

60. $\sqrt{2x - 4} = x - 14$
$\{20\}$

61. $w = \dfrac{\sqrt{w + 1}}{2}$
$\left\{\dfrac{1 + \sqrt{17}}{8}\right\}$

62. $y - 1 = \dfrac{\sqrt{y + 1}}{2}$
$\left\{\dfrac{9 + \sqrt{33}}{8}\right\}$

63. $\dfrac{t}{t - 2} = \dfrac{2t - 3}{t}$
$\{1, 6\}$

64. $\dfrac{z}{z + 3} = \dfrac{3z}{5z - 1}$
$\{0, 5\}$

65. $\dfrac{2}{x^2} + \dfrac{4}{x} + 1 = 0$
$\{-2 - \sqrt{2}, -2 + \sqrt{2}\}$

66. $\dfrac{1}{x^2} + \dfrac{3}{x} + 1 = 0$
$\left\{\dfrac{-3 - \sqrt{5}}{2}, \dfrac{-3 + \sqrt{5}}{2}\right\}$

Find the complex solutions to each equation. See Example 10.

67. $x^2 + 2x + 5 = 0$
$\{-1 - 2i, -1 + 2i\}$

68. $x^2 + 4x + 5 = 0$
$\{-2 - i, -2 + i\}$

69. $x^2 + 12 = 0$
$\{-2i\sqrt{3}, 2i\sqrt{3}\}$

70. $-3x^2 - 21 = 0$
$\{-i\sqrt{7}, i\sqrt{7}\}$

71. $5z^2 - 4z + 1 = 0$
$\left\{\dfrac{2 \pm i}{5}\right\}$

72. $2w^2 - 3w + 2 = 0$
$\left\{\dfrac{3 \pm i\sqrt{7}}{4}\right\}$

Find all real or imaginary solutions to each equation. Use the method of your choice.

73. $4x^2 + 25 = 0$ $\left\{-\dfrac{5}{2}i, \dfrac{5}{2}i\right\}$

74. $5w^2 - 3 = 0$ $\left\{-\dfrac{\sqrt{15}}{5}, \dfrac{\sqrt{15}}{5}\right\}$

75. $\left(p + \dfrac{1}{2}\right)^2 = \dfrac{9}{4}$ $\{-2, 1\}$

76. $\left(y - \dfrac{2}{3}\right)^2 = \dfrac{4}{9}$ $\left\{0, \dfrac{4}{3}\right\}$

77. $5t^2 + 4t - 3 = 0$ $\left\{\dfrac{-2 - \sqrt{19}}{5}, \dfrac{-2 + \sqrt{19}}{5}\right\}$

78. $3v^2 + 4v - 1 = 0$ $\left\{\dfrac{-2 - \sqrt{7}}{3}, \dfrac{-2 + \sqrt{7}}{3}\right\}$

79. $m^2 + 2m - 24 = 0$ $\{-6, 4\}$

80. $q^2 + 6q - 7 = 0$ $\{-7, 1\}$

81. $\left(a + \dfrac{2}{3}\right)^2 = -\dfrac{32}{9}$ $\left\{\dfrac{-2 - 4i\sqrt{2}}{3}, \dfrac{-2 + 4i\sqrt{2}}{3}\right\}$

82. $\left(w + \dfrac{1}{2}\right)^2 = -6$ $\left\{\dfrac{-1 - 2i\sqrt{6}}{2}, \dfrac{-1 + 2i\sqrt{6}}{2}\right\}$

83. $-x^2 + x + 6 = 0$ $\{-2, 3\}$

84. $-x^2 + x + 12 = 0$ $\{-3, 4\}$

85. $x^2 - 6x + 10 = 0$ $\{3 - i, 3 + i\}$

86. $x^2 - 8x + 17 = 0$ $\{4 - i, 4 + i\}$

87. $2x - 5 = \sqrt{7x + 7}$ $\{6\}$

88. $\sqrt{7x + 29} = x + 3$ $\{5\}$

89. $\dfrac{1}{x} + \dfrac{1}{x - 1} = \dfrac{1}{4}$ $\left\{\dfrac{9 - \sqrt{65}}{2}, \dfrac{9 + \sqrt{65}}{2}\right\}$

90. $\dfrac{1}{x} - \dfrac{2}{1 - x} = \dfrac{1}{2}$ $\left\{\dfrac{7 - \sqrt{41}}{2}, \dfrac{7 + \sqrt{41}}{2}\right\}$

If the solution to an equation is imaginary or irrational, it takes a bit more effort to check. Replace x by each given number to verify each statement.

91. Both $2 + \sqrt{3}$ and $2 - \sqrt{3}$ satisfy $x^2 - 4x + 1 = 0$.

92. Both $1 + \sqrt{2}$ and $1 - \sqrt{2}$ satisfy $x^2 - 2x - 1 = 0$.

93. Both $1 + i$ and $1 - i$ satisfy $x^2 - 2x + 2 = 0$.

94. Both $2 + 3i$ and $2 - 3i$ satisfy $x^2 - 4x + 13 = 0$.

Solve each problem.

95. *Approach speed.* The formula $1211.1L = CA^2S$ is used to determine the approach speed for landing an aircraft, where L is the gross weight of the aircraft in pounds, C is the

coefficient of lift, S is the surface area of the wings in square feet (ft^2), and A is approach speed in feet per second. Find A for the Piper Cheyenne, which has a gross weight of 8,700 lb, a coefficient of lift of 2.81, and wing surface area of 200 ft^2.   136.9 ft/sec

96. *Time to swing.* The period T (time in seconds for one complete cycle) of a simple pendulum is related to the length L (in feet) of the pendulum by the formula $8T^2 = \pi^2 L$. If a child is on a swing with a 10-foot chain, then how long does it take to complete one cycle of the swing?   3.5 sec

97. *Time for a swim.* Tropical Pools figures that its monthly revenue in dollars on the sale of x above-ground pools is given by $R = 1500x - 3x^2$, where x is less than 25. What number of pools sold would provide a revenue of $17,568?   12

98. *Pole vaulting.* In 1981 Vladimir Poliakov (USSR) set a world record of 19 ft $\frac{3}{4}$ in. for the pole vault (Doubleday Almanac). To reach that height, Poliakov obtained a speed of approximately 36 feet per second on the runway. The function $h = -16t^2 + 36t$ gives his height t seconds after leaving the ground.

a) Use the formula to find the exact values of t for which his height was 18 feet.

b) Use the accompanying graph to estimate the value of t for which he was at his maximum height.

c) Approximately how long was he in the air?
   a) 0.75 sec and 1.5 sec   b) 1.125 sec
   c) 2.25 sec

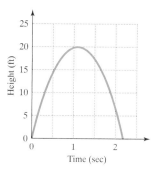

FIGURE FOR EXERCISE 98

GETTING MORE INVOLVED

99. *Discussion.* Which of the following equations is not a quadratic equation?
   a) $\pi x^2 - \sqrt{5}x - 1 = 0$ b) $3x^2 - 1 = 0$
   c) $4x + 5 = 0$ d) $0.009x^2 = 0$
   c

100. *Exploration.* Solve $x^2 - 4x + k = 0$ for $k = 0, 4, 5,$ and 10.

a) When does the equation have only one solution?

b) For what values of k are the solutions real?

c) For what values of k are the solutions imaginary?
   a) $k = 4$ b) $k < 4$ c) $k > 4$

101. *Cooperative learning.* Write a quadratic equation of each of the following types, then trade your equations with those of a classmate. Solve the equations and verify that they are of the required types.
  a) a single rational solution
  b) two rational solutions
  c) two irrational solutions
  d) two imaginary solutions

 102. *Exploration.* In the next section we will solve $ax^2 + bx + c = 0$ for x by completing the square. Try it now without looking ahead.

 GRAPHING CALCULATOR EXERCISES

For each equation, find approximate solutions rounded to two decimal places.

103. $x^2 - 7.3x + 12.5 = 0$ $\{4.56, 2.74\}$

104. $1.2x^2 - \pi x + \sqrt{2} = 0$ $\{2.04, 0.58\}$

105. $2x - 3 = \sqrt{20 - x}$ $\{3.53\}$

106. $x^2 - 1.3x = 22.3 - x^2$ $\{-3.03, 3.68\}$

8.2 **THE QUADRATIC FORMULA**

- Developing the Formula
- Using the Formula
- Number of Solutions
- Applications

Completing the square from Section 8.1 can be used to solve any quadratic equation. Here we apply this method to the general quadratic equation to get a formula for the solutions to any quadratic equation.

Developing the Formula

Start with the general form of the quadratic equation,

$$ax^2 + bx + c = 0.$$

Assume a is positive for now, and divide each side by a:

$$\frac{ax^2 + bx + c}{a} = \frac{0}{a}$$

$$x^2 + \frac{b}{a}x + \frac{c}{a} = 0$$

$$x^2 + \frac{b}{a}x = -\frac{c}{a} \quad \text{Subtract } \frac{c}{a} \text{ from each side.}$$

One-half of $\frac{b}{a}$ is $\frac{b}{2a}$, and $\frac{b}{2a}$ squared is $\frac{b^2}{4a^2}$:

$$x^2 + \frac{b}{a}x + \frac{b^2}{4a^2} = -\frac{c}{a} + \frac{b^2}{4a^2}$$

Factor the left-hand side and get a common denominator for the right-hand side:

$$\left(x + \frac{b}{2a}\right)^2 = \frac{b^2}{4a^2} - \frac{4ac}{4a^2} \qquad \frac{c(4a)}{a(4a)} = \frac{4ac}{4a^2}$$

$$\left(x + \frac{b}{2a}\right)^2 = \frac{b^2 - 4ac}{4a^2}$$

$$x + \frac{b}{2a} = \pm\sqrt{\frac{b^2 - 4ac}{4a^2}} \qquad \text{Even-root property}$$

$$x = \frac{-b}{2a} \pm \frac{\sqrt{b^2 - 4ac}}{2a} \qquad \text{Because } a > 0, \sqrt{4a^2} = 2a.$$

$$x = \frac{-b \pm \sqrt{b^2 - 4ac}}{2a}$$

We assumed a was positive so that $\sqrt{4a^2} = 2a$ would be correct. If a is negative, then $\sqrt{4a^2} = -2a$, and we get

$$x = \frac{-b}{2a} \pm \frac{\sqrt{b^2 - 4ac}}{-2a}.$$

However, the negative sign can be omitted in $-2a$ because of the $\pm$ symbol preceding it. For example, the results of $5 \pm (-3)$ and 5 ± 3 are the same. So when a is negative, we get the same formula as when a is positive. It is called the **quadratic formula.**

> ### The Quadratic Formula
>
> The solution to $ax^2 + bx + c = 0$, with $a \neq 0$, is given by the formula
>
> $$x = \frac{-b \pm \sqrt{b^2 - 4ac}}{2a}.$$

Using the Formula

The quadratic formula can be used to solve any quadratic equation.

E X A M P L E   1   **Two rational solutions**

Solve $x^2 + 2x - 15 = 0$ using the quadratic formula.

Solution

To use the formula, we first identify the values of a, b, and c:

$$1x^2 + 2x - 15 = 0$$
$$\uparrow \qquad \uparrow \qquad \uparrow$$
$$a \qquad \quad b \qquad \ c$$

The coefficient of x^2 is 1, so $a = 1$. The coefficient of $2x$ is 2, so $b = 2$. The constant term is -15, so $c = -15$. Substitute these values into the quadratic formula:

$$x = \frac{-2 \pm \sqrt{2^2 - 4(1)(-15)}}{2(1)}$$

$$= \frac{-2 \pm \sqrt{4 + 60}}{2}$$

$$= \frac{-2 \pm \sqrt{64}}{2}$$

$$= \frac{-2 \pm 8}{2}$$

$$x = \frac{-2 + 8}{2} = 3 \qquad \text{or} \qquad x = \frac{-2 - 8}{2} = -5$$

Check 3 and -5 in the original equation. The solution set is $\{-5, 3\}$.   ■

c a l c u l a t o r

c l o s e - u p

Note that the two solutions to

$$x^2 + 2x - 15 = 0$$

correspond to the two x-intercepts for the graph of the function

$$y = x^2 + 2x - 15.$$

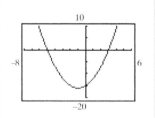

C A U T I O N   To identify a, b, and c for the quadratic formula, the equation must be in the standard form $ax^2 + bx + c = 0$. If it is not in that form, then you must first rewrite the equation.

EXAMPLE 2

One rational solution

Solve $4x^2 = 12x - 9$ by using the quadratic formula.

Solution

Rewrite the equation in the form $ax^2 + bx + c = 0$ before identifying a, b, and c:

$$4x^2 - 12x + 9 = 0$$

In this form we get $a = 4$, $b = -12$, and $c = 9$.

$$x = \frac{12 \pm \sqrt{(-12)^2 - 4(4)(9)}}{2(4)} \qquad \text{Because } b = -12, -b = 12.$$

$$= \frac{12 \pm \sqrt{144 - 144}}{8}$$

$$= \frac{12 \pm 0}{8} = \frac{12}{8} = \frac{3}{2}$$

Check $\frac{3}{2}$ in the original equation. The solution set is $\left\{\frac{3}{2}\right\}$. ■

Because the solutions to the equations in Examples 1 and 2 were rational numbers, these equations could have been solved by factoring. In the next example the solutions are irrational.

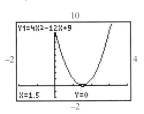

calculator

close-up

Note that the single solution to

$$4x^2 - 12x + 9 = 0$$

corresponds to the single x-intercept for the graph of the function

$$y = 4x^2 - 12x + 9.$$

EXAMPLE 3

Two irrational solutions

Solve $2x^2 + 6x + 3 = 0$.

Solution

Let $a = 2$, $b = 6$, and $c = 3$ in the quadratic formula:

$$x = \frac{-6 \pm \sqrt{(6)^2 - 4(2)(3)}}{2(2)}$$

$$= \frac{-6 \pm \sqrt{36 - 24}}{4} = \frac{-6 \pm \sqrt{12}}{4}$$

$$= \frac{-6 \pm 2\sqrt{3}}{4} = \frac{2(-3 \pm \sqrt{3})}{2 \cdot 2}$$

$$= \frac{-3 \pm \sqrt{3}}{2}$$

Check these values in the original equation. The solution set is $\left\{\frac{-3 \pm \sqrt{3}}{2}\right\}$. ■

EXAMPLE 4

Two imaginary solutions, no real solutions

Find the complex solutions to $x^2 + x + 5 = 0$.

Solution

Let $a = 1$, $b = 1$, and $c = 5$ in the quadratic formula:

$$x = \frac{-1 \pm \sqrt{(1)^2 - 4(1)(5)}}{2(1)} = \frac{-1 \pm \sqrt{-19}}{2} = \frac{-1 \pm i\sqrt{19}}{2}$$

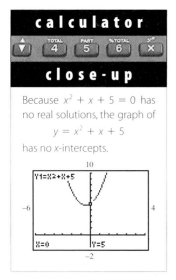

h e l p f u l   h i n t

If our only intent is to obtain the answer, then we would probably use a calculator that is programmed with the quadratic formula. However, by learning different methods, we gain insight into the problem and get valuable practice with algebra. So be sure you learn all of the methods.

Check these values in the original equation. The solution set is $\left\{\dfrac{-1 \pm i\sqrt{19}}{2}\right\}$. There are no real solutions to the equation. ■

You have learned to solve quadratic equations by four different methods: the even-root property, factoring, completing the square, and the quadratic formula. The even-root property and factoring are limited to certain special equations, but you should use those methods when possible. Any quadratic equation can be solved by completing the square or using the quadratic formula. Because the quadratic formula is usually faster, it is used more often than completing the square. However, completing the square is an important skill to learn. It will be used in the study of conic sections later in this text.

Methods for Solving $ax^2 + bx + c = 0$

Method	Comments	Examples
Even-root property	Use when $b = 0$.	$(x - 2)^2 = 8$ $x - 2 = \pm\sqrt{8}$
Factoring	Use when the polynomial can be factored.	$x^2 + 5x + 6 = 0$ $(x + 2)(x + 3) = 0$
Quadratic formula	Solves any quadratic equation	$x^2 + 5x + 3 = 0$ $x = \dfrac{-5 \pm \sqrt{25 - 4(3)}}{2}$
Completing the square	Solves any quadratic equation, but quadratic formula is faster	$x^2 - 6x + 7 = 0$ $x^2 - 6x + 9 = -7 + 9$ $(x - 3)^2 = 2$

M A T H   A T   W O R K

$x^2 + (x+1)^2 = 5^2$

Remodeling a kitchen can be an expensive undertaking. Choosing the correct style of cabinets, doors, and floor covering is just a small part of the process. Joe Prendergast, designer for Lee Kimball Kitchens, is involved in every step of creating a new kitchen for a client.

KITCHEN DESIGNER

The process begins with customers visiting the store to see what products are available. Once the client has decided on material and style, he or she fills out an extensive questionnaire so that Prendergast can get a sense of the client's lifestyle. Design plans are drawn using a scale of $\frac{1}{2}$ inch to 1 foot. Consideration is given to traffic patterns, doorways, and especially work areas where the cook would want to work unencumbered. Storage areas are a big consideration, as are lighting and color.

A new kitchen can take anywhere from 5 weeks to a few months or more, depending on how complicated the design and construction is. In Exercise 83 of this section we will find the dimensions of a border around a countertop.

Number of Solutions

The quadratic equations in Examples 1 and 3 had two real solutions each. In each of those examples the value of $b^2 - 4ac$ was positive. In Example 2 the quadratic equation had only one solution because the value of $b^2 - 4ac$ was zero. In Example 4 the quadratic equation had no real solutions because $b^2 - 4ac$ was negative. Because $b^2 - 4ac$ determines the kind and number of solutions to a quadratic equation, it is called the **discriminant.**

Number of Solutions to a Quadratic Equation

The quadratic equation $ax^2 + bx + c = 0$ with $a \neq 0$ has
    two real solutions if $b^2 - 4ac > 0$,
    one real solution if $b^2 - 4ac = 0$, and
    no real solutions (two imaginary solutions) if $b^2 - 4ac < 0$.

E X A M P L E   5

Using the discriminant

Use the discriminant to determine the number of real solutions to each quadratic equation.

a) $x^2 - 3x - 5 = 0$ **b)** $x^2 = 3x - 9$ **c)** $4x^2 - 12x + 9 = 0$

Solution

a) For $x^2 - 3x - 5 = 0$, use $a = 1$, $b = -3$, and $c = -5$ in $b^2 - 4ac$:

$$b^2 - 4ac = (-3)^2 - 4(1)(-5) = 9 + 20 = 29$$

Because the discriminant is positive, there are two real solutions to this quadratic equation.

b) For $x^2 - 3x + 9 = 0$, use $a = 1$, $b = -3$, and $c = 9$ in $b^2 - 4ac$:

$$b^2 - 4ac = (-3)^2 - 4(1)(9) = 9 - 36 = -27$$

Because the discriminant is negative, the equation has no real solutions. It has two imaginary solutions.

c) For $4x^2 - 12x + 9 = 0$, use $a = 4$, $b = -12$, and $c = 9$ in $b^2 - 4ac$:

$$b^2 - 4ac = (-12)^2 - 4(4)(9) = 144 - 144 = 0$$

Because the discriminant is zero, there is only one real solution to this quadratic equation.

Applications

With the quadratic formula we can easily solve problems whose solutions are irrational numbers. When the solutions are irrational numbers, we usually use a calculator to find rational approximations and to check.

E X A M P L E   6

Area of a tabletop

The area of a rectangular tabletop is 6 square feet. If the width is 2 feet shorter than the length, then what are the dimensions?

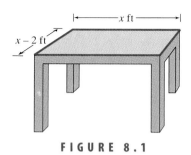

FIGURE 8.1

Solution

Let x be the length and $x - 2$ be the width, as shown in Fig. 8.1. Because the area is 6 square feet and $A = LW$, we can write the equation

$$x(x - 2) = 6$$

or

$$x^2 - 2x - 6 = 0.$$

Because this equation cannot be factored, we use the quadratic formula with $a = 1$, $b = -2$, and $c = -6$:

$$x = \frac{2 \pm \sqrt{(-2)^2 - 4(1)(-6)}}{2(1)}$$

$$= \frac{2 \pm \sqrt{28}}{2} = \frac{2 \pm 2\sqrt{7}}{2} = 1 \pm \sqrt{7}$$

Because $1 - \sqrt{7}$ is a negative number, it cannot be the length of a tabletop. If $x = 1 + \sqrt{7}$, then $x - 2 = 1 + \sqrt{7} - 2 = \sqrt{7} - 1$. Checking the product of $\sqrt{7} + 1$ and $\sqrt{7} - 1$, we get

$$(\sqrt{7} + 1)(\sqrt{7} - 1) = 7 - 1 = 6.$$

The exact length is $\sqrt{7} + 1$ feet, and the width is $\sqrt{7} - 1$ feet. Using a calculator, we find that the approximate length is 3.65 feet and the approximate width is 1.65 feet. ∎

WARM-UPS

True or false? Explain.

1. Completing the square is used to develop the quadratic formula.   True
2. For the equation $3x^2 = 4x - 7$, we have $a = 3$, $b = 4$, and $c = -7$. False
3. If $dx^2 + ex + f = 0$ and $d \neq 0$, then $x = \frac{-e \pm \sqrt{e^2 - 4df}}{2d}$.   True
4. The quadratic formula will not work on the equation $x^2 - 3 = 0$.   False
5. If $a = 2$, $b = -3$, and $c = -4$, then $b^2 - 4ac = 41$.   True
6. If the discriminant is zero, then there are no imaginary solutions.   True
7. If $b^2 - 4ac > 0$, then $ax^2 + bx + c = 0$ has two real solutions.   True
8. To solve $2x - x^2 = 0$ by the quadratic formula, let $a = -1$, $b = 2$, and $c = 0$.   True
9. Two numbers that have a sum of 6 can be represented by x and $x + 6$.   False
10. Some quadratic equations have one real and one imaginary solution.   False

8.2   EXERCISES

Reading and Writing   After reading this section, write out the answers to these questions. Use complete sentences.

1. What is the quadratic formula used for?
   The quadratic formula can be used to solve any quadratic equation.
2. When do you use the even-root property to solve a quadratic equation?
   The even-root property is used when $b = 0$.

3. When do you use factoring to solve a quadratic equation?
   Factoring is used when the quadratic polynomial is simple enough to factor.
4. When do you use the quadratic formula to solve a quadratic equation?
   The quadratic formula can be used on any quadratic equation, but generally we use factoring or the even-root property when applicable.

5. What is the discriminant?

The discriminant is $b^2 - 4ac$.

6. How many solutions are there to any quadratic equation in the complex number system?

In the complex number system any quadratic equation has either one or two solutions.

Solve each equation by using the quadratic formula. See Example 1.

7. $x^2 + 5x + 6 = 0$
$\{-3, -2\}$

8. $x^2 - 7x + 12 = 0$
$\{3, 4\}$

9. $y^2 + y = 6$
$\{-3, 2\}$

10. $m^2 + 2m = 8$
$\{-4, 2\}$

11. $6z^2 - 7z - 3 = 0$
$\left\{-\dfrac{1}{3}, \dfrac{3}{2}\right\}$

12. $8q^2 + 2q - 1 = 0$
$\left\{-\dfrac{1}{2}, \dfrac{1}{4}\right\}$

Solve each equation by using the quadratic formula. See Example 2.

13. $4x^2 - 4x + 1 = 0$
$\left\{\dfrac{1}{2}\right\}$

14. $4x^2 - 12x + 9 = 0$
$\left\{\dfrac{3}{2}\right\}$

15. $9x^2 - 6x + 1 = 0$
$\left\{\dfrac{1}{3}\right\}$

16. $9x^2 - 24x + 16 = 0$
$\left\{\dfrac{4}{3}\right\}$

17. $9 + 24x + 16x^2 = 0$
$\left\{-\dfrac{3}{4}\right\}$

18. $4 + 20x = -25x^2$
$\left\{-\dfrac{2}{5}\right\}$

Solve each equation by using the quadratic formula. See Example 3.

19. $v^2 + 8v + 6 = 0$
$\{-4 \pm \sqrt{10}\}$

20. $p^2 + 6p + 4 = 0$
$\{-3 \pm \sqrt{5}\}$

21. $-x^2 - 5x + 1 = 0$
$\left\{\dfrac{-5 \pm \sqrt{29}}{2}\right\}$

22. $-x^2 - 3x + 5 = 0$
$\left\{\dfrac{-3 \pm \sqrt{29}}{2}\right\}$

23. $2t^2 - 6t + 1 = 0$
$\left\{\dfrac{3 \pm \sqrt{7}}{2}\right\}$

24. $3z^2 - 8z + 2 = 0$
$\left\{\dfrac{4 \pm \sqrt{10}}{3}\right\}$

Solve each equation by using the quadratic formula. See Example 4.

25. $2t^2 - 6t + 5 = 0$
$\left\{\dfrac{3 \pm i}{2}\right\}$

26. $2y^2 + 1 = 2y$
$\left\{\dfrac{1 \pm i}{2}\right\}$

27. $-2x^2 + 3x = 6$
$\left\{\dfrac{3 \pm i\sqrt{39}}{4}\right\}$

28. $-3x^2 - 2x - 5 = 0$
$\left\{\dfrac{-1 \pm i\sqrt{14}}{3}\right\}$

29. $\dfrac{1}{2}x^2 + 13 = 5x$
$\{5 \pm i\}$

30. $\dfrac{1}{4}x^2 + \dfrac{17}{4} = 2x$
$\{4 \pm i\}$

Find $b^2 - 4ac$ and the number of real solutions to each equation. See Example 5.

31. $x^2 - 6x + 2 = 0$
28, 2

32. $x^2 + 6x + 9 = 0$
0, 1

33. $2x^2 - 5x + 6 = 0$
$-23, 0$

34. $-x^2 + 3x - 4 = 0$
$-7, 0$

35. $4m^2 + 25 = 20m$
0, 1

36. $v^2 = 3v + 5$
29, 2

37. $y^2 - \dfrac{1}{2}y + \dfrac{1}{4} = 0$
$-\dfrac{3}{4}, 0$

38. $\dfrac{1}{2}w^2 - \dfrac{1}{3}w + \dfrac{1}{4} = 0$
$-\dfrac{7}{18}, 0$

39. $-3t^2 + 5t + 6 = 0$
97, 2

40. $9m^2 + 16 = 24m$
0, 1

41. $9 - 24z + 16z^2 = 0$
0, 1

42. $12 - 7x + x^2 = 0$
1, 2

43. $5x^2 - 7 = 0$
140, 2

44. $-6x^2 - 5 = 0$
$-120, 0$

45. $x^2 = x$
1, 2

46. $-3x^2 + 7x = 0$
49, 2

Solve each equation by the method of your choice.

47. $\dfrac{1}{3}x^2 + \dfrac{1}{2}x = \dfrac{1}{3}$
$\left\{-2, \dfrac{1}{2}\right\}$

48. $\dfrac{1}{2}x^2 + x = 1$
$\{-1 \pm \sqrt{3}\}$

49. $\dfrac{w}{w-2} = \dfrac{w}{w-3}$
$\{0\}$

50. $\dfrac{y}{3y-4} = \dfrac{2}{y+4}$
$\{1 \pm i\sqrt{7}\}$

51. $\dfrac{9(3x-5)^2}{4} = 1$
$\left\{\dfrac{13}{9}, \dfrac{17}{9}\right\}$

52. $\dfrac{25(2x+1)^2}{9} = 0$
$\left\{-\dfrac{1}{2}\right\}$

53. $1 + \dfrac{20}{x^2} = \dfrac{8}{x}$
$\{4 \pm 2i\}$

54. $\dfrac{34}{x^2} = \dfrac{6}{x} - 1$
$\{3 \pm 5i\}$

55. $(x-8)(x+4) = -42$
$\{2 \pm i\sqrt{6}\}$

56. $(x-10)(x-2) = -20$
$\{6 \pm 2i\}$

57. $y = \dfrac{3(2y+5)}{8(y-1)}$
$\left\{-\dfrac{3}{4}, \dfrac{5}{2}\right\}$

58. $z = \dfrac{7z-4}{12(z-1)}$
$\left\{\dfrac{1}{4}, \dfrac{4}{3}\right\}$

 Use the quadratic formula and a calculator to solve each equation. Round answers to three decimal places and check your answers.

59. $x^2 + 3.2x - 5.7 = 0$ $\{-4.474, 1.274\}$

60. $x^2 + 7.15x + 3.24 = 0$ $\{-6.664, -0.486\}$

61. $x^2 - 7.4x + 13.69 = 0$ $\{3.7\}$

62. $1.44x^2 + 5.52x + 5.29 = 0$ $\{-1.917\}$

63. $1.85x^2 + 6.72x + 3.6 = 0$ $\{-2.979, -0.653\}$

64. $3.67x^2 + 4.35x - 2.13 = 0$ $\{-1.558, 0.373\}$

65. $3x^2 + 14{,}379x + 243 = 0$ $\{-4792.983, -0.017\}$

66. $x^2 + 12{,}347x + 6{,}741 = 0$ $\{-12{,}346.454, -0.546\}$

67. $x^2 + 0.00075x - 0.0062 = 0$ $\{-0.079, 0.078\}$

68. $4.3x^2 - 9.86x - 3.75 = 0$ $\{-0.332, 2.625\}$

Solve each problem. See Example 6.

69. Missing numbers. Find two positive real numbers that differ by 1 and have a product of 16.

$$\frac{1 + \sqrt{65}}{2} \text{ and } \frac{-1 + \sqrt{65}}{2}$$

70. Missing numbers. Find two positive real numbers that differ by 2 and have a product of 10.

$1 + \sqrt{11}$ and $-1 + \sqrt{11}$

71. More missing numbers. Find two real numbers that have a sum of 6 and a product of 4.

$3 + \sqrt{5}$ and $3 - \sqrt{5}$

72. More missing numbers. Find two real numbers that have a sum of 8 and a product of 2.

$4 - \sqrt{14}$ and $4 + \sqrt{14}$

73. Bulletin board. The length of a bulletin board is one foot more than the width. The diagonal has a length of $\sqrt{3}$ feet (ft). Find the length and width of the bulletin board.

Width $\dfrac{-1 + \sqrt{5}}{2}$ ft, length $\dfrac{1 + \sqrt{5}}{2}$ ft

74. Diagonal brace. The width of a rectangular gate is 2 meters (m) larger than its height. The diagonal brace measures $\sqrt{6}$ m. Find the width and height.

Width $-1 + \sqrt{2}$ m, length $1 + \sqrt{2}$ m

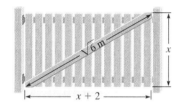

FIGURE FOR EXERCISE 74

75. Area of a rectangle. The length of a rectangle is 4 ft longer than the width, and its area is 10 square feet (ft^2). Find the length and width.

Width $-2 + \sqrt{14}$ ft, length $2 + \sqrt{14}$ ft

76. Diagonal of a square. The diagonal of a square is 2 m longer than a side. Find the length of a side.

$2 + 2\sqrt{2}$ m

If an object is given an initial velocity of v_0 feet per second from a height of s_0 feet, then its height S after t seconds is given by the formula $S = -16t^2 + v_0 t + s_0$.

77. Projected pine cone. If a pine cone is projected upward at a velocity of 16 ft/sec from the top of a 96-foot pine tree, then how long does it take to reach the earth?

3 sec

78. Falling pine cone. If a pine cone falls from the top of a 96-foot pine tree, then how long does it take to reach the earth?

$\sqrt{6}$ sec

 79. Penny tossing. If a penny is thrown downward at 30 ft/sec from the bridge at Royal Gorge, Colorado, how long does it take to reach the Arkansas River 1,000 ft below?

7.02 sec

 80. Foul ball. Suppose Charlie O'Brian of the Braves hits a baseball straight upward at 150 ft/sec from a height of 5 ft.

   a) Use the formula to determine how long it takes the ball to return to the earth.

   b) Use the accompanying graph to estimate the maximum height reached by the ball?

   a) 9.408 sec   **b)** 356.6 ft

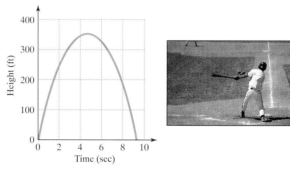

FIGURE FOR EXERCISE 80

In Exercises 81–83, solve each problem.

81. Recovering an investment. The manager at Cream of the Crop bought a load of watermelons for $200. She priced the melons so that she would make $1.50 profit on each melon. When all but 30 had been sold, the manager had recovered her initial investment. How many did she buy originally?

80

82. Sharing cost. The members of a flying club plan to share equally the cost of a $200,000 airplane. The members want to find five more people to join the club so that the cost per person will decrease by $2,000. How many members are currently in the club?

20

83. Kitchen countertop. A 30 in. by 40 in. countertop for a work island is to be covered with green ceramic tiles, except for a border of uniform width as shown in the figure. If the area covered by the green tiles is 704 square inches (in.^{2}), then how wide is the border?

4 in.

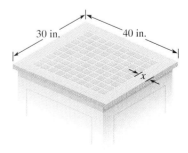

FIGURE FOR EXERCISE 83

GETTING MORE INVOLVED

 84. Discussion. Find the solutions to $6x^2 + 5x - 4 = 0$. Is the sum of your solutions equal to $-\frac{b}{a}$? Explain why the sum of the solutions to any quadratic equation is $-\frac{b}{a}$. (*Hint:* Use the quadratic formula.)

 85. Discussion. Use the result of Exercise 84 to check whether $\left\{\frac{2}{3}, \frac{1}{3}\right\}$ is the solution set to $9x^2 - 3x - 2 = 0$. If this solution set is not correct, then what is the correct solution set?

86. Discussion. What is the product of the two solutions to $6x^2 + 5x - 4 = 0$? Explain why the product of the solutions to any quadratic equation is $\frac{c}{a}$.

87. Discussion. Use the result of the previous exercise to check whether $\left\{\frac{9}{2}, -2\right\}$ is the solution set to $2x^2 - 13x + 18 = 0$. If this solution set is not correct, then what is the correct solution set?

 88. Cooperative learning. Work in a group to write a quadratic equation that has each given pair of solutions.
a) -4 and 5
b) $2 - \sqrt{3}$ and $2 + \sqrt{3}$
c) $5 + 2i$ and $5 - 2i$

 ### GRAPHING CALCULATOR EXERCISES

Determine the number of real solutions to each equation by examining the calculator graph of the corresponding function. Use the discriminant to check your conclusions.

89. $x^2 - 6.33x + 3.7 = 0$ $\{0.652, 5.678\}$
90. $1.8x^2 + 2.4x - 895 = 0$ $\{-22.975, 21.642\}$
91. $4x^2 - 67.1x + 344 = 0$ $\varnothing$
92. $-2x^2 - 403 = 0$ $\varnothing$
93. $-x^2 + 30x - 226 = 0$ $\varnothing$
94. $16x^2 - 648x + 6562 = 0$ $\varnothing$

8.3   MORE ON QUADRATIC EQUATIONS

In this section we use the ideas and methods of the previous sections to explore additional topics involving quadratic equations.

In this section

- Using the Discriminant in Factoring
- Writing a Quadratic with Given Solutions
- Equations Quadratic in Form
- Applications

Using the Discriminant in Factoring

Consider $ax^2 + bx + c$, where a, b, and c are integers with a greatest common factor of 1. If $b^2 - 4ac$ is a perfect square, then $\sqrt{b^2 - 4ac}$ is a whole number, and the solutions to $ax^2 + bx + c = 0$ are rational numbers. If the solutions to a quadratic equation are rational numbers, then they could be found by the factoring method. So if $b^2 - 4ac$ is a perfect square, then $ax^2 + bx + c$ factors. It is also true that if $b^2 - 4ac$ is not a perfect square, then $ax^2 + bx + c$ is prime.

EXAMPLE 1

Using the discriminant

Use the discriminant to determine whether each polynomial can be factored.
a) $6x^2 + x - 15$
b) $5x^2 - 3x + 2$

Solution

a) Use $a = 6$, $b = 1$, and $c = -15$ to find $b^2 - 4ac$:

$$b^2 - 4ac = 1^2 - 4(6)(-15) = 361$$

Because $\sqrt{361} = 19$, $6x^2 + x - 15$ can be factored. Using the ac method, we get

$$6x^2 + x - 15 = (2x - 3)(3x + 5).$$

b) Use $a = 5$, $b = -3$, and $c = 2$ to find $b^2 - 4ac$:

$$b^2 - 4ac = (-3)^2 - 4(5)(2) = -31$$

Because the discriminant is not a perfect square, $5x^2 - 3x + 2$ is prime. ■

Writing a Quadratic with Given Solutions

Not every quadratic equation can be solved by factoring, but the factoring method can be used (in reverse) to write a quadratic equation with any given solutions. For example, if the solutions to a quadratic equation are 5 and -3, we can reverse the steps in the factoring method as follows:

$$x = 5 \qquad \text{or} \qquad x = -3$$
$$x - 5 = 0 \qquad \text{or} \qquad x + 3 = 0$$
$$(x - 5)(x + 3) = 0 \qquad \text{Zero factor property}$$
$$x^2 - 2x - 15 = 0 \qquad \text{Multiply the factors.}$$

This method will produce the equation even if the solutions are irrational or imaginary.

E X A M P L E   2

Writing a quadratic given the solutions

Write a quadratic equation that has each given pair of solutions.

a) $4, -6$ **b)** $-\sqrt{2}, \sqrt{2}$

c) $-3i, 3i$

Solution

a) Reverse the factoring method using solutions 4 and -6:

$$x = 4 \qquad \text{or} \qquad x = -6$$
$$x - 4 = 0 \qquad \text{or} \qquad x + 6 = 0$$
$$(x - 4)(x + 6) = 0 \qquad \text{Zero factor property}$$
$$x^2 + 2x - 24 = 0 \qquad \text{Multiply the factors.}$$

b) Reverse the factoring method using solutions $-\sqrt{2}$ and $\sqrt{2}$:

$$x = -\sqrt{2} \qquad \text{or} \qquad x = \sqrt{2}$$
$$x + \sqrt{2} = 0 \qquad \text{or} \qquad x - \sqrt{2} = 0$$
$$\left(x + \sqrt{2}\right)\left(x - \sqrt{2}\right) = 0 \qquad \text{Zero factor property}$$
$$x^2 - 2 = 0 \qquad \text{Multiply the factors.}$$

c) Reverse the factoring method using solutions $-3i$ and $3i$:

$$x = -3i \qquad \text{or} \qquad x = 3i$$
$$x + 3i = 0 \qquad \text{or} \qquad x - 3i = 0$$
$$(x + 3i)(x - 3i) = 0 \qquad \text{Zero factor property}$$
$$x^2 - 9i^2 = 0 \qquad \text{Multiply the factors.}$$
$$x^2 + 9 = 0 \qquad \text{Note: } i^2 = -1$$

■

c a l c u l a t o r

c l o s e - u p

The graph of $y = x^2 + 2x - 24$ supports the conclusion in Example 2(a) because the graph crosses the x-axis at $(4, 0)$ and $(-6, 0)$.

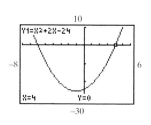

Equations Quadratic in Form

In a quadratic equation we have a variable and its square (x and x^2). An equation that contains an expression and the square of that expression is **quadratic in form** if substituting a single variable for that expression results in a quadratic equation. Equations that are quadratic in form can be solved by using methods for quadratic equations.

E X A M P L E   3

An equation quadratic in form
Solve $(x + 15)^2 - 3(x + 15) - 18 = 0$

Solution

Note that $x + 15$ and $(x + 15)^2$ both appear in the equation. Let $a = x + 15$ and substitute a for $x + 15$ in the equation:

$$(x + 15)^2 - 3(x + 15) - 18 = 0$$
$$a^2 - 3a - 18 = 0$$
$$(a - 6)(a + 3) = 0 \qquad \text{Factor.}$$

$$a - 6 = 0 \qquad \text{or} \qquad a + 3 = 0$$
$$a = 6 \qquad \text{or} \qquad a = -3$$
$$x + 15 = 6 \qquad \text{or} \quad x + 15 = -3 \quad \text{Replace } a \text{ by } x + 15.$$
$$x = -9 \quad \text{or} \qquad\qquad x = -18$$

Check in the original equation. The solution set is $\{-18, -9\}$. ■

In the next example we have a fourth-degree equation that is quadratic in form. Note that the fourth-degree equation has four solutions.

E X A M P L E   4

A fourth-degree equation
Solve $x^4 - 6x^2 + 8 = 0$.

Solution

Note that x^4 is the square of x^2. If we let $w = x^2$, then $w^2 = x^4$. Substitute these expressions into the original equation.

$$x^4 - 6x^2 + 8 = 0$$
$$w^2 - 6w + 8 = 0 \qquad \text{Replace } x^4 \text{ by } w^2 \text{ and } x^2 \text{ by } w.$$
$$(w - 2)(w - 4) = 0 \qquad \text{Factor.}$$

$$w - 2 = 0 \qquad \text{or} \qquad w - 4 = 0$$
$$w = 2 \qquad \text{or} \qquad w = 4$$
$$x^2 = 2 \qquad \text{or} \qquad x^2 = 4 \qquad \text{Substitute } x^2 \text{ for } w.$$
$$x = \pm\sqrt{2} \quad \text{or} \qquad x = \pm 2 \qquad \text{Even-root property}$$

Check. The solution set is $\{-2, -\sqrt{2}, \sqrt{2}, 2\}$. ■

> **helpful** **hint**
>
> The fundamental theorem of algebra says that the number of solutions to a polynomial equation is less than or equal to the degree of the polynomial. This famous theorem was proved by Carl Friederich Gauss when he was a young man.

C A U T I O N   If you replace x^2 by w, do not quit when you find the values of w. If the variable in the original equation is x, then you must solve for x.

E X A M P L E   5

A quadratic within a quadratic

Solve $(x^2 + 2x)^2 - 11(x^2 + 2x) + 24 = 0$.

Solution

Note that $x^2 + 2x$ and $(x^2 + 2x)^2$ appear in the equation. Let $a = x^2 + 2x$ and substitute.

$$a^2 - 11a + 24 = 0$$
$$(a - 8)(a - 3) = 0 \quad \text{Factor.}$$

$a - 8 = 0$ or $\qquad a - 3 = 0$

$\qquad a = 8$ or $\qquad\qquad a = 3$

$\quad x^2 + 2x = 8$ or $\qquad x^2 + 2x = 3 \quad$ Replace a by $x^2 + 2x$.

$x^2 + 2x - 8 = 0$ or $\quad x^2 + 2x - 3 = 0$

$(x - 2)(x + 4) = 0$ or $\quad (x + 3)(x - 1) = 0$

$x - 2 = 0$ or $x + 4 = 0$ or $x + 3 = 0$ or $x - 1 = 0$

$x = 2$ or $\quad x = -4$ or $\quad x = -3$ or $\quad x = 1$

Check. The solution set is $\{-4, -3, 1, 2\}$. ■

The next example involves a fractional exponent. To identify this type of equation as quadratic in form, recall how to square an expression with a fractional exponent. For example, $(x^{1/2})^2 = x$, $(x^{1/4})^2 = x^{1/2}$, and $(x^{1/3})^2 = x^{2/3}$.

c a l c u l a t o r

c l o s e - u p

The four x-intercepts on the graph of

$$y = (x^2 + 2x)^2$$
$$- 11(x^2 + 2x) + 24$$

support the conclusion in Example 5.

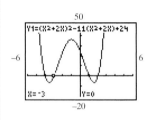

E X A M P L E   6

A fractional exponent

Solve $x - 9x^{1/2} + 14 = 0$.

Solution

Note that the square of $x^{1/2}$ is x. Let $w = x^{1/2}$; then $w^2 = (x^{1/2})^2 = x$. Now substitute w and w^2 into the original equation:

$$w^2 - 9w + 14 = 0$$
$$(w - 7)(w - 2) = 0$$

$w - 7 = 0$ or $w - 2 = 0$

$\quad w = 7$ or $\quad w = 2$

$\quad x^{1/2} = 7$ or $\quad x^{1/2} = 2 \quad$ Replace w by $x^{1/2}$.

$\quad\quad x = 49$ or $\quad\quad x = 4 \quad$ Square each side.

Because we squared each side, we must check for extraneous roots. First evaluate $x - 9x^{1/2} + 14$ for $x = 49$:

$$49 - 9 \cdot 49^{1/2} + 14 = 49 - 9 \cdot 7 + 14 = 0$$

Now evaluate $x - 9x^{1/2} + 14$ for $x = 4$:

$$4 - 9 \cdot 4^{1/2} + 14 = 4 - 9 \cdot 2 + 14 = 0$$

Because each solution checks, the solution set is $\{4, 49\}$. ■

C A U T I O N   An equation of quadratic form must have a term that is the square of another. Equations such as $x^4 - 5x^3 + 6 = 0$ or $x^{1/2} - 3x^{1/3} - 18 = 0$ are not quadratic in form and cannot be solved by substitution.

Applications

Applied problems often result in quadratic equations that cannot be factored. For such equations we use the quadratic formula to find exact solutions and a calculator to find decimal approximations for the exact solutions.

E X A M P L E   7

Changing area

Marvin's flower bed is rectangular in shape with a length of 10 feet and a width of 5 feet (ft). He wants to increase the length and width by the same amount to obtain a flower bed with an area of 75 square feet (ft^2). What should the amount of increase be?

Solution

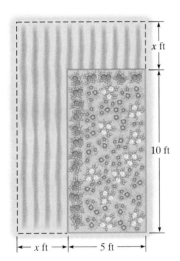

Let x be the amount of increase. The length and width of the new flower bed are $x + 10$ ft and $x + 5$ ft, as shown in Fig. 8.2. Because the area is to be 75 ft^2, we have

$$(x + 10)(x + 5) = 75.$$

Write this equation in the form $ax^2 + bx + c = 0$:

$$x^2 + 15x + 50 = 75$$
$$x^2 + 15x - 25 = 0 \quad \text{Get 0 on the right.}$$

$$x = \frac{-15 \pm \sqrt{225 - 4(1)(-25)}}{2(1)}$$

$$= \frac{-15 \pm \sqrt{325}}{2} = \frac{-15 \pm 5\sqrt{13}}{2}$$

Because the value of x must be positive, the exact increase is

$$\frac{-15 + 5\sqrt{13}}{2} \text{ ft.}$$

FIGURE 8.2

Using a calculator, we can find that x is approximately 1.51 ft. If $x = 1.51$ ft, then the new length is 11.51 ft, and the new width is 6.51 ft. The area of a rectangle with these dimensions is 74.93 ft^2. Of course, the approximate dimensions do not give exactly 75 ft^2. ■

E X A M P L E   8

Mowing the lawn

It takes Carla one hour longer to mow the lawn than it takes Sharon to mow the lawn. If they can mow the lawn in 5 hours working together, then how long would it take each girl by herself?

Solution

If Sharon can mow the lawn by herself in x hours, then she works at the rate of $\frac{1}{x}$ of the lawn per hour. If Carla can mow the lawn by herself in $x + 1$ hours, then she works at the rate of $\frac{1}{x + 1}$ of the lawn per hour. We can use a table to list all of the important quantities.

	Rate	Time	Work
Sharon	$\dfrac{1}{x}\ \dfrac{\text{lawn}}{\text{hr}}$	5 hr	$\dfrac{5}{x}$ lawn
Carla	$\dfrac{1}{x + 1}\ \dfrac{\text{lawn}}{\text{hr}}$	5 hr	$\dfrac{5}{x + 1}$ lawn

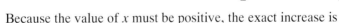

helpful hint

Note that the equation concerns the portion of the job done by each girl. We could have written an equation about the rates at which the two girls work. Because they can finish the lawn together in 5 hours, they are mowing together at the rate of $\frac{1}{5}$ lawn per hour. So

$$\frac{1}{x} + \frac{1}{x+1} = \frac{1}{5}.$$

Because they complete the lawn in 5 hours, the portion of the lawn done by Sharon and the portion done by Carla have a sum of 1:

$$\frac{5}{x} + \frac{5}{x+1} = 1$$

$$x(x+1)\frac{5}{x} + x(x+1)\frac{5}{x+1} = x(x+1)1 \quad \text{Multiply by the LCD.}$$

$$5x + 5 + 5x = x^2 + x$$

$$10x + 5 = x^2 + x$$

$$-x^2 + 9x + 5 = 0$$

$$x^2 - 9x - 5 = 0$$

$$x = \frac{9 \pm \sqrt{(-9)^2 - 4(1)(-5)}}{2(1)}$$

$$= \frac{9 \pm \sqrt{101}}{2}$$

Using a calculator, we find that $\frac{9 - \sqrt{101}}{2}$ is negative. So Sharon's time alone is

$$\frac{9 + \sqrt{101}}{2} \text{ hours.}$$

To find Carla's time alone, we add one hour to Sharon's time. So Carla's time alone is

$$\frac{9 + \sqrt{101}}{2} + 1 = \frac{9 + \sqrt{101}}{2} + \frac{2}{2} = \frac{11 + \sqrt{101}}{2} \text{ hours.}$$

Sharon's time alone is approximately 9.525 hours, and Carla's time alone is approximately 10.525 hours. ∎

WARM-UPS

True or false? Explain.

1. To solve $x^4 - 5x^2 + 6 = 0$ by substitution, we can let $w = x^2$.   True
2. We can solve $x^5 - 3x^3 - 10 = 0$ by substitution if we let $w = x^3$.   False
3. We always use the quadratic formula on equations of quadratic form.   False
4. If $w = x^{1/6}$, then $w^2 = x^{1/3}$.   True
5. To solve $x - 7\sqrt{x} + 10 = 0$ by substitution, we let $\sqrt{w} = x$.   False
6. If $y = 2^{1/2}$, then $y^2 = 2^{1/4}$.   False
7. If John paints a 100-foot fence in x hours, then his rate is $\frac{100}{x}$ of the fence per hour.   False
8. If Elvia drives 300 miles in x hours, then her rate is $\frac{300}{x}$ miles per hour (mph).   True
9. If Ann's boat goes 10 mph in still water, then against a 5-mph current, it will go 2 mph.   False
10. If squares with sides of length x inches are cut from the corners of an 11-inch by 14-inch rectangular piece of sheet metal and the sides are folded up to form a box, then the dimensions of the bottom will be $11 - x$ by $14 - x$.   False

8.3 EXERCISES

Reading and Writing *After reading this section, write out the answers to these questions. Use complete sentences.*

1. How can you use the discriminant to determine if a quadratic polynomial can be factored?

   If the coefficients are integers and the discriminant is a perfect square, then the quadratic polynomial can be factored.

2. What is the relationship between solutions to a quadratic equation and factors of a quadratic polynomial?

   The number k is a solution to a quadratic equation if and only if $x - k$ is a factor of the quadratic polynomial.

3. How do we write a quadratic equation with given solutions?

   If the solutions are a and b, then the quadratic equation $(x - a)(x - b) = 0$ has those solutions.

4. What is an equation quadratic in form?

   An equation of quadratic form is one that can be converted to a quadratic equation by making a substitution.

Use the discriminant to determine whether each quadratic polynomial can be factored, then factor the ones that are not prime. See Example 1.

5. $2x^2 - x + 4$
   Prime

6. $2x^2 + 3x - 5$
   $(2x + 5)(x - 1)$

7. $2x^2 + 6x - 5$
   Prime

8. $3x^2 + 5x - 1$
   Prime

9. $6x^2 + 19x - 36$
   $(3x - 4)(2x + 9)$

10. $8x^2 + 6x - 27$
    $(2x - 3)(4x + 9)$

11. $4x^2 - 5x - 12$
    Prime

12. $4x^2 - 27x + 45$
    $(4x - 15)(x - 3)$

13. $8x^2 - 18x - 45$
    $(4x - 15)(2x + 3)$

14. $6x^2 + 9x - 16$
    Prime

Write a quadratic equation that has each given pair of solutions. See Example 2.

15. $3, -7$
    $x^2 + 4x - 21 = 0$

16. $-8, 2$
    $x^2 + 6x - 16 = 0$

17. $4, 1$
    $x^2 - 5x + 4 = 0$

18. $3, 2$
    $x^2 - 5x + 6 = 0$

19. $\sqrt{5}, -\sqrt{5}$
    $x^2 - 5 = 0$

20. $-\sqrt{7}, \sqrt{7}$
    $x^2 - 7 = 0$

21. $4i, -4i$
    $x^2 + 16 = 0$

22. $-3i, 3i$
    $x^2 + 9 = 0$

23. $i\sqrt{2}, -i\sqrt{2}$
    $x^2 + 2 = 0$

24. $3i\sqrt{2}, -3i\sqrt{2}$
    $x^2 + 18 = 0$

25. $\dfrac{1}{2}, \dfrac{1}{3}$
    $6x^2 - 5x + 1 = 0$

26. $-\dfrac{1}{5}, -\dfrac{1}{2}$
    $10x^2 + 7x + 1 = 0$

Find all real solutions to each equation. See Example 3.

27. $(2a - 1)^2 + 2(2a - 1) - 8 = 0 \quad \left\{-\dfrac{3}{2}, \dfrac{3}{2}\right\}$

28. $(3a + 2)^2 - 3(3a + 2) = 10 \quad \left\{-\dfrac{4}{3}, 1\right\}$

29. $(w - 1)^2 + 5(w - 1) + 5 = 0 \quad \left\{\dfrac{-3 \pm \sqrt{5}}{2}\right\}$

30. $(2x - 1)^2 - 4(2x - 1) + 2 = 0 \quad \left\{\dfrac{3 \pm \sqrt{2}}{2}\right\}$

Find all real solutions to each equation. See Example 4.

31. $x^4 - 14x^2 + 45 = 0$
    $\{\pm\sqrt{5}, \pm 3\}$

32. $x^4 + 2x^2 = 15$
    $\{\pm\sqrt{3}\}$

33. $x^6 + 7x^3 = 8$
    $\{-2, 1\}$

34. $a^6 + 6a^3 = 16$
    $\{-2, \sqrt[3]{2}\}$

Find all real solutions to each equation. See Example 5.

35. $(x^2 + 2x)^2 - 7(x^2 + 2x) + 12 = 0 \quad \{-1 \pm \sqrt{5}, -3, 1\}$

36. $(x^2 + 3x)^2 + (x^2 + 3x) - 20 = 0 \quad \{-4, 1\}$

37. $(y^2 + y)^2 - 8(y^2 + y) + 12 = 0 \quad \{-3, -2, 1, 2\}$

38. $(w^2 - 2w)^2 + 24 = 11(w^2 - 2w) \quad \{-2, -1, 3, 4\}$

Find all real solutions to each equation. See Example 6.

39. $x^{1/2} - 5x^{1/4} + 6 = 0$
    $\{16, 81\}$

40. $2x - 5\sqrt{x} + 2 = 0$
    $\left\{\dfrac{1}{4}, 4\right\}$

41. $2x - 5x^{1/2} - 3 = 0$
    $\{9\}$

42. $x^{1/4} + 2 = x^{1/2}$
    $\{16\}$

Find all real solutions to each equation.

43. $x^{-2} + x^{-1} - 6 = 0$
    $\left\{-\dfrac{1}{3}, \dfrac{1}{2}\right\}$

44. $x^{-2} - 2x^{-1} = 8$
    $\left\{-\dfrac{1}{2}, \dfrac{1}{4}\right\}$

45. $x^{1/6} - x^{1/3} + 2 = 0$
    $\{64\}$

46. $x^{2/3} - x^{1/3} - 20 = 0$
    $\{-64, 125\}$

47. $\left(\dfrac{1}{y - 1}\right)^2 + \left(\dfrac{1}{y - 1}\right) = 6 \quad \left\{\dfrac{2}{3}, \dfrac{3}{2}\right\}$

48. $\left(\dfrac{1}{w + 1}\right)^2 - 2\left(\dfrac{1}{w + 1}\right) - 24 = 0 \quad \left\{-\dfrac{5}{6}, -\dfrac{5}{4}\right\}$

49. $2x^2 - 3 - 6\sqrt{2x^2 - 3} + 8 = 0 \quad \left\{\pm\dfrac{\sqrt{14}}{2}, \pm\dfrac{\sqrt{38}}{2}\right\}$

50. $x^2 + x + \sqrt{x^2 + x} - 2 = 0 \quad \left\{\dfrac{-1 \pm \sqrt{5}}{2}\right\}$

51. $x^{-2} - 2x^{-1} - 1 = 0$
    $\{-1 + \sqrt{2}, -1 - \sqrt{2}\}$

52. $x^{-2} - 6x^{-1} + 6 = 0$
    $\left\{\dfrac{3 + \sqrt{3}}{6}, \dfrac{3 - \sqrt{3}}{6}\right\}$

Find the exact solution to each problem. If the exact solution is an irrational number, then also find an approximate decimal solution. See Examples 7 and 8.

53. **Country singers.** Harry and Gary are traveling to Nashville to make their fortunes. Harry leaves on the train at 8:00 A.M. and Gary travels by car, starting at 9:00 A.M. To complete the 300-mile trip and arrive at the same time as Harry, Gary travels 10 miles per hour (mph) faster than the train. At what time will they both arrive in Nashville?
    2 P.M.

54. *Gone fishing.* Debbie traveled by boat 5 miles upstream to fish in her favorite spot. Because of the 4-mph current, it took her 20 minutes longer to get there than to return. How fast will her boat go in still water? $2\sqrt{34}$ or 11.662 mph

55. *Cross-country cycling.* Erin was traveling across the desert on her bicycle. Before lunch she traveled 60 miles (mi); after lunch she traveled 46 mi. She put in one hour more after lunch than before lunch, but her speed was 4 mph slower than before. What was her speed before lunch and after lunch?

Before $-5 + \sqrt{265}$ or 11.3 mph, after $-9 + \sqrt{265}$ or 7.3 mph

FIGURE FOR EXERCISE 55

56. *Extreme hardship.* Kim starts to walk 3 mi to school at 7:30 A.M. with a temperature of 0°F. Her brother Bryan starts at 7:45 A.M. on his bicycle, traveling 10 mph faster than Kim. If they get to school at the same time, then how fast is each one traveling?

Kim $-5 + \sqrt{145}$ or 7.042 mph, Bryan $5 + \sqrt{145}$ or 17.042 mph

57. *American pie.* John takes 3 hours longer than Andrew to peel 500 pounds (lb) of apples. If together they can peel 500 lb of apples in 8 hours, then how long would it take each one working alone?

Andrew $\dfrac{13 + \sqrt{265}}{2}$ or 14.6 hours, John $\dfrac{19 + \sqrt{265}}{2}$ or 17.6 hours

58. *On the half shell.* It takes Brent one hour longer than Calvin to shuck a sack of oysters. If together they shuck a sack of oysters in 45 minutes, then how long would it take each one working alone?

Brent $\dfrac{5 + \sqrt{13}}{4}$ or 2.151 hours, Calvin $\dfrac{1 + \sqrt{13}}{4}$ or 1.151 hours

59. *The growing garden.* Eric's garden is 20 ft by 30 ft. He wants to increase the length and width by the same amount to have a 1000-ft² garden. What should be the new dimensions of the garden?

Length $5 + 5\sqrt{41}$ or 37.02 ft, width $-5 + 5\sqrt{41}$ or 27.02 ft

60. *Open-top box.* Thomas is going to make an open-top box by cutting equal squares from the four corners of an 11 inch by 14 inch sheet of cardboard and folding up the sides. If the area of the base is to be 80 square inches, then what size square should be cut from each corner?

$\dfrac{25 - \sqrt{329}}{4}$ or 1.715 inches

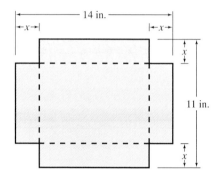

FIGURE FOR EXERCISE 60

61. *Pumping the pool.* It takes pump A 2 hours less time than pump B to empty a certain swimming pool. Pump A is started at 8:00 A.M., and pump B is started at 11:00 A.M. If the pool is still half full at 5:00 P.M., then how long would it take pump A working alone?

$14 + 2\sqrt{58}$ or 29.2 hours

62. *Time off for lunch.* It usually takes Eva 3 hours longer to do the monthly payroll than it takes Cicely. They start working on it together at 9:00 A.M. and at 5:00 P.M. they have 90% of it done. If Eva took a 2-hour lunch break while Cicely had none, then how much longer will it take for them to finish the payroll working together?

0.788 hour or 47 minutes

63. *Golden Rectangle.* One principle used by the ancient Greeks to get shapes that are pleasing to the eye in art and architecture was the Golden Rectangle. If a square is removed from one end of a Golden Rectangle, as shown in the figure on the next page, the sides of the remaining rectangle are proportional to the original rectangle. So the length and width of the original rectangle satisfy

$$\frac{L}{W} = \frac{W}{L - W}.$$

If the length of a Golden Rectangle is 10 meters, then what is its width?

$-5 + 5\sqrt{5}$ or 6.2 meters

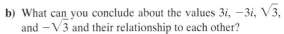

b) What can you conclude about the values $3i$, $-3i$, $\sqrt{3}$, and $-\sqrt{3}$ and their relationship to each other?

 65. *Cooperative learning.* Work with a group to write a quadratic equation that has each given pair of solutions.

a) $3 + \sqrt{5}, 3 - \sqrt{5}$ **b)** $4 - 2i, 4 + 2i$

c) $\dfrac{1 + i\sqrt{3}}{2}, \dfrac{1 - i\sqrt{3}}{2}$

 GRAPHING CALCULATOR EXERCISES

Solve each equation by locating the x-intercepts on the graph of a corresponding function. Round approximate answers to two decimal places.

66. $(5x - 7)^2 - (5x - 7) - 6 = 0$
$\{1, 2\}$

67. $x^4 - 116x^2 + 1600 = 0$
$\{-10, -4, 4, 10\}$

68. $(x^2 + 3x)^2 - 7(x^2 + 3x) + 9 = 0$
$\{-4.25, -3.49, 0.49, 1.25\}$

69. $x^2 - 3x^{1/2} - 12 = 0$
$\{4.27\}$

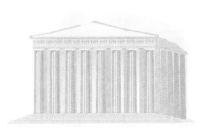

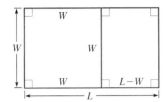

FIGURE FOR EXERCISE 63

GETTING MORE INVOLVED

 **64. *Exploration*.**

a) Given that $P(x) = x^4 + 6x^2 - 27$, find $P(3i)$, $P(-3i)$, $P(\sqrt{3})$, and $P(-\sqrt{3})$.

 8.4   QUADRATIC AND RATIONAL INEQUALITIES

We first solved inequalities in Chapter 2. In this section we solve inequalities involving quadratic polynomials. We use a new technique based on the rules for multiplying real numbers.

In this section

- Solving Quadratic Inequalities with a Sign Graph
- Solving Rational Inequalities with a Sign Graph
- Quadratic Inequalities That Cannot Be Factored
- Applications

Solving Quadratic Inequalities with a Sign Graph

An inequality involving a quadratic polynomial is called a **quadratic** inequality.

> **Quadratic Inequality**
>
> A quadratic inequality is an inequality of the form
>
> $$ax^2 + bx + c > 0,$$
>
> where a, b, and c are real numbers with $a \neq 0$. The inequality symbols $<$, $\leq$, and $\geq$ may also be used.

If we can factor a quadratic inequality, then the inequality can be solved with a **sign graph,** which shows where each factor is positive, negative, or zero.

E X A M P L E   1

Solving a quadratic inequality

Use a sign graph to solve the inequality $x^2 + 3x - 10 > 0$.

Solution

Because the left-hand side can be factored, we can write the inequality as

$$(x + 5)(x - 2) > 0.$$

This inequality says that the product of $x + 5$ and $x - 2$ is positive. If both factors are negative or both are positive, the product is positive. To analyze the signs of each factor, we make a sign graph as follows. First consider the possible values of the factor $x + 5$:

Value	Where	On the number line
$x + 5 = 0$	if $x = -5$	Put a 0 above -5.
$x + 5 > 0$	if $x > -5$	Put $+$ signs to the right of -5.
$x + 5 < 0$	if $x < -5$	Put $-$ signs to the left of -5.

The sign graph shown in Fig. 8.3 for the factor $x + 5$ is made from the information in the preceding table.

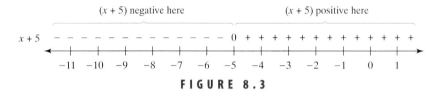

FIGURE 8.3

Now consider the possible values of the factor $x - 2$:

Value	Where	On the number line
$x - 2 = 0$	if $x = 2$	Put a 0 above 2.
$x - 2 > 0$	if $x > 2$	Put $+$ signs to the right of 2.
$x - 2 < 0$	if $x < 2$	Put $-$ signs to the left of 2.

We put the information for the factor $x - 2$ on the sign graph for the factor $x + 5$ as shown in Fig. 8.4. We can see from Fig. 8.4 that the product is positive if $x < -5$ and the product is positive if $x > 2$. The solution set for the quadratic inequality is shown in Fig. 8.5. Note that -5 and 2 are not included in the graph because for those values of x the product is zero. The solution set is $(-\infty, -5) \cup (2, \infty)$.

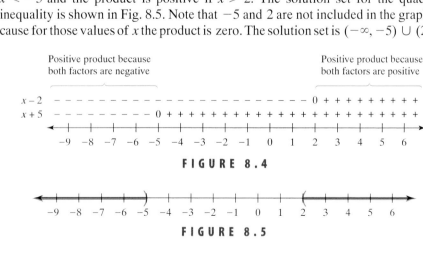

FIGURE 8.4

FIGURE 8.5

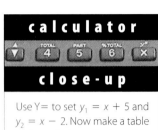

Use Y= to set $y_1 = x + 5$ and $y_2 = x - 2$. Now make a table and scroll through the table. The table numerically supports the sign graph in Fig. 8.4.

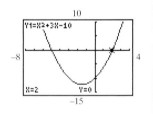

Note that the graph of $y = x^2 + 3x - 10$ is above the x-axis when $x < -5$ or when $x > 2$.

In the next example we will make the procedure from Example 1 a bit more efficient.

EXAMPLE 2

calculator

close-up

Use Y= to set $y_1 = 2x - 1$ and $y_2 = x + 3$. The table of values for y_1 and y_2 supports the sign graph in Fig. 8.6.

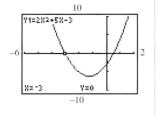

Note that the graph of $y = 2x^2 + 5x - 3$ is below the x-axis when x is between -3 and $\frac{1}{2}$.

Solving a quadratic inequality

Solve $2x^2 + 5x \le 3$ and graph the solution set.

Solution

Rewrite the inequality with 0 on one side:

$$2x^2 + 5x - 3 \le 0$$
$$(2x - 1)(x + 3) \le 0 \quad \text{Factor.}$$

Examine the signs of each factor:

$$2x - 1 = 0 \text{ if } x = \frac{1}{2}$$

$$2x - 1 > 0 \text{ if } x > \frac{1}{2}$$

$$2x - 1 < 0 \text{ if } x < \frac{1}{2}$$

$$x + 3 = 0 \text{ if } x = -3$$
$$x + 3 > 0 \text{ if } x > -3$$
$$x + 3 < 0 \text{ if } x < -3$$

Make a sign graph as shown in Fig. 8.6. The product of the factors is negative between -3 and $\frac{1}{2}$, when one factor is negative and the other is positive. The product is 0 at -3 and at $\frac{1}{2}$. So the solution set is the interval $\left[-3, \frac{1}{2}\right]$. The graph of the solution set is shown in Fig. 8.7.

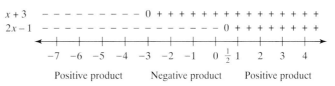

FIGURE 8.6

FIGURE 8.7

We summarize the strategy used for solving a quadratic inequality as follows.

Strategy for Solving a Quadratic Inequality with a Sign Graph

1. Write the inequality with 0 on the right.
2. Factor the quadratic polynomial on the left.
3. Make a sign graph showing where each factor is positive, negative, or zero.
4. Use the rules for multiplying signed numbers to determine which regions satisfy the original inequality.

Solving Rational Inequalities with a Sign Graph

The inequalities

$$\frac{x + 2}{x - 3} \le 2, \qquad \frac{2x - 3}{x + 5} \le 0 \qquad \text{and} \qquad \frac{2}{x + 4} \ge \frac{1}{x + 1}$$

are called **rational inequalities.** When we solve *equations* that involve rational expressions, we usually multiply each side by the LCD. However, if we multiply each side of any inequality by a negative number, we must reverse the inequality, and when we multiply by a positive number, we do not reverse the inequality. For this reason we generally *do not multiply inequalities by expressions involving variables.* The values of the expressions might be positive or negative. The next two examples show how to use a sign graph to solve rational inequalities that have variables in the denominator.

E X A M P L E   3

Solving a rational inequality

Solve $\frac{x + 2}{x - 3} \le 2$ and graph the solution set.

Solution

We *do not* multiply each side by $x - 3$. Instead, subtract 2 from each side to get 0 on the right:

$$\frac{x + 2}{x - 3} - 2 \le 0$$

$$\frac{x + 2}{x - 3} - \frac{2(x - 3)}{x - 3} \le 0 \quad \text{Get a common denominator.}$$

$$\frac{x + 2}{x - 3} - \frac{2x - 6}{x - 3} \le 0 \quad \text{Simplify.}$$

$$\frac{x + 2 - 2x + 6}{x - 3} \le 0 \quad \text{Subtract the rational expressions.}$$

$$\frac{-x + 8}{x - 3} \le 0 \quad \begin{array}{l}\text{The quotient of } -x + 8 \text{ and } x - 3 \text{ is}\\ \text{less than or equal to 0.}\end{array}$$

Examine the signs of the numerator and denominator:

$$\begin{array}{ll} x - 3 = 0 \text{ if } x = 3 & -x + 8 = 0 \text{ if } x = 8 \\ x - 3 > 0 \text{ if } x > 3 & -x + 8 > 0 \text{ if } x < 8 \\ x - 3 < 0 \text{ if } x < 3 & -x + 8 < 0 \text{ if } x > 8 \end{array}$$

Make a sign graph as shown in Fig. 8.8. Using the rule for dividing signed numbers and the sign graph, we can identify where the quotient is negative or zero. The solution set is $(-\infty, 3) \cup [8, \infty)$. Note that 3 is not in the solution set because the quotient is undefined if $x = 3$. The graph of the solution set is shown in Figure 8.9.

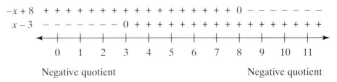

FIGURE 8.8

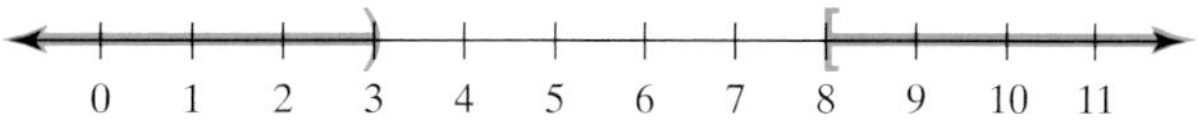

FIGURE 8.9

CAUTION   Remember to reverse the inequality sign when multiplying or dividing by a negative number. For example, $x - 3 > 0$ is equivalent to $x > 3$. But $-x + 8 > 0$ is equivalent to $-x > -8$, or $x < 8$.

E X A M P L E   4

Solving a rational inequality

Solve $\dfrac{2}{x + 4} \geq \dfrac{1}{x + 1}$ and graph the solution set.

Solution

We do not multiply by the LCD as we do in solving equations. Instead, subtract $\dfrac{1}{x + 1}$ from each side:

$$\frac{2}{x + 4} - \frac{1}{x + 1} \geq 0$$

$$\frac{2(x + 1)}{(x + 4)(x + 1)} - \frac{1(x + 4)}{(x + 1)(x + 4)} \geq 0 \quad \text{Get a common denominator.}$$

$$\frac{2x + 2 - x - 4}{(x + 1)(x + 4)} \geq 0 \quad \text{Simplify.}$$

$$\frac{x - 2}{(x + 1)(x + 4)} \geq 0$$

```
x + 1    – – – – – – – – – 0 + + + + + + + + + + + +
x + 4    – – – 0 + + + + + + + + + + + + + + + + + +
x – 2    – – – – – – – – – – – – – – – 0 + + + + + + +
         ┼───┼───┼───┼───┼───┼───┼───┼───┼───┼───┼→
        –5  –4  –3  –2  –1   0   1   2   3   4   5
```

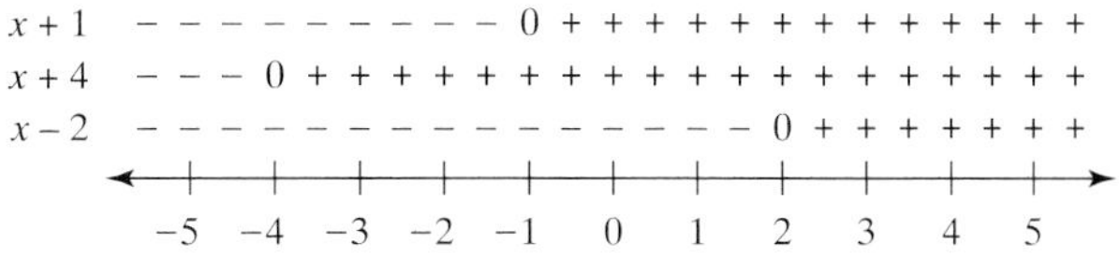

FIGURE 8.10

Make a sign graph as shown in Fig. 8.10. The computation of

$$\frac{x - 2}{(x + 1)(x + 4)}$$

involves multiplication and division. The result of this computation is positive if all of the three binomials are positive or if only one is positive and the other two are negative. The sign graph shows that this rational expression will have a positive value when x is between -4 and -1 and again when x is larger than 2. The solution set is $(-4, -1) \cup [2, \infty)$. Note that -1 and -4 are not in the solution set because they make the denominator zero. The graph of the solution set is shown in Fig. 8.11.

```
←──┼───(───┼───┼───┼───┼───┼───[───┼───┼───┼──→
  –5  –4  –3  –2  –1   0   1   2   3   4   5
```

FIGURE 8.11

Solving rational inequalities with a sign graph is summarized below.

> **Strategy for Solving a Rational Inequality with a Sign Graph**
>
> 1. Rewrite the inequality with 0 on the right-hand side.
> 2. Use only addition and subtraction to get an equivalent inequality.
> 3. Factor the numerator and denominator if possible.
> 4. Make a sign graph showing where each factor is positive, negative, or zero.
> 5. Use the rules for multiplying and dividing signed numbers to determine the regions that satisfy the original inequality.

Another method for solving quadratic and rational inequalities will be shown in Example 5. This method, called the **test point method,** can be used instead of the sign graph to solve the inequalities of Examples 1, 2, 3, and 4.

Quadratic Inequalities That Cannot Be Factored

The following example shows how to solve a quadratic inequality that involves a prime polynomial.

E X A M P L E 5

Solving a quadratic inequality using the quadratic formula

Solve $x^2 - 4x - 6 > 0$ and graph the solution set.

Solution

The quadratic polynomial is prime, but we can solve $x^2 - 4x - 6 = 0$ by the quadratic formula:

$$x = \frac{4 \pm \sqrt{16 - 4(1)(-6)}}{2(1)} = \frac{4 \pm \sqrt{40}}{2} = \frac{4 \pm 2\sqrt{10}}{2} = 2 \pm \sqrt{10}$$

As in the previous examples, the solutions to the equation divide the number line into the intervals $(-\infty, 2 - \sqrt{10})$, $(2 - \sqrt{10}, 2 + \sqrt{10})$, and $(2 + \sqrt{10}, \infty)$ on which the quadratic polynomial has either a positive or negative value. To determine which, we select an arbitrary **test point** in each interval. Because $2 + \sqrt{10} \approx 5.2$ and $2 - \sqrt{10} \approx -1.2$, we choose a test point that is less than -1.2, one between -1.2 and 5.2, and one that is greater than 5.2. We have selected -2, 0, and 7 for test points, as shown in Fig. 8.12. Now evaluate $x^2 - 4x - 6$ at each test point.

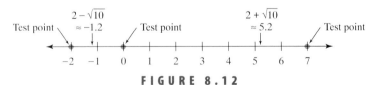

FIGURE 8.12

Test point	Value of $x^2 - 4x - 6$ at the test point	Sign of $x^2 - 4x - 6$ in interval of test point
-2	6	Positive
0	-6	Negative
7	15	Positive

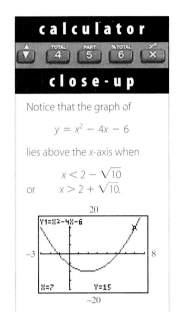

c a l c u l a t o r

c l o s e - u p

Notice that the graph of

$$y = x^2 - 4x - 6$$

lies above the *x*-axis when

$$x < 2 - \sqrt{10}$$
or $$x > 2 + \sqrt{10}.$$

Because $x^2 - 4x - 6$ is positive at the test points -2 and 7, it is positive at every point in the intervals containing those test points. So the solution set to the inequality $x^2 - 4x - 6 > 0$ is

$$(-\infty, 2 - \sqrt{10}) \cup (2 + \sqrt{10}, \infty),$$

and its graph is shown in Fig. 8.13.

FIGURE 8.13

The test point method used in Example 5 can be used also on inequalities that do factor. We summarize the strategy for solving inequalities using test points in the following box.

> ### Strategy for Solving Quadratic Inequalities Using Test Points
>
> 1. Rewrite the inequality with 0 on the right.
> 2. Solve the quadratic equation that results from replacing the inequality symbol with the equals symbol.
> 3. Locate the solutions to the quadratic equation on a number line.
> 4. Select a test point in each interval determined by the solutions to the quadratic equation.
> 5. Test each point in the original quadratic inequality to determine which intervals satisfy the inequality.

Applications

The following example shows how a quadratic inequality can be used to solve a problem.

E X A M P L E   6   **Making a profit**

Charlene's daily profit P (in dollars) for selling x magazine subscriptions is determined by the formula

$$P = -x^2 + 80x - 1500.$$

For what values of x is her profit positive?

Solution

We can find the values of x for which $P > 0$ by solving a quadratic inequality:

$$-x^2 + 80x - 1500 > 0$$
$$x^2 - 80x + 1500 < 0 \quad \text{Multiply each side by } -1.$$
$$(x - 30)(x - 50) < 0 \quad \text{Factor.}$$

Make a sign graph as shown in Fig. 8.14 on the next page. The product of the two factors is negative for x between 30 and 50. Because the last inequality is equivalent to the first, the profit is positive when the number of magazine subscriptions sold is greater than 30 and less than 50.

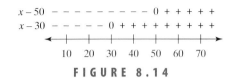

FIGURE 8.14

8.4   EXERCISES

Reading and Writing   After reading this section, write out the answers to these questions. Use complete sentences.

1. What is a quadratic inequality?
   A quadratic inequality has the form $ax^2 + bx + c > 0$. In place of $>$ we can also use $<$, $\leq$, or $\geq$.

2. What is a sign graph?
   A sign graph shows signs of the factors for all possible values of x.

3. What is a rational inequality?
   A rational inequality is an inequality involving a rational expression.

4. Why don't we usually multiply each side of an inequality by an expression involving a variable?
   Multiplying each side by a positive number does not change the direction of the inequality, but multiplying by a negative number does. So if we multiply by a variable, it is difficult to know which way the inequality goes.

Solve each inequality. State the solution set using interval notation and graph the solution set. See Examples 1 and 2.

5. $x^2 + x - 6 < 0$ $(-3, 2)$

6. $x^2 - 3x - 4 \geq 0$
   $(-\infty, -1] \cup [4, \infty)$

7. $y^2 - 4 > 0$
   $(-\infty, -2) \cup (2, \infty)$

8. $z^2 - 16 < 0$
   $(-4, 4)$

9. $2u^2 + 5u \geq 12$
   $(-\infty, -4] \cup \left[\frac{3}{2}, \infty\right)$

10. $2v^2 + 7v < 4$

$\left(-4, \dfrac{1}{2}\right)$

11. $4x^2 - 8x \geq 0$

$(-\infty, 0] \cup [2, \infty)$

12. $x^2 + x > 0$

$(-\infty, -1) \cup (0, \infty)$

13. $5x - 10x^2 < 0$

$(-\infty, 0) \cup \left(\dfrac{1}{2}, \infty\right)$

14. $3x - x^2 > 0$

$(0, 3)$

15. $x^2 + 6x + 9 \geq 0$ $(-\infty, \infty)$

16. $x^2 + 25 < 10x$ $\varnothing$

Solve each rational inequality. State and graph the solution set. See Examples 3 and 4.

17. $\dfrac{x}{x - 3} > 0$

$(-\infty, 0) \cup (3, \infty)$

18. $\dfrac{a}{a + 2} > 0$

$(-\infty, -2) \cup (0, \infty)$

19. $\dfrac{x + 2}{x} \leq 0$

$[-2, 0)$

20. $\dfrac{w - 6}{w} \leq 0$

$(0, 6]$

21. $\dfrac{t - 3}{t + 6} > 0$

$(-\infty, -6) \cup (3, \infty)$

22. $\dfrac{x - 2}{2x + 5} < 0$

$\left(-\dfrac{5}{2}, 2\right)$

23. $\dfrac{x}{x + 2} > -1$

$(-\infty, -2) \cup (-1, \infty)$

24. $\dfrac{x + 3}{x} \leq -2$

$[-1, 0)$

25. $\dfrac{2}{x - 5} > \dfrac{1}{x + 4}$

$(-13, -4) \cup (5, \infty)$

26. $\dfrac{3}{x + 2} > \dfrac{2}{x - 1}$

$(-2, 1) \cup (7, \infty)$

27. $\dfrac{m}{m - 5} + \dfrac{3}{m - 1} > 0$

$(-\infty, -5) \cup (1, 3) \cup (5, \infty)$

28. $\dfrac{p}{p - 16} + \dfrac{2}{p - 6} \leq 0$

$[-4, 6) \cup [8, 16)$

29. $\dfrac{x}{x - 3} \leq \dfrac{-8}{x - 6}$

$[-6, 3) \cup [4, 6)$

30. $\dfrac{x}{x + 20} > \dfrac{2}{x + 8}$ $(-\infty, -20) \cup (-10, -8) \cup (4, \infty)$

Solve each inequality. State and graph the solution set. See Example 5.

31. $x^2 - 2x - 4 > 0$ $(-\infty, 1 - \sqrt{5}) \cup (1 + \sqrt{5}, \infty)$

32. $x^2 - 2x - 5 \leq 0$ $[1 - \sqrt{6}, 1 + \sqrt{6}]$

33. $2x^2 - 6x + 3 \geq 0$ $\left(-\infty, \dfrac{3 - \sqrt{3}}{2}\right] \cup \left[\dfrac{3 + \sqrt{3}}{2}, \infty\right)$

34. $2x^2 - 8x + 3 < 0$ $\left(\dfrac{4 - \sqrt{10}}{2}, \dfrac{4 + \sqrt{10}}{2}\right)$

35. $y^2 - 3y - 9 \leq 0$ $\left[\dfrac{3 - 3\sqrt{5}}{2}, \dfrac{3 + 3\sqrt{5}}{2}\right]$

36. $z^2 - 5z - 7 < 0$ $\left(\dfrac{5 - \sqrt{53}}{2}, \dfrac{5 + \sqrt{53}}{2}\right)$

In Exercises 37–60, solve each inequality. State the solution set using interval notation.

37. $x^2 \leq 9$ $[-3, 3]$

38. $x^2 \geq 36$ $(-\infty, -6] \cup [6, \infty)$

39. $16 - x^2 > 0$ $(-4, 4)$

40. $9 - x^2 < 0$ $(-\infty, -3) \cup (3, \infty)$

41. $x^2 - 4x \geq 0$ $(-\infty, 0] \cup [4, \infty)$

42. $4x^2 - 9 > 0$ $\left(-\infty, -\dfrac{3}{2}\right) \cup \left(\dfrac{3}{2}, \infty\right)$

43. $3(2w^2 - 5) < w$ $\left(-\dfrac{3}{2}, \dfrac{5}{3}\right)$

44. $6(y^2 - 2) + y < 0$ $\left(-\dfrac{3}{2}, \dfrac{4}{3}\right)$

45. $z^2 \geq 4(z + 3)$ $(-\infty, -2] \cup [6, \infty)$

46. $t^2 < 3(2t - 3)$ $\varnothing$

47. $(q + 4)^2 > 10q + 31$ $(-\infty, -3) \cup (5, \infty)$

48. $(2p + 4)(p - 1) < (p + 2)^2$ $(-2, 4)$

49. $\dfrac{1}{2}x^2 \geq 4 - x$ $(-\infty, -4] \cup [2, \infty)$

50. $\dfrac{1}{2}x^2 \leq x + 12$ $[-4, 6]$

51. $\dfrac{x - 4}{x + 3} \leq 0$ $(-3, 4]$

52. $\dfrac{2x - 1}{x + 5} \geq 0$ $(-\infty, -5) \cup \left[\dfrac{1}{2}, \infty\right)$

53. $(x - 2)(x + 1)(x - 5) \geq 0$ $[-1, 2] \cup [5, \infty)$

54. $(x - 1)(x + 2)(2x - 5) < 0$ $(-\infty, -2) \cup (1, 2.5)$

55. $x^3 + 3x^2 - x - 3 < 0$ $(-\infty, -3) \cup (-1, 1)$

56. $x^3 + 5x^2 - 4x - 20 \geq 0$ $[-5, -2] \cup [2, \infty)$

57. $0.23x^2 + 6.5x + 4.3 < 0$ $(-27.58, -0.68)$

58. $0.65x^2 + 3.2x + 5.1 > 0$ $(-\infty, \infty)$

59. $\dfrac{x}{x - 2} > \dfrac{-1}{x + 3}$
$(-\infty, -2 - \sqrt{6}) \cup (-3, -2 + \sqrt{6}) \cup (2, \infty)$

60. $\dfrac{x}{3 - x} > \dfrac{2}{x + 5}$ $\left(\dfrac{-7 - \sqrt{73}}{2}, -5\right) \cup \left(\dfrac{-7 + \sqrt{73}}{2}, 3\right)$

Solve each problem by using a quadratic inequality. See Example 6.

61. *Positive profit.* The monthly profit P (in dollars) that Big Jim makes on the sale of x mobile homes is determined by the formula $P = x^2 + 5x - 50$. For what values of x is his profit positive? $6, 7, 8, \ldots$

62. *Profitable fruitcakes.* Sharon's revenue R (in dollars) on the sale of x fruitcakes is determined by the formula $R = 50x - x^2$. Her cost C (in dollars) for producing x fruit cakes is given by the formula $C = 2x + 40$. For what values of x is Sharon's profit positive? (Profit = revenue − cost.) $1, 2, 3, \ldots, 47$

If an object is given an initial velocity straight upward of v_0 feet per second from a height of s_0 feet, then its altitude S after t seconds is given by the formula

$$S = -16t^2 + v_0 t + s_0.$$

63. *Flying high.* An arrow is shot straight upward with a velocity of 96 feet per second (ft/sec) from an altitude of 6 feet. For how many seconds is this arrow more than 86 feet high?
4 seconds

 64. *Putting the shot.* In 1978 Udo Beyer (East Germany) set a world record in the shot-put of 72 ft 8 in. If Beyer had projected the shot straight upward with a velocity of 30 ft/sec from a height of 5 ft, then for what values of t would the shot be under 15 ft high?
$t < 0.43$ second or $t > 1.44$ seconds

If a projectile is fired at a 45° angle from a height of s_0 feet with initial velocity v_0 ft/sec, then its altitude S in feet after t seconds is given by

$$S = -16t^2 + \dfrac{v_0}{\sqrt{2}}t + s_0.$$

65. *Siege and garrison artillery.* An 8-inch mortar used in the Civil War fired a 44.5-lb projectile from ground level a distance of 3,600 ft when aimed at a 45° angle (Harold R. Peterson, *Notes on Ordinance of the American Civil War*). The accompanying graph shows the altitude of the projectile when it is fired with a velocity of $240\sqrt{2}$ ft/sec.
  a) Use the graph to estimate the maximum altitude reached by the projectile.
  900 ft
  b) Use the graph to estimate approximately how long the altitude of the projectile was greater than 864 ft.
  3 seconds
  c) Use the formula to determine the length of time for which the projectile had an altitude of more than 864 ft.
  3 seconds

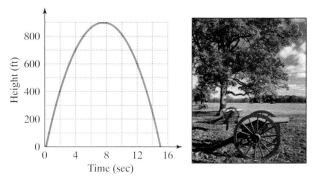

FIGURE FOR EXERCISE 65

 66. *Seacoast artillery.* The 13-inch mortar used in the Civil War fired a 220-lb projectile a distance of 12,975 ft when aimed at a 45° angle. If the 13-inch mortar was fired from a hill 100 ft above sea level with an initial velocity of 644 ft/sec, then for how long was the projectile more than 800 ft above sea level?
25.2 seconds

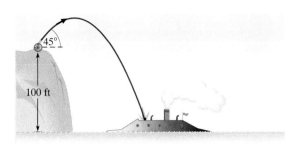

FIGURE FOR EXERCISE 66

GETTING MORE INVOLVED

67. *Cooperative learning.* Work in a small group to solve each inequality for x, given that h and k are real numbers with $h < k$.

a) $(x - h)(x - k) < 0$
(h, k)

b) $(x - h)(x - k) > 0$
$(-\infty, h) \cup (k, \infty)$

c) $(x + h)(x + k) < 0$
$(-k, -h)$

d) $(x + h)(x + k) \geq 0$
$(-\infty, -k] \cup [-h, \infty)$

e) $\dfrac{x - h}{x - k} \geq 0$
$(-\infty, h] \cup (k, \infty)$

f) $\dfrac{x + h}{x + k} \leq 0$
$(-k, -h]$

68. *Cooperative learning.* Work in a small group to solve $ax^2 + bx + c > 0$ for x in each case.

a) $b^2 - 4ac = 0$ and $a > 0$
$(-\infty, -b/(2a)) \cup (-b/(2a), \infty)$

b) $b^2 - 4ac = 0$ and $a < 0$ $\varnothing$

c) $b^2 - 4ac < 0$ and $a > 0$ $(-\infty, \infty)$

d) $b^2 - 4ac < 0$ and $a < 0$ $\varnothing$

e) $b^2 - 4ac > 0$ and $a > 0$
$$\left(-\infty, \frac{-b - \sqrt{b^2 - 4ac}}{2a}\right) \cup \left(\frac{-b + \sqrt{b^2 - 4ac}}{2a}, \infty\right)$$

f) $b^2 - 4ac > 0$ and $a < 0$
$$\left(\frac{-b - \sqrt{b^2 - 4ac}}{2a}, \frac{-b + \sqrt{b^2 - 4ac}}{2a}\right)$$

GRAPHING CALCULATOR EXERCISES

Match the given inequalities with their solution sets (a through d) by examining a table or a graph.

69. $x^2 - 2x - 8 < 0$ c

a. $(-2, 2) \cup (8, \infty)$

70. $x^2 - 3x > 54$ d

b. $(2, 4)$

71. $\dfrac{x}{x - 2} > 2$ b

c. $(-2, 4)$

72. $\dfrac{3}{x - 2} < \dfrac{5}{x + 2}$ a

d. $(-\infty, -6) \cup (9, \infty)$

COLLABORATIVE ACTIVITIES

Building a Room Addition

Leslie decides to build a rectangular studio on the south side of his house. The studio is attached so that the north side of the studio will be a portion of the current south side of the house. Because Leslie lives in the southwest, he decides to make the room from rammed earth, thus making the walls of the studio 2 feet thick. Because of the cost of building materials and labor, Leslie decides to limit the total area of the studio. He also decides to make the studio's inside south wall twice as long as its inside west wall.

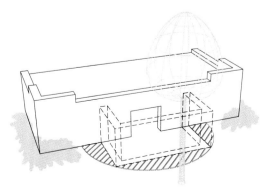

Grouping: Two to four students per group
Topic: Applications of the quadratic formula

Leslie would also like to build a semicircular patio around the studio (sliding glass doors will open from the studio out onto the patio). The studio will be circumscribed by the semicircle. A large tree, which he would like to keep for the summer shade is 20 feet from the south side of the house. See the accompanying sketch Leslie made of the studio and patio.

1. Working in your groups, finish the drawing of the house and studio, adding the given information. Define your variable(s) and find the internal dimensions of the rectangular studio if Leslie wants the *external* area of the studio to be 400 square feet. Round your answers to the nearest foot.

2. Determine how far the semicircular patio extends past the house at the farthest point. Round your answer to the nearest tenth of a foot.

3. Will Leslie be able to build the patio without removing the tree? Give a reason for your answer.

W R A P - U P

C H A P T E R 8

S U M M A R Y

Quadratic Equations

Examples

Quadratic equation

An equation of the form
$ax^2 + bx + c = 0$,
where a, b, and c are real numbers, with $a \neq 0$

$$x^2 = 11$$
$$(x - 5)^2 = 99$$
$$x^2 + 3x - 20 = 0$$

Methods for solving quadratic equations

Factoring:
Factor the quadratic polynomial, then set each factor equal to 0.

$$x^2 + x - 6 = 0$$
$$(x + 3)(x - 2) = 0$$
$$x + 3 = 0 \text{ or } x - 2 = 0$$

The even-root property:
If $x^2 = k$ $(k > 0)$, then $x = \pm\sqrt{k}$.
If $x^2 = 0$, then $x = 0$.
There are no real solutions to $x^2 = k$ for $k < 0$.

$$(x - 5)^2 = 10$$
$$x - 5 = \pm\sqrt{10}$$

Completing the square:
Take one-half of middle term, square it, then add it to each side.

$$x^2 + 6x = -4$$
$$x^2 + 6x + 9 = -4 + 9$$
$$(x + 3)^2 = 5$$

Quadratic formula:
Solves $ax^2 + bx + c = 0$ with $a \neq 0$:
$$x = \frac{-b \pm \sqrt{b^2 - 4ac}}{2a}$$

$$2x^2 + 3x - 5 = 0$$

$$x = \frac{-3 \pm \sqrt{3^2 - 4(2)(-5)}}{2(2)}$$

Number of solutions

Determined by the discriminant $b^2 - 4ac$:
$b^2 - 4ac > 0$ 2 real solutions

$$x^2 + 6x - 12 = 0$$
$$6^2 - 4(1)(-12) > 0$$

$b^2 - 4ac = 0$ 1 real solution

$$x^2 + 10x + 25 = 0$$
$$10^2 - 4(1)(25) = 0$$

$b^2 - 4ac < 0$ no real solutions, 2 imaginary solutions

$$x^2 + 2x + 20 = 0$$
$$2^2 - 4(1)(20) < 0$$

Factoring

The quadratic polynomial $ax^2 + bx + c$ (with integral coefficients) can be factored if and only if $b^2 - 4ac$ is a perfect square.

$$2x^2 - 11x + 12$$
$$b^2 - 4ac = 25$$
$$(2x - 3)(x - 4)$$

Writing equations

To write an equation with given solutions, reverse the steps in solving an equation by factoring.

$$x = 2 \text{ or } x = -3$$
$$(x - 2)(x + 3) = 0$$
$$x^2 + x - 6 = 0$$

Equations quadratic in form

Use substitution to convert to a quadratic.

$$x^4 + 3x^2 - 10 = 0$$
Let $a = x^2$
$$a^2 + 3a - 10 = 0$$

Quadratic and Rational Inequalities

Examples

| Quadratic inequality | An inequality involving a quadratic polynomial | $2x^2 - 7x + 6 \geq 0$
 $x^2 - 4x - 5 < 0$ |

| Rational inequality | An inequality involving a rational expression | $\dfrac{1}{x - 1} < \dfrac{3}{x - 2}$ |

Solving quadratic and rational inequalities

Get 0 on one side and express the other side as a product and/or quotient of linear factors. Make a sign graph showing the signs of the factors.

Use test points if the quadratic polynomial is prime.

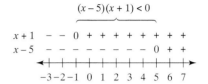

ENRICHING YOUR MATHEMATICAL WORD POWER

For each mathematical term, choose the correct meaning.

1. **quadratic equation**
   a. $ax + b = c$ with $a \neq 0$
   b. $ax^2 + bx + c = 0$ with $a \neq 0$
   c. $ax + b = 0$ with $a \neq 0$
   d. $a/x^2 + b/x = c$ with $x \neq 0$ b

2. **perfect square trinomial**
   a. a trinomial of the form $a^2 + 2ab + b^2$
   b. a trinomial of the form $a^2 + b^2$
   c. a trinomial of the form $a^2 + ab + b^2$
   d. a trinomial of the form $a^2 - 2ab - b^2$ a

3. **completing the square**
   a. drawing a perfect square
   b. evaluating $(a + b)^2$
   c. drawing the fourth side when given three sides of a square
   d. finding the third term of a perfect square trinomial   d

4. **quadratic formula**
   a. $x = \dfrac{-b \pm \sqrt{b^2 - 4ac}}{2}$
   b. $x = -b \pm \dfrac{\sqrt{b^2 - 4ac}}{2a}$
   c. $x = \dfrac{-b \pm \sqrt{b^2 - 4ac}}{2a}$
   d. $x = \dfrac{b \pm \sqrt{b^2 - 4ac}}{2a}$ c

5. **discriminant**
   a. the vertex of a parabola
   b. the radicand in the quadratic formula
   c. the leading coefficient in $ax^2 + bx + c$
   d. to treat unfairly   b

6. **quadratic in form**
   a. $ax^2 + bx + c = 0$
   b. a parabola
   c. an equation that is quadratic after a substitution
   d. having four equal sides   c

7. **quadratic inequality**
   a. $ax^2 + bx + c > 0$ with $a \neq 0$ or with $\geq, <,$ or $\leq$
   b. $ax + b > 0$ with $a \neq 0$ or with $\geq, <,$ or $\leq$
   c. completing the square
   d. the Pythagorean theorem   a

8. **sign graph**
   a. a graph showing the sign of x
   b. a sign on which a graph is drawn
   c. a number line showing the signs of factors
   d. to graph in sign language   c

9. **rational inequality**
   a. an inequality involving a rational expression(s)
   b. a quadratic inequality
   c. an inequality with rational exponents
   d. an inequality that compares two fractions   a

10. **test point**
    a. the end of a chapter
    b. to check if a point is in the right location
    c. a number that is used to check if an inequality is satisfied
    d. a positive integer   c

REVIEW EXERCISES

8.1 *Solve by factoring.*

1. $x^2 - 2x - 15 = 0$
$\{-3, 5\}$

2. $x^2 - 2x - 24 = 0$
$\{-4, 6\}$

3. $2x^2 + x = 15$
$\left\{-3, \dfrac{5}{2}\right\}$

4. $2x^2 + 7x = 4$
$\left\{-4, \dfrac{1}{2}\right\}$

5. $w^2 - 25 = 0$
$\{-5, 5\}$

6. $a^2 - 121 = 0$
$\{-11, 11\}$

7. $4x^2 - 12x + 9 = 0$
$\left\{\dfrac{3}{2}\right\}$

8. $x^2 - 12x + 36 = 0$
$\{6\}$

Solve by using the even-root property.

9. $x^2 = 12$
$\{\pm 2\sqrt{3}\}$

10. $x^2 = 20$
$\{\pm 2\sqrt{5}\}$

11. $(x - 1)^2 = 9$
$\{-2, 4\}$

12. $(x + 4)^2 = 4$
$\{-6, -2\}$

13. $(x - 2)^2 = \dfrac{3}{4}$
$\left\{\dfrac{4 \pm \sqrt{3}}{2}\right\}$

14. $(x - 3)^2 = \dfrac{1}{4}$
$\left\{\dfrac{5}{2}, \dfrac{7}{2}\right\}$

15. $4x^2 = 9$
$\left\{\pm\dfrac{3}{2}\right\}$

16. $2x^2 = 3$
$\left\{\pm\dfrac{\sqrt{6}}{2}\right\}$

Solve by completing the square.

17. $x^2 - 6x + 8 = 0$
$\{2, 4\}$

18. $x^2 + 4x + 3 = 0$
$\{-3, -1\}$

19. $x^2 - 5x + 6 = 0$
$\{2, 3\}$

20. $x^2 - x - 6 = 0$
$\{-2, 3\}$

21. $2x^2 - 7x + 3 = 0$
$\left\{\dfrac{1}{2}, 3\right\}$

22. $2x^2 - x = 6$
$\left\{-\dfrac{3}{2}, 2\right\}$

23. $x^2 + 4x + 1 = 0$
$\{-2 \pm \sqrt{3}\}$

24. $x^2 + 2x - 2 = 0$
$\{-1 \pm \sqrt{3}\}$

8.2 *Solve by the quadratic formula.*

25. $x^2 - 3x - 10 = 0$
$\{-2, 5\}$

26. $x^2 - 5x - 6 = 0$
$\{-1, 6\}$

27. $6x^2 - 7x = 3$
$\left\{-\dfrac{1}{3}, \dfrac{3}{2}\right\}$

28. $6x^2 = x + 2$
$\left\{-\dfrac{1}{2}, \dfrac{2}{3}\right\}$

29. $x^2 + 4x + 2 = 0$
$\{-2 \pm \sqrt{2}\}$

30. $x^2 + 6x = 2$
$\{-3 \pm \sqrt{11}\}$

31. $3x^2 + 1 = 5x$
$\left\{\dfrac{5 \pm \sqrt{13}}{6}\right\}$

32. $2x^2 + 3x - 1 = 0$
$\left\{\dfrac{-3 \pm \sqrt{17}}{4}\right\}$

Find the value of the discriminant and the number of real solutions to each equation.

33. $25x^2 - 20x + 4 = 0$
$0, 1$

34. $16x^2 + 1 = 8x$
$0, 1$

35. $x^2 - 3x + 7 = 0$
$-19, 0$

36. $3x^2 - x + 8 = 0$
$-95, 0$

37. $2x^2 + 1 = 5x$
$17, 2$

38. $-3x^2 + 6x - 2 = 0$
$12, 2$

Find the complex solutions to the quadratic equations.

39. $2x^2 - 4x + 3 = 0$
$\left\{\dfrac{2 \pm i\sqrt{2}}{2}\right\}$

40. $2x^2 - 6x + 5 = 0$
$\left\{\dfrac{3 \pm i}{2}\right\}$

41. $2x^2 + 3 = 3x$
$\left\{\dfrac{3 \pm i\sqrt{15}}{4}\right\}$

42. $x^2 + x + 1 = 0$
$\left\{\dfrac{-1 \pm i\sqrt{3}}{2}\right\}$

43. $3x^2 + 2x + 2 = 0$
$\left\{\dfrac{-1 \pm i\sqrt{5}}{3}\right\}$

44. $x^2 + 2 = 2x$
$\{1 \pm i\}$

45. $\dfrac{1}{2}x^2 + 3x + 8 = 0$
$\{-3 \pm i\sqrt{7}\}$

46. $\dfrac{1}{2}x^2 - 5x + 13 = 0$
$\{5 \pm i\}$

8.3 *Use the discriminant to determine whether each quadratic polynomial can be factored, then factor the ones that are not prime.*

47. $8x^2 - 10x - 3$
$(4x + 1)(2x - 3)$

48. $18x^2 + 9x - 2$
$(6x - 1)(3x + 2)$

49. $4x^2 - 5x + 2$
Prime

50. $6x^2 - 7x - 4$
Prime

51. $8y^2 + 10y - 25$
$(4y - 5)(2y + 5)$

52. $25z^2 - 15z - 18$
$(5z + 3)(5z - 6)$

Write a quadratic equation that has each given pair of solutions.

53. $-3, -6$
$x^2 + 9x + 18 = 0$

54. $4, -9$
$x^2 + 5x - 36 = 0$

55. $-5\sqrt{2}, 5\sqrt{2}$
$x^2 - 50 = 0$

56. $-2i\sqrt{3}, 2i\sqrt{3}$
$x^2 + 12 = 0$

Find all real solutions to each equation.

57. $x^6 + 7x^3 - 8 = 0$ $\{-2, 1\}$

58. $8x^6 + 63x^3 - 8 = 0$ $\left\{-2, \dfrac{1}{2}\right\}$

59. $x^4 - 13x^2 + 36 = 0$ $\{\pm 2, \pm 3\}$

60. $x^4 + 7x^2 + 12 = 0$ $\varnothing$

61. $(x^2 + 3x)^2 - 28(x^2 + 3x) + 180 = 0$ $\{-6, -5, 2, 3\}$

62. $(x^2 + 1)^2 - 8(x^2 + 1) + 15 = 0$ $\{\pm 2, \pm\sqrt{2}\}$

63. $x^2 - 6x + 6\sqrt{x^2 - 6x} - 40 = 0$ $\{-2, 8\}$

64. $x^2 - 3x - 3\sqrt{x^2 - 3x} + 2 = 0$ $\left\{-1, 4, \dfrac{3 \pm \sqrt{13}}{2}\right\}$

65. $t^{-2} + 5t^{-1} - 36 = 0$ $\left\{-\dfrac{1}{9}, \dfrac{1}{4}\right\}$

66. $a^{-2} + a^{-1} - 6 = 0$ $\left\{-\dfrac{1}{3}, \dfrac{1}{2}\right\}$

67. $w - 13\sqrt{w} + 36 = 0$ $\{16, 81\}$

68. $4a - 5\sqrt{a} + 1 = 0$ $\left\{\dfrac{1}{16}, 1\right\}$

8.4 *Solve each inequality. State the solution set using interval notation and graph it.*

69. $a^2 + a > 6$
$(-\infty, -3) \cup (2, \infty)$

70. $x^2 - 5x + 6 > 0$
$(-\infty, 2) \cup (3, \infty)$

71. $x^2 - x - 20 \le 0$
$[-4, 5]$

72. $a^2 + 2a \le 15$
$[-5, 3]$

73. $w^2 - w < 0$
$(0, 1)$

74. $x - x^2 \le 0$
$(-\infty, 0] \cup [1, \infty)$

75. $\dfrac{x - 4}{x + 2} \ge 0$
$(-\infty, -2) \cup [4, \infty)$

76. $\dfrac{x - 3}{x + 5} < 0$
$(-5, 3)$

77. $\dfrac{x - 2}{x + 3} < 1$
$(-3, \infty)$

78. $\dfrac{x - 3}{x + 4} > 2$
$(-11, -4)$

79. $\dfrac{3}{x + 2} > \dfrac{1}{x + 1}$
$(-2, -1) \cup \left(-\dfrac{1}{2}, \infty\right)$

80. $\dfrac{1}{x + 1} < \dfrac{1}{x - 1}$
$(-\infty, -1) \cup (1, \infty)$

MISCELLANEOUS

In Exercises 81–92, find all real or imaginary solutions to each equation.

81. $144x^2 - 120x + 25 = 0$ $\left\{\dfrac{5}{12}\right\}$

82. $49x^2 + 9 = 42x$ $\left\{\dfrac{3}{7}\right\}$

83. $(2x + 3)^2 + 7 = 12$ $\left\{\dfrac{-3 \pm \sqrt{5}}{2}\right\}$

84. $6x = -\dfrac{19x + 25}{x + 1}$ $\left\{-\dfrac{5}{3}, -\dfrac{5}{2}\right\}$

85. $1 + \dfrac{20}{9x^2} = \dfrac{8}{3x}$ $\left\{\dfrac{4 \pm 2i}{3}\right\}$

86. $\dfrac{x - 1}{x + 2} = \dfrac{2x - 3}{x + 4}$ $\{1 \pm \sqrt{3}\}$

87. $\sqrt{3x^2 + 7x - 30} = x$ $\left\{\dfrac{5}{2}\right\}$

88. $\dfrac{x^4}{3} = x^2 + 6$ $\{\pm\sqrt{6}, \pm i\sqrt{3}\}$

89. $2(2x + 1)^2 + 5(2x + 1) = 3$ $\left\{-2, -\dfrac{1}{4}\right\}$

90. $(w^2 - 1)^2 + 2(w^2 - 1) = 15$ $\{\pm 2i, \pm 2\}$

91. $x^{1/2} - 15x^{1/4} + 50 = 0$ $\{625, 10,000\}$

92. $x^{-2} - 9x^{-1} + 18 = 0$ $\left\{\dfrac{1}{6}, \dfrac{1}{3}\right\}$

Find exact and approximate solutions to each problem.

93. *Missing numbers.* Find two positive real numbers that differ by 4 and have a product of 4.
$-2 + 2\sqrt{2}$ and $2 + 2\sqrt{2}$, or 0.83 and 4.83

94. *One on one.* Find two positive real numbers that differ by 1 and have a product of 1.
$\dfrac{-1 + \sqrt{5}}{2}$ and $\dfrac{1 + \sqrt{5}}{2}$, or 0.62 and 1.62

95. *Big screen TV.* On a 19-inch diagonal measure television picture screen, the height is 4 inches less than the width. Find the height and width.
Width $\dfrac{4 + \sqrt{706}}{2}$ or 15.3 inches, height $\dfrac{-4 + \sqrt{706}}{2}$ or 11.3 inches

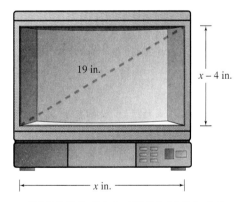
19 in.

$x - 4$ in.

x in.

FIGURE FOR EXERCISE 95

96. *Boxing match.* A boxing ring is in the shape of a square, 20 ft on each side. How far apart are the fighters when they are in opposite corners of the ring?
$20\sqrt{2}$ or 28.284 ft

97. *Students for a Clean Environment.* A group of environmentalists plans to print a message on an 8 inch by 10 inch paper. If the typed message requires 24 square inches of paper and the group wants an equal border on all sides, then how wide should the border be?

2 inches

FIGURE FOR EXERCISE 97

98. *Winston works faster.* Winston can mow his dad's lawn in one hour less than it takes his brother Willie. If they take 2 hours to mow it when working together, then how long would it take Winston working alone?

$\dfrac{3 + \sqrt{17}}{2}$ or 3.562 hours

99. *Ping Pong.* The table used for table tennis is 4 ft longer than it is wide and has an area of 45 ft^2. What are the dimensions of the table?

Width 5 ft, length 9 ft

FIGURE FOR EXERCISE 99

100. *Swimming pool design.* An architect has designed a motel pool within a rectangular area that is fenced on three sides as shown in the figure. If she uses 60 yards of fencing to enclose an area of 352 square yards, then what are the dimensions marked L and W in the figure? Assume L is greater than W.

$L = 22$ yards and $W = 16$ yards

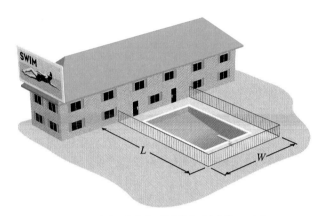

FIGURE FOR EXERCISE 100

101. *Decathlon champion.* For 1989 and 1990 Dave Johnson had the highest decathlon score in the world. When Johnson reached a speed of 32 ft/sec on the pole vault runway, his height above the ground t seconds after leaving the ground was given by $h = -16t^2 + 32t$. (The elasticity of the pole converts the horizontal speed into vertical speed.) Find the value of t for which his height was 12 ft.

0.5 second and 1.5 seconds

102. *Time of flight.* Use the information from Exercise 101 to determine how long Johnson was in the air. For how long was he more then 14 ft in the air?

2 seconds, $0.5\sqrt{2}$ or 0.707 seconds

CHAPTER 8 TEST

Calculate the value of $b^2 - 4ac$, and state how many real solutions each equation has.

1. $2x^2 - 3x + 2 = 0$ $-7, 0$

2. $-3x^2 + 5x - 1 = 0$ $13, 2$

3. $4x^2 - 4x + 1 = 0$ $0, 1$

Solve by using the quadratic formula.

4. $2x^2 + 5x - 3 = 0$ $\left\{-3, \dfrac{1}{2}\right\}$

5. $x^2 + 6x + 6 = 0$ $\{-3 \pm \sqrt{3}\}$

Solve by completing the square.

6. $x^2 + 10x + 25 = 0$ $\{-5\}$

7. $2x^2 + x - 6 = 0$ $\left\{-2, \dfrac{3}{2}\right\}$

Solve by any method.

8. $x(x + 1) = 12$ $\{-4, 3\}$

9. $a^4 - 5a^2 + 4 = 0$ $\{\pm 1, \pm 2\}$

10. $x - 2 - 8\sqrt{x - 2} + 15 = 0$ $\{11, 27\}$

Find the complex solutions to the quadratic equations.

11. $x^2 + 36 = 0$ $\{\pm 6i\}$

12. $x^2 + 6x + 10 = 0$ $\{-3 \pm i\}$

13. $3x^2 - x + 1 = 0$ $\left\{\dfrac{1 \pm i\sqrt{11}}{6}\right\}$

Write a quadratic equation that has each given pair of solutions.

14. $-4, 6$ $x^2 - 2x - 24 = 0$

15. $-5i, 5i$ $x^2 + 25 = 0$

Solve each inequality. State and graph the solution set.

16. $w^2 + 3w < 18$ $(-6, 3)$

17. $\dfrac{2}{x - 2} < \dfrac{3}{x + 1}$ $(-1, 2) \cup (8, \infty)$

Find the exact solution to each problem.

18. The length of a rectangle is 2 ft longer than the width. If the area is 16 ft^2, then what are the length and width?
Width $-1 + \sqrt{17}$ ft, length $1 + \sqrt{17}$ ft

19. A new computer can process a company's monthly payroll in one hour less time than the old computer. To really save time, the manager used both computers and finished the payroll in 3 hours. How long would it take the new computer to do the payroll by itself?
$\dfrac{5 + \sqrt{37}}{2}$ or 5.5 hours

Solve each equation.

1. $2x - 15 = 0$ $\left\{\dfrac{15}{2}\right\}$ **2.** $2x^2 - 15 = 0$ $\left\{\pm\dfrac{\sqrt{30}}{2}\right\}$

3. $2x^2 + x - 15 = 0$ $\left\{-3, \dfrac{5}{2}\right\}$

4. $2x^2 + 4x - 15 = 0$ $\left\{\dfrac{-2 \pm \sqrt{34}}{2}\right\}$

5. $|4x + 11| = 3$ $\left\{-\dfrac{7}{2}, -2\right\}$

6. $|4x^2 + 11x| = 3$ $\left\{-3, \dfrac{1}{4}, \dfrac{-11 \pm \sqrt{73}}{8}\right\}$

7. $\sqrt{x} = x - 6$ $\{9\}$

8. $(2x - 5)^{2/3} = 4$ $\left\{-\dfrac{3}{2}, \dfrac{13}{2}\right\}$

Solve each inequality.

9. $1 - 2x < 5 - x$ $(-4, \infty)$

10. $(1 - 2x)(5 - x) \le 0$ $\left[\dfrac{1}{2}, 5\right]$

11. $\dfrac{1 - 2x}{5 - x} \le 0$ $\left[\dfrac{1}{2}, 5\right)$

12. $|5 - x| < 3$ $(2, 8)$

13. $3x - 1 < 5$ and $-3 \le x$ $[-3, 2)$

14. $x - 3 < 1$ or $2x \ge 8$ $(-\infty, \infty)$

Solve each equation for y.

15. $2x - 3y = 9$ $y = \dfrac{2}{3}x - 3$

16. $\dfrac{y - 3}{x + 2} = -\dfrac{1}{2}$ $y = -\dfrac{1}{2}x + 2$

17. $3y^2 + cy + d = 0$ $y = \dfrac{-c \pm \sqrt{c^2 - 12d}}{6}$

18. $my^2 - ny = w$ $y = \dfrac{n \pm \sqrt{n^2 + 4mw}}{2m}$

19. $\dfrac{1}{3}x - \dfrac{2}{5}y = \dfrac{5}{6}$ $y = \dfrac{5}{6}x - \dfrac{25}{12}$

20. $y - 3 = -\dfrac{2}{3}(x - 4)$ $y = -\dfrac{2}{3}x + \dfrac{17}{3}$

Let $m = \dfrac{y_2 - y_1}{x_2 - x_1}$. Find the value of m for each of the following choices of x_1, x_2, y_1, and y_2.

21. $x_1 = 2, x_2 = 5, y_1 = 3, y_2 = 7$ $\dfrac{4}{3}$

22. $x_1 = -3, x_2 = 4, y_1 = 5, y_2 = -6$ $-\dfrac{11}{7}$

23. $x_1 = 0.3, x_2 = 0.5, y_1 = 0.8, y_2 = 0.4$ -2

24. $x_1 = \dfrac{1}{2}, x_2 = \dfrac{1}{3}, y_1 = \dfrac{3}{5}, y_2 = -\dfrac{4}{3}$ $\dfrac{58}{5}$

Solve each problem.

25. *Ticket prices.* In the summer of 1994 the rock group Pearl Jam testified before a congressional committee that Ticketmaster was unfairly raising the prices of the group's concert tickets. One member of the group stated that fans should not have to pay more than \$20 to see Pearl Jam. Of course, for any concert, as ticket prices rise, the number of tickets sold decreases, as shown in the figure. If you use the formula $n = 48{,}000 - 400p$ to predict the number sold depending on the price p, then how many will be sold at \$20 per ticket? How many will be sold at \$25 per ticket? Use the bar graph to estimate the price if 35,000 tickets were sold.   40,000, 38,000, \$32.50

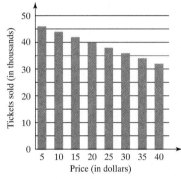

FIGURE FOR EXERCISE 25

26. *Increasing revenue.* Even though the number of tickets sold for a concert decreases with increasing price, the revenue generated does not necessarily decrease. Use the formula $R = p(48{,}000 - 400p)$ to determine the revenue when the price is \$20 and when the price is \$25. What price would produce a revenue of \$1.28 million? Use the graph to find the price that determines the maximum revenue.   \$800,000, \$950,000, \$40 or \$80, \$60

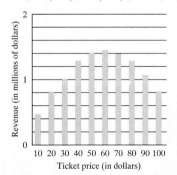

FIGURE FOR EXERCISE 26

CHAPTER 9

Additional Function Topics

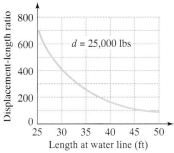

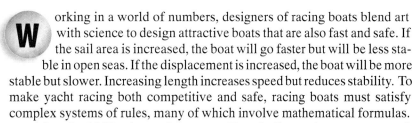

Working in a world of numbers, designers of racing boats blend art with science to design attractive boats that are also fast and safe. If the sail area is increased, the boat will go faster but will be less stable in open seas. If the displacement is increased, the boat will be more stable but slower. Increasing length increases speed but reduces stability. To make yacht racing both competitive and safe, racing boats must satisfy complex systems of rules, many of which involve mathematical formulas.

After the 1988 mismatch between Dennis Conner's catamaran and New Zealander Michael Fay's 133-foot monohull, an international group of yacht designers rewrote the America's Cup rules to ensure the fairness of the race. In addition to hundreds of pages of other rules, every yacht must satisfy the basic inequality

$$\frac{L + 1.25\sqrt{S} - 9.8\sqrt[3]{D}}{0.679} \le 24.000,$$

which balances the length L, the sail area S, and the displacement D.

In the 1979 Fastnet Race 15 sailors lost their lives. After *Exide Challenger*'s carbon-fiber keel snapped off Tony Bullimore spent 4 days inside the overturned hull before being rescued by the Australian navy. Yacht racing is a dangerous sport. To determine the general performance and safety of a yacht, designers calculate the displacement-length ratio, the sail area-displacement ratio, the ballast-displacement ratio, and the capsize screening value. In Exercises 73 and 74 of Section 9.1 we will see how composition of functions is used to define the displacement-length ratio and the sail area-displacement ratio.

9.1 COMBINING FUNCTIONS

In Sections 3.5 and 3.6 you learned the fundamental ideas about functions and their graphs. In this section you will learn how to combine functions to obtain new functions.

Basic Operations with Functions

An entrepreneur plans to rent a stand at a farmers market for $25 per day to sell strawberries. If she buys x flats of berries for $5 per flat and sells them for $9 per flat, then her daily cost in dollars can be written as a function of x:

$$C(x) = 5x + 25$$

Assuming she sells as many flats as she buys, her revenue in dollars is also a function of x:

$$R(x) = 9x$$

Because profit is revenue minus cost, we can find a function for the profit by subtracting the functions for cost and revenue:

$$P(x) = R(x) - C(x)$$
$$= 9x - (5x + 25)$$
$$= 4x - 25$$

The function $P(x) = 4x - 25$ expresses the daily profit as a function of x. Since $P(6) = -1$ and $P(7) = 3$, the profit is negative if 6 or fewer flats are sold and positive if 7 or more flats are sold.

In the example of the entrepreneur we subtracted two functions to find a new function. In other cases we may use addition, multiplication, or division to combine two functions. For any two given functions we can define the sum, difference, product, and quotient functions as follows.

Sum, Difference, Product, and Quotient Functions

Given two functions f and g, the functions $f + g$, $f - g$, $f \cdot g$, and $\frac{f}{g}$ are defined as follows:

Sum function: $(f + g)(x) = f(x) + g(x)$

Difference function: $(f - g)(x) = f(x) - g(x)$

Product function: $(f \cdot g)(x) = f(x) \cdot g(x)$

Quotient function: $\left(\dfrac{f}{g}\right)(x) = \dfrac{f(x)}{g(x)}$ provided that $g(x) \neq 0$

The domain of the function $f + g$, $f - g$, $f \cdot g$, or $\frac{f}{g}$ is the intersection of the domain of f and the domain of g. For the function $\frac{f}{g}$ we also rule out any values of x for which $g(x) = 0$.

EXAMPLE 1   **Operations with functions**

Let $f(x) = 4x - 12$ and $g(x) = x - 3$. Find the following.

a) $(f + g)(x)$ **b)** $(f - g)(x)$ **c)** $(f \cdot g)(x)$ **d)** $\left(\dfrac{f}{g}\right)(x)$

Solution

helpful **hint**

Note that we use $f + g, f - g,$ $f \cdot g,$ and f/g to name these functions only because there is no application in mind here. We generally use a single letter to name functions after they are combined as we did when using P for the profit function rather than $R - C$.

a) $(f + g)(x) = f(x) + g(x)$
$$= 4x - 12 + x - 3$$
$$= 5x - 15$$

b) $(f - g)(x) = f(x) - g(x)$
$$= 4x - 12 - (x - 3)$$
$$= 3x - 9$$

c) $(f \cdot g)(x) = f(x) \cdot g(x)$
$$= (4x - 12)(x - 3)$$
$$= 4x^2 - 24x + 36$$

d) $\left(\dfrac{f}{g}\right)(x) = \dfrac{f(x)}{g(x)} = \dfrac{4x - 12}{x - 3} = \dfrac{4(x - 3)}{x - 3} = 4$ for $x \neq 3$.

E X A M P L E   2

Evaluating a sum function

Let $f(x) = 4x - 12$ and $g(x) = x - 3$. Find $(f + g)(2)$.

Solution

In Example 1(a) we found a general formula for the function $f + g$, namely, $(f + g)(x) = 5x - 15$. If we replace x by 2, we get

$$(f + g)(2) = 5(2) - 15$$
$$= -5.$$

We can also find $(f + g)(2)$ by evaluating each function separately and then adding the results. Because $f(2) = -4$ and $g(2) = -1$, we get

$$(f + g)(2) = f(2) + g(2)$$
$$= -4 + (-1)$$
$$= -5.$$

Composition

helpful **hint**

The difference between the first four operations with functions and composition is like the difference between parallel and series in electrical connections. Components connected in parallel operate simultaneously and separately. If components are connected in series, then electricity must pass through the first component to get to the second component.

A salesperson's monthly salary is a function of the number of cars he sells: $1000 plus $50 for each car sold. If we let S be his salary and n be the number of cars sold, then S in dollars is a function of n:

$$S = 1000 + 50n$$

Each month the dealer contributes $100 plus 5% of his salary to a profit-sharing plan. If P represents the amount put into profit sharing, then P (in dollars) is a function of S:

$$P = 100 + 0.05S$$

Now P is a function of S, and S is a function of n. Is P a function of n? The value of n certainly determines the value of P. In fact, we can write a formula for P in terms of n by substituting one formula into the other:

$$P = 100 + 0.05S$$
$$= 100 + 0.05(1000 + 50n) \quad \text{Substitute } S = 1000 + 50n.$$
$$= 100 + 50 + 2.5n \quad\quad \text{Distributive property}$$
$$= 150 + 2.5n$$

Now P is written as a function of n, bypassing S. We call this idea **composition of functions.**

E X A M P L E   3

The composition of two functions

Given that $y = x^2 - 2x + 3$ and $z = 2y - 5$, write z as a function of x.

Solution

Replace y in $z = 2y - 5$ by $x^2 - 2x + 3$:

$$z = 2y - 5$$
$$= 2(x^2 - 2x + 3) - 5 \quad \text{Replace } y \text{ by } x^2 - 2x + 3.$$
$$= 2x^2 - 4x + 1$$

The equation $z = 2x^2 - 4x + 1$ expresses z as a function of x.   ■

The composition of two functions using f-notation is defined as follows.

Composition of Functions

The **composition** of f and g is denoted $f \circ g$ and is defined by the equation

$$(f \circ g)(x) = f(g(x)),$$

provided that $g(x)$ is in the domain of f.

The notation $f \circ g$ is read as "the composition of f and g" or "f compose g." The diagram in Fig. 9.1 shows a function g pairing numbers in its domain with numbers in its range. If the range of g is contained in or equal to the domain of f, then f pairs the second coordinates of g with numbers in the range of f. The composition function $f \circ g$ is a rule for pairing numbers in the domain of g directly with numbers in the range of f, bypassing the middle set. The domain of the function $f \circ g$ is the domain of g (or a subset of it) and the range of $f \circ g$ is the range of f (or a subset of it).

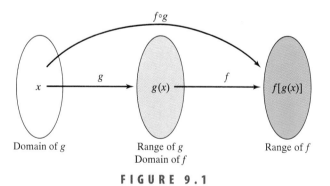

F I G U R E   9 . 1

CAUTION   The order in which functions are written is important in composition. For the function $f \circ g$ the function f is applied to $g(x)$. For the function $g \circ f$ the function g is applied to $f(x)$. The function closest to the variable x is applied first.

E X A M P L E   4

Composition of functions

Let $f(x) = 3x - 2$ and $g(x) = x^2 + 2x$. Find the following.

a) $(g \circ f)(2)$

b) $(f \circ g)(2)$

c) $(g \circ f)(x)$

d) $(f \circ g)(x)$

Solution

a) Because $(g \circ f)(2) = g(f(2))$, we first find $f(2)$:

$$f(2) = 3 \cdot 2 - 2 = 4$$

Because $f(2) = 4$, we have

$$(g \circ f)(2) = g(f(2)) = g(4) = 4^2 + 2 \cdot 4 = 24.$$

So $(g \circ f)(2) = 24$.

b) Because $(f \circ g)(2) = f(g(2))$, we first find $g(2)$:

$$g(2) = 2^2 + 2 \cdot 2 = 8$$

Because $g(2) = 8$, we have

$$(f \circ g)(2) = f(g(2)) = f(8) = 3 \cdot 8 - 2 = 22.$$

Thus $(f \circ g)(2) = 22$.

c) $(g \circ f)(x) = g(f(x))$

$$= g(3x - 2)$$

$$= (3x - 2)^2 + 2(3x - 2)$$

$$= 9x^2 - 12x + 4 + 6x - 4 = 9x^2 - 6x$$

So $(g \circ f)(x) = 9x^2 - 6x$.

d) $(f \circ g)(x) = f(g(x))$

$$= f(x^2 + 2x)$$

$$= 3(x^2 + 2x) - 2 = 3x^2 + 6x - 2$$

So $(f \circ g)(x) = 3x^2 + 6x - 2$.

calculator

close-up

Set $y_1 = 3x - 2$ and $y_2 = x^2 + 2x$. You can find the composition for Examples 4(a) and 4(b) by evaluating $y_2(y_1(2))$ and $y_1(y_2(2))$. Note that the order in which you evaluate the functions is critical.

```
Y₂(Y₁(2))
                    24
Y₁(Y₂(2))
                    22
```

helpful hint

A composition of functions can be viewed as two function machines where the output of the first is the input of the second.

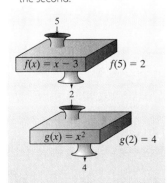

Notice that in Example 4(a) and (b), $(g \circ f)(2) \neq (f \circ g)(2)$. In Example 4(c) and (d) we see that $(g \circ f)(x)$ and $(f \circ g)(x)$ have different formulas defining them. In general, $f \circ g \neq g \circ f$. However, in Section 9.2 we will see some functions for which the composition in either order results in the same function.

It is often useful to view a complicated function as a composition of simpler functions. For example, the function $Q(x) = (x - 3)^2$ consists of two operations, subtracting 3 and squaring. So Q can be described as a composition of the functions $f(x) = x - 3$ and $g(x) = x^2$. To check this, we find $(g \circ f)(x)$:

$$(g \circ f)(x) = g(f(x))$$

$$= g(x - 3)$$

$$= (x - 3)^2$$

We can express the fact that Q is the same as the composition function $g \circ f$ by writing $Q = g \circ f$ or $Q(x) = (g \circ f)(x)$.

E X A M P L E   5

Expressing a function as a composition of simpler functions

Let $f(x) = x - 2$, $g(x) = 3x$, and $h(x) = \sqrt{x}$. Write each of the following functions as a composition, using f, g, and h.

a) $F(x) = \sqrt{x - 2}$

b) $H(x) = x - 4$

c) $K(x) = 3x - 6$

Solution

a) The function F consists of first subtracting 2 from x and then taking the square root of that result. So $F = h \circ f$. Check this result by finding $(h \circ f)(x)$:

$$(h \circ f)(x) = h(f(x)) = h(x - 2) = \sqrt{x - 2}$$

b) Subtracting 4 from x can be accomplished by subtracting 2 from x and then subtracting 2 from that result. So $H = f \circ f$. Check by finding $(f \circ f)(x)$:

$$(f \circ f)(x) = f(f(x)) = f(x - 2) = x - 2 - 2 = x - 4$$

c) Notice that $K(x) = 3(x - 2)$. The function K consists of subtracting 2 from x and then multiplying the result by 3. So $K = g \circ f$. Check by finding $(g \circ f)(x)$:

$$(g \circ f)(x) = g(f(x)) = g(x - 2) = 3(x - 2) = 3x - 6$$

CAUTION   In Example 5(a) we have $F = h \circ f$ because in F we subtract 2 before taking the square root. If we had the function $G(x) = \sqrt{x} - 2$, we would take the square root before subtracting 2. So $G = f \circ h$. Notice how important the order of operations is here.

In the next example we see functions for which the composition is the identity function. Each function undoes what the other function does. We will study functions of this type further in Section 9.2.

E X A M P L E   6

Composition of functions

Show that $(f \circ g)(x) = x$ for each pair of functions.

a) $f(x) = 2x - 1$ and $g(x) = \dfrac{x + 1}{2}$

b) $f(x) = x^3 + 5$ and $g(x) = (x - 5)^{1/3}$

Solution

a) $(f \circ g)(x) = f(g(x)) = f\left(\dfrac{x + 1}{2}\right)$

$$= 2\left(\dfrac{x + 1}{2}\right) - 1$$

$$= x + 1 - 1$$

$$= x$$

b) $(f \circ g)(x) = f(g(x)) = f((x - 5)^{1/3})$

$$= ((x - 5)^{1/3})^3 + 5$$

$$= x - 5 + 5$$

$$= x$$

WARM-UPS

True or false? Explain your answer.

1. If $f(x) = x - 2$ and $g(x) = x + 3$, then $(f - g)(x) = -5$.   True
2. If $f(x) = x + 4$ and $g(x) = 3x$, then $\left(\frac{f}{g}\right)(2) = 1$.   True
3. The functions $f \circ g$ and $g \circ f$ are always the same.   False
4. If $f(x) = x^2$ and $g(x) = x + 2$, then $(f \circ g)(x) = x^2 + 2$.   False
5. The functions $f \circ g$ and $f \cdot g$ are always the same.   False
6. If $f(x) = \sqrt{x}$ and $g(x) = x - 9$, then $g(f(x)) = f(g(x))$ for every x.   False
7. If $f(x) = 3x$ and $g(x) = \frac{x}{3}$, then $(f \circ g)(x) = x$.   True
8. If $a = 3b^2 - 7b$, and $c = a^2 + 3a$, then c is a function of b.   True
9. The function $F(x) = \sqrt{x - 5}$ is a composition of two functions.   True
10. If $F(x) = (x - 1)^2$, $h(x) = x - 1$, and $g(x) = x^2$, then $F = g \circ h$.   True

9.1   EXERCISES

Reading and Writing   *After reading this section, write out the answers to these questions. Use complete sentences.*

1. What are the basic operations with functions?
   The basic operations of functions are addition, subtraction, multiplication, and addition.

2. How do we perform the basic operations with functions?
   We perform the operation with functions by adding, subtracting, multiplying, or dividing the expressions that define the functions.

3. What is the composition of two functions?
   In the composition function the second function is evaluated on the result of the first function.

4. How is the order of operations related to composition of functions?
   Since each operation is a function, the order of operations determines the order in which the functions are composed.

Let $f(x) = 4x - 3$, and $g(x) = x^2 - 2x$. Find the following. See Examples 1 and 2.

5. $(f + g)(x)$
   $x^2 + 2x - 3$

6. $(f - g)(x)$
   $-x^2 + 6x - 3$

7. $(f \cdot g)(x)$

   $4x^3 - 11x^2 + 6x$

8. $\left(\frac{f}{g}\right)(x)$

   $\dfrac{4x - 3}{x^2 - 2x}$

9. $(f + g)(3)$
   12

10. $(f + g)(2)$
    5

11. $(f - g)(-3)$
    -30

12. $(f - g)(-2)$
    -19

13. $(f \cdot g)(-1)$
    -21

14. $(f \cdot g)(-2)$
    -88

15. $\left(\dfrac{f}{g}\right)(4)$ $\dfrac{13}{8}$

16. $\left(\dfrac{f}{g}\right)(-2)$ $-\dfrac{11}{8}$

For Exercises 17–24, use the two functions to write y as a function of x. See Example 3.

17. $y = 3a - 2, a = 2x - 6$ $y = 6x - 20$
18. $y = 2c + 3, c = -3x + 4$ $y = -6x + 11$
19. $y = 2d + 1, d = \dfrac{x + 1}{2}$ $y = x + 2$
20. $y = -3d + 2, d = \dfrac{2 - x}{3}$ $y = x$
21. $y = m^2 - 1, m = x + 1$ $y = x^2 + 2x$
22. $y = n^2 - 3n + 1, n = x + 2$ $y = x^2 + x - 1$
23. $y = \dfrac{a - 3}{a + 2}, a = \dfrac{2x + 3}{1 - x}$ $y = x$
24. $y = \dfrac{w + 2}{w - 5}, w = \dfrac{5x + 2}{x - 1}$ $y = x$

Let $f(x) = 2x - 3$, $g(x) = x^2 + 3x$, and $h(x) = \dfrac{x + 3}{2}$. Find the following. See Example 4.

25. $(g \circ f)(1)$ -2
26. $(f \circ g)(-2)$ -7
27. $(f \circ g)(1)$ 5
28. $(g \circ f)(-2)$ 28
29. $(f \circ f)(4)$ 7
30. $(h \circ h)(3)$ 3
31. $(h \circ f)(5)$ 5
32. $(f \circ h)(0)$ 0
33. $(f \circ h)(5)$ 5
34. $(h \circ f)(0)$ 0
35. $(g \circ h)(-1)$ 4
36. $(h \circ g)(-1)$ $\dfrac{1}{2}$
37. $(f \circ g)(2.36)$ 22.2992
38. $(h \circ f)(23.761)$ 23.761

39. $(g \circ f)(x)$

$4x^2 - 6x$

40. $(g \circ h)(x)$

$\dfrac{x^2 + 12x + 27}{4}$

41. $(f \circ g)(x)$

$2x^2 + 6x - 3$

42. $(h \circ g)(x)$

$\dfrac{x^2 + 3x + 3}{2}$

43. $(h \circ f)(x)$

x

44. $(f \circ h)(x)$

x

45. $(f \circ f)(x)$

$4x - 9$

46. $(g \circ g)(x)$

$x^4 + 6x^3 + 12x^2 + 9x$

47. $(h \circ h)(x)$

$\dfrac{x + 9}{4}$

48. $(f \circ f \circ f)(x)$

$8x - 21$

Let $f(x) = \sqrt{x}$, $g(x) = x^2$, and $h(x) = x - 3$. Write each of the following functions as a composition using f, g, or h. See Example 5.

49. $F(x) = \sqrt{x - 3}$

$F = f \circ h$

50. $N(x) = \sqrt{x} - 3$

$N = h \circ f$

51. $G(x) = x^2 - 6x + 9$

$G = g \circ h$

52. $P(x) = x$ for $x \geq 0$

$P = f \circ g$

53. $H(x) = x^2 - 3$

$H = h \circ g$

54. $M(x) = x^{1/4}$

$M = f \circ f$

55. $J(x) = x - 6$

$J = h \circ h$

56. $R(x) = \sqrt{x^2 - 3}$

$R = f \circ h \circ g$

57. $K(x) = x^4$

$K = g \circ g$

58. $Q(x) = \sqrt{x^2 - 6x + 9}$

$Q = f \circ g \circ h$

Show that $(f \circ g)(x) = x$ and $(g \circ f)(x) = x$ for each given pair of functions. See Example 6.

59. $f(x) = 3x + 5$, $g(x) = \dfrac{x - 5}{3}$

60. $f(x) = 3x - 7$, $g(x) = \dfrac{x + 7}{3}$

61. $f(x) = x^3 - 9$, $g(x) = \sqrt[3]{x + 9}$

62. $f(x) = x^3 + 1$, $g(x) = \sqrt[3]{x - 1}$

63. $f(x) = \dfrac{x - 1}{x + 1}$, $g(x) = \dfrac{x + 1}{1 - x}$

64. $f(x) = \dfrac{x + 1}{x - 3}$, $g(x) = \dfrac{3x + 1}{x - 1}$

65. $f(x) = \dfrac{1}{x}$, $g(x) = \dfrac{1}{x}$

66. $f(x) = 2x^3$, $g(x) = \left(\dfrac{x}{2}\right)^{1/3}$

Solve each problem.

67. *Color monitor.* The CTX CMS-1561 color monitor has a square viewing area that has a diagonal measure of 15 inches (Midwest Micro Catalog). Find the area of the viewing area in square inches (in.^{2}). Write a formula for the area of a square as a function of the length of its diagonal.

112.5 in.^{2}, $A = \dfrac{d^2}{2}$

68. *Perimeter.* Write a formula for the perimeter of a square as a function of its area.

$P = 4\sqrt{A}$.

69. *Profit function.* A plastic bag manufacturer has determined that the company can sell as many bags as it can produce each month. If it produces x thousand bags in a month, the revenue is $R(x) = x^2 - 10x + 30$ dollars, and the cost is $C(x) = 2x^2 - 30x + 200$ dollars. Use the fact that profit is revenue minus cost to write the profit as a function of x.

$P(x) = -x^2 + 20x - 170$

70. *Area of a sign.* A sign is in the shape of a square with a semicircle of radius x adjoining one side and a semicircle of diameter x removed from the opposite side. If the sides of the square are length $2x$, then write the area of the sign as a function of x.

$A = \dfrac{(32 + 3\pi)x^2}{8}$

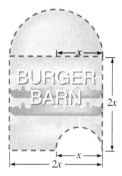

FIGURE FOR EXERCISE 70

71. *Junk food expenditures.* Suppose the average family spends 25% of its income on food, $F = 0.25I$, and 10% of each food dollar on junk food, $J = 0.10F$. Write J as a function of I.

$J = 0.025I$

72. *Area of an inscribed circle.* A pipe of radius r must pass through a square hole of area M as shown in the figure. Write the cross-sectional area of the pipe A as a function of M.

$A = \pi\dfrac{M}{4}$

FIGURE FOR EXERCISE 72

73. *Displacement-length ratio.* To find the displacement-length ratio D for a sailboat, first find x, where $x = (L/100)^3$ and L is the length at the water line in feet

(*Sail,* September 1997). Next find D, where $D = (d/2240)/x$ and d is the displacement in pounds.

a) For the Pacific Seacraft 40, $L = 30$ ft 3 in. and $d = 24{,}665$ pounds. Find D.

b) For a boat with a displacement of 25,000 pounds, write D as a function of L.

c) The graph for the function in part (b) is shown in the accompanying figure. For a fixed displacement, does the displacement-length ratio increase or decrease as the length increases?

a) 397.8 **b)** $D = \dfrac{1.116 \times 10^7}{L^3}$ **c)** decreases

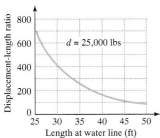

FIGURE FOR EXERCISE 73

74. Sail area-displacement ratio. To find the sail area-displacement ratio S, first find y, where $y = (d/64)^{2/3}$ and d is the displacement in pounds. Next find S, where $S = A/y$ and A is the sail area in square feet.

a) For the Pacific Seacraft 40, $A = 846$ square feet (ft^2) and $d = 24{,}665$ pounds. Find S.

b) For a boat with a sail area of 900 ft^2, write S as a function of d.

c) For a fixed sail area, does S increase or decrease as the displacement increases?

a) 15.97 **b)** $S = 14{,}400d^{-2/3}$ **c)** decreases

GETTING MORE INVOLVED

75. Discussion. Let $f(x) = \sqrt{x} - 4$ and $g(x) = \sqrt{x}$. Find the domains of f, g, and $g \circ f$.
$[0, \infty), [0, \infty), [16, \infty)$

76. Discussion. Let $f(x) = \sqrt{x - 4}$ and $g(x) = \sqrt{x - 8}$. Find the domains of f, g, and $f + g$.
$[4, \infty), [8, \infty), [8, \infty)$

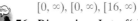

GRAPHING CALCULATOR EXERCISES

77. Graph $y_1 = x$, $y_2 = \sqrt{x}$, and $y_3 = x + \sqrt{x}$ in the same screen. Find the domain and range of $y_3 = x + \sqrt{x}$ by examining its graph. (On some graphing calculators you can enter y_3 as $y_3 = y_1 + y_2$.)
$[0, \infty), [0, \infty)$

78. Graph $y_1 = |x|$, $y_2 = |x - 3|$, and $y_3 = |x| + |x - 3|$. Find the domain and range of $y_3 = |x| + |x - 3|$ by examining its graph.
$(-\infty, \infty), [3, \infty)$

9.2 INVERSE FUNCTIONS

In Section 9.1 we introduced the idea of a pair of functions such that $(f \circ g)(x) = x$ and $(g \circ f)(x) = x$. Each function reverses what the other function does. In this section we explore that idea further.

Inverse of a Function

You can buy a 6-, 7-, or 8-foot conference table in the K-LOG Catalog for $299, $329, or $349, respectively. The set

$$f = \{(6, 299), (7, 329), (8, 349)\}$$

gives the price as a function of the length. We use the letter f as a name for this set or function, just as we use the letter f as a name for a function in the f-notation. In the function f, lengths in the domain $\{6, 7, 8\}$ are paired with prices in the range $\{299, 329, 349\}$. The **inverse** of the function f, denoted f^{-1}, is a function whose ordered pairs are obtained from f by interchanging the x- and y-coordinates:

$$f^{-1} = \{(299, 6), (329, 7), (349, 8)\}$$

We read f^{-1} as "f inverse." The domain of f^{-1} is $\{299, 329, 349\}$, and the range of f^{-1} is $\{6, 7, 8\}$. The inverse function reverses what the function does: it pairs prices in the range of f with lengths in the domain of f. For example, to find the cost of a

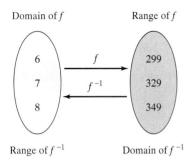

Domain of f Range of f

Range of f^{-1} Domain of f^{-1}

FIGURE 9.2

7-foot table, we use the function f to get $f(7) = 329$. To find the length of a table, that costs $349, we use the function f^{-1} to get $f^{-1}(349) = 8$. Of course, we could find the length of a $349 table by looking at the function f, but f^{-1} is a function whose input is price and whose output is length. In general, *the domain of f^{-1} is the range of f, and the range of f^{-1} is the domain of f.* See Fig. 9.2.

CAUTION   The -1 in f^{-1} is not read as an exponent. It does not mean $\frac{1}{f}$.

The cost per ribbon for Apple Imagewriter ribbons is a function of the number of boxes purchased:

$$g = \{(1, 4.85), (2, 4.60), (3, 4.60), (4, 4.35)\}$$

If we interchange the first and second coordinates in the ordered pairs of this function, we get

$$\{(4.85, 1), (4.60, 2), (4.60, 3), (4.35, 4)\}.$$

This set of ordered pairs is not a function because it contains ordered pairs with the same first coordinates and different second coordinates. So g does not have an inverse function. A function is **invertible** if you obtain a function when the coordinates of all ordered pairs are reversed. So f is invertible and g is not invertible. The function g is not invertible because the definition of function allows more than one number of the domain to be paired with the same number in the range. Of course, when this pairing is reversed, the definition of function is violated.

One-to-One Function

If a function is such that no two ordered pairs have different x-coordinates and the same y-coordinate, then the function is called a **one-to-one** function.

In a one-to-one function each member of the domain corresponds to just one member of the range, and each member of the range corresponds to just one member of the domain. *Functions that are one-to-one are invertible functions.*

Inverse Function

The inverse of a one-to-one function f is the function f^{-1}, which is obtained from f by interchanging the coordinates in each ordered pair of f.

helpful hint

Consider the universal product codes (UPC) and the prices for all of the items in your favorite grocery store. The price of an item is a function of the UPC because every UPC determines a price. This function is not invertible because you cannot determine the UPC from a given price.

E X A M P L E   1

Identifying invertible functions

Determine whether each function is invertible. If it is invertible, then find the inverse function.

a) $f = \{(2, 4), (-2, 4), (3, 9)\}$

b) $g = \left\{\left(2, \frac{1}{2}\right), \left(5, \frac{1}{5}\right), \left(7, \frac{1}{7}\right)\right\}$

c) $h = \{(3, 5), (7, 9)\}$

Solution

a) Since $(2, 4)$ and $(-2, 4)$ have the same y-coordinate, this function is not one-to-one, and it is not invertible.

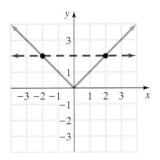

FIGURE 9.3

b) This function is one-to-one, and so it is invertible.

$$g^{-1} = \left\{\left(\frac{1}{2}, 2\right), \left(\frac{1}{5}, 5\right), \left(\frac{1}{7}, 7\right)\right\}$$

c) This function is invertible, and $h^{-1} = \{(5, 3), (9, 7)\}$.

You learned to use the vertical-line test in Section 3.6 to determine whether a graph is the graph of a function. The **horizontal-line test** is a similar visual test for determining whether a function is invertible. If a horizontal line crosses a graph two (or more) times, as in Fig. 9.3, then there are two points on the graph, say (x_1, y) and (x_2, y), that have different x-coordinates and the same y-coordinate. So the function is not one-to-one, and the function is not invertible.

Horizontal-Line Test

A function is invertible if and only if no horizontal line crosses its graph more than once.

EXAMPLE 2 **Using the horizontal-line test**

Determine whether each function is invertible by examining its graph.

a)

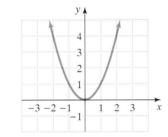

b)

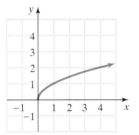

Solution

a) This function is not invertible because a horizontal line can be drawn so that it crosses the graph at $(2, 4)$ and $(-2, 4)$.

b) This function is invertible because every horizontal line that crosses the graph crosses it only once. ■

Identifying Inverse Functions

Consider the one-to-one function $f(x) = 3x$. The inverse function must reverse the ordered pairs of the function. Because division by 3 undoes multiplication by 3, we could guess that $g(x) = \frac{x}{3}$ is the inverse function. To verify our guess, we can use the following rule for determining whether two given functions are inverses of each other.

Identifying Inverse Functions

Functions f and g are inverses of each other if and only if
$$(g \circ f)(x) = x \text{ for every number } x \text{ in the domain of } f \text{ and}$$
$$(f \circ g)(x) = x \text{ for every number } x \text{ in the domain of } g.$$

In the next example we verify that $f(x) = 3x$ and $g(x) = \frac{x}{3}$ are inverses.

helpful **hint**

Tests such as the vertical-line test and the horizontal-line test are certainly not accurate in all cases. We discuss these tests to get a visual idea of what graphs of functions and invertible functions look like.

E X A M P L E   3

Identifying inverse functions

Determine whether the functions f and g are inverses of each other.

a) $f(x) = 3x$ and $g(x) = \dfrac{x}{3}$ **b)** $f(x) = 2x - 1$ and $g(x) = \dfrac{1}{2}x + 1$

c) $f(x) = x^2$ and $g(x) = \sqrt{x}$

Solution

a) Find $g \circ f$ and $f \circ g$:

$$(g \circ f)(x) = g(f(x)) = g(3x) = \frac{3x}{3} = x$$

$$(f \circ g)(x) = f(g(x)) = f\left(\frac{x}{3}\right) = 3 \cdot \frac{x}{3} = x$$

Because each of these equations is true for any real number x, f and g are inverses of each other. We write $g = f^{-1}$ or $f^{-1}(x) = \frac{x}{3}$.

b) Find the composition of g and f:

$$(g \circ f)(x) = g(f(x))$$

$$= g(2x - 1) = \frac{1}{2}(2x - 1) + 1 = x + \frac{1}{2}$$

So f and g are not inverses of each other.

c) If x is any real number, we can write

$$(g \circ f)(x) = g(f(x))$$

$$= g(x^2) = \sqrt{x^2} = |x|.$$

The domain of f is $(-\infty, \infty)$, and $|x| \neq x$ if x is negative. So g and f are not inverses of each other. Note that $f(x) = x^2$ is not a one-to-one function, since both $(3, 9)$ and $(-3, 9)$ are ordered pairs of this function. Thus $f(x) = x^2$ does not have an inverse. ∎

Switch-and-Solve Strategy

If an invertible function is defined by a list of ordered pairs, as in Example 1, then the inverse function is found by simply interchanging the coordinates in the ordered pairs. If an invertible function is defined by a formula, then the inverse function must reverse or undo what the function does. Because the inverse function interchanges the roles of x and y, we interchange x and y in the formula and then solve the new formula for y to undo what the original function did. This **switch-and-solve** strategy is illustrated in the next two examples.

E X A M P L E   4

The switch-and-solve strategy

Find the inverse of $h(x) = 2x + 1$.

Solution

First write the function as $y = 2x + 1$, then interchange x and y:

$$y = 2x + 1$$
$$x = 2y + 1 \quad \text{Interchange } x \text{ and } y.$$
$$x - 1 = 2y \quad \text{Solve for } y.$$
$$\frac{x - 1}{2} = y$$
$$h^{-1}(x) = \frac{x - 1}{2} \quad \text{Replace } y \text{ by } h^{-1}(x).$$

We can verify that h and h^{-1} are inverses by using composition:

$$(h^{-1} \circ h)(x) = h^{-1}(h(x)) = h^{-1}(2x + 1)$$

$$= \frac{2x + 1 - 1}{2} = \frac{2x}{2} = x$$

$$(h \circ h^{-1})(x) = h(h^{-1}(x)) = h\left(\frac{x - 1}{2}\right)$$

$$= 2 \cdot \frac{x - 1}{2} + 1 = x - 1 + 1 = x$$

E X A M P L E   5   **The switch-and-solve strategy**

If $f(x) = \dfrac{x + 1}{x - 3}$, find $f^{-1}(x)$.

Solution

Replace $f(x)$ by y, interchange x and y, then solve for y:

$$y = \frac{x + 1}{x - 3} \qquad \text{Use } y \text{ in place of } f(x).$$

$$x = \frac{y + 1}{y - 3} \qquad \text{Switch } x \text{ and } y.$$

$$x(y - 3) = y + 1 \qquad \text{Multiply each side by } y - 3.$$

$$xy - 3x = y + 1 \qquad \text{Distributive property}$$

$$xy - y = 3x + 1$$

$$y(x - 1) = 3x + 1 \qquad \text{Factor out } y.$$

$$y = \frac{3x + 1}{x - 1} \qquad \text{Divide each side by } x - 1.$$

$$f^{-1}(x) = \frac{3x + 1}{x - 1} \qquad \text{Replace } y \text{ by } f^{-1}(x).$$

You should check that $(f \circ f^{-1})(x) = x$ and $(f^{-1} \circ f)(x) = x$.

The strategy for finding the inverse of a function $f(x)$ is summarized as follows.

Switch-and-Solve Strategy for Finding f^{-1}

1. Replace $f(x)$ by y.
2. Interchange x and y.
3. Solve the equation for y.
4. Replace y by $f^{-1}(x)$.

If we use the switch-and-solve strategy to find the inverse of $f(x) = x^3$, then we get $f^{-1}(x) = x^{1/3}$. For $h(x) = 6x$ we have $h^{-1}(x) = \frac{x}{6}$. The inverse of $k(x) = x - 9$ is $k^{-1}(x) = x + 9$. For each of these functions there is an appropriate operation of arithmetic that undoes what the function does.

helpful hint

You should know from memory the inverses of simple functions that involve one or two operations. For example, the inverse of $f(x) = x + 99$ is $f^{-1}(x) = x - 99$. The inverse of $f(x) = x/33 + 22$ is $f^{-1}(x) = 33(x - 22)$.

If a function involves two operations, the inverse function undoes those operations in the opposite order from which the function does them. For example, the function $g(x) = 3x - 5$ multiplies x by 3 and then subtracts 5 from that result. To undo these operations, we add 5 and then divide the result by 3. So

$$g^{-1}(x) = \frac{x + 5}{3}.$$

Note that $g^{-1}(x) \neq \frac{x}{3} + 5$.

Even Roots or Even Powers

We need to use special care in finding inverses for functions that involve even roots or even powers. We saw in Example 3(c) that $f(x) = x^2$ is not the inverse of $g(x) = \sqrt{x}$. However, because $g(x) = \sqrt{x}$ is a one-to-one function, it has an inverse. The domain of g is $[0, \infty)$, and the range is $[0, \infty)$. So the inverse of g must have domain $[0, \infty)$ and range $[0, \infty)$. See Fig. 9.4. The only reason that $f(x) = x^2$ is not the inverse of g is that it has the wrong domain. So to write the

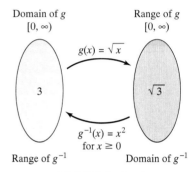

FIGURE 9.4

M A T H   A T   W O R K $x^2 + (x+1)^2 = 5^2$

When a serious automobile accident occurs in Massachusetts, Stephen Benanti, the Commanding Officer of the Accident Reconstruction Section of the State Police, may be called to examine the physical evidence at the scene. Physical evidence can consist of debris, scrapes, and gouges on the road; damage to fixed objects such as utility poles, trees, or guardrails; and the final resting position of and damage to the vehicles.

STATE POLICE OFFICER

One critical type of evidence that is sometimes found are skid marks on the road. Sergeant Benanti can use the lengths of the skid marks to calculate the speeds of accident vehicles. Using a sophisticated laser measuring device, Benanti can store data that later can be downloaded into a computer to reconstruct the accident scene. He must also conduct tests on the road surface to calculate the drag factor, which is the resistance between the tire and the road surface. A smooth, icy surface yields a much lower drag factor than a dry asphalt surface. The minimum speed formula that state troopers use to determine vehicles' speeds is a function of the drag factor and the skid distance: $S = \sqrt{30DF}$, where S = speed, D = distance skidded, and F = drag factor.

In Exercise 73 of this section you will use this minimum speed formula with a given length of skid marks to determine the speed of a vehicle.

inverse function, we must use the appropriate domain:

$$g^{-1}(x) = x^2 \quad \text{for} \quad x \geq 0$$

Note that by restricting the domain of g^{-1} to $[0, \infty)$, g^{-1} is one-to-one. With this restriction it is true that $(g \circ g^{-1})(x) = x$ and $(g^{-1} \circ g)(x) = x$ for every nonnegative number x.

EXAMPLE   6

Inverse of a function with an even exponent

Find the inverse of the function $f(x) = (x - 3)^2$ for $x \geq 3$.

Solution

Because of the restriction $x \geq 3$, f is a one-to-one function with domain $[3, \infty)$ and range $[0, \infty)$. The domain of the inverse function is $[0, \infty)$, and its range is $[3, \infty)$. Use the switch-and-solve strategy to find the formula for the inverse:

$$y = (x - 3)^2$$
$$x = (y - 3)^2$$
$$y - 3 = \pm\sqrt{x}$$
$$y = 3 \pm \sqrt{x}$$

Because the inverse function must have range $[3, \infty)$, we use the formula $f^{-1}(x) = 3 + \sqrt{x}$. Because the domain of f^{-1} is assumed to be $[0, \infty)$, no restriction is required on x.

Graphs of f and f^{-1}

Consider $f(x) = x^2$ for $x \geq 0$ and $f^{-1}(x) = \sqrt{x}$. Their graphs are shown in Fig. 9.5. Notice the symmetry. If we folded the paper along the line $y = x$, the two graphs would coincide.

If a point (a, b) is on the graph of the function f, then (b, a) must be on the graph of $f^{-1}(x)$. See Fig. 9.6. The points (a, b) and (b, a) lie on opposite sides of the diagonal line $y = x$ and are the same distance from it. For this reason the graphs of f and f^{-1} are symmetric with respect to the line $y = x$.

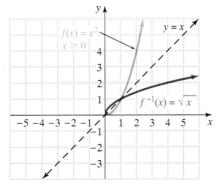

FIGURE 9.5

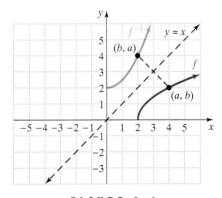

FIGURE 9.6

E X A M P L E 7

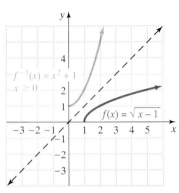

FIGURE 9.7

Inverses and their graphs

Find the inverse of the function $f(x) = \sqrt{x - 1}$ and graph f and f^{-1} on the same pair of axes.

Solution

To find f^{-1}, first switch x and y in the formula $y = \sqrt{x - 1}$:

$$x = \sqrt{y - 1}$$
$$x^2 = y - 1 \quad \text{Square both sides.}$$
$$x^2 + 1 = y$$

Because the range of f is the set of nonnegative real numbers $[0, \infty)$, we must restrict the domain of f^{-1} to be $[0, \infty)$. Thus $f^{-1}(x) = x^2 + 1$ for $x \geq 0$. The two graphs are shown in Fig. 9.7. ■

WARM-UPS

True or false? Explain your answer.

1. The inverse of $\{(1, 3), (2, 5)\}$ is $\{(3, 1), (2, 5)\}$ False
2. The function $f(x) = 3$ is a one-to-one function.   False
3. If $g(x) = 2x$, then $g^{-1}(x) = \frac{1}{2x}$.   False
4. Only one-to-one functions are invertible.   True
5. The domain of g is the same as the range of g^{-1}.   True
6. The function $f(x) = x^4$ is invertible.   False
7. If $f(x) = -x$, then $f^{-1}(x) = -x$.   True
8. If h is invertible and $h(7) = -95$, then $h^{-1}(-95) = 7$.   True
9. If $k(x) = 3x - 6$, then $k^{-1}(x) = \frac{1}{3}x + 2$.   True
10. If $f(x) = 3x - 4$, then $f^{-1}(x) = x + 4$.   False

9.2   EXERCISES

Reading and Writing *After reading this section, write out the answers to these questions. Use complete sentences.*

1. What is the inverse of a function?
   The inverse of a function is a function with the same ordered pairs except that the coordinates are reversed.

2. What is the domain of f^{-1}?
   The domain of f^{-1} is the range of f.

3. What is the range of f^{-1}?
   The range of f^{-1} is the domain of f.

4. What does the -1 in f^{-1} mean?
   The -1 in f^{-1} is not treated as an exponent. It is simply a notation for the inverse of the function f.

5. What is a one-to-one function?
   A function is one-to-one if no two ordered pairs have the same second coordinate with different first coordinates.

6. What is the horizontal-line test?
   The horizontal-line test says that if a horizontal line can be drawn to cross the graph of a function more than once, then the function is not one-to-one.

7. What is the switch-and-solve strategy for?
   The switch-and-solve strategy is used to find a formula for an inverse function.

8. How are the graphs of f and f^{-1} related?
   The graphs of f and f^{-1} are symmetric with respect to the line $y = x$.

Determine whether each function is invertible. If it is invertible, then find the inverse. See Example 1.

9. $\{(-3, 3), (-2, 2), (0, 0), (2, 2)\}$ No
10. $\{(1, 1), (2, 8), (3, 27)\}$ Yes, $\{(1, 1), (8, 2), (27, 3)\}$
11. $\{(16, 4), (9, 3), (0, 0)\}$ Yes, $\{(4, 16), (3, 9), (0, 0)\}$
12. $\{(-1, 1), (-3, 81), (3, 81)\}$ No
13. $\{(0, 5), (5, 0), (6, 0)\}$ No
14. $\{(3, -3), (-2, 2), (1, -1)\}$ Yes, $\{(-3, 3), (2, -2), (-1, 1)\}$
15. $\{(0, 0), (2, 2), (9, 9)\}$ Yes, $\{(0, 0), (2, 2), (9, 9)\}$
16. $\{(9, 1), (2, 1), (7, 1), (0, 1)\}$ No

Determine whether each function is invertible by examining the graph of the function. See Example 2.

17. No

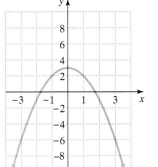

18. No

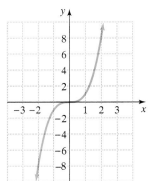

19. Yes

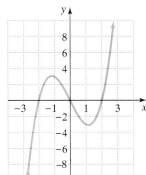

20. Yes

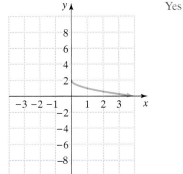

Determine whether each pair of functions f and g are inverses of each other. See Example 3.

21. $f(x) = 2x$ and $g(x) = 0.5x$ Yes

22. $f(x) = 3x$ and $g(x) = 0.33x$ No

23. $f(x) = 2x - 10$ and $g(x) = \frac{1}{2}x + 5$ Yes

24. $f(x) = 3x + 7$ and $g(x) = \frac{x - 7}{3}$ Yes

25. $f(x) = -x$ and $g(x) = -x$ Yes

26. $f(x) = \frac{1}{x}$ and $g(x) = \frac{1}{x}$ Yes

27. $f(x) = x^4$ and $g(x) = x^{1/4}$ No

28. $f(x) = |2x|$ and $g(x) = \left|\frac{x}{2}\right|$ No

Determine f^{-1} for each function by using the switch-and-solve strategy. Check that $(f \circ f^{-1})(x) = x$ and $(f^{-1} \circ f)(x) = x$. See Examples 4 and 5.

29. $f(x) = 5x$
$f^{-1}(x) = \dfrac{x}{5}$

30. $h(x) = -3x$
$h^{-1}(x) = -\dfrac{1}{3}x$

31. $g(x) = x - 9$
$g^{-1}(x) = x + 9$

32. $j(x) = x + 7$
$j^{-1}(x) = x - 7$

33. $k(x) = 5x - 9$
$k^{-1}(x) = \dfrac{x + 9}{5}$

34. $r(x) = 2x - 8$
$r^{-1}(x) = \dfrac{x + 8}{2}$

35. $m(x) = \dfrac{2}{x}$
$m^{-1}(x) = \dfrac{2}{x}$

36. $s(x) = \dfrac{-1}{x}$
$s^{-1}(x) = -\dfrac{1}{x}$

37. $f(x) = \sqrt[3]{x - 4}$
$f^{-1}(x) = x^3 + 4$

38. $f(x) = \sqrt[3]{x + 2}$
$f^{-1}(x) = x^3 - 2$

39. $f(x) = \dfrac{3}{x - 4}$
$f^{-1}(x) = \dfrac{3}{x} + 4$

40. $f(x) = \dfrac{2}{x + 1}$
$f^{-1}(x) = \dfrac{2}{x} - 1$

41. $f(x) = \sqrt[3]{3x + 7}$
$f^{-1}(x) = \dfrac{x^3 - 7}{3}$

42. $f(x) = \sqrt[3]{7 - 5x}$
$f^{-1}(x) = \dfrac{-x^3 + 7}{5}$

43. $f(x) = \dfrac{x + 1}{x - 2}$
$f^{-1}(x) = \dfrac{2x + 1}{x - 1}$

44. $f(x) = \dfrac{1 - x}{x + 3}$
$f^{-1}(x) = \dfrac{1 - 3x}{x + 1}$

45. $f(x) = \dfrac{x + 1}{3x - 4}$
$f^{-1}(x) = \dfrac{1 + 4x}{3x - 1}$

46. $g(x) = \dfrac{3x + 5}{2x - 3}$
$g^{-1}(x) = \dfrac{3x + 5}{2x - 3}$

Find the inverse of each function. See Example 6.

47. $p(x) = \sqrt[4]{x}$ $p^{-1}(x) = x^4$ for $x \geq 0$

48. $v(x) = \sqrt[6]{x}$ $v^{-1}(x) = x^6$ for $x \geq 0$

49. $f(x) = (x - 2)^2$ for $x \geq 2$ $f^{-1}(x) = 2 + \sqrt{x}$

50. $g(x) = (x + 5)^2$ for $x \geq -5$ $g^{-1}(x) = -5 + \sqrt{x}$

51. $f(x) = x^2 + 3$ for $x \geq 0$ $f^{-1}(x) = \sqrt{x - 3}$

52. $f(x) = x^2 - 5$ for $x \geq 0$ $f^{-1}(x) = \sqrt{x + 5}$

53. $f(x) = \sqrt{x + 2}$ $f^{-1}(x) = x^2 - 2$ for $x \geq 0$

54. $f(x) = \sqrt{x - 4}$ $f^{-1}(x) = x^2 + 4$ for $x \geq 0$

In Exercises 55–64, find the inverse of each function and graph f and f⁻¹ on the same pair of axes. See Example 7.

55. $f(x) = 2x + 3$

$f^{-1}(x) = \dfrac{1}{2}x - \dfrac{3}{2}$

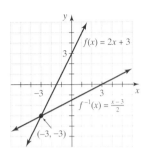

56. $f(x) = -3x + 2$

$f^{-1}(x) = -\dfrac{1}{3}x + \dfrac{2}{3}$

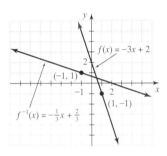

57. $f(x) = x^2 - 1$ for $x \geq 0$

$f^{-1}(x) = \sqrt{x + 1}$

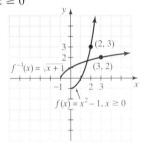

58. $f(x) = x^2 + 3$ for $x \geq 0$

$f^{-1}(x) = \sqrt{x - 3}$

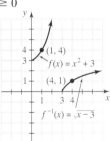

59. $f(x) = 5x$

$f^{-1}(x) = \dfrac{x}{5}$

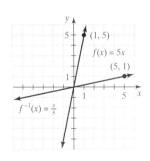

60. $f(x) = \dfrac{x}{4}$

$f^{-1}(x) = 4x$

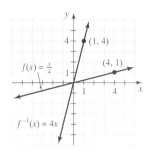

61. $f(x) = x^3$

$f^{-1}(x) = \sqrt[3]{x}$

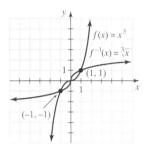

62. $f(x) = 2x^3$

$f^{-1}(x) = \sqrt[3]{\dfrac{x}{2}}$

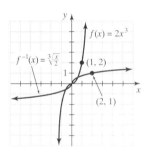

63. $f(x) = \sqrt{x - 2}$

$f^{-1}(x) = x^2 + 2$ for $x \geq 0$

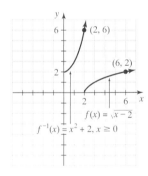

64. $f(x) = \sqrt{x + 3}$

$f^{-1}(x) = x^2 - 3$ for $x \geq 0$

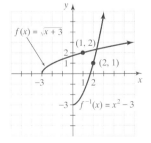

For each pair of functions, find $(f^{-1} \circ f)(x)$

65. $f(x) = x^3 - 1$ and $f^{-1}(x) = \sqrt[3]{x + 1}$ $(f^{-1} \circ f)(x) = x$

66. $f(x) = 2x^3 + 1$ and $f^{-1}(x) = \sqrt[3]{\dfrac{x - 1}{2}}$ $(f^{-1} \circ f)(x) = x$

67. $f(x) = \dfrac{1}{2}x - 3$ and $f^{-1}(x) = 2x + 6$ $(f^{-1} \circ f)(x) = x$

68. $f(x) = 3x - 9$ and $f^{-1}(x) = \dfrac{1}{3}x + 3$ $(f^{-1} \circ f)(x) = x$

69. $f(x) = \dfrac{1}{x} + 2$ and $f^{-1}(x) = \dfrac{1}{x - 2}$ $(f^{-1} \circ f)(x) = x$

70. $f(x) = 4 - \dfrac{1}{x}$ and $f^{-1}(x) = \dfrac{1}{4 - x}$ $(f^{-1} \circ f)(x) = x$

71. $f(x) = \dfrac{x + 1}{x - 2}$ and $f^{-1}(x) = \dfrac{2x + 1}{x - 1}$ $(f^{-1} \circ f)(x) = x$

72. $f(x) = \dfrac{3x - 2}{x + 2}$ and $f^{-1}(x) = \dfrac{2x + 2}{3 - x}$ $(f^{-1} \circ f)(x) = x$

Solve each problem.

 73. *Accident reconstruction.* The distance that it takes a car to stop is a function of the speed and the drag factor. The drag factor is a measure of the resistance between the tire and the road surface. The formula $S = \sqrt{30LD}$ is used to determine the minimum speed S [in miles per hour (mph)] for a car that has left skid marks of length L feet (ft) on a surface with drag factor D.

a) Find the minimum speed for a car that has left skid marks of length 50 ft where the drag factor is 0.75.
   33.5 mph

b) Does the drag factor increase or decrease for a road surface when it gets wet?
   decreases

c) Write L as a function S for a road surface with drag factor 1 and graph the function.

$$L = \dfrac{S^2}{30}$$

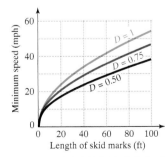

FIGURE FOR EXERCISE 73

74. *Area of a circle.* Let x be the radius of a circle and $h(x)$ be the area of the circle. Write a formula for $h(x)$ in terms of x. What does x represent in the notation $h^{-1}(x)$? Write a formula for $h^{-1}(x)$.

$h(x) = \pi x^2$, the area of the circle, $h^{-1}(x) = \sqrt{\dfrac{x}{\pi}}$

75. *Vehicle cost.* At Bill Hood Ford in Hammond a sales tax of 9% of the selling price x and a \$125 title and license fee are added to the selling price to get the total cost of a vehicle. Find the function $T(x)$ that the dealer uses to get the total cost as a function of the selling price x. Citizens National Bank will not include sales tax or fees in a loan. Find the function $T^{-1}(x)$ that the bank can use to get the selling price as a function of the total cost x.

$T(x) = 1.09x + 125$, $T^{-1}(x) = \dfrac{x - 125}{1.09}$

76. *Carpeting cost.* At the Windrush Trace apartment complex all living rooms are square, but the length of x feet may vary. The cost of carpeting a living room is \$18 per square yard plus a \$50 installation fee. Find the function $C(x)$ that gives the total cost of carpeting a living room of length x. The manager has an invoice for the total cost of a living room carpeting job but does not know in which apartment it was done. Find the function $C^{-1}(x)$ that gives the length of a living room as a function of the total cost of the carpeting job x.

$C(x) = 2x^2 + 50$, $C^{-1}(x) = \sqrt{\dfrac{x - 50}{2}}$

GETTING MORE INVOLVED

 77. *Discussion.* Let $f(x) = x^n$ for n a positive integer. For which values of n is f an invertible function? Explain.
An odd positive integer

 78. *Discussion.* Suppose f is a function with range $(-\infty, \infty)$ and g is a function with domain $(0, \infty)$. Is it possible that g and f are inverse functions? Explain.   No

 ## GRAPHING CALCULATOR EXERCISES

79. Most graphing calculators can form compositions of functions. Let $f(x) = x^2$ and $g(x) = \sqrt{x}$. To graph the composition $g \circ f$, let $y_1 = x^2$ and $y_2 = \sqrt{y_1}$. The graph of y_2 is the graph of $g \circ f$. Use the graph of y_2 to determine whether f and g are inverse functions.
Not inverses

80. Let $y_1 = x^3 - 4$, $y_2 = \sqrt[3]{x + 4}$, and $y_3 = \sqrt[3]{y_1} + 4$. The function y_3 is the composition of the first two functions. Graph all three functions on the same screen. What do the graphs indicate about the relationship between y_1 and y_2?
They are inverse functions.

9.3 VARIATION

If $y = 3x$, then as x varies so does y. Certain functions are customarily expressed in terms of variation. In this section you will learn to write formulas for those functions from verbal descriptions of the functions.

Direct Variation

In a community with an 8% sales tax rate, the amount of tax, t (in dollars), is a function of the amount of the purchase, a (in dollars). This function is expressed by the formula

$$t = 0.08a.$$

If the amount increases, then the tax increases. If a decreases, then t decreases. In this situation we say that t *varies directly with a, or t is directly proportional to a.* The constant tax rate, 0.08, is called the **variation constant** or **proportionality constant.** Notice that t is just a simple linear function of a. We are merely introducing some new terms to express an old idea.

> **Direct Variation**
>
> The statement **y varies directly as x,** or **y is directly proportional to x,** means that
>
> $$y = kx$$
>
> for some constant, k. The constant, k, is a fixed nonzero real number.

Finding the Proportionality Constant

If y varies directly as x and we know corresponding values for x and y, then we can find the proportionality constant.

E X A M P L E   1

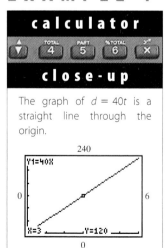

calculator

close-up

The graph of $d = 40t$ is a straight line through the origin.

Finding the proportionality constant

Joyce is traveling by car, and the distance she travels, d, varies directly with the amount of time, t, that she drives. In 3 hours she drove 120 miles. Find the proportionality constant and write d as a function of t.

Solution

Because d varies directly as t, we must have a constant k such that

$$d = kt.$$

Because $d = 120$ when $t = 3$, we can write

$$120 = k \cdot 3,$$

or

$$40 = k.$$

So the proportionality constant is 40 mph, and $d = 40t$. ∎

E X A M P L E   2

Direct variation

In a downtown office building the monthly rent for an office is directly proportional to the size of the office. If a 420-square-foot office rents for $1260 per month, then what is the rent for a 900-square-foot office?

Solution

Because the rent, R, varies directly with the area of the office, A, we have

$$R = kA.$$

Because a 420-square-foot office rents for $1260, we can substitute to find k:

$$1260 = k \cdot 420$$
$$3 = k$$

Now that we know the value of k, we can write

$$R = 3A.$$

To get the rent for a 900-square-foot office, insert 900 into this formula:

$$R = 3 \cdot 900$$
$$= 2700$$

So a 900-square-foot office rents for $2700 per month.

Inverse Variation

In making a 500-mile trip by car, the time it takes is a function of the speed of the car. The greater the speed, the less time it will take. If you decrease the speed, the time increases. We say that the time is *inversely proportional* to the speed. Using the formula $D = RT$ or $T = \dfrac{D}{R}$, we can write

$$T = \frac{500}{R}.$$

In general, we make the following definition.

> **Inverse Variation**
>
> The statement **y varies inversely as x,** or **y is inversely proportional to x,** means that
>
> $$y = \frac{k}{x}$$
>
> for some nonzero constant, k.

C A U T I O N   Be sure to understand the difference between direct and inverse variation. If y varies directly as x (with $k > 0$), then as x increases, y increases. If y varies inversely as x (with $k > 0$), then as x increases, y decreases.

E X A M P L E   3

Inverse variation

Suppose a is inversely proportional to b, and when $b = 5$, $a = \frac{1}{2}$. Find a when $b = 12$.

Solution

Because a is inversely proportional to b, we have

$$a = \frac{k}{b}$$

for some constant, k. Because $a = \frac{1}{2}$ when $b = 5$, we can find k by substituting these values into the formula:

$$\frac{1}{2} = \frac{k}{5}$$

$$\frac{5}{2} = k \qquad \text{Multiply each side by 5.}$$

Now to find a when $b = 12$, we use the formula with k replaced by $\frac{5}{2}$:

$$a = \frac{\frac{5}{2}}{b}$$

$$a = \frac{\frac{5}{2}}{12} = \frac{5}{2} \cdot \frac{1}{12} = \frac{5}{24}$$

Joint Variation

On a deposit of \$5000 in a savings account, the interest earned, I, depends on the rate, r, and the time, t. Assuming the interest is simple interest, we can use the formula $I = Prt$ to write

$$I = 5000rt.$$

The variable I is a function of two independent variables, r and t. In this case we say that I *varies jointly* as r and t.

Joint Variation

The statement y **varies jointly as x and z,** or y **is jointly proportional to x and z,** means that

$$y = kxz$$

for some nonzero constant, k.

E X A M P L E   4   **Joint variation**

Suppose y varies jointly with x and z, and $y = 12$ when $x = 5$ and $z = 2$. Find y when $x = 10$ and $z = -3$.

Solution

Because y varies jointly with x and z, we can write

$$y = kxz$$

for some constant, k. Now substitute $y = 12$, $x = 5$, and $z = 2$, and solve for k:

$$12 = k \cdot 5 \cdot 2$$

$$12 = 10k$$

$$\frac{6}{5} = k$$

Now that we know the value of k, we can rewrite the equation as

$$y = \frac{6}{5}xz.$$

To find y when $x = 10$ and $z = -3$, substitute into the equation:

$$y = \frac{6}{5}(10)(-3)$$

$$y = -36$$

More Variation

We frequently combine the ideas of direct, inverse, and joint variation with powers and roots. A combination of direct and inverse variation is referred to as **combined variation.** Study the examples that follow.

More Variation Examples

Statement	Formula
y varies directly as the square root of x.	$y = k\sqrt{x}$
y is directly proportional to the cube of x.	$y = kx^3$
y is inversely proportional to x^2.	$y = \dfrac{k}{x^2}$
y varies inversely as the square root of x.	$y = \dfrac{k}{\sqrt{x}}$
y varies jointly as x and the square of z.	$y = kxz^2$
y varies directly with x and inversely with the square root of z (combined variation).	$y = \dfrac{kx}{\sqrt{z}}$

CAUTION The variation terms never signify addition or subtraction. We always use multiplication unless we see the word "inversely." In that case we divide.

EXAMPLE 5 **Newton's law of gravity**

According to Newton's law of gravity, the gravitational attraction F between two objects with masses m_1 and m_2 is directly proportional to the product of their masses and inversely proportional to the square of the distance r between their centers. Write a formula for Newton's law of gravity.

Solution

Letting k be the constant of proportionality, we have

$$F = \frac{km_1m_2}{r^2}.$$

EXAMPLE 6 **House framing**

The time t that it takes to frame a house varies directly with the size of the house s in square feet and inversely with the number of framers n working on the job. If three framers can complete a 2,500-square-foot house in 6 days, then how long will it take six framers to complete a 4,500-square-foot house?

Solution

Because t varies directly with s and inversely with n, we have

$$t = \frac{ks}{n}.$$

Substitute $t = 6$, $s = 2500$, and $n = 3$ into this equation to find k:

$$6 = \frac{k \cdot 2500}{3}$$

$$18 = 2500k$$

$$0.0072 = k$$

Now use $k = 0.0072$, $s = 4500$, and $n = 6$ to find t:

$$t = \frac{0.0072 \cdot 4500}{6}$$

$$t = 5.4$$

So six framers can frame a 4500-square-foot house in 5.4 days.

WARM-UPS

True or false? Explain your answer.

1. If a varies directly as b, then $a = kb$.   True
2. If a is inversely proportional to b, then $a = bk$.   False
3. If a is jointly proportional to b and c, then $a = bc$.   False
4. If a is directly proportional to the square root of c, then $a = k\sqrt{c}$.   True
5. If b is directly proportional to a, then $b = ka^2$.   False
6. If a varies directly as b and inversely as c, then $a = \frac{kb}{c}$.   True
7. If a is jointly proportional to c and the square of b, then $a = \frac{kc}{b^2}$.   False
8. If a varies directly as c and inversely as the square root of b, then $a = \frac{kc}{b}$.   False
9. If b varies directly as a and inversely as the square of c, then $b = ka\sqrt{c}$.   False
10. If b varies inversely with the square of c, then $b = \frac{k}{c^2}$.   True

9.3   EXERCISES

Reading and Writing   *After reading this section, write out the answers to these questions. Use complete sentences.*

1. What does it mean that y varies directly as x?
   If y varies directly as x, then $y = kx$ for some constant k.
2. What is the constant of proportionality in a direct variation?
   The constant of proportionality in $y = kx$ is k.
3. What does it mean that y is inversely proportional to x?
   If y is inversely proportional to x, then $y = k/x$.
4. What is the difference between direct and inverse variation?
   In direct variation $y = kx$, whereas in inverse variation $y = k/x$.

5. What does it mean that y is jointly proportional to x and z?
   If y is jointly proportional to x and z, then $y = kxz$ for some constant k.
6. What is the difference between varies directly and directly proportional?
   Varies directly is the same as directly proportional.

Write a formula that expresses the relationship described by each statement. Use k as a constant of variation. See Examples 1–6.

7. a varies directly as m. $a = km$
8. w varies directly with P. $w = kP$
9. d varies inversely with e. $d = k/e$

10. y varies inversely as x. $y = k/x$

11. I varies jointly as r and t.
$I = krt$

12. q varies jointly as w and v.
$q = kwv$

13. m is directly proportional to the square of p.
$m = kp^2$

14. g is directly proportional to the cube of r.
$g = kr^3$

15. B is directly proportional to the cube root of w.
$B = k\sqrt[3]{w}$

16. F is directly proportional to the square of m.
$F = km^2$

17. t is inversely proportional to the square of x.
$t = \dfrac{k}{x^2}$

18. y is inversely proportional to the square root of z.
$y = \dfrac{k}{\sqrt{z}}$

19. v varies directly as m and inversely as n.
$v = \dfrac{km}{n}$

20. b varies directly as the square of n and inversely as the square root of v.
$b = \dfrac{kn^2}{\sqrt{v}}$

Find the proportionality constant and write a formula that expresses the indicated variation. See Example 1.

21. y varies directly as x, and $y = 6$ when $x = 4$.
$y = \dfrac{3}{2}x$

22. m varies directly as w, and $m = \frac{1}{3}$ when $w = \frac{1}{4}$.
$m = \dfrac{4}{3}w$

23. A varies inversely as B, and $A = 10$ when $B = 3$.
$A = \dfrac{30}{B}$

24. c varies inversely as d, and $c = 0.31$ when $d = 2$.
$c = \dfrac{0.62}{d}$

25. m varies inversely as the square root of p, and $m = 12$ when $p = 9$.
$m = \dfrac{36}{\sqrt{p}}$

26. s varies inversely as the square root of v, and $s = 6$ when $v = \frac{3}{2}$.
$s = \dfrac{3\sqrt{6}}{\sqrt{v}}$

27. A varies jointly as t and u, and $A = 6$ when $t = 5$ and $u = 3$.
$A = \dfrac{2}{5}tu$

28. N varies jointly as the square of p and the cube of q, and $N = 72$ when $p = 3$ and $q = 2$.
$N = p^2q^3$

29. y varies directly as x and inversely as z, and $y = 2.37$ when $x = \pi$ and $z = \sqrt{2}$.
$y = \dfrac{1.067x}{z}$

30. a varies directly as the square root of m and inversely as the square of n, and $a = 5.47$ when $m = 3$ and $n = 1.625$.
$a = \dfrac{8.339\sqrt{m}}{n^2}$

Solve each variation problem. See Examples 2–6.

31. If y varies directly as x, and $y = 7$ when $x = 5$, find y when $x = -3$.
$-\dfrac{21}{5}$

32. If n varies directly as p, and $n = 0.6$ when $p = 0.2$, find n when $p = \sqrt{2}$.
$3\sqrt{2}$

33. If w varies inversely as z, and $w = 6$ when $z = 2$, find w when $z = -8$.
$-\dfrac{3}{2}$

34. If p varies inversely as q, and $p = 5$ when $q = \sqrt{3}$, find p when $q = 5$.
$\sqrt{3}$

35. If A varies jointly as F and T, and $A = 6$ when $F = 3\sqrt{2}$ and $T = 4$, find A when $F = 2\sqrt{2}$ and $T = \frac{1}{2}$.
$\dfrac{1}{2}$

36. If j varies jointly as the square of r and the cube of v, and $j = -3$ when $r = 2\sqrt{3}$ and $v = \frac{1}{2}$, find j when $r = 3\sqrt{5}$ and $v = 2$.
-720

37. If D varies directly with t and inversely with the square of s, and $D = 12.35$ when $t = 2.8$ and $s = 2.48$, find D when $t = 5.63$ and $s = 6.81$.
3.293

38. If M varies jointly with x and the square of v, and $M = 39.5$ when $x = \sqrt{10}$ and $v = 3.87$, find M when $x = \sqrt{30}$ and $v = 7.21$.
237.469

Determine whether each equation represents direct, inverse, joint, or combined variation.

39. $y = \dfrac{78}{x}$ Inverse

40. $y = \dfrac{\pi}{x}$ Inverse

41. $y = \dfrac{1}{2}x$ Direct

42. $y = \dfrac{x}{4}$ Direct

43. $y = \dfrac{3x}{w}$ Combined

44. $y = \dfrac{4t^2}{\sqrt{x}}$ Combined

45. $y = \dfrac{1}{3}xz$ Joint

46. $y = 99qv$ Joint

In Exercises 47–61, solve each problem.

47. Lawn maintenance. At Larry's Lawn Service the cost of lawn maintenance varies directly with the size of the lawn. If the monthly maintenance on a 4,000-square-foot lawn is $280, then what is the maintenance fee for a 6,000-square-foot lawn? $420

48. Weight of the iguana. The weight of an iguana is directly proportional to its length. If a 4-foot iguana weighs 30 pounds, then how much should a 5-foot iguana weigh? 37.5 pounds

49. Gas laws. The volume of a gas in a cylinder at a fixed temperature is inversely proportional to the weight on the piston. If the gas has a volume of 6 cubic centimeters (cm³) for a weight of 30 kilograms (kg), then what would the volume be for a weight of 20 kg?   9 cm³

50. Selling software. A software vendor sells a software package at a price that is inversely proportional to the number of packages sold per month. When they are selling 900 packages per month, the price is $80 each. If they sell 1000 packages per month, then what should the new price be? $72 each

51. Costly culvert. The price of an aluminum culvert is jointly proportional to its radius and length. If a 12-foot culvert with a 6-inch radius costs $324, then what is the price of a 10-foot culvert with an 8-inch radius? $360

52. Pricing plastic. The cost of a piece of PVC water pipe varies jointly as its diameter and length. If a 20-foot pipe with a diameter of 1 inch costs $6.80, then what will be the cost of a 10-foot pipe with a $\frac{3}{4}$-inch diameter? $2.55

53. Reinforcing rods. The price of a steel rod varies jointly as the length and the square of the diameter. If an 18-foot rod with a 2-inch diameter costs $12.60, then what is the cost of a 12-foot rod with a 3-inch diameter? $18.90

54. Pea soup. The weight of a cylindrical can of pea soup varies jointly with the height and the square of the radius. If a 4-inch-high can with a 1.5-inch radius weighs 16 ounces, then what is the weight of a 5-inch-high can with a radius of 3 inches?   80 ounces

55. Falling objects. The distance an object falls in a vacuum varies directly with the square of the time it is falling. In the first 0.1 second after an object is dropped, it falls 0.16 feet.

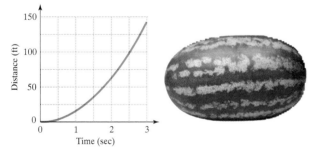

FIGURE FOR EXERCISE 55

a) Find the formula that expresses the distance d an object falls as a function of the time it is falling t.

b) How far does an object fall in the first 0.5 second after it is dropped?

c) How long does it take for a watermelon to reach the ground when dropped from a height of 100 feet?
   a) $d = 16t^2$ **b)** 4 feet   **c)** 2.5 seconds

56. Making Frisbees. The cost of material used in making a Frisbee varies directly with the square of the diameter. If it costs the manufacturer $0.45 for the material in a Frisbee with a 9-inch diameter, then what is the cost for the material in a 12-inch-diameter Frisbee? $0.80

57. Using leverage. The basic law of leverage is that the force required to lift an object is inversely proportional to the length of the lever. If a force of 2000 pounds applied 2 feet from the pivot point would lift a car, then what force would be required at 10 feet to lift the car?   400 pounds

58. Resistance. The resistance of a wire varies directly with the length and inversely as the square of the diameter. If a wire of length 20 feet and diameter 0.1 inch has a resistance of 2 ohms, then what is the resistance of a 30-foot wire with a diameter of 0.2 inch?   0.75 ohm

59. Computer programming. The time t required to complete a programming job varies directly with the complexity of the job and inversely with the number n of programmers working on the job. The complexity c is an arbitrarily assigned number between 1 and 10, with 10 being the most complex. It takes 8 days for a team of three programmers to complete a job with complexity 6. How long will it take five programmers to complete a job with complexity 9? 7.2 days

60. Shock absorbers. The volume of gas in a gas shock absorber varies directly with the temperature and inversely with the pressure. The volume is 10 cubic centimeters (cm³) when the temperature is 20°C and the pressure is 40 kg. What is the volume when the temperature is 30°C and the pressure is 25 kg?   24 cm³

61. Bicycle gear ratio. A bicycle's gear ratio G varies jointly with the number of teeth on the chain ring N (by the pedals) and the diameter of the wheel d, and inversely with the number of teeth on the cog c (on the rear wheel). A bicycle with 27-inch-diameter wheels, 26 teeth on the cog, and 52 teeth on the chain ring has a gear ratio of 54.

a) Find a formula that expresses the gear ratio as a function of N, d, and c. $G = \dfrac{Nd}{c}$

b) What is the gear ratio for a bicycle with 26-inch-diameter wheels, 42 teeth on the chain ring, and 13 teeth on the cog?   84

c) A five-speed bicycle with 27-inch-diameter wheels and 44 teeth on the chain ring has gear ratios of 52, 59, 70, 79, and 91. Find the number of teeth on the cog for each gear ratio.   23, 20, 17, 15, 13

d) For a fixed wheel size and chain ring, does the gear ratio increase or decrease as the number of teeth on the cog increases?   decreases

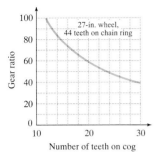

FIGURE FOR EXERCISE 61

 GRAPHING CALCULATOR EXERCISES

62. To see the difference between direct and inverse variation, graph $y_1 = 2x$ and $y_2 = \frac{2}{x}$ using $0 \leq x \leq 5$ and $0 \leq y \leq 10$. Which of these functions is increasing and which is decreasing?
y_1 increasing, y_2 decreasing

63. Graph $y_1 = 2\sqrt{x}$ and $y_2 = \frac{2}{\sqrt{x}}$ by using $0 \leq x \leq 5$ and $0 \leq y \leq 10$. At what point in the first quadrant do the curves cross? Which function is increasing and which is decreasing? Which represents direct variation and which represents inverse variation?
(1, 1), y_1 increasing, y_2 decreasing, y_1 direct variation, y_2 inverse variation

9.4 THE FACTOR THEOREM

In Chapter 5 you learned to add, subtract, multiply, divide, and factor polynomials. In this section we study functions defined by polynomials and learn to solve some higher-degree polynomial equations.

The Factor Theorem

Consider the polynomial function

$$P(x) = x^2 + 2x - 15.$$

The values of x for which $P(x) = 0$ are called the **zeros** or **roots** of the function. We can find the zeros of the function by solving the equation $P(x) = 0$:

$$x^2 + 2x - 15 = 0$$
$$(x + 5)(x - 3) = 0$$
$$x + 5 = 0 \qquad \text{or} \qquad x - 3 = 0$$
$$x = -5 \qquad \text{or} \qquad x = 3$$

helpful hint

Note that the zeros of the polynomial function are factors of the constant term 15.

Because $x + 5$ is a factor of $x^2 + 2x - 15$, -5 is a solution to the equation $x^2 + 2x - 15 = 0$ and a zero of the function $P(x) = x^2 + 2x - 15$. We can check that -5 is a zero of $P(x) = x^2 + 2x - 15$ as follows:

$$P(-5) = (-5)^2 + 2(-5) - 15$$
$$= 25 - 10 - 15$$
$$= 0$$

Because $x - 3$ is a factor of the polynomial, 3 is also a solution to the equation $x^2 + 2x - 15 = 0$ and a zero of the polynomial function. Check that $P(3) = 0$:

$$P(3) = 3^2 + 2 \cdot 3 - 15$$
$$= 9 + 6 - 15$$
$$= 0$$

Every linear factor of the polynomial corresponds to a zero of the polynomial function, and every zero of the polynomial function corresponds to a linear factor.

Now suppose $P(x)$ represents an arbitrary polynomial. If $x - c$ is a factor of the polynomial $P(x)$, then c is a solution to the equation $P(x) = 0$, and so $P(c) = 0$. If we divide $P(x)$ by $x - c$ and the remainder is 0, we must have

$$P(x) = (x - c)(\text{quotient}).$$ Dividend equals the divisor times the quotient.

If the remainder is 0, then $x - c$ is a factor of $P(x)$.

The **factor theorem** summarizes these ideas.

The Factor Theorem

The following statements are equivalent for any polynomial $P(x)$.

1. The remainder is zero when $P(x)$ is divided by $x - c$.

2. $x - c$ is a factor of $P(x)$.

3. c is a solution to $P(x) = 0$.

4. c is a zero of the function $P(x)$, or $P(c) = 0$.

To say that statements are equivalent means that the truth of any one of them implies that the others are true.

According to the factor theorem, if we want to determine whether a given number c is a zero of a polynomial function, we can divide the polynomial by $x - c$. The remainder is zero if and only if c is a zero of the polynomial function. The quickest way to divide by $x - c$ is to use synthetic division from Section 5.5.

E X A M P L E   1

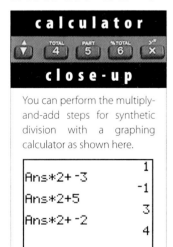

You can perform the multiply-and-add steps for synthetic division with a graphing calculator as shown here.

Using the factor theorem

Use synthetic division to determine whether 2 is a zero of $P(x) = x^3 - 3x^2 + 5x - 2$.

Solution

By the factor theorem, 2 is a zero of the function if and only if the remainder is zero when $P(x)$ is divided by $x - 2$. We can use synthetic division to determine the remainder. If we divide by $x - 2$, we use 2 on the left in synthetic division along with the coefficients $1, -3, 5, -2$ from the polynomial:

$$
\begin{array}{r|rrrr}
2 & 1 & -3 & 5 & -2 \\
  &   & 2 & -2 & 6 \\
\hline
  & 1 & -1 & 3 & 4
\end{array}
$$

Because the remainder is 4, 2 is not a zero of the function. ■

E X A M P L E   2

Using the factor theorem

Use synthetic division to determine whether -4 is a solution to the equation $2x^4 - 28x^2 + 14x - 8 = 0$.

Solution

By the factor theorem, -4 is a solution to the equation if and only if the remainder is zero when $P(x)$ is divided by $x + 4$. When dividing by $x + 4$, we use -4 in the synthetic division:

$$\begin{array}{r|rrrrr} -4 & 2 & 0 & -28 & 14 & -8 \\ & & -8 & 32 & -16 & 8 \\ \hline & 2 & -8 & 4 & -2 & 0 \end{array}$$

Because the remainder is zero, -4 is a solution to $2x^4 - 28x^2 + 14x - 8 = 0$. ∎

In the next example we use the factor theorem to determine whether a given binomial is a factor of a polynomial.

E X A M P L E   3

Using the factor theorem

Use synthetic division to determine whether $x + 4$ is a factor of $x^3 + 3x^2 + 16$.

Solution

According to the factor theorem, $x + 4$ is a factor of $x^3 + 3x^2 + 16$ if and only if the remainder is zero when the polynomial is divided by $x + 4$. Use synthetic division to determine the remainder:

$$\begin{array}{r|rrrr} -4 & 1 & 3 & 0 & 16 \\ & & -4 & 4 & -16 \\ \hline & 1 & -1 & 4 & 0 \end{array}$$

Because the remainder is zero, $x + 4$ is a factor, and the polynomial can be written as

$$x^3 + 3x^2 + 16 = (x + 4)(x^2 - x + 4).$$

Because $x^2 - x + 4$ is a prime polynomial, the factoring is complete. ∎

Solving Polynomial Equations

The techniques used to solve polynomial equations of degree 3 or higher are not as straightforward as those used to solve linear equations and quadratic equations. The next example shows how the factor theorem can be used to solve a third-degree polynomial equation.

E X A M P L E   4

Solving a third-degree equation

Suppose the equation $x^3 - 4x^2 - 17x + 60 = 0$ is known to have a solution that is an integer between -3 and 3 inclusive. Find the solution set.

Solution

Because one of the numbers $-3, -2, -1, 0, 1, 2,$ and 3 is a solution to the equation, we can use synthetic division with these numbers until we discover which one is a solution. We arbitrarily select 1 to try first:

$$\begin{array}{r|rrrr} 1 & 1 & -4 & -17 & 60 \\ & & 1 & -3 & -20 \\ \hline & 1 & -3 & -20 & 40 \end{array}$$

Because the remainder is 40, 1 is not a solution to the equation. Next try 2:

$$\begin{array}{r|rrrr} 2 & 1 & -4 & -17 & 60 \\ & & 2 & -4 & -42 \\ \hline & 1 & -2 & -21 & 18 \end{array}$$

Because the remainder is not zero, 2 is not a solution to the equation. Next try 3:

$$3 \begin{array}{|rrrr} 1 & -4 & -17 & 60 \\ & 3 & -3 & -60 \\ \hline 1 & -1 & -20 & 0 \end{array}$$

The remainder is zero, so 3 is a solution to the equation, and $x - 3$ is a factor of the polynomial. (If 3 had not produced a remainder of zero, then we would have tried $-3, -2, -1$, and 0.) The other factor is the quotient, $x^2 - x - 20$.

$$x^3 - 4x^2 - 17x + 60 = 0$$
$$(x - 3)(x^2 - x - 20) = 0 \quad \text{Use the results of synthetic division to factor.}$$
$$(x - 3)(x - 5)(x + 4) = 0 \quad \text{Factor completely.}$$
$$x - 3 = 0 \quad \text{or} \quad x - 5 = 0 \quad \text{or} \quad x + 4 = 0$$
$$x = 3 \quad \text{or} \quad x = 5 \quad \text{or} \quad x = -4$$

Check each of these solutions in the original equation. The solution set is $\{3, 5, -4\}$.

WARM-UPS

True or false? Explain your answers.

1. To divide $x^3 - 4x^2 - 3$ by $x - 5$, use 5 in the synthetic division.   True
2. To divide $5x^4 - x^3 + x - 2$ by $x + 7$, use -7 in the synthetic division.   True
3. The number 2 is a zero of $P(x) = 3x^3 - 5x^2 - 2x + 2$.   False
4. If $x^3 - 8$ is divided by $x - 2$, then $R = 0$.   True
5. If $R = 0$ when $x^4 - 1$ is divided by $x - a$, then $x - a$ is a factor of $x^4 - 1$.   True
6. If -2 satisfies $x^4 + 8x = 0$, then $x + 2$ is a factor of $x^4 + 8x$.   True
7. The binomial $x - 1$ is a factor of $x^{35} - 3x^{24} + 2x^{18}$.   True
8. The binomial $x + 1$ is a factor of $x^3 - 3x^2 + x + 5$.   True
9. If $x^3 - 5x + 4$ is divided by $x - 1$, then $R = 0$.   True
10. If $R = 0$ when $P(x) = x^3 - 5x - 2$ is divided by $x + 2$, then $P(-2) = 0$.   True

9.4 EXERCISES

Reading and Writing   After reading this section, write out the answers to these questions. Use complete sentences.

1. What is a zero of a function?
   A zero of the function f is a number a such that $f(a) = 0$.

2. What is a root of a function?
   A root of a function is the same as a zero.

3. What does it mean that statements are equivalent?
   Two statements are equivalent means that they are either both true or both false.

4. What is the quickest way to divide a polynomial by $x - c$?
   To divide by $x - c$ quickly, use synthetic division.

5. If the remainder is zero when you divide $P(x)$ by $x - c$, then what can you say about $P(c)$?
   If the remainder is zero when $P(x)$ is divided by $x - c$, then $P(c) = 0$.

6. What are two ways to determine whether c is a zero of a polynomial?
   The number c is a zero of a polynomial if the remainder in synthetic division is zero or if directly evaluating the polynomial at $x = c$ gives a value of zero.

Determine whether each given value of x is a zero of the given function. See Example 1.

7. $x = 1$, $P(x) = x^3 - x^2 + x - 1$ Yes

8. $x = -2$, $P(x) = -2x^3 - 5x^2 + 3x + 10$ Yes

9. $x = -3$, $P(x) = -x^4 - 3x^3 - 2x^2 + 18$ Yes

10. $x = 4$, $P(x) = x^4 - x^2 - 8x - 16$ No

11. $x = 2$, $P(x) = 2x^3 - 4x^2 - 5x + 9$ No

12. $x = -3$, $P(x) = x^3 + 5x^2 + 2x + 1$ No

Use synthetic division to determine whether each given value of x is a solution to the given equation. See Example 2.

13. $x = -3$, $x^3 + 5x^2 + 2x - 12 = 0$ Yes

14. $x = -5$, $x^2 - 3x - 40 = 0$ Yes

15. $x = -2$, $x^4 + 3x^3 - 5x^2 - 10x + 5 = 0$ No

16. $x = -3$, $-x^3 - 4x^2 + x + 12 = 0$ Yes

17. $x = 4$, $-2x^4 + 30x^2 + 5x + 12 = 0$ Yes

18. $x = 6$, $x^4 + x^3 - 40x^2 - 72 = 0$ Yes

19. $x = 3$, $0.8x^2 - 0.3x - 6.3 = 0$ Yes

20. $x = 5$, $6.2x^2 - 28.2x - 41.7 = 0$ No

Use synthetic division to determine whether the first polynomial is a factor of the second. If it is, then factor the polynomial completely. See Example 3.

21. $x - 3$, $x^3 - 6x - 9$ $(x - 3)(x^2 + 3x + 3)$

22. $x + 2$, $x^3 - 6x - 4$ $(x + 2)(x^2 - 2x - 2)$

23. $x + 5$, $x^3 + 9x^2 + 23x + 15$ $(x + 5)(x + 3)(x + 1)$

24. $x - 3$, $x^4 - 9x^2 + x - 7$ No

25. $x - 2$, $x^3 - 8x^2 + 4x - 6$ No

26. $x + 5$, $x^3 + 125$ $(x + 5)(x^2 - 5x + 25)$

27. $x + 1$, $x^4 + x^3 - 8x - 8$ $(x + 1)(x - 2)(x^2 + 2x + 4)$

28. $x - 2$, $x^3 - 6x^2 + 12x - 8$ $(x - 2)^3$

29. $x - 0.5$, $2x^3 - 3x^2 - 11x + 6$ $(2x - 1)(x - 3)(x + 2)$

30. $x - \dfrac{1}{3}$, $3x^3 - 10x^2 - 27x + 10$ $(3x - 1)(x - 5)(x + 2)$

Solve each equation, given that at least one of the solutions to each equation is an integer between -5 and 5. See Example 4.

31. $x^3 - 13x + 12 = 0$ $\{-4, 1, 3\}$

32. $x^3 + 2x^2 - 5x - 6 = 0$ $\{-3, -1, 2\}$

33. $2x^3 - 9x^2 + 7x + 6 = 0$ $\left\{-\dfrac{1}{2}, 2, 3\right\}$

34. $6x^3 + 13x^2 - 4 = 0$ $\left\{-\dfrac{2}{3}, -2, \dfrac{1}{2}\right\}$

35. $2x^3 - 3x^2 - 50x - 24 = 0$ $\left\{-4, -\dfrac{1}{2}, 6\right\}$

36. $x^3 - 7x^2 + 2x + 40 = 0$ $\{-2, 4, 5\}$

37. $x^3 + 5x^2 + 3x - 9 = 0$ $\{-3, 1\}$

38. $x^3 + 6x^2 + 12x + 8 = 0$ $\{-2\}$

39. $x^4 - 4x^3 + 3x^2 + 4x - 4 = 0$ $\{-1, 1, 2\}$

40. $x^4 + x^3 - 7x^2 - x + 6 = 0$ $\{-3, -1, 1, 2\}$

GETTING MORE INVOLVED

41. *Exploration.* We can find the zeros of a polynomial function by solving a polynomial equation. We can also work backward to find a polynomial function that has given zeros.

a) Write a first-degree polynomial function whose zero is -2. $f(x) = x + 2$

b) Write a second-degree polynomial function whose zeros are 5 and -5. $f(x) = x^2 - 25$

c) Write a third-degree polynomial function whose zeros are 1, -3, and 4. $f(x) = (x - 1)(x + 3)(x - 4)$

d) Is there a polynomial function with any given number of zeros? What is its degree?
yes, the degree is the same as the number of zeros

GRAPHING CALCULATOR EXERCISES

42. The x-coordinate of each x-intercept on the graph of a polynomial function is a zero of the polynomial function. Find the zeros of each function from its graph. Use synthetic division to check that the zeros found on your calculator really are zeros of the function.

a) $P(x) = x^3 - 2x^2 - 5x + 6$ $\{-2, 1, 3\}$

b) $P(x) = 12x^3 - 20x^2 + x + 3$ $\left\{-\dfrac{1}{3}, \dfrac{1}{2}, \dfrac{3}{2}\right\}$

43. With a graphing calculator an equation can be solved without the kind of hint that was given for Exercises 31–40. Solve each of the following equations by examining the graph of a corresponding function. Use synthetic division to check.

a) $x^3 - 4x^2 - 7x + 10 = 0$ $\{-2, 1, 5\}$

b) $8x^3 - 20x^2 - 18x + 45 = 0$ $\left\{-\dfrac{3}{2}, \dfrac{3}{2}, \dfrac{5}{2}\right\}$

COLLABORATIVE ACTIVITIES

Betting on Rockets

Gretchen and Rafael are launching model rockets in the park. They have a bet to see whose rocket can go the highest. Both have taken this math course and have some clues on how to use algebra to find out whose went the highest. Below is the formula that they will use to determine whose rocket has gone the highest:

$$h = -16t^2 + v_0t,$$

where h is the distance (feet) of the rocket above the ground, t is the time passed (seconds) since the rocket was launched, and v_0 is the speed of the rocket when launched (initial velocity).

They have decided to do three trials and to see whose is the highest two out of the three times.

From the following list, have each member of your group choose a different trial.

- **First Trial:** On the first launch Gretchen's rocket has a time in the air of 14 seconds, and Rafael's rocket has a time of 10 seconds.

Grouping: Three students per group
Topic: Applications of maximum and minimum of a parabola

- **Second Trial:** On the second launch Gretchen's rocket was in the air 11.5 seconds, and Rafael's rocket was in the air 12 seconds.

- **Third Trial:** On the third launch Gretchen's rocket was in the air 11.25 seconds, and Rafael's rocket was in the air 11.3 seconds.

Individually, complete the following for the trial you selected.

1. Find the initial velocity for each rocket using the formula $h = -16t^2 + v_0t$.

2. Graph the two quadratic functions. What is the h-value for each vertex? Whose rocket went the highest?

Together in your groups, compare your results and decide who wins the bet.

WRAP-UP   CHAPTER 9

SUMMARY

Combining Functions

		Examples
Sum	$(f + g)(x) = f(x) + g(x)$	For $f(x) = x^2$ and $g(x) = x + 1$ $(f + g)(x) = x^2 + x + 1$
Difference	$(f - g)(x) = f(x) - g(x)$	$(f - g)(x) = x^2 - x - 1$
Product	$(f \cdot g)(x) = f(x) \cdot g(x)$	$(f \cdot g)(x) = x^3 + x^2$
Quotient	$\left(\dfrac{f}{g}\right)(x) = \dfrac{f(x)}{g(x)}$	$\left(\dfrac{f}{g}\right)(x) = \dfrac{x^2}{x + 1}$
Composition of functions	$(g \circ f)(x) = g(f(x))$ $(f \circ g)(x) = f(g(x))$	$(g \circ f)(x) = g(x^2) = x^2 + 1$ $(f \circ g)(x) = f(x + 1)$ $\qquad = x^2 + 2x + 1$

Inverse Functions

		Examples
One-to-one function	A function in which no two ordered pairs have different x-coordinates and the same y-coordinate.	$f = \{(2, 20), (3, 30)\}$

Inverse function	The inverse of a one-to-one function f is the function f^{-1}, which is obtained from f by interchanging the coordinates in each ordered pair of f. The domain of f^{-1} is the range of f, and the range of f^{-1} is the domain of f.	$f^{-1} = \{(20, 2), (30, 3)\}$
Horizontal-line test	If there is a horizontal line that crosses the graph of a function more than once, then the function is not invertible.	
f-notation for inverse	Two functions f and g are inverses of each other if and only if both of the following conditions are met. 1. $(g \circ f)(x) = x$ for every number x in the domain of f. 2. $(f \circ g)(x) = x$ for every number x in the domain of g.	$f(x) = x^3 + 1$ $f^{-1}(x) = \sqrt[3]{x - 1}$
Switch-and-solve strategy for finding f^{-1}	1. Replace $f(x)$ by y. 2. Interchange x and y. 3. Solve for y. 4. Replace y by $f^{-1}(x)$.	$y = x^3 + 1$ $x = y^3 + 1$ $x - 1 = y^3$ $y = \sqrt[3]{x - 1}$ $f^{-1}(x) = \sqrt[3]{x - 1}$
Graphs of f and f^{-1}	Graphs of inverse functions are symmetric with respect to the line $y = x$.	

The Language of Variation

		Examples
Direct	y varies directly as x, $y = kx$	$z = 5m$
Inverse	y varies inversely as x, $y = \dfrac{k}{x}$	$a = \dfrac{1}{c}$
Joint	y varies jointly as x and z, $y = kxz$	$V = 6LW$
Combined	y varies directly as x and inversely as z, $y = \dfrac{kx}{z}$	$S = \dfrac{3A}{B}$

The Factor Theorem

		Examples
Factor theorem	The following are equivalent for $P(x)$ a polynomial in x. 1. The remainder is zero when $P(x)$ is divided by $x - c$. 2. $x - c$ is a factor of $P(x)$. 3. c is a solution to $P(x) = 0$. 4. c is a zero of the function $P(x)$, or $P(c) = 0$.	$P(x) = x^2 - x - 2$ $P(x) = (x - 2)(x + 1)$ $P(-1) = 0$, $P(2) = 0$ -1 and 2 both satisfy $x^2 - x - 2 = 0$

ENRICHING YOUR MATHEMATICAL WORD POWER

For each mathematical term, choose the correct meaning.

1. **composition of f and g**
   a. the function $f \circ g$ where $(f \circ g)(x) = f(g(x))$
   b. the function $f \circ g$ where $(f \circ g)(x) = g(f(x))$
   c. the function $f \cdot g$ where $(f \cdot g)(x) = f(x) \cdot g(x)$
   d. a diagram showing f and g a

2. **sum of f and g**
   a. the function $f \cdot g$ where $(f \cdot g)(x) = f(x) \cdot g(x)$
   b. the function $f + g$ where $(f + g)(x) = f(x) + g(x)$
   c. the function $f \circ g$ where $(f \circ g)(x) = g(f(x))$
   d. the function obtained by adding the domains of f and g b

3. **inverse of the function f**
   a. a function with the same ordered pairs as f
   b. the opposite of the function f
   c. the function $1/f$
   d. a function in which the ordered pairs of f are reversed   d

4. **one-to-one function**
   a. a constant function
   b. a function that pairs 1 with 1
   c. a function in which no two ordered pairs have the same first coordinate and different second coordinates
   d. a function in which no two ordered pairs have the same second coordinate and different first coordinates   d

5. **vertical-line test**
   a. a visual method for determining whether a graph is a graph of a function
   b. a visual method for determining whether a function is one-to-one
   c. using a vertical line to check a graph
   d. a test on vertical lines   a

6. **horizontal-line test**
   a. a test that horizontal lines must pass
   b. a visual method for determining whether a function is one-to-one
   c. a graph that does not cross the x-axis
   d. a visual method for determining whether a graph is a graph of a function   b

7. **proportionality constant**
   a. a direct variation
   b. a constant proportion
   c. a ratio that is constant
   d. the constant k in $y = kx$ d

8. **y varies directly as x**
   a. $y = kx^2$ where k is a constant
   b. $y = mx + b$ where m and b are nonzero constants
   c. $y = kx$ where k is a nonzero constant
   d. $y = k/x$ where k is a nonzero constant   c

9. **y varies inversely as x**
   a. $y = x/k$ where k is a nonzero constant
   b. $y = -x$
   c. $y = kx$ where k is a nonzero constant
   d. $y = k/x$ where k is a nonzero constant   d

10. **zero of a function**
    a. a number a such that $f(0) = a$
    b. a number a such that $f(a) = 0$
    c. the value of $f(0)$
    d. the number zero   b

REVIEW EXERCISES

9.1 Let $f(x) = 3x + 5$, $g(x) = x^2 - 2x$, and $h(x) = \dfrac{x - 5}{3}$. Find the following.

1. $f(-3)$
   -4

2. $h(-4)$
   -3

3. $(h \circ f)(\sqrt{2})$
   $\sqrt{2}$

4. $(f \circ h)(\pi)$
   π

5. $(g \circ f)(2)$
   99

6. $(g \circ f)(x)$
   $9x^2 + 24x + 15$

7. $(f + g)(3)$
   17

8. $(f - g)(x)$
   $-x^2 + 5x + 5$

9. $(f \cdot g)(x)$
   $3x^3 - x^2 - 10x$

10. $\left(\dfrac{f}{g}\right)(1)$
    -8

11. $(f \circ f)(0)$
    20

12. $(f \circ f)(x)$
    $9x + 20$

Let $f(x) = |x|$, $g(x) = x + 2$, and $h(x) = x^2$. Write each of the following functions as a composition of functions, using f, g, or h.

13. $F(x) = |x + 2|$
    $F = f \circ g$

14. $G(x) = |x| + 2$
    $G = g \circ f$

15. $H(x) = x^2 + 2$
    $H = g \circ h$

16. $K(x) = x^2 + 4x + 4$
    $K = h \circ g$

17. $I(x) = x + 4$
    $I = g \circ g$

18. $J(x) = x^4 + 2$
    $J = g \circ h \circ h$

9.2 Determine whether each function is invertible. If it is invertible, find the inverse.

19. $\{(-2, 4), (2, 4)\}$
    No

20. $\{(1, 1), (3, 3)\}$
    Yes, $\{(1, 1), (3, 3)\}$

21. $f(x) = 8x$
    Yes, $f^{-1}(x) = x/8$

22. $i(x) = -\dfrac{x}{3}$
    Yes, $i^{-1}(x) = -3x$

23. $g(x) = 13x - 6$

Yes, $g^{-1}(x) = \dfrac{x + 6}{13}$

24. $h(x) = \sqrt[3]{x - 6}$

Yes, $h^{-1}(x) = x^3 + 6$

25. $j(x) = \dfrac{x + 1}{x - 1}$

Yes, $j^{-1}(x) = \dfrac{x + 1}{x - 1}$

26. $k(x) = |x| + 7$

No

27. $m(x) = (x - 1)^2$

No

28. $n(x) = \dfrac{3}{x}$

Yes, $n^{-1}(x) = \dfrac{3}{x}$

Find the inverse of each function, and graph f and f^{-1} on the same pair of axes.

29. $f(x) = 3x - 1$

$f^{-1}(x) = \dfrac{1}{3}x + \dfrac{1}{3}$

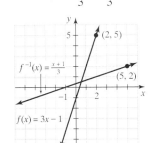

30. $f(x) = 2 - x^2$ for $x \geq 0$

$f^{-1}(x) = \sqrt{2 - x}$

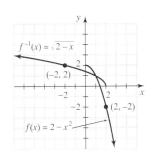

31. $f(x) = \dfrac{x^3}{2}$

$f^{-1}(x) = \sqrt[3]{2x}$

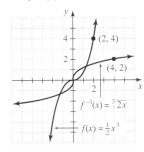

32. $f(x) = -\dfrac{1}{4}x$

$f^{-1}(x) = -4x$

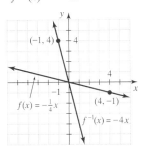

9.3 *Solve each variation problem.*

33. If y varies directly as m and $y = -3$ when $m = \dfrac{1}{4}$, find y when $m = -2$.   24

34. If a varies inversely as b and $a = 6$ when $b = -3$, find a when $b = 4$. $-\dfrac{9}{2}$

35. If c varies directly as m and inversely as n, and $c = 20$ when $m = 10$ and $n = 4$, find c when $m = 6$ and $n = -3$. -16

36. If V varies jointly as h and the square of r, and $V = 32$ when $h = 6$ and $r = 3$, find V when $h = 3$ and $r = 4$. $\dfrac{256}{9}$

9.4 *Given that either 2 or -2 is a solution to each of the following equations, solve each equation.*

37. $x^3 - 4x^2 - 11x + 30 = 0$ $\{-3, 2, 5\}$

38. $x^3 - 2x^2 - 6x + 12 = 0$ $\{-\sqrt{6}, \sqrt{6}, 2\}$

39. $x^3 - 5x^2 - 2x + 24 = 0$ $\{-2, 3, 4\}$

40. $x^3 + 7x^2 + 4x - 12 = 0$ $\{-6, -2, 1\}$

Determine whether the first polynomial is a factor of the second. If it is, then factor the polynomial completely.

41. $x - 3$, $x^3 + 4x^2 - 11x - 30$ $(x - 3)(x + 2)(x + 5)$

42. $x + 4$, $x^3 + x^2 - 10x + 8$ $(x + 4)(x - 2)(x - 1)$

43. $x + 2$, $x^3 - 5x - 6$ No

44. $x - 1$, $2x^3 - 5x^2 + 2x - 9$ No

MISCELLANEOUS

45. *Falling object.* If a ball is dropped from a tall building, then the distance traveled by the ball in t seconds varies directly as the square of the time t. If the ball travels 144 feet (ft) in 3 seconds, then how far does it travel in 4 seconds? 256 ft

46. *Studying or partying.* Evelyn's grade on a math test varies directly with the number of hours spent studying and inversely with the number of hours spent partying during the 24 hours preceding the test. If she scored a 90 on a test after she studied 10 hours and partied 2 hours, then what should she score after studying 4 hours and partying 6 hours? 12

47. *Inscribed square.* Given that B is the area of a square inscribed in a circle of radius r and area A, write B as a function of A.

$B = \dfrac{2A}{\pi}$

48. *Area of a window.* A window is in the shape of a square of side s, with a semicircle of diameter s above it. Write a

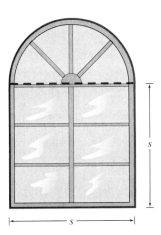

FIGURE FOR EXERCISE 48

function that expresses the total area of the window as a function of s.

$$A = \frac{(8 + \pi)s^2}{8}$$

49. **Composition of functions.** Given that $a = 3k + 2$ and $k = 5w - 6$, write a as a function of w.
$a = 15w - 16$

50. **Volume of a cylinder.** The volume of a cylinder with a fixed height of 10 centimeters (cm) is given by $V = 10\pi r^2$, where r is the radius of the circular base. Write the volume as a function of the area of the base, A. $V = 10A$

51. **Square formulas.** Write the area of a square A as a function of the length of a side of the square s. Write the length of a side of a square as a function of the area.
$A = s^2, s = \sqrt{A}$

52. **Circle formulas.** Write the area of a circle A as a function of the radius of the circle r. Write the radius of a circle as a function of the area of the circle. Write the area as a function of the diameter d.

$$A = \pi r^2, r = \sqrt{\frac{A}{\pi}}, A = \frac{\pi d^2}{4}$$

CHAPTER 9 TEST

Let $f(x) = -2x + 5$ and $g(x) = x^2 + 4$. Find the following.

1. $f(-3)$
   11

2. $(g \circ f)(-3)$
   125

3. $f^{-1}(11)$
   -3

4. $f^{-1}(x)$
   $\dfrac{x - 5}{-2}$

5. $(g + f)(x)$
   $x^2 - 2x + 9$

6. $(f \cdot g)(1)$
   15

7. $(f^{-1} \circ f)(1776)$
   1776

8. $(f/g)(2)$
   $\dfrac{1}{8}$

9. $(f \circ g)(x)$
   $-2x^2 - 3$

10. $(g \circ f)(x)$
    $4x^2 - 20x + 29$

Let $f(x) = x - 7$ and $g(x) = x^2$. Write each of the following functions as a composition of functions using f and g.

11. $H(x) = x^2 - 7$ $H = f \circ g$

12. $W(x) = x^2 - 14x + 49$ $W = g \circ f$

Determine whether each function is invertible. If it is invertible, find the inverse.

13. $\{(2, 3), (4, 3)\}$ Not invertible

14. $f(x) = \sqrt[3]{x} + 9$ $f^{-1}(x) = (x - 9)^3$

Solve each problem.

15. Find the inverse of $f(x) = \dfrac{2x + 1}{x - 1}$. $f^{-1}(x) = \dfrac{x + 1}{x - 2}$

16. The volume of a sphere varies directly as the cube of the radius. If a sphere with radius 3 feet (ft) has a volume of 36π cubic feet (ft³), then what is the volume of a sphere with a radius of 2 ft?
    $\dfrac{32\pi}{3}$ ft³

17. Suppose y varies directly as x and inversely as the square root of z. If $y = 12$ when $x = 7$ and $z = 9$, then what is the proportionality constant?
    $\dfrac{36}{7}$

18. The cost of a Persian rug varies jointly as the length and width of the rug. If the cost is \$2,256 for a 6 foot by 8 foot rug, then what is the cost of a 9 foot by 12 foot rug?
    \$5,076

19. Is $x + 1$ a factor of $2x^4 - 5x^3 + 3x^2 + 6x - 4$? Explain.
    Yes

20. Given that all of the solutions to the equation $x^3 - 12x^2 + 47x - 60 = 0$ are positive integers smaller than 10, find the solution set.
    $\{3, 4, 5\}$

Simplify each expression.

1. $125^{-2/3}$ $\dfrac{1}{25}$

2. $\left(\dfrac{8}{27}\right)^{-1/3}$ $\dfrac{3}{2}$

3. $\sqrt{18} - \sqrt{8}$ $\sqrt{2}$

4. $x^5 \cdot x^3$ x^8

5. $16^{1/4}$ 2

6. $\dfrac{x^{12}}{x^3}$ x^9

Find the real solution set to each equation.

7. $x^2 = 9$
$\{\pm 3\}$

8. $x^2 = 8$
$\{\pm 2\sqrt{2}\}$

9. $x^2 = x$
$\{0, 1\}$

10. $x^2 - 4x - 6 = 0$
$\{2 \pm \sqrt{10}\}$

11. $x^{1/4} = 3$
$\{81\}$

12. $x^{1/6} = -2$
$\varnothing$

13. $|x| = 8$
$\{\pm 8\}$

14. $|5x - 4| = 21$
$\left\{-\dfrac{17}{5}, 5\right\}$

15. $x^3 = 8$
$\{2\}$

16. $(3x - 2)^3 = 27$
$\left\{\dfrac{5}{3}\right\}$

17. $\sqrt{2x - 3} = 9$
$\{42\}$

18. $\sqrt{x - 2} = x - 8$
$\{11\}$

Sketch the graph of each set.

19. $\{(x, y) \mid y = 5\}$

20. $\{(x, y) \mid y = 2x - 5\}$

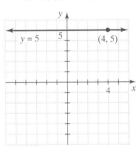

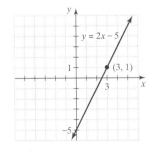

21. $\{(x, y) \mid x = 5\}$

22. $\{(x, y) \mid 3y = x\}$

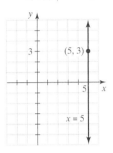

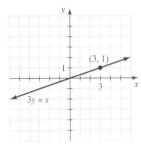

23. $\{(x, y) \mid y = 5x^2\}$

24. $\{(x, y) \mid y = -2x^2\}$

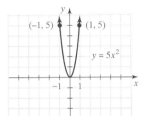

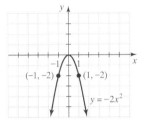

Find the missing coordinates in each ordered pair so that the ordered pair satisfies the given equation.

25. $(2, \ ), (3, \ ), (\ , 2), (\ , 16), \quad 2^x = y$
$(2, 4), (3, 8), (1, 2), (4, 16)$

26. $\left(\dfrac{1}{2}, \ \right), (-1, \ ), (\ , 16), (\ , 1), \quad 4^x = y$
$\left(\dfrac{1}{2}, 2\right), \left(-1, \dfrac{1}{4}\right), (2, 16), (0, 1)$

Find the domain of each expression.

27. $\sqrt{x}$ $[0, \infty)$

28. $\sqrt{6 - 2x}$ $(-\infty, 3]$

29. $\dfrac{5x - 3}{x^2 + 1}$ $(-\infty, \infty)$

30. $\dfrac{x - 3}{x^2 - 10x + 9}$ $(-\infty, 1) \cup (1, 9) \cup (9, \infty)$

Solve each problem.

31. ***Capital cost and operating cost.*** To decide when to replace company cars, an accountant looks at two cost components: capital cost and operating cost. The capital cost C (the difference between the original cost and the salvage value) for a certain car is $3,000 plus $0.12 for each mile that the car is driven.

a) Write the capital cost C as a linear function of x, the number of miles that the car is driven.

$$C = 0.12x + 3000$$

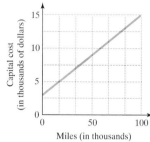 

FIGURE FOR EXERCISE 31(a)

b) The operating cost P is $0.15 per mile initially and increases linearly to $0.25 per mile when the car reaches 100,000 miles. Write P as a function of x, the number of miles that the car is driven.

$$P = 1 \times 10^{-6}x + 0.15$$

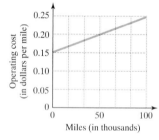 

FIGURE FOR EXERCISE 31(b)

32. ***Total cost.*** The accountant in the previous exercise uses the function $T = \frac{C}{x} + P$ to find the total cost per mile.

a) Find T for $x = 20,000$, $30,000$, and $90,000$.

$0.44, $0.40, $0.39

b) Sketch a graph of the total cost function.

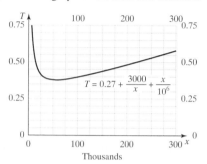

$$T = 0.27 + \frac{3000}{x} + \frac{x}{10^6}$$

c) The accountant has decided to replace the car when T reaches $0.38 for the second time. At what mileage will the car be replaced?

60,000 miles

d) For what values of x is T less than or equal to $0.38?

[50,000, 60,000]

CHAPTER 10

Exponential and Logarithmic Functions

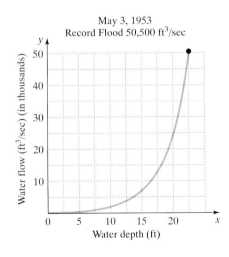

May 3, 1953
Record Flood 50,500 ft³/sec

Water flow (ft³/sec) (in thousands) — vertical axis: 10, 20, 30, 40, 50

Water depth (ft) — horizontal axis: 0, 5, 10, 15, 20

Water is one of the essentials of life, yet it is something that most of us take for granted. Among other things, the U.S. Geological Survey (U.S.G.S.) studies freshwater. For over 50 years the Water Resources Division of the U.S.G.S. has been gathering basic data about the flow of both freshwater and saltwater from streams and groundwater surfaces. This division collects, compiles, analyzes, verifies, organizes, and publishes data gathered from groundwater data collection networks in each of the 50 states, Puerto Rico, and the Trust Territories. Records of stream flow, groundwater levels, and water quality provide hydrological information needed by local, state, and federal agencies as well as the private sector.

There are many instances of the importance of the data collected by the U.S.G.S. For example, before 1987 the Tangipahoa River in Louisiana was used extensively for swimming and boating. In 1987 data gathered by the U.S.G.S. showed that fecal coliform levels in the river exceeded safe levels. Consequently, Louisiana banned recreational use of the river. Other studies by the Water Resources Division include the results of pollutants on salt marsh environments and the effect that salting highways in winter has on our drinking water supply.

In Exercises 85 and 86 of Section 10.2 you will see how data from the U.S.G.S. is used in a logarithmic function to measure water quality.

10.1   EXPONENTIAL FUNCTIONS

We have studied functions such as

$$f(x) = x^2, \qquad g(x) = x^3, \qquad \text{and} \qquad h(x) = x^{1/2}.$$

For these functions the variable is the base. In this section we discuss functions that have a variable as an exponent. These functions are called *exponential functions.*

Definition

Some examples of exponential functions are

$$f(x) = 2^x, \qquad f(x) = \left(\frac{1}{2}\right)^x, \qquad \text{and} \qquad f(x) = 3^x.$$

> **Exponential Function**
>
> An **exponential function** is a function of the form
> $$f(x) = a^x,$$
> where $a > 0$ and $a \neq 1$.

We rule out the base 1 in the definition because $f(x) = 1^x$ is the same as the constant function $f(x) = 1$. Zero is not used as a base because $0^x = 0$ for any positive x and nonpositive powers of 0 are undefined. Negative numbers are not used as bases because an expression such as $(-4)^x$ is not a real number if $x = \frac{1}{2}$.

In this section

- Definition
- Domain
- Graphing Exponential Functions
- Exponential Equations
- Applications

helpful hint

It is essential that you have a calculator for this chapter. The most modern calculators are the graphing calculators. They cost a bit more but are worth the price.

EXAMPLE 1   Evaluating exponential functions

Let $f(x) = 2^x$, $g(x) = \left(\frac{1}{4}\right)^{1-x}$, and $h(x) = -3^x$. Find the following.

a) $f\left(\frac{3}{2}\right)$ **b)** $f(-3)$ **c)** $g(3)$ **d)** $h(2)$

Solution

a) $f\left(\frac{3}{2}\right) = 2^{3/2} = \sqrt{2^3} = \sqrt{8} = 2\sqrt{2}$

b) $f(-3) = 2^{-3} = \dfrac{1}{2^3} = \dfrac{1}{8}$

c) $g(3) = \left(\frac{1}{4}\right)^{1-3} = \left(\frac{1}{4}\right)^{-2} = 4^2 = 16$

d) $h(2) = -3^2 = -9$ Note that $-3^2 \neq (-3)^2$. ■

For many applications of exponential functions we use base 10 or another base called e. The number e is an irrational number that is approximately 2.718. We will see how e is used in compound interest in Example 10 of this section. Base 10 will be used in the next section. Base 10 is called the **common base,** and base e is called the **natural base.**

EXAMPLE 2

Base 10 and base e

Let $f(x) = 10^x$ and $g(x) = e^x$. Find the following and round approximate answers to four decimal places.

a) $f(3)$ **b)** $f(1.51)$ **c)** $g(0)$ **d)** $g(2)$

Solution

a) $f(3) = 10^3 = 1000$

b) $f(1.51) = 10^{1.51} \approx 32.3594$ Use the 10^x key on a calculator.

c) $g(0) = e^0 = 1$

d) $g(2) = e^2 \approx 7.3891$ Use the e^x key on a calculator. ◼

Domain

In the definition of an exponential function no restrictions were placed on the exponent x because the domain of an exponential function is the set of all real numbers. So both rational and irrational numbers can be used as the exponent. We have been using rational numbers for exponents since Chapter 7, but we have not yet seen an irrational number as an exponent. Even though we do not formally define irrational exponents in this text, an irrational number such as π can be used as an exponent, and you can evaluate an expression such as 2^π by using a calculator. Try it:

$$2^\pi \approx 8.824977827$$

Graphing Exponential Functions

Even though the domain of an exponential function is the set of all real numbers, we can graph an exponential function by evaluating it for just a few integers.

EXAMPLE 3

Exponential functions with base greater than 1

Sketch the graph of each function.

a) $f(x) = 2^x$ **b)** $g(x) = 3^x$

Solution

a) We first make a table of ordered pairs that satisfy $f(x) = 2^x$:

x	-2	-1	0	1	2	3
$f(x) = 2^x$	$\frac{1}{4}$	$\frac{1}{2}$	1	2	4	8

As x increases, 2^x increases and 2^x is always positive. Because the domain of the function is $(-\infty, \infty)$, we draw the graph in Fig. 10.1 as a smooth curve through these points. From the graph we can see that the range is $(0, \infty)$.

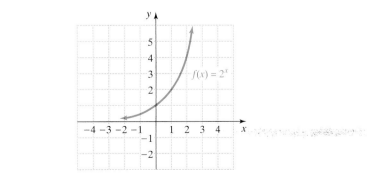

FIGURE 10.1

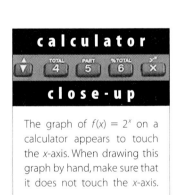
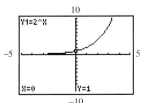

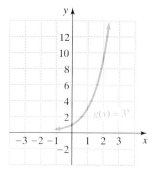

FIGURE 10.2

b) Make a table of ordered pairs that satisfy $g(x) = 3^x$:

x	-2	-1	0	1	2	3
$g(x) = 3^x$	$\frac{1}{9}$	$\frac{1}{3}$	1	3	9	27

As x increases, 3^x increases and 3^x is always positive. The graph is shown in Fig. 10.2. From the graph we see that the range is $(0, \infty)$. ■

Because $e \approx 2.718$, the graph of $f(x) = e^x$ lies between the graphs of $f(x) = 2^x$ and $g(x) = 3^x$, as shown in Fig. 10.3. Note that all three functions have the same domain and range and the same y-intercept. In general, the function $f(x) = a^x$ for $a > 1$ has the following characteristics:

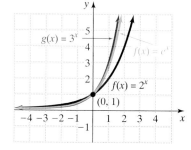

FIGURE 10.3

1. The y-intercept of the curve is $(0, 1)$.

2. The domain is $(-\infty, \infty)$, and the range is $(0, \infty)$.

3. The curve approaches the negative x-axis but does not touch it.

4. The y-values are increasing as we go from left to right along the curve.

EXAMPLE 4

Exponential functions with base between 0 and 1

Graph each function.

a) $f(x) = \left(\frac{1}{2}\right)^x$

b) $f(x) = 4^{-x}$

Solution

a) First make a table of ordered pairs that satisfy $f(x) = \left(\frac{1}{2}\right)^x$:

x	-2	-1	0	1	2	3
$f(x) = \left(\frac{1}{2}\right)^x$	4	2	1	$\frac{1}{2}$	$\frac{1}{4}$	$\frac{1}{8}$

As x increases, $\left(\frac{1}{2}\right)^x$ decreases, getting closer and closer to 0. Draw a smooth curve through these points as shown in Fig. 10.4.

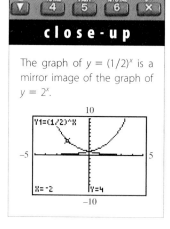
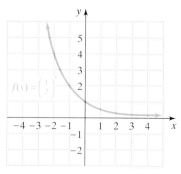

FIGURE 10.4

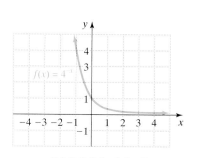

FIGURE 10.5

b) Because $4^{-x} = \left(\frac{1}{4}\right)^x$, we make a table for $f(x) = \left(\frac{1}{4}\right)^x$:

x	-2	-1	0	1	2	3
$f(x) = \left(\frac{1}{4}\right)^x$	16	4	1	$\frac{1}{4}$	$\frac{1}{16}$	$\frac{1}{64}$

As x increases, $\left(\frac{1}{4}\right)^x$, or 4^{-x}, decreases, getting closer and closer to 0. Draw a smooth curve through these points as shown in Fig. 10.5. ■

Notice the similarities and differences between the exponential function with $a > 1$ and with $0 < a < 1$. The function $f(x) = a^x$ for $0 < a < 1$ has the following characteristics:

1. The y-intercept of the curve is $(0, 1)$.
2. The domain is $(-\infty, \infty)$, and the range is $(0, \infty)$.
3. The curve approaches the positive x-axis but does not touch it.
4. The y-values are decreasing as we go from left to right along the curve.

CAUTION An exponential function can be written in more than one form. For example, $f(x) = \left(\frac{1}{2}\right)^x$ is the same as $f(x) = \frac{1}{2^x}$, or $f(x) = 2^{-x}$.

Although exponential functions have the form $f(x) = a^x$, other functions that have similar forms are also called exponential functions. Notice how changing the form $f(x) = a^x$ in the next two examples changes the shape and location of the graph.

EXAMPLE 5

Changing the shape and location

Sketch the graph of $f(x) = 3^{2x-1}$.

Solution

Make a table of ordered pairs:

x	-1	0	$\frac{1}{2}$	1	2
$f(x) = 3^{2x-1}$	$\frac{1}{27}$	$\frac{1}{3}$	1	3	27

The graph through these points is shown in Fig. 10.6. ■

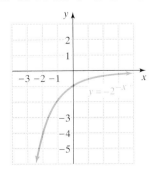

FIGURE 10.6

EXAMPLE 6

Changing the shape and location

Sketch the graph of $y = -2^{-x}$.

Solution

Because $-2^{-x} = -(2^{-x})$, all y-coordinates are negative. Make a table of ordered pairs:

x	-2	-1	0	1	2
$f(x) = -2^{-x}$	-4	-2	-1	$-\frac{1}{2}$	$-\frac{1}{4}$

The graph through these points is shown in Fig. 10.7. ■

FIGURE 10.7

Exponential Equations

In Chapter 9 we used the horizontal-line test to determine whether a function is one-to-one. Because no horizontal line can cross the graph of an exponential function more than once, exponential functions are one-to-one functions. For an exponential function one-to-one means that *if two exponential expressions with the same base are equal, then the exponents are equal.*

One-to-One Property of Exponential Functions

For $a > 0$ and $a \neq 1$,

$$\text{if} \quad a^m = a^n, \quad \text{then} \quad m = n.$$

In the next example we use the one-to-one property to solve equations involving exponential functions.

E X A M P L E   7

c a l c u l a t o r

| ▲ | TOTAL | PART | %TOTAL | yˣ |
| ▼ | 4 | 5 | 6 | × |

c l o s e - u p

You can see the solution to $2^{2x-1} = 8$ by graphing $y_1 = 2^{2x-1}$ and $y_2 = 8$. The x-coordinate of the point of intersection is the solution to the equation.

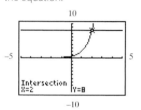

c a l c u l a t o r

| ▲ | TOTAL | PART | %TOTAL | yˣ |
| ▼ | 4 | 5 | 6 | × |

c l o s e - u p

The equation $9^{|x|} = 3$ has two solutions because the graphs of $y_1 = 9^{|x|}$ and $y_2 = 3$ intersect twice.

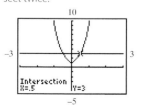

Using the one-to-one property

Solve each equation.

a) $2^{2x-1} = 8$ **b)** $9^{|x|} = 3$ **c)** $\dfrac{1}{8} = 4^x$

Solution

a) Because 8 is 2^3, we can write each side as a power of the same base, 2:

$$
\begin{aligned}
2^{2x-1} &= 8 && \text{Original equation} \\
2^{2x-1} &= 2^3 && \text{Write each side as a power of the same base.} \\
2x - 1 &= 3 && \text{One-to-one property} \\
2x &= 4 \\
x &= 2
\end{aligned}
$$

Check: $2^{2 \cdot 2 - 1} = 2^3 = 8$. The solution set is $\{2\}$.

b) Because $9 = 3^2$, we can write each side as a power of 3:

$$
\begin{aligned}
9^{|x|} &= 3 && \text{Original equation} \\
(3^2)^{|x|} &= 3^1 \\
3^{2|x|} &= 3^1 && \text{Power of a power rule} \\
2|x| &= 1 && \text{One-to-one property} \\
|x| &= \frac{1}{2} \\
x &= \pm\frac{1}{2}
\end{aligned}
$$

Check $x = \pm\frac{1}{2}$ in the original equation. The solution set is $\left\{-\frac{1}{2}, \frac{1}{2}\right\}$.

c) Because $\dfrac{1}{8} = 2^{-3}$ and $4 = 2^2$, we can write each side as a power of 2:

$$
\begin{aligned}
\frac{1}{8} &= 4^x && \text{Original equation} \\
2^{-3} &= (2^2)^x && \text{Write each side as a power of 2.} \\
2^{-3} &= 2^{2x} && \text{Power of a power rule} \\
2x &= -3 && \text{One-to-one property} \\
x &= -\frac{3}{2}
\end{aligned}
$$

Check $x = -\dfrac{3}{2}$ in the original equation. The solution set is $\left\{-\frac{3}{2}\right\}$.

The one-to-one property is also used to find the first coordinate when given the second coordinate of an exponential function.

E X A M P L E   8

Finding the x-coordinate in an exponential function

Let $f(x) = 2^x$ and $g(x) = \left(\frac{1}{2}\right)^{1-x}$. Find x if:

a) $f(x) = 32$

b) $g(x) = 8$

Solution

a) Because $f(x) = 2^x$ and $f(x) = 32$, we can find x by solving $2^x = 32$:

$$2^x = 32$$
$$2^x = 2^5 \quad \text{Write both sides as a power of the same base.}$$
$$x = 5 \quad \text{One-to-one property}$$

b) Because $g(x) = \left(\frac{1}{2}\right)^{1-x}$ and $g(x) = 8$, we can find x by solving $\left(\frac{1}{2}\right)^{1-x} = 8$:

$$\left(\frac{1}{2}\right)^{1-x} = 8$$
$$(2^{-1})^{1-x} = 2^3 \quad \text{Because } \tfrac{1}{2} = 2^{-1} \text{ and } 8 = 2^3$$
$$2^{x-1} = 2^3 \quad \text{Power of a power rule}$$
$$x - 1 = 3 \quad \text{One-to-one property}$$
$$x = 4$$

Applications

Exponential functions are used to describe phenomena such as population growth, radioactive decay, and compound interest. Here we discuss compound interest. If an investment is earning **compound interest,** then interest is periodically paid into the account and the interest that is paid also earns interest. If a bank pays 6% compounded quarterly on an account, then the interest is computed four times per year (every 3 months) at 1.5% (one-quarter of 6%). Suppose an account has \$5000 in it at the beginning of a quarter. We can apply the simple interest formula $A = P + Prt$, with $r = 6\%$ and $t = \frac{1}{4}$, to find how much is in the account at the end of the first quarter.

$$A = P + Prt$$
$$= P(1 + rt) \quad \text{Factor.}$$
$$= 5000\left(1 + 0.06 \cdot \frac{1}{4}\right) \quad \text{Substitute.}$$
$$= 5000(1.015)$$
$$= \$5075$$

To repeat this computation for another quarter, we multiply \$5075 by 1.015. If A represents the amount in the account at the end of n quarters, we can write A as an exponential function of n:

$$A = \$5000(1.015)^n$$

In general, the amount A is given by the following formula.

Compound Interest Formula

If P represents the principal, i the interest rate per period, n the number of periods, and A the amount at the end of n periods, then

$$A = P(1 + i)^n.$$

E X A M P L E   9

calculator

close-up

Graph $y = 350(1.01)^x$ to see the growth of the $350 deposit in Example 9 over time. After 360 months it is worth $12,582.37.

Compound interest formula

If $350 is deposited in an account paying 12% compounded monthly, then how much is in the account at the end of 6 years and 6 months?

Solution

Interest is paid 12 times per year, so the account earns $\frac{1}{12}$ of 12%, or 1% each month, for 78 months. So $i = 0.01$, $n = 78$, and $P = \$350$:

$$A = P(1 + i)^n$$
$$A = \$350(1.01)^{78}$$
$$= \$760.56$$

If we shorten the length of the time period (yearly, quarterly, monthly, daily, hourly, etc.), the number of periods n increases while the interest rate for the period decreases. As n increases, the amount A also increases but will not exceed a certain amount. That certain amount is the amount obtained from *continuous compounding* of the interest. It is shown in more advanced courses that the following formula gives the amount when interest is compounded continuously.

Continuous-Compounding Formula

If P is the principal or beginning balance, r is the annual percentage rate compounded continuously, t is the time in years, and A is the amount or ending balance, then

$$A = Pe^{rt}.$$

helpful hint

Compare Examples 9 and 10 to see the difference between compounded monthly and compounded continuously. Although there is not much difference to an individual investor, there could be a large difference to the bank. Rework Examples 9 and 10 using $50 million as the deposit.

C A U T I O N   The value of t in the continuous-compounding formula must be in years. For example, if the time is 1 year and 3 months, then $t = 1.25$ years. If the time is 3 years and 145 days, then

$$t = 3 + \frac{145}{365}$$
$$\approx 3.3973 \text{ years.}$$

E X A M P L E   1 0   **Continuous-compounding formula**

If $350 is deposited in an account paying 12% compounded continuously, then how much is in the account after 6 years and 6 months?

Solution

Use $r = 12\%$, $t = 6.5$ years, and $P = \$350$ in the formula for compounding interest continuously:

$$A = Pe^{rt}$$
$$= 350e^{(0.12)(6.5)}$$
$$= 350e^{0.78}$$
$$= \$763.52 \quad \text{Use the } e^x \text{ key on a scientific calculator.}$$

Note that compounding continuously amounts to a few dollars more than compounding monthly did in Example 9. ∎

calculator close-up

Graph $y = 350e^{0.12x}$ to see the growth of the $350 deposit in Example 10 over time. After 30 years it is worth $12,809.38.

MATH   AT   WORK

GEOPHYSICIST

Neal Driscoll, a geophysicist at the Lamont-Doherty Earth Observatory of Columbia University, explores both the ocean and the continents to understand the processes that shape the earth. What he finds fascinating is the interaction between the ocean and the land—not just at the shoreline, but underneath the sea as well.

To get a preliminary picture of the ocean floor, Dr. Driscoll spent part of the past summer working with the U.S.G.S., studying the effects of storms on beaches and underwater landscapes. The results of these studies can be used as a baseline to provide help to coastal planners who are building waterfront homes. Other information obtained can be used to direct transporters of dredged material to places where the material is least likely to affect plant, fish, and human life. The most recent study found many different types of ocean floor, ranging from sand and mud to large tracts of algae.

Imaging the seafloor is a difficult problem. It can be a costly venture, and there are numerous logistical problems. Recently developed technology, such as towable undersea cameras and satellite position systems, has made the task easier. Dr. Driscoll and his team use this new technology and sound reflection to gather data about how the sediment on the ocean floor changes in response to storm events. This research is funded by the Office of Naval Research (ONR).

In Exercise 28 of the Making Connections exercises you will see how a geophysicist uses sound to measure the depth of the ocean.

WARM-UPS

True or false? Explain your answer.

1. If $f(x) = 4^x$, then $f\left(-\dfrac{1}{2}\right) = -2$.   False

2. If $f(x) = \left(\dfrac{1}{3}\right)^x$, then $f(-1) = 3$.   True

3. The function $f(x) = x^4$ is an exponential function.   False

4. The functions $f(x) = \left(\dfrac{1}{2}\right)^x$ and $g(x) = 2^{-x}$ have the same graph.   True

5. The function $f(x) = 2^x$ is invertible.   True

6. The graph of $y = \left(\dfrac{1}{3}\right)^x$ has an x-intercept.   False

7. The y-intercept for $f(x) = e^x$ is $(0, 1)$.   True

8. The expression $2^{\sqrt{2}}$ is undefined.   False

9. The functions $f(x) = 2^{-x}$ and $g(x) = \dfrac{1}{2^x}$ have the same graph.   True

10. If \$500 earns 6% compounded monthly, then at the end of 3 years the investment is worth $500(1.005)^3$ dollars.   False

10.1   EXERCISES

Reading and Writing   After reading this section, write out the answers to these questions. Use complete sentences.

1. What is an exponential function?
   An exponential function has the form $f(x) = a^x$ where $a > 0$ and $a \neq 1$.

2. What is the domain of every exponential function?
   The domain of an exponential function is all real numbers.

3. What are the two most popular bases?
   The two most popular bases are e and 10.

4. What is the one-to-one property of exponential functions?
   The one-to-one property states that if $a^m = a^n$, then $m = n$.

5. What is the compound interest formula?
   The compound interest formula is $A = P(1 + i)^n$.

6. What does compounded continuously mean?
   When money is compounded continuously, we use the formula $A = Pe^{rt}$.

Let $f(x) = 4^x$, $g(x) = \left(\dfrac{1}{3}\right)^{x+1}$, and $h(x) = -2^x$. Find the following. See Example 1.

7. $f(2)$ 16

8. $f(-1)$ $\dfrac{1}{4}$

9. $f\left(\dfrac{1}{2}\right)$ 2

10. $f\left(-\dfrac{3}{2}\right)$ $\dfrac{1}{8}$

11. $g(-2)$ 3

12. $g(1)$ $\dfrac{1}{9}$

13. $g(0)$ $\dfrac{1}{3}$

14. $g(-3)$ 9

15. $h(0)$ -1

16. $h(3)$ -8

17. $h(-2)$ $-\dfrac{1}{4}$

18. $h(-4)$ $-\dfrac{1}{16}$

 Let $h(x) = 10^x$ and $j(x) = e^x$. Find the following. Use a calculator as necessary and round approximate answers to four decimal places. See Example 2.

19. $h(0)$ 1

20. $h(-1)$ 0.1

21. $h(2)$ 100

22. $h(3.4)$ 2511.886

23. $j(1)$ 2.718

24. $j(3.5)$ 33.115

25. $j(-2)$ 0.135

26. $j(0)$ 1

Sketch the graph of each function. See Examples 3 and 4.

27. $f(x) = 4^x$

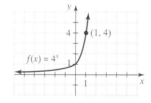

28. $g(x) = 5^x$

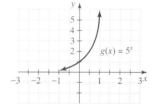

29. $h(x) = \left(\dfrac{1}{3}\right)^x$

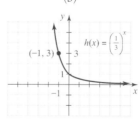

30. $i(x) = \left(\dfrac{1}{5}\right)^x$

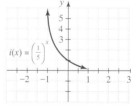

31. $y = 10^x$

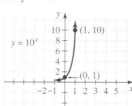

32. $y = (0.1)^x$

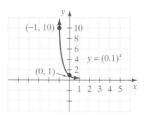

Sketch the graph of each function. See Examples 5 and 6.

33. $y = 10^{x+2}$

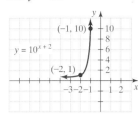

34. $y = 3^{2x+1}$

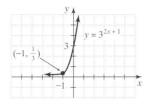

35. $f(x) = -2^x$

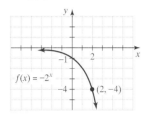

36. $k(x) = -2^{x-2}$

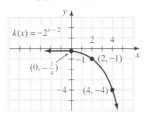

37. $g(x) = 2^{-x}$

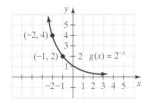

38. $A(x) = 10^{1-x}$

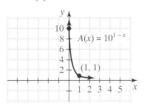

39. $f(x) = -e^x$

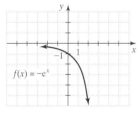

40. $g(x) = e^{-x}$

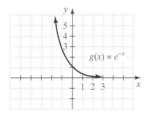

41. $H(x) = 10^{|x|}$

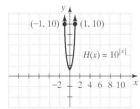

42. $s(x) = 2^{(x^2)}$

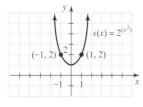

43. $P = 5000(1.05)^t$

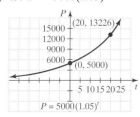

44. $d = 800 \cdot 10^{-4t}$

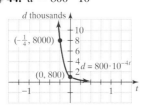

Solve each equation. See Example 7.

45. $2^x = 64$ $\{6\}$ **46.** $3^x = 9$ $\{2\}$

47. $10^x = 0.001$ $\{-3\}$ **48.** $10^{2x} = 0.1$ $\left\{-\dfrac{1}{2}\right\}$

49. $2^x = \dfrac{1}{4}$ $\{-2\}$ **50.** $3^x = \dfrac{1}{9}$ $\{-2\}$

51. $\left(\dfrac{2}{3}\right)^{x-1} = \dfrac{9}{4}$ $\{-1\}$ **52.** $\left(\dfrac{1}{4}\right)^{3x} = 16$ $\left\{-\dfrac{2}{3}\right\}$

53. $5^{-x} = 25$ $\{-2\}$ **54.** $10^{-x} = 0.01$ $\{2\}$

55. $-2^{1-x} = -8$ $\{-2\}$ **56.** $-3^{2-x} = -81$ $\{-2\}$

57. $10^{|x|} = 1000$ $\{-3, 3\}$ **58.** $3^{|2x-5|} = 81$ $\left\{\dfrac{1}{2}, \dfrac{9}{2}\right\}$

Let $f(x) = 2^x$, $g(x) = \left(\dfrac{1}{3}\right)^x$, and $h(x) = 4^{2x-1}$. Find x in each case. See Example 8.

59. $f(x) = 4$ 2   **60.** $f(x) = \dfrac{1}{4}$ -2

61. $f(x) = 4^{2/3}$ $\dfrac{4}{3}$ **62.** $f(x) = 1$ 0

63. $g(x) = 9$ -2 **64.** $g(x) = \dfrac{1}{9}$ 2

65. $g(x) = 1$ 0   **66.** $g(x) = \sqrt{3}$ $-\dfrac{1}{2}$

67. $h(x) = 16$ $\dfrac{3}{2}$ **68.** $h(x) = \dfrac{1}{2}$ $\dfrac{1}{4}$

69. $h(x) = 1$ $\dfrac{1}{2}$ **70.** $h(x) = \sqrt{2}$ $\dfrac{5}{8}$

   Solve each problem. See Example 9.

71. **Compounding quarterly.** If \$6,000 is deposited in an account paying 5% compounded quarterly, then what amount will be in the account after 10 years?
\$9,861.72

72. **Compounding quarterly.** If \$400 is deposited in an account paying 10% compounded quarterly, then what amount will be in the account after 7 years?
\$798.60

73. **Outstanding performance.** The top stock fund over 10 years was Fidelity Select-Home Finance because it

FIGURE FOR EXERCISE 73

returned an average of 27.6% annually for 10 years (Money's 1998 Guide to Mutual Funds, www.money.com).

a) How much was an investment of $10,000 in this fund in 1988 worth in 1998?
$114,421.26

b) Use the accompanying graph to estimate the year in which the $10,000 investment was worth $75,000.
1996

74. Second place. The Kaufman fund was the second best fund over 10 years with an average annual return of 26.5% (Money's 1998 Guide to Mutual Funds, www.money.com). How much was an investment of $10,000 in this fund in 1988 worth in 1998?
$104,931.35

75. Depreciating knowledge. The value of a certain textbook seems to decrease according to the formula $V = 45 \cdot 2^{-0.9t}$, where V is the value in dollars and t is the age of the book in years. What is the book worth when it is new? What is it worth when it is 2 years old?
$45, $12.92

76. Mosquito abatement. In a Minnesota swamp in the springtime the number of mosquitoes per acre appears to grow according to the formula $N = 10^{0.1t+2}$, where t is the number of days since the last frost. What is the size of the mosquito population at times $t = 10$, $t = 20$, and $t = 30$?
1,000, 10,000, 100,000

 In Exercises 77–82, solve each problem. See Example 10.

77. Compounding continuously. If $500 is deposited in an account paying 7% compounded continuously, then how much will be in the account after 3 years?
$616.84

78. Compounding continuously. If $7,000 is deposited in an account paying 8% compounded continuously, then what will it amount to after 4 years?
$9,639.89

79. One year's interest. How much interest will be earned the first year on $80,000 on deposit in an account paying 7.5% compounded continuously?
$6,230.73

80. Partial year. If $7,500 is deposited in an account paying 6.75% compounded continuously, then how much will be in the account after 5 years and 215 days?
$10,937.13

81. Radioactive decay. The number of grams of a certain radioactive substance present at time t is given by the formula $A = 300 \cdot e^{-0.06t}$, where t is the number of years. Find the amount present at time $t = 0$. Find the amount present

after 20 years. Use the accompanying graph to estimate the number of years that it takes for one-half of the substance to decay. Will the substance ever decay completely?
300 grams, 90.4 grams, 12 years, no

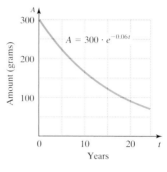

FIGURE FOR EXERCISE 81

82. Population growth. The population of a certain country appears to be growing according to the formula $P = 20 \cdot e^{0.1t}$, where P is the population in millions and t is the number of years since 1980. What was the population in 1980? What will the population be in the year 2000?
20 million, 147.8 million

GETTING MORE INVOLVED

 83. Exploration. An approximate value for e can be found by adding the terms in the following infinite sum:

$$1 + \frac{1}{1} + \frac{1}{2 \cdot 1} + \frac{1}{3 \cdot 2 \cdot 1} + \frac{1}{4 \cdot 3 \cdot 2 \cdot 1} + \cdots$$

Use a calculator to find the sum of the first four terms. Find the difference between the sum of the first four terms and e. (For e, use all of the digits that your calculator gives for e^1.) What is the difference between e and the sum of the first eight terms?
2.66666667, 0.0516, 2.8×10^{-5}

GRAPHING CALCULATOR EXERCISES

84. Graph $y_1 = 2^x$, $y_2 = e^x$, and $y_3 = 3^x$ on the same coordinate system. Which point do all three graphs have in common?
(0, 1)

85. Graph $y_1 = 3^x$, $y_2 = 3^{x-1}$, and $y_3 = 3^{x-2}$ on the same coordinate system. What can you say about the graph of $y = 3^{x-k}$ for any real number k?
The graph of $y = 3^{x-k}$ lies k units to the right of $y = 3^x$ when $k > 0$ and $|k|$ units to the left of $y = 3^x$ when $k < 0$.

10.2 LOGARITHMIC FUNCTIONS

In this section

- Definition
- Domain and Range
- Graphing Logarithmic Functions
- Logarithmic Equations
- Applications

In Section 10.1 you learned that exponential functions are one-to-one functions. Because they are one-to-one functions, they have inverse functions. In this section we study the inverses of the exponential functions.

Definition

Consider the base-2 exponential function $f(x) = 2^x$. Because any exponential function is one-to-one, f has an inverse function. The inverse function is called the **base-2 logarithm function.** We write $f^{-1}(x) = \log_2(x)$ and read $\log_2(x)$ as "the base-2 logarithm of x." The inverse function undoes what the function does. For example, because $f(5) = 2^5 = 32$, we have $f^{-1}(32) = \log_2(32) = 5$. See Fig. 10.8. So $\log_2(32)$ is the exponent that is used on the base 2 to obtain 32. In general, the inverse of the base-a exponential function is called the **base-a logarithm function** and *$\log_a(x)$ is the exponent that is used on the base a to obtain x.* The expression $\log_a(x)$ is called a **logarithm.**

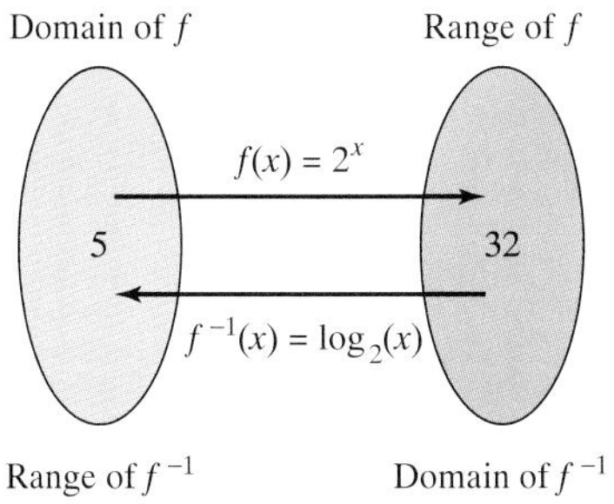
Domain of f
Range of f
$f(x) = 2^x$
5
32
$f^{-1}(x) = \log_2(x)$
Range of f^{-1}
Domain of f^{-1}

FIGURE 10.8

$\log_a(x)$

For any $a > 0$ and $a \neq 1$,

$$y = \log_a(x) \quad \text{if and only if} \quad a^y = x.$$

The definition of base-a logarithm says that the logarithmic equation $y = \log_a(x)$ and the exponential equation $a^y = x$ are equivalent.

EXAMPLE 1 **Using the definition of logarithm**

Write each logarithmic equation as an exponential equation and each exponential equation as a logarithmic equation.

a) $\log_5(125) = 3$ **b)** $6 = \log_{1/4}(x)$

c) $\left(\frac{1}{2}\right)^m = 8$ **d)** $7 = 3^z$

Solution

a) "The base-5 logarithm of 125 equals 3" means that 3 is the exponent on 5 that produces 125. So $5^3 = 125$.

b) The equation $6 = \log_{1/4}(x)$ is equivalent to $\left(\frac{1}{4}\right)^6 = x$ by the definition of logarithm.

c) The equation $\left(\frac{1}{2}\right)^m = 8$ is equivalent to $\log_{1/2}(8) = m$.

d) The equation $7 = 3^z$ is equivalent to $\log_3(7) = z$. ■

The definition of logarithm is also used to evaluate logarithmic functions.

EXAMPLE 2 **Finding logarithms**

Evaluate each logarithm.

a) $\log_5(25)$ **b)** $\log_2\left(\frac{1}{8}\right)$

c) $\log_{1/2}(4)$ **d)** $\log_{10}(0.001)$

Solution

a) The number $\log_5(25)$ is the exponent that is used on the base 5 to obtain 25. Because $25 = 5^2$, we have $\log_5(25) = 2$.

b) The number $\log_2\left(\frac{1}{8}\right)$ is the power of 2 that gives us $\frac{1}{8}$. Because $\frac{1}{8} = 2^{-3}$, we have $\log_2\left(\frac{1}{8}\right) = -3$.

c) The number $\log_{1/2}(4)$ is the power of $\frac{1}{2}$ that produces 4. Because $4 = \left(\frac{1}{2}\right)^{-2}$, we have $\log_{1/2}(4) = -2$.

d) Because $0.001 = 10^{-3}$, we have $\log_{10}(0.001) = -3$. ∎

There are two bases for logarithms that are used more frequently than the others: They are 10 and e. The base-10 logarithm is called the **common logarithm** and is usually written as $\log(x)$. The base-e logarithm is called the **natural logarithm** and is usually written as $\ln(x)$. Most scientific calculators have function keys for $\log(x)$ and $\ln(x)$. The simplest way to obtain a common or natural logarithm is to use a scientific calculator. However, a table of common logarithms can be found in Appendix C of this text.

In the next example we find natural and common logarithms of certain numbers without a calculator or a table.

E X A M P L E   3   **Finding common and natural logarithms**

Evaluate each logarithm.

a) $\log(1000)$

b) $\ln(e)$

c) $\log\left(\dfrac{1}{10}\right)$

Solution

a) Because $10^3 = 1000$, we have $\log(1000) = 3$.

b) Because $e^1 = e$, we have $\ln(e) = 1$.

c) Because $10^{-1} = \frac{1}{10}$, we have $\log\left(\frac{1}{10}\right) = -1$. ∎

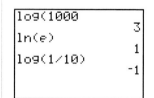

A graphing calculator has keys for the common logarithm (LOG) and the natural logarithm (LN).

```
log(1000
                3
ln(e)
                1
log(1/10)
               -1
```

Domain and Range

The domain of the exponential function $y = 2^x$ is $(-\infty, \infty)$, and its range is $(0, \infty)$. Because the logarithmic function $y = \log_2(x)$ is the inverse of $y = 2^x$, the domain of $y = \log_2(x)$ is $(0, \infty)$, and its range is $(-\infty, \infty)$.

CAUTION   Because the domain of $y = \log_a(x)$ is $(0, \infty)$ for any $a > 0$ and $a \neq 1$, expressions such as $\log_2(-4)$, $\log_{1/3}(0)$, and $\ln(-1)$ are undefined.

Graphing Logarithmic Functions

In Chapter 9 we saw that the graphs of a function and its inverse function are symmetric about the line $y = x$. Because the logarithm functions are inverses of exponential functions, their graphs are also symmetric about $y = x$.

E X A M P L E 4

calculator

close-up

The graphs of $y = \ln(x)$ and $y = e^x$ are symmetric with respect to the line $y = x$. Logarithmic functions with bases other than e and 10 will be graphed on a calculator in Section 10.4.

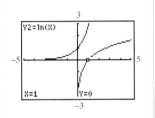

A logarithmic function with base greater than 1

Sketch the graph of $g(x) = \log_2(x)$ and compare it to the graph of $y = 2^x$.

Solution

Make a table of ordered pairs for $g(x) = \log_2(x)$ using positive numbers for x:

x	$\frac{1}{4}$	$\frac{1}{2}$	1	2	4	8
$g(x) = \log_2(x)$	-2	-1	0	1	2	3

Draw a curve through these points as shown in Fig. 10.9. The graph of the inverse function $y = 2^x$ is also shown in Fig. 10.9 for comparison. Note the symmetry of the two curves about the line $y = x$.

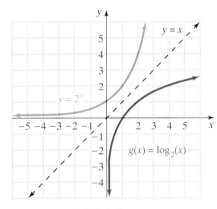

FIGURE 10.9

All logarithmic functions with the base greater than 1 have graphs that are similar to the one in Fig. 10.9. In general, the graph of $f(x) = \log_a(x)$ for $a > 1$ has the following characteristics (see Fig. 10.10):

1. The x-intercept of the curve is $(1, 0)$.
2. The domain is $(0, \infty)$, and the range is $(-\infty, \infty)$.
3. The curve approaches the negative y-axis but does not touch it.
4. The y-values are increasing as we go from left to right along the curve.

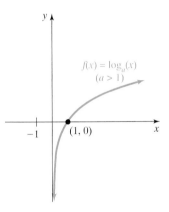

FIGURE 10.10

E X A M P L E 5

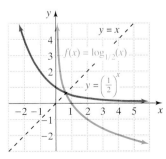

FIGURE 10.11

A logarithmic function with base less than 1

Sketch the graph of $f(x) = \log_{1/2}(x)$ and compare it to the graph of $y = \left(\frac{1}{2}\right)^x$.

Solution

Make a table of ordered pairs for $f(x) = \log_{1/2}(x)$ using positive numbers for x:

x	$\frac{1}{4}$	$\frac{1}{2}$	1	2	4	8
$f(x) = \log_{1/2}(x)$	2	1	0	-1	-2	-3

The curve through these points is shown in Fig. 10.11. The graph of the inverse function $y = \left(\frac{1}{2}\right)^x$ is also shown in Fig. 10.11 for comparison. Note the symmetry with respect to the line $y = x$.

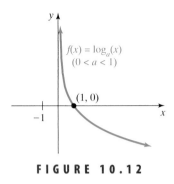

FIGURE 10.12

All logarithmic functions with the base between 0 and 1 have graphs that are similar to the one in Fig. 10.11. In general, the graph of $f(x) = \log_a(x)$ for $0 < a < 1$ has the following characteristics (see Fig. 10.12):

1. The x-intercept of the curve is $(1, 0)$.

2. The domain is $(0, \infty)$, and the range is $(-\infty, \infty)$.

3. The curve approaches the positive y-axis but does not touch it.

4. The y-values are decreasing as we go from left to right along the curve.

Figures 10.9 and 10.11 illustrate the fact that $y = \log_a(x)$ and $y = a^x$ are inverse functions for any base a. For any given exponential or logarithmic function the inverse function can be easily obtained from the definition of logarithm.

E X A M P L E   6 **Inverses of logarithmic and exponential functions**

Find the inverse of each function.

a) $f(x) = 10^x$ **b)** $g(x) = \log_3(x)$

Solution

a) The inverse of $f(x)$ is $f^{-1}(x) = \log_{10}(x)$ or $y = \log(x)$.

b) The inverse of $g(x) = \log_3(x)$ is $g^{-1}(x) = 3^x$ or $y = 3^x$. ■

Logarithmic Equations

In Section 10.1 we learned that the exponential functions are one-to-one functions. Because logarithmic functions are inverses of exponential functions, they are one-to-one functions also. For a base-a logarithmic function *one-to-one means that if the base-a logarithms of two numbers are equal, then the numbers are equal.*

One-to-One Property of Logarithms

For $a > 0$ and $a \neq 1$,

$$\text{if} \quad \log_a(m) = \log_a(n), \quad \text{then} \quad m = n.$$

The one-to-one property of logarithms and the definition of logarithms are the two basic tools that we use to solve equations involving logarithms. We use these tools in the next example.

E X A M P L E   7 **Logarithmic equations**

Solve each equation.

a) $\log_3(x) = -2$ **b)** $\log_x(8) = -3$ **c)** $\log(x^2) = \log(4)$

Solution

a) Use the definition of logarithms to rewrite the logarithmic equation as an equivalent exponential equation:

$$\log_3(x) = -2$$
$$3^{-2} = x \quad \text{Definition of logarithm}$$
$$\frac{1}{9} = x$$

The solution set is $\left\{\frac{1}{9}\right\}$.

b) Use the definition of logarithms to rewrite the logarithmic equation as an equivalent exponential equation:

$$\log_x(8) = -3$$
$$x^{-3} = 8 \qquad \text{Definition of logarithm}$$
$$(x^{-3})^{-1} = 8^{-1} \qquad \text{Raise each side to the } -1 \text{ power.}$$
$$x^3 = \frac{1}{8}$$
$$x = \sqrt[3]{\frac{1}{8}} = \frac{1}{2} \qquad \text{Odd-root property}$$

The solution set is $\left\{\frac{1}{2}\right\}$.

c) To write an equation equivalent to $\log(x^2) = \log(4)$, we use the one-to-one property of logarithms:

$$\log(x^2) = \log(4)$$
$$x^2 = 4 \qquad \text{One-to-one property of logarithms}$$
$$x = \pm 2 \qquad \text{Even-root property}$$

The solution set is $\{-2, 2\}$. ◼

CAUTION   If we have equality of two logarithms with the same base, we use the one-to-one property to eliminate the logarithms. If we have an equation with only one logarithm, such as $\log_a(x) = y$, we use the definition of logarithm to write $a^y = x$ and to eliminate the logarithm.

Applications

When money earns interest compounded continuously, the formula

$$t = \frac{1}{r} \ln\left(\frac{A}{P}\right)$$

expresses the relationship between the time in years t, the annual interest rate r, the principal P, and the amount A. This formula is used to determine how long it takes for a deposit to grow to a specific amount.

E X A M P L E   8

Finding the time for a specified growth

How long does it take $80 to grow to $240 at 12% compounded continuously?

Solution

Use $r = 0.12$, $P = \$80$, and $A = \$240$ in the formula, and use a calculator to evaluate the logarithm:

$$t = \frac{1}{0.12} \ln\left(\frac{240}{80}\right)$$
$$= \frac{\ln(3)}{0.12}$$
$$\approx 9.155$$

It takes approximately 9.155 years, or 9 years and 57 days. ◼

WARM-UPS

True or false? Explain.

1. The equation $a^3 = 2$ is equivalent to $\log_a(2) = 3$.   True
2. If (a, b) satisfies $y = 8^x$, then (a, b) satisfies $y = \log_8(x)$.   False
3. If $f(x) = a^x$ for $a > 0$ and $a \neq 1$, then $f^{-1}(x) = \log_a(x)$.   True
4. If $f(x) = \ln(x)$, then $f^{-1}(x) = e^x$.   True
5. The domain of $f(x) = \log_6(x)$ is $(-\infty, \infty)$.   False
6. $\log_{25}(5) = 2$ False
7. $\log(-10) = 1$ False
8. $\log(0) = 0$ False
9. $5^{\log_5(125)} = 125$ True
10. $\log_{1/2}(32) = -5$ True

10.2   EXERCISES

Reading and Writing   *After reading this section, write out the answers to these questions. Use complete sentences.*

1. What is the inverse function for the function $f(x) = 2^x$?
   If $f(x) = 2^x$, then $f^{-1}(x) = \log_2(x)$.

2. What is $\log_a(x)$?
   The expression $\log_a(x)$ is the exponent of a that produces x. So $a^{\log_a(x)} = x$.

3. What is the difference between the common logarithm and the natural logarithm?
   The common logarithm uses the base 10 and the natural logarithm uses base e.

4. What is the domain of $f(x) = \log_a(x)$?
   The domain of $f(x) = \log_a(x)$ is $(0, \infty)$.

5. What is the one-to-one property of logarithmic functions?
   The one-to-one property for logarithmic functions states that if $\log_a(m) = \log_a(n)$, then $m = n$.

6. What is the relationship between the graphs of $f(x) = a^x$ and $f^{-1}(x) = \log_a(x)$ for $a > 0$ and $a \neq 1$?
   The graphs of $f(x) = a^x$ and $f^{-1}(x) = \log_a(x)$ are symmetric about the line $y = x$.

Write each exponential equation as a logarithmic equation and each logarithmic equation as an exponential equation. See Example 1.

7. $\log_2(8) = 3$ $2^3 = 8$
8. $\log_{10}(10) = 1$ $10^1 = 10$
9. $10^2 = 100$ $\log(100) = 2$
10. $5^3 = 125$ $\log_5(125) = 3$
11. $y = \log_5(x)$ $5^y = x$
12. $m = \log_b(N)$ $b^m = N$
13. $2^a = b$ $\log_2(b) = a$
14. $a^3 = c$ $\log_a(c) = 3$
15. $\log_3(x) = 10$ $3^{10} = x$
16. $\log_c(t) = 4$ $c^4 = t$
17. $e^3 = x$ $\ln(x) = 3$
18. $m = e^x$ $\ln(m) = x$

Evaluate each logarithm. See Examples 2 and 3.

19. $\log_2(4)$ 2
20. $\log_2(1)$ 0
21. $\log_2(16)$ 4
22. $\log_4(16)$ 2
23. $\log_2(64)$ 6
24. $\log_8(64)$ 2
25. $\log_4(64)$ 3
26. $\log_{64}(64)$ 1
27. $\log_2\left(\dfrac{1}{4}\right)$ -2
28. $\log_2\left(\dfrac{1}{8}\right)$ -3
29. $\log(100)$ 2
30. $\log(1)$ 0
31. $\log(0.01)$ -2
32. $\log(10{,}000)$ 4
33. $\log_{1/3}\left(\dfrac{1}{3}\right)$ 1
34. $\log_{1/3}\left(\dfrac{1}{9}\right)$ 2
35. $\log_{1/3}(27)$ -3
36. $\log_{1/3}(1)$ 0
37. $\ln(e^3)$ 3
38. $\ln(1)$ 0
39. $\ln(e^2)$ 2
40. $\ln\left(\dfrac{1}{e}\right)$ -1

 Use a calculator to evaluate each logarithm. Round answers to four decimal places.

41. $\log(5)$ 0.6990
42. $\log(0.03)$ -1.5229
43. $\ln(6.238)$ 1.8307
44. $\ln(0.23)$ -1.4697

Sketch the graph of each function. See Examples 4 and 5.

45. $f(x) = \log_3(x)$
46. $g(x) = \log_{10}(x)$

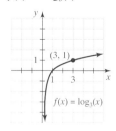

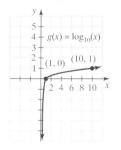

47. $y = \log_4(x)$

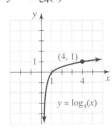

48. $y = \log_5(x)$

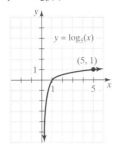

49. $h(x) = \log_{1/4}(x)$

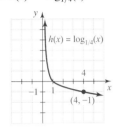

50. $y = \log_{1/3}(x)$

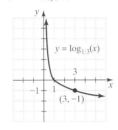

51. $y = \log_{1/5}(x)$

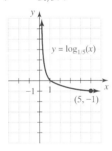

52. $y = \log_{1/6}(x)$

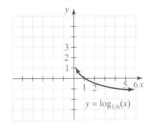

Find the inverse of each function. See Example 6.

53. $f(x) = 6^x$
$f^{-1}(x) = \log_6(x)$

54. $f(x) = 4^x$
$f^{-1}(x) = \log_4(x)$

55. $f(x) = \ln(x)$
$f^{-1}(x) = e^x$

56. $f(x) = \log(x)$
$f^{-1}(x) = 10^x$

57. $f(x) = \log_{1/2}(x)$
$f^{-1}(x) = \left(\dfrac{1}{2}\right)^x$

58. $f(x) = \log_{1/4}(x)$
$f^{-1}(x) = \left(\dfrac{1}{4}\right)^x$

Solve each equation. See Example 7.

59. $x = \left(\dfrac{1}{2}\right)^{-2}$ {4}

60. $x = 3^2$ {9}

61. $5 = 25^x$ $\left\{\dfrac{1}{2}\right\}$

62. $0.1 = 10^x$ {−1}

63. $\log(x) = -3$ {0.001}

64. $\log(x) = 5$ {100,000}

65. $\log_x(36) = 2$ {6}

66. $\log_x(100) = 2$ {10}

67. $\log_x(5) = -1$ $\left\{\dfrac{1}{5}\right\}$

68. $\log_x(16) = -2$ $\left\{\dfrac{1}{4}\right\}$

69. $\log(x^2) = \log(9)$
{±3}

70. $\ln(2x − 3) = \ln(x + 1)$
{4}

 Use a calculator to solve each equation. Round answers to four decimal places.

71. $3 = 10^x$ **72.** $10^x = 0.03$ **73.** $10^x = \dfrac{1}{2}$
{0.4771}   {−1.5229}   {−0.3010}

74. $75 = 10^x$ **75.** $e^x = 7.2$ **76.** $e^{3x} = 0.4$
{1.8751}   {1.9741}   {−0.3054}

Solve each problem. See Example 8. Use a calculator as necessary.

77. Double your money. How long does it take $5,000 to grow to $10,000 at 12% compounded continuously? 5.776 years

78. Half the rate. How long does it take $5,000 to grow to $10,000 at 6% compounded continuously?   11.552 years

79. Earning interest. How long does it take to earn $1,000 in interest on a deposit of $6,000 at 8% compounded continuously?   1.9269 years

80. Lottery winnings. How long does it take to earn $1,000 interest on a deposit of one million dollars at 9% compounded continuously?   4.054 days

The annual growth rate for an investment that is growing continuously is given by

$$r = \dfrac{1}{t}\ln\left(\dfrac{A}{P}\right),$$

where P is the principal and A is the amount after t years.

81. Top stock. An investment of $10,000 in Dell Computer stock in 1995 grew to $231,800 in 1998.
a) Assuming the investment grew continuously, what was the annual growth rate?
b) If Dell continues to grow at the same rate, then what will the $10,000 investment be worth in 2002?
 a) 104.8%   **b)** $15,320,208

82. Chocolate bars. An investment of $10,000 in 1980 in Hershey stock was worth $563,000 in 1998. Assuming the investment grew continuously, what was the annual growth rate?   22.39%

In chemistry the pH of a solution is defined by

$$\text{pH} = -\log_{10}[H+],$$

where H+ is the hydrogen ion concentration of the solution in moles per liter. Distilled water has a pH of approximately 7. A solution with a pH under 7 is called an acid, and one with a pH over 7 is called a base.

83. Tomato juice. Tomato juice has a hydrogen ion concentration of $10^{-4.1}$ mole per liter (mol/L). Find the pH of tomato juice.   4.1

84. Stomach acid. The gastric juices in your stomach have a hydrogen ion concentration of 10^{-1} mol/L. Find the pH of your gastric juices.   1

85. Neuse River pH. The pH of a water sample is one of the many measurements of water quality done by the U.S. Geological Survey. The hydrogen ion concentration of the water in the Neuse River at New Bern, North Carolina, was

1.58×10^{-7} mol/L on July 9, 1998 (Water Resources for North Carolina, wwwnc.usgs.gov). What was the pH of the water at that time?   6.8

86. *Roanoke River* **pH.** On July 9, 1998 the hydrogen ion concentration of the water in the Roanoke River at Janesville, North Carolina, was 1.995×10^{-7} mol/L (Water Resources for North Carolina, wwwnc.usgs.gov). What was the pH of the water at that time?   6.7

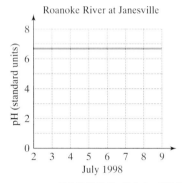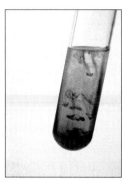

FIGURE FOR EXERCISE 86

Solve each problem.

87. *Sound level.* The level of sound in decibels (db) is given by the formula

$$L = 10 \cdot \log(I \times 10^{12}),$$

where I is the intensity of the sound in watts per square meter. If the intensity of the sound at a rock concert is 0.001 watt per square meter at a distance of 75 meters from the stage, then what is the level of the sound at this point in the audience?   90 db

88. *Logistic growth.* If a rancher has one cow with a contagious disease in a herd of 1,000, then the time in days t for n of the cows to become infected is modeled by

$$t = -5 \cdot \ln\left(\frac{1000 - n}{999n}\right).$$

Find the number of days that it takes for the disease to spread to 100, 200, 998, and 999 cows. This model, called

a *logistic growth model,* describes how a disease can spread very rapidly at first and then very slowly as nearly all of the population has become infected.   23.5, 27.6, 65.6, 69.1

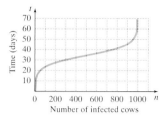

FIGURE FOR EXERCISE 88

GETTING MORE INVOLVED

 89. *Discussion.* Use the switch-and-solve method from Chapter 8 to find the inverse of the function $f(x) = 5 + \log_2(x - 3)$. State the domain and range of the inverse function. $f^{-1}(x) = 2^{x-5} + 3, (-\infty, \infty), (3, \infty)$

90. *Discussion.* Find the inverse of the function $f(x) = 2 + e^{x+4}$. State the domain and range of the inverse function.
$f^{-1}(x) = \ln(x - 2) - 4, (2, \infty), (-\infty, \infty)$

GRAPHING CALCULATOR EXERCISES

91. *Composition of inverses.* Graph the functions $y = \ln(e^x)$ and $y = e^{\ln(x)}$. Explain the similarities and differences between the graphs.
$y = \ln(e^x) = x$ for $-\infty < x < \infty,$ $y = e^{\ln(x)} = x$ for $0 < x < \infty$

92. *The population bomb.* The population of the earth is growing continuously with an annual rate of about 1.6%. If the present population is 6 billion, then the function $y = 6e^{0.016x}$ gives the population in billions x years from now. Graph this function for $0 \le x \le 200$. What will the population be in 100 years and in 200 years?
29.7 billion, 147.2 billion

10.3 PROPERTIES OF LOGARITHMS

The properties of logarithms are very similar to the properties of exponents because *logarithms are exponents.* In this section we use the properties of exponents to write some properties of logarithms. The properties will be used in solving logarithmic equations in Section 10.4.

Product Rule for Logarithms

If $M = a^x$ and $N = a^y$, we can use the product rule for exponents to write

$$MN = a^x \cdot a^y = a^{x+y}.$$

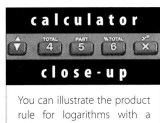
The equation $MN = a^{x+y}$ is equivalent to

$$\log_a(MN) = x + y.$$

Because $M = a^x$ and $N = a^y$ are equivalent to $x = \log_a(M)$ and $y = \log_a(N)$, we can replace x and y in $\log_a(MN) = x + y$ to get

$$\log_a(MN) = \log_a(M) + \log_a(N).$$

So *the logarithm of a product is the sum of the logarithms,* provided that all of the logarithms are defined. This rule is called the **product rule for logarithms.**

Product Rule for Logarithms

$$\log_a(MN) = \log_a(M) + \log_a(N)$$

E X A M P L E   1

Using the product rule for logarithms

Write each expression as a single logarithm.

a) $\log_2(7) + \log_2(5)$

b) $\ln(\sqrt{2}) + \ln(\sqrt{3})$

helpful   hint

The product rule for logs comes from the product rule for exponents. Since you add exponents when you multiply exponential expressions (and logarithms are exponents), it is not too surprising that the log of a product is the sum of the logs.

Solution

a) $\log_2(7) + \log_2(5) = \log_2(35)$ Product rule for logarithms

b) $\ln(\sqrt{2}) + \ln(\sqrt{3}) = \ln(\sqrt{6})$ Product rule for logarithms

Quotient Rule for Logarithms

If $M = a^x$ and $N = a^y$, we can use the quotient rule for exponents to write

$$\frac{M}{N} = \frac{a^x}{a^y} = a^{x-y}.$$

By the definition of logarithm, $\frac{M}{N} = a^{x-y}$ is equivalent to

$$\log_a\left(\frac{M}{N}\right) = x - y.$$

Because $x = \log_a(M)$ and $y = \log_a(N)$, we have

$$\log_a\left(\frac{M}{N}\right) = \log_a(M) - \log_a(N).$$

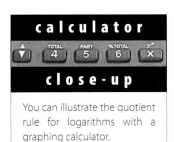
So *the logarithm of a quotient is equal to the difference of the logarithms,* provided that all logarithms are defined. This rule is called the **quotient rule for logarithms.**

Quotient Rule for Logarithms

$$\log_a\left(\frac{M}{N}\right) = \log_a(M) - \log_a(N)$$

E X A M P L E   2

Using the quotient rule for logarithms

Write each expression as a single logarithm.

a) $\log_2(3) - \log_2(7)$

b) $\ln(w^8) - \ln(w^2)$

Solution

a) $\log_2(3) - \log_2(7) = \log_2\left(\dfrac{3}{7}\right)$ Quotient rule for logarithms

b) $\ln(w^8) - \ln(w^2) = \ln\left(\dfrac{w^8}{w^2}\right)$ Quotient rule for logarithms

  $= \ln(w^6)$ Quotient rule for exponents

Power Rule for Logarithms

If $M = a^x$, we can use the power rule for exponents to write

$$M^N = (a^x)^N = a^{Nx}.$$

By the definition of logarithms, $M^N = a^{Nx}$ is equivalent to

$$\log_a(M^N) = Nx.$$

Because $x = \log_a(M)$, we have

$$\log_a(M^N) = N \cdot \log_a(M).$$

So *the logarithm of a power of a number is the power times the logarithm of the number,* provided that all logarithms are defined. This rule is called the **power rule for logarithms.**

Power Rule for Logarithms

$$\log_a(M^N) = N \cdot \log_a(M)$$

E X A M P L E   3

Using the power rule for logarithms

Rewrite each logarithm in terms of $\log(2)$.

a) $\log(2^{10})$ b) $\log(\sqrt{2})$ c) $\log\left(\dfrac{1}{2}\right)$

Solution

a) $\log(2^{10}) = 10 \cdot \log(2)$ Power rule for logarithms

b) $\log(\sqrt{2}) = \log(2^{1/2})$ Write $\sqrt{2}$ as a power of 2.

  $= \dfrac{1}{2}\log(2)$ Power rule for logarithms

c) $\log\left(\dfrac{1}{2}\right) = \log(2^{-1})$ Write $\dfrac{1}{2}$ as a power of 2.

  $= -1 \cdot \log(2)$ Power rule for logarithms

  $= -\log(2)$

calculator

close-up

Because logarithmic and exponential functions are inverses, applying one and then the other returns the original number.

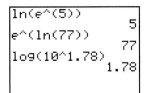

```
ln(e^(5))
                    5
e^(ln(77))
                   77
log(10^1.78)
                 1.78
```

Inverse Properties

An exponential function and logarithmic function with the same base are inverses of each other. For example, the logarithm of 32 base 2 is 5 and the fifth power of 2 is 32. In symbols, we have

$$2^{\log_2(32)} = 2^5 = 32.$$

If we raise 3 to the fourth power, we get 81; and if we find the base-3 logarithm of 81, we get 4. In symbols, we have

$$\log_3(3^4) = \log_3(81) = 4.$$

We can state the inverse relationship between exponential and logarithm functions in general with the following inverse properties.

Inverse Properties

1. $\log_a(a^M) = M$ **2.** $a^{\log_a(M)} = M$

E X A M P L E   4

Using the inverse properties

Simplify each expression.

a) $\ln(e^5)$ **b)** $2^{\log_2(8)}$

Solution

a) Using the first inverse property, we get $\ln(e^5) = 5$.

b) Using the second inverse property, we get $2^{\log_2(8)} = 8$. ■

   Note that there is more than one way to simplify the expressions in Example 4. Using the power rule for logarithms and the fact that $\ln(e) = 1$, we have $\ln(e^5) = 5 \cdot \ln(e) = 5$. Using $\log_2(8) = 3$, we have $2^{\log_2(8)} = 2^3 = 8$.

Using the Properties

We have already seen many properties of logarithms. There are three properties that we have not yet formally stated. Because $a^1 = a$ and $a^0 = 1$, we have $\log_a(a) = 1$ and $\log_a(1) = 0$ for any positive number a. If we apply the quotient rule to $\log_a(1/N)$, we get

$$\log_a\left(\frac{1}{N}\right) = \log_a(1) - \log_a(N) = 0 - \log_a(N) = -\log_a(N).$$

So $\log_a\left(\frac{1}{N}\right) = -\log_a(N)$. These three new properties along with all of the other properties of logarithms are summarized as follows.

Properties of Logarithms

If M, N, and a are positive numbers, $a \neq 1$, then

1. $\log_a(a) = 1$ **2.** $\log_a(1) = 0$

3. $\log_a(a^M) = M$ Inverse properties **4.** $a^{\log_a(M)} = M$

5. $\log_a(MN) = \log_a(M) + \log_a(N)$ Product rule

6. $\log_a\left(\frac{M}{N}\right) = \log_a(M) - \log_a(N)$ Quotient rule

7. $\log_a\left(\frac{1}{N}\right) = -\log_a(N)$ **8.** $\log_a(M^N) = N \cdot \log_a(M)$ Power rule

We have already seen several ways in which to use the properties of logarithms. In the next three examples we see more uses of the properties. First we use the rules of logarithms to write the logarithm of a complicated expression in terms of logarithms of simpler expressions.

E X A M P L E   5

close-up

Examine the values of $\log(9/2)$, $\log(9) - \log(2)$, and $\log(9)/\log(2)$.

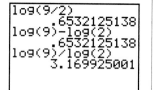

Using the properties of logarithms

Rewrite each expression in terms of $\log(2)$ and/or $\log(3)$.

a) $\log(6)$ **b)** $\log(16)$

c) $\log\left(\dfrac{9}{2}\right)$ **d)** $\log\left(\dfrac{1}{3}\right)$

Solution

a) $\log(6) = \log(2 \cdot 3)$

$\qquad\quad = \log(2) + \log(3)$ Product rule

b) $\log(16) = \log(2^4)$

$\qquad\qquad = 4 \cdot \log(2)$ Power rule

c) $\log\left(\dfrac{9}{2}\right) = \log(9) - \log(2)$ Quotient rule

$\qquad\qquad = \log(3^2) - \log(2)$

$\qquad\qquad = 2 \cdot \log(3) - \log(2)$ Power rule

d) $\log\left(\dfrac{1}{3}\right) = -\log(3)$ Property 7

CAUTION   Do not confuse $\dfrac{\log(9)}{\log(2)}$ with $\log\left(\dfrac{9}{2}\right)$. We can use the quotient rule to write $\log\left(\dfrac{9}{2}\right) = \log(9) - \log(2)$, but $\dfrac{\log(9)}{\log(2)} \neq \log(9) - \log(2)$. The expression $\dfrac{\log(9)}{\log(2)}$ means $\log(9) \div \log(2)$. Use your calculator to verify these two statements.

The properties of logarithms can be used to combine several logarithms into a single logarithm (as in Examples 1 and 2) or to write a logarithm of a complicated expression in terms of logarithms of simpler expressions.

E X A M P L E   6

Using the properties of logarithms

Rewrite each expression as a sum or difference of multiples of logarithms.

a) $\log\left(\dfrac{xz}{y}\right)$ **b)** $\log_3\left(\dfrac{(x-3)^{2/3}}{\sqrt{x}}\right)$

Solution

a) $\log\left(\dfrac{xz}{y}\right) = \log(xz) - \log(y)$ Quotient rule

$\qquad\qquad = \log(x) + \log(z) - \log(y)$ Product rule

b) $\log_3\left(\dfrac{(x-3)^{2/3}}{\sqrt{x}}\right) = \log_3((x-3)^{2/3}) - \log_3(x^{1/2})$ Quotient rule

$\qquad\qquad\qquad = \dfrac{2}{3}\log_3(x-3) - \dfrac{1}{2}\log_3(x)$ Power rule

In the next example we use the properties of logarithms to convert expressions involving several logarithms into a single logarithm. The skills we are learning here will be used to solve logarithmic equations in Section 10.4.

EXAMPLE 7   **Combining logarithms**

Rewrite each expression as a single logarithm.

a) $\dfrac{1}{2} \log(x) - 2 \cdot \log(x + 1)$ **b)** $3 \cdot \log(y) + \dfrac{1}{2} \log(z) - \log(x)$

Solution

a) $\dfrac{1}{2} \log(x) - 2 \cdot \log(x + 1) = \log(x^{1/2}) - \log((x + 1)^2)$ Power rule

$$= \log\!\left(\frac{\sqrt{x}}{(x + 1)^2} \right) \qquad \text{Quotient rule}$$

b) $3 \cdot \log(y) + \dfrac{1}{2} \log(z) - \log(x) = \log(y^3) + \log(\sqrt{z}) - \log(x)$ Power rule

$$= \log(y^3 \cdot \sqrt{z}) - \log(x) \qquad \text{Product rule}$$

$$= \log\!\left(\frac{y^3 \cdot \sqrt{z}}{x} \right) \qquad \text{Quotient rule}$$

WARM-UPS

True or false? Explain.

1. $\log_2\!\left(\dfrac{x^2}{8} \right) = \log_2(x^2) - 3$ True

2. $\dfrac{\log(100)}{\log(10)} = \log(100) - \log(10)$ False

3. $\ln(\sqrt{2}) = \dfrac{\ln(2)}{2}$ True

4. $3^{\log_3(17)} = 17$ True

5. $\log_2\!\left(\dfrac{1}{8} \right) = \dfrac{1}{\log_2(8)}$ False

6. $\ln(8) = 3 \cdot \ln(2)$ True

7. $\ln(1) = e$ False

8. $\dfrac{\log(100)}{10} = \log(10)$ False

9. $\dfrac{\log_2(8)}{\log_2(2)} = \log_2(4)$ False

10. $\ln(2) + \ln(3) - \ln(7) = \ln\!\left(\dfrac{6}{7} \right)$ True

---

10.3   EXERCISES

Reading and Writing   *After reading this section, write out the answers to these questions. Use complete sentences.*

1. What is the product rule for logarithms?
The product rule for logarithms states that $\log_a(MN) = \log_a(M) + \log_a(N)$.

2. What is the quotient rule for logarithms?
The quotient rule for logarithms states that $\log_a(M/N) = \log_a(M) - \log_a(N)$.

3. What is the power rule for logarithms?
The power rule for logarithms states that $\log_a(M^N) = N \cdot \log_a(M)$.

4. Why is it true that $\log_a(a^M) = M$?
Since $\log_a(a^M)$ is the exponent used on a to obtain a^M, we have $\log_a(a^M) = M$.

5. Why is it true that $a^{\log_a(M)} = M$?
Since $\log_a(M)$ is the exponent you would use on a to obtain M, using $\log_a(M)$ as the exponent produces M: $a^{\log_a(M)} = M$.

6. Why is it true that $\log_a(1) = 0$ for $a > 0$ and $a \neq 1$?
We have $\log_a(1) = 0$ because $a^0 = 1$.

Assume all variables involved in logarithms represent numbers for which the logarithms are defined.

Write each expression as a single logarithm and simplify. See Example 1.

7. $\log(3) + \log(7)$
$\log(21)$

8. $\ln(5) + \ln(4)$
$\ln(20)$

9. $\log_3(\sqrt{5}) + \log_3(\sqrt{x})$
$\log_3(\sqrt{5x})$

10. $\ln(\sqrt{x}) + \ln(\sqrt{y})$
$\ln(\sqrt{xy})$

11. $\log(x^2) + \log(x^3)$
$\log(x^5)$

12. $\ln(a^3) + \ln(a^5)$
$\ln(a^8)$

13. $\ln(2) + \ln(3) + \ln(5)$ $\ln(30)$

14. $\log_2(x) + \log_2(y) + \log_2(z)$ $\log_2(xyz)$

15. $\log(x) + \log(x + 3)$ $\log(x^2 + 3x)$

16. $\ln(x - 1) + \ln(x + 1)$ $\ln(x^2 - 1)$

17. $\log_2(x - 3) + \log_2(x + 2)$ $\log_2(x^2 - x - 6)$

18. $\log_3(x - 5) + \log_3(x - 4)$ $\log_3(x^2 - 9x + 20)$

Write each expression as a single logarithm. See Example 2.

19. $\log(8) - \log(2)$ $\log(4)$

20. $\ln(3) - \ln(6)$ $\ln\left(\frac{1}{2}\right)$

21. $\log_2(x^6) - \log_2(x^2)$
$\log_2(x^4)$

22. $\ln(w^9) - \ln(w^3)$
$\ln(w^6)$

23. $\log(\sqrt{10}) - \log(\sqrt{2})$ $\log(\sqrt{5})$

24. $\log_3(\sqrt{6}) - \log_3(\sqrt{3})$ $\log_3(\sqrt{2})$

25. $\ln(4h - 8) - \ln(4)$ $\ln(h - 2)$

26. $\log(3x - 6) - \log(3)$ $\log(x - 2)$

27. $\log_2(w^2 - 4) - \log_2(w + 2)$ $\log_2(w - 2)$

28. $\log_3(k^2 - 9) - \log_3(k - 3)$ $\log_3(k + 3)$

29. $\ln(x^2 + x - 6) - \ln(x + 3)$ $\ln(x - 2)$

30. $\ln(t^2 - t - 12) - \ln(t - 4)$ $\ln(t + 3)$

Write each expression in terms of $\log(3)$. See Example 3.

31. $\log(27)$ $3\log(3)$

32. $\log\left(\frac{1}{9}\right)$ $-2\log(3)$

33. $\log(\sqrt{3})$ $\frac{1}{2}\log(3)$

34. $\log(\sqrt[4]{3})$ $\frac{1}{4}\log(3)$

35. $\log(3^x)$ $x\log(3)$

36. $\log(3^{-99})$ $-99\log(3)$

Simplify each expression. See Example 4.

37. $\log_2(2^{10})$ 10   **38.** $\ln(e^9)$ 9   **39.** $5^{\log_5(19)}$ 19

40. $10^{\log(2.3)}$ 2.3   **41.** $\log(10^8)$ 8   **42.** $\log_4(4^5)$ 5

43. $e^{\ln(4.3)}$ 4.3   **44.** $3^{\log_3(5.5)}$ 5.5

Rewrite each expression in terms of $\log(3)$ and/or $\log(5)$. See Example 5.

45. $\log(15)$
$\log(3) + \log(5)$

46. $\log(9)$
$2\log(3)$

47. $\log\left(\frac{5}{3}\right)$
$\log(5) - \log(3)$

48. $\log\left(\frac{3}{5}\right)$
$\log(3) - \log(5)$

49. $\log(25)$
$2\log(5)$

50. $\log\left(\frac{1}{27}\right)$
$-3\log(3)$

51. $\log(75)$
$2 \cdot \log(5) + \log(3)$

52. $\log(0.6)$
$\log(3) - \log(5)$

53. $\log\left(\frac{1}{3}\right)$
$-\log(3)$

54. $\log(45)$
$2 \cdot \log(3) + \log(5)$

55. $\log(0.2)$
$-\log(5)$

56. $\log\left(\frac{9}{25}\right)$
$2\log(3) - 2 \cdot \log(5)$

Rewrite each expression as a sum or a difference of multiples of logarithms. See Example 6.

57. $\log(xyz)$ $\log(x) + \log(y) + \log(z)$

58. $\log(3y)$ $\log(3) + \log(y)$

59. $\log_2(8x)$ $3 + \log_2(x)$ **60.** $\log_2(16y)$ $4 + \log_2(y)$

61. $\ln\left(\frac{x}{y}\right)$ $\ln(x) - \ln(y)$ **62.** $\ln\left(\frac{z}{3}\right)$ $\ln(z) - \ln(3)$

63. $\log(10x^2)$ $1 + 2 \cdot \log(x)$

64. $\log(100\sqrt{x})$ $2 + \frac{1}{2}\log(x)$

65. $\log_5\left(\frac{(x - 3)^2}{\sqrt{w}}\right)$ $2\log_5(x - 3) - \frac{1}{2}\log_5(w)$

66. $\log_3\left(\frac{(y + 6)^3}{y - 5}\right)$ $3\log_3(y + 6) - \log_3(y - 5)$

67. $\ln\left(\frac{yz\sqrt{x}}{w}\right)$ $\ln(y) + \ln(z) + \frac{1}{2}\ln(x) - \ln(w)$

68. $\ln\left(\frac{(x - 1)\sqrt{w}}{x^3}\right)$ $\ln(x - 1) + \frac{1}{2}\ln(w) - 3\ln(x)$

Rewrite each expression as a single logarithm. See Example 7.

69. $\log(x) + \log(x - 1)$ $\log(x^2 - x)$

70. $\log_2(x - 2) + \log_2(5)$ $\log_2(5x - 10)$

71. $\ln(3x - 6) - \ln(x - 2)$ $\ln(3)$

72. $\log_3(x^2 - 1) - \log_3(x - 1)$ $\log_3(x + 1)$

73. $\ln(x) - \ln(w) + \ln(z)$ $\ln\left(\frac{xz}{w}\right)$

74. $\ln(x) - \ln(3) - \ln(7)$ $\ln\left(\frac{x}{21}\right)$

75. $3 \cdot \ln(y) + 2 \cdot \ln(x) - \ln(w)$ $\ln\left(\frac{x^2y^3}{w}\right)$

76. $5 \cdot \ln(r) + 3 \cdot \ln(t) - 4 \cdot \ln(s)$ $\ln\left(\frac{r^5t^3}{s^4}\right)$

77. $\frac{1}{2}\log(x - 3) - \frac{2}{3}\log(x + 1)$ $\log\left(\frac{(x - 3)^{1/2}}{(x + 1)^{2/3}}\right)$

78. $\frac{1}{2}\log(y - 4) + \frac{1}{2}\log(y + 4)$ $\log(\sqrt{y^2 - 16})$

79. $\frac{2}{3}\log_2(x - 1) - \frac{1}{4}\log_2(x + 2)$ $\log_2\left(\frac{(x - 1)^{2/3}}{(x + 2)^{1/4}}\right)$

80. $\frac{1}{2}\log_3(y + 3) + 6 \cdot \log_3(y)$ $\log_3\left(y^6\sqrt{y + 3}\right)$

Determine whether each equation is true or false.

81. $\log(56) = \log(7) \cdot \log(8)$
False

82. $\log\left(\frac{5}{9}\right) = \frac{\log(5)}{\log(9)}$
False

83. $\log_2(4^2) = (\log_2(4))^2$
True

84. $\ln(4^2) = (\ln(4))^2$
False

85. $\ln(25) = 2 \cdot \ln(5)$
True

86. $\ln(3e) = 1 + \ln(3)$
True

87. $\frac{\log_2(64)}{\log_2(8)} = \log_2(8)$
False

88. $\frac{\log_2(16)}{\log_2(4)} = \log_2(4)$
True

89. $\log\left(\frac{1}{3}\right) = -\log(3)$
True

90. $\log_2(8 \cdot 2^{59}) = 62$
True

91. $\log_2(16^5) = 20$
True

92. $\log_2\left(\frac{5}{2}\right) = \log_2(5) - 1$
True

93. $\log(10^3) = 3$
True

94. $\log_3(3^7) = 7$
True

95. $\log(100 + 3) = 2 + \log(3)$
False

96. $\frac{\log_7(32)}{\log_7(8)} = \frac{5}{3}$
True

Solve each problem.

97. *Growth rate.* The annual growth rate for continuous growth is given by

$$r = \frac{\ln(A) - \ln(P)}{t},$$

where P is the initial investment and A is the amount after t years.
a) Rewrite the formula using a single logarithm.
b) In 1998 a share of Microsoft stock was worth 27 times what it was worth in 1990. What was the annual growth rate for that period?
  a) $r = \ln((A/P)^{1/t})$ **b)** 41.2%

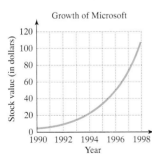

Growth of Microsoft

FIGURE FOR EXERCISE 97

98. *Diversity index.* The U.S.G.S. measures the quality of a water sample by using the diversity index d, given by

$$d = -[p_1 \cdot \log_2(p_1) + p_2 \cdot \log_2(p_2) + \cdots + p_n \cdot \log_2(p_n)],$$

where n is the number of different taxons (biological classifications) represented in the sample and p_1 through p_n are the percentages of organisms in each of the n taxons. The value of d ranges from 0 when all organisms in the water sample are the same to some positive number when all organisms in the sample are different. If two-thirds of the organisms in a water sample are in one taxon and one-third of the organisms are in a second taxon, then $n = 2$ and

$$d = -\left[\frac{2}{3}\log_2\left(\frac{2}{3}\right) + \frac{1}{3}\log_2\left(\frac{1}{3}\right)\right].$$

Use the properties of logarithms to write the expression on the right-hand side as $\log_2\left(\frac{3\sqrt[3]{2}}{2}\right)$. (In Section 10.4 you will learn how to evaluate a base-2 logarithm using a calculator.)

GETTING MORE INVOLVED

99. *Discussion.* Which of the following equations is an identity? Explain.
a) $\ln(3x) = \ln(3) \cdot \ln(x)$
b) $\ln(3x) = \ln(3) + \ln(x)$
c) $\ln(3x) = 3 \cdot \ln(x)$
d) $\ln(3x) = \ln(x^3)$
  b

100. *Discussion.* Which of the following expressions is not equal to $\log(5^{2/3})$? Explain.
a) $\frac{2}{3}\log(5)$ **b)** $\frac{\log(5) + \log(5)}{3}$
c) $(\log(5))^{2/3}$ **d)** $\frac{1}{3}\log(25)$
  c

GRAPHING CALCULATOR EXERCISES

101. Graph the functions $y_1 = \ln(\sqrt{x})$ and $y_2 = 0.5 \cdot \ln(x)$ on the same screen. Explain your results.
The graphs are the same because
$$\ln(\sqrt{x}) = \ln(x^{1/2}) = \frac{1}{2}\ln(x).$$

102. Graph the functions $y_1 = \log(x)$, $y_2 = \log(10x)$, $y_3 = \log(100x)$, and $y_4 = \log(1000x)$ using the viewing window $-2 \le x \le 5$ and $-2 \le y \le 5$. Why do these curves appear as they do?
Because $\log(10x) = 1 + \log(x)$, $\log(100x) = 2 + \log(x)$, and $\log(1000x) = 3 + \log(x)$; the graphs lie 1, 2, and 3 units above $y = \log(x)$.

103. Graph the function $y = \log(e^x)$. Explain why the graph is a straight line. What is its slope?
The graph is a straight line because $\log(e^x) = x \log(e) \approx 0.434x$. The slope is $\log(e)$ or approximately 0.434.

10.4 SOLVING EQUATIONS

We solved some equations involving exponents and logarithms in Sections 10.1 and 10.2. In this section we use the properties of exponents and logarithms to solve more complex equations.

Logarithmic Equations

The main tool that we have for solving logarithmic equations is the definition of logarithms: $y = \log_a(x)$ if and only if $a^y = x$. We can use the definition to rewrite any equation that has only one logarithm as an equivalent exponential equation.

EXAMPLE 1

A logarithmic equation with only one logarithm

Solve $\log(x + 3) = 2$.

Solution

Write the equivalent exponential equation:

$$\log(x + 3) = 2 \qquad \text{Original equation}$$
$$10^2 = x + 3 \qquad \text{Definition of logarithm}$$
$$100 = x + 3$$
$$97 = x$$

Check: $\log(97 + 3) = \log(100) = 2$. The solution set is $\{97\}$.

In the next example we use the product rule for logarithms to write a sum of two logarithms as a single logarithm.

EXAMPLE 2

Using the product rule to solve an equation

Solve $\log_2(x + 3) + \log_2(x - 3) = 4$.

Solution

Rewrite the sum of the logarithms as the logarithm of a product:

$$\log_2(x + 3) + \log_2(x - 3) = 4 \qquad \text{Original equation}$$
$$\log_2[(x + 3)(x - 3)] = 4 \qquad \text{Product rule}$$
$$\log_2[x^2 - 9] = 4 \qquad \text{Multiply the binomials.}$$
$$x^2 - 9 = 2^4 \qquad \text{Definition of logarithm}$$
$$x^2 - 9 = 16$$
$$x^2 = 25$$
$$x = \pm 5 \qquad \text{Even-root property}$$

To check, first let $x = -5$ in the original equation:

$$\log_2(-5 + 3) + \log_2(-5 - 3) = 4$$
$$\log_2(-2) + \log_2(-8) = 4 \qquad \text{Incorrect}$$

Because the domain of any logarithm function is the set of positive real numbers, these logarithms are undefined. Now check $x = 5$ in the original equation:

$$\log_2(5 + 3) + \log_2(5 - 3) = 4$$
$$\log_2(8) + \log_2(2) = 4$$
$$3 + 1 = 4 \quad \text{Correct}$$

The solution set is $\{5\}$.

C A U T I O N   Always check that your solutions to a logarithmic equation do not produce undefined logarithms in the original equation.

E X A M P L E   3

Using the one-to-one property of logarithms
Solve $\log(x) + \log(x - 1) = \log(8x - 12) - \log(2)$.

Solution

Apply the product rule to the left-hand side and the quotient rule to the right-hand side to get a single logarithm on each side:

$$\log(x) + \log(x - 1) = \log(8x - 12) - \log(2).$$
$$\log[x(x - 1)] = \log\left(\frac{8x - 12}{2}\right) \quad \text{Product rule; quotient rule}$$
$$\log(x^2 - x) = \log(4x - 6) \quad \text{Simplify.}$$
$$x^2 - x = 4x - 6 \quad \text{One-to-one property of logarithms}$$
$$x^2 - 5x + 6 = 0$$
$$(x - 2)(x - 3) = 0$$
$$x - 2 = 0 \quad \text{or} \quad x - 3 = 0$$
$$x = 2 \quad \text{or} \quad x = 3$$

Neither $x = 2$ nor $x = 3$ produces undefined terms in the original equation. Use a calculator to check that they both satisfy the original equation. The solution set is $\{2, 3\}$.

C A U T I O N   The product rule, quotient rule, and power rule do not eliminate logarithms from equations. To do so, we use the definition to change $y = \log_a(x)$ into $a^y = x$ or the one-to-one property to change $\log_a(m) = \log_a(n)$ into $m = n$.

Exponential Equations

If an equation has a single exponential expression, we can write the equivalent logarithmic equation.

E X A M P L E   4

A single exponential expression
Find the exact solution to $2^x = 10$.

Solution

The equivalent logarithmic equation is

$$x = \log_2(10).$$

The solution set is $\{\log_2(10)\}$. The number $\log_2(10)$ is the exact solution to the equation. Later in this section you will learn how to use the base-change formula to find an approximate value for an expression of this type.

---

c a l c u l a t o r

▲	TOTAL	PART	%TOTAL	▶*
▼	4	5	6	×

c l o s e - u p

Graph
$$y_1 = \log(x) + \log(x - 1)$$
and
$$y_2 = \log(8x - 12) - \log(2)$$
to see the two solutions to the equation in Example 3.

Intersection
X=3   Y=.77815125

In Section 10.1 we solved some exponential equations by writing each side as a power of the same base and then applying the one-to-one property of exponential functions. We review that method in the next example.

E X A M P L E 5

Powers of the same base

Solve $2^{(x^2)} = 4^{3x-4}$.

Solution

We can write each side as a power of the same base:

$$2^{(x^2)} = (2^2)^{3x-4} \qquad \text{Because } 4 = 2^2$$
$$2^{(x^2)} = 2^{6x-8} \qquad \text{Power of a power rule}$$
$$x^2 = 6x - 8 \qquad \text{One-to-one property of exponential functions}$$

$$x^2 - 6x + 8 = 0$$
$$(x - 4)(x - 2) = 0$$
$$x - 4 = 0 \quad \text{or} \quad x - 2 = 0$$
$$x = 4 \quad \text{or} \quad x = 2$$

Check $x = 2$ and $x = 4$ in the original equation. The solution set is $\{2, 4\}$. ■

For some exponential equations we cannot write each side as a power of the same base as we did in Example 5. In this case we take a logarithm of each side and simplify, using the rules for logarithms.

E X A M P L E 6

Exponential equation with two different bases

Find the exact and approximate solution to $2^{x-1} = 3^x$.

Solution

We first take the base-10 logarithm of each side:

$$2^{x-1} = 3^x \qquad \text{Original equation}$$
$$\log(2^{x-1}) = \log(3^x) \qquad \text{Take log of each side.}$$
$$(x - 1)\log(2) = x \cdot \log(3) \qquad \text{Power rule}$$
$$x \cdot \log(2) - \log(2) = x \cdot \log(3) \qquad \text{Distributive property}$$
$$x \cdot \log(2) - x \cdot \log(3) = \log(2) \qquad \text{Get all } x\text{-terms on one side.}$$
$$x[\log(2) - \log(3)] = \log(2) \qquad \text{Factor out } x.$$
$$x = \frac{\log(2)}{\log(2) - \log(3)} \qquad \text{Exact solution}$$
$$x \approx -1.7095 \qquad \text{Approximate solution}$$

You can use a calculator to check -1.7095 in the original equation. As the first step of the solution, we could have taken the logarithm of each side using any base. We chose base 10 so that we could use a calculator to find an approximate solution from the exact solution. ■

Changing the Base

Scientific calculators have an x^y key for computing any power of any base, in addition to the function keys for computing 10^x and e^x. For logarithms we have the keys ln and log, but there are no function keys for logarithms using other bases. To solve this problem, we develop a formula for expressing a base-a logarithm in terms of base-b logarithms.

If $y = \log_a(M)$, then $a^y = M$. Now we solve $a^y = M$ for y, using base-b logarithms:

$$a^y = M$$

$$\log_b(a^y) = \log_b(M) \quad \text{Take the base-}b\text{ logarithm of each side.}$$

$$y \cdot \log_b(a) = \log_b(M) \quad \text{Power rule}$$

$$y = \frac{\log_b(M)}{\log_b(a)} \quad \text{Divide each side by } \log_b(a).$$

Because $y = \log_a(M)$, we can write $\log_a(M)$ in terms of base-b logarithms.

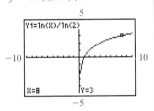

The base-change formula allows you to graph logarithmic functions with bases other than e and 10. For example, to graph $y = \log_2(x)$, graph $y = \ln(x)/\ln(2)$.

Base-Change Formula

If a and b are positive numbers not equal to 1 and M is positive, then

$$\log_a(M) = \frac{\log_b(M)}{\log_b(a)}.$$

In words, we take the logarithm with the new base and divide by the logarithm of the old base. The most important use of the base-change formula is to find base-a logarithms using a calculator. If the new base is 10 or e, then

$$\log_a(M) = \frac{\log(M)}{\log(a)} = \frac{\ln(M)}{\ln(a)}.$$

E X A M P L E   7

Using the base-change formula

Find $\log_7(99)$ to four decimal places.

Solution

Use the base-change formula with $a = 7$ and $b = 10$:

$$\log_7(99) = \frac{\log(99)}{\log(7)} \approx 2.3614$$

Check by finding $7^{2.3614}$ with your calculator. Note that we also have

$$\log_7(99) = \frac{\ln(99)}{\ln(7)} \approx 2.3614.$$

■

Strategy for Solving Equations

There is no formula that will solve every equation in this section. However, we have a strategy for solving exponential and logarithmic equations. The list on the next page summarizes the ideas that we need for solving these equations.

Solving Exponential and Logarithmic Equations

1. If the equation has a single logarithm or a single exponential expression, rewrite the equation using the definition $y = \log_a(x)$ if and only if $a^y = x$.
2. Use the properties of logarithms to combine logarithms as much as possible.
3. Use the one-to-one properties:
   a) If $\log_a(m) = \log_a(n)$, then $m = n$.
   b) If $a^m = a^n$, then $m = n$.
4. To get an approximate solution of an exponential equation, take the common or natural logarithm of each side of the equation.

Applications

In compound interest problems, logarithms are used to find the time it takes for money to grow to a specified amount.

E X A M P L E   8

Finding the time

If $500 is deposited into an account paying 8% compounded quarterly, then in how many quarters will the account have $1000 in it?

Solution

helpful hint

When we get $2 = (1.02)^n$, we can use the definition of log as in Example 8 or take the natural log of each side:

$$\ln(2) = \ln(1.02^n)$$
$$\ln(2) = n \cdot \ln(1.02)$$
$$n = \frac{\ln(2)}{\ln(1.02)}$$

In either way we arrive at the same solution.

We use the compound interest formula $A = P(1 + i)^n$ with a principal of $500, an amount of $1000, and an interest rate of 2% each quarter:

$$A = P(1 + i)^n$$
$$1000 = 500(1.02)^n \quad \text{Substitute.}$$
$$2 = (1.02)^n \quad \text{Divide each side by 500.}$$
$$n = \log_{1.02}(2) \quad \text{Definition of logarithm}$$
$$= \frac{\ln(2)}{\ln(1.02)} \quad \text{Base-change formula}$$
$$\approx 35.0028 \quad \text{Use a calculator.}$$

It takes approximately 35 quarters, or 8 years and 9 months, for the initial investment to be worth $1000. Note that we could also solve $2 = (1.02)^n$ by taking the common or natural logarithm of each side. Try it. ∎

WARM-UPS

True or false? Explain.

1. If $\log(x - 2) + \log(x + 2) = 7$, then $\log(x^2 - 4) = 7$.　True
2. If $\log(3x + 7) = \log(5x - 8)$, then $3x + 7 = 5x - 8$.　True
3. If $e^{x-6} = e^{x^2-5x}$, then $x - 6 = x^2 - 5x$.　True
4. If $2^{3x-1} = 3^{5x-4}$, then $3x - 1 = 5x - 4$.　False
5. If $\log_2(x^2 - 3x + 5) = 3$, then $x^2 - 3x + 5 = 8$.　True
6. If $2^{2x-1} = 3$, then $2x - 1 = \log_2(3)$.　True
7. If $5^x = 23$, then $x \cdot \ln(5) = \ln(23)$.　True
8. $\log_3(5) = \dfrac{\ln(3)}{\ln(5)}$　False
9. $\dfrac{\ln(2)}{\ln(6)} = \dfrac{\log(2)}{\log(6)}$　True
10. $\log(5) = \ln(5)$　False

10.4   EXERCISES

Reading and Writing   *After reading this section, write out the answers to these questions. Use complete sentences.*

1. What exponential equation is equivalent to $\log_a(x) = y$?
   The exponential equation $a^y = x$ is equivalent to $\log_a(x) = y$.

2. How can you find a logarithm with a base other than 10 or e using a calculator?
   According to the base-change formula,
   $\log_a(x) = \ln(x)/\ln(a)$

Solve each equation. See Examples 1 and 2.

3. $\log_2(x + 1) = 3$ $\{7\}$

4. $\log_3(x^2) = 4$ $\{\pm 9\}$

5. $\log(x) + \log(5) = 1$ $\{2\}$

6. $\ln(x) + \ln(3) = 0$ $\left\{\dfrac{1}{3}\right\}$

7. $\log_2(x - 1) + \log_2(x + 1) = 3$ $\{3\}$

8. $\log_3(x - 4) + \log_3(x + 4) = 2$ $\{5\}$

9. $\log_2(x - 1) - \log_2(x + 2) = 2$ $\varnothing$

10. $\log_4(8x) - \log_4(x - 1) = 2$ $\{2\}$

11. $\log_2(x - 4) + \log_2(x + 2) = 4$ $\{6\}$

12. $\log_6(x + 6) + \log_6(x - 3) = 2$ $\{6\}$

Solve each equation. See Example 3.

13. $\ln(x) + \ln(x + 5) = \ln(x + 1) + \ln(x + 3)$ $\{3\}$

14. $\log(x) + \log(x + 5) = 2 \cdot \log(x + 2)$ $\{4\}$

15. $\log(x + 3) + \log(x + 4) = \log(x^3 + 13x^2) - \log(x)$ $\{2\}$

16. $\log(x^2 - 1) - \log(x - 1) = \log(6)$ $\{5\}$

17. $2 \cdot \log(x) = \log(20 - x)$ $\{4\}$

18. $2 \cdot \log(x) + \log(3) = \log(2 - 5x)$ $\left\{\dfrac{1}{3}\right\}$

Solve each equation. See Examples 4 and 5.

19. $3^x = 7$
    $\{\log_3(7)\}$

20. $2^{x-1} = 5$
    $\{1 + \log_2(5)\}$

21. $2^{3x+4} = 4^{x-1}$
    $\{-6\}$

22. $9^{2x-1} = 27^{1/2}$
    $\left\{\dfrac{7}{8}\right\}$

23. $\left(\dfrac{1}{3}\right)^x = 3^{1+x}$
    $\left\{-\dfrac{1}{2}\right\}$

24. $4^{3x} = \left(\dfrac{1}{2}\right)^{1-x}$
    $\left\{-\dfrac{1}{5}\right\}$

 Find the exact solution and approximate solution to each equation. See Example 6.

25. $2^x = 3^{x+5}$ $\dfrac{5 \ln(3)}{\ln(2) - \ln(3)}$, -13.548

26. $e^x = 10^x$ 0

27. $5^{x+2} = 10^{x-4}$ $\dfrac{4 + 2 \log(5)}{1 - \log(5)}$, 17.932

28. $3^{2x} = 6^{x+1}$ $\dfrac{\ln(6)}{\ln(9) - \ln(6)}$, 4.419

29. $8^x = 9^{x-1}$ $\dfrac{\ln(9)}{\ln(9) - \ln(8)}$, 18.655

30. $5^{x+1} = 8^{x-1}$ $\dfrac{\ln(5) + \ln(8)}{\ln(8) - \ln(5)}$, 7.849

 Use the base-change formula to find each logarithm to four decimal places. See Example 7.

31. $\log_2(3)$ 1.5850

32. $\log_3(5)$ 1.4650

33. $\log_3\left(\dfrac{1}{2}\right)$ -0.6309

34. $\log_5(2.56)$ 0.5841

35. $\log_{1/2}(4.6)$ -2.2016

36. $\log_{1/3}(3.5)$ -1.1403

37. $\log_{0.1}(0.03)$ 1.5229

38. $\log_{0.2}(1.06)$ -0.0362

 For each equation, find the exact solution and an approximate solution when appropriate. Round approximate answers to three decimal places.

39. $x \cdot \ln(2) = \ln(7)$ $\dfrac{\ln(7)}{\ln(2)}$, 2.807

40. $x \cdot \log(3) = \log(5)$ $\dfrac{\log(5)}{\log(3)}$, 1.465

41. $3x - x \cdot \ln(2) = 1$ $\dfrac{1}{3 - \ln(2)}$, 0.433

42. $2x + x \cdot \log(5) = \log(7)$ $\dfrac{\log(7)}{2 + \log(5)}$, 0.313

43. $3^x = 5$ $\dfrac{\ln(5)}{\ln(3)}$, 1.465

44. $2^x = \dfrac{1}{3}$ $-\dfrac{\ln(3)}{\ln(2)}$, -1.585

45. $2^{x-1} = 9$ $1 + \dfrac{\ln(9)}{\ln(2)}$, 4.170

46. $10^{x-2} = 6$ $2 + \log(6)$, 2.778

47. $3^x = 20$ $\log_3(20)$, 2.727

48. $2^x = 128$ 7

49. $\log_3(x) + \log_3(5) = 1$ $\dfrac{3}{5}$

50. $\log(x) - \log(3) = \log(6)$ 18

51. $8^x = 2^{x+1}$ $\dfrac{1}{2}$

52. $2^x = 5^{x+1}$ $\dfrac{\ln(5)}{\ln(2) - \ln(5)}$, -1.756

 In Exercises 53–64, solve each problem. See Example 8.

53. **Finding the time.** How many months does it take for $1,000 to grow to $1,500 in an account paying 12% compounded monthly?   41 months

54. Finding the time. How many years does it take for $25 to grow to $100 in an account paying 8% compounded annually?
18 years

55. Going with the flow. The flow [in cubic feet per second (ft^3/sec)] of the Tangipahoa River at Robert, Louisiana, is a linear function of its depth (in feet) when the river is at normal levels (Exercise 87, Section 3.3). When flood stages are included, the flow y is better modeled by the exponential function $y = 114.308e^{0.265x}$, where x is the depth. Find the flow when the depth is 15.8 feet.
7,524 ft^3/sec

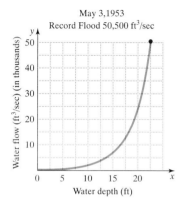
May 3, 1953
Record Flood 50,500 ft^3/sec

FIGURE FOR EXERCISES 55 AND 56

56. Record flood. Use the formula of the previous exercise to find the depth of the Tangipahoa River at Robert, Louisiana, on May 3, 1953 when the flow reached an all-time record of 50,500 ft^3/sec (U.S.G.S., waterdata.usgs.gov).
22.98 ft

57. Above the poverty level. In a certain country the number of people above the poverty level is currently 28 million and growing 5% annually. Assuming the population is growing continuously, the population P (in millions), t years from now, is determined by the formula $P = 28e^{0.05t}$. In how many years will there be 40 million people above the poverty level?
7.1 years

58. Below the poverty level. In the same country as in Exercise 57, the number of people below the poverty level is currently 20 million and growing 7% annually. This population (in millions), t years from now, is determined by the formula $P = 20e^{0.07t}$. In how many years will there be 40 million people below the poverty level?
9.9 years

59. Fifty-fifty. For this exercise, use the information given in Exercises 57 and 58. In how many years will the number of people above the poverty level equal the number of people below the poverty level?
16.8 years

60. Golden years. In a certain country there are currently 100 million workers and 40 million retired people. The population of workers is decreasing according to the formula $W = 100e^{-0.01t}$, where t is in years and W is in millions. The population of retired people is increasing according to the formula $R = 40e^{0.09t}$, where t is in years and R is in millions. In how many years will the number of workers equal the number of retired people?
9.2 years

61. Ions for breakfast. Orange juice has a pH of 3.7. What is the hydrogen ion concentration of orange juice? (See Exercises 83–86 of Section 10.2.)
2.0×10^{-4}

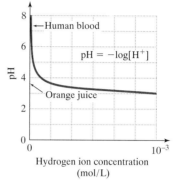

FIGURE FOR EXERCISES 61 AND 62

62. Ions in your veins. Normal human blood has a pH of 7.4. What is the hydrogen ion concentration of normal human blood?
4.0×10^{-8}

63. Diversity index. In Exercise 98 of Section 10.3 we expressed the diversity index d for a certain water sample as

$$d = \log_2\left(\frac{3\sqrt[3]{2}}{2}\right).$$

Use the base-change formula and a calculator to calculate the value of d. Round the answer to four decimal places.
0.9183

64. Quality water. In a certain water sample, 5% of the organisms are in one taxon, 10% are in a second taxon, 20% are in a third taxon, 15% are in a fourth taxon, 23% are in a fifth taxon, and the rest are in a sixth taxon. Use the formula given in Exercise 98 of Section 10.3 with $n = 6$ to find the diversity index of the water sample.
2.42

GETTING MORE INVOLVED

65. Exploration. Logarithms were designed to solve equations that have variables in the exponents, but logarithms can be used to solve certain polynomial equations. Consider the following example:

$$x^5 = 88$$

$$5 \cdot \ln(x) = \ln(88)$$

$$\ln(x) = \frac{\ln(88)}{5} \approx 0.895467$$

$$x = e^{0.895467} \approx 2.4485$$

Solve $x^3 = 12$ by taking the natural logarithm of each side. Round the approximate solution to four decimal places. Solve $x^3 = 12$ without using logarithms and compare with your previous answer.

2.2894

 **66.** *Discussion.* Determine whether each logarithm is positive or negative without using a calculator. Explain your answers.

a) $\log_2(0.45)$ b) $\ln(1.01)$
c) $\log_{1/2}(4.3)$ d) $\log_{1/3}(0.44)$
    a) negative   b) positive
    c) negative   d) positive

GRAPHING CALCULATOR EXERCISES

67. Graph $y_1 = 2^x$ and $y_2 = 3^{x-1}$ on the same coordinate system. Use the intersect feature of your calculator to find the point of intersection of the two curves. Round to two decimal places.   (2.71, 6.54)

68. Bob invested $1,000 at 6% compounded continuously. At the same time Paula invested $1,200 at 5% compounded monthly. Write two functions that give the amounts of Bob's and Paula's investments after x years. Graph these functions on a graphing calculator. Use the intersect feature of your graphing calculator to find the approximate value of x for which the investments are equal in value.
$y = 1000e^{0.06x}$, $y = 1200(1 + 0.05/12)^{12x}$, 18.0 years

69. Graph the functions $y_1 = \log_2(x)$ and $y_2 = 3^{x-4}$ on the same coordinate system and use the intersect feature to find the points of intersection of the curves. Round to two decimal places. (*Hint:* To graph $y = \log_2(x)$, use the base-change formula to write the function as $y = \ln(x)/\ln(2)$.)
(1.03, 0.04), (4.73, 2.24)

COLLABORATIVE ACTIVITIES

In How Much Space Could We Live?

The formula for population growth is

$$P(t) = P_0 e^{kt},$$

where $P(t)$ is the population after t years, P_0 is the population initially, k is the growth rate per year, and t is the number of years elapsed. We will find out how long it would take to cover the earth completely if the human population were growing exponentially. For all of your calculations, round your answers to one decimal place unless directed otherwise.

1. The population of the world in 1994 was approximately 5.5×10^9 people. If the population was 2.75×10^9 people in 1964, then what is the current growth rate (express as a percent)? Round your answer to the nearest tenth of a percent.

The earth has a total surface area of approximately 5.1×10^{14} square meters. Seventy percent of this surface area is rock, ice, sand, and open ocean. Another 8% of the total surface area is tundra, lakes and streams, continental shelves, algal beds and reefs, and estuaries. We will consider the remaining area to be suitable for growing food and for living space.

2. Determine the surface area available for growing food and for living space.

Grouping: Two students per group
Topic: Exponential and logarithmic functions

3. If each person needs 100 square meters of the earth's surface for living space and growing food, then in how many years after 1994 will the livable surface of the earth be used up? (Use the rate and the 1994 population from Question 1 and the surface area from Question 2.)

4. With your partner, think about the following questions and report your conclusions.

- Does 100 square meters per person for living space and growing food seem reasonable? Take into account that many people live in tall apartment buildings and how that translates into surface area used per person.

- How much room do you think it takes to grow animals for food (cows, chickens, pigs, etc.)? What about grains, vegetables, nuts, and fruit? Would food grow as well in desert areas, mountainous areas, or jungle areas?

- Would there be any space left for wild animals or natural plant life? Would there be any space left for shopping malls, movie theaters, concert halls, factories, office buildings, or parking lots?

WRAP-UP CHAPTER 10

SUMMARY

Exponential and Logarithmic Functions		Examples
Exponential function	A function of the form $f(x) = a^x$ for $a > 0$ and $a \neq 1$	$f(x) = 3^x$
Logarithm function	A function of the form $f(x) = \log_a(x)$ for $a > 0$ and $a \neq 1$	$f(x) = \log_2(x)$
	$y = \log_a(x)$ if and only if $a^y = x$.	$\log_3(8) = x \leftrightarrow 3^x = 8$
Common logarithm	Base-10: $f(x) = \log(x)$	$\log(100) = 2$ because $100 = 10^2$.
Natural logarithm	Base-e: $f(x) = \ln(x)$ $e \approx 2.718$	$\ln(e) = 1$ because $e^1 = e$.
Inverse functions	$f(x) = a^x$ and $g(x) = \log_a(x)$ are inverse functions.	If $f(x) = e^x$, then $f^{-1}(x) = \ln(x)$.

Properties		Examples
M, N, and a are positive numbers with $a \neq 1$.	$\log_a(a) = 1 \qquad \log_a(1) = 0$	$\log_5(5) = 1, \log_5(1) = 0$
Inverse properties	$\log_a(a^M) = M \qquad a^{\log_a(M)} = M$	$\log(10^7) = 7, e^{\ln(3.4)} = 3.4$
Product rule	$\log_a(MN) = \log_a(M) + \log_a(N)$	$\ln(3x) = \ln(3) + \ln(x)$
Quotient rule	$\log_a\left(\dfrac{M}{N}\right) = \log_a(M) - \log_a(N)$	$\ln\left(\dfrac{2}{3}\right) = \ln(2) - \ln(3)$
	$\log_a\left(\dfrac{1}{N}\right) = -\log_a(N)$	$\ln\left(\dfrac{1}{3}\right) = -\ln(3)$
Power rule	$\log_a(M^N) = N \cdot \log_a(M)$	$\log(x^3) = 3 \cdot \log(x)$
Base-change formula	$\log_a(M) = \dfrac{\log_b(M)}{\log_b(a)}$	$\log_3(5) = \dfrac{\ln(5)}{\ln(3)}$

Equations Involving Logarithms and Exponents		Examples
Strategy	1. If there is a single logarithm or a single exponential expression, rewrite the equation using the definition of logarithms: $y = \log_a(x)$ if and only if $a^y = x$.	$2^x = 3$ and $x = \log_2(3)$ are equivalent.

2. Use the properties of logarithms to combine logarithms as much as possible.

3. Use the one-to-one properties:
   a) If $\log_a(m) = \log_a(n)$, then $m = n$.
   b) If $a^m = a^n$, then $m = n$.

4. To get an approximate solution, take the common or natural logarithm of each side of an exponential equation.

$\log(x) + \log(x - 3) = 1$
$\log(x^2 - 3x) = 1$

$\ln(x) = \ln(5 - x),$
$x = 5 - x$
$2^{3x} = 2^{5x-7}, \ 3x = 5x - 7$
$2^x = 3, \ \ln(2^x) = \ln(3)$
$x \cdot \ln(2) = \ln(3)$
$x = \dfrac{\ln(3)}{\ln(2)}$

ENRICHING YOUR MATHEMATICAL WORD POWER

For each mathematical term, choose the correct meaning.

1. **exponential function**
   a. $f(x) = a^x$ where $a > 0$ and $a \neq 1$
   b. $f(x) = ax^2$ where $a \neq 0$
   c. $f(x) = ax + b$ where $a \neq 0$
   d. $f(x) = x^n$ where n is an integer   a

2. **common base**
   a. base 2
   b. base e
   c. base π
   d. base 10   d

3. **natural base**
   a. base 2
   b. base e
   c. base π
   d. base 10   b

4. **domain**
   a. the range
   b. the set of second coordinates of a relation
   c. the independent variable
   d. the set of first coordinates of a relation   d

5. **compound interest**
   a. simple interest
   b. $A = Prt$
   c. an irrational interest rate
   d. interest is periodically paid into the account and the interest earns interest   d

6. **continuous compounding**
   a. compound interest
   b. using $A = Pe^{rt}$ to compute the amount
   c. frequent compounding
   d. using $A = P(1 + i)^n$ to compute the amount   b

7. **base-a logarithm of x**
   a. the exponent that is used on the base a to obtain x
   b. the exponent that is used on x to obtain a
   c. the power of 10 that produces x
   d. the power of e that produces a a

8. **base-a logarithm function**
   a. $f(x) = a^x$ where $a > 0$ and $a \neq 1$
   b. $f(x) = \log_a(x)$ where $a > 0$ and $a \neq 1$
   c. $f(x) = \log_x(a)$ where $a > 0$ and $a \neq 1$
   d. $f(x) = \log(x)$ where $x > 0$ b

9. **common logarithm**
   a. $\log_2(x)$
   b. $\log(x)$
   c. $\ln(x)$
   d. $\log_3(x)$ b

10. **natural logarithm**
    a. $\log_2(x)$
    b. $\log(x)$
    c. $\ln(x)$
    d. $\log_3(x)$ c

REVIEW EXERCISES

10.1 *Use* $f(x) = 5^x$, $g(x) = 10^{x-1}$, *and* $h(x) = \left(\frac{1}{4}\right)^x$ *for Exercises 1–28. Find the following.*

1. $f(-2)$ $\dfrac{1}{25}$

2. $f(0)$ 1

3. $f(3)$ 125

4. $f(4)$ 625

5. $g(1)$ 1

6. $g(-1)$ $\dfrac{1}{100}$

7. $g(0)$ $\dfrac{1}{10}$

8. $g(3)$ 100

9. $h(-1)$ 4

10. $h(2)$ $\dfrac{1}{16}$

11. $h\left(\dfrac{1}{2}\right)$ $\dfrac{1}{2}$

12. $h\left(-\dfrac{1}{2}\right)$ 2

Find x in each case.

13. $f(x) = 25$ 2

14. $f(x) = -\dfrac{1}{125}$ No solution

15. $g(x) = 1000$ 4

16. $g(x) = 0.001$ -2

17. $h(x) = 32$ $-\dfrac{5}{2}$

18. $h(x) = 8$ $-\dfrac{3}{2}$

19. $h(x) = \dfrac{1}{16}$ 2

20. $h(x) = 1$ 0

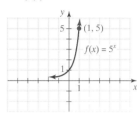 *Find the following.*

21. $f(1.34)$ 8.6421

22. $f(-3.6)$ 0.00305

23. $g(3.25)$ 177.828

24. $g(4.87)$ 7413.102

25. $h(2.82)$ 0.02005

26. $h(\pi)$ 0.01284

27. $h(\sqrt{2})$ 0.1408

28. $h\left(\dfrac{1}{3}\right)$ 0.6300

Sketch the graph of each function.

29. $f(x) = 5^x$

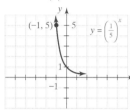

30. $g(x) = e^x$

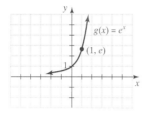

31. $y = \left(\dfrac{1}{5}\right)^x$

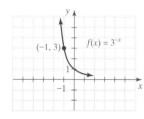

32. $y = e^{-x}$

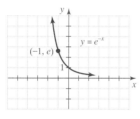

33. $f(x) = 3^{-x}$

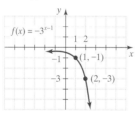

34. $f(x) = -3^{x-1}$

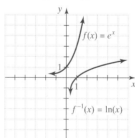

35. $y = 1 + 2^x$

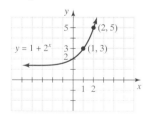

36. $y = 1 - 2^x$

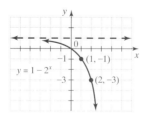

10.2 *Write each exponential equation as a logarithmic equation and each logarithmic equation as an exponential equation.*

37. $10^m = n$ $\log(n) = m$

38. $b = a^5$ $\log_a(b) = 5$

39. $h = \log_k(t)$ $k^h = t$

40. $\log_v(5) = u$ $v^u = 5$

Let $f(x) = \log_2(x)$, $g(x) = \log(x)$, and $h(x) = \log_{1/2}(x)$. Find the following.

41. $f\left(\dfrac{1}{8}\right)$
-3

42. $f(64)$
6

43. $g(0.1)$
-1

44. $g(1)$
0

45. $g(100)$
2

46. $h\left(\dfrac{1}{8}\right)$
3

47. $h(1)$
0

48. $h(4)$
-2

49. x, if $f(x) = 8$
256

50. x, if $g(x) = 3$
1000

51. $f(77)$
6.267

52. $g(88.4)$
1.946

53. $h(33.9)$
-5.083

54. $h(0.05)$
4.322

55. x, if $f(x) = 2.475$
5.560

56. x, if $g(x) = 1.426$
26.669

For each function f, find f^{-1} and sketch the graphs of f and f^{-1} on the same set of axes.

57. $f(x) = 10^x$
$f^{-1}(x) = \log(x)$

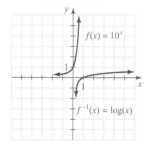

58. $f(x) = \log_8(x)$
$f^{-1}(x) = 8^x$

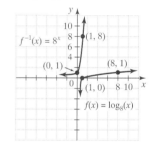

59. $f(x) = e^x$
$f^{-1}(x) = \ln(x)$

60. $f(x) = \log_3(x)$
$f^{-1}(x) = 3^x$

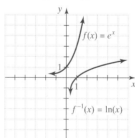

10.3 *Rewrite each expression as a sum or a difference of multiples of logarithms.*

61. $\log(x^2y)$
$2\log(x) + \log(y)$

62. $\log_3(x^2 + 2x)$
$\log_3(x) + \log_3(x + 2)$

63. $\ln(16)$ $4\ln(2)$

64. $\log\left(\dfrac{y}{\sqrt{x}}\right)$ $\log(y) - \dfrac{1}{2}\log(x)$

65. $\log_5\left(\dfrac{1}{x}\right)$

$-\log_5(x)$

66. $\ln\left(\dfrac{xy}{z}\right)$

$\ln(x) + \ln(y) - \ln(z)$

Rewrite each expression as a single logarithm.

67. $\dfrac{1}{2}\log(x + 2) - 2 \cdot \log(x - 1)$ $\log\left(\dfrac{\sqrt{x + 2}}{(x - 1)^2}\right)$

68. $3 \cdot \ln(x) + 2 \cdot \ln(y) - \dfrac{1}{3}\ln(z)$ $\ln\left(\dfrac{x^3 y^2}{\sqrt[3]{z}}\right)$

10.4 *Find the exact solution to each equation.*

69. $\log_2(x) = 8$ $\{256\}$

70. $\log_3(x) = 0.5$ $\{\sqrt{3}\}$

71. $\log_2(8) = x$ $\{3\}$

72. $3^x = 8$ $\{\log_3(8)\}$

73. $x^3 = 8$ $\{2\}$

74. $3^2 = x$ $\{9\}$

75. $\log_x(27) = 3$ $\{3\}$

76. $\log_x(9) = -\dfrac{1}{3}$ $\left\{\dfrac{1}{729}\right\}$

77. $x \cdot \ln(3) - x = \ln(7)$ $\left\{\dfrac{\ln(7)}{\ln(3) - 1}\right\}$

78. $x \cdot \log(8) = x \cdot \log(4) + \log(9)$ $\left\{\dfrac{\log(9)}{\log(2)}\right\}$

79. $3^x = 5^{x-1}$ $\left\{\dfrac{\ln(5)}{\ln(5) - \ln(3)}\right\}$

80. $5^{(2x^2)} = 5^{3-5x}$ $\left\{-3, \dfrac{1}{2}\right\}$

81. $4^{2x} = 2^{x+1}$ $\left\{\dfrac{1}{3}\right\}$

82. $\log(12) = \log(x) + \log(7 - x)$ $\{3, 4\}$

83. $\ln(x + 2) - \ln(x - 10) = \ln(2)$ $\{22\}$

84. $2 \cdot \ln(x + 3) = 3 \cdot \ln(4)$ $\{5\}$

85. $\log(x) - \log(x - 2) = 2$ $\left\{\dfrac{200}{99}\right\}$

86. $\log_2(x) = \log_2(x + 16) - 1$ $\{16\}$

 Use a calculator to find an approximate solution to each of the following. Round your answers to four decimal places.

87. $6^x = 12$ $\{1.3869\}$ **88.** $5^x = 8^{3x+2}$ $\{-0.8985\}$

89. $3^{x+1} = 5$ $\{0.4650\}$ **90.** $\log_3(x) = 2.634$ $\{18.0608\}$

MISCELLANEOUS

 Solve each problem.

91. *Compounding annually.* What does $10,000 invested at 11.5% compounded annually amount to after 15 years?
$51,182.68

92. *Doubling time.* How many years does it take for an investment to double at 6.5% compounded annually?
11.007 years

93. *Decaying substance.* The amount, A, of a certain radioactive substance remaining after t years, is given by the formula $A = A_0 e^{-0.0003t}$, where A_0 is the initial amount. If we have 218 grams of this substance today, then how much of it will be left 1,000 years from now?
161.5 grams

94. *Wildlife management.* The number of white-tailed deer in the Hiawatha National Forest is believed to be growing according to the function

$$P = 517 + 10 \cdot \ln(8t + 1),$$

where t is the time in years from the year 2000.
a) What is the size of the population in 2000?
b) In what year will the population reach 600?
c) Does the population as shown on the accompanying graph appear to be growing faster during the period 2000 to 2005 or during the period 2005 to 2010?
d) What is the average rate of change of the population for each period in part (c)?
a) 517   b) 2503   c) faster for 2000 to 2005
d) 7.4 deer per year, 1.4 deer per year

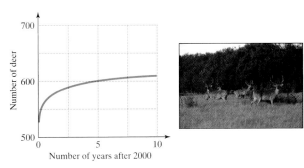

FIGURE FOR EXERCISE 94

95. *Comparing investments.* Melissa deposited $1,000 into an account paying 5% annually; on the same day Frank deposited $900 into an account paying 7% compounded continuously. Find the number of years that it will take for the amounts in the accounts to be equal.
5 years

96. *Imports and exports.* The value of imports for a small Central American country is believed to be growing according to the function

$$I = 15 \cdot \log(16t + 33),$$

and the value of exports appears to be growing according to the function

$$E = 30 \cdot \log(t + 3),$$

where I and E are in millions of dollars and t is the number of years after 2000.
a) What are the values of imports and exports in 2000?
$22.8 million, $14.3 million

b) Use the accompanying graph to estimate the year in which imports will equal exports.   2010
c) Algebraically find the year in which imports will equal exports.   2012

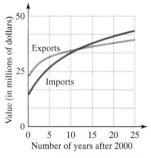

FIGURE FOR EXERCISE 96

97. Finding river flow. The U.S.G.S. measures the water height h (in feet above sea level) for the Tangipahoa River at Robert, Louisiana, and then finds the flow y [in cubic feet per second (ft^3/sec)], using the formula

$$y = 114.308e^{0.265(h-6.87)}.$$

Find the flow when the river at Robert is 20.6 ft above sea level.
4,347.5 ft^3/sec

98. Finding the height. Rewrite the formula in Exercise 97 to express h as a function of y. Use the new formula to find the water height above sea level when the flow is 10,000 ft^3/sec.

$$h = \frac{\ln(y/114.308)}{0.265} + 6.87, \quad 23.74 \text{ ft}$$

CHAPTER 10 TEST

Let $f(x) = 5^x$ and $g(x) = \log_5(x)$. Find the following.

1. $f(2)$ 25

2. $f(-1)$ $\dfrac{1}{5}$

3. $f(0)$ 1

4. $g(125)$ 3

5. $g(1)$ 0

6. $g\left(\dfrac{1}{5}\right)$ -1

Sketch the graph of each function.

7. $y = 2^x$

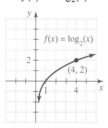

8. $f(x) = \log_2(x)$

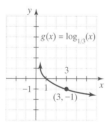

9. $y = \left(\dfrac{1}{3}\right)^x$

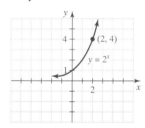

10. $g(x) = \log_{1/3}(x)$

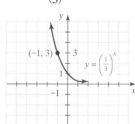

Suppose $\log_a(M) = 6$ and $\log_a(N) = 4$. Find the following.

11. $\log_a(MN)$ 10

12. $\log_a\left(\dfrac{M^2}{N}\right)$ 8

13. $\dfrac{\log_a(M)}{\log_a(N)}$ $\dfrac{3}{2}$

14. $\log_a(a^3 M^2)$ 15

15. $\log_a\left(\dfrac{1}{N}\right)$ -4

Find the exact solution to each equation.

16. $3^x = 12$ $\{\log_3(12)\}$

17. $\log_3(x) = \dfrac{1}{2}$ $\{\sqrt{3}\}$

18. $5^x = 8^{x-1}$ $\left\{\dfrac{\ln(8)}{\ln(8) - \ln(5)}\right\}$

19. $\log(x) + \log(x + 15) = 2$ $\{5\}$

20. $2 \cdot \ln(x) = \ln(3) + \ln(6 - x)$ $\{3\}$

 Use a scientific calculator to find an approximate solution to each of the following. Round your answers to four decimal places.

21. Solve $20^x = 5$. $\{0.5372\}$

22. Solve $\log_3(x) = 2.75$. $\{20.5156\}$

23. The number of bacteria present in a culture at time t is given by the formula $N = 10e^{0.4t}$, where t is in hours. How many bacteria are present initially? How many are present after 24 hours?   10; 147,648

24. How many hours does it take for the bacteria population of Problem 23 to double?   1.733 hours

Find the exact solution to each equation.

1. $(x - 3)^2 = 8$ $\{3 \pm 2\sqrt{2}\}$

2. $\log_2(x - 3) = 8$ $\{259\}$

3. $2^{x-3} = 8$ $\{6\}$

4. $2x - 3 = 8$ $\left\{\dfrac{11}{2}\right\}$

5. $|x - 3| = 8$ $\{-5, 11\}$

6. $\sqrt{x - 3} = 8$ $\{67\}$

7. $\log_2(x - 3) + \log_2(x) = \log_2(18)$ $\{6\}$

8. $2 \cdot \log_2(x - 3) = \log_2(5 - x)$ $\{4\}$

9. $\dfrac{1}{2}x - \dfrac{2}{3} = \dfrac{3}{4}x + \dfrac{1}{5}$ $\left\{-\dfrac{52}{15}\right\}$

10. $3x^2 - 6x + 2 = 0$ $\left\{\dfrac{3 \pm \sqrt{3}}{3}\right\}$

Find the inverse of each function.

11. $f(x) = \dfrac{1}{3}x$
$f^{-1}(x) = 3x$

12. $g(x) = \log_3(x)$
$g^{-1}(x) = 3^x$

13. $f(x) = 2x - 4$
$f^{-1}(x) = \dfrac{x + 4}{2}$

14. $h(x) = \sqrt{x}$
$h^{-1}(x) = x^2$ for $x \geq 0$

15. $j(x) = \dfrac{1}{x}$
$j^{-1}(x) = \dfrac{1}{x}$

16. $k(x) = 5^x$
$k^{-1}(x) = \log_5(x)$

17. $m(x) = e^{x-1}$
$m^{-1}(x) = 1 + \ln(x)$

18. $n(x) = \ln(x)$
$n^{-1}(x) = e^x$

Sketch the graph of each equation.

19. $y = 2x$

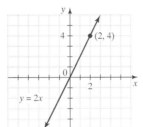

20. $y = 2^x$

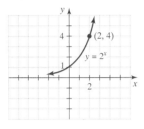

21. $y = x^2$

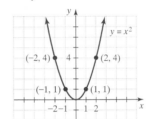

22. $y = \log_2(x)$

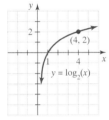

23. $y = \dfrac{1}{2}x - 4$

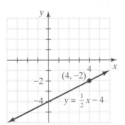

24. $y = |2 - x|$

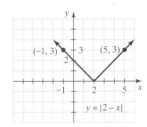

25. $y = 2 - x^2$

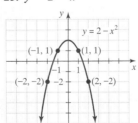

26. $y = e^2$

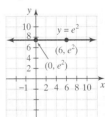

Solve each problem.

27. Civilian labor force. Using data from January 1988 through January 1998, the number of workers in the civilian labor force can be modeled by the linear function

$$n(t) = 1.53t + 113.82$$

or by the exponential function

$$n(t) = 114.0e^{t/80},$$

where t is the number of years since January of 1988 and $n(t)$ is in millions of workers, (Bureau of Labor Statistics, stats.bls.gov).

a) Graph both functions on the same coordinate system for $0 \le t \le 15$.

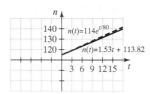

b) What does each model predict for the value of n in January of 1999?
linear 130.7 million, exponential 130.8 million

c) Use the Internet or your library to find the actual size of the civilian labor force in January of 1999.

d) Which model's prediction is closer to the actual number for January of 1999?

e) Answer parts (b), (c), and (d) for the present year.

 28. Measuring ocean depths. In this exercise you will see how a geophysicist uses sound reflection to measure the depth of the ocean. Let v be the speed of sound through the water and d_1 be the depth of the ocean below the ship, as shown in the accompanying figure.

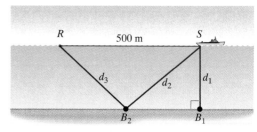

FIGURE FOR EXERCISE 28

a) The time it takes for sound to travel from the ship at point S straight down to the ocean floor at point B_1 and back to point S is 0.270 second. Write d_1 as a function of v.
$d_1 = 0.135v$

b) It takes 0.432 second for sound to travel from point S to point B_2 and then to a receiver at R, which is towed 500 meters behind the ship. Assuming $d_2 = d_3$, write d_2 as a function of v.
$d_2 = 0.216v$

c) Use the Pythagorean theorem to find v. Then find the ocean depth d_1.
$d_1 = 200.2$ meters

Nonlinear Systems and the Conic Sections

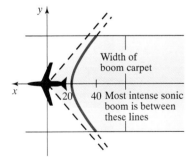

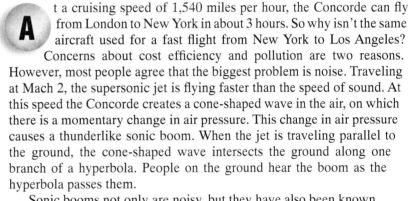

Width of boom carpet

20 40 Most intense sonic boom is between these lines

At a cruising speed of 1,540 miles per hour, the Concorde can fly from London to New York in about 3 hours. So why isn't the same aircraft used for a fast flight from New York to Los Angeles? Concerns about cost efficiency and pollution are two reasons. However, most people agree that the biggest problem is noise. Traveling at Mach 2, the supersonic jet is flying faster than the speed of sound. At this speed the Concorde creates a cone-shaped wave in the air, on which there is a momentary change in air pressure. This change in air pressure causes a thunderlike sonic boom. When the jet is traveling parallel to the ground, the cone-shaped wave intersects the ground along one branch of a hyperbola. People on the ground hear the boom as the hyperbola passes them.

Sonic booms not only are noisy, but they have also been known to cause physical destruction such as broken windows and cracked plaster. For this reason supersonic jets are restricted from flying over land areas in the United States and much of the world. Some engineers believe that changing the silhouette of the plane can lessen the sonic boom, but most agree that it is impossible to eliminate the noise altogether.

In this chapter we discuss curves, including the hyperbola, that occur when a geometric plane intersects a cone. In Exercise 54 of Section 11.4 you will see how the altitude of the aircraft is related to the width of the area where the sonic boom is heard.

In this

section

• Solving by Elimination

• Applications

11.1 NONLINEAR SYSTEMS OF EQUATIONS

We studied systems of linear equations in Chapter 4. In this section we turn our attention to nonlinear systems of equations.

Solving by Elimination

Equations such as

$$y = x^2, \qquad y = \sqrt{x}, \qquad y = |x|, \qquad y = 2^x, \qquad \text{and} \qquad y = \log_2(x)$$

are **nonlinear equations** because their graphs are not straight lines. We say that a system of equations is nonlinear if at least one equation in the system is nonlinear. We solve a nonlinear system just like a linear system, by elimination of variables. However, because the graphs of nonlinear equations may intersect at more than one point, there may be more than one ordered pair in the solution set to the system.

EXAMPLE 1   **A parabola and a line**

Solve the system of equations and draw the graph of each equation on the same coordinate system:

$$y = x^2 - 1$$
$$x + y = 1$$

Solution

We can eliminate y by substituting $y = x^2 - 1$ into $x + y = 1$:

$$x + y = 1$$
$$x + (x^2 - 1) = 1 \qquad \text{Substitute } x^2 - 1 \text{ for } y.$$
$$x^2 + x - 2 = 0$$
$$(x - 1)(x + 2) = 0$$
$$x - 1 = 0 \qquad \text{or} \qquad x + 2 = 0$$
$$x = 1 \qquad \text{or} \qquad x = -2$$

Replace x by 1 and -2 in $y = x^2 - 1$ to find the corresponding values of y:

$$y = (1)^2 - 1 \qquad y = (-2)^2 - 1$$
$$y = 0 \qquad\qquad y = 3$$

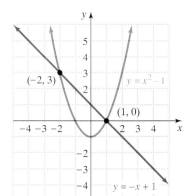

FIGURE 11.1

Check that each of the points $(1, 0)$ and $(-2, 3)$ satisfies both of the original equations. The solution set is $\{(1, 0), (-2, 3)\}$. If we solve $x + y = 1$ for y, we get $y = -x + 1$. The line $y = -x + 1$ has y-intercept $(0, 1)$ and slope -1. The graph of $y = x^2 - 1$ is a parabola with vertex $(0, -1)$. Of course, $(1, 0)$ and $(-2, 3)$ are on both graphs. The two graphs are shown in Fig. 11.1. ■

Graphing is not an accurate method for solving any system of equations. However, the graphs of the equations in a nonlinear system help us to understand how many solutions we should have for the system. It is not necessary to graph a system to solve it. Even when the graphs are too difficult to sketch, we can solve the system.

EXAMPLE 2   **Solving a system algebraically**

Solve the system: $x^2 + y^2 + 2y = 3$
$$x^2 - y = 5$$

Solution

If we substitute $y = x^2 - 5$ into the first equation to eliminate y, we will get a fourth-degree equation to solve. Instead, we can eliminate the variable x by writing $x^2 - y = 5$ as $x^2 = y + 5$. Now replace x^2 by $y + 5$ in the first equation:

$$x^2 + y^2 + 2y = 3$$
$$(y + 5) + y^2 + 2y = 3$$
$$y^2 + 3y + 5 = 3$$
$$y^2 + 3y + 2 = 0$$
$$(y + 2)(y + 1) = 0 \quad \text{Solve by factoring.}$$
$$y + 2 = 0 \quad \text{or} \quad y + 1 = 0$$
$$y = -2 \quad \text{or} \quad y = -1$$

Let $y = -2$ in the equation $x^2 = y + 5$ to find the corresponding x:

$$x^2 = -2 + 5$$
$$x^2 = 3$$
$$x = \pm\sqrt{3}$$

Now let $y = -1$ in the equation $x^2 = y + 5$ to find the corresponding x:

$$x^2 = -1 + 5$$
$$x^2 = 4$$
$$x = \pm 2$$

Check these values in the original equations. The solution set is

$$\{(\sqrt{3}, -2), (-\sqrt{3}, -2), (2, -1), (-2, -1)\}.$$

The graphs of these two equations intersect at four points.

E X A M P L E   3

Solving a system algebraically

Solve the system:
$$(1) \quad \frac{2}{x} + \frac{1}{y} = \frac{1}{5}$$

$$(2) \quad \frac{1}{x} - \frac{3}{y} = \frac{1}{3}$$

Solution

Usually with equations involving rational expressions we first multiply by the least common denominator (LCD), but this would make the given system more complicated. So we will just use the addition method to eliminate y:

$$\frac{6}{x} + \frac{3}{y} = \frac{3}{5} \qquad \text{Eq. (1) multiplied by 3}$$

$$\frac{1}{x} - \frac{3}{y} = \frac{1}{3} \qquad \text{Eq. (2)}$$

$$\frac{7}{x} \qquad = \frac{14}{15} \qquad \frac{3}{5} + \frac{1}{3} = \frac{14}{15}$$

$$14x = 7 \cdot 15$$

$$x = \frac{7 \cdot 15}{14} = \frac{15}{2}$$

To find y, substitute $x = \frac{15}{2}$ into Eq. (1):

$$\frac{2}{\frac{15}{2}} + \frac{1}{y} = \frac{1}{5}$$

$$\frac{4}{15} + \frac{1}{y} = \frac{1}{5} \qquad \frac{2}{\frac{15}{2}} = 2 \cdot \frac{2}{15} = \frac{4}{15}$$

$$15y \cdot \frac{4}{15} + 15y \cdot \frac{1}{y} = 15y \cdot \frac{1}{5} \qquad \text{Multiply each side by the LCD, } 15y.$$

$$4y + 15 = 3y$$

$$y = -15$$

Check that $x = \frac{15}{2}$ and $y = -15$ satisfy both original equations. The solution set is $\left\{\left(\frac{15}{2}, -15\right)\right\}$. ∎

A system of nonlinear equations might involve exponential or logarithmic functions. To solve such systems, you will need to recall some facts about exponents and logarithms.

E X A M P L E   4

A system involving logarithms

Solve the system
$$y = \log_2(x + 28)$$
$$y = 3 + \log_2(x)$$

Solution

Eliminate y by substitution $\log_2(x + 28)$ for y in the second equation:

$$\log_2(x + 28) = 3 + \log_2(x) \qquad \text{Eliminate } y.$$

$$\log_2(x + 28) - \log_2(x) = 3 \qquad \text{Subtract } \log_2(x) \text{ from each side.}$$

$$\log_2\left(\frac{x + 28}{x}\right) = 3 \qquad \text{Quotient rule for logarithms}$$

$$\frac{x + 28}{x} = 8 \qquad \text{Definition of logarithm}$$

$$x + 28 = 8x \qquad \text{Multiply each side by } x.$$

$$28 = 7x \qquad \text{Subtract } x \text{ from each side.}$$

$$4 = x \qquad \text{Divide each side by 7.}$$

If $x = 4$, then $y = \log_2(4 + 28) = \log_2(32) = 5$. Check $(4, 5)$ in both equations. The solution to the system is $\{(4, 5)\}$. ∎

Applications

The next example shows a geometric problem that can be solved with a system of nonlinear equations.

E X A M P L E   5

Nonlinear equations in applications

A 15-foot ladder is leaning against a wall so that the distance from the bottom of the ladder to the wall is one-half of the distance from the top of the ladder to the ground. Find the distance from the top of the ladder to the ground.

Solution

Let x be the number of feet from the bottom of the ladder to the wall and y be the number of feet from the top of the ladder to the ground (see Fig. 11.2). We can write two equations involving x and y:

$$x^2 + y^2 = 15^2 \quad \text{Pythagorean theorem}$$
$$y = 2x$$

Solve by substitution:

$$x^2 + (2x)^2 = 225 \quad \text{Replace } y \text{ by } 2x.$$
$$x^2 + 4x^2 = 225$$
$$5x^2 = 225$$
$$x^2 = 45$$
$$x = \pm\sqrt{45} = \pm 3\sqrt{5}$$

Because x represents distance, x must be positive. So $x = 3\sqrt{5}$. Because $y = 2x$, we get $y = 6\sqrt{5}$. The distance from the top of the ladder to the ground is $6\sqrt{5}$ feet.

FIGURE 11.2

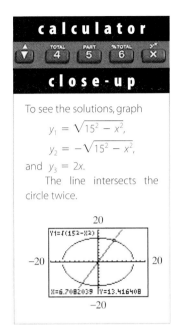

The next example shows how a nonlinear system can be used to solve a problem involving work.

E X A M P L E   6

Nonlinear equations in applications

A large fish tank at the Gulf Aquarium can usually be filled in 10 minutes using pumps A and B. However, pump B can pump water in or out at the same rate. If pump B is inadvertently run in reverse, then the tank will be filled in 30 minutes. How long would it take each pump to fill the tank by itself?

Solution

Let a represent the number of minutes that it takes pump A to fill the tank alone and b represent the number of minutes it takes pump B to fill the tank alone. The rate at which pump A fills the tank is $\frac{1}{a}$ of the tank per minute, and the rate at which pump B fills the tank is $\frac{1}{b}$ of the tank per minute. Because the work completed is the product of the rate and time, we can make the following table when the pumps work together to fill the tank:

	Rate	Time	Work
Pump A	$\dfrac{1}{a} \dfrac{\text{tank}}{\text{min}}$	10 min	$\dfrac{10}{a}$ tank
Pump B	$\dfrac{1}{b} \dfrac{\text{tank}}{\text{min}}$	10 min	$\dfrac{10}{b}$ tank

Note that each pump fills a fraction of the tank and those fractions have a sum of 1:

$$(1) \qquad \frac{10}{a} + \frac{10}{b} = 1$$

In the 30 minutes in which pump B is working in reverse, A puts in $\frac{30}{a}$ of the tank whereas B takes out $\frac{30}{b}$ of the tank. Since the tank still gets filled, we can write the following equation:

$$(2) \quad \frac{30}{a} - \frac{30}{b} = 1$$

Multiply Eq. (1) by 3 and add the result to Eq. (2) to eliminate b:

$$\frac{30}{a} + \frac{30}{b} = 3 \quad \text{Eq. (1) multiplied by 3}$$

$$\frac{30}{a} - \frac{30}{b} = 1 \quad \text{Eq. (2)}$$

$$\frac{60}{a} \qquad = 4$$

$$4a = 60$$

$$a = 15$$

Use $a = 15$ in Eq. (1) to find b:

$$\frac{10}{15} + \frac{10}{b} = 1$$

$$\frac{10}{b} = \frac{1}{3} \quad \text{Subtract } \frac{10}{15} \text{ from each side.}$$

$$b = 30$$

So pump A fills the tank in 15 minutes working alone, and pump B fills the tank in 30 minutes working alone. ■

WARM-UPS

True or false? Explain your answer.

1. The graph of $y = x^2$ is a parabola.   True
2. The graph of $y = |x|$ is a straight line.   False
3. The point $(3, -4)$ satisfies both $x^2 + y^2 = 25$ and $y = \sqrt{5x + 1}$.   False
4. The graphs of $y = \sqrt{x}$ and $y = -x - 2$ do not intersect.   True
5. Substitution is the only method for eliminating a variable when solving a nonlinear system.   False
6. If Bob paints a fence in x hours, then he paints $\frac{1}{x}$ of the fence per hour.   True
7. In a triangle whose angles are $30°$, $60°$, and $90°$, the length of the side opposite the $30°$ angle is one-half the length of the hypotenuse.   True
8. The formula $V = LWH$ gives the volume of a rectangular box in which the sides have lengths L, W, and H.   True
9. The surface area of a rectangular box is $2LW + 2WH + 2LH$.   True
10. The area of a right triangle is one-half the product of the lengths of its legs.   True

11.1   EXERCISES

Reading and Writing   *After reading this section, write out the answers to these questions. Use complete sentences.*

1. Why are some equations called nonlinear?

If the graph of an equation is not a straight line, then it is called nonlinear.

2. Why do we graph the equations in a nonlinear system?

With a graph we can see the approximate value of the solutions and the number of solutions.

3. Why don't we solve systems by graphing?

Graphing is not an accurate method for solving a system and the graphs might be difficult to draw.

4. What techniques do we use to solve nonlinear systems?

We generally use substitution and addition to solve nonlinear systems.

Solve each system and graph both equations on the same set of axes. See Example 1.

5. $y = x^2$
$x + y = 6$
$\{(2, 4), (-3, 9)\}$

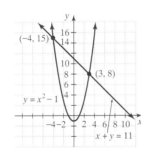

6. $y = x^2 - 1$
$x + y = 11$
$\{(-4, 15), (3, 8)\}$

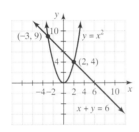

7. $y = |x|$
$2y - x = 6$

$\{(-2, 2), (6, 6)\}$

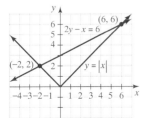

8. $y = |x|$
$3y = x + 6$

$\left\{\left(-\dfrac{3}{2}, \dfrac{3}{2}\right), (3, 3)\right\}$

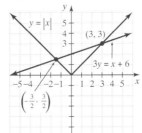

9. $y = \sqrt{2x}$
$x - y = 4$
$\{(8, 4)\}$

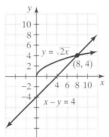

10. $y = \sqrt{x}$
$x - y = 6$
$\{(9, 3)\}$

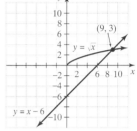

11. $4x - 9y = 9$
$xy = 1$

$\left\{\left(-\dfrac{3}{4}, -\dfrac{4}{3}\right), \left(3, \dfrac{1}{3}\right)\right\}$

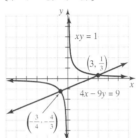

12. $2x + 2y = 3$
$xy = -1$

$\left\{\left(-\dfrac{1}{2}, 2\right), \left(2, -\dfrac{1}{2}\right)\right\}$

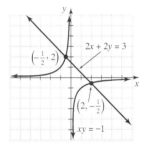

13. $y = -x^2 + 1$
$y = x^2$

$\left\{\left(\dfrac{\sqrt{2}}{2}, \dfrac{1}{2}\right), \left(-\dfrac{\sqrt{2}}{2}, \dfrac{1}{2}\right)\right\}$

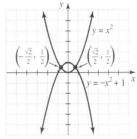

14. $y = x^2$
$y = \sqrt{x}$

$\{(0, 0), (1, 1)\}$

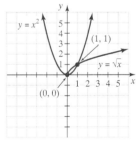

Solve each system. See Examples 2 and 3.

15. $x^2 + y^2 = 25$
$y = x^2 - 5$
$\{(0, -5), (3, 4), (-3, 4)\}$

16. $x^2 + y^2 = 25$
$y = x + 1$
$\{(-4, -3), (3, 4)\}$

17. $xy - 3x = 8$
$y = x + 1$
$\{(4, 5), (-2, -1)\}$

18. $xy + 2x = 9$
$x - y = 2$
$\{(3, 1), (-3, -5)\}$

19. $xy - x = 8$
$xy + 3x = -4$

$\left\{\left(-3, -\dfrac{5}{3}\right)\right\}$

20. $2xy - 3x = -1$
$xy + 5x = -7$

$\{(-1, 2)\}$

21. $\dfrac{1}{x} - \dfrac{1}{y} = 5$

$\dfrac{2}{x} + \dfrac{1}{y} = -3$

$\left\{\left(\dfrac{3}{2}, -\dfrac{3}{13}\right)\right\}$

22. $\dfrac{2}{x} - \dfrac{3}{y} = \dfrac{1}{2}$

$\dfrac{3}{x} + \dfrac{1}{y} = \dfrac{1}{2}$

$\left\{\left(\dfrac{11}{2}, -22\right)\right\}$

23. $x^2y = 20$

$xy + 2 = 6x$

$\left\{\left(-\dfrac{5}{3}, \dfrac{36}{5}\right), (2, 5)\right\}$

24. $y^2x = 3$

$xy + 1 = 6x$

$\left\{\left(\dfrac{1}{12}, -6\right), \left(\dfrac{1}{3}, 3\right)\right\}$

25. $x^2 + y^2 = 8$

$x^2 - y^2 = 2$

$\{(\sqrt{5}, \sqrt{3}), (\sqrt{5}, -\sqrt{3}), (-\sqrt{5}, \sqrt{3}), (-\sqrt{5}, -\sqrt{3})\}$

26. $x^2 + 2y^2 = 8$

$2x^2 - y^2 = 1$

$\{(\sqrt{2}, \sqrt{3}), (\sqrt{2}, -\sqrt{3}), (-\sqrt{2}, \sqrt{3}), (-\sqrt{2}, -\sqrt{3})\}$

27. $x^2 + xy - y^2 = -11$

$x + y = 7$

$\{(2, 5), (19, -12)\}$

28. $x^2 + xy + y^2 = 3$

$y = 2x - 5$

$\left\{\left(\dfrac{11}{7}, -\dfrac{13}{7}\right), (2, -1)\right\}$

29. $3y - 2 = x^4$

$y = x^2$

$\{(\sqrt{2}, 2), (-\sqrt{2}, 2), (1, 1), (-1, 1)\}$

30. $y - 3 = 2x^4$

$y = 7x^2$

$\left\{\left(\dfrac{\sqrt{2}}{2}, \dfrac{7}{2}\right), \left(-\dfrac{\sqrt{2}}{2}, \dfrac{7}{2}\right), (\sqrt{3}, 21), (-\sqrt{3}, 21)\right\}$

Solve the following systems involving logarithmic and exponential functions. See Example 4.

31. $y = \log_2(x - 1)$

$y = 3 - \log_2(x + 1)$

$\{(3, 1)\}$

32. $y = \log_3(x - 4)$

$y = 2 - \log_3(x + 4)$

$\{(5, 0)\}$

33. $y = \log_2(x - 1)$

$y = 2 + \log_2(x + 2)$

$\varnothing$

34. $y = \log_4(8x)$

$y = 2 + \log_4(x - 1)$

$\{(2, 2)\}$

35. $y = 2^{3x+4}$

$y = 4^{x-1}$

$\{(-6, 4^{-7})\}$

36. $y = 4^{3x}$

$y = \left(\dfrac{1}{2}\right)^{1-x}$

$\left\{\left(-\dfrac{1}{5}, 4^{-3/5}\right)\right\}$

Solve each problem by using a system of two equations in two unknowns. See Examples 5 and 6.

37. **Known hypotenuse.** Find the lengths of the legs of a right triangle whose hypotenuse is $\sqrt{15}$ feet and whose area is 3 square feet.

$\sqrt{3}$ ft and $2\sqrt{3}$ ft

38. **Known diagonal.** A small television is advertised to have a picture with a diagonal measure of 5 inches and a viewing area of 12 square inches (in.²). What are the length and width of the screen?

3 inches by 4 inches

FIGURE FOR EXERCISE 38

39. **House of seven gables.** Vincent has plans to build a house with seven gables. The plans call for an attic vent in the shape of an isosceles triangle in each gable. Because of the slope of the roof, the ratio of the height to the base of each triangle must be 1 to 4. If the vents are to provide a total ventilating area of 3,500 in.², then what should be the height and base of each triangle?

Height $5\sqrt{10}$ inches, base $20\sqrt{10}$ inches

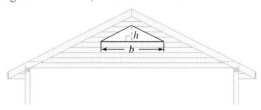

FIGURE FOR EXERCISE 39

40. **Known perimeter.** Find the lengths of the sides of a triangle whose perimeter is 6 feet (ft) and whose angles are 30°, 60°, and 90° (see Appendix A).

$3 - \sqrt{3}$ ft, $6 - 2\sqrt{3}$ ft, $-3 + 3\sqrt{3}$ ft

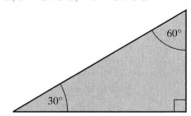

FIGURE FOR EXERCISE 40

41. **Filling a tank.** Pump A can either fill a tank or empty it in the same amount of time. If pump A and pump B are working together, the tank can be filled in 6 hours. When pump A was inadvertently left in the drain position while pump B was trying to fill the tank, it took 12 hours to fill the tank. How long would it take either pump working alone to fill the tank?

Pump A 24 hours, pump B 8 hours

42. **Cleaning a house.** Roxanne either cleans the house or messes it up at the same rate. When Roxanne is cleaning with her mother, they can clean up a completely messed up house in 6 hours. If Roxanne is not cooperating, it takes her mother 9 hours to clean the house, with Roxanne continually messing it up. How long would it take her mother to clean the entire house if Roxanne were sent to her grandmother's house?

$\dfrac{36}{5}$ hours

43. *Cleaning fish.* Jan and Beth work in a seafood market that processes 200 pounds of catfish every morning. On Monday, Jan started cleaning catfish at 8:00 A.M. and finished cleaning 100 pounds just as Beth arrived. Beth then took over and finished the job at 8:50 A.M. On Tuesday they both started at 8 A.M. and worked together to finish the job at 8:24 A.M. On Wednesday, Beth was sick. If Jan is the faster worker, then how long did it take Jan to complete all of the catfish by herself?

40 minutes

FIGURE FOR EXERCISE 43

44. *Building a patio.* Richard has already formed a rectangular area for a flagstone patio, but his wife Susan is unsure of the size of the patio they want. If the width is increased by 2 ft, then the area is increased by 30 square feet (ft²). If the width is increased by 1 ft and the length by 3 ft, then the area is increased by 54 ft². What are the dimensions of the rectangle that Richard has already formed?

12 ft by 15 ft

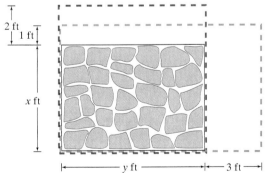

FIGURE FOR EXERCISE 44

45. *Fencing a rectangle.* If 34 ft of fencing are used to enclose a rectangular area of 72 ft², then what are the dimensions of the area?

8 ft by 9 ft

46. *Real numbers.* Find two numbers that have a sum of 8 and a product of 10.

$4 - \sqrt{6}$ and $4 + \sqrt{6}$

47. *Imaginary numbers.* Find two complex numbers whose sum is 8 and whose product is 20.

$4 - 2i$ and $4 + 2i$

48. *Imaginary numbers.* Find two complex numbers whose sum is -6 and whose product is 10.

$-3 + i$ and $-3 - i$

49. *Making a sign.* Rico's Sign Shop has a contract to make a sign in the shape of a square with an isosceles triangle on top of it, as shown in the figure. The contract calls for a total height of 10 ft with an area of 72 ft². How long should Rico make the side of the square and what should be the height of the triangle?

Side 8 ft, height of triangle 2 ft

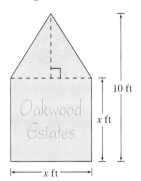

FIGURE FOR EXERCISE 49

50. *Designing a box.* Angelina is designing a rectangular box of 120 cubic inches that is to contain new Eaties breakfast cereal. The box must be 2 inches thick so that it is easy to hold. It must have 184 square inches of surface area to provide enough space for all of the special offers and coupons. What should be the dimensions of the box?

6 inches by 10 inches by 2 inches

 GRAPHING CALCULATOR EXERCISES

51. Solve each system by graphing each pair of equations on a graphing calculator and using the intersect feature to estimate the point of intersection. Find the coordinates of each intersection to the nearest hundredth.

a) $y = e^x - 4$
   $y = \ln(x + 3)$

b) $3^{y-1} = x$
   $y = x^2$

c) $x^2 + y^2 = 4$
   $y = x^3$

   a) $(1.71, 1.55)$, $(-2.98, -3.95)$
   b) $(1, 1)$, $(0.40, 0.16)$
   c) $(1.17, 1.62)$, $(-1.17, -1.62)$

11.2 THE PARABOLA

The parabola is one of four different curves that can be obtained by intersecting a cone and a plane as in Fig. 11.3. These curves, called **conic sections,** are the parabola, circle, ellipse, and hyperbola. You studied parabolas in Section 3.6, but in this section you will study parabolas in more detail. You will see how the parabola and the other conic sections are used in applications such as satellite dishes, telescopes, spotlights, and navigation.

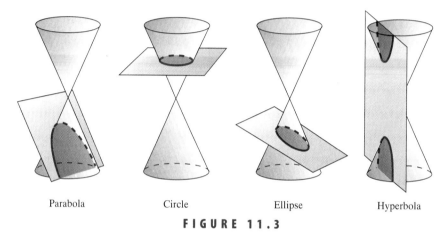

Parabola   Circle   Ellipse   Hyperbola

FIGURE 11.3

Developing the Equation

In Section 3.6 we called the graph of $y = ax^2 + bx + c$ a parabola. This equation is the **standard equation** of a parabola. In this section you will see that the following geometric definition describes the same curve as the equation.

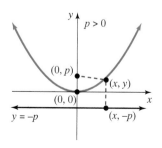

FIGURE 11.4

> **Parabola**
>
> Given a line (the **directrix**) and a point not on the line (the **focus**), the set of all points in the plane that are equidistant from the point and the line is called a **parabola.**

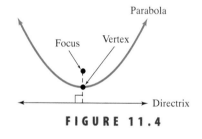

FIGURE 11.5

The **vertex** of the parabola is the midpoint of the line segment joining the focus and the directrix, perpendicular to the directrix. See Fig. 11.4.

The focus of a parabola is important in applications. When parallel rays of light travel into a parabolic reflector, they are reflected toward the focus as in Fig. 11.5. This property is used in telescopes to see the light from distant stars. If the light source is at the focus, as in a searchlight, the light is reflected off the parabola and projected outward in a narrow beam. This reflecting property is also used in camera lenses, satellite dishes, and eavesdropping devices.

To develop an equation for a parabola, given the focus and directrix, choose the point $(0, p)$, where $p > 0$ as the focus and the line $y = -p$ as the directrix, as shown in Fig. 11.6. The vertex of this parabola is $(0, 0)$. For an arbitrary point (x, y) on the parabola the distance to the directrix is the distance from (x, y) to $(x, -p)$.

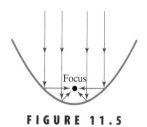

FIGURE 11.6

The distance to the focus is the distance between (x, y) and $(0, p)$. We use the fact that these distances are equal to write the equation of the parabola:

$$\sqrt{(x - 0)^2 + (y - p)^2} = \sqrt{(x - x)^2 + (y - (-p))^2}$$

To simplify the equation, first remove the parentheses inside the radicals:

$$\sqrt{x^2 + y^2 - 2py + p^2} = \sqrt{y^2 + 2py + p^2}$$

$$x^2 + y^2 - 2py + p^2 = y^2 + 2py + p^2 \quad \text{Square each side.}$$

$$x^2 = 4py \quad\quad \text{Subtract } y^2 \text{ and } p^2 \text{ from each side.}$$

$$y = \frac{1}{4p}x^2$$

$$y = ax^2 \quad\quad \text{Let } a = \frac{1}{4p}.$$

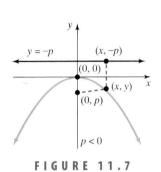

$y = -p$ $(x, -p)$
$(0, 0)$
(x, y) x
$(0, p)$

$p < 0$

FIGURE 11.7

So a curve that satisfies the geometric definition of parabola has an equation of the form $y = ax^2 + bx + c$.

If the focus is at $(0, p)$ with $p < 0$ and the directrix is $y = -p$, then the parabola opens downward as shown in Fig. 11.7. Deriving the equation from the distances (as was just done for $p > 0$) again yields $y = \frac{1}{4p}x^2$, which is of the form $y = ax^2$ for $a = \frac{1}{4p}$. Note that a and p have the same sign because $a = \frac{1}{4p}$.

If a parabola has vertex $(0, 0)$ and opens up or down, then its equation is of the form $y = ax^2$. In general, if (h, k) is the vertex, $(h, k + p)$ is the focus, and $y = k - p$ is the directrix, then we can develop the equation $y = a(x - h)^2 + k$ for the parabola just as we developed the equation $y = ax^2$. Graphs of $y = a(x - h)^2 + k$ are shown in Fig. 11.8.

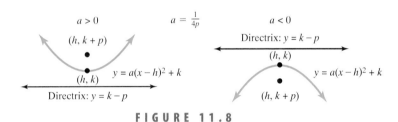

$a > 0$ $a = \frac{1}{4p}$ $a < 0$

$(h, k + p)$ Directrix: $y = k - p$

(h, k) $y = a(x - h)^2 + k$ (h, k) $y = a(x - h)^2 + k$

Directrix: $y = k - p$ $(h, k + p)$

FIGURE 11.8

Parabolas in the Form $y = a(x - h)^2 + k$

The graph of the equation $y = a(x - h)^2 + k$ $(a \neq 0)$ is a parabola with vertex (h, k), focus $(h, k + p)$, and directrix $y = k - p$, where $a = \frac{1}{4p}$. If $a > 0$, the parabola opens upward; if $a < 0$, the parabola opens downward.

Note that the locations of the focus and directrix determine the value of a and the shape and opening of the parabola.

CAUTION   For a parabola that opens upward, $p > 0$, and the focus $(h, k + p)$ is above the vertex (h, k). For a parabola that opens downward, $p < 0$, and the focus $(h, k + p)$ is below the vertex (h, k). In either case the distance from the vertex to the focus and the vertex to the directrix is $|p|$.

EXAMPLE 1

Finding the vertex, focus, and directrix, given an equation

Find the vertex, focus, and directrix for the parabola $y = x^2$.

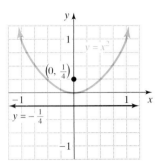

FIGURE 11.9

Solution

Compare $y = x^2$ to the general formula $y = a(x - h)^2 + k$. We see that $h = 0$, $k = 0$, and $a = 1$. So the vertex is $(0, 0)$. Because $a = 1$, we can use $a = \frac{1}{4p}$ to get

$$1 = \frac{1}{4p},$$

or $p = \frac{1}{4}$. Use $(h, k + p)$ to get the focus $\left(0, \frac{1}{4}\right)$. Use the equation $y = k - p$ to get $y = -\frac{1}{4}$ as the equation of the directrix. See Fig. 11.9. ■

EXAMPLE 2

Finding an equation, given a focus and directrix

Find the equation of the parabola with focus $(-1, 4)$ and directrix $y = 3$.

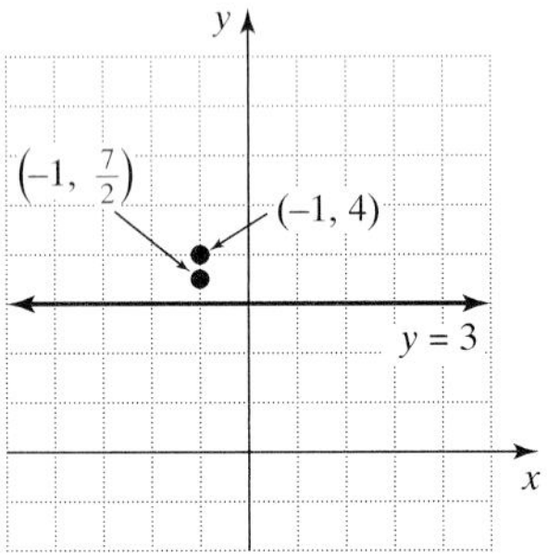

Solution

Because the vertex is halfway between the focus and directrix, the vertex is $\left(-1, \frac{7}{2}\right)$. See Fig. 11.10. The distance from the vertex to the focus is $\frac{1}{2}$. Because the focus is above the vertex, p is positive. So $p = \frac{1}{2}$, and $a = \frac{1}{4p} = \frac{1}{2}$. The equation is

$$y = \frac{1}{2}(x - (-1))^2 + \frac{7}{2}.$$

Simplify to get the equation $y = \frac{1}{2}x^2 + x + 4$. ■

FIGURE 11.10

The equation of a parabola can be written in two different forms. To change the form $y = a(x - h)^2 + k$ to the form $y = ax^2 + bx + c$, we square the binomial and combine like terms, as in Example 2. To change from $y = ax^2 + bx + c$ to $y = a(x - h)^2 + k$, we complete the square.

EXAMPLE 3

Converting $y = ax^2 + bx + c$ to $y = a(x - h)^2 + k$

Write $y = 2x^2 - 4x + 5$ in the form $y = a(x - h)^2 + k$ and identify the vertex, focus, and directrix of the parabola.

Solution

Use completing the square to rewrite the equation:

$$y = 2(x^2 - 2x) + 5$$
$$y = 2(x^2 - 2x + 1 - 1) + 5 \quad \text{Complete the square.}$$
$$y = 2(x^2 - 2x + 1) - 2 + 5 \quad \text{Move } 2(-1) \text{ outside the parentheses.}$$
$$y = 2(x - 1)^2 + 3$$

The vertex is $(1, 3)$. Because $a = \frac{1}{4p}$, we have

$$\frac{1}{4p} = 2,$$

The graphs of

$$y_1 = 2x^2 - 4x + 5$$

and

$$y_2 = 2(x - 1)^2 + 3$$

appear to be identical. This supports the conclusion that the equations are equivalent.

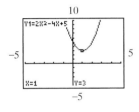

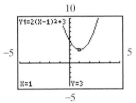

and $p = \frac{1}{8}$. Because the parabola opens upward, the focus is $\frac{1}{8}$ unit above the vertex at $\left(1, 3\frac{1}{8}\right)$, or $\left(1, \frac{25}{8}\right)$, and the directrix is the horizontal line $\frac{1}{8}$ unit below the vertex, $y = 2\frac{7}{8}$ or $y = \frac{23}{8}$.   ◼

C A U T I O N   Be careful when you complete a square within parentheses as in Example 3. For another example, consider the equivalent equations

$$y = -3(x^2 + 4x),$$
$$y = -3(x^2 + 4x + 4 - 4),$$

and

$$y = -3(x + 2)^2 + 12.$$

Identifying the Vertex from Standard Form

If the equation of a parabola is in the form $y = a(x - h)^2 + k$, then its vertex is (h, k). If we complete the square on the standard equation of the parabola $y = ax^2 + bx + c$ to get it into the form $y = a(x - h)^2 + k$, we will find the vertex in terms of a, b, and c:

$$y = ax^2 + bx + c$$

$$y = a\left(x^2 + \frac{bx}{a}\right) + c \qquad \text{Factor } a \text{ out of the first two terms.}$$

$$y = a\left(x^2 + \frac{bx}{a} + \frac{b^2}{4a^2} - \frac{b^2}{4a^2}\right) + c \qquad \frac{1}{2} \cdot \frac{b}{a} = \frac{b}{2a} \ \text{ and } \ \left(\frac{b}{2a}\right)^2 = \frac{b^2}{4a^2}$$

$$y = a\left(x^2 + \frac{bx}{a} + \frac{b^2}{4a^2}\right) - \frac{b^2}{4a} + c \qquad \text{Move } a\left(-\frac{b^2}{4a^2}\right) \text{ outside the parentheses.}$$

$$y = a\left(x^2 + \frac{bx}{a} + \frac{b^2}{4a^2}\right) - \frac{b^2}{4a} + \frac{c \cdot 4a}{4a} \qquad \text{Build up the denominator.}$$

$$y = a\left(x + \frac{b}{2a}\right)^2 + \frac{4ac - b^2}{4a} \qquad \text{Simplify.}$$

The equation is now in the form $y = a(x - h)^2 + k$, and the vertex is

$$\left(\frac{-b}{2a}, \frac{4ac - b^2}{4a}\right).$$

We summarize these results as follows.

Parabolas in the Form $y = ax^2 + bx + c$

The graph of $y = ax^2 + bx + c$ (for $a \neq 0$) is a parabola opening upward if $a > 0$ and downward if $a < 0$. The x-coordinate of the vertex is $x = \frac{-b}{2a}$.

You can use $\frac{4ac - b^2}{4a}$ to get the y-coordinate of the vertex, but it is usually easier to get it from $y = ax^2 + bx + c$ for $x = \frac{-b}{2a}$. We use the second approach in Example 4.

E X A M P L E   4

Finding the features of a parabola from standard form

Find the vertex, focus, and directrix of the parabola $y = -3x^2 + 9x - 5$, and determine whether the parabola opens upward or downward.

Solution

The x-coordinate of the vertex is

$$x = \frac{-b}{2a} = \frac{-9}{2(-3)} = \frac{-9}{-6} = \frac{3}{2}.$$

To find the y-coordinate of the vertex, let $x = \frac{3}{2}$ in $y = -3x^2 + 9x - 5$:

$$y = -3\left(\frac{3}{2}\right)^2 + 9\left(\frac{3}{2}\right) - 5 = -\frac{27}{4} + \frac{27}{2} - 5 = \frac{7}{4}$$

The vertex is $\left(\frac{3}{2}, \frac{7}{4}\right)$. Because $a = -3$, the parabola opens downward. To find the focus, use $-3 = \frac{1}{4p}$ to get $p = -\frac{1}{12}$. The focus is $\frac{1}{12}$ of a unit below the vertex at $\left(\frac{3}{2}, \frac{7}{4} - \frac{1}{12}\right)$ or $\left(\frac{3}{2}, \frac{5}{3}\right)$. The directrix is the horizontal line $\frac{1}{12}$ of a unit above the vertex, $y = \frac{7}{4} + \frac{1}{12}$ or $y = \frac{11}{6}$.   ∎

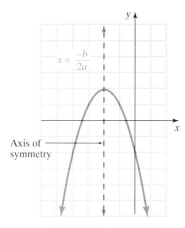

calculator

close-up

A calculator graph can be used to check the vertex and opening of a parabola.

Symmetry and Intercepts

The graph of $y = x^2$ shown in Fig. 11.9 is said to be **symmetric about the y-axis** because the two halves of the parabola would coincide if the paper were folded on the y-axis. In general, the vertical line through the vertex is called the **axis of symmetry** for the parabola. Because the x-coordinate of the vertex for $y = ax^2 + bx + c$ is $\frac{-b}{2a}$, the axis of symmetry is $x = \frac{-b}{2a}$. See Fig. 11.11.

The parabola $y = ax^2 + bx + c$ has exactly one y-intercept. If $x = 0$ in the equation, then $y = a(0)^2 + b(0) + c = c$. So the y-intercept is $(0, c)$.

To find the x-intercepts, we let $y = 0$ in the equation $y = ax^2 + bx + c$ and solve the quadratic equation $ax^2 + bx + c = 0$. The number of x-intercepts may be 0, 1, or 2, depending on the number of solutions to the quadratic equation. See Fig. 11.12.

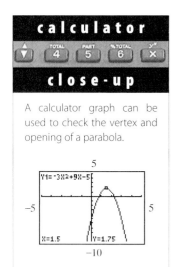

FIGURE 11.11

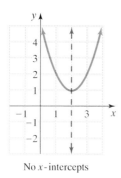

No x-intercepts

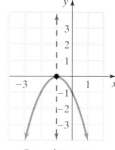

One x-intercept

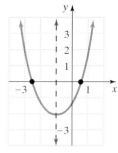

Two x-intercepts

FIGURE 11.12

Graphing a Parabola

When graphing a parabola, we can use the features that we have discussed to improve accuracy and understanding. These features are summarized in the following box as a strategy for graphing parabolas.

> **Graphing the Parabola $y = ax^2 + bx + c$**
>
> To graph the parabola $y = ax^2 + bx + c$, use the following facts:
>
> **1.** The parabola opens upward if $a > 0$ and opens downward if $a < 0$.
> **2.** The x-coordinate of the vertex is $\frac{-b}{2a}$.
> **3.** The graph is symmetric about the vertical line $x = \frac{-b}{2a}$.
> **4.** The y-intercept is $(0, c)$.
> **5.** The x-intercepts are found by solving $ax^2 + bx + c = 0$.

CAUTION   The focus and directrix are important features of a parabola, but they are not part of the curve itself. So we do not usually find the focus and directrix when graphing a parabola from an equation.

EXAMPLE 5

helpful hint

When drawing a parabola such as the one in Fig. 11.13 by hand, use your hand like a compass and draw it in two steps. If you are right-handed, sketch the left half of the parabola first. Then turn your paper upside down to sketch the right half.

Graphing a parabola

Determine whether the parabola $y = -x^2 - x + 6$ opens upward or downward, and find the vertex, axis of symmetry, x-intercepts, and y-intercept. Find several additional points on the parabola, and sketch the graph.

Solution

Because $a = -1$, this parabola opens downward. The x-coordinate of the vertex is

$$x = \frac{-b}{2a} = \frac{-(-1)}{2(-1)} = -\frac{1}{2}.$$

If $x = -\frac{1}{2}$, then

$$y = -\left(-\frac{1}{2}\right)^2 - \left(-\frac{1}{2}\right) + 6 = -\frac{1}{4} + \frac{1}{2} + 6 = \frac{25}{4}.$$

So the vertex is $\left(-\frac{1}{2}, \frac{25}{4}\right)$. The axis of symmetry is the vertical line $x = -\frac{1}{2}$. To find the x-intercepts, let $y = 0$ in the original equation:

$$-x^2 - x + 6 = 0$$
$$x^2 + x - 6 = 0$$
$$(x + 3)(x - 2) = 0$$
$$x + 3 = 0 \quad \text{or} \quad x - 2 = 0$$
$$x = -3 \quad \text{or} \quad x = 2$$

The x-intercepts are $(-3, 0)$ and $(2, 0)$. The y-intercept is $(0, 6)$. Using all of this information and the additional points $(1, 4)$, $(-2, 4)$, and $(-1, 6)$, we get the graph shown in Fig. 11.13. ■

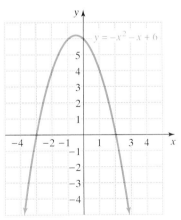

FIGURE 11.13

Maximum or Minimum Value of y

On a parabola that opens upward, the minimum value that y attains is the y-coordinate of the vertex. On a parabola that opens downward, the maximum value of y is again the y-coordinate of the vertex. We use the fact that the x-coordinate of the vertex is $\frac{-b}{2a}$ to obtain the maximum or minimum value of y.

E X A M P L E   6

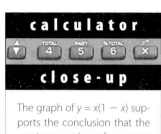

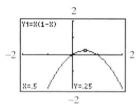

The graph of $y = x(1 - x)$ supports the conclusion that the maximum value of y occurs when $x = 1/2$.

Maximizing a product

Find two numbers that have the maximum product, subject to the condition that their sum is 1.

Solution

Because their sum is 1, we can let x represent one number and $1 - x$ represent the other. If y represents their product, we can write the equation

$$y = x(1 - x)$$

or

$$y = -x^2 + x.$$

This equation is the equation of a parabola that opens downward. Its highest point, which is the maximum value of y, occurs when

$$x = \frac{-b}{2a}$$
$$= \frac{-1}{2(-1)}$$
$$= \frac{1}{2}.$$

If $x = \frac{1}{2}$, then $1 - x = \frac{1}{2}$. The two numbers are $\frac{1}{2}$ and $\frac{1}{2}$. Among the numbers with a sum of 1, they give the maximum product. So no two numbers with a sum of 1 can have a product larger than $\frac{1}{4}$. ■

CAUTION Be sure to answer the question that is asked in a maximum-minimum problem. If $y = ax^2 + bx + c$, then the maximum (or minimum) value of y is the y-coordinate of the vertex. The value of x that causes y to reach its maximum (or minimum) value is the x-coordinate of the vertex.

E X A M P L E   7

Maximizing revenue

A manufacturer of in-line skates uses the formula $R = 360x - 2x^2$ to predict the weekly revenue in dollars that will be produced when the skates are priced at x dollars per pair. Find the price that will produce the maximum revenue, and find the maximum possible revenue.

Solution

The graph of $R = 360x - 2x^2$ is a parabola that opens downward as shown in Fig. 11.14 on the next page. To find the vertex of the parabola, use $a = -2$ and $b = 360$ in $x = \frac{-b}{2a}$:

$$x = \frac{-360}{2(-2)} = 90$$

Now use $x = 90$ in $R = 360x - 2x^2$ to find R:

$$R = 360(90) - 2 \cdot 90^2 = 16{,}200$$

So if the skates are priced at $90 per pair, then the manufacturer will receive the maximum weekly revenue $16,200.

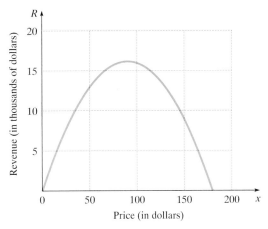

FIGURE 11.14

11.2   EXERCISES

Reading and Writing   After reading this section, write out the answers to these questions. Use complete sentences.

1. What is the definition of a parabola given in this section?
A parabola is the set of all points in a plane that are equidistant from a given line and a fixed point not on the line.

2. What is the location of the vertex?
The vertex is the midpoint of the line segment joining the focus and directrix, perpendicular to the directrix.

3. What are the two forms of the equation of a parabola?
A parabola can be written in the forms $y = ax^2 + bx + c$ or $y = a(x - h)^2 + k$.

4. What is the distance from the focus to the vertex in any parabola of the form $y = ax^2 + bx + c$?
The distance from the focus to the directrix is $|2p|$, where $a = 1/(4p)$.

5. How do we convert an equation of the form $y = ax^2 + bx + c$ into the form $y = a(x - h)^2 + k$?
We use completing the square to convert $y = ax^2 + bx + c$ into $y = a(x - h)^2 + k$.

6. How do we convert an equation of the form $y = a(x - h)^2 + k$ into the form $y = ax^2 + bx + c$?
To convert $y = a(x - h)^2 + k$ into the form $y = ax^2 + bx + c$, square the binomial, multiply by a, then add like terms.

Find the vertex, focus, and directrix for each parabola. See Example 1.

7. $y = 2x^2$ Vertex $(0, 0)$, focus $\left(0, \frac{1}{8}\right)$, directrix $y = -\frac{1}{8}$.

8. $y = \frac{1}{2}x^2$ Vertex $(0, 0)$, focus $\left(0, \frac{1}{2}\right)$, directrix $y = -\frac{1}{2}$.

9. $y = -\dfrac{1}{4}x^2$ Vertex $(0, 0)$, focus $(0, -1)$, directrix $y = 1$

10. $y = -\dfrac{1}{12}x^2$ Vertex $(0, 0)$, focus $(0, -3)$, directrix $y = 3$

11. $y = \dfrac{1}{2}(x - 3)^2 + 2$

Vertex $(3, 2)$, focus $(3, 2.5)$, directrix $y = 1.5$

12. $y = \dfrac{1}{4}(x + 2)^2 - 5$

Vertex $(-2, -5)$, focus $(-2, -4)$, directrix $y = -6$

13. $y = -(x + 1)^2 + 6$

Vertex $(-1, 6)$, focus $(-1, 5.75)$, directrix $y = 6.25$

14. $y = -3(x - 4)^2 + 1$

Vertex $(4, 1)$, focus $\left(4, \dfrac{11}{12}\right)$, directrix $y = \dfrac{13}{12}$

Find the equation of the parabola with the given focus and directrix. See Example 2.

15. Focus $(0, 2)$, directrix $y = -2$ $y = \dfrac{1}{8}x^2$

16. Focus $(0, -3)$, directrix $y = 3$ $y = -\dfrac{1}{12}x^2$

17. Focus $\left(0, -\dfrac{1}{2}\right)$, directrix $y = \dfrac{1}{2}$ $y = -\dfrac{1}{2}x^2$

18. Focus $\left(0, \dfrac{1}{8}\right)$, directrix $y = -\dfrac{1}{8}$ $y = 2x^2$

19. Focus $(3, 2)$, directrix $y = 1$ $y = \dfrac{1}{2}x^2 - 3x + 6$

20. Focus $(-4, 5)$, directrix $y = 4$ $y = \dfrac{1}{2}x^2 + 4x + \dfrac{25}{2}$

21. Focus $(1, -2)$, directrix $y = 2$ $y = -\dfrac{1}{8}x^2 + \dfrac{1}{4}x - \dfrac{1}{8}$

22. Focus $(2, -3)$, directrix $y = 1$ $y = -\dfrac{1}{8}x^2 + \dfrac{1}{2}x - \dfrac{3}{2}$

23. Focus $(-3, 1.25)$, directrix $y = 0.75$ $y = x^2 + 6x + 10$

24. Focus $\left(5, \dfrac{17}{8}\right)$, directrix $y = \dfrac{15}{8}$ $y = 2x^2 - 20x + 52$

Write each equation in the form $y = a(x - h)^2 + k$. Identify the vertex, focus, and directrix of each parabola. See Example 3.

25. $y = x^2 - 6x + 1$

$y = (x - 3)^2 - 8$, vertex $(3, -8)$, focus $(3, -7.75)$, directrix $y = -8.25$

26. $y = x^2 + 4x - 7$

$y = (x + 2)^2 - 11$, vertex $(-2, -11)$, focus $(-2, -10.75)$, directrix $y = -11.25$

27. $y = 2x^2 + 12x + 5$

$y = 2(x + 3)^2 - 13$, vertex $(-3, -13)$, focus $(-3, -12.875)$, directrix $y = -13.125$

28. $y = 3x^2 + 6x - 7$

$y = 3(x + 1)^2 - 10$, vertex $(-1, -10)$, focus $\left(-1, -9\dfrac{11}{12}\right)$, directrix $y = -10\dfrac{1}{12}$

29. $y = -2x^2 + 16x + 1$

$y = -2(x - 4)^2 + 33$, vertex $(4, 33)$, focus $\left(4, 32\dfrac{7}{8}\right)$, directrix $y = 33\dfrac{1}{8}$

30. $y = -3x^2 - 6x + 7$

$y = -3(x + 1)^2 + 10$, vertex $(-1, 10)$, focus $\left(-1, 9\dfrac{11}{12}\right)$, directrix $y = 10\dfrac{1}{12}$

31. $y = 5x^2 + 40x$

$y = 5(x + 4)^2 - 80$, vertex $(-4, -80)$, focus $\left(-4, -79\dfrac{19}{20}\right)$, directrix $y = -80\dfrac{1}{20}$

32. $y = -2x^2 + 10x$

$y = -2\left(x - \dfrac{5}{2}\right)^2 + \dfrac{25}{2}$, vertex $\left(\dfrac{5}{2}, \dfrac{25}{2}\right)$, focus $\left(\dfrac{5}{2}, \dfrac{99}{8}\right)$, directrix $y = \dfrac{101}{8}$

Find the vertex, focus, and directrix of each parabola (without completing the square), and determine whether the parabola opens upward or downward. See Example 4.

33. $y = x^2 - 4x + 1$

Vertex $(2, -3)$, focus $\left(2, -2\dfrac{3}{4}\right)$, directrix $y = -3\dfrac{1}{4}$, upward

34. $y = x^2 - 6x - 7$

Vertex $(3, -16)$, focus $\left(3, -15\dfrac{3}{4}\right)$, directrix $y = -16\dfrac{1}{4}$, upward

35. $y = -x^2 + 2x - 3$

Vertex $(1, -2)$, focus $\left(1, -2\dfrac{1}{4}\right)$, directrix $y = -1\dfrac{3}{4}$, downward

36. $y = -x^2 + 4x + 9$

Vertex $(2, 13)$, focus $\left(2, 12\dfrac{3}{4}\right)$, directrix $y = 13\dfrac{1}{4}$, downward

37. $y = 3x^2 - 6x + 1$

Vertex $(1, -2)$, focus $\left(1, -1\dfrac{11}{12}\right)$, directrix $y = -2\dfrac{1}{12}$, upward

38. $y = 2x^2 + 4x - 3$

Vertex $(-1, -5)$, focus $\left(-1, -4\dfrac{7}{8}\right)$, directrix $y = -5\dfrac{1}{8}$, upward

39. $y = -x^2 - 3x + 2$

Vertex $\left(-\dfrac{3}{2}, \dfrac{17}{4}\right)$, focus $\left(-\dfrac{3}{2}, 4\right)$, directrix $y = \dfrac{9}{2}$, downward

40. $y = -x^2 + 3x - 1$

Vertex $\left(\dfrac{3}{2}, \dfrac{5}{4}\right)$, focus $\left(\dfrac{3}{2}, 1\right)$, directrix $y = \dfrac{3}{2}$, downward

41. $y = 3x^2 + 5$

Vertex $(0, 5)$, focus $\left(0, 5\dfrac{1}{12}\right)$, directrix $y = 4\dfrac{11}{12}$, upward

42. $y = -2x^2 - 6$

Vertex $(0, -6)$, focus $\left(0, -6\dfrac{1}{8}\right)$, directrix $y = -5\dfrac{7}{8}$, downward

Find the vertex, axis of symmetry, x-intercepts, and y-intercept for each parabola. Find several additional points on the parabola, and then sketch its graph. See Example 5.

43. $y = x^2 - 3x + 2$

Vertex $\left(\dfrac{3}{2}, -\dfrac{1}{4}\right)$ axis of

symmetry $x = \dfrac{3}{2}$, intercepts

$(0, 2), (1, 0), (2, 0)$

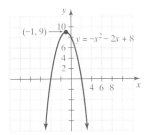

44. $y = x^2 + 6x + 8$

Vertex $(-3, -1)$, axis of symmetry $x = -3$, intercepts $(0, 8), (-4, 0), (-2, 0)$

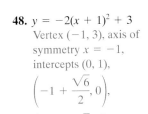

45. $y = -x^2 - 2x + 8$

Vertex $(-1, 9)$, axis of symmetry $x = -1$, intercepts $(0, 8), (-4, 0), (2, 0)$

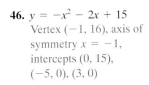

46. $y = -x^2 - 2x + 15$

Vertex $(-1, 16)$, axis of symmetry $x = -1$, intercepts $(0, 15), (-5, 0), (3, 0)$

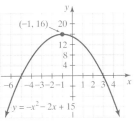

47. $y = (x + 2)^2 + 1$

Vertex $(-2, 1)$, axis of symmetry $x = -2$, intercept $(0, 5)$

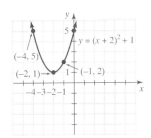

48. $y = -2(x + 1)^2 + 3$

Vertex $(-1, 3)$, axis of symmetry $x = -1$, intercepts $(0, 1)$,

$\left(-1 + \dfrac{\sqrt{6}}{2}, 0\right)$,

$\left(-1 - \dfrac{\sqrt{6}}{2}, 0\right)$

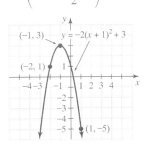

49. $y = x^2 + 2x + 1$

Vertex $(-1, 0)$, axis of symmetry $x = -1$, intercepts $(0, 1), (-1, 0)$

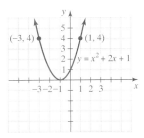

50. $y = x^2 - 6x + 9$

Vertex $(3, 0)$, axis of symmetry $x = 3$, intercepts $(0, 9), (3, 0)$

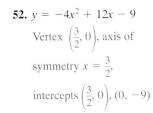

51. $y = -4x^2 + 4x - 1$

Vertex $\left(\dfrac{1}{2}, 0\right)$, axis of

symmetry $x = \dfrac{1}{2}$,

intercepts $(0, -1), \left(\dfrac{1}{2}, 0\right)$

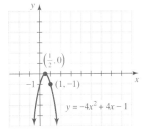

52. $y = -4x^2 + 12x - 9$

Vertex $\left(\dfrac{3}{2}, 0\right)$, axis of

symmetry $x = \dfrac{3}{2}$,

intercepts $\left(\dfrac{3}{2}, 0\right), (0, -9)$

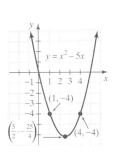

53. $y = x^2 - 5x$

Vertex is $\left(\dfrac{5}{2}, -\dfrac{25}{4}\right)$, axis of

symmetry $x = \dfrac{5}{2}$,

intercepts $(0, 0), (5, 0)$

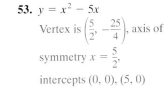

54. $y = 3x^2 - 9x$

Vertex $\left(\dfrac{3}{2}, -\dfrac{27}{4}\right)$, axis of

symmetry $x = \dfrac{3}{2}$,

intercepts $(0, 0), (3, 0)$

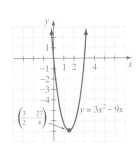

55. $y = 3x^2 + 5$

Vertex $(0, 5)$, axis of symmetry $x = 0$, intercept $(0, 5)$

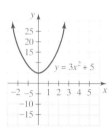

56. $y = -2x^2 + 3$

Vertex $(0, 3)$, axis of symmetry $x = 0$, intercepts $(0, 3)$, $\left(-\dfrac{\sqrt{6}}{2}, 0\right), \left(\dfrac{\sqrt{6}}{2}, 0\right)$

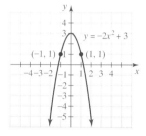

57. $y = x^2 - 2x - 1$

Vertex $(1, -2)$, axis of symmetry $x = 1$, intercepts $(0, -1)$, $(1 + \sqrt{2}, 0), (1 - \sqrt{2}, 0)$

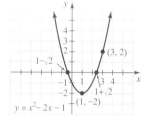

58. $y = x^2 - 4x + 1$

Vertex $(2, -3)$, axis of symmetry $x = 2$, intercepts $(0, 1), (2 - \sqrt{3}, 0)$, $(2 + \sqrt{3}, 0)$

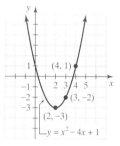

59. $y = (x - 5)^2$

Vertex $(5, 0)$, axis of symmetry $x = 5$, intercepts $(0, 25), (5, 0)$

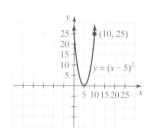

60. $y = 3(x - 1)^2 - 4$

Vertex $(1, -4)$, axis of symmetry $x = 1$, intercepts $(0, -1)$, $\left(1 + \dfrac{2\sqrt{3}}{3}, 0\right),$ $\left(1 - \dfrac{2\sqrt{3}}{3}, 0\right)$

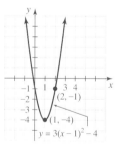

Solve each problem. See Examples 6 and 7.

61. Maximum product. Find two numbers that have the maximum possible product among numbers that have a sum of 8.

4 and 4

62. Maximum product. What is the maximum product that can be obtained by two numbers that a have a sum of -6?

9

63. Maximum area. A gardener plans to enclose a rectangular area with 160 feet (ft) of fencing. What dimensions for the length and width would maximize the area?

40 ft by 40 ft

64. Maximum area. Another gardener has a 4-foot-wide gate for her rectangular garden in place. If she has 100 ft of fencing in addition to the gate, then what dimensions for the length and width would maximize the area?

26 ft by 26 ft

65. Minimum hypotenuse. If the total length of the legs of a right triangle is 6 ft, then what lengths for the legs would minimize the square of the hypotenuse?

3 ft each

66. Maximum revenue. A concert promoter uses the formula $R = 7500p - 250p^2$ to calculate the total revenue for a concert if the ticket price is p dollars each. What ticket price will maximize the revenue? See the accompanying figure.

$15

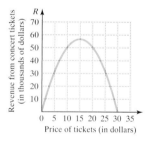

FIGURE FOR EXERCISE 66

67. Maximum height. If a punter kicks a football straight up at a velocity of 128 feet per second (ft/sec) from a height of 6 ft, then the football's distance above the earth after t seconds is given by the formula $s = -16t^2 + 128t + 6$. What is the maximum height reached by the ball? How long does the ball take to reach its maximum height? Could the punter

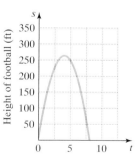

FIGURE FOR EXERCISE 67

hit the roof of the New Orleans Superdome, which is 273 ft above the playing field?   262 ft, 4 sec, no

68. Maximum height. If a baseball is hit straight upward at 150 ft/sec from a height of 5 ft, then its distance above the earth after t seconds is given by the formula $s = -16t^2 + 150t + 5$. What is the maximum height attained by the ball? How long does the ball take to reach its maximum height?   356.5625 ft, 4.6875 sec

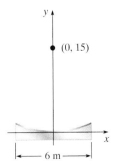

$v = 150$ ft/sec

5 ft

FIGURE FOR EXERCISE 68

69. Minimum cost. It costs Acme Manufacturing C dollars per hour to operate its golf ball division. An analyst has determined that C is related to the number of golf balls produced per hour, x, by the equation $C = 0.009x^2 - 1.8x + 100$. What number of balls per hour should Acme produce to minimize the cost per hour of manufacturing these golf balls?   100

70. Maximum profit. A chain store manager has been told by the main office that daily profit, P, is related to the number of clerks working that day, x, according to the equation $P = -25x^2 + 300x$. What number of clerks will maximize the profit, and what is the maximum possible profit?   6 clerks, $900

71. World's largest telescope. The largest reflecting telescope in the world is the 6-meter (m) reflector on Mount Pastukhov in Russia. The accompanying figure shows a cross section of a parabolic mirror 6 m in diameter with the vertex at the origin and the focus at (0, 15). Find the equation of the parabola.

$$y = \frac{1}{60}x^2$$

(0, 15)

6 m

FIGURE FOR EXERCISE 71

72. Arecibo Observatory. The largest radio telescope in the world uses a 1,000-ft parabolic dish, suspended in a valley in Arecibo, Puerto Rico. The antenna hangs above the vertex of the dish on cables stretching from two towers. The accompanying figure shows a cross section of the parabolic dish and the towers. Assuming the vertex is at (0, 0), find the equation for the parabola. Find the distance from the vertex to the antenna located at the focus.
$y = 0.0008x^2$, 312.5 ft

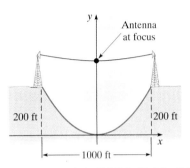

Antenna
at focus

200 ft   200 ft

1000 ft

FIGURE FOR EXERCISE 72

Graph both equations of each system on the same coordinate axes. Use elimination of variables to find all points of intersection.

73. $y = -x^2 + 3$
    $y = x^2 + 1$
    $\{(-1, 2), (1, 2)\}$

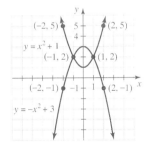

74. $y = x^2 - 3$
    $y = -x^2 + 5$
    $\{(2, 1), (-2, 1)\}$

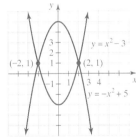

75. $y = x^2 - 2$
    $y = 2x - 3$
    $\{(1, -1)\}$

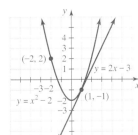

76. $y = x^2 + x - 6$
    $y = 7x - 15$
    $\{(3, 6)\}$

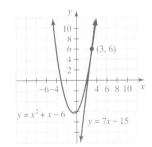

77. $y = x^2 + 3x - 4$
$y = -x^2 - 2x + 8$
$\left\{\left(\dfrac{3}{2}, \dfrac{11}{4}\right), (-4, 0)\right\}$

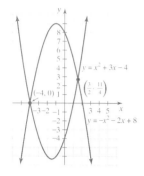

78. $y = x^2 + 2x - 8$
$y = -x^2 - x + 12$
$\left\{\left(\dfrac{5}{2}, \dfrac{13}{4}\right), (-4, 0)\right\}$

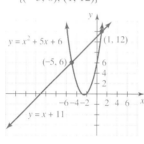

79. $y = x^2 + 3x - 4$
$y = 2x + 2$
$\{(-3, -4), (2, 6)\}$

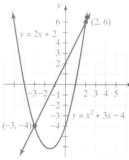

80. $y = x^2 + 5x + 6$
$y = x + 11$
$\{(-5, 6), (1, 12)\}$

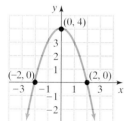

Write an equation in the form $y = ax^2 + bx + c$ for each given parabola.

81. $y = -x^2 + 4$

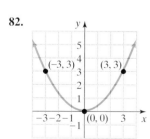

82. $y = \dfrac{1}{3}x^2$

83.

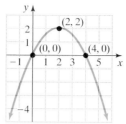

$y = -\dfrac{1}{2}x^2 + 2x$

84.

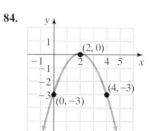

$y = -\dfrac{3}{4}x^2 + 3x - 3$

GETTING MORE INVOLVED

85. *Exploration.* Consider the parabola with focus $(p, 0)$ and directrix $x = -p$ for $p > 0$. Let (x, y) be an arbitrary point on the parabola. Write an equation expressing the fact that the distance from (x, y) to the focus is equal to the distance from (x, y) to the directrix. Rewrite the equation in the form $x = ay^2$, where $a = \dfrac{1}{4p}$.

86. *Exploration.* In general, the graph of $x = a(y - h)^2 + k$ for $a \neq 0$ is a parabola opening left or right with vertex at (k, h).
a) For which values of a does the parabola open to the right, and for which values of a does it open to the left?
b) What is the equation of its axis of symmetry?
c) Sketch the graphs $x = 2(y - 3)^2 + 1$ and
$x = -(y + 1)^2 + 2$.
   a) Right for $a > 0$ and left for $a < 0$ **b)** $y = h$

GRAPHING CALCULATOR EXERCISES

87. Graph $y = x^2$ using the viewing window with $-1 \leq x \leq 1$ and $0 \leq y \leq 1$. Next graph $y = 2x^2 - 1$ using the viewing window $-2 \leq x \leq 2$ and $-1 \leq y \leq 7$. Explain what you see.
The graphs have identical shapes.

88. Graph $y = x^2$ and $y = 6x - 9$ in the viewing window $-5 \leq x \leq 5$ and $-5 \leq y \leq 20$. Does the line appear to be tangent to the parabola? Solve the system $y = x^2$ and $y = 6x - 9$ to find all points of intersection for the parabola and the line.
Intersection $(3, 9)$

11.3   THE CIRCLE

In this section we continue the study of the conic sections with a discussion of the circle.

Developing the Equation

A circle is obtained by cutting a cone, as was shown in Fig. 11.3. We can also define a circle using points and distance, as we did for the parabola.

Circle

A **circle** is the set of all points in a plane that lie a fixed distance from a given point in the plane. The fixed distance is called the **radius,** and the given point is called the **center.**

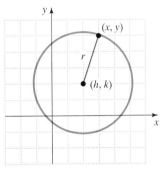

FIGURE 11.15

We can use the distance formula of Section 7.5 to write an equation for the circle with center (h, k) and radius r, shown in Fig. 11.15. If (x, y) is a point on the circle, its distance from the center is r. So

$$\sqrt{(x - h)^2 + (y - k)^2} = r.$$

We square both sides of this equation to get the **standard form** for the equation of a circle.

Standard Equation for a Circle

The graph of the equation

$$(x - h)^2 + (y - k)^2 = r^2$$

with $r > 0$, is a circle with center (h, k) and radius r.

Note that a circle centered at the origin with radius r ($r > 0$) has the standard equation

$$x^2 + y^2 = r^2.$$

EXAMPLE 1

Finding the equation, given the center and radius

Write the standard equation for the circle with the given center and radius.

a) Center $(0, 0)$, radius 2

b) Center $(-1, 2)$, radius 4

Solution

a) The center at $(0, 0)$ means that $h = 0$ and $k = 0$ in the standard equation. So the equation is $(x - 0)^2 + (y - 0)^2 = 2^2$, or $x^2 + y^2 = 4$. The circle with radius 2 centered at the origin is shown in Fig. 11.16.

b) The center at $(-1, 2)$ means that $h = -1$ and $k = 2$. So

$$[x - (-1)]^2 + [y - 2]^2 = 4^2.$$

Simplify this equation to get

$$(x + 1)^2 + (y - 2)^2 = 16.$$

FIGURE 11.16

The circle with center $(-1, 2)$ and radius 4 is shown in Fig. 11.17.

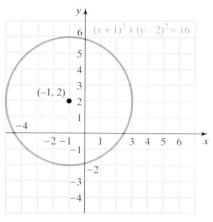

FIGURE 11.17

CAUTION   The equations $(x - 1)^2 + (y + 3)^2 = -9$ and $(x - 1)^2 + (y + 3)^2 = 0$ might look like equations of circles, but they are not. The first equation is not satisfied by any ordered pair of real numbers because the left-hand side is nonnegative for any x and y. The second equation is satisfied only by the point $(1, -3)$.

E X A M P L E   2

Finding the center and radius, given the equation

Determine the center and radius of the circle $x^2 + (y + 5)^2 = 2$.

Solution

We can write this equation as

$$(x - 0)^2 + [y - (-5)]^2 = (\sqrt{2})^2.$$

In this form we see that the center is $(0, -5)$ and the radius is $\sqrt{2}$.

E X A M P L E   3

Graphing a circle

Find the center and radius of $(x - 1)^2 + (y + 2)^2 = 9$, and sketch the graph.

Solution

The graph of this equation is a circle with center at $(1, -2)$ and radius 3. See Fig. 11.18 for the graph.

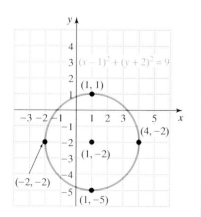

FIGURE 11.18

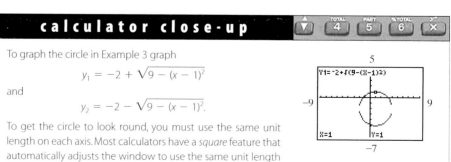

calculator close-up

To graph the circle in Example 3 graph

$$y_1 = -2 + \sqrt{9 - (x - 1)^2}$$

and

$$y_2 = -2 - \sqrt{9 - (x - 1)^2}.$$

To get the circle to look round, you must use the same unit length on each axis. Most calculators have a *square* feature that automatically adjusts the window to use the same unit length on each axis.

Equations Not in Standard Form

It is not easy to recognize that $x^2 - 6x + y^2 + 10y = -30$ is the equation of a circle, but it is. In the next example we convert this equation into the standard form for a circle by completing the squares for the variables x and y.

E X A M P L E   4

Converting to standard form

Find the center and radius of the circle given by the equation

$$x^2 - 6x + y^2 + 10y = -30.$$

Solution

helpful hint

What do circles and lines have in common? They are the two simplest graphs to draw. We have compasses to make our circles look good and rulers to make our lines look good.

To complete the square for $x^2 - 6x$, we add 9, and for $y^2 + 10y$, we add 25. To get an equivalent equation, we must add on both sides:

$$x^2 - 6x + y^2 + 10y = -30$$

$$x^2 - 6x + 9 + y^2 + 10y + 25 = -30 + 9 + 25 \qquad \text{Add 9 and 25 to both sides.}$$

$$(x - 3)^2 + (y + 5)^2 = 4 \qquad \text{Factor the trinomials on the left-hand side.}$$

From the standard form we see that the center is $(3, -5)$ and the radius is 2. ■

Systems of Equations

We first solved systems of nonlinear equations in two variables in Section 11.1. We found the points of intersection of two graphs without drawing the graphs. Here we will solve systems involving circles, parabolas, and lines. In the next example we find the points of intersection of a line and a circle.

E X A M P L E   5

Intersection of a line and a circle

Graph both equations of the system

$$(x - 3)^2 + (y + 1)^2 = 9$$
$$y = x - 1$$

on the same coordinate axes, and solve the system by elimination of variables.

Solution

The graph of the first equation is a circle with center at $(3, -1)$ and radius 3. The graph of the second equation is a straight line with slope 1 and y-intercept $(0, -1)$. Both graphs are shown in Fig. 11.19. To solve the system by elimination, we substitute $y = x - 1$ into the equation of the circle:

$$(x - 3)^2 + (x - 1 + 1)^2 = 9$$
$$(x - 3)^2 + x^2 = 9$$
$$x^2 - 6x + 9 + x^2 = 9$$
$$2x^2 - 6x = 0$$
$$x^2 - 3x = 0$$
$$x(x - 3) = 0$$

$$x = 0 \qquad \text{or} \qquad x = 3$$
$$y = -1 \qquad \qquad y = 2 \qquad \text{Because } y = x - 1$$

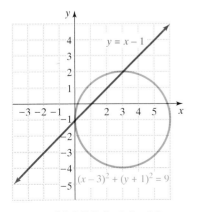

FIGURE 11.19

Check $(0, -1)$ and $(3, 2)$ in the original system and with the graphs in Fig. 11.19. The solution set is $\{(0, -1), (3, 2)\}$. ■

M A T H   A T   W O R K $x^2 + (x+1)^2 = 5^2$

Friedrich von Huene, a flautist and recorder player, has been crafting woodwind instruments in his family business for over 30 years. Because it is best to play music of earlier centuries on the instruments of their time, von Huene is using many originals as models for his flutes, recorders, and oboes of different sizes.

EARLY MUSICAL INSTRUMENT MAKER

Because museum instruments have many different pitch standards, their dimensions frequently have to be changed to accommodate pitch standards that musicians use today. For a lower pitch the length of the instrument as well as the inside diameter must be increased. For a higher pitch the length has to be shortened and the diameter decreased. However, the factor for changing the length will be different from the factor for changing the diameter. A row of organ pipes demonstrates that the larger and longer pipes are proportionately more slender than the shorter high-pitched pipes. A pipe an octave higher in pitch is about half as long as the pipe an octave lower, but its diameter will be about 0.6 as large as the diameter of the lower pipe.

When making a very large recorder, von Huene carefully chooses the length, the position of the tone holes, and the bore to get the proper volume of air inside the instrument. In Exercises 57 and 58 of this section you will make the kinds of calculations von Huene makes when he crafts a modern reproduction of a Renaissance flute.

WARM-UPS

True or false? Explain your answer.

1. The radius of a circle can be any nonzero real number.   False

2. The coordinates of the center must satisfy the equation of the circle.   False

3. The circle $x^2 + y^2 = 4$ has its center at the origin.   True

4. The graph of $x^2 + y^2 = 9$ is a circle centered at $(0, 0)$ with radius 9.   False

5. The graph of $(x - 2)^2 + (y - 3)^2 + 4 = 0$ is a circle of radius 2.   False

6. The graph of $(x - 3) + (y + 5) = 9$ is a circle of radius 3.   False

7. There is only one circle centered at $(-3, -1)$ passing through the origin.   True

8. The center of the circle $(x - 3)^2 + (y - 4)^2 = 10$ is $(-3, -4)$.   False

9. The center of the circle $x^2 + y^2 + 6y - 4 = 0$ is on the y-axis.   True

10. The radius of the circle $x^2 - 3x + y^2 = 4$ is 2.   False

11.3   EXERCISES

Reading and Writing   *After reading this section, write out the answers to these questions. Use complete sentences.*

1. What is the definition of a circle?

A circle is the set of all points in a plane that lie at a fixed distance from a fixed point.

2. What is the standard equation of a circle?

The equation $(x - h)^2 + (y - k)^2 = r^2$ is the standard equation of a circle with center (h, k) and radius r (for $r > 0$).

Write the standard equation for each circle with the given center and radius. See Example 1.

3. Center $(0, 3)$, radius 5 $x^2 + (y - 3)^2 = 25$

4. Center $(2, 0)$, radius 3 $(x - 2)^2 + y^2 = 9$

5. Center $(1, -2)$, radius 9 $(x - 1)^2 + (y + 2)^2 = 81$

6. Center $(-3, 5)$, radius 4 $(x + 3)^2 + (y - 5)^2 = 16$

7. Center $(0, 0)$, radius $\sqrt{3}$ $x^2 + y^2 = 3$

8. Center $(0, 0)$, radius $\sqrt{2}$ $x^2 + y^2 = 2$

9. Center $(-6, -3)$, radius $\dfrac{1}{2}$ $(x + 6)^2 + (y + 3)^2 = \dfrac{1}{4}$

10. Center $(-3, -5)$, radius $\dfrac{1}{4}$ $(x + 3)^2 + (y + 5)^2 = \dfrac{1}{16}$

11. Center $\left(\frac{1}{2}, \frac{1}{3}\right)$, radius 0.1 $\left(x - \dfrac{1}{2}\right)^2 + \left(y - \dfrac{1}{3}\right)^2 = 0.01$

12. Center $\left(-\frac{1}{2},\ 3\right)$, radius 0.2 $\left(x + \dfrac{1}{2}\right)^2 + (y - 3)^2 = 0.04$

Find the center and radius for each circle. See Example 2.

13. $(x - 3)^2 + (y - 5)^2 = 2$ $(3, 5), \sqrt{2}$

14. $(x + 3)^2 + (y - 7)^2 = 6$ $(-3, 7), \sqrt{6}$

15. $x^2 + \left(y - \dfrac{1}{2}\right)^2 = \dfrac{1}{2}$ $\left(0, \dfrac{1}{2}\right), \dfrac{\sqrt{2}}{2}$

16. $5x^2 + 5y^2 = 5$ $(0, 0), 1$

17. $4x^2 + 4y^2 = 9$ $(0, 0), \dfrac{3}{2}$

18. $9x^2 + 9y^2 = 49$ $(0, 0), \dfrac{7}{3}$

19. $3 - y^2 = (x - 2)^2$ $(2, 0), \sqrt{3}$

20. $9 - x^2 = (y + 1)^2$ $(0, -1), 3$

Sketch the graph of each equation. See Example 3.

21. $x^2 + y^2 = 9$

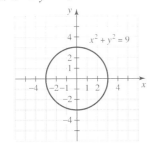

22. $x^2 + y^2 = 16$

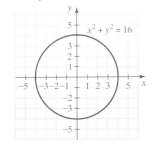

23. $x^2 + (y - 3)^2 = 9$

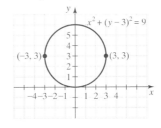

24. $(x - 4)^2 + y^2 = 16$

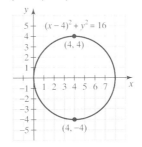

25. $(x + 1)^2 + (y - 1)^2 = 2$

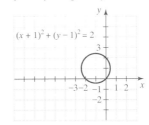

26. $(x - 2)^2 + (y + 2)^2 = 8$

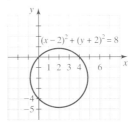

27. $(x - 4)^2 + (y + 3)^2 = 16$

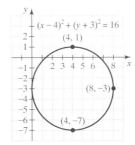

28. $(x - 3)^2 + (y - 7)^2 = 25$

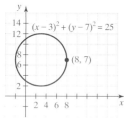

29. $\left(x - \dfrac{1}{2}\right)^2 + \left(y + \dfrac{1}{2}\right)^2 = \dfrac{1}{4}$

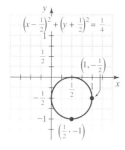

30. $\left(x + \dfrac{1}{3}\right)^2 + y^2 = \dfrac{1}{9}$

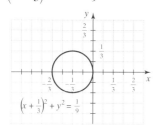

Rewrite each equation in the standard form for the equation of a circle, and identify its center and radius. See Example 4.

31. $x^2 + 4x + y^2 + 6y = 0$
$(x + 2)^2 + (y + 3)^2 = 13, (-2, -3), \sqrt{13}$

32. $x^2 - 10x + y^2 + 8y = 0$
$(x - 5)^2 + (y + 4)^2 = 41, (5, -4), \sqrt{41}$

33. $x^2 - 2x + y^2 - 4y - 3 = 0$
$(x - 1)^2 + (y - 2)^2 = 8, (1, 2), 2\sqrt{2}$

34. $x^2 - 6x + y^2 - 2y + 9 = 0$
$(x - 3)^2 + (y - 1)^2 = 1, (3, 1), 1$

35. $x^2 + y^2 = 8y + 10x - 32$
$(x - 5)^2 + (y - 4)^2 = 9, (5, 4), 3$

36. $x^2 + y^2 = 8x - 10y$
$(x - 4)^2 + (y + 5)^2 = 41, (4, -5), \sqrt{41}$

37. $x^2 - x + y^2 + y = 0$
$\left(x - \dfrac{1}{2}\right)^2 + \left(y + \dfrac{1}{2}\right)^2 = \dfrac{1}{2}, \left(\dfrac{1}{2}, -\dfrac{1}{2}\right), \dfrac{\sqrt{2}}{2}$

38. $x^2 - 3x + y^2 = 0$
$\left(x - \dfrac{3}{2}\right)^2 + y^2 = \dfrac{9}{4}, \left(\dfrac{3}{2}, 0\right), \dfrac{3}{2}$

39. $x^2 - 3x + y^2 - y = 1$
$\left(x - \dfrac{3}{2}\right)^2 + \left(y - \dfrac{1}{2}\right)^2 = \dfrac{7}{2}, \left(\dfrac{3}{2}, \dfrac{1}{2}\right), \dfrac{\sqrt{14}}{2}$

40. $x^2 - 5x + y^2 + 3y = 2$
$\left(x - \dfrac{5}{2}\right)^2 + \left(y + \dfrac{3}{2}\right)^2 = \dfrac{21}{2}, \left(\dfrac{5}{2}, -\dfrac{3}{2}\right), \dfrac{\sqrt{42}}{2}$

41. $x^2 - \dfrac{2}{3}x + y^2 + \dfrac{3}{2}y = 0$
$\left(x - \dfrac{1}{3}\right)^2 + \left(y + \dfrac{3}{4}\right)^2 = \dfrac{97}{144}, \left(\dfrac{1}{3}, -\dfrac{3}{4}\right), \dfrac{\sqrt{97}}{12}$

42. $x^2 + \dfrac{1}{3}x + y^2 - \dfrac{2}{3}y = \dfrac{1}{9}$
$\left(x + \dfrac{1}{6}\right)^2 + \left(y - \dfrac{1}{3}\right)^2 = \dfrac{1}{4}, \left(-\dfrac{1}{6}, \dfrac{1}{3}\right), \dfrac{1}{2}$

Graph both equations of each system on the same coordinate axes. Solve the system by elimination of variables to find all points of intersection of the graphs. See Example 5.

43. $x^2 + y^2 = 10$
$y = 3x$
$\{(1, 3), (-1, -3)\}$

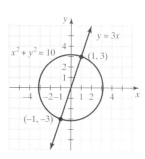

44. $x^2 + y^2 = 4$
$y = x - 2$
$\{(0, -2), (2, 0)\}$

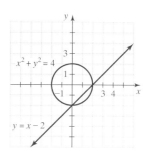

45. $x^2 + y^2 = 9$
$y = x^2 - 3$
$\{(0, -3), (\sqrt{5}, 2), (-\sqrt{5}, 2)\}$

46. $x^2 + y^2 = 4$
$y = x^2 - 2$
$\{(0, -2), (-\sqrt{3}, 1), (\sqrt{3}, 1)\}$

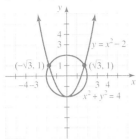

47. $(x - 2)^2 + (y + 3)^2 = 4$
$y = x - 3$
$\{(0, -3), (2, -1)\}$

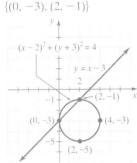

48. $(x + 1)^2 + (y - 4)^2 = 17$
$y = x + 2$
$\{(3, 5), (-2, 0)\}$

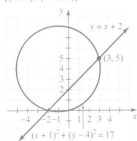

In Exercises 49–58, solve each problem.

49. Determine all points of intersection of the circle $(x - 1)^2 + (y - 2)^2 = 4$ with the y-axis.
$(0, 2 + \sqrt{3})$ and $(0, 2 - \sqrt{3})$

50. Determine the points of intersection of the circle $x^2 + (y - 3)^2 = 25$ with the x-axis.
$(-4, 0)$ and $(4, 0)$

51. Find the radius of the circle that has center $(2, -5)$ and passes through the origin.
$\sqrt{29}$

52. Find the radius of the circle that has center $(-2, 3)$ and passes through $(3, -1)$.
$\sqrt{41}$

53. Determine the equation of the circle that is centered at $(2, 3)$ and passes through $(-2, -1)$.
$(x - 2)^2 + (y - 3)^2 = 32$

54. Determine the equation of the circle that is centered at $(3, 4)$ and passes through the origin.
$(x - 3)^2 + (y - 4)^2 = 25$

55. Find all points of intersection of the circles $x^2 + y^2 = 9$ and $(x - 5)^2 + y^2 = 9$.
$\left(\dfrac{5}{2}, -\dfrac{\sqrt{11}}{2}\right)$ and $\left(\dfrac{5}{2}, \dfrac{\sqrt{11}}{2}\right)$

56. A donkey is tied at the point $(2, -3)$ on a rope of length 12. Turnips are growing at the point $(6, 7)$. Can the donkey reach them?
Yes

57. *Volume of a flute.* The volume of air in a flute is a critical factor in determining its pitch. A cross section of a Renaissance flute in C is shown in the accompanying figure. If the length of the flute is 2,874 millimeters, then what is the volume of air in the flute (to the nearest cubic millimeter (mm^3))? (*Hint:* Use the formula for the volume of a cylinder.)
755,903 mm^3

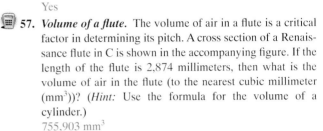

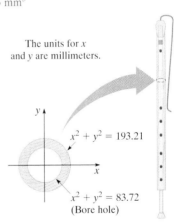

The units for x and y are millimeters.

$x^2 + y^2 = 193.21$

$x^2 + y^2 = 83.72$
(Bore hole)

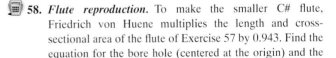

FIGURE FOR EXERCISES 57 AND 58

58. *Flute reproduction.* To make the smaller C# flute, Friedrich von Huene multiplies the length and cross-sectional area of the flute of Exercise 57 by 0.943. Find the equation for the bore hole (centered at the origin) and the volume of air in the C# flute.
$x^2 + y^2 = 78.95$, 672,186 mm^3

Graph each equation.

59. $x^2 + y^2 = 0$
$(0, 0)$ only

60. $x^2 - y^2 = 0$

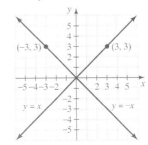

61. $y = \sqrt{1 - x^2}$

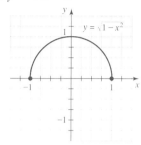

62. $y = -\sqrt{1 - x^2}$

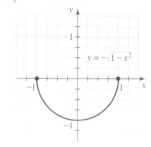

GETTING MORE INVOLVED

63. *Cooperative learning.* The equation of a circle is a special case of the general equation $Ax^2 + Bx + Cy^2 + Dy = E$, where $A, B, C, D,$ and E are real numbers. Working in small groups, find restrictions that must be placed on $A, B, C, D,$

and E so that the graph of this equation is a circle. What does the graph of $x^2 + y^2 = -9$ look like?

B and D can be any real numbers, but A must equal C, and $4AE + B^2 + D^2 > 0$. No ordered pairs satisfy $x^2 + y^2 = -9$.

64. *Discussion.* Suppose lighthouse A is located at the origin and lighthouse B is located at coordinates $(0, 6)$. The captain of a ship has determined that the ship's distance from lighthouse A is 2 and its distance from lighthouse B is 5. What are the possible coordinates for the location of the ship?

$$\left(-\frac{\sqrt{39}}{4}, \frac{5}{4}\right) \text{ and } \left(\frac{\sqrt{39}}{4}, \frac{5}{4}\right)$$

GRAPHING CALCULATOR EXERCISES

Graph each relation on a graphing calculator by solving for y and graphing two functions.

65. $x^2 + y^2 = 4$
   $y = \pm\sqrt{4 - x^2}$

66. $(x - 1)^2 + (y + 2)^2 = 1$
   $y = -2 \pm \sqrt{1 - (x - 1)^2}$

67. $x = y^2$
   $y = \pm\sqrt{x}$

68. $x = (y + 2)^2 - 1$
   $y = -2 \pm \sqrt{x + 1}$

69. $x = y^2 + 2y + 1$
   $y = -1 \pm \sqrt{x}$

70. $x = 4y^2 + 4y + 1$
   $y = \dfrac{-1 \pm \sqrt{x}}{2}$

11.4   THE ELLIPSE AND HYPERBOLA

In this section

• The Ellipse

• The Hyperbola

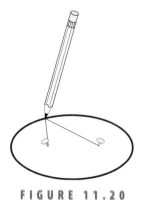

FIGURE 11.20

In this section we study the remaining two conic sections: the ellipse and the hyperbola.

The Ellipse

An ellipse can be obtained by intersecting a plane and a cone, as was shown in Fig. 11.3. We can also give a definition of an ellipse in terms of points and distance.

> **Ellipse**
>
> An **ellipse** is the set of all points in a plane such that the sum of their distances from two fixed points is a constant. Each fixed point is called a **focus** (plural: foci).

An easy way to draw an ellipse is illustrated in Fig. 11.20. A string is attached at two fixed points, and a pencil is used to take up the slack. As the pencil is moved around the paper, the sum of the distances of the pencil point from the two fixed points remains constant. Of course, the length of the string is that constant. You may wish to try this.

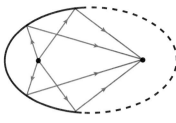

FIGURE 11.21

Like the parabola, the ellipse also has interesting reflecting properties. All light or sound waves emitted from one focus are reflected off the ellipse to concentrate at the other focus (see Fig. 11.21). This property is used in light fixtures where a concentration of light at a point is desired or in a whispering gallery such as Statuary Hall in the U.S. Capitol Building.

The orbits of the planets around the sun and satellites around the earth are elliptical. For the orbit of the earth around the sun, the sun is at one focus. For the elliptical path of an earth satellite the earth is at one focus and a point in space is the other focus.

Figure 11.22 shows an ellipse with foci $(c, 0)$ and $(-c, 0)$. The origin is the center of this ellipse. In general, the **center** of an ellipse is a point midway between the foci. The ellipse in Fig. 11.22 has x-intercepts at $(a, 0)$ and $(-a, 0)$ and y-intercepts at $(0, b)$ and $(0, -b)$. The distance formula can be used to write the following equation for this ellipse. (See Exercise 55.)

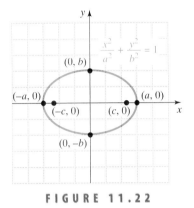

FIGURE 11.22

Equation of an Ellipse Centered at the Origin

An ellipse centered at $(0, 0)$ with foci at $(\pm c, 0)$ and constant sum $2a$ has equation

$$\frac{x^2}{a^2} + \frac{y^2}{b^2} = 1,$$

where a, b, and c are positive real numbers with $c^2 = a^2 - b^2$.

To draw a "nice-looking" ellipse, we would locate the foci and use string as shown in Fig. 11.20. We can get a rough sketch of an ellipse centered at the origin by using the x- and y-intercepts only.

EXAMPLE 1

Graphing an ellipse

Find the x- and y-intercepts for the ellipse and sketch its graph.

$$\frac{x^2}{9} + \frac{y^2}{4} = 1$$

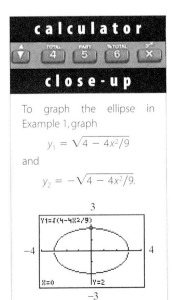

Solution

To find the y-intercepts, let $x = 0$ in the equation:

$$\frac{0}{9} + \frac{y^2}{4} = 1$$

$$\frac{y^2}{4} = 1$$

$$y^2 = 4$$

$$y = \pm 2$$

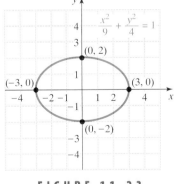

FIGURE 11.23

To find x-intercepts, let $y = 0$. We get $x = \pm 3$. The four intercepts are $(0, 2)$, $(0, -2)$, $(3, 0)$, and $(-3, 0)$. Plot the intercepts and draw an ellipse through them as in Fig. 11.23.

Ellipses, like circles, may be centered at any point in the plane. To get the equation of an ellipse centered at (h, k), we replace x by $x - h$ and y by $y - k$ in the equation of the ellipse centered at the origin.

Equation of an Ellipse Centered at (h, k)

An ellipse centered at (h, k) has equation

$$\frac{(x - h)^2}{a^2} + \frac{(y - k)^2}{b^2} = 1,$$

where a and b are positive real numbers.

E X A M P L E   2   **An ellipse with center (h, k)**

Sketch the graph of the ellipse:

$$\frac{(x - 1)^2}{9} + \frac{(y + 2)^2}{4} = 1$$

Solution

The graph of this ellipse is exactly the same size and shape as the ellipse

$$\frac{x^2}{9} + \frac{y^2}{4} = 1,$$

which was graphed in Example 1. However, the center for

$$\frac{(x - 1)^2}{9} + \frac{(y + 2)^2}{4} = 1$$

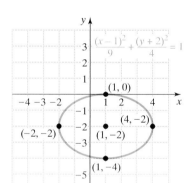

FIGURE 11.24

is $(1, -2)$. The denominator 9 is used to determine that the ellipse passes through points that are three units to the right and three units to the left of the center: $(4, -2)$ and $(-2, -2)$. See Fig. 11.24. The denominator 4 is used to determine that the ellipse passes through points that are two units above and two units below the center: $(1, 0)$ and $(1, -4)$. We draw an ellipse using these four points, just as we did for an ellipse centered at the origin.

The Hyperbola

A hyperbola is the curve that occurs at the intersection of a cone and a plane, as was shown in Fig. 11.3 in Section 11.2. A hyperbola can also be defined in terms of points and distance.

Hyperbola

A **hyperbola** is the set of all points in the plane such that the difference of their distances from two fixed points (foci) is constant.

Like the parabola and the ellipse, the hyperbola also has reflecting properties. If a light ray is aimed at one focus, it is reflected off the hyperbola and goes to the

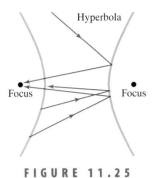

FIGURE 11.25

other focus, as shown in Fig. 11.25. Hyperbolic mirrors are used in conjunction with parabolic mirrors in telescopes.

The definitions of a hyperbola and an ellipse are similar, and so are their equations. However, their graphs are very different. Figure 11.26 shows a hyperbola in which the distance from a point on the hyperbola to the closer focus is N and the distance to the farther focus is M. The value $M - N$ is the same for every point on the hyperbola.

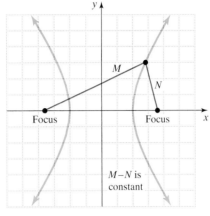

FIGURE 11.26

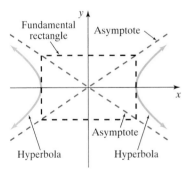

FIGURE 11.27

A hyperbola has two parts called **branches.** These branches look like parabolas, but they are not parabolas. The branches of the hyperbola shown in Fig. 11.27 get closer and closer to the dashed lines, called **asymptotes,** but they never intersect them. The asymptotes are used as guidelines in sketching a hyperbola. The asymptotes are found by extending the diagonals of the **fundamental rectangle,** shown in Fig. 11.27. The key to drawing a hyperbola is getting the fundamental rectangle and extending its diagonals to get the asymptotes. You will learn how to find the fundamental rectangle from the equation of a hyperbola. The hyperbola in Fig. 11.27 opens to the left and right.

If we start with foci at $(\pm c, 0)$ and a positive number a, then we can use the definition of a hyperbola to derive the following equation of a hyperbola in which the constant difference between the distances to the foci is $2a$.

Equation of a Hyperbola Centered at (0, 0)

A hyperbola centered at $(0, 0)$ with foci $(c, 0)$ and $(-c, 0)$ and constant difference $2a$ has equation

$$\frac{x^2}{a^2} - \frac{y^2}{b^2} = 1,$$

where a, b, and c are positive real numbers such that $c^2 = a^2 + b^2$.

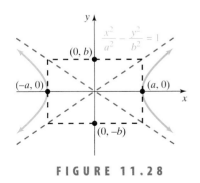

FIGURE 11.28

The graph of a general equation for a hyperbola is shown in Fig. 11.28. Notice that the fundamental rectangle extends to the x-intercepts along the x-axis and extends b units above and below the origin along the y-axis. The facts necessary for graphing a hyperbola centered at the origin and opening to the left and to the right are listed as follows.

Graphing a Hyperbola Centered at the Origin, Opening Left and Right

To graph the hyperbola $\frac{x^2}{a^2} - \frac{y^2}{b^2} = 1$:

1. Locate the x-intercepts at $(a, 0)$ and $(-a, 0)$.
2. Draw the fundamental rectangle through $(\pm a, 0)$ and $(0, \pm b)$.
3. Draw the extended diagonals of the rectangle to use as asymptotes.
4. Draw the hyperbola to the left and right approaching the asymptotes.

E X A M P L E   3

A hyperbola opening left and right

Sketch the graph of $\frac{x^2}{36} - \frac{y^2}{9} = 1$, and find the equations of its asymptotes.

Solution

The x-intercepts are $(6, 0)$ and $(-6, 0)$. Draw the fundamental rectangle through these x-intercepts and the points $(0, 3)$ and $(0, -3)$. Extend the diagonals of the fundamental rectangle to get the asymptotes. Now draw a hyperbola passing through the x-intercepts and approaching the asymptotes as shown in Fig. 11.29. From the graph in Fig. 11.29 we see that the slopes of the asymptotes are $\frac{1}{2}$ and $-\frac{1}{2}$. Because the y-intercept for both asymptotes is the origin, their equations are $y = \frac{1}{2}x$ and $y = -\frac{1}{2}x$.

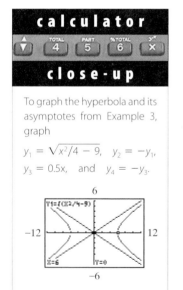

calculator

close-up

To graph the hyperbola and its asymptotes from Example 3, graph

$y_1 = \sqrt{x^2/4 - 9}$, $y_2 = -y_1$,
$y_3 = 0.5x$,   and $y_4 = -y_3$.

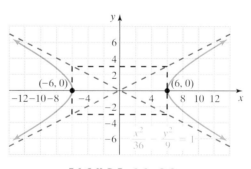

FIGURE 11.29

A hyperbola may open up and down. In this case the graph intersects only the y-axis. The facts necessary for graphing a hyperbola that opens up and down are summarized on the next page.

helpful hint

We could include here general formulas for the equations of the asymptotes, but that is not necessary. It is easier first to draw the asymptotes as suggested and then to figure out their equations by looking at the graph.

Graphing a Hyperbola Centered at the Origin, Opening Up and Down

To graph the hyperbola $\frac{y^2}{b^2} - \frac{x^2}{a^2} = 1$:

1. Locate the y-intercepts at $(0, b)$ and $(0, -b)$.
2. Draw the fundamental rectangle through $(0, \pm b)$ and $(\pm a, 0)$.
3. Draw the extended diagonals of the rectangle to use as asymptotes.
4. Draw the hyperbola opening up and down approaching the asymptotes.

EXAMPLE 4

A hyperbola opening up and down

Graph the hyperbola $\frac{y^2}{9} - \frac{x^2}{4} = 1$ and find the equations of its asymptotes.

Solution

If $y = 0$, we get

$$-\frac{x^2}{4} = 1$$
$$x^2 = -4.$$

Because this equation has no real solution, the graph has no x-intercepts. Let $x = 0$ to find the y-intercepts:

$$\frac{y^2}{9} = 1$$
$$y^2 = 9$$
$$y = \pm 3$$

The y-intercepts are $(0, 3)$ and $(0, -3)$, and the hyperbola opens up and down. From $a^2 = 4$ we get $a = 2$. So the fundamental rectangle extends to the intercepts $(0, 3)$ and $(0, -3)$ on the y-axis and to the points $(2, 0)$ and $(-2, 0)$ along the x-axis. We extend the diagonals of the rectangle and draw the graph of the hyperbola as shown in Fig. 11.30. From the graph in Fig. 11.30 we see that the asymptotes have slopes $\frac{3}{2}$ and $-\frac{3}{2}$. Because the y-intercept for both asymptotes is the origin, their equations are $y = \frac{3}{2}x$ and $y = -\frac{3}{2}x$. ■

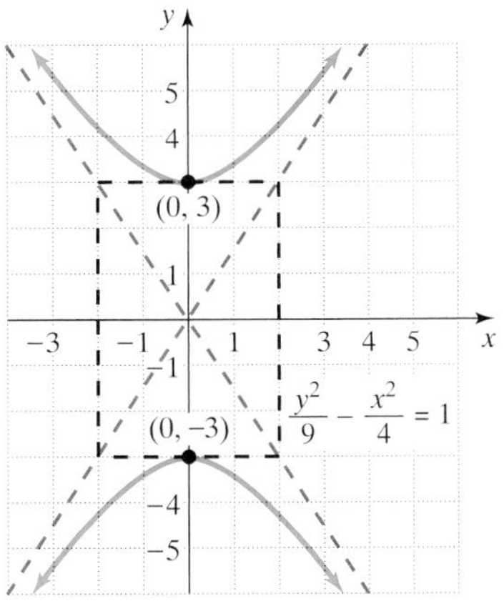

FIGURE 11.30

EXAMPLE 5

A hyperbola not in standard form

Sketch the graph of the hyperbola $4x^2 - y^2 = 4$.

Solution

First write the equation in standard form. Divide each side by 4 to get

$$x^2 - \frac{y^2}{4} = 1.$$

There are no y-intercepts. If $y = 0$, then $x = \pm 1$. The hyperbola opens left and right with x-intercepts at $(1, 0)$ and $(-1, 0)$. The fundamental rectangle extends to the intercepts along the x-axis and to the points $(0, 2)$ and $(0, -2)$ along the y-axis. We extend the diagonals of the rectangle for the asymptotes and draw the graph as shown in Fig. 11.31. ■

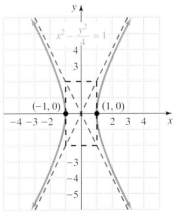

FIGURE 11.31

WARM-UPS

True or false? Explain your answer.

1. The x-intercepts of the ellipse $\frac{x^2}{36} + \frac{y^2}{25} = 1$ are $(5, 0)$ and $(-5, 0)$.   False

2. The graph of $\frac{x^2}{9} + \frac{y}{4} = 1$ is an ellipse.   False

3. If the foci of an ellipse coincide, then the ellipse is a circle.   True

4. The graph of $2x^2 + y^2 = 2$ is an ellipse centered at the origin.   True

5. The y-intercepts of $x^2 + \frac{y^2}{3} = 1$ are $(0, \sqrt{3})$ and $(0, -\sqrt{3})$.   True

6. The graph of $\frac{x^2}{9} + \frac{y}{4} = 1$ is a hyperbola.   False

7. The graph of $\frac{x^2}{25} - \frac{y^2}{16} = 1$ has y-intercepts at $(0, 4)$ and $(0, -4)$.   False

8. The hyperbola $\frac{y^2}{9} - x^2 = 1$ opens up and down.   True

9. The graph of $4x^2 - y^2 = 4$ is a hyperbola.   True

10. The asymptotes of a hyperbola are the extended diagonals of a rectangle.   True

11.4   EXERCISES

Reading and Writing   *After reading this section, write out the answers to these questions. Use complete sentences.*

1. What is the definition of an ellipse?
   An ellipse is the set of all points in a plane such that the sum of their distances from two fixed points is constant.

2. How can you draw an ellipse with a pencil and string?
   Attach a string to two thumbtacks and use a pencil to take up the slack as shown in the text.

3. Where is the center of an ellipse?
   The center of an ellipse is the point that is midway between the foci.

4. What is the equation of an ellipse centered at the origin?
   The equation of an ellipse centered at the origin is $\frac{x^2}{a^2} + \frac{y^2}{b^2} = 1$.

5. What is the equation of an ellipse centered at (h, k)?
   The equation of an ellipse centered at (h, k) is $\frac{(x - h)^2}{a^2} + \frac{(y - k)^2}{b^2} = 1$.

6. What is the definition of a hyperbola?
   A hyperbola is the set of all points in a plane such that the difference of their distances from two fixed points is constant.

7. How do you find the asymptotes of a hyperbola?
   The asymptotes of a hyperbola are the extended diagonals of the fundamental rectangle.

8. What is the equation of a hyperbola centered at the origin and opening left and right?
   The equation of a hyperbola centered at the origin and opening left and right is $\frac{x^2}{a^2} - \frac{y^2}{b^2} = 1$.

Sketch the graph of each ellipse. See Example 1.

9. $\frac{x^2}{9} + \frac{y^2}{4} = 1$

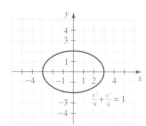

10. $\frac{x^2}{9} + \frac{y^2}{16} = 1$

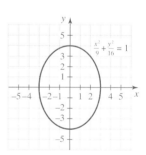

11. $\frac{x^2}{9} + y^2 = 1$

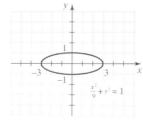

12. $x^2 + \frac{y^2}{4} = 1$

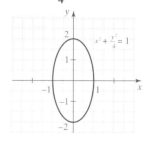

13. $\dfrac{x^2}{36} + \dfrac{y^2}{25} = 1$

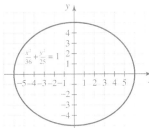

14. $\dfrac{x^2}{25} + \dfrac{y^2}{49} = 1$

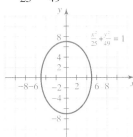

15. $\dfrac{x^2}{24} + \dfrac{y^2}{5} = 1$

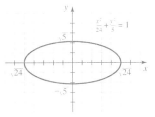

16. $\dfrac{x^2}{6} + \dfrac{y^2}{17} = 1$

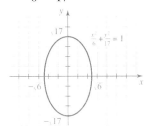

17. $9x^2 + 16y^2 = 144$

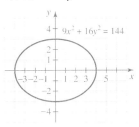

18. $9x^2 + 25y^2 = 225$

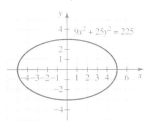

19. $25x^2 + y^2 = 25$

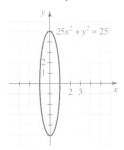

20. $x^2 + 16y^2 = 16$

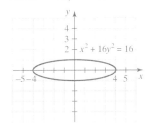

21. $4x^2 + 9y^2 = 1$

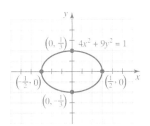

22. $25x^2 + 16y^2 = 1$

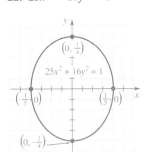

Sketch the graph of each ellipse. See Example 2.

23. $\dfrac{(x-3)^2}{4} + \dfrac{(y-1)^2}{9} = 1$

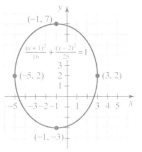

24. $\dfrac{(x+5)^2}{49} + \dfrac{(y-2)^2}{25} = 1$

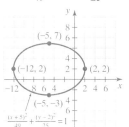

25. $\dfrac{(x+1)^2}{16} + \dfrac{(y-2)^2}{25} = 1$

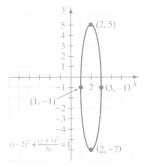

26. $\dfrac{(x-3)^2}{36} + \dfrac{(y+4)^2}{64} = 1$

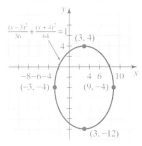

27. $(x-2)^2 + \dfrac{(y+1)^2}{36} = 1$

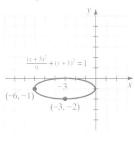

28. $\dfrac{(x+3)^2}{9} + (y+1)^2 = 1$

Sketch the graph of each hyperbola and write the equations of its asymptotes. See Examples 3–5.

29. $\dfrac{x^2}{4} - \dfrac{y^2}{9} = 1$

$y = \pm\dfrac{3}{2}x$

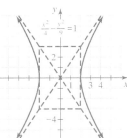

30. $\dfrac{x^2}{16} - \dfrac{y^2}{9} = 1$

$y = \pm\dfrac{3}{4}x$

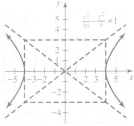

31. $\dfrac{y^2}{4} - \dfrac{x^2}{25} = 1$

$y = \pm\dfrac{2}{5}x$

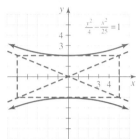

32. $\dfrac{y^2}{9} - \dfrac{x^2}{16} = 1$

$y = \pm\dfrac{3}{4}x$

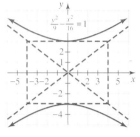

33. $\dfrac{x^2}{25} - y^2 = 1$

$y = \pm\dfrac{1}{5}x$

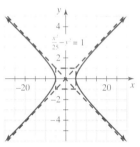

34. $x^2 - \dfrac{y^2}{9} = 1$

$y = \pm 3x$

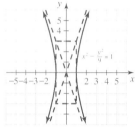

35. $x^2 - \dfrac{y^2}{25} = 1$

$y = \pm 5x$

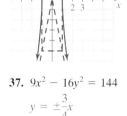

36. $\dfrac{x^2}{9} - y^2 = 1$

$y = \pm\dfrac{1}{3}x$

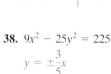

37. $9x^2 - 16y^2 = 144$

$y = \pm\dfrac{3}{4}x$

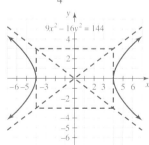

38. $9x^2 - 25y^2 = 225$

$y = \pm\dfrac{3}{5}x$

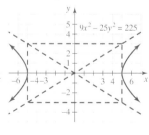

39. $x^2 - y^2 = 1$

$y = \pm x$

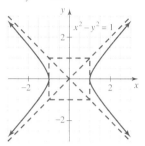

40. $y^2 - x^2 = 1$

$y = \pm x$

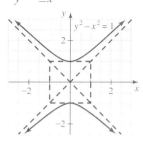

Graph both equations of each system on the same coordinate axes. Use elimination of variables to find all points of intersection.

41. $\dfrac{x^2}{4} + \dfrac{y^2}{9} = 1$

$x^2 - \dfrac{y^2}{9} = 1$

$\left(\dfrac{2\sqrt{10}}{5}, \dfrac{3\sqrt{15}}{5}\right),$

$\left(\dfrac{2\sqrt{10}}{5}, -\dfrac{3\sqrt{15}}{5}\right),$

$\left(-\dfrac{2\sqrt{10}}{5}, \dfrac{3\sqrt{15}}{5}\right),$

$\left(-\dfrac{2\sqrt{10}}{5}, -\dfrac{3\sqrt{15}}{5}\right)$

42. $x^2 - \dfrac{y^2}{4} = 1$

$\dfrac{x^2}{9} + \dfrac{y^2}{4} = 1$

$\left(\dfrac{3\sqrt{5}}{5}, \dfrac{4\sqrt{5}}{5}\right),$

$\left(\dfrac{3\sqrt{5}}{5}, -\dfrac{4\sqrt{5}}{5}\right),$

$\left(-\dfrac{3\sqrt{5}}{5}, \dfrac{4\sqrt{5}}{5}\right),$

$\left(-\dfrac{3\sqrt{5}}{5}, -\dfrac{4\sqrt{5}}{5}\right).$

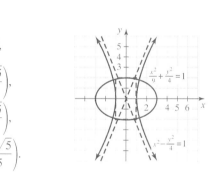

43. $\dfrac{x^2}{4} + \dfrac{y^2}{16} = 1$

$x^2 + y^2 = 1$

No points of intersection

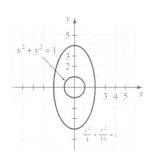

44. $x^2 + \dfrac{y^2}{9} = 1$

$x^2 + y^2 = 4$

$\left(\dfrac{\sqrt{10}}{4}, \dfrac{3\sqrt{6}}{4}\right),$

$\left(\dfrac{\sqrt{10}}{4}, -\dfrac{3\sqrt{6}}{4}\right),$

$\left(-\dfrac{\sqrt{10}}{4}, \dfrac{3\sqrt{6}}{4}\right),$

$\left(-\dfrac{\sqrt{10}}{4}, -\dfrac{3\sqrt{6}}{4}\right)$

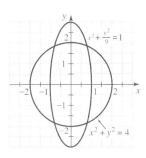

45. $x^2 + y^2 = 4$

$x^2 - y^2 = 1$

$\left(\dfrac{\sqrt{10}}{2}, \dfrac{\sqrt{6}}{2}\right), \left(\dfrac{\sqrt{10}}{2}, -\dfrac{\sqrt{6}}{2}\right),$

$\left(-\dfrac{\sqrt{10}}{2}, \dfrac{\sqrt{6}}{2}\right),$

$\left(-\dfrac{\sqrt{10}}{2}, -\dfrac{\sqrt{6}}{2}\right)$

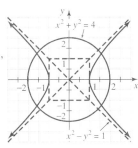

46. $x^2 + y^2 = 16$

$x^2 - y^2 = 4$

$(\sqrt{10}, \sqrt{6}),$

$(\sqrt{10}, -\sqrt{6}),$

$(-\sqrt{10}, \sqrt{6}),$

$(-\sqrt{10}, -\sqrt{6})$

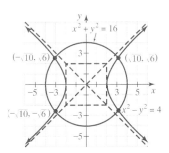

47. $x^2 + 9y^2 = 9$

$x^2 + y^2 = 4$

$\left(\dfrac{3\sqrt{6}}{4}, \dfrac{\sqrt{10}}{4}\right),$

$\left(\dfrac{3\sqrt{6}}{4}, -\dfrac{\sqrt{10}}{4}\right),$

$\left(-\dfrac{3\sqrt{6}}{4}, \dfrac{\sqrt{10}}{4}\right),$

$\left(-\dfrac{3\sqrt{6}}{4}, -\dfrac{\sqrt{10}}{4}\right)$

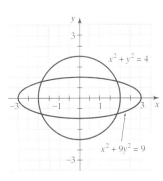

48. $x^2 + y^2 = 25$

$x^2 + 25y^2 = 25$

$(5, 0), (-5, 0)$

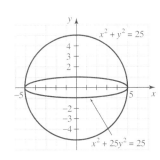

49. $x^2 + 9y^2 = 9$

$y = x^2 - 1$

$\left(\dfrac{\sqrt{17}}{3}, \dfrac{8}{9}\right), \left(-\dfrac{\sqrt{17}}{3}, \dfrac{8}{9}\right),$

$(0, -1)$

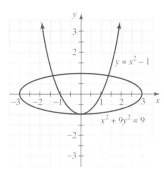

50. $4x^2 + y^2 = 4$

$y = 2x^2 - 2$

$(-1, 0), (1, 0), (0, -2)$

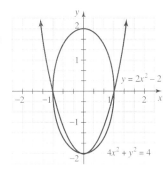

51. $9x^2 - 4y^2 = 36$

$2y = x - 2$

$(2, 0), \left(-\dfrac{5}{2}, -\dfrac{9}{4}\right)$

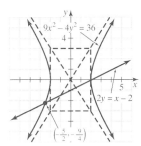

52. $25y^2 - 9x^2 = 225$

$y = 3x + 3$

$(0, 3), \left(-\dfrac{25}{12}, -\dfrac{13}{4}\right)$

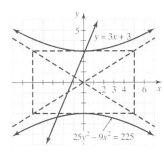

Solve each problem.

53. *Marine navigation.* The loran (long-range navigation) system is used by boaters to determine their location at sea. The loran unit on a boat measures the difference in time that it takes for radio signals from pairs of fixed points to reach the boat. The unit then finds the equations of two hyperbolas that pass through the location of the boat. Suppose a boat is located in the first quadrant at the intersection of $x^2 - 3y^2 = 1$ and $4y^2 - x^2 = 1$.

a) Use the graph on the next page to approximate the location of the boat.

b) Algebraically find the exact location of the boat.

    a) $(2.5, 1.5)$ **b)** $(\sqrt{7}, \sqrt{2})$

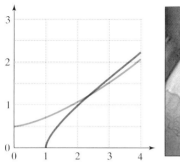

FIGURE FOR EXERCISE 53

54. Sonic boom. An aircraft traveling at supersonic speed creates a cone-shaped wave that intersects the ground along a hyperbola, as shown in the accompanying figure. A thunderlike sound is heard at any point on the hyperbola. This sonic boom travels along the ground, following the aircraft. The area where the sonic boom is most noticeable is called the *boom carpet*. The width of the boom carpet is roughly five times the altitude of the aircraft. Suppose the equation of the hyperbola in the figure is

$$\frac{x^2}{400} - \frac{y^2}{100} = 1,$$

where the units are miles and the width of the boom carpet is measured 40 miles behind the aircraft. Find the altitude of the aircraft.

$4\sqrt{3}$ or 6.9 miles

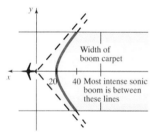

FIGURE FOR EXERCISE 54

GETTING MORE INVOLVED

55. Cooperative learning. Let (x, y) be an arbitrary point on an ellipse with foci $(c, 0)$ and $(-c, 0)$ for $c > 0$. The following equation expresses the fact that the distance from (x, y) to $(c, 0)$ plus the distance from (x, y) to $(-c, 0)$ is the constant value $2a$ (for $a > 0$):

$$\sqrt{(x - c)^2 + (y - 0)^2} + \sqrt{(x - (-c))^2 + (y - 0)^2} = 2a$$

Working in groups, simplify this equation. First get the radicals on opposite sides of the equation, then square both sides twice to eliminate the square roots. Finally, let $b^2 = a^2 - c^2$ to get the equation

$$\frac{x^2}{a^2} + \frac{y^2}{b^2} = 1.$$

56. Cooperative learning. Let (x, y) be an arbitrary point on a hyperbola with foci $(c, 0)$ and $(-c, 0)$ for $c > 0$. The following equation expresses the fact that the distance from (x, y) to $(c, 0)$ minus the distance from (x, y) to $(-c, 0)$ is the constant value $2a$ (for $a > 0$):

$$\sqrt{(x - c)^2 + (y - 0)^2} - \sqrt{(x - (-c))^2 + (y - 0)^2} = 2a$$

Working in groups, simplify the equation. You will need to square both sides twice to eliminate the square roots. Finally, let $b^2 = c^2 - a^2$ to get the equation

$$\frac{x^2}{a^2} - \frac{y^2}{b^2} = 1.$$

GRAPHING CALCULATOR EXERCISES

57. Graph $y_1 = \sqrt{x^2 - 1}$, $y_2 = -\sqrt{x^2 - 1}$, $y_3 = x$, and $y_4 = -x$ to get the graph of the hyperbola $x^2 - y^2 = 1$ along with its asymptotes. Use the viewing window $-3 \le x \le 3$ and $-3 \le y \le 3$. Notice how the branches of the hyperbola approach the asymptotes.

58. Graph the same four functions in Exercise 57, but use $-30 \le x \le 30$ and $-30 \le y \le 30$ as the viewing window. What happened to the hyperbola?

11.5 SECOND-DEGREE INEQUALITIES

In this section we graph second-degree inequalities and systems of inequalities involving second-degree inequalities.

Graphing a Second-Degree Inequality

A second-degree inequality is an inequality involving squares of at least one of the variables. Changing the equal sign to an inequality symbol for any of the equations of the conic sections gives us a second-degree inequality. Second-degree inequalities are graphed in the same manner as linear inequalities.

EXAMPLE 1

A second-degree inequality

Graph the inequality $y < x^2 + 2x - 3$.

Solution

We first graph $y = x^2 + 2x - 3$. This parabola has x-intercepts at $(1, 0)$ and $(-3, 0)$, y-intercept at $(0, -3)$, and vertex at $(-1, -4)$. The graph of the parabola is drawn with a dashed line, as shown in Fig. 11.32. The graph of the parabola divides the plane into two regions. Every point on one side of the parabola satisfies the inequality $y < x^2 + 2x - 3$, and every point on the other side satisfies the inequality $y > x^2 + 2x - 3$. To determine which side is which, we test a point that is not on the parabola, say $(0, 0)$. Because

$$0 < 0^2 + 2 \cdot 0 - 3$$

is false, the region not containing the origin is shaded, as in Fig. 11.32.

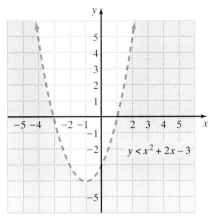

FIGURE 11.32

EXAMPLE 2

A second-degree inequality

Graph the inequality $x^2 + y^2 < 9$.

Solution

The graph of $x^2 + y^2 = 9$ is a circle of radius 3 centered at the origin. The circle divides the plane into two regions. Every point in one region satisfies $x^2 + y^2 < 9$, and every point in the other region satisfies $x^2 + y^2 > 9$. To identify the regions, we pick a point and test it. Select $(0, 0)$. The inequality

$$0^2 + 0^2 < 9$$

is true. Because $(0, 0)$ is inside the circle, all points inside the circle satisfy the inequality $x^2 + y^2 < 9$, as shown in Fig. 11.33. The points outside the circle satisfy the inequality $x^2 + y^2 > 9$.

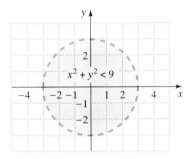

FIGURE 11.33

EXAMPLE 3

A second-degree inequality

Graph the inequality $\dfrac{x^2}{4} - \dfrac{y^2}{9} > 1$.

Solution

First graph the hyperbola $\frac{x^2}{4} - \frac{y^2}{9} = 1$. Because the hyperbola shown in Fig. 11.34 divides the plane into three regions, we select a test point in each region and check to see whether it satisfies the inequality. Testing the points $(-3, 0)$, $(0, 0)$, and $(3, 0)$ gives us the inequalities

$$\frac{(-3)^2}{4} - \frac{0^2}{9} > 1, \qquad \frac{0^2}{4} - \frac{0^2}{9} > 1, \qquad \text{and} \qquad \frac{3^2}{4} - \frac{0^2}{9} > 1.$$

Because only the first and third inequalities are correct, we shade only the regions containing $(3, 0)$ and $(-3, 0)$, as shown in Fig. 11.34.

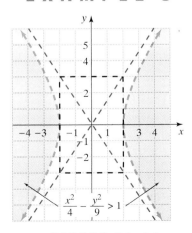

FIGURE 11.34

Systems of Inequalities

A point is in the solution set to a system of inequalities if it satisfies all inequalities of the system. We graph a system of inequalities by first determining the graph of each inequality and then finding the intersection of the graphs.

E X A M P L E   4

Systems of second-degree inequalities

Graph the system of inequalities:

$$\frac{y^2}{4} - \frac{x^2}{9} > 1$$

$$\frac{x^2}{9} + \frac{y^2}{16} < 1$$

Solution

Figure 11.35(a) shows the graph of the first inequality. In Fig. 11.35(b) we have the graph of the second inequality. In Fig. 11.35(c) we have shaded only the points that satisfy both inequalities. Figure 11.35(c) is the graph of the system.

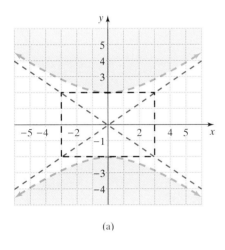

(a)

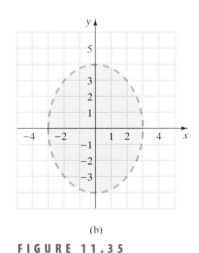

(b)

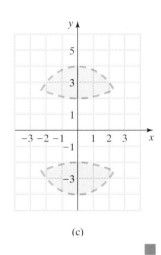

(c)

F I G U R E   1 1 . 3 5

WARM-UPS

True or false? Explain your answer.

1. The graph of $x^2 + y = 4$ is a circle of radius 2.   False
2. The graph of $x^2 + 9y^2 = 9$ is an ellipse.   True
3. The graph of $y^2 = x^2 + 1$ is a hyperbola.   True
4. The point $(0, 0)$ satisfies the inequality $2x^2 - y < 3$.   True
5. The graph of the inequality $y > x^2 - 3x + 2$ contains the origin.   False
6. The origin should be used as a test point for graphing $x^2 > y$.   False
7. The solution set to $x^2 + 3x + y^2 + 8y + 3 < 0$ includes the origin.   False
8. The graph of $x^2 + y^2 < 4$ is the region inside a circle of radius 2.   True
9. The point $(0, 4)$ satisfies $x^2 - y^2 < 1$ and $y > x^2 - 2x + 3$.   True
10. The point $(0, 0)$ satisfies $x^2 + y^2 < 1$ and $y < x^2 + 1$.   True

11.5   EXERCISES

Graph each inequality. See Examples 1–3.

1. $y > x^2$

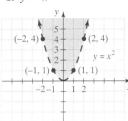

2. $y \leq x^2 + 1$

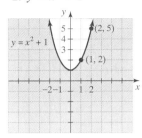

3. $y < x^2 - x$

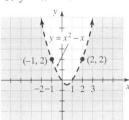

4. $y > x^2 + x$

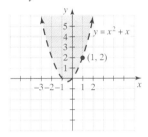

5. $y > x^2 - x - 2$

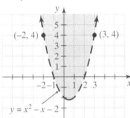

6. $y < x^2 + x - 6$

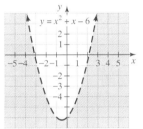

7. $x^2 + y^2 \leq 9$

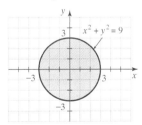

8. $x^2 + y^2 > 16$

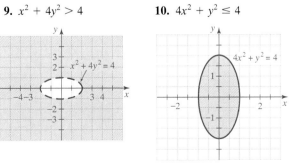

9. $x^2 + 4y^2 > 4$

10. $4x^2 + y^2 \leq 4$

11. $4x^2 - 9y^2 < 36$

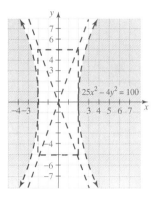

12. $25x^2 - 4y^2 > 100$

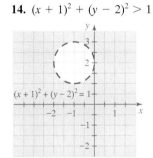

13. $(x - 2)^2 + (y - 3)^2 < 4$

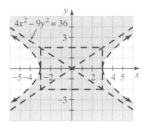

14. $(x + 1)^2 + (y - 2)^2 > 1$

15. $x^2 + y^2 > 1$

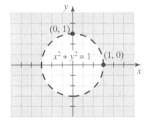

16. $x^2 + y^2 < 25$

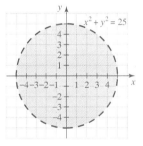

17. $4x^2 - y^2 > 4$

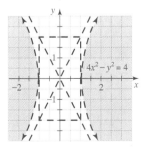

18. $x^2 - 9y^2 \leq 9$

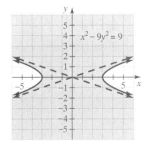

19. $y^2 - x^2 \leq 1$

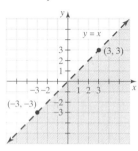

20. $x^2 - y^2 > 1$

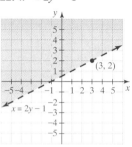

27. $y > x^2 + x$
$y < 5$

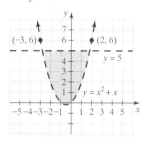

28. $y > x^2 + x - 6$
$y < x + 3$

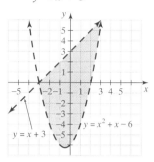

21. $x > y$

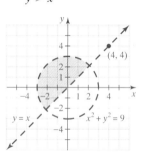

22. $x < 2y - 1$

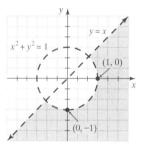

29. $y \geq x + 2$
$y \leq 2 - x$

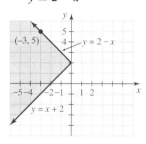

30. $y \geq 2x - 3$
$y \leq 3 - 2x$

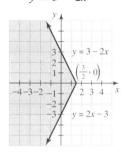

Graph the solution set to each system of inequalities. See Example 4.

23. $x^2 + y^2 < 9$
$y > x$

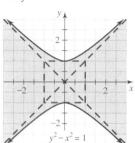

24. $x^2 + y^2 > 1$
$x > y$

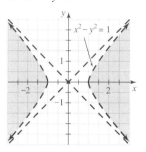

31. $4x^2 - y^2 < 4$
$x^2 + 4y^2 > 4$

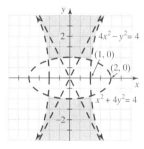

32. $x^2 - 4y^2 < 4$
$x^2 + 4y^2 > 4$

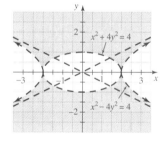

25. $x^2 - y^2 > 1$
$x^2 + y^2 < 4$

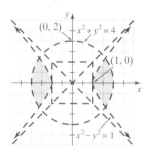

26. $y^2 - x^2 < 1$
$x^2 + y^2 > 9$

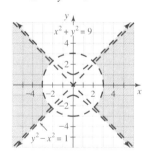

33. $x - y < 0$
$y + x^2 < 1$

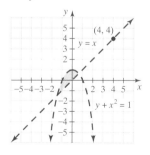

34. $y + 1 > x^2$
$x + y < 2$

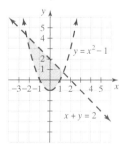

35. $y < 5x - x^2$
$x^2 + y^2 < 9$

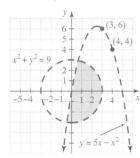

36. $y < x^2 + 5x$
$x^2 + y^2 < 16$

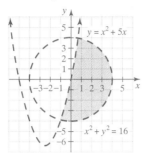

37. $y \geq 3$
$x \leq 1$

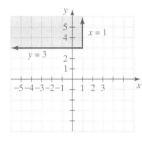

38. $x > -3$
$y < 2$

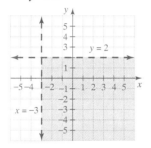

39. $4y^2 - 9x^2 < 36$
$x^2 + y^2 < 16$

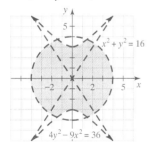

40. $25y^2 - 16x^2 < 400$
$x^2 + y^2 > 4$

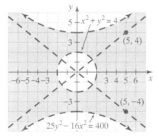

41. $y < x^2$
$x^2 + y^2 < 1$

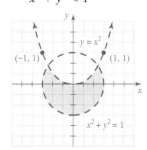

42. $y > x^2$
$4x^2 + y^2 < 4$

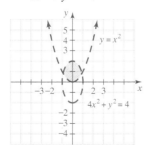

Solve the problem.

43. ***Buried treasure.*** An old pirate on his deathbed gave the following description of where he had buried some treasure on a deserted island: "Starting at the large palm tree, I walked to the north and then to the east, and there I buried the treasure. I walked at least 50 paces to get to that spot, but I was not more than 50 paces, as the crow flies, from the large palm tree. I am sure that I walked farther in the northerly direction than in the easterly direction." With the large palm tree at the origin and the positive y-axis pointing to the north, graph the possible locations of the treasure.

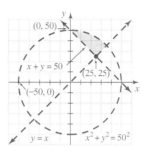

FIGURE FOR EXERCISE 43

GRAPHING CALCULATOR EXERCISES

44. Use graphs to find an ordered pair that is in the solution set to the system of inequalities:

$$y > x^2 - 2x + 1$$
$$y < -1.1(x - 4)^2 + 5$$

Verify that your answer satisfies both inequalities.

45. Use graphs to find the solution set to the system of inequalities:

$$y > 2x^2 - 3x + 1$$
$$y < -2x^2 - 8x - 1$$

No solution

COLLABORATIVE ACTIVITIES

Focus on Comets

Conic sections are used to model many different things in the natural world. Astronomers use mirrors in the shape of parabolas in telescopes. They have learned that planets can have elliptical as well as circular orbits around the sun and that comets may have orbits that resemble hyperbolas, parabolas, or ellipses. In this activity we will consider comets with these three types of orbits.

The orbit that a comet will take as it approaches the sun depends on its velocity (as well as other factors). If it has enough velocity to escape from the pull of the sun, it may take either a parabolic orbit or a hyperbolic orbit. If it doesn't have enough velocity, then it will take an elliptical orbit. Of course, a comet that has a parabolic or hyperbolic orbit will not come back again around our sun. Only comets with elliptical orbits do we see again.

For the problems below, round all your answers to two decimal places.

1. Halley's comet is in an elliptical orbit about the sun with the sun at one of the foci. In this problem we will make a scale model of the orbit of Halley's comet about the sun. We will choose our coordinate system so that the ellipse is centered at (0, 0). Halley's comet comes within 0.6 astronomical unit[1] from the sun at its closest point and 35 astronomical units at its farthest point. The position of the comet at these points

[1]An astronomical unit is the distance of the earth from the sun. The earth's orbit is almost circular and so is about the same distance from the sun at any point in its orbit.

Grouping: Two students per group
Topic: Conic sections

will correspond to the horizontal vertices. Find the equation for the ellipse. Determine where the foci should be. On a piece of cardboard, put thumbtacks at the foci. Determine the length of string you will need to draw the ellipse using the scale 1 centimeter = 1 astronomical unit (AU). Draw the comet's orbit, indicating which focus is the sun.

2. If the velocity of a comet equals the escape velocity (it is going just fast enough to get away), then its orbit will be parabolic. The sun will be at the focus of the parabola. We will place the vertex at the point (0, 0); in this case the vertex will be where the comet is the closest to the sun. Suppose we have a comet that is 0.75 AU from the sun at its closest point. Find the equation for the parabola. Graph the parabola.

3. If the velocity of a comet is greater than the escape velocity (it can easily escape from the sun's gravitational pull), its path will resemble one-half of a hyperbola with the sun as one focus. Assuming we have a left- and a right-opening hyperbola, we can model the path of such a comet along one of its branches. We will center the hyperbola at (0, 0), and place the sun at the left focus and draw the path of the comet as it approaches from the left. Suppose that the comet will be 1.5 AUs from the sun at its closest point. Assume the sun is at the point (−3, 0). Draw a sketch of this scenario. What is the equation for the hyperbola? Graph the hyperbola.

Extension: Graph all three equations on a graphing calculator or computer.

WRAP-UP   CHAPTER 11

SUMMARY

Nonlinear Systems

Nonlinear systems in two variables	Use substitution or addition to eliminate variables. Nonlinear systems may have several points in the solution set.	

Examples

$y = x^2$
$x^2 + y^2 = 4$
Substitution: $y + y^2 = 4$

Parabola

$y = a(x - h)^2 + k$

Opens upward for $a > 0$, downward for $a < 0$
Vertex at (h, k)
To find focus and directrix, use $a = \dfrac{1}{4p}$.
Distance from vertex to focus or directrix is $|p|$.

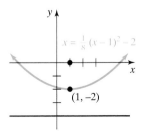
$x = \frac{1}{8}(x - 1)^2 - 2$

$(1, -2)$

$y = ax^2 + bx + c$

Opens upward for $a > 0$, downward for $a < 0$

The x-coordinate of the vertex is $\frac{-b}{2a}$.

Find the y-coordinate of the vertex by evaluating

$y = ax^2 + bx + c$ for $x = \frac{-b}{2a}$.

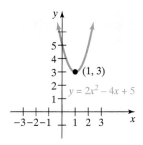

Circle

Examples

$(x - h)^2 + (y - k)^2 = r^2$

Center (h, k)

Radius r (for $r > 0$)

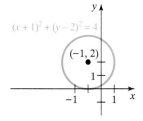

Centered at origin

$x^2 + y^2 = r^2$

Center $(0, 0)$

Radius r (for $r > 0$)

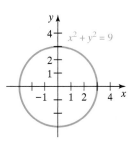

Ellipse

Examples

Centered at origin

$\dfrac{x^2}{a^2} + \dfrac{y^2}{b^2} = 1$

Center: $(0, 0)$

x-intercepts: $(a, 0)$ and $(-a, 0)$

y-intercepts: $(0, b)$ and $(0, -b)$

Foci: $(\pm c, 0)$ if $a^2 > b^2$ and $c^2 = a^2 - b^2$

$(0, \pm c)$ if $b^2 > a^2$ and $c^2 = b^2 - a^2$

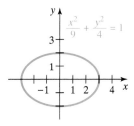

Arbitrary center

$\dfrac{(x - h)^2}{a^2} + \dfrac{(y - k)^2}{b^2} = 1$

Center: (h, k)

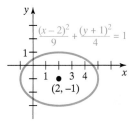

Hyperbola

Centered at origin
Opening left and right

Center $(0, 0)$
x-intercepts: $(a, 0)$ and $(-a, 0)$

$$\frac{x^2}{a^2} - \frac{y^2}{b^2} = 1$$

y-intercepts: none

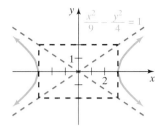

Centered at origin
Opening up and down

Centered at $(0, 0)$
x-intercepts: none

$$\frac{y^2}{b^2} - \frac{x^2}{a^2} = 1$$

y-intercepts: $(0, b)$ and $(0, -b)$

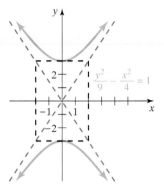

Second-Degree Inequalities

Examples

Solution set for
a single inequality

Graph the boundary curve obtained by replacing
the inequality symbol by the equal sign.
Use test points to determine which regions satisfy
the inequality.

$x^2 + y^2 < 16$

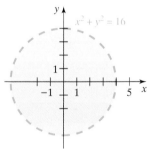

Solution set for a system
of inequalities

Graph the boundary curves. Then select a test
point in each region. Shade only the regions for
which the test point satisfies all inequalities of
the system.

$x^2 + y^2 < 16$
$y > x^2 - 1$

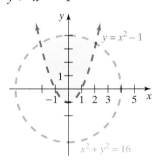

ENRICHING YOUR MATHEMATICAL WORD POWER

For each mathematical term, choose the correct meaning.

1. **nonlinear equation**
   a. an equation that is not lined up
   b. an equation whose graph is a straight line
   c. an equation whose graph is not a straight line
   d. an exponential equation   c

2. **parabola**
   a. the points in a plane that are equidistant from a point and a line
   b. the points in a plane that are a fixed distance from a fixed point
   c. the points in a plane that are equidistant from two fixed points
   d. the points in a plane the sum of whose distances from two fixed points is a constant   a

3. **directrix**
   a. the line $y = x$
   b. the line of symmetry of a parabola
   c. the x-axis
   d. the fixed line in the definition of parabola   d

4. **vertex of a parabola**
   a. the midpoint of the line segment joining the focus and directrix perpendicular to the directrix
   b. the focus
   c. the x-intercept
   d. the endpoint   a

5. **conic sections**
   a. the two halves of a cone
   b. the vertex and focus
   c. the curves obtained at the intersection of a cone and a plane
   d. the asymptotes   c

6. **axis of symmetry**
   a. the x-axis
   b. the y-axis
   c. the directrix
   d. the line of symmetry of a parabola   d

7. **circle**
   a. the points in a plane that are equidistant from a point and a line
   b. the points in a plane that are a fixed distance from a fixed point
   c. the points in a plane that are equidistant from two fixed points
   d. the points in a plane the sum of whose distances from two fixed points is a constant   b

8. **ellipse**
   a. the points in a plane that are equidistant from a point and a line
   b. the points in a plane that are a fixed distance from a fixed point
   c. the points in a plane that are equidistant from two fixed points
   d. the points in a plane such that the sum of their distances from two fixed points is constant   d

9. **hyperbola**
   a. the points in a plane that are equidistant from a point and a line
   b. the points in a plane that are a fixed distance from a fixed point
   c. the points in a plane such that the difference of their distances from two fixed points is constant
   d. the points in a plane such that the sum of their distances from two fixed points is a constant   c

10. **asymptotes**
    a. lines approached by a hyperbola
    b. lines approached by parabolas
    c. tangent lines to a circle
    d. lines that pass through the vertices of an ellipse   a

REVIEW EXERCISES

11.1 *Graph both equations on the same set of axes, then determine the points of intersection of the graphs by solving the system.*

1. $y = x^2$
   $y = -2x + 15$
   $\{(3, 9), (-5, 25)\}$

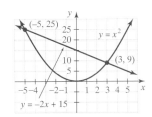

2. $y = \sqrt{x}$
   $y = \dfrac{1}{3}x$
   $\{(0, 0), (9, 3)\}$

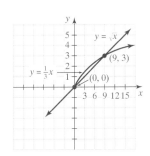

3. $y = 3x$

$y = \dfrac{1}{x}$

$\left\{\left(\dfrac{\sqrt{3}}{3}, \sqrt{3}\right), \left(-\dfrac{\sqrt{3}}{3}, -\sqrt{3}\right)\right\}$

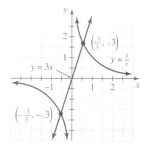

4. $y = |x|$

$y = -3x + 5$

$\left\{\left(\dfrac{5}{4}, \dfrac{5}{4}\right)\right\}$

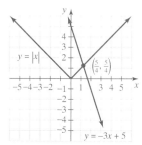

Solve each system.

5. $x^2 + y^2 = 4$

$y = \dfrac{1}{3}x^2$

$\{(\sqrt{3}, 1), (-\sqrt{3}, 1)\}$

6. $12y^2 - 4x^2 = 9$

$x = y^2$

$\left\{\left(\dfrac{3}{2}, -\dfrac{\sqrt{6}}{2}\right), \left(\dfrac{3}{2}, \dfrac{\sqrt{6}}{2}\right)\right\}$

7. $x^2 + y^2 = 34$

$y = x + 2$

$\{(-5, -3), (3, 5)\}$

8. $y = 2x + 1$

$xy - y = 5$

$\left\{\left(-\dfrac{3}{2}, -2\right), (2, 5)\right\}$

9. $y = \log(x - 3)$

$y = 1 - \log(x)$

$\{(5, \log(2))\}$

10. $y = \left(\dfrac{1}{2}\right)^x$

$y = 2^{x-1}$

$\left\{\left(\dfrac{1}{2}, \dfrac{\sqrt{2}}{2}\right)\right\}$

11. $x^4 = 2(12 - y)$

$y = x^2$

$\{(2, 4), (-2, 4)\}$

12. $x^2 + 2y^2 = 7$

$x^2 - 2y^2 = -5$

$\{(1, \sqrt{3}), (-1, \sqrt{3}),$
$(1, -\sqrt{3}), (-1, -\sqrt{3})\}$

11.2 *Sketch the graph of each parabola. Determine the x- and y-intercepts, vertex, focus, directrix, and axis of symmetry for each.*

13. $y = x^2 + 3x - 18$

Intercepts $(0, -18)$, $(-6, 0)$, $(3, 0)$, vertex $\left(-\dfrac{3}{2}, -\dfrac{81}{4}\right)$, axis of symmetry $x = -\dfrac{3}{2}$, focus $\left(-\dfrac{3}{2}, -20\right)$, directrix $y = -\dfrac{41}{2}$

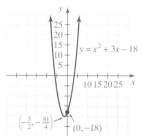

14. $y = x - x^2$

Intercepts $(0, 0)$, $(1, 0)$, vertex $\left(\dfrac{1}{2}, \dfrac{1}{4}\right)$, axis of symmetry $x = \dfrac{1}{2}$, focus $\left(\dfrac{1}{2}, 0\right)$, directrix $y = \dfrac{1}{2}$

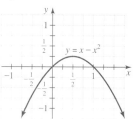

15. $y = x^2 + 3x + 2$

Intercepts $(0, 2)$, $(-2, 0)$, $(-1, 0)$, vertex $\left(-\dfrac{3}{2}, -\dfrac{1}{4}\right)$, axis of symmetry $x = -\dfrac{3}{2}$, focus $\left(-\dfrac{3}{2}, 0\right)$, directrix $y = -\dfrac{1}{2}$

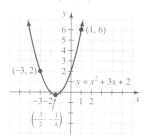

16. $y = -x^2 - 3x + 4$

Intercepts $(0, 4)$, $(-4, 0)$, $(1, 0)$, vertex $\left(-\dfrac{3}{2}, \dfrac{25}{4}\right)$, axis of symmetry $x = -\dfrac{3}{2}$, focus $\left(-\dfrac{3}{2}, 6\right)$, directrix $y = \dfrac{13}{2}$

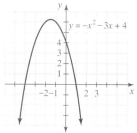

17. $y = -\dfrac{1}{2}(x - 2)^2 + 3$

Intercepts $(0, 1)$, $(2 \pm \sqrt{6}, 0)$, vertex $(2, 3)$ axis of symmetry $x = 2$, focus $\left(2, \dfrac{5}{2}\right)$, directrix $y = \dfrac{7}{2}$

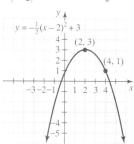

18. $y = \frac{1}{4}(x + 1)^2 - 2$

Intercepts $\left(0, -\frac{7}{4}\right)$, $\left(-1 \pm 2\sqrt{2}, 0\right)$, vertex $(-1, -2)$, axis of symmetry $x = -1$, focus $(-1, -1)$, directrix $y = -3$

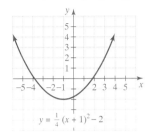

Write each equation in the form $y = a(x - h)^2 + k$, and identify the vertex of the parabola.

19. $y = 2x^2 - 8x + 1$ $y = 2(x - 2)^2 - 7, (2, -7)$

20. $y = -2x^2 - 6x - 1$ $y = -2\left(x + \frac{3}{2}\right)^2 + \frac{7}{2}, \left(-\frac{3}{2}, \frac{7}{2}\right)$

21. $y = -\frac{1}{2}x^2 - x + \frac{1}{2}$ $y = -\frac{1}{2}(x + 1)^2 + 1, (-1, 1)$

22. $y = \frac{1}{4}x^2 + x - 9$ $y = \frac{1}{4}(x + 2)^2 - 10, (-2, -10)$

Solve each problem.

23. Maximum product. Find two numbers that have the largest possible product among numbers that have a sum of 6.
3 and 3

24. Minimum product. Find two numbers that have the smallest possible product among numbers that have a difference of 10.
5 and -5

25. Maximum area. Otto has 90 feet of fencing that he plans to use to enclose a rectangular area. What dimensions will maximize the area of the rectangle?
22.5 ft by 22.5 ft

26. Maximum area. Suppose Otto already has a 6-foot-wide gate in place and he wants to use the gate and an additional 90 feet of fencing to enclose a rectangular area. What dimensions will maximize the area of the rectangle?
24 ft on each side

11.3 *Determine the center and radius of each circle, and sketch its graph.*

27. $x^2 + y^2 = 100$
(0, 0), 10

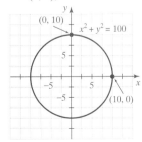

28. $x^2 + y^2 = 20$
$(0, 0), 2\sqrt{5}$

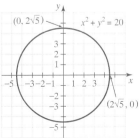

29. $(x - 2)^2 + (y + 3)^2 = 81$
$(2, -3), 9$

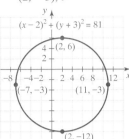

30. $x^2 + 2x + y^2 = 8$
$(-1, 0), 3$

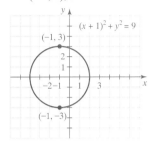

31. $9y^2 + 9x^2 = 4$
$(0, 0), \frac{2}{3}$

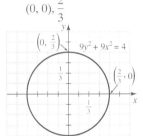

32. $x^2 + 4x + y^2 - 6y - 3 = 0$
$(-2, 3), 4$

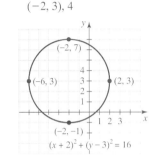

Write the standard equation for each circle with the given center and radius.

33. Center $(0, 3)$, radius 6 $x^2 + (y - 3)^2 = 36$

34. Center $(0, 0)$, radius $\sqrt{6}$ $x^2 + y^2 = 6$

35. Center $(2, -7)$, radius 5 $(x - 2)^2 + (y + 7)^2 = 25$

36. Center $\left(\frac{1}{2}, -3\right)$, radius $\frac{1}{2}$ $\left(x - \frac{1}{2}\right)^2 + (y + 3)^2 = \frac{1}{4}$

11.4 *Sketch the graph of each ellipse.*

37. $\frac{x^2}{36} + \frac{y^2}{49} = 1$

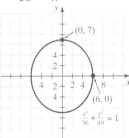

38. $\frac{x^2}{25} + y^2 = 1$

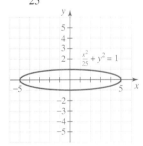

39. $25x^2 + 4y^2 = 100$

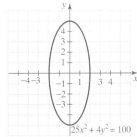

40. $6x^2 + 4y^2 = 24$

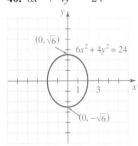

Sketch the graph of each hyperbola.

41. $\dfrac{x^2}{49} - \dfrac{y^2}{36} = 1$

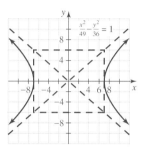

42. $\dfrac{y^2}{25} - \dfrac{x^2}{49} = 1$

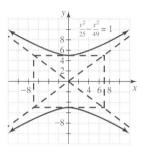

43. $4x^2 - 25y^2 = 100$

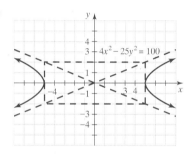

44. $6y^2 - 16x^2 = 96$

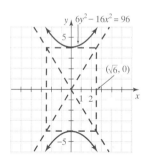

11.5 *Graph each inequality.*

45. $4x - 2y > 3$

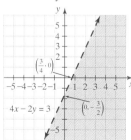

46. $y < x^2 - 3x$

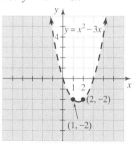

47. $y^2 < x^2 - 1$

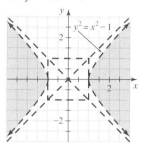

48. $y^2 < 1 - x^2$

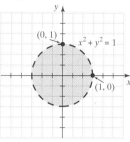

49. $4x^2 + 9y^2 > 36$

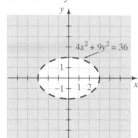

50. $x^2 + y > 2x - 1$

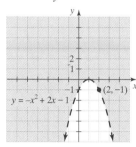

Graph the solution set to each system of inequalities.

51. $y < 3x - x^2$
$x^2 + y^2 < 9$

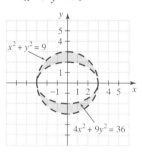

52. $x^2 - y^2 < 1$
$y < 1$

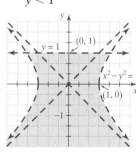

53. $4x^2 + 9y^2 > 36$
$x^2 + y^2 < 9$

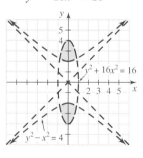

54. $y^2 - x^2 > 4$
$y^2 + 16x^2 < 16$

MISCELLANEOUS

Identify each equation as the equation of a straight line, parabola, circle, hyperbola, or ellipse. Try to do these without rewriting the equations.

55. $x^2 = y^2 + 1$
Hyperbola

56. $x = y + 1$
Line

57. $x^2 = 1 - y^2$
Circle

58. $x^2 = y + 1$
Parabola

59. $x^2 + x = 1 - y^2$
Circle

60. $(x - 3)^2 + (y + 2)^2 = 7$
Circle

61. $x^2 + 4x = 6y - y^2$
Circle

62. $4x + 6y = 1$
Line

63. $\dfrac{x^2}{3} - \dfrac{y^2}{5} = 1$
Hyperbola

64. $x^2 + \dfrac{y^2}{3} = 1$
Ellipse

65. $4y^2 - x^2 = 8$
Hyperbola

66. $9x^2 + y = 9$
Parabola

Sketch the graph of each equation.

67. $x^2 = 4 - y^2$

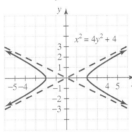

68. $x^2 = 4y^2 + 4$

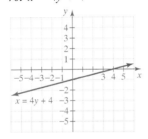

69. $x^2 = 4y + 4$

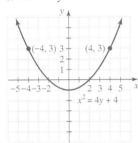

70. $x = 4y + 4$

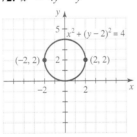

71. $x^2 = 4 - 4y^2$

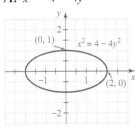

72. $x^2 = 4y - y^2$

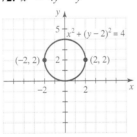

73. $x^2 = 4 - (y - 4)^2$

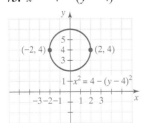

74. $(x - 2)^2 + (y - 4)^2 = 4$

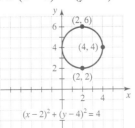

Write the equation of the circle with the given features.

75. Centered at the origin and passing through $(3, 4)$
$x^2 + y^2 = 25$

76. Centered at $(2, -3)$ and passing through $(-1, 4)$
$(x - 2)^2 + (y + 3)^2 = 58$

77. Centered at $(-1, 5)$ with radius 6
$(x + 1)^2 + (y - 5)^2 = 36$

78. Centered at $(0, -3)$ and passing through the origin
$x^2 + (y + 3)^2 = 9$

Write the equation of the parabola with the given features.

79. Focus $(1, 4)$ and directrix $y = 2$
$y = \frac{1}{4}(x - 1)^2 + 3$

80. Focus $(-2, 1)$ and directrix $y = 5$
$y = -\frac{1}{8}(x + 2)^2 + 3$

81. Vertex $(0, 0)$ and focus $\left(0, \frac{1}{4}\right)$
$y = x^2$

82. Vertex $(1, 2)$ and focus $\left(1, \frac{3}{2}\right)$
$y = -\frac{1}{2}(x - 1)^2 + 2$

83. Vertex $(0, 0)$, passing through $(3, 2)$, and opening upward
$y = \frac{2}{9}x^2$

84. Vertex $(1, 3)$, passing through $(0, 0)$, and opening downward
$y = -3(x - 1)^2 + 3$

Solve each system of equations.

85. $x^2 + y^2 = 25$
$y = -x + 1$
$\{(4, -3), (-3, 4)\}$

86. $x^2 - y^2 = 1$
$x^2 + y^2 = 7$
$\left\{(2, \sqrt{3}), (2, -\sqrt{3}), (-2, \sqrt{3}), (-2, -\sqrt{3})\right\}$

87. $4x^2 + y^2 = 4$
$x^2 - y^2 = 21$
$\varnothing$

88. $y = x^2 + x$
$y = -x^2 + 3x + 12$
$\{(3, 12), (-2, 2)\}$

Solve each problem.

89. *Perimeter of a rectangle.* A rectangle has a perimeter of 16 feet and an area of 12 square feet. Find its length and width.
6 ft, 2 ft

90. *Tale of two circles.* Find the radii of two circles such that the difference in areas of the two is 10π square inches and the difference in radii of the two is 2 inches.
$\dfrac{7}{2}$ in., $\dfrac{3}{2}$ in.

Sketch the graph of each equation.

1. $x^2 + y^2 = 25$

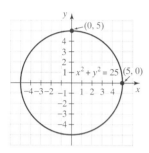

2. $\dfrac{x^2}{16} - \dfrac{y^2}{25} = 1$

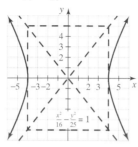

3. $y^2 + 4x^2 = 4$

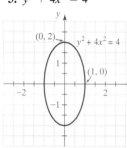

4. $y = x^2 + 4x + 4$

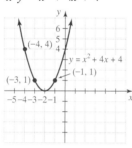

5. $y^2 - 4x^2 = 4$

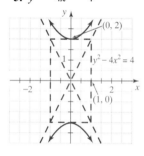

6. $y = -x^2 - 2x + 3$

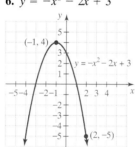

Sketch the graph of each inequality.

7. $x^2 - y^2 < 9$

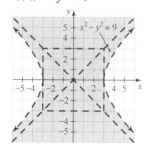

8. $x^2 + y^2 > 9$

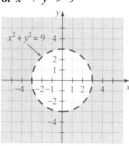

9. $y > x^2 - 9$

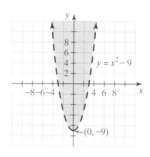

Graph the solution set to each system of inequalities.

10. $x^2 + y^2 < 9$
$x^2 - y^2 > 1$

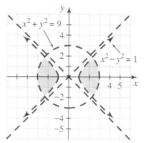

11. $y < -x^2 + x$
$y < x - 4$

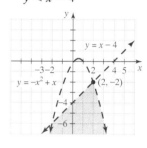

Solve each system of equations.

12. $y = x^2 - 2x - 8$
$y = 7 - 4x$
$\{(-5, 27), (3, -5)\}$

13. $x^2 + y^2 = 12$
$y = x^2$
$\left\{(\sqrt{3}, 3), (-\sqrt{3}, 3)\right\}$

Solve each problem.

14. Find the center and radius of the circle
$x^2 + 2x + y^2 + 10y = 10$.
$(-1, -5), 6$

15. Find the vertex, focus, and directrix of the parabola $y = x^2 + x + 3$. State the axis of symmetry and whether the parabola opens up or down.

Vertex $\left(-\dfrac{1}{2}, \dfrac{11}{4}\right)$, focus $\left(-\dfrac{1}{2}, 3\right)$, directrix $y = \dfrac{5}{2}$, axis of symmetry $x = -\dfrac{1}{2}$, upward

16. Write the equation $y = \dfrac{1}{2}x^2 - 3x - \dfrac{1}{2}$ in the form $y = a(x - h)^2 + k$.

$y = \dfrac{1}{2}(x - 3)^2 - 5$

17. If a ball is thrown straight up with a velocity of 64 feet per second from a height of 20 feet (ft), then its height s above the ground at time t is given by $s = -16t^2 + 64t + 20$. What is the maximum height reached by the ball? 84 ft

18. Write the equation of a circle with center $(-1, 3)$ that passes through $(2, 5)$. $(x + 1)^2 + (y - 3)^2 = 13$

19. Find the length and width of a rectangular room that has an area of 108 square feet and a perimeter of 42 ft. 12 ft, 9 ft

Sketch the graph of each equation.

1. $y = 9x - x^2$

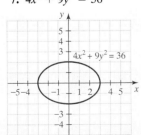

2. $y = 9x$

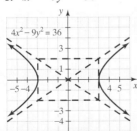

3. $y = (x - 9)^2$

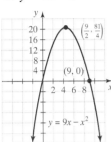

4. $y^2 = 9 - x^2$

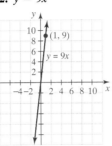

5. $y = 9x^2$

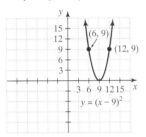

6. $y = |9x|$

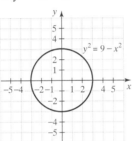

7. $4x^2 + 9y^2 = 36$

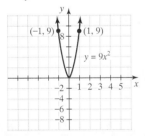

8. $4x^2 - 9y^2 = 36$

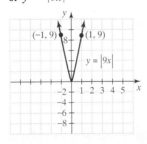

9. $y = 9 - x$

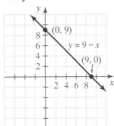

10. $y = 9^x$

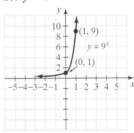

Find the following products.

11. $(x + 2y)^2$ $x^2 + 4xy + 4y^2$

12. $(x + y)(x^2 + 2xy + y^2)$ $x^3 + 3x^2y + 3xy^2 + y^3$

13. $(a + b)^3$ $a^3 + 3a^2b + 3ab^2 + b^3$

14. $(a - 3b)^2$ $a^2 - 6ab + 9b^2$

15. $(2a + 1)(3a - 5)$ $6a^2 - 7a - 5$

16. $(x - y)(x^2 + xy + y^2)$ $x^3 - y^3$

Solve each system of equations.

17. $2x - 3y = -4$
$x + 2y = 5$
$\{(1, 2)\}$

18. $x^2 + y^2 = 25$
$x + y = 7$
$\{(3, 4), (4, 3)\}$

19. $2x - y + z = 7$
$x - 2y - z = 2$
$x + y + z = 2$
$\{(1, -2, 3)\}$

20. $y = x^2$
$y - 2x = 3$
$\{(-1, 1), (3, 9)\}$

Solve each formula for the specified variable.

21. $ax + b = 0$, for x $x = -\dfrac{b}{a}$

22. $wx^2 + dx + m = 0$, for x $x = \dfrac{-d \pm \sqrt{d^2 - 4wm}}{2w}$

23. $A = \dfrac{1}{2}h(B + b)$, for B $B = \dfrac{2A - bh}{h}$

24. $\dfrac{1}{x} + \dfrac{1}{y} = \dfrac{1}{2}$, for x $x = \dfrac{2y}{y - 2}$

25. $L = m + mxt$, for m $m = \dfrac{L}{1 + xt}$

26. $y = 3a\sqrt{t}$, for t $t = \dfrac{y^2}{9a^2}$

Solve each problem.

27. Write the equation of the line in slope-intercept form that goes through the points $(2, -3)$ and $(-4, 1)$.

$$y = -\frac{2}{3}x - \frac{5}{3}$$

28. Write the equation of the line in slope-intercept form that contains the origin and is perpendicular to the line $2x - 4y = 5$.
$y = -2x$

29. Write the equation of the circle that has center $(2, 5)$ and passes through the point $(-1, -1)$.
$(x - 2)^2 + (y - 5)^2 = 45$

30. Find the center and radius of the circle $x^2 + 3x + y^2 - 6y = 0$.
$\left(-\dfrac{3}{2}, 3\right), \dfrac{3\sqrt{5}}{2}$

Perform the computations with complex numbers.

31. $2i(3 + 5i)$
$-10 + 6i$

32. i^6
-1

33. $(2i - 3) + (6 - 7i)$
$3 - 5i$

34. $\left(3 + i\sqrt{2}\right)^2$
$7 + 6i\sqrt{2}$

35. $(2 - 3i)(5 - 6i)$
$-8 - 27i$

36. $(3 - i) + (-6 + 4i)$
$-3 + 3i$

37. $(5 - 2i)(5 + 2i)$
29

38. $(2 - 3i) \div (2i)$
$-\dfrac{3}{2} - i$

39. $(4 + 5i) \div (1 - i)$
$-\dfrac{1}{2} + \dfrac{9}{2}i$

40. $\dfrac{4 - \sqrt{-8}}{2}$
$2 - i\sqrt{2}$

Solve.

41. *Going bananas.* Salvadore has observed that when bananas are \$0.30 per pound (lb), he sells 250 lb per day, and when bananas are \$0.40 per lb, he sells only 200 lb per day.

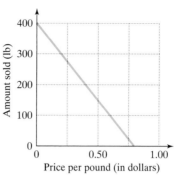

FIGURE FOR EXERCISE 41

a) Assume the number of pounds sold, q, is a linear function of the price per pound, x, and find that function.
b) Salvadore's daily revenue in dollars is the product of the number of pounds sold and the price per pound. Write the revenue as a function of x.
c) Graph the revenue function.
d) What price per pound maximizes his revenue?
e) What is his maximum possible revenue?
  a) $q = -500x + 400$ b) $R = -500x^2 + 400x$
  c)

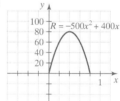

  d) \$0.40 per pound   e) \$80

12

Sequences and Series

Everyone realizes the importance of investing for the future. Some people go to great pains to study the markets and to make wise investment decisions. Some stay away from investing because they do not want to take chances. However, the most important factor in investing is making regular investments (*Money,* www.money.com). According to *Money,* if you had invested $5,000 in the stock market every year at the market high for that year (the worst time to invest) for the last 40 years, your investment would be worth $2.8 million today.

A sequence of periodic investments earning a fixed rate of interest can be thought of as a geometric sequence. In this chapter you will learn how to find the sum of a geometric sequence and to calculate the future value of a sequence of periodic investments. In Exercise 58 of Section 12.4 you will see how *Money* magazine calculated the value of $5,000 invested each year for 10 years in Fidelity's Magellan fund.

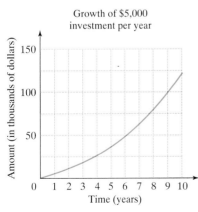

Growth of $5,000 investment per year

(12.1) SEQUENCES

The word "sequence" is a familiar word. We may speak of a sequence of events or say that something is out of sequence. In this section we give the mathematical definition of a sequence.

Definition

In mathematics we think of a sequence as a list of numbers. Each number in the sequence is called a **term** of the sequence. There is a first term, a second term, a third term, and so on. For example, the daily high temperature readings in Minot, North Dakota, for the first 10 days in January can be thought of as a finite sequence with 10 terms:

$$-9, -2, 8, -11, 0, 6, 14, 1, -5, -11$$

The set of all positive even integers,

$$2, 4, 6, 8, 10, 12, 14, \ldots,$$

can be thought of as an infinite sequence.

To give a precise definition of sequence, we use the terminology of functions. The list of numbers is the range of the function.

> **Sequence**
>
> A **finite sequence** is a function whose domain is the set of positive integers less than or equal to some fixed positive integer. An **infinite sequence** is a function whose domain is the set of all positive integers.

When the domain is apparent, we will refer to either a finite sequence or an infinite sequence simply as a sequence. For the independent variable of the function we will usually use n (for natural number) rather than x. For the dependent variable we write a_n (read "a sub n") rather than y. We call a_n the **nth term,** or the **general term** of the sequence. Rather than use the $f(x)$ notation for functions, we will define sequences with formulas. When n is used as a variable, we will assume it represents natural numbers only.

EXAMPLE 1

Listing terms of a finite sequence

List all of the terms of each finite sequence.

a) $a_n = n^2$ for $1 \leq n \leq 5$

b) $a_n = \dfrac{1}{n + 2}$ for $1 \leq n \leq 4$

Solution

a) Using the natural numbers from 1 through 5 in $a_n = n^2$, we get

$$a_1 = 1^2 = 1,$$
$$a_2 = 2^2 = 4,$$
$$a_3 = 3^2 = 9,$$
$$a_4 = 4^2 = 16,$$

and

$$a_5 = 5^2 = 25.$$

The five terms of this sequence are 1, 4, 9, 16, and 25. We often refer to the listing of the terms of the sequence as the sequence.

b) Using the natural numbers from 1 through 4 in $a_n = \dfrac{1}{n+2}$, we get the terms

$$a_1 = \frac{1}{1+2} = \frac{1}{3},$$

$$a_2 = \frac{1}{2+2} = \frac{1}{4},$$

$$a_3 = \frac{1}{5},$$

and

$$a_4 = \frac{1}{6}.$$

The four terms of the sequence are $\frac{1}{3}, \frac{1}{4}, \frac{1}{5},$ and $\frac{1}{6}$.

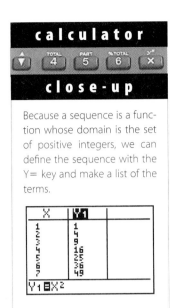

calculator

close-up

Because a sequence is a function whose domain is the set of positive integers, we can define the sequence with the Y= key and make a list of the terms.

EXAMPLE   2

Listing terms of an infinite sequence

List the first three terms of the infinite sequence whose nth term is

$$a_n = \frac{(-1)^n}{2^{n+1}}.$$

Solution

Using the natural numbers 1, 2, and 3 in the formula for the nth term yields

$$a_1 = \frac{(-1)^1}{2^{1+1}} = -\frac{1}{4}, \qquad a_2 = \frac{(-1)^2}{2^{2+1}} = \frac{1}{8}, \qquad \text{and} \qquad a_3 = \frac{(-1)^3}{2^{3+1}} = -\frac{1}{16}.$$

We write the sequence as follows:

$$-\frac{1}{4}, \frac{1}{8}, -\frac{1}{16}, \ldots$$

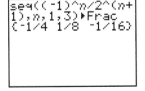

calculator

close-up

Some calculators have a sequence feature that allows you to specify the formula and which terms to evaluate. We can even get the terms as fractions.

```
seq((-1)^n/2^(n+
1),n,1,3)▶Frac
{-1/4 1/8 -1/16}
```

Finding a Formula for the nth Term

We often know the terms of a sequence and want to write a formula that will produce those terms. To write a formula for the nth term of a sequence, examine the terms and look for a pattern. Each term is a function of the term number. The first term corresponds to $n = 1$, the second term corresponds to $n = 2$, and so on.

EXAMPLE   3

A familiar sequence

Write the general term for the infinite sequence

$$3, 5, 7, 9, 11, \ldots.$$

Solution

The even numbers are all multiples of 2 and can be represented as $2n$. Because each odd number is 1 more than an even number, a formula for the nth term might be

$$a_n = 2n + 1.$$

helpful hint

Finding a formula for a sequence could be extremely difficult. For example, there is no known formula that will produce the sequence of prime numbers:

2, 3, 5, 7, 11, 13, 17, 19, ...

To be sure, we write out a few terms using the formula:

$$a_1 = 2(1) + 1 = 3$$
$$a_2 = 2(2) + 1 = 5$$
$$a_3 = 2(3) + 1 = 7$$

So the general term is $a_n = 2n + 1$.

CAUTION There can be more than one formula that produces the given terms of a sequence. For example, the sequence

$$1, 2, 4, \ldots$$

could have nth term $a_n = 2^{n-1}$ or $a_n = \frac{1}{2}n^2 - \frac{1}{2}n + 1$. The first three terms for both of these sequences are identical, but their fourth terms are different.

EXAMPLE 4

A sequence with alternating signs

Write the general term for the infinite sequence

$$1, -\frac{1}{4}, \frac{1}{9}, -\frac{1}{16}, \ldots.$$

Solution

To obtain the alternating signs, we use powers of -1. Because any even power of -1 is positive and any odd power of -1 is negative, we use $(-1)^{n+1}$. The denominators are the squares of the positive integers. So the nth term of this infinite sequence is given by the formula

$$a_n = \frac{(-1)^{n+1}}{n^2}.$$

Check this sequence by using this formula to find the first four terms.

In the next example we use a sequence to model a physical situation.

EXAMPLE 5

The bouncing ball

Suppose a ball always rebounds $\frac{2}{3}$ of the height from which it falls and the ball is dropped from a height of 6 feet. Write a sequence whose terms are the heights from which the ball falls. What is a formula for the nth term of this sequence?

Solution

On the first fall the ball travels 6 feet (ft), as shown in Fig. 12.1. On the second fall it travels $\frac{2}{3}$ of 6, or 4 ft. On the third fall it travels $\frac{2}{3}$ of 4, or $\frac{8}{3}$ ft, and so on. We write the sequence as follows:

$$6, 4, \frac{8}{3}, \frac{16}{9}, \frac{32}{27}, \ldots$$

The nth term can be written by using powers of $\frac{2}{3}$:

$$a_n = 6\left(\frac{2}{3}\right)^{n-1}$$

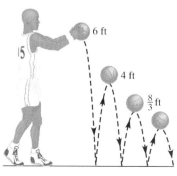

FIGURE 12.1

M A T H   A T   W O R K $x^2 + (x + 1)^2 = 5^2$

Most of us find an upholstered chair, sink into it, and remark on the comfort. Before design consultant Audrey Jordan sits down, she often looks at the fabric to observe the color and texture—and especially to see whether the fabric is one of her original designs. Fabric design is more than just an idea that is printed on a piece of cloth. Consideration must be given to the end product, which could be anything

**FABRIC
DESIGNER**

from a handbag to a large sofa. Colors and themes must be chosen with both current trends and styles in mind. Sometimes a design will be an overall or nondirectional pattern, such as polka dots, which can be cut randomly. More often, it will have a specific theme, such as fruit, which can be cut and sewn in only one direction.

For all products one of the main considerations is the vertical repeat. A good portion of textile machinery is standardized for vertical repeats of 27 inches or fractions thereof. For example, the vertical repeat could be every $13\frac{1}{2}$ inches or every 9 inches. Even though the horizontal repeat can vary, Ms. Jordan must consider both the horizontal and vertical repeats for a particular end product.

In Exercise 45 of this section you will find the standard vertical repeats for a textile machine.

WARM-UPS

True or false? Explain your answer.

1. The nth term of the sequence 2, 4, 6, 8, 10, . . . is $a_n = 2n$.   True
2. The nth term of the sequence 1, 3, 5, 7, 9, . . . is $a_n = 2n - 1$.   True
3. A sequence is a function.   True
4. The domain of a finite sequence is the set of positive integers.   False
5. The nth term of $-1, 4, -9, 16, -25, . . .$ is $a_n = (-1)^{n+1}n^2$.   False
6. For the infinite sequence $b_n = \frac{1}{n}$, the independent variable is $\frac{1}{n}$.   False
7. For the sequence $c_n = n^3$, the dependent variable is c_n.   True
8. The sixth term of the sequence $a_n = (-1)^{n+1}2^n$ is 64.   False
9. The symbol a_n is used for the dependent variable of a sequence.   True
10. The tenth term of the sequence 2, 4, 8, 16, 32, 64, 128, . . . is 1024.   True

12.1   EXERCISES

Reading and Writing   *After reading this section, write out the answers to these questions. Use complete sentences.*

1. What is a sequence?
   A sequence is a list of numbers.

2. What is a term of a sequence?
   Each number in the sequence is called a "term" of the sequence.

3. What is a finite sequence?
   A finite sequence is a function whose domain is the set of positive integers less than or equal to some fixed positive integer.

4. What is an infinite sequence?
   An infinite sequence is a function whose domain is the set of all positive integers.

List all terms of each finite sequence. See Example 1.

5. $a_n = n^2$ for $1 \le n \le 8$

1, 4, 9, 16, 25, 36, 49, 64

6. $a_n = -n^2$ for $1 \le n \le 4$

$-1, -4, -9, -16$

7. $b_n = \dfrac{(-1)^n}{n}$ for $1 \le n \le 10$

$-1, \dfrac{1}{2}, -\dfrac{1}{3}, \dfrac{1}{4}, -\dfrac{1}{5}, \dfrac{1}{6}, -\dfrac{1}{7}, \dfrac{1}{8}, -\dfrac{1}{9}, \dfrac{1}{10}$

8. $b_n = \dfrac{(-1)^{n+1}}{n}$ for $1 \le n \le 6$

$1, -\dfrac{1}{2}, \dfrac{1}{3}, -\dfrac{1}{4}, \dfrac{1}{5}, -\dfrac{1}{6}$

9. $c_n = (-2)^{n-1}$ for $1 \le n \le 5$

$1, -2, 4, -8, 16$

10. $c_n = (-3)^{n-2}$ for $1 \le n \le 5$

$-\dfrac{1}{3}, 1, -3, 9, -27$

11. $a_n = 2^{-n}$ for $1 \le n \le 6$

$\dfrac{1}{2}, \dfrac{1}{4}, \dfrac{1}{8}, \dfrac{1}{16}, \dfrac{1}{32}, \dfrac{1}{64}$

12. $a_n = 2^{-n+2}$ for $1 \le n \le 5$

$2, 1, \dfrac{1}{2}, \dfrac{1}{4}, \dfrac{1}{8}$

13. $b_n = 2n - 3$ for $1 \le n \le 7$

$-1, 1, 3, 5, 7, 9, 11$

14. $b_n = 2n + 6$ for $1 \le n \le 7$

8, 10, 12, 14, 16, 18, 20

15. $c_n = n^{-1/2}$ for $1 \le n \le 5$

$1, \dfrac{\sqrt{2}}{2}, \dfrac{\sqrt{3}}{3}, \dfrac{1}{2}, \dfrac{\sqrt{5}}{5}$

16. $c_n = n^{1/2} 2^{-n}$ for $1 \le n \le 4$

$\dfrac{1}{2}, \dfrac{\sqrt{2}}{4}, \dfrac{\sqrt{3}}{8}, \dfrac{1}{8}$

Write the first four terms of the infinite sequence whose nth term is given. See Example 2.

17. $a_n = \dfrac{1}{n^2 + n}$

$\dfrac{1}{2}, \dfrac{1}{6}, \dfrac{1}{12}, \dfrac{1}{20}$

18. $b_n = \dfrac{1}{(n + 1)(n + 2)}$

$\dfrac{1}{6}, \dfrac{1}{12}, \dfrac{1}{20}, \dfrac{1}{30}$

19. $b_n = \dfrac{1}{2n - 5}$

$-\dfrac{1}{3}, -1, 1, \dfrac{1}{3}$

20. $a_n = \dfrac{4}{2n + 5}$

$\dfrac{4}{7}, \dfrac{4}{9}, \dfrac{4}{11}, \dfrac{4}{13}$

21. $c_n = (-1)^n(n - 2)^2$

$-1, 0, -1, 4$

22. $c_n = (-1)^n(2n - 1)^2$

$-1, 9, -25, 49$

23. $a_n = \dfrac{(-1)^{2n}}{n^2}$

$1, \dfrac{1}{4}, \dfrac{1}{9}, \dfrac{1}{16}$

24. $a_n = (-1)^{2n+1} 2^{n-1}$

$-1, -2, -4, -8$

Write a formula for the general term of each infinite sequence. See Examples 3 and 4.

25. 1, 3, 5, 7, 9, . . . $a_n = 2n - 1$

26. 5, 7, 9, 11, 13, . . . $a_n = 2n + 3$

27. 1, -1, 1, -1, . . . $a_n = (-1)^{n+1}$

28. -1, 1, -1, 1, . . . $a_n = (-1)^n$

29. 0, 2, 4, 6, 8, . . . $a_n = 2n - 2$

30. 4, 6, 8, 10, 12, . . . $a_n = 2n + 2$

31. 3, 6, 9, 12, . . . $a_n = 3n$

32. 4, 8, 12, 16, . . . $a_n = 4n$

33. 4, 7, 10, 13, . . . $a_n = 3n + 1$

34. 3, 7, 11, 15, . . . $a_n = 4n - 1$

35. -1, 2, -4, 8, -16, . . . $a_n = (-1)^n 2^{n-1}$

36. 1, -3, 9, -27, . . . $a_n = (-3)^{n-1}$

37. 0, 1, 4, 9, 16, . . . $a_n = (n - 1)^2$

38. 0, 1, 8, 27, 64, . . . $a_n = (n - 1)^3$

Solve each problem. See Example 5.

39. *Football penalties.* A football is on the 8-yard line, and five penalties in a row are given that move the ball half the distance to the (closest) goal. Write a sequence of five terms that specify the location of the ball after each penalty.

$4, 2, 1, \dfrac{1}{2}, \dfrac{1}{4}$ yard line

40. *Infestation.* Leona planted 9 acres of soybeans, but by the end of each week, insects had destroyed one-third of the acreage that was healthy at the beginning of the week. How many acres does she have left after 6 weeks?

$\dfrac{64}{81}$ acre

41. *Constant rate of increase.* The MSRP for the 1999 Ford F-250 Lariat 4WD Super Duty Super Cab was $32,535

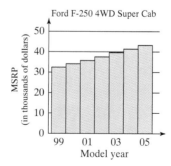

FIGURE FOR EXERCISE 41

(Edmund's New Car Prices, www.edmunds.com). Suppose the price of this truck increases by 5% each year. Find the prices to the nearest dollar for the 2000 through 2005 models.
$34,162, $35,870, $37,663, $39,546, $41,524, $43,600

42. Constant increase. The MSRP for a new 1999 Mercury Cougar was $21,455 (Edmund's New Car Prices, www.edmunds.com). Suppose the price of this car increases by $1,000 each year. Find the prices of the 2000 through 2005 models.
$22,455, $23,455, $24,455, $25,455, $26,455, $27,455

43. Economic impact. To assess the economic impact of a factory on a community, economists consider the annual amount the factory spends in the community, then the portion of the money that is respent in the community, then the portion of the respent money that is respent in the community, and so on. Suppose a garment manufacturer spends $1 million annually in its community and 80% of all money received in the community is respent in the community. Find the first four terms of the economic impact sequence.
$1,000,000, $800,000, $640,000, $512,000

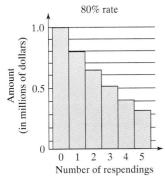

FIGURE FOR EXERCISE 43

44. Less impact. The rate at which money is respent in a community varies from community to community. Find the first four terms of the economic impact sequence for the manufacturer in Exercise 43, assuming only 50% of money received in the community is respent in the community.
$1,000,000, $500,000, $250,000, $125,000

45. Fabric design. A fabric designer must take into account the capability of textile machines to produce material with vertical repeats. A textile machine can be set up for a vertical repeat every $\frac{27}{n}$ inches (in.), where n is a natural number. Write the first five terms of the sequence $a_n = \frac{27}{n}$,

which gives the possible vertical repeats for a textile machine.
27 in., 13.5 in., 9 in., 6.75 in., 5.4 in.

46. Musical tones. The note middle C on a piano is tuned so that the string vibrates at 262 cycles per second, or 262 Hertz (Hz). The C note one octave higher is tuned to 524 Hz. The tuning for the 11 notes in between using the method called *equal temperament* is determined by the sequence $a_n = 262 \cdot 2^{n/12}$. Find the tuning for the 11 notes in between.
278, 294, 312, 330, 350, 371, 393, 416, 441, 467, 495 Hz.

GETTING MORE INVOLVED

47. Discussion. Everyone has two (biological) parents, four grandparents, eight great-grandparents, 16 great-great-grandparents, and so on. If we put the word "great" in front of the word "grandparents" 35 times, then how many of this type of relative do you have? Is this more or less than the present population of the earth? Give reasons for your answers.
137,438,953,500, larger

48. Discussion. If you deposit 1 cent into your piggy bank on September 1 and each day thereafter deposit twice as much as on the previous day, then how much will you be depositing on September 30? The total amount deposited for the month can be found without adding up all 30 deposits. Look at how the amount on deposit is increasing each day and see whether you can find the total for the month. Give reasons for your answers.
$5,368,709.12, $10,737,418.23

49. Cooperative learning. Working in groups, have someone in each group make up a formula for a_n, the nth term of a sequence, but do not show it to the other group members. Write the terms of the sequence on a piece of paper one at a time. After each term is given, ask whether anyone knows the next term. When the group can correctly give the next term, ask for a formula for the nth term.

50. Exploration. Find a real-life sequence in which all of the terms are the same. Find one in which each term after the first is one larger than the previous term. Find out what the sequence of fines is on your campus for your first, second, third, and fourth parking ticket.

51. Exploration. Consider the sequence whose nth term is $a_n = (0.999)^n$.
a) Calculate a_{100}, a_{1000}, and $a_{10,000}$.
0.9048, 0.3677, 0.00004517
b) What happens to a_n as n gets larger and larger?
a_n goes to zero

12.2 SERIES

If you make a sequence of bank deposits, then you might be interested in the total value of the terms of the sequence. Of course, if the sequence has only a few terms, you can simply add them. In Sections 12.3 and 12.4 we will develop formulas that give the sum of the terms for certain finite and infinite sequences. In this section you will first learn a notation for expressing the sum of the terms of a sequence.

Summation Notation

To describe the sum of the terms of a sequence, we use **summation notation.** The Greek letter Σ (sigma) is used to indicate sums. For example, the sum of the first five terms of the sequence $a_n = n^2$ is written as

$$\sum_{n=1}^{5} n^2.$$

You can read this notation as "the sum of n^2 for n between 1 and 5, inclusive." To find the sum, we let n take the values 1 through 5 in the expression n^2:

$$\sum_{n=1}^{5} n^2 = 1^2 + 2^2 + 3^2 + 4^2 + 5^2$$
$$= 1 + 4 + 9 + 16 + 25$$
$$= 55$$

In this context the letter n is the **index of summation.** Other letters may also be used. For example, the expressions

$$\sum_{n=1}^{5} n^2, \quad \sum_{j=1}^{5} j^2, \quad \text{and} \quad \sum_{i=1}^{5} i^2$$

all have the same value. Note that i is used as a variable here and not as an imaginary number.

EXAMPLE 1

Evaluating a sum in summation notation

Find the value of the expression

$$\sum_{i=1}^{3} (-1)^i (2i + 1).$$

Solution

Replace i by 1, 2, and 3, and then add the results:

$$\sum_{i=1}^{3} (-1)^i (2i + 1) = (-1)^1[2(1) + 1] + (-1)^2[2(2) + 1] + (-1)^3[2(3) + 1]$$
$$= -3 + 5 - 7$$
$$= -5$$

Series

The sum of the terms of the sequence 1, 4, 9, 16, 25 is written as

$$1 + 4 + 9 + 16 + 25.$$

This expression is called a *series*. It indicates that we are to add the terms of the given sequence. The sum, 55, is the sum of the series.

Series

The indicated sum of the terms of a sequence is called a **series.**

Just as a sequence may be finite or infinite, a series may be finite or infinite. In this section we discuss finite series only. In Section 12.4 we will discuss one type of infinite series.

Summation notation is a convenient notation for writing a series.

E X A M P L E   2

Converting to summation notation

Write the series in summation notation:

$$2 + 4 + 6 + 8 + 10 + 12 + 14$$

Solution

The general term for the sequence of positive even integers is $2n$. If we let n take the values from 1 through 7, then $2n$ ranges from 2 through 14. So

$$2 + 4 + 6 + 8 + 10 + 12 + 14 = \sum_{n=1}^{7} 2n.$$

■

E X A M P L E   3

Converting to summation notation

Write the series

$$\frac{1}{2} - \frac{1}{3} + \frac{1}{4} - \frac{1}{5} + \frac{1}{6} - \frac{1}{7} + \cdots + \frac{1}{50}$$

in summation notation.

helpful **hint**

A series is called an *indicated sum* because the addition is indicated by not actually being performed. The sum of a series is the real number obtained by actually performing the indicated addition.

Solution

For this series we let n be 2 through 50. The expression $(-1)^n$ produces alternating signs. The series is written as

$$\sum_{n=2}^{50} \frac{(-1)^n}{n}.$$

■

Changing the Index

In Example 3 we saw the index go from 2 through 50, but this is arbitrary. A series can be written with the index starting at any given number.

E X A M P L E   4

Changing the index

Rewrite the series

$$\sum_{i=1}^{6} \frac{(-1)^i}{i^2}$$

with an index j, where j starts at 0.

Solution

Because i starts at 1 and j starts at 0, we have $i = j + 1$. Because i ranges from 1 through 6 and $i = j + 1$, j must range from 0 through 5. Now replace i by $j + 1$ in the summation notation:

$$\sum_{j=0}^{5} \frac{(-1)^{j+1}}{(j + 1)^2}$$

Check that these two series have exactly the same six terms.

■

WARM-UPS

True or false? Explain your answer.

1. A series is the indicated sum of the terms of a sequence.   True

2. The sum of a series can never be negative.   False

3. There are eight terms in the series $\sum_{i=2}^{10} i^3$.   False

4. The series $\sum_{i=1}^{9} (-1)^i i^2$ and $\sum_{j=0}^{8} (-1)^j (j+1)^2$ have the same sum.   False

5. The ninth term of the series $\sum_{i=1}^{100} \dfrac{(-1)^i}{(i+1)(i+2)}$ is $\dfrac{1}{110}$.   False

6. $\sum_{i=1}^{2} (-1)^i 2^i = 2$ True

7. $\sum_{i=1}^{5} 3i = 3\left(\sum_{i=1}^{5} i\right)$ True

8. $\sum_{i=1}^{5} 4 = 20$ True

9. $\sum_{i=1}^{5} 2i + \sum_{i=1}^{5} 7i = \sum_{i=1}^{5} 9i$ True

10. $\sum_{i=1}^{3} (2i+1) = \left(\sum_{i=1}^{3} 2i\right) + 1$ False

12.2   EXERCISES

Reading and Writing *After reading this section, write out the answers to these questions. Use complete sentences.*

1. What is summation notation?
   Summation notation provides a way to write a sum without writing out all of the terms.

2. What is the index of summation?
   The index of summation is the variable used in summation notation.

3. What is a series?
   A series is the indicated sum of the terms of a sequence.

4. What is a finite series?
   A finite series is the indicated sum of the terms of a finite sequence.

Find the sum of each series. See Example 1.

5. $\sum_{i=1}^{4} i^2$

   30

6. $\sum_{j=0}^{3} (j+1)^2$

   30

7. $\sum_{j=0}^{5} (2j-1)$

   24

8. $\sum_{i=1}^{6} (2i-3)$

   24

9. $\sum_{i=1}^{5} 2^{-i}$

   $\dfrac{31}{32}$

10. $\sum_{i=1}^{5} (-2)^{-i}$

   $-\dfrac{11}{32}$

11. $\sum_{i=1}^{10} 5i^0$

   50

12. $\sum_{j=1}^{20} 3$

   60

13. $\sum_{i=1}^{3} (i-3)(i+1)$

   -7

14. $\sum_{i=0}^{5} i(i-1)(i-2)(i-3)$

   144

15. $\sum_{j=1}^{10} (-1)^j$

   0

16. $\sum_{j=1}^{11} (-1)^j$

   -1

Write each series in summation notation. Use the index i, and let i begin at 1 in each summation. See Examples 2 and 3.

17. $1 + 2 + 3 + 4 + 5 + 6$ $\sum_{i=1}^{6} i$

18. $2 + 4 + 6 + 8 + 10$ $\sum_{i=1}^{5} 2i$

19. $-1 + 3 - 5 + 7 - 9 + 11$ $\sum_{i=1}^{6} (-1)^i (2i-1)$

20. $1 - 3 + 5 - 7 + 9$ $\sum_{i=1}^{5} (-1)^{i+1}(2i-1)$

21. $1 + 4 + 9 + 16 + 25 + 36$ $\sum_{i=1}^{6} i^2$

22. $1 + 8 + 27 + 64 + 125$ $\sum_{i=1}^{5} i^3$

23. $\dfrac{1}{3} + \dfrac{1}{4} + \dfrac{1}{5} + \dfrac{1}{6}$ $\sum_{i=1}^{4} \dfrac{1}{2+i}$

24. $1 - \dfrac{1}{2} + \dfrac{1}{3} - \dfrac{1}{4} + \dfrac{1}{5} - \dfrac{1}{6}$ $\displaystyle\sum_{i=1}^{6} \dfrac{(-1)^{i+1}}{i}$

25. $\ln(2) + \ln(3) + \ln(4)$ $\displaystyle\sum_{i=1}^{3} \ln(i+1)$

26. $e^1 + e^2 + e^3 + e^4$ $\displaystyle\sum_{i=1}^{4} e^i$

27. $a_1 + a_2 + a_3 + a_4$ $\displaystyle\sum_{i=1}^{4} a_i$

28. $a^2 + a^3 + a^4 + a^5$ $\displaystyle\sum_{i=1}^{4} a^{i+1}$

29. $x_3 + x_4 + x_5 + \cdots + x_{50}$ $\displaystyle\sum_{i=1}^{48} x_{i+2}$

30. $y_1 + y_2 + y_3 + \cdots + y_{30}$ $\displaystyle\sum_{i=1}^{30} y_i$

31. $w_1 + w_2 + w_3 + \cdots + w_n$ $\displaystyle\sum_{i=1}^{n} w_i$

32. $m_1 + m_2 + m_3 + \cdots + m_k$ $\displaystyle\sum_{i=1}^{k} m_i$

Complete the rewriting of each series using the new index as indicated. See Example 4.

33. $\displaystyle\sum_{i=1}^{5} i^2 = \sum_{j=0}$

$\displaystyle\sum_{j=0}^{4} (j+1)^2$

34. $\displaystyle\sum_{i=1}^{6} i^3 = \sum_{j=0}$

$\displaystyle\sum_{j=0}^{5} (j+1)^3$

35. $\displaystyle\sum_{i=0}^{12} (2i-1) = \sum_{j=1}$

$\displaystyle\sum_{j=1}^{13} (2j-3)$

36. $\displaystyle\sum_{i=1}^{3} (3i+2) = \sum_{j=0}$

$\displaystyle\sum_{j=0}^{2} (3j+5)$

37. $\displaystyle\sum_{i=4}^{8} \dfrac{1}{i} = \sum_{j=1}$

$\displaystyle\sum_{j=1}^{5} \dfrac{1}{j+3}$

38. $\displaystyle\sum_{i=5}^{10} 2^{-i} = \sum_{j=1}$

$\displaystyle\sum_{j=1}^{6} 2^{-j-4}$

39. $\displaystyle\sum_{i=1}^{4} x^{2i+3} = \sum_{j=0}$

$\displaystyle\sum_{j=0}^{3} x^{2j+5}$

40. $\displaystyle\sum_{i=0}^{2} x^{3-2i} = \sum_{j=1}$

$\displaystyle\sum_{j=1}^{3} x^{5-2j}$

41. $\displaystyle\sum_{i=1}^{n} x^i = \sum_{j=0}$

$\displaystyle\sum_{j=0}^{n-1} x^{j+1}$

42. $\displaystyle\sum_{i=0}^{n} x^{-i} = \sum_{j=1}$

$\displaystyle\sum_{j=1}^{n+1} x^{-j+1}$

Write out the terms of each series.

43. $\displaystyle\sum_{i=1}^{6} x^i$ $x + x^2 + x^3 + x^4 + x^5 + x^6$

44. $\displaystyle\sum_{i=1}^{5} (-1)^i x^{i-1}$ $-1 + x - x^2 + x^3 - x^4$

45. $\displaystyle\sum_{j=0}^{3} (-1)^j x_j$ $x_0 - x_1 + x_2 - x_3$

46. $\displaystyle\sum_{j=1}^{5} \dfrac{1}{x_j}$ $\dfrac{1}{x_1} + \dfrac{1}{x_2} + \dfrac{1}{x_3} + \dfrac{1}{x_4} + \dfrac{1}{x_5}$

47. $\displaystyle\sum_{i=1}^{3} i x^i$ $x + 2x^2 + 3x^3$

48. $\displaystyle\sum_{i=1}^{5} \dfrac{x}{i}$ $x + \dfrac{x}{2} + \dfrac{x}{3} + \dfrac{x}{4} + \dfrac{x}{5}$

A series can be used to model the situation in each of the following problems.

49. Leap frog. A frog with a vision problem is 1 yard away from a dead cricket. He spots the cricket and jumps halfway to the cricket. After the frog realizes that he has not reached the cricket, he again jumps halfway to the cricket. Write a series in summation notation to describe how far the frog has moved after nine such jumps.

$\displaystyle\sum_{i=1}^{9} 2^{-i}$

50. Compound interest. Cleo deposited $1,000 at the beginning of each year for 5 years into an account paying 10% interest compounded annually. Write a series using summation notation to describe how much she has in the account at the end of the fifth year. Note that the first $1,000 will receive interest for 5 years, the second $1,000 will receive interest for 4 years, and so on.

$\displaystyle\sum_{i=1}^{5} 1000(1.1)^i$

51. Total economic impact. In Exercise 43 of Section 12.1 we described a factory that spends $1 million annually in a community in which 80% of all money received in the community is respent in the community. Use summation notation to write the sum of the first four terms of the economic impact sequence for the factory.

$\displaystyle\sum_{i=1}^{4} 1,000,000(0.8)^{i-1}$

52. Total spending. Suppose you earn $1 on January 1, $2 on January 2, $3 on January 3, and so on. Use summation notation to write the sum of your earnings for the entire month of January.

$\displaystyle\sum_{i=1}^{31} i$

GETTING MORE INVOLVED

53. Discussion. What is the difference between a sequence and a series?

A sequence is basically a list of numbers. A series is the indicated sum of the terms of a sequence.

54. Discussion. For what values of n is $\displaystyle\sum_{i=1}^{n} \dfrac{1}{i} > 4$?

$n \geq 31$

12.3 ARITHMETIC SEQUENCES AND SERIES

We defined sequences and series in Sections 12.1 and 12.2. In this section you will study a special type of sequence known as an arithmetic sequence. You will also study the series corresponding to this sequence.

Arithmetic Sequences

Consider the following sequence:

$$5, 9, 13, 17, 21, \ldots$$

This sequence is called an arithmetic sequence because of the pattern for the terms. Each term is 4 larger than the previous term.

helpful hint

Arithmetic used as an adjective (ar-ith-met'-ic) is pronounced differently from arithmetic used as a noun (a-rith'-me-tic). Arithmetic (the adjective) is accented similarly to geometric.

Arithmetic Sequence

A sequence in which each term after the first is obtained by adding a fixed amount to the previous term is called an **arithmetic sequence.**

The fixed amount is called the **common difference** and is denoted by the letter d. If a_1 is the first term, then the second term is $a_1 + d$. The third term is $a_1 + 2d$, the fourth term is $a_1 + 3d$, and so on.

Formula for the nth Term of an Arithmetic Sequence

The nth term, a_n, of an arithmetic sequence with first term a_1 and common difference d is

$$a_n = a_1 + (n - 1)d.$$

EXAMPLE 1

The nth term of an arithmetic sequence

Write a formula for the nth term of the arithmetic sequence

$$5, 9, 13, 17, 21, \ldots.$$

Solution

Each term of the sequence after the first is 4 more than the previous term. Because the common difference is 4 and the first term is 5, the nth term is given by

$$a_n = 5 + (n - 1)4.$$

We can simplify this expression to get

$$a_n = 4n + 1.$$

In the next example the common difference is negative.

EXAMPLE 2

An arithmetic sequence of decreasing terms

Write a formula for the nth term of the arithmetic sequence

$$4, 1, -2, -5, -8, \ldots.$$

Solution

Each term is 3 less than the previous term, so $d = -3$. Because $a_1 = 4$, we can write the nth term as

$$a_n = 4 + (n - 1)(-3),$$

or

$$a_n = -3n + 7.$$

In the next example we find some terms of an arithmetic sequence using a given formula for the nth term.

E X A M P L E   3

Writing terms of an arithmetic sequence

Write the first five terms of the sequence in which $a_n = 3 + (n - 1)6$.

Solution

Let n take the values from 1 through 5, and find a_n:

$$a_1 = 3 + (1 - 1)6 = 3$$
$$a_2 = 3 + (2 - 1)6 = 9$$
$$a_3 = 3 + (3 - 1)6 = 15$$
$$a_4 = 3 + (4 - 1)6 = 21$$
$$a_5 = 3 + (5 - 1)6 = 27$$

Notice that $a_n = 3 + (n - 1)6$ gives the general term for an arithmetic sequence with first term 3 and common difference 6. Because each term after the first is 6 more than the previous term, the first five terms that we found are correct.

The formula $a_n = a_1 + (n - 1)d$ involves four variables: a_1, a_n, n, and d. If we know the values of any three of these variables, we can find the fourth.

E X A M P L E   4

Finding a missing term of an arithmetic sequence

Find the twelfth term of the arithmetic sequence whose first term is 2 and whose fifth term is 14.

Solution

Before finding the twelfth term, we use the given information to find the missing common difference. Let $n = 5$, $a_1 = 2$, and $a_5 = 14$ in the formula $a_n = a_1 + (n - 1)d$ to find d:

$$14 = 2 + (5 - 1)d$$
$$14 = 2 + 4d$$
$$12 = 4d$$
$$3 = d$$

Now use $a_1 = 2$, $d = 3$ and $n = 12$ in $a_n = a_1 + (n - 1)d$ to find a_{12}:

$$a_{12} = 2 + (12 - 1)3$$
$$a_{12} = 35$$

Arithmetic Series

The indicated sum of an arithmetic sequence is called an **arithmetic series.** For example, the series

$$2 + 4 + 6 + 8 + 10 + \cdots + 54$$

is an arithmetic series because there is a common difference of 2 between the terms.

We can find the actual sum of this arithmetic series without adding all of the terms. Write the series in increasing order, and below that write the series in decreasing order. We then add the corresponding terms:

$$
\begin{array}{rcl}
S &=& 2 + 4 + 6 + 8 + \cdots + 52 + 54 \\
S &=& 54 + 52 + 50 + 48 + \cdots + 4 + 2 \\
\hline
2S &=& 56 + 56 + 56 + 56 + \cdots + 56 + 56
\end{array}
$$

Now, how many times does 56 appear in the sum on the right? Because

$$2 + 4 + 6 + \cdots + 54 = 2 \cdot 1 + 2 \cdot 2 + 2 \cdot 3 + \cdots + 2 \cdot 27,$$

there are 27 terms in this sum. Because 56 appears 27 times on the right, we have $2S = 27 \cdot 56$, or

$$S = \frac{27 \cdot 56}{2} = 27 \cdot 28 = 756.$$

If $S_n = a_1 + a_2 + a_3 + \cdots + a_n$ is any arithmetic series, then we can find its sum using the same technique. Rewrite S_n as follows:

$$
\begin{array}{rcl}
S_n &=& a_1 + (a_1 + d) + (a_1 + 2d) + \cdots + a_n \\
S_n &=& a_n + (a_n - d) + (a_n - 2d) + \cdots + a_1 \\
\hline
2S_n &=& (a_1 + a_n) + (a_1 + a_n) + (a_1 + a_n) + \cdots + (a_1 + a_n) \quad \text{Add.}
\end{array}
$$

Because $(a_1 + a_n)$ appears n times on the right, we have $2S_n = n(a_1 + a_n)$. Divide each side by 2 to get the following formula.

Sum of an Arithmetic Series

The sum, S_n, of the first n terms of an arithmetic series with first term a_1 and nth term a_n, is given by

$$S_n = \frac{n}{2}(a_1 + a_n).$$

E X A M P L E   5

The sum of an arithmetic series

Find the sum of the positive integers from 1 to 100 inclusive.

Solution

The described series, $1 + 2 + 3 + \cdots + 100$, has 100 terms. So we can use $n = 100$, $a_1 = 1$, and $a_n = 100$ in the formula for the sum of an arithmetic series:

$$S_n = \frac{n}{2}(a_1 + a_n)$$

$$S_{100} = \frac{100}{2}(1 + 100)$$

$$= 50(101)$$

$$= 5050$$

helpful **hint**

Legend has it that Carl F. Gauss knew this formula when he was in grade school. Gauss's teacher told him to add up the numbers from 1 through 100 for busy work. He immediately answered 5050.

E X A M P L E   6

The sum of an arithmetic series

Find the sum of the series

$$12 + 16 + 20 + \cdots + 84.$$

Solution

This series is an arithmetic series with $a_n = 84$, $a_1 = 12$, and $d = 4$. To get the number of terms, n, we use $a_n = a_1 + (n - 1)d$:

$$84 = 12 + (n - 1)4$$
$$84 = 8 + 4n$$
$$76 = 4n$$
$$19 = n$$

Now find the sum of these 19 terms:

$$S_{19} = \frac{19}{2}(12 + 84) = 912$$

WARM-UPS

True or false? Explain your answer.

1. The arithmetic sequence 3, 1, −1, −3, −5, . . . has common difference 2.
   False
2. The sequence 2, 5, 9, 14, 20, 27, . . . is an arithmetic sequence.   False
3. The sequence 2, 4, 2, 0, 2, 4, 2, 0, . . . is an arithmetic sequence.   False
4. The nth term of an arithmetic sequence with first term a_1 and common difference d is given by the formula $a_n = a_1 + nd$.   False
5. If $a_1 = 5$ and $a_3 = 10$ in an arithmetic sequence, then $a_4 = 15$.   False
6. If $a_1 = 6$ and $a_3 = 2$ in an arithmetic sequence, then $a_2 = 10$.   False
7. An arithmetic series is the indicated sum of an arithmetic sequence.   True
8. The series $\sum_{i=1}^{5}(3 + 2i)$ is an arithmetic series.   True
9. The sum of the first n counting numbers is $\frac{n(n + 1)}{2}$.   True
10. The sum of the even integers from 8 through 28 inclusive is $5(8 + 28)$.
    False

12.3   EXERCISES

Reading and Writing   *After reading this section, write out the answers to these questions. Use complete sentences.*

1. What is an arithmetic sequence?

   An arithmetic sequence is one in which each term after the first is obtained by adding a fixed amount to the previous term.

2. What is the nth term of an arithmetic sequence?

   The nth term of an arithmetic sequence is $a_1 + (n - 1)d$, where a_1 is the first term.

3. What is an arithmetic series?

   An arithmetic series is an indicated sum of an arithmetic sequence.

4. What is the formula for the sum of the first n terms of an arithmetic series?

   The formula for the sum of the first n terms of an arithmetic series is $\frac{n}{2}(a_1 + a_n)$.

Write a formula for the nth term of each arithmetic sequence. See Examples 1 and 2.

5. 0, 6, 12, 18, 24, . . . $a_n = 6n - 6$
6. 0, 5, 10, 15, 20, . . . $a_n = 5n - 5$
7. 7, 12, 17, 22, 27, . . . $a_n = 5n + 2$
8. 4, 15, 26, 37, 48, . . . $a_n = 11n - 7$
9. −4, −2, 0, 2, 4, . . . $a_n = 2n - 6$

10. $-3, 0, 3, 6, 9, \ldots$ $a_n = 3n - 6$

11. $5, 1, -3, -7, -11, \ldots$ $a_n = -4n + 9$

12. $8, 5, 2, -1, -4, \ldots$ $a_n = -3n + 11$

13. $-2, -9, -16, -23, \ldots$ $a_n = -7n + 5$

14. $-5, -7, -9, -11, -13, \ldots$ $a_n = -2n - 3$

15. $-3, -2.5, -2, -1.5, -1, \ldots$ $a_n = 0.5n - 3.5$

16. $-2, -1.25, -0.5, 0.25, \ldots$ $a_n = 0.75n - 2.75$

17. $-6, -6.5, -7, -7.5, -8, \ldots$ $a_n = -0.5n - 5.5$

18. $1, 0.5, 0, -0.5, -1, \ldots$ $a_n = -0.5n + 1.5$

In Exercises 19–32, write the first five terms of the arithmetic sequence whose nth term is given. See Example 3.

19. $a_n = 9 + (n - 1)4$ $9, 13, 17, 21, 25$

20. $a_n = 13 + (n - 1)6$ $13, 19, 25, 31, 37$

21. $a_n = 7 + (n - 1)(-2)$ $7, 5, 3, 1, -1$

22. $a_n = 6 + (n - 1)(-3)$ $6, 3, 0, -3, -6$

23. $a_n = -4 + (n - 1)3$ $-4, -1, 2, 5, 8$

24. $a_n = -19 + (n - 1)12$ $-19, -7, 5, 17, 29$

25. $a_n = -2 + (n - 1)(-3)$ $-2, -5, -8, -11, -14$

26. $a_n = -1 + (n - 1)(-2)$ $-1, -3, -5, -7, -9$

27. $a_n = -4n - 3$ $-7, -11, -15, -19, -23$

28. $a_n = -3n + 1$ $-2, -5, -8, -11, -14$

29. $a_n = 0.5n + 4$ $4.5, 5, 5.5, 6, 6.5$

30. $a_n = 0.3n + 1$ $1.3, 1.6, 1.9, 2.2, 2.5$

31. $a_n = 20n + 1000$ $1020, 1040, 1060, 1080, 1100$

32. $a_n = -600n + 4000$ $3400, 2800, 2200, 1600, 1000$

Find the indicated part of each arithmetic sequence. See Example 4.

33. Find the eighth term of the sequence that has a first term of 9 and a common difference of 6.

51

34. Find the twelfth term of the sequence that has a first term of -2 and a common difference of -3.

-35

35. Find the common difference if the first term is 6 and the twentieth term is 82.

4

36. Find the common difference if the first term is -8 and the ninth term is -64.

-7

37. If the common difference is -2 and the seventh term is 14, then what is the first term?

26

38. If the common difference is 5 and the twelfth term is -7, then what is the first term?

-62

39. Find the sixth term of the sequence that has a fifth term of 13 and a first term of -3.

17

40. Find the eighth term of the sequence that has a sixth term of -42 and a first term of 3.

-60

Find the sum of each given series. See Examples 5 and 6.

41. $1 + 2 + 3 + \cdots + 48$ 1176

42. $1 + 2 + 3 + \cdots + 12$ 78

43. $8 + 10 + 12 + \cdots + 36$ 330

44. $9 + 12 + 15 + \cdots + 72$ 891

45. $-1 + (-7) + (-13) + \cdots + (-73)$ -481

46. $-7 + (-12) + (-17) + \cdots + (-72)$ -553

47. $-6 + (-1) + 4 + 9 + \cdots + 64$ 435

48. $-9 + (-1) + 7 + \cdots + 103$ 705

49. $20 + 12 + 4 + (-4) + \cdots + (-92)$ -540

50. $19 + 1 + (-17) + \cdots + (-125)$ -477

51. $\sum_{i=1}^{12} (3i - 7)$ 150

52. $\sum_{i=1}^{7} (-4i + 6)$ -70

53. $\sum_{i=1}^{11} (-5i + 2)$ -308

54. $\sum_{i=1}^{19} (3i - 5)$ 475

Solve each problem using the ideas of arithmetic sequences and series.

55. *Increasing salary.* If a lab technician has a salary of $22,000 her first year and is due to get a $500 raise each year, then what will her salary be in her seventh year?

$25,000

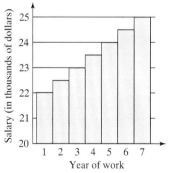

FIGURE FOR EXERCISE 55

56. *Seven years of salary.* What is the total salary for 7 years of work for the lab technician of Exercise 55?

$164,500

57. *Light reading.* On the first day of October an English teacher suggests to his students that they read five pages of a novel and every day thereafter increase their daily reading by two pages. If his students follow this suggestion, then how many pages will they read during October?

1,085

58. *Heavy penalties.* If an air-conditioning system is not completed by the agreed upon date, the contractor pays a penalty of $500 for the first day that it is overdue, $600 for the second day, $700 for the third day, and so on. If the system is completed 10 days late, then what is the total amount of the penalties that the contractor must pay?

$9,500

GETTING MORE INVOLVED

 59. *Discussion.* Which of the following sequences is not an arithmetic sequence? Explain your answer.

a) $\dfrac{1}{2}, 1, \dfrac{3}{2}, \ldots$ b) $\dfrac{1}{2}, \dfrac{1}{3}, \dfrac{1}{4}, \ldots$

c) $5, 0, -5, \ldots$ d) $2, 3, 4, \ldots$

b

 60. *Discussion.* What is the smallest value of n for which

$$\sum_{i=1}^{n} \dfrac{i}{2} > 50? \quad 14$$

 12.4   GEOMETRIC SEQUENCES AND SERIES

In Section 12.3 you studied the arithmetic sequences and series. In this section you will study sequences in which each term is a *multiple* of the term preceding it. You will also learn how to find the sum of the corresponding series.

Geometric Sequences

Consider the following sequence:

$$3, 6, 12, 24, 48, \ldots$$

Unlike an arithmetic sequence, these terms do not have a common difference, but there is a simple pattern to the terms. Each term after the first is twice the term preceding it. Such a sequence is called a geometric sequence.

Geometric Sequence

A sequence in which each term after the first is obtained by multiplying the preceding term by a constant is called a **geometric sequence.**

The constant is denoted by the letter r and is called the **common ratio.** If a_1 is the first term, then the second term is $a_1 r$. The third term is $a_1 r^2$, the fourth term is $a_1 r^3$, and so on. We can write a formula for the nth term of a geometric sequence by following this pattern.

Formula for the nth Term of a Geometric Sequence

The nth term, a_n, of a geometric sequence with first term a_1 and common ratio r is

$$a_n = a_1 r^{n-1}.$$

The first term and the common ratio determine all of the terms of a geometric sequence.

E X A M P L E   1

Finding the nth term

Write a formula for the nth term of the geometric sequence

$$6, 2, \frac{2}{3}, \frac{2}{9}, \ldots.$$

Solution

We can obtain the common ratio by dividing any term after the first by the term preceding it. So

$$r = 2 \div 6 = \frac{1}{3}.$$

Because each term after the first is $\frac{1}{3}$ of the term preceding it, the nth term is given by

$$a_n = 6\left(\frac{1}{3}\right)^{n-1}.$$

■

E X A M P L E   2

Finding the nth term

Find a formula for the nth term of the geometric sequence

$$2, -1, \frac{1}{2}, -\frac{1}{4}, \ldots.$$

Solution

We obtain the ratio by dividing a term by the term preceding it:

$$r = -1 \div 2 = -\frac{1}{2}$$

Each term after the first is obtained by multiplying the preceding term by $-\frac{1}{2}$. The formula for the nth term is

$$a_n = 2\left(-\frac{1}{2}\right)^{n-1}.$$

■

In the next example we use the formula for the nth term to write some terms of a geometric sequence.

E X A M P L E   3

Writing the terms

Write the first five terms of the geometric sequence whose nth term is

$$a_n = 3(-2)^{n-1}.$$

Solution

Let n take the values 1 through 5 in the formula for the nth term:

$$a_1 = 3(-2)^{1-1} = 3$$
$$a_2 = 3(-2)^{2-1} = -6$$
$$a_3 = 3(-2)^{3-1} = 12$$
$$a_4 = 3(-2)^{4-1} = -24$$
$$a_5 = 3(-2)^{5-1} = 48$$

Notice that $a_n = 3(-2)^{n-1}$ gives the general term for a geometric sequence with first term 3 and common ratio -2. Because every term after the first can be obtained by multiplying the previous term by -2, the terms 3, -6, 12, -24, and 48 are correct. ■

The formula for the nth term involves four variables: a_n a_1, r, and n. If we know the value of any three of them, we can find the value of the fourth.

E X A M P L E   4

Finding a missing term

Find the first term of a geometric sequence whose fourth term is 8 and whose common ratio is $\frac{1}{2}$.

Solution

Let $a_4 = 8, r = \frac{1}{2}$, and $n = 4$ in the formula $a_n = a_1 r^{n-1}$:

$$8 = a_1 \left(\frac{1}{2}\right)^{4-1}$$

$$8 = a_1 \cdot \frac{1}{8}$$

$$64 = a_1$$

So the first term is 64. ■

Finite Geometric Series

Consider the following series:

$$1 + 2 + 4 + 8 + 16 + \cdots + 512$$

The terms of this series are the terms of a finite geometric sequence. The indicated sum of a geometric sequence is called a **geometric series.**

We can find the actual sum of this finite geometric series by using a technique similar to the one used for the sum of an arithmetic series. Let

$$S = 1 + 2 + 4 + 8 + \cdots + 256 + 512.$$

Because the common ratio is 2, multiply each side by -2:

$$-2S = -2 - 4 - 8 - \cdots - 512 - 1024$$

Adding the last two equations eliminates all but two of the terms on the right:

$$
\begin{aligned}
S &= 1 + 2 + 4 + 8 + \cdots + 256 + 512 \\
-2S &= -2 - 4 - 8 - \cdots - 512 - 1024 \\
\hline
-S &= 1 - 1024 \quad \text{Add.} \\
-S &= -1023 \\
S &= 1023
\end{aligned}
$$

If $S_n = a_1 + a_1 r + a_1 r^2 + \cdots + a_1 r^{n-1}$ is any geometric series, we can find the sum in the same manner. Multiplying each side of this equation by $-r$ yields

$$-rS_n = -a_1 r - a_1 r^2 - a_1 r^3 - \cdots - a_1 r^n.$$

If we add S_n and $-rS_n$, all but two of the terms on the right are eliminated:

$$S_n = a_1 + a_1r + a_1r^2 + \cdots \qquad\qquad + a_1r^{n-1}$$
$$-rS_n = \qquad - a_1r - a_1r^2 - a_1r^3 - \cdots \qquad\qquad - a_1r^n$$

$$\begin{aligned} S_n - rS_n &= a_1 \qquad\qquad\qquad\qquad\qquad\quad - a_1r^n \qquad \text{Add.}\\ (1-r)S_n &= a_1(1-r^n) \qquad\qquad\qquad\qquad \begin{array}{l}\text{Factor out}\\ \text{common factors.}\end{array}\end{aligned}$$

Now divide each side of this equation by $1 - r$ to get the formula for S_n.

Sum of n Terms of a Geometric Series

If S_n represents the sum of the first n terms of a geometric series with first term a_1 and common ratio r $(r \neq 1)$, then

$$S_n = \frac{a_1(1 - r^n)}{1 - r}.$$

E X A M P L E   5   **The sum of a finite geometric series**

Find the sum of the series

$$\frac{1}{3} + \frac{1}{9} + \frac{1}{27} + \cdots + \frac{1}{729}.$$

Solution

The first term is $\frac{1}{3}$, and the common ratio is $\frac{1}{3}$. So the nth term can be written as

$$a_n = \frac{1}{3}\left(\frac{1}{3}\right)^{n-1}.$$

We can use this formula to find the number of terms in the series:

$$\frac{1}{729} = \frac{1}{3}\left(\frac{1}{3}\right)^{n-1}$$

$$\frac{1}{729} = \left(\frac{1}{3}\right)^{n}$$

Because $3^6 = 729$, we have $n = 6$. (Of course, you could use logarithms to solve for n.) Now use the formula for the sum of six terms of this geometric series:

$$S_6 = \frac{\dfrac{1}{3}\left[1 - \left(\dfrac{1}{3}\right)^6\right]}{1 - \dfrac{1}{3}} = \frac{\dfrac{1}{3}\left[1 - \dfrac{1}{729}\right]}{\dfrac{2}{3}}$$

$$= \frac{1}{3} \cdot \frac{728}{729} \cdot \frac{3}{2}$$

$$= \frac{364}{729}$$

E X A M P L E   6   **The sum of a finite geometric series**

Find the sum of the series

$$\sum_{i=1}^{12} 3(-2)^{i-1}.$$

Solution

This series is geometric with first term 3, ratio -2, and $n = 12$. We use the formula for the sum of the first 12 terms of a geometric series:

$$S_{12} = \frac{3[1 - (-2)^{12}]}{1 - (-2)} = \frac{3[-4095]}{3} = -4095$$

Infinite Geometric Series

Consider how a very large value of n affects the formula for the sum of a finite geometric series,

$$S_n = \frac{a_1(1 - r^n)}{1 - r}.$$

If $|r| < 1$, then the value of r^n gets closer and closer to 0 as n gets larger and larger. For example, if $r = \frac{2}{3}$ and $n = 10, 20$, and 100, then

$$\left(\frac{2}{3}\right)^{10} \approx 0.0173415, \quad \left(\frac{2}{3}\right)^{20} \approx 0.0003007, \quad \text{and} \quad \left(\frac{2}{3}\right)^{100} \approx 2.460 \times 10^{-18}.$$

Because r^n is approximately 0 for large values of n, $1 - r^n$ is approximately 1. If we replace $1 - r^n$ by 1 in the expression for S_n, we get

$$S_n \approx \frac{a_1}{1 - r}.$$

So as n gets larger and larger, the sum of the first n terms of the infinite geometric series

$$a_1 + a_1 r + a_1 r^2 + \cdots$$

gets closer and closer to $\frac{a_1}{1 - r}$, provided that $|r| < 1$. Therefore we say that $\frac{a_1}{1 - r}$ is the sum of *all* of the terms of the infinite geometric series.

Sum of an Infinite Geometric Series

If $a_1 + a_1 r + a_1 r^2 + \cdots$ is an infinite geometric series, with $|r| < 1$, then the sum S of all of the terms of this series is given by

$$S = \frac{a_1}{1 - r}.$$

calculator close-up

Experiment with your calculator to see what happens to r^n as n gets larger and larger.

```
.99^100
       .3660323413
.99^1000
    4.317124741E-5
.99^10000
    2.24877485E-44
```

EXAMPLE 7

helpful hint

You can imagine this series in a football game. The Bears have the ball on the Lions' 1 yard line. The Lions continually get penalties that move the ball one-half of the distance to the goal. Theoretically, the ball will never reach the goal, but the total distance it moves will get closer and closer to 1 yard.

Sum of an infinite geometric series

Find the sum

$$\frac{1}{2} + \frac{1}{4} + \frac{1}{8} + \frac{1}{16} + \cdots.$$

Solution

This series is an infinite geometric series with $a_1 = \frac{1}{2}$ and $r = \frac{1}{2}$. Because $r < 1$, we have

$$S = \frac{\dfrac{1}{2}}{1 - \dfrac{1}{2}} = 1.$$

For an infinite series the index of summation i takes the values 1, 2, 3, and so on, without end. To indicate that the values for i keep increasing without bound, we say that *i takes the values from* 1 *through* ∞ (infinity). Note that the symbol "∞" does not represent a number. Using the ∞ symbol, we can write the indicated sum of an infinite geometric series (with $|r| < 1$) by using summation notation as follows:

$$a_1 + a_1 r + a_1 r^2 + \cdots = \sum_{i=1}^{\infty} a_1 r^{i-1}$$

E X A M P L E 8 **Sum of an infinite geometric series**

Find the value of the sum

$$\sum_{i=1}^{\infty} 8 \left(\frac{3}{4} \right)^{i-1}.$$

Solution

This series is an infinite geometric series with first term 8 and ratio $\frac{3}{4}$. So

$$S = \frac{8}{1 - \dfrac{3}{4}} = 8 \cdot \frac{4}{1} = 32.$$

E X A M P L E 9 **Follow the bouncing ball**

Suppose a ball always rebounds $\frac{2}{3}$ of the height from which it falls and the ball is dropped from a height of 6 feet. Find the total distance that the ball travels.

Solution

The ball falls 6 feet (ft) and rebounds 4 ft, then falls 4 ft and rebounds $\frac{8}{3}$ ft. The following series gives the total distance that the ball falls:

$$F = 6 + 4 + \frac{8}{3} + \frac{16}{9} + \cdots$$

The distance that the ball rebounds is given by the following series:

$$R = 4 + \frac{8}{3} + \frac{16}{9} + \cdots$$

Each of these series is an infinite geometric series with ratio $\frac{2}{3}$. Use the formula for an infinite geometric series to find each sum:

$$F = \frac{6}{1 - \dfrac{2}{3}} = 6 \cdot \frac{3}{1} = 18 \text{ ft}, \qquad R = \frac{4}{1 - \dfrac{2}{3}} = 4 \cdot \frac{3}{1} = 12 \text{ ft}$$

The total distance traveled by the ball is the sum of F and R, 30 ft.

Annuities

One of the most important applications of geometric series is in calculating the value of an annuity. An **annuity** is a sequence of periodic payments. The payments might be loan payments or investments.

EXAMPLE 10 **Value of an annuity**

A deposit of $1,000 is made at the beginning of each year for 30 years and earns 6% interest compounded annually. What is the value of this annuity at the end of the 30th year?

Solution

The last deposit earns interest for only one year. So at the end of the 30th year it amounts to $1000(1.06)$. The next to last deposit earns interest for 2 years and amounts to $1000(1.06)^2$. The first deposit earns interest for 30 years and amounts to $1000(1.06)^{30}$. So the value of the annuity at the end of the 30th year is the sum of the finite geometric series

$$1000(1.06) + 1000(1.06)^2 + 1000(1.06)^3 + \cdots + 1000(1.06)^{30}.$$

Use the formula for the sum of 30 terms of a finite geometric series with $a_1 = 1000(1.06)$ and $r = 1.06$:

$$S_{30} = \frac{1000(1.06)(1 - (1.06)^{30})}{1 - 1.06} = \$83{,}801.68$$

So 30 annual deposits of $1,000 each amount to $83,801.68.

WARM-UPS

True or false? Explain your answer.

1. The sequence 2, 6, 24, 120, . . . is a geometric sequence.   False
2. For $a_n = 2^n$ there is a common difference between adjacent terms.   False
3. The common ratio for the geometric sequence $a_n = 3(0.5)^{n-1}$ is 0.5.   True
4. If $a_n = 3(2)^{-n+3}$, then $a_1 = 12$.   True
5. In the geometric sequence $a_n = 3(2)^{-n+3}$ we have $r = \frac{1}{2}$.   True
6. The terms of a geometric series are the terms of a geometric sequence.   True
7. To evaluate $\sum_{i=1}^{10} 2^i$, we must list all of the terms.   False

8. $\sum_{i=1}^{5} 6\left(\frac{3}{4}\right)^{i-1} = \dfrac{9\left[1 - \left(\frac{3}{4}\right)^5\right]}{1 - \frac{3}{4}}$ False   9. $10 + 5 + \frac{5}{2} + \cdots = \dfrac{10}{1 - \frac{1}{2}}$ True

10. $2 + 4 + 8 + 16 + \cdots = \dfrac{2}{1 - 2}$ False

12.4   EXERCISES

Reading and Writing   *After reading this section, write out the answers to these questions. Use complete sentences.*

1. What is a geometric sequence?

A geometric sequence is one in which each term after the first is obtained by multiplying the preceding term by a constant.

2. What is the nth term of a geometric sequence?

The nth term of a geometric sequence is $a_1 r^{n-1}$, where a_1 is the first term and r is the common ratio.

3. What is a geometric series?

A geometric series is an indicated sum of a geometric sequence.

4. What is the formula for the sum of the first n terms of a geometric series?

The sum of the first n terms of a geometric series is given by $S_n = \frac{a_1(1 - r^n)}{1 - r}$.

5. What is the approximate value of r^n when n is large and $|r| < 1$?

The approximate value of r^n when n is large and $|r| < 1$ is 0.

6. What is the formula for the sum of an infinite geometric series?

The sum of an infinite geometric series is given by $S = \frac{a_1}{1 - r}$, provided $|r| < 1$.

Write a formula for the nth term of each geometric sequence. See Examples 1 and 2.

7. $\frac{1}{3}, 1, 3, 9, \ldots$

$a_n = \frac{1}{3}(3)^{n-1}$

8. $\frac{1}{4}, 2, 16, \ldots$

$a_n = \frac{1}{4}(8)^{n-1}$

9. $64, 8, 1, \ldots$

$a_n = 64\left(\frac{1}{8}\right)^{n-1}$

10. $100, 10, 1, \ldots$

$a_n = 100\left(\frac{1}{10}\right)^{n-1}$

11. $8, -4, 2, -1, \ldots$

$a_n = 8\left(-\frac{1}{2}\right)^{n-1}$

12. $-9, 3, -1, \ldots$

$a_n = -9\left(-\frac{1}{3}\right)^{n-1}$

13. $2, -4, 8, -16, \ldots$

$a_n = 2(-2)^{n-1}$

14. $-\frac{1}{2}, 2, -8, 32, \ldots$

$a_n = -\frac{1}{2}(-4)^{n-1}$

15. $-\frac{1}{3}, -\frac{1}{4}, -\frac{3}{16}, \ldots$

$a_n = -\frac{1}{3}\left(\frac{3}{4}\right)^{n-1}$

16. $-\frac{1}{4}, -\frac{1}{5}, -\frac{4}{25}, \ldots$

$a_n = -\frac{1}{4}\left(\frac{4}{5}\right)^{n-1}$

Write the first five terms of the geometric sequence with the given nth term. See Example 3.

17. $a_n = 2\left(\frac{1}{3}\right)^{n-1}$

$2, \frac{2}{3}, \frac{2}{9}, \frac{2}{27}, \frac{2}{81}$

18. $a_n = -5\left(\frac{1}{2}\right)^{n-1}$

$-5, -\frac{5}{2}, -\frac{5}{4}, -\frac{5}{8}, -\frac{5}{16}$

19. $a_n = (-2)^{n-1}$

$1, -2, 4, -8, 16$

20. $a_n = \left(-\frac{1}{3}\right)^{n-1}$

$1, -\frac{1}{3}, \frac{1}{9}, -\frac{1}{27}, \frac{1}{81}$

21. $a_n = 2^{-n}$

$\frac{1}{2}, \frac{1}{4}, \frac{1}{8}, \frac{1}{16}, \frac{1}{32}$

22. $a_n = 3^{-n}$

$\frac{1}{3}, \frac{1}{9}, \frac{1}{27}, \frac{1}{81}, \frac{1}{243}$

23. $a_n = (0.78)^n$

$0.78, 0.6084, 0.4746,$
$0.3702, 0.2887$

24. $a_n = (-0.23)^n$

$-0.23, 0.0529, -0.0122,$
$0.0028, -0.0006$

Find the required part of each geometric sequence. See Example 4.

25. Find the first term of the geometric sequence that has fourth term 40 and common ratio 2.

5

26. Find the first term of the geometric sequence that has fifth term 4 and common ratio $\frac{1}{2}$.

64

27. Find r for the geometric sequence that has $a_1 = 6$ and $a_4 = \frac{2}{9}$.

$\frac{1}{3}$

28. Find r for the geometric sequence that has $a_1 = 1$ and $a_4 = -27$.

-3

29. Find a_4 for the geometric sequence that has $a_1 = -3$ and $r = \frac{1}{3}$.

$-\frac{1}{9}$

30. Find a_5 for the geometric sequence that has $a_1 = -\frac{2}{3}$ and $r = -\frac{2}{3}$.

$-\frac{32}{243}$

Find the sum of each geometric series. See Examples 5 and 6.

31. $\frac{1}{2} + \frac{1}{4} + \frac{1}{8} + \cdots + \frac{1}{512}$ $\frac{511}{512}$

32. $1 + \frac{1}{3} + \frac{1}{9} + \cdots + \frac{1}{81}$ $\frac{121}{81}$

33. $\frac{1}{2} - \frac{1}{4} + \frac{1}{8} - \frac{1}{16} + \frac{1}{32}$ $\frac{11}{32}$

34. $3 - 1 + \frac{1}{3} - \frac{1}{9} + \frac{1}{27} - \frac{1}{81}$ $\frac{182}{81}$

35. $30 + 20 + \frac{40}{3} + \cdots + \frac{1280}{729}$ $\frac{63,050}{729}$

36. $9 - 6 + 4 - \cdots - \frac{128}{243}$ $\frac{1261}{243}$

37. $\sum_{i=1}^{10} 5(2)^{i-1}$ 5115

38. $\sum_{i=1}^{7} (10,000)(0.1)^{i-1}$ 11,111.11

39. $\sum_{i=1}^{6} (0.1)^i$ 0.111111

40. $\sum_{i=1}^{5} (0.2)^i$ 0.24992

41. $\sum_{i=1}^{6} 100(0.3)^i$ 42.8259

42. $\sum_{i=1}^{7} 36(0.5)^i$ 35.71875

Find the sum of each infinite geometric series. See Examples 7 and 8.

43. $\dfrac{1}{8} + \dfrac{1}{16} + \dfrac{1}{32} + \cdots$

$\dfrac{1}{4}$

44. $\dfrac{1}{9} + \dfrac{1}{27} + \dfrac{1}{81} + \cdots$

$\dfrac{1}{6}$

45. $3 + 2 + \dfrac{4}{3} + \cdots$

9

46. $2 + 1 + \dfrac{1}{2} + \cdots$

4

47. $4 - 2 + 1 - \dfrac{1}{2} + \cdots$

$\dfrac{8}{3}$

48. $16 - 12 + 9 - \dfrac{27}{4} + \cdots$

$\dfrac{64}{7}$

49. $\displaystyle\sum_{i=1}^{\infty} (0.3)^i$

$\dfrac{3}{7}$

50. $\displaystyle\sum_{i=1}^{\infty} (0.2)^i$

$\dfrac{1}{4}$

51. $\displaystyle\sum_{i=1}^{\infty} 3(0.5)^{i-1}$

6

52. $\displaystyle\sum_{i=1}^{\infty} 7(0.4)^{i-1}$

$\dfrac{35}{3}$

53. $\displaystyle\sum_{i=1}^{\infty} 3(0.1)^i$

$\dfrac{1}{3}$

54. $\displaystyle\sum_{i=1}^{\infty} 6(0.1)^i$

$\dfrac{2}{3}$

55. $\displaystyle\sum_{i=1}^{\infty} 12(0.01)^i$

$\dfrac{4}{33}$

56. $\displaystyle\sum_{i=1}^{\infty} 72(0.01)^i$

$\dfrac{8}{11}$

Use the ideas of geometric series to solve each problem. See Examples 9 and 10.

 57. *Retirement fund.* Suppose a deposit of $2,000 is made at the beginning of each year for 45 years into an account paying 12% compounded annually. What is the amount of this annuity at the end of the 45th year? $3,042,435.27

 58. *World's largest mutual fund.* If you had invested $5,000 at the beginning of each year for the last 10 years in Fidelity's

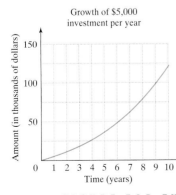

Growth of $5,000 investment per year

Amount (in thousands of dollars)

150

100

50

0   1 2 3 4 5 6 7 8 9 10

Time (years)

FIGURE FOR EXERCISE 58

Magellan fund you would have earned 18.97% compounded annually (Fidelity Investments, www.fidelity.com). Find the amount of this annuity at the end of the tenth year.
$146,763.44

 59. *Big saver.* Suppose you deposit one cent into your piggy bank on the first day of December and, on each day of December after that, you deposit twice as much as on the previous day. How much will you have in the bank after the last deposit?
$21,474,836.47

 60. *Big family.* Consider yourself, your parents, your grand-parents, your great-grandparents, your great-greatgrand-parents, and so on, back to your grandparents with the word "great" used in front 40 times. What is the total number of people you are considering?
8.796×10^{12}

61. *Total economic impact.* In Exercise 43 of Section 12.1 we described a factory that spends $1 million annually in a community in which 80% of the money received is respent in the community. Economists assume the money is respent again and again at the 80% rate. The total economic impact of the factory is the total of all of this spending. Find an approximation for the total by using the formula for the sum of an infinite geometric series with a rate of 80%.
$5,000,000

62. *Less impact.* Repeat Exercise 61, assuming money is respent again and again at the 50% rate.
$2,000,000

GETTING MORE INVOLVED

 63. *Discussion.* Which of the following sequences is not a geometric sequence? Explain your answer.
a) $1, 2, 4, \ldots$
b) $0.1, 0.01, 0.001, \ldots$
c) $-1, 2, -4, \ldots$
d) $2, 4, 6, \ldots$
   d

 **64.** *Discussion.* The repeating decimal number $0.44444 \ldots$ can be written as

$$\frac{4}{10} + \frac{4}{100} + \frac{4}{1000} + \cdots,$$

an infinite geometric series. Find the sum of this geometric series.
$\dfrac{4}{9}$

65. *Discussion.* Write the repeating decimal number $0.24242424 \ldots$ as an infinite geometric series. Find the sum of the geometric series.
$\dfrac{8}{33}$

12.5 BINOMIAL EXPANSIONS

In Chapter 5 you learned how to square a binomial. In this section you will study higher powers of binomials.

Some Examples

We know that $(x + y)^2 = x^2 + 2xy + y^2$. To find $(x + y)^3$, we multiply $(x + y)^2$ by $x + y$:

$$(x + y)^3 = (x^2 + 2xy + y^2)(x + y)$$
$$= (x^2 + 2xy + y^2)x + (x^2 + 2xy + y^2)y$$
$$= x^3 + 2x^2y + xy^2 + x^2y + 2xy^2 + y^3$$
$$= x^3 + 3x^2y + 3xy^2 + y^3$$

The sum $x^3 + 3x^2y + 3xy^2 + y^3$ is called the **binomial expansion** of $(x + y)^3$. If we again multiply by $x + y$, we will get the binomial expansion of $(x + y)^4$. This method is rather tedious. However, if we examine these expansions, we can find a pattern and learn how to find binomial expansions without multiplying.

Consider the following binomial expansions:

$$(x + y)^0 = 1$$
$$(x + y)^1 = x + y$$
$$(x + y)^2 = x^2 + 2xy + y^2$$
$$(x + y)^3 = x^3 + 3x^2y + 3xy^2 + y^3$$
$$(x + y)^4 = x^4 + 4x^3y + 6x^2y^2 + 4xy^3 + y^4$$
$$(x + y)^5 = x^5 + 5x^4y + 10x^3y^2 + 10x^2y^3 + 5xy^4 + y^5$$

Observe that the exponents on the variable x are decreasing, whereas the exponents on the variable y are increasing, as we read from left to right. Also notice that the sum of the exponents in each term is the same for that entire line. For instance, in the fourth expansion the terms x^4, x^3y, x^2y^2, xy^3, and y^4 all have exponents with a sum of 4. If we continue the pattern, the expansion of $(x + y)^6$ will have seven terms containing x^6, x^5y, x^4y^2, x^3y^3, x^2y^4, xy^5, and y^6. Now we must find the pattern for the coefficients of these terms.

Obtaining the Coefficients

If we write out only the coefficients of the expansions that we already have, we can easily see a pattern. This triangular array of coefficients for the binomial expansions is called **Pascal's triangle.**

$$
\begin{array}{ccccccccccc}
 & & & & & 1 & & & & & \\
 & & & & 1 & & 1 & & & & \\
 & & & 1 & & 2 & & 1 & & & \\
 & & 1 & & 3 & & 3 & & 1 & & \\
 & 1 & & 4 & & 6 & & 4 & & 1 & \\
1 & & 5 & & 10 & & 10 & & 5 & & 1
\end{array}
$$

$(x + y)^0 = 1$
$(x + y)^1 = 1x + 1y$
$(x + y)^2 = 1x^2 + 2xy + 1y^2$
$(x + y)^3 = 1x^3 + 3x^2y + 3xy^2 + 1y^3$
$(x + y)^4 = 1x^4 + 4x^3y + 6x^2y^2 + 4xy^3 + 1y^4$
Coefficients in $(x + y)^5$

Notice that each line starts and ends with a 1 and that each entry of a line is the sum of the two entries above it in the previous line. For instance, $4 = 3 + 1$, and $10 = 6 + 4$. Following this pattern, the sixth and seventh lines of coefficients are

$$1 \quad 6 \quad 15 \quad 20 \quad 15 \quad 6 \quad 1$$
$$1 \quad 7 \quad 21 \quad 35 \quad 35 \quad 21 \quad 7 \quad 1$$

Pascal's triangle gives us an easy way to get the coefficients for the binomial expansion with small powers, but it is impractical for larger powers. For larger powers we use a formula involving **factorial notation.**

$n!$ (n factorial)

If n is a positive integer, $n!$ (read "n factorial") is defined to be the product of all of the positive integers from 1 through n.

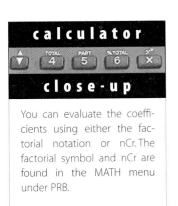

For example, $3! = 3 \cdot 2 \cdot 1 = 6$, and $5! = 5 \cdot 4 \cdot 3 \cdot 2 \cdot 1 = 120$. We also define $0!$ to be 1.

Before we state a general formula, consider how the coefficients for $(x + y)^4$ are found by using factorials:

$$\frac{4!}{4!\,0!} = \frac{4 \cdot 3 \cdot 2 \cdot 1}{4 \cdot 3 \cdot 2 \cdot 1 \cdot 1} = 1 \qquad \text{Coefficient of } x^4 \text{ (or } x^4 y^0\text{)}$$

$$\frac{4!}{3!\,1!} = \frac{4 \cdot 3 \cdot 2 \cdot 1}{3 \cdot 2 \cdot 1 \cdot 1} = 4 \qquad \text{Coefficient of } 4x^3 y$$

$$\frac{4!}{2!\,2!} = \frac{4 \cdot 3 \cdot 2 \cdot 1}{2 \cdot 1 \cdot 2 \cdot 1} = 6 \qquad \text{Coefficient of } 6x^2 y^2$$

$$\frac{4!}{1!\,3!} = \frac{4 \cdot 3 \cdot 2 \cdot 1}{1 \cdot 3 \cdot 2 \cdot 1} = 4 \qquad \text{Coefficient of } 4xy^3$$

$$\frac{4!}{0!\,4!} = \frac{4 \cdot 3 \cdot 2 \cdot 1}{1 \cdot 4 \cdot 3 \cdot 2 \cdot 1} = 1 \qquad \text{Coefficient of } y^4 \text{ (or } x^0 y^4\text{)}$$

Note that each expression has $4!$ in the numerator, with factorials in the denominator corresponding to the exponents on x and y.

The Binomial Theorem

We now summarize these ideas in the **binomial theorem.**

The Binomial Theorem

In the expansion of $(x + y)^n$ for a positive integer n, there are $n + 1$ terms, given by the following formula:

$$(x + y)^n = \frac{n!}{n!\,0!}x^n + \frac{n!}{(n-1)!\,1!}x^{n-1}y + \frac{n!}{(n-2)!\,2!}x^{n-2}y^2 + \cdots + \frac{n!}{0!\,n!}y^n$$

The notation $\binom{n}{r}$ is often used in place of $\dfrac{n!}{(n-r)!\,r!}$ in the binomial expansion. Using this notation, we write the expansion as

$$(x + y)^n = \binom{n}{0}x^n + \binom{n}{1}x^{n-1}y + \binom{n}{2}x^{n-2}y^2 + \cdots + \binom{n}{n}y^n.$$

Another notation for $\dfrac{n!}{(n-r)!r!}$ is $_nC_r$. Using this notation, we have

$$(x+y)^n = {_nC_0}x^n + {_nC_1}x^{n-1}y + {_nC_2}x^{n-2}y^2 + \cdots + {_nC_n}y^n.$$

E X A M P L E   1

Using the binomial theorem

Write out the first three terms of $(x+y)^9$.

Solution

$$(x+y)^9 = \frac{9!}{9!\,0!}x^9 + \frac{9!}{8!\,1!}x^8y + \frac{9!}{7!\,2!}x^7y^2 + \cdots$$
$$= x^9 + 9x^8y + 36x^7y^2 + \cdots$$

E X A M P L E   2

Using the binomial theorem

Write the binomial expansion for $(x^2 - 2a)^5$.

Solution

We expand a difference by writing it as a sum and using the binomial theorem:

$$(x^2 - 2a)^5 = (x^2 + (-2a))^5$$
$$= \frac{5!}{5!\,0!}(x^2)^5 + \frac{5!}{4!\,1!}(x^2)^4(-2a)^1 + \frac{5!}{3!\,2!}(x^2)^3(-2a)^2 + \frac{5!}{2!\,3!}(x^2)^2(-2a)^3$$
$$+ \frac{5!}{1!\,4!}(x^2)(-2a)^4 + \frac{5!}{0!\,5!}(-2a)^5$$
$$= x^{10} - 10x^8a + 40x^6a^2 - 80x^4a^3 + 80x^2a^4 - 32a^5$$

E X A M P L E   3

Finding a specific term

Find the fourth term of the expansion of $(a+b)^{12}$.

Solution

The variables in the first term are $a^{12}b^0$, those in the second term are $a^{11}b^1$, those in the third term are $a^{10}b^2$, and those in the fourth term are a^9b^3. So

$$\frac{12!}{9!\,3!}a^9b^3 = 220a^9b^3.$$

The fourth term is $220a^9b^3$.

calculator

close-up

Because $_nC_r = \dfrac{n!}{(n-r)!r!}$, we have

$$_{12}C_9 = \frac{12!}{3!\,9!} \quad \text{and} \quad _{12}C_3 = \frac{12!}{9!\,3!}.$$

So there is more than one way to compute $12!/(9!\,3!)$:

```
12!/(9!3!)
                220
12 nCr 9
                220
12 nCr 3
                220
```

Using the ideas of Example 3, we can write a formula for any term of a binomial expansion.

Formula for the *k*th Term of $(x+y)^n$

For k ranging from 1 to $n+1$, the kth term of the expansion of $(x+y)^n$ is given by the formula

$$\frac{n!}{(n-k+1)!(k-1)!}x^{n-k+1}y^{k-1}.$$

E X A M P L E   4

Finding a specific term

Find the sixth term of the expansion of $(a^2 - 2b)^7$.

Solution

Use the formula for the kth term with $k = 6$ and $n = 7$:

$$\frac{7!}{(7 - 6 + 1)!(6 - 1)!}(a^2)^2(-2b)^5 = 21a^4(-32b^5) = -672a^4b^5 \qquad \blacksquare$$

We can think of the binomial expansion as a finite series. Using summation notation, we can write the binomial theorem as follows.

The Binomial Theorem (Using Summation Notation)

For any positive integer n,

$$(x + y)^n = \sum_{i=0}^{n} \frac{n!}{(n - i)!i!}x^{n-i}y^i \qquad \text{or} \qquad (x + y)^n = \sum_{i=0}^{n} \binom{n}{i}x^{n-i}y^i.$$

E X A M P L E   5

Using summation notation

Write $(a + b)^5$ using summation notation.

Solution

Use $n = 5$ in the binomial theorem:

$$(a + b)^5 = \sum_{i=0}^{5} \frac{5!}{(5 - i)!i!}a^{5-i}b^i \qquad \blacksquare$$

WARM-UPS

True or false? Explain your answer.

1. There are 12 terms in the expansion of $(a + b)^{12}$.   False
2. The seventh term of $(a + b)^{12}$ is a multiple of a^5b^7.   False
3. For all values of x, $(x + 2)^5 = x^5 + 32$.   False
4. In the expansion of $(x - 5)^8$ the signs of the terms alternate.   True
5. The eighth line of Pascal's triangle is

$$1 \ 8 \ 28 \ 56 \ 70 \ 56 \ 28 \ 8 \ 1. \quad \text{True}$$

6. The sum of the coefficients in the expansion of $(a + b)^4$ is 2^4.   True
7. $(a + b)^3 = \sum_{i=0}^{3} \frac{3!}{(3 - i)!\,i!}a^{3-i}b^i$ True
8. The sum of the coefficients in the expansion of $(a + b)^n$ is 2^n.   True
9. $0! = 1!$ True
10. $\dfrac{7!}{5!\,2!} = 21$ True

12.5 EXERCISES

Reading and Writing *After reading this section, write out the answers to these questions. Use complete sentences.*

1. What is a binomial expansion?

The sum obtained for a power of a binomial is called a binomial expansion.

2. What is Pascal's triangle and how do you make it?

Pascal's triangle gives the coefficients for $(a + b)^n$ for $n = 1, 2, 3$, and so on. Each row starts and ends with a 1. The other terms are obtained by adding the closest two terms in the preceding row.

3. What does $n!$ mean?

The expression $n!$ is the product of the positive integers from 1 through n.

4. What is the binomial theorem?

The binomial theorem gives the expansion of $(a + b)^n$.

Evaluate each expression.

5. $\dfrac{5!}{2!\,3!}$ 10

6. $\dfrac{6!}{5!\,1!}$ 6

7. $\dfrac{8!}{5!\,3!}$ 56

8. $\dfrac{9!}{2!\,7!}$ 36

Use the binomial theorem to expand each binomial. See Examples 1 and 2.

9. $(r + t)^5$ $r^5 + 5r^4t + 10r^3t^2 + 10r^2t^3 + 5rt^4 + t^5$

10. $(r + t)^6$ $r^6 + 6r^5t + 15r^4t^2 + 20r^3t^3 + 15r^2t^4 + 6rt^5 + t^6$

11. $(m - n)^3$ $m^3 - 3m^2n + 3mn^2 - n^3$

12. $(m - n)^4$ $m^4 - 4m^3n + 6m^2n^2 - 4mn^3 + n^4$

13. $(x + 2a)^3$ $x^3 + 6ax^2 + 12a^2x + 8a^3$

14. $(a + 3b)^4$ $a^4 + 12a^3b + 54a^2b^2 + 108ab^3 + 81b^4$

15. $(x^2 - 2)^4$ $x^8 - 8x^6 + 24x^4 - 32x^2 + 16$

16. $(x^2 - a^2)^5$ $x^{10} - 5a^2x^8 + 10a^4x^6 - 10a^6x^4 + 5a^8x^2 - a^{10}$

17. $(x - 1)^7$ $x^7 - 7x^6 + 21x^5 - 35x^4 + 35x^3 - 21x^2 + 7x - 1$

18. $(x + 1)^6$ $x^6 + 6x^5 + 15x^4 + 20x^3 + 15x^2 + 6x + 1$

Write out the first four terms in the expansion of each binomial. See Examples 1 and 2.

19. $(a - 3b)^{12}$ $a^{12} - 36a^{11}b + 594a^{10}b^2 - 5940a^9b^3$

20. $(x - 2y)^{10}$ $x^{10} - 20x^9y + 180x^8y^2 - 960x^7y^3$

21. $(x^2 + 5)^9$ $x^{18} + 45x^{16} + 900x^{14} + 10{,}500x^{12}$

22. $(x^2 + 1)^{20}$ $x^{40} + 20x^{38} + 190x^{36} + 1140x^{34}$

23. $(x - 1)^{22}$ $x^{22} - 22x^{21} + 231x^{20} - 1540x^{19}$

24. $(2x - 1)^8$ $256x^8 - 1024x^7 + 1792x^6 - 1792x^5$

25. $\left(\dfrac{x}{2} + \dfrac{y}{3}\right)^{10}$ $\dfrac{x^{10}}{1024} + \dfrac{5x^9y}{768} + \dfrac{5x^8y^2}{256} + \dfrac{5x^7y^3}{144}$

26. $\left(\dfrac{a}{2} + \dfrac{b}{5}\right)^8$ $\dfrac{a^8}{256} + \dfrac{a^7b}{80} + \dfrac{7a^6b^2}{400} + \dfrac{7a^5b^3}{500}$

Find the indicated term of the binomial expansion. See Examples 3 and 4.

27. $(a + w)^{13}$, 6th term
$1287a^8w^5$

28. $(m + n)^{12}$, 7th term
$924m^6n^6$

29. $(m - n)^{16}$, 8th term
$-11{,}440m^9n^7$

30. $(a - b)^{14}$, 6th term
$-2002a^9b^5$

31. $(x + 2y)^8$, 4th term
$448x^5y^3$

32. $(3a + b)^7$, 4th term
$2835a^4b^3$

33. $(2a^2 - b)^{20}$, 7th term
$635{,}043{,}840a^{28}b^6$

34. $(a^2 - w^2)^{12}$, 5th term
$495a^{16}w^8$

Write each expansion using summation notation. See Example 5.

35. $(a + m)^8$

$\displaystyle\sum_{i=0}^{8} \frac{8!}{(8 - i)!\,i!} a^{8-i}m^i$

36. $(z + w)^{13}$

$\displaystyle\sum_{i=0}^{13} \frac{13!}{(13 - i)!\,i!} z^{13-i}w^i$

37. $(a - 2x)^5$

$\displaystyle\sum_{i=0}^{5} \frac{5!(-2)^i}{(5 - i)!\,i!} a^{5-i}x^i$

38. $(w - 3m)^7$

$\displaystyle\sum_{i=0}^{7} \frac{7!(-3)^i}{(7 - i)!\,i!} w^{7-i}m^i$

GETTING MORE INVOLVED

 39. *Discussion.* Find the trinomial expansion for $(a + b + c)^3$ by using $x = a$ and $y = b + c$ in the binomial theorem.
$a^3 + b^3 + c^3 + 3a^2b + 3a^2c + 3ab^2 + 3ac^2 + 3b^2c + 3bc^2 + 6abc$

 40. *Discussion.* What problem do you encounter when trying to find the fourth term in the binomial expansion for $(x + y)^{120}$? How can you overcome this problem? Find the fifth term in the binomial expansion for $(x - 2y)^{100}$.
$280{,}840x^{117}y^3$, $62{,}739{,}600x^{96}y^4$

COLLABORATIVE ACTIVITIES

Lotteries Are Series(ous)

Roberto and his brother-in-law Horatio each have a child who will be graduating from high school in 5 years. Each would like to buy his child a car for a graduation present. Horatio decides to buy two lottery tickets each week for the next 5 years, hoping to win and buy a new car for his child. The lottery tickets are $2.00 each at the local convenience store. Roberto, who doesn't

Grouping: Two to four students per group
Topic: Sequences and series

believe in lotteries, decides to set aside $4.00 each week and to deposit this money in a savings account each quarter for the next 5 years to buy a used car. He finds a bank that will pay 5% yearly interest compounded quarterly.

1. Write a series that represents how much money Horatio will spend on lottery tickets over a 5-year period. Assume there are 52 weeks in a year. Compute the total amount of money spent on lottery tickets.

2. Find the percent paid quarterly on the account. Find the number of compounding periods in 5 years. Find the amount of money Roberto will deposit each quarter.

3. Write a series to show how much money Roberto will have in his savings account at the end of 5 years.

4. Discuss in your groups the chances of Horatio winning the lottery compared to the sure savings Roberto has. Discuss the fact that Horatio would have to pay taxes on his lottery winnings and Roberto pays tax only on the interest earned. Do Horatio's chances increase the longer he buys lottery tickets?

WRAP-UP   CHAPTER 12

SUMMARY

Sequences and Series

Examples

| Sequence | Finite — A function whose domain is the set of positive integers less than or equal to a fixed positive integer | $3, 5, 7, 9, 11$
$a_n = 2n + 1$
$1 \le n \le 5$ |

| | Infinite — A function whose domain is the set of positive integers | $2, 4, 6, 8, \ldots$
$a_n = 2n$ |

| Series | The indicated sum of a sequence | $2 + 4 + 6 + \cdots + 50$ |

| Summation notation | $\displaystyle\sum_{i=1}^{n} a_i = a_1 + a_2 + a_3 + \cdots + a_n$ | $\displaystyle\sum_{i=1}^{25} 2i = 2 + 4 + \cdots + 50$ |

Arithmetic Sequences and Series

Examples

| Arithmetic sequence | Each term after the first is obtained by adding a fixed amount to the previous term. | $6, 11, 16, 21, \ldots$
Fixed amount, d, is 5. |

| nth term | The nth term of an arithmetic sequence is
$a_n = a_1 + (n - 1)d$. | If $a_1 = 6$ and $d = 5$, then
$a_n = 6 + (n - 1)5$. |

| Arithmetic series | The sum of an arithmetic sequence | $6 + 11 + 16 + 21$ |

| Sum of first n terms | $S_n = \dfrac{n}{2}(a_1 + a_n)$ | $S_4 = \dfrac{4}{2}(6 + 21) = 54$ |

Geometric Sequences and Series

Examples

| Geometric sequence | Each term after the first is obtained by multiplying the preceding term by a constant. | $2, 6, 18, 54, \ldots$
Constant, r, is 3. |

| nth term | The nth term of a geometric sequence is
$$a_n = a_1 r^{n-1}.$$ | $a_1 = 2, r = 3$
$a_n = 2 \cdot 3^{n-1}$ |

Geometric series (finite)	The indicated sum of a finite geometric sequence. $a_1 + a_1r + a_1r^2 + \cdots + a_1r^{n-1}$	$2 + 6 + 18 + 54 + 162$
Sum of first n terms	$S_n = \dfrac{a_1(1 - r^n)}{1 - r}$	$a_1 = 2, r = 3, n = 5$ $S_5 = \dfrac{2(1 - 3^5)}{1 - 3} = 242$
Geometric series (infinite)	$a_1 + a_1r + a_1r^2 + a_1r^3 + \cdots$	$8 + 4 + 2 + 1 + \dfrac{1}{2} + \cdots$
Sum of an infinite geometric series	$S = \dfrac{a_1}{1 - r}$, provided that $\lvert r \rvert < 1$	$a_1 = 8, r = \dfrac{1}{2}$ $S = \dfrac{8}{1 - \dfrac{1}{2}} = 16$
Factorial notation	The notation $n!$ represents the product of the positive integers from 1 through n.	$5! = 5 \cdot 4 \cdot 3 \cdot 2 \cdot 1 = 120$
Binomial theorem	$(x + y)^n = \dfrac{n!}{n!\,0!}x^n + \dfrac{n!}{(n - 1)!\,1!}x^{n-1}y$ $+ \dfrac{n!}{(n - 2)!\,2!}x^{n-2}y^2 + \cdots + \dfrac{n!}{0!\,n!}y^n$ Using summation notation: $(x + y)^n = \displaystyle\sum_{i=0}^{n} \dfrac{n!}{(n - i)!\,i!}x^{n-i}y^i = \sum_{i=0}^{n} \binom{n}{i}x^{n-i}y^i$	$(x + y)^3 = x^3 + 3x^2y$ $+ 3xy^2 + y^3$
kth term of $(x + y)^n$	$\dfrac{n!}{(n - k + 1)!(k - 1)!}x^{n-k+1}y^{k-1}$	Third term of $(a + b)^{10}$ is $\dfrac{10!}{8!\,2!}a^8b^2 = 45a^8b^2.$

ENRICHING YOUR MATHEMATICAL WORD POWER

For each mathematical term, choose the correct meaning.

1. **sequence**
   a. a list of numbers
   b. a procedure for getting the answer
   c. events that happen in order
   d. a linear function   a

2. **finite sequence**
   a. a short sequence
   b. a sequence of whole numbers
   c. a function whose domain is the set of positive integers

   d. a function whose domain is the set of positive integers less than or equal to a fixed positive integer   d

3. **infinite sequence**
   a. a short sequence
   b. a sequence of whole numbers
   c. a function whose domain is the set of positive integers
   d. a function whose domain is the set of positive integers less than or equal to a fixed positive integer   c

4. **series**
   a. a special sequence
   b. the indicated sum of the terms of a sequence

c. a sequence of positive numbers
d. a show with many episodes   b

5. arithmetic sequence
a. a sequence in which each term after the first is obtained by adding a fixed amount to the previous term
b. a sequence of fractions
c. a sequence found in arithmetic
d. a finite sequence   a

6. geometric sequence
a. a sequence of rectangles
b. a sequence of geometric formulas
c. a sequence in which each term after the first is obtained by multiplying the preceding term by a constant
d. a sequence in which the terms are geometric   c

7. geometric series
a. a series of geometric shapes
b. the indicated sum of an arithmetic sequence
c. a series of ratios
d. the indicated sum of a geometric sequence   d

8. binomial expansion
a. the trinomial obtained when a binomial is stretched
b. the expression obtained from raising a binomial to a whole number power
c. the coefficients of a binomial
d. the various powers of a binomial   b

9. Pascal's triangle
a. an equilateral triangle
b. a triangle formed by the graphs of three linear equations
c. the right triangle in the Pythagorean theorem
d. a triangular array of coefficients for the binomial expansions   d

10. $n!$
a. the product of the positive integers from 1 through n
b. the binomial coefficients
c. the n vertices of Pascal's triangle
d. 3.141592654   a

REVIEW EXERCISES

12.1 *List all terms of each finite sequence.*

1. $a_n = n^3$ for $1 \le n \le 5$
1, 8, 27, 64, 125

2. $b_n = (n - 1)^4$ for $1 \le n \le 4$
0, 1, 16, 81

3. $c_n = (-1)^n(2n - 3)$ for $1 \le n \le 6$
1, 1, −3, 5, −7, 9

4. $d_n = (-1)^{n-1}(3 - n)$ for $1 \le n \le 7$
2, −1, 0, 1, −2, 3, −4

Write the first three terms of the infinite sequence whose nth term is given.

5. $a_n = -\dfrac{1}{n}$

$-1, -\dfrac{1}{2}, -\dfrac{1}{3}$

6. $b_n = \dfrac{(-1)^n}{n^2}$

$-1, \dfrac{1}{4}, -\dfrac{1}{9}$

7. $b_n = \dfrac{(-1)^{2n}}{2n + 1}$

$\dfrac{1}{3}, \dfrac{1}{5}, \dfrac{1}{7}$

8. $a_n = \dfrac{-1}{2n - 3}$

$1, -1, -\dfrac{1}{3}$

9. $c_n = \log_2(2^{n+3})$
4, 5, 6

10. $c_n = \ln(e^{2n})$
2, 4, 6

12.2 *Find the sum of each series.*

11. $\displaystyle\sum_{i=1}^{3} i^3$ 36

12. $\displaystyle\sum_{i=0}^{4} 6$ 30

13. $\displaystyle\sum_{n=1}^{5} n(n - 1)$ 40

14. $\displaystyle\sum_{j=0}^{3} (-2)^j$ −5

Write each series in summation notation. Use the index i, and let i begin at 1.

15. $\dfrac{1}{4} + \dfrac{1}{6} + \dfrac{1}{8} + \cdots$ $\displaystyle\sum_{i=1}^{\infty} \dfrac{1}{2(i + 1)}$

16. $\dfrac{1}{3} + \dfrac{1}{4} + \dfrac{1}{5} + \cdots$ $\displaystyle\sum_{i=1}^{\infty} \dfrac{1}{i + 2}$

17. $0 + 1 + 4 + 9 + 16 + \cdots$ $\displaystyle\sum_{i=1}^{\infty} (i - 1)^2$

18. $-1 + 2 - 3 + 4 - 5 + 6 - \cdots$ $\displaystyle\sum_{i=1}^{\infty} i(-1)^i$

19. $x_1 - x_2 + x_3 - x_4 + \cdots$ $\displaystyle\sum_{i=1}^{\infty} (-1)^{i+1}x_i$

20. $-x^2 + x^3 - x^4 + x^5 - \cdots$ $\displaystyle\sum_{i=1}^{\infty} (-1)^i x^{i+1}$

12.3 *Write the first four terms of the arithmetic sequence with the given nth term.*

21. $a_n = 6 + (n - 1)5$ 6, 11, 16, 21

22. $a_n = -7 + (n - 1)4$ −7, −3, 1, 5

23. $a_n = -20 + (n - 1)(-2)$ −20, −22, −24, −26

24. $a_n = 10 + (n - 1)(-2.5)$ 10, 7.5, 5, 2.5

25. $a_n = 1000n + 2000$ 3000, 4000, 5000, 6000

26. $a_n = -500n + 5000$ 4500, 4000, 3500, 3000

Write a formula for the nth term of each arithmetic sequence.

27. $\dfrac{1}{3}, \dfrac{2}{3}, 1, \dfrac{4}{3}, \ldots$

$a_n = \dfrac{n}{3}$

28. 10, 6, 2, −2, ...

$a_n = -4n + 14$

29. $2, 4, 6, 8, \ldots$

$a_n = 2n$

30. $20, 10, 0, -10, \ldots$

$a_n = -10n + 30$

Find the sum of each arithmetic series.

31. $1 + 2 + 3 + \cdots + 24$ 300

32. $-5 + (-2) + 1 + 4 + \cdots + 34$ 203

33. $\dfrac{1}{6} + \dfrac{1}{2} + \dfrac{5}{6} + \dfrac{7}{6} + \cdots + \dfrac{11}{2}$ $\dfrac{289}{6}$

34. $-3 - 6 - 9 - 12 - \cdots - 36$ -234

35. $\displaystyle\sum_{i=1}^{7}(2i - 3)$ 35

36. $\displaystyle\sum_{i=1}^{6}[12 + (i - 1)5]$ 147

12.4 *Write the first four terms of the geometric sequence with the given nth term.*

37. $a_n = 3\left(\dfrac{1}{2}\right)^{n-1}$

$3, \dfrac{3}{2}, \dfrac{3}{4}, \dfrac{3}{8}$

38. $a_n = 6\left(-\dfrac{1}{3}\right)^{n}$

$-2, \dfrac{2}{3}, -\dfrac{2}{9}, \dfrac{2}{27}$

39. $a_n = 2^{1-n}$

$1, \dfrac{1}{2}, \dfrac{1}{4}, \dfrac{1}{8}$

40. $a_n = 5(10)^{n-1}$

$5, 50, 500, 5{,}000$

41. $a_n = 23(10)^{-2n}$

$0.23, 0.0023, 0.000023,$
0.00000023

42. $a_n = 4(10)^{-n}$

$0,4, 0.04, 0.004, 0.0004$

Write a formula for the nth term of each geometric sequence.

43. $\dfrac{1}{2}, 3, 18, \ldots$

$a_n = \dfrac{1}{2}(6)^{n-1}$

44. $-6, 2, -\dfrac{2}{3}, \dfrac{2}{9}, \ldots$

$a_n = -6\left(-\dfrac{1}{3}\right)^{n-1}$

45. $\dfrac{7}{10}, \dfrac{7}{100}, \dfrac{7}{1000}, \ldots$

$a_n = 0.7(0.1)^{n-1}$

46. $2, 2x, 2x^2, 2x^3, \ldots$

$a_n = 2x^{n-1}$

Find the sum of each geometric series.

47. $\dfrac{1}{3} + \dfrac{1}{9} + \dfrac{1}{27} + \dfrac{1}{81}$ $\dfrac{40}{81}$

48. $2 + 4 + 8 + 16 + \cdots + 512$ 1022

49. $\displaystyle\sum_{i=1}^{10} 3(10)^{-i}$ 0.3333333333

50. $\displaystyle\sum_{i=1}^{5}(0.1)^{i}$ 0.11111

51. $\dfrac{1}{4} + \dfrac{1}{12} + \dfrac{1}{36} + \dfrac{1}{108} + \cdots$ $\dfrac{3}{8}$

52. $12 + (-6) + 3 + \left(-\dfrac{3}{2}\right) + \cdots$ 8

53. $\displaystyle\sum_{i=1}^{\infty} 18\left(\dfrac{2}{3}\right)^{i-1}$ 54

54. $\displaystyle\sum_{i=1}^{\infty} 9(0.1)^{i}$ 1

12.5 *Use the binomial theorem to expand each binomial.*

55. $(m + n)^5$ $m^5 + 5m^4n + 10m^3n^2 + 10m^2n^3 + 5mn^4 + n^5$

56. $(2m - y)^4$ $16m^4 - 32m^3y + 24m^2y^2 - 8my^3 + y^4$

57. $(a^2 - 3b)^3$ $a^6 - 9a^4b + 27a^2b^2 - 27b^3$

58. $\left(\dfrac{x}{2} + 2a\right)^5$ $\dfrac{x^5}{32} + \dfrac{5x^4a}{8} + 5x^3a^2 + 20x^2a^3 + 40xa^4 + 32a^5$

Find the indicated term of the binomial expansion.

59. $(x + y)^{12}$, 5th term
$495x^8y^4$

60. $(x - 2y)^9$, 5th term
$2016x^5y^4$

61. $(2a - b)^{14}$, 3rd term
$372{,}736a^{12}b^2$

62. $(a + b)^{10}$, 4th term
$120a^7b^3$

Write each expression in summation notation.

63. $(a + w)^7$

$\displaystyle\sum_{i=0}^{7} \dfrac{7!}{(7 - i)!\, i!} a^{7-i}w^i$

64. $(m - 3y)^9$

$\displaystyle\sum_{i=0}^{9} \dfrac{9!(-3)^i}{(9 - i)!\, i!} m^{9-i}y^i$

MISCELLANEOUS

Identify each sequence as an arithmetic sequence, a geometric sequence, or neither.

65. $1, 3, 6, 10, 15, \ldots$
Neither

66. $9, 12, 16, \dfrac{64}{3}, \ldots$
Geometric

67. $9, 12, 15, 18, \ldots$
Arithmetic

68. $2, 4, 8, 16, \ldots$
Geometric

69. $0, 2, 4, 6, 8, \ldots$
Arithmetic

70. $0, 3, 9, 27, 81, \ldots$
Neither

Solve each problem.

71. Find the common ratio for the geometric sequence with first term 6 and fourth term $\dfrac{1}{30}$.
$\dfrac{1}{\sqrt[3]{180}}$

72. Find the common difference for an arithmetic sequence with first term 6 and fourth term 36.
10

73. Write out all of the terms of the series
$$\sum_{i=1}^{5} \dfrac{(-1)^i}{i!}.$$
$-1 + \dfrac{1}{2} - \dfrac{1}{6} + \dfrac{1}{24} - \dfrac{1}{120}$

74. Write out the first eight rows of Pascal's triangle.

75. Write out all of the terms of the series
$$\sum_{i=0}^{5} \dfrac{5!}{(5 - i)!\, i!} a^{5-i}b^i.$$
$a^5 + 5a^4b + 10a^3b^2 + 10a^2b^3 + 5ab^4 + b^5$

76. Write out all of the terms of the series
$$\sum_{i=0}^{8} \dfrac{8!}{(8 - i)!\, i!} x^{8-i}y^i.$$
$x^8 + 8x^7y + 28x^6y^2 + 56x^5y^3 + 70x^4y^4 + 56x^3y^5 +$
$28x^2y^6 + 8xy^7 + y^8$

77. How many terms are there in the expansion of $(a + b)^{25}$?
26

78. Calculate $\dfrac{12!}{8!\,4!}$.   495

 79. If \$3,000 is deposited at the beginning of each year for 16 years into an account paying 10% compounded annually, then what is the value of the annuity at the end of the 16th year?   \$118,634.11

 80. If \$3,000 is deposited at the beginning of each year for 8 years into an account paying 10% compounded annually,

then what is the value of the annuity at the end of the eighth year? How does the value of the annuity in this exercise compare to that of Exercise 79?
\$37,738.43. The time is half as much, and the amount is less than one-third as much.

 81. If one deposit of \$3,000 is made into an account paying 10% compounded annually, then how much will be in the account at the end of 16 years? Note that a single deposit is not an annuity.
\$13,784.92

CHAPTER 12 TEST

List the first four terms of the sequence whose nth term is given.

1. $a_n = -10 + (n - 1)6$
$-10, -4, 2, 8$

2. $a_n = 5(0.1)^{n-1}$
$5, 0.5, 0.05, 0.005$

3. $a_n = \dfrac{(-1)^n}{n!}$
$-1, \dfrac{1}{2}, -\dfrac{1}{6}, \dfrac{1}{24}$

4. $a_n = \dfrac{2n - 1}{n^2}$
$1, \dfrac{3}{4}, \dfrac{5}{9}, \dfrac{7}{16}$

Write a formula for the nth term of each sequence.

5. $7, 4, 1, -2, \ldots$ $a_n = 10 - 3n$

6. $-25, 5, -1, \dfrac{1}{5}, \ldots$ $a_n = -25\left(-\dfrac{1}{5}\right)^{n-1}$

7. $2, -4, 6, -8, 10, -12, \ldots$ $a_n = (-1)^{n-1}2n$

8. $1, 4, 9, 16, 25, \ldots$ $a_n = n^2$

Write out all of the terms of each series.

9. $\displaystyle\sum_{i=1}^{5}(2i + 3)$ $5 + 7 + 9 + 11 + 13$

10. $\displaystyle\sum_{i=1}^{6}5(2)^{i-1}$ $5 + 10 + 20 + 40 + 80 + 160$

11. $\displaystyle\sum_{i=0}^{4}\dfrac{4!}{(4 - i)!\,i!}m^{4-i}q^i$ $m^4 + 4m^3q + 6m^2q^2 + 4mq^3 + q^4$

Find the sum of each series.

12. $\displaystyle\sum_{i=1}^{20}(6 + 3i)$ 750

13. $\displaystyle\sum_{i=1}^{5}10\left(\dfrac{1}{2}\right)^{i-1}$ $\dfrac{155}{8}$

14. $\displaystyle\sum_{i=1}^{\infty}0.35(0.93)^{i-1}$ 5

15. $2 + 4 + 6 + \cdots + 200$ 10,100

16. $\dfrac{1}{4} + \dfrac{1}{8} + \dfrac{1}{16} + \cdots$ $\dfrac{1}{2}$

17. $2 + 1 + \dfrac{1}{2} + \dfrac{1}{4} + \cdots + \dfrac{1}{128}$ $\dfrac{511}{128}$

Solve each problem.

18. Find the common ratio for the geometric sequence that has first term 3 and fifth term 48.
± 2

19. Find the common difference for the arithmetic sequence that has first term 1 and twelfth term 122.
11

20. Find the fifth term in the expansion of $(r - t)^{15}$.
$1365r^{11}t^4$

21. Find the fourth term in the expansion of $(a^2 - 2b)^8$.
$-448a^{10}b^3$

22. If \$800 is deposited at the beginning of each year for 25 years into an account earning 10% compounded annually, then what is the value of this annuity at the end of the 25th year?
\$86,545.41

Let $f(x) = x^2 - 3$, $g(x) = 2x - 1$, $h(x) = 2^x$, and $m(x) = \log_2(x)$. Find the following.

1. $f(3)$

6

2. $f(n)$

$n^2 - 3$

3. $f(x + h)$

$x^2 + 2xh + h^2 - 3$

4. $f(x) - g(x)$

$x^2 - 2x - 2$

5. $g(f(3))$

11

6. $(f \circ g)(2)$

6

7. $m(16)$

4

8. $(h \circ m)(32)$

32

9. $h(-1)$

$\dfrac{1}{2}$

10. $h^{-1}(8)$

3

11. $m^{-1}(0)$

1

12. $(m \circ h)(x)$

x

Solve each variation problem.

13. If y varies directly as x, and $y = -6$ when $x = 4$, find y when $x = 9$. $-\dfrac{27}{2}$

14. If a varies inversely as b, and $a = 2$ when $b = -4$, find a when $b = 3$. $-\dfrac{8}{3}$

15. If y varies directly as w and inversely as t, and $y = 16$ when $w = 3$ and $t = -4$, find y when $w = 2$ and $t = 3$. $-\dfrac{128}{9}$

16. If y varies jointly as h and the square of r, and $y = 12$ when $h = 2$ and $r = 3$, find y when $h = 6$ and $r = 2$.   16

Sketch the graph of each inequality or system of inequalities.

17. $x > 3$ and $x + y < 0$

18. $|x - y| \geq 2$

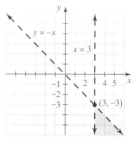

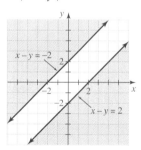

19. $y < -2x + 3$ and $y > 2^x$

20. $|y + 2x| < 1$

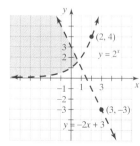

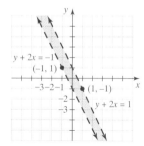

21. $x^2 + y^2 < 4$

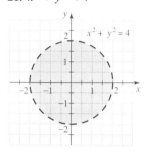

22. $x^2 - y^2 < 1$

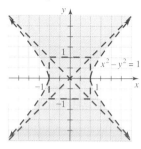

23. $y < \log_2(x)$

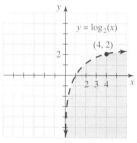

24. $x^2 + 2y < 4$

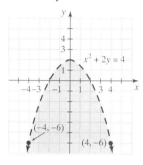

25. $\dfrac{x^2}{4} + \dfrac{y^2}{9} < 1$ and $y > x^2$

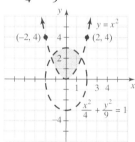

Perform the indicated operation and simplify. Write answers with positive exponents.

26. $\dfrac{a}{b} + \dfrac{b}{a}$

$\dfrac{a^2 + b^2}{ab}$

27. $1 - \dfrac{3}{y}$

$\dfrac{y - 3}{y}$

28. $\dfrac{x - 2}{x^2 - 9} - \dfrac{x - 4}{x^2 - 2x - 3}$

$\dfrac{10}{(x - 3)(x + 3)(x + 1)}$

29. $\dfrac{x^2 - 16}{2x + 8} \cdot \dfrac{4x^2 + 16x + 64}{x^3 - 16}$

$\dfrac{2(x^3 - 64)}{x^3 - 16}$

30. $\dfrac{(a^2 b)^3}{(ab^2)^4} \cdot \dfrac{ab^3}{a^{-4}b^2}$

$\dfrac{a^7}{b^4}$

31. $\dfrac{x^2 y}{(xy)^3} \div \dfrac{xy^2}{x^2 y^4}$

1

Simplify.

32. $8^{2/3}$ 4

33. $16^{-5/4}$ $\dfrac{1}{32}$

34. $-4^{1/2}$ -2

35. $27^{-2/3}$ $\dfrac{1}{9}$

36. -2^{-3} $-\dfrac{1}{8}$

37. $2^{-3/5} \cdot 2^{-7/5}$ $\dfrac{1}{4}$

38. $5^{-2/3} \div 5^{1/3}$ $\dfrac{1}{5}$

39. $(9^{1/2} + 4^{1/2})^2$ 25

Solve.

40. *Predicting heights of preschoolers.* A popular model in pediatrics for predicting the height of preschoolers is the JENNS model. According to this model, if $h(x)$ is the height [in centimeters(cm)] at age x (in years) for $0.25 \leq x \leq 6$, then

$$h(x) = 79.041 + 6.39x - e^{(3.261 - 0.993x)}.$$

a) Find the predicted height in inches for a child of age 4 years, 3 months.

b) If you have a graphing calculator, graph the function as shown in the accompanying figure.

c) Use your graphing calculator to find the age to the nearest tenth of a year for a child who has a height of 80 cm.

    a) 105.8 cm

    c) 1.3 years

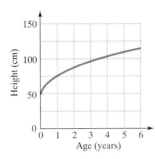

FIGURE FOR EXERCISE 40

Geometric Figures and Formulas

Triangle: A three-sided figure

Area: $A = \frac{1}{2}bh$, Perimeter: $P = a + b + c$

Sum of the measures of the angles is 180°.

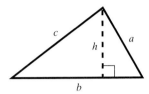

30-60-90 Right Triangle: The side opposite 30° is one-half the length of the hypotenuse.

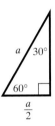

Trapezoid: A four-sided figure with one pair of parallel sides

Area: $A = \frac{1}{2}h(b_1 + b_2)$

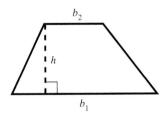

Right Triangle: A triangle with a 90° angle

Area $= \frac{1}{2}ab$, Perimeter: $P = a + b + c$

Pythagorean Theorem: $c^2 = a^2 + b^2$

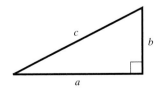

Parallelogram: A four-sided figure with opposite sides parallel

Area: $A = bh$

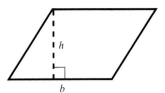

Rhombus: A four-sided figure with four equal sides

Perimeter: $P = 4a$

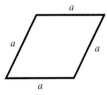

Rectangle: A four-sided figure with four right angles

Area: $A = LW$

Perimeter: $P = 2L + 2W$

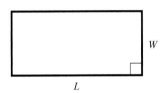

Circle

Area: $A = \pi r^2$

Circumference: $C = 2\pi r$

Diameter: $d = 2r$

Right Circular Cone

Volume: $V = \frac{1}{3}\pi r^2 h$

Lateral Surface Area: $S = \pi r\sqrt{r^2 + h^2}$

Rectangular Solid
Volume: $V = LWH$
Surface Area: $A = 2LW + 2WH + 2LH$

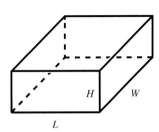

Square: A four-sided figure with four equal sides and four right angles
Area: $A = s^2$
Perimeter: $P = 4s$

Sphere

Volume: $V = \frac{4}{3}\pi r^3$

Surface Area: $S = 4\pi r^2$

Right Circular Cylinder
Volume: $V = \pi r^2 h$
Lateral Surface Area: $S = 2\pi rh$

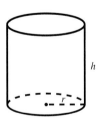

Geometric Terms

An **angle** is a union of two rays with a common endpoint.

A **right angle** is an angle with a measure of 90°.

Two angles are **complementary** if the sum of their measures is 90°.

An **isosceles triangle** is a triangle that has two equal sides.

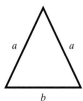

Similar triangles are triangles that have the same shape. Their corresponding angles are equal and corresponding sides are proportional:

$$\frac{a}{d} = \frac{b}{e} = \frac{c}{f}$$

An **acute angle** is an angle with a measure between 0° and 90°.

An **obtuse angle** is an angle with a measure between 90° and 180°.

Two angles are **supplementary** if the sum of their measures is 180°.

An **equilateral triangle** is a triangle that has three equal sides.

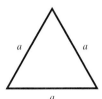

n	n^2	$\sqrt{n}$	n	n^2	$\sqrt{n}$	n	n^2	$\sqrt{n}$
1	1	1.0000	41	1681	6.4031	81	6561	9.0000
2	4	1.4142	42	1764	6.4807	82	6724	9.0554
3	9	1.7321	43	1849	6.5574	83	6889	9.1104
4	16	2.0000	44	1936	6.6332	84	7056	9.1652
5	25	2.2361	45	2025	6.7082	85	7225	9.2195
6	36	2.4495	46	2116	6.7823	86	7396	9.2736
7	49	2.6458	47	2209	6.8557	87	7569	9.3274
8	64	2.8284	48	2304	6.9282	88	7744	9.3808
9	81	3.0000	49	2401	7.0000	89	7921	9.4340
10	100	3.1623	50	2500	7.0711	90	8100	9.4868
11	121	3.3166	51	2601	7.1414	91	8281	9.5394
12	144	3.4641	52	2704	7.2111	92	8464	9.5917
13	169	3.6056	53	2809	7.2801	93	8649	9.6437
14	196	3.7417	54	2916	7.3485	94	8836	9.6954
15	225	3.8730	55	3025	7.4162	95	9025	9.7468
16	256	4.0000	56	3136	7.4833	96	9216	9.7980
17	289	4.1231	57	3249	7.5498	97	9409	9.8489
18	324	4.2426	58	3364	7.6158	98	9604	9.8995
19	361	4.3589	59	3481	7.6811	99	9801	9.9499
20	400	4.4721	60	3600	7.7460	100	10000	10.0000
21	441	4.5826	61	3721	7.8102	101	10201	10.0499
22	484	4.6904	62	3844	7.8740	102	10404	10.0995
23	529	4.7958	63	3969	7.9373	103	10609	10.1489
24	576	4.8990	64	4096	8.0000	104	10816	10.1980
25	625	5.0000	65	4225	8.0623	105	11025	10.2470
26	676	5.0990	66	4356	8.1240	106	11236	10.2956
27	729	5.1962	67	4489	8.1854	107	11449	10.3441
28	784	5.2915	68	4624	8.2462	108	11664	10.3923
29	841	5.3852	69	4761	8.3066	109	11881	10.4403
30	900	5.4772	70	4900	8.3666	110	12100	10.4881
31	961	5.5678	71	5041	8.4261	111	12321	10.5357
32	1024	5.6569	72	5184	8.4853	112	12544	10.5830
33	1089	5.7446	73	5329	8.5440	113	12769	10.6301
34	1156	5.8310	74	5476	8.6023	114	12996	10.6771
35	1225	5.9161	75	5625	8.6603	115	13225	10.7238
36	1296	6.0000	76	5776	8.7178	116	13456	10.7703
37	1369	6.0828	77	5929	8.7750	117	13689	10.8167
38	1444	6.1644	78	6084	8.8318	118	13924	10.8628
39	1521	6.2450	79	6241	8.8882	119	14161	10.9087
40	1600	6.3246	80	6400	8.9443	120	14400	10.9545

This table gives the common logarithms for numbers between 1 and 10. The common logarithms for other numbers can be found by using scientific notation and the properties of logarithms. For example, to find log(1230) we write

$$\log(1230) = \log(1.23 \times 10^3) = \log(1.23) + \log(10^3)$$
$$= 0.0899 + 3 = 3.0899$$

n	0	1	2	3	4	5	6	7	8	9
1.0	.0000	.0043	.0086	.0128	.0170	.0212	.0253	.0294	.0334	.0374
1.1	.0414	.0453	.0492	.0531	.0569	.0607	.0645	.0682	.0719	.0755
1.2	.0792	.0828	.0864	.0899	.0934	.0969	.1004	.1038	.1072	.1106
1.3	.1139	.1173	.1206	.1239	.1271	.1303	.1335	.1367	.1399	.1430
1.4	.1461	.1492	.1523	.1553	.1584	.1614	.1644	.1673	.1703	.1732
1.5	.1761	.1790	.1818	.1847	.1875	.1903	.1931	.1959	.1987	.2014
1.6	.2041	.2068	.2095	.2122	.2148	.2175	.2201	.2227	.2253	.2279
1.7	.2304	.2330	.2355	.2380	.2405	.2430	.2455	.2480	.2504	.2529
1.8	.2553	.2577	.2601	.2625	.2648	.2672	.2695	.2718	.2742	.2765
1.9	.2788	.2810	.2833	.2856	.2878	.2900	.2923	.2945	.2967	.2989
2.0	.3010	.3032	.3054	.3075	.3096	.3118	.3139	.3160	.3181	.3201
2.1	.3222	.3243	.3263	.3284	.3304	.3324	.3345	.3365	.3385	.3404
2.2	.3424	.3444	.3464	.3483	.3502	.3522	.3541	.3560	.3579	.3598
2.3	.3617	.3636	.3655	.3674	.3692	.3711	.3729	.3747	.3766	.3784
2.4	.3802	.3820	.3838	.3856	.3874	.3892	.3909	.3927	.3945	.3962
2.5	.3979	.3997	.4014	.4031	.4048	.4065	.4082	.4099	.4116	.4133
2.6	.4150	.4166	.4183	.4200	.4216	.4232	.4249	.4265	.4281	.4298
2.7	.4314	.4330	.4346	.4362	.4378	.4393	.4409	.4425	.4440	.4456
2.8	.4472	.4487	.4502	.4518	.4533	.4548	.4564	.4579	.4594	.4609
2.9	.4624	.4639	.4654	.4669	.4683	.4698	.4713	.4728	.4742	.4757
3.0	.4771	.4786	.4800	.4814	.4829	.4843	.4857	.4871	.4886	.4900
3.1	.4914	.4928	.4942	.4955	.4969	.4983	.4997	.5011	.5024	.5038
3.2	.5051	.5065	.5079	.5092	.5105	.5119	.5132	.5145	.5159	.5172
3.3	.5185	.5198	.5211	.5224	.5237	.5250	.5263	.5276	.5289	.5302
3.4	.5315	.5328	.5340	.5353	.5366	.5378	.5391	.5403	.5416	.5428
3.5	.5441	.5453	.5465	.5478	.5490	.5502	.5514	.5527	.5539	.5551
3.6	.5563	.5575	.5587	.5599	.5611	.5623	.5635	.5647	.5658	.5670
3.7	.5682	.5694	.5705	.5717	.5729	.5740	.5752	.5763	.5775	.5786
3.8	.5798	.5809	.5821	.5832	.5843	.5855	.5866	.5877	.5888	.5899
3.9	.5911	.5922	.5933	.5944	.5955	.5966	.5977	.5988	.5999	.6010
4.0	.6021	.6031	.6042	.6053	.6064	.6075	.6085	.6096	.6107	.6117
4.1	.6128	.6138	.6149	.6160	.6170	.6180	.6191	.6201	.6212	.6222
4.2	.6232	.6243	.6253	.6263	.6274	.6284	.6294	.6304	.6314	.6325
4.3	.6335	.6345	.6355	.6365	.6375	.6385	.6395	.6405	.6415	.6425
4.4	.6435	.6444	.6454	.6464	.6474	.6484	.6493	.6503	.6513	.6522
4.5	.6532	.6542	.6551	.6561	.6571	.6580	.6590	.6599	.6609	.6618
4.6	.6628	.6637	.6646	.6656	.6665	.6675	.6684	.6693	.6702	.6712
4.7	.6721	.6730	.6739	.6749	.6758	.6767	.6776	.6785	.6794	.6803
4.8	.6812	.6821	.6830	.6839	.6848	.6857	.6866	.6875	.6884	.6893
4.9	.6902	.6911	.6920	.6928	.6937	.6946	.6955	.6964	.6972	.6981
n	0	1	2	3	4	5	6	7	8	9

n	0	1	2	3	4	5	6	7	8	9
5.0	.6990	.6998	.7007	.7016	.7024	.7033	.7042	.7050	.7059	.7067
5.1	.7076	.7084	.7093	.7101	.7110	.7118	.7126	.7135	.7143	.7152
5.2	.7160	.7168	.7177	.7185	.7193	.7202	.7210	.7218	.7226	.7235
5.3	.7243	.7251	.7259	.7267	.7275	.7284	.7292	.7300	.7308	.7316
5.4	.7324	.7332	.7340	.7348	.7356	.7364	.7372	.7380	.7388	.7396
5.5	.7404	.7412	.7419	.7427	.7435	.7443	.7451	.7459	.7466	.7474
5.6	.7482	.7490	.7497	.7505	.7513	.7520	.7528	.7536	.7543	.7551
5.7	.7559	.7566	.7574	.7582	.7589	.7597	.7604	.7612	.7619	.7627
5.8	.7634	.7642	.7649	.7657	.7664	.7672	.7679	.7686	.7694	.7701
5.9	.7709	.7716	.7723	.7731	.7738	.7745	.7752	.7760	.7767	.7774
6.0	.7782	.7789	.7796	.7803	.7810	.7818	.7825	.7832	.7839	.7846
6.1	.7853	.7860	.7868	.7875	.7882	.7889	.7896	.7903	.7910	.7917
6.2	.7924	.7931	.7938	.7945	.7952	.7959	.7966	.7973	.7980	.7987
6.3	.7993	.8000	.8007	.8014	.8021	.8028	.8035	.8041	.8048	.8055
6.4	.8062	.8069	.8075	.8082	.8089	.8096	.8102	.8109	.8116	.8122
6.5	.8129	.8136	.8142	.8149	.8156	.8162	.8169	.8176	.8182	.8189
6.6	.8195	.8202	.8209	.8215	.8222	.8228	.8235	.8241	.8248	.8254
6.7	.8261	.8267	.8274	.8280	.8287	.8293	.8299	.8306	.8312	.8319
6.8	.8325	.8331	.8338	.8344	.8351	.8357	.8363	.8370	.8376	.8382
6.9	.8388	.8395	.8401	.8407	.8414	.8420	.8426	.8432	.8439	.8445
7.0	.8451	.8457	.8463	.8470	.8476	.8482	.8488	.8494	.8500	.8506
7.1	.8513	.8519	.8525	.8531	.8537	.8543	.8549	.8555	.8561	.8567
7.2	.8573	.8579	.8585	.8591	.8597	.8603	.8609	.8615	.8621	.8627
7.3	.8633	.8639	.8645	.8651	.8657	.8663	.8669	.8675	.8681	.8686
7.4	.8692	.8698	.8704	.8710	.8716	.8722	.8727	.8733	.8739	.8745
7.5	.8751	.8756	.8762	.8768	.8774	.8779	.8785	.8791	.8797	.8802
7.6	.8808	.8814	.8820	.8825	.8831	.8837	.8842	.8848	.8854	.8859
7.7	.8865	.8871	.8876	.8882	.8887	.8893	.8899	.8904	.8910	.8915
7.8	.8921	.8927	.8932	.8938	.8943	.8949	.8954	.8960	.8965	.8971
7.9	.8976	.8982	.8987	.8993	.8998	.9004	.9009	.9015	.9020	.9025
8.0	.9031	.9036	.9042	.9047	.9053	.9058	.9063	.9069	.9074	.9079
8.1	.9085	.9090	.9096	.9101	.9106	.9112	.9117	.9122	.9128	.9133
8.2	.9138	.9143	.9149	.9154	.9159	.9165	.9170	.9175	.9180	.9186
8.3	.9191	.9196	.9201	.9206	.9212	.9217	.9222	.9227	.9232	.9238
8.4	.9243	.9248	.9253	.9258	.9263	.9269	.9274	.9279	.9284	.9289
8.5	.9294	.9299	.9304	.9309	.9315	.9320	.9325	.9330	.9335	.9340
8.6	.9345	.9350	.9355	.9360	.9365	.9370	.9375	.9380	.9385	.9390
8.7	.9395	.9400	.9405	.9410	.9415	.9420	.9425	.9430	.9435	.9440
8.8	.9445	.9450	.9455	.9460	.9465	.9469	.9474	.9479	.9484	.9489
8.9	.9494	.9499	.9504	.9509	.9513	.9518	.9523	.9528	.9533	.9538
9.0	.9542	.9547	.9552	.9557	.9562	.9566	.9571	.9576	.9581	.9586
9.1	.9590	.9595	.9600	.9605	.9609	.9614	.9619	.9624	.9628	.9633
9.2	.9638	.9643	.9647	.9652	.9657	.9661	.9666	.9671	.9675	.9680
9.3	.9685	.9689	.9694	.9699	.9703	.9708	.9713	.9717	.9722	.9727
9.4	.9731	.9736	.9741	.9745	.9750	.9754	.9759	.9763	.9768	.9773
9.5	.9777	.9782	.9786	.9791	.9795	.9800	.9805	.9809	.9814	.9818
9.6	.9823	.9827	.9832	.9836	.9841	.9845	.9850	.9854	.9859	.9863
9.7	.9868	.9872	.9877	.9881	.9886	.9890	.9894	.9899	.9903	.9908
9.8	.9912	.9917	.9921	.9926	.9930	.9934	.9939	.9943	.9948	.9952
9.9	.9956	.9961	.9965	.9969	.9974	.9978	.9983	.9987	.9991	.9996
n	0	1	2	3	4	5	6	7	8	9

ANSWERS

CHAPTER 1

Section 1.1 Warm-ups F F F F T T T F T T

1. A set is a collection of objects.
2. A finite set has a fixed number of elements and an infinite set does not.
3. A Venn diagram is used to illustrate relationships between sets.
4. The intersection of two sets consists of elements that are in both sets, whereas the union of two sets consists of elements that are in one, in the other, or in both sets.
5. Every member of set A is also a member of set B.
6. The empty set is a subset of every set.
7. F **8.** F **9.** T **10.** F **11.** T **12.** F **13.** F
14. T **15.** F **16.** F **17.** F **18.** F **19.** $\varnothing$
20. $\{1, 2, 3, 4, 5, 6, 7, 8, 9\}$ **21.** $\{1, 3, 5\}$
22. $\{1, 2, 3, 4, 5, 7, 9\}$ **23.** $\{1, 2, 3, 4, 5, 6, 8\}$ **24.** $\{2, 4\}$
25. A **26.** B **27.** $\varnothing$ **28.** $\varnothing$ **29.** A **30.** N
31. $=$ **32.** $\neq$ **33.** $\cup$ **34.** $\cap$ **35.** $\cap$ **36.** $\cup$
37. $\notin$ **38.** $\in$ **39.** $\in$ **40.** $\in$ **41.** T **42.** T
43. T **44.** F **45.** T **46.** T **47.** T **48.** T
49. F **50.** F **51.** T **52.** T **53.** $\{2, 3, 4, 5, 6, 7, 8\}$
54. $\varnothing$ **55.** $\{3, 5\}$ **56.** $\{1, 2, 3, 4, 5, 7\}$
57. $\{1, 2, 3, 4, 5, 6, 8\}$ **58.** $\{2, 4\}$ **59.** $\{2, 3, 4, 5\}$
60. $\{2, 4\}$ **61.** $\{2, 3, 4, 5, 7\}$ **62.** $\{2, 3, 4, 5, 7\}$
63. $\{2, 3, 4, 5\}$ **64.** $\{2, 4\}$ **65.** $\{2, 3, 4, 5, 7\}$
66. $\{2, 3, 4, 5, 7\}$ **67.** $\subseteq$ **68.** $=$ **69.** $\in$ **70.** $\subseteq$
71. $\cap$ **72.** $\subseteq$ **73.** $\subseteq$ **74.** $\cap$ **75.** $\cap$ **76.** $\cup$
77. $\cup$ **78.** $\cap$ **79.** $\{2, 4, 6, \ldots, 18\}$ **80.** $\{7, 8, 9, \ldots\}$
81. $\{13, 15, 17, \ldots\}$ **82.** $\{1, 3, 5, \ldots, 13\}$
83. $\{6, 8, 10, \ldots, 78\}$ **84.** $\{13, 15, 17, \ldots, 55\}$
85. $\{x \mid x$ is a natural number between 2 and 7$\}$
86. $\{x \mid x$ is an odd natural number less than 8$\}$
87. $\{x \mid x$ is an odd natural number greater than 4$\}$
88. $\{x \mid x$ is a natural number greater than 3$\}$
89. $\{x \mid x$ is an even natural number between 5 and 83$\}$
90. $\{x \mid x$ is an odd natural number between 8 and 52$\}$
91. No **92.** $A = B, A = B$
93. **a)** $3 \in \{1, 2, 3\}$ **b)** $\{3\} \subseteq \{1, 2, 3\}$ **c)** $\varnothing \neq \{\varnothing\}$
94. **a)** $\varnothing, \{1\}, \{2\}, \{1, 2\}$ **b)** 8 **c)** 16 **d)** 2^n

Section 1.2 Warm-ups F T F F T F T T F T

1. The integers consist of the positive and negative counting numbers and zero.
2. The rational numbers consist of all numbers that can be expressed as a ratio of integers.
3. The repeating or terminating decimal numbers are rational numbers.
4. Decimals that neither repeat nor terminate are irrational numbers.
5. The set of real numbers is the union of the rational and irrational numbers.

6. The ratio of the circumference and diameter of any circle is π.
7. T **8.** T **9.** F **10.** T **11.** T **12.** F **13.** T **14.** F
15. $\{0, 1, 2, 3, 4, 5\}$ **16.** $\{1, 2, 3, 4, 5, 6\}$

17. $\{-4, -3, -2, -1, 0, 1, \ldots\}$

18. $\{3, 4, 5, 6, 7, 8, 9, 10, 11\}$ **19.** $\{1, 2, 3, 4\}$

20. $\{1, 2, 3, 4, 5, \ldots\}$ **21.** $\{-2, -1, 0, 1, 2, 3, 4\}$

22. $\{-3, -2, -1, 0, 1, 2, 3, 4, 5, 6\}$

23. All **24.** $\left\{\dfrac{8}{2}\right\}$ **25.** $\left\{0, \dfrac{8}{2}\right\}$ **26.** $\left\{-3, 0, \dfrac{8}{2}\right\}$
27. $\left\{-3, -\dfrac{5}{2}, -0.025, 0, 3\dfrac{1}{2}, \dfrac{8}{2}\right\}$ **28.** $\left\{-\sqrt{10}, \sqrt{2}\right\}$
29. T **30.** F **31.** F **32.** T **33.** T **34.** F **35.** F
36. T **37.** F **38.** T **39.** F **40.** T **41.** T **42.** F
43. $\subseteq$ **44.** $\subseteq$ **45.** $\not\subseteq$ **46.** $\not\subseteq$ **47.** $\subseteq$ **48.** $\subseteq$
49. $\subseteq$ **50.** $\subseteq$ **51.** $\subseteq$ **52.** $\subseteq$ **53.** $\in$ **54.** $\in$
55. $\in$ **56.** $\in$ **57.** $\in$ **58.** $\in$ **59.** $\notin$ **60.** $\in$
61. $\subseteq$ **62.** $\not\subseteq$ **63.** $\subseteq$ **64.** $\not\subseteq$

Section 1.3 Warm-ups T T F T T F F F F T

1. The absolute value of a number is the number's distance from 0 on the number line.
2. Add their absolute values, then affix the sign of the original numbers.
3. Subtract their absolute values and use the sign of the number with the larger absolute value.
4. The difference $a - b$ is defined as $a + (-b)$.
5. Multiply their absolute values, then affix a positive sign if the original numbers have the same sign and a negative sign if the original numbers have opposite signs.
6. The quotient $a \div b$ is defined as $a \cdot \dfrac{1}{b}$.
7. 34 **8.** 17 **9.** 0 **10.** 15 **11.** 0 **12.** 0 **13.** -9
14. -3 **15.** 9 **16.** 8 **17.** -3 **18.** -2 **19.** 4
20. 7 **21.** -7 **22.** -26 **23.** -2 **24.** -10
25. -10 **26.** 5 **27.** -26 **28.** 0 **29.** -2 **30.** -12
31. -5 **32.** -13 **33.** 0 **34.** 2 **35.** $\dfrac{1}{10}$ **36.** $-\dfrac{1}{4}$
37. $-\dfrac{1}{6}$ **38.** $\dfrac{5}{4}$ **39.** -14.98 **40.** -0.85 **41.** -2.71
42. -0.2 **43.** 2.803 **44.** -1.72143 **45.** -0.2649
46. -5.99 **47.** -3 **48.** -11 **49.** -11 **50.** -17

51. 13   **52.** 12   **53.** −6   **54.** −10   **55.** −9
56. −9   **57.** 23   **58.** 120   **59.** 1   **60.** $-\frac{1}{8}$ **61.** $-\frac{1}{2}$
62. $-\frac{1}{6}$ **63.** 1.97   **64.** −3.02   **65.** 7.3   **66.** −4.23
67. −50.73   **68.** −17.964   **69.** 1.27   **70.** 3.5
71. −75   **72.** −35   **73.** $\frac{1}{6}$ **74.** $\frac{3}{7}$ **75.** −0.09
76. 0.05   **77.** 0.2   **78.** −0.125   **79.** $\frac{1}{20}$ or 0.05
80. $-\frac{1}{5}$ or −0.2   **81.** $-\frac{5}{6}$ **82.** −8   **83.** $-\frac{10}{3}$ **84.** 8
85. −2   **86.** −42   **87.** −37.5   **88.** 54   **89.** −0.08
90. −14   **91.** 0.25   **92.** 20   **93.** −12   **94.** $\frac{8}{15}$
95. $\frac{3}{5}$ **96.** $-\frac{1}{5}$ **97.** −91.25   **98.** −50   **99.** 17,000
100. −16.52   **101.** −49   **102.** −49   **103.** −7
104. −52   **105.** 15   **106.** −75   **107.** −342
108. −41   **109.** $\frac{3}{4}$ **110.** $\frac{3}{8}$ **111.** −3   **112.** 0
113. 20   **114.** 52   **115.** 0   **116.** $\frac{15}{2}$ **117.** −180
118. −165   **119.** $-\frac{1}{6}$ **120.** $-\frac{1}{2}$ **121.** −55   **122.** 7
123. −1   **124.** $-\frac{4}{3}$ **125.** $-\frac{39}{2}$ **126.** $-\frac{67}{3}$
127. 27.99   **128.** 54.9   **129.** −29.3   **130.** −0.541
131. −0.7   **132.** −0.57   **133.** 2   **134.** 25
135. $44,400   **136.** $800,000, −$2.9 million   **137.** 20°
138. 11°C   **139.** 1,014 feet   **140.** 65,229 feet

Section 1.4 Warm-ups F T T F F F T T F F
1. An arithmetic expression is the result of writing numbers in a meaningful combination with the ordinary operations of arithmetic.
2. An expression is called a sum, a difference, a product, or a quotient if the last operation to be performed is addition, subtraction, multiplication, or division, respectively.
3. Grouping symbols are used to indicate the order in which operations are to be performed.
4. An exponential expression is an expression of the form a^n.
5. The order of operations tells us the order in which to perform operations when we omit grouping symbols.
6. The value of -3^2 is −9 and the value of $(-3)^2$ is 9.
7. −22   **8.** 1   **9.** −8   **10.** −36   **11.** −14
12. 0   **13.** 32   **14.** 81   **15.** 1   **16.** −1   **17.** $\frac{1}{9}$
18. $\frac{1}{64}$ **19.** 7   **20.** 10   **21.** 10   **22.** 4   **23.** 8
24. 13   **25.** −8   **26.** −19   **27.** 17   **28.** 14   **29.** $\frac{1}{24}$
30. $-\frac{1}{16}$ **31.** 58   **32.** −63   **33.** −25   **34.** 125
35. −200   **36.** 0   **37.** 40   **38.** 216   **39.** −25
40. −16   **41.** −7.5   **42.** 540   **43.** −22.4841   **44.** 3.69
45. −1.9602   **46.** −32.157432   **47.** −276.48
48. 8.86875   **49.** −2   **50.** 3   **51.** −1   **52.** 2

53. −6   **54.** 0   **55.** 0   **56.** 3   **57.** −7   **58.** 7
59. $-\frac{4}{3}$ **60.** $\frac{7}{4}$ **61.** −8   **62.** −15   **63.** 5   **64.** 0
65. $-\frac{5}{2}$ **66.** 3   **67.** 4   **68.** 1   **69.** $\frac{10}{9}$ **70.** 0
71. $\frac{3}{4}$ **72.** $\frac{7}{4}$ **73.** −2.67   **74.** 18   **75.** 41   **76.** 99
77. 27   **78.** 0   **79.** 1   **80.** 2   **81.** 9   **82.** 4
83. −1   **84.** $-\frac{5}{2}$ **85.** 26   **86.** −8   **87.** $-\frac{3}{2}$ **88.** 0
89. −2   **90.** −1   **91.** $\frac{5}{12}$ **92.** $\frac{1}{4}$ **93.** 17   **94.** −9
95. −46   **96.** 23   **97.** 3   **98.** 4   **99.** −8   **100.** 100
101. 26 beats per minute, age 43   **102.** 150, 127.5, age 67
103. 104 feet   **104.** 36,000 ft² (square feet)
105. a) $60,000   b) $60,776.47   **106.** $74,557.71
107. $5,500, $5,441.96   **108.** $62,222   **109.** $32,721

Section 1.5 Warm-ups T F F F F F T T F T
1. The commutative property of addition says that $a + b = b + a$ and the commutative property of multiplication says that $a \cdot b = b \cdot a$.
2. The associative property of addition says that $(a + b) + c = a + (b + c)$.
3. The commutative property of addition says that you get the same result when you add two numbers in either order. The associative property of addition deals with which two numbers are added first when adding three numbers.
4. The distributive property says that $a(b + c) = ab + ac$.
5. Zero is the additive identity because adding zero to a number does not change the number.
6. One is the multiplicative identity because multiplying a number by 1 does not change the number.
7. 1   **8.** −2   **9.** −14   **10.** 1   **11.** −24   **12.** −33
13. −1.7   **14.** 8.7   **15.** −19.8   **16.** −4.03
17. $4x - 24$ **18.** $5a - 5$ **19.** $2(m + 5)$ **20.** $3(y + 3)$
21. $3a + at$ **22.** $by + bw$ **23.** $-2w + 10$
24. $-4m + 28$ **25.** $-6 + 2y$ **26.** $-20 + 5p$
27. $5(x - 1)$ **28.** $3(y + 1)$ **29.** $2x + y$ **30.** $4y + w$
31. $6w + 9y$ **32.** $4x + 24$ **33.** $3(y - 5)$ **34.** $5(x + 2)$
35. $3(a + 3)$ **36.** $7(b - 7)$ **37.** $2x + 4$ **38.** $x + 2$
39. $-x + 2$ **40.** $-3x + 1$ **41.** 2   **42.** 3   **43.** 1
44. −1   **45.** $\frac{1}{6}$ **46.** $\frac{1}{8}$ **47.** 4   **48.** $\frac{4}{3}$ **49.** $-\frac{10}{7}$
50. $-\frac{10}{9}$ **51.** $-\frac{5}{9}$ **52.** $-\frac{5}{13}$ **53.** 0.6200
54. −0.1433   **55.** 0.7326   **56.** 0.0639   **57.** 1,450 mph
58. 2,193 mph   **59.** 217 buttons per hour
60. 0.06 house per hour
61. Commutative property of addition
62. Commutative property of multiplication
63. Distributive property
64. Associative property of multiplication
65. Associative property of multiplication
66. Distributive property
67. Multiplicative inverse property

68. Commutative property of addition
69. Commutative property of multiplication
70. Multiplication property of zero
71. Multiplicative identity property
72. Multiplicative inverse property
73. Distributive property   **74.** Additive identity property
75. Additive inverse property
76. Multiplicative identity property
77. Multiplication property of zero
78. Distributive property   **79.** Distributive property
80. Distributive property   **81.** $w + 5$ **82.** $2(x + 1)$
83. $(5x)y$ **84.** $\frac{1}{2} + x$ **85.** $\frac{1}{2}(x - 1)$ **86.** $3x - 21$
87. $3(2x + 3)$ **88.** $x + (7 + 3)$ **89.** 1   **90.** $-a + 3$
91. 0   **92.** 1   **93.** 4   **94.** 45

Section 1.6 Warm-ups T F T T F T F F F T
1. A term is a single number or a product of a number and one or more variables.
2. Like terms contain the same variables with the same powers.
3. The coefficient of a term is the number preceding the variables.
4. The distributive property is used to combine like terms.
5. You can multiply and divide unlike terms.
6. You can remove parentheses preceded by a negative sign by taking the opposite of every term in the parentheses.
7. 9,000   **8.** 7,500   **9.** 1   **10.** 1   **11.** 527   **12.** 294
13. 470   **14.** 530   **15.** 38   **16.** 34   **17.** 48,000
18. 45,000   **19.** 0   **20.** 456   **21.** 398   **22.** 185
23. 1   **24.** 1   **25.** 1,700   **26.** -330 **27.** 374
28. 764   **29.** 0   **30.** 0   **31.** $2n$ **32.** $12a$ **33.** $7w$
34. $10b$ **35.** $-11mw^2$ **36.** $-14b^2x$ **37.** $-3x$
38. $-11 - 7t$ **39.** $-4 - 7z$ **40.** $-16m$ **41.** $9t^2$
42. $5a + 4a^2$ **43.** $-4ab + 3a^2b$ **44.** $-2x^2y$ **45.** $8mn$
46. $2cm$ **47.** $-2x^3y$ **48.** $-4s^4t$ **49.** $-2kz^6$ **50.** 0
51. $28t$ **52.** $-12r$ **53.** $10x^2$ **54.** $21h^2$ **55.** h^2
56. $-x^2$ **57.** $-28w$ **58.** $5t$ **59.** $-x + x^2$
60. $-p^2 + p$ **61.** $25k^2$ **62.** $16y^2$ **63.** y **64.** z^2
65. $2y$ **66.** y **67.** $3x^3$ **68.** $-2x^2$ **69.** $x^2y + 5x$
70. $3xy^2 - 4w$ **71.** $-x + 2$ **72.** $2x + 3$ **73.** $\frac{1}{2}xt - 5$
74. $\frac{1}{2}xt^2 - 2$ **75.** $-3a + 1$ **76.** $3x + 7$ **77.** $10 - x$
78. $14 - w$ **79.** $3m + 1$ **80.** $1 - 9t$ **81.** $-12b + at$
82. $3x^2 - 2y$ **83.** $2t^2 - 3w$ **84.** $2n^2 - 4m$ **85.** $y^2 + z$
86. $3xy + zy - w$ **87.** $9x + 8$ **88.** $7x + 17$
89. $2x - 2$ **90.** $2x + 3$ **91.** $-2a^2 + 2c$ **92.** $2x^2 + 1$
93. $7t^2 - 9w$ **94.** $11xy^2 + 17$ **95.** $m - 12$
96. $5m - 12$ **97.** $-7k^3 - 17$ **98.** $4k^3 + 1$
99. $0.96x - 2$ **100.** $0.97x - 15$ **101.** $0.06x - 1.5$
102. $0.2x + 28$ **103.** $-4k + 16$ **104.** $3w - 8$
105. $-4.5x + 17.83$ **106.** $0.735x - 2.576$
107. $-4xy - 22$ **108.** $-18x^2 - 12$ **109.** $-29w^2$
110. $16w^3 - 2w^2$ **111.** $-6a^2w^2$ **112.** $-11a^2w - 3aw^2$
113. $2x^2y + \frac{1}{3}$ **114.** $\frac{1}{2}abc - 2bc$ **115.** $\frac{1}{4}m^2 - m$
116. $-3wyt + wy$ **117.** $4t^3 + 3t^2 - 1$ **118.** $-x^3 + 2$

119. $2xyz + xy - 3z$ **120.** $-2a^2b^4 - 1$ **121.** $3s + 6$ ft
122. $4w + 100$ ft, no   **123.** $\frac{13}{3}x$ m, $\frac{7}{6}x^2$ m^2
124. $4x$ in., x^2 in.^{2}

Enriching Word Power
1. a   **2.** c   **3.** a   **4.** d   **5.** a   **6.** c   **7.** a   **8.** d
9. b   **10.** a   **11.** b   **12.** a   **13.** b   **14.** c

Chapter 1 Review
1. T   **2.** F   **3.** F   **4.** T   **5.** T   **6.** F   **7.** F   **8.** F
9. T   **10.** T   **11.** T   **12.** F   **13.** F   **14.** F   **15.** T
16. T   **17.** F   **18.** F   **19.** T   **20.** T   **21.** $\{0, 1, 31\}$
22. $\{1, 31\}$ **23.** $\{-1, 0, 1, 31\}$
24. $\left\{-1, 0, 1, 1.732, \frac{22}{7}, 31\right\}$ **25.** $\{-\sqrt{2}, \sqrt{3}, \pi\}$
26. All   **27.** F   **28.** F   **29.** F   **30.** F   **31.** F
32. F   **33.** T   **34.** F   **35.** F   **36.** F   **37.** 5
38. -8 **39.** -12 **40.** -16 **41.** -24 **42.** 42
43. 2   **44.** -5 **45.** $-\frac{1}{6}$ **46.** $\frac{5}{12}$ **47.** 10   **48.** -5
49. 9.96   **50.** -3.05 **51.** -4 **52.** -0.16 **53.** 0
54. -0.2 **55.** -4 **56.** -0.18 **57.** 0   **58.** 3
59. -3 **60.** 2   **61.** 39   **62.** 55   **63.** 121   **64.** 53
65. 23   **66.** 7   **67.** 2   **68.** -19 **69.** 23   **70.** 16
71. -96 **72.** 2   **73.** 5   **74.** 12   **75.** -1 **76.** 2
77. 0.76   **78.** 4.98   **79.** 1   **80.** 21   **81.** -3 **82.** 7
83. 4   **84.** $\frac{1}{2}$ **85.** 1   **86.** 4   **87.** -8 **88.** -5
89. 1   **90.** 25   **91.** -35 **92.** 19   **93.** 2   **94.** $\frac{1}{2}$
95. 5   **96.** 5   **97.** -1 **98.** -1
99. Commutative property of addition
100. Multiplication property of zero
101. Distributive property   **102.** Additive inverse property
103. Associative property of multiplication
104. Commutative property of addition
105. Multiplicative identity property
106. Multiplicative inverse property
107. Multiplicative inverse property
108. Multiplicative identity property
109. Multiplication property of zero
110. Commutative property of addition
111. Additive identity property   **112.** Distributive property
113. Additive inverse property
114. Associative property of addition
115. $3(x - a)$ **116.** $5(x - y)$ **117.** $3w + 3$
118. $2m + 28$ **119.** $7(x + 1)$ **120.** $3(w + 1)$
121. $5x - 25$ **122.** $13b - 39$ **123.** $-6x + 15$
124. $-10 + 8x$ **125.** $p(1 - t)$ **126.** $b(a + 1)$
127. $7a + 2$ **128.** $3m + 4$ **129.** $-t - 2$ **130.** 0
131. $-2a + 4$ **132.** $-4w + 4y$ **133.** $4x + 33$
134. $-3x + 28$ **135.** $-0.8x - 0.48$
136. $-0.9x - 0.12$ **137.** $-0.05x - 1.85$
138. $-0.18x - 8$ **139.** $\frac{1}{4}x + 4$ **140.** $\frac{5}{4}x - \frac{1}{4}$
141. $-3x^2 - 2x + 1$ **142.** 2
143. 0, additive inverse, multiplication property of zero

144. 4, multiplicative inverse, multiplicative identity
145. 7680, distributive　　**146.** 280, distributive
147. 48, associative property of addition, additive inverse
148. 42, associative property of addition, additive inverse, additive identity
149. 0, distributive, additive inverse
150. 0, distributive, additive inverse, multiplication property of zero
151. 47, associative property of multiplication, multiplicative inverse
152. 90, distributive
153. -24, commutative property of multiplication, associative property of multiplication
154. 4, commutative property of multiplication, associative property of multiplication, multiplicative inverse
155. 0, additive inverse, multiplication property of zero
156. 0, multiplication property of zero
157. Seven shingles per minute　　**158.** $20x^2 + 4x$ dollars
159. \$36,761, 2006　　**160.** 10,179 gallons

Chapter 1 Test

1. $\{2, 3, 4, 5, 6, 7, 8, 10\}$　　**2.** $\{6, 7\}$　　**3.** $\{4, 6, 8, 10\}$
4. $\{0, 8\}$　　**5.** $\{-4, 0, 8\}$　　**6.** $\left\{-4, -\frac{1}{2}, 0, 1.65, 8\right\}$
7. $\{-\sqrt{3}, \sqrt{5}, \pi\}$　　**8.** -9　　**9.** 8　　**10.** -11　　**11.** -1.98
12. -2　　**13.** -4　　**14.** $\frac{1}{18}$　　**15.** -12　　**16.** 7　　**17.** -3
18. 0　　**19.** 4,780　　**20.** 240　　**21.** -30　　**22.** 40
23. 7　　**24.** 0　　**25.** Distributive property
26. Commutative property of multiplication
27. Associative property of addition
28. Additive inverse property
29. Commutative property of multiplication
30. Multiplication property of zero　　**31.** $11m - 3$
32. $0.95x + 2.9$　　**33.** $\frac{3}{4}x - \frac{5}{4}$　　**34.** $5x^2$
35. $3x^2 + 2x - 1$　　**36.** $-7xy$　　**37.** $5(x - 8)$
38. $7(t - 1)$　　**39.** 16 trees per hour
40. $4x - 8$, $x^2 - 4x$, 28 feet, 45 square feet
41. 12.6 billion

CHAPTER 2

Section 2.1 Warm-ups F T F T T T T T T T
1. An equation is a sentence that expresses the equality of two algebraic expressions.
2. A number satisfies the equation if the equation is true when the variable is replaced by the number.
3. Equivalent equations are equations that have the same solution set.
4. A linear equation in one variable is an equation of the form $ax + b = 0$.
5. If the equation involves fractions then multiply each side by the LCD.
6. An identity is an equation that is satisfied by all values of the variable for which both sides are defined.
7. A conditional equation is an equation that has at least one solution but is not an identity.
8. An inconsistent equation is an equation that has no solution.
9. Yes　　**10.** Yes　　**11.** Yes　　**12.** Yes　　**13.** No
14. Yes　　**15.** No　　**16.** Yes　　**17.** $\{-87\}$　　**18.** $\{60\}$
19. $\{-24\}$　　**20.** $\{13\}$　　**21.** $\left\{\frac{3}{2}\right\}$　　**22.** $\left\{-\frac{7}{5}\right\}$　　**23.** $\{-1\}$
24. $\{-5\}$　　**25.** $\{-3\}$　　**26.** $\{2\}$　　**27.** $\left\{\frac{5}{2}\right\}$　　**28.** $\{-7\}$
29. $\{18\}$　　**30.** $\{-47\}$　　**31.** $\{18\}$　　**32.** $\{24\}$　　**33.** $\{0\}$
34. $\{0\}$　　**35.** $\left\{-\frac{28}{3}\right\}$　　**36.** $\left\{-\frac{12}{5}\right\}$　　**37.** $\left\{-\frac{28}{5}\right\}$
38. $\left\{\frac{50}{3}\right\}$　　**39.** $\{2\}$　　**40.** $\left\{\frac{9}{10}\right\}$　　**41.** $\{12\}$　　**42.** $\{20\}$
43. $\{-7\}$　　**44.** $\{11\}$　　**45.** $\{12\}$　　**46.** $\{15\}$　　**47.** $\{6\}$
48. $\{8\}$　　**49.** $\{90\}$　　**50.** $\{70\}$　　**51.** $\{1000\}$　　**52.** $\{800\}$
53. $\{800\}$　　**54.** $\{200\}$　　**55.** $\varnothing$, inconsistent
56. $\{0\}$, conditional　　**57.** R, identity　　**58.** R, identity
59. $\{1\}$, conditional　　**60.** $\{5\}$, conditional　　**61.** R, identity
62. R, identity　　**63.** R, identity　　**64.** $\{x \mid x \neq 0\}$, identity
65. $\varnothing$, inconsistent　　**66.** $\varnothing$, inconsistent　　**67.** R, identity
68. R, identity　　**69.** $\{-4\}$, conditional　　**70.** $\{7\}$, conditional
71. $\{1\}$　　**72.** $\left\{\frac{1}{9}\right\}$　　**73.** $\left\{\frac{5}{18}\right\}$　　**74.** $\left\{\frac{13}{19}\right\}$　　**75.** $\left\{\frac{121}{48}\right\}$
76. $\{0\}$　　**77.** $\left\{\frac{19}{13}\right\}$　　**78.** $\left\{\frac{5}{2}\right\}$　　**79.** $\left\{\frac{10}{29}\right\}$　　**80.** $\left\{-\frac{650}{19}\right\}$
81. $\{3\}$　　**82.** $\left\{-\frac{1}{5}\right\}$　　**83.** $\{16\}$　　**84.** $\{-6\}$　　**85.** $\left\{\frac{3}{4}\right\}$
86. $\left\{-\frac{6}{5}\right\}$　　**87.** $\{-15\}$　　**88.** $\{-14\}$　　**89.** $\{6\}$　　**90.** $\{-8\}$
91. $\{6\}$　　**92.** $\{2\}$　　**93.** $\{-2\}$　　**94.** $\{-4\}$　　**95.** $\{53,191.49\}$
96. $\{842.11\}$　　**97.** $\{4.7\}$　　**98.** $\{1.7713\}$
99. a) 42.2 million　　**b)** 2010　　**c)** increasing
100. a) \$32,396　　**b)** 2007　　**101. a)** 1964　　**b)** 2005
102. 2022

Section 2.2 Warm-ups F F F T F T F T T F
1. A formula is an equation involving two or more variables.
2. A formula is used to show a relationship between two or more variable quantities.
3. Solving for a variable means to rewrite the formula by isolating the indicated variable.
4. If a variable occurs twice, then we usually use the distributive property to isolate it.
5. To find the value of a variable, solve for that variable, then replace all other variables with given numbers.
6. The formula for the volume of a rectangular solid is $V = LWH$.
7. $t = \dfrac{I}{Pr}$　　**8.** $r = \dfrac{d}{t}$　　**9.** $C = \dfrac{5}{9}(F - 32)$　　**10.** $h = \dfrac{2A}{b}$
11. $W = \dfrac{A}{L}$　　**12.** $r = \dfrac{C}{2\pi}$　　**13.** $b_1 = 2A - b_2$
14. $b_2 = 2A - b_1$　　**15.** $y = -\dfrac{2}{3}x + 3$　　**16.** $y = -\dfrac{5}{4}x + 2$
17. $y = x - 4$　　**18.** $y = x + 6$　　**19.** $y = \dfrac{3}{2}x - 6$

20. $y = \frac{4}{3}x - 4$ **21.** $y = \frac{1}{2}x + \frac{1}{2}$ **22.** $y = \frac{1}{3}x + \frac{5}{3}$

23. $t = \frac{A - P}{Pr}$ **24.** $r = \frac{A - P}{Pt}$ **25.** $a = \frac{1}{b + 1}$

26. $y = \frac{m}{1 - w}$ **27.** $y = \frac{12}{1 - x}$ **28.** $x = \frac{2}{y - 1}$

29. $x = \frac{6}{w^2 - y^2 - z^2}$ **30.** $x = \frac{5}{z^2 + w^2 - y^2}$

31. $R_1 = \frac{RR_2}{R_2 - R}$ **32.** $a = \frac{2b}{b - 2}$ **33.** 12.472

34. 4.957   **35.** 34.932   **36.** 2.395   **37.** 0.539

38. 0.176   **39.** $\frac{1}{3}$ **40.** $-\frac{13}{4}$ **41.** $\frac{13}{2}$ **42.** -4

43. -1 **44.** -15 **45.** 4   **46.** -3 **47.** -4.4507

48. 2.7773   **49.** $\frac{5}{8}$ **50.** $-\frac{1}{3}$ **51.** $-\frac{7}{2}$ **52.** 1

53. -10 **54.** $-\frac{4}{5}$ **55.** $-\frac{8}{3}$ **56.** 1   **57.** 4   **58.** $\frac{4}{3}$

59. $A = \pi r^2$ **60.** $C = \pi d$ **61.** $r = \frac{C}{2\pi}$ **62.** $d = \frac{C}{\pi}$

63. $W = \frac{P - 2L}{2}$ **64.** $L = \frac{A}{W}$ **65.** 15%

66. One-half year   **67.** 5.75 yards   **68.** $\frac{55}{7}$ meters

69. 7.2 feet   **70.** 2 meters   **71.** 5 feet   **72.** 2 feet
73. 15 feet   **74.** 8 meters   **75.** 14 inches
76. 10 centimeters   **77.** 168 feet   **78.** 160 feet
79. 1.5 meters   **80.** 12 inches   **81.** 3,979 miles
82. 159 miles   **83.** 9.55 inches   **84.** 5.57 meters
85. 95,232 pounds   **86.** 107,442 pounds, strong enough
87. a) 2,457   **b)** 420   **88. a)** 25.1%   **b)** 2024   **c)** 2024
89. 200 feet   **90.** $500   **91.** $1,200   **92.** 81 feet

Section 2.3 Warm-ups F T F F T T F T F F
1. Three unknown consecutive integers are represented by x, $x + 1$, and $x + 2$.
2. In either case we use x, $x + 2$, and $x + 4$, but for odd integers x represents an odd integer and for even integers x represents an even integer.
3. The formula $P = 2L + 2W$ expresses the perimeter in terms of length and width.
4. Addition can be indicated by the words, "sum," "more than," or "plus."
5. The commission is a percentage of the selling price.
6. Uniform motion is motion at a constant rate.
7. $x, x + 2$ **8.** $x, x + 2$ **9.** $x, 10 - x$ **10.** $x, x + 3$
11. $0.85x$ **12.** $3x$ **13.** $3x$ miles   **14.** $\frac{100}{x + 5}$ hours
15. $4x + 10$ **16.** $10 - x$ meters   **17.** 27, 28, 29
18. 56, 57, 58   **19.** 82, 84, 86   **20.** 26, 28, 30
21. 63, 65   **22.** 11, 13, 15, 17
23. Width 46 meters, length 93 meters
24. Width 1.5 feet, length 3.5 feet
25. 161 feet, 312 feet, 211 feet   **26.** 17.5 inches
27. Width 11.25 feet, length 27.5 feet
28. 3 feet wide, 7 feet high   **29.** $1,500 at 6%, $2,500 at 10%

30. Brother $2, sister $1   **31.** $80,000   **32.** $120,000
33. $\frac{40}{3}$ gallons   **34.** 18 liters   **35.** 1.36 ounces
36. 1.0596 ounces   **37.** 40 mph   **38.** 8 mph
39. 15 mph   **40.** 100 mph   **41.** $86,957   **42.** $43.60
43. $8450   **44.** $3,333.33   **45.** 18 feet   **46.** 27 feet
47. Packers 35, Chiefs 10
48. 16% Wal-Mart, 20% Toys "R" Us   **49.** $8.67 per pound
50. $1.65 per ounce   **51.** 30 pounds   **52.** 40 pounds
53. $\frac{40}{7}$ quarts   **54.** $\frac{16}{3}$ quarts   **55.** $300.6 billion
56. 5.96 children per woman
57. Brian $14,400, Daniel $7,200, Raymond $3,800
58. Lauren $88,880, Lisa $44,440, Lena $22,220, lawyer $8,888
59. 22, 23   **60.** 9, 11   **61.** 7.5 hours   **62.** 15 hours
63. 20 meters by 20 meters   **64.** 9 feet by 9 feet
65. $500 at 8%, $2,500 at 10%
66. $1,000 at 6%, $7,000 at 9%
67. 2 gallons of 5% solution, 3 gallons of 10% solution
68. 5 liters of 12% alcohol, 1 liter of water
69. Todd 46, Darla 32   **70.** Al 21, Bart 42, Carl 37

Section 2.4 Warm-ups F F T F F F T T T T
1. An inequality is a sentence that expresses inequality between two algebraic expressions.
2. To express inequality we use the symbols $<$, $\leq$, $>$, and $\geq$.
3. If a is less than b, then a lies to the left of b on the number line.
4. A linear inequality is an inequality of the form $ax + b > 0$ or with any of the other inequality symbols used in place of $>$.
5. When you multiply or divide by a negative number, the inequality symbol is reversed.
6. We can verbally indicate inequality with words like "less than," "at least," "greater than," and "at most."
7. F   **8.** F   **9.** T   **10.** T   **11.** T   **12.** F   **13.** T
14. T   **15.** Yes   **16.** No   **17.** No   **18.** Yes   **19.** No
20. Yes
21. $(-\infty, -1]$ **22.** $[-7, \infty)$

23. $(20, \infty)$ **24.** $(-\infty, 30)$

25. $[3, \infty)$ **26.** $(-\infty, -2)$

27. $(-\infty, 2.3)$ **28.** $(-\infty, 4.5]$

29. $(1, \infty)$ **30.** $(-\infty, 3)$ **31.** $(-\infty, -3]$ **32.** $[-2, \infty)$
33. $(-\infty, 5)$ **34.** $(-7, \infty)$ **35.** $[-4, \infty)$ **36.** $(-\infty, -9]$
37. $>$ **38.** $\leq$ **39.** $>$ **40.** $\leq$ **41.** $\leq$ **42.** $<$
43. $>$ **44.** $\leq$

45. $(-2, \infty)$

46. $(-\infty, -2]$

47. $[-4, \infty)$

48. $(-\infty, 3)$

49. $(5, \infty)$

50. $\left(-\infty, \dfrac{8}{3}\right)$

51. $[-3, \infty)$

52. $\left(-\infty, -\dfrac{7}{2}\right]$

53. $(13, \infty)$

54. $\left(\dfrac{27}{2}, \infty\right)$

55. $[-1, \infty)$

56. $[1, \infty)$

57. $(-\infty, 4]$

58. $(-\infty, 9)$

59. $\left(\dfrac{2}{3}, \infty\right)$
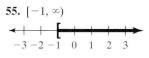
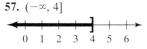

60. $(-\infty, -2)$

61. $\left(\dfrac{13}{3}, \infty\right)$
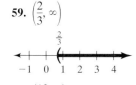

62. $\left(\dfrac{23}{2}, \infty\right)$
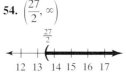

63. $(-\infty, \infty)$
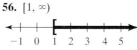
64. $\varnothing$

65. $\varnothing$　　**66.** $(-\infty, \infty)$
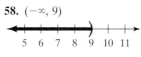

67. $(1, \infty)$
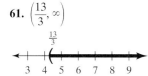

68. $\left(-\infty, \dfrac{1}{2}\right)$
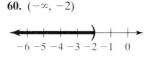

69. $\left(-\infty, \dfrac{49}{30}\right]$
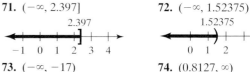
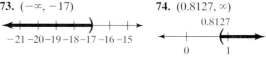

70. $\left(-\infty, \dfrac{33}{151}\right)$

71. $(-\infty, 2.397]$

72. $(-\infty, 1.52375)$

73. $(-\infty, -17)$

74. $(0.8127, \infty)$

75. $x =$ Tony's height, $x > 6$ feet
76. $a =$ Glenda's age, $a < 60$ years
77. $s =$ Wilma's salary, $s < \$80{,}000$
78. $w =$ Bubba's weight, $w > 80$ pounds
79. $v =$ speed of the Concorde, $v \le 1{,}450$ mph
80. $s =$ minimum speed, $s \ge 45$ mph
81. $a =$ amount Julie can afford, $a \le \$400$
82. $a =$ Fred's grade point average, $a \ge 3.2$
83. $b =$ Burt's height, $b \le 5$ feet
84. $r =$ Ernie's speed, $r \le 10$ mph
85. $t =$ Tina's hourly wage, $t \le \$8.20$
86. $s =$ selling price, $s \ge \$12{,}000$
87. $x =$ price of car, $x < \$9{,}100$
88. $x =$ price of sewing machine, $x \le \$636.36$
89. $x =$ price of truck, $x \ge \$9{,}100.92$
90. $x =$ Curly's contribution, $x \ge \$183.33$
91. a) Increasing　**b)** 2005　**92.** 2031
93. $x =$ final exam score, $x \ge 77$
94. $x =$ final exam score, $x \ge 98$
95. $x =$ the price of A-Mart jeans, $x < \$16.67$
96. $x =$ Al's rate, $x < 30$ mph　**97. a)** $[8, \infty)$
　b) $(-\infty, -6)$　**c)** $(2, \infty)$　**d)** $(-\infty, -12)$　**e)** $(2, \infty)$

Section 2.5 Warm-ups T T F T T T F T F T
1. A compound inequality consists of two inequalities joined with the words "and" or "or."
2. A compound inequality using and is true only when both simple inequalities are true.
3. A compound inequality using or is true when either one or the other or both inequalities is true.
4. Solve each simple inequality and then either union or intersect the solution sets.
5. The inequality $a < b < c$ means that $a < b$ and $b < c$.
6. The inequality $5 < x > 7$ has no meaning. All inequality symbols must point in the same direction in this notation.
7. No　**8.** Yes　**9.** Yes　**10.** Yes　**11.** No　**12.** Yes
13. No　**14.** Yes　**15.** Yes　**16.** Yes　**17.** Yes　**18.** Yes
19.　　　　　　　　　　**20.**

21.

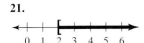

(number line with bracket at 2, shaded right) 0 1 2 3 4 5 6

22.

(number line, shaded left with parenthesis at 4) −1 0 1 2 3 4 5

23.

(number line, entire shaded) −3 −2 −1 0 1 2 3

24.

(number line, parenthesis at −2, bracket at 4) −2 −1 0 1 2 3 4

25. ∅

26.

(number line, entire shaded) −3 −2 −1 0 1 2 3

27.

(number line, bracket at 6, parenthesis at 9) 4 5 6 7 8 9 10 11

28. ∅

29. ∅

30.

(number line) −4 −3 −2 −1 0 1 2 3 4

31. $(-\infty, 1) \cup (10, \infty)$

(number line) −1 0 1 2 3 4 5 6 7 8 9 10 11

32. $(-\infty, -2) \cup (11, \infty)$

11
(number line) −4 −2 0 2 4 6 8 10 12 14

33. $(9, \infty)$

(number line, shaded right of 9) 7 8 9 10 11 12 13

34. $(10, \infty)$

(number line, shaded right of 10) 8 9 10 11 12 13 14

35. $(-6, \infty)$

(number line) −8 −7 −6 −5 −4 −3 −2

36. $(-8, \infty)$

(number line) −10 −9 −8 −7 −6 −5 −4

37. $(1, 4]$

(number line) 0 1 2 3 4 5

38. ∅

39. $(-\infty, \infty)$

(number line) −3 −2 −1 0 1 2 3

40. $\left(\dfrac{8}{15}, 4\right)$

$\frac{8}{15}$
(number line) 0 1 2 3 4 5

41. ∅   **42.** $(-\infty, \infty)$

(number line) −3 −2 −1 0 1 2 3

43. $(4, 7)$

(number line) 3 4 5 6 7 8

44. $(-1, 3)$

(number line) −2 −1 0 1 2 3 4

45. $[-3, 2)$

(number line) −3 −2 −1 0 1 2

46. $(-4, 2]$

(number line) −4 −3 −2 −1 0 1 2

47. $\left(-\dfrac{7}{3}, 3\right]$

$-\frac{7}{3}$
(number line) −3 −2 −1 0 1 2 3

48. $\left(-\dfrac{7}{2}, \dfrac{3}{2}\right]$

$-\frac{7}{2}$ $\frac{3}{2}$
(number line) −4 −3 −2 −1 0 1 2

49. $(-1, 5)$

(number line) −1 0 1 2 3 4 5

50. $(-4, 11)$

11
(number line) −4 −2 0 2 4 6 8 10 12

51. $[2, 3]$

(number line) 0 1 2 3 4 5

52. $[-4, 12]$

(number line) −4 −2 0 2 4 6 8 10 12

53. $(2, \infty)$ **54.** $(-6, \infty)$ **55.** $(-\infty, 5)$ **56.** $(-\infty, -2)$
57. $[2, 4]$ **58.** $[3, 8)$ **59.** $(-\infty, \infty)$
60. $(-\infty, -2] \cup (2, \infty)$ **61.** ∅   **62.** ∅   **63.** $[4, 5)$
64. $(0, 4]$ **65.** $[1, 6]$ **66.** $(0, 5)$ **67.** $x > 2$
68. $x \le 5$ **69.** $x < 3$ **70.** $x < -4$ or $x > 3$
71. $x > 2$ or $x \le -1$ **72.** $-1 < x < 2$
73. $-2 \le x < 3$ **74.** $x < 2$ **75.** $x \ge -3$
76. $x \le 0$ or $x > 1$ **77.** $(5, 7)$ **78.** $(-3, \infty)$
79. $(-1, 1] \cup (10, \infty)$ **80.** ∅   **81.** $(-3, 3)$

82. $\left(-\dfrac{14}{3}, \dfrac{1}{2}\right)$ **83.** x = final exam score, $73 \le x \le 86.5$

84. x = final exam score, $82 \le x \le 109$
85. x = price of truck, $\$11{,}033 \le x \le \$13{,}811$
86. x = selling price, $\$14{,}560 \le x \le \$14{,}900$
87. x = number of cigarettes on the run, $4 \le x \le 18$
88. w = width, $10 < w < 15$
89. a) 1,144,700   **b)** 2002   **c)** 2013   **d)** 2005
90. a) 33.15 million   **b)** 2000   **c)** 2004
91. $-b < x < -a$ provided $a < b$
92. Notation is used correctly in (a) and (e).
93. a) $(12, 32)$ **b)** $(-20, 10]$ **c)** $(0, 9)$ **d)** $[-3, -1]$
94. a) (s, t) if $s < t$, no solution if $t < s$
    b) (s, ∞) if $s > t$, (t, ∞) if $t < s$

Section 2.6 Warm-ups T F F T F T F F T F
1. Absolute value of a number is the number's distance from 0 on the number line.
2. Only 0 is 0 units from 0 on the number line.
3. Since both 4 and -4 are four units from 0, $|x| = 4$ has two solutions.
4. Since $|x| \ge 0$ for every real number x, $|x| = -3$ is impossible.
5. Since the distance from 0 for every number on the number line is greater than or equal to 0, $|x| \ge 0$.
6. Since $|x| \ge 0$ for all x, $|x| < -3$ is impossible.
7. $\{-5, 5\}$ **8.** $\{-2, 2\}$ **9.** $\{2, 4\}$ **10.** $\{3, 7\}$

11. $\{-3, 9\}$ **12.** $\{1, 13\}$ **13.** $\left\{-\dfrac{8}{3}, \dfrac{16}{3}\right\}$ **14.** ∅

15. $\{12\}$ **16.** $\left\{\dfrac{11}{3}, \dfrac{13}{3}\right\}$ **17.** $\{-20, 80\}$ **18.** $\{50\}$

19. ∅   **20.** $\{-10.5, 4.5\}$ **21.** $\{0, 5\}$ **22.** $\left\{\dfrac{5}{3}, -\dfrac{7}{3}\right\}$

23. $\{0.143, 1.298\}$ **24.** $\{-2.092, 7.382\}$
25. $\{-2, 2\}$ **26.** $\{-7, 7\}$ **27.** $\{-11, 5\}$ **28.** $\{-2, 6\}$

29. $\{0, 3\}$ **30.** $\{4, 12\}$ **31.** $\left\{-6, \dfrac{4}{3}\right\}$ **32.** $\{-3, 3\}$

33. $\{1, 3\}$ **34.** $\left\{-1, \dfrac{2}{3}\right\}$ **35.** $(-\infty, \infty)$ **36.** $(-\infty, \infty)$

37. $|x| < 2$ **38.** $|x| \le 5$ **39.** $|x| > 3$ **40.** $|x| \ge 6$
41. $|x| \le 1$ **42.** $|x| < 1$ **43.** $|x| \ge 2$ **44.** $|x| > 4$
45. No   **46.** No   **47.** Yes   **48.** Yes   **49.** No   **50.** No
51. Yes   **52.** Yes
53. $(-\infty, -6) \cup (6, \infty)$ **54.** $(-\infty, -3) \cup (3, \infty)$

(number line) −8 −6 −4 −2 0 2 4 6 8

(number line) −4 −3 −2 −1 0 1 2 3 4

55. $(-3, 3)$

56. $(-7, 7)$

57. $(-\infty, -1] \cup [5, \infty)$

58. $(-\infty, 4] \cup [6, \infty)$

59. $\left(-\frac{1}{2}, \frac{9}{2}\right)$

60. $(-1, 2)$

61. $[-2, 12]$

62. $[5, 7]$

63. $\left(-\infty, -\frac{9}{2}\right] \cup \left[\frac{15}{2}, \infty\right)$

64. $\left(-\infty, -\frac{5}{2}\right] \cup \left[\frac{15}{2}, \infty\right)$

65. $(-\infty, 2) \cup (2, \infty)$

66. $(-\infty, \infty)$ **67.** $(-\infty, \infty)$ **68.** $(-\infty, \infty)$ **69.** $\varnothing$

70. $\varnothing$ **71.** $(-\infty, \infty)$

72. $(-\infty, 5) \cup (5, \infty)$

73. $(-\infty, -3) \cup (-1, \infty)$ **74.** $[-1, 9]$ **75.** $(-4, 4)$

76. $(-\infty, -10] \cup [10, \infty)$ **77.** $(-1, 1)$

78. $(-\infty, -4) \cup (4, \infty)$ **79.** $(0.255, 0.847)$

80. $(-\infty, 1.0125] \cup [1.90625, \infty)$ **81.** $\varnothing$ **82.** $\left(-\infty, \frac{2}{3}\right]$

83. $(4, 5)$ **84.** $(1, 4) \cup (8, 9)$ **85.** 1401 or 1429

86. July 5 or July 27   **87.** Between 121 and 133 pounds

88. Greater than 125 or less than 95

89. a) 1 second   **b)** 1 second   **c)** $0.5 < t < 1.5$

90. 0.25 second

91. a) $(-\infty, \infty)$ **b)** $(-\infty, \infty)$ **c)** all reals except $n = 0$

92. b) $|m + n| \le |m| + |n|$

Enriching Word Power

1. c   **2.** b   **3.** d   **4.** c   **5.** c   **6.** d   **7.** d   **8.** a
9. d   **10.** d   **11.** d   **12.** c   **13.** d

Chapter 2 Review

1. $\{8\}$ **2.** $\{9\}$ **3.** $\left\{-\frac{3}{2}\right\}$ **4.** $\{5\}$ **5.** R **6.** $\{0\}$

7. $\varnothing$ **8.** R **9.** $\{0\}$ **10.** $\left\{-\frac{4}{3}\right\}$ **11.** $\{20\}$ **12.** $\{6\}$

13. $\{5\}$ **14.** $\{-5\}$ **15.** $\{5\}$ **16.** $\{4\}$ **17.** $x = \frac{-b}{a}$

18. $x = \frac{d - c}{m}$ **19.** $x = \frac{2}{c - a}$ **20.** $x = \frac{3}{m + 1}$

21. $x = \frac{P}{mw}$ **22.** $x = \frac{2}{yz}$ **23.** $x = \frac{2}{2w - 1}$

24. $x = \frac{a}{2a - 1}$ **25.** $y = \frac{3}{2}x + 3$ **26.** $y = \frac{4}{3}x + 3$

27. $y = -\frac{1}{3}x + 4$ **28.** $y = \frac{1}{2}x - 8$ **29.** $y = 2x - 20$

30. $y = \frac{2}{3}x + \frac{5}{4}$ **31.** Length 8.5 inches, width 14 inches

32. 10 feet   **33.** Wife \$27,000, Roy \$35,000

34. Duane \$16,000, wife \$17,000   **35.** \$9,500   **36.** \$620

37. 11 nickels, 4 dimes   **38.** 2 quarters, 12 nickels, 5 dimes

39. 15 miles   **40.** 200 mph

41. $(-3, \infty)$

42. $(-\infty, -5)$

43. $(0, \infty)$

44. $(0, \infty)$

45. $(-\infty, -8]$

46. $[-6, \infty)$

47. $\left(-\infty, \frac{11}{2}\right)$

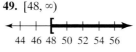

48. $(5, \infty)$

49. $[48, \infty)$

50. $[60, \infty)$

51. $(-\infty, -4) \cup (1, \infty)$

52. $(-\infty, 1) \cup (7, \infty)$

53. $(0, 9)$

54. $(-3, 0]$

55. $(0, \infty)$

56. $(-\infty, 5)$

57. $(-\infty, 4)$

58. $(8, \infty)$

59. $\varnothing$ **60.** $(-\infty, \infty)$ **61.** $(-\infty, \infty)$ **62.** $\varnothing$

63. $\left[-\frac{17}{2}, \frac{13}{2}\right]$

64. $\left(-2, \frac{19}{3}\right)$

65. $[1, \infty)$ **66.** $(-1, \infty)$ **67.** $(3, 6)$ **68.** $[0, 3]$

69. $(-\infty, \infty)$ **70.** $(-\infty, \infty)$ **71.** $[-2, -1]$ **72.** $\varnothing$

73.

74.

75. ∅

76.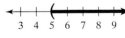
-21 -14 -7 0 7 14 21

77.

-8 -4 0 4 8 12 16

78.

3 4 5 6 7 8 9

79. ∅ **80.** ∅ **81.**
-12 -8 -4 0 4 8 12
  82. ∅

83. R **84.** R **85.**
-1 0 1 2 3 4 5

86.
$\frac{5}{2}$ $\frac{19}{2}$
2 3 4 5 6 7 8 9 10

87. x = rental price, $3 \le x \le \$5$ **88.** $[0, 7.5]$
89. $(40.2, 53.6)$ **90. a)** 43 cm **b)** $F > 47.0$
91. 81 or 91 **92.** Greater than 88 or less than 56
93. \$50,000 accountant, \$60,000 employees **94.** 25%
95. Washington County 1,200, Cade County 2,400
96. At least \$220,000 **97.** $x > 1$ **98.** $x \le 2$
99. $x = 2$ **100.** $3 \le x < 5$ **101.** $|x| = 3$
102. $x = 1$ **103.** $x \le -1$ **104.** $|x| > 2$
105. $|x| \le 2$ **106.** $|x| = 5$ **107.** $x \le 2$ or $x \ge 7$
108. $|x| \le 1$ **109.** $|x| > 3$ **110.** $x > 3$ or $x < -1$
111. $5 < x < 7$ **112.** $|x| > 4$ **113.** $|x| > 0$
114. $-6 \le x < 6$

Chapter 2 Test
1. $\{-4\}$ **2.** R **3.** $\{-6, 6\}$ **4.** $\{2, 5\}$ **5.** $y = \frac{2}{5}x - 4$
6. $y = \frac{5}{1 - 3x}$ **7.** $[4, 8]$

3 4 5 6 7 8 9

8. $(-\infty, -7) \cup (13, \infty)$

-15 -10 -5 0 5 10 15

9. $(5, \infty)$

3 4 5 6 7 8 9

10. $\left(-8, -\frac{1}{2}\right)$

$-\frac{1}{2}$
-8 -7 -6 -5 -4 -3 -2 -1 0

11. $[-5, 3)$

-5 -4 -3 -2 -1 0 1 2 3

12. $(-\infty, 15)$

11 12 13 14 15 16 17

13. $\{2, 5\}$ **14.** $(-\infty, \infty)$ **15.** ∅ **16.** $\{2.5\}$ **17.** ∅
18. R **19.** ∅ **20.** R **21.** $\{100\}$ **22.** 13 meters
23. 14 inches **24.** \$300 **25.** 30 liters
26. $|x - 28,000| > 3,000$, Brenda makes more than \$31,000 or less than \$25,000.

Making Connections Chapters 1–2
1. $11x$ **2.** $30x^2$ **3.** $3x + 1$ **4.** $4x - 3$ **5.** 899
6. 961 **7.** 841 **8.** 25 **9.** 13 **10.** -25 **11.** 5
12. -4 **13.** $-2x + 13$ **14.** 60 **15.** 72 **16.** -9
17. $-3x^3$ **18.** 1 **19.** $\{0\}$ **20.** R **21.** $\{0\}$ **22.** $\{1\}$

23. $\left\{-\frac{1}{3}\right\}$ **24.** $\{1\}$ **25.** R **26.** $\{1000\}$ **27.** $\left\{-\frac{17}{5}, 1\right\}$
28. a) 87,500 **b)** $C_r = 4,500 + 0.06x$, $C_b = 8,000 + 0.02x$
  c) 87,500 **d)** Buying is \$1,300 cheaper.
  e) $(75,000, 100,000)$

CHAPTER 3

Section 3.1 Warm-ups F F F T T T T T F T
1. The origin is the point where the x-axis and y-axis intersect.
2. An ordered pair is a pair of real numbers in which there is a first number and a second one.
3. Intercepts are points where a graph crosses the axes.
4. The graph of an equation of the type $y = k$ where k is a fixed number is a horizontal line.
5. The graph of an equation of the type $x = k$ where k is a fixed number is a vertical line.
6. The variable typically associated with the vertical axis is the dependent variable.
7. $(2, 0), (3, -3)$ **8.** $\left(-1, \frac{3}{2}\right), (4, 4)$
9. $(-4, -33), (22, 6)$ **10.** $\left(3, \frac{1}{3}\right), (1, -1)$
11. I, Graph for 11–23 odd **12.** II, Graph for 12–24 even

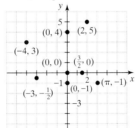

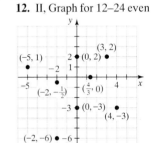

13. III **14.** III **15.** y-axis **16.** y-axis **17.** IV
18. x-axis **19.** II **20.** y-axis **21.** x-axis **22.** I
23. y-axis **24.** IV
25.   **26.**

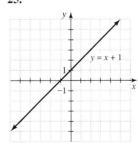   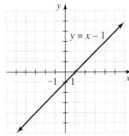

27.   **28.**

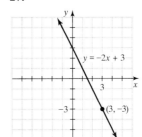   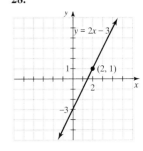

29.

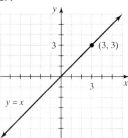

30.

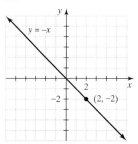

39.

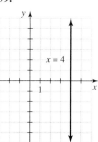

40.

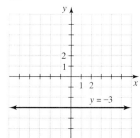

31.

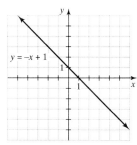

32.

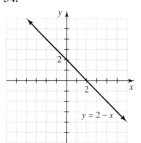

41.

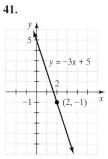

42.

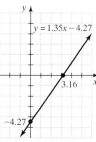

33.

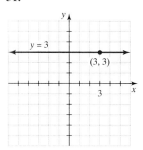

34.

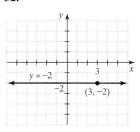

43.

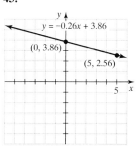

44.

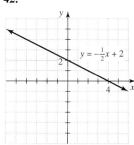

35.

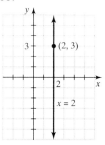

36.

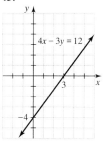

45.

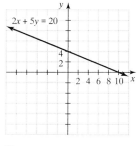

46.

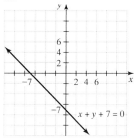

37.

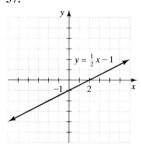

38.

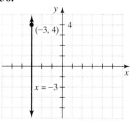

47.

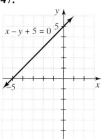

48.

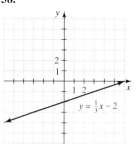

49.

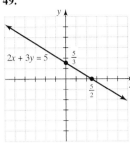

50.

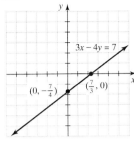

51.

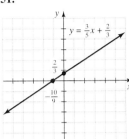

52.

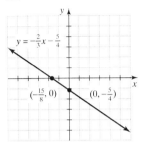

53. a) \$22,353 **b)** \$269

c)

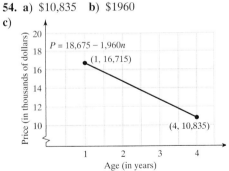

54. a) \$10,835 **b)** \$1960

c)

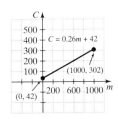

55. \$146

56. 31%

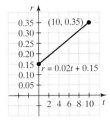

57. a) \$11.45 is the cost of a five-topping pizza.
   b) 11 is the number of toppings on a \$14.45 pizza.
58. a) \$16.95 **b)** 185 **c)** $(0, 4.95)$ **d)** $(-49.5, 0)$
59. a) Her weekly cost, revenue, and profit are \$517.50, \$1,275, and \$757.50.
   b) 1,100. She had a profit of \$995 on selling 1,100 roses.
   c) 995. The difference between revenue and cost is \$995, which is her profit.
60. a) 24 ft/sec, -8 ft/sec, going down
   b) 2.75 second, at maximum height
   c) $(0, 88)$ indicates that at $t = 0$ second the velocity was 88 ft/sec, $(2.75, 0)$ indicates that at $t = 2.75$ second the velocity was 0 ft/sec **d)** -88 ft/sec
61. a) $(3, 4)$ **b)** $(2, -3)$ **c)** $\left(\dfrac{5}{12}, \dfrac{3}{8}\right)$
62. a) $(3, 4), (4, 0), (1, 2)$ **c)** $\left(\dfrac{8}{3}, 2\right)$
63.

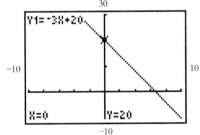

64.

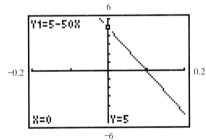

65.

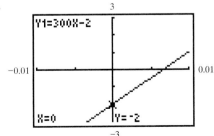

66.

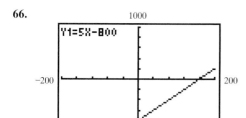

67.

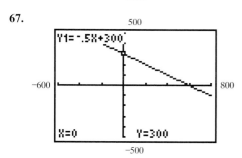

68.

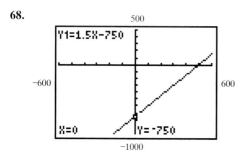

Section 3.2 Warm-ups T F F T F F F T F F

1. Slope measures the steepness of a line.
2. The rise is the change in y-coordinates and run is the change in x-coordinates.
3. A horizontal line has zero slope because it has no rise.
4. Slope is undefined for vertical lines because the run is zero and division by zero is undefined.
5. If m_1 and m_2 are the slopes of perpendicular lines, then $m_1 = -\dfrac{1}{m_2}$.
6. If m_1 and m_2 are the slopes of parallel lines, then $m_1 = m_2$.
7. $\dfrac{2}{3}$ 8. $-\dfrac{2}{3}$ 9. Undefined 10. 0 11. -1 12. 1
13. $\dfrac{3}{2}$ 14. 3 15. -1 16. $-\dfrac{1}{2}$ 17. $-\dfrac{5}{3}$ 18. 2
19. $\dfrac{4}{7}$ 20. 2 21. 5 22. $-\dfrac{11}{14}$ 23. $-\dfrac{5}{3}$ 24. $\dfrac{1}{2}$
25. $-\dfrac{3}{5}$ 26. $-\dfrac{10}{3}$ 27. $-\dfrac{2}{5}$ 28. 6 29. -3 30. $\dfrac{9}{2}$
31. 0 32. 0 33. Undefined 34. Undefined 35. 0.169
36. -2.767 37. -1.273 38. -1.910 39. $\dfrac{8}{3}$ 40. $\dfrac{7}{3}$
41. $\dfrac{3}{7}$ 42. $\dfrac{4}{7}$ 43. $-\dfrac{5}{4}$ 44. $\dfrac{1}{5}$ 45. Yes 46. No

47. No **48.** Yes **49.** No **50.** Yes
51. a) 204 **b)** $22,500 **c)** $22,563
52. a) $12,000 **b)** -1250 **c)** $12,345
53. $(3, 5), (0, -7)$ **54.** $4, 6, 6\dfrac{2}{3}$ **55.** $-\dfrac{5}{2}$ **56.** -3 or 1
57. -4.049 **58.** 1.726
59. A horizontal line has zero slope and a vertical line has undefined slope.
60. Every line goes through at least two quadrants. A nonhorizontal, nonvertical line that misses quadrant II or IV or both has a positive slope. A nonhorizontal, nonvertical line that misses quadrant I or III or both has a negative slope.
61. $-2, \dfrac{1}{2}$, perpendicular **62.** $2, -\dfrac{1}{2}$, perpendicular
63. Increasing m makes the graph increase faster. The slopes of these lines are 1, 2, 3, and 4.
64. Decreasing m makes the graph decrease faster. The slopes of these lines are $-1, -2, -3$, and -4.

Section 3.3 Warm-ups T F F T T F T F F T
1. Point-slope form is $y - y_1 = m(x - x_1)$, where m is the slope and (x_1, y_1) is a point on the line.
2. Slope-intercept form is $y = mx + b$, where m is the slope and $(0, b)$ is the y-intercept.
3. To write an equation of a line, we need the slope and a point on the line.
4. Standard form is $Ax + By = C$, where A, B, and C are real numbers with A and B not both zero.
5. To find the slope from standard form solve the equation for y to get the form $y = mx + b$, where m is the slope.
6. To graph a line knowing the slope and y-intercept, start at the y-intercept and count off the rise and run to locate a second point. Then draw a line through the y-intercept and your second point.
7. $y = 2x - 7$ 8. $y = 6x + 17$ 9. $y = -\dfrac{1}{2}x + 2$
10. $y = \dfrac{2}{3}x + 3$ 11. $y = -\dfrac{3}{7}x + \dfrac{27}{7}$ 12. $y = -x - 1$
13. $y = 4x - 8$ 14. $y = -\dfrac{5}{6}x$ 15. $y = -\dfrac{3}{4}x$
16. $y = -\dfrac{1}{2}x - 3$ 17. $y = \dfrac{1}{2}x + 2$ 18. $y = -\dfrac{2}{3}x - 2$
19. $x = 1$ 20. $y = -2$ 21. $y = -x$ 22. $y = x$
23. $y = \dfrac{3}{2}x - 3$ 24. $y = 3x + 3$ 25. $y = -x + 2$
26. $y = -\dfrac{1}{2}x + 2$ 27. $x - 3y = 6$ 28. $x - 2y = -14$
29. $x - 2y = -13$ 30. $x - 4y = 2$ 31. $2x - 6y = 11$
32. $3x - 12y = 13$ 33. $5x + 6y = 890$
34. $3x - 7y = 200$ 35. $y = -\dfrac{2}{5}x + \dfrac{1}{5}, -\dfrac{2}{5}, \left(0, \dfrac{1}{5}\right)$
36. $y = x - \dfrac{2}{3}, 1, \left(0, -\dfrac{2}{3}\right)$ 37. $y = 3x - 2, 3, (0, -2)$
38. $y = -\dfrac{1}{2}x + \dfrac{5}{2}, -\dfrac{1}{2}, \left(0, \dfrac{5}{2}\right)$ 39. $y = 2, 0, (0, 2)$

40. $y = 9, 0, (0, 9)$ **41.** $y = 3x - 1, 3, (0, -1)$

42. $y = -2x + 6, -2, (0, 6)$ **43.** $y = \frac{3}{2}x + 11, \frac{3}{2}, (0, 11)$

44. $y = -\frac{3}{5}x + \frac{36}{5}, -\frac{3}{5}, \left(0, \frac{36}{5}\right)$ **45.** $y = \frac{1}{3}x + \frac{7}{12}, \frac{1}{3}, \left(0, \frac{7}{12}\right)$

46. $y = -\frac{1}{2}x + \frac{11}{24}, -\frac{1}{2}, \left(0, \frac{11}{24}\right)$

47. $y = 0.01x + 6057, 0.01, (0, 6057)$
48. $y = 0.05x + 4900.5, 0.05, (0, 4900.5)$
49. $x - 2y = -10$ **50.** $10x - 2y = -1$
51. $2x + y = 4$ **52.** $5x + 4y = 20$ **53.** $2x - y = -5$
54. $3x + y = 0$ **55.** $x + 2y = 7$ **56.** $3x - 5y = -14$
57. $2x + y = 3$ **58.** $x - 3y = 5$ **59.** $3x - y = -9$
60. $2x + 3y = 25$ **61.** $y = 5$ **62.** $x = -1$

63. 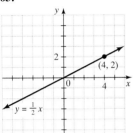   **64.**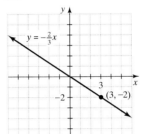

65. 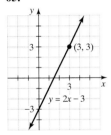   **66.**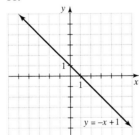

67. 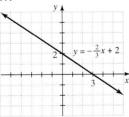   **68.**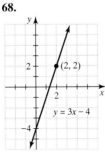

69. 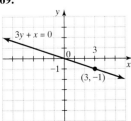   **70.**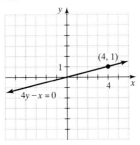

71. 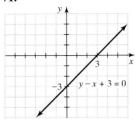   **72.**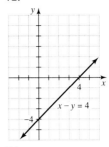

73. 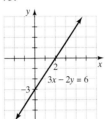   **74.**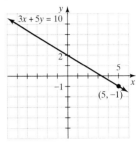

75. 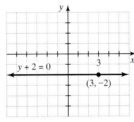   **76.**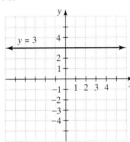

77. 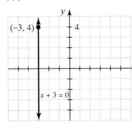   **78.**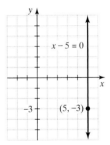

79. Perpendicular   **80.** Perpendicular   **81.** Parallel
82. Parallel   **83.** Perpendicular   **84.** Parallel
85. $t = \frac{7}{6}s + 60, 95°F$ **86.** $C = \frac{1}{2}n + 1000, \$1000.50$
87. a) $y = 0.4x - 774$ **b)** 28 billion tons
88. a) $y = 0.1x - 3.5$ **b)** 7 billion tons
89. a) $w = 304.5d - 1493.5$ **b)** 884.6 ft³/sec   **c)** increasing
90. a) $58.25x + 47.50y = 5,031,250$ **b)** 63,000
   c) (0, 105,921.1), (86,373.4, 0), The intercepts give the number of shares if all money was spent on only one type of stock.
   d) decrease
94. The lines are perpendicular and will appear so in a window in which the length of one unit on the x-axis is equal to the length of one unit on the y-axis.
95. The lines intersect at (50, 97).

Section 3.4 Warm-ups T F T F T T F T F T

1. A linear inequality is an inequality of the form $Ax + By \leq C$ (or using $<$, $>$, or $\geq$), where A, B, and C are real numbers and A and B are not both zero.

2. The solution set to a linear inequality in two variables is usually illustrated with a graph.

3. If the inequality includes equality then the line should be solid.

4. We shade the side on which the inequality is satisfied.

5. The test point method is used to determine which side of the boundary line to shade.

6. To graph a compound inequality, we find either the union or intersection of the regions determined by each simple inequality.

7.

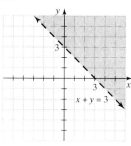

8.

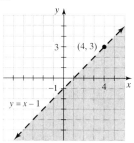

9.

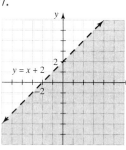

10.

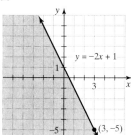

11.

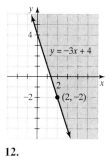

12.

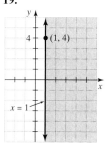

13.

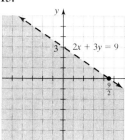

14.

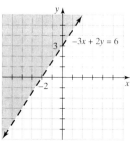

15.

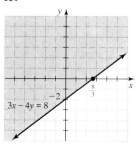

16.

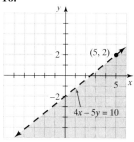

17.

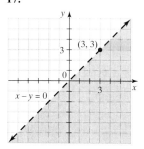

18.

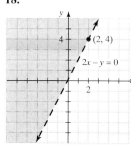

19.

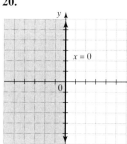

20.

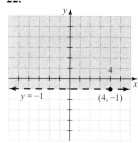

21.

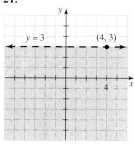

22.

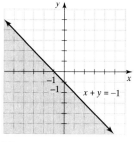

23.

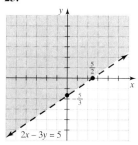

24.

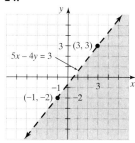

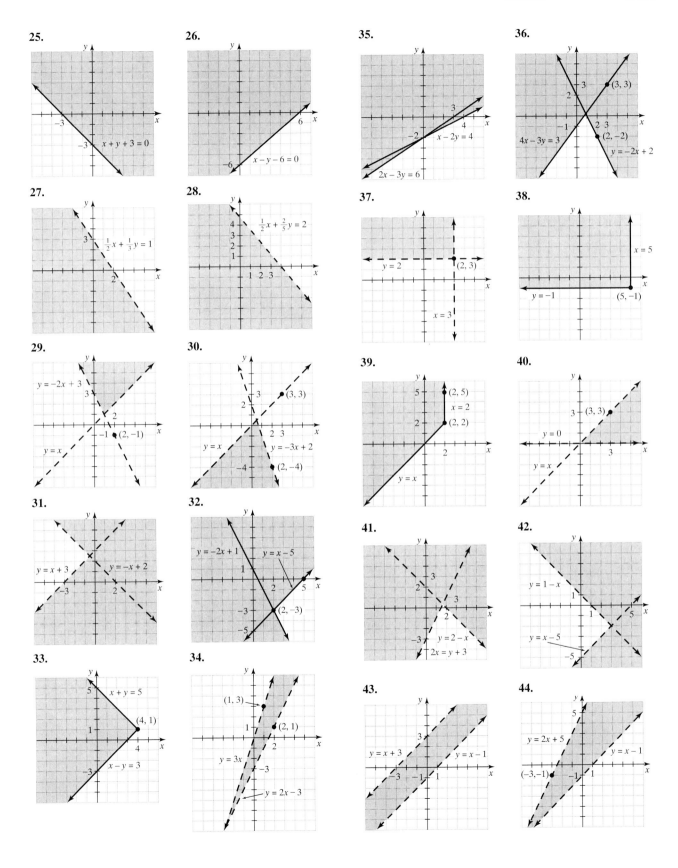

45.

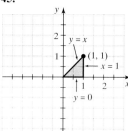

46.

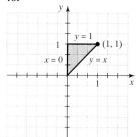

55.

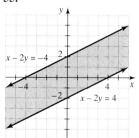

56.

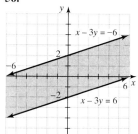

47.

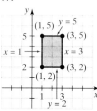

48.

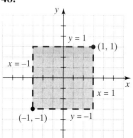

57.

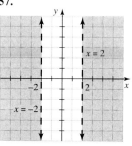

58.

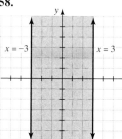

49.

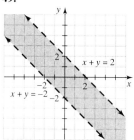

50.

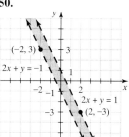

59.

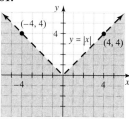

60.

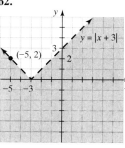

51.

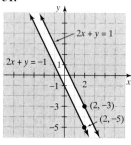

52.

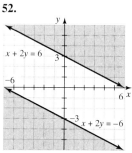

61.

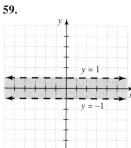

62.

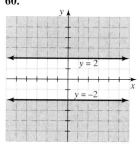

53.

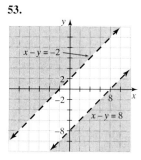

54.

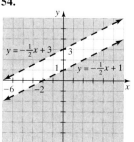

63.

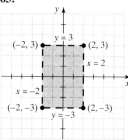

64.

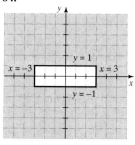

65.

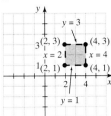

66.

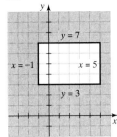

67.

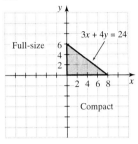

68.

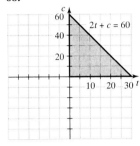

69.

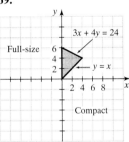

70.

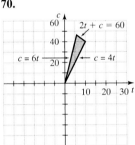

71.

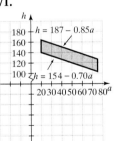

72.

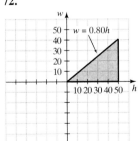

73.

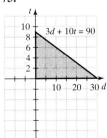

74. a) $3r + t \le 330$ **b)** no **c)** yes

Section 3.5 Warm-ups F T T F F F T T F T
1. A function is a set of ordered pairs in which no two have the same first coordinate and different second coordinates.
2. A relation is any set of ordered pairs.
3. The domain of a relation is the set of all possible first coordinates.
4. The range of relation is the set of all possible second coordinates.
5. The *f*-notation is the notation in which we use $f(x)$ rather than y as the dependent variable.
6. The average rate of change of the function $f(x)$ on the interval $[a, b]$ is $(f(b) - f(a))/(b - a)$.
7. Yes 8. No 9. No 10. Yes 11. No 12. No
13. Yes 14. Yes 15. Yes 16. Yes 17. No 18. No
19. Yes 20. Yes 21. Yes 22. Yes 23. No 24. Yes
25. Yes 26. Yes 27. No 28. No 29. Yes 30. Yes
31. No 32. Yes 33. $\{2\}, \{3, 5, 7\}$ 34. $\{3, 4, 5\}, \{1\}$
35. R, $[0, \infty)$ 36. R, R 37. $[0, \infty), [0, \infty)$
38. $[0, \infty), [1, \infty)$ 39. 10 40. 298 41. 12 42. 20
43. 1 44. 17 45. $\dfrac{7}{3}$ 46. 17 47. -5 or 1
48. 5 or -9 49. $4a - 1$ 50. $4a + 3$ 51. $4x + 7$
52. $4x + 4h - 1$ 53. $\dfrac{1}{x + 5}$ 54. $\dfrac{1}{x}$ 55. $\dfrac{1}{x + h + 2}$
56. $\dfrac{1}{a}$ 57. 4 58. 2 59. 21 60. $\dfrac{1}{5}$ 61. $-\dfrac{1}{8}$
62. $-\dfrac{3}{2}$ 63. 2.337 64. 18.591 65. 66.75
66. 0.894 67. -15 68. $3h$ 69. -10 70. $-5h$
71. 3 72. 3 73. $A = s^2$ 74. $P = 4s$
75. $C = 3.98y$ 76. $P = 14.5h$ 77. $C = 0.50n + 14.95$
78. $C = 10n + 30$ 79. $h = 4t + 70$
80. $C = 0.02p + 0.60$ 81. $-\$1,006.25$ per year
82. $-\$5,313$ per year 83. 7.5 mph per second
84. $2.25 billion per year per month
85. -92.8 ft/sec, -94.4 ft/sec, -95.84 ft/sec
86. -95.984 ft/sec, -95.9984 ft/sec, -96 ft/sec

Section 3.6 Warm-ups T T T F T F T T F T
1. A linear function is a function of the form $f(x) = mx + b$, where m and b are real numbers with $m \ne 0$.
2. A constant function is a function of the form $f(x) = k$, where k is a real number.
3. The graph of a constant function is a horizontal line.
4. The absolute value function has a V-shaped graph.
5. The graph of a quadratic function is a parabola.
6. If there is a vertical line that crosses a graph more than once, the graph is not the graph of a function.
7. R, $\{-2\}$ 8. R, $\{4\}$

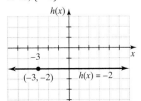

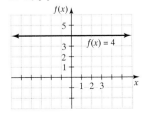

9. R, R

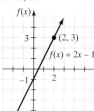

10. R, R

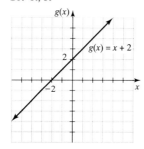

11. R, R

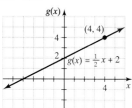

12. R, R

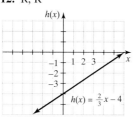

13. R, R

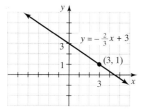

14. R, R

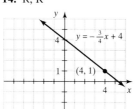

15. R, R

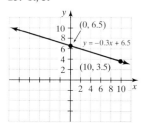

16. R, R

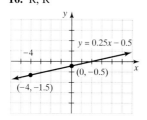

17. $(-\infty, \infty), [1, \infty)$

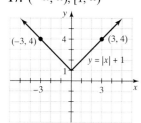

18. $(-\infty, \infty), [-3, \infty)$

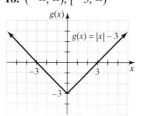

19. $(-\infty, \infty), [0, \infty)$

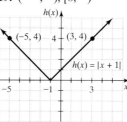

20. $(-\infty, \infty), [0, \infty)$

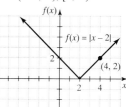

21. $(-\infty, \infty), [0, \infty)$

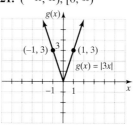

22. $(-\infty, \infty), [0, \infty)$

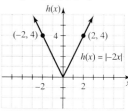

23. $(-\infty, \infty), [0, \infty)$

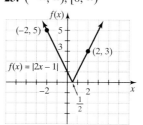

24. $(-\infty, \infty), [0, \infty)$

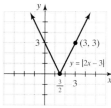

25. $(-\infty, \infty), [1, \infty)$

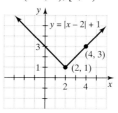

26. $(-\infty, \infty), [2, \infty)$

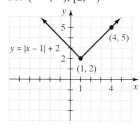

27. $(-\infty, \infty), [2, \infty)$

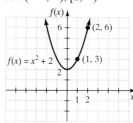

28. $(-\infty, \infty), [-4, \infty)$

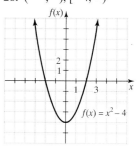

29. $(-\infty, \infty)$, $[0, \infty)$

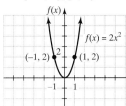

30. $(-\infty, \infty)$, $(-\infty, 0]$

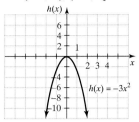

39. $[1, \infty)$, $[0, \infty)$

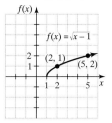

40. $[-1, \infty)$, $[0, \infty)$

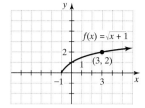

31. $(-\infty, \infty)$, $(-\infty, 6]$

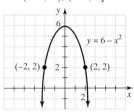

32. $(-\infty, \infty)$, $(-\infty, 3]$

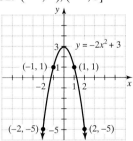

41. $[0, \infty)$, $(-\infty, 0]$

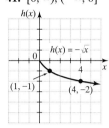

42. $[1, \infty)$, $(-\infty, 0]$

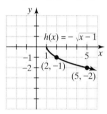

33. $(-\infty, \infty)$, $(-\infty, 2]$

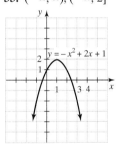

34. $(-\infty, \infty)$, $[-3, \infty)$

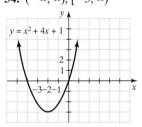

43. $[0, \infty)$, $[2, \infty)$

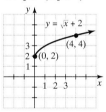

44. $[0, \infty)$, $[1, \infty)$

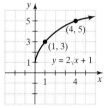

35. $(-\infty, \infty)$, $[-1, \infty)$

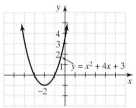

36. $(-\infty, \infty)$, $(-\infty, 1]$

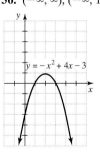

45. $[0, \infty)$, $(-\infty, \infty)$

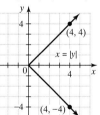

46. $(-\infty, 0]$, $(-\infty, \infty)$

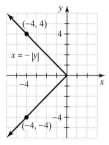

37. $[0, \infty)$, $[0, \infty)$

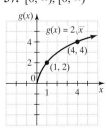

38. $[0, \infty)$, $[-1, \infty)$

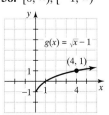

47. $(-\infty, 0]$, $(-\infty, \infty)$

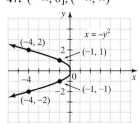

48. $(-\infty, 1]$, $(-\infty, \infty)$

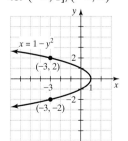

49. $\{5\}$, $(-\infty, \infty)$

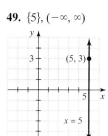

50. $\{-3\}$, $(-\infty, \infty)$

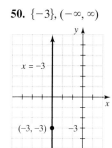

65. $(-\infty, \infty)$, $[-1, \infty)$

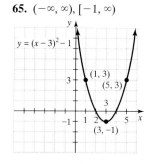

66. $(-\infty, \infty)$, $[-4, \infty)$

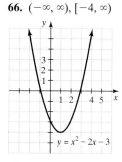

51. $[-9, \infty)$, $(-\infty, \infty)$

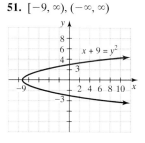

52. $[-3, \infty)$, $(-\infty, \infty)$

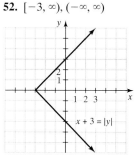

67. $(-\infty, \infty)$, $[1, \infty)$

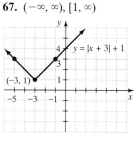

68. $(-\infty, \infty)$, $(-\infty, \infty)$

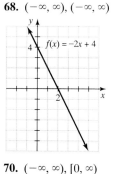

53. $[0, \infty)$, $[0, \infty)$

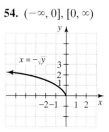

54. $(-\infty, 0]$, $[0, \infty)$

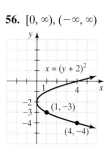

69. $[0, \infty)$, $[-3, \infty)$

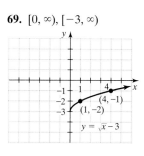

70. $(-\infty, \infty)$, $[0, \infty)$

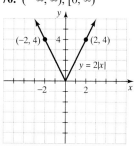

55. $[0, \infty)$, $(-\infty, \infty)$

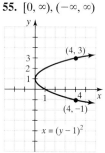

56. $[0, \infty)$, $(-\infty, \infty)$

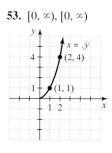

71. $(-\infty, \infty)$, $(-\infty, \infty)$

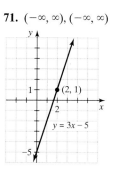

72. $(-\infty, \infty)$, $[0, \infty)$

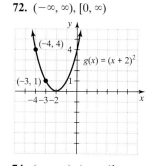

57. No   **58.** No   **59.** Yes
60. Yes   **61.** No   **62.** Yes
63. $(-\infty, \infty)$, $(-\infty, 1]$

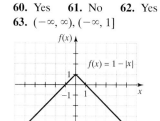

64. $[3, \infty)$, $[0, \infty)$

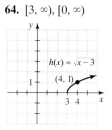

73. $(-\infty, \infty)$, $(-\infty, 0]$

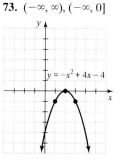

74. $(-\infty, \infty)$, $(-\infty, 4]$

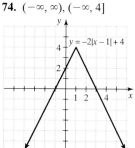

75. 64 feet

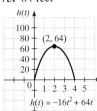

76. 16 feet

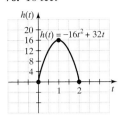

77. a) 2 P.M.   **b)** 170 parts per million
78. b) 150   **c)** $62.50

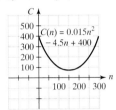

79. a) 0.2134 mg/mm^2 **b)** 1925
80. 12.6 cm, 2,827.4 ft^2
81. The graph of $f(x) = \sqrt{x^2}$ is the same as the graph of $f(x) = |x|$.
82. $(-\infty, 0) \cup (0, \infty)$, $(-\infty, 0) \cup (0, \infty)$
83. For large values of k the graph gets narrower and for smaller values of k the graph gets broader.
84. The graph of $y = x^2 + k$ moves upward for $k > 0$ and downward for $k < 0$.
85. The graph of $y = (x - k)^2$ moves to the right for $k > 0$ and to the left for $k < 0$.
86. The equation $x = y^2$ is equivalent to $y = \sqrt{x}$ or $y = -\sqrt{x}$.
87. The graph of $y = f(x - k)$ lies to the right of the graph of $y = f(x)$ when $k > 0$.

Enriching Word Power
1. d   **2.** a   **3.** a   **4.** b   **5.** c   **6.** a   **7.** b   **8.** b
9. a   **10.** c   **11.** b   **12.** c   **13.** b   **14.** a   **15.** d
16. b   **17.** d

Chapter 3 Review
1. III   **2.** y-axis   **3.** x-axis   **4.** II   **5.** y-axis   **6.** I
7. IV   **8.** IV   **9.** $(0, 2), \left(\frac{2}{3}, 0\right), (4, -10), \left(\frac{5}{3}, -3\right)$
10. $\left(0, \frac{5}{3}\right), \left(\frac{5}{2}, 0\right), \left(-6, \frac{17}{3}\right), (-5, 5)$ **11.** 1   **12.** $-\frac{11}{5}$
13. $\frac{3}{7}$ **14.** $\frac{1}{2}$ **15.** $\frac{3}{8}$ **16.** 0   **17.** $\frac{7}{11}$ **18.** 0
19. $-3, (0, 4)$ **20.** $\frac{3}{2}, \left(0, -\frac{1}{2}\right)$ **21.** $\frac{2}{3}, \left(0, \frac{7}{3}\right)$
22. 0, (0, 8)   **23.** $2x - 3y = 12$ **24.** $5x + 100y = 26$
25. $x - 2y = -5$ **26.** $6x - 4y = 3$ **27.** $x - 2y = 7$
28. $3x - y = -2$ **29.** $3x + 4y = 18$ **30.** $x - 4y = 0$
31. $y = 5$ **32.** $x + y = 0$

33.

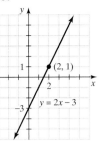

34.

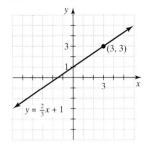

35.

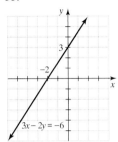

36.

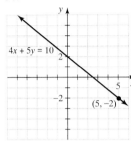

37.

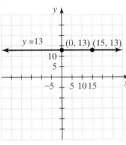

38.

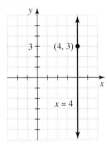

39.

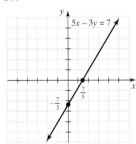

40.

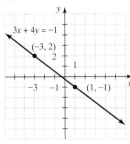

41.

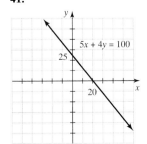

42.

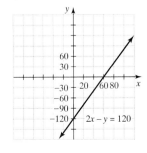

43.

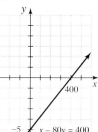

$x - 80y = 400$

44.

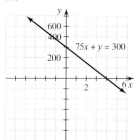

$75x + y = 300$

53.

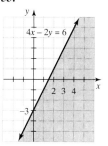

$4x - 2y = 6$

54.

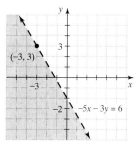

$(-3, 3)$

$-5x - 3y = 6$

45.

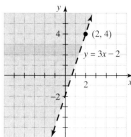

$(2, 4)$

$y = 3x - 2$

46.

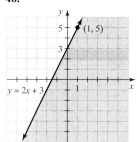

$(1, 5)$

$y = 2x + 3$

55.

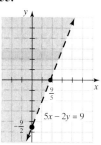

$\frac{9}{5}$

$5x - 2y = 9$

$-\frac{9}{2}$

56.

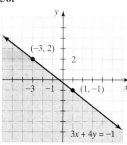

$(-3, 2)$

$(1, -1)$

$3x + 4y = -1$

47.

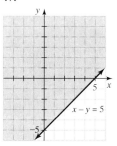

$x - y = 5$

48.

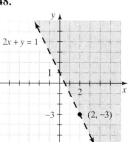

$2x + y = 1$

$(2, -3)$

57.

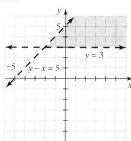

$y = 3$

$y - x = 5$

58.

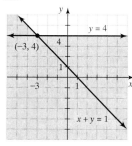

$y = 4$

$(-3, 4)$

$x + y = 1$

49.

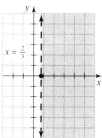

$x = \frac{2}{3}$

50.

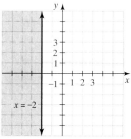

$x = -2$

59.

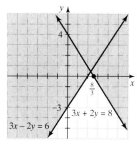

$\frac{8}{3}$

$3x + 2y = 8$

$3x - 2y = 6$

60.

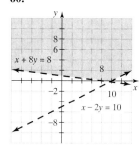

$x + 8y = 8$

$x - 2y = 10$

51.

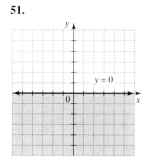

$y = 0$

52.

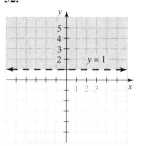

$y = 1$

61.

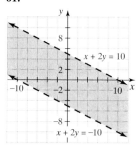

$x + 2y = 10$

$x + 2y = -10$

62.

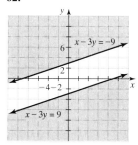

$x - 3y = -9$

$x - 3y = 9$

63.

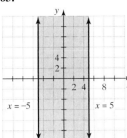

64.

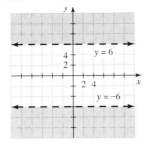

99. $(-\infty, \infty), [0, \infty)$

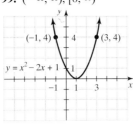

100. $(-\infty, \infty), [-16, \infty)$

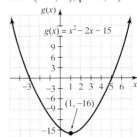

65.

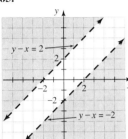

66.

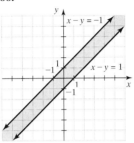
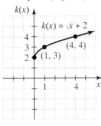

101. $[0, \infty), [2, \infty)$

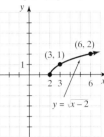

102. $[2, \infty), [0, \infty)$

67. No **68.** Yes **69.** Yes **70.** No **71.** No **72.** Yes
73. $\{1, 2\}, \{3\}$ **74.** $(-\infty, \infty), (-\infty, 0]$
75. $(-\infty, \infty), (-\infty, \infty)$ **76.** $[0, \infty), [0, \infty)$
77. $(-\infty, \infty), [0, \infty)$ **78.** $[0, \infty), [-1, \infty)$ **79.** -5

80. -11 **81.** -6 **82.** -4 **83.** $-\dfrac{21}{4}$ **84.** $-\dfrac{25}{4}$

85. $2a - 5$ **86.** $2x + 1$ **87.** $2a - 7$ **88.** $2x + 2h - 5$

89. 3 **90.** $\dfrac{5}{2}$ **91.** 200 **92.** $\dfrac{1}{10}$ **93.** $-\dfrac{1}{24}$ **94.** 3

95. $(-\infty, \infty), (-\infty, \infty)$ **96.** $(-\infty, \infty), (-\infty, \infty)$

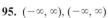
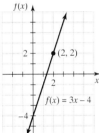
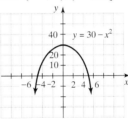

103. $(-\infty, \infty), (-\infty, 30]$

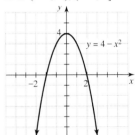

104. $(-\infty, \infty), (-\infty, 4]$

97. $(-\infty, \infty), [-2, \infty)$

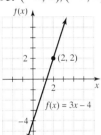

98. $(-\infty, \infty), [0, \infty)$

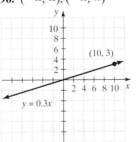
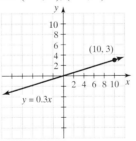

105. $\{2\}, (-\infty, \infty)$

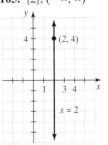

106. $[-1, \infty), (-\infty, \infty)$

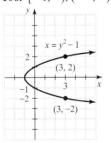

107. $[1, \infty), (-\infty, \infty)$

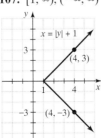

108. $[0, \infty), [1, \infty)$

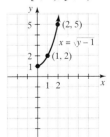

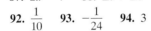
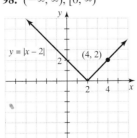
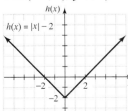

109. $3x - y = 6$ **110.** $x + 2y = 4$ **111.** $x + 2y = 7$
112. $y = -3$ **113.** $x = 2$ **114.** $4x + 7y = 30$
115. $y = 0$ **116.** $x - 3y = 11$ **117.** $2x - y = -6$
118. $x = 2$ **119.** $3x - y = -6$ **120.** $x + y = 0$
121. $y = 5$ **122.** $x = -3$ **126.** Yes
127. a) $h = 220 - a$ **b)** 180 **c)** decreases
128. $h = 3.5d + 71.5$, 106.5
129. 62 days

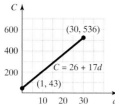

130. a) $w \leq 0.95h$ **b)** No   **c)** 40 inches

Chapter 3 Test

1. $(0, 5), \left(\frac{5}{2}, 0\right), \left(\frac{13}{2}, -8\right)$ **2.** $-\frac{6}{5}$

3. $\frac{8}{5}, (0, 2)$ **5.** $V = -2{,}000a + 22{,}000$

6. $x + 2y = 6$ **7.** $4x + y = -7$
8. $5x + 3y = 19$ **9.** $x - 2y = -4$

10.

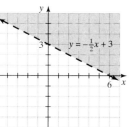

11.

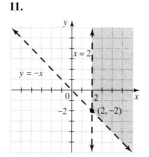

12.

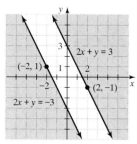

13.

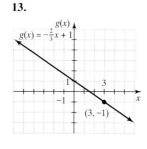

14.

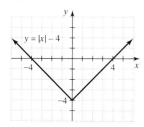

15.

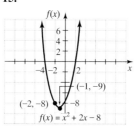

16.

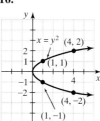

17.

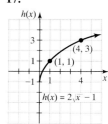

18. Yes   **19.** $11, -2a + 3$ **20.** 15   **21.** $(-\infty, \infty), [0, \infty)$
22. $S = 0.50n + 3$ **23.** 1 second

Making Connections Chapters 1–3

1. 128   **2.** 64   **3.** 49   **4.** -29 **5.** -5 **6.** -5
7. $12t^2$ **8.** $7t$ **9.** $x + 2$ **10.** $7y$ **11.** $7x - 32$
12. $-21x^2 + 8x$ **13.** $\{27\}$ **14.** $\{200\}$ **15.** $\left\{\frac{7}{3}\right\}$
16. $\left\{\frac{4}{3}, \frac{10}{3}\right\}$ **17.** $\varnothing$ **18.** $\left\{0, \frac{14}{3}\right\}$

19.

```
  ←+——+——(——+——+——+——+——+→
   2   3   4   5   6   7   8
```

20.

```
  ←+——+——[——+——+——+——+→
   0   1   2   3   4   5   6
```

21.

```
  ←————————+——+——+——+→
   -1   0   1   2   3   4   5
```

22.

```
        -3
  ←——[——+——+——+——+——(——→
   -4  -2   0   2   4   6   8
```

23.

```
  ←+——(——+——+——+——)——+→
   0   1   2   3   4   5   6
```

24.

```
  ←——[——+——+——+——+——+——+——[——→
   -4 -3 -2 -1  0  1  2  3  4  5
```

25.

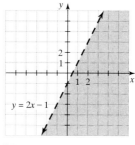

26.

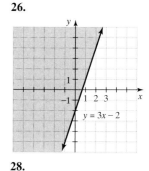

27.

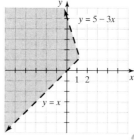

28.

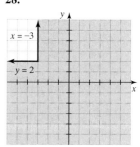

29. a) $b = 500a - 24{,}000$, $b = 667a - 34{,}689$,
$b = 800a - 43{,}600$ **b)** \$8{,}000   **c)** 69
d) The slopes 500, 667, and 800 indicate the additional amount
per year received beyond the basic amount in each category.

CHAPTER 4

Section 4.1 Warm-ups T F F T T T T T T T

1. The intersection point of the graphs is the solution to an independent system.
2. The lines do not intersect if the system has no solution.
3. The graphing method can be very inaccurate.
4. For substitution we eliminate a variable by substituting one equation into the other.
5. If the equation you get after substituting turns out to be incorrect, like $0 = 9$, then the system has no solution.
6. If the substitution results in an identity, then the system is dependent.
7. $\{(1, 2)\}$ 8. $\{(2, -1)\}$ 9. $\{(0, -1)\}$ 10. $\{(-1, -1)\}$
11. $\{(2, -1)\}$ 12. $\{(-1, 3)\}$ 13. $\varnothing$ 14. $\varnothing$
15. $\{(x, y) \mid x + 2y = 8\}$ 16. $\{(x, y) \mid 2x - 3y = 6\}$
17. c   18. d   19. b   20. a   21. $\{(8, 3)\}$, independent
22. $\{(3, 7)\}$, independent   23. $\{(-3, 2)\}$, independent
24. $\{(-2, -5)\}$, independent   25. $\varnothing$, inconsistent
26. $\{(x, y) \mid 2x - y = 3\}$, dependent
27. $\{(x, y) \mid y = 2x - 5\}$, dependent
28. $\left\{\left(\frac{13}{9}, -\frac{1}{9}\right)\right\}$, independent   29. $\{(5, -1)\}$, independent
30. $\{(6, 2)\}$, independent   31. $\{(7, 7)\}$, independent
32. $\{(4, 5)\}$, independent   33. $\{(15, 25)\}$, independent
34. $\{(30, 20)\}$, independent   35. $\{(20, 10)\}$, independent
36. $\{(8, 4)\}$, independent   37. $\left\{\left(\frac{9}{2}, -\frac{1}{2}\right)\right\}$, independent
38. $\{(0, -3)\}$, independent   39. $\varnothing$, inconsistent
40. $\{(x, y) \mid 3x - y = 12\}$, dependent
41. $\{(x, y) \mid 3y - 2x = -3\}$, dependent   42. $\varnothing$, inconsistent
43. $\{(0.75, 1.125)\}$, independent
44. $\{(4.75, 11.375)\}$, independent
45. Width 13 feet, length 28 feet
46. Hsu $35,929, Alkena $48,397
47. $14,000 at 5%, $16,000 at 10%
48. Marysville 13,000, Springfield 12,000   49. -12 and 14
50. -4 and -12 51. 94 toasters, 6 vacation coupons
52. 103 student tickets, 206 adult tickets
53. State tax $3,553, federal tax $28,934
54. State tax $11,066, federal tax $115,574
55. $20,000
56. a) $(35,000, 15,000)$ b) bonus $13,043, taxes $34,783
57. a) $500,000   b) $300,000   c) 20,000   d) $400,000
58. a)

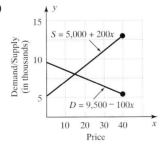

$S = 5,000 + 200x$

$D = 9,500 - 100x$

b) The supply increases.   c) The demand decreases.
d) Equilibrium price is $15.
59. a   60. c   61. a) $(2.8, 2.6)$ b) $(1.0, -0.2)$

Section 4.2 Warm-ups T F T T F T T F T T
1. In this section we learned the addition method.
2. We try to eliminate a variable by adding the equations.
3. In some cases we multiply one or both of the equations on each side to change the coefficients of the variable that we are trying to eliminate.
4. If a false equation, such as $3 = 4$, results from addition of the equations, then the equations are inconsistent.
5. If an identity, such as $0 = 0$, results from addition of the equations, then the equations are dependent.
6. Addition is usually easier to use when the equations are in the same form.
7. $\{(8, -1)\}$ 8. $\{(1, -2)\}$ 9. $\{(5, -7)\}$ 10. $\{(1, 1)\}$
11. $\{(-1, 3)\}$ 12. $\{(3, -4)\}$ 13. $\{(-1, -3)\}$
14. $\{(1, -2)\}$ 15. $\{(-2, -5)\}$ 16. $\{(-3, 8)\}$
17. $\{(22, 26)\}$ 18. $\{(8, 5)\}$ 19. $\varnothing$, inconsistent
20. $\varnothing$, inconsistent   21. $\{(x, y) \mid 5x - y = 1\}$, dependent
22. $\{(x, y) \mid 4x + 3y = 2\}$, dependent
23. $\left\{\left(\frac{5}{2}, 0\right)\right\}$, independent   24. $\{(0, 4)\}$, independent
25. $\{(12, 6)\}$ 26. $\{(4, -6)\}$ 27. $\{(-8, 6)\}$ 28. $\{(2, 3)\}$
29. $\{(16, 12)\}$ 30. $\{(63, 0)\}$ 31. $\{(12, 7)\}$
32. $\{(30, 100)\}$ 33. $\{(400, 800)\}$ 34. $\{(600, 500)\}$
35. $\{(1.5, 1.25)\}$ 36. $\{(1.125, -1.5)\}$ 37. $1.00
38. Books $0.25 each, magazines $0.15 each
39. 1,380 students   40. 462 boys, 924 girls
41. 31 dimes, 4 nickels   42. 17 nickels, 35 pennies
43. a) $(20, 30)$ b) 20 pounds chocolate, 30 pounds peanut butter
44. 20 pounds regular, 40 pounds no-fat   45. 4 hours
46. 147 miles   47. 80%   48. 90%
49. Width 150 meters, length 200 meters
50. Darren 432 feet, Douglas 360 feet

Section 4.3 Warm-ups F F T F F T T F F F
1. A linear equation in three variables is an equation of the form $Ax + By + Cz = D$ where A, B, and C cannot all be zero.
2. An order triple is a collection of three numbers [written as (a, b, c)] in which the order of the numbers is important.
3. A solution to a system of linear equations in three variables is an ordered triple that satisfies all of the equations in the system.
4. We solve systems in three variables by using addition or substitution to eliminate variables.
5. The graph of a linear equation in three variables is a plane in a three-dimensional coordinate system.
6. For an inconsistent system at least two of the planes are parallel.
7. $\{(1, 2, -1)\}$ 8. $\{(2, -1, 3)\}$ 9. $\{(1, 3, 2)\}$
10. $\{(3, 2, 1)\}$ 11. $\{(1, -5, 3)\}$ 12. $\{(1, 2, -3)\}$
13. $\{(-1, 2, -1)\}$ 14. $\{(3, -2, 2)\}$ 15. $\{(-1, -2, 4)\}$
16. $\{(1, -1, -2)\}$ 17. $\{(1, 3, 5)\}$ 18. $\{(2, 4, 6)\}$
19. $\{(3, 4, 5)\}$ 20. $\{(-2, 4, 6)\}$ 21. $\varnothing$
22. $\{(x, y, z) \mid -x + 2y - 3z = -6\}$
23. $\{(x, y, z) \mid 3x - y + z = 5\}$ 24. $\varnothing$ 25. $\varnothing$ 26. $\varnothing$
27. $\{(x, y, z) \mid 5x + 4y - 2z = 150\}$
28. $\{(x, y, z) \mid 3x - 2y + 50z = 300\}$

29. $\{(0.1, 0.3, 2)\}$ **30.** $\{(0.36, -0.12, 0.06)\}$
31. $1,500 stocks, $4,500 bonds, $6,000 mutual fund
32. $16,000 at 5%, $12,000 at 6%, $32,000 at 7%
33. Coffee $0.40, doughnut $0.35, tip $0.55
34. Anna 108 pounds, Bob 118 pounds, Chris 92 pounds
35. $0.95   **36.** Edwin 24, father 51, grandfather 84
37. Soup 14 ounces, tuna 8 ounces, error 2 ounces
38. 137
39. $24,000 teaching, $18,000 painting, $6,000 royalties
40. 15 nickels, 10 dimes, 2 quarters

Section 4.4 Warm-ups T T T F T F F F T F
1. A matrix is a rectangular array of numbers.
2. A row runs horizontally and a column runs vertically.
3. The order of a matrix is the number of rows and columns.
4. An element of a matrix is a number that occupies a position in the matrix.
5. An augmented matrix is a matrix where the entries in the first column are the coefficients of x, the entries in the second column are the coefficients of y, and the entries in the third column are the constants from a system of two linear equations in two unknowns.
6. The goal of Gaussian elimination is to get ones on the diagonal.
7. 2×2 **8.** 2×3 **9.** 3×2 **10.** 3×3

11. 3×1 **12.** 1×3 **13.** $\begin{bmatrix} 2 & -3 & 9 \\ -3 & 1 & -1 \end{bmatrix}$

14. $\begin{bmatrix} 1 & -1 & 4 \\ 2 & 1 & 3 \end{bmatrix}$ **15.** $\begin{bmatrix} 1 & -1 & 1 & 1 \\ 1 & 1 & -2 & 3 \\ 0 & 1 & -3 & 4 \end{bmatrix}$

16. $\begin{bmatrix} 1 & 1 & 0 & 2 \\ 0 & 1 & -3 & 5 \\ -3 & 0 & 2 & 8 \end{bmatrix}$ **17.** $5x + y = -1$
                                  $2x - 3y = 0$
18. $x = 4$
      $y = -3$

19. $x = 6$ **20.** $x + 4z = 3$
      $-x + z = -3$ $2y + z = -1$
      $x + y = 1$ $x + y + z = 1$

21. $R_1 \leftrightarrow R_2$ **22.** $-3R_1 + R_2 \to R_2$ **23.** $\frac{1}{5}R_2 \to R_2$
24. $R_2 + R_1 \to R_1$ **25.** $\{(1, 2)\}$ **26.** $\{(3, 4)\}$
27. $\{(4, 5)\}$ **28.** $\{(1, 1)\}$ **29.** $\{(1, -1)\}$ **30.** $\{(-1, 1)\}$
31. $\{(7, 6)\}$ **32.** $\{(3, 5)\}$ **33.** $\varnothing$ **34.** $\varnothing$
35. $\{(x, y) \mid x + 2y = 1\}$ **36.** $\{(x, y) \mid 2x - 3y = 1\}$
37. $\{(1, 2, 3)\}$ **38.** $\{(3, 2, 1)\}$ **39.** $\{(1, 1, 1)\}$
40. $\{(1, 2, 1)\}$ **41.** $\{(1, 2, 0)\}$ **42.** $\{(2, 1, 1)\}$
43. $\{(1, 0, 1)\}$ **44.** $\{(4, 3, 2)\}$
45. $\{(x, y, z) \mid x - y + z = 1\}$
46. $\{(x, y, z) \mid 2x - y + z = 1\}$ **47.** $\varnothing$ **48.** $\varnothing$

Section 4.5 Warm-ups T F F T T T T T F F
1. A determinant is a real number associated with a square matrix.
2. Cramer's rule can be used to solve systems of linear equations.
3. Cramer's rule works on systems that have exactly one solution.

4. For inconsistent and dependent systems the determinant of the matrix of coefficients is zero.
5. -1 **6.** -1 **7.** -3 **8.** 0   **9.** -14 **10.** 16
11. 0.4   **12.** 16   **13.** $\{(1, -3)\}$ **14.** $\{(-2, 5)\}$
15. $\{(1, 1)\}$ **16.** $\{(-2, -3)\}$ **17.** $\left\{\left(\frac{23}{13}, \frac{9}{13}\right)\right\}$
18. $\{(3, 4)\}$ **19.** $\{(10, 15)\}$ **20.** $\{(20, 12)\}$
21. $\left\{\left(\frac{27}{4}, \frac{13}{2}\right)\right\}$ **22.** $\{(-8, 12)\}$ **23.** $\varnothing$
24. $\{(x, y) \mid -x + 3y = 6\}$ **25.** $\left\{\left(\frac{14}{3}, \frac{2}{3}\right)\right\}$ **26.** $\{(1, -5)\}$
27. $\{(x, y) \mid 4x - y = 6\}$ **28.** $\varnothing$
29. $\{(x, y) \mid y = 3x - 12\}$ **30.** $\{(4, -3)\}$ **31.** $\{(2, -5)\}$
32. $\{(-2, -3)\}$ **33.** $\{(480, 24)\}$ **34.** $\{(76, 22)\}$
35. a) $(9, 11)$ **b)** 9 servings peas, 11 servings beets
36. Cornies 6 servings, Oaties 3 servings   **37.** 37   **38.** 48
39. Milk $2.40, magazine $2.25
40. 50 washing machines, 40 refrigerators
41. 12 singles, 10 doubles
42. 16 box wrenches, 12 open-end wrenches
43. Gary 39, Harry 34   **44.** $29°$ and $61°$
45. Square 10 feet, triangle $\frac{40}{3}$ feet   **46.** $0.75
47. 10 gallons of 10% solution, 20 gallons of 25% solution
48. Emily 62 mph, Camille 56 mph

Section 4.6 Warm-ups T F T F F T T T F F
1. A minor for an element in a 3×3 matrix is the determinant of a 2×2 matrix.
2. A minor for an element is obtained by deleting the row and column of the element and finding the determinant of the 2×2 matrix that remains.
3. The sign array tells what signs to use in the expansion by minors.
4. Cramer's rule solves only those systems that have a unique solution.
5. 11   **6.** -24 **7.** 4   **8.** -18 **9.** 3   **10.** 1   **11.** 1
12. -1 **13.** -7 **14.** -1 **15.** -1 **16.** -5 **17.** 9
18. -26 **19.** 5   **20.** 9   **21.** 22   **22.** 3   **23.** 6
24. -6 **25.** 70   **26.** -28 **27.** 25   **28.** -1
29. $\{(1, 2, 3)\}$ **30.** $\{(1, -2, 3)\}$ **31.** $\{(-1, 1, 2)\}$
32. $\{(2, -1, -2)\}$ **33.** $\{(-3, 2, 1)\}$ **34.** $\{(6, -2, 3)\}$
35. $\left\{\left(\frac{3}{2}, \frac{1}{2}, 2\right)\right\}$ **36.** $\left\{\left(\frac{1}{2}, \frac{5}{2}, -2\right)\right\}$ **37.** $\{(0, 1, -1)\}$
38. $\{(1, 0, -1)\}$ **39.** $\{(x, y, z) \mid 2x - y + z = 1\}$ **40.** $\varnothing$
41. $\varnothing$ **42.** $\{(x, y, z) \mid x - y + z = 5\}$ **43.** $\{(1, 3, -6)\}$
44. $\varnothing$
45. Mimi 36 pounds, Mitzi 32 pounds, Cassandra 107 pounds
46. 20 nickels, 15 dimes, 6 quarters
47. $39°, 51°, 90°$ **48.** $20°, 40°, 120°$

Section 4.7 Warm-ups F F F F F T F T F T
1. A constraint is an inequality that restricts the values of the variables.
2. Linear programming is the process used to maximize or minimize a linear function subject to linear constraints.
3. Constraints may be limitations on the amount of available supplies, money, or other resources.

4. A linear function of two variables is a function of the form $f(x, y) = Ax + By + C$.

5. The maximum or minimum of a linear function subject to linear constraints occurs at a vertex of the region determined by the constraints.

6. Write the constraints, graph the region that they determine, locate each vertex, then evaluate the function at each vertex, and identify the maximum or minimum.

7.

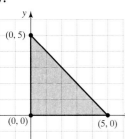

8.

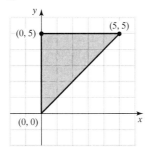

9.

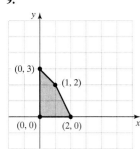

10.

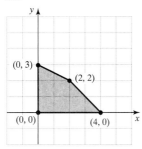

11.

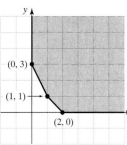

12.

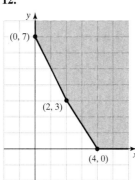

13.

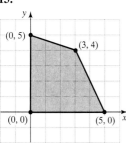

14.

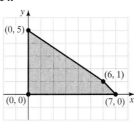

15.

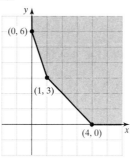

16.

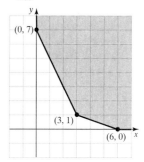

17. **a)** 0, 320,000, 510,000, 450,000
   b) 30 TV ads and 60 radio ads
18. **a)** 0, 160,000, 390,000, 450,000
   b) 50 TV ads and 0 radio ads
19. 6 doubles, 4 triples   **20.** 16 chairs, 0 swings
21. 0 doubles, 8 triples   **22.** 8 chairs, 12 swings
23. 1.75 cups Doggie Dinner, 5.5 cups Puppie Power
24. 3 supervisors, 3 helpers
25. 10 cups Doggie Dinner, 0 cups Puppie Power
26. 0 supervisors, 6 helpers
27. Laundromat $8,000, carwash $16,000
28. 0 economy models, 60 deluxe models

Enriching Word Power

1. c   **2.** a   **3.** a   **4.** d   **5.** b   **6.** c   **7.** a   **8.** c
9. d   **10.** b   **11.** a   **12.** d

Chapter 4 Review

1. $\{(1, 1)\}$, independent   **2.** $\varnothing$, inconsistent
3. $\{(1, -1)\}$, independent   **4.** $\{(2, 3)\}$, independent
5. $\{(-3, 2)\}$, independent   **6.** $\{(-3, -6)\}$, independent
7. $\varnothing$, inconsistent   **8.** $\{(x, y) \mid 2x - y = 3\}$, dependent
9. $\{(-1, 5)\}$, independent   **10.** $\{(-2, -3)\}$, independent
11. $\{(x, y) \mid 3x - 2y = 12\}$, dependent   **12.** $\varnothing$, inconsistent
13. $\{(1, -3, 2)\}$ **14.** $\{(-2, 1, 5)\}$ **15.** $\varnothing$
16. $\{(x, y, z) \mid x - y + z = 1\}$ **17.** $\{(2, -4)\}$ **18.** $\{(2, -4)\}$
19. $\{(1, 1, 2)\}$ **20.** $\{(2, 1, 3)\}$ **21.** 2   **22.** 1   **23.** -0.2
24. $\frac{1}{60}$ **25.** $\{(-1, -2)\}$ **26.** $\{(2, -4)\}$ **27.** $\{(2, 1)\}$
28. $\{(x, y) \mid 2x - 5y = -1\}$ **29.** $\varnothing$ **30.** $\{(4, 3)\}$
31. 58   **32.** -10 **33.** -30 **34.** -7 **35.** $\{(1, 2, -3)\}$
36. $\varnothing$ **37.** $\{(-1, 2, 4)\}$ **38.** $\{(x, y, z) \mid x - y - z = 1\}$
39. (0, 0), (0, 3), (4, 1), (5, 0)   **40.** (0, 6), (2, 3), (8, 0)

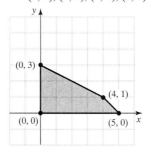

41. 30   **42.** 36   **43.** 78   **44.** 53   **45.** 36 minutes
46. 16 years old to drive, 12 years old at present
47. 4 liters of 30% solution A, 8 liters of 20% solution B, 8 liters of 60% solution C
48. a) (13,000, 7,000)   **b)** $13,200 in Dell, $6,800 in bonds
49. Three servings of each

Chapter 4 Test

1. $\{(1, 3)\}$ **2.** $\left\{\left(\frac{5}{2}, -3\right)\right\}$ **3.** $\{(x, y) \mid y = x - 5\}$

4. $\{(-1, 3)\}$ **5.** $\varnothing$ **6.** Inconsistent   **7.** Dependent
8. Independent   **9.** $\{(1, -2, -3)\}$ **10.** $\{(2, 5)\}$
11. $\{(3, 1, 1)\}$ **12.** -18 **13.** -2 **14.** $\{(-1, 2)\}$
15. $\{(2, -2, 1)\}$ **16.** Singles $18, doubles $25
17. Jill 17 hours, Karen 14 hours, Betsy 62 hours   **18.** 44

Making Connections Chapters 1–4

1. -81 **2.** 7   **3.** 73   **4.** 5.94   **5.** $-t - 3$
6. $-0.9x + 0.9$ **7.** $3x^2 + 2x - 1$ **8.** y **9.** $y = \frac{3}{5}x - \frac{7}{5}$
10. $y = \frac{C}{D}x - \frac{W}{D}$ **11.** $y = \frac{K}{W - C}$ **12.** $y = \frac{bw - 2A}{b}$
13. $\{(4, -1)\}$ **14.** $\{(500, 700)\}$ **15.** $\{(x, y) \mid x + 17 = 5y\}$
16. $\varnothing$ **17.** $y = \frac{5}{9}x + 55$ **18.** $y = -\frac{11}{6}x + \frac{2}{3}$
19. $y = 5x + 26$ **20.** $y = \frac{1}{2}x + 5$ **21.** $y = 5$
22. $x = -7$
23. a) Machine A
   b) Machine B $0.04 per copy, machine A $0.03 per copy
   c) The slopes 0.04 and 0.03 are the per copy cost for each machine
   d) B: $y = 0.04x + 2000$, A: $y = 0.03x + 4000$
   e) 200,000

CHAPTER 5

Section 5.1   Warm-ups T F F T T F T T T F

1. An exponential expression is an expression of the form a^n.
2. The expression $a^{-n} = 1/a^n$ for any positive integer n and $a \neq 0$.
3. The product rule says that $a^m a^n = a^{m+n}$.
4. The quotient rule says that $a^m/a^n = a^{m-n}$.
5. To convert a number in scientific notation to standard notation, move the decimal point n places to the left if the exponent on 10 is $-n$ or move the decimal point n places to the right if the exponent on 10 is n, assuming that n is a positive integer.
6. To convert from standard notation to scientific notation, count the number of decimal places, n, required to move the decimal point so that there is one nonzero digit to the left of the decimal point. Use 10^n if you moved the decimal point to the left and use 10^{-n} if you moved the decimal point to the right.

7. 16   **8.** 27   **9.** $\frac{1}{16}$ **10.** $\frac{1}{27}$ **11.** 16   **12.** $\frac{5}{3}$
13. $\frac{16}{9}$ **14.** 10,000   **15.** -18 **16.** -2 **17.** 0.003
18. 50,000   **19.** 16   **20.** 16   **21.** 2^{17} **22.** $6x^5$

23. $\frac{2}{y^2}$ **24.** 1   **25.** $\frac{30}{x^7}$ **26.** $\frac{4}{a^{17}}$ **27.** -2 **28.** 1
29. 1   **30.** 1   **31.** 2   **32.** $\frac{5}{y}$ **33.** 32   **34.** 5000
35. $\frac{100}{3}$ **36.** $\frac{6}{y^2}$ **37.** $\frac{8y^2}{5x^2}$ **38.** $\frac{x^3}{75y^3}$ **39.** $\frac{1}{48x^2}$
40. $\frac{24}{y^5}$ **41.** x^2 **42.** a^5 **43.** 3^8 **44.** 6^7 **45.** $\frac{1}{3a^3}$
46. $\frac{a}{7}$ **47.** $\frac{-3}{w^8}$ **48.** $\frac{-5}{x^8}$ **49.** $3^8 w^6$ **50.** $\frac{w^9}{2^8}$ **51.** $\frac{xy}{2}$
52. $\frac{1}{5r^8}$ **53.** 9   **54.** 16   **55.** -14 **56.** -162
57. $\frac{1}{16}$ **58.** $\frac{1}{27}$ **59.** $\frac{4}{9}$ **60.** 1   **61.** $-15a^4$ **62.** $\frac{10}{x^3}$
63. $8a^3$ **64.** $\frac{10x^2}{y^5}$ **65.** a **66.** $3b$ **67.** $\frac{-2x^2y^4}{3}$
68. $-x^2y^5$ **69.** $\frac{5}{3}$ **70.** 7   **71.** $\frac{6}{25}$ **72.** $-\frac{2}{3}$ **73.** 3
74. 3   **75.** -2 **76.** -3 **77.** -4 **78.** -4 **79.** -3
80. 4   **81.** 486,000,000   **82.** 380   **83.** 0.00000237
84. 0.00162   **85.** 4,000,000   **86.** 496,000   **87.** 0.000005
88. 0.048   **89.** 3.2×10^5 **90.** 4.3298×10^7
91. 7.1×10^{-7} **92.** 8.94×10^{-6} **93.** 7×10^{-5}
94. 8.2951×10^6 **95.** 2.35×10^7 **96.** 4.3×10^{-10}
97. 7.5×10^{-1} **98.** 4×10^{-3} **99.** 3×10^{22}
100. 2×10^7 **101.** -1.2×10^{13} **102.** 1×10^{-16}
103. 1.578×10^5 **104.** 2.036×10^{12} **105.** 9.187×10^{-5}
106. 9.652×10^6 **107.** 5.961×10^{157} **108.** 3.645×10^{140}
109. 3.828×10^{30} **110.** 7.048   **111.** 4.910×10^{11} feet
112. 8.3 minutes   **113.** 3.83×10^7 seconds
114. 7.480×10^{-9} meters   **115.** 2.6 lb/person/day
116. a) 4.4 lb/person/day in 1998
   b) 5.0 lb/person/day in 2010

Section 5.2   Warm-ups F T T F F F T T T T

1. The power of a power rule says that $(a^m)^n = a^{mn}$.
2. The power of a product rule says that $(ab)^m = a^m b^m$.
3. The power of a quotient rule says that $(a/b)^m = a^m/b^m$.
4. Principal is the amount of money invested initially.
5. To compute the amount A when interest is compounded annually, use $A = P(1 + i)^n$ where P is the principal, i is the annual interest rate, and n is the number of years.
6. To compute the present value P for the amount A in n years at annual interest rate i, use $P = A(1 + i)^{-n}$.

7. 64   **8.** 81   **9.** y^{10} **10.** x^{12} **11.** $\frac{1}{x^8}$ **12.** $\frac{1}{x^{14}}$
13. m^{18} **14.** a^9 **15.** 1   **16.** $\frac{1}{m^5}$ **17.** $\frac{1}{x^2}$ **18.** a^2
19. $81y^2$ **20.** $-8a^3$ **21.** $25w^6$ **22.** $\frac{-8}{w^{15}}$ **23.** $\frac{x^9}{y^6}$
24. $\frac{a^4}{b^6}$ **25.** $\frac{b^2}{9a^2}$ **26.** $\frac{x^3}{8y^6}$ **27.** $\frac{6x^3}{y}$ **28.** $15a^2b$
29. $\frac{1}{8a^3b^4}$ **30.** $\frac{1}{81x^4y^6}$ **31.** $\frac{w^3}{8}$ **32.** $\frac{m^2}{25}$ **33.** $-\frac{27a^3}{64}$
34. $\frac{16}{81b^4}$ **35.** $\frac{x^2y^2}{4}$ **36.** $\frac{27}{8a^6b^3}$ **37.** $\frac{y^2}{9x^6}$ **38.** $-\frac{x^3}{8y^6}$

39. $\dfrac{25}{4}$ **40.** $\dfrac{16}{9}$ **41.** 4 **42.** $\dfrac{9}{4}$ **43.** $-\dfrac{27}{8x^3}$ **44.** $-\dfrac{c}{ab}$

45. $\dfrac{27y^3}{8x^6}$ **46.** a^2b^8 **47.** 5^{6t} **48.** 3 **49.** 2^{6w^2}

50. 6^{2x} **51.** 7^{m+3} **52.** 4^p **53.** 8^{5a+11} **54.** 5^{8-7y}

55. $6x^9$ **56.** $9x^8$ **57.** $-8x^6$ **58.** $\dfrac{6}{x^2}$ **59.** $\dfrac{3z}{x^2y}$

60. $\dfrac{x^2y^2}{2}$ **61.** $-\dfrac{3}{2}$ **62.** -5 **63.** $\dfrac{4x^6}{9}$ **64.** $-\dfrac{8y^{12}}{x^3}$

65. $-\dfrac{x^2}{2}$ **66.** $-\dfrac{27}{x^6}$ **67.** $\dfrac{y^3}{8x^3}$ **68.** $\dfrac{3y}{2x^2}$ **69.** $\dfrac{b^{14}}{5a^7}$

70. $\dfrac{16m^7}{n^{17}}$ **71.** $\dfrac{x^6}{16y^8}$ **72.** $\dfrac{y^2}{27x^{10}}$ **73.** $3ac^8$ **74.** $\dfrac{1}{7xy^3z^{11}}$

75. 2^{11} **76.** 2^{60} **77.** 2^{-4} **78.** 2^6 **79.** 2^{6n} **80.** 2^{n-5}

81. 2^{4m^2} **82.** 2^{-16n} **83.** 25 **84.** 6.25 **85.** 0.75

86. 2 **87.** 850.559 **88.** 18.700 **89.** 1.533

90. 505.080 **91.** $56,197.12

92. a) $23,673.64 **b)** $33,273.20

93. $2,958.64 **94.** $275,928.73

95. a) 75.1 years **b)** 81.4 years

96. a) 80.1 years **b)** 7.9 years **c)** At 80 both males and females can expect about 5 more years.

99. d **100. a)** True **b)** False

101. b) (6.116, 20,000) **c)** 6.116 years

102. b) (87.54, 86)
c) At 87.54 years of age you can expect to live until 86. The model fails here.

Section 5.3 Warm-ups F F F T F T F T T F

1. A term of a polynomial is a single number or the product of a number and one or more variables raised to whole number powers.

2. The number preceding the variable in each term is the coefficient of that variable.

3. A constant is simply a number.

4. A polynomial is a single term or a finite sum of terms.

5. The degree of a polynomial in one variable is the highest power of the variable in the polynomial.

6. To multiply a binomial and a trinomial, we use the distributive property.

7. Yes **8.** Yes **9.** No **10.** No **11.** Yes **12.** Yes

13. No **14.** No **15.** $4, -8$, binomial

16. $3, -1$, binomial **17.** 0, 0, monomial

18. 0, 0, monomial **19.** 7, 0, monomial

20. 4, 0, monomial **21.** 6, 1, trinomial

22. $3, \dfrac{1}{2}$, trinomial **23.** 80 **24.** 0 **25.** -29 **26.** 0

27. 0 **28.** 0 **29.** $3a + 2$ **30.** $3w - 1$

31. $5xy + 25$ **32.** $2ab + 1$ **33.** $2x - 9$

34. $3y^2 + y - 9$ **35.** $2x^3 - x^2 - 2x - 8$

36. $-x^2 + 5x - 7$ **37.** $11x^2 - 2x - 9$

38. $-x^2 - 8x + 9$ **39.** $x + 5$ **40.** $2x + 9$

41. $-6x^2 + 5x + 2$ **42.** $-4x^2 + 10x + 4$

43. $2x$ **44.** $w + 1$ **45.** $-15x^6$ **46.** $2a^3b^6$

47. $-3x + 2$ **48.** $x^2 - 3x + 9$ **49.** $15x^4y^4 - 20x^3y^3$

50. $24y^6z^3 - 9y^5z^2 + 6y^5z$ **51.** $x^2 - 4$ **52.** $x^2 - 1$

53. $2x^3 - x^2 + x - 6$ **54.** $x^3 - 7x^2 + 14x - 8$

55. $-10x^2 + 15x$ **56.** $-6a^4 + 10a^3 - 14a$

57. $x^2 + 10x + 25$ **58.** $a^2 - b^2$ **59.** $2x^2 + 9x - 18$

60. $6x^4 - 11x^2 - 10$ **61.** $x^3 - y^3$ **62.** $a^3 + b^3$

63. $2x - 5$ **64.** $x^3 + 6x - 8$ **65.** $4a^2 - 11a - 4$

66. $2w^2 - w - 1$ **67.** $4w^2 - 8w - 7$

68. $a^3 + 2a^2 - 2a - 1$ **69.** $x^3 - 8$ **70.** $a^3 - 27$

71. $xz - wz + 2xw - 2w^2$ **72.** $w^2t^2 - at^2 + 3w^2 - 3a$

73. $6x^2y^2 + 7xy - 5$ **74.** $3a^2b^2 + 20ab - 32$

75. $14.4375x^2 - 29.592x - 9.72$

76. $x^3 - 34.88x^2 + 30.888x - 5.936$

77. $15.369x^2 + 1.88x - 0.866$

78. $-19.51x^3 - 3.7x^2 + 1.94x + 15.4$ **79.** $\dfrac{3}{4}x + \dfrac{3}{2}$

80. $\dfrac{2}{3}x - \dfrac{1}{2}$ **81.** $-\dfrac{1}{2}x^2 + x$ **82.** $x^2 - \dfrac{4}{3}x - \dfrac{5}{6}$

83. $3x^2 - 21x + 1$ **84.** $-3x^3 + 18x^2 - 13x - 17$

85. $-4x^2 - 62x - 168$ **86.** $x^4 - 25x^2 + 20x - 4$

87. $x^4 - m^2 - 4m - 4$ **88.** $9x^4 + 11x^2 + 4$

89. $-6x^3 - 53x^2 - 8x$ **90.** $-3x^2 + 16x + 120$

91. $-4a^{2m} - 4a^m + 5$ **92.** $-3b^{3z} + b^{2z} + 1$

93. $x^{2n} + 2x^n - 3$ **94.** $8y^{2t} + 2y^t - 21$

95. $5z^{3w} - z^{1+w}$ **96.** $w^{3p} - 1$ **97.** $x^{6r} + y^3$

98. $4x^{2a} - z^2$ **99.** $75 **100.** $150 **101.** $49.35, $15.61

102. $37.30, $247.90, There is a greater difference in height between adjacent bars as the number of openers increases.

103. a) 6.2 years **b)** no **c)** 2037

104. a) 1969 **b)** $0.1726y - 268.445$

Section 5.4 Warm-ups T T F T F T T T F F

1. The distributive property is used in multiplying binomials.

2. FOIL stands for first, outer, inner, last.

3. The purpose of FOIL is to provide a fast way to find the product of two binomials.

4. The square of a sum is the square of the first term plus twice the product of the two terms plus the square of the last term.

5. The square of a difference is the square of the first term minus twice the product of the two terms plus the square of the last term.

6. The product of a sum and a difference of the same two terms is the square of the first term minus the square of the second term.

7. $x^2 + 2x - 8$ **8.** $x^2 + 2x - 15$ **9.** $2x^2 + 7x + 3$

10. $2y^2 + 7y + 6$ **11.** $2a^2 - 7a - 15$

12. $3x^2 - 13x - 30$ **13.** $4x^4 - 49$ **14.** $9y^6 - 64$

15. $2x^6 + 7x^3 - 4$ **16.** $6t^4 + t^2 - 12$

17. $w^2 + 5wz - 6z^2$ **18.** $w^2 + 2wy - 8y^2$

19. $9k^2 + 6kt - 8t^2$ **20.** $7a^2 + 5ax - 2x^2$

21. $xy - 3y + xw - 3w$ **22.** $yz - y + 2z - 2$

23. $m^2 + 6m + 9$ **24.** $a^2 + 4a + 4$ **25.** $a^2 - 8a + 16$

26. $b^2 - 6b + 9$ **27.** $4w^2 + 4w + 1$

28. $9m^2 + 24m + 16$ **29.** $9t^2 - 30tu + 25u^2$

30. $9w^2 - 12wx + 4x^2$ **31.** $x^2 + 2x + 1$

32. $d^2 + 10d + 25$ **33.** $a^2 - 6ay^3 + 9y^6$

34. $9m^2 - 30mn^3 + 25n^6$ **35.** $\dfrac{1}{4}x^2 - x + 1$

36. $\dfrac{4}{9}x^2 - \dfrac{2}{3}x + \dfrac{1}{4}$ **37.** $w^2 - 81$ **38.** $m^2 - 16$

39. $w^6 - y^2$ **40.** $a^6 - x^2$ **41.** $4x^2 - 49$ **42.** $25x^2 - 9$

43. $9x^4 - 4$ **44.** $16y^4 - 1$ **45.** $m^2 + 2mt + t^2 - 25$
46. $4x^2 + 12x + 9 - y^2$ **47.** $y^2 - r^2 - 10r - 25$
48. $x^2 - 9 + 6k - k^2$ **49.** $4y^2 - 4yt + t^2 + 12y - 6t + 9$
50. $u^2 - 6uv + 9v^2 - 8u + 24v + 16$
51. $9h^2 + 6hk - 6h + k^2 - 2k + 1$
52. $4p^2 - 12pq - 24p + 9q^2 + 36q + 36$
53. $x^2 + 3x - 54$ **54.** $6x^4 - x^2 - 12$ **55.** $25 - x^2$
56. $16 - a^2b^2$ **57.** $6x^2 + 7ax - 20a^2$ **58.** $x^{10} - 4$
59. $2t^2 + 2tw - 3t - 3w$ **60.** $5ax^2 - 9ax + 5bx - 9b$
61. $9x^4 + 12x^2y^3 + 4y^6$ **62.** $25a^8 - 20a^4b + 4b^2$
63. $6y^2 - 4y - 10$ **64.** $6b^2 + 3b - 9$ **65.** 2496
66. $9,999$ **67.** $49x^2 + 42x + 9$ **68.** $1 - 2pq + p^2q^2$
69. $36y^3 + 12y^2 + y$ **70.** $100y^3 - 20y^2 + y$
71. $2ah + h^2$ **72.** $2x + h$ **73.** $x^3 + 6x^2 + 12x + 8$
74. $a^4 + 4a^3 + 6a^2 + 4a + 1$ **75.** $y^3 - 9y^2 + 27y - 27$
76. $16b^4 + 32b^3 + 24b^2 + 8b + 1$
77. $16.32x^2 - 10.47x - 17.55$
78. $28.09x^2 - 97.52x + 84.64$
79. $12.96y^2 + 31.68y + 19.36$ **80.** $10.89a^2 - 62.41b^2$
81. $x^{3m} + 2x^{2m} + 3x^m + 6$ **82.** $a^{2n} - b^2$
83. $a^{3n+1} + a^{2n+1} - 3a^{n+1}$ **84.** $1 + 3x^{2b} + 5x^{3b}$
85. $a^{2m} + 2a^{m+n} + a^{2n}$ **86.** $x^{2w} - 2x^{w+t} + x^{2t}$
87. $15y^{3m} + 24y^{2m}z^k + 20y^m z^{3-k} + 32z^3$
88. $4x^{3a-4} + 3y^{b+5}x^{2a-3} - 8x^{a-1}y^{4-b} - 6y^9$
89. $x^2 + 4x + 3$ **90.** $x^2 + 14x - 120$ square feet
91. $4x^2 - 36x + 80, 66.24$ km^2
92. $x^3 + 12x^2 + 48x + 64$ in.^{3}
93. $4x^3 - 20x^2 + 24x, 5.9$ ft^3 **94.** $4.5x + 7$ dollars

Section 5.5 Warm-ups F F T T T F F T T T
1. If $a \div b = c$, then the divisor is b, the dividend is a, and the quotient is c.
2. For long division polynomials should be written with the exponents in descending order.
3. If the term x^n is missing in the dividend, insert the term $0 \cdot x^n$ for the missing term.
4. Stop the long division process when the remainder has a smaller degree than the divisor.
5. Synthetic division is used only for dividing by a binomial of the form $x - c$.
6. The remainder is zero if and only if the dividend is a factor of the divisor.
7. $12x^4$ **8.** $6x^2$ **9.** -2 **10.** $-2a$ **11.** $2b - 3$
12. $4x^2 - 3x$ **13.** $x + 2$ **14.** $x^2 - 2x + 4$
15. $-5x^2 + 4x - 3$ **16.** $3x^2 - 2x + 4$ **17.** $\frac{7}{2}x^2 - 2x$
18. $\frac{3}{2}x - \frac{5}{4}$ **19.** $x + 5, -2$ **20.** $x + 2, 1$ **21.** $x - 4, 8$
22. $3, 3$ **23.** $x^2 - 2x + 4, 0$ **24.** $y^2 + y + 1, 0$
25. $a^2 + 2a + 8, 11$ **26.** $w^2 + 3w + 6, 9$
27. $x^2 - 2x + 3, -6$ **28.** $a^2 - 3a + 7, -18$
29. $x^3 + 3x^2 + 6x + 11, 21$ **30.** $3x^3 - 6x^2 + 11x - 19, 44$
31. $-3x^2 - 1, x - 4$ **32.** $x^2 + x - 3, -3x + 7$
33. $\frac{1}{2}x - \frac{5}{4}, -\frac{7}{4}$ **34.** $\frac{1}{3}x - \frac{7}{3}, 15$ **35.** $\frac{2}{3}x + \frac{1}{9}, \frac{56}{9}$
36. $\frac{3}{2}x + \frac{5}{4}, -\frac{9}{4}$ **37.** $2 + \frac{10}{x - 5}$ **38.** $1 + \frac{1}{x - 1}$

39. $x - 1 + \frac{1}{x + 1}$ **40.** $x - 3 + \frac{18}{x + 3}$
41. $x^2 - 2x + 4 + \frac{-8}{x + 2}$ **42.** $x^2 + 2x + 4 + \frac{7}{x - 2}$
43. $x + \frac{2}{x}$ **44.** $x + \frac{3}{2x}$ **45.** $x - 6 + \frac{21}{x + 2}$
46. $x - 2 + \frac{-16}{x - 3}$ **47.** $3x^2 - x - 1 + \frac{6}{x - 1}$
48. $-2x^2 + 5x - 10 + \frac{17}{x + 2}$ **49.** $x^2 - 3x, -3$
50. $x^2 + 9x + 24, 67$ **51.** $2x - 6, 11$ **52.** $3x - 13, 30$
53. $3x^3 + 9x^2 + 12x + 43, 120$
54. $-2x^3 - 4x^2 - 5x - 10, -25$
55. $x^4 + x^3 + x^2 + x + 1, 0$
56. $x^5 - x^4 + x^3 - x^2 + x - 1, 0$
57. $x^2 - 2x - 1, 8$ **58.** $x^2 + 4x + 13, 45$
59. $2.3x + 0.596, 0.79072$ **60.** $1.6x - 6.38, 16.184$
61. No   **62.** No   **63.** Yes   **64.** No   **65.** Yes
66. No   **67.** Yes   **68.** Yes   **69.** Yes   **70.** Yes
71. Yes   **72.** Yes   **73.** $(x - 4)(x - 2)$
74. $(x + 8)(x - 5)$ **75.** $(w - 3)(w^2 + 3w + 9)$
76. $(w + 5)(w^2 - 5w + 25)$ **77.** $(x^2 - 2x + 2)(x - 2)$
78. $(2x^2 - 2x + 7)(x + 1)$ **79.** $(z + 3)(z + 3)$
80. $(2a - 5)(2a - 5)$ **81.** $(2y + 1)(3y + 1)$
82. $(4y - 3)(3y + 2)$ **83. a)** $AC(x) = 0.03x + 300$
b) no   **c)** Because $AC(x)$ is very close to 300 for x less than 15, the graph looks horizontal.
84. $AP(x) = 100 + 2x, \$124$ **85.** $x - 1$ feet
86. $h + 3$ **87.** $6,333.3$ cubic meters
88. $V = \frac{H(a^2 + ab + b^2)}{3}$

Section 5.6 Warm-ups T T T T F F F T F F
1. A prime number is a natural number greater than 1 that has no factors other than itself and 1.
2. A natural number is factored completely when it is expressed as a product of prime numbers.
3. The greatest common factor for the terms of a polynomial is a monomial that includes every number or variable that is a factor of all terms of the polynomial.
4. The greatest common factor can be factored out with a positive coefficient or a negative coefficient.
5. A linear polynomial is a polynomial of the form $ax + b$ with $a \neq 0$.
6. A quadratic polynomial is a polynomial of the form $ax^2 + bx + c$ with $a \neq 0$.
7. A prime polynomial is a polynomial that cannot be factored. Any monomial is also called a prime polynomial.
8. A polynomial is factored completely when it is expressed as a product of prime polynomials.
9. 12   **10.** $14a$ **11.** 3   **12.** $14x$ **13.** $6xy$ **14.** $20a^2b^2$
15. $x(x^2 - 5)$ **16.** $10(x^2 - 2y^3)$ **17.** $12w(4x + 3y)$
18. $14w(3z + 2a)$ **19.** $2x(x^2 - 2x + 3)$
20. $6x(x^2 - 2x + 3)$ **21.** $12a^3b^2(3b^4 - 2a + 5a^2b)$
22. $22x^6y^6z(2x^2 - 5y^3z)$ **23.** $(x - 6)(a + b)$
24. $(y - 4)(3 + b)$ **25.** $(y - 1)^2(y + z)$
26. $(w - 2)^2(w + 3)$ **27.** $2(x - y), -2(-x + y)$

28. $3(-x + 2)$, $-3(x - 2)$ **29.** $3x(2x - 1)$, $-3x(-2x + 1)$
30. $5x(2x + 1)$, $-5x(-2x - 1)$
31. $w^2(-w + 3)$, $-w^2(w - 3)$
32. $2w^3(-w + 3)$, $-2w^3(w - 3)$
33. $a(-a^2 + a - 7)$, $-a(a^2 - a + 7)$
34. $2a^2(-a^2 - 2a + 3)$, $-2a^2(a^2 + 2a - 3)$
35. $(x - 10)(x + 10)$ **36.** $(9 - y)(9 + y)$
37. $(2y - 7)(2y + 7)$ **38.** $(4b - 1)(4b + 1)$
39. $(3x - 5a)(3x + 5a)$ **40.** $(11a - b)(11a + b)$
41. $(12wz - 1)(12wz + 1)$ **42.** $(xy - 3c)(xy + 3c)$
43. $(x - 10)^2$ **44.** $(y + 5)^2$ **45.** $(2m - 1)^2$
46. $(3t + 5)^2$ **47.** $(w - t)^2$ **48.** $(2r + 5t)^2$
49. $(a - 1)(a^2 + a + 1)$ **50.** $(w + 1)(w^2 - w + 1)$
51. $(w + 3)(w^2 - 3w + 9)$ **52.** $(x - 4)(x^2 + 4x + 16)$
53. $(2x - 1)(4x^2 + 2x + 1)$ **54.** $(3x + 1)(9x^2 - 3x + 1)$
55. $(a + 2)(a^2 - 2a + 4)$ **56.** $(m - 2)(m^2 + 2m + 4)$
57. $2(x + 2)(x - 2)$ **58.** $3x(x - 3)(x + 3)$
59. $x(x + 5)^2$ **60.** $5a^2m(a - 3)(a + 3)$ **61.** $(2x + 1)^2$
62. $a(x - 4)^2$ **63.** $(x + 3)(x + 7)$ **64.** $(x - 2)(x - 5)$
65. $3y(2y + 1)$ **66.** $y(4y - 1)$ **67.** $(2x - 5)^2$
68. $ax(ax - 3)^2$ **69.** $2m(m - n)(m^2 + mn + n^2)$
70. $y^2(5x^3 - y^3)$ **71.** $(2x - 3)(x - 2)$
72. $(2x + 1)(x + 3)$ **73.** $a(3a + w)(3a - w)$
74. $2b(n - b)^2$ **75.** $-5(a - 3)^2$ **76.** $-2(x - 5)(x + 5)$
77. $2(2 - 3x)(4 + 6x + 9x^2)$
78. $x^2y(3 - 4y)(9 + 12y + 16y^2)$ **79.** $-3y(y + 3)^2$
80. $-2n(m + 2)^2$ **81.** $-7(ab + 1)(ab - 1)$
82. $-17a(a + 1)$ **83.** $(x^5 - 3)(x^5 + 3)$
84. $(y^4 - 2)(y^4 + 2)$ **85.** $(z^6 - 3)^2$ **86.** $(a^3 + 5)^2$
87. $2x(x^3 + 2)^2$ **88.** $x(x^6 - 3)^2$ **89.** $x(2x^2 + 1)^2$
90. $2(3x^3 + 2)^2$ **91.** $(x^2 - 2)(x^4 + 2x^2 + 4)$
92. $(y^2 - 3)(y^4 + 3y^2 + 9)$ **93.** $2(x^3 + 2)(x^6 - 2x^3 + 4)$
94. $x(x^4 + 1)(x^8 - x^4 + 1)$ **95.** $(a^n - 1)(a^n + 1)$
96. $(b^{2n} - 3)(b^{2n} + 3)$ **97.** $(a^r + 3)^2$ **98.** $(u^{3n} - 2)^2$
99. $(x^m - 2)(x^{2m} + 2x^m + 4)$ **100.** $(y^n + 1)(y^{2n} - y^n + 1)$
101. $(a^m - b)(a^{2m} + a^m b + b^2)$
102. $(r^m + 2t)(r^{2m} - 2r^m t + 4t^2)$ **103.** $k(k^w - 5)^2$
104. $a(2a^t + 1)^2$ **105.** $uv^{2k}(v^{2k} - u)^2$
106. $u^m v^n(u^m + v^n)^2$ **107.** 9   **108.** 16   **109.** 20
110. 42   **111.** 16   **112.** 25   **113.** 100   **114.** 25
115. a) $b - 3$ **b)** 4,050 cubic inches (in.^{3})   **c)** 30 inches
116. $3y + 6$ centimeters

Section 5.7 Warm-ups T F F F T F F F T T

1. To factor $x^2 + bx + c$, find two integers whose sum is b and whose product is c.
2. To factor $ax^2 + bx + c$, find two integers whose product is ac and whose sum is b and then use grouping.
3. Trial and error means simply to write down possible factors and then to use FOIL to check until you get the correct factors.
4. When factoring a polynomial, first factor out the greatest common factor.
5. $(x + 1)(x + 3)$ **6.** $(y + 2)(y + 3)$ **7.** $(a + 10)(a + 5)$
8. $(t + 8)(t + 3)$ **9.** $(y - 7)(y + 2)$ **10.** $(x - 6)(x + 3)$
11. $(x - 2)(x - 4)$ **12.** $(y - 10)(y - 3)$
13. $(a - 9)(a - 3)$ **14.** $(x - 6)(x + 5)$

15. $(a + 10)(a - 3)$ **16.** $(w + 30)(w - 1)$
17. $(3w + 1)(2w + 1)$ **18.** $(4x + 3)(x + 2)$
19. $(2x + 1)(x - 3)$ **20.** $(2a - 1)(a + 2)$
21. $(2x + 5)(2x + 3)$ **22.** $(2y + 3)(3y + 4)$
23. $(2x - 1)(3x - 1)$ **24.** $(3m + 4)(2m - 3)$
25. $(3y + 1)(4y - 1)$ **26.** $(4x - 1)(3x + 2)$
27. $(6a - 5)(a + 1)$ **28.** $(10b + 3)(3b - 1)$
29. $(2x - 1)(x + 8)$ **30.** $(3a + 2)(a + 6)$
31. $(3b + 5)(b - 7)$ **32.** $(2y - 3)(y - 7)$
33. $(3w - 4)(2w - 9)$ **34.** $(3x - 2)(5x + 3)$
35. $(4x - 1)(x - 1)$ **36.** $(4x + 3)(x + 1)$
37. $(5m - 2)(m + 3)$ **38.** $(5t + 1)(t - 2)$
39. $(3y + 4)(2y - 5)$ **40.** $(7u - 3)(u + 2)$
41. $(x^3 + 5)(x^3 - 7)$ **42.** $(x^2 + 10)(x^2 - 3)$
43. $(a^{10} - 10)^2$ **44.** $(b^8 + 11)^2$
45. $-2a(3a^2 + 1)(2a^2 + 1)$ **46.** $-b(2b^3 - 3)(2b^3 + 1)$
47. $(x^a + 5)(x^a - 3)$ **48.** $(y^b + 5)(y^b - 4)$
49. $(x^a - y^b)(x^a + y^b)$ **50.** $(w^{2m} - a)(w^{2m} + a)$
51. $(x^4 - 3)(x^4 + 2)$ **52.** $(m^5 - 6)(m^5 + 1)$
53. $x^a(x - 1)(x + 1)$ **54.** $y(y^a - 1)(y^a + 1)$
55. $(x^a + 3)^2$ **56.** $(x^a - y^b)^2$ **57.** $y(4yz^3 - 3)(yz^3 + 2)$
58. $u^2(2u^2v^3 - 3)(u^2v^3 + 4)$ **59.** $2(x + 5)^2$ **60.** $3(a + 1)^2$
61. $a(a - 6)(a + 6)$ **62.** $x(x - 6)(x + 1)$
63. $5(a + 6)(a - 1)$ **64.** $2(a - 7)(a + 6)$
65. $2(x + 8y)(x - 8y)$ **66.** $a(a - 3)^2$
67. $-3(x + 3)(x - 4)$ **68.** $x(y - 10)(y + 7)$
69. $m^3(m + 10)^2$ **70.** $4(a - 2)^2$ **71.** $(3x + 4)(2x + 5)$
72. $(2y - 1)(y - 6)$ **73.** $(y - 6)^2$ **74.** $(m - 1)^2$
75. $(3m - 5n)(3m + 5n)$ **76.** $n^2(m - n)^2$
77. $5(a + 6)(a - 2)$ **78.** $-3(y - 5)(y + 2)$
79. $-2(w - 10)(w + 1)$ **80.** $z(x + y)^2$
81. $x^2(w + 10)(w - 10)$ **82.** $(3x + 5)^2$
83. $(3x + 1)(3x - 1)$ **84.** $(3w + 4)(2w - 9)$
85. $(4x + 5)(2x - 3)$ **86.** $(2w + 3)^2$ **87.** $(2x - 5)^2$
88. $(3m - 11)(3m + 11)$ **89.** $(3a + 10)^2$
90. $10(w + 4)(w - 2)$ **91.** $4(a + 2)(a + 4)$
92. $(5x - 2)(4x - 3)$ **93.** $3m(m - 2)(m^2 + 2m + 4)$
94. $6z(w + 1)(w^2 - w + 1)$ **95.** a and b   **96.** b

Section 5.8 Warm-ups F T T F F T F T F T T F

1. Always factor out the greatest common factor first.
2. In factoring a binomial, look for a difference of two squares and a sum or difference of two cubes.
3. In factoring a trinomial, look for the perfect square trinomials.
4. In factoring a four-term polynomial, try to factor by grouping.
5. Prime   **6.** Not prime   **7.** Not prime   **8.** Prime
9. Not prime   **10.** Prime   **11.** Prime   **12.** Not prime
13. Prime   **14.** Prime   **15.** Prime   **16.** Prime
17. $(a^2 - 5)^2$ **18.** $(3y^2 + 2)^2$ **19.** $(x^2 - 2)(x - 2)(x + 2)$
20. $(x^3 + 3)(x - 1)(x^2 + x + 1)$ **21.** $3(x - 2)(3x - 4)$
22. $(2x - 1)(2x + 3)$
23. $2(y - 2)(y^2 + 2y + 4)(y + 2)(y^2 - 2y + 4)$
24. $6(1 - y)(1 + y + y^2)(1 + y)(1 - y + y^2)$
25. $2(4a^2 + 3)(4a^2 - 3)$ **26.** $2(a - 2)(a + 2)(a^2 + 4)$
27. $(x + 3)(x - 2)(x^2 - x + 6)$

28. $(y^2 - 2y - 1)(y + 1)^2$ **29.** $(m + 5)(m + 1)$
30. $4w(w - 4)$ **31.** $(3y - 7)(y + 4)$
32. $(2w + 3)(w + 5)$ **33.** $(y - 3)(y + 3)(y - 1)(y + 1)$
34. $(m - 4)(m + 4)(m - 2)(m + 2)$ **35.** $(a + b)(x + y)$
36. $(7 + k)(x + z)$ **37.** $(x - 3)(x + 3)(x + 1)$
38. $(x - 5)(x + 5)(x + 1)$ **39.** $(w - 3)(a - b)$
40. $(w + 2)(x - y)$ **41.** $(a + 3)^2(a^2 - 3a + 9)$
42. $(c - 5)(a - b)$ **43.** $(y - 5)(y + 2)(y^2 - 2y + 4)$
44. $(x - 3)(x^2 + a)$ **45.** $(d - w)(ay + 1)$
46. $(y + a)(x + b)$ **47.** $(a + y)(x - 1)(x + 1)$
48. $(c + d)(a - b)(a + b)$ **49.** $(y + b)(y + 1)(y^2 - y + 1)$
50. $(b + w^2)(a + m)$ **51.** $(3x - 4)^2$
52. $-3(x - 8)(x + 2)$ **53.** $(3x - 1)(4x - 3)$ **54.** Prime
55. $3a(a + 3)(a^2 - 3a + 9)$ **56.** $-a(a - 5)(a + 5)$
57. $2(x^2 + 16)$ **58.** $x(x + 2)^2$ **59.** Prime
60. $(x - 1)(x^2 + x + 1)(x + 2)$
61. $(x + y - 1)(x + y + 1)$ **62.** $x(x^2 + 9)$
63. $ab(a - b)(a + b)$ **64.** $2(m - 5)(m^2 + 5m + 25)$
65. $(x + 2)(x - 2)(x^2 + 2x + 4)$ **66.** $(x + 3)(x + 7)$
67. $n(m + n)^2$ **68.** $b(a - 3)^2$ **69.** $(2 + w)(m + n)$
70. $(w - 5)(a + b)$ **71.** $(2w + 3)(2w - 1)$
72. Prime   **73.** $(t^2 + 7)(t^2 - 3)$ **74.** $(m^2 + 1)(m^2 + 4)$
75. $-a(a + 10)(a - 3)$ **76.** $y^2(2y - 5)(y + 4)$
77. $(y + 2)(y + 6)$ **78.** $2(2t + 1)(t + 2)$
79. $-2(w - 5)(w + 5)(w^2 + 25)$
80. $5a(a - 1)(a + 1)(a^2 + 1)$ **81.** $8a(a^2 + 1)$
82. $ax(w + 1)$ **83.** $(w + 2)(w + 8)$ **84.** $(a - 7)(a - 5)$
85. $a(2w - 3)^2$ **86.** $an(3n - 2)(3n + 7)$ **87.** $(x - 3)^2$
88. $x(x + 6)^2$ **89.** $3x^2(x - 5)(x + 5)$
90. $3(x^2 + 3x + 4)$ **91.** $n(m - 1)(m^2 + m + 1)$
92. $m^2(m^2 + 16)$ **93.** $2(3x + 5)(2x - 3)$
94. $3(5x - 4)(6x + 5)$ **95.** $2(a^3 - 16)$
96. $(2x - 3)(6x - 5)$ **97.** $(a^m - 1)(a^{2m} + a^m + 1)$
98. $(x^{2a} + 2)(x^{4a} - 2x^{2a} + 4)$
99. $(a^w - b^{2n})(a^{2w} + a^w b^{2n} + b^{4n})$
100. $(x^n - 3)(x^n + 3)$ **101.** $(t^n - 2)(t^n + 2)(t^{2n} + 4)$
102. $a^2(a^n + 1)(a^{2n} - a^n + 1)$ **103.** $a(a^n - 5)(a^n + 3)$
104. $x^m(x^m + 3)(x^m - 2)$ **105.** $(a^n - 3)(a^n + b)$
106. $(z + x)(x^m + 5)$

Section 5.9 Warm-ups F T T F T F T T F F
1. The zero factor property says that if $a \cdot b = 0$ then either $a = 0$ or $b = 0$.
2. A quadratic equation is an equation of the form $ax^2 + bx + c = 0$ with $a \neq 0$.
3. The hypotenuse of a right triangle is the side opposite the right angle.
4. The legs of a right triangle are the sides that form the right angle.
5. The Pythagorean theorem says that a triangle is a right triangle if and only if the sum of the squares of the legs is equal to the square of the hypotenuse.
6. The diagonal of a rectangle is the line segment that joins two opposite vertices.

7. $\{-4, 5\}$ **8.** $\{-5, 6\}$ **9.** $\left\{\frac{5}{2}, -\frac{4}{3}\right\}$ **10.** $\left\{-\frac{8}{3}, \frac{3}{4}\right\}$
11. $\{-7, 2\}$ **12.** $\{-3, 9\}$ **13.** $\{0, 7\}$ **14.** $\{0, 5\}$

15. $\{-4, 5\}$ **16.** $\{-6, 7\}$ **17.** $\{-3, 4\}$ **18.** $\{-6, -2\}$
19. $\left\{-4, \frac{5}{2}\right\}$ **20.** $\left\{-3, -\frac{2}{3}\right\}$ **21.** $\{-2, 0, 2\}$ **22.** $\{-4, 0, 4\}$
23. $\{-5, -4, 5\}$ **24.** $\{-4, -2, 4\}$ **25.** $\{-1, 1, 2\}$
26. $\{-5, 1, 5\}$ **27.** $\{-3, -1, 1, 3\}$ **28.** $\{-5, -3, 3, 5\}$
29. $\{-8, -6, 4, 6\}$ **30.** $\{-7, -3, 1, 5\}$ **31.** $\{-4, -2, 0\}$
32. $\{-8, -4, 0\}$ **33.** $\{-7, -3, 1\}$ **34.** $\{-5, -3, 4, 6\}$
35. $\left\{-\frac{3}{2}, 2\right\}$ **36.** $\left\{-5, \frac{1}{3}\right\}$ **37.** $\{-6, -3, -2, 1\}$
38. $\{-8, -4, -2, 2\}$ **39.** $\{-6, 1\}$ **40.** $\{1\}$
41. $\{-5, 2\}$ **42.** $\{-7, 2\}$ **43.** $\{-5, -4, 0\}$
44. $\{-1, 0, 3\}$ **45.** $\{-3, 0, 3\}$ **46.** $\{-5, 0, 5\}$
47. $\left\{-3, \frac{1}{2}, 3\right\}$ **48.** $\{-3, 1, 9\}$ **49.** $\left\{\frac{3}{2}\right\}$ **50.** $\left\{-\frac{1}{4}\right\}$
51. $\{0, -b\}$ **52.** $\{-a, -b\}$ **53.** $\left\{-\frac{b}{a}, \frac{b}{a}\right\}$ **54.** $\left\{-\frac{a}{3}\right\}$
55. $\left\{-\frac{b}{2}\right\}$ **56.** $\{-b, b\}$ **57.** $\left\{-\frac{3}{a}, 1\right\}$ **58.** $\left\{-\frac{b}{a}\right\}$
59. Width 4 inches, length 6 inches
60. Width 9 yards, length 13 yards   **61.** 4 and 9
62. 2 and 4.5   **63.** Length 43 inches, width 22 inches
64. 6 feet by 8 feet
65. a) 4 seconds   **b)** 4 seconds   **c)** 64 feet   **d)** 2 seconds
66. 10 seconds   **67.** Width 5 feet, length 12 feet
68. 6 feet and 8 feet   **69.** Width 5 feet, length 12 feet
70. Length 8 inches, width 6 inches
71. 3 and 4, or -4 and -3 **72.** Length 15 feet, width 9 feet
73. Length 20 feet, width 12 feet   **74.** 8

Enriching Word Power
1. d   **2.** b   **3.** c   **4.** d   **5.** b   **6.** a   **7.** c   **8.** a
9. d   **10.** a   **11.** d   **12.** c   **13.** a   **14.** c   **15.** b
16. c   **17.** a   **18.** c

Chapter 5 Review
1. 2   **2.** 1   **3.** 36   **4.** 225   **5.** $-\frac{1}{27}$ **6.** $\frac{1}{4}$ **7.** 1
8. 3^{11} **9.** $\frac{8}{x^3}$ **10.** $-\frac{12}{a^7}$ **11.** $\frac{1}{y^2}$ **12.** w^6 **13.** a^7
14. m^{11} **15.** $\frac{2}{x^4}$ **16.** $-\frac{y^4}{x^{10}}$ **17.** 8,360,000
18. 34,000,000   **19.** 0.00057   **20.** 0.004
21. 8.07×10^6 **22.** 9×10^4 **23.** 7.09×10^{-4}
24. 5×10^{-7} **25.** 1×10^2 **26.** 6×10^{33} **27.** $\frac{1}{a}$
28. $\frac{x^8}{81y^4}$ **29.** $\frac{n^2}{m^{16}}$ **30.** $\frac{w^5 y}{x^7}$ **31.** $\frac{81}{16}$ **32.** $\frac{9}{a^8}$
33. $\frac{25}{36}$ **34.** 36   **35.** $-\frac{4}{3ab}$ **36.** $-\frac{5}{4x^5 y^3}$ **37.** $\frac{b^{14}}{a^7}$
38. $72x^{13}$ **39.** 5^{6w-1} **40.** 3^{7y} **41.** 7^{15a-40} **42.** 2^{12+6k}
43. $8w + 2$ **44.** $3xy - 4a$ **45.** $-6x + 3$
46. $-2x^2 + 3x + 1$ **47.** $x^3 - 4x^2 + 8x - 8$
48. $x^3 + 3x^2 + 50$ **49.** $-4xy + 22z$ **50.** $-4x + 19$
51. $5m^5 - m^3 + 2m^2$ **52.** $a^3 + 6a^2 + 12a + 8$
53. $x^2 + 4x - 21$ **54.** $k^2 - k - 20$ **55.** $z^2 - 25y^2$
56. $m^2 - 9$ **57.** $m^2 + 16m + 64$ **58.** $b^2 + 4ab + 4a^2$
59. $w^2 - 10xw + 24x^2$ **60.** $2w^2 + 9w - 18$
61. $k^2 - 6k + 9$ **62.** $n^2 - 10n + 25$ **63.** $m^4 - 25$

64. $6k^4 + 8k^2t - 30t^2$ **65.** $x^2 + 3x - 5, 0$
66. $2x^2 - x + 3, 0$ **67.** $m^3 - m^2 + m - 1, 0$
68. $x^3 + x^2 + x + 1, 0$ **69.** $a^6 + 2a^3 + 4, 0$
70. $a + b, 0$ **71.** $m^2 + 2m - 6, 0$ **72.** $-1, 0$
73. $x + 1 + \dfrac{-4}{x - 1}$ **74.** $x + \dfrac{2}{x + 3}$ **75.** $3 + \dfrac{6}{x - 2}$
76. $4 + \dfrac{20}{x - 5}$ **77.** Yes   **78.** Yes   **79.** Yes   **80.** No
81. No   **82.** No   **83.** Yes   **84.** Yes   **85.** $3(x - 2)$
86. $x(7x - 1)$ **87.** $-4(-a + 5)$ **88.** $-w(-w + 1)$
89. $-w(w - 3)$ **90.** $(-3)(2 - x)$ **91.** $(y - 9)(y + 9)$
92. $(rt - 3v)(rt + 3v)$ **93.** $(2x + 7)^2$ **94.** $(y - 10)^2$
95. $(t - 9)^2$ **96.** $(2w + s)^2$ **97.** $(t - 5)(t^2 + 5t + 25)$
98. $(2y + 1)(4y^2 - 2y + 1)$ **99.** $(x - 10)(x + 3)$
100. $(y + 8)(y - 4)$ **101.** $(w - 7)(w + 4)$
102. $(2t - 1)(3t - 1)$ **103.** $(2m + 7)(m - 1)$
104. $(4x - 3)(3x - 2)$ **105.** $m(m^3 - 5)(m^3 + 2)$
106. $w(3w^2 - 5)(2w^2 + 1)$ **107.** $5(x + 2)(x^2 - 2x + 4)$
108. $w(w - 3)^2$ **109.** $(3x + 2)(3x + 1)$
110. $a(x + 1)(x^2 - x + 1)$ **111.** $(x - 1)(x + 1)^2$
112. $2(4x + 1)(2x - 1)$ **113.** $-y(x - 4)(x + 4)$
114. $-5(m - 1)(m + 1)$ **115.** $-ab^2(a - 1)^2$
116. $-2(w + 4)^2$ **117.** $(x - 1)(x^2 + 9)$
118. $(w^2 + 3)(w - 1)(w + 1)$ **119.** $(x - 2)(x + 2)(x^2 + 3)$
120. $(2x - 1)(4x^2 + 2x + 1)$ **121.** $a^3(a - 1)(a^2 + a + 1)$
122. $(a + 2)(a - b)$ **123.** $-2(2m + 3)^2$
124. $-3(x + 5)(x - 2)$ **125.** $(2x - 7)(2x + 1)$
126. $(m - 10)(m - 3)$
127. $(x + 2)(x^2 - 2x + 4)(x - 1)(x^2 + x + 1)$
128. $2a(2a - 1)(2a + 1)(4a^2 + 1)$
129. $(a^2 - 11)(a^2 - 12)$ **130.** $(x + 1)(x - 3)(x + 3)$
131. $(x^k - 7)(x^k + 7)$
132. $(x^k - 1)(x^{2k} + x^k + 1)(x^k + 1)(x^{2k} - x^k + 1)$
133. $(m^a - 3)(m^a + 1)$ **134.** $(2y^n - 3)(y^n - 2)$
135. $(3z^k - 2)^2$ **136.** $(5z^{3m} + 2)^2$ **137.** $(y^a - b)(y^a + c)$
138. $x(y^b + 2)(x - 1)(x + 1)$ **139.** $\{0, 5\}$ **140.** $\{-3, -2\}$
141. $\{0, 5\}$ **142.** $\{-8, 7\}$ **143.** $\{-1/2, 5\}$
144. $\{-4, -3, 3\}$ **145.** $\{-5, -1, 1\}$ **146.** $\left\{-\dfrac{3}{4}, \dfrac{1}{3}\right\}$
147. $\{-3, -1, 1, 3\}$ **148.** $\{-2, -1, 4, 5\}$ **149.** 6 feet
150. $-5, -4, -3,$ or $3, 4, 5$ **151.** $12a + 4$ kilometers
152. $12x + 36$ ft^2 **153.** a) 68.7 years   b) 6.5 years
154. 75.9 years   **155.** a) 15%   b) \$12,196.46
156. \$433.16

Chapter 5 Test

1. $\dfrac{1}{9}$ **2.** 36   **3.** 8   **4.** $12x^7$ **5.** $4y^{12}$ **6.** $64a^6b^3$

7. $\dfrac{27}{x^6}$ **8.** $\dfrac{2a^3}{b^3}$ **9.** 3,240,000,000   **10.** 0.0008673
11. 2.4×10^{-5} **12.** 2×10^{-13} **13.** $3x^3 + 3x^2 - 2x + 3$
14. $-2x^2 - 8x - 3$ **15.** $x^3 - 5x^2 + 13x - 14$
16. $x^2 + 4x - 5$ **17.** $x^3 - 6x^2 + 12x - 8$ **18.** -1
19. $2x^2 - 11x - 21$ **20.** $x^2 - 12x + 36$

21. $4x^2 + 20x + 25$ **22.** $9y^4 - 25$ **23.** $5 + \dfrac{-15}{x + 3}$

24. $x + 5 + \dfrac{4}{x - 2}$ **25.** $(a - 6)(a + 4)$ **26.** $(2x + 7)^2$
27. $3(m - 2)(m^2 + 2m + 4)$ **28.** $2y(x - 4)(x + 4)$
29. $(a - 5)(2x + 3)$ **30.** $(x - 1)(x + 1)(x^2 + 4)$
31. $\left\{-5, \dfrac{3}{2}\right\}$ **32.** $\{-2, 0, 2\}$ **33.** $\{-4, -3, 2, 3\}$
34. Width 8 inches, height 6 inches   **35.** 38.0, 10.8, 7.9

Making Connections Chapters 1–5

1. 16   **2.** -8 **3.** $\dfrac{1}{16}$ **4.** 2   **5.** 1   **6.** $\dfrac{1}{6}$ **7.** $\dfrac{1}{12}$

8. 49   **9.** 64   **10.** 8   **11.** -29 **12.** $\dfrac{11}{30}$ **13.** $\{200\}$

14. $\left\{\dfrac{9}{5}\right\}$ **15.** $\left\{-9, \dfrac{3}{2}\right\}$ **16.** $\left\{-\dfrac{15}{2}, 0\right\}$ **17.** $\left\{2, \dfrac{8}{5}\right\}$

18. $\left\{\dfrac{9}{5}\right\}$ **19.** $\varnothing$ **20.** $\{-3, -2, 1, 2\}$ **21.** $\left\{-5, \dfrac{1}{2}\right\}$

22. $\left\{-\dfrac{2}{3}, \dfrac{4}{3}\right\}$ **23.** 5×10^{10} **24.** 2.05×10^5

25. a) \$20,000   b) \$2,500   c) \$20,000   d) 0

CHAPTER 6

Section 6.1 Warm-ups T T F T F F F T T F

1. A rational expression is a ratio of two polynomials with the denominator not equal to zero.
2. The domain of a rational expression is all real numbers except those that cause the denominator to be zero.
3. The basic principle of rational numbers says that $(ab)/(ac) = b/c$, provided a and c are not zero.
4. To reduce a rational expression, factor the numerator and denominator completely and then divide out the common factors.
5. We build up the denominator by multiplying the numerator and denominator by the same expression.
6. Average cost is total cost divided by the number of items.
7. $\{x \mid x \neq 1\}$ **8.** $\{x \mid x \neq -5\}$ **9.** $\{z \mid z \neq 0\}$
10. $\{z \mid z \neq 0\}$ **11.** $\{y \mid y \neq -2 \text{ and } y \neq 2\}$
12. $\{y \mid y \neq -3 \text{ and } y \neq 3\}$ **13.** $\{a \mid a \neq -2 \text{ and } a \neq -3\}$
14. $\left\{b \mid b \neq -\dfrac{1}{2} \text{ and } b \neq 4\right\}$ **15.** $\{x \mid x \neq -4 \text{ and } x \neq 0\}$
16. $\{x \mid x \neq -3 \text{ and } x \neq 0\}$
17. $\{x \mid x \neq -3 \text{ and } x \neq 0 \text{ and } x \neq 2\}$
18. $\{x \mid x \neq -1 \text{ and } x \neq 0 \text{ and } x \neq 1\}$
19. $\dfrac{2}{19}$ **20.** $\dfrac{2}{13}$ **21.** $\dfrac{1}{5}$ **22.** $\dfrac{11}{7}$ **23.** $\dfrac{x + 1}{2}$

24. $a + 1$ **25.** $-\dfrac{3}{5}$ **26.** -5 **27.** $\dfrac{b}{a^2}$ **28.** $\dfrac{2y}{3z}$

29. $\dfrac{-w}{3x^2y}$ **30.** $\dfrac{-3a^2b^8}{4c^4}$ **31.** $\dfrac{b^2}{1 + a}$ **32.** $\dfrac{b^3 - a}{a}$

33. $-\dfrac{1}{2}$ **34.** $-\dfrac{1}{2}$ **35.** $\dfrac{x + 2}{x}$ **36.** $\dfrac{x - 2}{x}$

37. $a^2 + ab + b^2$ **38.** $\dfrac{9x^2 - 3xy + y^2}{2}$ **39.** $\dfrac{x^2 - 1}{x^2 + 1}$

40. $\dfrac{a^2 - b^2}{a^2 + b^2}$ **41.** $\dfrac{x - 2}{2x - 6}$ **42.** $\dfrac{x + 3}{x - 2}$ **43.** $\dfrac{x^2 + 7x - 4}{x^2 - 16}$

44. $\dfrac{(x+2)(x^2+4)}{2}$ **45.** $\dfrac{b+3}{a+y}$ **46.** $\dfrac{x-3}{x+5}$ **47.** $\dfrac{10}{50}$

48. $\dfrac{6}{9}$ **49.** $\dfrac{3x}{3x^2}$ **50.** $\dfrac{3a^2b^3}{a^3b^5}$ **51.** $\dfrac{5x-5}{x^2-2x+1}$

52. $\dfrac{x^2+3x}{x^2-9}$ **53.** $\dfrac{-3}{-6x-6}$ **54.** $\dfrac{10}{15x-20}$ **55.** $\dfrac{5a}{a}$

56. $\dfrac{3a+3}{a+1}$ **57.** $\dfrac{x^2+x-2}{x^2+2x-3}$ **58.** $\dfrac{x^2+4x}{x^2-x-20}$

59. $\dfrac{-7}{1-x}$ **60.** $\dfrac{-2}{2b-2a}$ **61.** $\dfrac{3x^2-6x+12}{x^3+8}$

62. $\dfrac{x^3+2x^2+4x}{x^3-8}$ **63.** $\dfrac{2x^2+9x+10}{6x^2+13x-5}$

64. $\dfrac{2a^2-9a}{4a^2-16a-9}$ **65.** $\dfrac{4}{7}$ **66.** -1 **67.** $\dfrac{1}{10}$

68. $-\dfrac{22}{17}$ **69.** Undefined   **70.** Undefined   **71.** $\dfrac{7}{21}$

72. $\dfrac{12}{3}$ **73.** $\dfrac{10}{2}$ **74.** $\dfrac{12}{16}$ **75.** $\dfrac{3a}{a^2}$ **76.** $\dfrac{10}{2y}$

77. $\dfrac{-2}{b-a}$ **78.** $\dfrac{-3}{4-x}$ **79.** $\dfrac{2x+2}{x^2-1}$ **80.** $\dfrac{5x-15}{x^2-9}$

81. $\dfrac{-2}{3-w}$ **82.** $\dfrac{2}{x-5}$ **83.** $\dfrac{x+2}{3}$ **84.** $\dfrac{1}{2}$ **85.** $\dfrac{1}{x-4}$

86. $\dfrac{x+1}{x}$ **87.** $\dfrac{a+1}{a}$ **88.** $\dfrac{1}{x+3}$ **89.** $\dfrac{x^2+x+1}{x^3-1}$

90. $\dfrac{x^3-8}{x^2-4}$ **91.** x^a-2 **92.** $\dfrac{x^b-3}{x^b-6}$ **93.** $\dfrac{1+w}{x^a-m}$

94. x^a-2 **95.** $\dfrac{x^{2b}+x^b+1}{x^b+1}$ **96.** $\dfrac{x^a+1}{2x^a-1}$

97. $\dfrac{250}{x}$ mph   **98.** $\dfrac{24}{x+1}$ suits per day

99. a) $A(n)=\dfrac{0.50n+45}{n}$ dollars   **b)** 7.5 cents

c) decreases   **d)** increases

100. $A(n)=\dfrac{50{,}000+300n}{n}$ dollars, \$250 per person

101. a) $p(n)=\dfrac{0.053n^2-0.64n+6.71}{3.43n+87.24}$

b) 7.7%, 18.5%, 41.4%

102. a) $p(n)=\dfrac{25.2n+1069}{41.7n+1429}$ **b)** 69.5%

103. The value of $R(x)$ gets closer and closer to $\dfrac{1}{2}$.

104. The value of $H(x)$ gets closer and closer to $\dfrac{7}{3}$.

Section 6.2 Warm-ups F F T F T T T T T F
1. To multiply rational numbers, multiply the numerators and the denominators.
2. To multiply rational expressions, multiply the numerators and the denominators.
3. The expressions $a-b$ and $b-a$ are opposites.
4. To divide rational numbers, invert the divisor and multiply.

5. $\dfrac{5}{11}$ **6.** $\dfrac{5}{14}$ **7.** $\dfrac{ab}{4}$ **8.** $\dfrac{2y}{3}$ **9.** $\dfrac{1}{2}$ **10.** -2

11. $\dfrac{x+1}{x^2+1}$ **12.** $\dfrac{x^2+1}{x+1}$ **13.** $\dfrac{1}{a-b}$ **14.** $-\dfrac{1}{2}$

15. $\dfrac{a+2}{2}$ **16.** $w+1$ **17.** $-\dfrac{2}{3}$ **18.** $-\dfrac{19x}{3}$

19. $-7a-14$ **20.** $-2x^3$ **21.** $\dfrac{3}{2}$ **22.** 6   **23.** $\dfrac{63}{5}$

24. $\dfrac{21a^2}{2b^2c^7}$ **25.** $\dfrac{6b}{5c^8}$ **26.** $\dfrac{5x^5}{4yz}$ **27.** $\dfrac{w}{w-1}$ **28.** $\dfrac{-4}{a+3}$

29. $\dfrac{2}{x-y}$ **30.** 2   **31.** $\dfrac{2}{x}$ **32.** $\dfrac{1}{x+4}$ **33.** $2x-2y$

34. $\dfrac{2a-b}{2}$ **35.** $2x+10$ **36.** $\dfrac{x^2+1}{x+1}$ **37.** $\dfrac{a-b}{6}$

38. $\dfrac{2}{a+b}$ **39.** $3a-3b$ **40.** x^2+6x+9 **41.** $\dfrac{5x}{6}$

42. $\dfrac{x}{2a}$ **43.** 3   **44.** $\dfrac{1}{2}$ **45.** $\dfrac{1}{12}$ **46.** $\dfrac{b}{2a}$ **47.** $\dfrac{2x}{3}$

48. $\dfrac{2x}{y}$ **49.** -1 **50.** $b-a$ **51.** -2 **52.** $\dfrac{5}{x}$

53. $\dfrac{a+b}{a}$ **54.** $-\dfrac{1}{2}$ **55.** $2a+2b$ **56.** $3x+9$ **57.** $\dfrac{3x}{5y}$

58. $\dfrac{b^2-4a}{2a}$ **59.** $\dfrac{3a}{10b}$ **60.** $\dfrac{6}{a}$ **61.** $\dfrac{7x^2}{3x+2}$

62. $\dfrac{3x^2+9x}{x^2-4}$ **63.** $-\dfrac{a^6b^2}{8c^2}$ **64.** $-\dfrac{w^2y^4}{3}$ **65.** $\dfrac{2m^8n^2}{3}$

66. $\dfrac{r}{t^4}$ **67.** $\dfrac{2x-3}{8x-4}$ **68.** $\dfrac{3}{x}$ **69.** $\dfrac{k+m}{m-k}$ **70.** $\dfrac{a+b}{a-b}$

71. $\dfrac{y^b}{x^a}$ **72.** $x^{a+2}y^{b+7}$ **73.** $\dfrac{x^a-1}{x^a+2}$ **74.** 1   **75.** $\dfrac{m^k+1}{m^k+2}$

76. $\dfrac{m^k-1}{m^{2k-1}+m^{k-1}}$ **77.** 7.1%   **78.** 54%   **79.** $\dfrac{75}{x}$ miles

80. $\dfrac{5x}{4}$ mph   **81.** e   **82.** b

Section 6.3 Warm-ups F F T F F F T T T F
1. The sum of a/b and c/b is $(a+c)/b$.
2. The LCD is the least common multiple of the denominators.
3. The least common multiple (LCM) of some numbers is the smallest number that is a multiple of all of the numbers.
4. To find the LCM for some polynomials, first factor completely, then use the LCM of the coefficients and every other factor with the highest exponent that occurs on that factor in any of the polynomials.
5. To add rational expressions with different denominators, you must build up the expressions to equivalent expressions with the same denominator.
6. You do not need identical denominators for division and multiplication.

7. $4x$ **8.** $3x^2$ **9.** $\dfrac{-x+1}{x}$ **10.** $\dfrac{1-y}{y}$ **11.** $\dfrac{5}{2}$

12. a^2-ab+b^2 **13.** $\dfrac{2}{x-3}$ **14.** -1 **15.** 120

16. 396   **17.** $30x^3y$ **18.** $36a^3b^5$ **19.** $a^3b^5c^2$
20. $x^2y^6z^3$ **21.** $x(x+2)(x-2)$ **22.** $y(y-5)(y+2)$

23. $12a + 24$ **24.** $4a^2 - 6a$ **25.** $(x - 1)(x + 1)^2$

26. $(y - 5)(y + 3)^2$ **27.** $x(x - 4)(x + 4)(x + 2)$

28. $5z^2 - 125$ **29.** $\dfrac{17}{140}$ **30.** $\dfrac{1}{144}$ **31.** $\dfrac{1}{40}$ **32.** $\dfrac{109}{520}$

33. $\dfrac{3w + 5z}{w^2 z^2}$ **34.** $\dfrac{2b - 3a}{a^2 b^2}$ **35.** $\dfrac{2x - 1}{24}$ **36.** $\dfrac{-a - 9}{30}$

37. $\dfrac{11x}{10a}$ **38.** $-\dfrac{5x}{24y}$ **39.** $\dfrac{9 - 4xy}{4y}$ **40.** $\dfrac{b^2 - 4ac}{4a}$

41. $\dfrac{-2a - 14}{a(a + 2)}$ **42.** $\dfrac{-x - 3}{x(x + 1)}$ **43.** $\dfrac{3a - b}{(a - b)(a + b)}$

44. $\dfrac{8x - 4}{(x + 2)(x - 2)}$ **45.** $\dfrac{4x + 9}{(x + 3)(x - 3)}$

46. $\dfrac{6x + 25}{(x - 5)(x + 5)}$ **47.** 0

48. $\dfrac{-4}{x - 5}$ **49.** $\dfrac{11}{2x - 4}$ **50.** $\dfrac{25}{3x - 9}$

51. $\dfrac{-x + 3}{(x - 1)(x + 2)(x + 3)}$ **52.** $\dfrac{-3x}{(x - 2)(x + 2)(x - 5)}$

53. $\dfrac{x^2 + 7x - 18}{(x + 1)(x + 3)(x - 3)}$ **54.** $\dfrac{3x^2 + 4x - 22}{(x + 2)(x + 3)(x - 4)}$

55. $\dfrac{8x - 2}{x(x - 1)(x + 2)}$ **56.** $\dfrac{4a^2 + 8a - 2}{a(a + 1)(a - 1)}$

57. $\dfrac{7}{12}$ **58.** $\dfrac{17}{20}$ **59.** $-\dfrac{19}{40}$ **60.** $\dfrac{3a + 10}{6}$

61. $\dfrac{5x}{6}$ **62.** $-\dfrac{y}{12}$ **63.** $\dfrac{3a - 2b}{3b}$ **64.** $\dfrac{x + 27}{9x}$

65. $\dfrac{3a + 2}{3}$ **66.** $\dfrac{m + 3y}{3}$ **67.** $\dfrac{a + 3}{a}$ **68.** $\dfrac{x + 1}{x}$

69. $\dfrac{3}{x}$ **70.** $\dfrac{4a + 2}{a}$ **71.** $\dfrac{8x + 3}{12x}$ **72.** $\dfrac{x + 1}{5x}$ **73.** -3

74. $\dfrac{z + 3}{z + 1}$ **75.** $\dfrac{-x - 1}{(x + 2)(x + 3)}$ **76.** $\dfrac{-a - 7}{(a + 3)(a + 1)}$

77. $\dfrac{2}{(a^3 - 1)(a^3 + 1)}$ **78.** $\dfrac{2x^3}{(x^3 + 8)(x^3 - 8)}$ **79.** b

80. $\dfrac{a + b}{ab}$ **81.** $\dfrac{5x^2 + x + 4}{(x - 1)(x^2 + x + 1)}$ **82.** $\dfrac{a - 3}{(a + 1)^2}$

83. $\dfrac{x^2 + 25}{x(x - 5)}$ **84.** $\dfrac{1}{a - 2}$ **85.** $\dfrac{-w^2 + 2w + 8}{(w + 3)(w^2 - 3w + 9)}$

86. $\dfrac{-3a^2 + 6a - 5}{a^3 + 8}$ **87.** $\dfrac{a - 3}{a^2 + 2a + 4}$ **88.** $\dfrac{z + 2}{z^2 + 2z + 4}$

89. $\dfrac{-w^3 - 2w^2 - 5w + 6}{(w - 2)(w + 2)(w^2 + 2w + 4)}$

90. $\dfrac{-x^3 + 5x^2 - 10x - 6}{(x^2 - 9)(x^2 - 3x + 9)}$

91. $\dfrac{x^3 + x^2 + 2x + 1}{(x - 1)(x + 1)(x^2 + x + 1)}$ **92.** $\dfrac{x^3 - 4x - 6}{(x^3 - 1)(x + 1)}$

93. $\dfrac{16x + 8}{x^2 + x}$ **94.** $\dfrac{1200x - 3600}{x^2 - 6x}$ **95.** $\dfrac{3x + 60}{x}$

96. $\dfrac{x + 6}{3x}$ **97.** $\dfrac{300x + 500}{x^2 + 5x}$ hours   **98.** $\dfrac{7}{2x}$ hours

Section 6.4 Warm-ups F T T T F F F F T T

1. A complex fraction is a fraction that contains fractions in the numerator, denominator, or both.

2. One method is to perform the operations in the numerator and then in the denominator, and then divide the results. The other method is to multiply the numerator and the denominator by the LCD for all of the fractions.

3. $\dfrac{10}{3}$ **4.** $\dfrac{35}{22}$ **5.** -8 **6.** $\dfrac{3 - 5x}{9x + 21}$ **7.** $\dfrac{a^2 + ab}{a - b}$

8. $\dfrac{mn^3 - n^4}{m^2 - 3m}$ **9.** $\dfrac{a^2 b + 3a}{a + b^2}$ **10.** $\dfrac{m^2 n - 2m}{n - 3m}$ **11.** $\dfrac{x - 3y}{x + y}$

12. $\dfrac{8t + 12w}{w - t}$ **13.** $\dfrac{60m - 3m^2}{8m + 36}$ **14.** $\dfrac{42z - 12}{2 - z}$

15. $\dfrac{a^2 - ab}{b^2}$ **16.** $\dfrac{2xy + y}{2x}$ **17.** $\dfrac{xy + x^2}{y^2 - x^2 y^2}$

18. $\dfrac{3ab^3 - 6a^3}{a^2 b^2 + 2b^3}$ **19.** $\dfrac{x + 2}{x - 2}$ **20.** $\dfrac{x - 2}{x - 5}$

21. $\dfrac{y^2 - y - 2}{(y - 1)(3y + 4)}$ **22.** $\dfrac{2a^2 - 3a - 14}{4a^2 - a - 14}$ **23.** $\dfrac{4x - 10}{x - 4}$

24. $\dfrac{x - 10}{3x - 5}$ **25.** $\dfrac{6w - 3}{2w^2 + w - 4}$ **26.** $\dfrac{2x + 1}{2x^2 + 3x - 3}$

27. $\dfrac{2b - a}{a - 3b}$ **28.** $\dfrac{7x + 2}{-2x - 8}$ **29.** $\dfrac{3a - 7}{5a - 2}$

30. $\dfrac{2x^2 - 12x - 6}{x^2 - 3x - 12}$ **31.** $\dfrac{3m^2 - 12m + 12}{(m - 3)(2m - 1)}$

32. $\dfrac{-y^2 - 11y - 24}{3y^2 + 33y + 54}$ **33.** $\dfrac{2x^2 + 4x + 5}{4x^2 - 2x - 6}$

34. $\dfrac{-3a^2 + 2a + 8}{5a^2 - 13a + 22}$ **35.** $\dfrac{yz + wz}{wy + wz}$ **36.** $\dfrac{b - a}{b + a}$

37. $\dfrac{x}{x + 1}$ **38.** $\dfrac{2a + 1}{a}$ **39.** $\dfrac{a^2 + b^2}{ab^3}$ **40.** $\dfrac{n^3 + m^3}{m^4 n}$

41. $\dfrac{a - 1}{a}$ **42.** $\dfrac{a - m}{am}$ **43.** $\dfrac{1}{x^2 - x + 1}$ **44.** $\dfrac{x^2 + x + 1}{x + 1}$

45. $2m - 3$ **46.** $2x^2 - 3$ **47.** $-\dfrac{1}{ab}$ **48.** $-a^2 b^2$

49. $-x^3 y^3$ **50.** $\dfrac{a^2 b^3 - a^3 b^2}{a + b}$ **51.** $x - 2$ **52.** $a + 3$

53. $\dfrac{xy}{x + y}$ **54.** $\dfrac{a^2 b^2}{b^2 - 2ab + a^2}$ **55.** $-1.7333, -\dfrac{26}{15}$

56. $-0.1163, -\dfrac{5}{43}$ **57.** $0.1667, \dfrac{1}{6}$ **58.** $1.7391, \dfrac{40}{23}$

59. 47.4% **60.** $38.3\%, 65.2\%$ **61.** 49.5 mph

62. 109.3 cents per gallon

64. a) $\dfrac{5}{8}, \dfrac{11}{18}$ **b)** The denominator is larger than the numerator in the first fraction.

65. $\dfrac{2x + 1}{3x + 2}, x \neq 0, -1, -\dfrac{1}{2}, -\dfrac{2}{3}$

Section 6.5 Warm-ups T F F T F T F F F T

1. The first step is to multiply each side of the equation by the LCD.

2. A solution to the equation can cause 0 to appear in a denominator.

3. A proportion is an equation expressing equality of two rational expressions.

4. In $a/b = c/d$ the means are b and c.

5. In $a/b = c/d$ the extremes are a and d.

6. The extremes-means property says that if $a/b = c/d$ then $bc = ad$.

7. $\{-24\}$ **8.** $\{10\}$ **9.** $\left\{\dfrac{22}{15}\right\}$ **10.** $\{1\}$ **11.** $\{5\}$

12. $\left\{-\dfrac{1}{9}\right\}$ **13.** $\{1, 6\}$ **14.** $\{2, 15\}$ **15.** $\{20, 25\}$

16. $\{-60, 10\}$ **17.** $\varnothing$ **18.** $\varnothing$ **19.** $\{-2\}$ **20.** $\{-3\}$

21. $\{-5, 1\}$ **22.** $\left\{-\dfrac{9}{2}, 4\right\}$ **23.** $\left\{\dfrac{8}{3}\right\}$ **24.** $\left\{\dfrac{45}{7}\right\}$ **25.** $\left\{-\dfrac{3}{4}\right\}$

26. $\left\{-\dfrac{15}{7}\right\}$ **27.** $\left\{-\dfrac{14}{5}\right\}$ **28.** $\left\{-\dfrac{40}{3}\right\}$ **29.** $\{20\}$

30. $\{-5\}$ **31.** $\{-3, 3\}$ **32.** $\{-5, 5\}$ **33.** $\{-5, 6\}$

34. $\{-6, 1\}$ **35.** $\{-6\}$ **36.** $\varnothing$ **37.** $\left\{\dfrac{11}{2}\right\}$ **38.** $\{0\}$

39. $\{-6, 6\}$ **40.** $\{-9, 9\}$ **41.** $\{8\}$ **42.** $\{6\}$ **43.** $\{2, 4\}$

44. $\{-5, 2\}$ **45.** $\{-1\}$ **46.** $\{9\}$ **47.** $\{5\}$ **48.** $\{6\}$

49. $\{-3, 3\}$ **50.** $\{-5, 5\}$ **51.** $\{8\}$ **52.** $\left\{\dfrac{5}{6}\right\}$

53. $\left\{\dfrac{25}{2}\right\}$ **54.** $\left\{\dfrac{15}{2}\right\}$ **55.** $\{4\}$ **56.** $\{-4\}$ **57.** $\{-5, 2\}$

58. $\left\{\dfrac{14}{9}, 2\right\}$ **59.** \$166,666.67   **60.** \$200,000   **61.** 138

62. 763   **63.** Width 132 cm, length 154 cm

64. 426 cars, 284 pickups   **65.** 20%, 96%

66. 49.75%, 66%   **67. a)** \$17,142.86 **b)** \$57,142.86

68. a) 557 **b)** 7   **70. a)** 0, 1 **b)** 1 **c)** -1

Section 6.6 Warm-ups F T F T T T F T F F

1. $y = 5x - 7$ **2.** $y = -6x + 46$ **3.** $y = -\dfrac{1}{3}x + 1$

4. $y = -\dfrac{2}{3}x - \dfrac{17}{3}$ **5.** $y = mx - bm + a$

6. $y = ax - ak + h$ **7.** $y = -\dfrac{7}{3}x - \dfrac{29}{3}$

8. $y = -\dfrac{9}{4}x + \dfrac{3}{4}$ **9.** $f = \dfrac{F}{M}$ **10.** $A = P(1 + rt)$

11. $D^2 = \dfrac{4A}{\pi}$ **12.** $r^2 = \dfrac{V}{\pi h}$ **13.** $m_1 = \dfrac{Fr^2}{km_2}$

14. $v^2 = \dfrac{rF}{m}$ **15.** $q = \dfrac{pf}{p - f}$ **16.** $R_1 = \dfrac{RR_2}{R_2 - R}$

17. $a^2 = \dfrac{b^2}{1 - e^2}$ **18.** $b^2 = a^2 - a^2e^2$ **19.** $T_1 = \dfrac{P_1V_1T_2}{P_2V_2}$

20. $P_2 = \dfrac{P_1V_1T_2}{T_1V_2}$ **21.** $h = \dfrac{3V}{4\pi r^2}$ **22.** $S = 2\pi rh + 2\pi r^2$

23. $\dfrac{1}{2}$ **24.** 0.05   **25.** 24   **26.** $\dfrac{4}{3}$ **27.** $\dfrac{32}{3}$ **28.** $\dfrac{144}{5}$

29. -6.517 **30.** -0.046 **31.** 19.899   **32.** -62.033

33. 1.910   **34.** 93.133   **35.** 4 mph   **36.** 60 mph

37. Patrick 24 minutes, Guy 36 minutes, 100 mph

38. 30 mph   **39.** 10 mph   **40.** 7.5 mph   **41.** 6 hours

42. $\dfrac{120}{23}$ hours   **43.** 60 minutes   **44.** 6 hours

45. 30 minutes   **46.** $\dfrac{60}{7}$ hours

47. 10 pounds apples, 12 pounds oranges

48. Rabbit \$0.60 per pound, raccoon \$1.80 per pound

49. 6 ohms   **50.** $R_2 = 2$ ohms, $R_1 = 3$ ohms

51. a) 80 **b)** $C(n) = \dfrac{24{,}000}{n}$ **52.** 20   **53.** 25.255 days

54. 9.7706 minutes

Enriching Word Power

1. b   **2.** d   **3.** b   **4.** d   **5.** b   **6.** a   **7.** a   **8.** d

9. a   **10.** b   **11.** d   **12.** a   **13.** c   **14.** d

Chapter 6 Review

1. $\{x \mid x \neq 1\}$ **2.** $\{x \mid x \neq -5 \text{ and } x \neq 5\}$

3. $\{x \mid x \neq -1 \text{ and } x \neq 2\}$ **4.** $\{x \mid x \neq 0 \text{ and } x \neq 1\}$

5. $\dfrac{c^2}{a^2b}$ **6.** $\dfrac{x^2 + 1}{3}$ **7.** $\dfrac{4x^2}{3y}$ **8.** $x - 1$ **9.** $-a$

10. $2x^3 + 2x^2 + 2x$ **11.** $\dfrac{3}{2}$ **12.** $\dfrac{3x - 3y}{x + y}$ **13.** $6x(x - 2)$

14. $2(x - 2)(x + 2)(x + 4)(x^2 + 2x + 4)$ **15.** $12a^5b^3$

16. $(2x - 3)(2x + 3)^2$ **17.** $\dfrac{3x + 11}{2(x - 3)(x + 3)}$

18. $\dfrac{-2x + 27}{(x - 3)(x + 4)}$ **19.** $\dfrac{aw - 5b}{a^2b^2}$ **20.** $\dfrac{x^2 + 4x}{(x - 1)(x + 1)}$

21. $\dfrac{21}{10x - 60}$ **22.** $\dfrac{5x + 14}{4x - 4}$ **23.** $\dfrac{7 - 3y}{4y - 3}$ **24.** $\dfrac{a^4 - b^3}{a^4b + ab^3}$

25. $\dfrac{b^3 - a^2}{ab}$ **26.** $\dfrac{q^2 + p^2}{pq^2}$ **27.** $\left\{-\dfrac{16}{3}\right\}$ **28.** $\left\{\dfrac{9}{2}\right\}$

29. $\{10\}$ **30.** $\{6\}$ **31.** $y = mx + b$ **32.** $A = \dfrac{h}{2}(b_1 + b_2)$

33. $m = \dfrac{Fr}{v^2}$ **34.** $r = \dfrac{A - P}{Pt}$ **35.** $r = \dfrac{3A}{2\pi h}$ **36.** $b = \dfrac{2w^2}{a}$

37. $y = 2x - 17$ **38.** $y = -\dfrac{1}{2}x + 3$ **39.** $\dfrac{4}{3x}$ **40.** $\left\{\dfrac{1}{2}\right\}$

41. $\dfrac{10 + 7y}{6xy}$ **42.** $\dfrac{-x + 6}{x(x - 2)}$ **43.** $\dfrac{8a + 10}{(a - 5)(a + 5)}$

44. $\{7\}$ **45.** $\{-8, 8\}$ **46.** $\dfrac{2}{5}$ **47.** $-\dfrac{3}{2}$ **48.** $\{-10, 10\}$

49. $\{-9, 9\}$ **50.** $\dfrac{x + 2}{x^2 + 2x + 4}$ **51.** $\dfrac{1}{x - m}$ **52.** $\left\{-\dfrac{21}{5}\right\}$

53. $\dfrac{8a + 20}{(a - 5)(a + 5)(a + 1)}$ **54.** $\dfrac{w + 2}{(w + 1)(w - 1)}$

55. $\dfrac{-15a + 10}{2(a + 2)(a + 3)(a - 3)}$ **56.** $\dfrac{-8a - 1}{3(a - 2)(a + 2)(a - 1)}$

57. $\{2\}$ **58.** $\varnothing$ **59.** $\left\{-\dfrac{5}{2}\right\}$ **60.** $\dfrac{-3}{(a + 3)^2}$ **61.** $-\dfrac{1}{3}$

62. $\{-2, 3\}$ **63.** $\dfrac{x - 2}{3x + 3}$ **64.** $\dfrac{x^2 - x + 1}{2}$

65. $\dfrac{3a^2 + 7a + 16}{a^3 - 8}$ **66.** $\{-5, 3\}$ **67.** $\dfrac{3 - x}{(x + 1)(x + 3)}$

68. $\{-1, 6\}$ **69.** $\dfrac{a + w}{a + 4}$ **70.** $\dfrac{3}{2y + 4}$ **71.** $\{-6, 8\}$

72. $\{4\}$ **73.** $\dfrac{18}{3x}$ **74.** $\dfrac{2}{a}$ **75.** $\dfrac{-3}{b - a}$ **76.** $\dfrac{2}{x - a}$

77. $\dfrac{4x}{x}$ **78.** $\dfrac{5ab}{b}$ **79.** $10x$ **80.** $3a^2$ **81.** $\dfrac{1}{3}$ **82.** $\dfrac{1}{2x}$

83. $\dfrac{1}{a + 3}$ **84.** $\dfrac{x + 2}{x^2 - 4}$ **85.** $\dfrac{3}{10}$ **86.** $\dfrac{1}{20}$ **87.** $\dfrac{5a}{6}$

88. $\dfrac{8x}{15}$ **89.** $\dfrac{b - a}{ab}$ **90.** $\dfrac{3b - 2w}{bw}$ **91.** $\dfrac{a - 3}{3}$

92. $\dfrac{x - y}{y}$ **93.** $\dfrac{2a + 1}{a}$ **94.** $\dfrac{3x - 1}{x}$ **95.** 1 **96.** y

97. $1 - x$ **98.** -1 **99.** -1 **100.** $\dfrac{1}{2}$ **101.** $\dfrac{b}{6a}$

102. $2a - 6$ **103.** -2 **104.** $\dfrac{1}{w - 2}$

105. $\dfrac{3x^a + 4}{(x^a - 2)(x^a + 3)}$ **106.** $\dfrac{1}{x(x^k - 1)}$ **107.** $2x^k$

108. y^{3a} **109.** $448{,}483$ **110. a)** $6{,}943$ **b)** $530{,}410$
111. 10 hours   **112.** 7 mph   **113.** 400 hours
114. 12 minutes

Chapter 6 Test

1. $\left\{x \mid x \ne \dfrac{4}{3}\right\}$ **2.** $\{x \mid x \ne 3 \text{ and } x \ne -3\}$ **3.** $(-\infty, \infty)$

4. $\dfrac{3a^3}{2b}$ **5.** $\dfrac{-x - y}{2x - 2y}$ **6.** $-\dfrac{1}{36}$ **7.** $\dfrac{7y^2 + 3}{y}$

8. $\dfrac{5}{a - 9}$ **9.** $\dfrac{4a + 3b}{24a^2b^2}$ **10.** $\dfrac{a^3}{30b^2}$ **11.** $-\dfrac{3}{a + b}$

12. $\dfrac{1}{x^2 - 1}$ **13.** $\dfrac{-4x + 2}{(x + 2)(x - 2)(x - 5)}$ **14.** $\dfrac{m + 1}{3}$

15. $\left\{\dfrac{12}{7}\right\}$ **16.** $\{4, 10\}$ **17.** $\{-2, 2\}$ **18.** $t = \dfrac{a^2}{W}$

19. $b = \dfrac{2a}{a - 2}$ **20.** $\dfrac{16}{9 - 6x}$ **21.** $w - m$ **22.** $\dfrac{3a^2}{2}$

23. 24 minutes   **24.** 30 miles   **25.** 9

Making Connections Chapters 1–6

1. $\left\{\dfrac{15}{4}\right\}$ **2.** $\{-4, 4\}$ **3.** $\left\{\dfrac{12}{5}\right\}$ **4.** $\{-6, 3\}$ **5.** $\left\{\dfrac{1}{4}\right\}$

6. $\{6\}$ **7.** $\left\{\dfrac{1}{2}\right\}$ **8.** $\left\{\dfrac{3}{2}\right\}$ **9.** $\{-3, 3\}$ **10.** $\{-8, 3\}$

11. $\left\{\dfrac{1}{3}, \dfrac{2}{3}\right\}$ **12.** $\{-3, 9\}$ **13.** $\left\{-\dfrac{9}{2}, \dfrac{1}{2}\right\}$ **14.** $\varnothing$

15. $y = \dfrac{C - Ax}{B}$ **16.** $y = -\dfrac{1}{3}x + \dfrac{4}{3}$ **17.** $y = \dfrac{C}{A - B}$

18. $y = A$ or $y = -A$ **19.** $y = 2A - 2B$

20. $y = \dfrac{2AC}{2B + C}$ **21.** $y = \dfrac{3}{4}x - \dfrac{3}{2}$ **22.** $y = A$ or $y = 2$

23. $y = \dfrac{2A - BC}{B}$ **24.** $y = B$ or $y = -C$ **25.** $12x^{13}$

26. $3x^5 + 15x^8$ **27.** $25x^{12}$ **28.** $27a^9b^6$ **29.** $-4a^6b^6$

30. $\dfrac{1}{32x^{10}}$ **31.** $\dfrac{27x^{12}y^{15}}{8}$ **32.** $\dfrac{a^2}{4b^6c^2}$ **33.** $\dfrac{ab + a^2b^4}{b + a^2}$

34. $a + b$ **35. a)** 2,188 calories   **b)** increases
  c) $B = 9.56W + 918.7$ **d)** $B = 2328 - 4.68A$

CHAPTER 7

Section 7.1 Warm-ups T F T F T T T F F F
1. If $a^n = b$ then a is an nth root of b.
2. If $a^n = b$, then a is an even root of b provided n is even and a is an odd root of b provided n is odd.
3. The principal root is the positive even root of a positive number.
4. The nth root of b is written as $b^{1/n}$.
5. The nth root of 0 is 0.
6. The expression $a^{-m/n}$ represents the reciprocal of the nth root of the mth power of a.
7. 10 **8.** 13 **9.** 3 **10.** 2 **11.** -3 **12.** -2

13. $\dfrac{1}{2}$ **14.** $-\dfrac{1}{2}$ **15.** Not a real number

16. Not a real number   **17.** 10 **18.** 3 **19.** -4
20. -2 **21.** -1 **22.** -5 **23.** 8 **24.** 125 **25.** 9
26. -8 **27.** -125 **28.** -1000 **29.** Not a real number

30. Not a real number   **31.** $\dfrac{1}{2}$ **32.** $\dfrac{1}{3}$ **33.** $\dfrac{1}{16}$ **34.** $\dfrac{1}{8}$

35. $-\dfrac{1}{8}$ **36.** $\dfrac{1}{81}$ **37.** Not a real number

38. Not a real number   **39.** $3^{7/12}$ **40.** $2^{5/6}$ **41.** 1 **42.** 1

43. $\dfrac{1}{2}$ **44.** $\dfrac{1}{3}$ **45.** 2 **46.** $\dfrac{1}{3}$ **47.** 6 **48.** 4 **49.** 4

50. 9 **51.** 81 **52.** $\dfrac{1}{9}$ **53.** $\dfrac{1}{4}$ **54.** 25 **55.** $\dfrac{9}{8}$

56. $\dfrac{25}{27}$ **57.** $|x|$ **58.** $|y|$ **59.** a^4 **60.** $|b^5|$ **61.** y

62. w^3 **63.** $|3x^3y|$ **64.** $|2a^2b|$ **65.** $\left|\dfrac{3x^3}{y^5}\right|$ **66.** $\dfrac{4a^4}{|y^9|}$

67. $x^{3/4}$ **68.** $y^{2/3}$ **69.** $\dfrac{y^{3/2}}{x^{1/4}}$ **70.** $a^{3/2}b^{2/3}$ **71.** $\dfrac{1}{w^{8/3}}$

72. $\dfrac{1}{a^{3/2}}$ **73.** $\dfrac{y}{x^{1/2}}$ **74.** $\dfrac{y^{1/2}}{x^{2/3}}$ **75.** $12x^8$ **76.** $5a^{8/3}$

77. $\dfrac{x^{1/4}}{2y^{1/2}z^{1/4}}$ **78.** $\dfrac{3x^4z^6}{y^5}$ **79.** $\dfrac{a^2}{b}$ **80.** $\dfrac{64a^3}{b^2}$

81. 9　**82.** 4^8　**83.** $-\dfrac{1}{8}$　**84.** $-\dfrac{1}{125}$　**85.** $\dfrac{1}{625}$　**86.** $\dfrac{1}{9}$

87. $2^{1/4}$　**88.** $\dfrac{1}{3}$　**89.** 3　**90.** 4　**91.** 3　**92.** 9

93. $\dfrac{4}{9}$　**94.** $-\dfrac{3}{2}$　**95.** Not a real number　**96.** $\dfrac{4}{3}$

97. $3x^{9/2}$　**98.** $-3x^3$　**99.** $\dfrac{a^2}{27}$　**100.** $\dfrac{x}{25}$　**101.** $a^{5/4}b$

102. $m^{3/2}n^{5/2}$　**103.** $k^{9/2}m^4$　**104.** $tv^{13/6}$　**105.** 1.2599

106. 2.2361　**107.** -1.4142　**108.** -1.4422　**109.** 2

110. 6　**111.** 1.9862　**112.** 17　**113.** 2.5　**114.** 3.375

115. $a^{3m/4}$　**116.** $b^{n/6}$　**117.** $a^{2m/15}$　**118.** $b^{n/12}$

119. a^nb^m　**120.** $a^{3m}b^{2n}$　**121.** $a^{4m}b^{2n}$　**122.** $\dfrac{b^{1/n}}{a^{1/m}}$

123. 13 inches　**124.** 2 meters　**125.** 274.96 m^2

126. a) 1.52 AU　**b)** 27 years　**127.** 47.2%　**128.** 16.9%

129. 6.1%　**130.** 1.65%

Section 7.2 Warm-ups T F T F T F T F T T

1. The expression $\sqrt[n]{a}$ is called a radical.

2. The expressions $\sqrt[n]{a}$ and $a^{1/n}$ both represent the nth root of a.

3. The product rule for radicals says that $\sqrt[n]{a}\cdot\sqrt[n]{b}=\sqrt[n]{ab}$.

4. The product rule can be used to factor out a perfect square from the radicand as in $\sqrt{18}=\sqrt{9}\sqrt{2}=3\sqrt{2}$.

5. The quotient rule for radicals says that $\sqrt[n]{a}/\sqrt[n]{b}=\sqrt[n]{a/b}$.

6. The simplified form for a radical expression has no perfect nth powers as factors of the radicand, no fractions inside the radical, and no radicals in the denominator.

7. $27^{1/2}$　**8.** $(-27)^{1/3}$　**9.** $x^{5/2}$　**10.** $a^{3/2}$　**11.** $a^{12/3}$

12. $w^{-27/3}$　**13.** $-\sqrt[3]{5}$　**14.** $-\sqrt{7}$　**15.** $\sqrt[5]{2^2}$　**16.** $\sqrt[3]{3^2}$

17. $\sqrt[3]{x^{-2}}$　**18.** $\sqrt[5]{x^{-2}}$　**19.** 11　**20.** 8　**21.** -10

22. -1　**23.** Not a real number　**24.** 2　**25.** a^8　**26.** b^{18}

27. 4^8　**28.** w^2　**29.** w^6　**30.** a^4　**31.** $3\sqrt{w}$

32. $6\sqrt{m}$　**33.** $2\sqrt{5}$　**34.** $5\sqrt{2}$　**35.** $3\sqrt{5w}$　**36.** $4\sqrt{3t}$

37. $12\sqrt{2}$　**38.** $11\sqrt{2}$　**39.** $3\sqrt[3]{2}$　**40.** $-2\sqrt[3]{6}$

41. $2\sqrt[4]{2a}$　**42.** $2\sqrt[4]{5xy}$　**43.** $\dfrac{3}{10}$　**44.** $\dfrac{5}{2}$　**45.** $\dfrac{5\sqrt{2}}{3}$

46. $\dfrac{3\sqrt{2}}{5}$　**47.** $-\dfrac{5}{2}$　**48.** $\dfrac{2}{3}$　**49.** $\dfrac{2\sqrt[3]{2x}}{3}$　**50.** $-\dfrac{3\sqrt[3]{3a}}{10}$

51. $\dfrac{2\sqrt{5}}{5}$　**52.** $\dfrac{5\sqrt{3}}{3}$　**53.** $\dfrac{\sqrt{21}}{7}$　**54.** $\dfrac{\sqrt{30}}{5}$　**55.** $\dfrac{\sqrt[3]{2}}{2}$

56. $\dfrac{7\sqrt[3]{9}}{3}$　**57.** $\dfrac{\sqrt[3]{150}}{5}$　**58.** $\dfrac{\sqrt[4]{6}}{3}$　**59.** $\dfrac{\sqrt{15}}{6}$　**60.** $\dfrac{\sqrt{14}}{6}$

61. $\dfrac{1}{2}$　**62.** $\dfrac{1}{3}$　**63.** $\dfrac{\sqrt{2}}{2}$　**64.** $\dfrac{\sqrt{6}}{4}$　**65.** $\dfrac{\sqrt[3]{14}}{2}$

66. $\dfrac{\sqrt[4]{125}}{5}$　**67.** $2x^4\sqrt{3}$　**68.** $6x^5\sqrt{2}$　**69.** $2a^4b\sqrt{15ab}$

70. $3w^7z^3\sqrt[3]{7wz}$　**71.** $\dfrac{\sqrt{xy}}{y}$　**72.** $\dfrac{x\sqrt{a}}{a}$　**73.** $\dfrac{a\sqrt{ab}}{b^4}$

74. $\dfrac{w^2\sqrt{w}}{y^4}$　**75.** $2x^4\sqrt[3]{2x}$　**76.** $2x^5\sqrt[3]{3x^2}$　**77.** $x^2y\sqrt{xy^2}$

78. $w^3y\sqrt[4]{w^2y^3}$　**79.** $2x^4\sqrt[5]{2x^2}$　**80.** $x^2y\sqrt[5]{x^2z^3}$　**81.** $\dfrac{\sqrt[3]{ab^2}}{b}$

82. $\dfrac{\sqrt[4]{aw}}{w}$　**83.** 27　**84.** $\dfrac{1}{8}$　**85.** $\dfrac{1}{10}$

86. Not a real number　**87.** $\dfrac{2\sqrt{2x}}{7}$　**88.** $\dfrac{2\sqrt{3b}}{11}$

89. $\dfrac{2\sqrt[4]{2a}}{3}$　**90.** $\dfrac{3\sqrt[4]{2y}}{5}$　**91.** $-3x^3y^2\sqrt[3]{y^2}$　**92.** $2y^2z^2\sqrt[4]{2z^3}$

93. $\dfrac{\sqrt{b}}{a}$　**94.** $\dfrac{n^2}{m}$　**95.** $\dfrac{\sqrt[3]{18}}{6b^2}$　**96.** $\dfrac{\sqrt[3]{60x^2y}}{6x}$　**97.** 2.887

98. 0.272　**99.** 0.693　**100.** 3.826　**101.** 1.310　**102.** 1

103. 1　**104.** 1.2　**105.** $-4°F,\ -10°F$　**106.** Minneapolis

107. a) $t=\dfrac{\sqrt{h}}{4}$　**b)** $\dfrac{\sqrt{10}}{2}$ sec　**c)** 100 ft

108. 32.6 seconds　**109.** 5.8 knots　**110.** No, 1.84 m^2

111. 114.1 ft/sec, 77.8 mph

112. a) 102.3 ft/sec　**b)** 8,840 lb

Section 7.3 Warm-ups F T F F T F T F F T

1. Like radicals are radicals with the same index and the same radicand.

2. Like radicals are combined using the distributive property just as we combine like terms.

3. In the product rule the radicals must have the same index but do not have to have the same radicand.

4. To multiply radicals of different indices, we convert them to equivalent radicals with the same index.

5. $-\sqrt{3}$　**6.** $-2\sqrt{5}$　**7.** $9\sqrt{7x}$　**8.** $10\sqrt{6a}$　**9.** $5\sqrt[3]{2}$

10. $5\sqrt[3]{4}$　**11.** $4\sqrt{3}-2\sqrt{5}$　**12.** $-6\sqrt{2}+4\sqrt{3}$

13. $5\sqrt[3]{x}$　**14.** $-3\sqrt[3]{5y}+2\sqrt[3]{x}$　**15.** $2\sqrt{x}-\sqrt{2x}$

16. $2\sqrt[3]{ab}+6\sqrt{a}$　**17.** $2\sqrt{2}+2\sqrt{7}$　**18.** $2\sqrt{3}+2\sqrt{6}$

19. $-\sqrt{2}$　**20.** $-3\sqrt{5}$　**21.** $\dfrac{3\sqrt{2}}{2}$　**22.** $-\dfrac{2\sqrt{3}}{3}$

23. $\dfrac{21\sqrt{5}}{5}$　**24.** $\dfrac{9\sqrt{2}}{2}$　**25.** $x\sqrt{5x}+2x\sqrt{2}$

26. $4\sqrt{2x}-8x^2\sqrt{3x}$　**27.** $5\sqrt[3]{3}$　**28.** $7\sqrt[3]{3}$　**29.** $-\sqrt[4]{3}$

30. $3\sqrt[5]{2}$　**31.** $ty\sqrt[3]{2t}$　**32.** $8z\sqrt[3]{2w^2z^2}$　**33.** $\sqrt{15}$

34. $\sqrt{35}$　**35.** $30\sqrt{2}$　**36.** $-24\sqrt{5}$　**37.** $6a\sqrt{14}$

38. $50\sqrt{c}$　**39.** $3\sqrt[4]{3}$　**40.** $5\sqrt[4]{4}$　**41.** 12　**42.** 32

43. $\dfrac{2x\sqrt[3]{3x}}{3}$　**44.** $\dfrac{2x\sqrt[4]{5x}}{5}$　**45.** $6\sqrt{2}+18$

46. $2\sqrt{15}+30$　**47.** $5\sqrt{2}-2\sqrt{5}$　**48.** $3\sqrt{10}-\sqrt{6}$

49. $3\sqrt[3]{t^2}-t\sqrt[3]{3}$　**50.** $2\sqrt[3]{3x}-\sqrt[3]{4x}$　**51.** $-7-3\sqrt{3}$

52. $-7-4\sqrt{5}$　**53.** 2　**54.** $27+10\sqrt{2}$

55. $-8-6\sqrt{5}$　**56.** $12+2\sqrt{6}$　**57.** $-6+9\sqrt{2}$

58. $2\sqrt{6}+7$　**59.** $\sqrt[6]{3^5}$　**60.** $\sqrt[4]{27}$　**61.** $\sqrt[10]{5^7}$

62. $\sqrt[15]{2^8}$　**63.** $\sqrt[6]{500}$　**64.** $\sqrt[6]{864}$　**65.** $\sqrt[12]{432}$

66. $\sqrt[12]{648}$　**67.** -1　**68.** 46　**69.** 3　**70.** 1　**71.** 19

72. 2　**73.** 13　**74.** 5　**75.** $25-9x$　**76.** $16y-9z$

77. $11\sqrt{3}$　**78.** $6\sqrt{2}$　**79.** $10\sqrt{30}$　**80.** $30\sqrt{15}$

81. $8-\sqrt{7}$　**82.** 3　**83.** $16w$　**84.** $15m$

85. $3x^2\sqrt{2x}$　**86.** $2t^4\sqrt{5t}$　**87.** $\dfrac{5\sqrt{2}}{12}$　**88.** $\dfrac{-\sqrt{3}}{3}$

89. $28 + \sqrt{10}$ **90.** $3 + 7\sqrt{6}$ **91.** $\dfrac{8\sqrt{2}}{15}$

92. $\dfrac{5\sqrt{2} + 4\sqrt{3}}{20}$ **93.** 17 **94.** -19 **95.** $9 + 6\sqrt{x} + x$

96. $1 - 2\sqrt{x} + x$ **97.** $25x - 30\sqrt{x} + 9$

98. $9a + 12\sqrt{a} + 4$ **99.** $x + 3 + 2\sqrt{x + 2}$

100. $x + 2\sqrt{x - 1}$ **101.** $-\sqrt{w}$ **102.** $6\sqrt{m}$ **103.** $a\sqrt{a}$

104. $4w\sqrt{y}$ **105.** $7\sqrt{2a}$ **106.** $16\sqrt{2z}$ **107.** $3x^2\sqrt{x}$

108. $6x\sqrt{2x}$ **109.** $a - a^6$ **110.** $wz - 4y^8$

111. $13x\sqrt[3]{2x}$ **112.** $-xy^2\sqrt[3]{3x^2y}$ **113.** $\dfrac{y^2\sqrt[3]{2x^2y}}{2x}$

114. $\dfrac{2\sqrt[4]{9z}}{3z}$ **115.** $\dfrac{x^2\sqrt[4]{5}}{5}$ **116.** $a - a^2$ **117.** $\sqrt[6]{32x^5}$

118. $\sqrt[12]{128m^4n^3}$ **119.** $3\sqrt{2}$ square feet (ft²)

120. $3\sqrt{3}$ cubic meters (m³) **121.** $\dfrac{9\sqrt{2}}{2}$ ft²

122. $3\sqrt{5}$ square meters (m²) **123.** No **124.** a and d

Section 7.4 Warm-ups T T F T F T F T T T

1. $\sqrt{3}$ **2.** $\sqrt{2}$ **3.** $\dfrac{\sqrt{15}}{5}$ **4.** $\dfrac{\sqrt{35}}{7}$ **5.** $\dfrac{3\sqrt{2}}{10}$ **6.** $\dfrac{\sqrt{5}}{10}$

7. $\dfrac{\sqrt{2}}{3}$ **8.** $\dfrac{5\sqrt{2}}{4}$ **9.** $\sqrt[3]{10}$ **10.** 2 **11.** $x^2\sqrt[3]{4}$

12. $a^2\sqrt[4]{2}$ **13.** $2 + \sqrt{5}$ **14.** $2 + \sqrt{2}$ **15.** $1 - \sqrt{3}$

16. $1 - \sqrt{2}$ **17.** $\dfrac{1 + \sqrt{6} + \sqrt{2} + \sqrt{3}}{2}$

18. $\dfrac{3\sqrt{6} - 5\sqrt{2}}{4}$ **19.** $\dfrac{2\sqrt{3} - \sqrt{6}}{3}$ **20.** $\dfrac{5\sqrt{7} + 5\sqrt{5}}{2}$

21. $\dfrac{6\sqrt{6} + 2\sqrt{15}}{13}$ **22.** $\dfrac{15\sqrt{10} - 3\sqrt{30}}{44}$

23. $\dfrac{18\sqrt{5} + 3\sqrt{10} - 2\sqrt{6} - 12\sqrt{3}}{66}$

24. $-\dfrac{17\sqrt{3} + 49}{59}$ **25.** $128\sqrt{2}$ **26.** 729 **27.** $x^2\sqrt{x}$

28. $8y\sqrt{y}$ **29.** $-27x^4\sqrt{x}$ **30.** $16x^6$ **31.** $8x^5$

32. $32y^4$ **33.** $4\sqrt[3]{25}$ **34.** $18\sqrt[3]{2}$ **35.** x^4 **36.** $8y^2\sqrt[4]{y}$

37. $\dfrac{\sqrt{6} + 2\sqrt{2}}{2}$ **38.** $\sqrt{7}$ **39.** $2\sqrt{6}$

40. $\dfrac{3\sqrt{6} + 2\sqrt{10}}{12}$ **41.** $\dfrac{\sqrt{2}}{2}$ **42.** 2 **43.** $\dfrac{2}{3}$ **44.** $\dfrac{2\sqrt{a}}{3a}$

45. $\dfrac{2 - \sqrt{2}}{5}$ **46.** $\dfrac{2 - \sqrt{7}}{3}$ **47.** $\dfrac{1 + \sqrt{3}}{2}$ **48.** $\dfrac{1 + \sqrt{2}}{2}$

49. $a - 3\sqrt{a}$ **50.** $6m - 18\sqrt{m}$ **51.** $4a\sqrt{a} + 4a$

52. $3a\sqrt{b} + 3\sqrt{ab}$ **53.** $12m$ **54.** $36y$ **55.** $4xy^2z$

56. $25a^3b$ **57.** $m - m^2$ **58.** $w - w^2$ **59.** $5x\sqrt[3]{x}$

60. $3a\sqrt[3]{2a}$ **61.** $8m^4\sqrt[4]{8m^2}$ **62.** $-32t^6\sqrt[6]{32t^4}$

63. $2\sqrt{2} - 2$ **64.** $-2 - 2\sqrt{2}$ **65.** $2 + 8\sqrt{2}$

66. $\dfrac{2\sqrt{3}}{5}$ **67.** $\dfrac{3\sqrt{2} + 2\sqrt{3}}{6}$ **68.** $\dfrac{10\sqrt{3} + 3\sqrt{5}}{15}$

69. $7\sqrt{2} - 1$ **70.** $\dfrac{\sqrt{5} + 5\sqrt{3}}{2}$ **71.** $\dfrac{4x + 4\sqrt{x}}{x - 4}$

72. $\dfrac{3\sqrt{5} + y\sqrt{5} - 2\sqrt{5y}}{9 - y}$ **73.** $\dfrac{x + \sqrt{x}}{x - x^2}$

74. $\dfrac{x^2 + 8x\sqrt{x} - 45\sqrt{x}}{x^2 - 9x}$ **75.** $\dfrac{\sqrt{6}}{3}$ **76.** $\dfrac{2}{\sqrt{2}}$

77. $\dfrac{\sqrt{2} + 1}{1}$ **78.** $\dfrac{6 - 2\sqrt{6}}{2}$ **79.** $\dfrac{\sqrt{x} + 1}{x - 1}$

80. $\dfrac{15 + 5\sqrt{x}}{9 - x}$ **81.** $\dfrac{3\sqrt{2} - 3x}{2 - x^2}$ **82.** $\dfrac{8\sqrt{3} - 4a}{12 - a^2}$

83. 3.968 **84.** 4.882 **85.** 12.124 **86.** 12.124
87. 14.697 **88.** 14.697 **89.** 8.873 **90.** 8.873
91. 0.725 **92.** -1.725 **93.** -0.419 **94.** -3.581
95. a) $x^3 - 2$ **b)** $(x + \sqrt[3]{5})(x^2 - \sqrt[3]{5}x + \sqrt[3]{25})$ **c)** 3
   d) $(\sqrt[3]{a} + \sqrt[3]{b})(\sqrt[3]{a^2} - \sqrt[3]{ab} + \sqrt[3]{b^2})$
   $(\sqrt[3]{a} - \sqrt[3]{b})(\sqrt[3]{a^2} + \sqrt[3]{ab} + \sqrt[3]{b^2})$

96. b

Section 7.5 Warm-ups F T F F T F F T T T

1. The odd-root property says that if n is an odd positive integer, then $x^n = k$ is equivalent to $x = \sqrt[n]{k}$ for any real number k.

2. The even-root property says that if n is a positive even integer, then $x^n = k$ is equivalent for $x = \pm\sqrt[n]{k}$ for $k > 0$, $x = 0$ for $k = 0$, and has no solution for $k < 0$.

3. An extraneous solution is a solution that appears when solving an equation but does not satisfy the original equation.

4. Raising each side to an even power can produce an extraneous root because the even powers of both negative and positive numbers are positive. For example, if $\sqrt{x} = -2$, then squaring each side produces an extraneous root.

5. $\{-10\}$ **6.** $\{5\}$ **7.** $\left\{\dfrac{1}{2}\right\}$ **8.** $\left\{-\dfrac{1}{3}\right\}$ **9.** $\{1\}$ **10.** $\{0\}$

11. $\{-2\}$ **12.** $\{9\}$ **13.** $\{-5, 5\}$ **14.** $\{-6, 6\}$

15. $\{-2\sqrt{5}, 2\sqrt{5}\}$ **16.** $\{-2\sqrt{10}, 2\sqrt{10}\}$ **17.** $\varnothing$ **18.** $\varnothing$

19. $\{-1, 7\}$ **20.** $\{-3, 7\}$ **21.** $\{-1 - 2\sqrt{2}, -1 + 2\sqrt{2}\}$

22. $\{-3 - 2\sqrt{3}, -3 + 2\sqrt{3}\}$ **23.** $\{-\sqrt{10}, \sqrt{10}\}$

24. $\{\pm 3\sqrt{2}\}$ **25.** $\{3\}$ **26.** $\left\{\dfrac{3}{2}\right\}$ **27.** $\{-2, 2\}$

28. $\{-2, 2\}$ **29.** $\{52\}$ **30.** $\{37\}$ **31.** $\left\{\dfrac{9}{4}\right\}$ **32.** $\{3\}$

33. $\{9\}$ **34.** $\{10\}$ **35.** $\varnothing$ **36.** $\varnothing$ **37.** $\{3\}$ **38.** $\left\{\dfrac{2}{5}\right\}$

39. $\{-5, 3\}$ **40.** $\{-4, 5\}$ **41.** $\{1\}$ **42.** $\{5\}$ **43.** $\varnothing$

44. $\{3\}$ **45.** $\{6\}$ **46.** $\{0, 4\}$ **47.** $\{7\}$ **48.** $\{27\}$

49. $\{-5\}$ **50.** $\{2\}$ **51.** $\{-3\sqrt{3}, 3\sqrt{3}\}$

52. $\{-2\sqrt{2}, 2\sqrt{2}\}$ **53.** $\left\{-\dfrac{1}{27}, \dfrac{1}{27}\right\}$ **54.** $\left\{-\dfrac{1}{8}, \dfrac{1}{8}\right\}$

55. $\{512\}$ **56.** $\{19,683\}$ **57.** $\left\{\dfrac{1}{81}\right\}$ **58.** $\{16\}$ **59.** $\left\{0, \dfrac{2}{3}\right\}$

60. $\{0, 2\}$ **61.** $\left\{\dfrac{4 - \sqrt{2}}{4}, \dfrac{4 + \sqrt{2}}{4}\right\}$ **62.** $\{24\}$ **63.** $\varnothing$

64. $\varnothing$ **65.** $\sqrt{13}$ **66.** $2\sqrt{2}$ **67.** $2\sqrt{17}$ **68.** $\sqrt{58}$
69. $\sqrt{65}$ **70.** $\sqrt{58}$ **71.** $\{-\sqrt{2}, \sqrt{2}\}$ **72.** $\{-\sqrt{7}, \sqrt{7}\}$

73. $\{-5\}$ **74.** $\{10\}$ **75.** $\left\{-\dfrac{1}{3}, \dfrac{1}{3}\right\}$ **76.** $\left\{-\dfrac{1}{2}, \dfrac{1}{2}\right\}$ **77.** $\varnothing$

78. $\{2 + 2\sqrt[3]{2}\}$ **79.** $\{-9\}$ **80.** $\{-26\}$ **81.** $\left\{\dfrac{5}{4}\right\}$ **82.** $\{0\}$

83. $\{-9, 4\}$ **84.** $\varnothing$ **85.** $\left\{-\dfrac{2}{3}, 2\right\}$ **86.** $\left\{-\dfrac{3}{2}, 3\right\}$

87. $\{-2 - 2\sqrt[4]{2}, -2 + 2\sqrt[4]{2}\}$ **88.** $\{-1 - 2\sqrt[4]{3}, -1 + 2\sqrt[4]{3}\}$

89. $\{0\}$ **90.** $\{-2\}$ **91.** $\left\{\dfrac{1}{2}\right\}$ **92.** $\left\{\pm\dfrac{1}{2}\right\}$

93. $\left\{-\dfrac{\sqrt{3}}{3}, \dfrac{\sqrt{3}}{3}\right\}$ **94.** $\left\{-\dfrac{\sqrt{2}}{6}, \dfrac{\sqrt{2}}{6}\right\}$ **95.** $4\sqrt{2}$ feet

96. $4\sqrt{5}$ meters **97.** $5\sqrt{2}$ feet **98.** $2\sqrt[3]{10}$ feet
99. 50 feet **100.** 13 meters

101. a) 1.89 **b)** $d = \dfrac{64b^3}{C^3}$ **c)** $d > 19{,}683$ pounds

102. a) 15.9 **b)** $d = \left(\dfrac{16A}{S}\right)^{3/2}$ **103.** $\sqrt[6]{32}$ meters

104. $2\sqrt{2}$ cubic feet (ft^3) **105.** $\sqrt{73}$ kilometers (km)
106. $\sqrt{13}$ km **107.** $S = P(1 + r)^n, P = S(1 + r)^{-n}$
108. $2\sqrt{2}$ ft^3 **109.** 9.5 AU **110.** 0.61 year
111. $\{-1.8, 1.8\}$ **112.** $\{-2.042\}$ **113.** $\{4.993\}$
114. $\{55.653\}$ **115.** $\{-26.372, 26.372\}$ **116.** $\{1.074\}$

Section 7.6 Warm-ups T F F T T T T F T F
1. A complex number is a number of the form $a + bi$, where a and b are real numbers.
2. An imaginary number is a complex number in which $b \neq 0$.
3. The union of the real numbers and the imaginary numbers is the set of complex numbers.
4. Addition, subtraction, and multiplication of the complex number is done as if the complex numbers were binomials with i being a variable. When i^2 occurs, we replace it with -1.
5. The conjugate of $a + bi$ is $a - bi$.
6. To divide complex numbers, write the quotient as a fraction and multiply the numerator and denominator by the conjugate of the denominator.
7. $-2 + 8i$ **8.** $4 + 2i$ **9.** $-4 + 4i$ **10.** $-4 - i$
11. -2 **12.** -10 **13.** $-8 - 2i$ **14.** $-8 + 5i$
15. $6 + 15i$ **16.** $4 - 12i$ **17.** $-2 - 10i$ **18.** $18 + 6i$
19. $-4 - 12i$ **20.** $15 - 10i$ **21.** $-10 + 24i$
22. $2 + 11i$ **23.** $-1 + 3i$ **24.** $-4 - 19i$ **25.** $-5i$
26. 10 **27.** 29 **28.** $7 + 24i$ **29.** 2 **30.** 40 **31.** 20
32. 17 **33.** -9 **34.** -25 **35.** -25 **36.** -81
37. 16 **38.** $8i$ **39.** i **40.** 1 **41.** 34 **42.** 10
43. 5 **44.** 52 **45.** 5 **46.** 13 **47.** 7 **48.** 21
49. $\dfrac{12}{17} - \dfrac{3}{17}i$ **50.** $\dfrac{42}{53} + \dfrac{12}{53}i$ **51.** $\dfrac{4}{13} + \dfrac{7}{13}i$

52. $\dfrac{1}{5} + \dfrac{13}{5}i$ **53.** $3 - 4i$ **54.** $-2 - \dfrac{5}{3}i$ **55.** $1 + 3i$

56. $-\dfrac{3}{2} + \dfrac{1}{2}i$ **57.** $2 + 2i$ **58.** $3 + 3i$ **59.** $5 + 6i$

60. $2 + 12i$ **61.** $7 - i\sqrt{6}$ **62.** $3 + i\sqrt{5}$ **63.** $5i\sqrt{2}$
64. $i\sqrt{5}$ **65.** $1 + i\sqrt{3}$ **66.** $-2 - i\sqrt{2}$

67. $-1 - \dfrac{1}{2}i\sqrt{6}$ **68.** $-2 - \dfrac{1}{2}i\sqrt{5}$ **69.** $\{\pm 6i\}$ **70.** $\{\pm 2i\}$

71. $\{\pm 2i\sqrt{3}\}$ **72.** $\{\pm 5i\}$ **73.** $\left\{\pm\dfrac{i\sqrt{10}}{2}\right\}$ **74.** $\left\{\pm\dfrac{2i\sqrt{3}}{3}\right\}$

75. $\{\pm i\sqrt{2}\}$ **76.** $\{\pm i\}$ **77.** $18 - i$ **78.** 13 **79.** $5 + i$

80. $1 + 2i$ **81.** $-\dfrac{6}{25} - \dfrac{17}{25}i$ **82.** $\dfrac{2}{5} - \dfrac{1}{5}i$ **83.** $3 + 2i$

84. $12 + 3i$ **85.** -9 **86.** -64 **87.** $3i\sqrt{3}$ **88.** $2i$
89. $-5 - 12i$ **90.** $16 + 30i$ **91.** $-2 + 2i\sqrt{2}$

92. $\dfrac{1}{3} + \dfrac{1}{2}i\sqrt{3}$

Enriching Word Power
1. d **2.** b **3.** b **4.** b **5.** d **6.** b **7.** c **8.** a
9. d **10.** c **11.** a **12.** c **13.** d **14.** b

Chapter 7 Review
1. $\dfrac{1}{9}$ **2.** -125 **3.** 4 **4.** 5 **5.** $\dfrac{1}{1000}$ **6.** $\dfrac{1}{100}$

7. $27x^{1/2}$ **8.** $\dfrac{x^{1/2}z}{y^{5/2}}$ **9.** $a^{7/2}b^{7/2}$ **10.** $\dfrac{v^2}{t}$ **11.** $x^{3/4}y^{5/4}$

12. ab **13.** $6x^2\sqrt{2x}$ **14.** $3y^4z^2\sqrt{10y}$ **15.** $2x\sqrt[3]{9x^2}$

16. $3a^2b^3\sqrt[3]{3a^2}$ **17.** 8 **18.** $8\sqrt{2}$ **19.** $\dfrac{\sqrt{10}}{5}$

20. $\dfrac{\sqrt{6}}{6}$ **21.** $\dfrac{\sqrt[3]{18}}{3}$ **22.** $\dfrac{\sqrt[3]{3}}{3}$ **23.** $\dfrac{2\sqrt{3x}}{3x}$ **24.** $\dfrac{3\sqrt{2y}}{2y}$

25. $\dfrac{y\sqrt{15y}}{3}$ **26.** $\dfrac{x^2\sqrt{10x}}{4}$ **27.** $\dfrac{3\sqrt[3]{4a^2}}{2a}$ **28.** $\sqrt[3]{a}$

29. $\dfrac{5\sqrt[4]{27x^2}}{3x}$ **30.** $\dfrac{\sqrt[4]{a^2b}}{a}$ **31.** $2xy^3\sqrt[4]{3x}$ **32.** $2x^2y^2\sqrt[5]{y^2}$

33. 13 **34.** 14 **35.** $3\sqrt{5} - 2\sqrt{3}$ **36.** $2\sqrt{3} + \sqrt{2}$

37. $\dfrac{10\sqrt{3}}{3}$ **38.** $\dfrac{\sqrt{2}}{4}$ **39.** $30 - 21\sqrt{6}$ **40.** $-2a + 2ab^3$

41. $6 - 3\sqrt{3} + 2\sqrt{2} - \sqrt{6}$ **42.** $2x + \sqrt{xy} - y$

43. $\sqrt[3]{5}$ **44.** $5x\sqrt[3]{2x}$ **45.** $\dfrac{5\sqrt{2}}{2}$ **46.** $5\sqrt{3}$ **47.** 9

48. $-512x^4\sqrt{x}$ **49.** $1 - \sqrt{2}$ **50.** $\dfrac{1 + \sqrt{2}}{2}$

51. $\dfrac{-\sqrt{6} - 3\sqrt{2}}{2}$ **52.** $-2\sqrt{15} + 5\sqrt{3}$ **53.** $\dfrac{3\sqrt{2} + 2}{7}$

54. $\dfrac{-3\sqrt{y} + y}{9 - y}$ **55.** $256w^{10}$ **56.** m^{14} **57.** $\{-4, 4\}$

58. $\{-10, 10\}$ **59.** $\{3, 7\}$ **60.** $\{2, 12\}$

61. $\{-1 - \sqrt{5}, -1 + \sqrt{5}\}$ **62.** $\{-5 - \sqrt{3}, -5 + \sqrt{3}\}$

63. $\varnothing$ **64.** $\{-8, 0\}$ **65.** $\{10\}$ **66.** $\{11\}$ **67.** $\{9\}$

68. $\left\{\dfrac{17}{2}\right\}$ **69.** $\{-8, 8\}$ **70.** $\left\{-\dfrac{1}{8}, \dfrac{1}{8}\right\}$ **71.** $\{124\}$

72. $\left\{\dfrac{23}{8}, \dfrac{25}{8}\right\}$ **73.** $\{7\}$ **74.** $\{-5, 2\}$ **75.** $\{2, 3\}$ **76.** $\{5\}$

77. $\{9\}$ **78.** $\{1\}$ **79.** $\{4\}$ **80.** $\{9\}$ **81.** $5 + 25i$

82. $12 + i$ **83.** $7 - 3i$ **84.** $5 - 5i$ **85.** $-1 + 2i$

86. $2 - i$ **87.** $2 + i$ **88.** $2 + 3i$ **89.** $2 - i\sqrt{3}$

90. $2 + i\sqrt{2}$ **91.** $\dfrac{5}{17} - \dfrac{14}{17}i$ **92.** $\dfrac{3}{13} + \dfrac{11}{13}i$ **93.** $\{\pm 10i\}$

94. $\left\{\pm\dfrac{i\sqrt{3}}{5}\right\}$ **95.** $\left\{\pm\dfrac{3i\sqrt{2}}{2}\right\}$ **96.** $\left\{\pm\dfrac{2i\sqrt{6}}{3}\right\}$ **97.** False

98. True   **99.** True   **100.** False   **101.** True

102. True   **103.** False   **104.** True   **105.** False

106. True   **107.** False   **108.** False   **109.** False

110. True   **111.** True   **112.** True   **113.** False

114. True   **115.** True   **116.** False   **117.** False

118. True   **119.** False   **120.** True   **121.** True

122. False   **123.** True   **124.** False   **125.** $2\sqrt{58}$

126. $2\sqrt{17}$ **127.** $5\sqrt{30}$ seconds   **128.** 36 feet

129. $10\sqrt{7}$ feet   **130.** $113\dfrac{1}{3}$ yards   **131.** $200\sqrt{2}$ feet

132. $8\sqrt{3}$ feet   **133.** $26.4\sqrt[3]{25}$ ft^2 **134.** $2\sqrt[6]{432}$ ft

135. 6.7%   **136.** 1.04%   **137.** $V = \dfrac{29\sqrt{LCS}}{CS}$

138. 2,227 ft^3/sec, 5.5 ft

Chapter 7 Test

1. 4   **2.** $\dfrac{1}{8}$ **3.** $\sqrt{3}$ **4.** 30   **5.** $3\sqrt{5}$ **6.** $\dfrac{6\sqrt{5}}{5}$

7. 2   **8.** $6\sqrt{2}$ **9.** $\dfrac{\sqrt{15}}{6}$ **10.** $\dfrac{2 + \sqrt{2}}{2}$ **11.** $4 - 3\sqrt{3}$

12. $2ay^2\sqrt[4]{2a}$ **13.** $\dfrac{\sqrt[3]{4x}}{2x}$ **14.** $\dfrac{2a^4\sqrt{2ab}}{b^2}$ **15.** $-3x^3$

16. $2m\sqrt{5m}$ **17.** $x^{3/4}$ **18.** $3y^2x^{1/4}$ **19.** $2x^2\sqrt[3]{5x}$

20. $19 + 8\sqrt{3}$ **21.** $\dfrac{5 + \sqrt{3}}{11}$ **22.** $\dfrac{6\sqrt{2} - \sqrt{3}}{23}$

23. $22 + 7i$ **24.** $1 - i$ **25.** $\dfrac{1}{5} - \dfrac{7}{5}i$ **26.** $-\dfrac{3}{4} + \dfrac{1}{4}i\sqrt{3}$

27. $\{-5, 9\}$ **28.** $\left\{-\dfrac{7}{4}\right\}$ **29.** $\{-8, 8\}$ **30.** $\left\{\pm\dfrac{4}{3}i\right\}$ **31.** $\{3\}$

32. $\{5\}$ **33.** $2\sqrt{2}$ **34.** $\dfrac{3\sqrt{2}}{2}$ feet   **35.** 25 and 36

36. Length 6 ft, width 4 ft   **37.** 39.53 AU, 164.97 years

Making Connections Chapters 1–7

1. $\left\{-\dfrac{4}{7}\right\}$ **2.** $\left\{\dfrac{3}{2}\right\}$

3. $(-\infty, -3) \cup (-2, \infty)$
$-5\ -4\ -3\ -2\ -1\ \ 0$

4. $\left\{\dfrac{3}{2}\right\}$ **5.** $(-\infty, 1)$
$-3\ -2\ -1\ \ 0\ \ 1\ \ 2\ \ 3$

6. $\varnothing$ **7.** $\{9\}$ **8.** $\varnothing$ **9.** $\{-12, -2\}$ **10.** $\left\{\dfrac{1}{16}\right\}$

11. $(-6, \infty)$
$-8\ -7\ -6\ -5\ -4\ -3\ -2$

12. $\left\{-\dfrac{1}{64}, \dfrac{1}{64}\right\}$ **13.** $\left\{-\dfrac{\sqrt{3}}{3}, \dfrac{\sqrt{3}}{3}\right\}$ **14.** R

15. $\left(-\dfrac{1}{3}, 3\right)$
$-2\ -1\ \ 0\ \ 1\ \ 2\ \ 3\ \ 4$

16. $\left\{\dfrac{1}{3}\right\}$ **17.** $\{82\}$ **18.** $\left\{\dfrac{6}{5}, \dfrac{12}{5}\right\}$ **19.** $\{100\}$ **20.** R

21. $\{4\sqrt{30}\}$ **22.** $\{400\}$ **23.** $\left\{\dfrac{13 + 9\sqrt{2}}{3}\right\}$

24. $\{-3\sqrt{2}, 3\sqrt{2}\}$ **25.** $\{5\}$ **26.** $\{7 + 3\sqrt{6}\}$

27. $\{-2, 3\}$ **28.** $\{-5, 2\}$ **29.** $\{-2, 3\}$ **30.** $\left\{\dfrac{1}{2}, 3\right\}$

31. 3   **32.** -2 **33.** $\dfrac{1}{2}$ **34.** $\dfrac{1}{3}$ **35.** 48.5 cm^3

36. 14%, 56 cm^3

CHAPTER 8

Section 8.1 Warm-ups F F F F T F F F T F F

1. In this section, quadratic equations are solved by factoring, the even root property, and completing the square.

2. If $b = 0$ in $ax^2 + bx + c = 0$, then the equation can be solved by the even-root property.

3. The last term is the square of one-half the coefficient of the middle term.

4. If the leading coefficient is not 1, then the first step is to divide each side by the leading coefficient.

5. $\{-2, 3\}$ **6.** $\{-4, -2\}$ **7.** $\{-5, 3\}$ **8.** $\{-3, 5\}$

9. $\left\{-1, \dfrac{3}{2}\right\}$ **10.** $\left\{-\dfrac{3}{2}, \dfrac{5}{3}\right\}$ **11.** $\{-7\}$ **12.** $\{3\}$

13. $\{-4, 4\}$ **14.** $\left\{-\dfrac{5}{2}, \dfrac{5}{2}\right\}$ **15.** $\{-9, 9\}$ **16.** $\left\{-\dfrac{3}{2}, \dfrac{3}{2}\right\}$

17. $\left\{-\dfrac{4}{3}, \dfrac{4}{3}\right\}$ **18.** $\{-4\sqrt{2}, 4\sqrt{2}\}$ **19.** $\{-1, 7\}$

20. $\{-7, -3\}$ **21.** $\{-1 - \sqrt{5}, -1 + \sqrt{5}\}$

22. $\{2 - 2\sqrt{2}, 2 + 2\sqrt{2}\}$ **23.** $\left\{\dfrac{3 - \sqrt{7}}{2}, \dfrac{3 + \sqrt{7}}{2}\right\}$

24. $\left\{\dfrac{-2 - \sqrt{5}}{3}, \dfrac{-2 + \sqrt{5}}{3}\right\}$ **25.** $x^2 + 2x + 1$

26. $m^2 + 14m + 49$ **27.** $x^2 - 3x + \dfrac{9}{4}$ **28.** $w^2 - 5w + \dfrac{25}{4}$

29. $y^2 + \dfrac{1}{4}y + \dfrac{1}{64}$ **30.** $z^2 + \dfrac{3}{2}z + \dfrac{9}{16}$ **31.** $x^2 + \dfrac{2}{3}x + \dfrac{1}{9}$

32. $p^2 + \dfrac{6}{5}p + \dfrac{9}{25}$ **33.** $(x + 4)^2$ **34.** $(x - 5)^2$

35. $\left(y - \dfrac{5}{2}\right)^2$ **36.** $\left(w + \dfrac{1}{2}\right)^2$ **37.** $\left(z - \dfrac{2}{7}\right)^2$

38. $\left(m - \dfrac{3}{5}\right)^2$ **39.** $\left(t + \dfrac{3}{10}\right)^2$ **40.** $\left(h + \dfrac{3}{4}\right)^2$

41. $\{-3, 5\}$ **42.** $\{-1, 7\}$ **43.** $\{-10, 2\}$ **44.** $\{-1, -9\}$

45. $\{-5, 7\}$ **46.** $\{-2, 4\}$ **47.** $\{-4, 5\}$ **48.** $\{-2, 5\}$

49. $\{-7, 2\}$ **50.** $\{-2, 1\}$ **51.** $\left\{-1, \dfrac{3}{2}\right\}$ **52.** $\left\{-\dfrac{5}{2}, 3\right\}$

53. $\{-2 - \sqrt{10}, -2 + \sqrt{10}\}$

54. $\{-3 - \sqrt{17}, -3 + \sqrt{17}\}$

55. $\{-4 - 2\sqrt{5}, -4 + 2\sqrt{5}\}$

56. $\{-5 - 2\sqrt{7}, -5 + 2\sqrt{7}\}$

57. $\left\{\dfrac{-3 - \sqrt{41}}{4}, \dfrac{-3 + \sqrt{41}}{4}\right\}$

58. $\left\{\dfrac{-5 - \sqrt{33}}{4}, \dfrac{-5 + \sqrt{33}}{4}\right\}$ **59.** $\{4\}$ **60.** $\{20\}$

61. $\left\{\dfrac{1 + \sqrt{17}}{8}\right\}$ **62.** $\left\{\dfrac{9 + \sqrt{33}}{8}\right\}$ **63.** $\{1, 6\}$ **64.** $\{0, 5\}$

65. $\{-2 - \sqrt{2}, -2 + \sqrt{2}\}$ **66.** $\left\{\dfrac{-3 - \sqrt{5}}{2}, \dfrac{-3 + \sqrt{5}}{2}\right\}$

67. $\{-1 - 2i, -1 + 2i\}$ **68.** $\{-2 - i, -2 + i\}$

69. $\{-2i\sqrt{3}, 2i\sqrt{3}\}$ **70.** $\{-i\sqrt{7}, i\sqrt{7}\}$ **71.** $\left\{\dfrac{2 \pm i}{5}\right\}$

72. $\left\{\dfrac{3 \pm i\sqrt{7}}{4}\right\}$ **73.** $\left\{-\dfrac{5}{2}i, \dfrac{5}{2}i\right\}$ **74.** $\left\{-\dfrac{\sqrt{15}}{5}, \dfrac{\sqrt{15}}{5}\right\}$

75. $\{-2, 1\}$ **76.** $\left\{0, \dfrac{4}{3}\right\}$ **77.** $\left\{\dfrac{-2 - \sqrt{19}}{5}, \dfrac{-2 + \sqrt{19}}{5}\right\}$

78. $\left\{\dfrac{-2 - \sqrt{7}}{3}, \dfrac{-2 + \sqrt{7}}{3}\right\}$ **79.** $\{-6, 4\}$ **80.** $\{-7, 1\}$

81. $\left\{\dfrac{-2 - 4i\sqrt{2}}{3}, \dfrac{-2 + 4i\sqrt{2}}{3}\right\}$

82. $\left\{\dfrac{-1 - 2i\sqrt{6}}{2}, \dfrac{-1 + 2i\sqrt{6}}{2}\right\}$ **83.** $\{-2, 3\}$ **84.** $\{-3, 4\}$

85. $\{3 - i, 3 + i\}$ **86.** $\{4 - i, 4 + i\}$

87. $\{6\}$ **88.** $\{5\}$

89. $\left\{\dfrac{9 - \sqrt{65}}{2}, \dfrac{9 + \sqrt{65}}{2}\right\}$ **90.** $\left\{\dfrac{7 - \sqrt{41}}{2}, \dfrac{7 + \sqrt{41}}{2}\right\}$

95. 136.9 ft/sec **96.** 3.5 sec **97.** 12

98. a) 0.75 sec and 1.5 sec **b)** 1.125 sec **c)** 2.25 sec

99. c **100. a)** $k = 4$ **b)** $k < 4$ **c)** $k > 4$

103. $\{4.56, 2.74\}$ **104.** $\{2.04, 0.58\}$ **105.** $\{3.53\}$

106. $\{-3.03, 3.68\}$

Section 8.2 Warm-ups T F T F T T T T F F

1. The quadratic formula can be used to solve any quadratic equation.

2. The even-root property is used when $b = 0$.

3. Factoring is used when the quadratic polynomial is simple enough to factor.

4. The quadratic formula can be used on any quadratic equation, but generally we use factoring or the even-root property when applicable.

5. The discriminant is $b^2 - 4ac$.

6. In the complex number system any quadratic equation has either one or two solutions.

7. $\{-3, -2\}$ **8.** $\{3, 4\}$ **9.** $\{-3, 2\}$ **10.** $\{-4, 2\}$

11. $\left\{-\dfrac{1}{3}, \dfrac{3}{2}\right\}$ **12.** $\left\{-\dfrac{1}{2}, \dfrac{1}{4}\right\}$ **13.** $\left\{\dfrac{1}{2}\right\}$ **14.** $\left\{\dfrac{3}{2}\right\}$ **15.** $\left\{\dfrac{1}{3}\right\}$

16. $\left\{\dfrac{4}{3}\right\}$ **17.** $\left\{-\dfrac{3}{4}\right\}$ **18.** $\left\{-\dfrac{2}{5}\right\}$ **19.** $\{-4 \pm \sqrt{10}\}$

20. $\{-3 \pm \sqrt{5}\}$ **21.** $\left\{\dfrac{-5 \pm \sqrt{29}}{2}\right\}$ **22.** $\left\{\dfrac{-3 \pm \sqrt{29}}{2}\right\}$

23. $\left\{\dfrac{3 \pm \sqrt{7}}{2}\right\}$ **24.** $\left\{\dfrac{4 \pm \sqrt{10}}{3}\right\}$ **25.** $\left\{\dfrac{3 \pm i}{2}\right\}$

26. $\left\{\dfrac{1 \pm i}{2}\right\}$ **27.** $\left\{\dfrac{3 \pm i\sqrt{39}}{4}\right\}$ **28.** $\left\{\dfrac{-1 \pm i\sqrt{14}}{3}\right\}$

29. $\{5 \pm i\}$ **30.** $\{4 \pm i\}$ **31.** 28, 2 **32.** 0, 1

33. $-23, 0$ **34.** $-7, 0$ **35.** 0, 1 **36.** 29, 2 **37.** $-\dfrac{3}{4}, 0$

38. $-\dfrac{7}{18}, 0$ **39.** 97, 2 **40.** 0, 1 **41.** 0, 1 **42.** 1, 2

43. 140, 2 **44.** $-120, 0$ **45.** 1, 2 **46.** 49, 2

47. $\left\{-2, \dfrac{1}{2}\right\}$ **48.** $\{-1 \pm \sqrt{3}\}$ **49.** $\{0\}$ **50.** $\{1 \pm i\sqrt{7}\}$

51. $\left\{\dfrac{13}{9}, \dfrac{17}{9}\right\}$ **52.** $\left\{-\dfrac{1}{2}\right\}$ **53.** $\{4 \pm 2i\}$ **54.** $\{3 \pm 5i\}$

55. $\{2 \pm i\sqrt{6}\}$ **56.** $\{6 \pm 2i\}$ **57.** $\left\{-\dfrac{3}{4}, \dfrac{5}{2}\right\}$ **58.** $\left\{\dfrac{1}{4}, \dfrac{4}{3}\right\}$

59. $\{-4.474, 1.274\}$ **60.** $\{-6.664, -0.486\}$ **61.** $\{3.7\}$

62. $\{-1.917\}$ **63.** $\{-2.979, -0.653\}$

64. $\{-1.558, 0.373\}$ **65.** $\{-4792.983, -0.017\}$

66. $\{-12,346.454, -0.546\}$ **67.** $\{-0.079, 0.078\}$

68. $\{-0.332, 2.625\}$ **69.** $\dfrac{1 + \sqrt{65}}{2}$ and $\dfrac{-1 + \sqrt{65}}{2}$

70. $1 + \sqrt{11}$ and $-1 + \sqrt{11}$ **71.** $3 + \sqrt{5}$ and $3 - \sqrt{5}$

72. $4 - \sqrt{14}$ and $4 + \sqrt{14}$

73. Width $\dfrac{-1 + \sqrt{5}}{2}$ ft, length $\dfrac{1 + \sqrt{5}}{2}$ ft

74. Width $-1 + \sqrt{2}$ m, length $1 + \sqrt{2}$ m

75. Width $-2 + \sqrt{14}$ ft, length $2 + \sqrt{14}$ ft

76. $2 + 2\sqrt{2}$ m **77.** 3 sec **78.** $\sqrt{6}$ sec

79. 7.02 sec **80. a)** 9.408 sec **b)** 356.6 ft

81. 80 **82.** 20 **83.** 4 in.

89. $\{0.652, 5.678\}$ **90.** $\{-22.975, 21.642\}$

91. $\varnothing$ **92.** $\varnothing$ **93.** $\varnothing$ **94.** $\varnothing$

Section 8.3 Warm-ups T F F T F F F T F F

1. If the coefficients are integers and the discriminant is a perfect square, then the quadratic polynomial can be factored.

2. The number k is a solution to a quadratic equation if and only if $x - k$ is a factor of the quadratic polynomial.

3. If the solutions are a and b, then the quadratic equation $(x - a)(x - b) = 0$ has those solutions.

4. An equation of quadratic form is one that can be converted to a quadratic equation by making a substitution.

5. Prime   **6.** $(2x + 5)(x - 1)$ **7.** Prime   **8.** Prime

9. $(3x - 4)(2x + 9)$ **10.** $(2x - 3)(4x + 9)$ **11.** Prime

12. $(4x - 15)(x - 3)$ **13.** $(4x - 15)(2x + 3)$ **14.** Prime

15. $x^2 + 4x - 21 = 0$ **16.** $x^2 + 6x - 16 = 0$

17. $x^2 - 5x + 4 = 0$ **18.** $x^2 - 5x + 6 = 0$

19. $x^2 - 5 = 0$ **20.** $x^2 - 7 = 0$ **21.** $x^2 + 16 = 0$

22. $x^2 + 9 = 0$ **23.** $x^2 + 2 = 0$ **24.** $x^2 + 18 = 0$

25. $6x^2 - 5x + 1 = 0$ **26.** $10x^2 + 7x + 1 = 0$

27. $\left\{-\dfrac{3}{2}, \dfrac{3}{2}\right\}$ **28.** $\left\{-\dfrac{4}{3}, 1\right\}$ **29.** $\left\{\dfrac{-3 \pm \sqrt{5}}{2}\right\}$

30. $\left\{\dfrac{3 \pm \sqrt{2}}{2}\right\}$ **31.** $\{\pm\sqrt{5}, \pm 3\}$ **32.** $\{\pm\sqrt{3}\}$

33. $\{-2, 1\}$ **34.** $\{-2, \sqrt[3]{2}\}$ **35.** $\{-1 \pm \sqrt{5}, -3, 1\}$

36. $\{-4, 1\}$ **37.** $\{-3, -2, 1, 2\}$ **38.** $\{-2, -1, 3, 4\}$

39. $\{16, 81\}$ **40.** $\left\{\dfrac{1}{4}, 4\right\}$ **41.** $\{9\}$ **42.** $\{16\}$ **43.** $\left\{-\dfrac{1}{3}, \dfrac{1}{2}\right\}$

44. $\left\{-\dfrac{1}{2}, \dfrac{1}{4}\right\}$ **45.** $\{64\}$ **46.** $\{-64, 125\}$ **47.** $\left\{\dfrac{2}{3}, \dfrac{3}{2}\right\}$

48. $\left\{-\dfrac{5}{6}, -\dfrac{5}{4}\right\}$ **49.** $\left\{\pm\dfrac{\sqrt{14}}{2}, \pm\dfrac{\sqrt{38}}{2}\right\}$ **50.** $\left\{\dfrac{-1 \pm \sqrt{5}}{2}\right\}$

51. $\{-1 + \sqrt{2}, -1 - \sqrt{2}\}$ **52.** $\left\{\dfrac{3 + \sqrt{3}}{6}, \dfrac{3 - \sqrt{3}}{6}\right\}$

53. 2 P.M.   **54.** $2\sqrt{34}$ or 11.662 mph

55. Before $-5 + \sqrt{265}$ or 11.3 mph, after $-9 + \sqrt{265}$ or 7.3 mph

56. Kim $-5 + \sqrt{145}$ or 7.042 mph, Bryan $5 + \sqrt{145}$ or 17.042 mph

57. Andrew $\dfrac{13 + \sqrt{265}}{2}$ or 14.6 hours, John $\dfrac{19 + \sqrt{265}}{2}$ or 17.6 hours

58. Brent $\dfrac{5 + \sqrt{13}}{4}$ or 2.151 hours, Calvin $\dfrac{1 + \sqrt{13}}{4}$ or 1.151 hours

59. Length $5 + 5\sqrt{41}$ or 37.02 ft, width $-5 + 5\sqrt{41}$ or 27.02 ft

60. $\dfrac{25 - \sqrt{329}}{4}$ or 1.715 inches

61. $14 + 2\sqrt{58}$ or 29.2 hours   **62.** 0.788 hour or 47 minutes

63. $-5 + 5\sqrt{5}$ or 6.2 meters   **66.** $\{1, 2\}$

67. $\{-10, -4, 4, 10\}$ **68.** $\{-4.25, -3.49, 0.49, 1.25\}$

69. $\{4.27\}$

Section 8.4 Warm-ups F F F F T T T T T F

1. A quadratic inequality has the form $ax^2 + bx + c > 0$. In place of $>$ we can also use $<$, $\leq$, or $\geq$.

2. A sign graph shows signs of the factors for all possible values of x.

3. A rational inequality is an inequality involving a rational expression.

4. Multiplying each side by a positive number does not change the direction of the inequality, but multiplying by a negative number does. So if we multiply by a variable, it is difficult to know which way the inequality goes.

5. $(-3, 2)$

6. $(-\infty, -1] \cup [4, \infty)$

7. $(-\infty, -2) \cup (2, \infty)$

8. $(-4, 4)$

9. $(-\infty, -4] \cup \left[\dfrac{3}{2}, \infty\right)$

10. $\left(-4, \dfrac{1}{2}\right)$

11. $(-\infty, 0] \cup [2, \infty)$

12. $(-\infty, -1) \cup (0, \infty)$

13. $(-\infty, 0) \cup \left(\dfrac{1}{2}, \infty\right)$

14. $(0, 3)$

15. $(-\infty, \infty)$ **16.** $\varnothing$

17. $(-\infty, 0) \cup (3, \infty)$

18. $(-\infty, -2) \cup (0, \infty)$

19. $[-2, 0)$

20. $(0, 6]$

21. $(-\infty, -6) \cup (3, \infty)$

22. $\left(-\dfrac{5}{2}, 2\right)$

23. $(-\infty, -2) \cup (-1, \infty)$

24. $[-1, 0)$

25. $(-13, -4) \cup (5, \infty)$

26. $(-2, 1) \cup (7, \infty)$

27. $(-\infty, -5) \cup (1, 3) \cup (5, \infty)$

28. $[-4, 6) \cup [8, 16)$ **29.** $[-6, 3) \cup [4, 6)$

30. $(-\infty, -20) \cup (-10, -8) \cup (4, \infty)$

31. $\left(-\infty, 1 - \sqrt{5}\right) \cup \left(1 + \sqrt{5}, \infty\right)$

32. $\left[1 - \sqrt{6}, 1 + \sqrt{6}\right]$

33. $\left(-\infty, \dfrac{3 - \sqrt{3}}{2}\right] \cup \left[\dfrac{3 + \sqrt{3}}{2}, \infty\right)$

34. $\left(\dfrac{4 - \sqrt{10}}{2}, \dfrac{4 + \sqrt{10}}{2}\right)$

35. $\left[\dfrac{3 - 3\sqrt{5}}{2}, \dfrac{3 + 3\sqrt{5}}{2}\right]$

36. $\left(\dfrac{5 - \sqrt{53}}{2}, \dfrac{5 + \sqrt{53}}{2}\right)$

37. $[-3, 3]$ **38.** $(-\infty, -6] \cup [6, \infty)$ **39.** $(-4, 4)$
40. $(-\infty, -3) \cup (3, \infty)$ **41.** $(-\infty, 0] \cup [4, \infty)$
42. $\left(-\infty, -\dfrac{3}{2}\right) \cup \left(\dfrac{3}{2}, \infty\right)$ **43.** $\left(-\dfrac{3}{2}, \dfrac{5}{3}\right)$ **44.** $\left(-\dfrac{3}{2}, \dfrac{4}{3}\right)$
45. $(-\infty, -2] \cup [6, \infty)$ **46.** $\varnothing$ **47.** $(-\infty, -3) \cup (5, \infty)$
48. $(-2, 4)$ **49.** $(-\infty, -4] \cup [2, \infty)$ **50.** $[-4, 6]$
51. $(-3, 4]$ **52.** $(-\infty, -5) \cup \left[\dfrac{1}{2}, \infty\right)$
53. $[-1, 2] \cup [5, \infty)$ **54.** $(-\infty, -2) \cup (1, 2.5)$
55. $(-\infty, -3) \cup (-1, 1)$ **56.** $[-5, -2] \cup [2, \infty)$
57. $(-27.58, -0.68)$ **58.** $(-\infty, \infty)$
59. $\left(-\infty, -2 - \sqrt{6}\right) \cup \left(-3, -2 + \sqrt{6}\right) \cup (2, \infty)$
60. $\left(\dfrac{-7 - \sqrt{73}}{2}, -5\right) \cup \left(\dfrac{-7 + \sqrt{73}}{2}, 3\right)$

61. $6, 7, 8, \ldots$ **62.** $1, 2, 3, \ldots, 47$ **63.** 4 seconds
64. $t < 0.43$ second or $t > 1.44$ seconds
65. a) 900 ft **b)** 3 sec **c)** 3 sec   **66.** 25.2 seconds
67. a) (h, k) **b)** $(-\infty, h) \cup (k, \infty)$
   c) $(-k, -h)$ **d)** $(-\infty, -k] \cup [-h, \infty)$
   e) $(-\infty, h] \cup (k, \infty)$ **f)** $(-k, -h]$

68. a) $(-\infty, -b/(2a)) \cup (-b/(2a), \infty)$
   b) $\varnothing$ **c)** $(-\infty, \infty)$ **d)** $\varnothing$
   e) $\left(-\infty, \dfrac{-b - \sqrt{b^2 - 4ac}}{2a}\right) \cup \left(\dfrac{-b + \sqrt{b^2 - 4ac}}{2a}, \infty\right)$
   f) $\left(\dfrac{-b - \sqrt{b^2 - 4ac}}{2a}, \dfrac{-b + \sqrt{b^2 - 4ac}}{2a}\right)$
69. c   **70.** d   **71.** b   **72.** a

Enriching Word Power Chapter 8
1. b   **2.** a   **3.** d   **4.** c   **5.** b
6. c   **7.** a   **8.** c   **9.** a   **10.** c

Chapter 8 Review

1. $\{-3, 5\}$ **2.** $\{-4, 6\}$ **3.** $\left\{-3, \dfrac{5}{2}\right\}$ **4.** $\left\{-4, \dfrac{1}{2}\right\}$

5. $\{-5, 5\}$ **6.** $\{-11, 11\}$ **7.** $\left\{\dfrac{3}{2}\right\}$ **8.** $\{6\}$ **9.** $\{\pm 2\sqrt{3}\}$

10. $\{\pm 2\sqrt{5}\}$ **11.** $\{-2, 4\}$ **12.** $\{-6, -2\}$

13. $\left\{\dfrac{4 \pm \sqrt{3}}{2}\right\}$ **14.** $\left\{\dfrac{5}{2}, \dfrac{7}{2}\right\}$ **15.** $\left\{\pm \dfrac{3}{2}\right\}$ **16.** $\left\{\pm \dfrac{\sqrt{6}}{2}\right\}$

17. $\{2, 4\}$ **18.** $\{-3, -1\}$ **19.** $\{2, 3\}$ **20.** $\{-2, 3\}$

21. $\left\{\dfrac{1}{2}, 3\right\}$ **22.** $\left\{-\dfrac{3}{2}, 2\right\}$ **23.** $\left\{-2 \pm \sqrt{3}\right\}$

24. $\left\{-1 \pm \sqrt{3}\right\}$ **25.** $\{-2, 5\}$ **26.** $\{-1, 6\}$ **27.** $\left\{-\dfrac{1}{3}, \dfrac{3}{2}\right\}$

28. $\left\{-\dfrac{1}{2}, \dfrac{2}{3}\right\}$ **29.** $\left\{-2 \pm \sqrt{2}\right\}$ **30.** $\left\{-3 \pm \sqrt{11}\right\}$

31. $\left\{\dfrac{5 \pm \sqrt{13}}{6}\right\}$ **32.** $\left\{\dfrac{-3 \pm \sqrt{17}}{4}\right\}$ **33.** 0, 1

34. 0, 1   **35.** $-19, 0$ **36.** $-95, 0$ **37.** 17, 2   **38.** 12, 2

39. $\left\{\dfrac{2 \pm i\sqrt{2}}{2}\right\}$ **40.** $\left\{\dfrac{3 \pm i}{2}\right\}$ **41.** $\left\{\dfrac{3 \pm i\sqrt{15}}{4}\right\}$

42. $\left\{\dfrac{-1 \pm i\sqrt{3}}{2}\right\}$ **43.** $\left\{\dfrac{-1 \pm i\sqrt{5}}{3}\right\}$ **44.** $\{1 \pm i\}$

45. $\left\{-3 \pm i\sqrt{7}\right\}$ **46.** $\{5 \pm i\}$ **47.** $(4x + 1)(2x - 3)$
48. $(6x - 1)(3x + 2)$ **49.** Prime   **50.** Prime
51. $(4y - 5)(2y + 5)$ **52.** $(5z + 3)(5z - 6)$
53. $x^2 + 9x + 18 = 0$ **54.** $x^2 + 5x - 36 = 0$
55. $x^2 - 50 = 0$ **56.** $x^2 + 12 = 0$ **57.** $\{-2, 1\}$
58. $\left\{-2, \dfrac{1}{2}\right\}$ **59.** $\{\pm 2, \pm 3\}$ **60.** $\varnothing$ **61.** $\{-6, -5, 2, 3\}$
62. $\{\pm 2, \pm \sqrt{2}\}$ **63.** $\{-2, 8\}$ **64.** $\left\{-1, 4, \dfrac{3 \pm \sqrt{13}}{2}\right\}$
65. $\left\{-\dfrac{1}{9}, \dfrac{1}{4}\right\}$ **66.** $\left\{-\dfrac{1}{3}, \dfrac{1}{2}\right\}$ **67.** $\{16, 81\}$ **68.** $\left\{\dfrac{1}{16}, 1\right\}$
69. $(-\infty, -3) \cup (2, \infty)$ **70.** $(-\infty, 2) \cup (3, \infty)$

71. $[-4, 5]$ **72.** $[-5, 3]$

73. $(0, 1)$

74. $(-\infty, 0] \cup [1, \infty)$

75. $(-\infty, -2) \cup [4, \infty)$

76. $(-5, 3)$

77. $(-3, \infty)$

78. $(-11, -4)$

79. $(-2, -1) \cup \left(-\dfrac{1}{2}, \infty\right)$

80. $(-\infty, -1) \cup (1, \infty)$

81. $\left\{\dfrac{5}{12}\right\}$ **82.** $\left\{\dfrac{3}{7}\right\}$ **83.** $\left\{\dfrac{-3 \pm \sqrt{5}}{2}\right\}$ **84.** $\left\{-\dfrac{5}{3}, -\dfrac{5}{2}\right\}$

85. $\left\{\dfrac{4 \pm 2i}{3}\right\}$ **86.** $\{1 \pm \sqrt{3}\}$ **87.** $\left\{\dfrac{5}{2}\right\}$

88. $\{\pm\sqrt{6}, \pm i\sqrt{3}\}$ **89.** $\left\{-2, -\dfrac{1}{4}\right\}$ **90.** $\{\pm 2i, \pm 2\}$

91. $\{625, 10{,}000\}$ **92.** $\left\{\dfrac{1}{6}, \dfrac{1}{3}\right\}$

93. $-2 + 2\sqrt{2}$ and $2 + 2\sqrt{2}$, or 0.83 and 4.83

94. $\dfrac{-1 + \sqrt{5}}{2}$ and $\dfrac{1 + \sqrt{5}}{2}$, or 0.62 and 1.62

95. Width $\dfrac{4 + \sqrt{706}}{2}$ or 15.3 inches, height $\dfrac{-4 + \sqrt{706}}{2}$ or 11.3 inches

96. $20\sqrt{2}$ or 28.284 ft **97.** 2 inches

98. $\dfrac{3 + \sqrt{17}}{2}$ or 3.562 hours **99.** Width 5 ft, length 9 ft

100. $L = 22$ yards and $W = 16$ yards

101. 0.5 second and 1.5 seconds

102. 2 seconds, $0.5\sqrt{2}$ or 0.707 second

Chapter 8 Test

1. $-7, 0$ **2.** $13, 2$ **3.** $0, 1$ **4.** $\left\{-3, \dfrac{1}{2}\right\}$

5. $\{-3 \pm \sqrt{3}\}$ **6.** $\{-5\}$ **7.** $\left\{-2, \dfrac{3}{2}\right\}$ **8.** $\{-4, 3\}$

9. $\{\pm 1, \pm 2\}$ **10.** $\{11, 27\}$ **11.** $\{\pm 6i\}$ **12.** $\{-3 \pm i\}$

13. $\left\{\dfrac{1 \pm i\sqrt{11}}{6}\right\}$ **14.** $x^2 - 2x - 24 = 0$ **15.** $x^2 + 25 = 0$

16. $(-6, 3)$

17. $(-1, 2) \cup (8, \infty)$

18. Width $-1 + \sqrt{17}$ ft, length $1 + \sqrt{17}$ ft

19. $\dfrac{5 + \sqrt{37}}{2}$ or 5.5 hours

Making Connections Chapters 1–8

1. $\left\{\dfrac{15}{2}\right\}$ **2.** $\left\{\pm\dfrac{\sqrt{30}}{2}\right\}$ **3.** $\left\{-3, \dfrac{5}{2}\right\}$ **4.** $\left\{\dfrac{-2 \pm \sqrt{34}}{2}\right\}$

5. $\left\{-\dfrac{7}{2}, -2\right\}$ **6.** $\left\{-3, \dfrac{1}{4}, \dfrac{-11 \pm \sqrt{73}}{8}\right\}$ **7.** $\{9\}$

8. $\left\{-\dfrac{3}{2}, \dfrac{13}{2}\right\}$ **9.** $(-4, \infty)$ **10.** $\left[\dfrac{1}{2}, 5\right]$ **11.** $\left[\dfrac{1}{2}, 5\right)$

12. $(2, 8)$ **13.** $[-3, 2)$ **14.** $(-\infty, \infty)$ **15.** $y = \dfrac{2}{3}x - 3$

16. $y = -\dfrac{1}{2}x + 2$ **17.** $y = \dfrac{-c \pm \sqrt{c^2 - 12d}}{6}$

18. $y = \dfrac{n \pm \sqrt{n^2 + 4mw}}{2m}$ **19.** $y = \dfrac{5}{6}x - \dfrac{25}{12}$

20. $y = -\dfrac{2}{3}x + \dfrac{17}{3}$ **21.** $\dfrac{4}{3}$ **22.** $-\dfrac{11}{7}$ **23.** -2

24. $\dfrac{58}{5}$ **25.** 40,000, 38,000, $32.50

26. $800,000, $950,000, $40 or $80, $60

CHAPTER 9

Section 9.1 Warm-ups T T F F F F T T T T

1. The basic operations of functions are addition, subtraction, multiplication, and division.

2. We perform the operations with functions by adding, subtracting, multiplying, or dividing the expressions that define the functions.

3. In the composition of functions the second function is evaluated on the result of the first function.

4. Since each operation is a function, the order of operations determines the order in which the functions are composed.

5. $x^2 + 2x - 3$ **6.** $-x^2 + 6x - 3$ **7.** $4x^3 - 11x^2 + 6x$

8. $\dfrac{4x - 3}{x^2 - 2x}$ **9.** 12 **10.** 5 **11.** -30 **12.** -19

13. -21 **14.** -88 **15.** $\dfrac{13}{8}$ **16.** $-\dfrac{11}{8}$

17. $y = 6x - 20$ **18.** $y = -6x + 11$ **19.** $y = x + 2$

20. $y = x$ **21.** $y = x^2 + 2x$ **22.** $y = x^2 + x - 1$

23. $y = x$ **24.** $y = x$ **25.** -2 **26.** -7 **27.** 5

28. 28 **29.** 7 **30.** 3 **31.** 5 **32.** 0 **33.** 5 **34.** 0

35. 4 **36.** $\dfrac{1}{2}$ **37.** 22.2992 **38.** 23.761 **39.** $4x^2 - 6x$

40. $\dfrac{x^2 + 12x + 27}{4}$ **41.** $2x^2 + 6x - 3$ **42.** $\dfrac{x^2 + 3x + 3}{2}$

43. x **44.** x **45.** $4x - 9$ **46.** $x^4 + 6x^3 + 12x^2 + 9x$

47. $\dfrac{x + 9}{4}$ **48.** $8x - 21$ **49.** $F = f \circ h$ **50.** $N = h \circ f$

51. $G = g \circ h$ **52.** $P = f \circ g$ **53.** $H = h \circ g$

54. $M = f \circ f$ **55.** $J = h \circ h$ **56.** $R = f \circ h \circ g$

57. $K = g \circ g$ **58.** $Q = f \circ g \circ h$

67. 112.5 in.^{2}, $A = \dfrac{d^2}{2}$ **68.** $P = 4\sqrt{A}$.

69. $P(x) = -x^2 + 20x - 170$ **70.** $A = \dfrac{(32 + 3\pi)x^2}{8}$

71. $J = 0.025I$ **72.** $A = \pi\dfrac{M}{4}$

73. a) 397.8 **b)** $D = \dfrac{1.116 \times 10^7}{L^3}$ **c)** decreases

74. a) 15.97 **b)** $S = 14{,}400d^{-2/3}$ **c)** decreases

75. $[0, \infty), [0, \infty), [16, \infty)$ **76.** $[4, \infty), [8, \infty), [8, \infty)$

77. $[0, \infty), [0, \infty)$ **78.** $(-\infty, \infty), [3, \infty)$

Section 9.2 Warm-ups F F F T T F T T T F

1. The inverse of a function is a function with the same ordered pairs except that the coordinates are reversed.

2. The domain of f^{-1} is the range of f.

3. The range of f^{-1} is the domain of f.

4. The -1 in f^{-1} is not treated as an exponent. It is simply a notation for the inverse of the function f.

5. A function is one-to-one if no two ordered pairs have the same second coordinate with different first coordinates.

6. The horizontal-line test says that if a horizontal line can be drawn to cross the graph of a function more than once, then the function is not one-to-one.

7. The switch-and-solve strategy is used to find a formula for an inverse function.

8. The graphs of f and f^{-1} are symmetric with respect to the line $y = x$.

9. No   **10.** Yes, $\{(1, 1), (8, 2), (27, 3)\}$

11. Yes, $\{(4, 16), (3, 9), (0, 0)\}$

12. No   **13.** No

14. Yes, $\{(-3, 3), (2, -2), (-1, 1)\}$

15. Yes, $\{(0, 0), (2, 2), (9, 9)\}$

16. No   **17.** No   **18.** No

19. Yes   **20.** Yes   **21.** Yes   **22.** No   **23.** Yes

24. Yes   **25.** Yes   **26.** Yes   **27.** No   **28.** No

29. $f^{-1}(x) = \dfrac{x}{5}$ **30.** $h^{-1}(x) = -\dfrac{1}{3}x$

31. $g^{-1}(x) = x + 9$ **32.** $j^{-1}(x) = x - 7$

33. $k^{-1}(x) = \dfrac{x + 9}{5}$ **34.** $r^{-1}(x) = \dfrac{x + 8}{2}$

35. $m^{-1}(x) = \dfrac{2}{x}$ **36.** $s^{-1}(x) = -\dfrac{1}{x}$

37. $f^{-1}(x) = x^3 + 4$ **38.** $f^{-1}(x) = x^3 - 2$

39. $f^{-1}(x) = \dfrac{3}{x} + 4$ **40.** $f^{-1}(x) = \dfrac{2}{x} - 1$

41. $f^{-1}(x) = \dfrac{x^3 - 7}{3}$ **42.** $f^{-1}(x) = \dfrac{-x^3 + 7}{5}$

43. $f^{-1}(x) = \dfrac{2x + 1}{x - 1}$ **44.** $f^{-1}(x) = \dfrac{1 - 3x}{x + 1}$

45. $f^{-1}(x) = \dfrac{1 + 4x}{3x - 1}$ **46.** $g^{-1}(x) = \dfrac{3x + 5}{2x - 3}$

47. $p^{-1}(x) = x^4$ for $x \geq 0$ **48.** $v^{-1}(x) = x^6$ for $x \geq 0$

49. $f^{-1}(x) = 2 + \sqrt{x}$ **50.** $g^{-1}(x) = -5 + \sqrt{x}$

51. $f^{-1}(x) = \sqrt{x - 3}$ **52.** $f^{-1}(x) = \sqrt{x + 5}$

53. $f^{-1}(x) = x^2 - 2$ for $x \geq 0$

54. $f^{-1}(x) = x^2 + 4$ for $x \geq 0$

55. $f^{-1}(x) = \dfrac{1}{2}x - \dfrac{3}{2}$

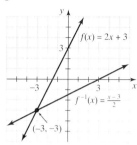

56. $f^{-1}(x) = -\dfrac{1}{3}x + \dfrac{2}{3}$

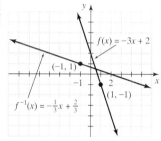

57. $f^{-1}(x) = \sqrt{x + 1}$ **58.** $f^{-1}(x) = \sqrt{x - 3}$

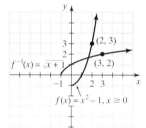   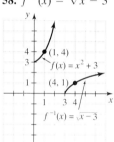

59. $f^{-1}(x) = \dfrac{x}{5}$ **60.** $f^{-1}(x) = 4x$

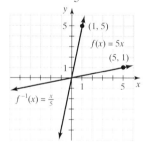   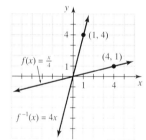

61. $f^{-1}(x) = \sqrt[3]{x}$ **62.** $f^{-1}(x) = \sqrt[3]{\dfrac{x}{2}}$

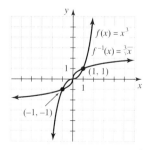   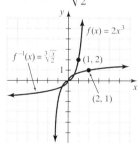

63. $f^{-1}(x) = x^2 + 2$ for $x \geq 0$

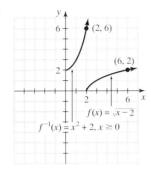

64. $f^{-1}(x) = x^2 - 3$ for $x \geq 0$

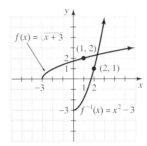

65. $(f^{-1} \circ f)(x) = x$ **66.** $(f^{-1} \circ f)(x) = x$
67. $(f^{-1} \circ f)(x) = x$ **68.** $(f^{-1} \circ f)(x) = x$
69. $(f^{-1} \circ f)(x) = x$ **70.** $(f^{-1} \circ f)(x) = x$
71. $(f^{-1} \circ f)(x) = x$ **72.** $(f^{-1} \circ f)(x) = x$
73. a) 33.5 mph   **b)** decreases   **c)** $L = \dfrac{S^2}{30}$
74. $h(x) = \pi x^2$, the area of the circle, $h^{-1}(x) = \sqrt{\dfrac{x}{\pi}}$
75. $T(x) = 1.09x + 125$, $T^{-1}(x) = \dfrac{x - 125}{1.09}$
76. $C(x) = 2x^2 + 50$, $C^{-1}(x) = \sqrt{\dfrac{x - 50}{2}}$
77. An odd positive integer   **78.** No
79. Not inverses   **80.** They are inverse functions.

Section 9.3 Warm-ups T F F T F T F F F T
1. If y varies directly as x, then $y = kx$ for some constant k.
2. The constant of proportionality in $y = kx$ is k.
3. If y is inversely proportional to x, then $y = k/x^2$.
4. In direct variation $y = kx$, whereas in inverse variation $y = k/x$.
5. If y is jointly proportional to x and z, then $y = kxz$ for some constant k.
6. Varies directly is the same as directly proportional.
7. $a = km$ **8.** $w = kP$ **9.** $d = k/e$ **10.** $y = k/x$
11. $I = krt$ **12.** $q = kwv$ **13.** $m = kp^2$ **14.** $g = kr^3$
15. $B = k\sqrt[3]{w}$ **16.** $F = km^2$ **17.** $t = \dfrac{k}{x^2}$ **18.** $y = \dfrac{k}{\sqrt{z}}$
19. $v = \dfrac{km}{n}$ **20.** $b = \dfrac{kn^2}{\sqrt{v}}$ **21.** $y = \dfrac{3}{2}x$ **22.** $m = \dfrac{4}{3}w$
23. $A = \dfrac{30}{B}$ **24.** $c = \dfrac{0.62}{d}$ **25.** $m = \dfrac{36}{\sqrt{p}}$ **26.** $s = \dfrac{3\sqrt{6}}{\sqrt{v}}$
27. $A = \dfrac{2}{5}tu$ **28.** $N = p^2 q^3$ **29.** $y = \dfrac{1.067x}{z}$

30. $a = \dfrac{8.339\sqrt{m}}{n^2}$ **31.** $-\dfrac{21}{5}$ **32.** $3\sqrt{2}$ **33.** $-\dfrac{3}{2}$
34. $\sqrt{3}$ **35.** $\dfrac{1}{2}$ **36.** -720 **37.** 3.293   **38.** 237.469
39. Inverse   **40.** Inverse   **41.** Direct   **42.** Direct
43. Combined   **44.** Combined   **45.** Joint   **46.** Joint
47. \$420   **48.** 37.5 pounds   **49.** 9 cm^3 **50.** \$72 each
51. \$360   **52.** \$2.55   **53.** \$18.90   **54.** 80 ounces
55. a) $d = 16t^2$ **b)** 4 feet   **c)** 2.5 seconds   **56.** \$0.80
57. 400 pounds   **58.** 0.75 ohm   **59.** 7.2 days   **60.** 24 cm^3
61. a) $G = \dfrac{Nd}{c}$ **b)** 84   **c)** 23, 20, 17, 15, 13   **d)** decreases
62. y_1 increasing, y_2 decreasing
63. (1, 1), y_1 increasing, y_2 decreasing, y_1 direct variation, y_2 inverse variation

Section 9.4 Warm-ups T T F T T T T T T T
1. A zero of the function f is a number a such that $f(a) = 0$.
2. A root of a function is the same as a zero.
3. Two statements are equivalent means that they are either both true or both false.
4. To divide by $x - c$ quickly, use synthetic division.
5. If the remainder is zero when $P(x)$ is divided by $x - c$, then $P(c) = 0$.
6. The number c is a zero of a polynomial if the remainder in synthetic division is zero or if directly evaluating the polynomial at $x = c$ gives a value of zero.
7. Yes   **8.** Yes   **9.** Yes   **10.** No   **11.** No   **12.** No
13. Yes   **14.** Yes   **15.** No   **16.** Yes   **17.** Yes
18. Yes   **19.** Yes   **20.** No   **21.** $(x - 3)(x^2 + 3x + 3)$
22. $(x + 2)(x^2 - 2x - 2)$ **23.** $(x + 5)(x + 3)(x + 1)$
24. No   **25.** No   **26.** $(x + 5)(x^2 - 5x + 25)$
27. $(x + 1)(x - 2)(x^2 + 2x + 4)$ **28.** $(x - 2)^3$
29. $(2x - 1)(x - 3)(x + 2)$ **30.** $(3x - 1)(x - 5)(x + 2)$
31. $\{-4, 1, 3\}$ **32.** $\{-3, -1, 2\}$ **33.** $\left\{-\dfrac{1}{2}, 2, 3\right\}$
34. $\left\{-\dfrac{2}{3}, -2, \dfrac{1}{2}\right\}$ **35.** $\left\{-4, -\dfrac{1}{2}, 6\right\}$ **36.** $\{-2, 4, 5\}$
37. $\{-3, 1\}$ **38.** $\{-2\}$ **39.** $\{-1, 1, 2\}$ **40.** $\{-3, -1, 1, 2\}$
41. a) $f(x) = x + 2$ **b)** $f(x) = x^2 - 25$
    c) $f(x) = (x - 1)(x + 3)(x - 4)$
    d) yes, the degree is the same as the number of zeros
42. a) $\{-2, 1, 3\}$ **b)** $\left\{-\dfrac{1}{3}, \dfrac{1}{2}, \dfrac{3}{2}\right\}$
43. a) $\{-2, 1, 5\}$ **b)** $\left\{-\dfrac{3}{2}, \dfrac{3}{2}, \dfrac{5}{2}\right\}$

Enriching Word Power
1. a   **2.** b   **3.** d   **4.** d   **5.** a   **6.** b   **7.** d   **8.** c
9. d   **10.** b

Chapter 9 Review
1. -4 **2.** -3 **3.** $\sqrt{2}$ **4.** π **5.** 99
6. $9x^2 + 24x + 15$ **7.** 17   **8.** $-x^2 + 5x + 5$
9. $3x^3 - x^2 - 10x$ **10.** -8 **11.** 20   **12.** $9x + 20$

13. $F = f \circ g$ **14.** $G = g \circ f$ **15.** $H = g \circ h$
16. $K = h \circ g$ **17.** $I = g \circ g$ **18.** $J = g \circ h \circ h$
19. No **20.** Yes, $\{(1, 1), (3, 3)\}$ **21.** Yes, $f^{-1}(x) = x/8$

22. Yes, $i^{-1}(x) = -3x$ **23.** Yes, $g^{-1}(x) = \dfrac{x + 6}{13}$

24. Yes, $h^{-1}(x) = x^3 + 6$ **25.** Yes, $j^{-1}(x) = \dfrac{x + 1}{x - 1}$

26. No **27.** No **28.** Yes, $n^{-1}(x) = \dfrac{3}{x}$

29. $f^{-1}(x) = \dfrac{1}{3}x + \dfrac{1}{3}$ **30.** $f^{-1}(x) = \sqrt{2 - x}$

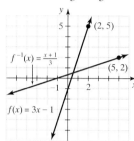
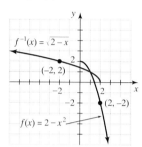

31. $f^{-1}(x) = \sqrt[3]{2x}$
32. $f^{-1}(x) = -4x$

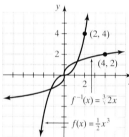
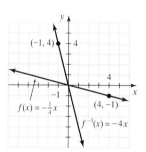

33. 24 **34.** $-\dfrac{9}{2}$ **35.** -16 **36.** $\dfrac{256}{9}$

37. $\{-3, 2, 5\}$ **38.** $\{-\sqrt{6}, \sqrt{6}, 2\}$ **39.** $\{-2, 3, 4\}$
40. $\{-6, -2, 1\}$ **41.** $(x - 3)(x + 2)(x + 5)$
42. $(x + 4)(x - 2)(x - 1)$ **43.** No **44.** No

45. 256 ft **46.** 12 **47.** $B = \dfrac{2A}{\pi}$ **48.** $A = \dfrac{(8 + \pi)s^2}{8}$

49. $a = 15w - 16$ **50.** $V = 10A$ **51.** $A = s^2, s = \sqrt{A}$

52. $A = \pi r^2, r = \sqrt{\dfrac{A}{\pi}}, A = \dfrac{\pi d^2}{4}$

Chapter 9 Test
1. 11 **2.** 125 **3.** -3 **4.** $\dfrac{x - 5}{-2}$ **5.** $x^2 - 2x + 9$

6. 15 **7.** 1776 **8.** $\dfrac{1}{8}$ **9.** $-2x^2 - 3$

10. $4x^2 - 20x + 29$ **11.** $H = f \circ g$ **12.** $W = g \circ f$
13. Not invertible **14.** $f^{-1}(x) = (x - 9)^3$

15. $f^{-1}(x) = \dfrac{x + 1}{x - 2}$ **16.** $\dfrac{32\pi}{3}$ ft^3 **17.** $\dfrac{36}{7}$ **18.** $5,076

19. Yes **20.** $\{3, 4, 5\}$

Making Connections Chapters 1–9
1. $\dfrac{1}{25}$ **2.** $\dfrac{3}{2}$ **3.** $\sqrt{2}$ **4.** x^8 **5.** 2 **6.** x^9 **7.** $\{\pm3\}$

8. $\{\pm2\sqrt{2}\}$ **9.** $\{0, 1\}$ **10.** $\{2 \pm \sqrt{10}\}$ **11.** $\{81\}$

12. $\varnothing$ **13.** $\{\pm8\}$ **14.** $\left[-\dfrac{17}{5}, 5\right]$ **15.** $\{2\}$ **16.** $\left\{\dfrac{5}{3}\right\}$

17. $\{42\}$ **18.** $\{11\}$

19. 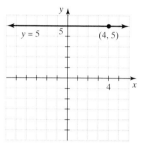 **20.**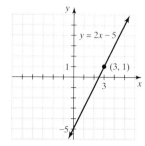

21. 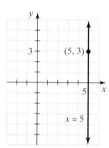 **22.**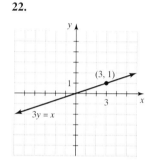

23. 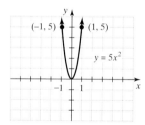 **24.**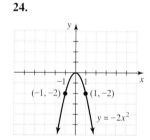

25. $(2, 4), (3, 8), (1, 2), (4, 16)$

26. $\left(\dfrac{1}{2}, 2\right), \left(-1, \dfrac{1}{4}\right), (2, 16), (0, 1)$ **27.** $[0, \infty)$ **28.** $(-\infty, 3]$

29. $(-\infty, \infty)$ **30.** $(-\infty, 1) \cup (1, 9) \cup (9, \infty)$
31. a) $C = 0.12x + 3000$ **b)** $P = 1 \times 10^{-6}x + 0.15$
32. a) $0.44, $0.40, $0.39
b)

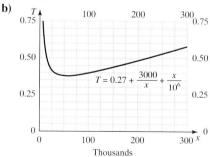

c) 60,000 miles **d)** $[50,000, 60,000]$

CHAPTER 10

Section 10.1 Warm-ups F T F T T F T F T F

1. An exponential function has the form $f(x) = a^x$ where $a > 0$ and $a \neq 1$.
2. The domain of an exponential function is all real numbers.
3. The two most popular bases are e and 10.
4. The one-to-one property states that if $a^m = a^n$, then $m = n$.
5. The compound interest formula is $A = P(1 + i)^n$.
6. When money is compounded continuously, we use the formula $A = Pe^{rt}$

7. 16 8. $\frac{1}{4}$ 9. 2 10. $\frac{1}{8}$ 11. 3 12. $\frac{1}{9}$ 13. $\frac{1}{3}$

14. 9 15. -1 16. -8 17. $-\frac{1}{4}$ 18. $-\frac{1}{16}$ 19. 1

20. 0.1 21. 100 22. 2511.886 23. 2.718 24. 33.115
25. 0.135 26. 1

27.

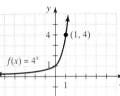

28.

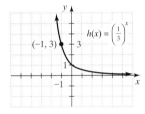

29.

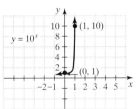

30.

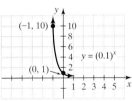

31.

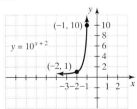

32.

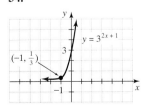

33.

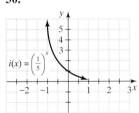

34.

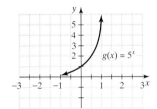

35.

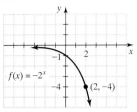

36.

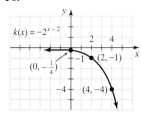

37.

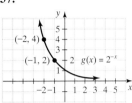

38.

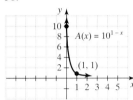

39.

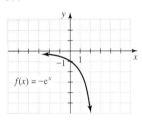

40.

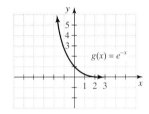

41.

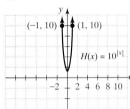

42.

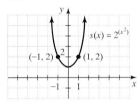

43.

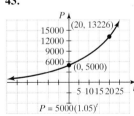

44.

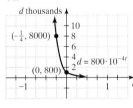

45. $\{6\}$ 46. $\{2\}$ 47. $\{-3\}$ 48. $\left\{-\frac{1}{2}\right\}$ 49. $\{-2\}$

50. $\{-2\}$ 51. $\{-1\}$ 52. $\left\{-\frac{2}{3}\right\}$ 53. $\{-2\}$ 54. $\{2\}$

55. $\{-2\}$ 56. $\{-2\}$ 57. $\{-3, 3\}$ 58. $\left\{\frac{1}{2}, \frac{9}{2}\right\}$ 59. 2

60. -2 61. $\frac{4}{3}$ 62. 0 63. -2 64. 2 65. 0

66. $-\frac{1}{2}$ 67. $\frac{3}{2}$ 68. $\frac{1}{4}$ 69. $\frac{1}{2}$ 70. $\frac{5}{8}$ 71. \$9,861.72

72. \$798.60 73. **a)** \$114,421.26 **b)** 1996

74. $104,931.35   **75.** $45, $12.92
76. 1,000, 10,000, 100,000   **77.** $616.84   **78.** $9,639.89
79. $6,230.73   **80.** $10,937.13
81. 300 grams, 90.4 grams, 12 years, no
82. 20 million, 147.8 million
83. 2.66666667, 0.0516, 2.8×10^{-5} **84.** (0, 1)
85. The graph of $y = 3^{x-k}$ lies k units to the right of $y = 3^x$ when $k > 0$ and $|k|$ units to the left of $y = 3^x$ when $k < 0$.

Section 10.2 Warm-ups T F T T F F F F T T

1. If $f(x) = 2^x$, then $f^{-1}(x) = \log_2(x)$.
2. The expression $\log_a(x)$ is the exponent of a that produces x. So $a^{\log_a(x)} = x$.
3. The common logarithm uses the base 10 and the natural logarithm uses base e.
4. The domain of $f(x) = \log_a(x)$ is $(0, \infty)$.
5. The one-to-one property for logarithmic functions states that if $\log_a(m) = \log_a(n)$, then $m = n$.
6. The graphs of $f(x) = a^x$ and $f^{-1}(x) = \log_a(x)$ are symmetric about the line $y = x$.
7. $2^3 = 8$ **8.** $10^1 = 10$ **9.** $\log(100) = 2$
10. $\log_5(125) = 3$ **11.** $5^y = x$ **12.** $b^m = N$
13. $\log_2(b) = a$ **14.** $\log_a(c) = 3$ **15.** $3^{10} = x$
16. $c^4 = t$ **17.** $\ln(x) = 3$ **18.** $\ln(m) = x$ **19.** 2
20. 0   **21.** 4   **22.** 2   **23.** 6   **24.** 2   **25.** 3   **26.** 1
27. -2 **28.** -3 **29.** 2   **30.** 0   **31.** -2 **32.** 4
33. 1   **34.** 2   **35.** -3 **36.** 0   **37.** 3   **38.** 0
39. 2   **40.** -1 **41.** 0.6990   **42.** -1.5229
43. 1.8307   **44.** -1.4697

45.
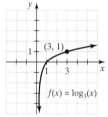

46.
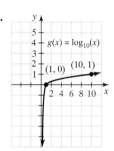

47.
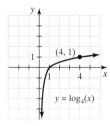

48.
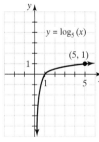

49.
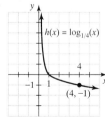

50.
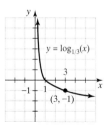

51.
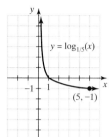

52.
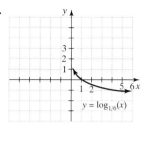

53. $f^{-1}(x) = \log_6(x)$ **54.** $f^{-1}(x) = \log_4(x)$
55. $f^{-1}(x) = e^x$ **56.** $f^{-1}(x) = 10^x$ **57.** $f^{-1}(x) = \left(\frac{1}{2}\right)^x$
58. $f^{-1}(x) = \left(\frac{1}{4}\right)^x$ **59.** {4}   **60.** {9}   **61.** $\left\{\frac{1}{2}\right\}$
62. $\{-1\}$ **63.** {0.001}   **64.** {100,000}   **65.** {6}
66. {10}   **67.** $\left\{\frac{1}{5}\right\}$ **68.** $\left\{\frac{1}{4}\right\}$ **69.** $\{\pm 3\}$ **70.** {4}
71. {0.4771}   **72.** $\{-1.5229\}$ **73.** $\{-0.3010\}$
74. {1.8751}   **75.** {1.9741}   **76.** $\{-0.3054\}$
77. 5.776 years   **78.** 11.552 years   **79.** 1.9269 years
80. 4.054 days   **81. a)** 104.8%   **b)** $15,320,208
82. 22.39%   **83.** 4.1   **84.** 1   **85.** 6.8   **86.** 6.7
87. 90 db   **88.** 23.5, 27.6, 65.6, 69.1
89. $f^{-1}(x) = 2^{x-5} + 3$, $(-\infty, \infty)$, $(3, \infty)$
90. $f^{-1}(x) = \ln(x - 2) - 4$, $(2, \infty)$, $(-\infty, \infty)$
91. $y = \ln(e^x) = x$ for $-\infty < x < \infty$, $y = e^{\ln(x)} = x$ for $0 < x < \infty$
92. 29.7 billion, 147.2 billion

Section 10.3 Warm-ups T F T T F T F F F T

1. The product rule for logarithms states that $\log_a(MN) = \log_a(M) + \log_a(N)$.
2. The quotient rule for logarithms states that $\log_a(M/N) = \log_a(M) - \log_a(N)$.
3. The power rule for logarithms states that $\log_a(M^N) = N \cdot \log_a(M)$.
4. Since $\log_a(a^M)$ is the exponent used on a to obtain a^M, we have $\log_a(a^M) = M$.
5. Since $\log_a(M)$ is the exponent you would use on a to obtain M, using $\log_a(M)$ as the exponent produces M: $a^{\log_a(M)} = M$.
6. We have $\log_a(1) = 0$ because $a^0 = 1$.
7. $\log(21)$ **8.** $\ln(20)$ **9.** $\log_3(\sqrt{5x})$ **10.** $\ln(\sqrt{xy})$
11. $\log(x^5)$ **12.** $\ln(a^8)$ **13.** $\ln(30)$ **14.** $\log_2(xyz)$
15. $\log(x^2 + 3x)$ **16.** $\ln(x^2 - 1)$ **17.** $\log_2(x^2 - x - 6)$
18. $\log_3(x^2 - 9x + 20)$ **19.** $\log(4)$ **20.** $\ln\left(\frac{1}{2}\right)$
21. $\log_2(x^4)$ **22.** $\ln(w^6)$ **23.** $\log(\sqrt{5})$ **24.** $\log_3(\sqrt{2})$
25. $\ln(h - 2)$ **26.** $\log(x - 2)$ **27.** $\log_2(w - 2)$
28. $\log_3(k + 3)$ **29.** $\ln(x - 2)$ **30.** $\ln(t + 3)$
31. $3\log(3)$ **32.** $-2\log(3)$ **33.** $\frac{1}{2}\log(3)$
34. $\frac{1}{4}\log(3)$ **35.** $x\log(3)$ **36.** $-99\log(3)$ **37.** 10
38. 9   **39.** 19   **40.** 2.3   **41.** 8   **42.** 5   **43.** 4.3

44. 5.5 **45.** $\log(3) + \log(5)$ **46.** $2\log(3)$
47. $\log(5) - \log(3)$ **48.** $\log(3) - \log(5)$ **49.** $2\log(5)$
50. $-3\log(3)$ **51.** $2\log(5) + \log(3)$
52. $\log(3) - \log(5)$ **53.** $-\log(3)$
54. $2\log(3) + \log(5)$ **55.** $-\log(5)$
56. $2\log(3) - 2\log(5)$ **57.** $\log(x) + \log(y) + \log(z)$
58. $\log(3) + \log(y)$ **59.** $3 + \log_2(x)$ **60.** $4 + \log_2(y)$
61. $\ln(x) - \ln(y)$ **62.** $\ln(z) - \ln(3)$ **63.** $1 + 2\log(x)$
64. $2 + \dfrac{1}{2}\log(x)$ **65.** $2\log_5(x-3) - \dfrac{1}{2}\log_5(w)$
66. $3\log_3(y+6) - \log_3(y-5)$
67. $\ln(y) + \ln(z) + \dfrac{1}{2}\ln(x) - \ln(w)$
68. $\ln(x-1) + \dfrac{1}{2}\ln(w) - 3\ln(x)$ **69.** $\log(x^2 - x)$
70. $\log_2(5x - 10)$ **71.** $\ln(3)$ **72.** $\log_3(x+1)$
73. $\ln\left(\dfrac{xz}{w}\right)$ **74.** $\ln\left(\dfrac{x}{21}\right)$ **75.** $\ln\left(\dfrac{x^2 y^3}{w}\right)$ **76.** $\ln\left(\dfrac{r^5 t^3}{s^4}\right)$
77. $\log\left(\dfrac{(x-3)^{1/2}}{(x+1)^{2/3}}\right)$ **78.** $\log\left(\sqrt{y^2 - 16}\right)$
79. $\log_2\left(\dfrac{(x-1)^{2/3}}{(x+2)^{1/4}}\right)$ **80.** $\log_3\left(y^6\sqrt{y+3}\right)$ **81.** False
82. False **83.** True **84.** False **85.** True **86.** True
87. False **88.** True **89.** True **90.** True **91.** True
92. True **93.** True **94.** True **95.** False **96.** True
97. $r = \ln\left((A/P)^{1/t}\right)$, 41.2% **99.** b **100.** c
101. The graphs are the same because
$$\ln(\sqrt{x}) = \ln(x^{1/2}) = \dfrac{1}{2}\ln(x).$$
102. Because $\log(10x) = 1 + \log(x)$, $\log(100x) = 2 + \log(x)$, and $\log(1000x) = 3 + \log(x)$; the graphs lie 1, 2, and 3 units above $y = \log(x)$.
103. The graph is a straight line because $\log(e^x) = x\log(e) \approx 0.434x$. The slope is $\log(e)$ or approximately 0.434.

Section 10.4 Warm-ups T T T F T T T F T F
1. The exponential equation $a^y = x$ is equivalent to $\log_a(x) = y$.
2. According to the base change formula, $\log_a(x) = \ln(x)/\ln(a)$.
3. $\{7\}$ **4.** $\{\pm 9\}$ **5.** $\{2\}$ **6.** $\left\{\dfrac{1}{3}\right\}$ **7.** $\{3\}$ **8.** $\{5\}$
9. $\varnothing$ **10.** $\{2\}$ **11.** $\{6\}$ **12.** $\{6\}$ **13.** $\{3\}$ **14.** $\{4\}$
15. $\{2\}$ **16.** $\{5\}$ **17.** $\{4\}$ **18.** $\left\{\dfrac{1}{3}\right\}$ **19.** $\{\log_3(7)\}$
20. $\{1 + \log_2(5)\}$ **21.** $\{-6\}$ **22.** $\left\{\dfrac{7}{8}\right\}$ **23.** $\left\{-\dfrac{1}{2}\right\}$
24. $\left\{-\dfrac{1}{5}\right\}$ **25.** $\dfrac{5\ln(3)}{\ln(2) - \ln(3)}$, -13.548 **26.** 0
27. $\dfrac{4 + 2\log(5)}{1 - \log(5)}$, 17.932 **28.** $\dfrac{\ln(6)}{\ln(9) - \ln(6)}$, 4.419
29. $\dfrac{\ln(9)}{\ln(9) - \ln(8)}$, 18.655 **30.** $\dfrac{\ln(5) + \ln(8)}{\ln(8) - \ln(5)}$, 7.849
31. 1.5850 **32.** 1.4650 **33.** -0.6309 **34.** 0.5841
35. -2.2016 **36.** -1.1403 **37.** 1.5229 **38.** -0.0362

39. $\dfrac{\ln(7)}{\ln(2)}$, 2.807 **40.** $\dfrac{\log(5)}{\log(3)}$, 1.465 **41.** $\dfrac{1}{3 - \ln(2)}$, 0.433
42. $\dfrac{\log(7)}{2 + \log(5)}$, 0.313 **43.** $\dfrac{\ln(5)}{\ln(3)}$, 1.465
44. $-\dfrac{\ln(3)}{\ln(2)}$, -1.585 **45.** $1 + \dfrac{\ln(9)}{\ln(2)}$, 4.170
46. $2 + \log(6)$, 2.778 **47.** $\log_3(20)$, 2.727 **48.** 7
49. $\dfrac{3}{5}$ **50.** 18 **51.** $\dfrac{1}{2}$ **52.** $\dfrac{\ln(5)}{\ln(2) - \ln(5)}$, -1.756
53. 41 months **54.** 18 years **55.** 7,524 ft^3/sec
56. 22.98 ft **57.** 7.1 years **58.** 9.9 years **59.** 16.8 years
60. 9.2 years **61.** 2.0×10^{-4} **62.** 4.0×10^{-8}
63. 0.9183 **64.** 2.42 **65.** 2.2894
66. a) negative **b)** positive **c)** negative **d)** positive
67. (2.71, 6.54)
68. $y = 1000e^{0.06x}$, $y = 1200(1 + 0.05/12)^{12x}$, 18.0 years
69. (1.03, 0.04), (4.73, 2.24)

Enriching Word Power
1. a **2.** d **3.** b **4.** d **5.** d **6.** b **7.** a **8.** b
9. b **10.** c

Chapter 10 Review
1. $\dfrac{1}{25}$ **2.** 1 **3.** 125 **4.** 625 **5.** 1 **6.** $\dfrac{1}{100}$ **7.** $\dfrac{1}{10}$
8. 100 **9.** 4 **10.** $\dfrac{1}{16}$ **11.** $\dfrac{1}{2}$ **12.** 2 **13.** 2
14. No solution **15.** 4 **16.** -2 **17.** $-\dfrac{5}{2}$ **18.** $-\dfrac{3}{2}$
19. 2 **20.** 0 **21.** 8.6421 **22.** 0.00305 **23.** 177.828
24. 7413.102 **25.** 0.02005 **26.** 0.01284 **27.** 0.1408
28. 0.6300
29.
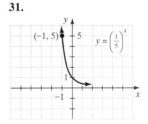
30.
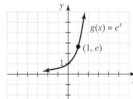
31.
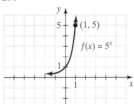
32.
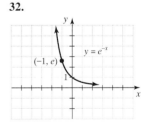
33.
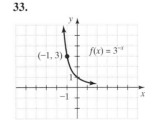
34.
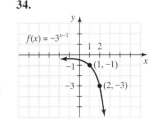

35.

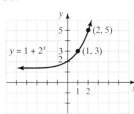

36.

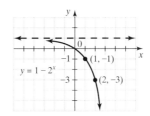

37. $\log(n) = m$ **38.** $\log_a(b) = 5$ **39.** $k^h = t$
40. $v^u = 5$ **41.** -3 **42.** 6   **43.** -1 **44.** 0
45. 2   **46.** 3   **47.** 0   **48.** -2 **49.** 256   **50.** 1000
51. 6.267   **52.** 1.946   **53.** -5.083
54. 4.322   **55.** 5.560   **56.** 26.669
57. $f^{-1}(x) = \log(x)$ **58.** $f^{-1}(x) = 8^x$

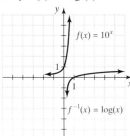

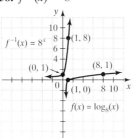

59. $f^{-1}(x) = \ln(x)$ **60.** $f^{-1}(x) = 3^x$

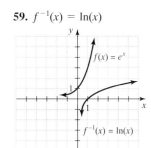

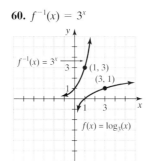

61. $2\log(x) + \log(y)$ **62.** $\log_3(x) + \log_3(x + 2)$
63. $4\ln(2)$ **64.** $\log(y) - \dfrac{1}{2}\log(x)$ **65.** $-\log_5(x)$
66. $\ln(x) + \ln(y) - \ln(z)$ **67.** $\log\left(\dfrac{\sqrt{x + 2}}{(x - 1)^2}\right)$
68. $\ln\left(\dfrac{x^3 y^2}{\sqrt[3]{z}}\right)$ **69.** $\{256\}$ **70.** $\{\sqrt{3}\}$ **71.** $\{3\}$

72. $\{\log_3(8)\}$ **73.** $\{2\}$ **74.** $\{9\}$ **75.** $\{3\}$ **76.** $\left\{\dfrac{1}{729}\right\}$
77. $\left\{\dfrac{\ln(7)}{\ln(3) - 1}\right\}$ **78.** $\left\{\dfrac{\log(9)}{\log(2)}\right\}$ **79.** $\left\{\dfrac{\ln(5)}{\ln(5) - \ln(3)}\right\}$
80. $\left\{-3, \dfrac{1}{2}\right\}$ **81.** $\left\{\dfrac{1}{3}\right\}$ **82.** $\{3, 4\}$ **83.** $\{22\}$ **84.** $\{5\}$
85. $\left\{\dfrac{200}{99}\right\}$ **86.** $\{16\}$ **87.** $\{1.3869\}$ **88.** $\{-0.8985\}$
89. $\{0.4650\}$ **90.** $\{18.0608\}$ **91.** \$51,182.68
92. 11.007 years   **93.** 161.5 grams

94. a) 517   **b)** 2503   **c)** faster for 2000 to 2005
    d) 7.4 deer per year, 1.4 deer per year
95. 5 years   **96. a)** \$22.8 million, \$14.3 million   **b)** 2010
    c) 12 years
97. 4,347.5 ft³/sec   **98.** $h = \dfrac{\ln(y/114.308)}{0.265} + 6.87,\ 23.74$ ft

Chapter 10 Test
1. 25   **2.** $\dfrac{1}{5}$ **3.** 1   **4.** 3   **5.** 0   **6.** -1
7.

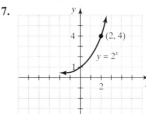

8.

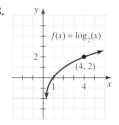

9.

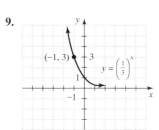

10.

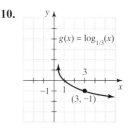

11. 10   **12.** 8   **13.** $\dfrac{3}{2}$ **14.** 15   **15.** -4
16. $\{\log_3(12)\}$ **17.** $\{\sqrt{3}\}$ **18.** $\left\{\dfrac{\ln(8)}{\ln(8) - \ln(5)}\right\}$
19. $\{5\}$ **20.** $\{3\}$ **21.** $\{0.5372\}$ **22.** $\{20.5156\}$
23. 10, 147,648   **24.** 1.733 hours

Making Connections Chapters 1–10
1. $\{3 \pm 2\sqrt{2}\}$ **2.** $\{259\}$ **3.** $\{6\}$ **4.** $\left\{\dfrac{11}{2}\right\}$ **5.** $\{-5, 11\}$
6. $\{67\}$ **7.** $\{6\}$ **8.** $\{4\}$ **9.** $\left\{-\dfrac{52}{15}\right\}$ **10.** $\left\{\dfrac{3 \pm \sqrt{3}}{3}\right\}$
11. $f^{-1}(x) = 3x$ **12.** $g^{-1}(x) = 3^x$ **13.** $f^{-1}(x) = \dfrac{x + 4}{2}$
14. $h^{-1}(x) = x^2$ for $x \geq 0$ **15.** $j^{-1}(x) = \dfrac{1}{x}$
16. $k^{-1}(x) = \log_5(x)$ **17.** $m^{-1}(x) = 1 + \ln(x)$
18. $n^{-1}(x) = e^x$
19.   **20.**

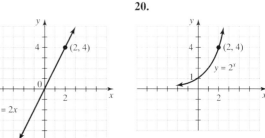

21.

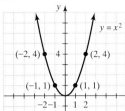

22.

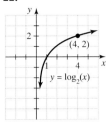

23.

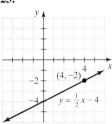

24.

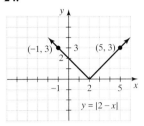

25.

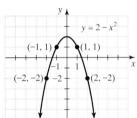

26.

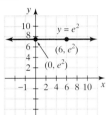

27. a)

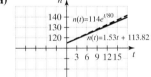

b) linear 130.7 million, exponential 130.8 million

28. a) $d_1 = 0.135v$ **b)** $d_2 = 0.216v$ **c)** $d_1 = 200.2$ meters

CHAPTER 11

Section 11.1 Warm-ups T F F T F T T T T T

1. If the graph of an equation is not a straight line, then it is called nonlinear.

2. With a graph we can see the approximate value of the solutions and the number of solutions.

3. Graphing is not an accurate method for solving a system and the graphs might be difficult to draw.

4. We generally use substitution and addition to solve nonlinear systems.

5. $\{(2, 4), (-3, 9)\}$

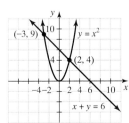

6. $\{(-4, 15), (3, 8)\}$

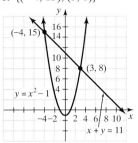

7. $\{(-2, 2), (6, 6)\}$

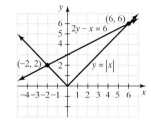

8. $\left\{\left(-\dfrac{3}{2}, \dfrac{3}{2}\right), (3, 3)\right\}$

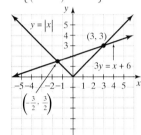

9. $\{(8, 4)\}$

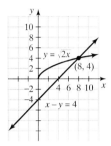

10. $\{(9, 3)\}$

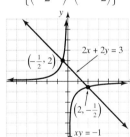

11. $\left\{\left(-\dfrac{3}{4}, -\dfrac{4}{3}\right), \left(3, \dfrac{1}{3}\right)\right\}$

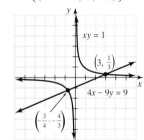

12. $\left\{\left(-\dfrac{1}{2}, 2\right), \left(2, -\dfrac{1}{2}\right)\right\}$

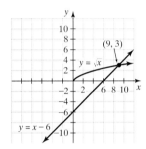

13. $\left\{\left(\dfrac{\sqrt{2}}{2}, \dfrac{1}{2}\right), \left(-\dfrac{\sqrt{2}}{2}, \dfrac{1}{2}\right)\right\}$

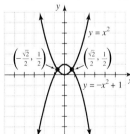

14. $\{(0, 0), (1, 1)\}$

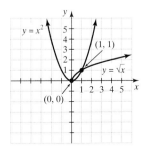

15. $\{(0, -5), (3, 4), (-3, 4)\}$ **16.** $\{(-4, -3), (3, 4)\}$

17. $\{(4, 5), (-2, -1)\}$ **18.** $\{(3, 1), (-3, -5)\}$

19. $\left\{\left(-3, -\dfrac{5}{3}\right)\right\}$ **20.** $\{(-1, 2)\}$

21. $\left\{\left(\dfrac{3}{2}, -\dfrac{3}{13}\right)\right\}$ **22.** $\left\{\left(\dfrac{11}{2}, -22\right)\right\}$

23. $\left\{\left(-\dfrac{5}{3}, \dfrac{36}{5}\right), (2, 5)\right\}$ **24.** $\left\{\left(\dfrac{1}{12}, -6\right), \left(\dfrac{1}{3}, 3\right)\right\}$

25. $\{(\sqrt{5}, \sqrt{3}), (\sqrt{5}, -\sqrt{3}), (-\sqrt{5}, \sqrt{3}), (-\sqrt{5}, -\sqrt{3})\}$

26. $\{(\sqrt{2}, \sqrt{3}), (\sqrt{2}, -\sqrt{3}), (-\sqrt{2}, \sqrt{3}), (-\sqrt{2}, -\sqrt{3})\}$

27. $\{(2, 5), (19, -12)\}$ **28.** $\left\{\left(\dfrac{11}{7}, -\dfrac{13}{7}\right), (2, -1)\right\}$

29. $\{(\sqrt{2}, 2), (-\sqrt{2}, 2), (1, 1), (-1, 1)\}$

30. $\left\{\left(\dfrac{\sqrt{2}}{2}, \dfrac{7}{2}\right), \left(-\dfrac{\sqrt{2}}{2}, \dfrac{7}{2}\right), (\sqrt{3}, 21), (-\sqrt{3}, 21)\right\}$

31. $\{(3, 1)\}$ **32.** $\{(5, 0)\}$ **33.** $\varnothing$ **34.** $\{(2, 2)\}$

35. $\{(-6, 4^{-7})\}$ **36.** $\left\{\left(-\dfrac{1}{5}, 4^{-3/5}\right)\right\}$ **37.** $\sqrt{3}$ ft and $2\sqrt{3}$ ft

38. 3 inches by 4 inches

39. Height $5\sqrt{10}$ inches, base $20\sqrt{10}$ inches

40. $3 - \sqrt{3}$ ft, $6 - 2\sqrt{3}$ ft, $-3 + 3\sqrt{3}$ ft

41. Pump A 24 hours, pump B 8 hours

42. $\dfrac{36}{5}$ hours   **43.** 40 minutes   **44.** 12 ft by 15 ft

45. 8 ft by 9 ft   **46.** $4 - \sqrt{6}$ and $4 + \sqrt{6}$

47. $4 - 2i$ and $4 + 2i$ **48.** $-3 + i$ and $-3 - i$

49. Side of square 8 ft, height of triangle 2 ft

50. 6 inches by 10 inches by 2 inches

51. a) $(1.71, 1.55), (-2.98, -3.95)$ **b)** $(1, 1), (0.40, 0.16)$
   c) $(1.17, 1.62), (-1.17, -1.62)$

Section 11.2 Warm-ups F T T F F T T F T T

1. A parabola is the set of all points in a plane that are equidistant from a given line and a fixed point not on the line.

2. The vertex is the midpoint of the line segment joining the focus and directrix, perpendicular to the directrix.

3. A parabola can be written in the forms $y = ax^2 + bx + c$ or $y = a(x - h)^2 + k$.

4. The distance from the focus to the directrix is $|2p|$, where $a = 1/(4p)$.

5. We use completing the square to convert $y = ax^2 + bx + c$ into $y = a(x - h)^2 + k$.

6. To convert $y = a(x - h)^2 + k$ into the form $y = ax^2 + bx + c$, square the binomial, multiply by a, then add like terms.

7. Vertex $(0, 0)$, focus $\left(0, \dfrac{1}{8}\right)$, directrix $y = -\dfrac{1}{8}$.

8. Vertex $(0, 0)$, focus $\left(0, \dfrac{1}{2}\right)$, directrix $y = -\dfrac{1}{2}$.

9. Vertex $(0, 0)$, focus $(0, -1)$, directrix $y = 1$

10. Vertex $(0, 0)$, focus $(0, -3)$, directrix $y = 3$

11. Vertex $(3, 2)$, focus $(3, 2.5)$, directrix $y = 1.5$

12. Vertex $(-2, -5)$, focus $(-2, -4)$, directrix $y = -6$

13. Vertex $(-1, 6)$, focus $(-1, 5.75)$, directrix $y = 6.25$

14. Vertex $(4, 1)$, focus $\left(4, \dfrac{11}{12}\right)$, directrix $y = \dfrac{13}{12}$

15. $y = \dfrac{1}{8}x^2$ **16.** $y = -\dfrac{1}{12}x^2$ **17.** $y = -\dfrac{1}{2}x^2$

18. $y = 2x^2$ **19.** $y = \dfrac{1}{2}x^2 - 3x + 6$

20. $y = \dfrac{1}{2}x^2 + 4x + \dfrac{25}{2}$ **21.** $y = -\dfrac{1}{8}x^2 + \dfrac{1}{4}x - \dfrac{1}{8}$

22. $y = -\dfrac{1}{8}x^2 + \dfrac{1}{2}x - \dfrac{3}{2}$ **23.** $y = x^2 + 6x + 10$

24. $y = 2x^2 - 20x + 52$

25. $y = (x - 3)^2 - 8$, vertex $(3, -8)$, focus $(3, -7.75)$, directrix $y = -8.25$

26. $y = (x + 2)^2 - 11$, vertex $(-2, -11)$, focus $(-2, -10.75)$, directrix $y = -11.25$

27. $y = 2(x + 3)^2 - 13$, vertex $(-3, -13)$, focus $(-3, -12.875)$, directrix $y = -13.125$

28. $y = 3(x + 1)^2 - 10$, vertex $(-1, -10)$, focus $\left(-1, -9\dfrac{11}{12}\right)$, directrix $y = -10\dfrac{1}{12}$

29. $y = -2(x - 4)^2 + 33$, vertex $(4, 33)$, focus $\left(4, 32\dfrac{7}{8}\right)$, directrix $y = 33\dfrac{1}{8}$

30. $y = -3(x + 1)^2 + 10$, vertex $(-1, 10)$, focus $\left(-1, 9\dfrac{11}{12}\right)$, directrix $y = 10\dfrac{1}{12}$

31. $y = 5(x + 4)^2 - 80$, vertex $(-4, -80)$, focus $\left(-4, -79\dfrac{19}{20}\right)$, directrix $y = -80\dfrac{1}{20}$

32. $y = -2\left(x - \dfrac{5}{2}\right)^2 + \dfrac{25}{2}$, vertex $\left(\dfrac{5}{2}, \dfrac{25}{2}\right)$, focus $\left(\dfrac{5}{2}, \dfrac{99}{8}\right)$, directrix $y = \dfrac{101}{8}$

33. Vertex $(2, -3)$, focus $\left(2, -2\dfrac{3}{4}\right)$, directrix $y = -3\dfrac{1}{4}$, upward

34. Vertex $(3, -16)$, focus $\left(3, -15\dfrac{3}{4}\right)$, directrix $y = -16\dfrac{1}{4}$, upward

35. Vertex $(1, -2)$, focus $\left(1, -2\dfrac{1}{4}\right)$, directrix $y = -1\dfrac{3}{4}$, downward

36. Vertex $(2, 13)$, focus $\left(2, 12\dfrac{3}{4}\right)$, directrix $y = 13\dfrac{1}{4}$, downward

37. Vertex $(1, -2)$, focus $\left(1, -1\dfrac{11}{12}\right)$, directrix $y = -2\dfrac{1}{12}$, upward

38. Vertex $(-1, -5)$, focus $\left(-1, -4\dfrac{7}{8}\right)$, directrix $y = -5\dfrac{1}{8}$, upward

39. Vertex $\left(-\dfrac{3}{2}, \dfrac{17}{4}\right)$, focus $\left(-\dfrac{3}{2}, 4\right)$, directrix $y = \dfrac{9}{2}$, downward

40. Vertex $\left(\dfrac{3}{2}, \dfrac{5}{4}\right)$, focus $\left(\dfrac{3}{2}, 1\right)$, directrix $y = \dfrac{3}{2}$, downward

41. Vertex $(0, 5)$, focus $\left(0, 5\dfrac{1}{12}\right)$, directrix $y = 4\dfrac{11}{12}$, upward

42. Vertex $(0, -6)$, focus $\left(0, -6\dfrac{1}{8}\right)$, directrix $y = -5\dfrac{7}{8}$, downward

43. Vertex $\left(\frac{3}{2}, -\frac{1}{4}\right)$, axis of symmetry $x = \frac{3}{2}$, intercepts $(0, 2)$, $(1, 0), (2, 0)$

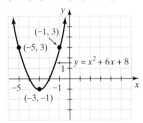

44. Vertex $(-3, -1)$, axis of symmetry $x = -3$, intercepts $(0, 8), (-4, 0), (-2, 0)$

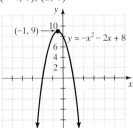

45. Vertex $(-1, 9)$, axis of symmetry $x = -1$, intercepts $(0, 8)$, $(-4, 0), (2, 0)$

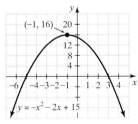

46. Vertex $(-1, 16)$, axis of symmetry $x = -1$, intercepts $(0, 15), (-5, 0), (3, 0)$

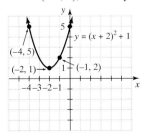

47. Vertex $(-2, 1)$, axis of symmetry $x = -2$, intercept $(0, 5)$

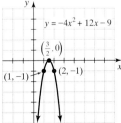

48. Vertex $(-1, 3)$, axis of symmetry $x = -1$, intercepts $(0, 1)$, $\left(-1 + \frac{\sqrt{6}}{2}, 0\right), \left(-1 - \frac{\sqrt{6}}{2}, 0\right)$

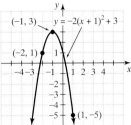

49. Vertex $(-1, 0)$, axis of symmetry $x = -1$, intercepts $(0, 1)$, $(-1, 0)$

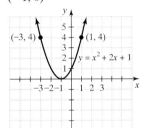

50. Vertex $(3, 0)$, axis of symmetry $x = 3$, intercepts $(0, 9), (3, 0)$

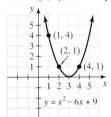

51. Vertex $\left(\frac{1}{2}, 0\right)$, axis of symmetry $x = \frac{1}{2}$, intercepts $(0, -1)$, $\left(\frac{1}{2}, 0\right)$

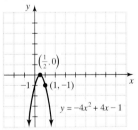

52. Vertex $\left(\frac{3}{2}, 0\right)$, axis of symmetry $x = \frac{3}{2}$, intercepts $\left(\frac{3}{2}, 0\right)$, $(0, -9)$

53. Vertex $\left(\frac{5}{2}, -\frac{25}{4}\right)$, axis of symmetry $x = \frac{5}{2}$, intercepts $(0, 0), (5, 0)$

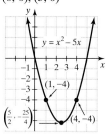

54. Vertex $\left(\frac{3}{2}, -\frac{27}{4}\right)$, axis of symmetry $x = \frac{3}{2}$, intercepts $(0, 0)$, $(3, 0)$

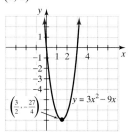

55. Vertex $(0, 5)$, axis of symmetry $x = 0$, intercept $(0, 5)$

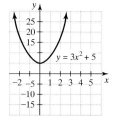

56. Vertex $(0, 3)$, axis of symmetry $x = 0$, intercepts $(0, 3)$, $\left(-\frac{\sqrt{6}}{2}, 0\right), \left(\frac{\sqrt{6}}{2}, 0\right)$

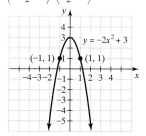

57. Vertex $(1, -2)$, axis of symmetry $x = 1$, intercepts $(0, -1), \left(1 + \sqrt{2}, 0\right), \left(1 - \sqrt{2}, 0\right)$

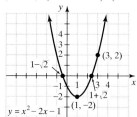

58. Vertex $(2, -3)$, axis of symmetry $x = 2$, intercepts $(0, 1)$, $\left(2 - \sqrt{3}, 0\right), \left(2 + \sqrt{3}, 0\right)$

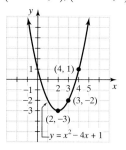

59. Vertex $(5, 0)$, axis of symmetry $x = 5$, intercepts $(0, 25)$, $(5, 0)$

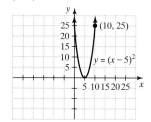

60. Vertex $(1, -4)$, axis of symmetry $x = 1$, intercepts $(0, -1), \left(1 + \frac{2\sqrt{3}}{3}, 0\right), \left(1 - \frac{2\sqrt{3}}{3}, 0\right)$

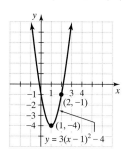

61. 4 and 4   **62.** 9   **63.** 40 ft by 40 ft
64. 26 ft by 26 ft   **65.** 3 ft each   **66.** $15
67. 262 ft, 4 sec, no   **68.** 356.5625 ft, 4.6875 sec
69. 100   **70.** 6 clerks, $900   **71.** $y = \frac{1}{60}x^2$
72. $y = 0.0008x^2$, 312.5 ft
73. $\{(-1, 2), (1, 2)\}$ **74.** $\{(2, 1), (-2, 1)\}$

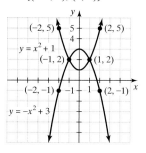
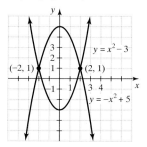

75. $\{(1, -1)\}$

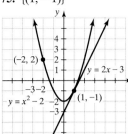

76. $\{(3, 6)\}$

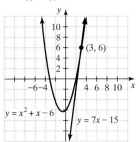

77. $\left\{\left(\dfrac{3}{2}, \dfrac{11}{4}\right), (-4, 0)\right\}$

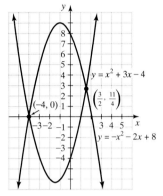

78. $\left\{\left(\dfrac{5}{2}, \dfrac{13}{4}\right), (-4, 0)\right\}$

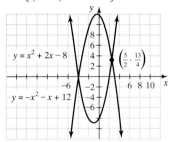

79. $\{(-3, -4), (2, 6)\}$ **80.** $\{(-5, 6), (1, 12)\}$

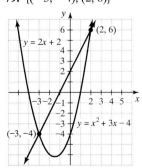
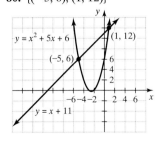

81. $y = -x^2 + 4$ **82.** $y = \dfrac{1}{3}x^2$ **83.** $y = -\dfrac{1}{2}x^2 + 2x$

84. $y = -\dfrac{3}{4}x^2 + 3x - 3$

86. a) Right for $a > 0$ and left for $a < 0$ **b)** $y = h$

87. The graphs have identical shapes.
88. Intersection $(3, 9)$

Section 11.3 Warm-ups F F T F F F T F T F
1. A circle is the set of all points in a plane that lie at a fixed distance from a fixed point.
2. The equation $(x - h)^2 + (y - k)^2 = r^2$ is the standard equation of a circle with center (h, k) and radius r (for $r > 0$).
3. $x^2 + (y - 3)^2 = 25$ **4.** $(x - 2)^2 + y^2 = 9$
5. $(x - 1)^2 + (y + 2)^2 = 81$
6. $(x + 3)^2 + (y - 5)^2 = 16$ **7.** $x^2 + y^2 = 3$
8. $x^2 + y^2 = 2$ **9.** $(x + 6)^2 + (y + 3)^2 = \dfrac{1}{4}$
10. $(x + 3)^2 + (y + 5)^2 = \dfrac{1}{16}$
11. $\left(x - \dfrac{1}{2}\right)^2 + \left(y - \dfrac{1}{3}\right)^2 = 0.01$
12. $\left(x + \dfrac{1}{2}\right)^2 + (y - 3)^2 = 0.04$ **13.** $(3, 5), \sqrt{2}$
14. $(-3, 7), \sqrt{6}$ **15.** $\left(0, \dfrac{1}{2}\right), \dfrac{\sqrt{2}}{2}$ **16.** $(0, 0), 1$
17. $(0, 0), \dfrac{3}{2}$ **18.** $(0, 0), \dfrac{7}{3}$ **19.** $(2, 0), \sqrt{3}$
20. $(0, -1), 3$
21.

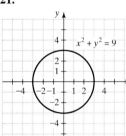

22.

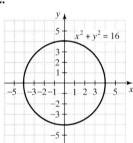

23.

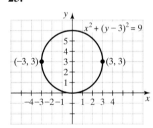

24.

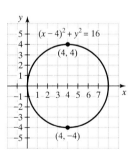

25.

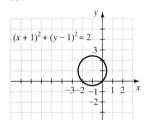

26.

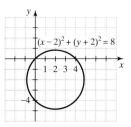

27.

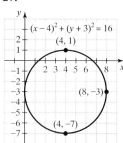

28.

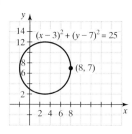

29.

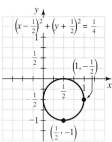

30.

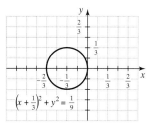

31. $(x + 2)^2 + (y + 3)^2 = 13, (-2, -3), \sqrt{13}$

32. $(x - 5)^2 + (y + 4)^2 = 41, (5, -4), \sqrt{41}$

33. $(x - 1)^2 + (y - 2)^2 = 8, (1, 2), 2\sqrt{2}$

34. $(x - 3)^2 + (y - 1)^2 = 1, (3, 1), 1$

35. $(x - 5)^2 + (y - 4)^2 = 9, (5, 4), 3$

36. $(x - 4)^2 + (y + 5)^2 = 41, (4, -5), \sqrt{41}$

37. $\left(x - \dfrac{1}{2}\right)^2 + \left(y + \dfrac{1}{2}\right)^2 = \dfrac{1}{2}, \left(\dfrac{1}{2}, -\dfrac{1}{2}\right), \dfrac{\sqrt{2}}{2}$

38. $\left(x - \dfrac{3}{2}\right)^2 + y^2 = \dfrac{9}{4}, \left(\dfrac{3}{2}, 0\right), \dfrac{3}{2}$

39. $\left(x - \dfrac{3}{2}\right)^2 + \left(y - \dfrac{1}{2}\right)^2 = \dfrac{7}{2}, \left(\dfrac{3}{2}, \dfrac{1}{2}\right), \dfrac{\sqrt{14}}{2}$

40. $\left(x - \dfrac{5}{2}\right)^2 + \left(y + \dfrac{3}{2}\right)^2 = \dfrac{21}{2}, \left(\dfrac{5}{2}, -\dfrac{3}{2}\right), \dfrac{\sqrt{42}}{2}$

41. $\left(x - \dfrac{1}{3}\right)^2 + \left(y + \dfrac{3}{4}\right)^2 = \dfrac{97}{144}, \left(\dfrac{1}{3}, -\dfrac{3}{4}\right), \dfrac{\sqrt{97}}{12}$

42. $\left(x + \dfrac{1}{6}\right)^2 + \left(y - \dfrac{1}{3}\right)^2 = \dfrac{1}{4}, \left(-\dfrac{1}{6}, \dfrac{1}{3}\right), \dfrac{1}{2}$

43. $\{(1, 3), (-1, -3)\}$

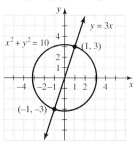

44. $\{(0, -2), (2, 0)\}$

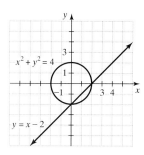

45. $\left\{(0, -3), (\sqrt{5}, 2), (-\sqrt{5}, 2)\right\}$

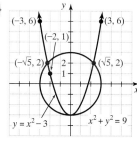

46. $\left\{(0, -2), (-\sqrt{3}, 1), (\sqrt{3}, 1)\right\}$

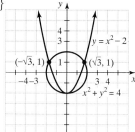

47. $\{(0, -3), (2, -1)\}$

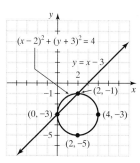

48. $\{(3, 5), (-2, 0)\}$

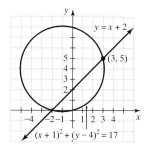

49. $\left(0, 2 + \sqrt{3}\right)$ and $\left(0, 2 - \sqrt{3}\right)$ **50.** $(-4, 0)$ and $(4, 0)$

51. $\sqrt{29}$ **52.** $\sqrt{41}$ **53.** $(x - 2)^2 + (y - 3)^2 = 32$

54. $(x - 3)^2 + (y - 4)^2 = 25$

55. $\left(\dfrac{5}{2}, -\dfrac{\sqrt{11}}{2}\right)$ and $\left(\dfrac{5}{2}, \dfrac{\sqrt{11}}{2}\right)$

56. Yes   **57.** $755{,}903$ mm^3

58. $x^2 + y^2 = 78.95$, $672{,}186$ mm^3 **59.** $(0, 0)$ only

60. 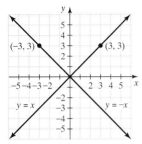   **61.**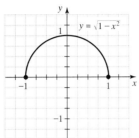

62.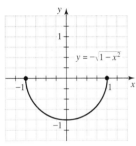

63. B and D can be any real numbers, but A must equal C, and $4AE + B^2 + D^2 > 0$.

64. $\left(-\dfrac{\sqrt{39}}{4}, \dfrac{5}{4}\right)$ and $\left(\dfrac{\sqrt{39}}{4}, \dfrac{5}{4}\right)$ **65.** $y = \pm\sqrt{4 - x^2}$

66. $y = -2 \pm \sqrt{1 - (x - 1)^2}$ **67.** $y = \pm\sqrt{x}$

68. $y = -2 \pm \sqrt{x + 1}$ **69.** $y = -1 \pm \sqrt{x}$

70. $y = \dfrac{-1 \pm \sqrt{x}}{2}$

Section 11.4 Warm-ups F F T T T F F T T T

1. An ellipse is the set of all points in a plane such that the sum of their distances from two fixed points is constant.

2. Attach a string to two thumb tacks and use a pencil to take up the slack as shown in the text.

3. The center of an ellipse is the point that is midway between the foci.

4. The equation of an ellipse centered at the origin is $\dfrac{x^2}{a^2} + \dfrac{y^2}{b^2} = 1$.

5. The equation of an ellipse centered at (h, k) is $\dfrac{(x - h)^2}{a^2} + \dfrac{(y - k)^2}{b^2} = 1$.

6. A hyperbola is the set of all points in a plane such that the difference of their distances from two fixed points is constant.

7. The asymptotes of a hyperbola are the extended diagonals of the fundamental rectangle.

8. The equation of a hyperbola centered at the origin and opening left and right is of the form $\dfrac{x^2}{a^2} - \dfrac{y^2}{b^2} = 1$.

9.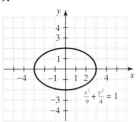

10.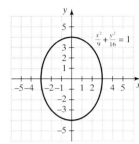

11.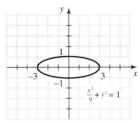

12.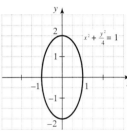

13.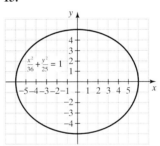

14.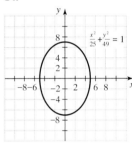

15.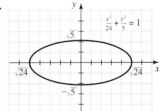

16.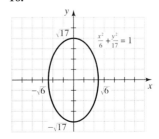

17.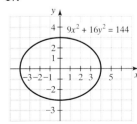

18.

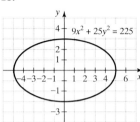

19.

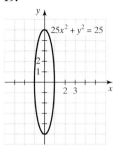

28.

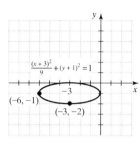

29. $y = \pm\dfrac{3}{2}x$

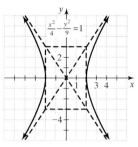

20.

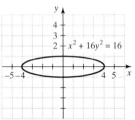

21.

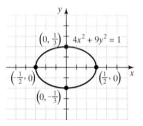

30. $y = \pm\dfrac{3}{4}x$

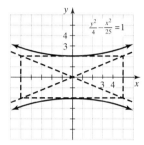

31. $y = \pm\dfrac{2}{5}x$

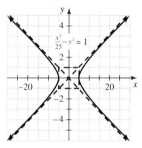

22.

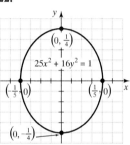

23.

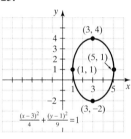

32. $y = \pm\dfrac{3}{4}x$

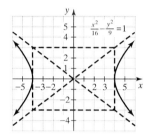

33. $y = \pm\dfrac{1}{5}x$

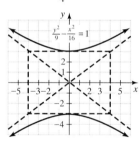

24.

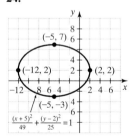

25.

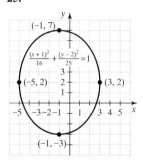

34. $y = \pm 3x$

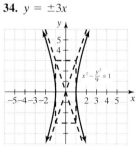

35. $y = \pm 5x$

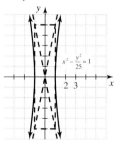

26.

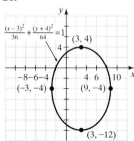

27.

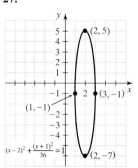

36. $y = \pm\dfrac{1}{3}x$

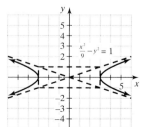

37. $y = \pm\dfrac{3}{4}x$

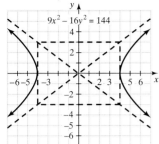

38. $y = \pm\dfrac{3}{5}x$

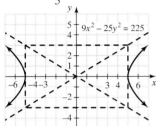

39. $y = \pm x$

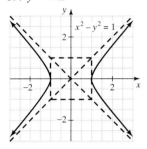

40. $y = \pm x$

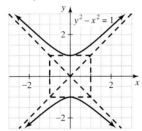

41. $\left(\dfrac{2\sqrt{10}}{5}, \dfrac{3\sqrt{15}}{5}\right),$

$\left(\dfrac{2\sqrt{10}}{5}, -\dfrac{3\sqrt{15}}{5}\right),$

$\left(-\dfrac{2\sqrt{10}}{5}, \dfrac{3\sqrt{15}}{5}\right),$

$\left(-\dfrac{2\sqrt{10}}{5}, -\dfrac{3\sqrt{15}}{5}\right)$

42. $\left(\dfrac{3\sqrt{5}}{5}, \dfrac{4\sqrt{5}}{5}\right),$

$\left(\dfrac{3\sqrt{5}}{5}, -\dfrac{4\sqrt{5}}{5}\right),$

$\left(-\dfrac{3\sqrt{5}}{5}, \dfrac{4\sqrt{5}}{5}\right),$

$\left(-\dfrac{3\sqrt{5}}{5}, -\dfrac{4\sqrt{5}}{5}\right).$

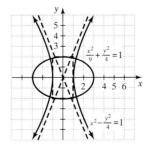

43. No points of intersection

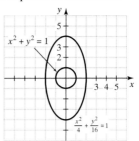

44. $\left(\dfrac{\sqrt{10}}{4}, \dfrac{3\sqrt{6}}{4}\right), \left(\dfrac{\sqrt{10}}{4}, -\dfrac{3\sqrt{6}}{4}\right),$

$\left(-\dfrac{\sqrt{10}}{4}, \dfrac{3\sqrt{6}}{4}\right), \left(-\dfrac{\sqrt{10}}{4}, -\dfrac{3\sqrt{6}}{4}\right)$

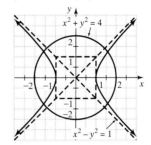

45. $\left(\dfrac{\sqrt{10}}{2}, \dfrac{\sqrt{6}}{2}\right), \left(\dfrac{\sqrt{10}}{2}, -\dfrac{\sqrt{6}}{2}\right),$

$\left(-\dfrac{\sqrt{10}}{2}, \dfrac{\sqrt{6}}{2}\right), \left(-\dfrac{\sqrt{10}}{2}, -\dfrac{\sqrt{6}}{2}\right)$

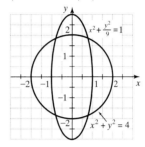

46. $\left(\sqrt{10}, \sqrt{6}\right), \left(\sqrt{10}, -\sqrt{6}\right), \left(-\sqrt{10}, \sqrt{6}\right),$

$\left(-\sqrt{10}, -\sqrt{6}\right)$

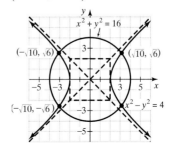

47. $\left(\dfrac{3\sqrt{6}}{4}, \dfrac{\sqrt{10}}{4}\right), \left(\dfrac{3\sqrt{6}}{4}, -\dfrac{\sqrt{10}}{4}\right),$

$\left(-\dfrac{3\sqrt{6}}{4}, \dfrac{\sqrt{10}}{4}\right), \left(-\dfrac{3\sqrt{6}}{4}, -\dfrac{\sqrt{10}}{4}\right)$

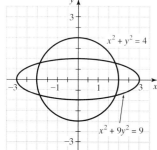

48. $(5, 0), (-5, 0)$

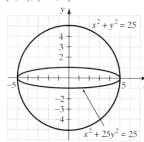

49. $\left(\dfrac{\sqrt{17}}{3}, \dfrac{8}{9}\right), \left(-\dfrac{\sqrt{17}}{3}, \dfrac{8}{9}\right), (0, -1)$

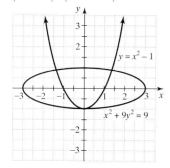

50. $(-1, 0), (1, 0), (0, -2)$

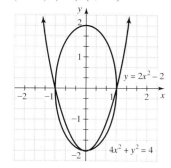

51. $(2, 0), \left(-\dfrac{5}{2}, -\dfrac{9}{4}\right)$

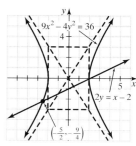

52. $(0, 3), \left(-\dfrac{25}{12}, -\dfrac{13}{4}\right)$

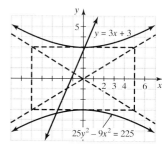

53. a) $(2.5, 1.5)$ **b)** $\left(\sqrt{7}, \sqrt{2}\right)$ **54.** $4\sqrt{3}$ or 6.9 miles

Section 11.5 Warm-ups F T T T F F F T T T

1.

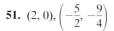

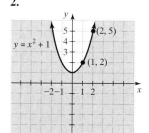

2.

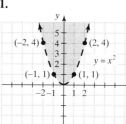

3.

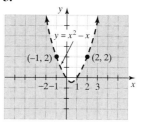

4.

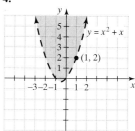

5.

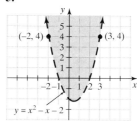

6.

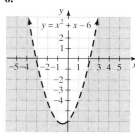

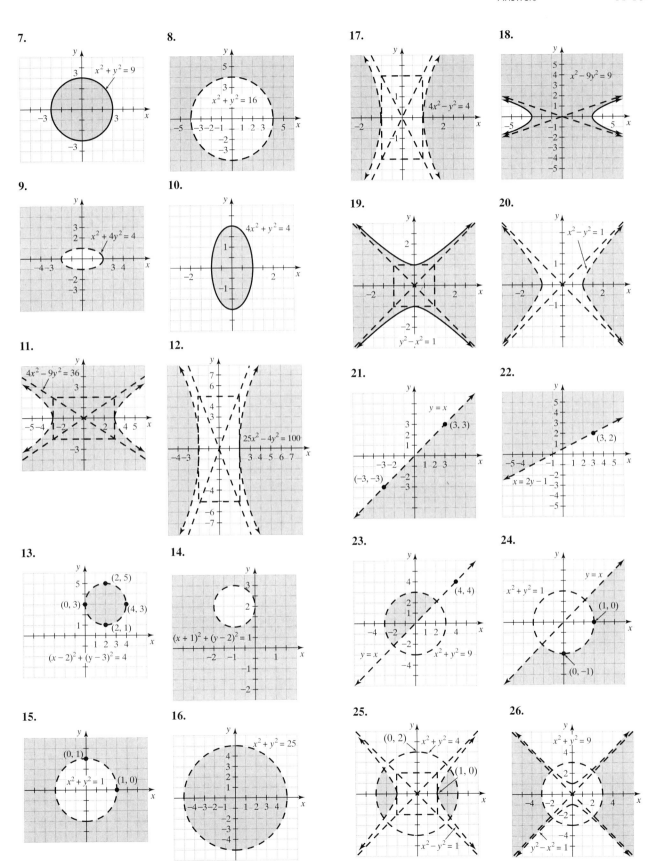

27.

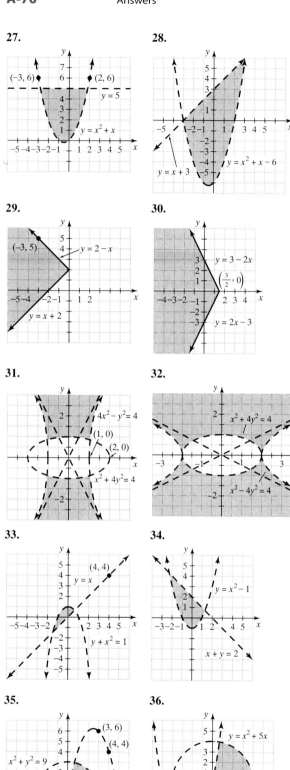

28.

29.

30.

31.

32.

33.

34.

35.

36.

37.

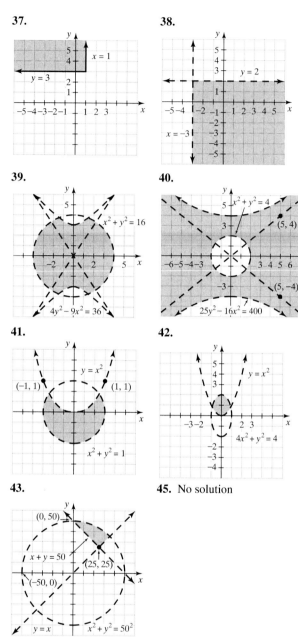

38.

39.

40.

41.

42.

43.

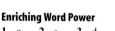

45. No solution

Enriching Word Power

1. c **2.** a **3.** d **4.** a **5.** c **6.** d **7.** b **8.** d
9. c **10.** a

Chapter 11 Review

1. $\{(3, 9), (-5, 25)\}$

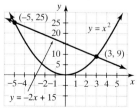

2. $\{(0, 0), (9, 3)\}$

3. $\left\{\left(\frac{\sqrt{3}}{3}, \sqrt{3}\right), \left(-\frac{\sqrt{3}}{3}, -\sqrt{3}\right)\right\}$

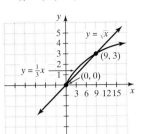

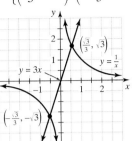

15. Intercepts $(0, 2)$, $(-2, 0)$, $(-1, 0)$, vertex $\left(-\frac{3}{2}, -\frac{1}{4}\right)$, axis of symmetry $x = -\frac{3}{2}$, focus $\left(-\frac{3}{2}, 0\right)$, directrix $y = -\frac{1}{2}$

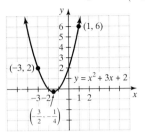

4. $\left\{\left(\frac{5}{4}, \frac{5}{4}\right)\right\}$

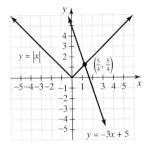

16. Intercepts $(0, 4)$, $(-4, 0)$, $(1, 0)$, vertex $\left(-\frac{3}{2}, \frac{25}{4}\right)$, axis of symmetry $x = -\frac{3}{2}$, focus $\left(-\frac{3}{2}, 6\right)$, directrix $y = \frac{13}{2}$

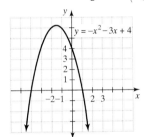

5. $\{(\sqrt{3}, 1), (-\sqrt{3}, 1)\}$ **6.** $\left\{\left(\frac{3}{2}, -\frac{\sqrt{6}}{2}\right), \left(\frac{3}{2}, \frac{\sqrt{6}}{2}\right)\right\}$

7. $\{(-5, -3), (3, 5)\}$ **8.** $\left\{\left(-\frac{3}{2}, -2\right), (2, 5)\right\}$

9. $\{(5, \log(2))\}$ **10.** $\left\{\left(\frac{1}{2}, \frac{\sqrt{2}}{2}\right)\right\}$ **11.** $\{(2, 4), (-2, 4)\}$

12. $\{(1, \sqrt{3}), (-1, \sqrt{3}), (1, -\sqrt{3}), (-1, -\sqrt{3})\}$

17. Intercepts $(0, 1)$, $(2 \pm \sqrt{6}, 0)$, vertex $(2, 3)$, axis of symmetry $x = 2$, focus $\left(2, \frac{5}{2}\right)$, directrix $y = \frac{7}{2}$

13. Intercepts $(0, -18)$, $(-6, 0)$, $(3, 0)$, vertex $\left(-\frac{3}{2}, -\frac{81}{4}\right)$, axis of symmetry $x = -\frac{3}{2}$, focus $\left(-\frac{3}{2}, -20\right)$, directrix $y = -\frac{41}{2}$

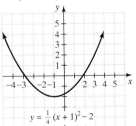

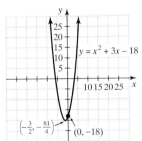

18. Intercepts $\left(0, -\frac{7}{4}\right)$, $(-1 \pm 2\sqrt{2}, 0)$, vertex $(-1, -2)$, axis of symmetry $x = -1$, focus $(-1, -1)$, directrix $y = -3$

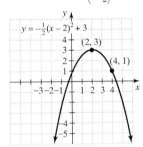

14. Intercepts $(0, 0)$, $(1, 0)$, vertex $\left(\frac{1}{2}, \frac{1}{4}\right)$, axis of symmetry $x = \frac{1}{2}$, focus $\left(\frac{1}{2}, 0\right)$, directrix $y = \frac{1}{2}$

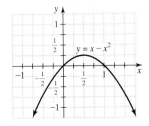

19. $y = 2(x - 2)^2 - 7$, $(2, -7)$

20. $y = -2\left(x + \frac{3}{2}\right)^2 + \frac{7}{2}$, $\left(-\frac{3}{2}, \frac{7}{2}\right)$

21. $y = -\frac{1}{2}(x + 1)^2 + 1$, $(-1, 1)$

22. $y = \frac{1}{4}(x + 2)^2 - 10$, $(-2, -10)$

23. 3 and 3   **24.** 5 and -5 **25.** 22.5 ft by 22.5 ft

26. 24 ft on each side

27. $(0, 0)$, 10

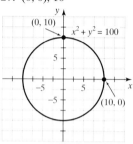

28. $(0, 0)$, $2\sqrt{5}$

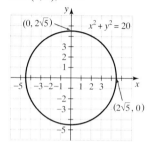

29. $(2, -3)$, 9

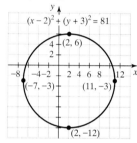

30. $(-1, 0)$, 3

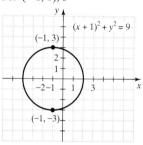

31. $(0, 0)$, $\frac{2}{3}$

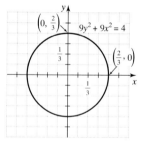

32. $(-2, 3)$, 4

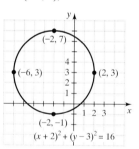

33. $x^2 + (y - 3)^2 = 36$

34. $x^2 + y^2 = 6$

35. $(x - 2)^2 + (y + 7)^2 = 25$

36. $\left(x - \frac{1}{2}\right)^2 + (y + 3)^2 = \frac{1}{4}$

37.

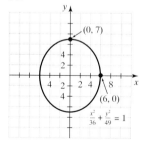

38.

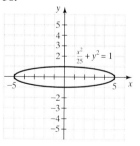

39.

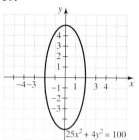

40.

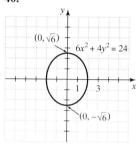

41.

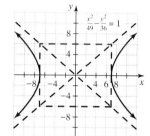

42.

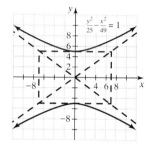

43.

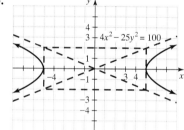

44.

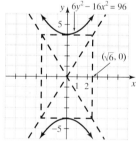

45.

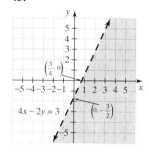

46.

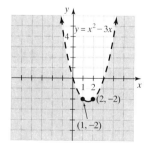

47.

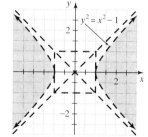

48.

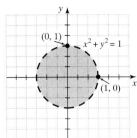

49.

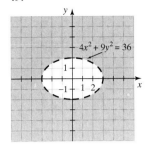

50.

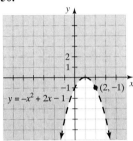

51.

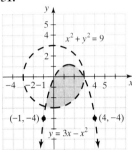

52.

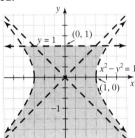

53.

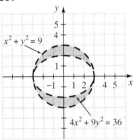

54.

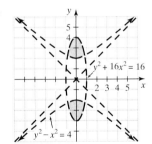

55. Hyperbola **56.** Line **57.** Circle **58.** Parabola
59. Circle **60.** Circle **61.** Circle **62.** Line
63. Hyperbola **64.** Ellipse **65.** Hyperbola **66.** Parabola

67.

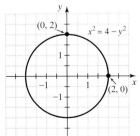

68.

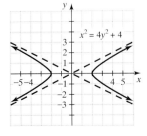

69.

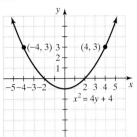

70.

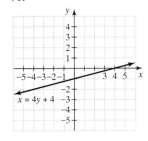

71.

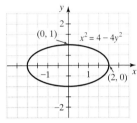

72.

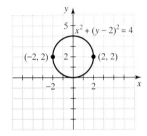

73.

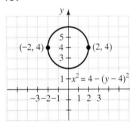

74.

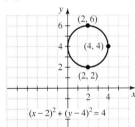

75. $x^2 + y^2 = 25$ **76.** $(x - 2)^2 + (y + 3)^2 = 58$
77. $(x + 1)^2 + (y - 5)^2 = 36$ **78.** $x^2 + (y + 3)^2 = 9$
79. $y = \frac{1}{4}(x - 1)^2 + 3$ **80.** $y = -\frac{1}{8}(x + 2)^2 + 3$
81. $y = x^2$ **82.** $y = -\frac{1}{2}(x - 1)^2 + 2$ **83.** $y = \frac{2}{9}x^2$
84. $y = -3(x - 1)^2 + 3$ **85.** $\{(4, -3), (-3, 4)\}$
86. $\{(2, \sqrt{3}), (2, -\sqrt{3}), (-2, \sqrt{3}), (-2, -\sqrt{3})\}$
87. $\varnothing$ **88.** $\{(3, 12), (-2, 2)\}$ **89.** 6 ft, 2 ft
90. $\frac{7}{2}$ in., $\frac{3}{2}$ in.

Chapter 11 Test

1.

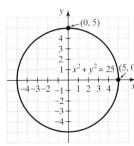

2.

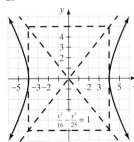

3.

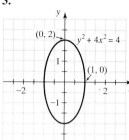

4.

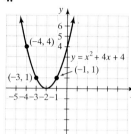

16. $y = \frac{1}{2}(x - 3)^2 - 5$ **17.** 84 ft

18. $(x + 1)^2 + (y - 3)^2 = 13$ **19.** 12 ft, 9 ft

Making Connections Chapters 1–11

1.

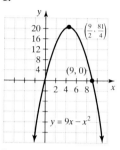

2.

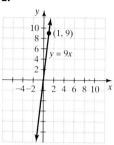

5.

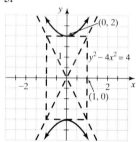

6.

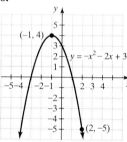

3.

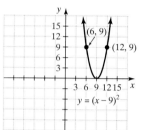

4.

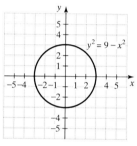

7.

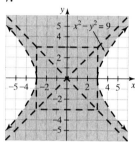

8.

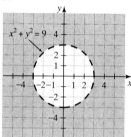

5.

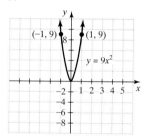

6.

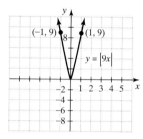

9.

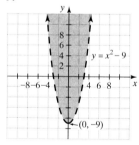

10.

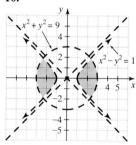

7.

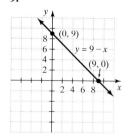

8.

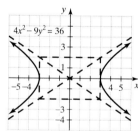

11.

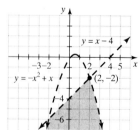

12. $\{(-5, 27), (3, -5)\}$ **13.** $\{(\sqrt{3}, 3), (-\sqrt{3}, 3)\}$

14. $(-1, -5)$, 6

15. Vertex $\left(-\frac{1}{2}, \frac{11}{4}\right)$, focus $\left(-\frac{1}{2}, 3\right)$, directrix $y = \frac{5}{2}$, axis of symmetry $x = -\frac{1}{2}$, upward

9.

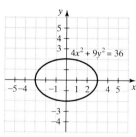

10.

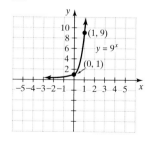

11. $x^2 + 4xy + 4y^2$ **12.** $x^3 + 3x^2y + 3xy^2 + y^3$
13. $a^3 + 3a^2b + 3ab^2 + b^3$ **14.** $a^2 - 6ab + 9b^2$
15. $6a^2 - 7a - 5$ **16.** $x^3 - y^3$ **17.** $\{(1, 2)\}$
18. $\{(3, 4), (4, 3)\}$ **19.** $\{(1, -2, 3)\}$ **20.** $\{(-1, 1), (3, 9)\}$

21. $x = -\dfrac{b}{a}$ **22.** $x = \dfrac{-d \pm \sqrt{d^2 - 4wm}}{2w}$

23. $B = \dfrac{2A - bh}{h}$ **24.** $x = \dfrac{2y}{y - 2}$ **25.** $m = \dfrac{L}{1 + xt}$

26. $t = \dfrac{y^2}{9a^2}$ **27.** $y = -\dfrac{2}{3}x - \dfrac{5}{3}$ **28.** $y = -2x$

29. $(x - 2)^2 + (y - 5)^2 = 45$ **30.** $\left(-\dfrac{3}{2}, 3\right), \dfrac{3\sqrt{5}}{2}$

31. $-10 + 6i$ **32.** -1 **33.** $3 - 5i$ **34.** $7 + 6i\sqrt{2}$
35. $-8 - 27i$ **36.** $-3 + 3i$ **37.** 29 **38.** $-\dfrac{3}{2} - i$
39. $-\dfrac{1}{2} + \dfrac{9}{2}i$ **40.** $2 - i\sqrt{2}$

41. **a)** $q = -500x + 400$ **b)** $R = -500x^2 + 400x$
c)

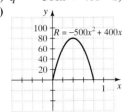

d) \$0.40 per pound   **e)** \$80

CHAPTER 12

Section 12.1 Warm-ups T T T F F F T F T T
1. A sequence is a list of numbers.
2. Each number in the sequence is called a "term" of the sequence.
3. A finite sequence is a function whose domain is the set of positive integers less than or equal to some fixed positive integer.
4. An infinite sequence is a function whose domain is the set of all positive integers.
5. $1, 4, 9, 16, 25, 36, 49, 64$ **6.** $-1, -4, -9, -16$
7. $-1, \dfrac{1}{2}, -\dfrac{1}{3}, \dfrac{1}{4}, -\dfrac{1}{5}, \dfrac{1}{6}, -\dfrac{1}{7}, \dfrac{1}{8}, -\dfrac{1}{9}, \dfrac{1}{10}$

8. $1, -\dfrac{1}{2}, \dfrac{1}{3}, -\dfrac{1}{4}, \dfrac{1}{5}, -\dfrac{1}{6}$ **9.** $1, -2, 4, -8, 16$

10. $-\dfrac{1}{3}, 1, -3, 9, -27$ **11.** $\dfrac{1}{2}, \dfrac{1}{4}, \dfrac{1}{8}, \dfrac{1}{16}, \dfrac{1}{32}, \dfrac{1}{64}$

12. $2, 1, \dfrac{1}{2}, \dfrac{1}{4}, \dfrac{1}{8}$ **13.** $-1, 1, 3, 5, 7, 9, 11$

14. $8, 10, 12, 14, 16, 18, 20$ **15.** $1, \dfrac{\sqrt{2}}{2}, \dfrac{\sqrt{3}}{3}, \dfrac{1}{2}, \dfrac{\sqrt{5}}{5}$

16. $\dfrac{1}{2}, \dfrac{\sqrt{2}}{4}, \dfrac{\sqrt{3}}{8}, \dfrac{1}{8}$ **17.** $\dfrac{1}{2}, \dfrac{1}{6}, \dfrac{1}{12}, \dfrac{1}{20}$ **18.** $\dfrac{1}{6}, \dfrac{1}{12}, \dfrac{1}{20}, \dfrac{1}{30}$

19. $-\dfrac{1}{3}, -1, 1, \dfrac{1}{3}$ **20.** $\dfrac{4}{7}, \dfrac{4}{9}, \dfrac{4}{11}, \dfrac{4}{13}$ **21.** $-1, 0, -1, 4$

22. $-1, 9, -25, 49$ **23.** $1, \dfrac{1}{4}, \dfrac{1}{9}, \dfrac{1}{16}$ **24.** $-1, -2, -4, -8$

25. $a_n = 2n - 1$ **26.** $a_n = 2n + 3$ **27.** $a_n = (-1)^{n+1}$
28. $a_n = (-1)^n$ **29.** $a_n = 2n - 2$ **30.** $a_n = 2n + 2$
31. $a_n = 3n$ **32.** $a_n = 4n$ **33.** $a_n = 3n + 1$
34. $a_n = 4n - 1$ **35.** $a_n = (-1)^n 2^{n-1}$
36. $a_n = (-3)^{n-1}$ **37.** $a_n = (n - 1)^2$
38. $a_n = (n - 1)^3$ **39.** $4, 2, 1, \dfrac{1}{2}, \dfrac{1}{4}$ **40.** $\dfrac{64}{81}$ acre
41. \$34,162, \$35,870, \$37,663, \$39,546, \$41,524, \$43,600
42. \$22,455, \$23,455, \$24,455, \$25,455, \$26,455, \$27,455
43. \$1,000,000, \$800,000, \$640,000, \$512,000
44. \$1,000,000, \$500,000, \$250,000, \$125,000
45. 27 in., 13.5 in., 9 in., 6.75 in., 5.4 in.
46. 278, 294, 312, 330, 350, 371, 393, 416, 441, 467, 495 Hz.
47. 137,438,953,500, larger
48. \$5,368,709.12, \$10,737,418.23
51. **a)** 0.9048, 0.3677, 0.00004517   **b)** a_n goes to zero

Section 12.2 Warm-ups T F F F F T T T T F
1. Summation notation provides a way to write a sum without writing out all of the terms.
2. The index of summation is the variable used in summation notation.
3. A series is the indicated sum of the terms of a sequence.
4. A finite series is the indicated sum of the terms of a finite sequence.
5. 30   **6.** 30   **7.** 24   **8.** 24   **9.** $\dfrac{31}{32}$ **10.** $-\dfrac{11}{32}$

11. 50   **12.** 60   **13.** -7 **14.** 144   **15.** 0   **16.** -1

17. $\displaystyle\sum_{i=1}^{6} i$ **18.** $\displaystyle\sum_{i=1}^{5} 2i$ **19.** $\displaystyle\sum_{i=1}^{6} (-1)^i(2i - 1)$

20. $\displaystyle\sum_{i=1}^{5} (-1)^{i+1}(2i - 1)$ **21.** $\displaystyle\sum_{i=1}^{6} i^2$ **22.** $\displaystyle\sum_{i=1}^{5} i^3$

23. $\displaystyle\sum_{i=1}^{4} \dfrac{1}{2 + i}$ **24.** $\displaystyle\sum_{i=1}^{6} \dfrac{(-1)^{i+1}}{i}$ **25.** $\displaystyle\sum_{i=1}^{3} \ln(i + 1)$

26. $\displaystyle\sum_{i=1}^{4} e^i$ **27.** $\displaystyle\sum_{i=1}^{4} a_i$ **28.** $\displaystyle\sum_{i=1}^{4} a^{i+1}$ **29.** $\displaystyle\sum_{i=1}^{48} x_{i+2}$

30. $\displaystyle\sum_{i=1}^{30} y_i$ **31.** $\displaystyle\sum_{i=1}^{n} w_i$ **32.** $\displaystyle\sum_{i=1}^{k} m_i$ **33.** $\displaystyle\sum_{j=0}^{4} (j + 1)^2$

34. $\displaystyle\sum_{j=0}^{5} (j + 1)^3$ **35.** $\displaystyle\sum_{j=1}^{13} (2j - 3)$ **36.** $\displaystyle\sum_{j=0}^{2} (3j + 5)$

37. $\displaystyle\sum_{j=1}^{5} \dfrac{1}{j + 3}$ **38.** $\displaystyle\sum_{j=1}^{6} 2^{-j-4}$ **39.** $\displaystyle\sum_{j=0}^{3} x^{2j+5}$ **40.** $\displaystyle\sum_{j=1}^{3} x^{5-2j}$

41. $\displaystyle\sum_{j=0}^{n-1} x^{j+1}$ **42.** $\displaystyle\sum_{j=1}^{n+1} x^{-j+1}$ **43.** $x + x^2 + x^3 + x^4 + x^5 + x^6$

44. $-1 + x - x^2 + x^3 - x^4$ **45.** $x_0 - x_1 + x_2 - x_3$

46. $\dfrac{1}{x_1} + \dfrac{1}{x_2} + \dfrac{1}{x_3} + \dfrac{1}{x_4} + \dfrac{1}{x_5}$ **47.** $x + 2x^2 + 3x^3$

48. $x + \dfrac{x}{2} + \dfrac{x}{3} + \dfrac{x}{4} + \dfrac{x}{5}$ **49.** $\displaystyle\sum_{i=1}^{9} 2^{-i}$ **50.** $\displaystyle\sum_{i=1}^{5} 1000(1.1)^i$

51. $\displaystyle\sum_{i=1}^{4} 1,000,000(0.8)^{i-1}$ **52.** $\displaystyle\sum_{i=1}^{31} i$

53. A sequence is basically a list of numbers. A series is the indicated sum of the terms of a sequence.

54. $n \geq 31$

Section 12.3 Warm-ups F F F F F F T T T F

1. An arithmetic sequence is one in which each term after the first is obtained by adding a fixed amount to the previous term.

2. The nth term of an arithmetic sequence is $a_1 + (n - 1)d$, where a_1 is the first term.

3. An arithmetic series is an indicated sum of an arithmetic sequence.

4. The formula for the sum of the first n terms of an arithmetic series is $\frac{n}{2}(a_1 + a_n)$.

5. $a_n = 6n - 6$ **6.** $a_n = 5n - 5$ **7.** $a_n = 5n + 2$
8. $a_n = 11n - 7$ **9.** $a_n = 2n - 6$ **10.** $a_n = 3n - 6$
11. $a_n = -4n + 9$ **12.** $a_n = -3n + 11$
13. $a_n = -7n + 5$ **14.** $a_n = -2n - 3$
15. $a_n = 0.5n - 3.5$ **16.** $a_n = 0.75n - 2.75$
17. $a_n = -0.5n - 5.5$ **18.** $a_n = -0.5n + 1.5$
19. 9, 13, 17, 21, 25 **20.** 13, 19, 25, 31, 37
21. 7, 5, 3, 1, -1 **22.** 6, 3, 0, -3, -6
23. -4, -1, 2, 5, 8 **24.** -19, -7, 5, 17, 29
25. -2, -5, -8, -11, -14 **26.** -1, -3, -5, -7, -9
27. -7, -11, -15, -19, -23
28. -2, -5, -8, -11, -14 **29.** 4.5, 5, 5.5, 6, 6.5
30. 1.3, 1.6, 1.9, 2.2, 2.5 **31.** 1020, 1040, 1060, 1080, 1100
32. 3400, 2800, 2200, 1600, 1000 **33.** 51 **34.** -35
35. 4 **36.** -7 **37.** 26 **38.** -62 **39.** 17 **40.** -60
41. 1176 **42.** 78 **43.** 330 **44.** 891 **45.** -481
46. -553 **47.** 435 **48.** 705 **49.** -540 **50.** -477
51. 150 **52.** -70 **53.** -308 **54.** 475 **55.** \$25,000
56. \$164,500 **57.** 1085 **58.** \$9,500 **59.** b **60.** 14

Section 12.4 Warm-ups F F T T T T F F T F

1. A geometric sequence is one in which each term after the first is obtained by multiplying the preceding term by a constant.

2. The nth term of a geometric sequence is $a_1 r^{n-1}$, where a_1 is the first term and r is the common ratio.

3. A geometric series is an indicated sum of a geometric sequence.

4. The sum of the first n terms of a geometric series is given by $S_n = \frac{a_1(1 - r^n)}{1 - r}$.

5. The approximate value of r^n when n is large and $|r| < 1$ is 0.

6. The sum of an infinite geometric series is given by $S = \frac{a_1}{1 - r}$, provided $|r| < 1$.

7. $a_n = \frac{1}{3}(3)^{n-1}$ **8.** $a_n = \frac{1}{4}(8)^{n-1}$ **9.** $a_n = 64\left(\frac{1}{8}\right)^{n-1}$

10. $a_n = 100\left(\frac{1}{10}\right)^{n-1}$ **11.** $a_n = 8\left(-\frac{1}{2}\right)^{n-1}$

12. $a_n = -9\left(-\frac{1}{3}\right)^{n-1}$ **13.** $a_n = 2(-2)^{n-1}$

14. $a_n = -\frac{1}{2}(-4)^{n-1}$ **15.** $a_n = -\frac{1}{3}\left(\frac{3}{4}\right)^{n-1}$

16. $a_n = -\frac{1}{4}\left(\frac{4}{5}\right)^{n-1}$ **17.** $2, \frac{2}{3}, \frac{2}{9}, \frac{2}{27}, \frac{2}{81}$

18. $-5, -\frac{5}{2}, -\frac{5}{4}, -\frac{5}{8}, -\frac{5}{16}$ **19.** 1, -2, 4, -8, 16

20. $1, -\frac{1}{3}, \frac{1}{9}, -\frac{1}{27}, \frac{1}{81}$ **21.** $\frac{1}{2}, \frac{1}{4}, \frac{1}{8}, \frac{1}{16}, \frac{1}{32}$

22. $\frac{1}{3}, \frac{1}{9}, \frac{1}{27}, \frac{1}{81}, \frac{1}{243}$

23. 0.78, 0.6084, 0.4746, 0.3702, 0.2887
24. -0.23, 0.0529, -0.0122, 0.0028, -0.0006 **25.** 5

26. 64 **27.** $\frac{1}{3}$ **28.** -3 **29.** $-\frac{1}{9}$ **30.** $-\frac{32}{243}$

31. $\frac{511}{512}$ **32.** $\frac{121}{81}$ **33.** $\frac{11}{32}$ **34.** $\frac{182}{81}$ **35.** $\frac{63050}{729}$

36. $\frac{1261}{243}$ **37.** 5115 **38.** 11,111.11 **39.** 0.111111

40. 0.24992 **41.** 42.8259 **42.** 35.71875 **43.** $\frac{1}{4}$

44. $\frac{1}{6}$ **45.** 9 **46.** 4 **47.** $\frac{8}{3}$ **48.** $\frac{64}{7}$ **49.** $\frac{3}{7}$ **50.** $\frac{1}{4}$

51. 6 **52.** $\frac{35}{3}$ **53.** $\frac{1}{3}$ **54.** $\frac{2}{3}$ **55.** $\frac{4}{33}$ **56.** $\frac{8}{11}$

57. \$3,042,435.27 **58.** \$146,763.44 **59.** \$21,474,836.47
60. 8.796×10^{12} **61.** \$5,000,000 **62.** \$2,000,000

63. d **64.** $\frac{4}{9}$ **65.** $\frac{8}{33}$

Section 12.5 Warm-ups F F F T T T T T T T

1. The sum obtained for a power of a binomial is called a binomial expansion.

2. Pascal's triangle gives the coefficients for $(a + b)^n$ for $n = 1, 2, 3$, and so on. Each row starts and ends with a 1. The other terms are obtained by adding the closest two terms in the preceding row.

3. The expression $n!$ is the product of the positive integers from 1 through n.

4. The binomial theorem gives the expansion of $(a + b)^n$.

5. 10 **6.** 6 **7.** 56 **8.** 36
9. $r^5 + 5r^4t + 10r^3t^2 + 10r^2t^3 + 5rt^4 + t^5$
10. $r^6 + 6r^5t + 15r^4t^2 + 20r^3t^3 + 15r^2t^4 + 6rt^5 + t^6$
11. $m^3 - 3m^2n + 3mn^2 - n^3$
12. $m^4 - 4m^3n + 6m^2n^2 - 4mn^3 + n^4$
13. $x^3 + 6ax^2 + 12a^2x + 8a^3$
14. $a^4 + 12a^3b + 54a^2b^2 + 108ab^3 + 81b^4$
15. $x^8 - 8x^6 + 24x^4 - 32x^2 + 16$
16. $x^{10} - 5a^2x^8 + 10a^4x^6 - 10a^6x^4 + 5a^8x^2 - a^{10}$
17. $x^7 - 7x^6 + 21x^5 - 35x^4 + 35x^3 - 21x^2 + 7x - 1$
18. $x^6 + 6x^5 + 15x^4 + 20x^3 + 15x^2 + 6x + 1$
19. $a^{12} - 36a^{11}b + 594a^{10}b^2 - 5940a^9b^3$
20. $x^{10} - 20x^9y + 180x^8y^2 - 960x^7y^3$
21. $x^{18} + 45x^{16} + 900x^{14} + 10500x^{12}$
22. $x^{40} + 20x^{38} + 190x^{36} + 1140x^{34}$
23. $x^{22} - 22x^{21} + 231x^{20} - 1540x^{19}$

24. $256x^8 - 1024x^7 + 1792x^6 - 1792x^5$

25. $\dfrac{x^{10}}{1024} + \dfrac{5x^9y}{768} + \dfrac{5x^8y^2}{256} + \dfrac{5x^7y^3}{144}$

26. $\dfrac{a^8}{256} + \dfrac{a^7b}{80} + \dfrac{7a^6b^2}{400} + \dfrac{7a^5b^3}{500}$

27. $1287a^8w^5$ **28.** $924m^6n^6$ **29.** $-11440m^9n^7$

30. $-2002a^9b^5$ **31.** $448x^5y^3$ **32.** $2835a^4b^3$

33. $635{,}043{,}840a^{28}b^6$ **34.** $495a^{16}w^8$

35. $\displaystyle\sum_{i=0}^{8} \dfrac{8!}{(8-i)!\,i!}a^{8-i}m^i$ **36.** $\displaystyle\sum_{i=0}^{13} \dfrac{13!}{(13-i)!\,i!}z^{13-i}w^i$

37. $\displaystyle\sum_{i=0}^{5} \dfrac{5!(-2)^i}{(5-i)!\,i!}a^{5-i}x^i$ **38.** $\displaystyle\sum_{i=0}^{7} \dfrac{7!(-3)^i}{(7-i)!\,i!}w^{7-i}m^i$

39. $a^3 + b^3 + c^3 + 3a^2b + 3a^2c + 3ab^2 + 3ac^2 + 3b^2c + 3bc^2 + 6abc$

40. $280{,}840x^{117}y^3,\ 62{,}739{,}600x^{96}y^4$

Enriching Word Power

1. a **2.** d **3.** c **4.** b **5.** a **6.** c **7.** d **8.** b
9. d **10.** a

Chapter 12 Review

1. $1, 8, 27, 64, 125$ **2.** $0, 1, 16, 81$ **3.** $1, 1, -3, 5, -7, 9$

4. $2, -1, 0, 1, -2, 3, -4$ **5.** $-1, -\dfrac{1}{2}, -\dfrac{1}{3}$ **6.** $-1, \dfrac{1}{4}, -\dfrac{1}{9}$

7. $\dfrac{1}{3}, \dfrac{1}{5}, \dfrac{1}{7}$ **8.** $1, -1, -\dfrac{1}{3}$ **9.** $4, 5, 6$ **10.** $2, 4, 6$

11. 36 **12.** 30 **13.** 40 **14.** -5

15. $\displaystyle\sum_{i=1}^{\infty} \dfrac{1}{2(i+1)}$ **16.** $\displaystyle\sum_{i=1}^{\infty} \dfrac{1}{i+2}$ **17.** $\displaystyle\sum_{i=1}^{\infty} (i-1)^2$

18. $\displaystyle\sum_{i=1}^{\infty} i(-1)^i$ **19.** $\displaystyle\sum_{i=1}^{\infty} (-1)^{i+1}x_i$ **20.** $\displaystyle\sum_{i=1}^{\infty} (-1)^i x^{i+1}$

21. $6, 11, 16, 21$ **22.** $-7, -3, 1, 5$

23. $-20, -22, -24, -26$ **24.** $10, 7.5, 5, 2.5$

25. $3000, 4000, 5000, 6000$ **26.** $4500, 4000, 3500, 3000$

27. $a_n = \dfrac{n}{3}$ **28.** $a_n = -4n + 14$ **29.** $a_n = 2n$

30. $a_n = -10n + 30$ **31.** 300 **32.** 203 **33.** $\dfrac{289}{6}$

34. -234 **35.** 35 **36.** 147 **37.** $3, \dfrac{3}{2}, \dfrac{3}{4}, \dfrac{3}{8}$

38. $-2, \dfrac{2}{3}, -\dfrac{2}{9}, \dfrac{2}{27}$ **39.** $1, \dfrac{1}{2}, \dfrac{1}{4}, \dfrac{1}{8}$ **40.** $5, 50, 500, 5{,}000$

41. $0.23, 0.0023, 0.000023, 0.00000023$

42. $0.4, 0.04, 0.004, 0.0004$

43. $a_n = \dfrac{1}{2}(6)^{n-1}$ **44.** $a_n = -6\left(-\dfrac{1}{3}\right)^{n-1}$

45. $a_n = 0.7(0.1)^{n-1}$ **46.** $a_n = 2x^{n-1}$ **47.** $\dfrac{40}{81}$ **48.** 1022

49. 0.3333333333 **50.** 0.11111 **51.** $\dfrac{3}{8}$ **52.** 8 **53.** 54

54. 1 **55.** $m^5 + 5m^4n + 10m^3n^2 + 10m^2n^3 + 5mn^4 + n^5$

56. $16m^4 - 32m^3y + 24m^2y^2 - 8my^3 + y^4$

57. $a^6 - 9a^4b + 27a^2b^2 - 27b^3$

58. $\dfrac{x^5}{32} + \dfrac{5x^4a}{8} + 5x^3a^2 + 20x^2a^3 + 40xa^4 + 32a^5$

59. $495x^8y^4$ **60.** $2016x^5y^4$

61. $372{,}736a^{12}b^2$ **62.** $120a^7b^3$

63. $\displaystyle\sum_{i=0}^{7} \dfrac{7!}{(7-i)!\,i!}a^{7-i}w^i$ **64.** $\displaystyle\sum_{i=0}^{9} \dfrac{9!(-3)^i}{(9-i)!\,i!}m^{9-i}y^i$

65. Neither **66.** Geometric **67.** Arithmetic

68. Geometric **69.** Arithmetic **70.** Neither

71. $\dfrac{1}{\sqrt[3]{180}}$ **72.** 10 **73.** $-1 + \dfrac{1}{2} - \dfrac{1}{6} + \dfrac{1}{24} - \dfrac{1}{120}$

75. $a^5 + 5a^4b + 10a^3b^2 + 10a^2b^3 + 5ab^4 + b^5$

76. $x^8 + 8x^7y + 28x^6y^2 + 56x^5y^3 + 70x^4y^4 + 56x^3y^5 + 28x^2y^6 + 8xy^7 + y^8$

77. 26 **78.** 495 **79.** \$118,634.11

80. \$37,738.43, the time is half as much, and the amount is less than one-third as much.

81. \$13,784.92

Chapter 12 Test

1. $-10, -4, 2, 8$ **2.** $5, 0.5, 0.05, 0.005$ **3.** $-1, \dfrac{1}{2}, -\dfrac{1}{6}, \dfrac{1}{24}$

4. $1, \dfrac{3}{4}, \dfrac{5}{9}, \dfrac{7}{16}$ **5.** $a_n = 10 - 3n$ **6.** $a_n = -25\left(-\dfrac{1}{5}\right)^{n-1}$

7. $a_n = (-1)^{n-1}2n$ **8.** $a_n = n^2$ **9.** $5 + 7 + 9 + 11 + 13$

10. $5 + 10 + 20 + 40 + 80 + 160$

11. $m^4 + 4m^3q + 6m^2q^2 + 4mq^3 + q^4$

12. 750 **13.** $\dfrac{155}{8}$ **14.** 5 **15.** $10{,}100$ **16.** $\dfrac{1}{2}$

17. $\dfrac{511}{128}$ **18.** ± 2 **19.** 11 **20.** $1365r^{11}t^4$

21. $-448a^{10}b^3$ **22.** \$86,545.41

Making Connections Chapters 1–12

1. 6 **2.** $n^2 - 3$ **3.** $x^2 + 2xh + h^2 - 3$ **4.** $x^2 - 2x - 2$

5. 11 **6.** 6 **7.** 4 **8.** 32 **9.** $\dfrac{1}{2}$ **10.** 3 **11.** 1

12. x **13.** $-\dfrac{27}{2}$ **14.** $-\dfrac{8}{3}$ **15.** $-\dfrac{128}{9}$ **16.** 16

17.

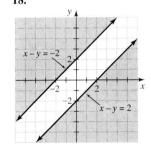

18.

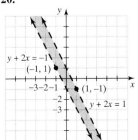

19.

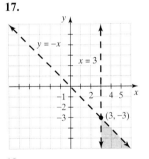

20.

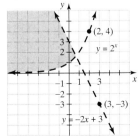

21.

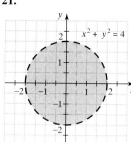

22.

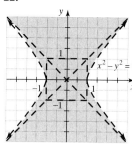

25.

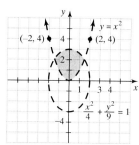

23.

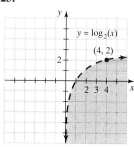

24.

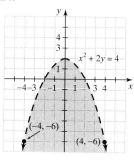

26. $\dfrac{a^2 + b^2}{ab}$ **27.** $\dfrac{y - 3}{y}$

28. $\dfrac{10}{(x - 3)(x + 3)(x + 1)}$ **29.** $\dfrac{2(x^3 - 64)}{x^3 - 16}$

30. $\dfrac{a^7}{b^4}$ **31.** 1 **32.** 4 **33.** $\dfrac{1}{32}$

34. -2 **35.** $\dfrac{1}{9}$ **36.** $-\dfrac{1}{8}$

37. $\dfrac{1}{4}$ **38.** $\dfrac{1}{5}$ **39.** 25

40. a) 105.8 cm   **c)** 1.3 years

INDEX